11th International Conference on Magnet Technology
(MT-11)

Volume 1

Proceedings of the 11th International Conference on Magnet Technology (MT-11) held in Tsukuba, Japan, 28 August to 1 September 1989

The complete proceedings are available in 2 volumes.

11th International Conference on Magnet Technology
(MT-11)

Edited by

T. SEKIGUCHI

Yokohama National University

and

S. SHIMAMOTO

Japan Atomic Energy Research Institute

Volume 1

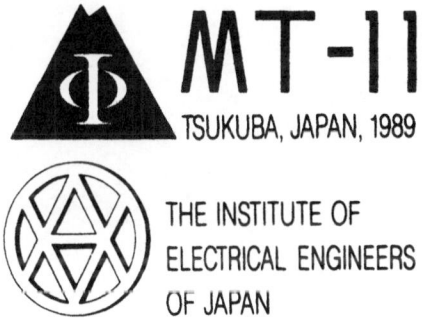

MT-11 TSUKUBA, JAPAN, 1989

THE INSTITUTE OF
ELECTRICAL ENGINEERS
OF JAPAN

ELSEVIER APPLIED SCIENCE
LONDON and NEW YORK

ELSEVIER SCIENCE PUBLISHERS LTD
Crown House, Linton Road, Barking, Essex IG11 8JU, England

Sole Distributor in the USA and Canada
ELSEVIER SCIENCE PUBLISHING CO., INC.
655 Avenue of the Americas, New York, NY 10010, USA

WITH 231 TABLES AND 865 ILLUSTRATIONS

© 1990 THE INSTITUTE OF ELECTRICAL ENGINEERS OF JAPAN

Softcover reprint of the hardcover 1st edition 1990

British Library Cataloguing in Publication Data

International Conference on Magnet Technology *(11th: 1989: Tsukuba, Japan)*
 11th International Conference on Magnet Technology (MT-11).
 1. Electrical engineering components: Magnets
 I. Title   II. Sekiguchi, T. (Tadashi)   III. Shimamoto, S.
 621.34

ISBN-13: 978-94-010-6832-1   e-ISBN-13: 978-94-009-0769-0
DOI: 10.1007/978-94-009-0769-0

Library of Congress CIP data applied for

Preface

Over the years the aim of the International Conference on Magnet Technology has been the exchange of information on the design, construction and operation of magnets for a variety of applications, such as high energy physics, fusion, electrical machinery and others. The aim has included advances in materials for magnet conductors, insulators and supporting structures. Since its inception the focus of the International Conference on Magnet Technology has gradually shifted to superconducting magnets. Now almost all papers are related to superconductivity.

The 11th International Conference on Magnet Technology (MT-11) was organized by the combined efforts of the Institute of Electrical Engineers of Japan, the Association for Promotion of Electrical, Electronic and Information Engineering, and the Tokyo Section of the IEEE. The Conference was held at the Tsukuba University Hall, Tsukuba, Japan, from 28 August to 1 September 1989, courtesy of the University of Tsukuba. The Tsukuba University Hall was large enough to host invited talks, parallel sessions, poster sessions and industrial exhibitions. 461 participants from 19 countries registered for MT-11, and 280 invited and contributed papers were presented. The papers were reviewed not only by the Program Committee but also by foreign participants.

Working sessions and social events were characterized by a truly international atmosphere. Scientific as well as cultural excursions were organized so that foreign visitors could experience the spirit of modern Japan. 26 companies, of which 8 were from Western countries, participated in the industrial exhibition which featured diverse products and services of interest to the magnet community.

On behalf of the International Organizing Committee I wish good luck to Professor V. A. Glukhikh, the Chairman of MT-12, the next International Conference on Magnet Technology, to be held in Leningrad, USSR, in June 1991.

TADASHI SEKIGUCHI
Chairman of MT-11

Acknowledgments

As Chairman of the MT-11 Program Committee, I wish to express my appreciation to the participants who contributed final papers for the MT-11 proceedings. Unfortunately, certain papers were not received by the Program Committee before the deadline for printing. We apologize for not being able to publish these papers in the MT-11 proceedings, nevertheless we hope that they will be published elsewhere.

The MT-11 Program Committee and I would like to thank the industrial sponsors who supported this conference. Finally, we would like to express our hearty gratitude to the many individuals without whose dedication MT-11 would not have progressed as smoothly as it did.

S. SHIMAMOTO
Secretary General of MT-11
Program Chairman of MT-11

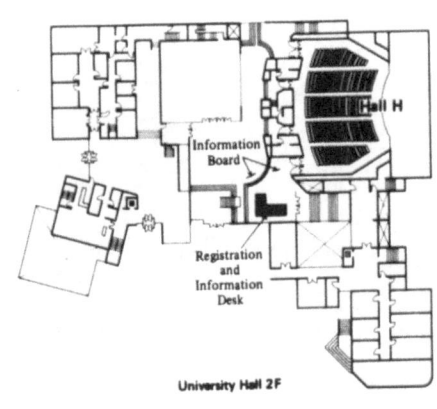

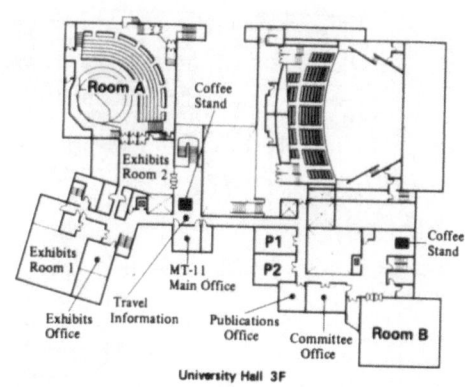

University Hall 3F

Reception

Flower Arrangement

Sumie(Chinese Ink Painting)

Excursions

Special Activities

Banquet

Contents

VOLUME 1

Plenary Talks

High Energy Physics I

High Energy Physics II

High Energy Physics III

High Energy Physics IV

Energy Storage, Power Applications and MHD Magnets

SC Generators and Motors

xix

VOLUME 2

Fusion Magnets

Conductor Development

Characteristics at Pulsed Operation

Stability and Reliability

Instrumentation and Measurements

Imaging Magnets

Magnet Technologies for Researches

Cryogenic System and Components

High T_c Superconductors

xxx

ORGANIZERS

The Institute of Electrical Engineers of Japan
Association for Promotion of Electrical, Electronic and Information Engineering
IEEE Tokyo Section

in conjunction with:

The University of Tsukuba
Japan Atomic Energy Research Institute
National Laboratory for High Energy Physics
National Research Institute for Metals
Electrotechnical Laboratory

COLLABORATING ORGANIZERS AND SOCIETIES

Atomic Energy Society of Japan
Institute for Materials Research, Tohoku University
Institute of Plasma Physics, Nagoya University
Japan Atomic Industrial Forum Incorporated
Japanese Association of Refrigeration
Japanese Society of Radiation Chemistry
Physical Society of Japan
The Chemical Society of Japan
The Cryogenic Association of Japan
The Institute of Electronic, Information and Communication Engineers
The Iron and Steel Institute of Japan
The Japan Institute of Metals
The Japan Society for Composite Materials
The Japan Society of Applied Physics
The Japan Society of Mechanical Engineers
The Japan Society of Plasma Science and Nuclear Fusion Research
The Society of Chemical Engineers, Japan
The Society of Polymer Science, Japan
The Magnetics Society of Japan

INTERNATIONAL ORGANIZING COMMITTEE

Chairman:
  T. Sekiguchi   Japan

Committee:
G. Bronca	France
N.A. Chernoplekov	Union of Soviet Socialist Republics
H. Collan	Finland
J.H. Coupland	United Kingdom
F.R. de Boer	The Netherlands
S. Han	People's Republic of China
I. Hlasnik	Czechoslovakia
P. Komarek	Federal Republic of Germany
F. Lange	German Democratic Republic
M. Morpurgo	Switzerland
P.A. Reeve	Canada
N. Sacchetti	Italy
S. Shimamoto	Japan
K. Trojnar	Poland
G. Vecsey	Switzerland
J.E.C. Williams	United States of America

NATIONAL ORGANIZING COMMITTEE

Honorary Chairman:
S. Yamamura   Professor Emeritus, The University of Tokyo

Chairman:
T. Sekiguchi   Yokohama National University
       Professor Emeritus, The University of Tokyo

Secretary General:
S. Shimamoto   Japan Atomic Energy Research Institute

Vice-Secretary:
H. Hirabayashi   National Laboratory for High Energy Physics

Advisers:

T. Anayama	Tohoku University
F. Irie	Kinki University
Y. Kyotani	Technova Inc.
T. Mitsui	The Tokyo Electrical Power Co., Inc.
S. Mori	Japan Atomic Energy Research Institute
Y. Muto	Tohoku University
K. Narimatsu	The Kansai Electric Power Co., Inc.
T. Nishikawa	National Laboratory for High Energy Physics
K. Oshima	Professor Emeritus, The University of Tokyo (deceased)
Y. Sekine	The Institute of Electrical Engineers of Japan
T. Sugano	Association for Promotion of Electrical, Electronic and Information Engineering
K. Tachikawa	Tokai University
H. Takahashi	The Chubu Electric Power Co., Inc.
S. Tanaka	The Chubu Electric Power Co., Inc.
T. Uchida	Institute of Plasma Physics, Nagoya University
T. Utsunomiya	IEEE Tokyo Section
K. Yamamoto	Japan Atomic Industrial Forum Inc.
M. Yamamoto	Takushoku University

Inspectors:

H. Kaminosono	Central Research Institute of Electric Power Industry
T. Takasuna	Hitachi Ltd.

Committee:

Y. Aiyama	Electrotechnical Laboratory
A. Furuya	The Institute of Electrical Engineers of Japan
K. Hida	Kobe Steel, Ltd.
H. Hirabayashi	National Laboratory for High Energy Physics
T. Hiraki	Kawasaki Heavy Industries, Ltd.
T. Ichino	The Institute of Electrical Engineers of Japan
T. Inoue	Nippon Steel Corporation
K. Ishikawa	Mitsubishi Heavy Industries, Ltd.
T. Isobe	The Furukawa Electric Co., Ltd.
T. Itoh	Mitsubishi Electric Corporation
H. Kamitsubo	The Institute of Physical and Chemical Research
M. Katsurai	The University of Tokyo
Y. Kito	The Institute of Electrical Engineers of Japan
H. Kohda	The Institute of Electrical Engineers of Japan
K. Kondo	The University of Tsukuba
K. Koyama	Electrotechnical Laboratory
M. Kubo	Sumitomo Heavy Industries, Ltd.
Y. Miki	The Institute of Electrical Engineers of Japan
Y. Miki	The Institute of Electrical Engineers of Japan
Y. Miyashita	NKK Corporation

S. Mori	The University of Tsukuba
K. Nagai	Toshiba Corporation
K. Nagamine	The University of Tokyo
Y. Nakagawa	Tohoku University
M. Nakahara	Kobe Steel, Ltd.
T. Nakahara	Sumitomo Electric Industries, Ltd.
Y. Nakazato	Fuji Electric Co., Ltd.
H. Nomura	Kawasaki Steel Corporation
T. Oboshi	The Institute of Electrical Engineers of Japan
T. Okada	Osaka University
T. Ogasawara	Nihon University
T. Ogura	Mitsubishi Cable Industries, Ltd.
K. Onishi	The Japan Steel Works, Ltd.
Y. Saiga	Ishikawajima-Harima Heavy Industries Co., Ltd.
T. Sekine	Mitsubishi Cable Industries, Ltd.
S. Shimamoto	Japan Atomic Energy Research Institute
S. Tanaka	Fujikura Ltd.
F. Terasaki	Sumitomo Metal Industries, Ltd.
S. Tokunaga	Showa Electric Wire & Cable Co., Ltd.
S. Tomiyama	Electrotechnical Laboratory
O. Tsukamoto	Yokohama National University
K. Yamafuji	Kyushu University
B. Yoda	Hitachi Cable Ltd.

TREASURY COMMITTEE

Chairman:
S. Mori	Japan Atomic Energy Research Institute

Secretaries:
H. Hirabayashi	National Laboratory for High Energy Physics
T. Mitsui	The Tokyo Electric Power Co., Inc.
T. Nakahara	Sumitomo Electric Industries, Ltd.
T. Sekiguchi	Yokohama National University
S. Shimamoto	Japan Atomic Energy Research Institute
T. Shintomi	National Laboratory for High Energy Physics
T. Takasuna	Hitachi Ltd.

Committee:
Y. Aiyama	Electrotechnical Laboratory
T. Anayama	Tohoku University
T. Inoue	Nippon Steel Corporation
F. Irie	Kinki University
T. Itoh	Mitsubishi Electric Corporation
T. Kasahara	Hokkaido University
K. Kohra	Foundation for High Energy Accelerator Science
Y. Kyotani	Technova Inc.
K. Nagai	Toshiba Corporation
Y. Nakazato	Fuji Electric Co., Ltd.
K. Narimatsu	The Kansai Electric Power Co., Inc.
T. Nishikawa	National Laboratory for High Energy Physics
O. Ohara	Teisan KK
Y. Saiga	Ishikawajima-Harima Heavy Industries Co., Ltd.
H. Takahashi	The Chubu Electric Power Co., Inc.
S. Tanaka	The Chubu Electric Power Co., Inc.
I. Todoriki	Japan Industrial Technology Association

EXECUTIVE COMMITTEE

Chairman:
S. Shimamoto   Japan Atomic Energy Research Institute

Co-chairman:
H. Hirabayashi   National Laboratory for High Energy Physics

Program Committee

Chairman:
S. Shimamoto   Japan Atomic Energy Research Institute

Co-chairman:
H. Tsuji   Japan Atomic Energy Research Institute

Committee:
T. Ando   Japan Atomic Energy Research Institute
T. Hamajima   Toshiba Corporation
H. Hirabayashi   National Laboratory for High Energy Physics
T. Kato   Japan Atomic Energy Research Institute
H. Kobayashi   Nihon University
K. Koizumi   Japan Atomic Energy Research Institute
N. Maki   Hitachi Ltd.
T. Matsushita   Kyushu University
O. Motojima   National Institute for Fusion Science
H. Nakajima   Japan Atomic Energy Research Institute
M. Nishi   Japan Atomic Energy Research Institute
T. Nitta   Kyoto University
K. Noto   Iwate University
M. Ohta   Japan Atomic Energy Research Institute
K. Okuno   Japan Atomic Energy Research Institute
T. Onishi   Electrotechnical Laboratory
F. Sumiyoshi   Kagoshima University
M. Suzuki   Tohoku University
E. Tada   Japan Atomic Energy Research Institute
T. Takagi   The University of Tokyo
Y. Takahashi   Japan Atomic Energy Research Institute
M. Takeo   Kyushu University
J. Tani   Tohoku University
K. Togano   National Research Institute for Metals
H. Wada   National Research Institute for Metals
T. Yamada   Mitsubishi Electric Corporation
J. Yamamoto   Kyoto University

Accounts Committee

Chairman:
T. Shintomi   National Laboratory for High Energy Physics

Committee:
T. Haruyama   National Laboratory for High Energy Physics
T. Isono   Japan Atomic Energy Research Institute
S. Mitsunobu   National Laboratory for High Energy Physics
E. Tada   Japan Atomic Energy Research Institute
Y. Takada   The University of Tsukuba

Local Executive Committee

Chairman:

H. Hirabayashi · · · · · National Laboratory for High Energy Physics

Committee:

K. Arai	Electrotechnical Laboratory
K. Endo	National Laboratory for High Energy Physics
T. Haruyama	National Laboratory for High Energy Physics
K. Hosoyama	National Laboratory for High Energy Physics
K. Inoue	National Research Institute for Metals
T. Isono	Japan Atomic Energy Research Institute
K. Kaiho	Electrotechnical Laboratory
K. Kawano	Japan Atomic Energy Research Institute
K. Maehata	National Laboratory for High Energy Physics
Y. Makida	National Laboratory for High Energy Physics
T. Mito	Kyoto University
S. Mitsunobu	National Laboratory for High Energy Physics
I. Muta	Saga University
H. Nakai	National Laboratory for High Energy Physics
T. Ogitsu	National Laboratory for High Energy Physics
N. Ohuchi	National Laboratory for High Energy Physics
K. Okuno	Japan Atomic Energy Research Institute
T. Shintomi	National Laboratory for High Energy Physics
M. Sugimoto	Japan Atomic Energy Research Institute
E. Tada	Osaka University
A. Terashima	National Laboratory for High Energy Physics
K. Togano	National Research Institute for Metals
K. Tsuchiya	National Laboratory for High Energy Physics
H. Wada	National Research Institute for Metals
M. Wake	National Laboratory for High Energy Physics
A. Yamamoto	National Laboratory for High Energy Physics
K. Yoshida	Japan Atomic Energy Research Institute

Publication Staff

Chairman:

K. Okuno · · · · · Japan Atomic Energy Research Institute

Staff:

T. Kato · · · · · Japan Atomic Energy Research Institute
K. Yoshida (*Secretary*)

SPONSORS

The National Organizing Committee wish to thank the many generous sponsors, listed below, who have contributed to the success of this conference:

Daido Sanso KK
Fuji Electric Co., Ltd.
Hitachi Ltd.
Hitachi Sanso KK
Hoxan Corporation
Ishikawajima-Harima Heavy Industries Co., Ltd.
Iwatani International Co., Ltd.
Iwatani Naoji Foundation
Japan Industrial Technology Association
Kawasaki Heavy Industries, Ltd.

Koike Sanso Kogyo Co., Ltd.
Maekawa Mfg. Co., Ltd.
Mitsubishi Electric Corporation
Mitsubishi Heavy Industries, Ltd.
Nippon Sanso KK
Osaka Sanso Kogyo, Ltd.
Showa Denko KK
Sumitomo Heavy Industries, Ltd.
Taiyo Sanso Co., Ltd.
Teisan KK
The Federation of Electric Power Companies
The Japan Iron and Steel Federation
The Japanese Electric Wire and Cable Makers' Association
Tokin Corporation
Tokyo Cryogenic Industries Corp., Ltd.
Tomoe Shokai Co., Ltd.
Toshiba Corporation
Toyo Sanso Co., Ltd.
Tsukuba Expo'85 Memorial Foundation

LIST OF EXHIBITORS

Ansaldo Componenti SpA (Italy)
Central Japan Railway Company (Japan)
Daiei Musen Denki Co., Ltd. (Japan)
The Furukawa Electric Co., Ltd. (Japan)
Hitachi, Ltd. (Japan)
Ishikawajima-Harima Heavy Industries Co., Ltd. (Japan)
Interatom GmbH (Federal Republic of Germany)
Intermagnetics General Corporation (USA)
Iwatani Cryo-Techno Corporation (Japan)
Kawasaki Heavy Industries, Ltd. (Japan)
Kawasaki Steel Corporation (Japan)
Kobe Steel, Ltd. (Japan)
Magnex Scientific Ltd. (UK)
Mitsubishi Electric Corporation (Japan)
Mitsubishi Heavy Industries, Ltd. (Japan)
Nippon Steel Corporation (Japan)
Outokumpu Cupper (Finland)
Sumitomo Electric Industries, Ltd. (Japan)
Sumitomo Heavy Industries, Ltd. (Japan)
Thevenet+Clerjoune (France)
Tokin Corporation (Japan)
Torisha Co., Ltd. (Japan)
Toshiba Corporation (Japan)
Vacuumschmelze GmbH (Federal Republic of Germany)
Vector Fields Ltd. (UK)

PROGRESS IN MR SYSTEMS FOR MEDICAL APPLICATIONS

Isamu Mano

Department of Radiology, Toshiba General Hospital
6-3-22 Higashiooi, Shinagawa-ku, Tokyo 140, Japan

ABSTRACT

Magnetic resonance imaging, or MRI for short, is a diagnostic
modality based on clinical images produced using the magnetic
resonance signals emitted from organ's protons. It was in 1980s
that MRI first proved to be really practicable for clinical
examinations. Since then, MRI has made such a remarkable progress
that it is now indispensable for routine examinations. MRI has
various advantages that other imaging modalities do not possess,
the most important of which are noninvasive nature, ability to
reveal soft tissues 3-dimensional imaging and suitability for
examinations of vessels. While MRI does have some disadvantages
and problems, some of these are gradually improved or solved.
Judging from the coming progress made in fast and ultra-fast
imaging and its successful application to organ functions, there
are good prospects of a fruitful future for MRI.

1. INTRODUCTION

Magnetic resonance
imaging (MRI) is a diagnostic
procedure based on clinical
images reconstructed using the
MR signals emitted from the
spin-oriented nuclei in the
tissues of the body.

MR phenomenon was first
discovered in 1946 by Bloch et
al[1] and Purcell et al[2].
From the viewpoint of medical
applications, it may be
briefly described as follows:

First, the patient is
placed in a static magnetic
field with a uniform specified
strength. Then, a high-
frequency pulsed magnetic
field, with a frequency
determined by the static
magnetic field strength, is
applied. The nuclei of the
hydrogen atoms (^{1}H) in the
tissue, for example, give rise
to magnetic resonance. That
is, the ^{1}H nuclei (or protons)
absorb a specific amount of
energy, making a transition to
an excited state. Removing
the applied field causes the
^{1}H nuclei to revert to their
original state, releasing

their energy in the form of electromagnetic waves, that is, as MR signals. Thus, the process leading to the emission of MR signals is referred to as MR phenomenon, and the imaging technique and diagnostic modality based on this phenomenon, MRI. The MR signals generated in this way contain useful medical information as to the biological tissues as discussed later.

2. PROGRESS OF MRI IN RECENT YEARS

It was in 1980[3] that MRI first proved to be a practical diagnostic modality, although it had already been studied a few years before by several workers who had recognized the MRI capabilities. Among those pioneering workers were Abe[4] from Japan who developed the magnetic field focus method, Damadian[5] who reported extended relaxation times in malignant tumors and Lauterbur[6] who presented the MR imaging technique, following such distinguished workers as Hinshaw[7], Mansfield, Garroway et al[8], Claw, Young Andrew et al[9] from Nottingham University, and Mallard et al[10] from Aberdeen University.

MRI has made remarkable progress in the 1980s, which could be compared to the rapid advancement of computed tomography (CT). Among the Western businesses which became involved in the development of the MRI scanner were GE, Technicare, Diasonics, Picker, Siemens and Philips. Then, Japanese manufactures as Toshiba, Hitachi, Shimadzu, Mitsubishi, Yokogawa Electric and Asahi Medical, joined in the search for improvements. A tech-

nological base was firmly established with the development of a series of innovations such as multislice, multiecho, oblique section and three-dimensional imaging. As the result of research efforts, the high signal-to-noise ratio, which has long been considered a major problem, has been steadily improved. Needless to say, these advances would have been done with the improvement of the superconducting magnet, RF coils, power supplies, and the computer hardwares, as well as softwares.

As noted above, MRI made remarkable progress in the middle of the 1980s, and since then has increasingly spreaded to the medical fields. From the clinical viewpoint at the present time, MRI has gained its role in the medicine as follows:

Regarding diagnostic value for the brain and spinal cord, MRI is superior to CT; for the bones, joints, muscles and abdominal organs, it is at least equivalent to CT. With the development of fast imaging techniques in recent years, MRI is proving more popular for the chest as well as for the abdomen (Fig. 1). The hardware of MRI has been continued to be more compact, lighter and cheaper. The successful development of high-performance refrigerators has greatly reduced maintenance costs (Fig. 2). As a result of these, the number of scanners installed is steadily increasing (450 in Japan and 1,550 in the USA, at the end of 1988), indicating their widespread use as routine tools for clinical examinations.

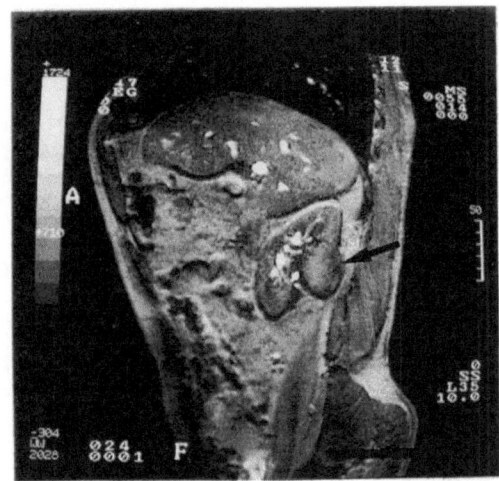

FIGURE 1

Renal cell carcinoma (arrow) taken by fast imaging technique. 1.5T, FE55/14

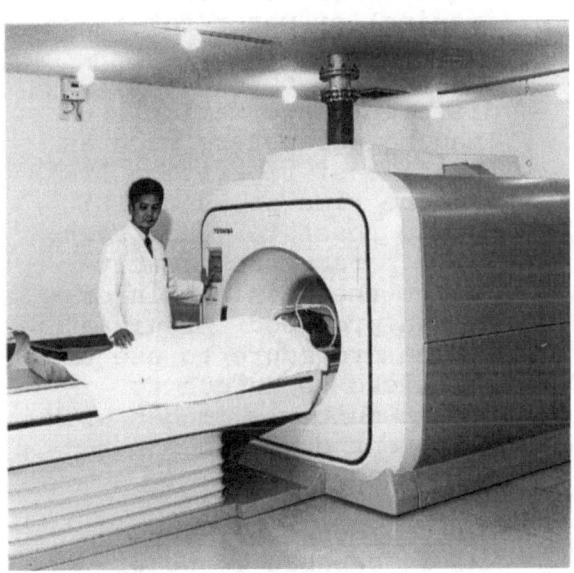

FIGURE 2

0.5T superconducting type MRI scanner. Very compact with low maintenance costs.

3. ADVANTAGES OF MRI

MRI has various advantages that other imaging modalities do not possess, the most important of which is its ability to reveal soft tissues.

(1) <u>A Noninvasive, Non-hazardous, Diagnostic Modality</u>
There are no harmful effects as there are for X-ray examinations (Fig. 3).

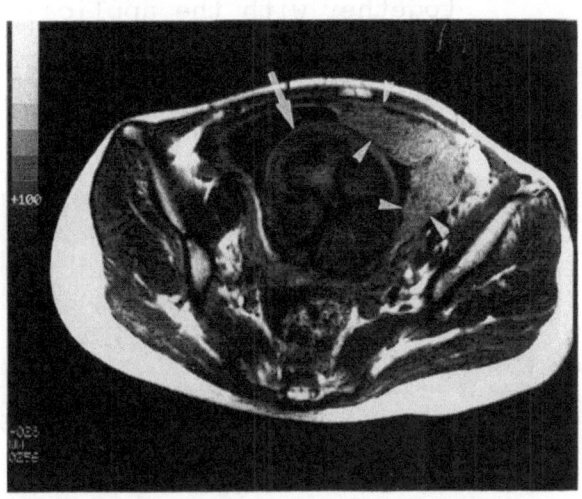

FIGURE 3

Fetus (arrow) MRI. 0.5T, SE500/40. Arrow heads show the placenta.

(2) <u>High Contrast in Soft Tissues</u>
For CT, the range of X-ray absorption values for soft tissues (except fat) is only a few percent of the difference between air and water values, whereas for MRI, the equivalent parameter has a value of several tens or more percent, provided that an appropriate pulse sequence is used.

(3) <u>High Spatial Resolution for Soft Tissues</u>
The spatial resolution of CT is determined primarily by the size of the aperture and the numbers of the X-

ray detector, whereas for MRI, it is by the strength of the applied magnetic field gradient. Therefore, the spatial resolution can be controlled voluntary, with the only disadvantage being that the image noise increases. However, the high contrast in soft tissues that MRI provides, together with the application of special coils technique, ultimately gives better spatial resolution than for CT (Fig. 4).

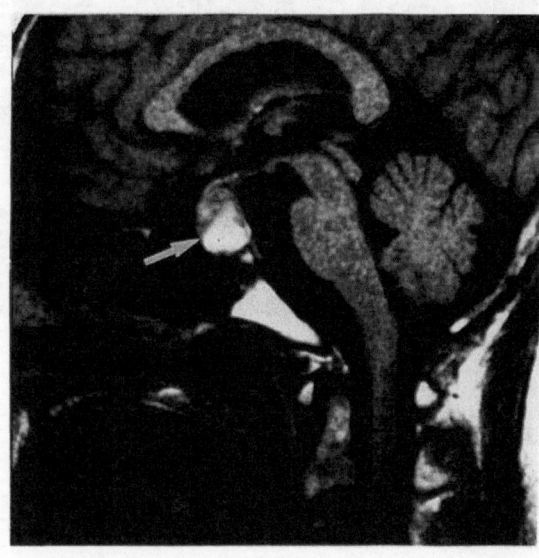

FIGURE 4

High resolution image of the pituitary gland tumor (arrow). 0.5T, SE500/40. Strong signal in the tumor means the hematoma.

(4) Fewer Artifacts
MRI is free from the artifacts caused by bone or air. Organ motion causes no streaking, only shading in most cases. However, in clinical use it must be kept in mind that MRI does have its own characteristic artifacts.

(5) Blood Flow Information, Making MR Angiography Possible
At the routine spin echo (SE)-sequence, MR signals are not emitted from arteries with normal blood flow. This results in black areas on the image. If the imaging parameters are properly selected, the sequence allows the speed and direction of blood flow to be determined. MR angiography is thus possible.

(6) Extensive Freedom in Selecting Imaging Modes and Parameters
MR imaging is generally performed using an SE-sequence. For special purposes, IR-sequence, FE-sequence (a kind of fast imaging) or water/fat image separation mode can be selected to obtain entirely different information. Imaging parameters such as the time of repetition (TR) and the time of echo (TE) can be reset for each scan. These parameters greatly affect the clinical information produced. It is a common procedure to use the multiecho technique to obtain multiple images with different echo times simultaneously.

(7) Three-dimensional Images
MRI is capable of multi-slice imaging. Selecting an SE-sequence with a longer TR, for example, permits ten slices or more to be imaged simultaneously. The ability to image slices in any direction, including coronal or sagittal sections, allows 3-dimensional morphology to be viewed. Recently, real 3-dimensional imaging, which permits the 3-

dimensional object data to be acquired at a single run, has become available.

(8) <u>High Imaging Efficiency</u>
MRI offers high imaging efficiency for long structures like the spinal cord or blood vessels, because any longitudinal section can be imaged. Selecting the multi-slice method with thin slices permits in everycases longitudinal sections of the organ to be imaged, using the "multi-planar reconstruction (MPR)" image processing technique.

(9) <u>Gated Scan to the Cardiac Cycle and Respiration</u>
MRI is performed entirely by electronic means, which makes it easy to synchronize the physiological signals. ECG-gated imaging has become indispensable for examinations of the cardiovascular and aortic systems. ECG-gated imaging, using the fast imaging technique, permits dynamic observation of cardiac cycles and blood flow.

(10) <u>Fast and Ultra-fast Imaging</u>
Fast imaging is now clinically available, taking 10 to 25 seconds per run. This technique, with applications in cardiac cine MRI, MR angiography and 3-dimensional imaging, has proven very useful. In addition, ultra-fast imaging has recently become clinically possible (Fig. 5), though some problems are yet to be solved. The imaging time is reduced to several tens of milliseconds, making clear cardiac images without ECG gating.

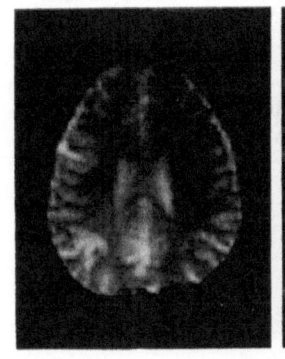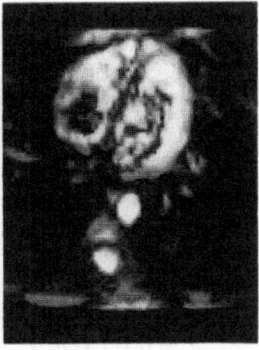

FIGURE 5

A head and a cardiac images taken by ultra-fast imaging mode. The scanning times are 33 msecond respectively.

(11) <u>Chemical Shift Information</u>
For ^{1}H-MRI, selecting the water/fat image separation mode permits water to be distinguished from fat. The same mode can be applied to ^{31}P, although in practice this nuclide does not give a sufficiently enough signal-to-noise ratio even in a magnetic field as high as 1.5 to 2 teslas. Therefore, in practice, the ^{31}P MR spectrum is acquired, using a square region of interest (ROI) at the ^{1}H images obtained, or with a surface coil used to localize the signal source.

(12) <u>Contrast Media</u>
Gd-DTPA is now clinically available, with almost the same distribution properties as the iodine contrast medium used in CT. Moreover, regarding the sensitivity and durability, Gd-DTPA is somewhat superior in most cases to the iodine contrast medium.

(13) <u>Few Machine Troubles and Flexibility to Upgrading.</u> The fault-free operation of the MRI systems is due to the absence of mechanical movable parts in the main body, in contrast to CT.

4. PROBLEMS OF MRI AND THEIR SOLUTIONS

While MRI offers a number of advantages as described above, it does have some disadvantages and problems. Some of these are at present being improved or solved. Eight items are discussed here.

(1) <u>Long Imaging Time</u>
MRI is far inferior to CT as for the patient through-put at present. However, in recent years, fast MR imaging has become available, which takes ten to several tens of second per run. In the future, ultra-fast imaging will be available, so there are good prospects of solving this problem.

(2) <u>Low Specificity</u>
In general, MRI is very sensitive, but not very specific for particular diseases. For example, it appears to be impossible, except in a few cases, to characterize tumor tissues. However, it may be possible to improve this specificity using contrast media.

(3) <u>Bad Data Compatibility</u>
This is because the imaging technique and the parameter setting depend on the user. Even if the same machine is considered, it is not possible to compare absolute values produced,

as the signal intensity is set for every run. Although T1 is approximately proportional to the static magnetic field strength, an exact proportionality does not hold. For this reason, diagnoses made with MRI are based on relative values.

(4) <u>No MR Signals from Calcification, Compact Bone, and Gas</u>
These substances offer important clues in the diagnosis of conventional X-ray imaging.

(5) <u>Not Available to All Patients</u>
When performing MR imaging, magnetic substances must be removed from the patient to the maximum possible extent, otherwise, not only will artifacts result, but there is a risk of injury and machine failure. Thus patients relying on respirators and patients with implanted pacemakers are excluded from MRI examinations. Great care must also be taken with patients with artificial organs and surgical clips.

(6) <u>Machine and Maintenance Costs</u>
At present, an MRI machine costs about one hundred million yen for the 0.5 tesla superconducting type, slightly more expensive than a CT machine. However, this is about one third, compared to the MRI machine price of a few years before. As for operating and maintenance costs of the superconducting machine, with its refrigerator, these amount to several million yen per year.

(7) <u>Some Restrictions on Suitable Locations for MRI Machines</u>
Electrical equipment sensitive to magnetic fields should be located by a certain distance away from the MRI machine, because of the high field it generates. In addition, the MRI machine should not be installed, for example, near elevators, because it requires an extremely uniform field. For these reasons, machines with magnetic shields are becoming more popular. An RF shield is also required to block out external electromagnetic noise.

(8) <u>Emergency Procedures</u>
Ordinary emergency lifesaving equipment should not be brought into the MRI room, as this is usually constructed of magnetic materials. In emergencies, the patient must always be brought out from the main magnet before emergency procedures are performed, because the magnetic field cut is difficult.

5. CONCLUSION

We have described a general account of MRI and its progress during this decade as well as its advantages and problems at the present time. MRI is spreading to more and more medical facilities, and still continues to make progress technologically. Particularly worthy of note is the progress made in fast imaging and ultra-fast imaging. This will not only permit whole organs to be imaged, but also allow 3-dimensional, dynamic and vascular images to be obtained. As contrast media are improved, and progress is made in imaging techniques based on chemical shift and making use of nuclides other than ^{1}H, more useful information about organ function will be provided. In addition, ^{31}P MR spectroscopy will gradually find clinical application. Under these circumstances, the MR future prospects would be splendid.

REFERENCES

1) Block F, Hansen WW, Packard M: Nuclear induction, Phys Rev 69: 127, 1946

2) Purcell EM, Torrey HC, Pound RV: Resonance absorption by nuclear magnetic moments in a solid, Phys Rev 69: 37-38, 1946

3) Kaufman L, Crooks LE, Margulis AR: Nuclear Magnetic Resonance in Medicine, Igaku-shoin, New York-Tokyo, 1981

4) Abe Z, Tanaka K, Hotta M, Imai M: Non-invasive measurements of biological information with application of nuclear magnetic resonance, (In) Biologic and clinical effects of low-frequency magnetic and electric fields, pp295-317, CC Thomas, Springfield, 1974

5) Damadian R: Tumor detection by nuclear magnetic resonance, Science 171: 1151-1153, 1971

6) Lauterbur PC: Image formation by induced local interactions: Example employing nuclear magnetic resonance, Nature 242: 190-

191, 1973

7) Hinshaw WS: Image formation
by nuclear magnetic
resonance: The sensitive-
point method, J Appl Phys
C47: 3709-3721, 1976

8) Garroway AN, Grannell PK,
Mansfield P: Image
formation in NMR by a
selective irradiative
process, J Phys C7: 457-
462, 1974

9) Gumby P: The new wave in
medicine: Nuclear magnetic
resonance, JAMA 247: 151-
159, 1982

10) Mallard J, Hutchison JMS,
Edelstein WA, et al: In
vivo n.m.r. imaging in
medicine: The Aberdeen
approach, both physical and
biological, Phil Trans R
Soc Lond B289: 519-533,
1980

Current Status of Maglev Transportation System in the World

Junji Fujie
Director of Maglev Laboratory
Railway Technical Research Institute
Japan

Abstract

At present, among maglevs driven by normal-conducting electro-magnetic suspension systems or superconducting electrodynamic suspension systems in the world, some have reached the stage of commercial operation, while others are still in the stage of test runs. Taking Birmingham, HSST, Transrapid, MLU and M-Bahn as examples, main features such as suspension, guidance and propulsion systems, air gap between vehicle and ground and the like as well as magnetic technology as a key factor in these systems are described, in addition to brief introduction of some proposed projects or concepts about new maglev systems.

1. Introduction

With increased attention drawn to congested intra-city and suburban transportation, and in enhanced expectation of low-pollution high-safety magnetic levitation system(maglev) as a superspeed railway suitable for a new era in the 21st century, the developments of various maglev systems for intra-city, airport access and high-speed intercity transportations are now promoted in major countries of the World.

This paper describes current status of development and planning for the maglev system now under test runs or in commercial operations as well as newly proposed maglev systems.

2. Electromagnetic Suspension(EMS) System

As is generally known, an EMS system is based on the principle of attracting on-board electromagnets to steel rails laid on the guideway. In this system an air gap between the ground surface and the vehicle bottom surface is about 10 to 15 mm, the attraction force of electromagnet must be controlled, and suspension can be kept even while standing. Low or medium speed systems under development use short stator type linear induction motor(LIM) as a drive system with traction power collected from third-rail. On the other hand, highspeed systems under development use long stator type linear synchronous motor(LSM) with on-board power collected through magnetic induction, which permits a maglev vehicle to run fully non-contacted. EMS systems other than those under development in USSR and Romania are as follows:

Fig. 1. Birmingham maglev vehicle
(England)

2.1 Birmingham System

Maglev development in Great Britain was started in 1973 with the initiative taken by the British Railways(BR). In 1975, a 100 m test track was laid on the compound of the BR Research Centre at Derby and test runs of a 2.7 t car were started(1). As a result of the development research, in May 1984 the maglev came into commercial service on a 620 m long track between Birmingham Airport and BR Birmingham station (Fig. 1).

This system is operated by two maglev trains shuttling on the parallel two tracks. Main features of the maglev vehicle are as follows: length 6 m; width 2.25 m; height 3.08 m; loaded weight 8 t; unloaded weight 5 t; speed 48 km/h.

Suspension is induced by eight E-core axial-flux magnets with aluminium conductors. For guidance of the vehicle, the magnets of each pair are laterally offset by 10 mm on opposite sides of the support-rail center line. Normal gap is 15 mm and magnetic flux density within air gap is 0.8 T. Steel support-rail laid on the guideway is laminated vertically to a width of 50 mm. The magnetic flux paths stretch in the running direction of the vehicle.

The power for suspension and control is about 3 kW/t. When air gap is 15 mm, magnet lift/weight ratio is 11.7 with current density of 1 A/mm^2 per magnet. In the meantime, magnets are equipped with temperature sensors to prevent overheating. Magnet driver uses a chopper module consisting of power transistors. Switching frequency is 1 kHz.

The drive system consists of single-sided axial-flux short stator type LIM with reaction plate laid at the central portion of the track and with air gap of 20 mm. The driving inverter is pulse width modulated(PWM) type with a continuous rating of 240 kVA and with a 60 seconds rating of 325 kVA.

Since inauguration of the first commercial operation in the world, Birmingham maglev has not suffered any major troubles these five years with operation of 20 hours per day.

2.2 HSST System

In order to improve convenience of air transportation by easing congestion of airport-access traffic, research and development of HSST system has been promoted by the Japan Air Lines since 1974 in Japan. After unmanned HSST-01 registered 307.8 km/h, runs of HSST were demonstrated in several international expositions. Now HSST-05 commercially operated with a business license issued on a limited conditions authorized by the Ministry of Transportation on a 515 m long single track line in the Yokohama Expo'89. HSST-05 is a 2-car unit. Its main features are as follows: length 36.3 m, width 3.0 m, height 3.6 m, tare weight 20 t per car, maximum permissible load 27 t and maximum design speed 60 km/h (2). Vehicle structure is designed in due consideration of 200 km/h operation. The 2-car unit has 160 seats(Fig. 2).

The aluminium alloy welded structure module has various built-in equipment for suspension, guidance, propulsion, braking etc. Accordingly, this module performs a set of functions assinged

Fig. 2. HSST-05 in the Yokohama Expo'89
(Japan)

to each equipment. A total of 8 modules consiting of 4 on each side are installed on the body through air springs.

Four magnets are installed within a module and they constitute 2 sets of 2 magnets in a staggered arrangement. The magnets in each set are horizontally offset by 5 mm on opposite sides of the rail center line. This arrangement generates lateral damping forces and causes one magnet to carry out its duties of both suspension and guidance. Each independently controlled module has 2 sets of gap sensors and 4 accelerometers. Two sets of magnet-drivers supply each magnet with electric current.

HSST unlike Birmingham allocates the poles of magnets on both sides of aluminium conductor coils. Consequently, the features of HSST are as follows: magnetic pole width is about 25 mm; the width of rail portion constituting magnetic pole is a half that of Birmingham; and non-laminated rail is in use.

The air gap of HSST-05 is 9 mm, while its mechanical gap is designed to be 7 mm. Meanwhile, HSST-03 has an air gap of 11 mm with a magnetic attraction force/weight ratio of 6.68 and it consumes power equal to about 1.7 kW/t. Therefore, the power consumption by HSST-05 is estimated at about 1.2 kW/t. This shows a smaller gap is more advantageous.

As for propulsion, short stator type LIM is contained within each module, generating a mechanical gap of 11 mm and a rated driving force of 2.6 kN/LIM.

2.3 Transrapid

The basic research for maglev in West Germany was started about 1966. In 1971, a test vehicle TR 01 driven by a short stator type LIM was manufactured and test runs were started. In 1977, a vapour-rocket driven "KOMET" successfully registered a speed of 400 km/h (3). And then, propulsion system has been changed to a long stator type LSM to facilitate speed up, decrease of vehicle weight, larger space for passenger rooms, non-contacting power collection and so on. The construction of a test track at Emsland was started in 1978 and test runs using a Transrapid(TR) 06 vehicle began in 1983(Fig. 3).

Unlike Birmingham, the TR is driven by a long stator type LSM system with 2 types of magnets: one intended for

Fig. 3. TR 06 in Emsland test track (West Germany)

both propulsion and suspension; the other exclusively for vehicle guidance. Other than these features, inductive power-colletion is available for the TR.

TR 06 is a 2-car unit with length 54.2 m, width 3,7 m, height 4.2 m, overall weight 122 t and 196 seats. In 1988, the speed of 412.6 km/h was successfully attained by the manned TR 06, which had already logged a total distance of 65000 km. Meanwhile, TR 07, an improved version of TR 06, has the following main features: length about 50 m, width 3.7 m, height 3.9 m and tare weight 50 t. TR 07 has just started running tests at Emsland.

The ground rails of TR serve for both propulsion and suspension and they are made of laminated core with a slot within which winding cables of 32.1 mm diameter and 150 mm^2 cross section are laid. The windings are electrically series-connected. The stator windings for three-phase are fixed by fastenings at their lower portions without protruding the fastenings out of iron core which constitutes a magnetic path. A piece of laminated core is 185 mm wide, 91.5 mm high and 1031 mm long with pole pitch of 258 mm, consequently generating a speed of 400 km/h when the supplied frequency is 215 Hz.

The on-board magnets for suspension are installed on the truck facing the rail. The magnetic poles of these magnets are alternately arranged. Six poles of these magnets constitute a module with dimensions of 232 mm width, 190 mm height and 1318 mm length. Four intermediate poles differ from 2 end poles. On the surface of magnet a slot is cut in which linear generator (LG) windings are embedded. When a maglev train runs, magnetic flux of suspension magnets is modulated by stator slot, inducing ac voltage to be rectified as an on-board power source.

The air gap between vehicle and ground is set at 8 mm. The design values of nominal coil current and energy consumption are respectively 25 A and 1.5 kW/t (4),(5).

Guidance magnets are installed on the truck opposite to guidance rail. Two long coils of the guidance magnets are wound in the direction of rail. The core has a length of 1533 mm and is longitudinally stretched.

The magnet weight including mechanical suspension and the nominal force are respectively 330 kg and 18.7 kN for suspension magnet, whereas for guidance magnet these values are respectively 270 kg and 9 kN. The net weight of a support system is about 32 t.

TR 07 is designed so as to decrease the height of pole and to lengthen a unit of magnet to 3022 mm for suspension magnet and to 3050 mm for guidance one with a whole magnet mass lowered, resulting in a force versus weight ratio of 10:1 (6),(7). A cross section of guidance magnet has changed from C-type for TR 06 to E-type for TR 07.

3. Eletrodynamic Suspension(EDS) System

An electrodynamic suspension(EDS) system suggested in 1966 by J. R. Powell and G. R. Damby is based on the principle of suspending a vehicle by repulsion force obtained in an electromagnetic induction phenomenon between on-board magnets and ground conductors during running. To apply this principle, in the U.S.A. the MIT-Raytheon joint team has started Magneplane Project and carried out test runs using a 1/25 scale model on a 120 m long guideway. Meanwhile, there was a Stanford Research Institute(SRI) Project using a test vehicle on U-shape aluminum guideway. In 1976, another group consisting of AEG, BBC and Siemens staff in West Germany registered 230 km/h using a test vehicle named EET 01/02 on a 880 m long loop track with a radius of 140 m. Also in Canada, an EDS system has been regarded as a future intercity transportation. The concept of this Canadian system has been published. Now, only Japan is developing a peculiar EDS system using test vehicles.

Fig. 4. 3-car unit MLU 001 in Miyazaki test track (Japan)

The features of an EDS system using superconducting magnets are as follows: although electric power to maintain a cryogenic temperature is needed, the power for suspension is unnecessary because superconducting magnet is used in a persistent current mode; a supension/guidance system is self-stabilized without control; the tolerance of guideway irregularity is set larger because an air gap between vehicle and ground is kept at an order of 10 cm; and so on. These features permit the maglev vehicle to be reduced in weight and to run at a speed exceeding 500 km/h.

3.1 MLU System

The Japanese National Railways(JNR) embarked on the research of superspeed railways in 1962. However, it was not until 1970 that an earnest effort to develop an EDS system was made by the JNR. In 1972 the suspension/propulsion performance of superconductivity was acertained in principle using test vehicles of LSM 200 and ML 100 in the compound of the Railway Technical Research Institute(RTRI) of Japan. In 1974, ML 100A test vehicle debuted, which meant the completion of a fundamental structure of an EDS system driven by a long stator type LSM. And in the same year fully non-contacting test runs were successfully carried out.

In 1977, test runs using ML 500 vehicle with tare weight of 10 t were started on the inverted-T type guideway at Miyazaki Maglev Test Center. In 1979, a world record of 517 km/h was attained. Thereafter, the guideway cross section has been converted to U type, and on that guideway test runs of 3-car unit MLU 001 and a 400.8 km/h manned run

Fig. 5. New test vehicle MLU 002
(Japan)

by a 2-car unit have been successfully carried out. No.1 vehicle of MLU 001
has logged a total distance exceeding
40000 km with a total of about 10000
demonstration riders(Fig. 4). In March
1987, a test vehicle MLU 002 was
completed moving a step further toward
a practical version. At present, test
runs using this vehicle are in progress
(Fig. 5).

The cross section of the Miyazaki
maglev test track is a U type reinforced
concrete guideway. On the inner wall
of the guideway, 1.1 m long and 0.7 m
wide ground coils for propulsion/guidance
made of aluminium conductor are arranged
with 1.4 m pitch, in three phases and
at 120 degrees, generating a speed of
500 km/h when a frequency of 33 Hz is
supplied. Meanwhile, aluminium suspension coils with a unit dimension of
0.45 m length and 0.33 m width are laid
with 0.7 m pitch on the guideway surface.

A 3-car unit MLU 001 is 28.8 m long,
3 m wide and 3.3 m high with 32 seats
and a total weight of 30 t. Superconducting magnets of MLU 001 distributive-
ly installed in a row consisting of
12 poles on each side of the body bottom
surface to a total of 24 poles in two
rows.

MLU 002, a one-car unit, has 22 m
overall length, 3 m width, 3.7 m height,
44 seats, 17 t total weight and 2 bogie
trucks. For MLU 002, 6 superconducting magnet poles on both sides with
3 poles on each side are installed on
the truck. A total of 12 poles are set
on 2 trucks, that is, leading truck
and trailing one, supporting the body.
Main design features of MLU 002 are
as follows: suspension force 196 kN

with an effective gap 110 mm, guidance
force 83.3 kN in case of 50 mm displacement with an effective gap more than
150 mm, propulsion force from 0 to 79.4
kN and maximum speed 420 km/h. A superconducting magnet of MLU 002 consists
of 3 separate magnetic parts including
a coil in each part. Upper portion of
a unit of these 3 parts is connected
with a liquid helium tank combined with
an on-board refrigerator(Fig. 6). Overall
dimensions of a 3-pole magnet including
all these parts are 6.1 m length, 0.45 m
width, 0.88 m height and about 1 t
weight. Outer vessel is made of
aluminium, and inner vessel made of
stainless steel. The columns made of
fiber reinforced plastics(FRP) are used
as load support. Carbon fiber reinforced plastics(CFRP) materials with low
thermal conductivity are in use at the
portions within the temperature ranges
from liquid helium to liquid nitrogen,
whereas glass fiber reinforced plastics
(GFRP) materials are at the portions
within the temperature range from liquid
nitrogen to the ambient temperature.
These two FRP materials together constitute a hybrid structure.

A superconducting coil is of a racetrack type with 2.1 m coil pitch, 1.7 m
length and 0.5 m width. The weight of
a coil is 75 kg, its cross section is
46 mm high and 72 mm wide, and the
magnetomotive force is 800 kA at its
maximum and 700 kA in operation. A coil
has a turn number of 1167 and a maximum
magnetic field of 5.92 T. The current
density for a coil is 243 A/mm^2. The
wire in use has a fine multi-filament
structure of niobium-titanium alloy

Fig. 6. Superconducting Magnet of
MLU 002 (Japan)

and copper. Copper ratio is 1 for MLU 002 against 2 for MLU 001. This lower copper ratio makes it possible to decrease the coil weight and to shorten the distance between the center of coil cross section and the ambient temperature portion. The wire has a cross section 1.0 mm high x 2.1 mm wide and is constituted of 2383 superconductor cores with a diameter of 23 μm (8),(9).

Persistent current switch is of thermal type with an off-state resistance of 50 ohms and an energizing speed from 5 to 10 A/sec., which makes it possible to make excitation in about one minute and to lower the consumption of liquid helium.

Two systems of on-board refrigerators are in use: Claude-cycle system for leading truck and Stirling-cycle system for trailing truck, both of which have the same refrigerating capacity of 5 W at 4.4 K.

In the meantime, development of a superconducting magnet more stable in reliability is now being pushed with reliability and endurance tests.

In order to lower air resistance and to construct tunnels with decreased cross sections, maglev vehicle is to be designed as low in height as possible. Furthermore, the design of a maglev train is proceeded along the lines that an articulated truck system is adopted with superconducting magnets located at car ends in order to decrease the magnetic field intensity within passenger rooms.

In order to have no magnetic drag peak at lower speed and a decreased magnetic drag at higher speed, a side wall suspension system in which null-flux connected pairs of suspension coils are installed on both side walls to suspend maglev vehicle is now being examined for adoption. The relevant running tests have been completed on the Miyazaki Maglev Test Track. As a result, its effectiveness has been confirmed.

On August 7 this year the Maglev Investigation Committee of Ministry of Transport selected a final candidate site for construction of a 40 to 45 km long test track in Yamanashi Prefecture which will be able to be taken as the origin for New Maglev Shinkansen extending from Tokyo to Osaka.

4. New Systems

Various new concepts of maglevs using permanent magnets already magnetically energized, Meissner effect of superconductivity, improved LIM other than normal-conducting or superconducting magnets have been proposed. It may be possible to use them to solve the problems faced by the present systems. Now, some of them are going to inaugurate commercial operation. Moreover in 1970s, has been developed a Romag system using LIM for propulsion, levitation and guidance in the U.S.A.

4.1 M-Bahn System

The research of this system was started in 1973. And in 1975, a 1400 m long test track was completed and two vehicles, one for 40 passengers and the other for 70 passengers, were manufactured for test runs. Now, in West Berlin continuous endurance tests are in progress on a 1.6 km long track including a 1 km long double track section and 3 turnouts; before long its commercial operation is to be inaugurated.

M-Bahn with on-board permanent magnets is driven by long stator type normally-conducting LSM system. Attraction force between propulsion permanent magnet and motor stator core spports most, or 95 %, of vehicle mass. In order to control suspension force, vertical guide roller mechanically varies air gap matching weight variations due to loading or unloading. Thus this system might better be called electromagnetic semi-levitation system(Fig. 7).

Fig. 7. M-Bahn vehicle (West Berlin)

Development target of the system is exclusively set at low speed version with no electrical control of suspension and with only on-board service power being supplied. Consequently, this maglev vehicle is lighter by about one half than the metro car approximately the same size as the maglev. Vehicle dimensions are as follows: 12 m length, 2.3 m width, 2.4 m height, 9 t unloaded weight, 18 t loaded weight with accommodation capacity of 80 passengers. Maximum speed and acceleration are respectively 80 km/h and 1.3 m/s^2. Support truck installed with permanent magnet made of samarium cobalt alloy series is connected through primary spring with intermediate truck whose secondary spring supports the carbody. Two vertical guide rollers made of urethane resin installed on the support truck which sandwiches the running track from top and bottom are connected with intermediate truck through a linkage. When the load is increased, the support truck installed with permanent magnet nears the bottom surface of the running track with the air gap changing from 24 mm in a normal state to 10 mm in a fully-loaded state. The carbody is guided by horizontal-guidance wheels. The main features of LSM are pole pitch of 0.12 m, core width of 0.15 m and conductor made of aluminium cable with a driving inverter output of 0 to 440 V, 550 A and 0 to 95 Hz.

4.2 STARLIM System

A U-type LIM developed since 1970s has demonstrated its superior characteristics through a rolling test rig at the INLETS Research Institute in France. In 1979, in Grenoble a full scale U-type LIM for 1 MW and 200 km/h was manufactured and tested for the first time in France. In 1982, it was agreed that development of a short stator type maglev be proceeded under French-West German cooperation[11]. Now, a project of STARLIM is in progress in which the Germans take charge of a suspension/guidance system, and the French a U-type LIM system.

An EMS system developed in West Germany is applied in the STARLIM suspension system, in which suspension and guidance are independently controlled by separate magnets. The structure of the U-type LIM as a driving system, unlike a single-sided system used for another short stator type LIM, is of three-phase type without coil ends whose core is wound on the outer round surface. The secondary conductor is in the shape of a U-type groove into which a motor is held keeping a certain air gap. In order to decrease leakage flux an aluminium or copper damping cage is installed on the upper portion of motor coil. Main features are: considerably smaller dimensions than single-sided LIMs; attraction force, which generates between secondary conductors, around both legs of U-type LIM offset by itself. In case of 10 mm air gap, efficiency by power factor makes about 0.5 to 0.55 within a highspeed range. In the meantime, the outer side of U-type secondary conductor is made of steel panel; both sides are also used as guidance rails for maglev vehicles.

The main features of a proposed vehicle are as follows: 2.9 m height, 2.6 m width, 30 m length, 164 seats, 150 to 200 km/h maximum speed, 700 to 1050 kW output of LIM. Bogie truck supporting the vehicle is installed with 1.5 m long U-type LIM at the central portion and 3.0 m long magnet for suspension/guidance at both sides[12].

4.3 Maglev System Applying New Concepts

Some new levitation systems with various structures other than those of already developed ones have been proposed.

(1) The concept of a Mixed-MU Levitation System is proposed in Great Britain. This system consists of a magnet system in which on-ground steel rail is sandwiched between on-board superconducting coils and on-board superconducting screen is located facing rail surface[13]. Steel rail and coils generate an attraction force which induces a suspension force with a problem of no vertical and lateral stabilities, characterizing this system. This problem is solved by the superconducting screen which shields magnetic field making use of Meissner effect. In this solution, because of a local reversal of the field gradient, the variation of a force with vertical and lateral displacement acts for safety. This system has the advantages that the suspension force is maintained during stoppage and a large air gap between vehicle and the ground is available. As for the problem of eddy-current loss suffered on the ground rail during running, it is proposed that rail be laminated to reduce

the loss.

(2) A Mixed-f concept has been suggested recently in West Germany(14). As with Mixed-mu Levitation System, the structure with superconducting screen is adopted also for this system. This proposal is feasible if multiple pole arrangement is taken for permanent magnets with the number of rail elements increased. However, a high temperature superconductor of which Tc is preferably 300 K must be used to gain a sufficient spring constant.

(3) A concept of Alternating-Gradient Attractive Levitation has been proposed recently in the U.S.A.(15). On-board superconducting magnets are installed so as to generate attraction and suspension from the bottom side during running. This system is said to have the advantanges of dynamic stability being kept under a certain condition, the same large air gap as EDS, low magnetic drag, no need for current control and the like.

(4) For the existing MLU systems, on-ground coils contribute to propulsion /guidance and levitation. Meanwhile, for a newly proposed system, side wall suspension divides these functions into propulsion and supension/guidance. Moreover, though practical problems are not yet straightened out, the author and others are working on a system in which the ground coil is exclusively responsible for all functions of propulsion, suspension and guidance.

(5) Other than superconductivity applications above mentioned, a hybrid system combining permanent magnets with electromagnets is proposed to decrease control power for electromagnets(14). Meanwhile in the U.S.A. a short stator linear synchronous unipolar motor on-board system which is available for propulsion, suspension and guidance has been proposed(16).

5. Conclusions

The development of maglev system is reaching practical stage in the transition period of 1990s to the 21st century. The development of magnet, regardless of superconducting or normally-conducting as a main factor of maglev system, seems to become increasingly important for improving its performance in terms of light weight, small size, low consumption of power

and high reliability. In addition, if high-temperature superconductors come to be applied, it can be foreseen that the new systems will be further promoted. I would like to expect continued growth of magnetic technology.

References

(1) Nenadovic, V and Riches, E.E.: "Maglev at Birmingham Airport; from system concept to successful operation", GEC Review, Vol.1,No.1(1985)

(2) Ohishi, A: "HSST-05 system; general and operational outline at YES'89", International Conference Maglev'89 (1989)

(3) Bohn, G and Alser, H.: "The magnetic train TRANSRAPID 06", International Conference on Maglev & Linear Drives (1986)

(4) Friedrich, R., Dreiman, K., Leistikow, R., Böhm, E. and Weller, A.: "Propulsion and power supply system of the TRANSRAPID 06 vehicle design and test results, Part I: Propulsion", International Conference on Maglev Transport '85 (1985)

(5) Bohn, G and Steinmetz, G.: "The electromagnetic suspension system of the magnetic train Transrapid", International Conference on Maglev Transport '85 (1985)

(6) Miller, L.: "The maglev transportion system TRANSRAPID and ULIMAS", International Conference on Maglev & Linear Drives 1986 (1986)

(7) Meins, J. and Ruoss, W.: "Test result of the levitation system (Transrapid vehicles TR 06 and TR 07)", Tenth International Conference on Magnetically Levitated Systems 1988 (1988)

(8) Jizo, Y., Fujiwara, S. and Nemoto, K.: "Superconducting magnet of new test vehicle MLU 002 or Japanese EDS system", Tenth International Conference on Magnetically Levitated System 1988 (1988)

(9) Tanaka, H.: "Application of Superconductivity to Transportation: Magnetically Levitated Train", 1st International Symposium on Superconductivity 1988 (1988)

(10) Dreimann, K.: "The M-Bahn maglev rapid transit system; Technology, Status, Experience", IEEE 1987 (1987)

(11) Fintescu, N.D. and Pascal, J.P.: "Tests results of full-scale 1 MW linear induction motor with PWM inverter", International Conference on Maglev & Linear Drives 1986 (1986)

(12) Projet de Coopération Rapport final, STARLIM (Source; Pascal, J.P., INRETS)

(13) Jones, D.I., Pattullo, A.W. and Paul, R.J.A.: "Assessment of eddy-current effect in the Mixed-MU Levitation System", Tenth International Conference on Magnetically Levitated System 1988 (1988)

(14) Weh, H.: "Magnetic levitation technology and its development potential", International Conference Maglev '89 (1989)

(15) Hull, J.R.: "Attractive levitation for high-speed ground transport with large guideway clearance and alternating-gradient stabilization", Intermag '89 (1989)

(16) Johnson, L.R., Rote, D.M., Hull, J.R., Coffey, H.T., Daley, J.G. and Geise, R.F.: "Maglev vehicles and superconductor technology; Integration of high-speed ground transportation into the air travel system", Argonne National Laboratory (1989)

PRESENT ACHIEVEMENTS AND PROSPECTS FOR SUPERCONDUCTING TOKAMAKS IN THE WORLD

P. Komarek
Kernforschungszentrum Karlsruhe GmbH
Institut für Technische Physik
Postfach 3640, D-7500 Karlsruhe, FRG

ABSTRACT

The operation of tokamaks with superconducting toroidal field coil systems is providing information how to integrate a large cryogenic component in the real environment of a tokamak and should demonstrate the reliable performance of superconducting magnet systems in all desired and even unexpected operation modes of a fusion device. While the large tokamaks Tore Supra and T 15 are just in different advanced stages of initial operation, the small tokamak TRIAM-1M is already fulfilling this purpose. Simultaneously, mission oriented technology development effort is carried out worldwide, partially coordinated and in any case unified in the strategy by the IAEA/ITER design work.

INTRODUCTION

The tokamak is the plasma confinement system studied most intensively, with good prospects for the first generation of nuclear devices, planned for the next decade, e.g. ITER, NET, FER and later reactors. Thus, it is obvious that the implementation of superconductivity for tokamak magnet systems is an issue since many years. Two simultaneous efforts can be reviewed, firstly the construction of small (T 7, TRIAM) and medium large tokamaks (T 15, TORE SUPRA) with superconducting toroidal field (TF) coils and the technological developments with demonstration of the magnet availability, directly for the generation of engineering test reactors. Both approaches complement each other for providing confidence in the use of superconductivity in fusion devices.

CONSTRAINTS FOR SUPERCONDUCTIVITY IN PLASMA PHYSICS EXPERIMENTS

The plasma physicists were rather cautious concerning the acceptance of superconducting magnets in their experiments in the past. Critical questions in comparison to copper magnets were
- sufficient high overall current density in spite of the additional space demand for the cryostat;
- stability of the superconducting state in case of fast changing magnetic field transients (e.g. plasma disruptions);
- sufficient space for diagnostics (access to the plasma chamber) in spite of the cryostat needs;
- cryogenic cooling needs;
- reliability and industrial availability of the s.c. magnet technology;
- reasonable costs.

Answers to these questions in the early seventies, especially for small machines with short plasma burn time and relatively few operation hours for the magnet system per year, were such that it was well justified to construct such machines with copper coils and flywheel generator power supplies. In the recent years, when larger tokamaks with longer plasma burn times became desirable, superconducting magnet systems started to become competitive or even preferable, for the TF-coils first. Furtheron, all designs for the next generation tokamaks show clearly the need for a construction with superconducting coils for the toroidal (TF) and poloidal field (PF) coil systems.

The different development approaches, for the tokamaks of the present generation (T 15, TORE SUPRA) on one hand and the large systems for the next decade on the other hand, show the broad spectrum and the evolution on design solutions for the superconducting systems, meeting the different needs. They are discussed in the following chapters.

THE DIFFERENT APPROACHES USED FOR SERVING PRESENT TOKAMAK EXPERIMENT NEEDS WITH S.C. MAGNETS

Tokamak experiments under operation now are based on designs developed 6 - 10 years ago. Thereby, two large devices with superconducting TF-coils have been constructed in the recent years, namely TORE SUPRA at Cadarache, France [1], and T 15 at the Kurchatov Institute at Moscow, USSR [2]. Beside these devices, a small but remarkable superconducting tokamak, TRIAM-1M [3] has been constructed and is under operation successfully at the Kyushu University in Japan. As this project and its achivements are discussed in a separate contribution [22] this article shall focus on the discussion of results from T15 and TORE S.

Design Approaches and Developments

Tore Supra and T 15 are tokamaks of very similar size and similar main plasma physics parameters. Therefore, it is very interesting to observe the different approaches which have been taken to meet the similar specifications for the TF-coils. Table 1 shows the machine parameters as far as they are relevant for the TF-coil system. For the problem to achieve a magnetic field of 9 $\div$ 9,3 T on the winding at an overall current density of 30 - 40 MA/m^2, two very different solutions were developed. The TORE SUPRA group based their design on the well available and known NbTi conductor, but had to implement the less established HeII cooling to achieve the goal, the T 15 group trusted the benefits of forced flow HeI cooling, but had to implement the so far less developed Nb_3Sn conductor technology. Both groups were aware about development needs and accompanied the design work with extended and careful development programmes. Table 2 shows an attempt to compare the arguments for and against each solution, based on the knowledge when both groups started the projects. While the main "risks" for the TORE SUPRA approach could be seen in the HeII cryogenics, for the T 15 approach it was the rather early development status of Nb_3Sn conductor and winding technology. Table 3 gives a survey on the selected conductor and winding parameters verified by the development programmes. Some characteristics of each solution can be seen. The stability margin in the case of TORE SUPRA is mainly based on the enthalpy content of the He and the large heat transfer area, in the case of T 15 it is based on the current margin and the large copper content.

The results of the developments have been presented on several earlier conferences and need not to be reviewed again [5,6].

Coil Tests during Construction

During the construction phase prototype coils and the coils delivered from the serial fabrication have been tested in large facilities built up for this purpose at CEN Saclay [7] and the Kurchatov Institute [4] respectively. All TORE SUPRA coils passed the tests in accordance to the specifications, in the T 15 coil tests beside some discharge voltage limitations due to noninsulated parts, some pecularities of the Nb_3Sn became visible. The observation was, that all coils reached the envisaged test current, but showed losses in the range of 20 - 80 W each, indicating some resistivity [8]. This was probably not a

TABLE 1

The main characteristics of TORE SUPRA, T 15 and TRIAM-1M in respect to the TF-coils [1,2,3].

	TORE SUPRA	T 15	TRIAM-1M
Plasma major radius R [m]	2.25	2.43	0.8
Plasma minor radius a [m]	0.70	0.70	0.12 x 0.18
Plasma current I_p [MA]	1.7	1.4 - 2.3	0.5
Typical plasma discharge duration [s]	30 - 120	26	0.5
Toroidal magnetic field at the plasma centre B_T [T]	4.5	3.5 ÷ 5.0	8
Maximum magnetic field at the winding B_m [T]	9.0	6.5 ÷ 9.3	11
Average current density in the winding $\langle j \rangle$ [MA/m^2]	40	33	57.4
Coil shape	circular	circular	D
Toroidal field coil mean diameter [m]	2.6	2.59	0.74 x 0.97
Number of coils	18	24	16
Total magnetic energy in the TF-coils [MJ]	600	380 - 790	76

big surprise, because in the meantime more detailed knowledge about reversible and irreversible strain degradation of Nb3Sn was accumulated by many laboratories around the world. In addition, the large Nb3Sn coil built and operated within the LCT project showed a similar behaviour [9]. The T 15 group explained the resistivity as determined by conductor defects, which appeared at the different stages of the manufacturing process of the conductor and the windings. To the author, this is a reasonable explanation but it is not necessarily the only one for reaching such a resistivity. From the extended measurements on Nb3Sn conductors worldwide, we also know now two principle facts which together can explain a certain remaining resistivity in Nb3Sn coils. Firstly, the critical current I_c shows a strong strain dependence under vertical or perpendicular forces. The overall I_c-increase or degradation is different for different conductor composite types, but in principle a degradation occures as soon as the filaments themselves are loaded by tension or compression [10]. The second effect concerns the voltage current (U-I) characteristic, where the transition around I_c

can be much less steep for Nb3Sn than for NbTi, with "measurable" voltages on short pieces already far below the defined one for I_c [18]. Figure 1 describes the situation qualitatively. In the simplest way the transition can be described by $U \propto I^n$. The "n-factor" for NbTi wires is usually rather high (50 ÷ 100), indicating a steep transition. For unstrained Nb3Sn-conductors the n-fac-

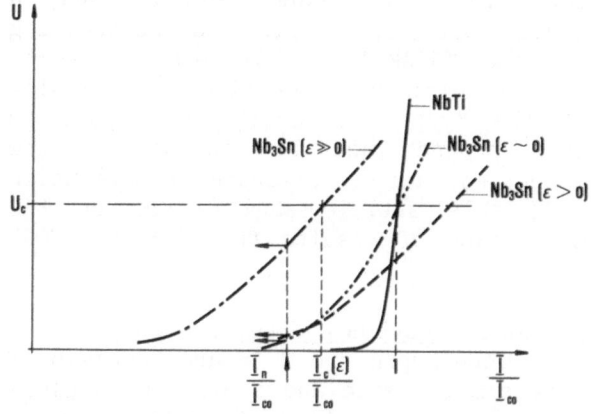

FIGURE 1 Differences in the voltage current characteristics of NbTi and Nb3Sn wires

TABLE 2
Comparison of arguments for and against the different design approaches selected for TORE SUPRA and T 15 coils respectively, based on the state of the art at the time of the beginning of the projects.

	TORE SUPRA	T 15
Conductor material	NbTi: well developed and available from industry in large quantities; performance well known.	Nb_3Sn: less developed and investigated in practical use; implies some uncertainties in the performance yet.
Conductor composite	tailoring and fabrication based on industrial standard	tailoring needs fabrication developments, involving time and costs.
Conductor handling	well known and with predictable risk	difficult and more risky
Winding manufacturing	as executed for many earlier magnets of pool boiling type	standard in electrical engineering of water cooled windings, but conductor treatment and connections demanding.
Cryogenics	Principles of pressurized HeII pool boiling known, but not on an engineering standard; → no information on reliability yet.	Forced flow HeI cooling known from a few earlier magnet systems, but two phase flow thermohydraulic needs additional care; → experience limited.
Cooling mode	All general advantages and disadvantages of pool boiling magnet systems implied.	All general advantages and disadvantages of forced flow cooled magnets implied.
Scalability of the design approach	It was aimed that the technological option NbTi + pressurized HeII at 1.8 K can be extrapolated up to larger fusion devices.	The higher development effort has been seen justified by the large potential of Nb_3Sn for all future tokamaks.

tor, especially in high field, is already lower ($\sim 20 \div 50$), but for a Nb_3Sn-conductor under strain, beside I_c also n is further degraded. Thus, if such a conductor is operated e.g. at 75 % of I_{c_0} (I_{c_0} ... unstrained short sample critical current) and this I_{c_0} is defined as usual e.g. at $U_c = 1 \mu V/cm$, a voltage of $0.1 \div 0.5 \mu V/cm$ could easily be present at the operation current. Let e.g. 200 m of such a conductor with 5000 A rated current be in the high field region, a resistive load of $10 \div 50$ W will be present. Thus, this effect can also contribute to the observed behaviour of the T 15 coils (and earlier of the Westinghouse LCT coil). This is supported by the observation at the T 15 coil tests, that at coils with low resistive losses (20 - 30 W) the voltage was well distributed over the whole length of the winding. Coils with higher losses (~ 80 W) showed an unequal voltage

distribution with peaks on certain pancakes, which indicated that really some defects due to manufacturing steps might be present too [19]. Of course, detailed knowledge about the strain behaviour of the T15 conductor and the actual strain distribution within the winding in loaded condition would be needed to clarify which contributions dominate the observed resistivity values.

Operation Experiences

In the meantime the TORE SUPRA tokamak started its operation and further experiences with the magnet system have been made. From March to July 1988, in an initial plasma operation mode, an operation current of 600 A was used to yield a 2 T field on the plasma axis. Within this period no superconductivity problems occurred, even at several discharges (on purpose) in semi-rapid mode. It was already plan-

TABLE 3

Characteristic data of the conductors, windings and cryogenic supplies of TORE
SUPRA and T 15 [1,2,4]

	TORE SUPRA	T 15
Superconductor	NbTi	Nb_3Sn
Stabilizer	Cu/CuNi	Cu
Dimensions [mm]	2.8 x 5.6	18 x 6.5
Copper to superconductor ratio	2:1	30:1
He fraction within the winding / conductor [%]	28.6	2.6
Rated current at 9/9,3 T [kA]	1.4	5.6
Critical current [kA] at field B [T] and temperature T [K]	1.4/9/4.2	10/8.5/4.2
Single conductor length in the winding [m]	622	340
Transient magnetic field in a disruption ΔB [T] τ [ms]	0.6 10 - 20	0.7 20
Number of turns in the winding	2028	456
Refrigeration power at 80 K [kW] at 4 K [W] at 1,75 K [W]	40 650 + 100 lHe/h 300	165 1200 + 70 lHe/h -

ned to go stepwise to the full rated current (1400 A), when in a fast discharge (not caused by a quench) from the 600 A level one coil became partially damaged. After careful evaluation, the only hypothesis which the group sees in spite of its very low probability, is the occurrence of a shortcircuit caused by a small metallic piece, followed by an arcing over one double pancake [20]. It was concluded from the measurements, that about 200 kJ had been dissipated in this arc, important enough for some damage, but without being totally disastrous. The effect is surprising because all coils had passed acceptance tests including discharges with voltages two times larger than those in the discharge mode. Thus, the metal piece - if one - must have been brought into this position through the He-supply. The event is of course a warning concerning some remaining risk on the continuous voltage capability of pool boiling coils in spite of good experiences with many such magnets in the past.

To avoid an early shut down before the initial plasma physics programme has been finished successfully, the TORE SUPRA group invented a unique operation scheme, demonstrating simultaneously a very stable performance of the HeII cooled magnets. While all other 17 coils remain in a continuous operation mode, the damaged coil is operated in a pulsed mode (as a resistive coil) with a 10 s plateau for the plasma operation. In such a cycle about 20 kJ are dissipated in this coil, increasing the HeII temperature from 1.8 K to 1.98 K only. This can be measured by careful temperature recording with sensors located in the cryogenic satellite. The available refrigeration power allows to operate in this pulsed mode with a repetition rate of 20 minutes. Many plasma runs had been performed in this mode until April 1989. In the scheduled shut down until September 1989 the damaged coil will be replaced by a spare coil and the damaged one will be investigated later on in all details to clarify finally the reason for damage.

The situation at T 15 is unfortunately not so well advanced. A first attempt to cool down the magnet system within the completed tokamak installation was carried out in December 1988.

It was not possible to go below 11 K, probably by limitations of the cryogenic vacuum, high heat losses in the He transfer lines and perhaps limited refrigeration power. Measures to improve these and further identified items, as high voltage insulation and compressor performance, need some time and a further cool down will not be possible before autumn 1989.

THE RUNNING DEVELOPMENTS FOR THE NEEDS OF NEXT GENERATION EXPERIMENTS

Development Strategy and Earlier Achievements

There exists a worldwide consensus since many years, that the next generation of large tokamaks which aim to work as engineering test devices, must be constructed with superconducting coils. For such machines national and international designs have been carried out (FED, OTR, INTOR) or are proceeding (NET, FER, ITER) and decisions about a construction can be expected for the first half of the next decade. These activities have been accompanied by extensive superconducting magnet development programmes from the beginning. Justification for that effort is, that as soon as the construction of such a tokamak will start, the magnet technology must be available on a proven status with sufficient know how for serial fabrication and design tools guaranting reliable margins for the magnet operation. The magnet system is one of the major components of the basic machine and must be considered as "semipermanently" installed, thus its reliability and availability must be very high already from the beginning of machine operation.

The national efforts in these developments became embedded in some international ones in the last decade and the strategy a rather common one in all nations. The successful completed IEA-Large Coil Task can be seen as an encouraging milestone and prerequisit for the present development tasks and test facilities [9]. Table 4 summarizes the development strategy.

Conductors

Many different conductor types have been designed and developed on a labora-

tory scale. Within the present designs of NET and ITER the variety on different designs has been narrowed down to a few basic types, as indicated in Figure 2 [11]. All conductors are forced flow cooled and have a thick steel conduit which acts not only as the helium pressure vessel, but also as a mechanical force transmission structure within the winding. Nb_3Sn is the reference superconducting material, but NbTi - if necessary forced flow HeII cooled - the back up, or for the outer ring coils, even an equally suited choice. For the inner ring coils (OH-coils) a high rated current with a rather small bending radius favour coils in wind and react technique with advanced successors of the earlier "cable in conduit" conductors or, as alternative solution, NbTi-cable conductors with HeII cooling. For the TF-coils in react and wind technique, rectangular conductors with a flat Nb_3Sn cable in the neutral bending zone, covered in a sophisticated manner by the stabilizer including cooling channels, are the common choice. The conductor designs differ only in the way how the stabilizer and the cooling channels are arranged, in accordance to different preferred fabrication techniques. For the case that a wind and react construction is considered for the TF-coils too, the same conductor type as for the OH-coil can be considered. Within the development the optimization of reliable fabrication on long length is a major and expensive task, which needs involvement of motivated industry. It might become a decisive issue for the conductor choice.

Test Facilities

Important are most relevant tests for these conductors to prove their capabilities. Beside experimental arrangements to measure j_c (B, ε), a.c. losses and stability, a few large facilities to test certain lengths of full size conductors are needed and are under construction. One can mention here especially the SULTAN III facility at PSI, Switzerland [12] and the FENIX facility at LLNL, USA [21].

Ultimate integral tests of long lengths of conductors and the winding technique should finally be done by

TABLE 4

The worldwide development strategy of superconducting magnets for the next
generation fusion devices (beside the operation of medium size tokamaks with s.c.
TF-coils)

Task	Goals	Status
LCT	Demonstrate reliable operation of large s.c. TF-coils and prove design and fabrication principles, by operating a compact torus.	Successfully completed.
Conductor development	Alternative designs and proto-type fabrication for NET, FER, INTOR, ITER and others.	Concentrated now on few full size prototypes for NET / ITER.
Conductor test facilities	Test of short / medium lengths of the developed conductors, mainly j_c $(B, \varepsilon, \dot{B}$; $Q_{disturbance})$	Several test stands available or under construction, up to large facilities as SULTAN III and FENIX.
Model and proto-type coil test facilities	Integral test of conductor and winding techniques under most relevant boundary conditions.	Medium size facilities under operation (SULTAN II, DPC) or construction (POLO); large facilities under design (e.g. TOSKA-Upgrade).

operating model or prototype coils in a
most relevant test environment. A
successful test stand for subsize TF
conductors was SULTAN II at PSI,
Switzerland [13]. For poloidal conductors the test facility DPC at JAERI,
Japan [14] and the POLO project of KfK,
Germany with CEA, France [15] can be
mentioned in this respect. Large facilities for coil testing are under
preparation at JAERI, Japan [14] and
KfK, Germany [16]. The TOSKA-Upgrade
facility at KfK e.g., will have a three
phase operation within the EURATOM/NET
technology programme. In stage 1 the
KfK-LCT coil will be reinstalled with
stronger reinforcement and a cooling
circuit for forced flow HeII. Goal is to
demonstrate the high field capability of
the well proven NbTi coil technology by
subcooling. If successful, this magnet
technology can be declared as fully
available for future tokamak systems
with fields up to 11 T at the coils. In
the second stage, test pancakes (model
coils) coming out of the NET coil development will be installed adjacent to
the LCT coil, which provides then the
background field for first testing of
the model coils. In the 3rd stage all
model coils will be mounted together,
forming a large 12 T solenoid for the
ultimate tests of all windings, also in
a pulsed mode (by transfering energy
from one coil part to another one).

CONCLUSIONS

The operation of tokamaks with
superconducting TF-coil systems is providing information how to integrate a
large cryogenic component in the real
environment of a tokamak and should
demonstrate the reliable performance of
superconducting magnet systems in all
desired and even unexpected operation
modes of a fusion device. While the
large tokamaks TORE SUPRA and T 15 are
just in differently advanced stages of
initial operation, the small tokamak
TRIAM-1M is already fulfilling its purpose [22].

Simultaneously, mission oriented
technology development effort is carried
out worldwide, partially coordinated and
in any case unified in the strategy by
the IAEA/ITER design work [17].

All these activities together form
a sound basis for the construction of
the magnet systems for the next generation of tokamaks, which aim to operate as
fusion engineering test facilities. The
magnet system as major component of the

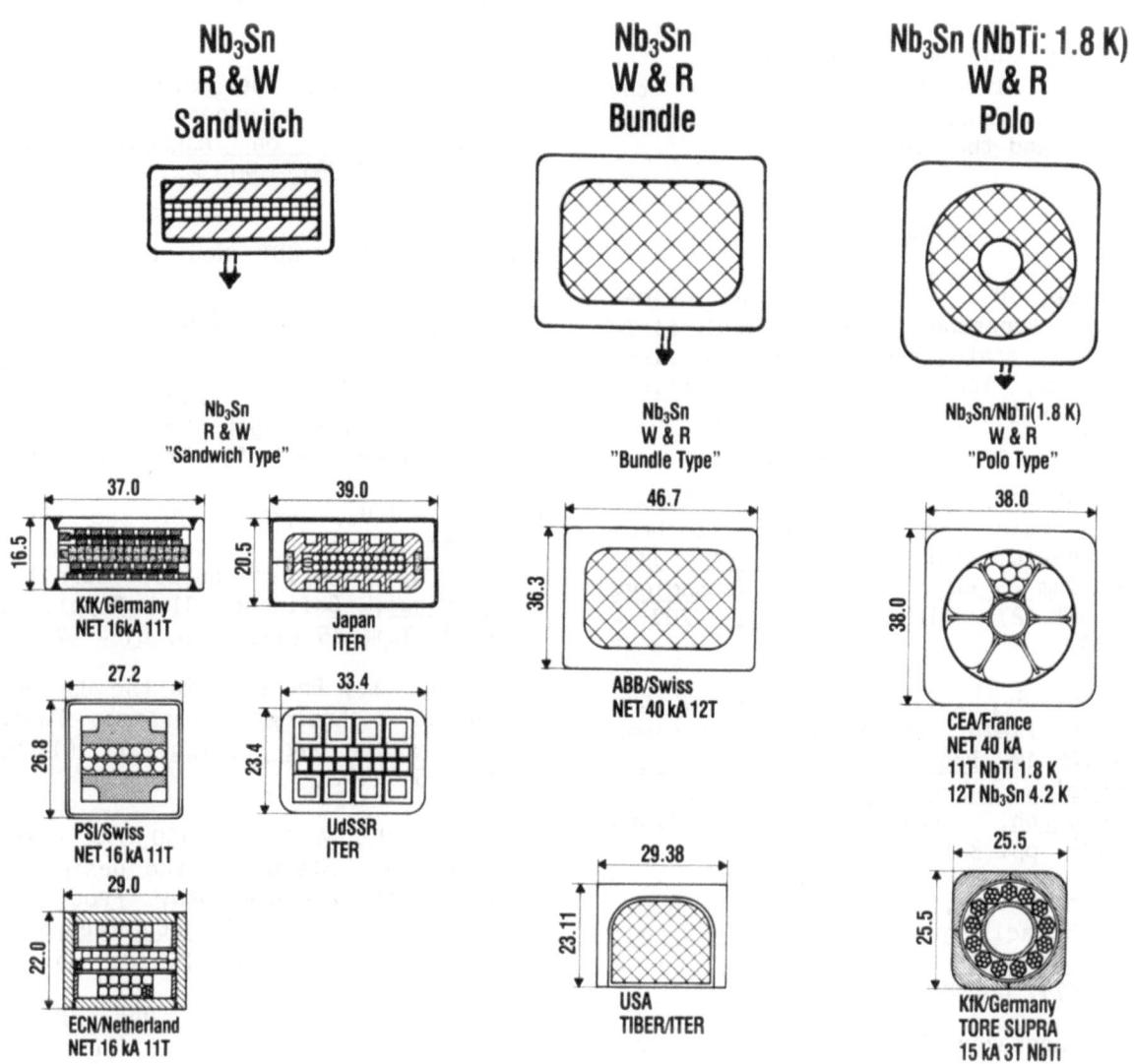

FIGURE 2 Conductors envisaged for NET/ITER

basic device must be available on a proven engineering basis from the beginning, which means in the first half of the next decade. This is a strong justification for all the effort reviewed in the previous chapters.

The already achieved advanced stage of the development is of benefit to other confinement approaches too and encouraged certainly the plan to built a new large helitron device near Nagoya, Japan, with superconducting magnets. Also the envisaged next large stellarator at IPP Garching, Germany, (W7X) will probably be constructed with superconducting coils, too.

The big challenge and the large resources needed, call for joint international tasks to share the capabilities and to optimize the chances for necessary progress in reasonable time. This has e.g. executed very successfully at the IEA Large Coil Task and is proceeding in several bilateral collaborations and in the coordinated ITER technology programme. Further international agreements are under discussion. This gives confidence that the required goals can be met in time.

ACKNOWLEDGEMENT

The author likes to thank very much Dr. B. Turck (CEN Cadarache) for providing many information about TORE SUPRA in an excellent prepared manner. Very valuable informations about T15 could be collected during a visite at the Kurchatov Institute, based on the scientific exchange agreement between the USSR and the F.R.G.

REFERENCES

1. Turck, B., TORE SUPRA: a tokamak with superconducting toroidal field coils; status report after the first plasma, IEEE Trans. on Magnetics, Vol. MAG-25, No 2 (1989), pp. 1473 - 1480.

2. Kadomtsev, B., T 15 installation. The main characteristics. Electromagnetic system, Proc. 12th Symposium on Fusion Technology (SOFT 12), Jülich 1982, pp. 207 - 218.

3. Itoh, S., Initial operation of the high field superconducting tokamak TRIAM-1M, Proc. Int. Conf. on Plasma Physics and Controlled Nucl. Fusion Research, IAEA-CN-47/H II-3, Kyoto 1986, pp. 321 - 331

4. Chernoplekov, N.A., Monoszon, N.A., T 15 facility and tests, IEEE Trans. on Magnetics, Vol. MAG-23, No 2 (1987), pp. 826 - 830.

5. Chernoplekov, N.A., Status and trends of s.c. magnets for thermonuclear research in USSR, IEEE Trans. on Magnetics, Vol. MAG-17, No 5 (1981), pp. 2158 - 2166.

6. Aymar, R., TORE SUPRA-Status report concerning the superconducting magnet after the qualifying development program, IEEE Trans. on Magnetics, Vol. MAG-17, No 5 (1981), pp. 1911 - 1914.

7. Ribaud, P., Tests of the toroidal field coils of TORE SUPRA, Proc. 14th Symp. on Fusion Technology (SOFT 14), Avignon 1986.

8. Cheverev, N.S., T-15 results testing of systems and parts, Proc. 15th Symp. on Fusion Technology (SOFT 15), Utrecht 1988.

9. Beard, D.S., The IEA Large Coil Task, Fusion Engineering and Design, 7 (1988), No 1,2.

10. Flükiger, R., The material aspects in advanced superconducting wires, IEEE Trans. on Magnetics, Vol. MAG-24 (1988), No 2, pp. 1019 - 1022.

11. Toschi, R., NET and the European Fusion Technology Programme, Fusion Engineering and Design, 11 (1989), pp. 47 - 62.

12. della Corte, A., The SULTAN III project, Proc. 15th Symp. on Fusion Technology (SOFT 15), Utrecht 1988.

13. Elen, J.D., Upgrade of the SULTAN superconducting test facility to 12 T by three A-15 coils, Journal de Physique, Colloque C1, Suppl. on No 1, Tome 45 (1984), p. C1 - 97.

14. Iijima, T., Progress in the Japanese Fusion Technology Programme, Fusion Engineering and Design, 11 (1989), pp. 35 - 46.

15. Jeske, U., A 2 MJ, 150 T/s pulsed ring coil. Status of the design and the test arrangements, Proc. 9th Int. Conf. on Magnet Technology (MT-9), Zürich 1985, pp. 32 - 35.

16. Hofmann, A., Further use of the Euratom LCT coil, Proc. 15th Symp. on Fusion Technology (SOFT 15), Utrecht 1988.

17. Clarke, J., The ITER activities, Proc. 12 Int. Conf. on Plasma Physics and Controlled Nucl. Fusion Research, IAEA-CN50, Nizza 1988.

18. Schneider, T. (KfK), unpublished.

19. Kostenko, A., private communication.

20. Turck, B. (CEN Cadarache), private communication.

21. Miller, J., unpublished information within the ITER workshops.

22. Nakamura, Y., Nagao, A., Hirahi, N., Itoh, S., Reliable and stable operation of the high field superconducting tokamak TRIAM-1M, this conference, paper MD-01.

WORLD'S ACHIEVEMENTS IN THE DEVELOPMENT OF SUPERCONDUCTING MATERIALS AT LIQUID HELIUM TEMPERATURE

K. Inoue
National Research Institute for Metals
1-2-1 Sengen, Tsukuba-shi, Ibaraki 305, Japan

ABSTRACT

Ductile Nb-Ti superconducting alloy is practically used for generating relatively low magnetic fields. On the other hand, brittle Nb_3Sn and V_3Ga superconducting compounds are practically used in high fields. Remarkable improvements have been achieved on the critical current densities, J_c, and the a.c. properties of the Nb-Ti multifilamentary wire. Nb-tube processed $(Nb,Ti)_3Sn$ multifilamentary conductor shows the best high-field J_c among the multifilamentary superconductors, although a similar high J_c is obtained for improved surface-diffusion V_3Ga tape at 17 - 20 T and 4.2 K. A Nb_3Al ultra-fine multifilamentary conductor developed recently is worth noticing as a promising candidate of the new practical superconductors.

INTRODUCTION

Since the discovery of high T_c oxide superconductors, the developing rate of metallic superconducting materials has become slow, because the interests of many material researchers shifted from the metallic superconductors to the oxide superconductors. Recently the boom of oxide superconductors has slowed down a little. Consequently R & D studies on metallic superconductors are being restored gradually. Therefore the recent achievements in the development of metallic superconductors are relatively poor. However some certain improvements of J_c have been achieved in several practical superconductors.

J_c's of 3000 - 3800 A/mm^2 at 4.2 K and 5 T have been attained for Nb-Ti multifilamentary conductor with Nb-diffusion barriers between the Nb-Ti filaments and the Cu matrix [1]. The desirable combination of Ti-addition and Nb-tube process have realized overall J_c's of 500 - 700 A/mm^2 at 4.2 K and 16 T without Cu in the Nb_3Sn multifilamentary conductor [2], which are the best high-field J_c's among those of the commercialized multifilamentary superconductors. Also similarly high J_c's are obtained for the improved surface-diffusion processed V_3Ga tapes at 17 - 20 T and 4.2 K [3].

Ultra-fine Nb-Ti multifilamentary wire

with Cu-Ni matrix is being developed as an a.c. superconductor [4]. The recent progress in lowering a.c. losses in superconductors should make such a.c. superconducting applications as superconducting transformers and superconducting armatures possible.

Ultrafine Nb_3Al multifilamentary wire has been fabricated by drawing a composite of Nb matrix and Al alloy cores, and it is worth being noted as the most promising candidate for a new practical superconductor [5]. The reason is that the wire shows excellent J_c, as high as that of practical Nb_3Sn multifilamentary wires, as well as a superior strain effect. The remarkable improvements of $\vec{J_c}$ above 20 T are achieved for the liquid-quenched Nb_3Al-based compounds [6] and the powder-processed $PbMo_6S_8$ compounds [7].

PROGRESS IN PRACTICAL NB-TI SUPERCONDUCTOR

Since the fine α-Ti precipitations, formed by annealing Nb-Ti alloy at about 400 ℃, were found to act as effective pinning centers, the Nb-Ti alloy has been the major practical superconductor for generating magnetic fields below 9 T. The reason is that the excellent mechanical properties of the Nb-Ti alloy make the engineering aspect of magnet fabrication much simpler than that of the other superconducting materials. The fabrication of Nb-Ti composites is much simpler also.

The enhancement of J_c has been quite successful during the past decade in Nb-Ti composites, and contributes to substantial savings in the cost of superconducting magnets. These improvements of J_c have been obtained mainly by optimizing both conditions of heat treatment at 350 - 450 ℃, and the process of cold-drawing between the heat treatments. The reasons are that the size, the shape, and the density of the α-Ti precipitations, which determine the J_c of the Nb-Ti alloy through the pinning mechanism, can be controlled by both the heat treatment conditions and the cold-drawing conditions. On the other hand the uniformity of the Nb-Ti filaments (not only shape but also composition) in the longitudinal direction of the wire is also one of the important factors for obtaining high J_c's in Nb-Ti multifilamentary conductors.

However it is very difficult to realize these two factors simultaneously in practice, because Cu-Ti intermetallic compounds are formed during the fabrication process of the Nb-Ti/Cu composite, and cause sausaging in the filaments which results in the degradation of the transport current and, in the extreme case, the breakage of the composite wire. Nb diffusion barriers between the Nb-Ti filaments and the Cu matrix is very effective in preventing the formation of Cu-Ti intermetallic compounds. The use of highly homogeneous Nb-Ti alloy ingot as a row material is also effective in obtaining high J_c's. It has been successful at several companies to fabricate Nb-Ti multifilamentary wires with Nb diffusion barriers which exhibit J_c's at 4.2 K of 3000 - 3800 A/mm^2 at 5 T and 1000 - 1300 A/mm^2 at 8 T [1], as shown in Fig. 1.

Owing to the recent development of low a.c. loss Nb-Ti superconductors, it has become possible to design or produce the primary models of the a.c. superconducting machines such as transformers, limiters,

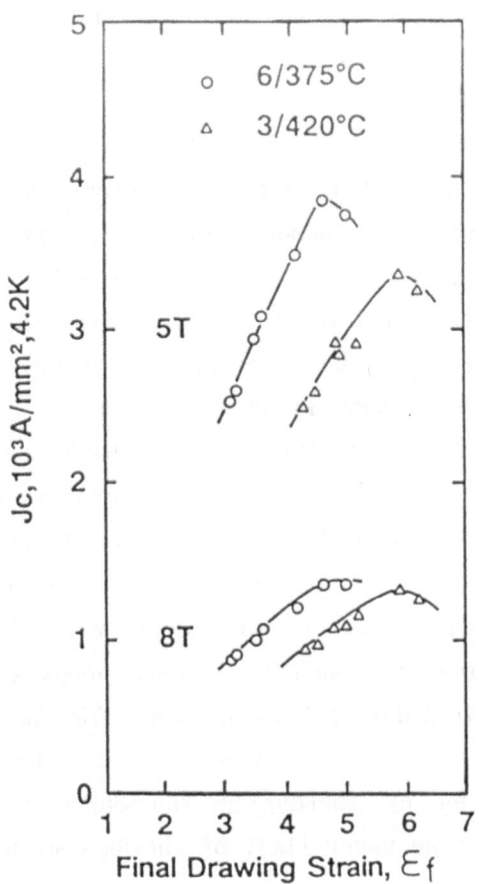

FIGURE 1 J_o of Nb-46.5 wt%Ti multifilamentary conductor with Nb diffusion barrier as a parameter of final drawing strain after several heat treatments [1].

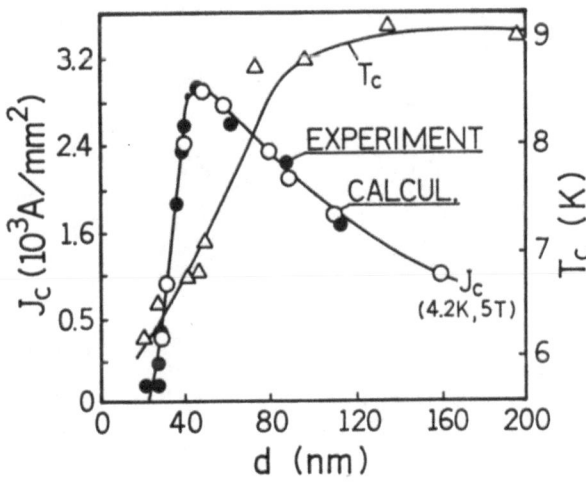

FIGURE 2 T_o and J_o vs filament diameter curves for ultra-fine Nb-Ti multifilamentary conductor [8].

armatures, and so on [4]. Lowering of the a.c. losses has been performed by reducing the Nb-Ti filament diameters to submicron size, decreasing the twist pitch to below 1 mm, and using highly resistive matrices such as Cu-Ni alloys. The 50/60 Hz losses in this kind of wires are reduced to about 100 kW/m³ at 1 T and 4.2 K.

In attempts to further reduce the a.c. losses by reducing the filamentary diameter, it was observed that proximity coupling of closely spaced Nb-Ti filaments led to an increase in losses. To eliminate the proximity coupling in the submicron Nb-Ti multifilamentary conductors with Cu-30%Ni matrix, the desirable filament spacing should be above 40 - 130 nm depending on the experimental results. The desirable filament spacing can be reduced by increasing the matrix resistivity or by the addition of magnetic impurities such as manganese. On the other hand, the proximity effects reduce the J_o and the T_o of Nb-Ti filaments, when their sizes are reduced below 45 nm and 100 nm, respectively, as shown in [8] Fig. 2.

Stabilization technology of a.c. superconductors has not yet been established. However, small a.c. coils, wound by a.c. Nb-Ti conductor with Cu-Ni matrix and without pure Cu, show relatively stabilized characteristics which seem to be explained by the fact that the size of the minimum propagating normal zone is comparable to those of Nb-Ti filaments [9]. Namely, the heat generated in the ultra-fine Nb-Ti filament may be absorbed rapidly by the surrounding Cu-Ni matrix. Rapid propagation rates of the normal zone have been also reported for the a.c. superconducting coils.

PROGRESS IN PRACTICAL Nb₃Sn AND V₃Ga SUPERCONDUCTORS

About 20 years ago, Cu was found to increase drastically the formation rate of V₃Ga and Nb₃Sn in the diffusion process at low temperatures. The short time/low temperature reactions lead to the formation of high J_c compounds with fine grains. It is generally believed that grain boundaries are the most effective pinning centers in A15 compounds such as Nb₃Sn and V₃Ga. The J_o's of surface diffusion-processed Nb₃Sn and V₃Ga tapes were improved at first by Cu addition. Then multifilamentary Nb₃Sn and V₃Ga conductors were developed by the bronze process and are used mainly for high-field magnets at present. Particularly bronze processed (Nb,Ti)₃Sn or (Nb,Ta)₃Sn multifilamentary conductors have become the major high field superconductors above 9 T [10], since Ti or Ta additions are found to improve the high-field superconducting properties of Nb₃Sn. On the other hand the V₃Ga conductors have been minor, because their fabrication cost is higher than that of Nb₃Sn conductor due to the expensive raw materials.

The bronze process has been industrially established technology but has some weak points. One of them is that many repetitions of the annealing process are necessary during cold-drawing, which raises the fabrication cost of Nb₃Sn conductors. Besides, the volume fraction of Nb₃Sn in the bronze processed conductor is relatively small, resulting in relatively small overall J_o. The reason is that the bronze used in the bronze process contains Sn of less than 8 at % and remains mostly as a non-superconducting and unnecessary portion in the conductor after the diffusion reaction.

To overcome these weak points in bronze process, several processes have been derived from bronze process and investigated as shown in Fig. 3. Among them [11] the internal Sn/modified jelly roll process, [12] the internal Sn process, and [2] the Nb tube process, are the most interesting ones by which to fabricate industrially multifilamentary Nb₃Sn conductors with high J_c. In these processes Cu and Sn instead of bronze are assembled into the starting composite. Both Cu and Sn can be cold drawn heavily without annealing, although the bronze needs to be annealed during cold-drawing. The Sn content in the composite can also be increased by increasing the Sn/Cu ratio, while the upper limit of the Sn content in

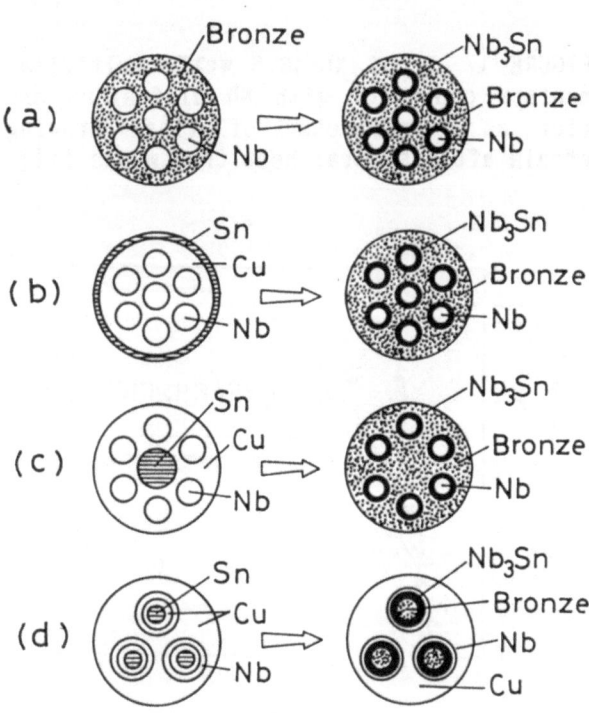

FIGURE 3 (a) Bronze process, and derived processes; (b) external Sn diffusion process, (c) internal Sn diffusion process, and (d) Nb tube process.

the bronze process is about 8 at % due to the poor cold-workability of Sn-rich bronze.

The desirable combination of Ti-addition and Nb-tube process has led to overall J_o(4.2 K, 16 T)'s of 500 - 700 A/mm^2 (without Cu) in Nb$_3$Sn multifilamentary conductors [2], which are the best high-field J_o's among the commercialized multifilamentary superconductors (Fig. 4). Typically, the Nb-tube process is as follows. Several single-core wires, consisting of a Nb-1%Ti tube with a Cu sheathed Sn core inside and a Cu tube outside, are bundled together into a Cu tube and extruded by hydrostatic extrusion. Then they are drawn to the final sizes without any intermediate annealing and finally reacted to form a (Nb,Ti)$_3$Sn layer inside each Nb tube filament. The Nb-tube processed (Nb,Ti)$_3$Sn multifilamentary con-

ductor was used for the innermost coil of a 20.1 T superconducting magnet operated at 1.8 K. The filament size in this wire is typically larger than 100 μm, for preventing the Nb tube breakage due to heavy cold-drawing. If the filament size can be reduced extremely in this process, many desirable improvements should be achieved in the J_o, a.c. losses, and mechanical properties.

Surface diffusion processed V$_3$Ga tapes are known to have large overall J_o's in high fields. The two-stage reaction, which consists of a first reaction at about 700 ℃ to form thick-enough V$_3$Ga layers and a subsequent second reaction at about 600 ℃ to achieve high H$_{c2}$, was found to improve the overall J_o(4.2 K) up to 300 A/mm^2 (without Cu) at 19 T [3]. The improved V$_3$Ga tape was used for the innermost coil of a 18.1 T superconducting magnet operated at 4.2 K.

Many other fabrication processes for Nb$_3$Sn and V$_3$Ga conductors have been also proposed and investigated; in situ process, infiltration process, powder metallurgy process and so on. Large overall J_o's and/or excellent mechanical properties have been reported for these superconductors. However more development is needed to produce industrial-scale stabilized superconductors through these processes.

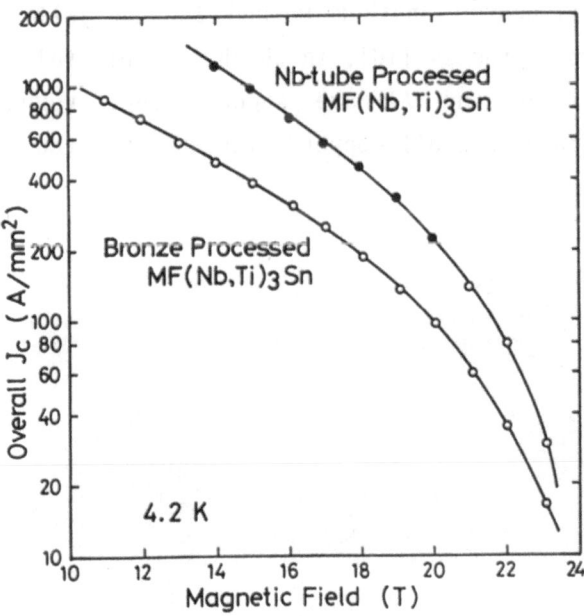

FIGURE 4 Overall J_o (without Cu) vs magnetic field curves at 4.2 K for the Nb-tube processed and the bronze processed (Nb,Ti)$_3$Sn multifilamentary conductors.

DEVELOPMENTS OF OTHER METALLIC SUPERCONDUCTING MATERIALS

Many metallic superconductors are known to have higher T$_c$'s or H$_{c2}$'s than those of Nb$_3$Sn and V$_3$Ga, e.g. Nb$_3$Ge, Nb-

₃Al, Nb₃(Al,Ge), Nb₃Ga, NbN, Nb(N,C), Pb-Mo₆S₈ and V₂(Hf,Zr). However, the formation rate of these compounds through diffusion processes is very slow at low temperatures compared to those of Nb₃Sn and V₃Ga. Therefore, R & D studies on these compounds concentrated on those fabricated by the physical vapour deposition (PVD), the chemical vapour deposition (CVD) and the liquid quench processes. Although the production of multifilamentary conductors is very difficult through these processes, relatively high J_c's can be obtained due to the formation of fine grain compounds.

The rapid quenching techniques from the liquid state are potentially useful for the preparation of Nb₃Al-based compounds. A supersaturated bcc phase of a Nb₃(Al,Ge) ternary alloy, prepared by liquid quenching, was transformed by annealing at 850 ℃ into the A15 phase with high J_c(4.2 K) of 2000 A/mm² at 20 T [13]. An amorphous ribbon of a Nb₃(Al,Si,B) alloy, prepared by liquid quenching, was annealed at 620 ℃ to be crystallized into a fine grained A15 phase ribbon with very high J_c(4.2 K),

of 20000 A/mm² at 20 T [6].

The fabrication of long/uniform conductor is very difficult through the liquid quenching process. Laser beam and electron beam irradiation processes [14] were proposed to overcome the weak point of the liquid quenching process. J_c(4.2 K) for the reacted area of about 200 A/mm² at 25 T has been obtained for a Nb-Al-Ge tape made by powder metallurgy process and then irradiated by an electron beam of 320 W.

The diffusion process is the most suitable method for producing multifilamentary conductors commercially, and it has been mainly investigated for the fabrication of Nb₃Al conductors by two different methods. One of them is the powder-metallurgy process [15], in which a Nb and Al powder mixture was packed into a copper alloy sheath, cold-worked into wire and finally heat treated to form Nb₃Al filaments. Another method is the jelly-roll process [16], in which two thin foils of Nb and Al are superimposed, wound around a small copper cylinder, inserted

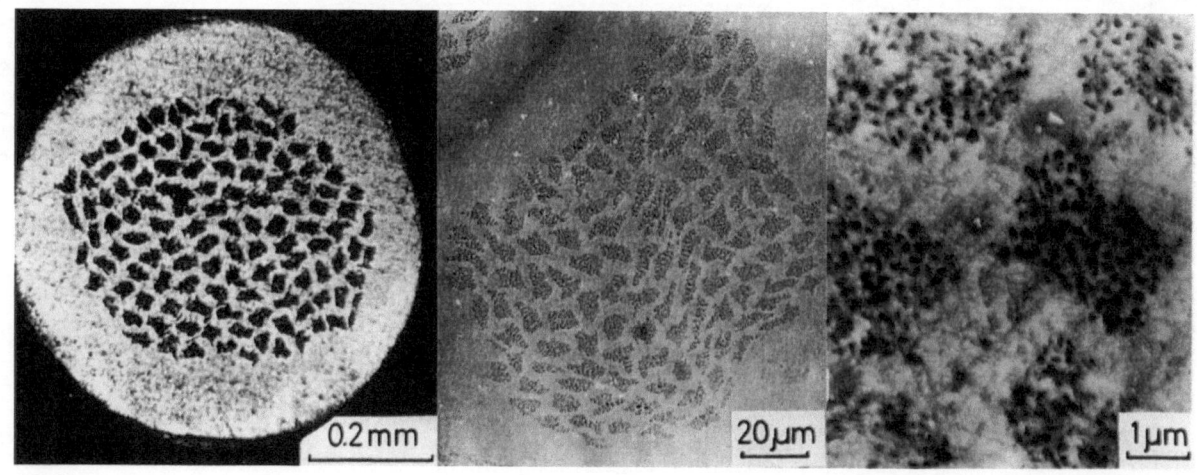

FIGURE 5   Typical OM and SEM photographs of the transverse cross section of the 1.8 million-core Nb/Al-7at%Mg composite wire of 0.7 mm φ.

into a copper sheath, cold-worked into a wire, and finally heat-treated to form a Nb_3Al layer. For obtaining high J_o's in these methods, it is necessary to form Nb_3Al layers by the short-distance Nb/Al diffusion reaction. However, it has been very difficult to realize the short-distance diffusion reaction in practical large-scale multifilamentary conductors due to the poor cold-workability of Nb/Al composites.

The cold-workability was found to improve very much by hardening the Al cores by alloying them with Mg, Ag, Cu and/or Zn. We have been successful in realizing the short-distance Nb/Al diffusion reaction and moreover in making a multifilamentary conductor with ultra-fine continuous Nb_3Al filaments (Fig. 5) through a newly-developed composite process [4], in which a multicore composite consisting of a niobium matrix and a large number of Al-based alloy core is cold-drawn into a wire with continuous ultra-fine Al-based alloy

filaments below 300 nm, and then heat-treated to form Nb_3Al filaments by the diffusion reaction.

When the Al-alloy core size is 30 - 90 nm, the conductor shows excellent super-conducting properties (Fig. 6), e.g. T_o of 15 - 17 K, $H_{o2}(4.2 K)$ of 21 - 25 T and an overall $J_o(4.2 K, 10 T)$ without Cu of 1000 - 1500 A/mm^2, which are comparable to those of the commercial Nb_3Sn multifilamentary conductors. The superior mechanical properties of this conductor have been reported as shown in Fig. 7. Low a.c. losses are also expected from this conductor. Therefore, the Nb_3Al ultra-fine multifilamentary conductor is worth noticing as the most promising candidate for the new practical superconductors at present.

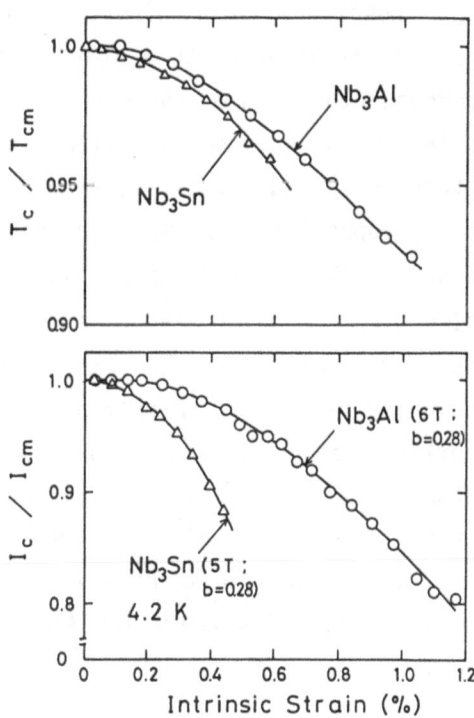

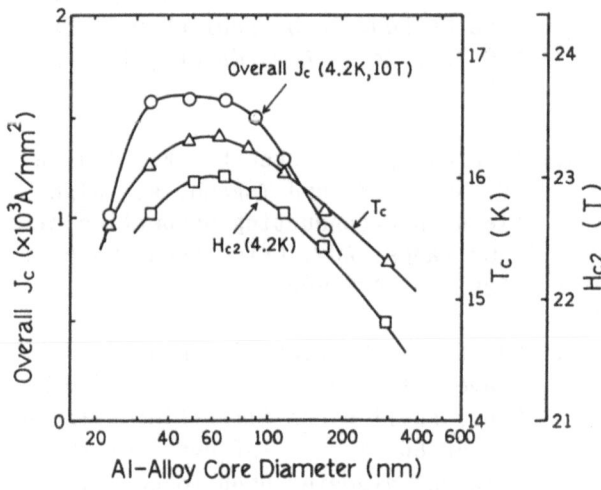

FIGURE 6 Dependence of overall $J_o(4.2 K, 10 T)$ without Cu, $H_{o2}(4.2 K)$ and T_o on the calculated Al-alloy core diameters for the Nb/Al-5at%Mg ultra-fine multifilamentary composites.

FIGURE 7 Typical T_c/T_{cm} vs intrinsic strain and I_c/I_{cm} vs intrinsic strain curves for ultra-fine Nb_3Al multifilamentary conductor and Nb_3Sn multifilamentary conductor [17].

The Chevrel phase compound $PbMo_6S_8$ has become of much interest since its H_{c2} exceeds 50 T at 4.2 K. Various fabrication processes have been studied for this superconductor to attain a practically sufficient J_c. Among them the powder metallurgy process is attractive from the industrial point of view. In this process a mixture of Mo, Pb and molybdenum-sulphide powders is packed in a Ta or Nb tube with a Cu alloy sheath, then cold-drawn to a final size and heat treated at about 1000 ℃ to form a $PbMo_6S_8$ core. The conductor shows a $J_c(4.2$ K) of 200 A/mm^2 at 20 T [7].

CVD processed Nb_3Ge tapes, reactive sputtered NbN films and composite-processed $V_2(Hf,Zr)$ multifilamentary conductors were also studied intensely because of their excellent superconducting properties. However greater efforts should be necessary to develop them into practical superconductors. The forecast of R & D studies on the advanced metallic superconductors is very difficult in relation with the development of high T_c oxide superconductors. Much improvement have been achieved on the J_c of high T_c oxide superconductors, although the many problems concerning with their anisotropy, chemical stability, electromagnetic stability and contact resistivity have not yet been solved. The Ag-sheath processed Bi-Sr-Ca-Cu oxide tapes have been reported to exhibit $J_c(4.2$ K)'s above 100 A/mm^2 at 20 - 30 T.

ACKNOWLEDGEMENTS

I am grateful to the R & D staffs of National Research Institute for Metals, Furukawa Electric Co., Ltd., Toshiba Co., Ltd., Showa Electric Wire & Cable Co. and Mitsubishi Electric Co., whose assistance made this review possible.

REFERENCES

1. Matsumoto, K. and Tanaka, Y., High critical current density in multifilamentary Nb-Ti superconducting wires. to be published in Proc. 6th US-Japan Workshop on High Field Superconductors (1989, Boulder, Colorado); Hong, S., Geschwindner, D., Mantone, A., Marencik, W., Zarek, S., and Zhou, R., High current density of Nb-Ti composite. IEEE Trans. on Magn., 1989, **25**, 1934-1936; Kanithi, H., Expectations and limitations of J_c in practical Nb-Ti conductors. Adv. Cryo. Engn. Mater., 1988, **34**, 951-958.

2. Shiraki, H., Nakayama, S., Tanaka, M., Murase, S., Aoki, N., Ichihara, M., Watanabe, K., Noto, K., and Muto, Y., High-field superconducting properties of Ti doped Nb_3Sn conductor by the Nb tube method. to be published in Proc. MRS Int. Meet. Adv. Mater., (1988, Tokyo).

3. Tachikawa, K., Takeuchi, K., Iijima, Y., Inoue, K. and Togano, K., High field superconducting properties of V_3Ga tapes. Adv. Cryo. Engn. Mater., 1988, **34**, 585-592.

4. Dubot, P., Fevrier, A., Renard, J.C., Taavergnier, J.P., Goyer, J., and Ky, G.H., Nb-Ti wires ultra-fine filaments for 50-60 Hz use: influence of the filament diameter upon losses. IEEE Trans. Magn., 1985, **MAG-21**, 177-180.

5. Takeuchi, T., Iijima, Y., Kosuge, M., Kuroda, T., Yuyama, M. and Inoue, K. Effects of additive elements on continuous ultra-fine Nb_3Al MF conductors. IEEE Trans. Magn., 1989, **25**,

2068-2075.

6. Clapp, M.T. and Shi, D. New processing technique for forming flexible A15 superconducting tapes with extremely high critical current densities and magnetic fields. Appl. Phys. Lett., 1986, **49**, 1305-1307.

7. Kubo, Y., Yoshizaki, K., Fujiwara, F., Noto, K. and Watanabe, K. Small coil tests of Chevrel-phase $PbMo_6S_8$ wires. to be published in Proc. MRS Int. Meet. Adv. Mater., (1988, Tokyo).

8. Hlasnik, I. Progress and problems in superconducting composites for ac applications. Proc. Int. Symp. Flux Pinning & Electromagnetic Properties in Superconductors., (Fukuoka, 1985) Matsukuma Press Co, p.247-253.

9. Shimizu, E. and Ito, D. Development of 50 kVA superconducting coil with Nb-Ti ultra-fine multifilamentary supercon- ductor. Cryogenic Engineering (in Japanese), 1988, **23**, 192-204.

10. Tachikawa, K., Itoh, K., Wada, H., Gould, D., Jones, H., Walter, C.R., Goodrich, L.F., Ekin, J.W. and Bray, L.S. VAMAS intercomparison of critical current measurement in Nb_3Sn wires. IEEE Trans. Magn. 1989, **25**, 2368-2374.

11. Smathers, D., O'Larey, P., Siddall, M., and McDonald, W. Status of the superconductor development program at Teledyne Wah Chang Albany. Adv. Cryo. Engn. Mater., 1988, **34**, 515-522.

12. Hazelton, D.W., and Ozeryansky, G.M. Development of internal tin Nb_3Sn conductor for high field magnet use. Adv. Cryo. Engn. Mater., 1988, **34**, 499-506.

13. Togano, K., Takeuchi, T. and Tachikawa K. A15 $Nb_3(Al,Ge)$ superconductor prepared by transformation from liquid quenched body-centered cubic phase. Appl. Phys. Lett. 1982, **41**, 199-201.

14. Kumakura, H., Togano, K., Iijima, Y. and Tachikawa, K. Critical current measurement and coil tests for Nb_3Al superconducting tapes fabricated by continuous laser and electron beam irradiation. Adv. Cryo. Engn. Mater., 1988, **34**, 469-475.

15. Thieme, C.L.H., Pourrahimi, S. and Foner, S. Nb_3Al wire produced by powder metallurgy and rapid quenching from high temperatures. IEEE Trans. Magn. 1989, **25**, 1992-1995.

16. Bruzzese, R., Sacchetti, N., Spadoni, M., Barani, G., Donati, G. and Ceresara, S. Improved critical current densities in Nb_3Al based conductors. IEEE Trans. Magn., 1987, **MAG-23**, 653-656.

17. Kuroda, T., Wada, H., Iijima, Y. and Inoue K. Strain effect on supercon- ducting properties in Nb_3Al multifila- mentary wires. J. Appl. Phys., 1989, **65**, 4445.

FIRST Nb₃Sn, 1m LONG SUPERCONDUCTING DIPOLE MODEL MAGNETS FOR LHC BREAK THE 10 Tesla FIELD THRESHOLD

A.Asner,R.Perin
C E R N
Geneva/Switzerland

S.Wenger,F.Zerobin
E L I N
Weiz/Austria

ABSTRACT

In late 1986 CERN and ELIN joined in a collaboration to develop a 1 m long, 50 mm aperture Nb₃Sn high field dipole model magnet for LHC. CERN provided the basic know-how and the cables and ELIN was doing design and manufacturing both of a mirror test dipole followed by a final dipole magnet. A winding technology has been developed based on the "wind and react" method. The excitation coils are wound of two different, 17 mm wide, Nb₃Sn cables with an inorganic insulation. After reaction the coils are epoxy vacuum impregnated and mounted into a mechanical support structure consisting of Al-collars, a split cold iron yoke and an outer aluminium cylinder. Design and technology were proven in a magnetic mirror dipole where a single pole was tested in February 1989: a maximum magnetic field of 10'2 T was reached at 17'4 kA and 4'3 K after a few quenches. The dipole magnet itself was successfully tested at the beginning of June 1989: a central bore field of 9'5 T was reached at 4'3K, the maximum field at the Nb₃Sn cable being about 10'05T.
The results achieved represent world record performance for high field dipole configurations and fully confirm the soundness of the chosen technological options.
Together with earlier successful testing of NbTi wound model dipoles working at 1'8K, CERN's high field collider magnet development is now solidly based on two very promising technologies. These technologies now are the basis for the development of 10 m long twin-aperture dipoles for the LHC.

INTRODUCTION

CERN is studying the implantation of a Large Hadron Collider in the already existing 27 km long LEP tunnel (1). The existence of this tunnel represents a great advantage for the project. Since the attainable beam energy however is directly proportional to the magnetic field of the magnets, it is important to push the technology of superconducting magnets towards the highest possible fields. The envisaged field level is in the 8 to 10T range. About 2000 dipole magnets of the "twin aperture design" will be required, each about 10 m long (1). Currently the developement of two magnet technologies is being pursued:

The first line makes use of superconducting NbTi cables for the excitation coils, cooled by superfluid helium at 1'8K. This solution requires a cryogenic system which is complex in comparison with a liquid helium system working at 4'2K. Following this technology, a 1m long single bore superconducting dipole magnet has been built (2) and successfully tested. A promising central field of 9'3T was obtained.

The second line makes use of Nb₃Sn cables for the excitation coils, the coils being manufactured according to the "wind and react" method (3) and working at about 4'2K. This method is the only one to be envisaged in view of the very small bending radii which are to be realized during winding of the coils.

Technological development was required to solve the problems related to the special insulation which has to withstand the Nb₃Sn reaction temperature of about 700°C. Due to the brittleness of Nb3Sn, the coils must be vacuum-impregnated after reaction.

GENERAL DESIGN

Although the dipole is called a model magnet, its transverse dimensions, aperture, coil cross-section, collars and iron structure correspond to future dipole magnets for LHC. The length, however, is reduced to 1m instead of 10 m and also the magnet has only one bore instead of the twin-aperture design envisaged for LHC (4).

Nevertheless the model magnet includes all main "problems" such as the high reaction temperature resistant cable insulation, the determination and shaping

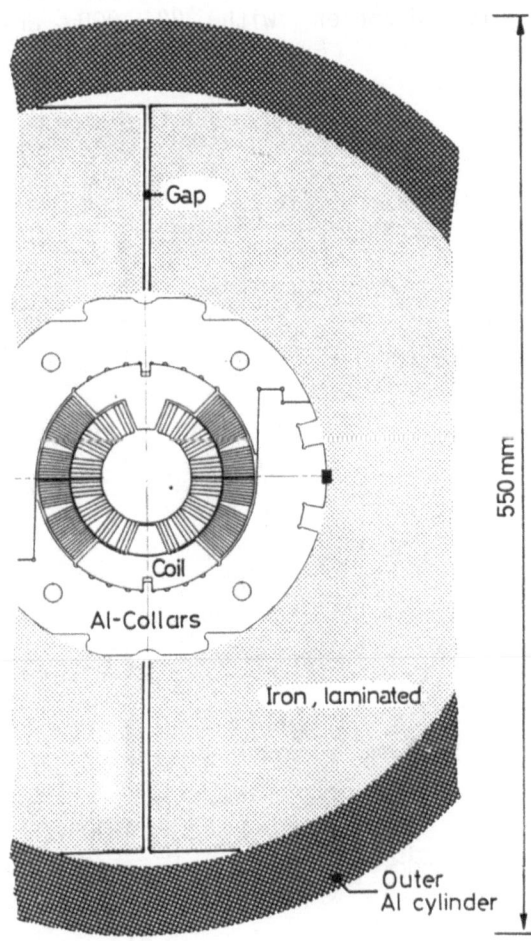

Fig.1: Schematic cross-section of the 1m long full aperture dipole magnet

of the coil ends, the necessarily low-ohmic splice connection between the two Nb₃Sn cables and between Nb₃Sn and the outgoing NbTi cables respectively, the tooling and many others.

The dipole cross-section is shown in figures 1 and 2: the inner and the outer layer of the excitation coils are wound with two different Nb₃Sn cables. Collars made of a special aluminium alloy are placed around the coils, surrounded by two vertically split iron halves and finally an outer retaining cylinder also made of a special aluminum alloy with high tensile strength. Figure 2 exhibits the distribution of the insulated cables into current blocks, the blocks separated by key-stoned copper wedges in order to obtain a dipole field with minimum harmonic content.

The mirror cross-section is shown in figure 3. The mirror was designed as a coil test facility. Thus only one coil system is inserted, the other coil system being replaced by a magnetic "mirror" iron insert. Comparing figures 1 and 3 one realizes that the mirror does not use collars, and its outer diameter is reduced.

The main parameters of the magnets, computed for a central field of 10T, are

```
Central field ...................10 Tesla
Nominal cooling temperature ....... 4'2 K
Nominal current (dipole) ........16'35 kA
Maximum field at inner cable...appr.10'5T
Maximum field at outer cable...appr. 8'8T
Magnetic forces,total ...... x  4600 kN/m
                            y -2400 kN/m
                            z   360 . kN
Stored energy (dipole) ........ 316 kJ/m
Stored energy (mirror) .........177 kJ/m
```

MECHANICAL STRUCTURE

As the electromagnetic forces at 10T are extremely high, a "hybrid" mechanical structure was chosen (4) (7). In such a structure as shown in figure 1, pre-stressing of the coil/collar assembly at room temperature provides only a part of the final coil pre-stress. During the cool-down the coil/collar assembly is further compressed due to the shrinking of the outer aluminum cylinder. This effects the closing of the (carefully computed and checked) vertical gap between the split iron halves. Once closed, the compressive forces generated on the gap planes more than counterbalance the outwards directed horizontal electromagnetic forces of the fully energized dipole magnet. This two-way compression of the coils as well as the whole mechanical structure has been verified on a full-cross-section 15cm mechanical model. The finite element computations have been confirmed with fair accuracy (6).

The stresses in the active parts remain within tolerable limits. The distribution of stresses and deformations of the coils had to be optimized in order to minimize the effect of disturbing the field harmonics. The whole structure must be designed in such a way that the coils are kept always and everywhere under compression. In chosing the initial prestress in the Nb_3Sn coils one has to be careful: on the one hand, permanent compression on all parts of the coil is indispensable, on the other hand measurements on Nb_3Sn strands and on

Nb_3Sn cables indicate a decrease of the critical current under transverse compression (5). Thus the total coil stress at 4'2K has to be chosen very carefully i.e. between the minimum tolerable compression of the coils at the most critical interface with their center posts and the maximum tolerable compression on the midplane between the coils (6).

THE Nb_3Sn CABLES

The dimensions of the two tapered Rutherford-type cables for the inner and for the outer layer are: 2'19/2'69 x 16'81 mm² for the inner layer and 1'47/1'79 x 16'81 mm² for the outer layer, respectively . The cables are insulated, 0'14 mm each side. The thicker cable consists of 24 strands of 1'38mm diameter with 50 000 Nb_3Sn filaments of 2'56μm in a bronze matrix; the Cu/non-Cu ratio is 0'38; a cross-section of the cable is shown in reference (6). The thin cable consists of 36 strands of 0'92mm diameter with 20 000 Nb_3Sn

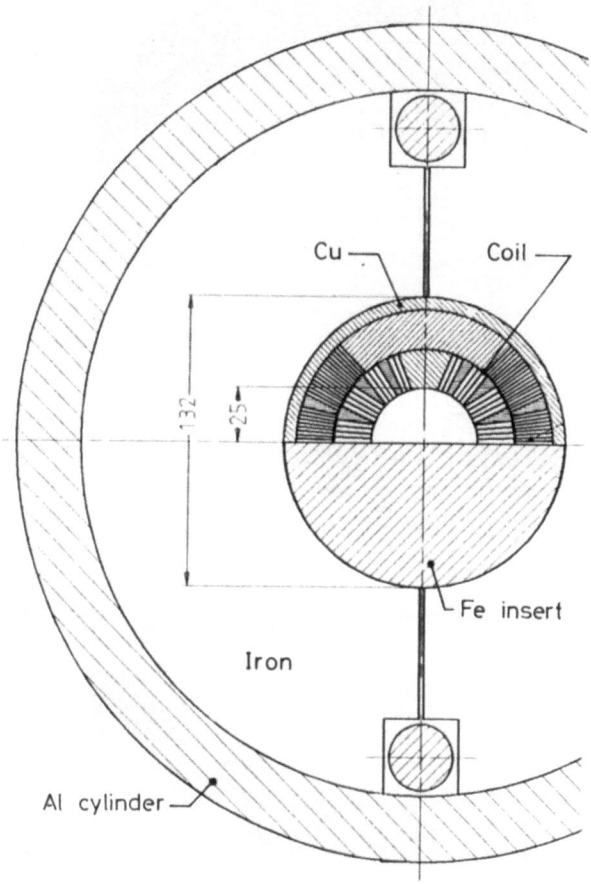

Fig.2: Cross-section model of collared coils

Fig.3: Schematic cross-section of the mirror dipole

filaments of 2'56µm each; here the stabilizing Cu/non-Cu ratio is 0'36. For the mirror, the inner cable had the same overall dimensions as the thick cable mentioned above, but with a reduced Cu/non-Cu ratio of 0'22 and thus a higher content of Nb_3Sn. All cables were delivered by Vacuumschmelze GbmH (Germany). Details concerning these cables and their successfull testing were exposed in reference (6). The cable performances as well as the calculated load lines are summarized in figure 4.

MAGNET MANUFACTURING

The various coil layers are wound with insulated (but non-reacted) Nb_3Sn cables on an appropriate winding machine. The machine is equipped with precisely machined mandrels, clamps and compression devices in order to ensure a correct geometry of the coils. The wound layers

are placed into a reaction oven to be reacted " in situ" in inert gas. The reaction temperature was 675°C and the duration 144 hours. As the coils are fabricated according to the "wind and react" process, only glass and/or mica tapes could be used for the interturn insulation. Preceeding detailed investigations concerned the performance of the insulation: reasonable mechanical strength both before and after thermal treatment, dimensions under compression, good impregnability,sufficient dielectric strength, etc.

After reaction, the internal coil splice between inner and outer layer is done. This Nb_3Sn-Nb_3Sn junction was specially developed in order to achieve a low-ohmic connection with $\leq$ 1 Nanoohm at 17'4 kA/10T/4'2K.

Each excitation coil, composed of one inner layer and one outer layer, is then covered by a coil-to-ground insulation. After brazing the Nb_3Sn-NbTi junction of the outgoing cables the coil is placed into a precisely machined impregnation mould, pre-tested and vacuum impregnated with epoxy resin.

A high standard of quality assurance is indispensable during coil fabrication; corrections after the vacuum impregnation are absolutely impossible.

After impregnation, the coils are pre-compressed and the elastic modulus is determined.After assembly of coils and collars the collars are compressed and locked. As the magnetic forces are high when the magnet is energized, rigid collars and appropriate clamping of the collars is indispensable. Different collar designs and clamping methods were investigated and experimentally proven. All tests were performed at room temperature as well as in liquid nitrogen (6).

The collared coil pair is placed into the laminated iron halves and the rigid end plates for the retention of the axial forces are tightened by strong longitudinal bolts. Finally the outer cylinder made of aluminium alloy is shrink fitted, the two coils are interconnected and the outgoing cables tightly fixed into their support structure. Instrumentation wiring is done followed by final inspections and tests.

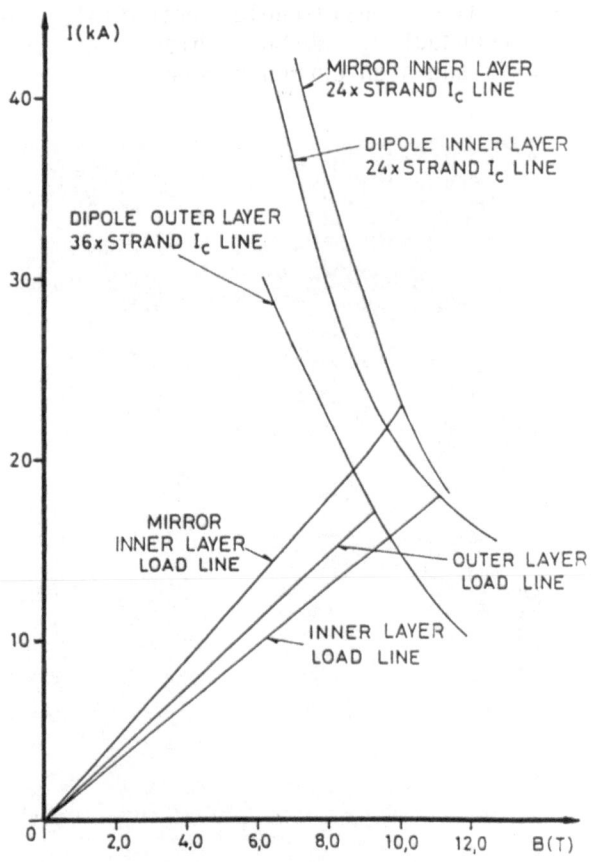

Fig.4: Cable I_c-lines and calculated coil load lines

RESULTS OF MIRROR TESTS AND MAIN DIPOLE TESTS

The lm long, 5cm-half-bore mirror dipole (cross section see figure 3) was tested at CERN in February 1989. Several test runs were made, varying both the current ramp rates and the operating temperature and also included thermal cycles. The quench history is shown in figure 5: After a few quenches, a maximum field of 10'2T was attained at 4'3K with an excitation current of 17'43kA. When cooled to 1'8K, the quenches were registered at essentially the same field which means that the mechanical limit of the mirror was reached. After one full thermal cycle the mirror magnet again reached the same field level mentioned above without re-training.

The lm long, 5cm-full-bore main dipole (cross section see figure 1) was tested at the beginning of June 1989 at CERN. Figure 6 shows the magnet, beeing prepared for testing. This magnet reached a central bore field of 9'5T at 4'3K, which is related to a maximum field at the innermost turn of the inner layer of about 10'05T. Thermal cycling of the magnet did not affect the field level, too; that means that practically no retraining occured.

The field level obtained in the mirror is slightly higher than in the main dipole. This is due to the fact that the inner mirror cable contains more Nb_3Sn than the inner cable of the main dipole. Additionally the magnetic forces and thus stress effects in the main dipole are higher than in the mirror.

CONCLUSION AND PROSPECTS

The successful development, fabrication and testing of the mirror device as well as of the lm long, 5cm aperture main dipole magnet fully confirmed the validity of the technological options of Nb_3Sn and of the "wind and react" technology. Both magnets were designed and manufactured in an industrial style and reached record field levels. To achieve this outstanding results, the ELIN-CERN collaboration was pursuing a vigorous technological effort related to all crucial components of the magnets. Also the considerable potentials of superconducting Nb_3Sn cables for high fields and high currents were confirmed.

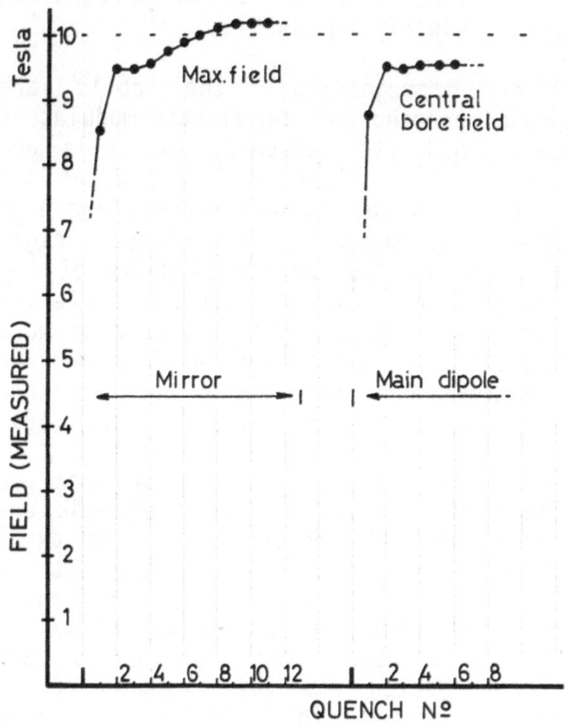

Fig.5: Quench history of the mirror dipole and of the main dipole

Fig.6: 1 m dipole magnet at the cryo-lab at CERN.

However, a continuous effort should follow this great initial success. It should aim at the development of advanced and/or less expensive Nb$_3$Sn cables and the development of better inorganic insulation materials, which would lead to a safer and quicker coil fabrication and thus reduce fabrication costs. Another concentrated effort is requested to clarify the effect of transverse compression on the critical current of Nb$_3$Sn cables in the presence of high magnetic field, this effort being necessary to exploit the possible design limits of Nb$_3$Sn magnets.

REFERENCES

1. G.Brianti, K.Hübner:"The Large Hadron Collider in the LEP tunnel", CERN report 87-05,1987

2. R.Perin, D.Leroy, G.Spigo:"The first industry made model magnet for the CERN Large Hadron Collider", IEEE Transactions on Magnetics, Vol. 25, pp. 1632 ... 1635, March 1989

3. A.Asner:" Development and testing of the first Nb$_3$Sn wound, in-situ reacted high field superconducting quadrupole of CERN", paper presented at the 1982 ASC, Knoxville, USA

4. D.Hagedorn,D.Leroy,R.Perin: "Towards the development of high field super-conducting magnets for a hadron collider in the LEP tunnel", Proc. of the MT 9, Zürich, 1985, pp 86..91

5. Pasztor, B.Jacob:"Effect of transverse compression stress on the critical current of cabled Nb$_3$Sn conductor", IEEE Transactions on Magnetics, Vol. 25, pp. 2379 ... 2381, March 1989

6. S.Wenger,F.Zerobin,A.Asner: "Towards a 1m long high field Nb$_3$Sn dipole magnet of the ELIN-CERN collaboration; development and technological aspects", IEEE Transactions on Magnetics, Vol. 25, pp. 1636 .. 1639, March 1989.

7. Perin,R: "Preliminary views on a 8 Tesla dipole magnet for the Large Hadron Collider". CERN.TIS/MC/85-22, LHC Note 32, August 1985.

ACKNOWLEDGEMENT

The authors wish to acknowledge the contributions and the participation of various colleagues from ELIN and CERN, in particular of Mr. Gottfried Simon from ELIN who designed the magnets. Fruitful technical discussions with the members of the CERN-EMA group and the members of ELIN-DEF are acknowledged, too. An important contribution was given by AMAG Aluminium Ranshofen, Austria, by providing all aluminium components. Last not least the authors acknowledge the testing of the magnets by F.Haug and L.Oberli, CERN.

FIRST RESULTS OF THE HIGH-FIELD MAGNET DEVELOPMENT
FOR THE LARGE HADRON COLLIDER

The CERN LHC Magnet Team, reported by R. Perin
European Organization for Nuclear Research (CERN),
CH-1211 Geneva 23 (Switzerland)

ABSTRACT

After a review of the present state of the R.&D. programme, the results of the tests of the first three model magnets are described.

The first magnet, designed and built as a joint venture by CERN and ANSALDO Componenti, Italy, is based on the NbTi technology. It was designed for 8 T nominal field, but it passed 8.5 T without any quench and attained a 9.3 T central field. Three test campaigns were performed, two at CERN and one at the CEA, Saclay, Laboratory (France). No re-training occurred even after having kept the magnet at room temperature for several months. A second magnet of the same design, but using a mechanically stronger conductor insulation, showed good reproducibility of results and reached a 9.45 T central field at 1.8 K.

The third magnet, based on the Nb_3Sn wind-and-react technology, was a joint project of CERN and ELIN-UNION, Austria. Design and technology were at first tested in a magnetic mirror device in which a full magnet pole was tested : 10.2 T central field was reached after a few quenches. The complete magnet was tested in June 1989 and reached a 9.5 T central field at 4.3 K.

INTRODUCTION

A high-energy proton-proton collider is proposed at CERN as the next facility to push further the frontiers of particle physics. This Large Hadron Collider (LHC) [1] would be installed in the tunnel of the Large Electron Positron Collider (LEP) which is now completed. The envisaged field level in the dipoles is in the range 8 to 10 T. To prepare for the LHC, an R.&D. programme has been launched for both presently possible routes to high-field magnets, i.e. using NbTi superconductor at about 2 K and Nb_3Sn at 4.5 K [2].

A Collaboration has been set up by CERN and several European Laboratories and Institutions [2]. Some co-operation

agreements have been officially established as shown in Table 1, while the other Institutions are collaborating, still in an informal way.

Industry is taking part in the R.&D. effort, not only in the development of components, but also in the design and construction of magnet models and prototypes. The most important common development projects with industrial firms are listed in Table 2. Other collaborations concern development of superconducting wires and cables, insulation systems and cryostat design.

The first magnet model (8TM1) has been designed and built as a joint venture of CERN and the firm ANSALDO Componenti, Italy [3]. A second model magnet (8TM2) of the same design was built by the same firm. A third magnet

TABLE 1

OFFICIAL CO-OPERATION AGREEMENTS WITH EUROPEAN INSTITUTIONS

Institute	Subject of common development
CEA (France)	– Test station for dipole magnets – 1.8 K helium circulation test installation – Mesurements of properties of cables – Heat transfer studies – Design and construction of a prototype quadrupole
INFN (Italy)	– Development of NbTi conductors with 5µm Ø filaments – Construction of 10 T, twin–aperture, magnets
FOM–UT–NIKHEF (Netherlands)	– Development of a 10 T twin–aperture dipole model magnet with Nb_3Sn conductor

model, using Nb_3Sn superconductor was designed and built [4,5] as a joint project of CERN and ELIN–UNION, Austria.

The single aperture configuration has been adopted in this initial development phase in order to limit investment on tooling and conductors, and the design field of the first two models was limited at 8 T because at that time cables with the width needed for a 10 T design (~ 17 mm) were not available. In the meantime 17 mm wide cables with the appropriate aspect (width/thickness) ratio were developed by European industry and a 10 T, twin-aperture, NbTi magnet was designed [6,7] (Fig. 1). Four 1 m long model magnets are being built by Ansaldo (I), Elin (A), Holec (NL) and Jeumont–Schneider (F).

To further explore the Nb_3Sn route, the development of a 10 T twin–aperture model magnet has been started in the Netherlands by a CERN–FOM–UT–NIKHEF Collaboration [8].

In parallel with the short model magnets, a 9 m long twin–aperture prototype [9] was designed at CERN and ordered to ABB, Germany. It uses HERA type coils in order to save time and money, but the rest of the structure corresponds to the LHC 10 T design. The cryostat is being made by FBM, Italy. The complete cryomagnet will be tested in CEA, Saclay, France, in spring 1990.

Design of a full size (~ 10 m long) 10 T, twin-aperture prototype dipole has been started in collaboration with INFN, Italy, and procurement of the necessary conductor and major tooling has been initiated.

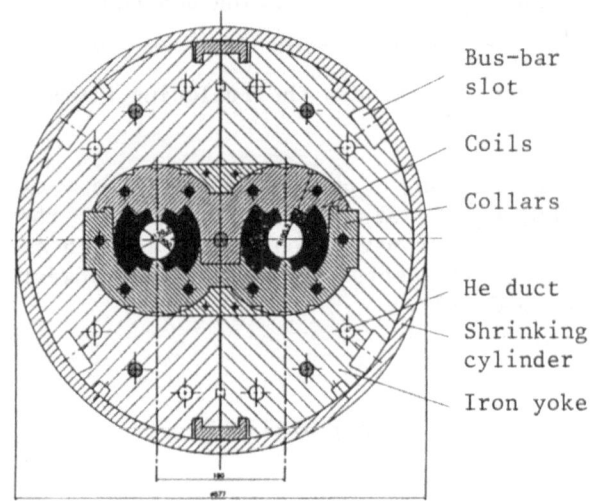

Bus–bar slot

Coils

Collars

He duct

Shrinking cylinder

Iron yoke

FIG. 1 Cross–section of the LHC 10 T magnet

RESULTS OF CABLE DEVELOPMENT

The cable for the inner layer, made of 26, 1.29 mm Ø, strands has the overall dimensions $2.06/2.50 \times 17 mm^2$. The cable for the outer layer made of 40, 0.8 mmØ strands, has $1.3/1.67 \times 17 mm^2$ cross-section. The mechanical stability of these unsoldered cables is now acceptable for winding. The current densities obtained in the outer cable are suitable for LHC : $1100 A/mm^2$ in the NbTi at 8 T, 4.2 K in 15 µm filaments. The fabrication process is now being optimized to obtain high current densities in strands of 1.29 mm Ø with 5 µm NbTi filaments.

TABLE 2

COMMON DEVELOPMENTS WITH INDUSTRIAL FIRMS

FIRM	Subject of common development	Starting year	State
ANSALDO Componenti (Italy)	Design and manufacture of a 1 m long, 8 T, NbTi model dipole (8TM1)	1986	Successfully concluded
ELIN-UNION (Austria)	Design and manufacture of a 1 m long, 9 ÷ 10 T, Nb$_3$Sn model dipole	1986	Successfully concluded
TESLA ENGINEERING assisted by RAL (U.K.)	Design and manufacture of a prototype sextupole/dipole corrector	1988	In progress
ABB (Switzerland)	High current diodes for magnet protection	1988	In progress
ACICA Consortium*) (Spain)	Design and manufacture of a prototype tuning quadrupole	1989	In progress

*) ACICA includes the following firms : ABENGOA, AME, CANZLER, CENEMESA and INDAR.

For Nb$_3$Sn, current densities of 1600 A/mm^2 at 11 T, 4.2 K, in the non-Cu part have been obtained in 15 to 20 μm filaments following the powder method [10].

TEST STATIONS

CERN

The 1 m long (1.35 m overall length) models have been tested in the CERN 1.8 K test station. It mainly consists of a vertical double-bath cryostat, a 100 ℓ/h refrigerator/liquefier, a 20 kA power supply and suitable cryogenic current leads. Bath temperature can be varied between 4.2 K and 1.6 K. Cooling capacity at 1.8 K is approximately 10 W [11].

CEN-SACLAY

A larger installation [12] has been built at CEN-SACLAY for testing model and prototype (10 m long) magnets for the LHC. The cooling power of the liquefier for the test of the first model (8TM1) was 200 W at 4.4 K. For the tests of the prototypes the cooling power is being increased to 450 W at 4.4 K and to ~ 40 W at 1.8 K.

8 TESLA MODEL MAGNETS

8TM1

This magnet model is an 8 T nominal field, single-aperture, 50 mm bore, 1 m long dipole using NbTi conductor and operating at superfluid helium temperatures. In its design [3,13] care has been taken to choose components that could be easily procured. So,the superconducting cable was designed such that use could be made of strands left over from the production of the cable for the DESY-HERA project [14]. The magnet cross-section is presented in Fig. 2 and the assembled magnet in its helium container is shown in Fig. 3. A list of the main parameters is presented in Table 3. The superconducting cable, supplied by Vacuumschmelze GmbH (Germany), is impregnated with a 3.5 x 10^{-8} Ωm (at 4.2 K) resistivity solder.

The design of the force containment structure was made for a 9.5 T central field. The coils are surrounded by 3 mm thick laminated aluminium alloy collars which are locked together by longitudinal stainless steel tie rods. The iron yoke is vertically split in two symmetric parts which are mounted around the collars under a pre-determined horizontal compressive force and with a

~ 0.5 mm vertical gap between them at room temperature. They are clamped by a stainless steel shrinking cylinder which pushes the iron halves to close their gap during cool-down. The functioning of this special mechanical structure, first proposed in 1985, has been described elsewhere [13]. Half of the axial e.m. force (~ 200 kN at B_o = 9.3 T) acting on the magnet ends is supported by thick stainless steel plates that are fixed to the shrinking cylinder.

Two test compaigns took place at CERN. In the first one the central field was voluntarily limited at 9.1 T. The magnet was subsequently warmed up to room temperature and then cooled down again for the second campaign of tests. A third test campaign was carried out at CEN, SACLAY, France, [15] in a horizontal cryostat after the magnet had stayed at room temperature during six months. The shrinking cylinder of the magnet formed the main part of the helium container in the horizontal cryostat (Fig. 3). The history of quenches can be seen in Fig. 4. The first quench occurred at 8.55 T central field. At the second cool-down 9.3 T were reached at the third quench. After the third cool-down two natural quenches occurred, the first one at 8.9 T and the second at ~ 9.35 T (10530 A) showing an excellent behaviour of the magnet after a half year at room temperature.

TABLE 3

MAIN PARAMETERS OF 8TM1 MODEL MAGNET

Coil inner diameter	50.0 mm
Coil outer diameter	102.6 mm
Yoke inner diameter	176.8 mm
Yoke outer diameter	396.0 mm
Shrinking cylinder outer Ø	412.0 mm
Coil straight length (inner)	800.0 mm
Overall coil length (outer)	1'060.0 mm
Length (without conn.)	1'265.0 mm
Overall magnet length	1'353.0 mm
Nominal induction B_o	8.0 T
Peak field in winding at B_o	8.3 T
Operation current at B_o	9'180 A
Turns per pole inner/outer shell	21/21
Stored energy at B_o	160 kJ
Stored energy at B_o = 9.3 T	216 kJ
Cable geometry	1.31,1.66 x 12.6 mm^2
No of strands	30
Strand diameter	0.84 mm
NbTi filament diameter	10 µm
Cu/Sc ratio	1.8
Insulation :	
2 x 0.025 mm Kapton 50 % overlapped + 0.12 mm glass fiber, B-stage epoxy helicoidally wound with 2 mm spacing	

Artificial quenches have then been initiated by a heater placed on the end of the coil outer layer, delaying the time of opening the power supply switch. With a delay of 96 ms, the maximum temperature in the coil was 107 K. The measured magnetic length was 951.5 mm at 9180 A, to be compared to the coil inner layer straight part of 800 mm.

8TM2

The 8TM2 model magnet is of the same design and has been built by the same manufacturer as the 8TM1. It had two purposes :
a) to test a new cable insulation.
b) To test reproducibility.
A weak point found in the fabrication of the windings of 8MT1 was the fragility of the cable insulation which led to frequent ruptures at the end bends. A new insulation was developed in collaboration with Isovolta (Austria) replacing the outer glass fiber tape by a new glass aramid fiber tape which was then applied to the conductor by the firm Alfacavi-Pirelli (Italy) (Fig. 5).

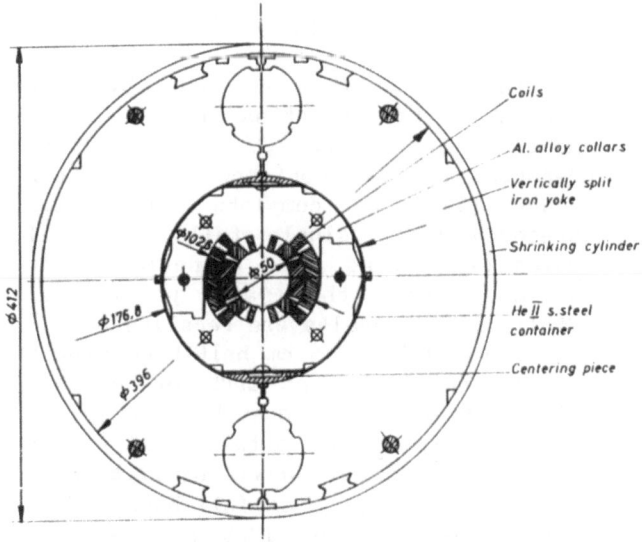

Coils
Al. alloy collars
Vertically split iron yoke
Shrinking cylinder
He II s.steel container
Centering piece

FIG. 2 Cross-section of the 8 T Model Magnet for the LHC

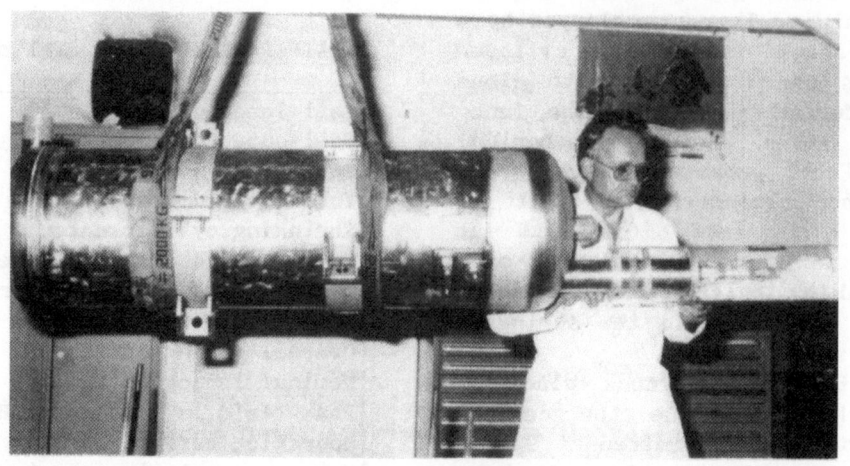

FIG. 3   8TM1 Magnet in its Helium Container

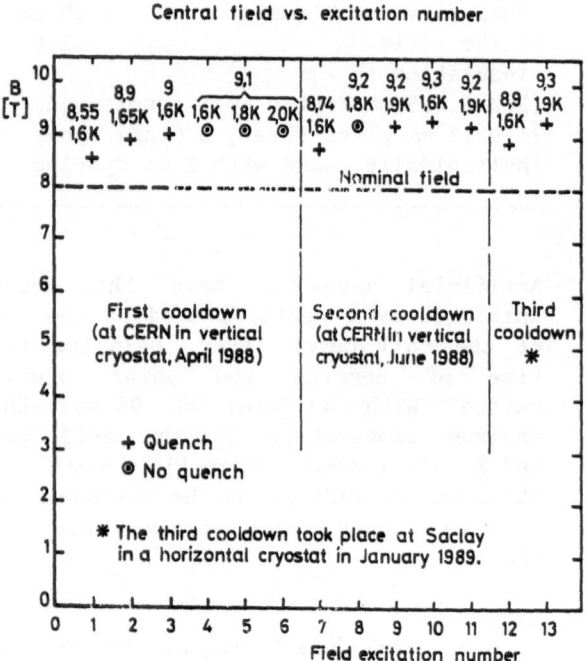

FIG. 4   Quench history of 8TM1

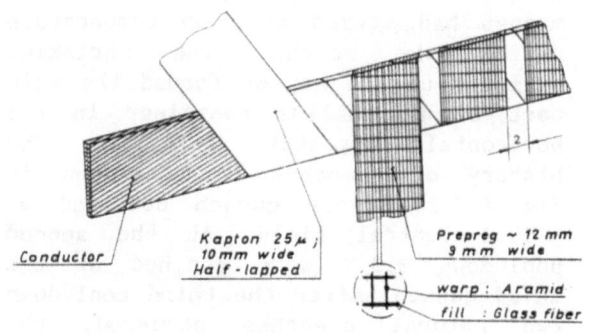

FIG. 5   Cable insulation for 8TM2 magnet

The magnet was tested in July 1989. It fulfilled the two above mentioned goals and behaved in a very similar way to 8TM1 reaching a slightly higher maximum central field, 9.45 T at 10660 A.

Nb₃Sn MODEL DIPOLE

The work carried out by the CERN-ELIN Collaboration on a Nb₃Sn model magnet and its results are reported at this Conference in another paper [5]. The complete magnet was preceded by a single full size coil which was tested in a magnetic mirror device. Only the main parameters, Table 4, and results are recalled here.

The 1 m long, 5 cm half-bore mirror dipole was tested at CERN in February 1989. After a few quenches, a 10.2 T maximum field was attained at 4.3 K with an excitation current of 17.43 kA.

The 1 m long, 5 cm full-bore dipole was tested in June 1989 at CERN. It reached a central bore field of 9.5 T at 4.3 K, which corresponds to about 10.05 T peak field at the conductor.

ment of high-field magnets for the LHC. The work must proceed with the construction and testing of long dipoles, quadrupoles, correction magnets and auxiliary equipment, in order to establish a sound technological basis for the construction of the LHC.

TABLE 4

MAIN PARAMETERS OF DIPOLE MAGNET AND MIRROR DEVICE

Central field	10 tesla
Nominal cooling temperature	4.2 K
Nominal current (dipole)	16.35 kA
Max. field at inner cable appr.	10.5 T
Max. field at outer cable appr.	8.8 T
Stored energy (dipole)	316 kJ/m
Stored energy (mirror)	177 kJ/m

CONCLUSIONS

Encouraging results have been obtained in the first phase of develop-

ACKNOWLEDGEMENTS

The important contribution of European Industry and Institutions is gratefully acknowledged. Several groups at CERN have contributed in the reported activities; the authors are particularly grateful to the SPS Div. Power Converters Group and the EF Div. Cryogenics Group for their assistance in the tests. Thanks are due to Mrs. Ch. Parthé for typing the manuscript.

REFERENCES

1. The LHC Working Group, Large Hadron Collider in the LEP Tunnel, CERN 87-05, May 1, 1987.
2. R. Perin, Progress on the Superconducting Magnets for the LHC, IEEE Trans. on Magnetics, Vol. 24, pages 734-740, March 1988.
3. R. Perin, D. Leroy, G. Spigo, The first industry made model magnet for the CERN LHC, IEEE Trans. on Magnetics, Vol. 25, pages 1632-1635, March 1989.
4. A. Asner, S. Wenger, F. Zerobin, Towards a 1 m long high-field Nb$_3$Sn dipole magnet of the ELIN-CERN Collaboration, IEEE Trans. on Magnets, Vol. 25, pages 1636-1639, March 1989.
5. A. Asner, R. Perin, S. Wenger, F. Zerobin, First Nb$_3$Sn, 1 m long superconducting model magnets for LHC break the 10 tesla field threshold, Paper presented at this Conference.
6. D. Hagedorn, D. Leroy, R. Perin, Towards the Dev. of high-field sup. Magnets for a Hadron Collider in the LEP tunnel, Proc. MT-9, Zurich, Sept. 9-13, 1985.
7. D. Leroy, R. Perin, G. de Rijk, W. Thomi, Design of a high-field twin-aperture superconducting dipole model, IEEE Trans. on Magnetics, Vol. 24, pages 1373-1376, March 1988.
8. H.H.J. ten Kate, C. Daum, L.J.M. van de Klundert, W. Hoogland, T.A. Roeterdink, F. van Overbeeke, R. Perin, Development of a 10 T Nb$_3$Sn twin-aperture model dipole magnet for the CERN LHC, Paper presented at this Conference.
9. D. Hagedorn, Ph. Lebrun, D. Leroy, R. Perin, J. Vlogaert, A. McInturff, Design and construction of a twin-aperture prototype magnet for the CERN LHC project, Paper presented at this Conference.
10. E.M. Hornsveld, J.D. Elen, L.A.M. van Hoogendam, Dev. of ECN-type NB$_3$Sn wire to smaller fil. size, Adv. in Cryog. Eng. Mat. Plenum Publ. 34, 493-8, 1987.
11. F. Haug, D. Hagedorn, L. Oberli, Cryogenics of the 1.8 K test station for 10 T superconducting magnet models, Proc. 12th Int. Cryog. Conf., Southampton, July 1215, 1988.
12. P. Chaumette, C. Cure, J. Deregel, P. Genevey, A large 1.8 K facility for magnet tests, Proc. ICEC/ICMC Conf., Los Angeles, July 24-28, 1989.
13. R. Perin, Preliminary views on a 8 tesla dipole magnet for the LHC, CERN TIS/MC/85-22, LHC Note 32, August 1985.
14. S. Wolff, The superconducting Magnet System of HERA, Proc. 8th Int. Conf. on Magnet Technology, Zürich, September 1985.
15. P. Genevey, Essai du dipole 8TM1, CEN-Saclay note 102.06R.000.TC3, 18-7-1989.

DESIGN AND CONSTRUCTION OF A TWIN-APERTURE PROTOTYPE MAGNET
FOR THE CERN LHC PROJECT

D. Hagedorn, Ph. Lebrun, D. Leroy, R. Perin, J. Vlogaert
European Organization for Nuclear Research (CERN),
CH-1211 Geneva 23 (Switzerland)
A. McInturff, FERMI National Accelerator Laboratory,
P.O. Box 500, Batavia IL 60510 (USA)

ABSTRACT

A twin-aperture, 10 m long, prototype magnet was designed by CERN and built by industry in order to gain experience in the construction of full-scale, twin-aperture, high-field magnets and with the superfluid helium operation of such magnets. In order to gain time and reduce the cost of development and tooling, the superconducting windings were chosen to be the same type and geometry as those of the dipoles of the HERA proton ring but operated at the lower temperature of 1.8 K.

The active part of the magnet consists of two dipole coils wound from a keystoned NbTi cable, clamped with aluminium collars. The collar/coil assembly is surrounded by a split low carbon steel yoke enclosed by an aluminium alloy shrinking cylinder.

The cryostat consists of a stainless steel liquid helium container, a radiation screen at 4.5 K, a liquid nitrogen screen, and a steel vacuum vessel.

A central bore field of 7.5 tesla is predicted for an operational temperature of 1.8 K.

INTRODUCTION

CERN is studying the possibility of a Large Hadron Collider (LHC) [1] being installed in the LEP tunnel. A development programme [2,3] has therefore been started for the required high-field superconducting magnets which guide the two counter-rotating proton beams in the quasi-circular tunnel.

In parallel with the 1 m long model magnet programme, which produces both single and dual aperature dipoles (8 to 10 tesla field range), it was decided to design and order from industry a twin-aperature 9 m long prototype magnet. The primary objective of which is to gain experience both in the construction of long two-in-one "mechanically hybrid"

structures and with superfluid helium operation in a long horizontal cryostat.

In order to gain time and save in the tooling costs, superconducting coils of the same type and geometry as those of the DESY HERA proton ring dipoles were used. The geometry of yoke and shrinking cylinder is the same as in the twin-aperture 10 T dipole.

MAGNET AND CRYOSTAT ASSEMBLY

The main features of the TAP magnet and cryostat are shown in Figure 1 (transverse cross-section) and Figure 2 (longitudinal cross-section). The major parameters are given in Table 1.

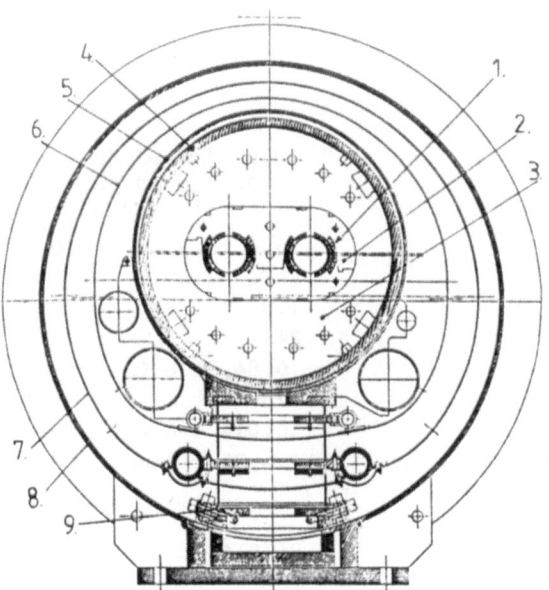

Fig 1 - Transverse cross-section

1. Excitation coil of the HERA type
2. Pair of collars
3. Split iron yoke
4. Shrinking cylinder
5. Helium vessel
6. Radiation screen
7. Liquid nitrogen screen
8. Vacuum vessel
9. Support post

ACTIVE PART

The excitation coils used in the TAP magnet are of the same type and geometry as those used in the dipoles of the DESY-HERA proton ring [4], but operated at 1.8 K. They are wound from a keystoned Rutherford type cable with 24 strands, each coated with a 0.005 mm thick silver-tin layer [5]. The cable is insulated first by a layer of 12 mm wide 0.025 mm thick Kapton tape wrapped with a 60 % overlap and then a 9 mm wide 0.12 mm thick glass fiber tape impregnated with 21 % (weight) of "B" stage epoxy is wrapped over the first. The latter tape has 3 mm gaps between wraps to allow free helium passage between adjacent turns. The excitation coil of a beam channel is composed of two identical half coils, having each a two shell structure. The inner shell has 32 turns per coil half, the outer 20 turns. The current density is uniform in both shells. The approximation of the cosine θ current distribution by the angular length and radial thickness of the shells is further improved with the insertion of an insulated copper wedge in each quadrant and shell after the 4th turn.

Two excitation coils are clamped with pairs of 3 mm thick collars of

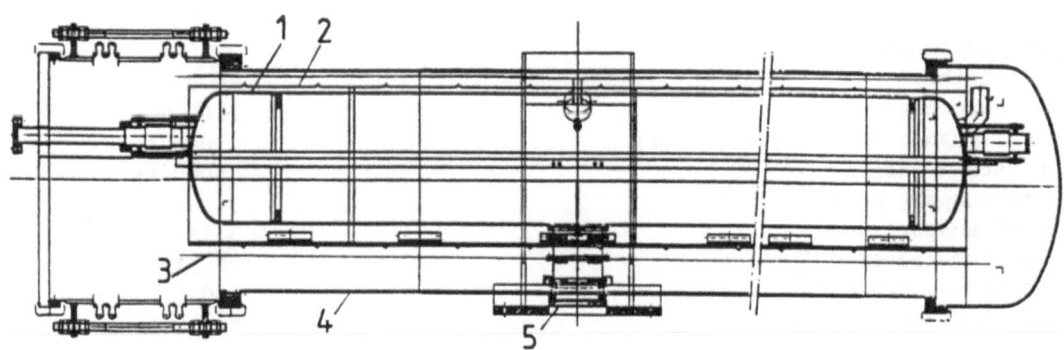

Fig 2.- Longitudinal cross-section (Test lay-out).

1. Liquid helium vessel
2. Radiation screen
3. Liquid nitrogen screen
4. Vacuum vessel
5. Support post

TABLE 1

Main parameters of the TAP magnet and cryostat

Cable Parameters

Cable width	10.0 ± 0.03 mm
Thin edge thickness	1.28 ± 0.02 mm
Thick edge thickness	1.67 ± 0.02 mm
Twist pitch	75.0 to 95.0 ± 2.0 mm
Superconductor	NbTi alloy
Copper to superconductor volume ratio	1.8
Number of strands	24
Diameter of strand before cabling	0.84 ± 0.01 mm
Number of filaments per strand	1230
Diameter of the filaments	0.014 ± 0.002 mm
Filament twist pitch	25 ± 3 mm
Minimum critical current measured with a magnetic field perpendicular to the large surface of the cable at 5.5 T and 4.6 K ($\rho(eff) = 10^{-14}$ Ω-m)	8610 A
Residual resistivity ratio of copper at zero field	73
Current density in SC at 7.5 T central field and 1.8 K	1859 A/mm^2

Magnet Parameters

Expected central field at 1.8 K	7.5 T
Excitation current at 7.5 T	8625 A
Stored energy at 7.5 T	4.06 MJ
Maximum field in winding at 7.5 T central field	8.05 T
Coil inner diameter	75 mm
Distance between aperture axes	180 mm
Resultant magnetic forces per quadrant	
horizontal component (inner and outer layer)	1.78 MN/m
vertical component (inner layer)	0.51 MN/m
(outer layer)	0.49 MN/m
Longitudinal magnetic force (total at either extremity)	0.55 MN
Overall length of magnet	9.15 m
Outer diameter of magnet	580 mm

Cryostat Parameters

Overall length	10.1 m
External diameter	1.0 m
Cold mass (magnet and helium vessel)	15000 kg
Mass of vacuum vessel	5000 kg
Number of support posts	3
Spacing of support posts	3.16 m
Liquid helium capacity	260 ℓ
Liquid nitrogen capacity	40 ℓ
Nominal temperature levels	80 K, 5 K, 2 K

AlMg 4.5 Mn G35 material. A pair of collars consists of two different types of mating collars, the difference being that the central part of the first type has a protruding tongue that fits into a groove of the second. Pairs of collars are reversed with respect to their neighbours and locked with stainless steel rods in three horizontal axial holes. Stainless steel shims are inserted at the poles in order to obtain the required prestress on the coils. A short press of 0.5 m length that can generate a force of over 5 MN is used to compress the collars around the coils in 20 successive steps. The continuity of the collar/coil assembly is obtained by longitudinally screwing the rods together. The magnet yoke is of the split iron type [6] with an about 1 mm average

non-parallel vertical gap between halves at room temperature. The yoke halves are assembled from precision punched low carbon steel laminations, which have been phosphatised to prevent oxidation. The laminations are preassembled into 0.6 m long stacks. "B"stage epoxy impregnated 0.28 mm thick polyamide sheets are inserted between the laminations of the stack. The whole stack is then cured in an oven for two hours at 160°C. The purpose of the polyamide sheet is to obtain the same longitudinal shrinkage for the yoke as for the coil and to make it impervious to liquid helium. The pre-assembled and cured stacks are aligned and 9 m long stainless steel rods are then inserted. Tapered cutouts are planned for the outer surface of the stack into which are fitted wedge-shaped stainless steel keys. These maintain the relative position of adjacent blocks.

The 0.6 m long stacks are terminated by special laminations with machined grooves allowing radial flow of the helium. The first and last stacks of the half yokes contain a number of austenitic stainless steel laminations to avoid magnetic field enhancement in the coil ends.

Low friction sheets are inserted between the coil/collar assembly and the half yokes and between the half yokes and the outer shrinkage cylinder to allow for relative tangential and longitudinal movements of the structure during thermal cycling.

The outer aluminium cylinder is composed of two preformed half shells made from AlMg 4.5 Mn alloy. The half shells are obtained by the successive bending of a 20 mm thick rolled plate. They are tightly fitted to the yoke, collar and coil assemblies. They are then welded together on both sides simultaneously by automatic MIG welding machines.

The azimuthal stresses introduced by weld shrinkage contribute to the required prestress of the assembly. The aluminium cylinder is terminated by two welded aluminium rings which retain the two aluminium end plates. At room temperature the aluminium cylinder is slightly longer than the yoke to permit free longitudinal shrinkage during cool down.

The combination of aluminium collars and shrinking cylinder was chosen in order to obtain the required prestress on the coils and sufficient contact pressure on the mating faces of the half yokes at operating temperature, while minimizing those stresses at room temperature.

THE CRYOSTAT

Although the main purpose of the cryostat is to contain the TAP magnet and permit its cryogenic operation and testing, it will also be used as a full-scale test bed to assess the preliminary design of future LHC cryostats. This will permit to study some of the mechanical and thermal problems associated with large-scale superfluid helium cryogenics, therefore establish an early technological reference (materials, components, manufacturing techniques, quality control) for the implementation of the industrial series production.

In order to be compatible with the magnet string configuration [1] and its associated cryogenic distribution scheme [7] in the LHC tunnel, the TAP cryostat has been designed for serial connection with similar units. As a consequence, its stand-alone operation requires the use of feed and end boxes to connect to the cryogenic test station.

The TAP magnet is contained in an integrally-welded austenitic stainless steel (X2 CrNi 19-11) helium vessel and operates in a static bath of "pressurized" 0.11 MPa (1.1 bar) superfluid helium at about 2 K. The magnet and helium vessel are cooled by longitudinal conduction in the static superfluid bath from a "cold source" (saturated superfluid helium) located in the test station. In order to safely sustain the pressure rise, due to the rapid energy release into the liquid helium from a resistive transition of the magnet, the design minimum pressure of the helium vessel was chosen to be 2 MPa (20 bar).

The magnet and helium vessel assembly rest inside the cylindrical vacuum vessel on three low-heat conduction support posts positioned longitudinally for equal loading. The vacuum vessel is made of construction steel (DIN 17102) for reasons of cost and magnetic shielding of stray flux. It has austenitic stainless steel flanges and connecting sleeve equipped with large diameter bellows. The vessel also serves as the vacuum insulation jacket for the four cryogenic distribution lines which, in the LHC, will provide the refrigeration power at 2 K and 5 K along the magnet string [7]. In the TAP cryostat, the 5 K lines provide an intermediate temperature shield to intercept heat leaks in steady state operation.

The magnet and cryogenic lines are surrounded by a low-emissivity screen made of aluminium alloy (Peraluman 250) and bridged thermally to the 5 K pipes. This geometry minimizes the residual heat leak into the 2 K zone. A second almost concentric screen is cooled by liquid nitrogen and covered with multilayer reflective insulation. This screen is made of industrial grade aluminium and is also bridged thermally to the support posts. It intercepts the largest fraction of the heat leak at a temperature level of 80 K. The designs of both screens ensure quasi-isothermal conditions in a selfsupporting mechanical structure. These screens must resist deformation when submitted to electromagnetic forces generated by induced eddy currents during a magnet quench.

The support posts subjected to a compressive load consist of epoxy-glass-fiber tubular columns with metallic base, top, and two intercept plates. The intercept plates are connected to the two thermal screens. Mechanical and thermal performance of the support posts require efficient and reliable joining techniques between the non-metallic columns and the metallic plates. This is achieved by epoxy glueing and proper choice of materials ensuring compatible shrinkage during cool down. The support posts fixed to the cold mass can allow for differential contractions between helium vessel, screens, and vacuum vessel by means of sliding supports and rollers, respectively. The roller-equipped base-plates also permit insertion of the cold assembly into the vessel during the assembly of the cryostat. In operation, the central support post is the fixed point between the cold mass and the external vessel. Inertial forces resulting from accelerations encountered during transport are contained by demountable shipping restraints.

INSTRUMENTATION AND PROTECTION

The magnet and cryostat assembly is extensively instrumented. This is in order to assess the device's performance and to facilitate it's operation. The instrumentation monitors voltages, temperatures, flow rates, pressures, stresses, and displacements. Thermal measurements of interest include cool down and steady state heat loads, heat intercept efficiencies, etc. The ability to monitor rapid pressure changes as a function of position along the length can help in the location of quenches. Stress measurements of interest include those of the collars, shrinking cylinder, helium vessel walls, and the helium vessel support posts. The displacements of concern are closure of the iron yoke and the cross checking of stresses in support posts and differential shrinkage coefficients of the materials used in the design.

The TAP magnet presents a particularly challenging problem in coil protection. The original HERA design parameters were 5 kA at 4.6 K. The lower temperature operation of the TAP magnet and the subsequent increase of operational current, magnetic field and stored energy necessitated a redesign of the protection heater contained in the insulation to ground . The heater must be able to warm up a sufficiently large coil volume rapidly enough above transition temperature to avoid a catastrophic overheating of the quenched volume. There are also other constraints, i.e. acceptable voltage and temperature distributions after and during a quench. The largest convenient heater that can be placed in the ground insulation is a factor of two larger than the standard HERA heater. The number of turns however that are heated by it are more than doubled with respect to the HERA dipoles due to a wedge in the outer winding. The maximum computed quenching coil temperature at the expected central field (7.5 T, 8625 A, 1.8 K), if the external current opening switch fails, is 255 K. If the switch works, it will be 203 K. If the heater fails, the calculated temperatures are twice as large and the possibility of damage or destruction of the coil is present. The maximum coil temperatures calculated at the intersection of the magnet load line and the best short sample current curve are 310 K when the switch fails and 240 K when the switch works.

These calculated temperatures indicate that it is advisable to ascertain the performance characteristics of the double heater first at a lower magnet current (3000 to 5000 A) where the external switch can protect the quenching coil from overheating. The procedure is to initiate a quench using the protection heaters and delay the external switch to a later but still conservatively safe time (less than 12×10^6 A^2 s). This approach enables the determination of the energy input necessary to the protection heater to conservatively insure its timely operation later on at higher magnet currents where the external switch alone cannot protect the windings in all circumstances.

STATE

Following competitive tendering, CERN has awarded ABB MANNHEIM (Germany) the contract for construction of the active part and F.B.M.-HUDSON Italiana of Milano (Italy) with the manufacturing and assembly of the cryostat. The superconducting cables were supplied by ABB, Switzerland.

The active part is presently being assembled and tested at room temperature and at liquid nitrogen temperature at ABB MANNHEIM. The active part will then be transported to F.B.M., where it will be assembled in the cryostat. The completed cryostat and magnet assembly will then be transported to CEN SACLAY (France) for cryogenic performance measurements and tests.

ACKNOWLEDGEMENTS

The authors want to express their gratitude to ABB MANNHEIM, especially to Mr. M. Förster, and to F.B.M., especially Dr. M. Mischiatti, and pay tribute to their colleagues at CERN, G. de Rijk, M. Bona, D. Perini, M. Granier and to A. Yamamoto from KEK (Japan) for their help.

REFERENCES

1. Brianti, G., and Hübner, K., (editors), The Large Hadron Collider in the LEP tunnel, CERN-87-05, 27 May 1987.

2. Perin, R., Progress on the superconducting magnets for the Large Hadron Collider, IEEE Transactions on Magnetics, Vol. 24, pages 734-740, March 1988 and CERN LHC Note 63, September 1987.

3. Leroy, D., Perin, R., de Rijk, G., Thomi, W., Design of a high-field twin-aperture superconducting dipole model, IEEE Transactions on Magnetics, Vol. 24, pages 1373-1376, March 1988 and CERN LHC Note 62, September 1987.

4. Wolff, S., The superconducting collared Coil for the Dipoles of the Proton Ring of HERA, DESY, HERA Specification, February 8, 1985, and revised on February 24, 1987.

5. Maix, R.K., Salathé, D., Wipf, S.L., Garber, M., Manufacture and Testing of 465 km superconducting Cable for the HERA Dipole Magnets, IEEE Trans. on Magnetics, Vol. 25, pages 1656-1659, March 1989.

6. Hagedorn, D., Leroy, D., Perin, R., Towards the Development of high-field superconducting Magnets for a Hadron Collider in the LEP Tunnel, Proc. of the 9th Int. Conf. on Magnet Technology, Zurich, September 9-13, 1985.

7. Claudet, G., Disdier, F., Gauthier, A., Lebrun, Ph., Morpurgo, M., and Schmid, J., Proc. ICEC 12, p. 497, Butterworth and Co., 1988.

COUPLING LOSSES IN A SUPERCONDUCTING MODEL MAGNET FOR THE L.H.C.

P. Tixador*, D. Leroy, L. Oberli

European Organization for Nuclear Research (CERN), CH-1211 Geneva , Switzerland.

* now : CNRS-CRTBT/LEG , B.P. 166 X - 38042 Grenoble Cedex, France.

ABSTRACT

In the framework of a future Large Hadron Collider (L.H.C.) at C.E.R.N. high field (8-10 T) magnet models are in development. We have studied and measured the strand coupling losses in a single aperture, 1 meter long dipole. This dipole works at superfluid Helium temperature with a NbTi soldered cable. The calculations are based on the intrinsic decay time constant of the induced currents. This parameter has been obtained experimentaly on a stack of magnet cables. The shape factor has been calculated for any dipole. Modelling the magnet as a transformer with a lossless primary and a secondary short circuited on a resistance representing the induced currents losses we have shown that the losses during a discharge may be calculated only from the simple measurement of the tension of the dipole after the current of the magnet becomes zero. Calculations and measurements are performed for discharges of magnet at high current rates (80 - 200 A/s). The results from calculations and measurements are in good agreement .

INTRODUCTION

A developement of high field (8-10 T) magnet models and prototypes is carried out for the Large Hadron Collider (L.H.C.) at C.E.R.N. [1][2]. This proposed superconducting collider is intented to be installed in the existing 27 km long tunnel of the L.E.P. (Large Electron Positron Colliding beam accelerator). The losses in the superconducting magnets during the ramping or the decay of the magnetic field are an important feature for the developement of superconducting cables and they have a major impact on the installed cryogenic power for L.H.C.

Subject of this report is the calculation and the measurements of the strand coupling losses for quick linear discharges in a model magnet.

MODEL MAGNET CHARACTERISTICS

This one meter long, 8 T nominal field, superconducting dipole has been designed and constructed in close collaboration with European Laboratories [3] and Industries. The superconducting coils are wounded with a NbTi cable and are cooled by superfluid Helium at 1.8K to 2 K.

A cross section of the magnet is presented fig. 1 whereas further informations may be found in reference [3]. The winding of the first dipole consists of two layers of the same superconducting flat cable.

55

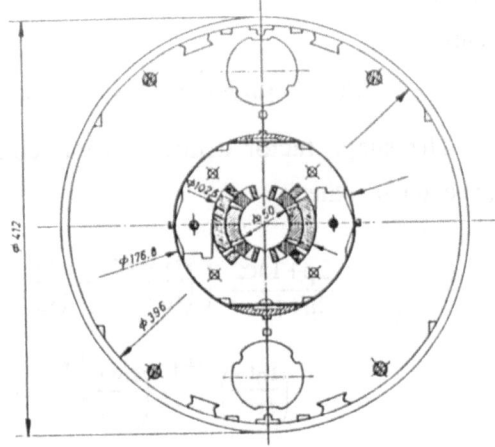

FIGURE 1 : cross-section of the magnet .

The main characteristics of the cable are listed in table 1. The cable of the model dipole is fully soldered. To get an high resistivity at low temperature ($\rho = 3.5\ \mu\Omega$cm) to decrease the losses, 4.5 % of Indium has been included into the solder.

TABLE 1 : Main characteristics of the cable.

Cable geometry	
- side heights (mm)	1.31/1.66
- width (mm)	12.6
Number of strands	30
Strand diameter (mm)	0.84
NbTi filament diameter (μm)	10
Number of filaments	2460
Cu/Sc ratio	1.8
Twist pitch of the strand (mm)	25
Transposition pitch (cable) (mm)	114

A fondamental parameter for the strand coupling losses of a cable is $\theta/1$-N [4], where θ is the decay time constant of the coupling currents and N the demagnetization factor of the winding. $\theta/1$-N is an intrinsic property of the cable whereas θ depends on the winding.

For a magnetic induction with following characteristics :

- perpendicular to the broad face of the cable,
- a linear time ramping of duration T from a steady state field down to zero,
- a linear spatial variation across the cable between the values B_1 and B_2 at t=0,

the expressions for the loss density w and the power density p are :

$$w = \frac{1}{\mu_0}\left(\frac{B_1+B_2}{2}\right)^2 \frac{\theta}{1-N}\frac{1}{T}\left(1 - \frac{\theta}{T}\left(1-e^{-T/\theta}\right)\right) \quad (1)$$

$t < T$:

$$p = \frac{1}{\mu_0}\left(\frac{B_1+B_2}{2\ T}\right)^2 \frac{\theta}{1-N}\left(1-e^{-t/\theta}\right)^2 \quad (2)$$

$t > T$:

$$p = \frac{1}{\mu_0}\left(\frac{B_1+B_2}{2\ T}\right)^2 \frac{\theta}{1-N}\left(1-e^{-T/\theta}\right)^2 e^{-2(t-T)/\theta} \quad (3)$$

For the cable $\theta/1$-N has been measured from a stack of short length cables and amounts to 4.7 s.

EVALUATION OF THE SHAPE FACTOR 1-N FOR A DIPOLE

Assuming that the tangential component of the dipole field is constant in a block of conductors, the inter-strand coupling currents in the cable are independant of the angular position. When the cables are submitted to a variable field, the induced currents through the 2c wide cable are represented by two current sheets K_1 and K_2 of opposite direction. They are located at $\pm$ c/2 from the center of the cable (fig. 2a) at two radius denoted respectively by r_1 and r_2. As the current is conservative K_1 and K_2 satisfy the relation :

$$K_1\ r_1 = K_2\ r_2$$

The magnetic induction created by the two current sheets in a block at the point A (fig. 2a) is given by [5] :

$$B(A) = \frac{\mu_0}{2} \sum_{n=1}^{\infty} K_1^n \left\{ \left(\frac{r_1}{r_0}\right)^{n+1} \left(1 - \left(\frac{r_0}{r_s}\right)^{2n}\right) + \frac{r_1}{r_2}\left(1 + \left(\frac{r_2}{r_s}\right)^{2n}\right)\left(\frac{r_0}{r_2}\right)^{n-1} \right\} \quad (6)$$

where n, odd (symmetry), is the range of the harmonic and K_1^n the n th harmonic of K_1.

$$n = 2p+1 \text{ with n integer} \; ; \; K_1^n = \frac{4}{\pi}\frac{\sin n\alpha}{n} K_1$$

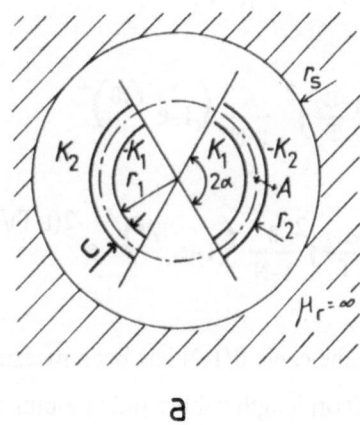

a

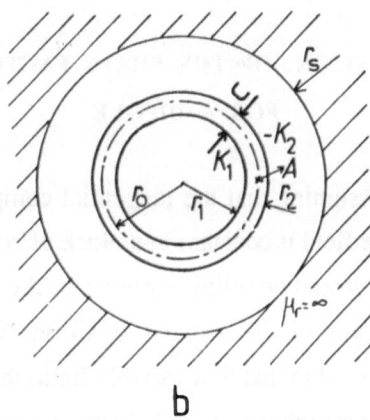

b

FIGURE 2 : modelling of the coupling currents in a dipole.

The shape factor verifies the relation :

$$\frac{1-N}{1-N'} = \frac{B}{B'} \quad (4)$$

where B and B' are the resulting inductions of the same coupling currents (equal representative current sheet) but for windings characterized respectively by the demagnetization factors N and N'. By analogy with a torus, the demagnetization factor is zero for

the winding shown by figure 2b while B'(A) amounts to :

$$B'(A) = \mu_0 K_1 \frac{r_1}{r_0} \quad (5)$$

The shape factor deduced from relations (4),(5) and (6) is then :

$$1-N = \frac{2}{\pi} \sum_{p=0}^{\infty} \frac{\sin(2p+1)\alpha}{2p+1} \left\{ \left(\frac{r_1}{r_0}\right)^{2p+1} + \left(\frac{r_0}{r_2}\right)^{2p+1} + \left(\frac{r_0 r_2}{r_s r_s}\right)^{2p+1} - \left(\frac{r_0 r_1}{r_s r_s}\right)^{2p+1} \right\}$$

The values of the shape factor are given for the various α corresponding to the conductor blocks in the dipole using the notations :

$$\frac{r_1}{r_0} = x_1 \qquad \frac{r_0}{r_2} = x_2 \qquad \frac{r_0 r_2}{r_s r_s} = x_3 \qquad \frac{r_0 r_1}{r_s r_s} = x_4$$

$$\alpha = \frac{\pi}{2} \quad 1-N = \frac{2}{\pi} \left\{ \sum_{i=1}^{3} \text{arctg } x_i - \text{arctg } x_4 \right\}$$

$$\alpha = \frac{\pi}{3} \quad 1-N = \frac{1}{\pi} \left\{ \sum_{i=1}^{3} \text{arctg}\frac{\sqrt{3}\,x_i}{1-x_i^2} - \text{arctg}\frac{\sqrt{3}\,x_4}{1-x_4^2} \right\}$$

$$\alpha = \frac{\pi}{4}$$

$$1-N = \frac{1}{\pi} \left\{ \sum_{i=1}^{3} \text{arctg}\frac{\sqrt{2}\,x_i}{1-x_i^2} - \text{arctg}\frac{\sqrt{2}\,x_4}{1-x_4^2} \right\}$$

$$\alpha = \frac{\pi}{6} \quad 1-N = \frac{2}{3\pi}$$

$$\left\{ \sum_{i=1}^{3} \text{arctg } x_i + \text{arctg } x_i^3 + \frac{1}{2}\text{arctg}\frac{x_i}{1-x_i^2} \right\}$$

$$- \left\{ \text{arctg } x_4 + \text{arctg } x_4^3 + \frac{1}{2}\text{arctg}\frac{x_4}{1-x_4^2} \right\}$$

For the studied model dipole the numerical values amount to :

$1-N_{\alpha=\pi/2} = 0.95 \qquad 1-N_{\alpha=\pi/3} = 0.94$

$1-N_{\alpha=\pi/4} = 0.92 \qquad 1-N_{\alpha=\pi/6} = 0.88$

THEORETICAL CALCULATION OF THE MAGNET LOSSES

The calculation deals only with the coupling current losses induced by the field perpendicular to the broad face of the cable in the straight parts of the magnet supposed to be one meter long. This losses are dominant in the measured dipole magnet due to the geometry of the cable and the fast variation of the field. We have then neglected the losses due to the field parallel to the broad face of the cable and the losses in the superconducting strands (magnetization and inter filament coupling losses).

Using expressions (1) (2) and (3) the losses are evaluated for each block of conductors listed in fig.3 for a linear discharge from 10 kA down to 0 in 80 s (dB/dt = 0.107 T/s).

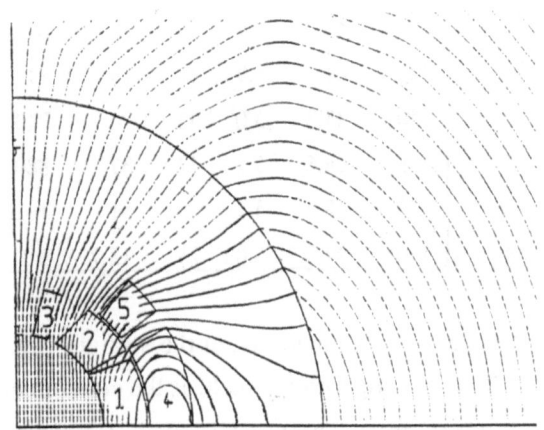

FIGURE 3 : flux lines distribution at I = 10 kA and numbers of the blocks.

Hypothesis for calculation :

- B_1 and B_2 : in each block B_1 and B_2 are supposed to be constant and equal to the mean values of the tangential induction component along respectively the inner and outer surfaces of the blocks given by the programme POISSON for I= 10 kA. A field distribution is given fig.3.

- 1-N : considering the relative values of the losses in the different blocks, 1-N is calculated only from the blocks 1 and 2. Coupling losses being proportionnal to the mean induction variation speed, the current sheets modeling these currents in the blocks 1 and 2 are in the ratio of the mean induction in both blocks. The field in the block 2 is 80% of the field in the block 1. To calculate the shape factor of the blocks 1 and 2 the magnetic field is subdivided in two parts : the first one which covers the two blocks $(-\pi/3 ; +\pi/3)$ amounts to 80% of the field, the second part amounts to 20% of the field and covers only the first block $(-\pi/6 ; +\pi/6)$.

$$1-N = 0.8 \, (1-N)_{\alpha=\pi/3} + 0.2 \, (1-N)_{\alpha=\pi/6}$$

$$1-N = 0.93$$

It appears that the shape factor in the tested model magnet is close to 1.

As $\theta/1-N$ is an intrinsic property of the cable which has been measured equal to 4.7 s, the decay time constant in the magnet is deduced to be :

4.7 s (1-N) = 4.4 s.

The losses calculation for each block are reported in table 2.

TABLE 2 : losses calculation for each block

n° block/ number of cables	B_1 (T)	B_2 (T)	Volume (cm³)	w MJ/m³	E (J)	P_{max} (W)
1/10	8.0	2.6	187	1.24	232	3.07
2/8	6.6	1.9	150	0.80	120	1.58
3/3	2.5	0.0	56	0.07	4	0.05
4/15	2.1	-2.4	280	<0.01	0.3	<0.01
5/6	1.1	-1.3	112	<0.01	<0.1	<0.01

The values for the dipole are :
- maximal dissipated power : P_{max} = 18.8 W
- total dissipated energy : E = 1425 J

EXPERIMENTAL LOSSES MEASUREMENT.

The magnet is seen as a transformer with a lossless primary and a secondary short-circuited on a resistance (fig. 4). The secondary current represents the coupling currents.

The equations governing these circuits are :

$$V = L\frac{dI}{dt} + M\frac{di}{dt}$$

$$0 = R\,i + 1\frac{di}{dt} + M\frac{dI}{dt}$$

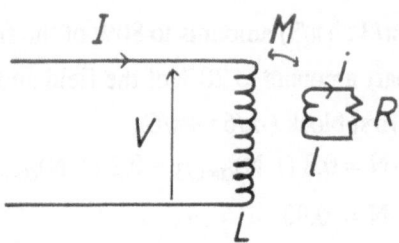

FIGURE 4 : electrical modelling of the magnet.

$$\left(\theta = \frac{1}{R}\right)$$

If I^0 represents the initial main current I and P_f the power dissipated in the resistance R the resolution of the equations leads to the following results :

$t < T$: linear magnet discharge $\frac{dI}{dt} = \frac{I^0}{T} = $ constant

$$V = L\frac{I^0}{T} - \frac{M^2}{T}\frac{I^0}{T}\exp\left(-\frac{t}{\theta}\right)$$

$$P_f = \frac{M^2}{R}\left(\frac{I^0}{T}\right)^2\left[1-e^{-t/\theta}\right]^2$$

$t > T$: discharge ended $I = 0$

$$V = \frac{M^2}{T}\frac{I^0}{T}\left[1-e^{-T/\theta}\right]\exp\left(-\frac{t-T}{\theta}\right)$$

$$P_f = \frac{M^2}{R}\left(\frac{I^0}{T}\right)^2\left[1-e^{-T/\theta}\right]^2\exp\left(-2\frac{t-T}{\theta}\right)$$

The total energy E is :

$$E = \frac{M^2}{R}\left(\frac{I^0}{T}\right)^2\left\{T-\theta\left[1-e^{-T/\theta}\right]\right\}$$

The voltage across the magnet when $I=0$ ($t = T^+$) is noted V^0, it amounts to :

$$V^0 = \frac{M^2}{T}\left(\frac{I^0}{T}\right)\left[1-e^{-T/\theta}\right]$$

P_f and E may be expressed using V^0 :

$$E = \theta\frac{I^0}{T}\,V^0\,\frac{T-\theta\left[1-e^{-T/\theta}\right]}{1-e^{-T/\theta}}$$

$t < T$: $$P_f = \theta\frac{I^0}{T}\,V^0\,\frac{\left[1-e^{-t/\theta}\right]^2}{1-e^{-T/\theta}}$$

$t > T$: $$P_f = \theta\frac{I^0}{T}\,V^0\left[1-e^{-T/\theta}\right]\exp\left(-2\frac{t-T}{\theta}\right)$$

The losses can be calculated from the measurements of θ and V^0 since T and I^0 are imposed.

V^0 and θ may be deduced from the decay of V when the discharge of the magnet is ended (logarithmic decrement method).

The Measurements are reported in table 3.

TABLE 3 : fast discharge measurements.

I^0 (kA)	dI/dt	T	θ	V^0	P_{max}	E
test number	(A/s)	(s)	(s)	(mV)	(W)	(J)
5/1	80	62.5	5.0	14	5.0	320
5/2	100	50	5.0	19	9.5	430
5/3	125	40	4.9	30	18	640
8/4	125	64	4.9	31	19	1120
5/5	150	33.3	5.0	35	26	740
5/6	175	28.6	5.1	42	37	880
5/7	200	25	5.1	50	51	1020
3/8	110	27.3	5.0	25	13.7	310

V^0 has been extrapolated from the straight line of slope θ in semi logarithmic scale as the current cut is not perfect. It can be verified that the quantity $\frac{V^0}{1-\exp(-T/\theta)}$ is approximately proportionnal to $\frac{I^0}{T}$ as theoretically expected (fig. 5).

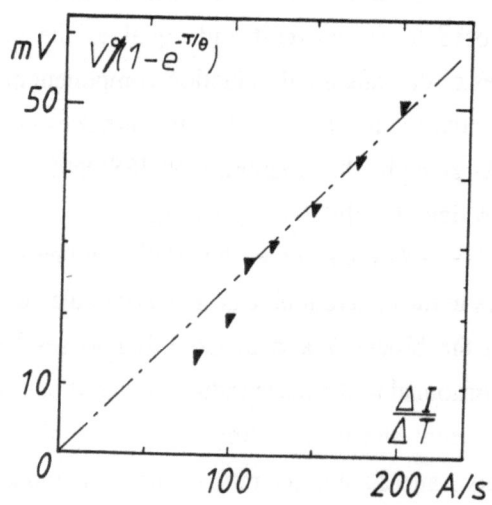

FIGURE 5 : variation of $\frac{V^0}{1-\exp(-T/\theta)}$ with $\frac{I^0}{T}$.

The value of θ amounts to 5s $\pm$ 2% for the eight tests. Compared with the theoretical value the relative error is 14 %.

Considering an initial current of 10kA we get from test n°4 the following results:

$$P_{max} = 19 \text{ W} \qquad E = 1420 \text{ J}$$

The theoretical results for this discharge are recalled:

$$P_{max} = 18.8 \text{ W} \qquad E = 1425 \text{ J}$$

The agreement is very good considering all the hypothesis done both for theory and measurement.

CONCLUSION

The measurement of the voltage across the magnet is a very simple method to get the coupling current losses in a magnet. The results on the model magnet indicate small differences (10%) with the theoretical calculations. The results have to be confirmed by tests on other magnets.

AKNOWLEDGEMENTS

The short sample measurements of the cable were performed using the equipement facilities of the C.E.A. at Cadarache in the group of B. Turck. The authors gratefully acknowledge him for his interest and support.

REFERENCES

1. G. Brianti, K. Hubner, The Large Hadron Collider in the L.E.P. tunnel. CERN 87-05
2. R. Perin, Progress on the superconducting magnets for the large Hadron collider. , IEEE Trans. on Magn. , 1988 , 24 , 734-740.
3. R. Perin, D. Leroy, G. Spigo, The first, industry made, model magnet for the Large Hadron Collider , IEEE Trans. on Magn. , 1989 , 25 , 1632-1635
4. A. M. Campbell, A general treatment of losses in multifilamentary superconductors, Cryogenics , 1982 , 3-16.
5. A. Hughes, T.J.E. Miller , Analysis of fields and inductances in air-cored and iron-cored synchronous machines. Proc. IEE , 1977 , 124 ,121-126.

CRITICAL CURRENT MEASUREMENTS OF PROTOTYPE CABLES FOR THE CERN LHC UP TO 50 kA AND BETWEEN 7 AND 13 TESLA USING A SUPERCONDUCTING TRANSFORMER CIRCUIT[*]

H.H.J. ten Kate, B. ten Haken, S. Wessel,
J.A. Eikelboom[+], E.M. Hornsveld[+]

University of Twente, Applied Superconductivity Centre,
P.O. Box 217, 7500 AE Enschede, and
(+) ECN, Petten, The Netherlands.

ABSTRACT

The current carrying properties of several prototype Nb_3Sn cables were investigated in the framework of the Netherlands part of the LHC magnet development programme. A facility in which LHC type of cables can be tested with a current up to 50 kA in a background field of 7 to 13 tesla was constructed and taken into operation. The sample current is provided by an inductive method using a superconducting transformer circuit by which the sample conductor is part of the secondary circuit. This method avoids the enormous helium loss in the 50 kA current leads and a costly low-ripple conventional 50 kA supply. The conductors under test are measured in the shape of a 4-turns spiral coil or a hair-pin. The investigated ECN type of Nb_3Sn cables have an excellent average critical current density in the strands of about 900 A/mm^2 at 10 T with 55% stabilizing copper. The paper reports on the measuring device and results of critical current measurements are presented.

INTRODUCTION

The investigation of the quality of a superconductor is an essential part of any magnet development programme especially when operating current and magnetic field are at a high level. As a contribution to the magnet development programme for the proposed Large Hadron Collider LHC [1], a 1 meter 10 tesla Nb_3Sn twin aperture model dipole magnet is developed in an UT-NIKHEF programme carried out in collaboration with ECN, SMIT WIRE and HOLEC [2].

The nominal current of these magnets is about 16 kA and the required test current to measure the critical current in the relevant field range of 7 to 11 tesla is 50 kA maximum for the best conductors.

The usual method of measuring the voltage-current characteristics in order to determine the critical current density and other qualifying parameters becomes unpractical when the current carrying capacity of the conductors go beyond a level of 10 kA and the conductors have a large cross-section. A direct current feeding of the sample conductor is, at these levels of current and magnetic field, time consuming and expensive. In order to overcome this we apply a superconducting power supply i.e. a superconducting transformer and switching circuit [3] (also called single

---

* These investigations in the programme of the Foundation for Fundamental Research on Matter FOM are supported, in part, by the Netherlands Technology Foundation STW.

step superconducting rectifier) to generate the test current in the cable. This
method of maximum current testing avoids
the enormous helium loss of 150 L/hr
associated with the 50 kA current
feedthroughs and the costly high quality,
low current ripple 50 kA power supply.

The method of current induction using
a superconducting transformer is reviewed
elsewhere [4]. The technique is based on a
ten years experience at our university
with this type of circuits [5,6,7]. At the
moment there exist worldwide only a few
facilities in which measurements on sections of cables in the current range of 10
to 20 kA at magnetic fields around 10
tesla can be performed.

In the following sections we report
on the presently available LHC type of
conductors, the method of measuring the
critical current, the superconducting
transformer circuit as well as the results
found with two Nb_3Sn cables.

Nb_3Sn LHC CONDUCTORS

The coils in the LHC main dipole magnets have a two layer geometry in which
large-aspect ratio cables are distributed
in a specific way in order to approximate
the ideal cosine distribution of current
around the beam channel. The LHC specification for the critical current density
required in these cables to attain 10 T is
1300 A/mm^2 at 11 T and 4.2 K assuming a
copper part for stabilization not less
than 50 %.

Up to now prototype Nb_3Sn LHC cables
have been made by two manufacturers both
using a different production method. Vacuumschmelze (VAC) made cables based on the
conventional bronze process but the copper
part had to be limited to about 38 % in
order to get a sufficiently high current
density. These conductors were used by
ELIN in the first 1 meter single aperture
LHC model magnet which recently was tested
up to a quench field of 9.45 T [8].

The second type of Nb_3Sn conductor is
a development of the Netherlands Energy
Research Foundation ECN. The "powder
method" conductor is characterized by the
application of Nb tubes filled with $NbSn_2$
powder and surrounded by pure copper. The
hexagonal elements are stacked, assembled
to a billet and then drawn to thin wire.
During the heat treatment the Nb_3Sn filaments grow from the inside of the Nb
tubes thus preventing a contamination of
the copper matrix, see figure 1. A wire

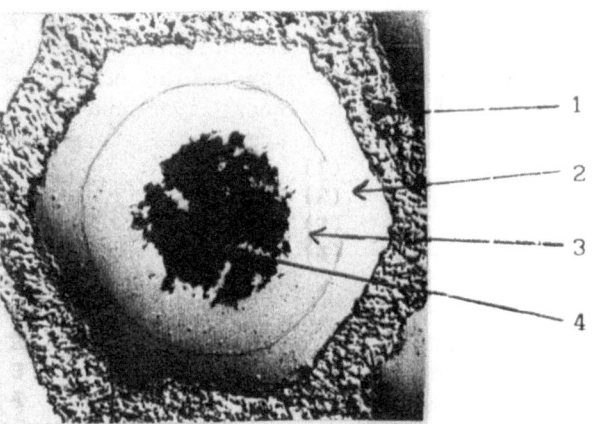

FIGURE 1. Detail of a hexagon in an ECN
type Nb_3Sn conductor showing: (1) the copper matrix, (2) Nb tube, (3) Nb_3Sn layer
and (4) the residues in the core.

with a pure Cu matrix is obtained without
the presence of bronze in the wire
section. The obvious advantage is the
average current density which is much
larger than in the bronze conductors. On
the other hand, it is hardly possible to
produce wires with sub-micron filaments
due to the grain size of the powders.
Wires with 36 and 192 filaments were made
in a relatively large quantity for the
SULTAN coils and NET model conductors.
Double stacked wires with 324, 684 and
1332 filaments were already made on a
laboratory scale [9].

Based on the 192 filaments conductor
a few LHC type of cables have been made.
The main characteristics of the VAC and
the ECN cables are given in Table 1. As an
example, the cross-section of the ECN(B)
cable, developed for the outer layer of
the LHC dipole coils, is shown in figure 2. The filament size in the 192 filaments conductor is to thick to be acceptable for the final LHC from the viewpoint of conductor magnetization. A next
objective is to reduce the filament size
to below 15 μm by further developing the
1332 filaments- and comparable conductors,
and by developing two stage cables.

METHOD OF MEASURING

The conductor to be tested is made
part of a closed circuit with a superconducting switch and the secondary windings
of the transformer as illustrated in
figure 3. The primary coil of the
transformer is connected to a 100A/10V
power supply. When the current amplification factor of the transformer is

TABLE 1.

Main characteristics of prototype Nb$_3$Sn LHC cables of VAC [8] and ECN for the inner and outer layer. The nominal current for all cables is about 16 kA at 10.2 T for the inner layer and 8.7 T for the outer coil layer. All of these are single stage keystoned Rutherford cables.

	inner layer cables		outer layer cables	
	VAC	ECN(A)	VAC	ECN(B)
size [mm^2]	2.19/2.69×16.8	2.40/2.70×16.4	1.47/1.79×16.8	1.45/1.77×16.4
Cu part [%]	29	55	25	55
strands [-]	24	24	36	36
str.diam.[mm]	1.37	1.35	0.92	0.90
filaments [-]	50,000	192	20,000	192
fil.diam.[µm]	2.6	47	2.5	33

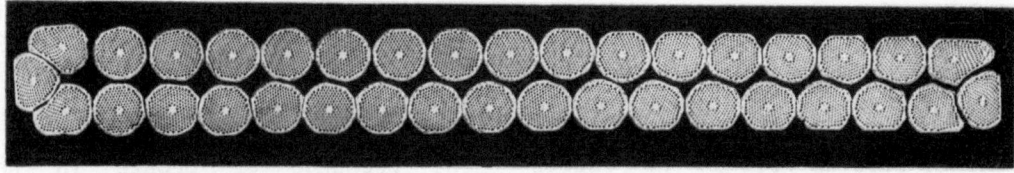

FIGURE 2. Cross-section of an ECN-type Nb$_3$Sn keystoned LHC cable, see Table 1, 36 strands of 192 filaments, 55 % Cu, size 1.45/1.77×16.4 mm^2.

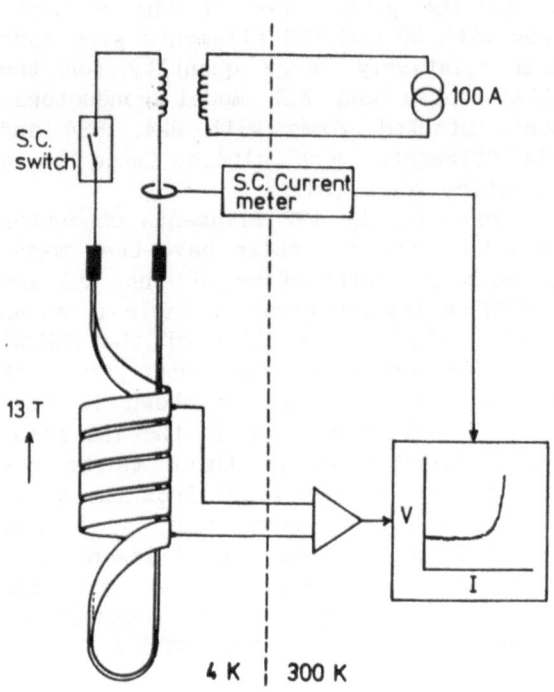

FIGURE 3. The superconducting transformer circuit connected to a test conductor having the shape of a small coil.

high, in this case 520, a very large secondary current can be induced provided the resistance in the secondary circuit is sufficiently low.

The measuring cycle is now as follows: First the primary current is swept to a certain value while the switch in the secondary circuit is open. After that the switch is closed and the primary current is swept thus inducing an increasing secondary current. The precise shape of this current can be prescribed when the transformer operates in a feedback mode.

The secondary current is measured using a superconducting current meter especially developed for this type of circuits [10]. The voltage coming up when the current approaches the critical value is measured with a Keithly nano-voltmeter.

TABLE 2.

Parameters of the superconducting transformer used for sample powering.

shape: two concentric solenoids		
primary inductance	[H]:	1.0
secondary inductance	[µH]:	0.33
mutual inductance	[mH]:	0.17
current ampl. factor	[-]:	520
primary turns	[-]:	3600
secondary turns	[-]:	1
inner diameter	[mm]:	80
outer diameter	[mm]:	145
length	[mm]:	70
maximum prim. current	[A]:	90
maximum sec. current	[kA]:	50

FIGURE 4. Photograph of the superconducting transformer circuit: (1) concentric primary and secondary coils, (2) sc. current meter, (3) insert of 13 T magnet with conductor sample (bore 60 mm).

The superconducting transformer circuit.

The main parameters of the transformer are collected in Table 2, and its practical realization is pictured in figure 4. The single secondary turn was made with two NbTi/Cu LHC cables connected in parallel. They were provided free of charge by Alsthom France. The test conductor is connected to the secondary turn by means of 15 cm long soldered joints. The time constant of the circuit

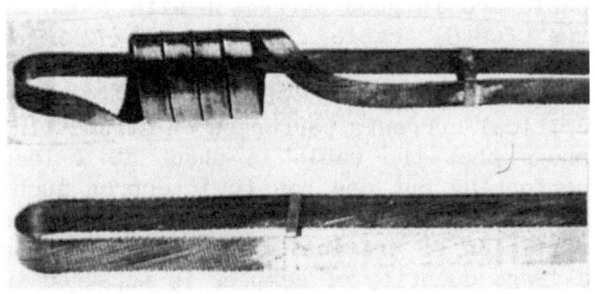

FIGURE 5. Photographs showing the two conductor geometries used: a 4-turns spiral coil and a hairpin.

is about 800 s which corresponds to a joint resistance of 0.2 nΩ, a value also required in the LHC magnets.

The transformer is mounted above the magnet that provides the external field at the sample.

The test conductors in the magnetic field.

Two basic geometries of the conductors in the external field were applied as indicated in figure 5. The hair-pin has the advantage of easy and quick mounting, but it is impossible to measure accurately the voltage below the critical current. The length of the conductor in the 4-turns spiral coil is about 60 cm which is provides better results when measuring in the low voltage regime. Moreover the direction of the magnetic field on the conductor is different in both cases. The inhomogeneity of the external magnetic field along the sample is ± 5 %. Note that the self field of the 4 turns also causes a relatively large field gradient over the conductor in the perpendicular direction.

The conductor samples are mounted on

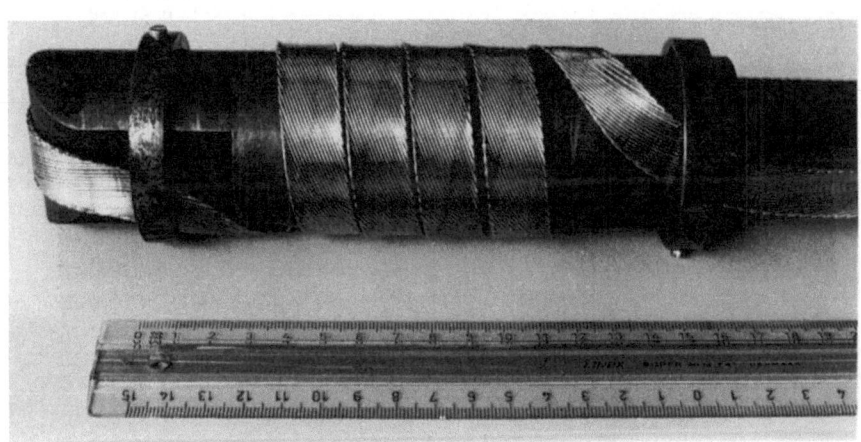

FIGURE 6. The 4-turns spiral coil on the sample holder before impregnation.

64

stainless steel tubes which are, after the Nb$_3$Sn forming reaction heat treatment, wrapped up glass tape and vacuum impregnated with STYCAST 2850FT. In this way the conductor is insulated in order to approximate the colling conditions in a bath cooled impregnated coil. Figure 6 shows a cable on the sample holder before wrapping and impregnation.

RESULTS

Results of the following measurements are presented:
(1) 4-turns coil test of the ECN(B) cable,
(2) hair-pin test of the ECN(B) cable,
(3) hair-pin test of the ECN(A) cable.
Both cables are specified in table 2. The critical current of both VAC cables will be measured in the coming months. The results are collected in table 3 and shown in figure 7. Note that the load lines of the inner and outer coils in the LHC dipoles are also indicated in the figure.

In the case of a hair-pin test the cable section between the voltage taps is so short (2 cm) that interpretation of the voltage is not meaningful. Therefore the quench current of the hair-pin samples is mentioned in table 3. The voltage measured on the 4-turns coil over 62 cm cable provides nice voltage-current traces which can be used to determine the critical current at the relevant criterion of $10^{-14}\Omega$m, see figure 8. For example the critical voltage at 10 kA in the ECN(B) cable is 2.7 nV. The quench current of the 4-turns coil generally occurs at a voltage

TABLE 3.
Results of: the quench (I_q) and critical currents (I_c) of the ECN A and B Nb$_3$Sn cables, $10^{-14}\Omega$m non copper critical current density of the B cable as calculated from I_c^B; all values are at 4.2 K.

B[T]	I_q^B[kA]	I_c^B[kA]	I_q^A[kA]	J_c^B[A/mm^2]
7	36.3			3380 *
8	30.9			2850 *
9	26.1		35	2390 *
10	22.0		29	1990 *
11	17.8	16.5	24	1700
12	14.4	13.5	19	1300
13	11.2	9.5	15	920

* calculated with (I_q^B-1.5).

level of 5 to 20 nV which implies that the quench current that can be expected lies about 0.5 to 1.5 kA above the critical current at about 10 kA, as was found in the measurements. Note that the samples are impregnated with epoxy which implies that the cables will quench in the vicinity of the $10^{-14}\Omega$m critical current.

The quench currents of the ECN(A) cable are in good agreement with those of the ECN(B) cable. Due to the larger section a factor 1.5 should be found.

A comparison with short sample critical currents performed on strand wire shows that the cable is about 10 % less performing but one has to interpret such results carefully since generally the variation of critical currents found when a large quantity of samples is measured is typically between 5 and 10 %.

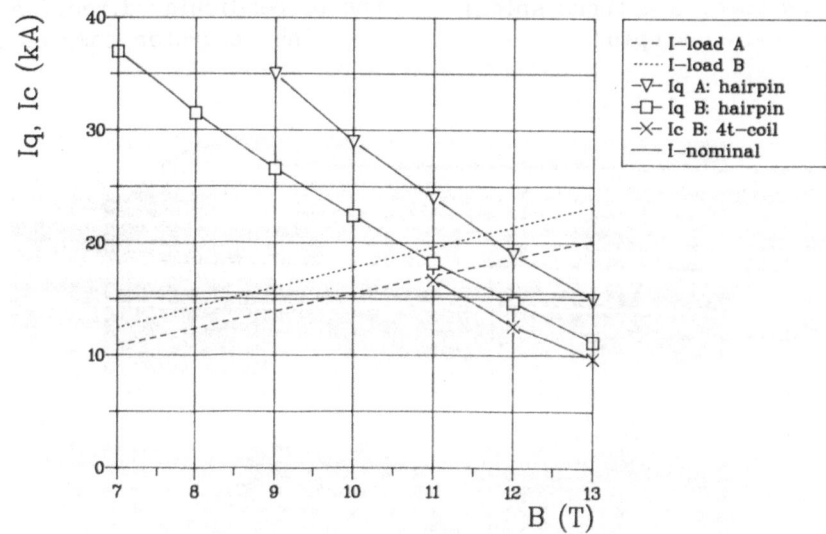

FIGURE 7. Measured quench- and critical currents of the 3 measured samples. The load lines of both coils in the LHC dipole magnet are also indicated.

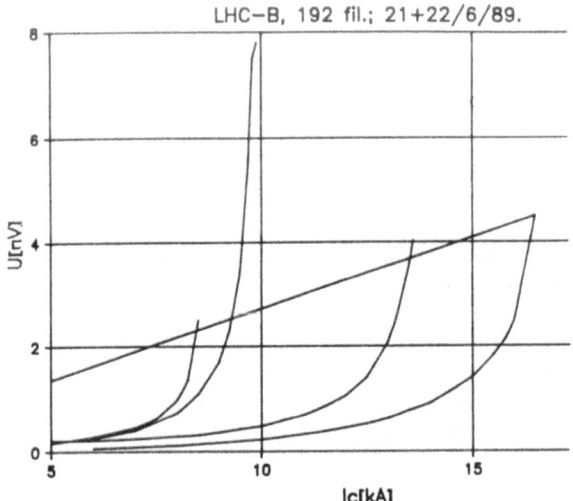

LHC-B, 192 fil.; 21+22/6/89..

FIGURE 8. Measured voltage versus current curves, Nb_3Sn LHC-B cable at 11,12,13 and 13.5 tesla.

CONCLUSIONS

Especially for the conductor development part of the LHC Nb_3Sn magnet program we designed and constructed a superconducting transformer circuit in order to measse critical currents. The circuit provides a maximum test current of 50 kA. It is a new facility to investigate the current carrying properties of medium size superconducting cables in the shape of either a hairpin or a 4-turns spiral coil in a background field up to 15 tesla..

The critical currents found in the experimental cables made by ECN are very high. A non-copper critical current density of 2000 A/mm^2 at 10 tesla and 4.2 K was found. The critical current at the working points of the LHC collider of 8.7 and 10.2 tesla for the outer and inner coils respectively, is about 90% above the operating current of about 16 kA. This means that when applying these conductors the stability margin at 10 tesla operation of the LHC is very high.

In principle the current density in the ECN type of Nb_3Sn conductors is large enough to attain 11 tesla operation provided the enhanced forces in the windings and structure could be handled.

The superconducting transformer circuit, as presented here, provides a very convenient, reliable and inexpensive method to measure critical currents in large conductors.

REFERENCES

1. Perin, P., Progress on the superconducting magnets for the Large Hadron Collider. IEEE Transaction on Magnetics, MAG-24, 734 (1988).
2. Ten Kate, H.H.J., Daum, C., Van de Klundert, L.J.M., Bode, H., Hoogland, W., Roeterdink, T.A., Van Overbeeke, F., Perin, R., On the development of a 1 meter "twin aperture" 10 T Nb_3Sn dipole magnet for the CERN LHC. Proc. IISSC, first Int. Industr. conf. on the SSC, New Orleans, Febr. 1989; and this MT11 conference paper JG-01.
3. Ten Kate, H. H. J., Superconducting Rectifiers. Thesis University of Twente, April 14, 1984.
4. Mulder, G., Ten Kate, H.H.J., Krooshoop, H.J.G., Van de Klundert, L.J.M, Inductive method for maximum current testing of superconducting cables. In Proc. MT11 conference, August 28 1989, Tsukuba, Japan.
5. Ten Kate, H.H.J., Pijper,H., Nijhuis, A., Van de Klundert, L.J.M., Maximum current and quench sensitivity test of 40 kA multistrand NbTi/CuNi conductor. In Proc. MT-9. SIN, Zurich, 1985, 584-7.
6. Ten Kate, H.H.J., Uyttewaal, W., Ten Haken, B. Van de Klundert, L.J.M., The Twente high-current conductor test facility, first results on critical current and propagation in two cables. In Advances in Cryogenic Engineering, vol. 33, ed. R.W. Fast, Plenum Press, New York, 1988, 211-18.
7. Mulder, G.B.J., Krooshoop, H.J.G, Nijhuis, A., Ten Kate, H.H.J. and Van de Klundert, L.J.M., A convenient method for testing high-current superconducting cables. Presented at CEC/ICMC-89, Los Angeles, July 1989.
8. Wenger, S., Zerobin, F., Asner, A., Towards a 1 m long high field Nb_3Sn dipole magnet of the ELIN-CERN collaboration for the LHC-project. IEEE Transaction on Magnetics, MAG-25 (1989) and this conference paper JA1.
9. Hornsveld, E.M., Elen,J.D., Van Beijnenand, L.A.M., Hoogendam, P., Development of ECN-type Niobium-tin wire towards smaller filament size. Adv. in Cryog. Eng., vol. 34, 493 (1987).
10. Ten Kate, H.H.J., Nederpelt, W., Juffermans, P., Van Overbeeke, F.,Van de Klundert, L.J.M, A new type of superconducting direct current meter for 25 kA. In Adv. in Cryog. Eng., vol. 31, Plenum Press, NY, 1986, 1309-12.

DEVELOPMENT OF A 10 T Nb$_3$Sn "TWIN APERTURE" MODEL DIPOLE MAGNET FOR THE CERN LHC*

H.H.J. ten Kate, C. Daum[+] L.J.M. van de Klundert, H. Bode[@] W. Hoogland[+]
J.D. Elen[#] F. van Overbeeke[&] R. Perin[o]

University of Twente, Applied Superconductivity Centre, POB 217,
7500 AE Enschede, (+) NIKHEF-H, Amsterdam; (#) ECN, Petten;
(&) HOLEC, Ridderkerk; (@) SMIT WIRE, Nijmegen, all in
the Netherlands; and (o) CERN, Geneva, Switzerland.

ABSTRACT

A colliding energy of 16 TeV as requested for the proposed new hadron collider LHC of CERN can only be attained if the techniques to construct 10 T dipole magnets are available since LHC should be installed in the existing LEP tunnel. At the moment a development programme is carried out at CERN in collaboration with European laboratories and industries to construct several 10 tesla model magnets following either the NbTi/2 K or the Nb$_3$Sn/4 K routes. Here we present a new initiative to explore the Nb$_3$Sn route. A 4 years R&D programme was started in 1988, in the framework of a co-operation agreement with CERN, by the Applied Superconductivity Centre at the University of Twente and NIKHEF, the National Institute for Nuclear and High Energy Physics in collaboration with ECN, HOLEC and SMIT WIRE. The paper reports on various design aspects and the present state of development.

INTRODUCTION

The proposed new Large Hadron Collider (LHC) providing proton-proton collisions up to an energy level of about 16 TeV is studied at CERN. It should provide a much higher collision energy as a next step towards new discoveries in, and better understanding of, particle physics.

A combination of LHC with the existing LEP tunnel and equipment would lead to an elegant and economical machine. Obviously, when using the LEP tunnel with its circumference of 27 km, the required magnetic field in the LHC dipole magnets is about 10 T in order to attain the 16 TeV

---

* These investigations in the programme of the Foundation for Fundamental Research on Matter FOM are supported, in part, by the Netherlands Technology Foundation STW.

centre-of-mass proton-proton collision energy. The restricted space available in the LEP tunnel forces to combine both beam channels into a single magnet instead of using two separate accelerator rings.

The so-called "2-in-1" or twin aperture magnet is illustrated in figure 1. Note that construction details as well as cryostat parts are not shown. The beam channels each having a set of coils are combined in a single force-retaining collaring structure and a yoke that links the magnetic fields in the beam channels.

Reduced space requirements and costs are obvious advantages but the magnetic coupling of the channels is a disadvantage of the "2-in-1" concept. The LHC rings consist of about 1860 main dipole magnets with a length of about 10 meter. More general information about LHC and the magnet technology can be found in references [1,2,3].

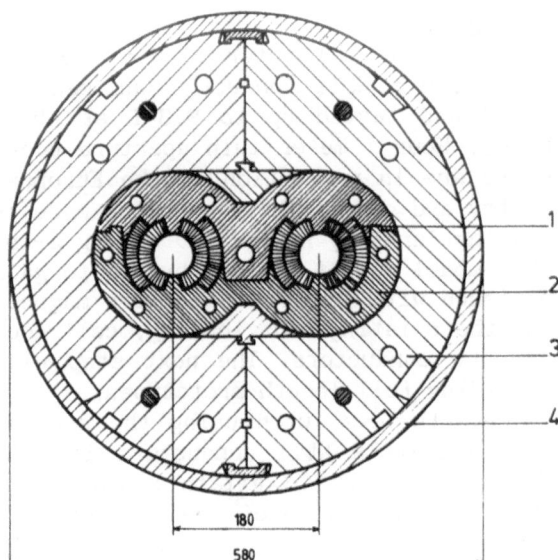

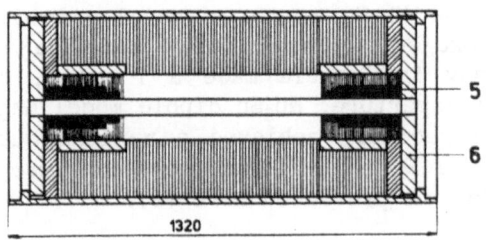

FIGURE 1. Schematic views of the cross and longitudinal sections of a 10 T twin aperture LHC dipole magnet showing the main parts: (1) coils, (2) collars, (3) cold iron yoke, (4) outer shrinking cylinder enclosing the cold magnet parts, (5) coil head sections, (6) end-flange, end-plates.

LHC DIPOLE MAGNET PARAMETERS

The criteria which have led to the present "2-in-1" geometry are the 10 T on the beam, the aperture size and the intra-beam distance restricted by the size of the LEP tunnel. The resulting design shows a 2-shell coil system with graded current density, surrounded by a laminated clamping structure, collars, and a vertically split and laminated cold iron yoke. This package is enclosed by an outer shrinking cylinder that handles the major part of the forces. The main parameters optimized for the requirements are given in Table 1.

The various model magnets have a 1:1 scale magnet cross-section, but a reduced length of 1 meter in order to keep down the costs of development.

The magnetic field is generated by two (almost) identical coil systems which

TABLE 1.
Main parameters of the LHC dipole magnets

nominal magnetic field	=	10.0 T
max. field in layer 1	=	10.2 T
max. field in layer 2	=	8.7 T
full magnetic length	=	9.5 m
coil inner diameter	=	50 mm
coil outer diameter	=	122 mm
distance between beams	=	180 mm
overall mass	=	18 tons
cold mass	=	15 tons
mass of superconductor	=	600 kg
stored energy at 10 T	=	684 kJ/m
operating current	=	16 kA
maximum Lorentz forces:		
in x-direction	=	4.6 MN/m
in y-direction	=	-1.2 MN/m
on coil head	=	0.17MN

are positioned around the two beam channels. They have a two-layer geometry and the 4 sub-coils are connected in series. A specific distribution of conductors is required to obtain a good field quality. An example of a possible conductor distribution is shown in figure 2. Two different conductors are applied, one for each layer. They have equal currents and width of 17 mm but a different thickness. This configuration provides multipoles less than the required 10^{-4} [4].

The coil support structure has to provide first, a sufficient rigidity and a sufficiently high dimensional accuracy to guarantee the field quality during the entire operating cycle, and second the

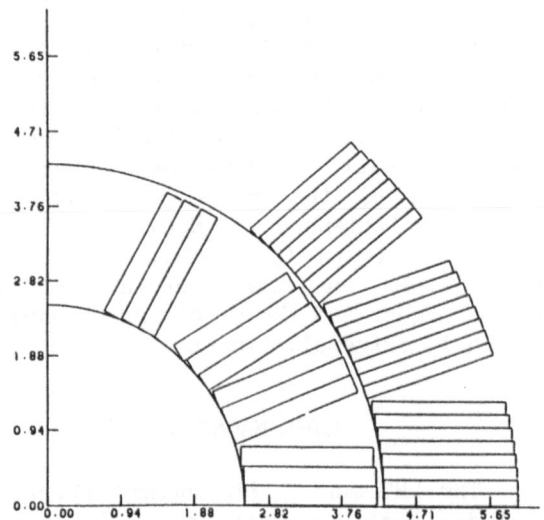

FIGURE 2. Example of a conductor distribution using a rectangular conductor.

coils have always to be under compression to avoid coil or conductor displacements that can initiate a quench which is fatal to accelerator operation. In contrast to the case of the common 5 and 6 T dipole magnets, the required pre-stress on the coils for 10 T operation is so large that it is not practical nor optimal to use only the collars to provide the pre-stress. Therefore, clamping of the coils in LHC dipole magnets is obtained by a combined action of collars, iron yoke and the outer cylinder in the "mechanically hybrid" support structure [4]. About 30% of the force F_x in the horizontal plane of 4.6 MN per meter at 10 T is taken by the collars and the major part of 70% is taken by the outer cylinder.

NbTi OR Nb$_3$Sn

A magnetic field of 10 T can be attained either by using a NbTi/Cu superconductor cooled with superfluid He at 1.8 K or a Nb$_3$Sn/Cu conductor operating at 4 K.

The major disadvantages of using NbTi are the complexity of the required large scale 2 K cooling system, the limited temperature margin, and the extremely low enthalpy at 2 K. In particular the latter property means that the magnet is extremely sensitive to energy releases caused by beam losses. This could initiate a quench and reduce the efficiency of operation of the accelerator.

Nb$_3$Sn has a critical temperature of 18 K instead of 9 K with NbTi and its critical field is, about 20 T at 4.2 K instead of the 13.7 T of NbTi at 1.8 K. Moreover, the enthalpy at 4.5 K is about a factor of 40 larger than of NbTi at 1.8 K.

These obvious advantages are partly compensated by the more difficult production procedures as the application of the "wind-and-react" process and vacuum impregnation of the coils. Moreover, the extreme brittleness of Nb$_3$Sn limits the allowable strain to a clear maximum of 0.2-0.3%. So, an analysis of the mechanics of the magnet and the resulting mechanical design is extremely important.

MODEL MAGNETS

The LHC magnet development programme includes both the NbTi and Nb$_3$Sn routes, but the major effort is concentrated on the NbTi/1.8 K version. In this case the magnet technology is more straightforward

and based on experiences with TEVATRON and HERA magnets.

Meanwhile a 1 meter single aperture model using a NbTi conductor was made by ANSALDO and tested up to 9.3 T at about 2 K [6]. Four 1 meter twin aperture model magnets are in production in the European industry and one of these is being made in the Netherlands by HOLEC. A 9 m long 2-in-1 NbTi model magnet is made by ABB to gain experience with long twin aperture magnets using the existing HERA tooling, coil geometry and conductors. The next step will be building full length prototypes.

Nevertheless, when using the commercially available NbTi conductors it will be very difficult to attain a clear and stable 10 T operation of the complete accelerator. Especially the sections in the accelerator which are exposed to an increased level of radiation are critical.

The Nb$_3$Sn route is developed within two projects. First, CERN and the ELIN company in Austria made a 1 meter, also single aperture, Nb$_3$Sn dipole model magnet which recently achieved a maximum quench field of 9.45 T [7] and the second project is the one presented here [8].

THE Nb$_3$Sn CONDUCTOR

In order to reach the 10 T nominal field the conductor for the inner and outer layer should have an average current density of at least 650 A/mm^2 at 11 T and a copper part for stabilization not less than 50 %. The preliminary parameters of the two cables as required for the inner and outer coils are given in Table 2.

At present three practical methods for Nb$_3$Sn wire production exist, the bron-

TABLE 2.
Parameters of Nb$_3$Sn cables for the inner- and outer layer respectively (1m magnet).

		inner,	outer layer
strand diam.	[mm]	0.90	1.35
percentage Cu	[%]	50-60	50-60
number of strands		24	36
twist length	[mm]	30	30
RRR value		>100	>100
bare width	[mm]	16.4	16.4
mean thickness	[mm]	2.43	1.61
cable pitch	[mm]	<120	<120
cable length	[m]	4 × 25	4 × 40
nominal current	[kA]	≈16@10.2 T	≈16@8.7 T
critical current	[kA]	>17@11 T	>18@8.5 T

ze, internal tin and powder method wires.

The "bronze" route conductors have the advantage of being available on an industrial scale but, inherent to the presence of bronze and Ta in the wire section, the wire is less performing in terms of the average current density. The LHC specification can only be reached by reducing the amount of copper in the wire section to 30 % instead of the 50 % The first Nb_3Sn single aperture LHC model magnet was made with this material [7].

With the "internal tin" method it is probably possible just to meet the 10 T specification with 50 % copper in the wire section, but at this state of developments no practical lengths are available. In any case, using these materials, it is not possible to go substantially beyond the specifications to further increase the magnetic field or the operational margin.

The third method is a development of The Netherlands Energy Research Foundation, ECN. It is characterized by the use of Nb tubes with a filling of $NbSn_2$ powder [9]. The residual part of the Nb tubes acts as a diffusion barrier protecting the pure copper matrix against contamination with tin during the reaction heat treatment. A pure Cu matrix is present without bronze.

As a consequence, the average current density in the ECN type powder method wire is, with the same amount of stabilizing copper, much higher than with the bronze process. A disadvantage of this type of wire production is that, it is hardly possible to make sub-micron filaments.

Wires with 36 and 192 filaments were made in a production quantity of about 500 kg for application in SULTAN coils and NET prototype conductors. Double stack wires with 324, 684 and 1332 filaments have already been produced on a laboratory scale. A few characteristic wire parameters are given in Table 3.

The non-copper critical current densities in the wires are in the range of 1600 to 2000 A/mm^2 at 11 T and 4.2 K.

TABLE 3.
Comparison of two ECN type Nb_3Sn wires with a wire diameter of 0.85 mm [9].

			192	1332
number of filaments			192	1332
percentage of copper	[%]		52	55
reaction time @675C^o	[hrs]		45	16
Nb tube diameter	[μm]		45	15
Nb_3Sn fil. diameter	[μm]		31	10

ECN is engaged in the development of wind and react cables for the LHC project of CERN. Recently critical current measurements were performed on two conductor samples of which one is shown in figure 3 [10]. This prototype cable for the outer coils shows an excellent performance and has a critical current of 30 kA at the working point of 8.7 T which is about 90 % over the nominal operating current. This result clearly demonstrates the superior critical current density of this material.

FIGURE 3. Photograph of an LHC Nb_3Sn cable developed by ECN; 36 strands 0.90 mm diameter, 192 filaments, 55% Cu matrix, size = 1.45/1.77×16.40 mm.^{2}

DEVELOPMENT PROGRAMME Nb_3Sn MODEL MAGNET

The research and development program for the Nb_3Sn LHC model magnet includes an integral design study of the LHC dipole magnet components in order to generate the best possible solution for the Nb_3Sn magnet within the boundaries set by the general LHC magnet specifications.

Important subjects and components which have to be developed are:

Design of the coil section. The use of rectangular Rutherford cables was studied since they are easier to produce, cheaper and more homogeneous [4].

The superconducting cable. The commercial availability of niobium-tin wires is presently limited to bronze type, not fully attaining the LHC specifications. A programme has been set up to investigate the use of the powder method Nb_3Sn conductor. The particular objective for LHC is to reduce the filament size ECN type conductors to below 15 μm in order to reduce filament magnetization. In the framework of the LHC project, the critical currents of LHC-type Nb_3Sn cables are measured [10].

Calculation of stationary and dynamic cable and filament magnetization. First results can be found in [11,12].

3-D calculation of magnetic fields and forces. An approximation was made using the method of image currents with saturated iron [13]. A more exact result is obtained with the FEM programme TOSCA which allows to calculate the contribution

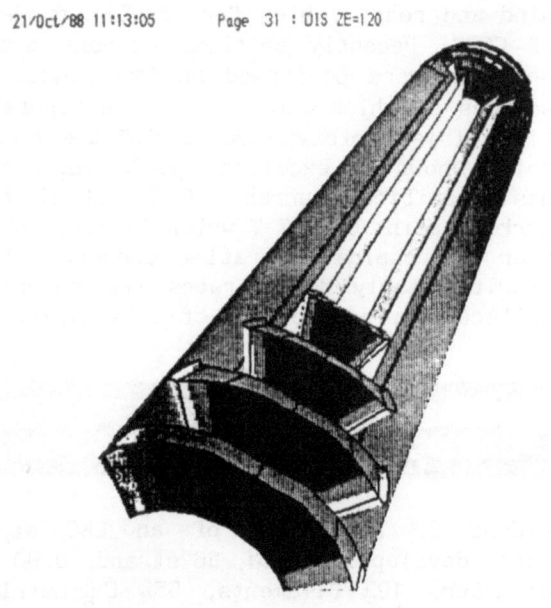

21/Oct/88 11:13:05   Page 31 : DIS ZE=120

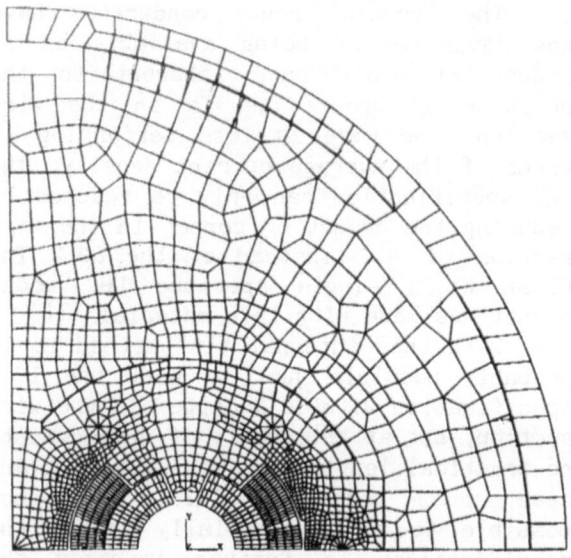

FIGURE 5. 2-D finite element model of the 2-in-1 dipole magnet.

FIGURE 4. 3-D impression of a half coil, 1 m LHC dipole magnet, made with TOSCA.

of the yoke at all fields [14]. In order to illustrate this a representation of a coil in TOSCA is shown in figure 4.

3-D computation of coil and conductor deformation, is in progress using the FEM programs ANSYS and SUPERTAB. A grid representing a quadrant of the 2-in-1 LHC dipole magnet is shown in figure 5.

Coil production, winding, pressing, heat treatment, vacuum impregnation, collaring and magnet assembly.

Effects of heat treatments and thermal contraction of all materials used.

Sensors and actuators. Strain, voltage and temperature sensors, and actuators to diagnose and to protect the magnet.

Magnet testing. Quench protection, powering, test procedures. A pre-test of the magnet at the Applied Superconductivity Centre is foreseen using a superconducting rectifier [15].

A new winding machine and collaring press have been constructed and installed, see figure 6. The obviously critical activity is the production of the superconducting cable. A first magnet test is sheduled for the end of 1990.

CONCLUSIONS

In the framework of the CERN LHC magnet development programme a Nb_3Sn dipole magnet is being designed and constructed in a collaboration between the University of Twente, NIKHEF, ECN, SMIT WIRE, HOLEC and CERN. It will be the first 1 meter 10 T Nb_3Sn twin aperture magnet and the second LHC Nb_3Sn dipole magnet.

The proposed application of ECN type of Nb_3Sn wire could lead to the highest possible field in LHC types of dipole magnets due to the superior current density in comparison to NbTi/1.8K or other commercially available Nb_3Sn materials.

The current carrying capacity is sufficient to attain 11 T operation provided the mechanical structure and the conductor can handle the enhanced forces. This possibility is the most striking advantage of using this Nb_3Sn conductor instead of NbTi or other Nb_3Sn conductors.

The magnet development programme is in progress and a first test of the magnet is scheduled for autumn 1990.

ACKNOWLEDGEMENTS

The authors wish to acknowledge the Netherlands Technology Foundation STW for the financial support and the colleagues Dick ter Avest, Bennie ten Haken, Sander Wessel, Rens Dubbeldam, Hans Geerinck and Wim van Emden, for their work and motivation which are essential to gain success.

REFERENCES

1.   Brianti, G., Hübner, K., The Large

FIGURE 6. Winding machine and press installed in a facility at HOLEC.

Hadron Collider in the LEP-tunnel. CERN report 87-5, 1987.

2. Perin, R., Progress on the superconducting magnets for the Large Hadron Collider. IEEE Transaction on Magnetics, MAG-24, 734 (1988).

3. Leroy, D., Perin, R., de Rijk, G., Thomi, W., Design of a high field twin-aperture superconducting dipole model, IEEE Transaction on Magnetics, MAG-24, 1373-1376, (1988).

4. Ter Avest, D., Solutions of dipole field optimization using conductors with a rectangular cross section. Internal project report 6LHC890420, UT-ASC, april 1989.

5. Hagedorn, D., Leroy, D., Perin, R., Towards the development of high field superconducting magnets for hadron collider in the LEP tunnel. Proc. MT-9 Zurich 9-13 Sept. 1985.

6. Perin, R., Leroy, D., Spigo, G., The first, industry made, model magnet for the CERN Large Hadron Collider. IEEE Trans. on Magn., MAG-25 (1989).

7. Wenger, S., Zerobin, F., Asner, A., Towards a 1 m long high field Nb_3Sn dipole magnet of the ELIN-CERN collaboration for the LHC-project. IEEE Transaction on Magnetics, MAG-25 (1989) and this conference paper JA1.

8. Ten Kate, H.H.J., Daum, C., Van de Klundert, L.J.M., Bode, H., Hoogland, W., Roeterdink, T.A., Van Overbeeke, F., Perin, R., On the development of a 1 meter "twin aperture" 10 T Nb_3Sn dipole magnet for the CERN LHC. Proc. IISSC, first Int. Industrial confer. on the SSC, New Orleans, Febr. 89.

9. Hornsveld, E.M., Elen,J.D., Van Beijnenand, L.A.M., Hoogendam, P., Development of ECN-type Niobium-tin wire towards smaller filament size. Adv. in Cryog. Eng., vol. 34, 493 (1987).

10. Ten Kate, H.H.J., Ten Haken, B., Wessel, S., Roeterdink, T.A., Hornsveld, E.M., Critical current measurement of prototype cables for the CERN LHC up to 50 kA and 13 tesla using a single step superconducting rectifier. Presented at MT-11 Japan, August 28, 1989, paper JF14.

11. Ter Avest, D., Van de Klundert, L.J.M., A new method to calculate conductor magnetization in accelerator dipoles. Proc. IISSC New Orleans, Feb. 8-10, 1989.

12. Ter Avest, D., Van de Klundert, L.J.M., Filament and cable magnetization in superconducting accelerator dipoles. Presented at MT-11 Japan, August 28, 1989, paper JF17.

13. Daum, C. and Ter Avest, D., Three-dimensional computation of magnetic fields and lorentz forces of an LHC dipole magnet using the method of image currents. Presented at MT-11 Japan, August 28, 1989, paper KG02, see also report LHC note no 94.

14. Ter Avest, D., Daum C. and Ten Kate, H.H.J., Magnetic field and lorentz forces in an LHC-dipole magnet. 3-D analysis using the fem programme TOSCA. Presented at MT-11 Japan, August 28, 1989, paper KJ01.

15. Ten Kate, H.H.J., Superconducting Rectifiers. Thesis University of Twente, April 14, 1984.

MAGNETIC FIELDS AND LORENTZ FORCES IN AN LHC DIPOLE MAGNET; 3-D ANALYSIS USING THE FEM PROGRAM TOSCA.

D. ter Avest, C. Daum[*] and H.H.J. ten Kate,
University of Twente, POB 217, 7500 AE ENSCHEDE, the Netherlands,
[*]NIKHEF-H, Amsterdam, the Netherlands.

ABSTRACT

This paper describes an investigation of magnetic fields and lorentz forces in the straight section and the coil end of an LHC-dipole magnet using the finite element program TOSCA. Starting from an existing design of the straight section the constant perimeter description was used to generate the coil end geometry. Results will be shown of the local magnetic field and lorentz force in the coil end.

INTRODUCTION

The design of the CERN-LHC dipole magnets is characterised by the effort to attain the highest possible magnetic field in the twin bores of the magnet (see fig. 1). In order to optimize the use of superconductor and to ensure a safe performance, the two layers of the coils should simultaneously reach their critical current as a function of the local maximum field. This requires a detailed study of magnetic fields in the conductor region of the magnet. Usually, the peak field of a dipole occurs in the magnet end, where the windings have a complex geometry and the tensile stress in the conductor has a maximum. The complex geometry of the ends inhibits proper support of the conductor. Therefore, it is the end part that limits the performance of the magnet. To prevent this, the maximum field occurring in the ends should be reduced. This can be achieved by an increment of the inner radius of the iron yoke in the end part of the magnet. A 3-D calculation of the magnetic field in the end part should reveal the reduction of the peak field due to this increment. Also, the lorentz forces on the conductors must be calculated in order to get the total force in the longitudinal direction of the magnet. The calculations were performed on the design of the 1 meter prototype LHC dipole magnet. For this we used the software package called TOSCA [1], a finite element program that solves electromagnetic problems in 3-D.

MODELING

The main advantage of using a finite element method is that a B(H) function can be specified in the program, which allows the study of magnetization in the iron at any value of the magnetic field. As a consequence, a non-linear B(I) relation will appear in contrast to analytical calculations where B/I is always constant due to the required assumption that μ_r is infinite. A second important benefit is the possibility of precise modeling of the iron yoke. For analytical calculations a symmetric structure must always be assumed in order to use the method of image currents [2]. Fig. 2 shows the geometry of the iron as used in the calculation. Considering the symmetry in the 2-in-1 magnet, only one 8th part of the volume of this magnet (one octant) needs to be considered. The geometry of the conductors

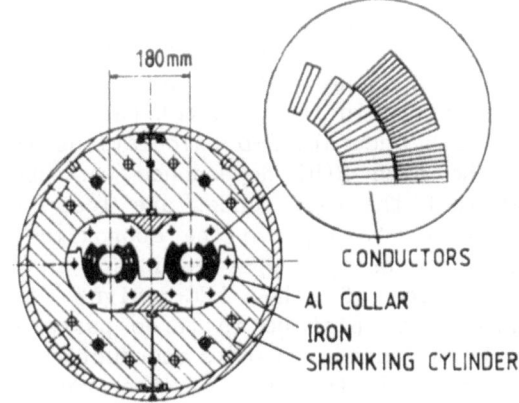

FIGURE 1 Cross section of the CERN-LHC 10 T prototype dipole magnet. The enlarged part shows the conductors.

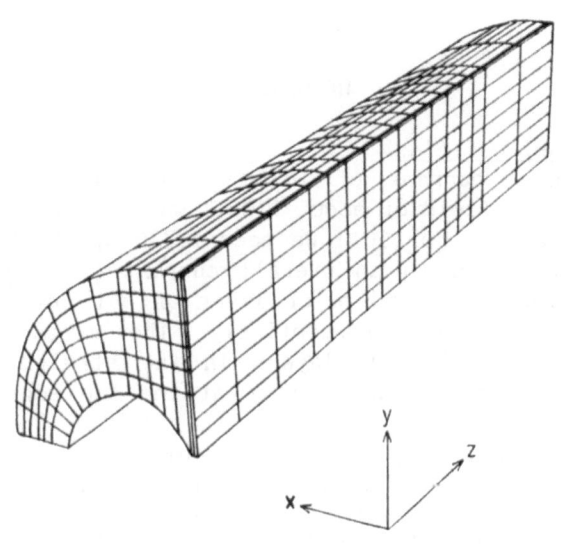

FIGURE 2 View of the iron as modeled by TOSCA. Only one 8th of the volume of the magnet needs to be studied.

in the magnet ends is very difficult to describe mathematically if keystoned Rutherford type cables are used, as is the case for the LHC model magnets. Therefore a convenient model must be applied to approximate the real shape. In our case the so-called constant perimeter description [3] was chosen. In this description all parts of the conductor have the same perimeter in the coil end. In TOSCA a module is present which approximates this constant perimeter model. This is due to the fact that in thick conductors on a relatively small bending radius the constant perimeter condition cannot be fulfilled for all parts of the conductor. We used the condition that all central points of the conductor, located on the central radius, have the same perimeter. The algorithm used in TOSCA to create the constant perimeter end is described briefly as follows. A milling cutter, which

maintains a constant angle to the magnet axis, is moved over the winding cylinder. The path described by the cutter gives the geometry of the conductor on the winding cylinder. The result is that a conductor always consists of two parallel lines in radial direction and two concentrical segments of circles. The path of the axis of the cutter, developed on the winding cylinder is in our case a half circle. As the algorithm uses a parameterisation for the conductors, the field induced by the current in a conductor can be obtained with a predefined accuracy by numerical integration. In order to reduce the CPU time mainly necessary for the integration, the number of conductors should be minimised. Therefore, a complete block of conductors in the magnet is modeled as one constant perimeter conductor, using the same current density as in the separate conductors. Although this may seem to give inaccurate results, one should realize that the local magnetic field is composed of the contributions of all the conductors in the magnet. Comparing the values of local magnetic field in the straight part of the magnet with accurate 2-D calculations [4], where the same B(H) dependence was used, showed that the differences are less than 2%. In order to improve the approximation of the shape of the upper conductor in the outer layer of the LHC-design, this conductor was split up in three parts and modeled as three separate conductors. A consequence of the use of the constant perimeter description in TOSCA is that the magnetic field inside the magnet cannot be analysed in terms of multipole components. This has been the subject of a separate study [5].

RESULTS

Magnetic fields

With a current of 14556 A per cable a central field of about 9.88 T was obtained. Therefore, for 10 T we need 14736 A, which corresponds to an overall current density in the inner and outer layer of 341 A/mm^2 and 526 A/mm^2 respectively. The maximum field in the two layers in the straight part of the magnet occurs at the top edge of the top conductors of the layers. For the inner layer the maximum is found exactly on the inside corner and amounts 10.22 T for a central field of 10 T. In the outer layer, the maximum is found at about 2 mm outward from the inside corner and amounts 8.66 T. Fig. 5 shows B_y along the magnet axis. The straight section of the magnet has a length of 0.68 m and the longest conductor has a length of 1.03 m. The calculated magnetic length of the

magnet, defined as

$$L_{mag} = \int_{-\infty}^{\infty} B_y(0,0,z)dz \; / \; B_y(0,0,0),$$

was found to be 0.87 m. The magnetic field in the outer conductors is not significantly changed if the length of the straight part of the conductors in the outer layer is increased by 1 cm. To check the sensitivity of the value of the magnetic field to small geometry changes the inclination angle of conductor 4 was changed. This angle which is the angle between milling cutter and winding cylinder was reduced by 10 degrees. This should alter the path of this conductor in the end. No significant change in the magnetic field value was found. It seems that the local magnetic field is not mainly determined by the local shape of a conductor, but by the added contributions of all conductors.

The maximum field in the magnet end is usually an absolute maximum and is found on the innermost conductor (closest to the winding cylinder top) just before the conductor starts the bending over the winding cylinder as indicated in fig. 3. The effect of an increment of the inner radius of the iron yoke in order to reduce the peak field in the magnet end is demonstrated in figs. 4 and 5. The inner radius of the iron in the end part was increased from 100 mm to 128 mm. For the inner conductor layer the maximum in the field plot disappears completely. In the outer layer only a small maximum is left which is less than 0.1 T greater than the field in the straight part of the magnet. Additionally, calculations were performed with an inner radius of the iron in the magnet end of 135 mm. Table 1 presents the effect of the increase of this radius. It

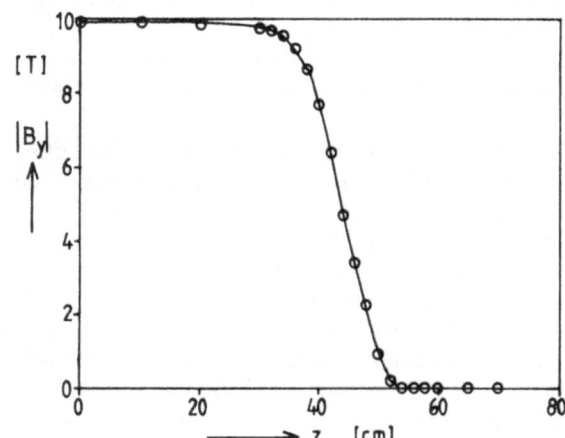

FIGURE 3 The dipole field B_y along the magnet axis.

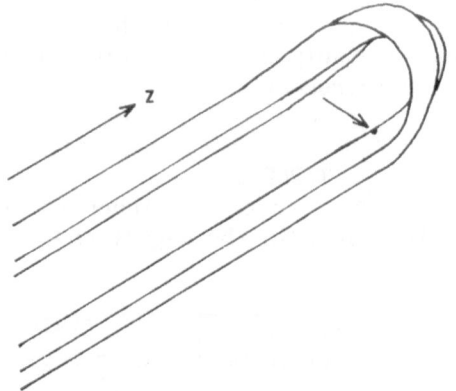

FIGURE 4 Position of
the maximum field in
the end part of the
magnet.
(inner conductor layer)

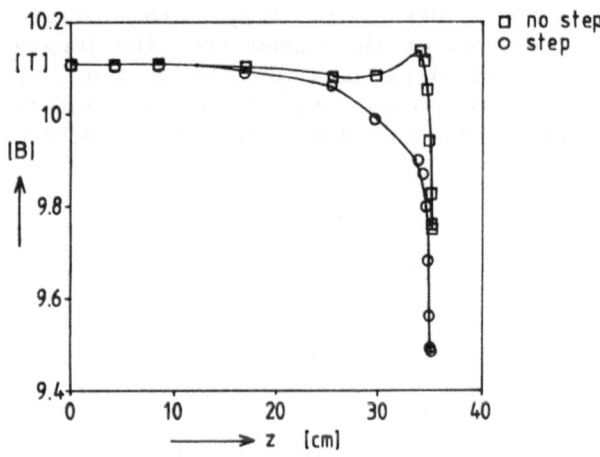

FIGURE 5 The effect of
an increase of the inner
radius of the iron
(step) on the max. field
occurring in the inner
conductor layer of the
magnet.

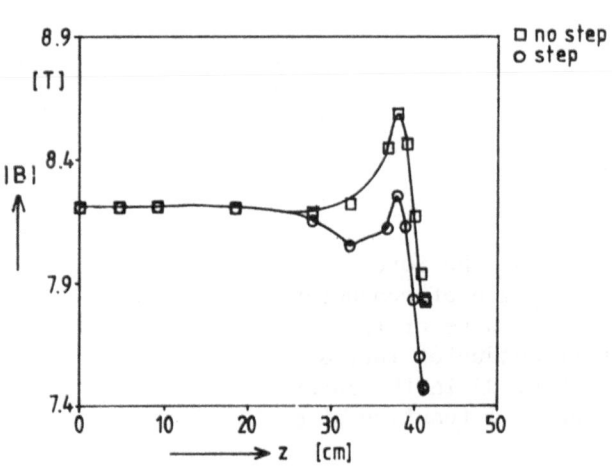

FIGURE 6 The effect of
an increase of the inner
radius of the iron
(step) on the max. field
occurring in the outer
conductor layer of the
magnet.

is clear that a satisfactory reduction of the peak field is already obtained by increasing the inner radius of the iron to 128 mm, and probably it is sufficient to use a smaller increment.

TABLE 1

Effect of the increase of the inner radius of the iron yoke in the magnet end.

inner radius [mm]	max. field inner layer [T]	max. field outer layer [T]
100	10.26	9.02
128	10.02	8.70
135	9.99	8.69

In order to study the field contribution of the magnet end, the length of the straight part of the magnet was varied. The load current for a central field of 10 T as a function of the length L

of the straight part of the magnet amounts:
I = 14736 A for L = 0.68 m, (model magnet),
     14778 A   L = 6.8 m and
     14792 A   L = 16.8 m.
It is clear from these data that the magnet end gives a larger contribution to the central field than the distant part of the straight section. Compared to the analytical calculation [2], the current surplus, or saturation factor, amounts 2.3%. This is in excellent agreement with the 2-D calculations [4], where a saturation factor of 2.2% was found.

Lorentz forces
Figs. 6,7 show for two conductors the force per unit of length of conductor in the magnet end as a function of the angle φ. This is the angle between milling cutter and the line in the ϕ-z plane where the conductor end part starts; $\varphi = 0$ at the beginning of the end part and $\varphi = 90$ on top of the winding cylinder. The force in the y-direction in conductors 4 and 8 is directed to the axis of the magnet. However, the total force is always directed

FIGURE 7 The force per unit length of conductor in the centre of the upper conductor (nr. 4, see Fig. 9) in the inner layer as a function of φ.

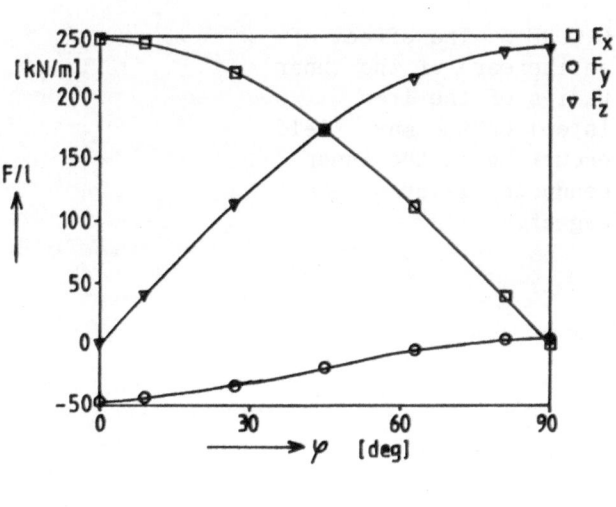

FIGURE 8 The force per unit length of conductor in the centre of the upper conductor (nr. 8, see Fig. 9) in the outer layer as a function of φ.

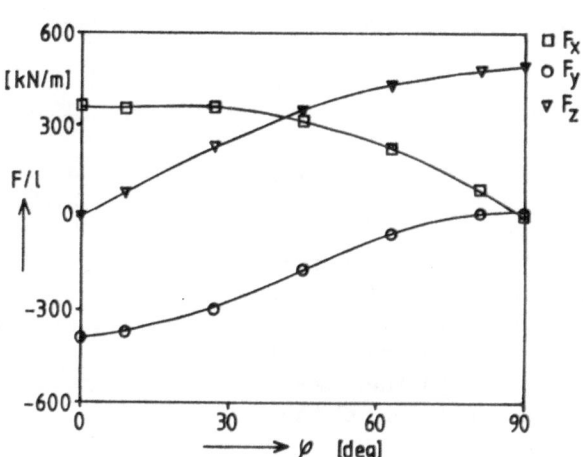

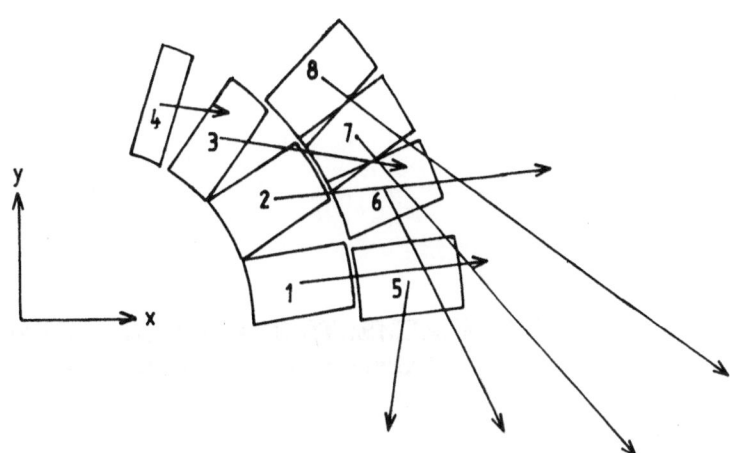

FIGURE 9 Plot of the total
force vectors per octant of
the magnet end part.

in the azimuthal direction of the
conductors or outwards. It was checked that
this is the case independent of the
z-position. Fig. 8 shows the direction of
the total force on the end part of one
octant of the magnet. Table 2 gives the
total force in the magnet end part.

TABLE 2

Total lorentz force per conductor in one
octant of the magnet end part, at a central
field of 10 T. Forces in kN.

cond	F_x	F_y	F_z	F_{tot}
1	8.5	0.8	4.5	9.6
2	12.4	1.0	8.6	15.1
3	8.1	-1.3	7.4	11.0
4	3.0	-0.4	2.9	4.2
5	-1.1	-6.0	5.7	8.3
6	5.3	-10.3	10.6	15.6
7	12.1	-13.4	17.3	25.0
8	17.8	-12.2	21.0	30.1

CONCLUSIONS

It was shown that an increase of the
inner radius of the iron yoke in the end
part to 128 mm reduces the local peak
field. The local value of the magnetic
field in the end part is not mainly
determined by the local shape of a
conductor but by the added contributions of
all conductors. Small geometry changes
should therefore not change the solution.
The force on the conductors is always
directed in the azimuthal direction or
outwards. Therefore, there is no need for
an inner supporting structure. The total
force in the z-direction per dipole amounts
311 kN at a central field of 10 T.

ACKNOWLEDGEMENTS

The authors wish to thank Dr. R. Perin for
the hospitality and support of the
CERN/SPS/EMA group, and R. Perin, D. Leroy
and G. de Rijk for many fruitful
discussions. These investigations are part
of the UT-NIKHEF programme in the
Netherlands to develop a 1 meter Nb_3Sn twin
aperture dipole model magnet for LHC at
CERN Geneva. This research in the programme
of the Foundation for Fundamental Research
on Matter (FOM) has been supported by the
Netherlands Technology Foundation (STW).

REFERENCES

1. TOSCA, Vector Fields Ltd, Oxford, UK
2. C. Daum and D. ter Avest. Three
   dimensional computation of magnetic
   fields and lorentz forces of an LHC
   dipole magnet using the method of image
   currents. Presented at this conference.
3. H.I. Rosten. A method for the
   calculation of the magnetostatic field
   from a wide class of current geometries.
   Chapter 4. RL-74-077.
4. G. de Rijk. Calculations with the 2-D
   FEM program POISSON.
   Private communications.
5. D. ter Avest. Solutions of dipole field
   optimization using conductors with a
   rectangular cross section. Internal
   report University of Twente, april 1989.
   To be published.

MEASUREMENTS OF QUENCH PROPAGATION VELOCITY ALONG A SUPER-STABILIZED CONDUCTOR*

A. Devred

SSC Central Design Group†

c/o Lawrence Berkeley Laboratory

One Cyclotron Road, MS 90-4040

Berkeley, CA 94720 USA

ABSTRACT

The transition of a superstabilized conductor to the normal resistive state is accompanied by a redistribution of current which, due to the size of the normal metal matrix, is long compared with the thermal propagation and greatly affects the propagation velocity. To simulate adiabatic propagation, a 2-m sample of ALEPH conductor, suspended in a vacuum and indirectly cooled at its extremities, was tested. Quenches were induced by heaters, and velocities were measured using voltage taps. Results are presented of tests at different currents and under different magnetic fields. These results are in good agreement with theoretical predictions.

INTRODUCTION

In composite, multifilament conventional superconductors, normal zone propagation is a strictly thermal phenomenon. It occurs close to the transition front, where a fraction of the power dissipated in the normal zone is transmitted by conduction to the superconducting zone, which in turn heats up and goes into transition. In superstabilized superconductors, like that developed for the ALEPH solenoid[1] where a conventional composite is enclosed in a large section of aluminum, an electromagnetic diffusion phenomenon is added. Indeed, close to the transition front, the current previously carried by the filaments is expelled toward the copper of the composite and toward the aluminum. Given the large section of

the aluminum, the current needs a certain amount of time and space to diffuse itself. The transition front thus leaves a wake of electromagnetic diffusion, along which the dissipated power density per unit volume progressively decreases.

The question then arises of how the current redistribution affects the propagation velocity. In previous papers[2,3,4] we have developed a theoretical model for the propagation of the normal zone along a layer of superstabilized conductors representative of the ALEPH solenoid shown in Figure 1. In this indirect cooling configuration, the annular ring and the layer insulation introduce a high thermal resistance between the helium and the conductors. Our model therefore assumes that, at the scale of the propagation phenomenon, the amount of heat transferred to the helium is negligible and thus that the layer of conductors can be considered adiabatic. Treating the composite material as a homogeneous medium, and assuming that the current diffusion through the copper of the composite is instantaneous at the

*This work is part of a Ph.D. thesis completed at the Commissariat à l'Energie aAtomique, CEN/Saclay, DPhPE/STIPE/STCM, 91191 Gif-sur-Yvette cédex (France)

†Operated by Universities Research Association, Inc., for the U.S. Department of Energy.

scale of the diffusion through the aluminium, we solve the equations of electromagnetic behavior and establish a general formula for the velocity. Reference 5 provides a summary of our theoretical results.

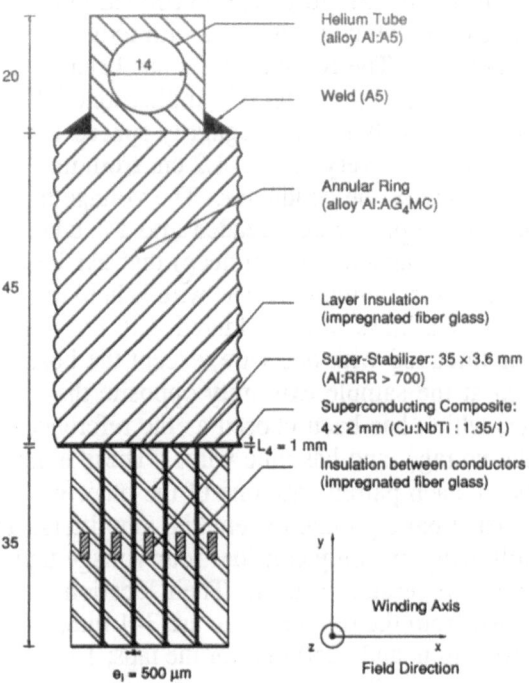

Figure 1. Cut away view of the ALEPH solenoid.

This paper reports the results of the experimental tests of our analytical formula. So far, propagation velocities along superstabilized conductors have only been measured in helium bath[6] or in a 1-m bore model solenoid where the transverse effects predominate.[7] Our goal was to recreate as closely as possible the assumptions of our theoretical model. Among them are (1) a constant-velocity propagation mode and (2) the hypothesis of adiabaticity. To be able to demonstrate the existence of a constant propagation velocity, we considered a 2-m-long sample of ALEPH conductor. To simulate adiabatic behavior, we operated the sample in a vacuum and reduced as much as possible the thermal contact between the conductor and the mechanical supports. The whole test setup was placed inside a large superconducting coil to be able to generate on the sample a magnetic field comparable to the one seen by the conductor in the real solenoid. The quenches were induced by heaters, and the propagation was monitored by voltage taps and temperature sensors regularly spaced along the sample.

Three kinds of influences were studied: the operating current, the external magnetic field, and the initial energy deposition. While the current and the field dramatically affect the velocity, the amount of energy put into the heater to induce the quench does not influence the subsequent propagation. As expected, the propagation velocity does not depend on the initial conditions. In the following, we limit our presentation to the first two influences.

DESCRIPTION OF THE EXPERIMENTAL SET UP

To constrain the magnetic forces, the conductor sample is wound around a stainless-steel mandrel of external diameter 0.72 m (determined by the bore of the external coil) and fastened by seven epoxy staples. Two layers of Kapton (each 2-mm thick) insulate the conductor from the mandrel. The question is now what is the thermal coupling between the conductor, the Kapton, and the mandrel, and does it influence the propagation velocity? When the external field is nil, the conductor loop tends to expand in its own field, only held by the staples; we then expect the conductor to behave adiabatically. Under an external magnetic field, the circulation of the currents in the external coil and in the sample have been designed so that the conductor loop contracts onto the mandrel. In such cases, it is more likely that part or all of the Kapton will be involved in the propagation process. For the mandrel, however, temperature sensors located in holes drilled at its periphery record no variation during quench testing. Adding this observation to the fact that it takes more than 100 hours to cool down the mandrel, while the conductor is cold in less than 50 hours, we conclude that the thermal resistance between the conductor and the mandrel is high enough to be considered as infinite at the scale of the propagation.

The sample is cooled by thermal conduction along the conductor using two helium exchangers. The location and the mounting of these exchangers are shown in Figure 2. Each consists of a copper tube (10 mm i.d., 12 mm o.d.), brazed on a 20-cm-long and 3-cm-thick copper plate. The copper plate is soldered on the previously tinned conductor. At both extremities of the sample, bending considerations have made it necessary to remove the aluminum from the conductor. The conductor is then soldered on a 2-cm-thick copper plate, which plays the role of stabilizer and improves the heat transfer between the sample and the exchangers (one extremity also includes a splice between composites). To avoid circulation of current through the helium pipes and to ensure electrical insulation to ground, insulating sleeves have been installed at appropriate locations in the refrigeration circuit. The mandrel has no specific refrigeration and only cools down through its weak thermal coupling to the conductor. The whole setup is surrounded by a shield at 20 K.

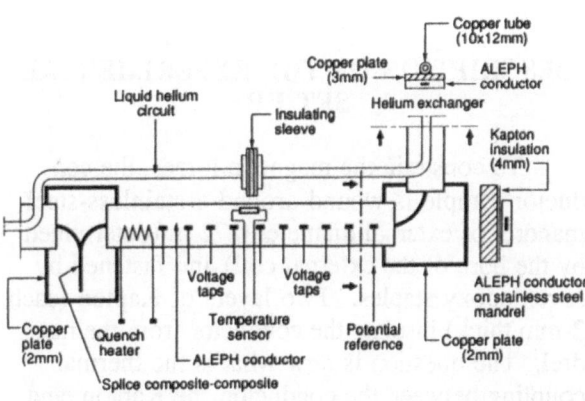

Figure 2. Experimental setup for measuring the propagation velocity along a sample of ALEPH conductor.

During testing, the operating pressure is regulated to 1.2 10^{-5} Pa corresponding to helium temperature of 4.4 K. Nevertheless, the sample temperature stabilizes around 5.1 K, 0.7 degrees higher than the cold source. This indicates residual thermal contributions, which can be estimated to be around 70 mK.[8] We have not been able to identify the origin of these residual thermal contributions, but we can verify that their influence on the velocity measurements is negligible. Indeed, we can bracket the temperature gradient induced along the sample by such thermal contributions. An upper limit is set by assuming that the 70 mK are punctually received in the middle of the sample, which leads to a gradient of 0.3 K. A lower limit is determined by assuming that the power is uniformly received along the sample, which leads to a value of 0.15 K. The temperature profiles corresponding to these gradients are very flat and thus should not invalidate our measurements.

As shown in Figure 2, the sample is equipped with a spot-heater, temperature sensors, and voltage taps. The heater, located at one extremity of the sample, consists of a flat bifilar coil (38 × 30 mm^2), wound with constantan wire, and glued to the conductor with Eccobond. The heater resistance is 38.4 Ω and hardly depends

on the temperature. The temperature sensors are calibrated Allen Bradley resistors, which have been chosen for their high sensitivity at low temperatures and their stability under low magnetic fields.[9] In order to decrease the response time, the carbon insulation of the sensors is filed to bare the active piece (graphite). The active piece is then inserted and glued with Eccobond into a copper sleeve, which is soldered with indium on the conductor. The response time has been measured to be of the order of 1 millisecond.[8] There are ten voltage taps regularly spaced along the sample (one every 20 cm) 1.2 cm from the superior edge of the conductor. The voltage taps consist of a copper wire threaded into a counter bore (2-mm diameter, 3-mm deep) that has been drilled into the aluminium. The wire is held in position by a brass screw soldered with tin. Each bore is threaded with one wire, except the bore located at the sample extremity opposite the heater, which has been chosen as the reference for the voltage, and has nine wires. These nine wires are each paired with one of the other wires; the twisted pairs go into differential amplifiers. In the following, we only consider four voltage taps and one temperature sensor. Their locations reckoned from the reference tap are 341 mm, 744 mm, 1046 mm, and 1250 mm for the taps; 1345 mm for the sensor. The taps are numbered starting with the one opposite the reference.

The external field is created by a superconducting magnet with a 1-m bore, equipped with an internal cryostat. The cryogenic circuits for the magnet and the sample are dissociated. The vacuum in the internal cryostat is maintained around 10^{-4} Pa.

TEST RESULTS

Unexpected behavior of the sample temperature appeared during testing. As we said, when the current is nil, the conductor temperature T_s stabilizes around 5.1 K. During the ramping of the current, eddy currents are induced in the stabilizer, which overheat the conductor. They disappear once a constant current has been established; the conductor then cools off and stabilizes again at a constant temperature. However, this constant temperature is higher than 5.1 K, increasing linearly with the current squared. The slope of the temperature increase also depends on the external field B_e. Expressions for T_s = f(I) can be found in Table 1. This shows the existence of additional heating, presumably of Joule origin. An evaluation of the resistance R_J producing this heating is given by

$$R_J = \frac{T_s(I) - T_s(I=0)}{R_{th}I^2},$$

where I is the current and R_{th} is the thermal resistance between the conductor and the exchangers. From the geometry of the exchangers, R_{th} can be evaluated around 10 K/W. The slope of T_s versus I when $B_e = 0$, then gives

$$R_J = 4.1\ 10^{-9}\ \Omega.$$

This value is typical of a weld between superconductors. We then conclude that the additional heat is generated in the splice located at the extremity of our sample (see Figure 2). The fact that the initial temperature of the sample is higher than expected is not a problem, since we can measure it. Of course, our comparison between experimental data and theoretical calculations will rely on these measured temperatures.

TABLE 1.
Selected Parameters for the Analytical Calculations

Critical temperatures (K)

$$T_C = 9.5 - 0.5\ B$$

$$T_{CI} = 9.5 - 0.5\left(B + \frac{1+(0.9-0.09B)B}{2435}I\right)$$

Field on the sample (T)

$$B_{max} = B_e + 10^{-4}\ I \qquad B_{av} = B_e + 5\ 10^{-5}\ I$$

Sample temperature (K)

$$T_s(0\ \text{T}) = 5.1 + 2.4\ 10^{-8}\ I^2$$
$$T_s(1\ \text{T}) = 5.3 + 4.3\ 10^{-8}\ I^2$$
$$T_s(1.55\ \text{T}) = 5.55 + 2.8\ 10^{-8}\ I^2$$
$$T_s(2.1\ \text{T}) = 5.25 + 4.75\ 10^{-8}\ I^2$$

Figures 3a and 3b feature typical records of a voltage channel (channel 1) and of the nearby temperature sensor. The current at quench is 2500 A, and the external field is nil. The heater is fired at $t_0 = 375$ ms, and the pulse lasts $\tau_h = 100$ ms. The temporal evolution of the voltage can clearly be divided into four phases that are delimited on Figure 3a: (1) For $t \leq t_1$, the voltage is nil; the quench has not yet reached tap 1, and the conductor between tap 1 and the reference tap is still superconducting. (2) For $t_1 \leq t \leq t_2$, the voltage rises rapidly; the quench has now reached

tap 1 and propagates towards the reference tap. (3) For $t_2 \leq t \leq t_3$, the voltage decreases progressively; the whole conductor has now switched to the normal resistive state, but the current has yet to complete its redistribution through the superstabilizer, causing the decay of the apparent resistance. (4) For $t \geq t_3$, the voltage starts to rise again, but very slowly; the current is now redistributed, but as a result of the Joule heating the aluminum resistivity increases, causing the increase in the apparent resistance.

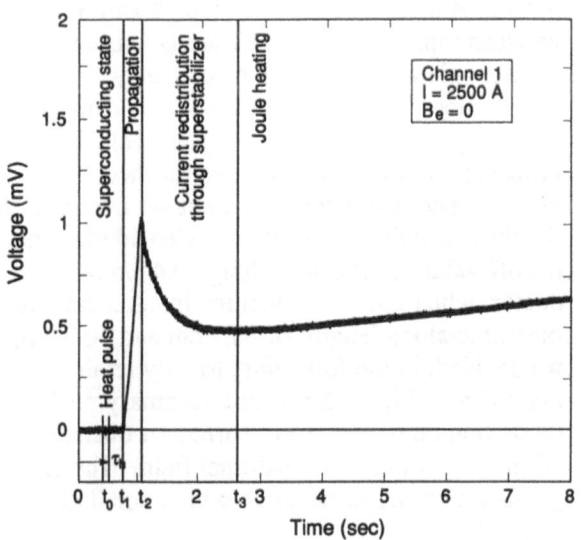

Figure 3a. Typical voltage evolution during a quench induced at 2500 A ($B_e = 0$ T).

These different phases can be correlated with the temporal evolution of the temperature shown in Figure 3b. For instance, at $t = t_1$, the temperature is about 9 K, which corresponds to the transition temperature of the conductor at 2500 A and thus correlates with the passage of the transition front near tap 1. Also, at $t = t_3$, the temperature is about 15 K, which, for the given purity of aluminum, corresponds to the end of the residual resistivity plateau; for temperatures greater than 15 K, the sample resistance thus increases.

From these data we conclude that thermal propagation is fast compared with electromagnetic diffusion, as predicted by the model. One can even verify that the times t_2 and t_3 are coherent with the characteristic times defined in Ref. 4 for these two phenomena.

To illustrate how the propagation velocity is determined, consider Figure 4. The current at quench is 1115 A, and the external field is nil. The four traces correspond to the voltages between the reference tap and the four taps mentioned earlier. They rise in sequence. As expected, the quench, initiated at one extremity by the heater, gently propagates along the sample successively hitting the different taps. The velocity is simply measured by dividing the length between two successive taps by the difference in time between the takeoffs of the two voltages. The four voltage traces allow three different velocity measurements and thus enable us to see its evolution along the sample. It appears that the velocities measured between taps 1 and 2 are systematically 15 to 20 percent lower than the velocities measured between taps 2 and 3 and between taps 3 and 4, which are consistent. We are not able to explain this discrepancy. The sample does not indicate any obvious damage between taps 1 and 2. If the constant-velocity propagation mode had not been reached, we would expect the velocity to increase over each of the three lengths of conductor. Considering that the off-value is measured along 20.4 cm of conductor, when the two velocities in agreement are measured along lengths of 30.2 cm and 40.3 cm, we decided, in the following, to only retain these two values. Figure 5 shows a summary plot of these velocities versus the current at quench for all the tests run in a nil external field. The data points lie on the same curve with a small dispersion.

The data are summarized in Figure 6 as a function of the current; the size of the plotting symbols represents the error bar. As in Figure 5, and for each field value, the data points lie on the same gentle curve.

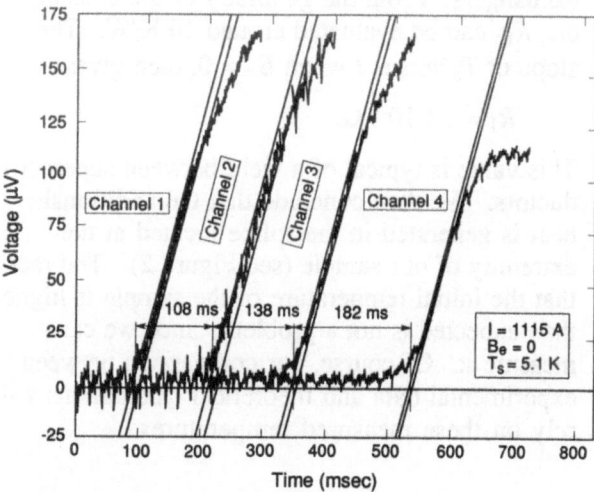

Figure 4. Example of determination of the propagation velocity for a quench induced at 1115 A (B_e = 0 T).

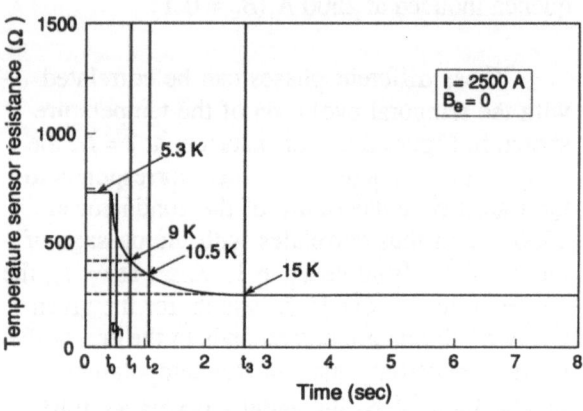

Figure 3b. Typical temperature evolution during a quench induced at 2500 A (B_e = 0 T).

Tests have been run for four external field values: 0 T, 1 T, 1.55 T, and 2.1 T. For each field value, the current in the sample was progressively increased from 1000 A to the critical current, in increments of 500 A. Each run, which provides two velocity measurements, was systematically doubled. This gives us about four hundred data points.

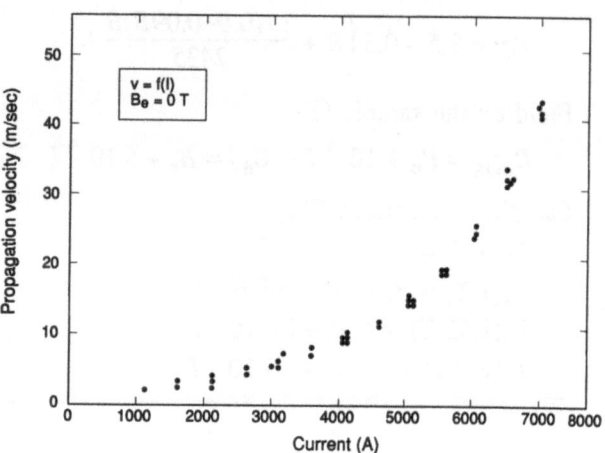

Figure 5. Summary plot of the velocity measurements (B_e = 0 T).

COMPARISON BETWEEN THEORY AND EXPERIMENT

The last step of our study is to compare the experimental data with the theoretical predictions. The details of our analytical formula can be found in Ref. 5. We nevertheless need to specify the geometrical and physical characteristics we use in our calculation. The overall size of the super-stabilized conductor is 34.15×3.45 mm^2. The composite is a 14-strand cable 5.75×1.36 mm^2. The strand diameter is 0.8 mm; the copper-to-superconductor ratio is 1.8 to 1. The copper and aluminum residual resistivity ratios are respectively 200 and 2200. Critical current measurements have been made at 4.2 K on six strands cut from the same conductor used in the experiment. The parametrization of T_{CI} given in Table 1 has been established assuming that the critical current is a linear function of the temperature. Also in Table 1 are formulas for the peak field, B_{max}, and the average field, B_{av}, on the conductor. The external field is given by the transfer function of the external coil; the own field is computed numerically. At a given current, the critical temperatures are calculated using the peak value, while the resistivities and the thermal conductivities of the materials are calculated using the average value. Lastly, the table shows linear approximations of $T_s = f(I)$ for the four values of B_e.

An obvious verification before analyzing the data of Figure 7 is to compare the critical currents actually measured in the experiment with the short sample predictions. For a given external field value, the short-sample current is the solution of the implicit equation

$$T_{CI}[I, B_{max}(I)] = T_s(I).$$

As shown in Table 2, these values are in good agreement.

With more confidence in our material properties, we turn to velocities. The data in Figure 6 show that for low currents, the values measured under external field are lower than the values without field. This is unexpected since T_C, T_{CI}, and the enthalpy per unit volume are decreasing functions of the magnetic field; the velocity should thus increase. (This is in fact the case when velocities under field are compared.) So, something, under the effect of the field, must slow down the propagation. One explanation is that, when the external field is nil, the conductor loop tends to expand under its own field, so that the thermal contact between the conductor and the Kapton insulation is poor, while under an external field, the loop contracts onto the mandrel, so that the contact between the conductor and the Kapton is greatly improved. This leads us to conclude that, in the absence of magnetic field, the Kapton does not interfere with the propagation process; under an external field, part or all of it is involved. In our analytical calculations, we then consider

the conductor as bare in the absence of external field, but associated with the 4 mm of Kapton in the presence of field. Computation results are also plotted in Figure 6. Predictions and measurements appear in good agreement.

TABLE 2.
Comparison Between Predicted and Measured Critical Currents

B_{max} (T)	Predicted Critical Current (A)	Measured Critical Current (A)
0	7750	7500
1	6050	6000
1.55	5300	5100
2.1	4700	4500

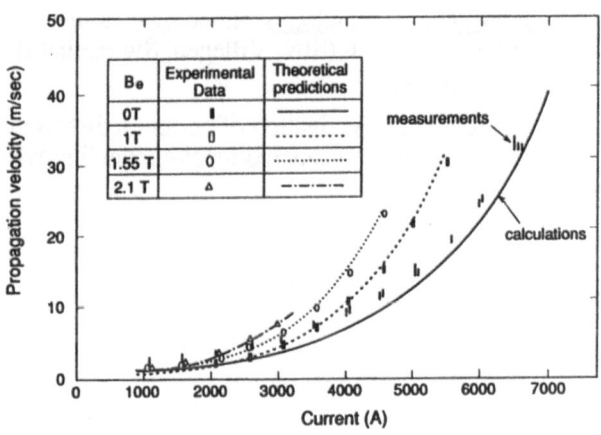

Figure 6. Comparison between predictions and measurements.

CONCLUSION

We have experimentally demonstrated the existence of a constant propagation velocity of the normal zone along superstabilized conductors, and we have verified that this velocity is in accord with the analytical law we had previously established. Our model of the current redistribution through the superstabilizer is thus justified. The next step towards a complete understanding of the quenching of the ALEPH solenoid would now be to investigate the transverse propagation.

ACKNOWLEDGMENTS

I am particularly grateful to the staff of the Sevice Technique de Cryogénie et de Magnétisme du CEN Saclay, whose skill and hard work made these tests possible. I want also to thank Katharine Metropolis for her comments on the manuscript, and Donna Matthews and Annie Calinog for their help in editing the text and preparing the figures.

REFERENCES

1. Desportes, H., LeBars, J., and Meuris, C., General design and conductor study for the ALEPH superconducting solenoid. (Proceedings of the 8th International Conference on Magnet Technology), J. Phys. (Paris), 1984, Colloque-45, C1–341.

2. Devred, A. and Meuris, C., Analytical solution for the propagation velocity of the normal zone in large matrix super-stabilized superconductor., (Proceedings of the 9th International Conference on Magnet Technology), edited by C. Marinucci and P. Weymuth (SIN, Villegen, Switzerland), 1985, p. 544.

3. Devred, A., Investigation of current redistribution in superstabilized superconducting winding when switching to the normal resistive state. J. Appl. Phys., 1989, 65 (10), 3963.

4. Devred, A., Investigation of the normal zone along a superstabilized superconducting solenoid. J. Appl. Phys. , 1989, 66 (6), 2689.

5. Devred, A., General formula for the adiabatic propagation velocity of the normal zone. IEEE Transaction on Magnetics, 1989, MAG-25 No. 2, 1968.

6. Scherer M., and Turowski, P., Investigation of the propagation velocity of a normal-conducting zone in technical superconductors. Cryogenics, 1978, 15, 515.

7. Hirabayashi, H., Morimoto, K., Make, M., Yamada, R., Yamamoto, A., Mori, S., Yoshizaki, R., Kawakami, H., Kondo, K., Aihara, K., Kazawa, Y., Kimura, H., Ogata, H., Saito, R., Suzuki, S., and Miyake, Y., Measurement of propagation velocities of the normal zone in a 1mϕ×1m superconducting solenoid magnet. Jpn. J. Appl. Phys., 1981, 20, 2243.

8. Lottin, J. Cl., private communication.

9. Rubin, L. G. and Brandt, B. L., Letter to the editor on low-temperature thermometry in high magnetic field. Cryogenics, 1987, 27, 269.

MECHANICAL ANALYSIS OF DIFFERENT YOKE CONFIGURATIONS FOR THE SSC DIPOLE

M. S. Chapman, J. M. Cortella, and R. I. Schermer
SSC Central Design Group*
Lawrence Berkeley Laboratory
MS 90/4040
One Cyclotron Road
Berkeley, California 94720

R. H. Wands
Fermi National Accelerator Laboratory*
P. O. Box 500
Batavia, Illinois 60510

ABSTRACT

Finite element methods are used to calculate the additional mechanical support offered by different configurations of the yoke and shell for the SSC dipoles. In this analysis, horizontally and vertically split yokes with different gaps between the yoke halves are evaluated in terms of collar deflections and coil stresses. The results show that the yoke offers significant additional support for the collars against the Lorentz forces. Collar deflections due to Lorentz forces can be reduced 50–75% by using the various yoke configurations studied here. Additionally, the analysis indicates that vertically split yokes are preferable to horizontally split yokes for maintaining a uniform stress state across the coil poles, and that for either horizontally or vertically split yokes, an open midplane gap between the yoke halves at 4.2 K results in a smaller coil stress loss during cooldown.

INTRODUCTION

Recent SSC dipoles have included two significant mechanical changes designed to better restrain the coils against Lorentz forces developed during energization. First, the yoke and skin have been redesigned to provide additional clamping of the coils, and second, the endplates of the magnet have been strengthened against axial deflection. These magnets have demonstrated improved training behavior over their predecessors, which relied entirely on the collars to clamp the coils and employed relatively flexible endplates [1,2]. Here, finite element methods are used to calculate the additional transverse support provided by the yoke and the shell for the present yoke design and for other possible configurations.

* Operated by the Universities Research Association, Inc., for the U. S. Department of Energy

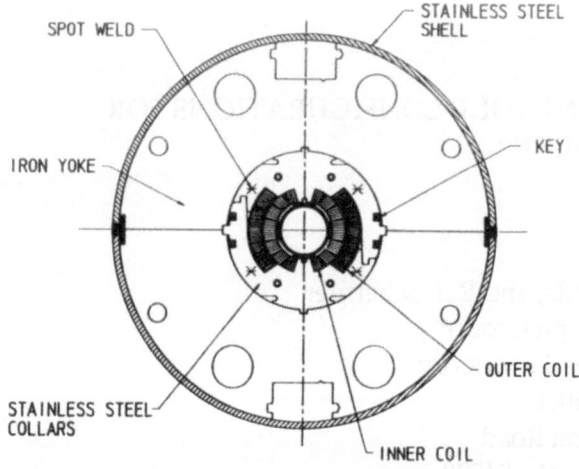

Figure 1. Cross section of the SSC Dipole

The cross section of the present design of the SSC dipole (designated C358D) is shown in Figure 1. Stainless steel, spot-welded collar laminations are tightly clamped around the coils and locked together with inserted keys. The collars azimuthally precompress the coils and support them against the Lorentz forces developed as the magnet is energized. After the coils are collared, the horizontally split carbon steel yokes and stainless steel shell are assembled around the collared coil, and the shell is welded. When the magnet is cooled down to 4.2 K, the collars and coils shrink more than the yoke and thus tend to separate from the yoke. However, yoke/collar contact along the region of the horizontal midplane is necessary in order for the yoke to be able to provide additional support to the coils during energization. To achieve this contact with the horizontally split yokes, a vertical interference fit is called for between the yoke and collar at 4.2 K such that there is a gap between the yoke halves. Tension developed in the shell due to weld shrinkage and thermal contraction tends to close the gap, and in doing so, loads the collar with a vertical force via the yoke. This force causes horizontal expansion of the collar and an inward bending of the yoke, which combine to produce yoke/collar contact along the horizontal midplane.

For the vertically split yokes, yoke/collar contact along the horizontal midplane is achieved in a more straightforward manner. As long as a sufficient horizontal interference fit is specified between the yokes and the collars, the tension developed in the shell will naturally draw the yokes in against the horizontal midplane of the collars.

It is possible to control the amount of shell tension transferred to the collars and coils by varying the yoke/collar interference fit (and thus the initial yoke midplane gap). If the specified interference is large enough so that the yoke midplane gap remains open throughout cooldown, all of the shell tension is transferred to the collars and the coils. If the yoke gap closes during cooldown, then only a portion of the developed shell tension is transferred to the collars and coils; the remainder is reacted by the closed yoke gap.

In this analysis, four yoke configurations are considered: two with horizontally split yokes and two with vertically split yokes. For each orientation of the yoke split, we consider either open or closed yoke midplane gaps at 4.2 K. In addtion to these four cases, a model with unsupported collars is considered for comparative purposes. A description of these cases is

Table 1. Yoke Configurations Analyzed

Case	Type of yoke	Yoke gap at 4.2 K
I	none	n/a
II	Horizontal	closed
III	Horizontal	open
IV	Vertical	closed
V	Vertical	open

presented in Table 1. Case II, with horizontally split yokes with a closed midplane gap at 4.2 K, represents the present SSC dipole design.

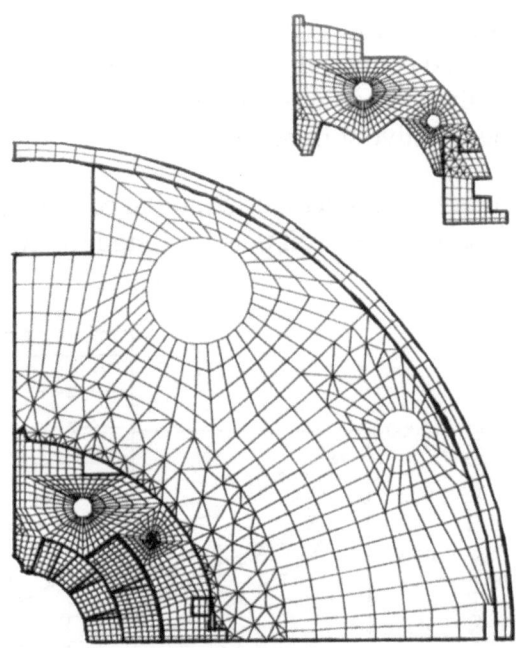

Figure 2. Finite element model of SSC dipole. Portions of collar shown separately for clarity.

THE MODEL

The ANSYS finite element code by Swanson Analysis Systems is used for this analysis. The 2–D finite element model of the magnet cross section is shown in Figure 2. The model includes the inner and outer coils surrounded by the collars, which are in turn surrounded by the yoke and shell. The collars are modeled in three separated pieces representing two laminations in depth, and constraint equations are used on the collar midplanes to enforce rotational symmetry. Additional constraints for the collars are provided by the inserted keys and spot-welds between successive pairs of laminations. Compression-only "gap" elements are used along interfaces that are permitted to separate. Horizontally split yokes are modeled with a symmetry boundary condition along the vertical midplane and a gapped boundary condition along the horizontal midplane that permits separation. For the vertically split yoke, these boundary conditions are reversed. Different initial yoke midplane gaps are specified using the gap elements along the appropriate midplane. The size of

the initial gap is chosen so that after cooldown, the gap will be either open or closed. For the unsupported collars, the yoke and shell are not included in the model.

The models are loaded with sequential load steps representing coil prestress assembly, welding of the shell, cooldown to 4.2 K, and energization to 6.6 T. The coil prestress loading is modeled with positive vertical displacements of the coil midplanes to produce average azimuthal compressive coil stresses of 8000 psi at the inner coil midplane and 6000 psi at the outer coil midplane. The welding of the shell, which produces azimuthal tension due to weld shrinkage, is represented by a negative displacement of the horizontal midplane of the shell sufficient to produce 30 kpsi in the shell. At present, it is difficult to determine the amount of tension developed in the shell during welding because the welds are performed manually, and yoke halves are only loosely clamped around the collars before welding. With improved tooling that incorporates automatic welding machines and a hydraulic press to compress the yoke halves together, it is hoped that the shell stresses can be more uniformly controlled. Cooldown to 4.2 K is modeled in one step using the integrated coefficients of thermal contractions for the various materials in the cross section.

Table 2. Material Properties

	Young's Modulus E (psi)	$\Delta l / l$ (293→4.2 K)
coils	1.5×10^6	−0.0040
collars	30×10^6	−0.0031
yokes	30×10^6	−0.0020
shell	30×10^6	−0.0031

These coefficients, along with the Young's moduli, are listed in Table 2. The Lorentz forces, which have been calculated using a magnetic version of this model as reported in [3], are applied as nodal forces on the coils for currents of 2000, 3000, 4000, 5000, 6000, 6400, and 7000 amps. The operating current of the dipole at 6.6 T is 6400 amps.

RESULTS

The calculated collar deflections due to the Lorentz forces acting on the coils at 6.6 T are shown in Figure 3. For all the yoke configurations studied, the additional support offered by the yoke reduces collar deflections significantly. Note that the yoke is effectively more stiff against collar motion when the yoke gap (horizontal or vertical) is closed during energization. This is because when the yoke gap is closed, the yoke has the same stiffness of an equivalent solid ring, whereas when the yoke gap is open, it has the lesser stiffness of an

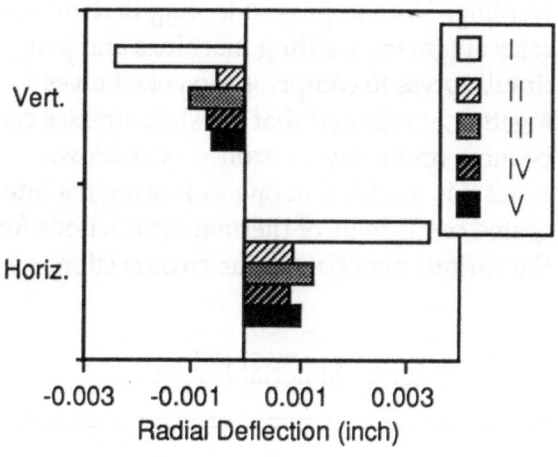

Figure 3. Radial collar deflections due to Lorentz forces only

equivalent split ring. It is surprising at first glance that the deflections of Case V, which has a vertically split yoke with an open gap, are so small. One might think that with the yoke gap open along the vertical midplane, the yoke would be unable to provide support against horizontal collar deflections. However, the yoke is preloaded in bending during the welding and cooldown steps such that it contacts

the collars along the top in addition to along the side. As the magnet is energized, the preloaded contact along the top of the collars partially unloads and contributes considerably to the effective stiffness of the yoke.

In Figure 4, the average azimuthal compressive stress at the inner coil pole is plotted for the

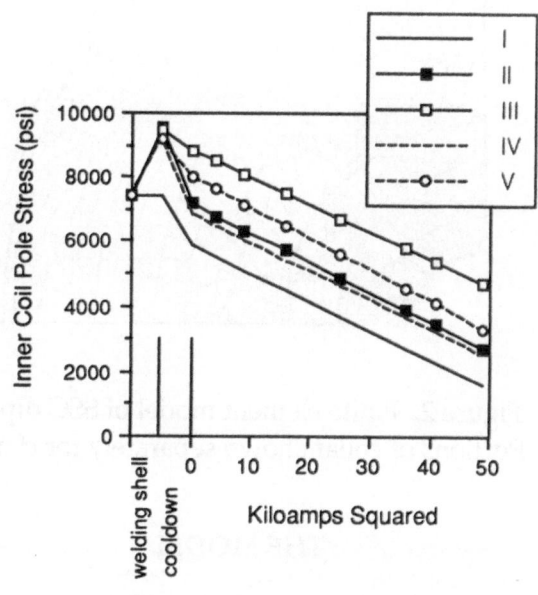

Figure 4. Average azimuthal stress (compressive) across inner coil pole face for welding of shell, cooldown, and energization.

different loadings. During welding, the yoke midplane gaps remain open for all cases, and the tension developed in the shell due to weld shrinkage is completely transferred into the coils and collars. This results in an increase of azimuthal stress in the inner coil of about 2000 psi. The different increase between the vertically split yokes and the horizontally split yokes can be attributed to the different load sharing between the collars and coils in the horizontal and vertical directions. For Case I, in which the yoke is not included in the model, the welding step is not included. The cooldown step produces different changes in coil stress depending on the particular yoke configuration. For the free-standing collars in Case I, the coil stress drops 1600 psi during cooldown because the coil shrinks more than the collars. For Cases III and V, the yoke midplane gap remains open during cooldown, and the drop in coil stress is minimized because of the increase in shell tension during this step. In cases II and IV, the yoke gap closes during cooldown and a portion of the shell stress is transferred into the now-closed yoke midplane gap. After the yoke gap closes, the collar/yoke contact begins to reduce due to greater thermal contraction of the collars and coils. However, at the end of cooldown, there is still significant contact between the yoke and collar, so the inner coil stress remains larger than the free-standing collar value. During energization to 6.6 T, the rate of inner coil stress loss versus current squared is linear and does not vary significantly from model to model.

Deflections of the collar and coils produce bending stresses in the coils such that the stress profile across the poles is non-uniform. The bending stresses are analogous to those developed in a thick cylindrical ring subjected to a

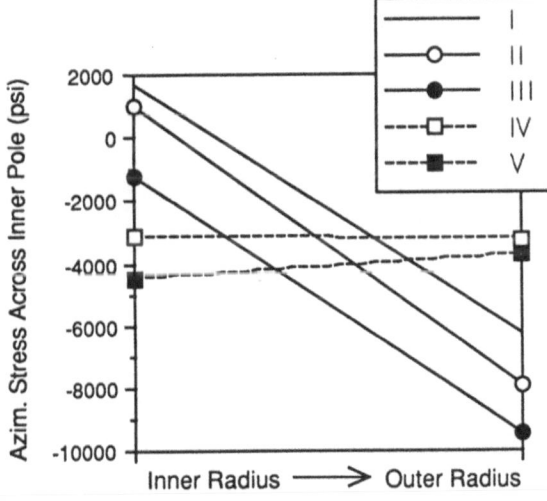

Figure 5. Linearized azimuthal stress profile across the inner coil pole at 4.2 K, 6.6 T assuming an initial coil prestress of -8000 psi. Positive stress values result from the linearization of the actual stress distribution across the pole; in the model, gap elements between the coil and the collar prevent tensile stress from developing.

transverse point load. The calculated stress profile across the inner coil pole at 6.6 T is shown in Figure 6 for the different cases. The deflection of the collar due to the Lorentz forces causes the inner radius of the coil pole to lose more stress than the outer radius. While reducing the collar deflections should, in general, reduce the stress gradient, this is not the case with the horizontally split yokes because the vertical clamping of the collar causes a significant stress gradient with the same sign as that produced during energization. The vertically split yokes, on the other hand, clamp the collars in the horizontal direction. This produces a stress gradient opposite to that produced during energization, and the two effects cancel. This cancellation produces a uniform stress profile across the coil pole, as indicated by the dashed lines in Figure 5.

Maintaining compressive stress across the poles is generally considered essential in limiting quench-causing conductor motion. It may be somewhat alarming, then, that the calculation predicts that the inner radius of the inner coil pole unloads completely at 6.6 T for both the unsupported collars and the present design. The calculation employs a relatively simple isotropic, linear, elastic model of the coil, however, so the calculated stress quantities are only qualitatively correct. Including the non-linear stress/strain behavior of the coil in the model would likely keep the pole from unloading because the coil becomes very soft at low stresses.

Although the cases with the open yoke gaps show the highest average stress at the inner pole (Figure 4), these configurations may not prove to be a practical design from a magnetic standpoint. With an open midplane gap, the final shape of the coils, and thus the magnetic field uniformity, is determined to some extent by the amount of tension in the shell. Because of the nature of the welding process, the final shell tension is hard to control, and therefore, an open midplane gap may result in significant magnet-to-magnet field variations. In addition to potentially affecting the field uniformity, an open gap at the yoke horizontal midplane reduces the magnitude of the field. The severities of these effects are not precisely known at this point and need to be better evaluated regardless of which yoke configuration is used.

CONCLUSION

From a mechanical point of view, the vertically split yoke with an open midplane gap after cooldown seems to be a promising configuration for the SSC dipole. It provides excellent support against collar deflections, minimal stress loss during cooldown, and a uniform stress profile across the inner coil pole at operating field. If an open yoke gap proves to adversely affect the magnetic field uniformity in an uncontrollable manner, then the vertically split yoke with a closed gap seems to be a good alternative.

The horizontally split yokes also provide excellent support against collar deflections and show small stress losses during cooldown. However, the bending stresses induced in the coils by the vertical clamping of the collars by the yokes produce a severe, undesirable stress gradient across the inner coil pole.

REFERENCES

1. Peoples, J. for the teams at BNL, Fermilab, LBL, and the SSC Central Design Group, Status of the SSC superconducting magnet program. In IEEE Trans. on Magnetics, vol. 25, no. 2, March 1989, p. 1444.

2. J. D. Jackson, ed., Conceptual Design of the Superconducting Super Collider, SSC–SR–2020, SSC Central Design Group, Lawrence Berkeley Laboratory, One Cyclotron Road, Berkeley, California, March 1986.

3. Wands, R. I. and Chapman, M. S., Finite element analysis of dipole magnets for the Superconducting Super Collider. In 1989 ANSYS Conference Proceedings, vol. II, May 1989, p. 7.42.

INVESTIGATION OF HEATER-INDUCED QUENCHES
IN A FULL-LENGTH SSC R&D DIPOLE

A. Devred, M. Chapman, J. Cortella, A. Desportes, J. DiMarco, J. Kaugerts,
R. Schermer, J. C. Tompkins, and J. Turner
SSC Central Design Group*
c/o Lawrence Berkeley Laboratory,
Berkeley, CA 94720 USA

J. G. Cottingham, P. Dahl, G. Ganetis, M. Garber, A. Ghosh, C. Goodzeit,
A. Greene, J. Herrera, S. Kahn, E. Kelly, G. Morgan, A. Prodell, E. P. Rohrer,
W. Sampson, R. Shutt, P. Thompson, P. Wanderer, and E. Willen
Brookhaven National Laboratory
Upton, NY 11973 USA

M. Bleadon, B. C. Brown, R. Hanft, M. Kuchnir, M. Lamm, P. Mantsch,
P. O. Mazur, D. Orris, J. Peoples, J. Strait, and G. Tool
Fermi National Accelerator Laboratory*
Batavia, IL 60510 USA

S. Caspi, W. Gilbert, C. Peters, J. Rechen, J. Royer,
R. Scanlan, C. Taylor, and J. Zbasnik
Lawrence Berkeley Laboratory †
Berkeley, CA 94720 USA

ABSTRACT

A 17-m-long SSC R&D dipole magnet instrumented with quench heaters and numerous voltage taps has been tested. These voltage taps enable (1) accurate localization of the quench start, (2) detailed studies of quench development, and (3) determination of coil temperature rise during a quench. The hot-spot temperature is determined by measuring the resistance of the conductor in the vicinity of the heater and is plotted versus number of MIITs. Measured temperatures are found to be in good agreement with predictions based on the assumption that the conductor is heated adiabatically. Finally, a limit to be imposed on the number of MIITs to operate the magnet safely is determined.

*Operated by Universities Research Association, Inc., for the U.S. Department of Energy.

†Work supported by the Office of Energy Research, Office of High-Energy Physics, High-Energy Physics Division, U.S. Department of Energy, under contract no. DE-AC03-76SF00098.

INTRODUCTION

A primary concern when building a 17-m-long superconducting dipole magnet is protection of its coil during a quench: How fast will the quench propagate? How much energy will be dissipated in the coil? How hot will the conductor get?

The first two questions were addressed in a previous paper,[1] which presented statistical studies of propagation velocities and of the number of MIITs* over the several long SSC dipole prototypes tested in the last two years.[2,3] Even though the quenches developed much faster than was ever seen or predicted before, the mechanism was reproducible from magnet to magnet,[1] depending only on the fraction of short sample. We also established a neat correlation between the number of MIITs and the inverse of the propagation velocity, illustrating the benefit of this fast quench development in reducing the overall energy dissipated in the coil. This paper addresses the third question and establishes the correlation between the number of MIITs and the maximum temperature reached by the conductor during a quench, also called the *hot-spot* temperature.

The temperature increase during a quench is central to the issue of safety. The temperature has to be limited to avoid failure of the Kapton insulation and degradation of the superconductor critical current. These effects both occur at a temperature of about 1000 K.[4] The SSC prototypes are currently operated with a maximum allowance of 800 K.

MEASURING THE HOT-SPOT TEMPERATURE

The best way to measure the temperature of a copper-stabilized conductor is to use the conductor itself as a temperature sensor. The resistivity of copper, ρ_{Cu}, is a well-tabulated function of three parameters: the copper residual-resistivity ratio, RRR; the temperature, T; and the magnetic field, B.[5] Let us consider a length L of conductor, carrying a current I. Once it has switched to the normal resistive state, the conductor sample exhibits an apparent resistance R_a

$$R_a = \frac{U}{I},$$

where U is the voltage across the sample. If the copper RRR and the magnetic field are known, an estimation of the conductor temperature, T, assumed to be uniform along the length L of the

sample, is obtained by solving the implicit equation in T

$$\rho_{Cu}\,(RRR,T,B) = \frac{r_{CuS}}{1+r_{CuS}}\,\frac{R_a S}{L}, \qquad (1)$$

where r_{CuS} is the copper-to-superconductor ratio, and S is the conductor cross-sectional area. Monitoring U and I during a quench then enables us to determine the temporal evolution of T.

Since we are interested in the hot-spot temperature, we have to monitor a voltage across a length of conductor surrounding the hot spot. Under normal conditions, the maximum temperature of a quenching coil is reached at the point where the quench originated. The quench start locations are not known in advance for spontaneous quenches, but they are well defined for spot-heater induced quenches. A simple experiment is to equip a magnet coil with spot heaters closely surrounded by two voltage taps. This has been done on a few full-length SSC dipole prototypes, including DD0017, which we will now discuss.

Figure 1 shows a cross section of the Brookhaven-design collared coil. The coil consists of four separately wound parts joined during assembly: two inner (upper and lower) and two outer (upper and lower) quarter coils. The inner quarter coils contain sixteen turns and three copper wedges; the outer quarter coils contain twenty turns and only one wedge. (Turns are counted starting from the midplane of the coil.) The characteristics of the inner- and outer-layer cables wound in DD0017 are given in Table 1.

TABLE 1.
Selected DD0017 Cable Characteristics

Quarter Coil	$S(m^2)$	r_{CuS}	RRR
Lower inner	11.79 10^{-6}	1.59	81
Lower outer	9.89 10^{-6}	1.62	83

*Between 10 and 295 K.

To establish the temperature-versus-MIITs correlations, the coil was equipped with six sets (spot heater + two voltage taps): two on the pole turn and two on the midplane turn of the lower inner quarter coil (one about 2 m from each end), and two on the pole turn of the lower outer coil (one at each end, in the middle of the curved sections). The length L between the two voltage taps is 25.4 cm for the inner-coil spot-heaters and 11.5 cm for the outer-coil spot-heaters. The RRR measurements provided in Table 1 were

---

*The number of MIITs is the integral over time t

$$MIITs = \frac{1}{10^6}\int_{\infty}^{0} dt\, I^2(t), \quad (A^2 sec)$$

where I is the current.

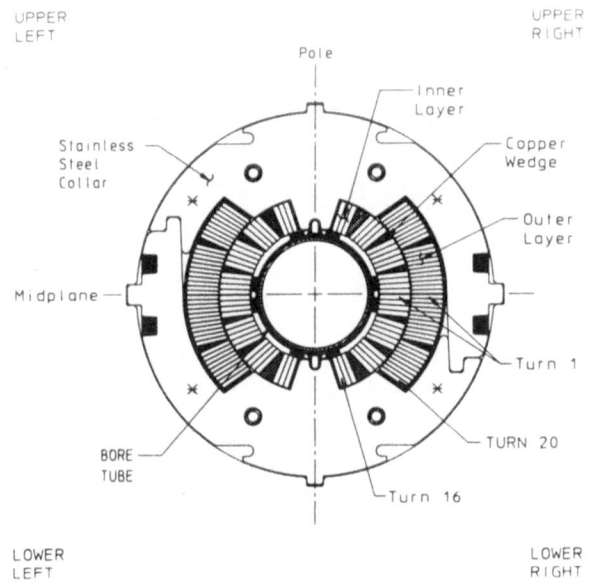

Figure 1. SSC Dipole cross section (C358).

made after testing, by warming up the whole magnet to a temperature of 12–13 K and circulating a current of about 10 A. The measurements were highly reproducible from one set of spot-heater taps to another. The magnetic field on the conductor, of course, depends on the coil turn and layer, and on the current. Numerical computations of the transfer functions $B = f(I)$ for the turns of interest are given in Table 2.

TABLE 2.
Transfer functions for the turns of interest

Inner coil turn 16	$B = 0.7505 + 0.9470 \ 10^{-3} I$
Inner coil turn 1	$B = 0.7183 + 0.9064 \ 10^{-3} I$
Outer coil turn 20	$B = 0.6266 + 0.7555 \ 10^{-3} I$

An important part of magnet DD0017 testing was devoted to spot-heater quenching, varying the current at quench, and successively firing different heaters. Figures 2 and 3 show typical records of the current I and the voltage U monitored during a quench induced at 6500 A by one of the inner-coil turn-1 heaters. The apparent resistance can then be calculated for any time, and the temperature can be estimated by solving Eq. (1), where the correct value of B for the given value of I has been introduced. The time, t_n, result is plotted as a continuous curve in Figure 4. As expected, it appears that after about 300 milliseconds the temperature reaches a kind of plateau. (The undulation for times greater than 400 milliseconds can be attri-

buted to calculation errors, since current and voltage then are both fairly small.) This plateau value defines the maximum temperature, T_{max}, reached by the coil during the quench. In our example, the number of MIITs is 7.39 and the maximum temperature is 157 K.

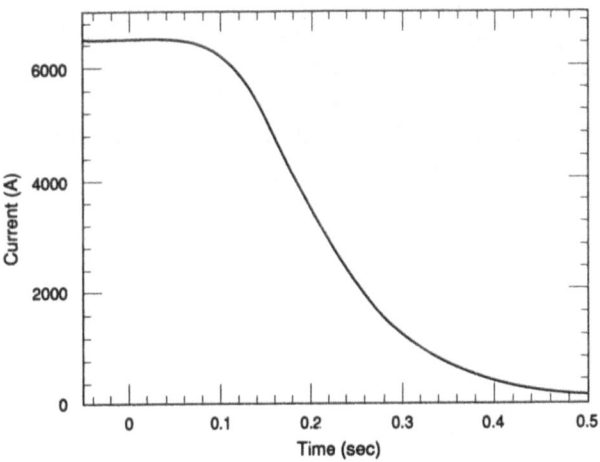

Figure 2. Current decay during a quench induced at 6500 A on turn 1 of the inner coil.

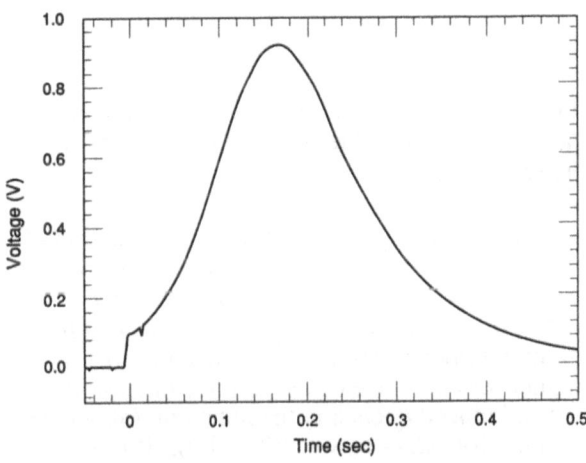

Figure 3. Voltage evolution across the spot-heater taps during a quench induced at 6500 A on turn 1 of the inner-coil.

The continuous curve of Figure 5 shows the result of the temperature computation for a quench induced at 6500 A on turn 20 of the outer coil. The shape of the curve is similar to Figure 4: after about 250 milliseconds, the temperature reaches a very stable plateau. Nevertheless, the value of this plateau is much higher than for the inner layer. At the same current as in the previous example, the number of MIITs is 8.39, and the maximum temperature is 289 K. The higher value of MIITs tells us that the quench propagates slower in the outer coil than in the inner coil, which is understandable because the field on the outer layer is much less than the field on the inner layer. More MIITs give, of course, a higher temperature. Also contributing to the higher temperature is the fact that the outer-layer conductor has about 15 percent less copper than the inner layer conductor, but, as it carries the same current, it dissipates 22.5 percent more Joule power. The interesting conclusion of this comparison is that, from the safety point of view, the limiting conductor is in the outer layer.

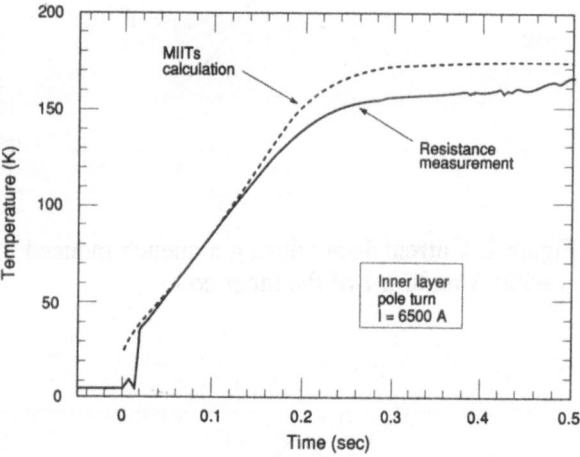

Figure 4. Evolution of the hot-spot temperature during a quench induced at 6500 A on turn 1 of the inner coil.

In Figure 6 we have plotted, as a function of the number of MIITs, the maximum temperatures measured for all the spot-heater quenches induced in DD0017. This plot confirms that for a given number of MIITs the temperature rise is higher for an induced quench in the outer-coil than for an induced quench in the inner coil (about 100 K more for MIITs between 8 and 9). It also appears that, for quenches in the inner coil, pole-turn and midplane-turn induced quenches are indistinguishable. This tells us that the 4 percent lower field on the turn-1 conductor does not greatly affect the overall power dissipation seen by this conductor as compared with that seen in the pole turn. (There is indeed a slight difference in the number of MIITs as a function of current for these two spot-heater locations—at a given current, the number of MIITs produced by a turn-1 induced quench is a few percent higher than for a turn-16 induced quench—showing that the propagation of a quench initiated in the midplane turn is presumably slower than the propagation of a quench initiated in the pole turn.) Nonetheless, the main

conclusion that can be drawn from Figure 6 is that for both conductor layers the temperature rise during a quench remains well below the limit of 800 K and that the safety of the magnet coil is ensured for MIITs numbers less than 9.

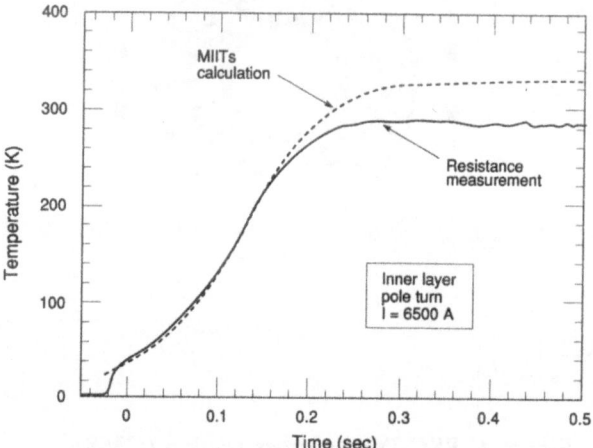

Figure 5. Evolution of the hot-spot temperature during a quench induced at 6500 A on the outer coil turn 20.

PREDICTIONS AND MEASUREMENTS COMPARED

To analytically predict the correlation of temperature-MIITs, we write the heat-balance equation for a small volume of conductor near the hot spot

$$C(T)\frac{\partial T}{\partial t} = \frac{1+r_{CuS}}{r_{CuS}} \rho_{Cu}(RRR,T,B)\left(\frac{I}{S}\right)^2 - H(t), \quad (2)$$

where C is the specific heat per unit volume of conductor and H is the power transferred to the surrounding medium, either by thermal conduction along the conductor or to the conductor insulation and the helium. An overestimation of the hot-spot temperature is given by neglecting H, e.g., by considering that the conductor near the hot spot behaves adiabatically,[6] and by assuming that the magnetic field remains constant, equals to its value B_0 at $t = 0$. Integrating Eq. (2) over time yields an implicit equation in T_{max}

$$\int_{T_{max}}^{T_0} dT \frac{r_{CuS}}{1+r_{CuS}} \frac{C(T)}{\rho_{Cu}(RRR,T,B_0)} = \cdots$$

$$\frac{1}{10^6}\int_{\infty}^{0} dt\, I^2(t) = \text{MIITs}, \quad (3)$$

where T_0 is the initial temperature. In fact, since we know I as a function of t, we can solve Eq. (3) for each value of time and thus predict the temporal evolution of the hot-spot temperature.

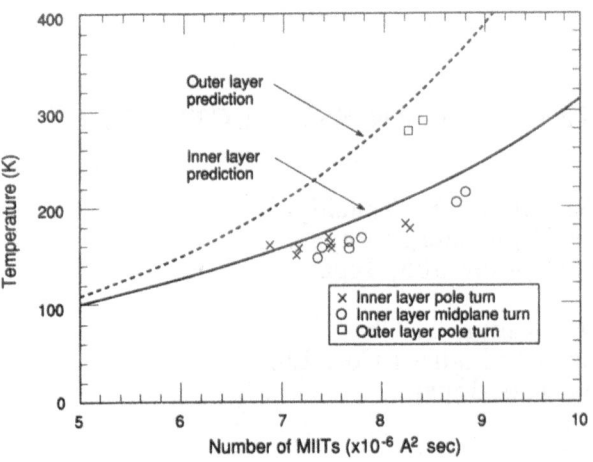

Figure 6. Temperature-versus-MIITs correlations

The dashed curves in Figures 4 and 5 show two examples of computation compared with the measured temperatures already presented. For the induced quench in the inner layer, B_0 has been chosen to be constant, equal to 7 T. The outer-layer computation requires more care. As the outer-layer spot-heaters are located at the ends of the coil, in the middle of the turn-20 curved sections, they sit outside of the iron yoke and see a very reduced field. In our computation, we have thus chosen to take $B_0 = 0$. In both cases, predictions and measurements are in good agreement. This implies that the hypothesis of justified, at least at the observed time scale. In other words, the amount of heat transferred to the surrounding media is small compared with the Joule heating, or the time required by the heat to be transferred to (or to be diffused through) the surrounding medium is large compared with 300 milliseconds. (The existence of a plateau in the temporal evolution of the temperature was already a clue that H could be neglected.)

With more confidence in our hypotheses, we return to Figure 6. The continuous and dashed curves are the temperature-versus-MIITs correlations, for the inner- and the outer-layer spot-conductors, as predicted by Eq. (3), using the same values of B_0 as above. The data points and the analytical predictions are in fairly good agreement. Equation (3) thus furnishes a reliable basis for predicting temperature-versus-MIITs correlations. Such agreements between hot-spot temperature measurements and predictions have also been seen on a short HERA model dipole.[7]

To conclude our safety analysis, it just remains to determine how the limit of 800 K on the peak temperature translates into a MIITs number. Iterations of Eq. (3) give 14 for the inner-coil turn-16 conductor (with $B_0 = 7$ T) and 10 for the outer-coil turn 20 (with $B_0 = 5.6$ T). A limit of 10 on the number of MIITs thus ensures that the conductor never gets higher than 800 K, wherever the quench occurs.

CONCLUSION

The use of spot heaters with two close voltage taps has enabled us to accurately measure the coil temperature increase during a quench. The measurements appear in good agreement with estimations assuming adiabatic heating of the conductor. We extrapolated that the number of MIITs has to be limited to ten to limit the peak temperature to 800 K.

REFERENCES

1. Devred, A., for the SSC Magnet R&D Collaboration, Quench characteristics of full-length SSC R&D dipole magnets. Presented at the 1989 Cryogenic Engineering Conference, Los Angeles, California, July 24–26,1989, to appear in the proceedings.

2. Strait, J., for the SSC Magnet R&D Collaboration, Test of full scale SSC R&D dipole magnets. IEEE Trans. Magn., 1989, 25, 1455–1458.

3. Tompkins, J., for the SSC Magnet R&D Collaboration, Performance of full length SSC model dipoles: results from 1988 tests. Presented at the First International Industrial Symposium on the Super Collider, New Orleans, Louisiana, February 8–10,1989, to appear in the proceedings.

4. Stiening, R., private communication.

5. Simon, N. G. and Reed, R. P., Materials for Superconducting System. NBS, March, 1987.

6. Wilson, N. M., Superconducting magnets. Clarendon Press, Oxford, 1983, p. 200.

7. Bonmann, D., Meß, K., Otterpohl, U., Schmüser, P. and Schweiger, M., Investigations on heater induced quenches in a superconducting test dipole coil for the HERA proton accelerator. DESY HERA 87-13, June 1987.

DESIGN OF A VERY HIGH FIELD MAGNET FOR AN SSC DECTECTOR

H. Hirabayashi, A. Maki, S. Terada, K. Tsuchiya
KEK, National Laboratory for High Energy Physics,
1-1 Oho, Tsukuba-shi, Ibaraki-ken 305, Japan
and
T. Akiyama, T. Doi, H. Kakui, T. Oba
IHI, Ishikawajima-Harima Heavy Industries Co., Ltd.,
Chiyoda-ku, Tokyo 100, Japan

ABSTRACT

The feasibility of a very high field superconducting solenoidal magnet for high energy collider experiments at SSC is investigated. We present a design for a magnet that is 2 m in diameter, 2.5 m in length with a central field of 6 Tesla. The magnet coil has a multi-layer structure in which every layer is wound on its own aluminum bobbin. These bobbin-coil layers are shrink-fit together into a single coil. This structure withstands the strong magnetic pressure while keeping the coil plus cryostat thickness less than 0.5 nuclear interaction lengths thus minimizing its effect on the performance of hadron calorimeters located outside the coil. An evaluation of the mechanical and thermal properties of the coil is given.

INTRODUCTION

Solenoidal magnetic detectors are frequently used in high energy physics experiments at colliding beam accelerators. Particles produced in beam-beam collisions are momentum analyzed by means of tracking devices located inside of the magnetic field. As the beam energy increases, the demands on the tracking system becomes more severe. The general tendency has been to meet these demands by increasing the size of the system. Another approach, which avoids the excessive growth of the detector size with increasing beam energy, is to exploit recent developments in the technology associated with superconductivity and use very high magnetic fields. A successful example of this approach is the AMY experiment[1] at TRISTAN, KEK, in Japan. In spite of the small size of the AMY magnet[2] (2.5 m in diameter, 1.5 m in length), its 3 T central

field provides a charged particle momentum resolution of $\Delta p_t/p_t=0.6\%$ x $p_t(GeV)$, comparable to that of larger, more conventional detectors. The small size of the detector minimizes muon backgrounds arising from the decay-in-flight of hadrons and readily accommodates low-beta configurations of the accelerator. The high field has enabled a new scheme of identifying electrons that is being realized in the AMY experiment utilizing the fact that secondary electrons radiate detectable synchrotron X-rays of tens of kev energy as they bend in the high magnetic field. Moreover, the experience with AMY has demonstrated that a small size can result in a substantial reduction in cost without compromising the detector's capabilities. We have aggressively extended the concept of the AMY experiment and investigated the feasibility of the very high field 6 T superconducting solenoid[3] suitable for use in experiments involving very high enegy (20 TeV + 20 TeV) proton-proton collisions at the Superconducting Super-Collider (SSC).

MAGNET DESIGN

Requirements

For purposes of design, we assume an SSC detector[4] with tracking devices and electromagnetic calorimeters located inside the magnet and with hadron calorimeters and muon detection systems placed outside the coil. In such a configuration, the magnet can be kept relatively small if it provides a high field. We studied the design of a solenoid that is 2.0 m in diameter and 2.5 m in length. By choosing a field strength of 6 T at the magnet center, we achieve a similar value of Bl^2 as larger designs based on more conventional field strengths[5]. Since the hadron calorimeters are located outside of the magnet coil and thick regions of uninstrumented material in front of them may degrade their performance, the thickness of the magnet must be kept thinner than a half of an absorption length, which is equivalent to about 20 cm of aluminum. In order to minimize regions of inactive material and gaps in the detector, the cryostat for the coil must be made as small as possible.

Magnetic properties

The magnetic field and forces were estimated using the computer programs POISSON[6] and JMAG[7]. We assumed the geometry shown in Fig. 1; the coil is surrounded by an iron yoke that is far removed from the coil's proximity and is, presumably, part of the muon detection system. In this configuration, the magnet is very nearly an air core solenoid; it requires 1.57×10^7 ampere turns to produce 6.0 T at the coil center. The field increases with radius, reaching a maximum of 6.6 T at the inner surface of the coil at its midplane. The stored energy in the magnetic field is 155 MJ.

The choice of the magnet current is the essential parameter which drives the rest of the design of the coil. We selected this to be 15 kA after considering a number of factors including: the number of turns and coil layers, the length and cross section of the conductor, the flexibility of the conductor in coil winding, the inductance and resistance of the coil, the characteristics of the current leads and power supply, etc. A large current value has the advantage of simplifying the

mechanical structure by reducing the number of turns and layers. In addition, the fewer number of turns reduces the coil's inductance and resistance, making the quench protection simpler. On the other hand, a large current requires a thick conductor, large current leads and a large power supply. The main parameters of the coil are listed in Table 1.

TABLE 1

Parameters of the Solenoid

Current	15,000	A
Central field	6.0	T
Maximum field	6.6	T
Coil inner diameter	2.0	m
Coil outer diameter	2.2	m
Coil length	2.5	m
Coil winding		
number of layers	6	
number of turns in a layer	174	
total turns of the winding	1,044	
Total ampere turns	15,660,000	A
Inductance	1.37	H
Stored energy	155.3	MJ
Coil weight	6.9	ton
Cryostat inner diameter	1.892	m
Cryostat outer diameter	2.376	m
Cryostat length	2.96	m
Cryostat weight	1.5	ton

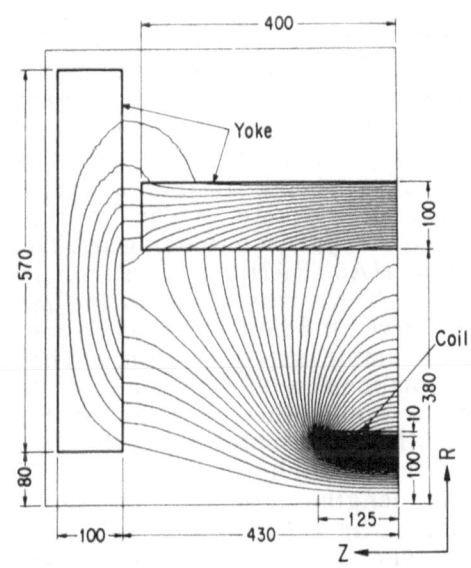

FIG. 1. Quadrant of the coil and return yoke configuration. The magnetic flux lines at 1.57×10^7 ampere turns as calculated by POISSON.

Coil conductor

The conductor is an aluminum-stabilized NbTi/Cu superconductor insulated with a 0.1 mm thick layer of Kapton, as shown in Fig. 2. At the rated current of 15kA, the current density in the NbTi superconductor is $750A/mm^2$, less than half of the critical current (at 4.4 K and 6.6 T) for the best NbTi/Cu superconducting wire[8] that is currently available.

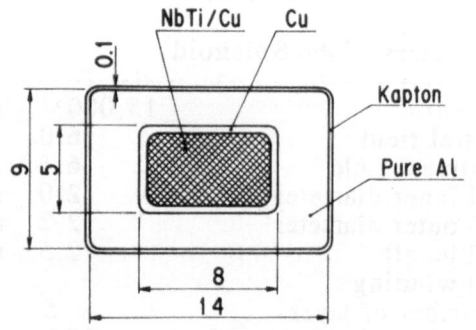

FIG. 2. Cross-section of the conductor.

Coil structure

The interaction between the strong magnetic field and the high current in the coil generates a large expansive magnetic force of 124.2 MN, equivalent to the pressure of about 150 MPa, on the coil structure. In order to sustain this large pressure, a multi-layer structure, similar to that used for high pressure gas vessels, was adopted. As shown in Fig. 3, the coil consists of six layers with each layer supported by its own outer aluminum bobbin. Since the magnetic pressure is strongest at the inner layer and weakest at the outer layer, the bobbin thicknesses vary accordingly. All layers are assembled by means of a shrink-fit into a single tight mechanical structure. The results of a structual analysis, done using the computer program NASTRAN[9], is summarized in Table 2. The stresses due to the magnetic forces were all below the allowed maximum with an ample safety margin. The radial field components at the ends of the coil generate a compressive magnetic force of 57.37 MN at the midplane of the coil; this contracts the conductor layers by about 2 mm in length. This contraction could introduce slip between the bobbins and the coil layers during magnet excitation. However, the results of the structual analysis indicate that this slip can be easily prevented by simply gluing the coil winding on its outer bobbin and by assembling the layers with a modest prestress by means of a shrink-fit.

TABLE 2

Stresses on the bobbin and conductor

Layer No.*)	Thickness (mm)	Max. stress **) (MPa)
1. Conductor	9	303.2
Bobbin	12	180.3
2. Conductor	9	298.0
Bobbin	10	175.5
3. Conductor	9	292.3
Bobbin	8	171.3
4. Conductor	9	287.5
Bobbin	6	171.3
5. Conductor	9	283.2
Bobbin	5	164.1
6. Conductor	9	278.3
Bobbin	5	159.8

*) Layers are counted from inner most layer
**) The maximum stress obtains at the center of the layer. The indicated stress should be compared with the maximum allowed values of 428 MPa and 228 MPa for the coil conductor and the aluminum bobbin, respectively.

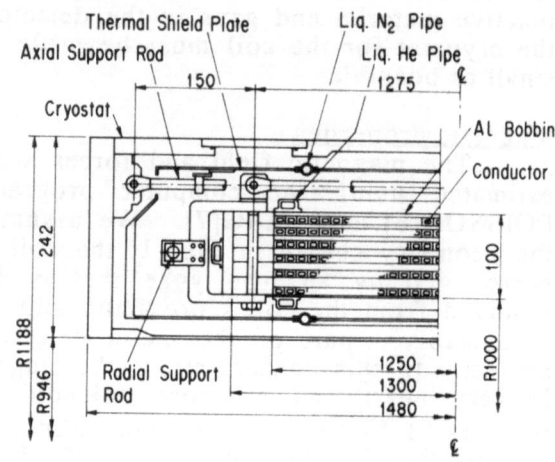

FIG. 3. Structure of the coil.

The superconducting coil must be suspended by a support system in a cryostat with a minimal heat leak. In the design of the support structure, the magnetic decentering forces caused by possible misalignment of the coil relative to the yoke and the weight of the coil have to be taken into account. However, since the yoke is far away from the coil, the decentering forces between the coil and the yoke are small; they are estimated to be 1470 N and 735 N per mm of misalignment in axial and radial direction, respectively. Thus, the support system essentially only has to counter the coil weight of 6.9 tons, which can be done by a simple radial support. A thin axial support rod between the endplate of the cryostat and the other end of the coil braces against possible axial movements of the coil. With this support structure, the cryostat can be made relatively short, while keeping the radial thickness as thin. The radial thickness of the magnet is 15.7 cm measured in aluminum equivalent, thinner than the required maximum allowed thickness of 20 cm. The major components contributing to the magnet thickness are listed in Table 3.

TABLE 3

The radial thickness of the magnet coil and cryostat (aluminum equivalent).

Cryostat	
inner wall	2.96 mm
outer wall	11.82 mm
liq. N2 thermal shield	4.0 mm
Coil bobbin	48.39 mm
Conductor	
NbTi filament (average)	24.69 mm
Copper (average)	28.22 mm
Al stabilizer (average)	36.86 mm
Total	156.94 mm

Thermal properties

The coil is cooled by the flow of two-phase helium through an aluminum coolant pipe of cross section 15 x 18 mm^2, welded on the inner and outer surfaces of the coil as illustrated in Fig. 4. The temperatures at various places in the coil were calculated using NASTRAN. In the calculation, the thermal contact between the coolant pipe and the coil surface was assumed to be 20 %. Heat leaks through the coil support rods and radiation through the superinsulation layers in the cryostat were taken into account. The calculation shows a highest temperature in the coil of 4.32 K. The pressure drop of the two-phase helium flow in the coolant pipe and helium transfer lines was studied for various helium flow rates. It is 0.012 bar at the optimum helium flow rate of 2.5 g/s. This pressure drop gives a very small temperature rise of 0.01K above 4.2K. From these calculations, it appears certain that this relatively simple cooling scheme is adequate. The total heat load of the magnet is moderate. This is estimated to be 32.4 W for 4.2 K helium circuit and 192 W for 80 K nitrogen circuit. In addition, the 15 kA

current leads are estimated to consume 44 l/h of liquid helium.

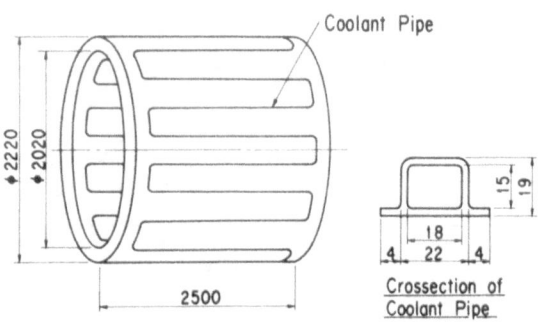

FIG. 4. Arrangement of the coolant pipes.

The cool-down of the coil from room temperature was also simulated using the computer codes PHENICS[10] and ISTRAN/HEAT[11]. As shown in Fig. 5, the temperature of the coolant helium is gradually lowered from 200 K to 4.2 K in order to avoid the thermal stress due to a temperature gradient in the coil. The flow rate of the helium, also shown in Fig. 5, is adjusted from 25 to 200 l/h in order to optimize the helium pressure in the coolant pipe while avoiding large temperature gradients in the coil. The helium pressure and the coil temperature are shown in Figs. 6 and 7, respectively. The coil can be cooled within a week; during which time the maximum helium pressure and the maximum temperature difference in the coil are 9.0 bar and 56.8 K, respectively.

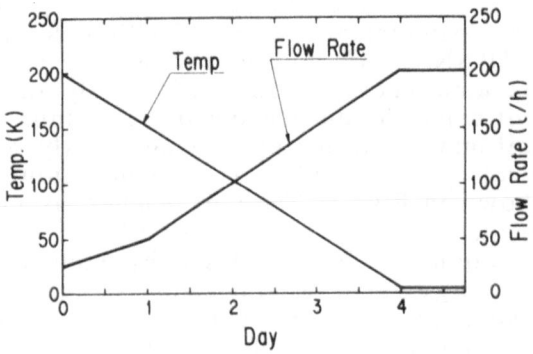

FIG. 5. The temperature and flow rate of the coolant helium as a function of time during the cool-down procedure.

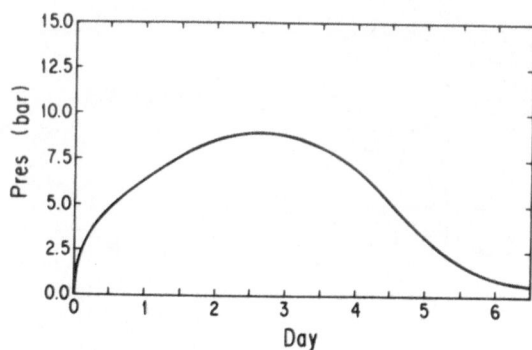

FIG. 6. Pressure of the coolant helium during the cool-down procedure.

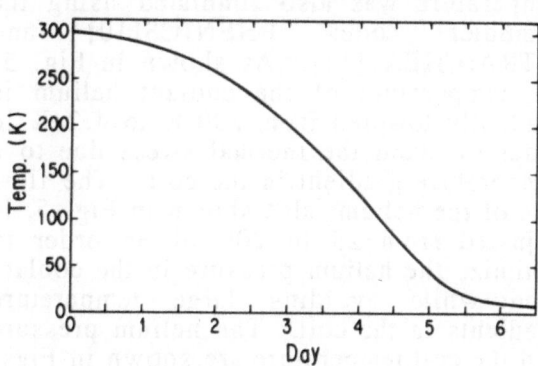

FIG. 7. The temperature of the coil during the cool-down procedure.

Quench properties

Although the coil conductor is stabilized with high purity aluminum, it must be assured that the coil is safe from quench. The behaviour during a quench was simulated using PHOENICS and ISTRAN/HEAT. The 155 MJ of stored magnetic energy is released rapidly in the coil when the quench occurs, and thus the coil may be heated up to dangerously high levels. The results of the simulation are shown in Fig. 8. The coil reaches its peak temperature of 132 K in 12 s. This situation can be moderated by connecting an energy dump resistor across the coil; the resistance of which should be relatively large compared with the coil resistance. In that case, most of the stored energy is dissipated in the resistor. This technique is commonly used for protecting super-conducting coils during a quench. However, the resistance of the dump resistor is limited by the ability for the coil to support high voltage differences, since a larger dump resistor results in higher voltages across the coil. In order to extract

the stored energy as much as possible, it is effective to increase the resistance of the dump resistor in inverse proportion to the decaying magnet current so as to keep the voltage constantly high during the quench. With the resistance of the dump resistor adjusted so that the voltage drop across it was kept at 2 kV, the temperature rise in the coil is reduced to a maximum value of 64 K.

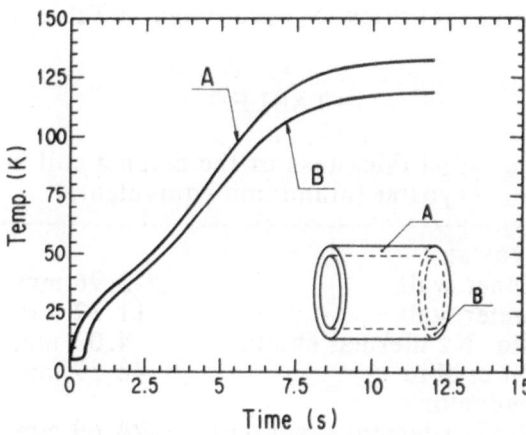

FIG. 8. The temperature rise of the coil after the quench that originates at point A.

CONCLUSION

A compact (2.0 m in diameter x 2.5 m in length) very high field (6 T) superconducting solenoidal magnet was designed for use in an SSC experiment. In the design study, the magnet showed reasonable mechanical, thermal and quench properties and there appear to be no major practical problems associated with its fabrication and operation. The magnet coil plus cryostat can be made with an amount of material corresponding to less than one-half of a nuclear absoption length of material.

ACKNOWLEADGEMENT

This work has been supported by the Japan-US Collaboration Program in High Energy Physics.

REFERENCES

1. C. Back et al., "*AMY-Chan, a High Resolution Lepton Detector for TRISTAN*", TRISTAN-EXP-003 KEK (1984); H. Sagawa et al., Phys. Rev. Lett. **60** (1988) 93.

2. Y. Doi et al., Nucl. Instr. and Meth. **A274** (1989) 95.

3. S. Terada et al, KEK Preprint 88-68 (1988), contributed to the 1988 Summer Study on High Energy Physics in the 1990's, Snowmass, Colorado, June 27 - July 15, 1988

4. J. Kirkby, T. Kondo and S.L. Olsen, *"Report of the Compact Detector Subgroup"*, in Proceeding of the Workshop on Experiments, Detectors, and Experimental Areas for the Supercollider, July 7-17, 1987, Berkeley, California.

5. G.G. Hanson, S. Mori, L.G. Pondrom and H.H. Williams, *"Report of the Larger Solenoid Detector Group"*, ibid.

6. M. Cyr and C. Iselin, POISSON, CERN (1976).

7. Developed by Japan Information Service Ltd., Tokyo, Japan.

8. S. Hong, D. Geschwindner, A. Mantone, W. Marancik, S. Zalek and R. Zhou, IEEE Trans. Magn. **MAG-25** (1989) 1934.

9. NASTRAN, National Aeronautics Space Administration, USA.

10. Developed by H.I. Rosten and D. Spalding, Concentration Heat and Momentum Ltd., London, England.

11. Developed by IHI, Ishikawajima-Harima Heavy Industries Co., Ltd., Tokyo, Japan.

STUDY OF THE CURRENT CARRYING ELEMENT OF THE SUPERCONDUCTING DIPOLE MAGNET FOR THE UNK

A.N.Baidakov, K.P.Myznikov, L.S.Shirshov, V.P.Snitko, A.N.Surkov, V.V.Sytnik,
N.V.Taran, V.A.Vasiliev, M.G.Vybornov, A.V.Zlobin
Institute for High Energy Physics, 142284, Serpukhov, Moscow region, USSR

ABSTRACT

The current carrying element of the superconducting dipole magnet of the Accelerating and Storage Complex (UNK) is a flat transposed Rutherford-type cable consisting of 19 strands, 9 of which are coated by Sn+5% Ag alloy. The cable is manufactured from $\emptyset$ 0.85 mm composite wires containing 8910 Nb-Ti filaments embedded into a copper matrix. The critical current density in the 5 T field at 4.2 K is at least $2.3 \cdot 10^5$ A/cm^2 which enables one to attain a bore field exceeding 6.4 T at a coil temperature of 4.25 K. The paper presents the results on the study of the basic parameters of the current carrying element which determine the paramount operational characteristics of the superconducting magnets for the UNK.

INTRODUCTION

A superconducting (SC) dipole for the UNK should satisfy the following basic requirements:

- the operating fields in the $\emptyset$ 70 mm bore should vary from 0.67 T to 5 T and have the quality required;

- the magnet should possess a stable operational feature at a coolant temperature of 4.4-4.6 K under the conditions of ac losses and radiation heat depositions;

- the value of the ac losses in the coil and in the design elements in the UNK operational cycle as well as the static heat leaks should not exceed the limit specified by the feasibilities of the UNK cryogenic system;

- the maximum quench-induced coil heating should not exceed 200 K.

The above parameters are governed by the dipole design, production technology, by the properties of the current carrying element and structural materials.

The basic element of a UNK dipole is a two-shell coil with an inner diameter of 80 mm and a cold iron. Figure 1 shows the cross section of such a dipole and table 1 presents its basic parameters.

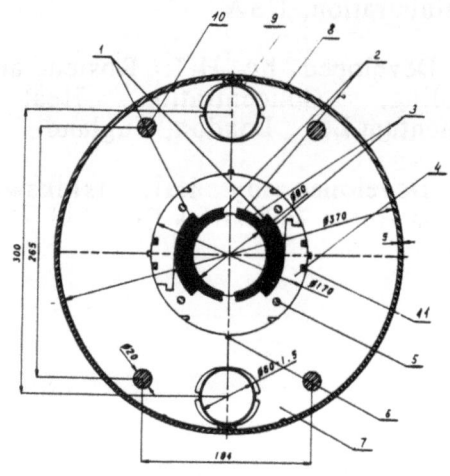

FIGURE 1

Cross section of a UNK dipole: 1 - outer layer of the coil, 2 - inner layer, 3 - spacers, 4 - collars, 5 - pin, 6 - rectangular lug, 7 - iron, 8 - helium vessel, 9 - two-phase helium pipe, 10 - stud, 11 - key.

Parameter	Inner layer	Outer layer
Number of turns	2x34	2x21
Turn width, mm	8.7	8.7
Mean turn thickness, mm	1.63	1.63
Angular layer dimension, deg	2x75.96	2x44.34
Angular spacer dimension, deg	4.25	7.24
Angular dimension of the median spacer, deg	0.2	1.05
Inner layer radius, mm	40.0	49.2
Ratio B_{max}/B_O	1.096	0.806
Quench temperature for the 5 T bore field, K	5.4	6.3

Table 1: The Basic Parameters of a UNK dipole

The current carrying element of a SC dipole for the UNK is a flat transposed Rutherford-type SC cable. The dipole is designed to use SC wires having Ø 6 μm filaments and a critical current density of 2.3×10^5 A/cm^2 in the 5 T field at 4.2 K. The industry has brought the production technology of such wires for SC magnets to a commercial level[1]. IHEP has designed and produced a series of machines and devices for the SC cable production[2].

This work presents some results on the study of the properties of the SC wires and cable which determine the paramount characteristics of the UNK SC dipole.

THE STUDY OF COMPOSITE SC WIRES
The basic characteristics of SC wires are given in Table 2.

Superconducting alloy	Nb+50%Ti (NT-50)
Wire diameter, mm	0.85+0.03
Number of SC filaments	8910
SC filament diameter, μm	6
Twist pitch, mm	10
Matrix material	copper
Copper-to-superconductor ratio (packing factor)	1.38+0.12 (0.42+0.02)
Copper residual resistivity ratio	⩾70
Critical current at 5 T at 4.2 K, A	⩾550
Critical current density at 5 T at 4.2 K, A/cm^2	⩾$2.3 \cdot 10^5$

Table 2: The Basic Characteristics of Composite SC Wires of the SC Magnets for the UNK.

An apprecible increase of the current density of SC wires with simultaneous decrease of the filament diameter has been accomplished by optimizing the conditions of thermal and mechanical treatment of SC wires and by increasing the longitudinal homogeneity of SC filaments using a thicker diffusion barrier at their surface.

The magnetization and hysteresis loss in SC wires have been studied in a transverse magnetic field at a temperature of boiling liquid helium. The measurement technique is described in paper[3]. Figure 2 presents the dependence of the hysteresis magnetization of SC filaments during field rise, $M_h^\uparrow$, and drop, $M_h^\downarrow$. It also shows for illustration purposes similar dependences for wires consisting of Ø 10 μm and Ø 6.5 μm filaments[4] which were used earlier for simulation of SC dipoles for the UNK. The critical current density was the same in all samples, $2.3 \cdot 10^5$ A/cm^2, at 5 T and 4.2 K. As is seen from the figure, a thicker diffusion barrier on the surface of SC filaments resulted in an increase of the hysteresis magnetization of the SC wire in fields below 2 T.

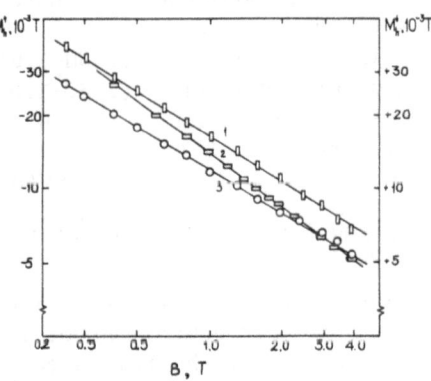

FIGURE 2

The hysteresis filament magnetization during rise, $M_h^\uparrow$, and fall, $M_h^\downarrow$, of external field: 1-d=10 μm, 2-d=6 μm, 3-d=6.5 μm.

Figure 3 presents the specific hysteresis loss in the SC wire versus the field amplitude in the triangular cycle. It also presents for comparison the hysteresis loss in a SC wire with Ø 10 μm SC filaments. As compared with this loss, the one in SC wires with Ø 6 μm filaments decreased almost proportionally to the filament diameter. The effect of a thicker diffusion barrier on the value of hysteresis loss in a SC wire with Ø 6 μm filaments does not exceed 10%.

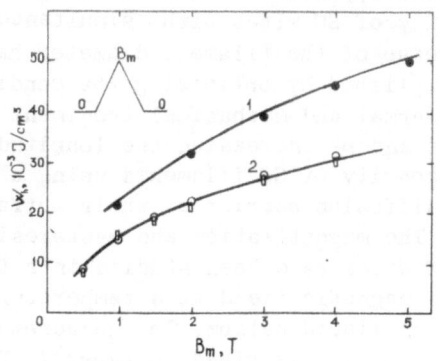

FIGURE 3

Specific hysteresis loss versus the field amplitude in the triangular cycle: 1-d= =10 μm, 2-d=6 μm.

THE STUDY OF TRANSPOSED SC CABLE

Table 3 presents the basic parameters of the SC cable of UNK magnets.

Number of strands		19
Cable type		ZEBRA
Strand coating		
	9 strands	Sn+5% Ag
	10 strands	bare
Cross section		
	inner layer, mm^2	(1.30-1.62)x8.5
	outer layer, mm^2	(1.33-1.59)x8.5
Transposition pitch, mm		62
Critical current at 5 T at 4.2 K, A		⩾9500

Table 3: The Basic Parameters of the SC Cable

To obtain the required bore field quality the geometric dimensions of the cable should satisfy the tolerances to a very good accuracy, not worse than +30 μm in the width and not worse than ±5 μm in the thickness. To attain the required accuracy in producing the transverse cross of the SC cable, a special forming device without the force closure of the forming rollers has been designed. Besides, the dimensions are controlled cyclewise with the help of a special measuring unit providing an accuracy of measuring the median line of the cable of ±1 μm. The information obtained is used subsequently to regulate the forming device[5]. Figure 4 shows the histogram of the distribution for deviations of the medium thickness of the trapezoid - shaped cross section from the calculated value. As is seen from the figure, the r.m.s. deviation of the median thickness does not exceed ±3 μm. The r.m.s. variation in the cable width is ±10 μm.

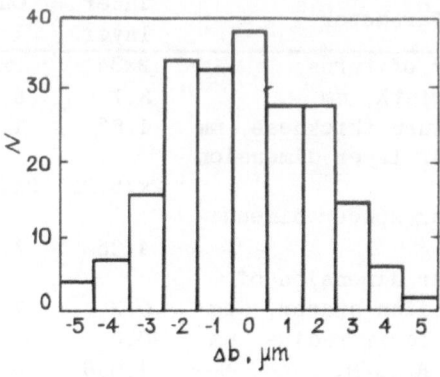

FIGURE 4

Histogram of the distribution of the deviations of the median cable thickness from the calculated value.

Figure 5 gives the results on measuring the dependence of the critical current of the cable on the field induction at 4.25 K. The measurements were carried out with the help of the device described in paper[6]. The critical current measured at longitudinal electric field of 1 μV/cm was 9750 A. This is 7% less than the sum of the critical currents of the initial underformed wires which is due to the effect of deformation of the transverse cable cross section. In this figure the load lines for the UNK dipole are denoted by numbers 1 and 2. As is seen from the figure, this critical current of the cable allows one to obtain a bore field of at least 6.4 T at 4.25 K.

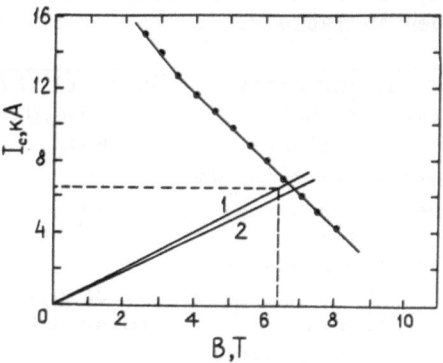

FIGURE 5

Critical current of the cable versus the induction of external field at 4.25 K: 1,2 - load lines for a UNK dipole.

The ac losses of package samples of the cables usedfor the UNK dipole have been measured in an varying external field normal to the cable plane under the stress on the samples close to that in the dipole coil using the technique described in paper[7]). The samples were cured at a temperature of 150°C during 5 hours as it is done with the dipole coil. Figure 6 shows the specific ac losses in the cable in the triangular cycle versus the field amplitude and ramp rate. Ac losses in the Rutherford-type SC cable consist of hysteresis loss in the SC filaments and eddy current losses which have cable and matrix components. For UNK ramp rates of 0.1 T/s the main contribution to the value of ac losses comes from the hysteresis and cable components. The matrix loss in SC wires with a filament twist pitch of not more that 10 mm does not exceed 10% of the eddy current losses.

Quench-induced cable heating has been studied. Figure 7 shows variations of the cable temperature with time measured for different values of the transport current. The values of the quench integral $\int I^2 dt$ in units $10^6 A^2 \cdot s$ at a given instant of time are specified on the curves of heating and are connected by the dashed lines. The calculated values of the temperature during adiabatic heating which correspond to the given quench integrals are plotted in the temperature axis. It is seen that for small values of time when the heating conditions approach adiabatic ones, the measurement results are in a good agreement with the calculation.

CONCLUSIONS

The studies carried out show that the developed composite SC wire with a critical current density above $2.3 \cdot 10^5$ A/cm^2 consisting of Ø 6 μm filaments possesses good mechanical and electric properties allowing transposition and precision moulding of the SC cable with a high packing factor of 90%, simultaneously retaining the current density of wires. The high current density of SC wires and cable allow one to obtain a bore field of at least 6.4 T at 4.25 K. Since the maximum operating field in the UNK cycle is accepted to be 5 T, the 25-30% reserve available will ensure a reliable operation of the superconducting magnet system of the accelerating and storage complex at the 4.4-4.6 K coolant temperature under the conditions of ac losses and radiation-induced heat depositions. The calculated addition to the sextupole nonlinearity brought about by the hysteresis magnetization of the current element with Ø 6 μm filaments is not more than $1.1 \cdot 10^{-4}$ in the central cross section of the dipole

at the 3.5 cm bore radius at the injection field 0.67 T. Taking into account the envisaged correction system, this is within the tolerances. Proceeding from the above results on the study of the ac losses in the cable of the chosen design, the mean power of ac losses in the coil in the acceleration cycle with a linear 40-s field rise in the bore from 0.67 T to 5 T, 38-s flattop and linear 40-s field drop from 5 T to 0.67 T will be 0.6 W per meter of the dipole length, 0.46 W/m being the hysteresis loss in the filaments and 0.1 W/m being the eddy current loss in the cable.

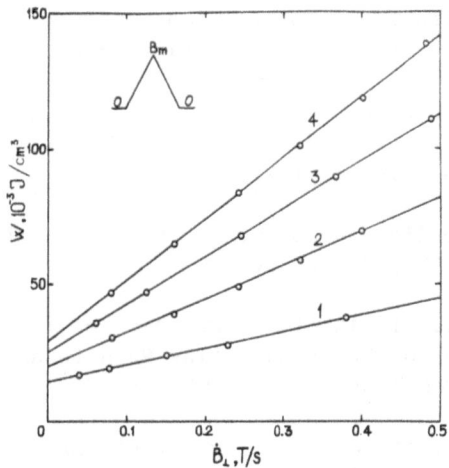

FIGURE 6
Specific as losses in the cable versus the field amplitude and ramp rate in the triangular cycle: 1 - B_m = 1,T, 2 - B_m = 3 T, 4 - B_m = 4 T.

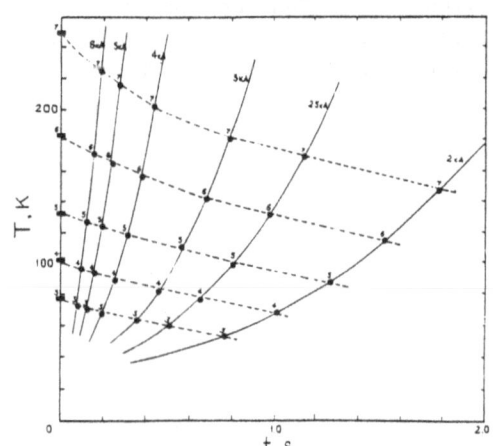

FIGURE 7
Time variation in the cable temperature after quench for different values of the transport current. Dashed lines correspond to $\int I^2 dt$=const.

As it follows from the study, with internal strip heaters placed on the outer layer of the coil, the value of the quench integral during dissipation of the stored energy in the coil does not exceed $6.5 \cdot 10^6$ $A^2 \cdot s$. This means that the maximum quench - induced coil heating will not exceed 200 K.

REFERENCES

1. G.K.Zelensky et al. "Development of Composite Superconductors for Accelerator-Storage Magnets". Report at the Intern. Conf. on Cryogenic Materials (Structure, Applications and Properties). Shenuang, P.R. China, 1988.

2. V.P.Oshchepkov et al. Electrotechnical Industry. Series "Cable Engineering". 1983, issue 7(221), p. 23.

3. A.G.Aleksandrov et al. Preprint IHEP 87-197, Serpukhov, 1987.

4. V.Y.Fil'kin et al. Proceed. of Workshop on Superconducting Magnets and Cryogenics. - BNL, 1984, p. 56.

5. E.P.Borisov et al. Preprint IHEP 89-23, Serpukhov, 1989.

6. L.S.Shirshov and S.Enderlein. Cryogenics, 1985, v. 25, p. 527.

7. Yu.P.Dmitrevsky et al. Proceed. of II All-Union Conference on Technical Application of Superconductivity. - Leningrad, Leningrad Research Computing Centre, 1984, v. 1, p. 309.

HEAT LOAD ON A SUPERCONDUCTING UNK DIPOLE

A.I.Ageyev, Yu.G.Bozhko, E.M.Kashtanov, S.S.Kozub, K.P.Myznikov,
Ya.V.Shpakovich, V.V.Sytnik, A.V.Tarasov, A.A.Zolotov
Institute for High Energy Physics, 142284, Serpukhov, Moscow region, USSR

ABSTRACT

The development of superconducting (SC) dipole magnets for the UNK rests upon the cold-iron design. The support system of the helium vessel of a SC UNK dipole consists of four vertical suspensions and four horizontal extension rods placed in two cross sections across the magnet length and also of two anchor extension rods fixing the centre of the helium vessel in thermal cycles. To reduce the value of thermal radiation, use is made of the nitrogen shield fixed with vertical suspensions and of a multilayer insulation put onto the nitrogen shield and helium vessel.

The heat load and the heat leak through the supports to the helium vessel of the SC UNK dipole have been studied. The techniques applied to study these heat loads are presented. The measured value of the heat load is 5-6 W, 1 W falls at the supports. The experimental results obtained are compared with the calculational estimates and feasibilities for a further reduction of a heat load to the helium vessel of the magnet are discussed.

INTRODUCTION

Heat load on the superconducting magnets contributes essentially into the heat load on the cryogenic system of the IHEP Accelerating and Storage Complex[1]. The heat load on liquid helium is due to heat leaks induced by the residual gas, thermal bridges. The value of the residual pressure in the insulation area of the magnet is assumed not to exceed 10^{-4} Pa, in which case the heat leak through the residual gas is negligible. The thermal radiation to the helium vessel is determined by its surface area and reflecting capacity and also by the temperature of the surface radiating energy on the vessel. The main contribution into the heat leak across the thermal bridges comes from the helium vessel support system. This paper describes such a system for the UNK dipole and presents some results on measuring the heat load to the helium vessel and heat leak through the magnet supports. The results are compared with those for a warm-iron design[2].

MAGNET DESIGN

Figure 1 shows the cross-sectional view of a UNK dipole. The superconducting coil collared with stainless steel laminations is placed in the iron yoke inside the helium vessel. The magnet is cooled by a single-phase helium flow, a part of which passes through the coil and another one through the bypass channels to be cooled by a two-phase helium counterflow. The ∅ 0.38 m, 4.7 T 6 m long helium vessel is fixed to the vacuum vessel with the help of the support system (see fig.2). This system consists of four vertical suspensions and four horizontal extension rods placed in two cross sections of the magnet length and also of two anchor extension rods centering the helium vessel with respect to the vacuum one.

A suspension consists of two ∅ 12 mm studs 190 mm and 312 mm long and a horizontal extension rod consists of two ∅ 10 mm studs 175 and 322 mm long. The studs are connected by nut 14 (see fig.1) fixed to the nitrogen shield, the shorter stud

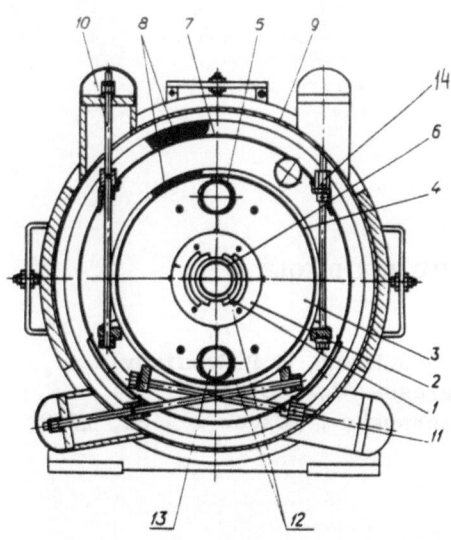

FIGURE 1
Cross section of a SC dipole for the UNK:
1 - coil, 2 - collars, 3 - yoke, 4 - helium vessel, 5 - two-phase helium pipe,
6 - beam pipe, 7 - nitrogen shield, 8 -
multilayer insulation, 9 - vacuum vessel,
10 - vertical suspension, 11 - horizontal
extension rod, 12 - single-phase helium,
13 - two-phase helium, 14 - nut.

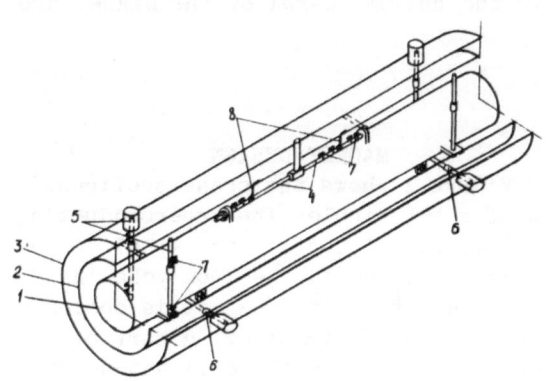

FIGURE 2
Support system of a SC dipole for the UNK:
1 - helium vessel, 2 - nitrogen shield,
3 - vacuum vessel, 4 - anchor extension
rods, 5 - vertical suspensions, 6 - horizontal extension rods, 7 - heat leak measuring thermometer, 8- heat sink strap.

being placed between the vacuum vessel
and nitrogen shield and the longer one
between the shield and helium vessel. This
design of suspensions and two-part extension rods enables one to simplify essentially the cryostat assembling. An anchor
extension rod is a Ø 12 mm stud 500 mm

long fixed with one end to the vacuum vessel and with another one to the helium
vessel. At a distance of 160 mm from the
attaching point to the vacuum vessel the
anchor extension rods have a thermal connection with the nitrogen shield.

The suspensions and extension rods
are manufactured from titanium alloy BT5-1.
The choice of this alloy for the load-bearing elements of the support system is
substantiated by its noticeable strength[3]
and a relatively low thermal conductivity
at low temperatures.

To reduce the thermal radiation, the
nitrogen shield fixed with vertical suspensions and a multilayer insulation wound
onto the nitrogen shield and helium vessel
is used. The superconducting dipole was
tested at a force circulating test facility[4] with a view to measure the heat load
on the helium vessel and the heat leak
through the support system.

HEAD LOAD ON THE HELIUM VESSEL

The scheme used to measure this load
is presented in fig. 3. For the known parameters of the subcooled helium flow at
the input and output points of the section
under measurements, the value of the heat
load on the helium vessel can be found
from the expression

$$Q_c = G(H''-H'), \qquad (1)$$

where G is the helium consumption rate,
H' and H'' is the enthalpy at the input
and output, respectively. The parameters
of the direct flow at the input and output (points 1,2) and those of the return
flow at the output (point 3) of the helium
vessel were determined with the help of
temperature-sensitive elements (resistors
TBO-0.125[5]) and pressure transducers. Pri-
or to their installation into the magnet,

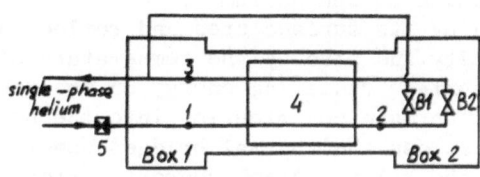

FIGURE 3
The scheme of measuring the heat load:
1,2,3 - the points where the helium flow
parameters were measured, 4 - helium vessel, 5 - Venturi nozzle.

the graduation marks of the resistors was checked from which the pairs having the minimum divergence of the readings were selected. The pressure was measured by "Sapphire" - type transducers, the liquid helium consumption rate was determined using a Venturi nozzle.

During the experiment valve B1 was open and B2 (see fig.3) was closed to determine the enthaply difference between points 1 and 2. The measurements were carried out for 4 values of liquid helium consumption for a magnet, which were in the 50-120 g/s range. The temperature drop between points 1 and 2 was in this case 0.04-0.02 K, and the relevant value of heat load was 10 ± 0.4 W. To pick out from this value the thermal radiation from the nitrogen shields of the test facility boxes, helium temperature shields were placed in front of the helium vessel ends. This complies with the conditions of heat transfer in a string of the UNK SC magnets, where the coils produce a regular structure in which the helium vessels turn to each other with their end parts. After these shields were installed, the heat load dropped down to 5.0 ± 0.5 W. To analyze the measured value of the heat load it is necessary to pick out of it the heat leak through the support system.

HEAT LEAK THROUGH THE SUPPORT SYSTEM

To measure this heat leak, TBO-0.125 resistors were placed on two vertical suspensions and anchor extension rods. They were used to determine temperatures T_1 and T_2 at two points of the suspension and extension rod between the helium vessel and nitrogen shield and also at two points between the nitrogen shield and vacuum vessel (see fig.2). The value of the heat leak through the suspension (extension rod) was found from the expression

$$Q_o = \overline{\lambda}(T_2 - T_1)F/L, \qquad (2)$$

where $\overline{\lambda}$ is the mean-integral thermal conductivity within the T_2-T_1 temperature range, F is the cross sectional area of the suspension (extension rod), L is the distance between the temperature-sensitive elements. They were placed as close to the helium vessel and nitrogen shield as possible in order to consider more completely the effect of thermal radiation exchange between the suspension (extension rod) and surrounding surfaces.

The $\overline{\lambda}$ was calculated from the results (see fig.4) on measuring the thermal conductivity of the BT5-1 titanium alloy at the setup[6] using the longitudinal axial flow technique. Figure 4 also presents the thermal conductivity of some other structural materials[7].

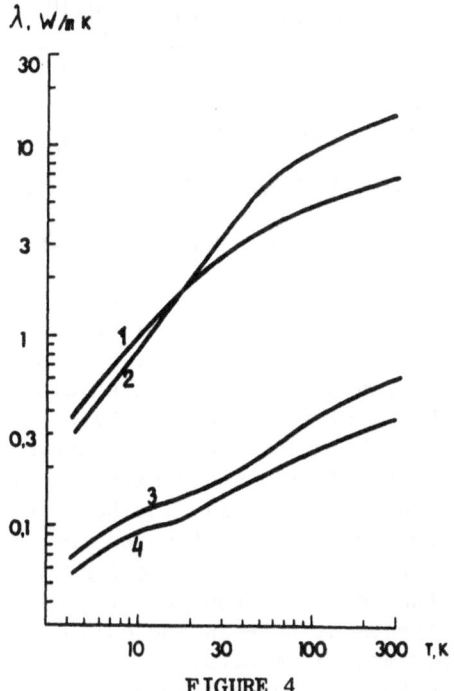

FIGURE 4

The thermal conductivity of structural materials versus temperature: 1 - titanium alloy BT5-1, 2 - stainless steel X18H10T, 3 - uniaxial glassfibre base laminate 27-63C along the fibres; 4 - glass-fabric base laminate CT9Φ -HT normal to the layers.

The results on measuring and calculating the heat leak through the suspensions and extension rods to the helium vessel and nitrogen shield are presented in the Table. As to the suspensions, the measured values of the heat leaks are in a good agreement between each other and with the calculated values. It should be noted that on cooling down the superconducting coil, the temperature field of the suspensions actually remained time-invariant.

The temperature field of the anchor extension rods connecting the nitrogen shield and helium vessel established very slowly. This was caused by a worse contact between the extension rods and helium vessel due to a larger contraction during cooldown of the vessel as compared with the extension rods. That is why the measured values of heat leaks across the anchor extension rods to the helium vessel appeared to be less than the calculated

Element of the support system	Element of the cryostat accepting heat leak	Calculated heat leak, W	N of suspension, extension rod	Length of the measured section, mm	Temperature drop, K	Measured heat leak, W
vertical suspension	helium vessel	0.080	1	40	12.6	0.087
			2	50	15.9	0.092
	nitrogen shield	0.90	1	15	31.5	1.00
			2	15	31.9	1.01
anchor extension rod	helium vessel	0.10	1	121	8.2	0.083
			2	120	1.6	0.023
	nitrogen shield	1.3	1	40	10.6	0.53
horizontal extension rod	helium vessel	0.065				
	nitrogen shield	0.70				

Table: Heat Leak Across the Support System of a UNK Dipole

ones and differ. The temperature field of the anchor extension connecting the vacuum vessel and nitrogen shield actually did not vary with time after cooldown was over. The measured value of heat leak across the extension rod to the nitrogen shield appeared to be lower than the calculated one due to the thermal radiation exchange of this section of the anchor extension rod with the nitrogen shield. The heat leak across the horizontal extension rod was not measured but since its design is similar to that of the anchor one there may be a good agreement between the calculated and experimental results.

The total value of the heat leak to the helium vessel was 1 W and that to the nitrogen shield was 12 W. These values coincide with the calculated ones. The analysis of the obtained data shows that in order to decrease the heat load of the SC dipole further, it is necessary to decrease the thermal radiation by raising the reflecting capacity of the outer surface of the multilayer insulation of the helium vessel and by a more efficient cooling of the sections of the nitrogen shield between the SC dipoles.

The obtained results allow one to find the value of heat load on a warm-iron SC dipole to a better accuracy. The measurements of such a magnet showed that the heat load was $Q_C = 12$ W[2] without the shields placed in front of the helium vessel ends. With the thermal radiation effect excluded, the value of the heat load is brought down to 7 W. On the other hand, the heat leak through the supports to the helium vessel of this magnet is 4 W[8]. Then, with account of about the same design of the helium vessels, and consequently, of the equal values of thermal radiation to them in cold and warm-iron dipoles the heat load on the latter should not exceed 8 W.

CONCLUSIONS

The performed study showed that the heat load on the helium vessel of the SC dipole for the UNK is 5 W, with 1 W falling at the support system. The total heat load on the liquid helium in SC magnets of the 2nd and 3d phases of the UNK is 26 kW which is about a half of the power of the cryogenic system for the machine. This means that the chosen design of the dipole for the UNK ensures the acceptable value of the heat load on its cryogenic system.

REFERENCES

1. V.I.Balbekov et al.// Proc. of XII Intern. Conf. on High Energy Accel. - Batavia. 1983. P. 40.

2. A.I.Ageyev et al.// Proc. of 1986 ICFA Workshop on Superconducting Magnets and Cryogenics.- Brookhaven 1986. P. 13.

3. Reference Book on Physical and Technical Foundations of Cryogenics. Edited by M.P.Malkov. M.: Energoatomizdat, 1985, p. 236.

4. A.I.Ageyev et al.// Proc. of the 10-th Intern. Cryogenic Engineering Conference. - Butterworth. U.K. 1984. P. 742.

5. V.I.Datskov. - Preprint JINR 8-83-717, Dubna, 1983.

6. U.Esher, S.S.Kozub. - Preprint IHEP 80-43, Serpukhov, 1980.

7. S.S.Kozub, Ya.V.Shpakovich. - Inzhenerno-Fizicheskiy Zhurnal, 1988. v. 54, N. 5, p. 852.

8. S.S.Kozub, Ya.V.Shpakovich. - Preprint IHEP 89-24, Serpukhov, 1989.

STUDY OF THE MODELS OF SUPERCONDUCTING DIPOLE MAGNET FOR THE UNK

A.I.Ageyev, N.I.Andreyev, V.I.Balbekov, V.I.Dolzhenkov, K.F.Gertsev, V.I.Gridasov,
K.P.Myznikov, A.P.Orlov, N.L.Smirnov, V.V.Sytnik, N.M.Tarakanov, L.M.Tkachenko,
Yu.V.Vanenkov, L.M.Vasiliev, V.A.Vasiliev, V.V.Yelistratov, A.V.Zlobin
Institute for High Energy Physics, 142284, Serpukhov, Moscow region, USSR

ABSTRACT

The basic element of the superconducting (SC) dipole magnet for the UNK is a shell-type two-layer coil with an inner diameter of 80 mm and cold iron. This design will use SC wires consisting of $\emptyset$ 6 μm filaments and having the minimum critical current density of $2.3 \cdot 10^5$ A/cm^2 in the 5 T field at 4.2 K. The current carrying element is a Rutherford-type SC cable consisting of 19 strands with 9 coated with the Sn+5% Ag alloy. This work presents some results on the study on training, mechanical stability, ac losses, static heat leaks, ramp rate characteristics of SC dipole models for the UNK with two types of cable insulation. Some results on measuring and calculating the transfer function and the bore field nonlinearities of short and full-scale models are also presented.

INTRODUCTION

To substantiate the choice of the serial design of SC magnets for the Accelerating and Storage Complex (UNK) IHEP has studied two types of SC dipoles models with a warm and cold iron yoke[1,2]. The analysis of the obtained results has shown that the basic characteristics of the dipoles of the two types satisfy the requirements imposed. But still, the cold-iron dipoles exhibited a number of advantages important for the serial production of SC magnets. The most important are a decrease of static heat leaks to the helium vessel, a cut in the amount of the superconductor used and a simpler cryostat design. With these considerations taken into account, the cold-iron design was favoured over the warm-iron one for the development of serial SC magnets for the UNK. The short and full-scale models of SC dipole magnet for the UNK has been designed and tested proceeding from the results on tests of SC dipole models with cold iron and reduced consumption of the superconductor[3].

MAGNET DESIGN

The basic element of the SC dipole magnet for the UNK is a 5800 mm long shell-type two-layer coil with an inner diameter of 80 mm and a cold iron 5600 mm long. Figure 1 shows the cross section of a SC dipole and Table 1 presents its basic characteristics.

Parameter	Inner layer	Outer layer
Number of turns	2x34	2x21
Turn width, mm	8.7	8.7
Mean turn thickness, mm	1.63	1.63
Angular layer dimension, deg	2x75.96	2x44.34
Angular spacer dimension, deg	4.25	7.24
Angular dimension of the median spacer, deg	2x0.10	2x0.5
Inner layer radius, mm	40	49.2
Ratio B_{max}/B_0	1.096	0.806
Quench temperature for the 5 T fore field, K	5.4	6.3

Table 1. The Basic Parameters of SC dipole for the UNK.

To improve the bore field quality, the inner and outer layer have spacers placed in each quadrant between the 29-th--30-th and 16-th-17-th turns, respectively. They are used to suppress even design nonlinearities of the 2nd, 4-th, 6-th and 8-th orders.

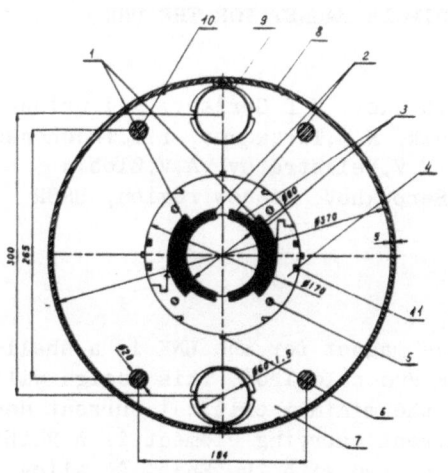

FIGURE 1

Cross section of a SC dipole for the UNK: 1 - outer layer, 2 - inner layer, 3 - spacers, 4 - collars, 5 - pin, 6 - rectangular lug, 7 - magnetic shield, 8 - helium vessel, 9 - two-phase helium pipe, 10 - stud, 11 - key.

To suppress the design nonlinearities of the 2nd and 4-th orders of the integral field and to reduce the value of the field in the end parts exceeding that in the rectilinear, these parts have a block-wise layout of turns.

The basic characteristics of the current carrying element of the SC dipole for the UNK are presented in Tables 2 and 3.

Superconducting alloy Nb+50%wt.Ti (NT-50)	
Conductor diameter, mm	0.85+0.03
Number of filaments	8910
Filament diameter, μm	6
Twist pitch, mm	10
Matrix material	Cu
Copper-to-superconductor ratio	1.38+0.12
(packing factor)	(0.42+0.02)
Copper residual resistivity ratio	$\geqslant$70
Critical current at 5 T and 4.2 K, A	$\geqslant$550
Critical current density at 5 T and 4.2 K, A/mm^2	$\geqslant$2300

Table 2. Conductor Specification

Number of strands	19
Cable type	ZEBRA
Strand coating	
9 strands	Sn+5% Ag
10 strands	bare
Cross section	
inner layer, mm^2	(1.30-1.62)x8.5
outer layer, mm^2	(1.33-1.59)x8.5
Cable twist pitch, mm	62
Cable critical current at 5 T and 4.2 K, A	$\geqslant$9500

Table 3. The Basic Parameters of the SC Cable.

The SC cable of the majority of the dipole models was insulated by 2 layers of 20 μm thick kapton tape and a layer of epoxy-impregnated fiberglass 100 μm thick and 10 mm wide wound with a 1 mm gap. Two short models were manufactured from the cable insulated by 2 layers of kapton tape 20 μm thick and a layer of kapton tape 40 μm thick and 10 mm wide with a binder applied on either side. The mean thickness of the cable without insulation was increased in this case in such a way that the total cable thickness should remain unchanged.

The fiberglass-insulated coils were moulded at a stress of 500-800 kg/cm^2 to be cured at 150°C during 5 hours. The kapton-insulated coils were moulded at the same stress and cured at 140°C during 2 hours.

The collars of the majority of the coils were manufactured from 1.5 mm thick sheets of stainless steel 12X18H10T. Its magnetic susceptibility measured at the helium temperature varied from $1.15 \cdot 10^{-2}$ at B=1 T to $1.0 \cdot 10^{-2}$ at B=5 T. One short model was manufactured using collars from 1.5 mm thick sheets of stainless steel 03X20H16АГ6 whose susceptibility at helium temperatures is constant, $0.75 \cdot 10^{-2}$, within the 0-5 T field range[4].

The prestress in the collared coil at ambient temperature in the linear part of the inner layer is at least 1000 kg/cm^2 and in that of the outer one is at least 600 kg/cm^2. To raise the mechanical stability of the end parts they are loaded axially with end bolted flanges. The prestress in the end parts is 500 kg/cm^2 at least. To counterbalance the prestress loss in the kapton-insulated coil during cooldown which is caused by a large shrinkage of the insulation the prestress at the ambient temperature was increased by 10%.

The magnetic shield for all models is manufactured from unannealed 3 mm thick electrical-sheet steel 20848. Its saturation magnetization and coercive force at the ambient temperature is $M_s=2.15$ T and $H_c=2.7$ Oe[5]. When choosing the geometric dimensions of the magnetic shield the dimensions of the grooves for rectangular lugs of the collars and the holes for liquid helium and the shield thickness were optimized accordingly. This was necessitated by the need to decrease the cold mass and reduce the shield saturation effect on the value of lower-order field nonlinearities in the central cross section.

Figure 2 shows the cross section of the SC dipole assembled in a force-circulating cryostat. The magnet is cooled by a single-phase helium flow a part of which goes through the circular channels in the coil and another one goes into the bypass channel in the shield to exchange heat with two-phase helium counterflow going through the pipe inside the bypass. The helium vessel is fixed to the warm vacuum shell with the help of titanium vertical suspensions and horizontal extension rods placed in two transverse cross sections over the magnet length. Longitudinal motions of the helium vessel are hampered by anchor extension rods fixing the central cross section and allowing a free motion of the magnet ends in thermal cycles.

To decrease the radiation-induced heat leaks to the helium vessel the cryostat is fitted with an aluminium nitrogen shield and a multilayer superinsulation applied on the nitrogen shield and helium vessel.

TEST RESULTS

Short 0.75 m long models were tested in a pooling cryostat in free boiling helium at 4.25 K. The fullscale model was tested at a force-circulating test facility cooling the magnets by single-phase helium at 4.3-4.4 K.

Figure 3 presents the results on training some short (DXB) and a full-scale (DDXB) SC dipole models. After the 1st quench of all magnets the field induction exceeded 5 T, the maximum operating field in the UNK cycle. The training process for short models with the kapton and fiberglass insulation (DXB1, DXB2, DXB4) and for the model with the kapton

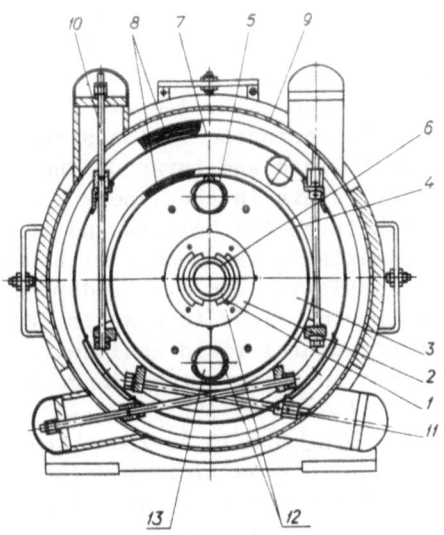

FIGURE 2

Cross section of a SC dipole assembled in a force-circulating cryostat: 1 - coil, 2 - collars, 3 - magnetic shield, 4 - helium vessel, 5 - two-phase helium pipe, 6 - beam pipe, 7 - nitrogen shield, 8 - superinsulation, 9 - vacuum vessel, 10 - suspension, 11 - extension rod, 12 - single-phase helium, 13 - two-phase helium.

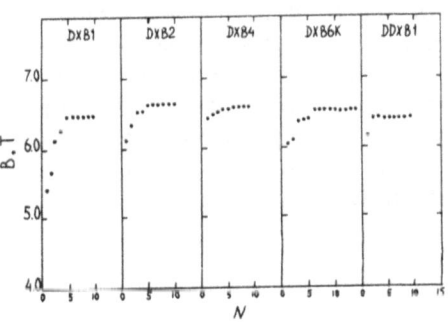

FIGURE 3

Training curves of short (DXB) and full-scale (DDXB) SC dipole models.

insulation (DXB6K) took not more than 5-6 quenches. After training the maximum bore field was 6.4-6.6 T and was determined by the critical current of the short sample with account of the temperature and field induction in the coil. The maximum bore field of the full-scale model, 6.4 T, was attained actually without training in the 2nd quench.

The mechanical stability of the dipole coil assembly was checked using the data on the collar deformation and on the inner stress variation caused by Lorentz forces and also the data on the bore field nonlinearity variations in the cycle during the measurements of short models without the iron. The coil deformation versus current is close to elastic. For the 6 kA current, the median dimension of the coil assembly increases by 140 μm. For currents up to 6 kA, variation in the stress between the collars and turns of the inner and outer layers in the linear part as well as between the collars and turns at the end parts in the models with two types of insulation had also a dependence on the current close to elastic. Variation in the structural field nonlinearities did not exceed in this current range the values $|\Delta C_3| = 1 \cdot 10^{-4}$, $|\Delta C_5| = 1.5 \cdot 10^{-4}$, $|\Delta C_7| = 1 \cdot 10^{-4}$, $|\Delta C_9| = 0.5 \cdot 10^{-4}$. The obtained results point to a sufficient value of the coil prestress and to a sufficient rigidity of the collars, ensuring a high mechanical stability of the SC dipole coil assembly.

Figure 4 shows the ramp rate characteristics of some short SC dipole models. Models DXB1, DXB2, DXB4 were manufactured from the cable insulated by kapton and fiberglass whereas the cable of model DXB6K was insulated by kapton only. In magnet DXB2 the interlayer helium channel was replaced by fiverglass spacers 0.5 mm thick. As is seen from the plots, a reduction in the SC dipole quench current with the ramp rate raised up to 500 A/s will not exceed 7%. The ramp rate characteristics of SC dipoles satisfy the imposed requirements both in the UNK operating cycle and during emergency removal of the energy stored in the magnet.

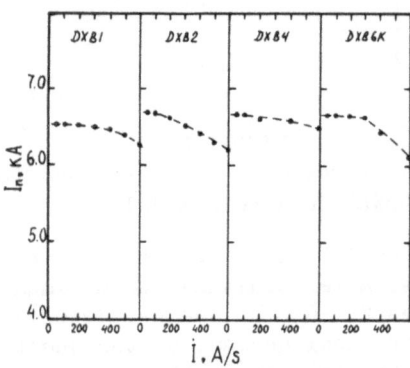

FIGURE 4
Ramp rate characteristics of a SC dipole.

Figure 5 shows the ac loss versus the current ramp rate and amplitude in the triangular cycle for a short model. The star denotes the ac loss in the UNK cycle with a 40-s linear growth of the bore field from 0.67 T to 5 T, 38-s field flattop and a 40-s linear field drop from

5 T to 0.67 T. This loss, if calculated per 1 m of the magnet length, is 130 J and makes the 1.1 W/m contribution into the cryogenic system of the UNK.

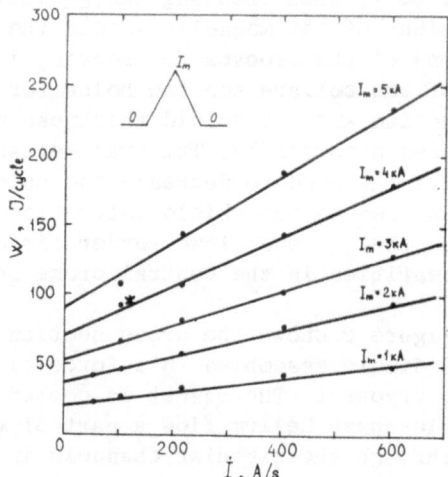

FIGURE 5
AC loss in a short model versus the current amplitude and ramp rate in the cycle: ✕ - UNK cycle.

Table 4 presents the calculated distribution on the basic constituents of ac loss in the coil per 1 m of the SC dipole.

	AC loss, J/m				Mean power in the cycle, W/m
	filaments	matrix	cable	total	
Inner layer	38	3	11	52	0.44
Outer layer	16	1	1	18	0.15
Total	54	4	12	70	0.59

Table 4. AC Loss in the SC dipole Coil in the UNK Operating Cycle.

The ac loss in the current carrying element is about a half of that in the magnet, whereas another half is related to the one in the magnetic shield. According to the measurements[5], 60 J/m dissipates as heat in the iron of SC magnets in the UNK cycle.

Static heat leaks in the helium vessel of SC dipoles have also been measured. The total static heat leaks were 5 W per magnet, and those on suspensions were 1 W.

The measured value of the transfer function caused by the coil geometry of short models having the 1.5 kA current varied from 0.9868 T/kA to 0.9882 T/kA and of the one with the 5 kA current varied from 0.9757 T/kA to 0.9759 T/kA. The design values obtained with account of the magnetic characteristics of the collars and yoke were 0.9868 T/kA and 0.9751 T/kA, respectively.

Table 5 presents the values of direct structural bore field nonlinearities in the central cross section of the magnet at a radius of 3.5 cm for the 2 kA coil current for short models with kapton-fiber-glass insulation (DXB1, DXB2) and for those with only kapton one (DXB6K, DXB7K).

	DXB1	DXB2	DXB6K	DXB7K	Mean value	Calcu-lation
C_2	-0.75	-0.55	+0.55	+0.15	-0.15	0
C_3	-7.9	-5.6	-6.0	-4.4	-6.0	-6.6
C_4	-0.75	+0.7	+0.2	-0.85	+0.2	0
C_5	+2.8	+4.5	+3.5	+3.5	+3.6	+2.5
C_6	+0.2	+0.8	+0.3	+0.5	+0.45	0
C_7	-1.5	-2.0	-2.6	-3.1	-2.6	-1.4
C_8	-0.2	0	0	+0.4	+0.05	0
C_9	+1.0	+1.8	+2.0	+3.0	+2.0	+0.9
C_{10}	+0.5	+0 2	+0.15	+1.0	+0.46	0

Table 5. Direct Structural Central Field Nonlinearities of Short Models with a Coil Current of 2 kA (in units 10^{-4}).

As is seen from these data, the values of odd-order nonlinearities are close to zero, with their spread not exceeding $+10^{-4}$. The nonzero values of even-order nonlinearities are in a good agreement with those calculated and are determined by the effect of magnetization of the collars. In the DXB3 magnet whose collars were manufactured from the stainless steel of type 03Х20Н16АГ6 , the absolute values of the transfer function and of even-order nonlinearities decreased due to a reduction of the magnet susceptibility of the collars. For further decrease of the effect of magnetization of the collars on the bore field quality the collars will be produced from austenite-stable stainless steel having a magnetic susceptibility of ~ 0.005. The difference in the values of even-order nonlinearities of magnets with the same insulation of the SC cable and of those with a different one is $(1-2)\cdot 10^4$. The spread in the magnetic characteristics of short models studied is within the tolerances, and the one of those of fullscale magnets will be studied further.

Figure 6 presents the results on measuring and calculating the transfer function of a full-scale SC dipole versus the coil current. The measured value is in a good agreement with the one calculated with account of real magnetic characteristics of the collars and iron within the whole range of the operating currents in

the UNK cycle. As was measured, the iron saturation in the region of high currents leads to a 1.03% reduction of the transfer function value for the maximum operating current whereas the calculated decrease of the transfer function value is 0.8%.

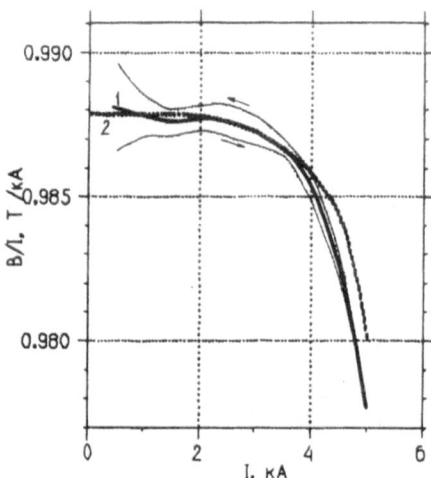

FIGURE 6
Transfer function of a SC dipole versus the coil current: 1 - measurements, 2 - calculation.

Figure 7 presents the results on measuring and calculating the integral values of C_3, C_5, C_7 and C_9 in the bore of full-scale dipole at the 3.5 cm radius versus the coil current. As is seen from the given curves, the effect of iron saturation on the value of variation in C_3 and C_5 field constituents in the range of large currents actually corresponds to the calculation. The nonzero integral values of C_3, C_5, C_7 and C_9 are explained by the effects of magnetization of the collars and by slight deviations of the coil geometry from the design one. The differences in the value of the transfer function of the SC dipole and in that of sextupole nonlinearity during current input and output are determined by the hysteresis magnetization of the current carrying element[6].

The analysis of the magnetic measurements shows that the values of gradient, sextupole and octupole nonlinearities can be corrected in the UNK operating cycle by the relevant field correction system. Higher-order and edge field nonlinearities are within the tolerances of the acceleration, stacking and extraction modes.

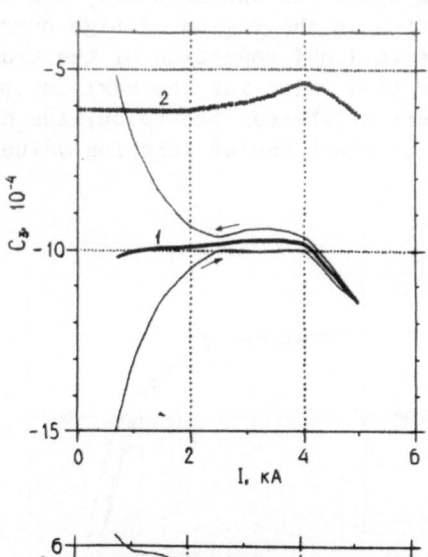

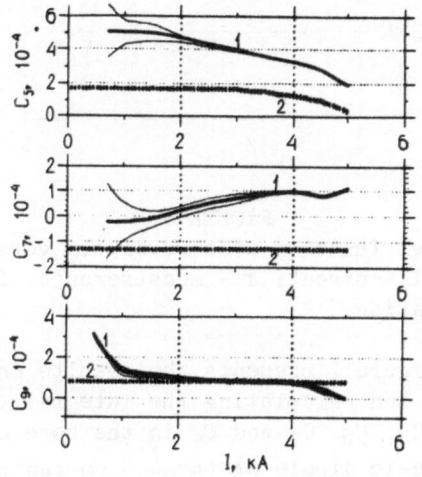

FIGURE 7
Bore field constituents C_3, C_5, C_7 and C_9 versus the coil current: 1 - measurements, 2 - calculation.

CONCLUSIONS

The study of the SC dipole for the UNK indicates that the chosen design satisfies the requirements imposed on the bore field value and quality, on the level of ac loss and static heat leaks. The available 25-30% reserve in the critical current will help ensure a reliable operation of the SC magnets in the UNK cycle at 4.4-4.6 K under the conditions of ac loss and radiation heat depositions.

IHEP is currently carrying on the work on preparation for the serial production of the SC dipoles for the UNK. The final choice of the insulation and structurual materials and correction of the coil dimensions will be made proceed-

ing from the results on the production and tests of a batch of the SC dipoles.

REFERENCES

1. A.I.Ageyev et al. Proceed. of XIII Intern. Conf. on High Energy Accel. - M.: Nauka (Siberian Branch), 1987. V. 2. P. 28.

2. A.I.Ageyev et al. Proceed. of Workshop on Superconducting Magnets and Cryogenics. - BNL, 1986, P. 199.

3. A.I.Ageyev et al. Preprint IHEP 89-27, Serpukhov, 1989.

4. A.N.Baidakov et al. Preprint IHEP 89-59, Serpukhov, 1989.

5. A.N.Baidakov et al. Preprint IHEP 89-26, Serpukhov, 1989.

6. K.F.Gertsev et al. IEEE Trans. on Magnetics, 1988, 24, N 20, P. 812.

EXPERIMENTAL STUDY OF COOLDOWN AND WARMUP OF A UNK SUPERCONDUCTING DIPOLE

A.I.Ageyev, Yu.G.Bozhko, K.P.Myznikov, A.N.Shamichev,
V.V.Sytnik, A.V.Tarasov, A.V.Zhirnov
Institute for High Energy Physics, 142284, Serpukhov, Moscow region, USSR

ABSTRACT

A number of calculational and experimental studied of cooldown to 4.5 K and warmup processes of the UNK superconducting dipoles have been carried out at IHEP. This paper presents some results on the experimental study of these processes, obtained for a 6-m cold-iron superconducting dipole. 232 temperature-sensitive elements arranged in 14 cross sections of the dipole were used to measure the temperature fields. Both routine technological processes and modes related to quenches were studied. Using the data on the detailed temperature distributions and the duration of the processes we could also conclude on the thermodynamic stresses in the helium cryostat pipes. The experimental results also made it possible to calibrate the three-dimensional model for calculational analysis of the thermal processes in the dipole.

For the model with the pipes connected to the cryostat flanges rigidly it took 12 hours to cool down the dipole and 24 hours to warm up it due to thermodynamic limitations. For the model with the aperture pipe having no such connections these times were 8.5 and 10 hours, respectively. This means that elimination of rigid connections, i.e. application of flexible compensating elements, allows one to avoid the inevitable adjustment of technological modes and to provide the required performance time both with the unit for mass calibration of dipoles and in a string of 110 dipoles of the UNK. The cooldown time of a SC dipole after an intentional quench at the experimental test facility did not exceed 10 minutes.

1. INTRODUCTION

To study the cooldown and warmup processes of superconducting (SC) magnets for the IHEP UNK, some calculational and experimental methods and models[1] have been developed. Their underlying idea was to study separately the longitudinal and cross sectional thermal processes in a string of 110 SC magnets. It could be made because the durations of these processes in a UNK string differed from each other essentially. However, longitudinal and transverse thermal processes taking place in a single SC dipole become comparable in time and therefore this approach was not suitable any longer. Therefore a three-dimensional calculational model has been developed which, in addition to nonstationary temperature fields, also enabled one to calculate the thermal and mechanical stresses in the helium cryostat pipes. To verify the reliability of this model and make a detailed experimental study of the processes, one of the SC dipoles was fitted with temperature-sensitive elements (TSE's). Below the specific features of measuring the temperature dipole fields are presented and the basic results obtained are discussed.

2. TEMPERATURE-SENSITIVE ELEMENTS IN SC DIPOLE AND THE CRYOGENIC TEST FACILITY

Temperature-sensitive elements were placed in 14 cross sections along the magnet (fig.1a). The layout of TSE's in each cross section is shown in figs.1b-d. The beginning, middle and end of the dipole had the greatest number of elements placed

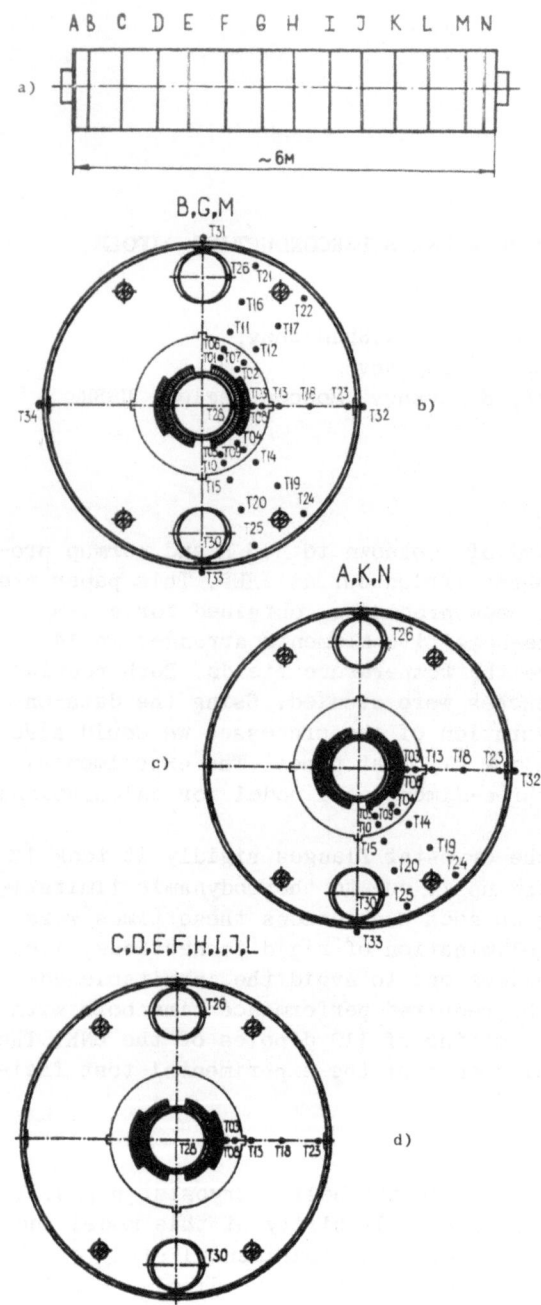

FIGURE 1

Layout of TSE's in SC dipole: a) longitudinal; b-d) in transferse cross sections.

there. As to other cross sections, the TSE's were put in specific places only. This layout of the elements and their comparatively small number allowed one to obtain a sufficient amount of data necessary for a detailed analysis of thermal

processes taking place in the SC dipole. The total number of the TSE's used is 232, including those placed on the nitrogen shield, the helium vessel suspensions and anchor tie-rods.

The dipole was tested at a test facility[2] for the full-scale UNK magnets which was modified to realize the required cooldown and warmup processes. The basic element of the test facility is the KGU-400/4.5 helium plant producing about 150 l of liquid helium per hour. At 80 K, the helium mass flow was about 0.04 kg/s and at 5 K it was about 0.014 kg/s. To prevent the liquid helium from finding its way into the compressor system, the magnet output is equipped with an electric heater having a power regulated up to 40 kW.

Resistors TBO-0.125[3] were used as temperature-sensitive elements in the dipole and in the helium flows of the test facility. Secondary blocks were 16-channel temperature indicators manufactured on the basis of an integrating-type digital-to-analog converter. A two-computer system incorporating an SM-4 computer and microcomputer "Electronika-60" was used for data accumulation and processing. The data obtained were transferred into the SM-4 computer to be used later for digital and graphic presentation and were also recorded onto output media. The data from about 70 TSE's were used for the current monitoring of the processes.

3. MEASUREMENT RESULTS

The magnet was cooled down in two stages: from 290 to 80 K and from 80 to 4.5 K. First, the magnet whose aperture pipe was connected rigidly (welded) to the helium cryostat flanges was studied. In this case, just at the 1st stage of cooldown the thermal and mechanical stresses developed in the aperture pipe, in contrast to those in the return flow pipes, reached admissible (safe) values (fig.2). This necessitated regulation of the cooldown mode, i.e. a decrease of the helium flow rate in this case. The values of the stresses developed were determined from mutual shrinkages of the cryostat pipes (fig.1) and of the outer helium cover with account of the real temperature distributions measured in them. The total cooldown time of the magnet was more than 12 hours while the 1st stage of cooldown lasted 8 hours. In the warmup mode, the helium cryostat pipes were undergoing contracting stresses, i.e. "were working" for steadiness. Under this condition, the maximum value of admissible stresses in the aperture pipe was accepted to be 5 kg/mm^2, which made the duration of the warmup processes 24 hours.

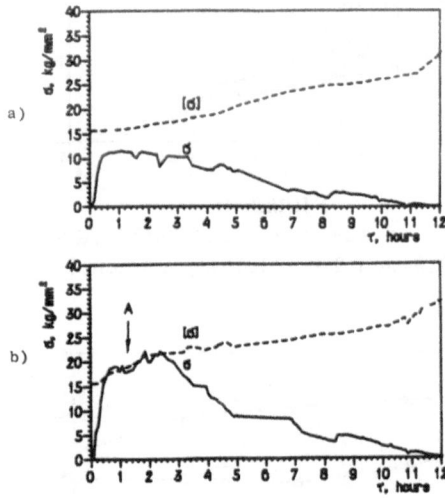

FIGURE 2

Variation of thermal and mechanical stresses in the rigid-coupled helium cryostat pipes in the cooldown mode; a) in the return flow pipe; b) in the aperture pipe; A - the area of the mode regulation.

To remove the limitations on the cooldown and warmup rate, the rigid coupling between the aperture pipe and cryostat flanges was eliminated. This brought the duration of magnet cooldown down to 8.5 hours while the 1st stage lasted about 5 hours (see fig.3). This figure also shows the calculational data obtained. In this case, the experimental values of the coefficients of convective heat transfer to helium found from the measured temperature distributions were used. As is seen, the agreement between the experimental and calculational values is fairly good. The calculational technique used one-dimensional equations for gas and a two-dimensional equation of thermal conductivity for the dipole cross section written with account of the relationship between the elements of the finite-element grid along the longitudinal axis. The values of the temperature stresses in the return flow pipes of the helium cryostat did not exceed the tolerable ones. Figure 4 shows the variation in the measured and calculated longitudinal temperature distributions for points T23 and T26 (see fig.1).

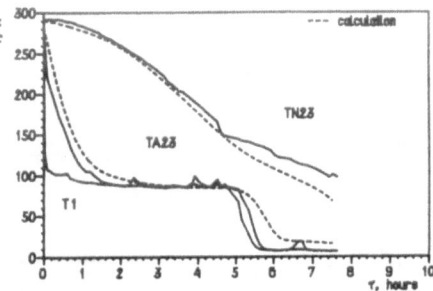

FIGURE 3

Time variation of temperatures TA23, TN23 and helium temperature T1 at the input.

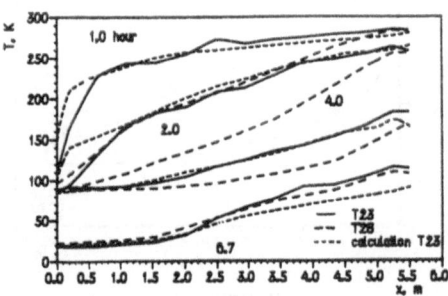

FIGURE 4

Longitudinal temperature distributions T23 and T26.

Figure 5 presents the data on warming up the "uncoupled" aperture pipe magnet from a temperature level of ~60 K, i.e. temperature T2 at the magnet output and those at points TA23 and TN23 (see fig.1). The variations in the temperature stresses in one of the return flow helium pipes for this mode is given in fig.6. As is seen, the values of the stresses developed did not exceed the tolerable ones, but still were much larger than 5 kg/mm^2. Figure 7 shows the longitudinal temperature distributions. The total duration of the process was about 10 hours.

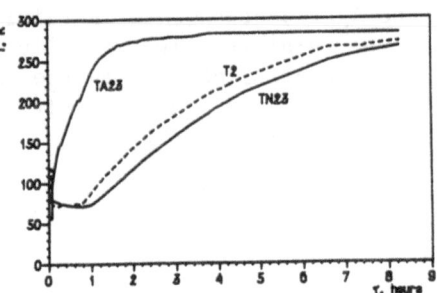

FIGURE 5

Variation of temperatures TA23, TN23 and helium temperature T2 at the output in the warmup mode.

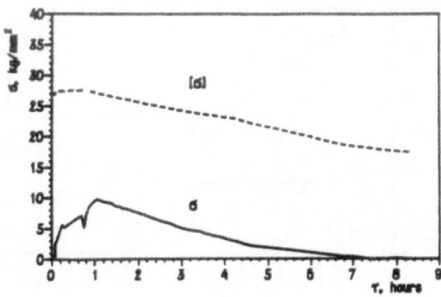

FIGURE 6
Thermal and mechanical stresses in the return flow pipe in the warmup mode.

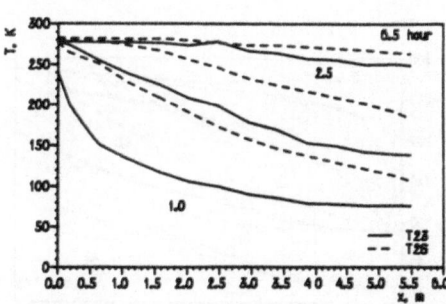

FIGURE 7
Longitudinal distributions of temperatures T26 and T23 during magnet warmup.

So, with the rigid coupling between the aperture pipe and the cryostat eliminated, this enables one to avoid damaging the SC dipole at high tenperature gradients and to ensure a reliable guidance of the technological modes required.

Quench-induced thermal processes were also studied. The time variation of temperature gradients TG03, TG08, TG13, TG18 and TG28 for one quench is shown in fig.8. In this case, the cooldown lasted about 10 minutes. The efficiency of energy removal from the SC dipole was about 60% for a 0.125 Ohm external load resistor. For the ramp rate characteristic, the warmup processes took about the same time since the I_c ($\dot{I}$)-dependence is fairly smooth. The relative value of helium heating in the central channel in the TSE TG28 (see fig.1) and, consequently, of the SC coil was about T=12-15 K.

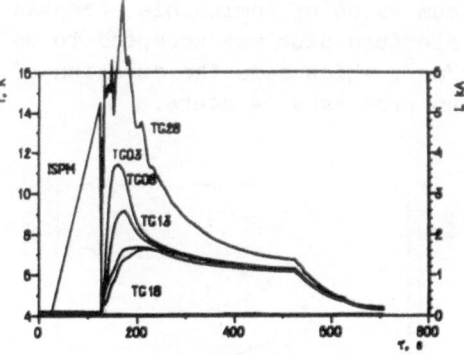

FIGURE 8
Variation of temperatures TG03, TG08, TG13, TG18 and TG28 during a quench.

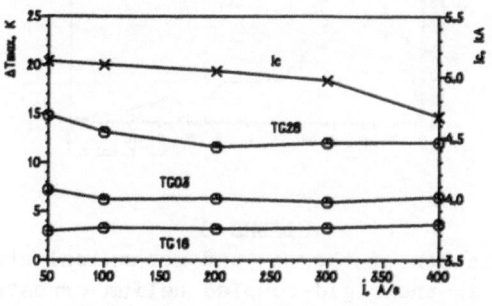

FIGURE 9
Magnet heating at points TG03, TG18, TG28 during a quench versus ramp rate.

CONCLUSIONS
The scope of the work done allowed us to carry out a detailed study of nonstationary temperature fields developed in a SC UNK dipole during cooldown and warmup and quenches. The operational presentation of the data on thermal stresses developed made it possible to ensure the control and monitoring of the modes as well as to prevent magnet damages.

REFERENCES
1. A.G.Abramov et al. Simulation of Cooldown and Warmup Processes of SC UNK Dipoles // Proceed. of X Al-Union High Energy Charged Part. Accel. - Dubna, 1986. - V. 2. -P. 296-299.

2. A.I.Ageyev et al. Test Facility for Full-Scale SC UNK Magnets // ICEC-10. - Finland, 1984. - P. 742-745.

3. V.I.Datskov. Technical Thermometers Based on Serial TBO - Type Resistors. - Dubna, 1983. - 8 p. (Preprint) JINR, 8-83-717.

MAGNETIC SYSTEM OF THE PULSE STRETCHER RING PSR-2000

P.I. Gladkikh, Yu.N. Grigor'ev, S.V. Efimov, A.Yu. Zelinskij,
I.M. Karnaukhov, S.G. Kononenko, N.I. Mocheshnikov, A.S. Tarasenko,
A.A. Shcherbakov
Kharkov Institute of Physics and Technology, Ukrainian SSR Academy of
Sciences, 310108 Kharkov, USSR

M.G. Nagaenko, A.V. Popov, B.V. Rozhdestvenskij, N.F. Shilkin
D.V. Efremov Scientific Research Institute for Electrophysical Equipment
188631 Leningrad, USSR

ABSTRACT
A 3 GeV electron stretcher ring designed at the Kharkov Institute
of Physics and Technology is intended for studies in nuclear and elementary particle physics. The magnet lattice is being optimized to produce
beams of synchrotron radiation. Main parameters of the magnet elements
are given along with the analysis of mechanical and alignment tolerances. Beam injection and extraction systems are briefly outlined, the
closed orbit correction system is also described.

A project of a pulse stretcher ring (PSR) /1/ for the Kharkov
electron linear accelerator /2/
is being developed (in cooperation with the D.V. Efremov Institute) to provide a quasicontinu-
ous electron beam which is necessary for experiments in nuclear and
particle physics, and also to make
a source of high-brilliance synchrotron radiation. This report
comprises the main characteristics
of the PSR magnetic system.

The magnet lattice of the PSR
is shown in Fig. 1. The principal
criteria of choosing this particular design stem from the requirements to the quality of the extracted beam imposed by the program of nuclear physics experiments and by the electron linac

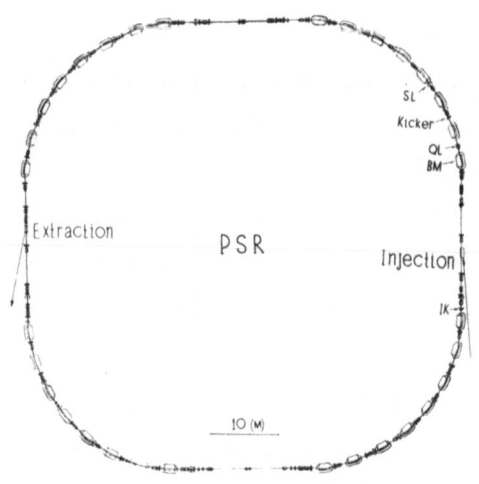

FIGURE 1. Magnet lattice of PSR-
2000. BM: dipole magnet, QL: quadrupole, IK: injection kicker,
SL: sextupole.

parameters /2/.

The magnet lattice of the PSR is a system involving four superperiods with separated bending and focusing functions. The arc is made up by 8 dipole magnets with a beam bending angle of 11.25°, and by 7 quadrupole magnets securing achromaticity in the straight section of the trajectory.

The straight section comprises 5 quadrupole magnets which ensure (together with the quadrupole magnets of the arc) the radial and vertical stability, and the PSR radial tune ν_x variation in the range from 5.33 to 5.5 at ν_z = 5.11. This allows the beam extraction at the third-order resonance or at the parametric resonance. To compensate negative chromaticity, the arc comprises four sextupole magnets, which are used in conjunction with four sextupole magnets installed in achromatic straight sections to produce the resonance harmonic of necessary amplitude and phase (for the extraction at ν_x = 16/3). Two pulsed quadrupole magnets installed in achromatic straight sections tune to resonance. For the parametric resonance extraction (ν_x = 5.5), the resonance-harmonic excitation and tuning to resonance were achieved by means of two pulsed quadrupole magnets and two octupole magnets positioned in achromatic straight sections. The amplitude and dispersion focusing

functions on one superperiod (ν_x = 5.33, ν_z = 5.11), as computed by the DECA program /3/ are shown in Fig. 2.

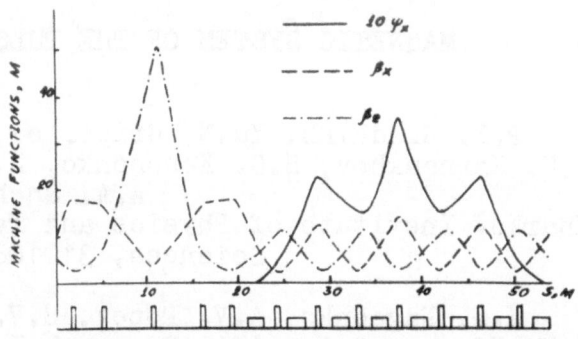

FIGURE 2. Machine functions.

The beam is injected by the use of a bump produced by three injection kickers with a strength of H·L = 0.02 (Tl·m) and an off time of 0.1μs. This ensures a two-turn injection for the M/3 resonance extraction, and a monoturn injection for the parametric resonance extraction. Two magnet septums MSI1 and MSI2 are used in the injection system, having the following respective characteristics: the peak fields of 0.9 Tl and 0.25 Tl, septum lengths 1.4 m and 1.15m, working volumes (height x width) 15mm x 60 mm and 20mm x 60 mm, septum thicknesses 17 mm and 1.5mm.

Besides the magnetic septums, the beam extraction system also comprises an electrostatic septum, 1.42 m long, with a highest field of 50 kV/cm and a septum thickness of 0.1 mm.

The yokes of all the magnets

are assembled from electrical steel laminations, the cross section of the dipole magnet being shown in Fig. 3.

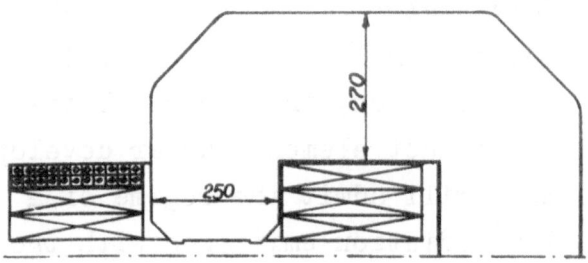

FIGURE 3. Dipole magnet cross section

The main parameters of the dipole magnet are:

magnetic field induction (Tl) 1.3
yoke length (m)   1.747
bending radius (m)   8.897
gap height (mm)   55.0
pole piece width (mm)   180.0
working region width (mm)   80.0
magnetic field inhomogeneity across the working volume (%) 0.3.

The dipole magnet winding consists of 132 turns of the water-cooled aluminium pipe.

As an example, Fig. 4 shows the field in the dipole magnet cross section computed with the

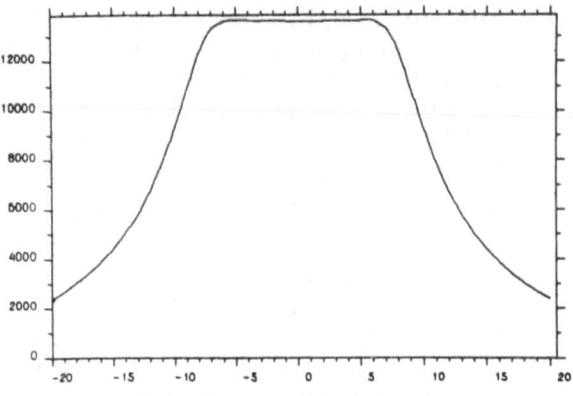

FIGURE 4. Magnetic field distribution in the gap of the dipole magnet.

TRESKA program /4/.

The cross sections of the quadrupole and sextupole magnets are shown in Figs. 5 and 6, respectively.

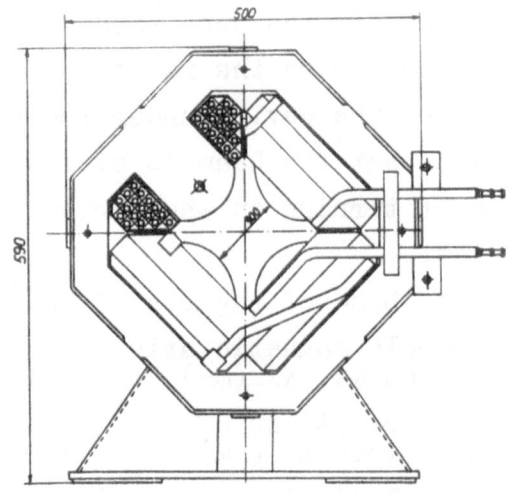

FIGURE 5. Quadrupole magnet cross section

The main parameters of the quadrupole magnet are:

magnetic field induction gradient (Tl/m)	4	7
aperture diameter (mm)	150	100
pole piece length (mm)	500	500.

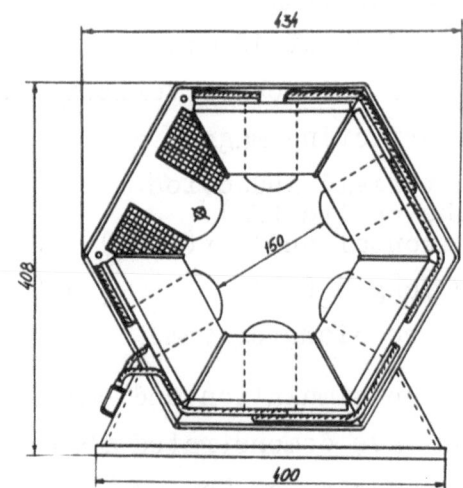

FIGURE 6. Sextupole magnet cross section

The main parameters of the sextupole magnet are:

magnetic field gradient (Tl/m^2) 40

aperture diameter (mm) 150

pole piece length (mm) 250

Simulation of equilibrium orbit distortions due to errors in manufacture and alignment of magnet elements has specified the requirements on the rms tolerances of the device under the condition that the beam is stored without turning on the correction system:

angular bend of magnets $2 \cdot 10^{-4}$

quadrupole magnetic axis deviation Δx, Δz (mm) 0.1

length deviations of magnet elements (mm) 1.0

azimuthal displacements of magnet elements (mm) 1.0

The correction system including 35 correcting two-coordinate dipole magnets ensures general correction of the equilibrium orbit over the whole ring perimeter to an accuracy of Δx, Δz ~1 to 2 mm for the original orbit distortions of about 10 mm.

The main characteristics of the correcting magnet are:

magnet field induction in the centre (Tl) (horizontal and vertical) 0.025

yoke length (mm) 150

aperture diameter (mm) 200

The simulation and calculations have demonstrated that the chosen design of the magnet lattice will ensure the following parameters of the extracted beam:

average current up to 30 A

energy range 0.5 to 3 GeV

emittance 0.1 mm mrad

extraction efficiency 98 %

duty factor 0.9

energy spread 0.1 %

Besides, the present lattice and magnet elements under development will admit the regime of a low radiation emittance with the pulse stretcher ring operating as a source of high-brilliance synchrotron radiation /5/.

REFERENCES

1. Boldyshev, V.F. et al., A design of 3 GeV CW electron accelerator facility, in Proceedings of the IEEE Particle Accelerator Conference, Washington, D.C., March 16-19, 1987, vol. 2, pp. 883-86.

2. Vishnyakov, V.A. et al., Electron linear accelerator as a PSR injector, in Proceedings of the 13th Linear Accelerator Conference, Stanford, June 2-6, pp. 576-80.

3. Gladkikh, P.I., Zelinskij, A.Yu. and Strelkov, M.A., DECA - a program packet for cyclic accelerator calculations (in Russian). KhFTI preprint (in publication), 1989.

4. Gorlovoj, M.V., Dajkovskij, A.G., Ershov, O.Yu. and Ryabov, A.A., On a new program packet for solving magnetostatics problems (in Russian). IFVEh preprint (in publication), 1989.

5. Boldyshev, V.F. et al., The Kharkov 3 GeV pulse stretcher and storage ring as a source of synchrotron radiation, in Proceedings of the 3rd International Conference on Synchrotron Radiation Instrumentation: SRI-88 (Tsukuba, 1988) p. A-009.

THE BENDING MAGNET SYSTEM OF LEP

M. Giesch and J.P. Gourber, CERN, Geneva, Switzerland

ABSTRACT

The main bending magnets of LEP are made of 3280 "steel-concrete" cores of 5.75 m length, disposed end-to-end generally in long units of six cores. Just prior to transportation into the tunnel, the cores are pre-assembled at the surface in pairs equipped with their four water-cooled aluminium bars and with their vacuum chamber. The bars are welded together in the tunnel by the MIG process so as to form a single powering circuit equivalent to one turn on each pole. Water-cooled cables are used to interconnect the bending magnets between the arcs. The paper recalls the main features of the cores and describes the excitation circuit in more detail; the criteria retained for the design of the bars and cables are presented together with the experience gained with the series production and with the assembly and interconnection of the magnets.

INTRODUCTION

The large electron positron collider (LEP) which is now being commissioned at CERN has a large circumference and a correspondingly low bending field (215 G at 20 GeV and only 1100 G at the maximum energy of 100 GeV) in order to limit the synchrotron radiation. This feature and the overall economical constraints have led to a specific design of the bending magnets based on the use of steel concrete cores excited by long excitation bars[1].

The regular cells in the arcs are 79 m long and comprise each two bending magnet units of six cores (Fig. 1) and two quadrupoles independently powered. At the ends of the arcs there are shorter cells with units of 4 or 2 cores. In order to reduce the synchrotron radiation at the experiments, the last half cell at each arc end incorporates a one tenth field magnet of 23.4 m length also connected in series with the main bending magnets.

THE STEEL CONCRETE CORES

The design, manufacture and performance of the steel-concrete cores have been described in other papers[2][3][4] and are only briefly recalled here. The 3280 cores of 5.75 m long and 4.6 t mass are made of stacks of low-carbon steel laminations, 1.5 mm thick, spaced by 4 mm and embedded in a cement mortar. Four longitudinal rods compress the core which behaves like a prestressed concrete beam (Fig. 2). This construction with a steel filling factor of only 0.27 results in a large saving with respect to a classical all steel magnet; furthermore, the field enhancement in the steel laminations is favourable for the magnetic performance of the magnet specially at injection level. The cores have been produced by two civil engineering firms over a period of 3 years. Upon delivery at CERN, the mechanical strength and the gap geometry have been measured on each individual core; these measurements served for the acceptance of the core and to determine the angular position of the median plane which was later used for the alignment in

FIGURE 1: Bending magnets in the LEP tunnel

the tunnel. The shrinkage of the mortar in the transverse plane gives rise to compression stresses in the steel laminations which reduce the permeability and thus the field in the gap. To alleviate this effect, one year after casting, each core has been submitted to a stress-relieving treatment and to a measurement of the excitation characteristic which results have been used to pair the cores in the tunnel; at the same time a flux loop was inserted in the lower pole of each core and calibrated. These flux loops connected in series allow the precise determination ($\pm$ 5 10^{-4}) of the

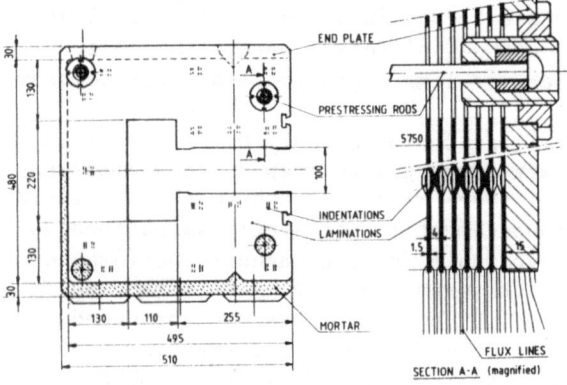

FIGURE 2: Steel concrete cores

field integral across the ring and thus of the beam energy while taking into

account any further ageing of the cores. The whole core production was then stored, waiting for bar assembly and installation in the tunnel.

THE EXCITATION CIRCUIT

The dipole cores are excited by means of four excitation bars which are connected in series in the tunnel so as to form a single powering circuit equivalent to one turn on each pole (Fig. 3). Using only the two inner bars connected in a two-turn circuit of 27 km circumference was considered at the beginning of the project, but discarded because of the stray field produced on the equipment in the tunnel and at the surface (perturbation of TV screens, aircraft compasses, etc.). The two inner excitation bars and one outer excitation bar are in one loop of the circuit, the other outer excitation bar is in the return loop. The excitation current is delivered by two power converters which are located in the same surface building and are inserted in the middle of each loop. The middle point of the power converters is earthed which allows the maximum voltage of the circuit with respect to ground to be only one fourth of the total voltage required. The excitation bars are held in

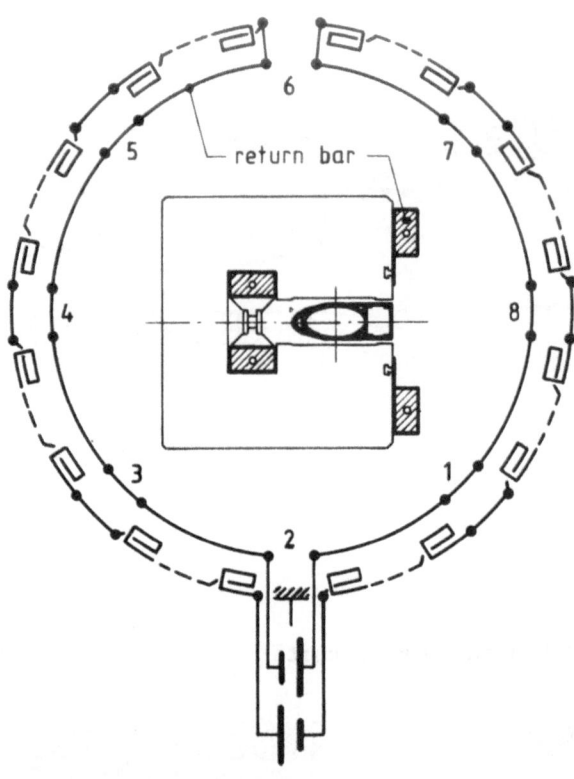

FIGURE 3: Excitation circuit of the bending magnets

position by stainless steel supports, which are dimensioned such that free movement of the bars, due to thermal elongation is guaranteed over the entire length of the magnet unit. The bars are fixed to the end-plate of the first core at one end of the magnet unit, movements due to thermal elongation are taken up by means of flexible connections at the other end (Fig. 4). Interconnection of dipoles behind the quadrupoles in the arcs is made by bars of square cross-section which at the same time provide the distribution of the cooling water to the excitation bars. Interconnection between the arcs is made by water-cooled cables.

Aluminium was imposed as conductor material by the demineralized water circuit common with that of the vacuum chamber. Electro-graded aluminium E-Al 99.5 was used throughout.

A cost optimization was done by minimizing the sum of the capital cost and electricity cost for 30000 hours of operation at 70 GeV. It gave an optimal current density around 0.8 A mm^{-2} which led to a cross-section of 44 x 90 mm^2 with a cooling hole of 11.5 mm for the excitation bars and 60 x 60 mm^2 with a

cooling hole of 15 mm for the interconnection bars. The excitation current at the maximum energy of 100 GeV is 4480 A and the total voltage 3600 V.

In order to save costs on components and installation, the excitation bars have been manufactured in lengths of about 12 m so as to equip pairs of cores.

The insulation of the bars is made from epoxy preimpregnated glass tape of 0.28 mm thickness. After bending, machining, welding and sandblasting of the bars, 4 layers of tape were applied with a half overlap, giving after curing a nominal insulation thickness of 2 mm. Curing of the insulation was made in an autoclave filled with asphalt compound under a pressure of 7 bar. The curing time was 8 hours at 160 °C. Each finished bar passed a high-voltage test at 8 kV r.m.s. for 1 min after a minimum of 6 hours immersion in water. The radiation resistance of the insulation is 5 10^7 Gy.

Four water-cooled cables are used to connect the dipoles between the arcs. A cross-section of the cable is shown in Fig. 5; it is made from 6 solid aluminium sectors with a section of 300 mm^2 each, spaced around the central cooling tube. The insulation between the conductor and the outer copper shield is a halogen-free EPR compound of 3.4 mm thickness. The nominal outer diameter is 67 mm, the bending radius about 1.5 m. The cable ends are welded to the excitation bars by means of simple connector pieces.

In order to avoid static pressure problems due to the 45 m shaft depth, conventional aluminium cables have been used for the part of the circuit between the converters at the surface and the tunnel; 12 aluminium cables of 630 mm^2 cooled by natural convection were needed for each branch of the circuit.

ASSEMBLY AND INSTALLATION

Preassembly of the dipole pairs was made in a dedicated surface building at a rate fixed by the installation in the tunnel (3 to 12 pairs of cores per day). The assembly comprises the following operations: alignment of the two cores, mounting of the four excitation bars using an automated frame, mounting of the vacuum chamber, glueing of the switches for thermal protection of the bars, insulation tests of the bars at 7 kV d.c.

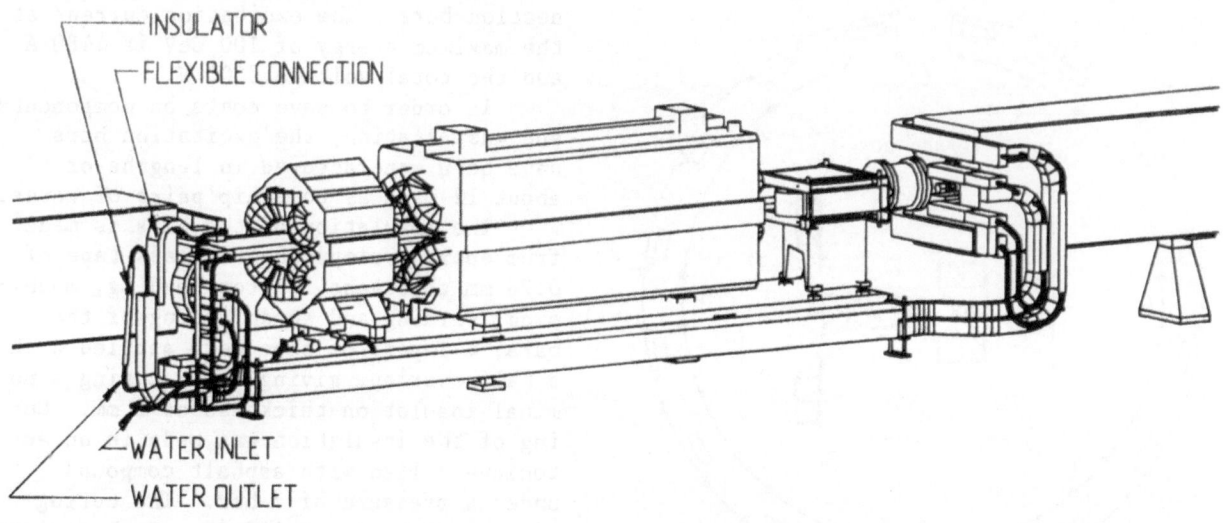

FIGURE 4: Interconnection of bending magnet units

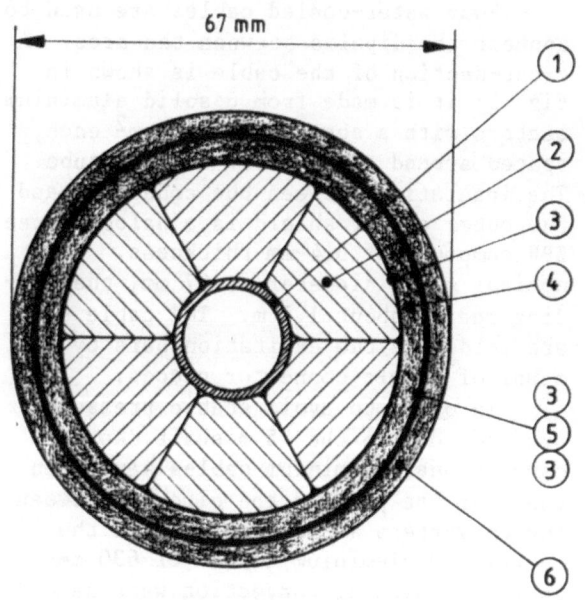

67 mm

① ② ③ ④ ③ ⑤ ③ ⑥

1 Cooling tube E-Al 99.5 Ø 15/18 mm
2 Conductor segments
  E-Al 99.5 6 x 300 mm^2
3 Separating tape
4 Insulation EPR 3.4 mm
5 Copper tape 0.1 mm
6 Sheath PE - copolymer 3.1 mm

FIGURE 5: Cross-section of water-cooled
cables

Fig. 6 shows the mounting of the inner
bars using the automated frame: the lower
inner bar being already installed, the
upper one is moved towards its final

FIGURE 6: Installation of bars

position. After assembly, the pairs
were clamped in special frames and transported by monorail into the tunnel.

There, after alignment of the dipole
pairs, the excitation bars were joined by
welding. The ends of the bars had been
machined such that when placed end to end
a V-shaped cavity is formed which was
filled with aluminium using the MIG-
welding technique. A stainless steel
tube of 12 mm inner diameter and 1 mm
thickness is embedded in the weld and as-
sures the continuity of the cooling hole

In order to avoid piercing this tube by the arc, it is shrink-fitted with a 3 mm thick aluminium tube which disperses the heat. A water-cooled jig was used for holding the ends in place during welding and to keep the temperature of the insulation, which starts at 100 mm from the end, at a temperature below 100°C. Fig. 7 shows the bars clamped with the junction tube prior to welding. After welding, the junctions are insulated by two U-shaped channels nested together so as to surround completely the bar. These channels are made from a non-hygroscopic polyester compound moulded under high pressure.

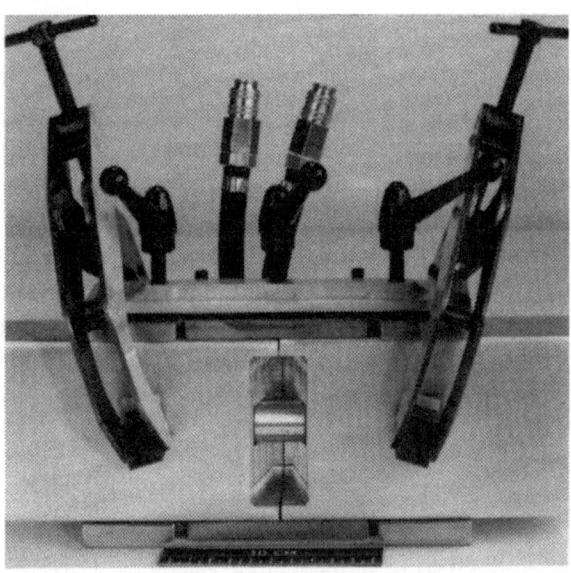

FIGURE 7: Welding jig

Interconnection of bending magnet units is made as shown in Fig. 4. The interconnection bars are welded directly to the excitation bars on one end (fixed end) and via flexible connections at the other end (floating end). The bending magnet units have two cooling circuits in parallel, each being constituted of one inner and one outer excitation bar in series. The water connections are made from aluminium tubes welded into holes, machined at the ends of the excitation and interconnection bars. Tubes which are between bars at different electrical potential are equipped with insulators. The end connections are enclosed in glass polyester covers bolted to the core end plates. During the interconnection work in the tunnel, a 30 bar pressure test and a flow test were performed on each half cell; an insulation test was also done at 5 kV d.c. prior to welding the two flexible connections.

The interconnection work in the tunnel lasted about 16 months without major difficulties. The whole circuit of about 90 km of bars showed an insulation resistance of about 1 M Ω.

ACKNOWLEDGMENTS

The successful completion of the project owes much to the know-how of the manufacturing firms. Cockerill-Sambre (B) and Styner & Bienz (CH) supplied the steel laminations; L'Entreprise Industrielle (F) and PORR AG (AU) manufactured the cores; Ansaldo (I) and Kabelmetal (D) supplied the bars and water-cooled cables respectively; Metareg (F) and Entrepose (F) performed the assembly and installation work.

The authors are also grateful to many colleagues at CERN who participated to this project from the design stage to the final commissioning.

REFERENCES

(1) Resegotti, L., The LEP magnet system. In Journal de Physique, Vol. 45, Colloque C1, 1984, pp. C1-233-39.

(2) Gourber, J.P., Billan, J., Laeger, H., Perrot, A., and Resegotti, L., On the way to the series production of steel-concrete cores for the LEP dipole magnets. In IEEE Trans. on Nucl. Sci., Vol. NS-30, No. 4, Part. II, 1983, pp. 3614-16.

(3) Billan, J., Gourber, J.P., Henrichsen, K.N., Laeger, H., and Resegotti, L., Influence of mortar-induced stresses on the magnetic characteristics of the LEP dipole cores. In IEEE Trans. on Magnetics, Vol. 24, No. 2, 1988, pp. 843-46.

(4) Billan, J., Gourber, J.P., Guignard, G., Henrichsen, K.N., Maugain, J.M., and Wolf, R., Magnetic performance of the LEP bending magnets. In Proc. 1989 Part. Acc. Conf., Chicago, March 1989.

CONSTRUCTION OF THE L3 MAGNET

F. Wittgenstein[1], A. Hervé[1], M. Feldmann[1], D. Luckey[2] and I. Vetlitsky[3]

1   CERN, European Organization for Nuclear Reseach, CH–1211 Geneva 23, Switzerland
2   Massachussetts Institute of Technology (MIT), Boston, MA 02115, USA
3   Institute of Theoretical and Experimental Physics (ITEP), Moscow 117259, USSR

ABSTRACT

The L3 detector, an international collaboration of 36 universities and 13 countries, is installing a new detector on the CERN LEP accelerator. This detector includes a huge magnet of 8000 t and 16 m in diameter. For a central field of .5 T, the rated power goes up to a level of 4 MW. The magnetized volume is of the order of 1300 m^3. The project has been accepted at the beginning of 1983 and operated for its first tests in October 1988. The helicoidal coil of 1100 t of aluminium plates has been manufactured on the site of CERN in two workshops, specially installed for this purpose. The plates, 60 mm thick, are welded together using the electron beam welding techniques. The yoke, ~ 4000 t of Fe–bars, has been machined in Novosibirsk. The poles, ~ 3000 t made of a supporting structure and filling plates, includes two moving doors to give access to the inside of the magnet and allow the installation of the different detectors. The total capital cost of the magnet amounts to 24 M$, plus 3 M$ for the detectors support structure. This paper describes the design, manufacturing, assembly and first runs of this magnet.

INTRODUCTION

The effort to build the L3 detector is the first major collaboration between scientists from the USA, Europe, the Soviet Union and the People's Republic of China.

A number of technological advances have been incorporated that make L3 uniquely capable of discovering new phenomena in elementary particle physics.

This experiment is housed in the tunnel of the Large Electron–Positron (LEP) accelerator run by 14 associated European countries at CERN. The tunnel, 27 km in circumference and buried between 50 m and 150 m underground accelerated electron and positrons in opposite directions. Every part of the detector involved collaboration. The magnet includes specially the collaboration of Switzerland, USA and USSR. Other countries participating in the delivery of major parts of the magnet are: France, East Germany, Sweden and the United Kingdom [1].

DESIGN PARAMETERS

All the detectors are mounted in the inside volume of a huge solenoid coil which is surrounded by an iron yoke and closed at its ends by two poles equipped with hinged doors to give access to the inner detectors. The purpose of this unmovable "magnetic cave" is to offer the maximum volume of magnetic field to the movable detectors. Figure 1 shows the general layout of the detector [2].

water circuits welded on the inner and outer edges of the sectors.

The cooling tubes are made of "Extrudal 050", an aluminium alloy with higher content of Si and Mg and with better qualities concerning corrosion.

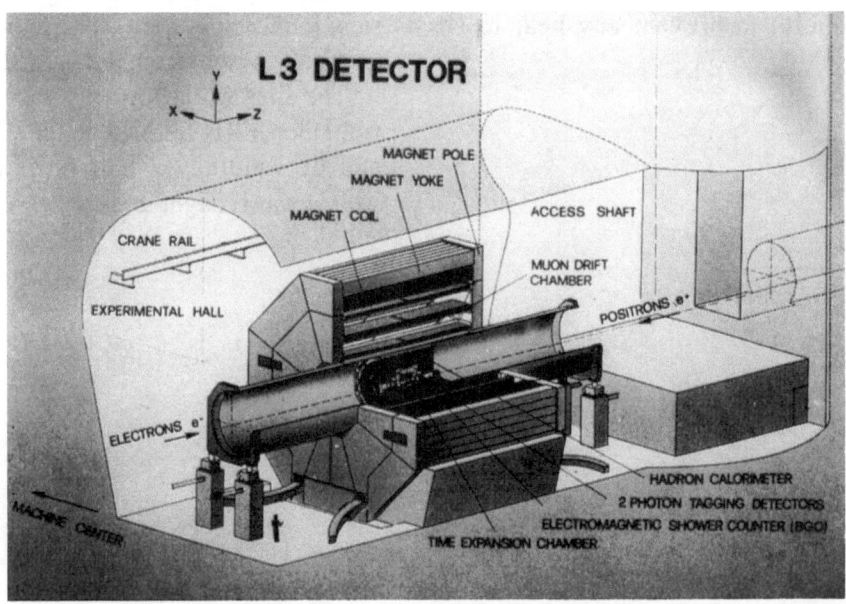

Fig. 1   Layout of the L3 detector

COIL DESIGN

To reduce the capitalization cost, the coil is made of semi–industrial plates (sectors) welded together in two steps on the CERN site. The manufacturing process of the coils explains the polygonal design of the magnet.

The selected material "Anticorodal 041", with a composition of at least 0.3% Si and 0.3% Mg, with heat treatment "71" has only 6% less conductivity than pure aluminium but present better mechanical strength and welding capabilities.

The rated current of 30 kA is a compromise between three parameters, i.e.:
• the production capabilities of the aluminium supplier concerning dimensions, plates thickness, flatness qualities;
• the investment in handling tools and manpower for manufacturing the coils;
• the difficulties inherent to the transport of high currents.(table 1).

The cooling is provided by two independent

Table 1 Main magnet parameters		
Inside radius	5930	mm
Width of the coil	890	mm
Outside radius	7900	mm
Total length	14000	mm
Power at the taps	4.2	MW
Central field	0.5	T
Coil contribution	0.36	T
Stored energy	150	MJ
Amper turns	5	MAt
Rated current	30	kA
Current density	55.5	A/cm^2
Cooling water	150	m^3/h
Coil weight (Al)	1100	t
Shielding weight	6700	t

The inter–turn insulation is provided by glass fiber plates (10 mm) with a superposition of mylar (0.2 mm).

The coil, 168 turns, is subdivided in 28 pancakes which are bolted together with 16 bolts. The mechanical rigidity of each pancake is given by 96 bolts. The bolts are triple-insulated with shrinkable shirts.

The free gaps between turns are closed with a special rubber joint to prevent any heat losses outside of the coil (figs 2–3).

Fig. 2 Storage of half turns

The pancakes are laid down on insulated bronze skates which can move, according to the thermal expansion, on two rails embedded in the lower part of the magnet yoke. The coil is pressed between the two poles and on one side through 15 electrical-insulated air springs to compensate the thermal expansion.

An active thermal shield is placed on the inner diameter of the coil to prevent any heat flux to disturb the internal volume.

COIL MANUFACTURING

The manufacturing procedure has been determined by the dimensions of the detector and imposed the installation of two local special dedicated workshops.

A new industrial process has been developed for the welding of the sectors using the 'electron beam welding technology (fig. 4).

Therefore, an important set of special tools was developed to support the quasi–industrial production centred on a welding gun of 45 kW at 50 kV. To operate under the best conditions a mobile vacuum chamber was specially built and successfully operated during four years. In a first step, four sectors were welded together bringing the unit basic weight from 0.8 t to 3.2 t.

The first workshop was moved after two years from the CERN SPS site to the P2 experimental site and by welding 12 half turns together, the unit weight was brought to 38 t.

Fig. 3 Complete magnet pancake

Each of the 28 pancakes has 4 cooling circuits; this design is a compromise between the ideal case of having two circuits per turn and an enormous amount of piping operation or 2 circuits per pancake and water pressure problems. The cooling circuits includes more than 6000 welding joints.

Many checks concerning dimensions, thermal, mechanical, electrical, pressure and corrosion behaviour have been conducted during the manufacturing period.

Fig. 4 Electron welding gun

MAGNETIC STRUCTURE

The pole magnetic structure are made of self–supporting skeletons comprising 1100 t of steel pieces giving the required rigidity and serving as a support and as a reference frame on which to mount the filling material. The filling material (5600 t) is there to provide the mass needed for the magnetic flux return, both in the two poles and in the return yoke (barrel). The filling material is coming from the USSR under the auspices of the Institute of Theoretical and Experimental Physics (ITEP). It is made of 50 and 40 mm thick soft iron plates cut to the desired shapes and tack–welded together to form individual masses of ~ 40 t for the barrel and 15 t for the poles. Each pole consists of two parts, the crown and the double doors. All this parts are made of open frames bolted together and positioned with expansion keys. The crown forms a complete ring and each door a half ring. The frame elements of the doors are joined by structural welds made on the spot. Two rails on each side of the open frames are used to guide the stacks of plates making up the filling material.

Each half door, when filled with plates, has a weight of 340 t and rests on grease skates positioned under the centre of gravity and rotates around huge hinges.

The hinges can be mechanically disconnected from the doors using self–blocking hydraulic jacks to prevent overstressing of the hinges due to the magnetic pressure on the poles (fig. 5) [2].

Fig. 5 General view of the assembly
showing two of the magnet doors

ENERGY SUPPLY

To simplify the magnet operation and restrict the underground required volume it was decided to dispose the power supply at the surface hall and to connect it to the magnet through the pit with a water–cooled bus bars system. The general features of the hollow bus bars are the following:

Diameters	140/110 Ø mm
Number of circuits/polarity	6
Circuit length (+ & –)	163 m
Resistance at 60C	5.8 μΩ/m
Max. installed power/m	160 W/m

To reduce the fringe field and the radiated electrical noise, both polarities of the bus bars are twisted.

The power supply is a thyristor power converter delivering a maximum current of 31.5 kA at 150 V. It consists of two HV transformers followed by six banks of water–cooled thyristors equipped with DC passive filters and free–wheel diodes. The magnet coil is earthed in the middle connection.

MONITORING

The magnet system includes 159 separate cooling circuits and 29 interior electrical connections. Therefore, many detectors have been embedded in the coils to monitor the magnet operations. In addition, potential and field monitoring devices, across the whole length of the magnet, water flowmeters and control valves should also be checked. Therefore about 2000 detectors are wired and brought to two concentrating racks just outside of the magnet barrel.

The local magnet monitoring system is placed on the plug of the pit and, from this point, general information concerning the magnet behaviour is directly fed into the detector slow control system.

DETECTORS SUPPORT STRUCTURE

The Support Tube (ST) could be described as a 32 m long, 50 mm thick, 4450 mm outer diameter heterogeneous tube with one flange support at each end to transmit the load to the ground. Its total mass is 247 t. The part of the ST which is inside the magnet (14.1 m) is in stainless steel with a 4.6 m long octogonal double–walled section in the middle. The remaining portion is in carbon steel. Each flange support is fitted on two servo–controlled mechanical jacks to allow continuous alignment of the ST with respect to the LEP beam.

Inside the double–walled central portion of the ST are located the inner detectors which have a total mass of 370 t. The outer layer is the Muon Filter (56 t) which is located inside the eight cells provided by the double–walled section.

Outside the ST, and inside the magnet, is situated the Muon Spectrometer. The muon chambers are arranged in half–length octant modules which are supported, using two Torque Tubes (TTs), on rails attached to the exterior of the ST.

Each TT, which is in stainless steel, has a mass of 29.5 t. The two TTs, which have been fitted over the cylindrical part of the ST before the welding of the last flange support, are now captive on the ST. The finished ST/TT unit has thus a mass of 306 t (fig. 6).

Fig. 6 Assembly of the detectors support structure in the magnet

ASSEMBLY

The assembly of the magnet progressed in four phases:

(a) First phase: the lower 3/8 of the barrel together with 5/8 of one of the poles has been assembled in position and aligned with respect to the LEP beam to form the coil cradle which is the foundation and reference support for the coil.

(b) Second phase: the 28 sets of coil sub-assemblies have been mounted in this cradle, aligned with respect to the LEP beam and electrically connected in series by welding. The 5/8 of the second pole, together with part of the two vertical walls of the barrel have then be mounted. Then the cooling pipe work for the coil has been completed.

(c) Third phase: the remaining part of the magnetic barrel and the doors structure.

(d) Fourth phase: the filling material of the two poles.

OPERATION

The magnet has been successfully operated for its acceptance tests in September 1988. Immediately afterwards, the ST/TT Unit was installed inside the magnet and the first detector run started in July 1989 [4].

AKNOWLEDGEMENTS

The authors are most thankful to the L3 Collaboration, led by Professor S.C.C. Ting, for their confidence along the project and more particularly to the support staff of CERN (Geneva), ETH (Zürich), ITEP (Moscow) and MIT (Cambridge, USA) who have largely contributed to the success of this huge project.

REFERENCES

[1] The L3 Administration and Communication Group, The L3 Experiment: Progress in Physics, Technology and International Collaboration, ed. S.C.Marks,.CERN (1988).

[2] L3 Collaboration, Technical proposal, May 1983.

[3] F. Wittgenstein, Construction d'un grand aimant de détecteur au CERN, Ingénieurs et architectes suisses 8 (1988) 111–114.

[4] L3 Collaboration, Construction of the L3 Detector (to be submitted to Nucl. Instr. and Methods).

CRITICAL CURRENT MEASUREMENTS OF SUPERCONDUCTING CABLES FOR HERA DIPOLE MAGNETS USING THE FACILITY MA.RI.S.A.

P.Fabbricatore, R.Musenich, R.Parodi, R.Vaccarone

I.N.F.N., Sezione di Genova, via Dodecaneso 33, I-16146, Genova, ITALY

ABSTRACT

The facility MA.RI.S.A., built up during 1986 at the I.N.F.N. laboratory in Genoa, was used during last years to measure the critical current of several superconducting cables (for HERA dipole magnets) up to 6.4 T field at 4.2 K. Since the first measurements, a comparison was made with the results obtained at Brookhaven National Laboratory (BNL) on samples of the same cable. A difference of about 600 A corresponding to 7-8 % was found (our measured critical currents were lower than the BNL ones). These results gave impulse to a research aiming to understand the influence of the several parameters involved in the Critical Current measurement in order to compare results obtained in different sets-up. The experience made during two years of tests is reported. Two parameters are introduced: (a) The Effective Critical Field at the cable; (b) The effect of the field inhomogeneity on the Critical Current degradation.

INTRODUCTION

During the last three year, at the I.N.F.N. laboratory in Genoa, critical current measurements of several superconducting (S/C) cables for HERA dipole magnets were performed. The measurements were done using the facility MA.RI.S.A. up to 6.4 Tesla at 4.2 K.

Since the first measurements, a comparison was made with the results obtained at Brookhaven National Laboratory (BNL) on samples of the same cable: our measured critical currents were always lower than BNL ones with an average difference of about 600 A corresponding to 7-8 % of the critical current (the last one being about 8000 A at 6 T and 4.6 K). As no experimental errors were found to explain such a difference we started a research aiming to understand the influence on Critical Current of the several parameters characteristic of the two sets-up. Generally Critical Current values are completely defined giving temperature and peak magnetic field (obtained by adding the external field and the maximum self field) but it is obvious that in this case the abovementioned parameters are not enough: we propose that two new parameters must be introduced. The former is

the Effective Critical Field at the cable (external field plus an average self field) widely discussed in ref.[1] that substitutes the Peak Field. The latter is the effect of the field inhomogeneity on the critical current degradation.

CRITICAL CURRENT MEASUREMENT ON HERA CABLE

Standard Criterium

Critical current measurements are made by applying a magnetic field B_{ext} normal to the wide face of a sample kept at temperature of boiling liquid helium (LHe). The electrical resistance is measured as function of the current; the critical current is defined as the current flowing through the conductor when an electrical resistivity of $10^{-14}\Omega$ m is measured. The critical field connected to the measured critical current is calculated by adding the maximum self field at the conductor ΔB_{sf}, (i.e. the field generated by the current flowing through the conductor), to the applied field B_{ext}.

The critical current is determined according to

the following steps:

I - I_c is measured using the aforementioned resistive criterium at applied field B_{ext} and at temperature T.

II - The field compensation is made by adding the maximum self field: $B = B_{ext} + I_c * \beta_s$ (where β_s is the slope of the maximum self field vs the current)

III -The measurements are referred to temperature T=4.6 K using the formulae found by Lubell[2].

IV -The results at three different fields are plotted and a best fit is performed, so that the critical currents at fields 5, 5.5, 6 Tesla are determined.

Other informations obtained in critical current measurements are the quench current and the n-value. The quench current is the maximum current flowing through the cable before complete transition to the normal state. The n-value is the slope of the curve LogV vs. LogI (V is the measured voltage drop at the sample, I is the current through the sample).

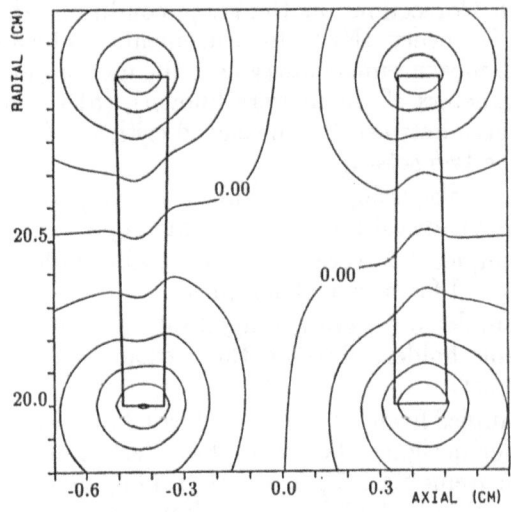

FIGURE 1 - Field map at the sample (INFN): two samples configuration, I = 8000 A, isolines 0.09 T, B_{max} = 5.46 T, B_{min} = 4.56 T

INFN-Genoa Set-up

In our set-up [3] the external magnetic field is given by a S/C solenoid (MARISA I) allowing measurements up to 6.4 T [4]. Several samples (from 2 to 6) are arranged inside the solenoid in turns of 0.41 m mean diameter. The sample length is about 1.2 m. To avoid destructive effects due to the magnetic forces the samples are connected in series not-inductively. The sample holder is composed by fiber-glass epoxy disks having suitable seats to host the samples and several holes allowing a good thermal ex-

change with the LHe bath. In the last version the disks are clamped together using stainless steel flanges and 16 bolts 20 mm diameter. This mechanical system allows to produce clamping forces up to 700 KN. For HERA cables the time for set-up arrangement was about two weeks.

The temperature of the sample is assumed to be the temperature of the LHe bath. The temperature measurement is done by measuring both the vapour pression above the LHe bath and the electrical resistance of a Carbon Glass sensor placed into the bath.

The temperature measurement is affected by an error of 0.05 K.

The magnetic field is measured directely at the sample by a Hall effect probe.

The current is measured by using a Zero Flux Transformer, with accuracy of at least 10^{-4}.

The residual ripple of the controlled rectifier of the DC power supply is 20 A p-p at frequency greater than 300 Hz.

These considerations lead to evaluate in about 200 A the maximum error on the measured critical current.

BNL Set-up

In BNL set-up [5] the magnetic field is generated by a short length S/C dipole magnet. Two samples are connected in series not-inductively and placed along the dipole axis. The maximum field available is 5.9 T.

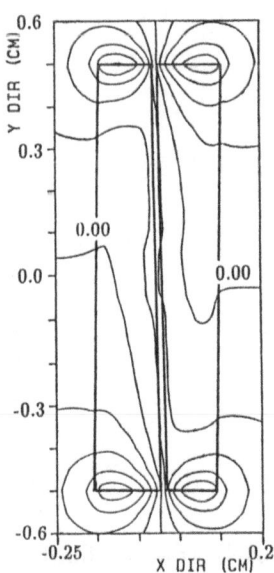

FIGURE 2 - Field map at the sample (BNL): two samples configuration, I = 8000 A, isolines 0.05 T, B_{max} = 5.27 T, B_{min} = 4.76 T

Comparison between the two Sets-up

There are several differences between INFN and BNL sets-up that can be the cause of different results. In order to compare the measurements obtained in the two systems, the following features must be taken into account:

I - The samples geometry is strongly different (a loop for INFN set-up, straigth cable for BNL); the influence of the mechanical pressure applied to the samples to hold the magnetic forces in the two geometries is not well understood. We observed in our set-up, applying the same pressure applied in BNL set-up (40 MPa), the occurrance of disturbances due to movements of the strands. As consequence more training quenches and generally premature quenching of the samples were observed well below the critical current values.

II - INFN set-up allows measurement at higher field level; measurements were made in the range of applied field 5.8 T - 6.4 T at currents of 9000-6000 A. Due to the disturbances no results were obtained at fields lower than 5.8 T (where Ic $\simeq$9000 A).

III - The magnetic field profile at the sample is strongly different in the two sets-up. This is mainly due to the spacing between nearest neighbour samples (10 mm for INFN , 0.25 mm for BNL). The field at the samples is shown in fig.1 for the INFN set-up and fig.2 for the BNL one in a configuration of a pair of samples; for both cases the external field is 5 T and the current flowing through the conductors 8000 A.

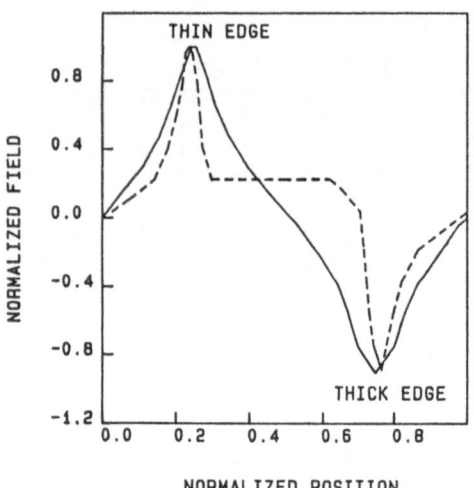

FIGURE 3 - BNL (dashed line) and INFN (continuous line) field profile at the sample

It is remarkable that in INFN set-up the Peak Field region is wider than in BNL one; this can be better seen in fig. 3, where the field at the sample (both for INFN and BNL set up) normalized at the maximum field is shown as function of the position moving around the cable .

That occurrance gave us the starting idea that the self field has more effect in our set-up compared to the BNL one. We developed a theory in order to correctly consider the self field effect [1]; the main feature of this theory is that the Critical Field does not coincide with the Peak Field but with an Effective Field, depending on several parameters (The maximum self field, the twist pitch of the strands, the sample geometry, the n-value, etc.).

EXPERIMENTAL RESULTS

Description

Table 1 shows the results of measurements on 24 different samples, performed using the facility MA.RI.S.A. in the 1987-1988 period. For each sample they are shown : (1) Identification (LMI and BBC are referred to the cables produced respectively by Europa Metalli and Brown Boveri), (2) Date of measurement, (3) Critical Current using the "Peak Field Correction" for INFN and BNL sets-up at 6 Tesla and 4.6 K, (4) Critical Current using the "Effective Field Correction" at the same conditions than (3) for every INFN measurements and some BNL measuremements (where the measurement parameters of which were known), (5) the differences between the measured critical currents in the two sets-up.

Before discussing the results, we want to add some historical notes to the data reported.

- Samples 1-2 were placed in a sample holder designed for a pair of samples.

- Samples 3-6 were arranged into an improved sample holder, allowing the measurement of 4 samples.

- Samples from 7 to 24 were hosted in a sample holder designed for 6 samples. Performing the measurements 1-12 a lot of training quenches were observed and the maximum voltage per unit length measured before the quench was 20 μV/m (some times 50 μV/m), being the critical voltage about 4-6 μV/m. In order to reduce the training and to perform measurements up to the real quench current, foreseen at higher voltage level, a new sample holder was designed allowing us to apply higher mechanical pressure.

- Samples 13-24 were measured using the new sample holder. Some problems occurred for the samples 13-18 when we were getting experience in adjustments of the sample holder.

- Because the mechanical disturbances did not disappear, limiting the measurement capacity of the apparatus (some times it was impossible to measure the critical current), samples 19-24 were impregnated with Woods-metal, so that movements of the strands were forbidden.

TABLE 1
Measurement results of Critical Current at B=6T, T=4.6 K

N.	Ident.	Date	I_c INFN-Ge		I_c BNL		Difference	
			Max	Eff	Max	Eff	Max	Eff
1	LMI32	Oct87	7700	7400	8350	*	-650	*
2	LMI38	Oct87	7780	7470	8330	*	-550	*
3	LMI173	Feb88	7330	6800	7920	*	-590	*
4	LMI118	Feb88	6900	6450	7560	6900	-660	-450
5	LMI181	Feb88	7330	6770	7960	7300	-630	-520
6	LMI177	Feb88	7630	7050	8050	*	-420	*
7	LMI189	Mar88	7300	6740	7950	7280	-650	-540
8	LMI190	Mar88	7600	7040	7800	7150	-200	-110
9	LMI194	Mar88	7800	7220	7950	*	-150	*
10	LMI214	Mar88	6900	6400	7430	*	-530	*
11	LMI216	Mar88	6730	6270	7460	6830	-690	-560
12	LMI199	Mar88	7830	7170	7850	7170	-20	0
13	LMI224	Apr88	7040	6400	8180	*	-1140	*
14	LMI225	Apr88	7080	6500	8110	*	-1020	*
15	LMI226	Apr88	6770	6150	7600	*	-830	*
16	LMI243	Jul88	7600	7050	8190	*	-590	*
17	LMI263	Jul88	7590	7010	8110	*	-520	*
18	LMI267	Jul88	7800	7150	8120	*	-320	*
19	LMI223	Dec88	6900	6350	8270	*	-1370	*
20	BBC33	Dec88	7200	6600	7840	7200	-640	-600
21	LMI222	Dec88	7000	6420	8230	*	-1230	*
22	BBC70	Dec88	8200	7540	8650	*	-450	*
23	LMI227	Dec88	7000	6430	7580	*	-580	*
24	LMI59	Dec88	7300	6700	8380	*	-680	*

Discussion

From the results of the measurements a simple consideration can be done:

In spite of some experimental parameters (mechanical pressure, sample holder configuration and cable cooling) were changed our measured values of critical current are always lower than BNL ones for amount of about 600 A (average value). This difference is slightly reduced if the "Effective Field", more correct in our opinion, is considered.

From the analisys of the self field in both the sets-up it results that for the INFN measurements the Effective Field Correction leads to increase the field of about 1000-1200 Gauss at 7000 A and for BNL measurements, at the same current, of 300-400 Gauss. These values are strongly different from the Peak Field ones (4000 Gauss for INFN, 2800 Gauss for BNL), however this consideration does not change very much the difference between INFN and BNL measured critical currents.

From this analisys it came out the idea that the difference between INFN and BNL measurements could be due to some physical effects connected to the different sets-up. Our attention was devoted again to the magnetic field; the analisys of the field at the samples in the two sets-up (already reported in fig.1 and 2), that first took us to develop the concept of Effective Field, suggested us the idea of a degradation effect.

On this kind of conductor (Rutherford of trapezoidal cross section) the cabling affects the critical current: a degradation occurs specially in the region of thin edge. On measuring the critical current in both the sets-up, the systems are adjusted in such a way that the Peak Field is at the thin edge. We made the hypothesys that in our set-up, due to the wider extension of the Peak Field region at the sample, the degradation effect becomes more evident (see fig.3).

In order to prove this assumption three experiments were planned:

-1- Some samples were measured again reversing the field so that the Peak Field is applied at the thick edge. The difference of measured critical current gives informations about the degradation. The results are shown in Table 2; difference from 500 A to about 900 A were measured. This large difference in critical current, measured reversing the field, was the first indication proving strong degradation effect.

-2- A sample measured in our set up was removed from the sample holder and measured again by BNL. The sample was LMI214; a critical current of about 600 A higher than our measurement was observed. This result gave the information that our set up did not cause degradation of the sample (due for istance to the mechanical stress).

TABLE 2
Difference between Critical Current values reversing the field

Id.	Date	Difference (A)
LMI223	Jan-89	500
LMI222	Jan-89	500
BBC33	Jan-89	770
LMI227	Jan-89	700
LMI59	Jan-89	870

-3- The last experiment we performed was aimed to reproduce in our set-up the same magnetic condition of BNL one.

Two samples, LMI177 and LMI181, already measured on february 1988, were placed in the sample holder connected not inductively and separed by a Kapton tape so that the distance was reduced from 10 mm to about 0.2 mm. Due to the strong modification required for the sample holder some difficulties were met performing the measurements and we were not able to measure the critical current of sample LMI177. The difficulties were mainly related to the low thermal exchange that, while prevented the measurements of LMI177, affected negatively the measurement of the other sample: in fact the temperature of the sample was higher than the helium bath one and not easy to evaluate. To avoid this effect a new sample holder had to be designed, but much time is required for its construction and we decided to go on by using the modified one.

As regard the sample LMI181, we measured a critical current value of 7600 A at 4.6 K and 6 T, about 300 A higher than the value obtained on february 1988. This value is still 350 A lower than the one measured by BNL (with maximum field correction) but the effect of the low thermal exchange must be taken into account.

In spite of a difference is still observed between our values and BNL ones the result of the experiment proved that the samples geometry has a great importance in critical current measurements due to the effect of the field inhomogeneity on the critical current degradation.

CONCLUSIONS

Two years of experimental work on critical cur-

rent measurements of large cables for High Energy Physics applications, showed that the critical current for this kind of cables has quite complex implications; it is not possible to transfer the concepts developed for small wires to these cable. We had experienced the importance of the field inhomogeneity at the sample. Both the "Effective Field" and the "Field Effect on the Degradation" are two important parameters to be taken into account.

The several measurements performed gave us the further indication that measurements on large cables using Round Simmetries (like in a solenoidal magnet) requires carefull and critical operations in the sample holder preparation.

REFERENCES

1. P.Fabbricatore, R.Musenich, R.Parodi, S.Pepe, R.Vaccarone, "Self field effects in the critical current measurements of superconducting wires and cables", to be published on Cryogenics.

2. M.S.Lubell, IEEE Trans.Magn., Mag-19, 1983, 3, pp.754-757

3. P.Fabbricatore, R.Parodi, A.Matrone, R.Vaccarone, "A multiple sample holder for J_c measurements on Hera cables" Proceeding ICEC 12, Butterworth, 1988, 903-907

4. P.Fabbricatore, A.Parodi, R.Parodi, C.Salvo, R.Vaccarone, "MARISA a test facility for research in applied superconductivity", Proceeding ICEC 12, Butterworth, 1988, pp. 879-882

5. W.B.Sampson, "Procedure for measuring the electrical properties of superconductors for accelerators magnets", Proceeding of Workshop on Superconducting Magnets and Cryogenics, B.N.L. May 12-16 1986, pagg. 153-156

TIME DEPENDENCE OF PERSISTENT CURRENT EFFECTS IN THE SUPERCONDUCTING HERA MAGNETS

H. BRÜCK, ZHENGKUAN JIAO, D. GALL, G. KNIES, J. KRZYWINSKI, R. MEINKE, H. PREISSNER
Deutsches Elektronen-Synchrotron DESY, Hamburg, Germany
P. SCHMÜSER
II. Institut für Experimental-Physik der Universität Hamburg, Germany

ABSTRACT
The higher multipole fields which are generated by persistent eddy currents in the HERA dipole and quadrupole magnets show a logarithmic time dependence for times between 200 s and 4000 s. The persistent current contribution to the main field of the magnets decreases as well. The logarithmic decay rates vary considerably from magnet to magnet and are quite different for superconductors from different manufacturers. The influence of various parameters is investigated.

INTRODUCTION

During the injection of the proton beam at an energy of 40 GeV, the superconducting HERA magnets are excited to only 5% of their nominal field. The field distortions from persistent currents are appreciable and require the use of correction coils, in particular for the sextupole which has a strong influence on the chromaticity of the accelerator. The time dependence of these field distortions, first observed at the Fermilab Tevatron [1], leads to additional complications. From measurements on preseries HERA magnets we were able to show [2] that the multipole fields decrease proportional to the logarithm of time, contrary to the exponential decay of eddy currents in circuits with inductive and resistive components. Flux creep in hard superconductors is known [3] to lead to a logarithmic time decrease of the critical current density, and this phenomenon is the presently accepted explanation of the time dependence of the multipole fields.

The accumulation of 210 proton bunches in HERA is expected to take about 30 minutes. A precise knowledge of the time dependence of the field distortions is needed to enable appropriate correction schemes. Unfortunately, the logarithmic decay rates show a large variation from magnet to magnet and appear to be quite different for different superconducting cables. For this reason time dependence measurements are performed for all dipole magnets and a large number of quadrupoles. The present status of the data will be discussed.

The magnetic measurement methods and results on quench currents, training behaviour and the analysis of multipole data of the magnets are reported in a separate contribution [4] to this conference.

MEASUREMENTS

The persistent eddy currents in the superconductor filaments generate multipole fields of all orders allowed by coil symmetry: $n = 1, 3, 5, \ldots$ in a dipole, $n = 2, 6, 10, \ldots$ in a quadrupole. Their contribution to the main dipole field will be denoted by $\tilde{B}_1$. For an increasing coil current ("up-ramp" branch of the hysteresis) $\tilde{B}_1$ is negative, i.e. the amplitude of the main field is reduced as might be expected from Lenz's rule; also the sextupole b_3 respectively the 12-pole b_6 are negative [4]. If the coil current is decreased ("down-ramp" branch) all signs are reversed.

Most of our time dependence measurements are performed on the "up-ramp" branch at a magnet current of 250 A, corresponding to the injection energy. To establish a well-defined initial condition the magnet is quenched and then the current is cycled 50 A $\rightarrow$ 6000 A $\rightarrow$ 50 A at a rate of 10 - 20 A/s. The minimum current of 50 A has been chosen to ensure a proper power supply regulation, which was found to be essential for obtaining reproducible results. The current of 250 A is usually approached with a low speed of 1 A/s so as to reduce the overshoot values to < 0.8 A.

In Fig. 1 the absolute value of the sextupole coefficient in a HERA dipole is plotted against the logarithm of time. As time origin we have chosen the instant at which the coil current reached the value of 250 A. The sextupole coefficient decreases by about 25% within 10 hours. Between 200 s and 4000 s the drop is almost linear; at larger times but also below 100 s the slope levels off. Figure 1b shows that the data points at small times can be described by a logarithmic fit as well if the time origin is shifted by an appropriate amount. The physical significance of the shifted time origin is not yet clear.

The absolute value of the persistent current contribution $\tilde{B}_1$ to the main dipole field is shown in Fig. 2. A similar logarithmic decrease is observed. Also the time variation of the 12-pole in a quadrupole (Fig. 3) resembles that of the sextupole.

The resolution of our "stretched-wire" system [4] is barely sufficient to determine the time dependence of the quadrupole gradient g. The relative change between 200 s and 2000 s, averaged over 3 quadrupoles, is $\Delta g/g = (1.2 \pm 0.6) \cdot 10^{-4}$ compared to an estimated $1.5 \cdot 10^{-4}$ from the observed time variation of the 12-pole.

Of great interest for the injection of protons into HERA is of course the average change in the dipole and multipole fields during the 30 minute long accumulation time of particle bunches. The injection will start a few minutes

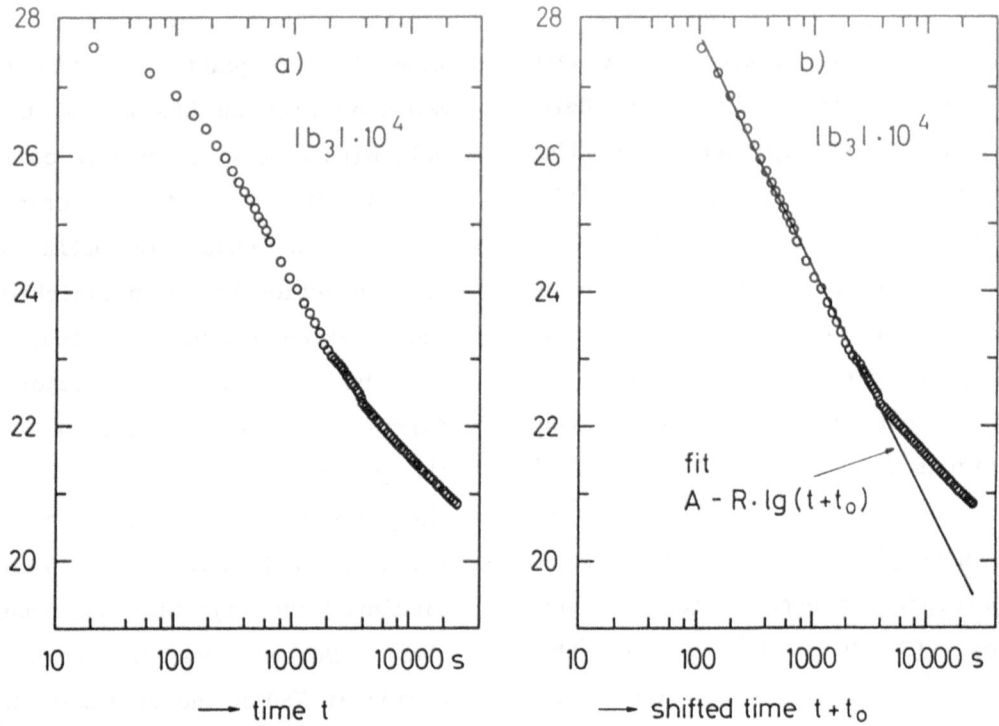

FIGURE 1

a) Time dependence of the absolute value   b) Same data plotted on a shifted time
   of the sextupole in a HERA dipole   scale $t+t_o$

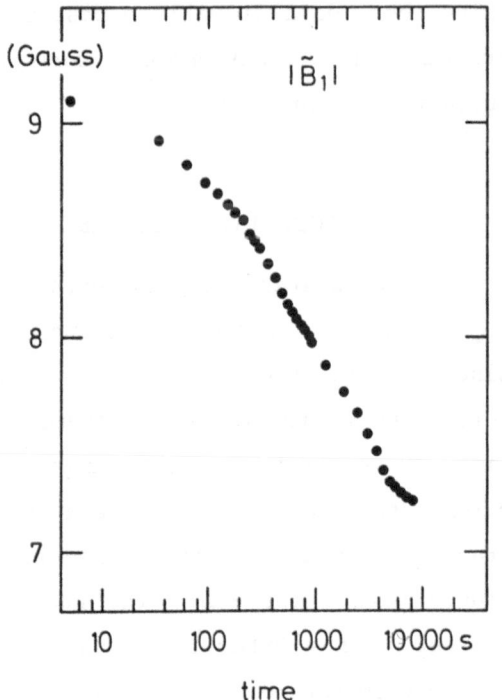

FIGURE 2
Time dependence of the eddy current
contribution $\tilde{B}_1$ to the main dipole field

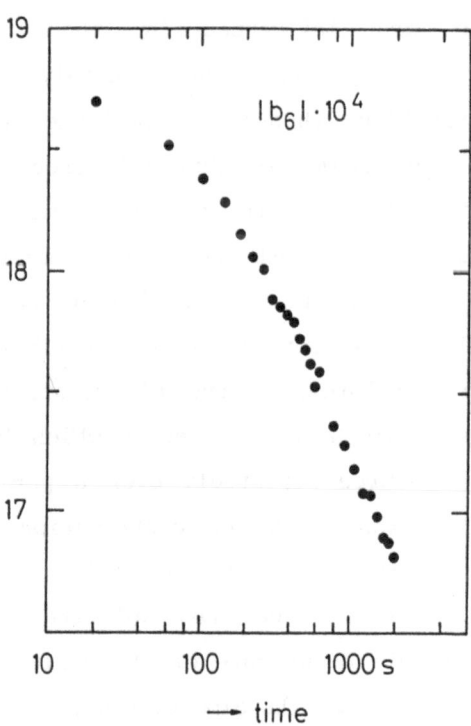

FIGURE 3
Time dependence of the 12-pole
in a quadrupole

after the magnet current has been stabilized. We determine the change of the field components between 200 s and 2000 s on the unshifted time scale. Here the data are well represented by the form A-R·lg(t) so this change is identical to the logarithmic decay rate R.

The decay rates of the sextupole in 51 dipoles and of the 12-pole in 88 quadrupoles are presented in Fig. 4. The first observation is that the distributions are rather wide. Secondly, within the limited statistics the average decay in the dipoles made with LMI superconductor appears to be twice as large $((3.7\pm0.6)\cdot10^{-4})$ as in the dipoles with ABB superconductor $((1.8\pm0.5)\cdot10^{-4})$. That a difference exists may not be too surprising since the rate of flux creep depends on the properties of the pinning centers and these are influenced by the wire production process. Unfortunately, no quantitative theory is available and even the parameters that influence the decay rate of the critical current density are unknown. Experimental uncertainties like insufficient power supply regulation may have contributed to a broadening of the distributions. However, for a single magnet which has been measured repeatedly over a period of 2 months, a narrow distribution is obtained (see insert to Fig. 4a).

In Fig. 5 we have plotted the logarithmic decay rate of the sextupole against that of the dipole. A very clear correlation is observed, supporting the hypothesis of a time dependent critical current density. Note that different experimental methods are used to determine the two quantities: the dipole is measured with an NMR probe, the sextupole with a rotating pickup coil.

A further study concerns the acceleration phase in HERA. When the beam injection has been finished and the acceleration starts new eddy currents will be induced in the superconductor filaments. The resulting sextupole changes in the dipoles may shift the chromaticity, as observed in the Tevatron [5]. In Fig. 6 we have simulated the injection and acceleration phase. During the 30 minute waiting period at 250 A the sextupole b_3 drifts away from the hysteresis curve. After that the dipole current is increased and b_3 moves on a smooth curve, joining the hysteresis curve again at about 280 A. Using these data it will be possible to program the sextupole correction currents appropriately.

SPECIAL INVESTIGATIONS

The persistent eddy currents in the superconductor reflect the complete history of the transport current variations in the magnet. Therefore the time decay rates might depend on parameters like the ramp speed of the transport current, the duration of pauses during the ramp, the number of current cycles after a quench and the helium temperature. Detailed studies are in progress and some first results will be presented.

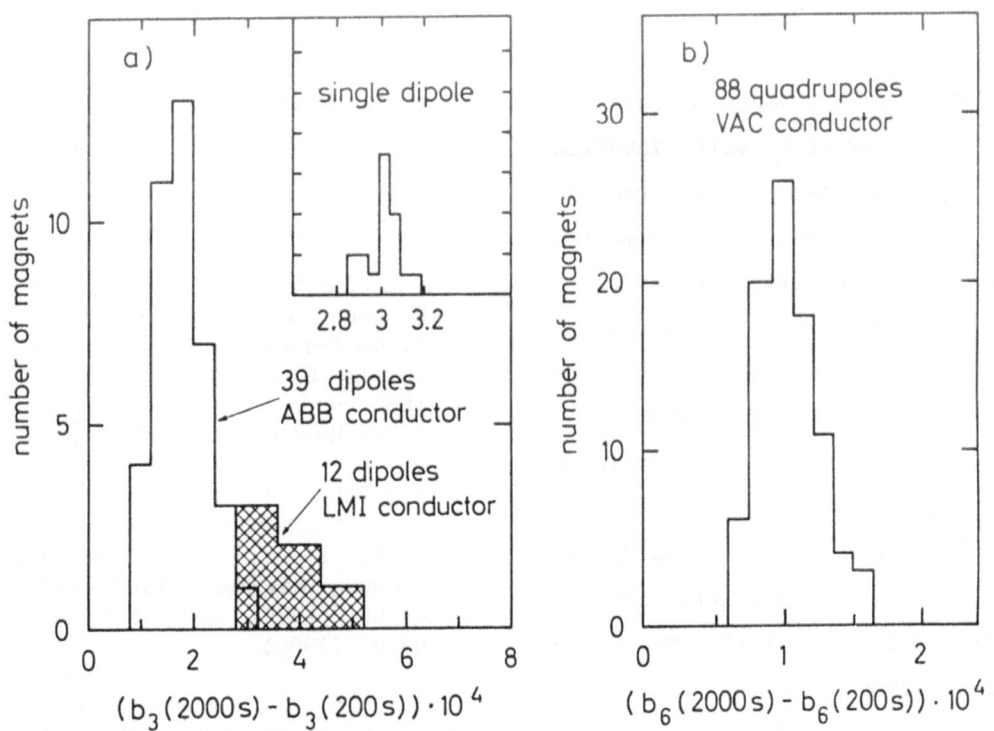

FIGURE 4
Decay rates (change between 200 s and 2000 s) of
a) sextupole in dipoles   b) 12-pole in quadrupoles

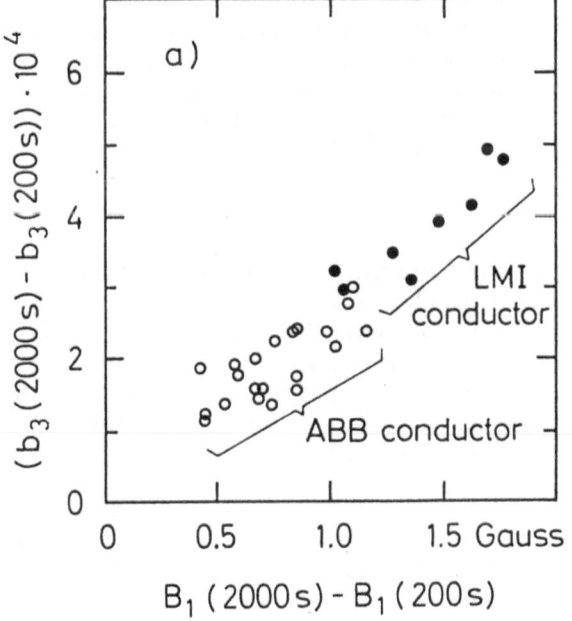

FIGURE 5
Correlation between the decay rates of
the sextupole and of the main field in
the dipole magnets

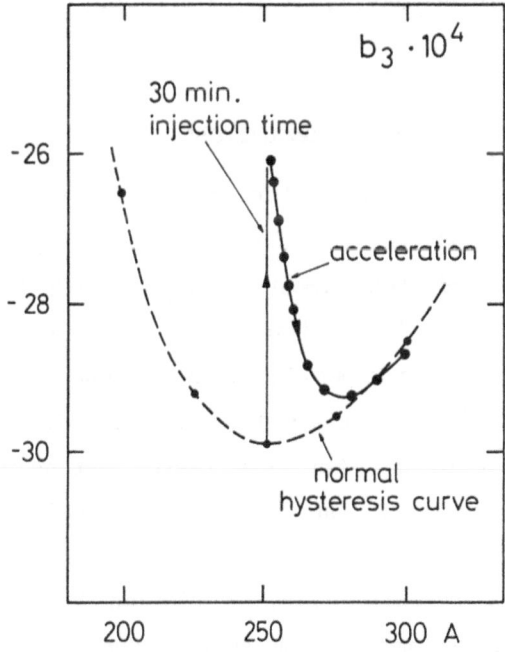

FIGURE 6
Time variation of the sextupole coefficient in a dipole during the injection
and acceleration phase in HERA

For a selected dipole the decay rate of the sextupole has been measured with one, two or three current cycles 50 A → 6000 A → 50 A made in advance. Contrary to experience with Tevatron magnets [5] the decay rate remains constant within our experimental resolution. This observation is very important for the HERA accelerator. During a beam loss several magnets may quench. Then the initial condition of the whole machine can be reestablished by simply going through a current cycle.

The ramp from 50 A to 250 A has been performed with different speeds, ranging form 1 A/s to 20 A/s. The decay rate stays almost constant, but the sextupole values at 20 A/s are lower than those at 1 A/s by about $1 \cdot 10^{-4}$.

A break of up to 30 minutes at the 6000 A point of the initial current cycle had no appreciable effect.

CONCLUSIONS

The field distortions from persistent currents are well understood and can be predicted almost quantitatively [6] provided the critical current density $J_c(B,T)$ is known. However, no quantitative theory exists for the time dependence of J_c. For this reason extensive measurements on the decay rates of multipoles are carried out. The present data reveal a large variation between magnets made from superconductors of different origin and even among the magnets made from the superconducting cable of a single manufacturer. By measuring the time dependence for all dipoles reliable information will be gained for the compensation of the field distortions in HERA.

REFERENCES

1 D.A.Finley, D.A.Edwards, R.W. Hanft, R. Johnson, A.D.McInturff, J. Strait, "Time Dependent Chromaticity Changes in the Tevatron," Proc. of the 1987 IEEE Part. Accel. Conf., Washington D.C., March 16-19, 1987, pp. 151-3

2 K.H. Meß, P. Schmüser, Lectures at the CERN-DESY School "Superconductivity in Particle Accelerators", Hamburg May/June 1988, CERN report 89-04 (1989) and DESY report HERA 89-01 (1989)

3 Y.B.Kim, C.F.Hempstead, A.R. Strnad, "Critical Persistent Currents in Hard Superconductors"; P.W. Anderson, "Theory of Flux Creep in Hard Superconductors", Phys. Rev. Letters 9,309 (1962)

4 H.R.Barton Jr.,R.Bouchard,Yanfang Bi, H.Brück, M.Dabrowska, Wenliang He, Zhuomin Chen,Zhengkuan Jiao, D. Gall, G. Knies, J, Krzywinski, J. Kulka, Liangzhen Lin,A. Makulski, R. Meinke, F. Müller, K. Nesteruk, J. Nogiec, H.Preißner, P.Schmüser, Z.Skotniczny, M. Surala, Wenlong Shi, contribution to this conference

5 D.A.Herrup, M.J.Syphers, D.E.Johnson, R.P.Johnson,A.V.Tollestrup,R.W.Hanft, B.C. Brown, M.J. Lamm, M. Kuchnir, A.D. McInturff, "Time Variations of Fields in Superconducting Magnets and their Effects on Accelerators"; R.W. Hanft, B.C. Brown, D.A. Herrup, M.J.Lamm, A.D.McInturff, M.J.Syphers, "Studies of Time Dependence of Fields in Tevatron Superconducting Dipole Magnets", IEEE Trans. Mag. 25, No. 2, 1643 and 1647 (1989)

6 H.Brück,R.Meinke,F.Müller, P.Schmüser "Field Distortions from Persistent Currents in the Superconducting HERA Magnets", DESY report 89-041 (1989)

PERFORMANCE OF THE SUPERCONDUCTING MAGNETS FOR THE HERA ACCELERATOR

H. R. Barton Jr., R. Bouchard, Yanfang Bi, H. Brück, M. Dabrowska,
D. Darvill, Wanliang He, Zhuomin Chen, Zhengkuan Jiao, D. Gall, G. Knies,
J. Krzywinski, J. Kulka, A. Ladage, R. Lange, Liangzhen Lin, A. Makulski,
R. Meinke, F. Müller, K. Nesteruk, J. Nogiec, H. Preissner, W. Rakoczy,
P. Schmüser, M. Surala, E. Schnacke, Z. Skotniczny, Wenlong Shi
Deutsches Elektronen-Synchroton DESY, 2000 Hamburg 52, Germany

ABSTRACT

The superconducting dipole and quadrupole magnets for the proton-electron collider HERA are subjected to extensive tests including measurements of heat loads, helium leak rates, maximum field capability, field integral, field direction and field uniformity. The magnets measured so far exceed the design performance by a safe margin.

INTRODUCTION

The proton storage ring of the proton-electron collider HERA [1], presently under construction at DESY in Hamburg, is equipped with 422 superconducting dipole [2] and 224 quadrupole [3] magnets.

For the nominal proton energy of 820 GeV the dipole fields of 4.68 T and quadrupole gradients of 90.2 T/m are achieved with a current of 5027 A. The coils are wound from a keystoned Rutherford-type cable with 24 (23) strands in the dipole (quadrupole). The magnetic length is 8.8 m for the dipole and 1.9 m for the quadrupole. At the low injection energy of 40 GeV the magnets suffer from severe field distortions caused by persistent eddy currents in the superconductor. Superconducting correction coils are mounted on the beam pipe of the magnets to compensate the sextupole and 10-pole in the dipole magnets and the 12-pole in the quadrupoles.

The industrial production of the magnets is underway and a large number have been delivered and tested.

MEASUREMENTS ON MAGNET PERFORMANCE

The magnets are first checked for all important mechanical tolerances and for electrical integrity. Then they are mounted on test stands, the helium supply is connected, and the vacuum vessel is leak-checked. After cooling to liquid helium temperature the helium leak rates are measured.

For several magnets the heat loads on the cryostat and the thermal shields have been determined. The observed values for dipoles (quadrupoles) are 5.5 W (8.5 W) at the liquid helium container and 25 W (33 W) at the shield at 60 K.

The leak rate of the cold helium system into the insulation vacuum is measured for each magnet. Typical leak rates for quadrupoles are 10^{-8} mbar l/sec, a factor of 100 below the required limit. After curing some initial problems the leak rates in the dipole cryostats are now also acceptable.

The following magnetic measurements are performed. For all dipoles, the field integral is determined with a precision of 0.01 % and the field direction with respect to gravity to within 0.2 mrad, using a detector which contains a nuclear magnetic resonance (NMR) probe, two orthogonal Hall probes and an electronic tilt sensor.

In the quadrupoles, the integrated gradient is measured with a precision of 0.05 %, the field direction to within 0.3 mrad and the magnet axis to within 0.2 mm by means of a "stretched-wire" system. A thin wire, stretched parallel to the quadrupole axis, is precisely moved in the horizontal or vertical direction and the induced voltage is recorded.

The multipole components of the magnets are measured with rotating pick-up coils with an internal compensation of the main field. The accuracy is 0.001 % relative to the main pole.

Maximum Field Capability

All HERA magnets are quenched a number of times to investigate their training properties and maximum current capability. The magnets show basically no training; more than 80% reach the critical current of the cable already in the first quench. The quench current distribution at a helium temperature of 4.75 K is shown in Fig. 1. The values measured in the dipoles are lower since there is a higher local field in the coils. Not a single quench has been observed below the nominal current of 5027 A. With one exception, all magnets measured so far would allow to operate the HERA proton ring up to an energy of 1 TeV.

Field Integral and Field Direction

The integrated dipole fields and quadrupole gradients, normalized to the magnet current, are shown in Fig. 2. Apart from a systematic length difference of 0.1 % between magnets from different manufacturers the distributions are quite narrow with rms deviations of 0.03 % for the dipoles and 0.06 % for the quadrupoles. Fig. 3 shows the variation of the field direction of a dipole along the magnet axis. For most magnets, the fluctuations are within the specified limits of ± 3 mrad.

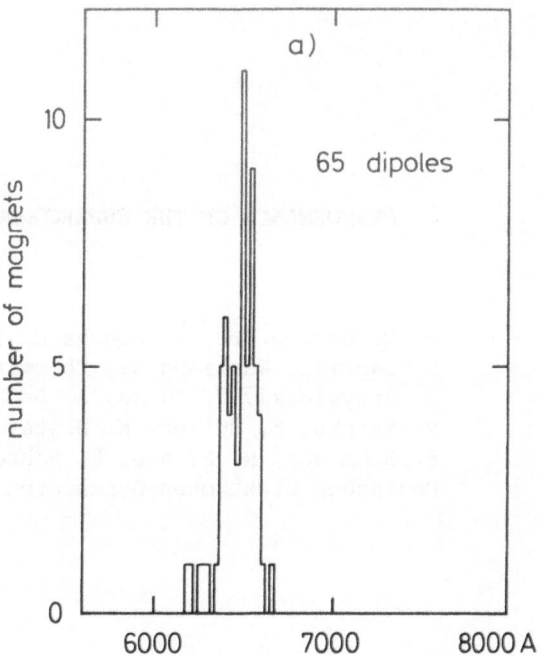

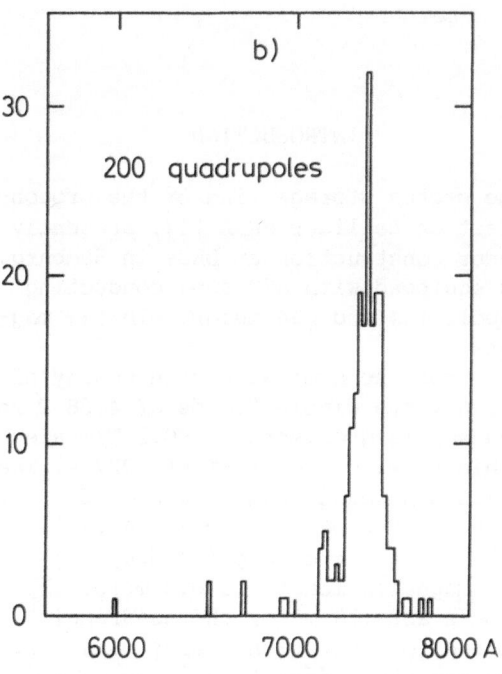

FIGURE 1   Quench current distribution for (a) dipoles, (b) quadrupoles at 4.75 K. The nominal operating current is 5027 A at a maximum helium temperature of 4.6 K.

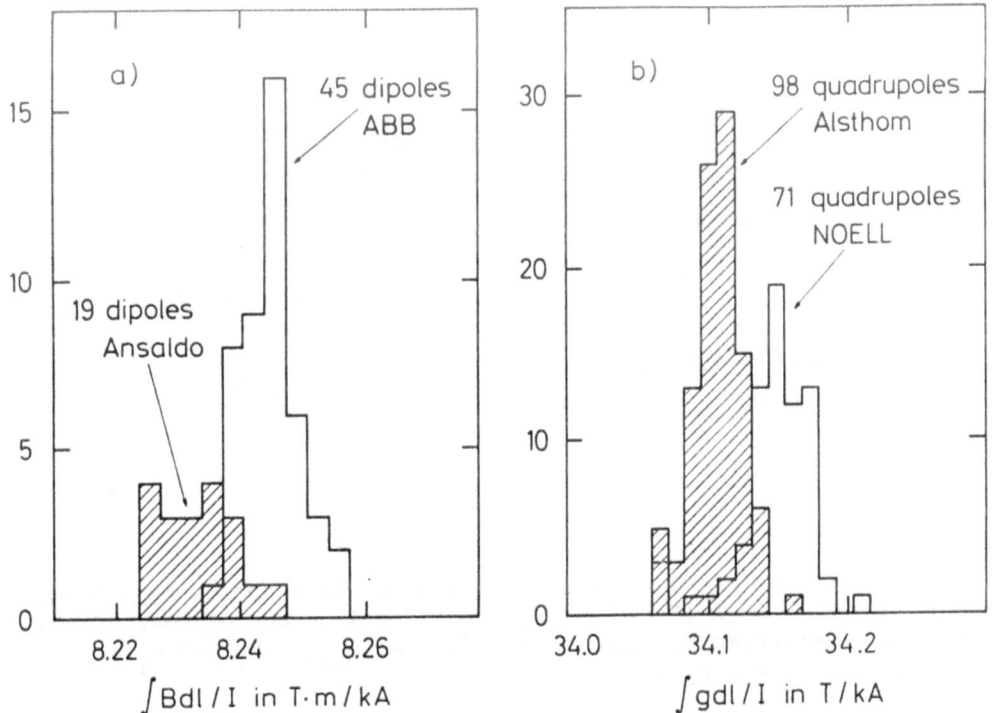

FIGURE 2   The integrated dipole fields (a) and quadrupole gradients (b), normalized to the current (measured at I = 2100 A in the dipoles and I = 5000 A in the quadrupoles).

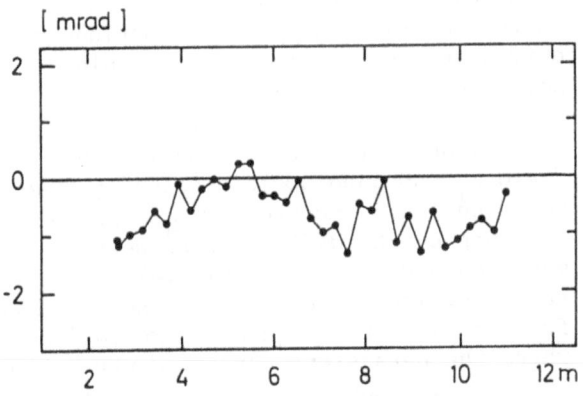

FIGURE 3   Field direction with respect to gravity along the axis of a dipole. The data points are reproducible to within 0.1 mrad, so the observed structure is a property of the magnet.

Harmonic Analysis of Field Uniformity

The field of an accelerator magnet is conveniently expressed in terms of a multipole expansion in a cylindrical coordinate system (r,θ,z)

$$B_\theta(r,\theta) = B_{ref} \sum_{n=1}^{\infty} \left(\frac{r}{r_o}\right)^{n-1} (b_n \cos(n\theta) + a_n \sin(n\theta))$$

and B_r correspondingly. Here r_o = 25 mm is the reference radius of the expansion and B_{ref} the amplitude of the main field component at the reference radius. The b_n and a_n are the normal and skew multipole coefficients.

The magnet coils without yoke are measured at room temperature during the industrial production process. The complete magnets are measured in the DESY test facility in the superconducting state.

The b_n multipoles from "warm" and "cold" measurements of dipoles and quadrupoles are shown in Fig. 4 together with their rms standard deviations. The observed field distortions are very

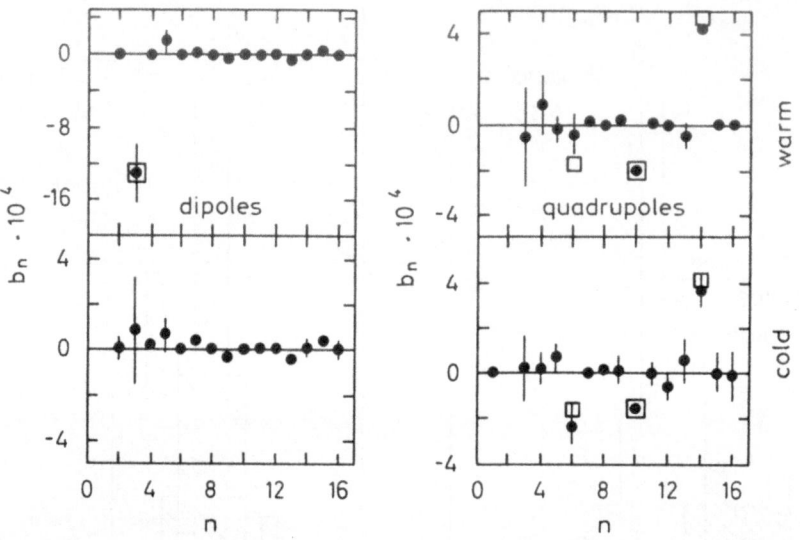

FIGURE 4   Multipole coefficients b_n of dipoles and quadrupoles from "warm" measurements and at 5000 A ("cold"). The theoretical values of nonvanishing coefficients are indicated by squares. For dipoles $B_{ref} = B_1$ and $b_1 = 1$; for quadrupoles, $B_{ref} = g\, r_o$ and $b_2 = 1$.

small and well within the allowed limits of $\pm 3 \cdot 10^{-4}$ for the dipoles. Fig. 5 demonstrates that a correlation exists between the sextupoles measured for warm coils and those for complete magnets in the superconducting state. A systematic difference of $13 \cdot 10^{-4}$, due to the contribution of the iron yoke, has been subtracted.

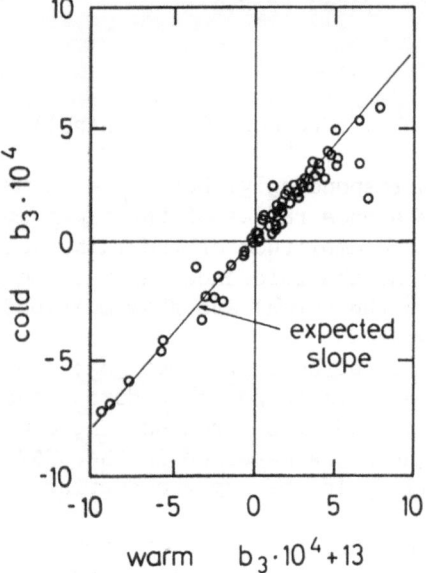

FIGURE 5   Correlation between the "warm" and "cold" measurements of the sextupole in the dipole magnets.

Effects from Persistent Currents

The persistent eddy currents which are induced in the niobium-titanium filaments during ramping have opposite direction for increasing and decreasing main field. The resulting hysteresis is clearly observed in the 6- and 10-pole coefficients of the dipoles and the 12- and 20-pole coefficients of the quadrupoles (Fig. 6). The predictions of an eddy current model [4], shown as continuous curves in Fig. 6, are in remarkable agreement with the data.

The persistent currents have also a significant influence on the main dipole field B_1 and quadrupole gradient g. Their contribution is denoted by $\tilde{B}_1$ resp. $\tilde{g}$ and is given by the difference between measured and computed field values at low currents. The latter ones are scaled down from measurements at higher excitation where the persistent current contribution is small and the iron yoke saturation still negligible. The data for $\tilde{B}_1/B_1$ and $\tilde{g}/g$ are shown in Fig. 7. Again a hysteresis and good agreement with the model predictions are observed.

The persistent current effects show a time dependence which is discussed in a separate contribution [5] to this conference.

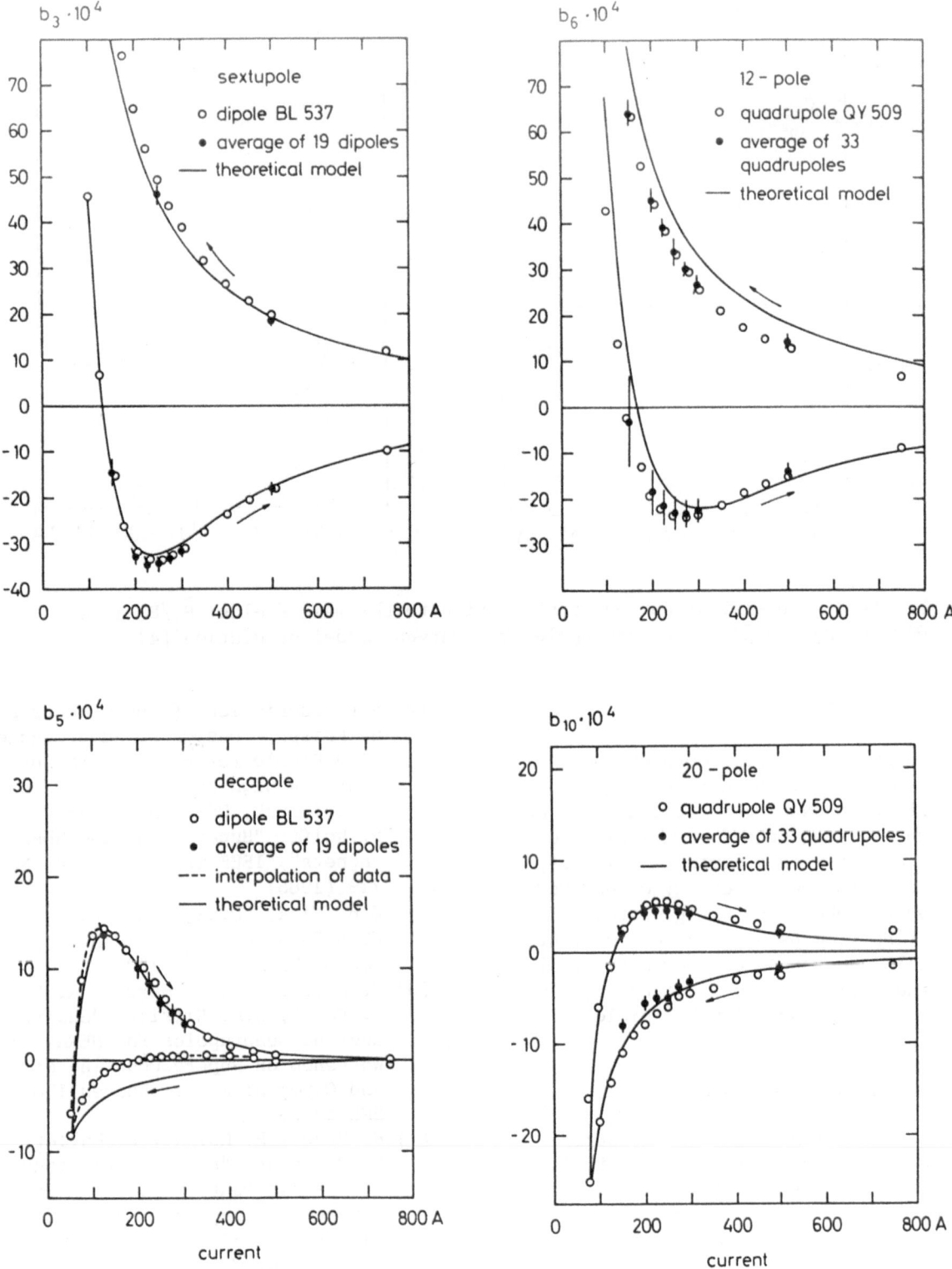

FIGURE 6   Hysteresis measurements of higher multipoles in dipoles and in quadrupoles. Open points: single magnet. Full points: average of 19 dipoles (33 quadrupoles). The curves are model calculations [4]. The ramp direction of the current is indicated by arrows.

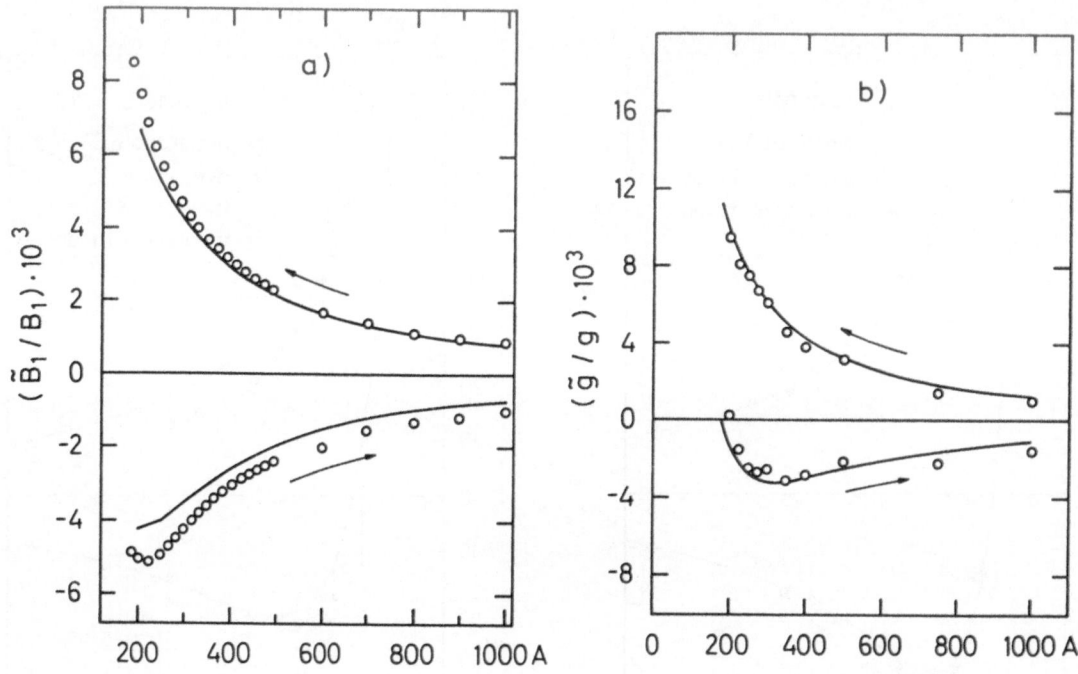

FIGURE 7 Persistent current contribution to the main fields: $\tilde{B}_1/B_1$ in a dipole (a) and $\tilde{g}/g$ in a quadrupole (b). Curves: model prediction[4].

SUMMARY

A large number of superconducting HERA magnets have been tested. The magnets show excellent field quality and quench training behaviour allowing to operate HERA with a comfortable safety margin at the design energy of 820 GeV and even higher. The field distortions resulting from persistent eddy currents are well understood both experimentally and theoretically, so a compensation with correction coils will be possible.

ACKNOWLEDGEMENTS

We thank all members of the measuring group for their efforts in data taking and the staff of the DESY refrigeration plant for the regular supply with liquid helium to operate the magnet test facility.

REFERENCES

[1] For a recent review of the HERA collider see B.H. Wiik, "HERA Status", Proceedings of the XXIV Intern. Conf. on High Energy Physics, Munich 1988

[2] For description of the magnets see: H. Kaiser, "Design of Superconducting Dipole for HERA", 13th Intern. Conf. on High Energy Accelerators; Novosibirsk, USSR, Aug. 1986 S. Wolff, "Superconducting HERA Magnets", IEEE Trans. Mag.24, No.2, 719 (1988) B.H. Wiik, "Design and Status of the HERA Superconducting Magnets", DESY report HERA 88-05 (1988)

[3] R. Auzolle, P. Le Marrec, A. Patoux, J. Perot, J.M. Rifflet, "Superconducting Quadrupoles for HERA", ICFA Workshop on Superconducting Magnets and Cryogenics, Brookhaven 1986, BNL 52006

[4] H. Brück, R. Meinke, F. Müller, P. Schmüser, "Field Distortions from Persistent Currents in the Superconducting HERA Magnets", DESY report 89-41 (1989), submitted to Z. Phys. C

[5] H. Brück, Zhengkuan Jiao, D. Gall, G. Knies, J. Krzywinski, R. Meinke, H. Preissner, P. Schmüser, "Time Dependence of Eddy Current Effects in the Superconducting HERA Magnets", this conference

INDUSTRIAL SERIES FABRICATION OF 722 SUPERCONDUCTING DIPOLE-AND SEXTUPOLE/QUADRUPOLE-CORRECTION MAGNETS FOR THE HERA PROTON STORAGE RING BY HOLEC, RIDDERKERK, THE NETHERLANDS.

H.J. Israël, P.A.M. Bracké, Holec Machines & Apparaten B.V., Ridderkerk,
The Netherlands.
P. Schmüser, II Institut für Experimentalphysik, Universität Hamburg, West-Germany.
C. Daum, NIKHEF-H, Amsterdam, The Netherlands.

ABSTRACT

The adjustment of the working point and the chromaticity correction of the HERA proton ring are provided by quadrupole and sextupole superconducting correction coils The orbit correction in the proton ring is realized by means of superferric correction dipoles. The layout, manufacture and performance of both types of correction magnets will be described and the experimental results on quench behaviour and field quality will be presented.

QUADRUPOLE/SEXTUPOLE CORRECTION COILS

The quadrupole and sextupole correction coils are mounted on the beam pipe inside the main dipole magnet (see Fig. 1).

Between the correction coils and the main dipole coil remains a 4 mm wide annular slit for the passage of single phase liquid helium of 4.35 to 4.60K in forced flow. Figures 2a, b, c show a cross section and an unwrapped view of the correction coil package.

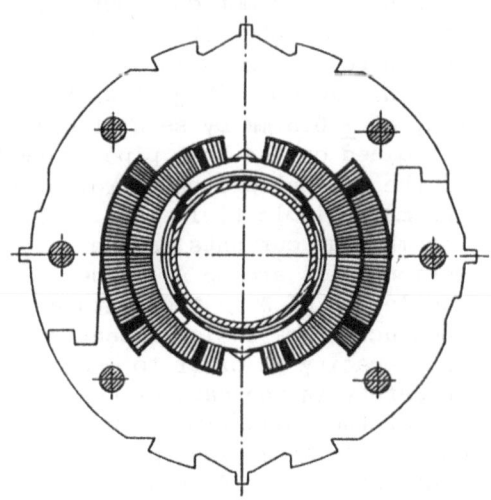

FIGURE 1

Cross section of the main dipole coil and the beam pipe with the correction coils. The centring is provided by G11 spacers at 45°, 135°, 225° and 315°. The spacers are 50 mm long and are glued on the beam pipe at seven locations.

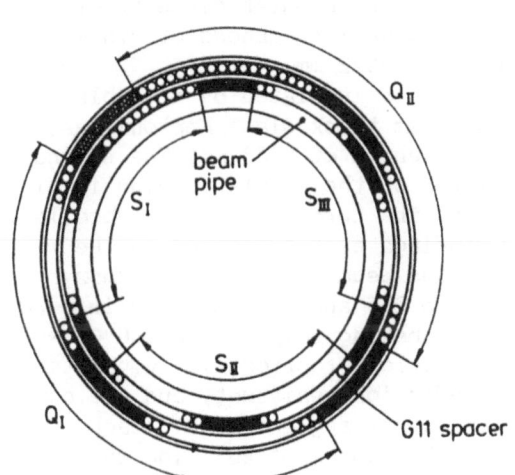

FIGURE 2a

Cross section of the beam pipe with the correction coils. S_I, S_{II}, S_{III} subcoils of the sextupole; Q_I, Q_{II} subcoils of the quadrupole.

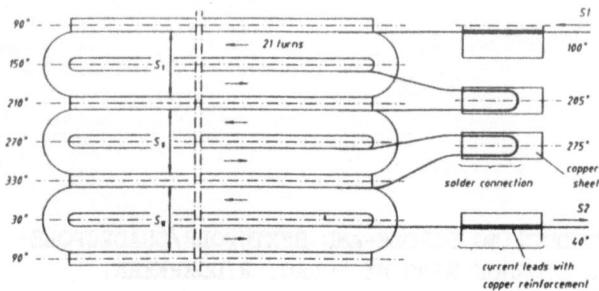

FIGURE 2b
Unwrapped view of the sextupole coil
showing the subcoils S_I, S_{II} and S_{III},
the internal solder connections and the
current leads. The direction of the
current flow is indicated by arrows.

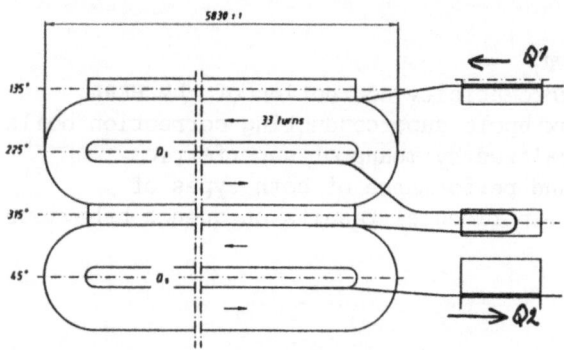

FIGURE 2c
Unwrapped view of the quadrupole coil
with subcoils Q_1, Q_{II}.

The beam pipe is a seamless stainless
steel tube (DIN 1.4429, equivalent to
316 LN) with an outer diameter of 60.3 ±
0.3 mm, a wall thickness of 2.5 mm and a
length of 9582 mm.
The pipe is insulated by a double layer
of 0.15 mm thick glass-Kapton-glass tape
over a length of 9268 mm. The nominal
diameter of the insulated pipe is 61.0 ±
0.1 mm. The three sextupole subcoils of
21 windings each are glued on the
insulated beam pipe with an expoxy
(Epikote 215 and Versamid 140 in a ratio
of 1:1, baking temperature 150°C) which
is suitable for cryogenic temperatures
and radiation resistant, and are covering
an azimuthal angle of 100° each, between
G11 centre cores of 20° wide which are
centred at azimuthal angles of 90°, 210°
and 330° and are glued on the beam pipe
with an angular accuracy of ± 0.2° over
the whole lenth of 5900 ± 1 mm including
the coil heads.

The sextupole subcoils are manufactured
with an angular accuracy of ± 0.3°. For
the nominal sextupole field of 0,020T at
ro = 25 mm a current of 46 A is needed.
The sextupole coil is insulated by one
layer of glass tape and provided by
compression wrapping of two layers of
glass fibre, glued with epoxy. The outer
diameter of this layer is 63.8 ± 0.1 mm.
The two quadrupole subcoils of 33
windings each are glued on this layer and
are covering an azimuthal angle of 150°
each, between G11 centre cores of 16.75 ±
0.05 mm wide (30°) and centred at 135° ±
0.2° and 315° ± 0.2°. Both coils are
mounted with an angular accuracy of ±
0.2° over the whole length of 5830 ± 1
mm. The quadrupole subcoils are manufactured with an angular accuracy of ±
0.3°. The nominal quadrupole field at ro
= 25 mm is B_2 = 0.040 T and is achieved
with a current of 77 A. Figures 2b and 2c
show the connections between the subcoils
and the current leads. The correction
coil package is surrounded with a glass
fibre compression wrapping with a
thickness of about 0.5 mm. The sextupole
is covered by two layers and the
quadrupole by a single layer wound from a
very strong glass fibre (0.31 mm^2
VETROTEX R glass, tensile strength 3600
N/mm^2) with a tension of 800 N/mm^2 and a
pitch of 2.5 mm. The compression wrapping
provides a high radial pressure on the
windings of about 6 N/mm^2 at T = 4K,
which together with the glue joints
inhibits conductor motion under the
influence of Lorentz forces. The
correction coil package is centered in
the main dipole coil (Fig.1) with an
accuracy of ± 0.3 mm by sets of G11
spacers glued on the beam pipe every 1.1
m. For the nominal proton energy of 820
GeV the main dipole field is 4.69 T
and the maximum currents in the
correction coils are 85 A in the
quadrupole and 65 A in the sextupole. The
superconductor has a significantly higher
current capacity in order to increase the
quench safety in the case of beam-induced
heating of the coils (image currents or
proton beam losses). The specified
critical current is 250 A at 4.60 K and
5.5 T. The superconductor is a single
strand wire of 0.70 mm diameter, a copper
to NbTi ratio of 1.8:1 and a filament
diameter of 15 μm. The wire is insulated
by 25 μm Kapton tape with 50% overlap and
two glass fibre layers with a B-stage
epoxy coating. In order to keep the

narrow tolerances on the azimuthal angles of the coils the insulated superconductor must have a nominal diameter of 1.03 mm with very small tolerances (less than 0.01 mm). The glass fibre layer of the wire is filled with such an amount of B-stage epoxy that the wire can be compressed in a heated mould to the design diameter of 1.03 mm. The coils are wound on a cylindrical mandrel with precise steel keys to serve as core. After the winding procedure the mandrel with coil is covered with a top mould. Along the sides of the coil and in between the mandrel and the top mould precise rulers are placed to determine the lateral dimension and the coils are compressed to their accurate dimensions and simultaneously baked at 150°C for two hours. For positioning of G11 cores and spacers and mounting the S-and Q-coils a well levelled table together with a special tool and an accurately machined triangular/rectangular alignment tool is used.

PERFORMANCE

The field measurement is done at room temperature. The Q- and S-coil is excited with an alternating current of 0.7 A and 11 Hz and the signal induced in a 2.05 m long pickup coil is registered by a lock-in amplifier and digitized. A full rotation of the pickup coil is divided into 100 steps. A micro computer controls the measurements and performs a Fourier transformation to compute the multipole coefficents. The procedure is repeated for both coil layers in three longitudinal positions of the pickup coil to cover the entire length of 5.9 m. Fig. 3 summarizes the results of the field measuremts of all 467 Q/S-coils. In the top figure is plotted the difference between the measured and expected normal multipole coefficients. The values are compatible with zero; the errors are much below the qualification limit of 0.01. For the skew multipole coefficients a_n a similar distribution is presented in the lower figure.

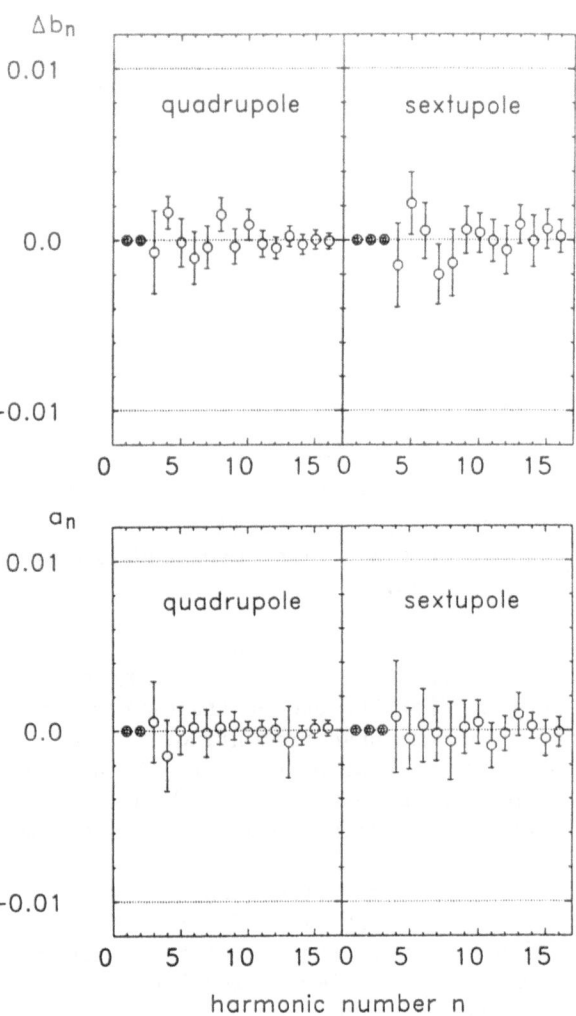

FIGURE 3
The normal and skew multipole coefficients of all 467 correction coils. The allowed limits are $\pm$ 0.01 for all a_n and $\blacktriangle b_n$.

The field measurement allows also to determine the misalignment angle between the Q- en S-coil. The average value is 1.0 $\pm$ 3.2 mrad wich again is well below the specified limits of $\pm$ 7 mrad.

number of magnets

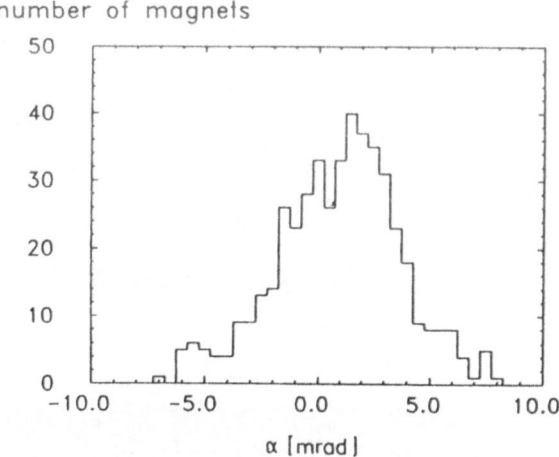

FIGURE 4

Distribution of the difference angle
between the Q- and S-coil in all 467
correction coils. The specified limits
are ± 7 mrad.

We conclude that the field quality is
much better than specified and more than
adequate for the use of the correction
coils in the HERA protonring.

A cryogenic test of the correction coils
is performed in a vertical bath cryostat
containing a dipole magnet with a maximum
field of 5.08T.

In Fig. 5 we show the quench current
distribution for all 467 correction
coils. The test results of the series
prodution were almost excellent.

number of magnets

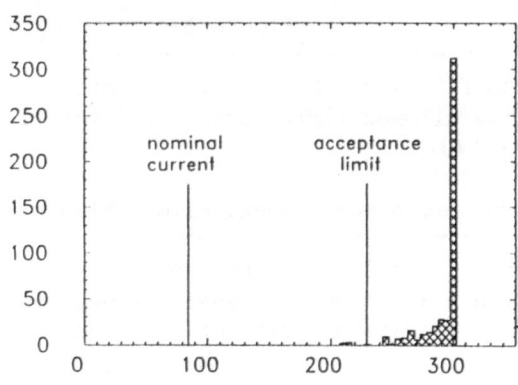

quench current [A]

FIGURE 5

The quench currents of all 467 correction
coils measured in an external field of
5.08 T and a helium bath temperature of
4.35 - 4.40 K. Plotted is the lower value
of the sextupole or quadrupole quench
currents. The specified critical current
of the superconductor corresponds to 290A

Almost 80% reached or exceeded the
specified critical current of the
superconductor. About 5% of the coils
were below the qualification level of 230
A, but even those coils were well above
the highest operating current. All
rejected correction coils have been
equipped with new subcoils and all have
passed the second test.

The number of traning steps is plotted in
Fig. 6.It is in general between 0 and 5.
One of the early correction coils, B17,
needed 11 training steps in the first
cryogenic test. When tested a second time
a few weeks later no further traning was
observed. Only one correction coil out of
460 had to be rejected because of
excessive and repeated training. All
other coils with some initial training
showed no further training with reversed
polarity or in a second test. More than
99% of the coils exceeded the nominal
operating current without any quench and
more than 90% went beyond twice the
nominal current without training.

number of magnets

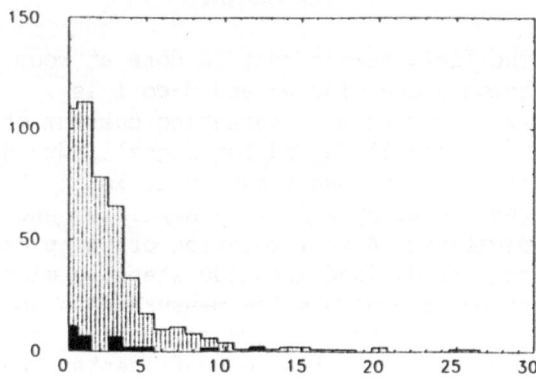

number of training steps

FIGURE 6

The number of training steps for all 467
correction coils (hatched area). Plotted
is the larger of the values found in
either the quadrupole or the sextupole
layer. Marked in black is the training
step distribution for those coils wich
initially failed the qualification level
of 230 A. These coils were repaired. They
have been retested and are included in
the hatched diagram.

One of the prototype coils was retested
more than a year later. It reached 300 A
without training. This demonstrates that
no fatigue occurs in the R glass
compression wrapping and the glue joints.

From the fact that the majority of the correction coils reached the critical current of the superconductor with very few if any training steps we can conclude that the conductors are well immobilized. Careful workmanship and a flawless superconductor are of course a necessary condition for such a good performance.

DIPOLE CORRECTION MAGNET

The correction dipole is a window frame magnet with an iron yoke of 610 mm length and 75 mm gap and two superconducting saddle coils. A cross section is shown in Fig. 7.

Since every dipole has to be powered individuallly, a low operating current (35 A for the nominal field of 1.17 T) and a correspondingly large number of turns (1000 per coil) are essential.

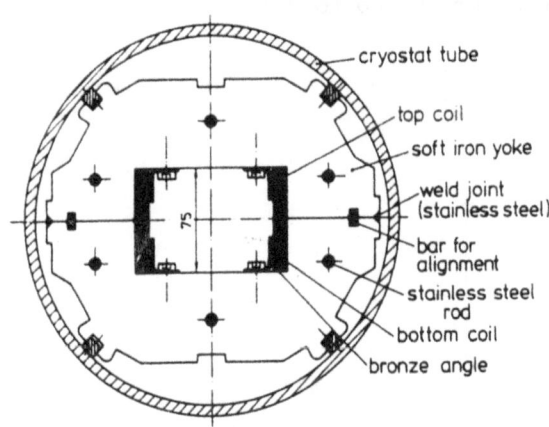

FIGURE 7
Cross section of correction dipole, mounted in the cryostat.

The superconductor has a diameter of 0.56 mm, a copper to NbTi ratio of 3.7:1 and contains 36 filaments of 45 μm diameter. The specified critical current is 250 A at 1.5 T. The wire is insulated by a polyesterimid varnish.

The coils are wound in a flat, racetrack-like winding mould with the wire passing through epoxy. When the winding is finished, the long straight sections and the arcs of the coil are constrained by precise compression bars. The center of the short straight section (the arc of the coil) is clamped in a fibre glass bracket. Then the long straight sections are folded up, allowing a natural bending of the unconstrained short sections into the saddle shape.

After bending the coil is baked in an oven at 150°C for two hours. This novel technique of producing saddle-shaped coils with many turns of thin conductor has proven to be quite reliable. The epoxy used for impregnation is a mixture of Epikote 215 and Versamid 140 in a ratio of 1:1. In the preseries and early series production the epoxy was used without filling material. Since in a number of magnets excessive training was observed, it was decided to add Al_2O_3 powder (40% by volume) to the epoxy, thereby reducing the differential shrinkage between the superconductor and the impregnation during cooldown. The magnets with Al_2O_3 filled epoxy show infact a considerable reduction in training steps. The coil is continually checked for damage to wires or lacquer insulation or loose wires in the coil heads. Besides mechanical and electrical measurements are made to ensure that all technical requirements and tolerances are met.

The iron yoke with a length of 610 + 0/-1 mm is split into two halves which are assembled from 5 mm thick precision-stamped soft iron laminations.

The coils are insulated with 125 μm Kapton and fixed in the half yokes by means of bronze angles. A "Z" shaped 0.15 mm stainless steel foil is placed between the insulated coil and the yoke to avoid friction of the coil on the rough inner surface of the yoke during cooldown. The bronze angles are fixed on the yoke by means of stainless knobs which are spot-welded against the laminations.

The laminations are held together by 8 mm thick stainless steel rods. These rods determine the longitudinal shrinkage during cooldown. The soft steel shrinks by 0.2% whereas the stainless steel rods and the superconducting coils contract by 0.3%. The stacking of the laminations is sufficiently loose to allow for a 0.3% shrinkage of the magnet. The two half yokes are mounted in a press and pushed together with a force of 5000 N. Two keys of accurately ground flat steel (5.9 mm wide, 10 mm high) are inserted into the slots of 5.9 mm width to ensure the horizontal alighnment of the yoke halves. A welding machine is making the two weld joints in the horizontal plane simultaneously.

158

The welding material is stainless steel with almost the same thermal contraction as the coils.
The field quality requires a straightness of the yokes of better than 0.2 mm, a twist angle of less than 3 mrad, and a parallelism of the pole faces of better than 0.5 mm. The coils have a height and width tolerance of 0.03 mm and 0.1 mm respectively. These requiremets are achieved using precision tooling.

PERFORMANCE

The cryogenic test of the correction dipoles comprises measurements of the quench current and the central field, using a Hall probe.
A multipole measurement with a rotating pickup coil is performed only for sample magnets since the field quality is largely determined by the accurate iron yoke and very reproducible from magnet to magnet. The observed distribution of quench currents is shown in Fig. 8. The specified qualification level is 75 A. (At this current, the lorentz forces acting on the windings are a factor of 4 larger than at the nominal current of 35 A). The majority of the magnets achieves quench currents above 100 A.

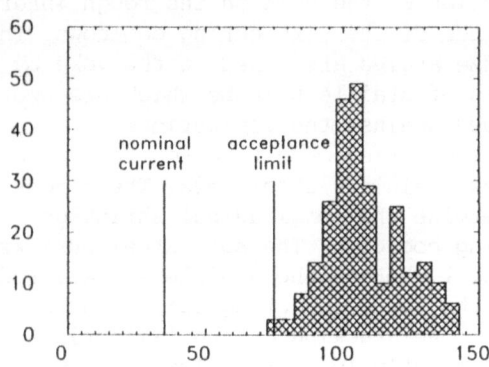

FIGURE 8
Final acceptance results of the quench currents of all 255 correction dipole-magnets for the HERA-project, manufactured by HOLEC B.V.

In normal accelerator operation the magnets have a large safety margin because the operating current at 1.5 T is less than 20% of the critical current. In Fig. 9a and 9b we summarize the field measurements done on the series magnet.

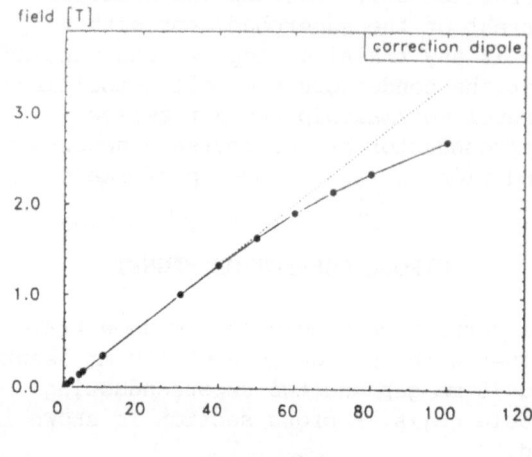

FIGURE 9a
Central field as a function of the current.

The field increases almost linearly with the current up to 2 T.

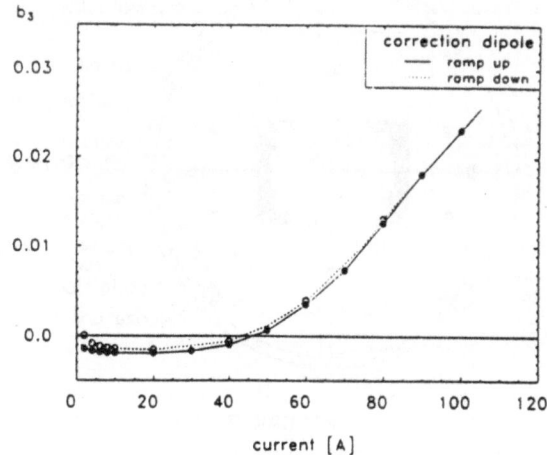

FIGURE 9b
Sextupole coefficient b_3 as a function of the current.

The only significant higher multipole is the normal sextupole. The coefficient b_3 is slightly negative for small currents, but rises steeply for currents above 50 A, where the iron yoke starts to saturate. The remanent field in the iron has been measured to be 5 Gauss.

ACKNOWLEDGEMENTS

We want to thank DESY, NIKHEF-H and the University Twente for their support in the development and testing of the D- and Q/S-correction coils.

STRUCTURAL ANALYSIS OF THE LHC 10 T TWIN-APERTURE DIPOLE

D. Leroy, R. Perin, D. Perini
European Organization for Nuclear Research (CERN),
CH-1211 Geneva 23 (Switzerland)
A. Yamamoto, Nat. Lab. for High Energy Phys. (KEK),
Tsukuba, Ibaraki-ken 305 (Japan)

ABSTRACT

The Large Hadron Collider (LHC) is proposed at CERN as the next facility to push further the frontier of high energy physics. The design of the superconducting dipole magnets is based upon the "two in one" concept in which the dipoles of the two rings of the collider have a common magnetic circuit. Twin aperture models (MTA1) designed for a 10 T field level are being developed in collaboration with European Industries. These high field magnets require a sophisticated mechanical design of the various parts as collars, split iron yoke and outer shrinking cylinder. A structural analysis of the MTA1 magnets has been made at CERN by using the FEM code "ANSYS". We report on the mechanical analysis and optimization of the magnet during assembly, cooldown and under the electromagnetic forces.

INTRODUCTION

The design of the superconducting magnets for a Large Hadron Collider (LHC), which is proposed at CERN [1] is based upon the "two in one" concept. In this configuration, the two dipole magnets of the two rings have a common magnetic circuit. This solution is economical in space and cost and should allow the installation of very high field dipoles in the existing LEP tunnel at CERN. The envisaged field level for LHC is in the range 8 to 10 T. A R. & D. programme is pursued at CERN to reach this goal [2]. To demonstrate the feasibility of the LHC machine, 10 T twin-aperture models and prototypes have to be constructed and tested.

Four one-meter long models are now under construction in collaboration with industries : ANSALDO (Italy), ELIN (Austria), HOLEC (Netherlands), JEUMONT-SCHNEIDER (France). The four models, which make use of NbTi superconducting cables working at 2 K, have various technical solutions and are based on the design report presented in Ref. [3]. The present report deals with more detailed analysis of the mechanical structure of the straight part of the 10 T twin-aperture superconducting dipole. The mechanical structure is designed to sustain an horizontal resultant of magnetic forces corresponding to 11.2 T field. The main parameters of the 10 T twin-aperture dipole are summarized in Table 1.

DESCRIPTION OF THE MECHANICAL STRUCTURE AND DESIGN PRINCIPLES

Fig. 1 shows the cross section of the magnet. The coil support structure is formed by :

A twin collar made of Al alloy in which the coils are assembled and prestressed at ~ 55 MPa at room temperature.

An iron yoke, split in two parts at the vertical symmetry plane of the twin-aperture magnet. The two iron inserts between collars and yoke at the vertical plane serve to reduce the quadrupole component due to iron saturation.

A stainless steel or aluminium alloy outer cylinder around the yoke.

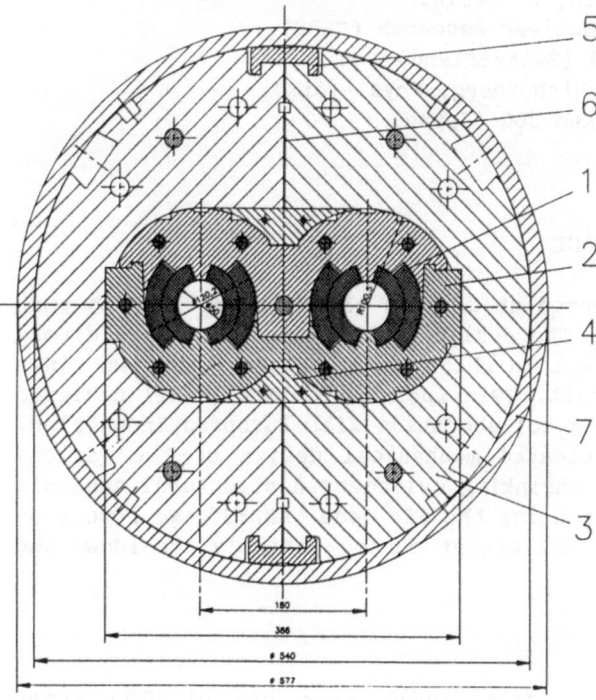

Fig. 1 - Cross section of MTA1 magnet.
1. Coils,   2. Al   collars,   3. Yoke,
4. Iron   insert,   5. Clamp,   6. Gap,
7. Outer shrinking cylinder.

The two parts of the yoke are separated by a gap at room temperature and so adjusted that, after cool down, two necessary conditions are fulfilled :

1. The collars and the yoke are in contact along the horizontal axis so that the horizontal component of the magnetic forces is well sustained by collars and yoke.

2. A required compressive force is produced on the mating faces of the yoke parts. The compressive force will be so that the gap stays closed at full excitation of the magnet.

TABLE 1

Dipole Parameters at 2 K	
Bore field [T]	10
Current [A]	15000
Turns per beam channel	
1st layer	2 x 13
2nd layer	2 x 24
Overall current density	
1st layer	357.5 A/mm^2
2nd layer	532.9 A/mm^2
Coil inner diameter	50 mm
Coil outer diameter	120.2 mm
Distance between aperture axes	180 mm
Collars outer dimensions	381x201 mm
Iron outer diameter	540 mm
Stored Energy for both combined channels	684 kJ/m
Resultant of magnetic forces in the first coil quadrant :	
Σ F_x	227.6 t/m
Σ F_y 1st layer	-23.4 t/m
Σ F_y 2nd layer	-98.0 t/m

In these conditions, the resulting mechanical structure is very rigid and has therefore smaller outward displacements under the large magnetic forces and consequently small stresses are induced in the coils by the bending moment effects. Moreover, when the inward displacement of the iron in the horizontal direction is larger than the corresponding shrinkage of the collared coils during cool down, an additional non-uniform prestress is introduced on the coils at low temperature.

The possibility to have, after cool down, a compressive force on the mating faces of the half-yokes and a closed contact between the collars and the yoke depends on the relative dimensions of the components of the structure, the choice of materials, the prestress in the outer shrinkage cylinder and the dimension of the gap between the two half-yokes.

The wanted gap is fixed and controlled at room temperature by means of clamps which maintain the two half-yokes at the desired position. After the assembly of the outer shrinking cylinder, the clamp force is transferred to the outer cylinder.

The difference between the required

stressing of the coils at operating conditions and the loading produced by the cool down of the structure determines the prestress which has to be provided at room temperature by coil collaring. The guide line for the stresses in the coils at full excitation is to have no region of the coil under tensile stresses and to keep the coils in contact with the collars [4]. Moreover, to limit the creep effects in the coils (flow of insulating materials) it is of great importance to have the prestress as small as possible at room temperature. If one assumes an infinitely rigid mechanical structure and neglects the stresses due to bending moments in the coils, a first idea of the minimum prestress in the coils at room temperature is given by :

$$\sigma_{\theta,0,coil} \simeq \Sigma \frac{F_y}{2w} - \Delta\alpha \cdot E_{coil} \approx 31 MPa$$

in which w = 17.3 mm is the width of the insulated cable E_{coil} = 20 GPa is the apparent Young's modulus of the coil, $\Delta\alpha$ is the difference of the integrated thermal contraction coefficient between 2 K and 300 K for coils and collars.

This relation ensures the contact between the coil and the collars at the upper angle of the coil layers.

OPTIMIZATION OF THE MECHANICAL STRUCTURE

Calculations of the mechanical structure have been made using the computer code Ansys[5]. Fig. 2 shows the general mesh for the LHC magnets used in the calculations. The collars are supported in the vertical and horizontal plane. There is a clearance of 0.5 mm between collars and iron yoke at the upper surface of the collars. The interferences between the various parts of the mechanical structure are simulated by "gap elements" provided in Ansys code.

The stresses and deformations have been calculated following the various sequences of the assembly : collaring, clamping, outer cylinder assembly, excitation at 11.2 T. The major results are summarized in Table 2 for the external cylinder made of Al alloy. The various points where stresses or dis-

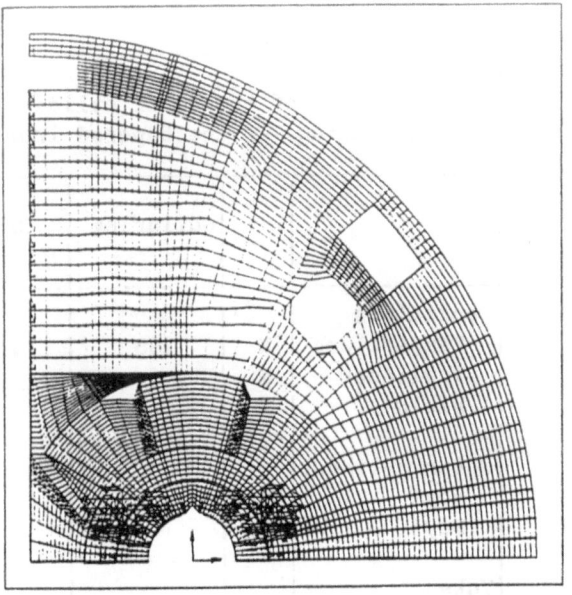

Fig. 2 - Mesh used in Ansys code.

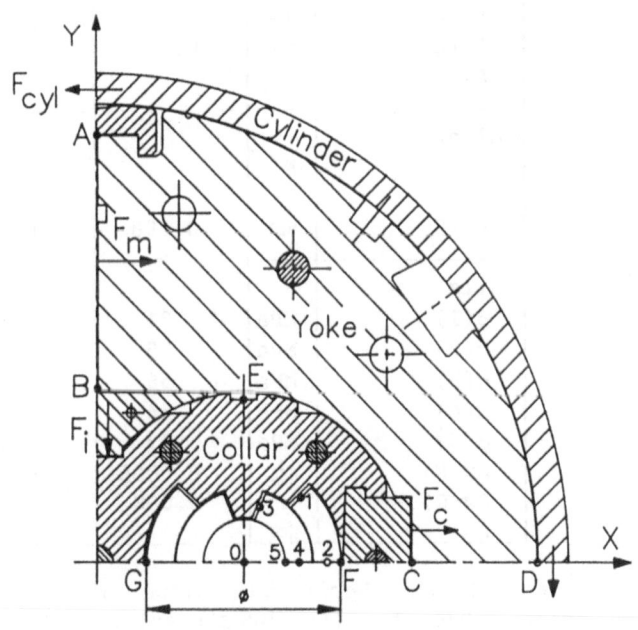

Fig. 3 - Sketch with the various points presented in Table 2 and Table 3.

placements are reported as indicated on Fig. 3. The presented results deal with the right part of the coils. The calculations refer to 1 mm of magnet length.

Coil Prestressing by collaring

The required prestressing obtained at room temperature by collaring the coils is simulated by the insertion of

TABLE 2 - SUMMARY OF THE MAIN RESULTS FOR MTA1

Material of the outer cylider :	Al alloy, 18.5 mm thick
Dimensions of the tapered iron half gap	0.51/0.44 mm

		Collaring	Clamp	Clamp + outer cylinder	Cool-down	Excitation
Conditions						
Temperature	K	293	293	293	2	2
Tcylinder	K			258	(− 33)	(− 33)
Field	T					11.2
Outer cylinder						
σcyl	Mpa			19	94	95
Fcyl	N			346	1739	1757
Yoke						
Fm	N				− 1505	− 180
ΔFm	N					+ 1325
Fi	N		− 327	− 328	− 485	− 1029
Δx(A)	mm		− 0.103	− 0.103	− 0.52	− 0.52
Δy(A)	mm		− 0.001	− 0.007	− 0.63	− 0.658
Δx(B)	mm		− 0.09	− 0.102	− 0.44	− 0.44
Δy(B)	mm		0.0001	0.001	− 0.318	− 0.337
Δx(C)	mm	− 0.013	− 0.07	− 0.077	− 0.803	− 0.725
Δx(D)	mm		− 0.078	− 0.086	− 0.964	− 0.899
Collar						
Fc	N		− 373	− 429	− 223	− 1599
ΔFc	N					− 1376
Δy(E)	mm	0.298	0.394	0.397	− 0.006	− 0.048
Coils						
σ(1)	MPa	− 31	− 32	− 32	− 41	− 7
σ(2)	MPa	− 35	− 35	− 35	− 45	− 99
σ(3)	MPa	− 56	− 59	− 60	− 76	− 29
σ(4)	MPa	− 54	− 56	− 56	− 70	− 96
σ(5)	MPa	− 77	− 67	− 66	− 83	− 132
Δx(F)	mm	− 0.0005	− 0.053	− 0.06	− 0.624	− 0.508
Δx(G)	mm	− 0.008	− 0.005	− 0.005	− 0.129	− 0.167
Δφ	mm					− 0.154

shims at the horizontal plane of the coils. Due to the prestressing, the collars are deformed. The interference between the collars and the iron yoke is checked so that the relative displacement of the various parts is not prevented during cool down.

Tapered shims with an angle of ~ 0.25° and 0.13° should allow more homogeneous distribution of the azimuthal stresses in the coil.

A separate model of the collars has been studied to see their behaviour in 3D. Fig. 3 and Fig. 4 show the deformation of collars and the stresses in coils after collaring. The stresses in the collars around the longitudinal rods are high and could reach locally the plastic deformation. This deformation will be taken into account in the determination of the shims between coils and collars. Special care has to be taken when the rods are inserted in the collars to avoid excessive overstressing of the coils during collaring.

Clamping of the Yoke

The two half-yokes are pressed on the collared coils at 370 N and precisely machined clamps are inserted. The

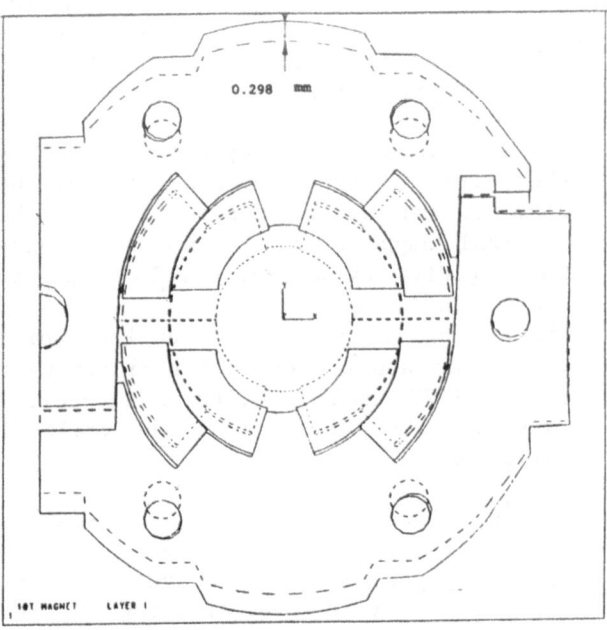

Fig. 4 Deformation of coils after collaring.

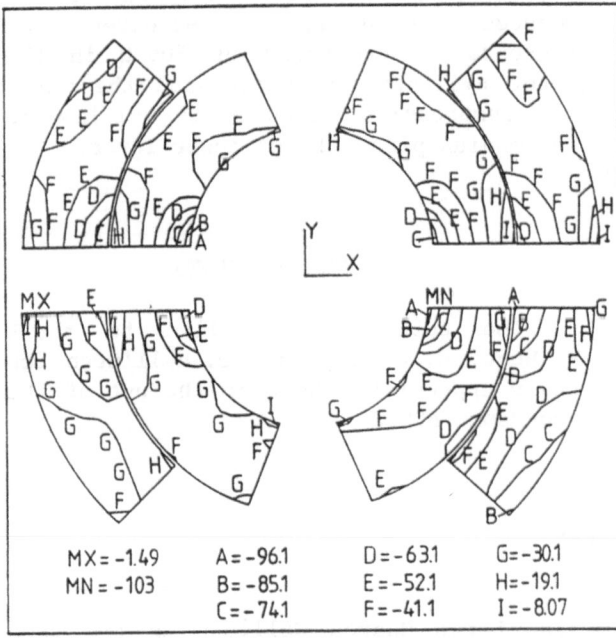

MX = -1.49	A = -96.1	D = -63.1	G = -30.1
MN = -103	B = -85.1	E = -52.1	H = -19.1
	C = -74.1	F = -41.1	I = -8.07

Fig. 5 - Stresses in coils after collaring (N/mm^2)

gap between the half-yokes can then be verified. In this procedure, the collars move by 0.07 mm and the half gap closes by 0.103 mm so that the initial tapered gap of 0.52/0.44 mm becomes 0.42/0.35 mm when the clamp is installed. The hori-

zontal radius of the yoke is reduced by 0.078 mm. The azimuthal stresses in the coils vary slightly.

Prestressing of the outer Cylinder around the Yoke

The required prestressing of 20 MPa in the outer cylinder at room temperature is simulated in Ansys by a shrink-fitting with a ΔT = 35 K, which corresponds to ΔR = - 0.13 mm. The ratio between the stiffness of the collars and the outer cylinder is so that ΔR is shared almost equally between the two elements. The force previously taken by the clamp is then transferred to the outer cylinder and no change occurs in the structure.

Cool-down at 2 K

At the cool-down, the outer cylinder drives the structure so that a mating force of 1505 N appears between the two half-yokes in the vertical plane and a compressive force on the collars of 223 N in the horizontal direction and 485 N in the vertical direction. The mean azimuthal stress in the coils is increased by 15 MPa on the inner layer and 10 MPa on the outer layer. During cool-down, the iron yoke rotates following the vertical contraction of the collar. This rotation enhances the displacement of the gap by 0.13 mm at the top of the gap and by 0.07 mm at the bottom of the gap. After cool-down, the magnet is no longer circular by 0.33 mm in radius. The stresses in collars reach 90 MPa at the upper angle of the outer layer.

Excitation at B = 11.2 T

It can be seen that the outward electromagnetic force (2940 N) is shared nearly equally between collars and iron structure (1325 and 1376 N). The radial deformation in the median plane of the collared coil at excitation is 0.078 mm. With collars only, the displacement would have been 0.289 mm. The displacement of the coils themselves is 0.116 mm for the right side and 0.038 mm for the left side so that the magnetic axis moves by 0.078 mm.

At the excitation of 11.2 T, the iron gap stays closed and, as foreseen, there is practically no change in the outer cylinder stresses.

It can be seen that the coils are still in a compressive state and the maximum stress is 132 MPa on the inner radius at the median plane.

Stainless steel outer cylinder

Calculations have been made when the outer cylinder is made of stainless steel, 12 mm thick. Due to the lower contraction, the prestress in the cylinder has to be increased so that the yoke stays in contact with the collars after cooldown. If the collars were infinitely rigid, the prestress would be ~250 MPa or ΔT = 120 K for the simulation in ANSYS. After assembly of the outer cylinder on the iron and collars, the prestress is 116 MPa at room temperature. When the clamp is installed with a force of 1400 N, the initial tapered half-gap of 0.52/0.44 mm becomes 0.20/0.20 mm. The collars are displaced by 0.2 mm and the stress distribution varies mainly in the inner layer. Compared with the Al alloy case, the compressive stress is increased by 9 MPa at the upper angle (point 3), but is decreased by 16 MPa at the mediam plane-inner radius (point 5). However, radial stresses of 50 MPa appear in the outer layer. After cool-down, the situation is similar to the one with an Al alloy cylinder. At 11.2 T, the stresses in the stainless steel cylinder do not exceed 162 MPa.

CONCLUSIONS

1. The mechanical calculations made with ANSYS on the straight part of the magnet indicate that the mechanical structure of the twin aperture 10T dipole performs well up to 11.2T.

2. The mechanical structure resulting from the assembly of collars/yoke/shrinking cylinder provides a deformation of the collared coils less than 0.08 mm at 11.2 T (0.06 mm at 10 T). The overall rigidity is increased by a factor 3.7 by the effect of the iron and outer cylinder. It has to be mentioned that the vertical support of the iron increases the rigidity by a factor 2.15. The gap stays closed up to a field of 11.8 T. The use of stainless steel collars which would increase the overall rigidity by a factor 1.75 would require a supplementary prestress of 8 MPa in the coils at collaring.

3. The stresses in collars are maximum at room temperature around the rods. The collaring procedure has to be well controlled to avoid overstressing of the coils. Fatigue tests have to be made on the collars.

4. The maximum compressive stress which is 130 MPa at 11.2 T (120 MPa at 10 T) could be reduced at 110 MPa for a magnet working at maximum 10 T if the coil prestressing at room temperature is reduced accordingly. Moreover an increase in prestress of 500 N in the outer cylinder would decrease the maximum compressive stresses by ~ 6 MPa in the median plane of the inner layer.

ACKNOWLEDGEMENTS

The authors would like to thank M. Bona, G. de Rijk and D. dell'Orco for their appreciated help in the use of the programme ANSYS.

REFERENCES

1. Brianti, G., and Hübner, K., (editors), The Large Hadron Collider in the LEP tunnel, CERN-87-05, 27 May 1987.
2. Perin, R., Progress on the superconducting magnets for the Large Hadron Collider, IEEE Transactions on Magnetics, Vol. 24, pages 734-740, March 1988 and CERN LHC Note 63, September 1987.
3. Leroy, D., Perin, R., de Rijk, G., Thomi, W., Design of a high-field twinaperture superconducting dipole model, IEEE Transactions on Magnetics, vol. 24, pages 1373-1376, March 1988 and CERN LHC Note 62, September 1987.
4. Hagedorn, D., Leroy, D., Perin, R., Towards the Development of high-field superconducting Magnets for a Hadron Collider in the LEP Tunnel, Proc. of the 9th Int. Conf. on Magnet Technology, Zurich, September 9-13, 1985.
5. ANSYS, Trademark from SWANSON Analysis Inc, Houston, U.S.A.

DEVELOPMENT OF SUPERCONDUCTING DIPOLE MAGNETS
FOR ACCELERATOR

K. Asano, T. Yamagiwa

Hitachi Ltd., Hitachi-shi, Ibaraki-ken, 317, Japan

and

T. Shintomi

KEK, National Laboratory for High Energy Physics,

1-1 Oho, Tsukuba-Shi Ibaraki-ken, 305, Japan

ABSTRACT

Small bore and high field dipole magnets are required for high energy hardron colliders to minimize cost of the facilities. In accordance with this requirements three 1.4 m - long dipoles with a 40 mm coil inner diameter and a field of 6.6T have been built and tested. The main features of the dipoles are a two layer cos θ superconducting coil, made from NbTi Rutherford type cables, clamped with 1.5 mm thick high manganese steel laminations and a cold laminated iron yoke surrounding the collared coil. The copper to superconductor ratio of inner coil cable, yoke structure and manufacturing method of the coil etc. are intentionally changed in the three magnets to compare the magnet performance.

INTRODUCTION

In order to survey the possibility of small bore and high field dipole magnets for accelerator, as a series of high field dipole development at KEK[1], KEK planned, developed and then arranged for prototype dipole magnets to be constructed by industrial firm. Hitachi Ltd. decided to manufacture three magnets and design and manufacturing conditions were intentionally changed in those magnets to compare the magnet performance.

The outline and some details on the magnet construction and testing are given in the followings.

DESCRIPTION OF THE MAGNET

General Design

A cross section through the coil and clamping structure are shown in Fig. 1. The coil cross section is a two layer cosθ winding with an aperture of 40 mm and an outer diameter of 80 mm. Those coils are tightly clamped by the collar with an outer diameter of 110 mm and are surrounded by a cold laminated iron yoke.

The magnet is designed to operate at a field of 6.6T in 4.2 K liquid helium with a current of 6504 A. The overall length of the coil is 1170 mm.

The main parameters are listed in Table 1.

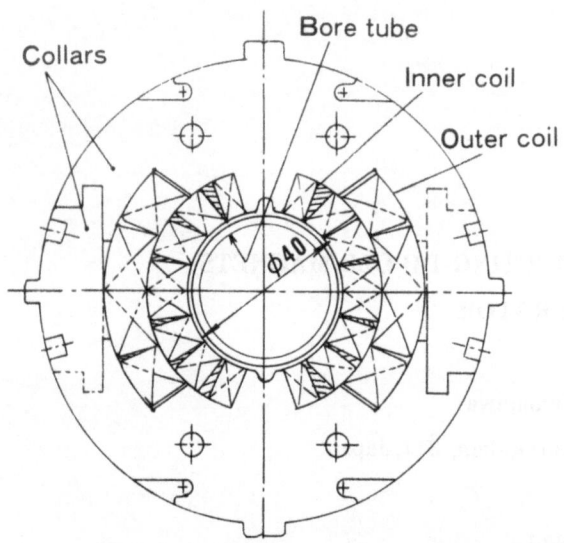

FIGURE 1.   Collared coil cross section

TABLE 1

Magnet parameters

Central field		6.6 T
Current		6504 A
Current density (inner coil)		450 A/mm²
Inductance		4.3 mH
Stored energy		90 KJ
Number of turns	inner	16×2
	outer	20×2
Overall length		1170 mm
Coil straight section length	inner	1000 mm
	outer	969 mm
Coil	inner diameter	40 mm
	outer diameter	79.6 mm
Collars	Materials	High manganese steel
	Thickness	1.52 mm
	Radial thickness	15 mm
Yoke	Materials	Low carbon steel
	Thickness	1.6 mm

Superconductor

Two kinds of keystoned cable of the Rutherford type, each having different dimensions and strands, are used for the inner and outer coil. It is composed of 23 strands of 0.81 mm diameter for the inner coil and 30 strands of 0.65 mm diameter for the outer coil. Each strand contains about 5µm superconducting filaments of high homogeneity NbTi in the copper matrix. The cable compaction factor is near 0.9.

The cable is insulated with Kapton tape and partially wrapped with a fiberglass tape. A first layer of 25µm thick Kapton tape is helically wrapped around the cable with half lap and a second layer of 0.1 mm thick fiberglass tape is also wrapped on the first layer with 10% gap. The second layer fiberglass tape is covered on the outside surface with B stage epoxy resin.

The main parameters of the cable are listed in Table 2. Each conductors used for three magnets have different characteristics on filament diameter, copper to superconductor ratio and critical current performance.

TABLE 2

Conductor parameters

	Inner	Outer
Cable width	9.30 mm	9.73 mm
Cable mid thickness	1.48 mm	1.17 mm
Keystone angle	1.6°	1.2°
Number of strands	23	30
Strand diameter	0.81 mm	0.65 mm
Filament diameter	5 µm	5 µm
Copper/super ratio	1.3~1.5	1.8
RRR of stabilizing copper	≧70	≧65
Cable twist pitch	79 mm	74 mm
Critical current (at 4.2K)	7167 A at 7 T	7860 A at 5.6 T

Coil and clamping structure

The two inner and two outer coil segments are wound separately on a winding machine and then they are cured to precise dimensions under relatively high temperature and pressure. Four coils are assembled around a 316 L stainless steel bore tube of 32 mm inner diameter. Kapton and Teflon sheets are inserted between the inner and outer coil layers and Kapton sheets are inserted at the mid-plane as well. Some G-10 spacers are settled on the bore tube in order to keep the accurate coil location and define helium cooling passages. Simple model of sextupole correction coils of 300 mm length are settled at the lead end side between the coil and the bore tube. Those are made of NbTi single wire of 0.3 mm diameter.

Structural support of the coils is provided by interlocking collars, 15 mm in radial thickness. Nonmagnetic high manganess steel (KHMN-30L)[2] is adopted as a collar material because of extremely low and stable magnetic permeability at liquid helium temperature and against deformations. However there is a difference in thermal contractions between coils and collars. In particular the integrated contraction of KHMN-30L is relatively smaller than that of stainless steel. Therefore, It is necessary to take into account the above when the collars are fastened together by means of 316LN stainless steel keys while the pressure is applied, so that the applied prestress remained the coils after the press is removed.

Cable connections between the inner and outer coil are made in the collar structure and cable connection between upper and lower coils are made at the outside of collars with oxygen-free copper block as a stabilizer.

External view of the completed collared coil is shown in Fig. 2.

Yoke and helium vessel

Sprit iron yoke of 267 mm diameter made of 1.6 mm thick low carbon steel laminations is equipped in this dipole magnet. This provides the magnetic function of a flux return path and acts as a shield to minimize the stray magnetic field. Helium vessel is formed by means that stainless steel support shells of 5 mm thickness are covered and welded outside of the yoke.

FIGURE 2.   Completed collared coil

MAGNET CONSTRUCTION

As this dipole is designed to operate near a conductor short sample critical current, it is important to study the relation between features of the magnets and quench performance. Some of the important valuable things in determining the number of quenches of the magnets are conductor performance, especially the ratio of copper to NbTi ratio in the inner coil conductor, collar key structure as a clamping technique, yoke end structure and the strength of the coil ends. Those factors are intentionally changed in the three magnets to compare the training performance and study those effects as shown in Table 3. Common features in this series of the magnets are: collar configurations except for key slot; coil end configurations; sliding collared coil assemblies; coil splice.

Superconducting cables for the inner coil with different ratio of copper, 1.6, 1.4 and 1.3, are used for the three magnets, respectively. Superconducting cables manufactured by Hitachi cable Ltd. are used for the first and second magnets and that manufactured by Furukawa Electric Co. is used for the last one. Tapered key structure of the collar is adopted and there are magnetic void in the yoke ends partially for the second and the last magnets. It is also important to mold the coil into a precise shape and high mechanical rigidity in order to avoid the disturbance caused by wire movements.

Actually the coils were molded to a fixed size with the maximum pressure applied during molding varied due to within-tolerance variation of the size of the materials used in the coil. The coils were cured in the pressure range 10-15 kg/mm^2, 4.5-8 kg/mm^2 and 6-8 kg/mm^2 for the first, second and last magnets, respectively.

The Young's modulus and the azimuthal size of the molded coils were measured. The Young's moduli of the inner coil ranged between 646-699 kg/mm^2, 632-941 kg/mm^2 and 822-1186 kg/mm^2, and also the oversizes of the coils, measured with an applied pressure of 5.5 kg/mm^2, were distributed between 0.20-0.32 mm, 0.57-0.65 mm and 0.17-0.45 mm, for the first, second and last magnets, respectively. The Young's modulus of the outer coils around 500 kg/mm^2, measured with an applied pressure of 4.5 kg/mm^2, were slightly lower than that of the inner coils.

The thickness of G-10 collaring shims, located between collars and coils, were adjusted to obtain the prescribed azimuthal coil prestress in accordance with those measured values.

TABLE 3.   Major difference between three dipoles

Magnet	Superconducting cable 1) Manufacturer 2) Cu: SC (Inner) 3) Critical current	Collar key structure	Yoke end structure	Coil end curing temperature
No. 1	1) Hitachi Cable Ltd.	Straight (Rectangular)	No void	130 °C Room temperature (additional)
	2) 1.59 : 1			
	3) 6946A at 7T			
No. 2	1) Hitachi Cable Ltd.	Tapered	Void (partially)	130 °C
	2) 1.41 : 1			
	3) 7406A at 7T			
No. 3	1) Furukawa Electric Co.	Tapered	Void (partially)	130 °C
	2) 1.30 : 1			
	3) 12600 A at 5T 4970 A at 8T			

MAGNET PERFORMANCE

Excitation test

After assembling of the magnets and tested in a vertical cryostat. A thyristorized power supply and the quench protection resistor of 50 mΩ was provided. The threshold level of the quench detection circuit was set for 500 mV for at least 0.5 sec. After the magnets were cooled down with liquid nitrogen and liquid helium, they were excited under the condition of the maximum current of 6800 A limited by test facilities.

Training history

The training performance and reached current of the three magnets, in liquid helium at 4.2 K, are shown in Fig. 3. there is a big difference obviously between the first magnet and the other magnets. The second and last ones exhibited similar behavior except the first quench current.

The first quenches of the magnets occurred at a current that is lower than the design value and its value seems to be improved with the increased magnet number. The second and the last one reached the design current after one training quenches, where the maximum field on the conductor was about 7 Tesla and the stored energy was 90 KJ. On the other hand, it took eight training quenches for the first one to reach the design current.

During the test, the voltage signals all the coils were monitored in order to determine which coil initiated quench. All quenches excepting the first quench of the first magnet, which occured in the outer coil, occured in the inner coils.

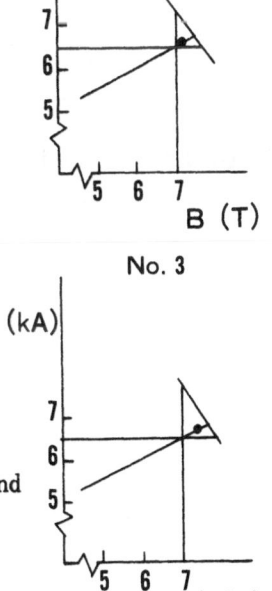

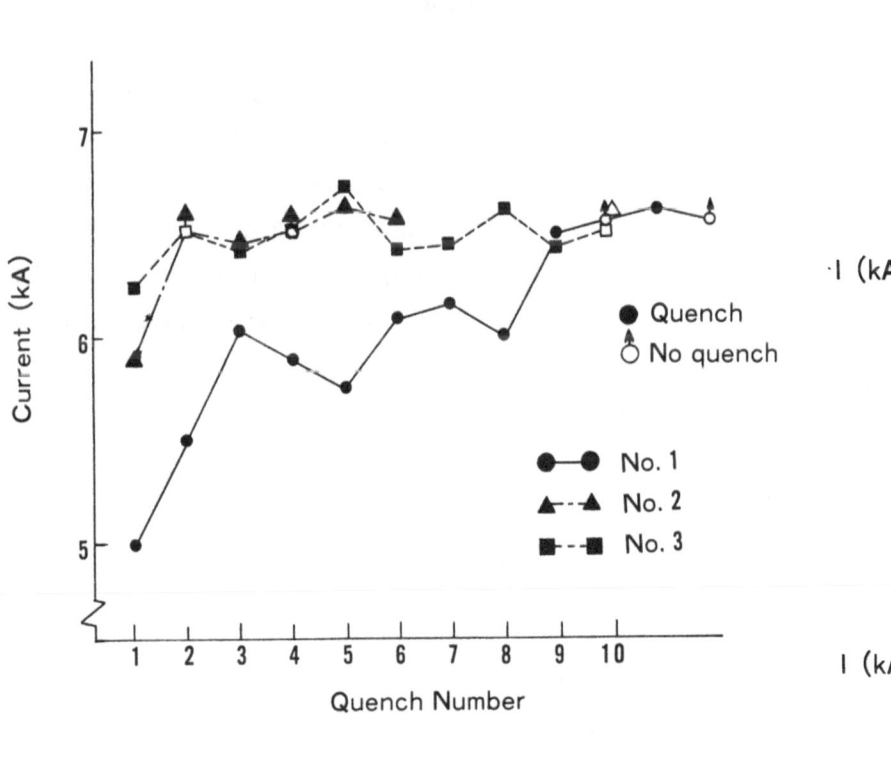

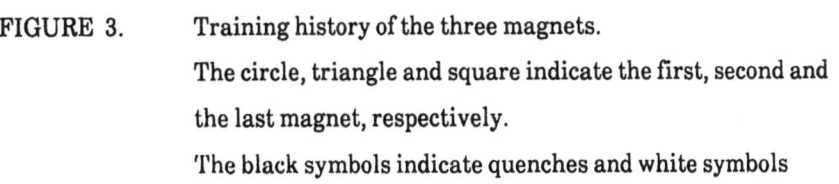

FIGURE 3.   Training history of the three magnets.
The circle, triangle and square indicate the first, second and the last magnet, respectively.
The black symbols indicate quenches and white symbols indicate no quenches.

Other properties

In this dipole magnet design, it is recognized that the magnet coils should be tightly clamped and compressed to minimize conductor motion under the action of the Lorentz forces during magnet excitation. For example in the azimuthal direction, this phenomena represents a reduced collar stress, when the coil is excited. It is important to know if the coils are still being held under compression at operation.

In order to monitor the azimuthal collar stress, straingauges were installed in the center collar block of the third magnet. In the case of inner pole of the collar, measured stress is reduced in proportion to the increase of the excitation current as shown in Fig. 4. This shows that the coil were still being held under compression at the design current.

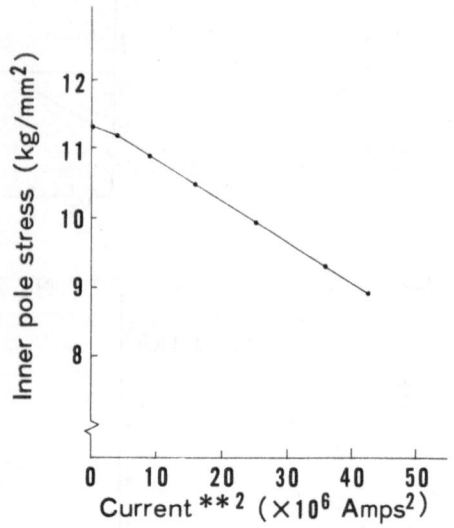

FIGURE 4.   The inner pole stress of the collar versus square of the current

CONCLUSION

To survey the possibility of small bore and high field dipole magnet, we successfully assembled and tested the three magnets and obtained the relation between some features of the magnets and quench performance. The three magnets trained and reached the design current and quench performance was improved in the second and the last ones. As a result of this development, the design and the manufacturing method used for those dipole is suitable and shows the construction possibility of small bore and high field accelerator dipole magnet.

ACKNOWLEDEMENT

The authors would like to express their gratitude to Professors T. Nishikawa, S. Ozaki and H. Hirabayashi for their continuous encouragement and support for this work.

REFERENCES

1.   Shintomi, T et al., The construction and test results of a 10T dipole. IEEE Trans. Nucl. Sci., NS-32, 1985, 3181

2.   Kawasaki Steel Co., Test results of KHMN trial products. (Private communication)

SUPERCONDUCTING QUADRUPOLE MAGNETS FOR THE TRISTAN LOW-BETA INSERTION

K. Tsuchiya, N. Ohuchi, A. Terashima, H. Nakayama, K. Egawa, S. Takasaki
KEK, National Laboratory for High Energy Physics,
1-1 Oho, Tsukuba-shi, Ibaraki-ken, 305, Japan
and
K. Asano, R. Hoshi, T. Yamagiwa
Hitachi Ltd., Hitachi-shi, Ibaraki-ken, 317, Japan

ABSTRACT

This paper describes the construction and test of the first pair of industrially fabricated superconducting quadrupole magnets intended for use in the TRISTAN low-beta insertions. These magnets are iron-free with dimensions of 140 mm coil inner diameter, 280 mm collar outer diameter and 1450 mm physical length. The design field gradient is 70 T/m and the good field aperture is 80 mm in diameter. During the first test in a vertical cryostat, these magnets achieved their design currents of 3405 A after only one training quench and reached 4000 A after three quenches. The quench characteristics and field quality of both magnets are very similar.

INTRODUCTION

In order to increase the luminosity of the TRISTAN electron-positron collider[1], eight high field-gradient superconducting quadrupole magnets, two quadrupoles for one collision point, are required. For this purpose, KEK has been developing an iron free quadrupole with a field gradient of 70 T/m. The fourth prototype made in KEK reached a short sample critical current after four training quenches and its field quality was good enough to use in the storage ring. Subsequently, KEK arranged for two quadrupole magnets to be constructed by an industrial firm, where KEK provides all scientific and technical know-how about the design and the construction method while the company does the actual fabrication. In the following section, we review the quadrupole design and construction, and report on the performance of industry-made magnets.

DESCRIPTION OF THE MAGNET

General design

A cross section of the collared coil is shown in Fig. 1. The coil cross section is a four-layer two-wedge design with an aperture of 140 mm and an outer diameter of 217.7 mm. The inner diameter of the collar is 218.7 mm; its outer diameter is 280 mm. The magnet is designed to operate at a field gradient of 70 T/m in 4.7 K single phase liquid helium[2] with a current of 3405 A. The effective length of the magnet is 1.17m; the physical length is 1.45 m. Using a two-dimensional field calculation based on an analytical formula[3], wedge positions that minimize the allowable higher harmonics were determined. The mean thickness of the conductor used for the design was 1.44 mm; this was determined by measuring a stack of 22 samples of insulated cable placed under the standard applied pressure of 3kg/mm^2.

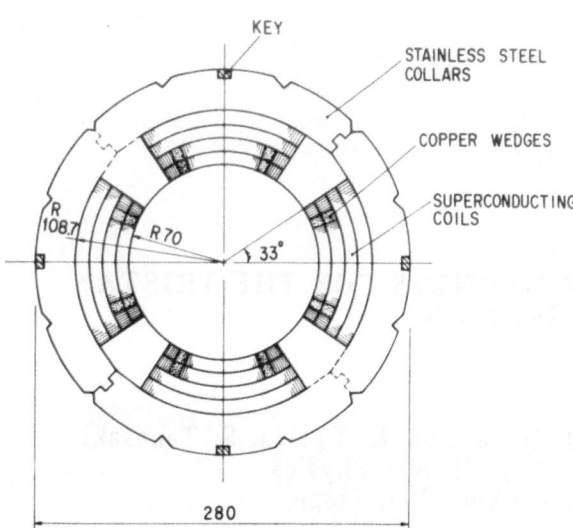

FIGURE 1. Collared coil cross section.

Special features of the design are as follows: (1) precisely machined fiber-reinforced-plastic (FRP) end spacers are used in order to make the end tight and to reduce the training; (2) double pancake winding method is adopted to reduce the number of electrical joints between the coils; (3) two kinds of cable, each having different cable lay directions, are used for the winding in order to avoid a twist of the assembled coil; a left-handed corkscrew cable was used for odd pole windings and a right-handed corkscrew cable was used for even pole windings. The main parameters of the quadrupole are listed in Table 1.

Conductor

The conductor is a keystoned cable of the Rutherford type as has been used in many accelerator magnets. It is composed of 27 strands of 0.68 mm diameter, each strand contains thin superconducting filaments of high homogeneity NbTi and is coated with 1 μm silver-tin solder. The ratio of copper to superconductor is 1.7 and the residual resistivity ratio of the stabilizing copper is about 180. The cable compaction factor is 0.89 so that 11 % of the cable area is expected to be filled with liquid helium when the magnet is cooled. For the right-handed corkscrew cables we used strand wire with a left-handed twist; for the left-handed corkscrew cables we did the opposite.

The insulation consists of a double layer of 25 μm Kapton foil and a layer of 50 μm Kapton with about 25 μm of B-stage epoxy on the outside surface. The 25 μm Kapton tape (12 mm width) is wrapped on the cable with a 50 % overlap and the 50 μm tape is wrapped on it with 1.25 mm gaps between adjacent layers.

The main parameters of the cable are listed in Table 2. The critical currents of the cable were measured at both Hitachi Laboratory and Brookhaven National Laboratory; the difference between the two measurements was about 2%. The critical currents of the cable (at a resistivity of 1 x 10^{-12} Ω cm) are shown in Table 2 and Fig. 4.

TABLE 1

Parameters of Superconducting Quadrupole

Field gradient	70	T/m
Current	3405	A
Overall length	1450	mm
Magnetic length	1170	mm
Inductance	58	mH
Stored energy	336	kJ
Field uniformity		
$\Delta B/B_2$	$\leq 5 \times 10^{-4}$	
Max. field on the conductor	6	Tesla
Coil (4-layer)		
Inner diameter	140	mm
Outer diameter	217.7	mm
Collars		
Materials	SUS316LN	
Radial thickness, nominal	30	mm

TABLE 2

Conductor Parameters

Cable dimensions		
height	9.09 ± 0.03	mm
small width	1.19 ± 0.01	mm
large width	1.35 ± 0.01	mm
Number of strands	27	
Strand diameter	0.683 ± 0.001	mm
Number of filaments	2264	
Filament diameter	8.7	μm
Copper/super ratio	1.7	
RRR of stabilizing copper	180 ± 20	
Cable twist pitch	70 ± 5	mm
Critical current in cable (at 4.2K)		
	8160 A	at 5T
	6425 A	at 6T
	4935 A	at 7T

Coil production

Eight separate coils, each consisting of a two-layer double-pancake winding, were made for each magnet. The coil production procedure was similar to that reported previously[4]; the first layer of the coil was wound on a convex mandrel by inserting a copper wedge and FRP end spacers in the appropriate positions and heat cured in a press at about 130 °C. After removal from the press, the second layer was wound onto the first layer in a similar manner, with an 0.5 mm thickness fishbone-type G-10 spacer between the first and second layers. The resulting two layer coil was then cured with an oversize of 0.4 mm inserted in the press. The third and fourth layer coils were made by the same process and oversized by 0.3 mm. In each curing process, an axial pressure was also applied to make the coil tight and to produce an uniform coil length. In Fig. 2 the deviations of the length of each layer from the design values are shown for both quadrupoles, H1 and H2. The Young's modulus and the azimuthal size of the cured coils were also measured. The Young's moduli ranged between 300~400 kg/mm^2 and the oversizes of the coils, measured with an applied pressure of 3 kg/mm^2, were distributed between 0.1 and 0.28mm.

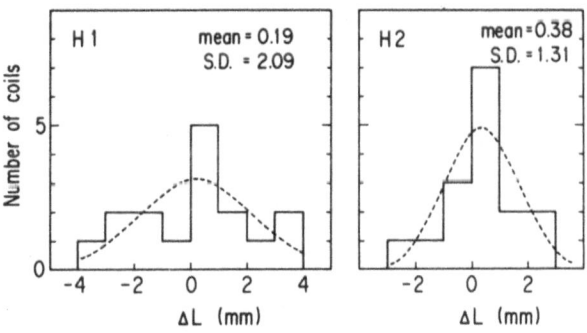

FIGURE 2. Histograms of the distribution of the deviations of the cured coil lengths from their design values.

Coil assembly

The eight double pancake coils were mounted around another cylindrical mandrel. This mandrel was not laminated since it had to be able to slide along the coil surface during removal after collaring. Ground insulation, made of two layers of 250 μm thick Mylar sheet, was placed on the outer surface of the coils over which

316LN stainless steel collars were stacked. The stacking was done so that two adjacent pairs of laminations were mounted perpendicular to each other so that two opposite windings were alternatively held by the central keys of the laminations. The collared magnet was then put into a collaring press which applied the correct pressure and displacement on the collars from four perpendicular radial directions. The press exerted a typical radial force on the collars of 7.0×10^5 kg. The collars were fastened together by means of 316LN stainless steel keys while the pressure was applied, so that the applied prestress remained on the coils after the press was removed. The mandrel was then pulled out from the collared coils and the electrical connections between the coils were established by means of a special soldering jig.

PERFORMANCE OF THE MAGNET

Training history

After the assembling of the magnets in a factory, they were transported to KEK and tested in a vertical cryostat. For the test, a transistorized power supply and a SCR switch were connected to the magnet. The quench protection resistor was 100 mΩ and the threshold level of the quench detection circuit was set for 1 V for at least 10 msec. During the test, the magnet was excited with a ramp rate of about 10 A/sec. The magnets were trained until the current reached 4000A, the maximum current guaranteed by the manufacturer.

The training performance of the two quadrupoles, in liquid helium at 4.2 K, is shown in Fig. 3. Both magnets exhibited very similar behavior. The first quenches of the magnets occurred at a current that is slightly lower than the design value but, in both cases, the second quenches were well above it. Both magnets reached 4000A after four training quenches, where the maximum field on the conductor was 7 Tesla and the stored energy in the magnet was 464 kJ.

The load line of the quadrupole and critical current of the superconducting cable are shown in Fig. 4.

During the test, the voltage signals of all the double pancake coils were monitored

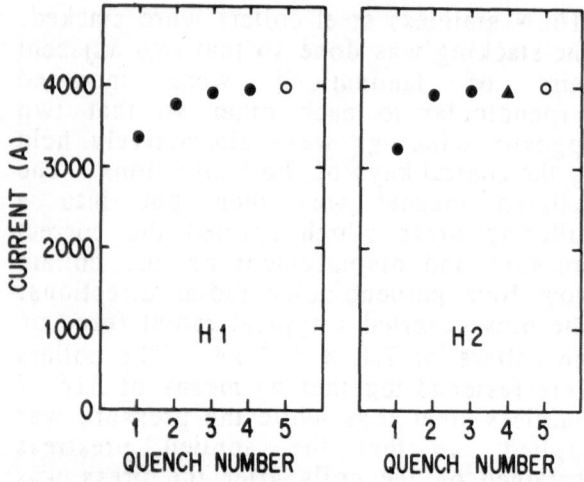

FIGURE 3. Training history of the two quadrupole magnets. The black circles indicate inner coil quenches, the white circle indicate no quenches and the black triangle indicates an outer coil quench.

in order to determine which coil winding initiated the quench. In the case of first quadrupole, H1, all the quenches occurred in the inner coil windings (1st and 2nd layer coils) of the magnet. In the second quadrupole, H2, the first three quenches were in inner coils but the fourth quench was in an outer coil (3rd and 4th layer coil).

Field quality

Field quality measurements were made with a 1.5 m long room temperature rotating coil of 39.43 mm radius inserted in the bore of the magnet. The integrated harmonic content is expressed as the ratio of the harmonic field strength to the quadrupole field at the radius of the measuring cylinder. The reproducibility of the measurements was 1×10^{-4}. Table 3 summarizes the allowed and unallowed harmonics, given by the integrated field strength ratio. As can be seen in the table, the current dependence of the multipoles is small, indicating that the mechanical support system of the quadrupole is withstanding the large electro-magnetic bursting forces of 9.4×10^4 kg/m and the coils are clamped with sufficient precompression. Only the normal dodecapole, B_6, displays the well known hysteresis effects caused by persistent currents at low magnet excitation, as is indicated in Fig. 5. The curves of two magnets are very similar; the slight shift between them may be due to a small difference in the coil cross section. The data indicate that the deviation from homogeneity, $\Delta B/B_2$ in the median plane and at the reference radius of 40 mm, was less than 3×10^{-4}.

The final field measurement, with the magnet in a horizontal cryostat and cooled to 4.7 K, are not yet complete. Preliminary

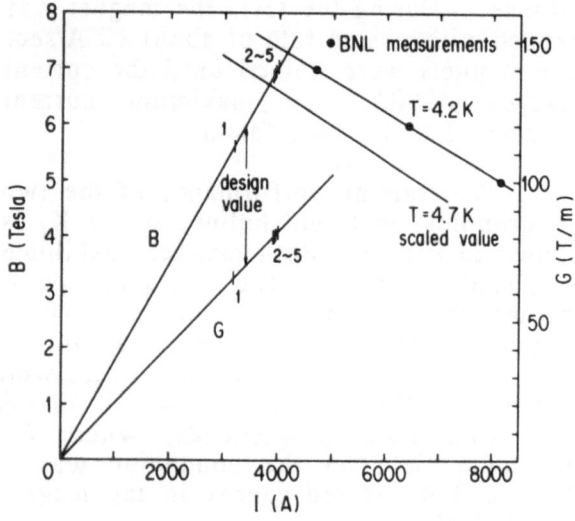

FIGURE 4. Critical currents of the superconducting cable compared to the load line of the quadrupole. The number on the load line shows the quench points of the second magnet, H2.

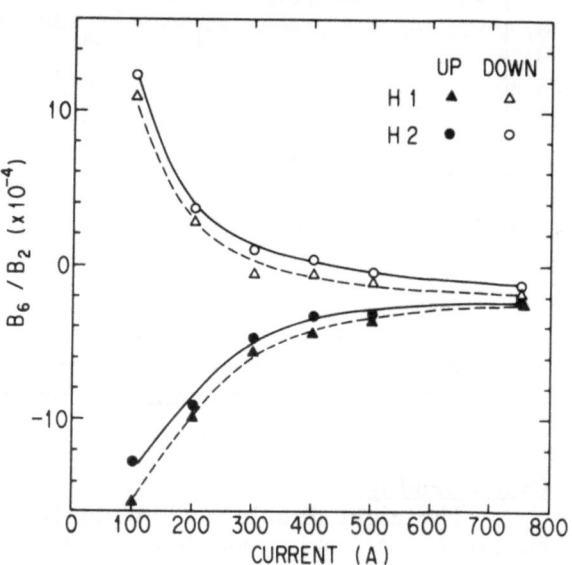

FIGURE 5. The hysteresis of the 12-pole component due to persistent currents in the 8.7 μm thick NbTi filaments.

TABLE 3
Harmonic content of the integrated magnetic field ($B_n/B_2 \times 10^4$).

	H1				H2			
	1000A	2000A	3000A	3500A	1000A	2000A	3000A	3500A
3 N	1.47	2.21	1.47	2.35	0.61	0.92	1.38	0.65
S	-6.08	-6.36	-6.48	-6.68	-1.76	-1.55	-1.41	-1.28
4 N	-1.43	-1.87	-2.38	-1.93	-0.11	-1.59	-0.25	-0.68
S	4.03	4.38	4.30	4.49	4.67	4.15	4.35	4.56
5 N	0.60	0.15	-0.34	0.63	0.88	0.81	0.48	0.02
S	-1.24	-0.69	-0.85	-0.13	-0.80	-1.04	-1.69	-1.20
6 N	-2.80	-2.66	-1.90	-1.89	-0.89	-1.11	-1.29	-0.82
S	5.88	0.95	0.53	0.72	-0.08	0.51	1.18	0.37
7 N	0	0	0	0	0	0.33	-0.63	0
S	0	0	0	0	0	0.37	-0.40	0
8 N	-0.89	-1.00	-0.65	-0.54	-0.31	-1.89	-0.26	0.42
S	1.05	0.71	0.65	0.57	1.40	0.41	-0.70	-0.16
9N	0	0	-0.09	-0.19	0.79	-0.38	0.44	-0.44
S	0	0	0.52	-0.41	-0.40	-0.78	-0.80	-0.65
10 N	-0.79	-0.50	0.04	0.19	-0.71	-0.22	-0.87	-0.03
S	-0.03	-0.50	0.74	-0.61	-0.35	0.45	0.28	0.34

data give a transfer function for both magnets of 0.0204 T/m/A and the effective magnetic length of 1.174 m. The difference of these values between the two magnets were within 1×10^{-3}.

CONCLUSION

The two quadrupoles constructed by Hitachi Ltd. reached the design current of 3405A with one quench and 4000A after three quenches. The field homogeneity is within a few 10^{-4} for most of the multipoles inside the useful aperture. Both magnets show very similar characteristics, indicating that the design and the manufacturing method used for the magnets are suitable for mass production. Production of another six quadrupoles is now proceeding at Hitachi Ltd. and will be completed by April 1990.

ACKNOWLEDGEMENT

We would like to express our gratitude to Professors Y. Kimura, H. Hirabayashi and K. Endo for their continuous encouragement and support. We are very grateful to Dr. W. Sampson and his group of Brookhaven National Laboratory for measuring the critical current of the cable. We would like to thank KEK machine shop group for their enthusiastic support.

REFERENCES

1. Isagawa, S., TRISTAN achievements and future plans. KEK Preprint 88-9 May 1988.

2. Tsuchiya, K., Ohuchi, N., Terashima, A. and Shinkai, K., Cryogenic system of the superconducting insertion quadrupole magnet for TRISTAN main ring. to be published in Advances in Cryogenic Engineering, Vol. 35.

3. Halbach, K., Field and first order perturbation effects in two-dimensional conductor dominated magnets, Nucl. Instrum. Methods., 1970, 78, 185.

4. Tsuchiya, K., Hosoyama, K., Ajima, Y., Egawa, K., Kumagai, N., Mitsunobu, S., Sakakibara, Y., Terashima, A. and Ogitsu, T., A prototype superconducting insertion quadrupole magnet for TRISTAN. Advances in Cryogenic Engineering, Vol. 31, Plenum Press, New York, 1986, 173.

TRAINING IN HIGH-PERFORMANCE SUPERCONDUCTING DIPOLE MAGNETS AND ROOM-TEMPERATURE ACOUSTIC SIGNAL ATTENUATION*

O. O. Ige[†] and Y. Iwasa[†]

Plasma Fusion Center

Massachusetts Institute of Technology, Cambridge MA 02139

ABSTRACT

High-performance superconducting dipole magnets such as those used in accelerators are susceptible to premature quenches due to mechanical disturbances, chiefly conductor motion and epoxy cracking. If indeed mechanical disturbances are the major cause of quenches, we can expect the structural integrity of a magnet to affect the 4.2-K performance of the magnet. We have measured the longitudinal attenuation of suitable acoustic signals in ten R-series Fermi research dipole magnets. We found that attenuation thus measured correlates well with the 4.2 K performance of the magnets: the higher the attenuation in a particular magnet, the greater the number of quenches it took to train the magnet. A simple explanation, based on the assumption that the primary premature quench-inducing mechanism in this class of magnets is frictional sliding of the conductor, is presented. This technique has the potential for use as a pre-4.2 K test screening of high-performance dipoles.

INTRODUCTION

Training has been a subject of much research since the early days of superconducting magnets [1]. The primary premature quench-inducing mechanisms in superconducting magnets, particularly high-field and high-current-density types, such as dipoles and quadrupoles used in particle accelerators, have been identified as frictional slip of the superconductor and epoxy debonding in epoxy-impregnated windings [2]. Earlier studies have shown that mechanical disturbances are the primary cause of premature quenches in this class of magnets [3].

The technique of evaluating the structural integrity of materials by acoustic methods is not new. Known variously as *pulse-echo*, *through line transmission*, or *ultrasonic evaluation*, these techniques were first used for detecting flaws in components; their use was later extended to characterizing material properties [4]. Briefly, the method consists of the generation of acoustic pulses in a series of specimens containing the range of material properties of interest, and the measurement of selected propagational acoustic parameters, usually attenuation and velocity. Correlations are then sought between material property and the selected acoustic parameter.

The motivation for the study presented here stemmed from our understanding of premature quenches. If, as widely believed, mechanical disturbances are indeed responsible for premature quenches (and hence training), then our aim is to identify mechanical properties which, when mea-

* This work was supported by the Division of High Energy Physics of the U.S. Department of Energy.

† Also the Department of Mechanical Engineering.

sured at room-temperature, will give us an indication of the relative performance of the magnets at 4.2 K. To this end, we measured the longitudinal attenuation characteristics of ten R-series, 0.86-m long, 7.5-cm bore Fermi research dipole magnets that had been previously quench-tested at 4.2 K. Other than in length, these magnets were very similar to the full-length Tevatron model. The cross-section is as shown in Fig. 1. They all had NbTi coils with copper-to-superconductor ratio of 1.8 (or 1.3 for a few). All but one had welded stainless-steel collars. The only exception, RG1001, had aluminium collars.

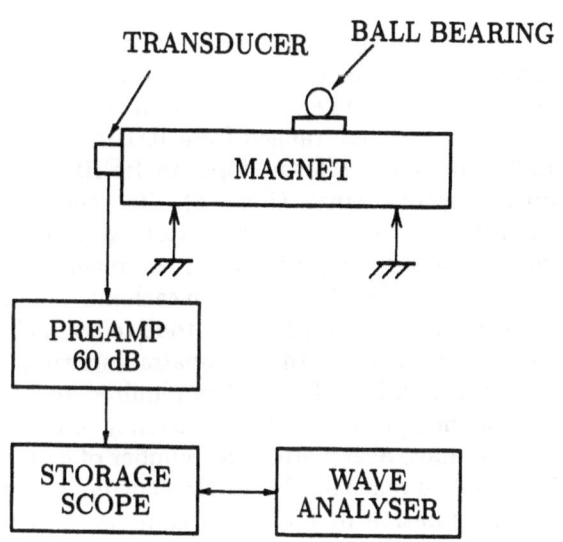

FIGURE 2. Block diagram of experimental set-up.

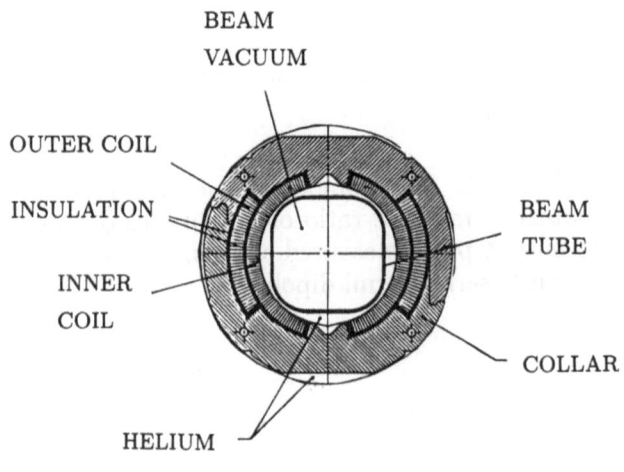

FIGURE 1. Fermi dipole cross-section.

positions were reversed.

The magnet is excited by dropping a steel ball bearing from a fixed height onto the collars. Amplitude and frequency data are then measured with the storage oscilloscope. A typical pulse is shown in Fig. 3.

EXPERIMENT

A schematic of the experimental arrangement is shown in Fig. 2. It consists mainly of a piezoelectric transducer, the output of which is amplified by 60 dB and then displayed on a storage oscilloscope. The oscilloscope also has software for spectral analysis.

In principle, the aim is to generate a pulse at one point on the magnet and measure the output (and hence attenuation) at various longitudinal positions. However, this is difficult to implement in practice because of the laminated collars which render coupling between them and the sensor uneven. Instead, the sensor is mounted at a fixed location (on the end plate) while the source is moved along the magnet. By the principle of reciprocity, the measured attenuation should be the same as though the source and sensor

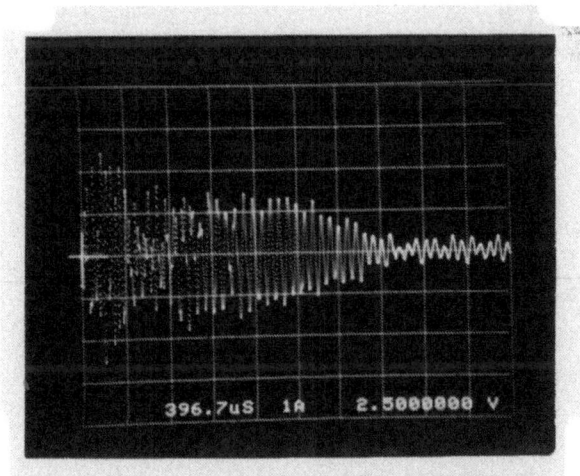

FIGURE 3. Oscillogram of a ball-bearing impact on dipole RD1001. The source is 5 cm from the receiver. Vertival sensitivity: 2.5 volts per division. Horizontal sensitivity: 400 μs per division.

RESULTS

In all, the amplitude ~ distance variation of ten R-series magnets were measured. We observed a wide range of variation in the attenuation curves. They ranged from RD1001, with a high, virtually uniform slope, to RF1001, with an asymptotic curve (Fig. 4). To relate such room-temperature-measured attenuation curves with the 4.2 K performance of the magnets, we have plotted the attenuation in each magnet versus the number of quenches it took to train (Fig. 5). Observe that, as the attenuaton increases, so does the number of quenches required to train the magnet. To check the consistency of this result, we have also plotted the number of quenches it took to reach 90 % of critical current against the attenuation in Fig. 6. Again, we find the same correlation.

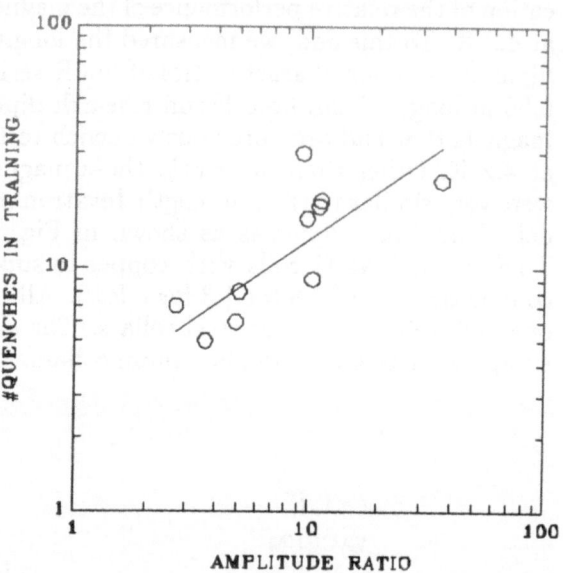

FIGURE 5. The number of quenches to reach plateau versus the ratio of amplitudes of input to output pulse measured at room temperature in ten R-series Fermi dipoles.

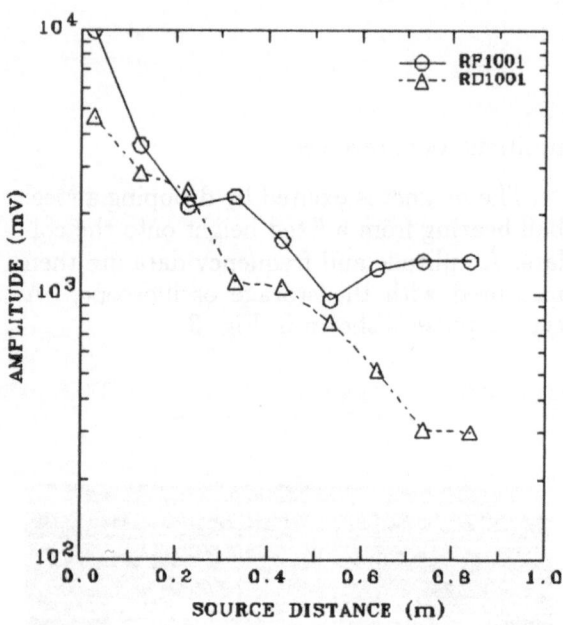

FIGURE 4. Longitudinal attenuation curves of a *good* (RF1001) and a *bad* (RD1001) magnet.

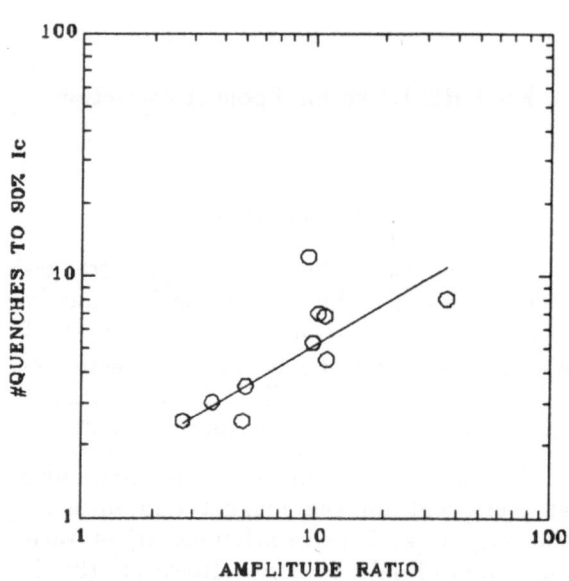

FIGURE 6. The number of quenches to 90 % of critical current versus the ratio of amplitudes of input to output pulse measured at room temperature in ten R-series Fermi dipoles.

DISCUSSION

In order to understand these results better, we conducted similar experiments on the different components that make up a collared coil assembly: collar packs and uncollared coils, this time using one of the SSC dipoles being assembled at Brookhaven National Laboratory (BNL). Although different in size, the SSC dipoles are similar in many ways to the Fermi dipoles. Note, from the plots in Fig. 7, that attenuation is high in the collar packs and low in the uncollared coils, with that for the collared coil somewhere in between. Although there are two coils, for simplicity in this discussion, we shall treat them as one.

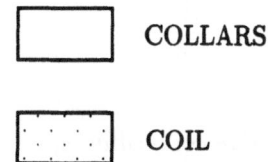

(a) "Full contact" interface

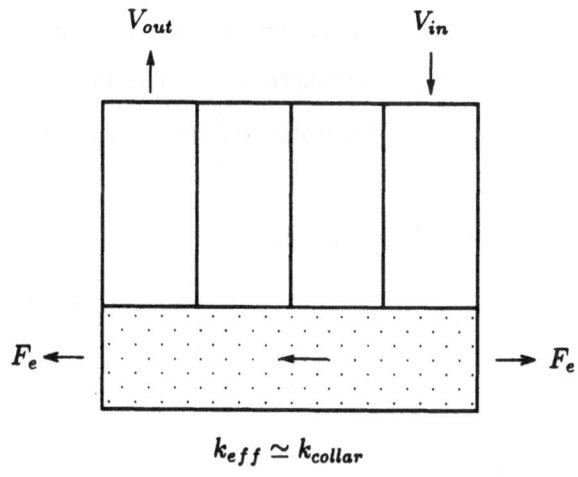

$$k_{eff} \simeq k_{collar}$$

(b) "No contact" interface

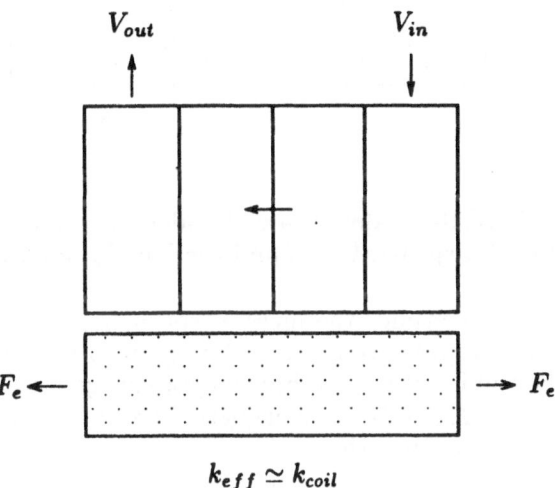

$$k_{eff} \simeq k_{coil}$$

FIGURE 8. Illustration of how collars ~ coil coupling affects effective coil stiffness. (a) perfect coupling, (b) poor coupling.

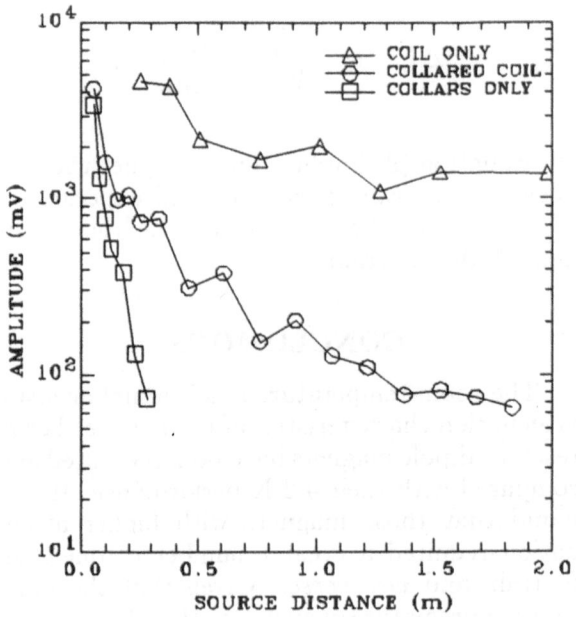

FIGURE 7. Amplitude versus source distance curves for the component elements of a collared coil compared with that for the collared coil assembly.

A diagrammatic explanation of what is happening is shown in Fig. 8. We believe that the measured attenuation for a given magnet depends largely on the effectiveness of the coupling between the collars and the coils. Figure 8 (a) illustrates the case of perfect coupling between the collars and the coils. In this case, acoustic waves generated in the collars are immediately transmitted to the coils where they travel with little attenuation before re-emerging in the collars at the end to be picked up by the transducer. The measured attenuation in this case will be low, reflecting mainly that of the uncollared coil. Noting that the collars are much stiffer than the coils, the effective stiffness of the coils, under axial electromagnetic force F_e will be close to that of the coil. Now consider the other extreme case in Fig. 8 (b), where coupling is so poor that the coil is practically isolated from the collars. In this case, the wave's only path is through the collars; the attenuation will be close to that of the collar pack, and the coils' effective stiffness to that of the uncollared coils. Assuming a linear stiffness ~ attenuation profile, these statements translate into the following equation:

$$\frac{k_{eff} - k_{coil}}{k_{collar} - k_{coil}} \approx \frac{A_{meas} - A_{collar}}{A_{coil} - A_{collar}} \quad , \quad (1)$$

where

k_{eff} is the effective coil stiffness,

k_{coil} is the intrinsic coil stiffness,

k_{collar} is the intrinsic collar stiffness,

A_{coil} is the attenuation in an uncollared coil,

A_{collar} is the attenuation in a collar pack, and

A_{meas} is the measured collared-coil attenuation.

This can be simplified to

$$k_{eff} \sim 1 - \frac{A_{meas}}{A_{collar}} \quad . \quad (2)$$

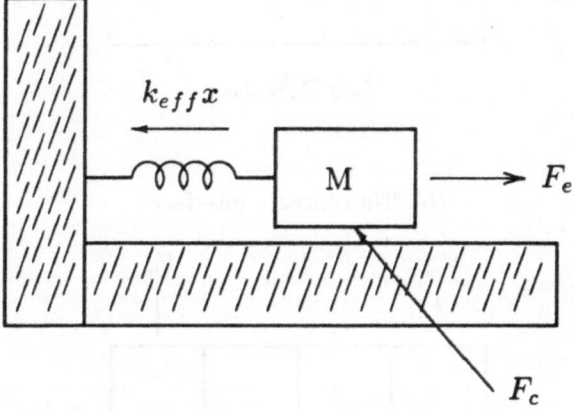

FIGURE 9. Model for a coil under electromagnetic, F_e, spring, $k_{eff}x$, and friction, F_c, forces.

Now consider the mass $\sim$ spring system in Fig. 9 representing a coil during energization. The coil is subject to three forces: spring, static friction, F_c, and electromagnetic force, F_e. Force balance requires that

$$F_e = k_{eff}x - F_c \quad (3)$$

or

$$\Delta F_e = k_{eff}\Delta x \quad . \quad (4)$$

In the vicinity of the critical surface, the energy of a slip event required to initiate a quench will be some value

$$\Delta E = \frac{1}{2}k_{eff}(\Delta x)^2 \quad . \quad (5)$$

Eliminating Δx from equations (4) and (5) gives

$$\Delta F_e = \sqrt{2k_{eff}\Delta E} \quad , \quad (6)$$

that is,

$$\Delta F_e \sim \sqrt{k_{eff}} \quad . \quad (7)$$

Since $F_e \sim I^2$, where I is the current in the coils,

$$\Delta I \sim \sqrt{k_{eff}} \quad . \quad (8)$$

Substituting for k_{eff} in equation (2) gives

$$\Delta I \sim \sqrt{1 - \frac{A_{meas}}{A_{collar}}} \quad . \quad (9)$$

Equation (9) implies that a magnet with low measured overall attenuation will experience a higher increment in current with each successive quench during training.

CONCLUSIONS

The room-temperature longitudinal acoustic attenuation characteristics of ten R-series Fermi research dipole magnets have been measured and compared with their 4.2 K performance. It was found that those magnets with higher attenuation required a greater number of quenches to train and *vice versa*. Given that the main cause of premature quenches in this class of magnets is conductor motion, this correlation may be explained in terms of the effective coil stiffness, which depends on the interface coupling between the collars and the coils. The procedure described here is promising as a technique for screening superconducting magnets at room temperature before testing them at 4.2 K.

REFERENCES

[1] Wilson, M.N., Superconducting Magnets, Clarendon Press, Oxford, 1983, chapter 5.

[2] Tsukamoto, O., Maguire, J.F., Bobrov, E.S., and Iwasa, Y., Identification of Quench Origins in a Superconductor with Acoustic Emission and Voltage Measurements. Appl. Phys. Lett., 1981, 39, p. 172.

[3] Ige, O. O., McInturff, A. D., and Iwasa, Y., Acoustic emission monitoring results from a Fermi dipole, Cryogenics, March 1986, vol. 26, pp. 131-140.

[4] Vary, A. and Lark, R.F., Correlation of fiber composite tensile strength with the ultrasonic stress wave factor, NASA Tech. Memo. April 1978, TM-78846.

DEVELOPMENT OF SUPERCONDUCTING DIPOLE MAGNET WITH

IDEAL ARCH STRUCTURE USING LARGE KEYSTONE ANGLE CABLE

T. Shintomi, A. Terashima and H. Hirabayashi

National Laboratory for High Energy Physics
Tsukuba-shi, Ibaraki, Japan

ABSTRACT

A superconducting dipole magnet with an ideal arch structure has been designed and developed for applications to big hadron collider accelerators. The dipole magnet has an aperture of 5 cm in inner coil diameter and the straight section of 1 m. The magnet can produce magnetic field of 6.6 tesla at the normal operation current of 5,890 A. We have developed a cable which has the large keystone angle of 3.1 degrees and fits to the ideal arch structure of 5 cm inner diameter of the magnet. The cable for the inner coil is composed of 23 strands whose diameter is 0.808 mm having filaments of 6 μm diameter. The fabrication of the cable has not met with serious problems and the degradation by cabling is less than 3 %. At the excitation test, the first R & D dipole magnet has several quenches up to the operation current, and reached to the I_c limit at the tenth quenches. The training will be reduced with tight collaring.

INTRODUCTION

A few projects of the high energy hadron collider accelerator have been planned and developed. Those accelerators require a large number of superconducting dipole magnets, for example about 8,000 for SSC and 4,000 for LHC in the current designs. It is very important to fabricate superconducting dipole magnets to have good performances. Especially the magnets without quench to I_c limit are required. Another important point is field quality. The dipole magnet for accelerator use is fabricated using a so called cos θ method with compacted strand cables. When the windings of the coil cross section are arranged to have an ideal arch structure, the cable should be made to have a proper keystone angle according to the arch structure. A smaller inner coil diameter requires a larger keystone angle cable. The superconducting dipole magnets of high energy colliders have a relatively small aperture, for example in the current designs, the inner diameters of the coil are 4 cm for the SSC dipole magnets[1] and 5 cm for LHC[2]. Since the angle of the existing cables has been limited below 2 degrees, the dipoles with such small aperture need some wedges inside the windings to fit the coil cross section to the arch structure. Those wedges sometimes cause difficulties for keeping magnet performances.

We have developed a superconducting dipole magnet of 5 cm inner diameter with an ideal arch structure using a large keystone angle cable of 3.1 degrees. In the fabricated dipole magnets it is not necessary to have any wedges inside the windings. The paper describes the design and development of the dipole magnet with the large keystone angle cable.

STRUCTURE OF DIPOLE MAGNETS

A few coil configurations have been considered to obtain uniform dipole field. In general, the cos θ method is adopted for high energy accelerators as shown in Fig. 1. The uniform dipole field is produced in the region where the two circles intersect each other.

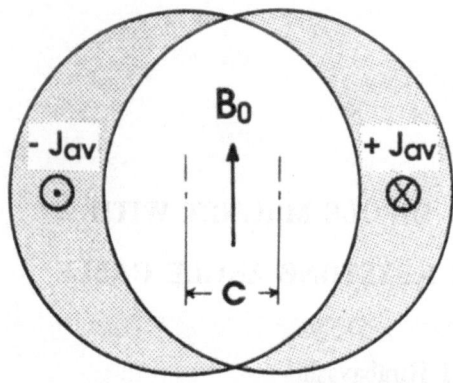

Fig. 1 An ideal cos θ method for dipole magnets.

The produced magnetic field B_0 is a perfect dipole field and given as follows.

$$B_0 = \mu_0 J_{av} c / 2$$

where J_{av} is average current density in the winding and c the separation of the circle center points.

As an approximation of cos θ, a double shell structure is usually used with a keystone angle, Rutherford type, cable. If the required aperture of the dipole becomes small, the keystone angle of cable should be increased. Some wedges are set inside the windings to fit the keystone angle to the coil arch structure because of the small aperture and the fabrication limit of the keystone angle. The wedges made of copper plate have different and high Young's modulus from that of the cable. This means non-uniform deformation under stresses by collaring and electromagnetic forces. It gives rise to the distortion of the field quality.[3] Moreover, the wedges are complex and cause of difficulty at the coil fabrication processes.

DEVELOPMENT OF LARGE KEYSTONE ANGLE CABLE

Necessary Condition of Keystone Angle

When a dipole magnet is wound to have an ideal arch structure, the coil inner diameter D is given as follows.

$$D = 2 (t / \theta - w)$$

where t and w are the thickness at the wide edge and the width of a keystone cable with insulation, respectively. θ is the keystone angle. In case of the

SSC dipole magnet, t, w and θ for the inner coil are 1.888 mm 9.600 mm and 1.6 degrees including insulation, respectively. To fabricate a dipole magnet of 40 mm inner diameter with the ideal arch structure, the keystone angle of the cable should be 3.6 degrees. In case of 50 mm inner diameter, it might be 3.1 degrees.

Fabrication Test of Large Keystone Cable

Fabrication test has been performed for various keystone angles of compacted strand cables to make the cable for a magnet which has the inner coil diameter of 50 mm.[4] To know the dependence of degradation of the critical current on the keystone angle and the packing factor, the cabling with several angles from 1.6 to 3.0 degrees and packing factors of ~90 and ~95 % has been tried. The existing compacted strand cables have usually been fabricated with the former packing factor. We have tried to do the cables with the latter one for test. The trial-fabricated cables are 9.3 mm wide without insulations using 23 strands whose diameter is 0.808 mm. The parameters of the cables are shown in Table 1.

Table 1 Parameters of the trial-fabricated cables.

Strand	
superconductor	NbTi
diameter [mm]	0.808
Cu/SC ratio	1.3 & 1.5
number of filament	12,800 & 7,300
filament diameter [μm]	4.8 & 6.0
filament spacing [μm]	0.6 & 1.5
twist pitch [mm]	25 & 12.5
J_c: 5 T, 4.2 K [A/mm^2]	2,590~2,740
Cable	
number of strands	23
cabling pitch [mm]	79
width [mm]	9.30
keystone angle [°]	1.6, 2.0, 2.4, 3.0
packing factor [%]	~90 & ~95

Cabling has been easily performed for the large keystone angle up to 3 degrees. Heavy deformation is occurred at the narrow edge for the larger keystone angle cable.

Degradation Measurement

Critical currents of the fabricated cables were measured by sampling three strands from each compacted strand cable under the magnetic field of 5 tesla. By comparing the measured values to that of the same strand before cabling, the degradation due to cabling has been estimated. The degradation curves of critical current according to the keystone angles and the packing factors are shown in Fig. 2.

As in the figure, the degradation of the cable with the packing factor of 90 ~ 92 % is almost linearly increased with the keystone angle. As for the cable of the packing factor of 94 ~ 95 %, it increases up to 5 % for 2.4 degrees but decreases down to 4 % for 3.0 degrees. Those are allowable and the same level as one of the existing compacted strand cables. Therefore, the large keystone angle cables are acceptable for dipole magnets.

To check the causes of the degradation, the break of the filaments has been checked and only a small amount of breaks were observed for all of the trial-fabricated cables. From the results, the degradation seems to be caused by the sausaging of filaments with heavy deformation at the narrow edge.

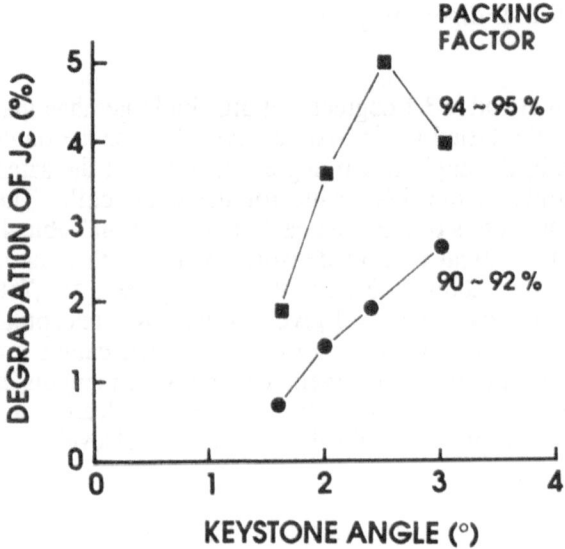

Fig. 2 Measured degradation of the critical current for the trial-fabricated cables with the key-stone angle and the packing factor of 90 ~ 95 % at 5 T.

DEVELOPMENT OF 5 CM ID DIPOLE MAGNET

Design
A model magnet with the ideal arch structure has been developed at KEK. We selected a coil which has an inner diameter of 5 cm, a length of the straight section of 1 m and central magnetic field of 6.6 tesla.

As mentioned before, dipole magnets for accelerators have to produce field uniformity better than 1×10^{-4}. To control the magnetic field in the aperture, there are a few adjustable parameters, *i.e.* cable width, current density and angles of inner and outer windings. Moreover, the grading of the windings should be taken into consideration for the coil design from an economical point of view.

We have fixed the cable widths of the inner and outer coils at 9.3 mm without insulations. The same width for both cables makes the coil fabrication processes simple. From those conditions, multipole components were calculated, and the angles of windings and average current densities were fixed to minimize the sextupole and decapole components. The magnet can produce the magnetic field of 6.6 tesla at the current of 5,890 A. The fixed size of the cross section and the drawings of the model dipole magnet are shown in Figs. 3 and 4. The parameters are also shown in Table 2.

Table 2 Parameters of the 5 cm R & D dipole magnet.

Central magnetic field [T]		6.6
Rated current [A]		5,890
Coil length straight section [m]		1.0
	overall [m]	1.35
Coil inner diameter [cm]		5
Number of turns	inner	24
	outer	22
Overall current density	inner [A/mm^2]	390
	outer [A/mm^2]	500

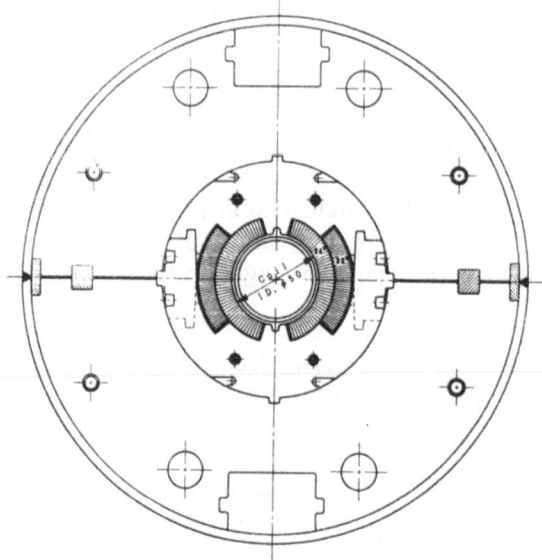

Fig. 3 Cross section of the 5 cm R & D dipole magnet with ideal arch structure using large keystone angle of 3.1 degrees.

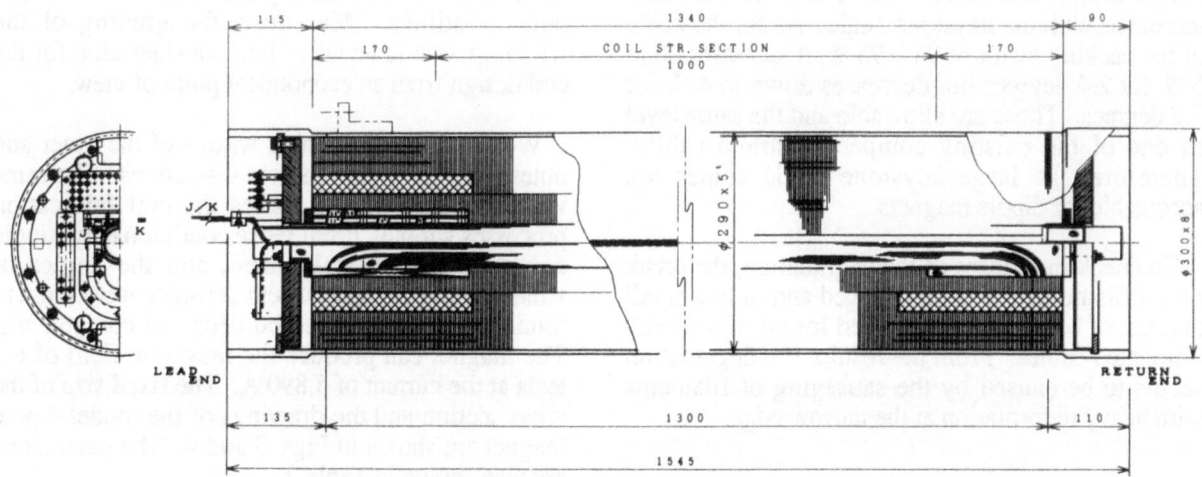

Fig. 4 Drawings of the 5 cm R & D dipole magnet.

Magnetic Field Calculation

The magnetic field calculations have been performed using the code "POISSON". The calculated harmonics at 1 cm radius are shown in Table 3.

Table 3 Calculated harmonics of the 5 cm R & D dipole magnet at 1 cm radius.

Transfer function		11.37 G/A
Sextupole	b_3	-1.57×10^{-4}
Decapole	b_5	0.10
14-pole	b_7	0.26
18-pole	b_9	- 0.10

The change of the harmonics according to the displacement of the coil at the excitation has been calculated. In the calculation the coil may be deformed from a circular form to elliptical one by expanding to the horizontal direction. Such deformation makes the magnetic field uniformity worse.

The coil deformation has been calculated using the finite element method and the maximum displacement will be 90 μm at the median plane.

The calculated maximum magnetic field is 6.95 tesla at the inner coil at the operation current. The load curves for the inner and outer coils are shown in Fig. 5.

Cables

According to the successful results of the large keystone angle cables, cables has been fabricated for the 1 m model dipole magnet with 5 cm inner coil diameter. The keystone angle of the cable for the

inner coil is 3.1 degrees and little bit larger than that of the tested cable. As for the cable for the outer coil, the angle is 1.6 degrees and almost the same angle as the SSC cable for the inner coil. The parameters of the used cable are shown in Table 4. The critical current densities of the both strands before cabling at 5 T and 4.2 K are more than 2,750 A and over the level given in the SSC conceptual design report. The J_c of the stranded cables are almost at the same level. The fabricated lengths of the both cables are 300 m and it was cleared that those cables are available in a commercial level.

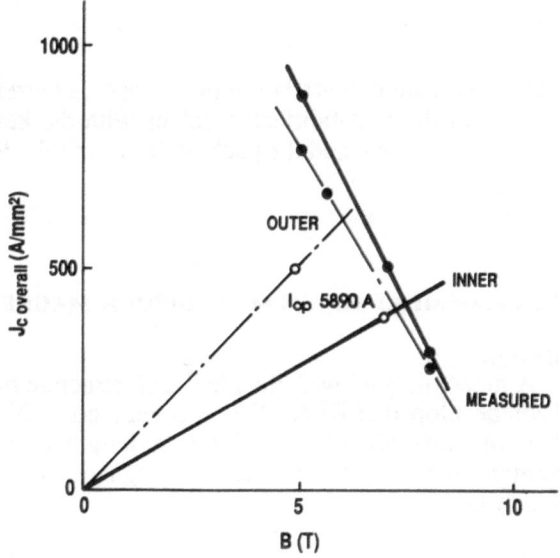

Fig. 5 Load curves for the inner and outer coils of the 5 cm R & D dipole magnet.

Table 4 Parameters of the large keystone angle cables for the 5 cm R & D dipole magnet.

Strand	Inner	Outer
diameter (mm)	0.808	0.614
Cu/SC ratio	1.4	1.8
number of filament	7,600	3,750
filament diameter (μm)	6.0	6.5
filament spacing (μm)	> 1.0	> 1.0
twist pitch (mm)	13.0~13.5	13.0~13.5
J_c: 5 T, 4.2 K (A/mm^2)	2,870	2,970
Cable		
number of strands	23	30
cabling pitch (mm)	79	74.5~75.0
thin edge thickness (mm)	1.14	0.91
thick edge thickness (mm)	1.64	1.19
width (mm)	9.34	9.30
keystone angle (°)	3.07	1.73
J_c: 5 T, 4.2 K (A/mm^2)	2,715	2,785
resistance ratio of Cu	> 110	> 110
fabricated length (m)	300	300

The cross section of the fabricated cable for the inner coil is shown in Fig. 6.

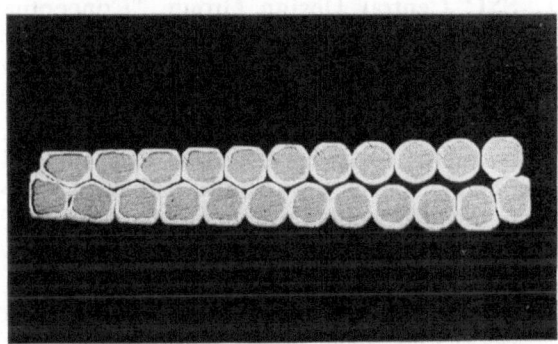

Fig. 6 Cross section of the keystone cable for the inner coil of the 5 cm R & D dipole magnet.

Fabrication and Excitation Test

One of the important features of the magnet is no quench performance up to the I_c limit. The mechanical clamping of the coil is a very important factor not to move the cable. The collaring is one of useful methods for clamping. A stainless steel, SS316LN, collar whose outer diameter is 130 mm was used. The tapered keys were adopted to fix the collar.

Spacers at the end sections are necessary to compensate the keystone angle of the cables. Two spacers were inserted for the inner coil and one for the outer coil. It is necessary for clamping that the spacers fit the curvature of the end windings completely. The spacers made of G-10 were worked upon to a most fitted shape using a computer controlled five-axis machine, and reshaped by trial winding test. Figure 7 shows the cured inner coil and the spacers.

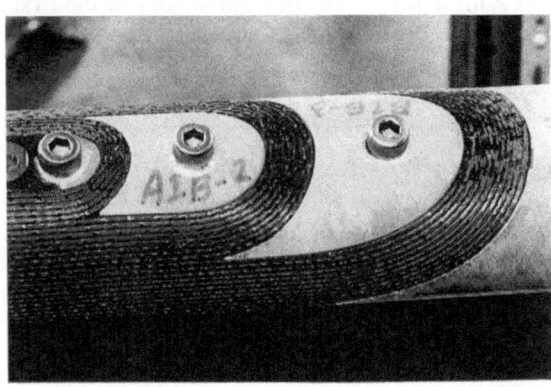

Fig. 7 Picture of the cured inner coil of the first 5 cm R & D dipole magnet.

The winding processes were very simple and easy, and finished in a relatively short time because of no wedges inside as shown in Fig. 8. The first R & D magnet has been developed in a very short period of time, only three months, including the design processes.

Fig. 8 Cutout of the collared coil of the first 5 cm R & D dipole magnet.

The excitation test of the finished 5 cm R & D dipole magnet has been performed. The magnet was cooled with pool boiling in a vertical cryostat. The current ramp rate was around 30 A/s. The first quench occurred at the current of 4,540 A and the second was at 5,303 A. After five quenches the magnet was excited almost at the design current of 5,890 A. At the tenth quench the magnet reached 99 % of the short sample critical current limit. In consideration of the calculation error, it may be the short sample limit. The quench history of the first 5 cm R & D dipole magnet is shown in Fig. 9.

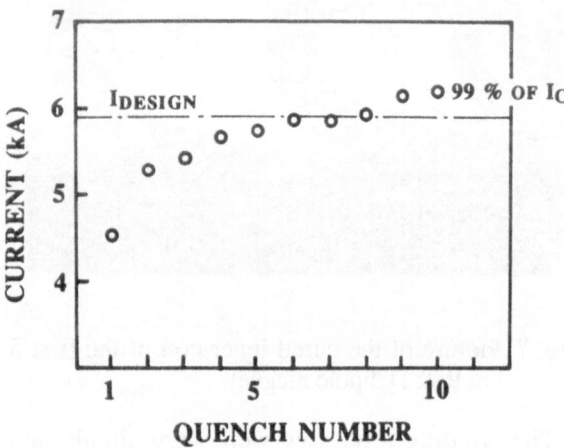

Fig. 9 Quench history of the R & D dipole magnet.

SUMMARY

The first 5 cm R & D dipole magnet has been developed in a very short period of time, only three months, including the design processes. There are no serious obstacle to fabricate them. The observed degradations of the large keystone cables up to 3 degrees with the packing factor of ~ 90 % are below the value of 3 % which is almost the same level in the usually fabricated compacted strand cables. The cable which has the keystone angle of 3.1 degrees and length of 300 m was fabricated for the first magnet of 1 m long. The inner coil diameter of the magnet is 5 cm. The critical current densities of the fabricated cables are more than 2,700 A/mm^2 without copper matrix and almost the same level

given in the SSC conceptual design report. To fix the coil tightly, the rigid spacers made of G-10 were used at the end sections. The excitation test of the R & D dipole magnet has been performed. After several quenches the magnet was excited at the design current and performed up to the short sample limit at the tenth quenches. The training will be reduced with tight and controlled clamping.

ACKNOWLEDGEMENT

The authors express their gratitude to Profs. T. Nishikawa and S. Ozaki for their continuous encouragement. They also express their thanks to the staff of the mechanical engineering center and the cryogenics center at KEK for their vigorous efforts for fabrication and test of the R & D dipole magnet. They are indebted to Mr. T. Kawaguchi of Mitsubishi Electric Corp., H. Miyazawa of Hitachi Ltd. and S. Murai of Toshiba Corp. for their useful discussions.

REFERENCES

1. SSC Central Design Group, "Conceptual Design of the Superconducting Supercollider," Lawrence Berkeley Laboratory, CDG SSC-SR-2020, March 1986

2. D. Leroy, R. Perin, G. de Rijk and W. Thomi, "Design of a High-Field Twin Aperture Superconducting Dipole Model," IEEE Trans. on Magnetics, MAG-24, 1373 (1988)

3. K. Ishibashi, et al., "Design Study on Future Accelerator Magnets Using Largely Keystoned Cables," IEEE Trans. on Mangetics, MAG-25, 1628 (1989)

4. T. Shintomi, et al., "Development of Large Keystone Angle Cable for Dipole Magnet with Ideal Arch Structure," Submitted to Adv. Cry. Eng., Vol. 36

THE TBA PROTOTYPE DIPOLE MAGNET OF SRRC

C. H. Chang, C. S. Hwang, J. C. Lee, W. C. Chou, J. H. Huang,

P.K. Tseng, C. S. Hsue, G. J. Hwang

Synchrotron Radiation Research Center, 8th Fl. No. 6, Roosevelt Road, Sec.1 Taipei, Taiwan, R.O.C.

ABSTRACT

A rectangular TBA (Triple Bending Achromat) combined function dipole magnet has been designed for the SRRC (Synchrotron Radiation Research Center) storage ring. The magnet design was calculated using the "MAGNET" program and measured by our Hall probe automatic mapping system.

The effective lengths with and without shims have been measured. The $\Delta B/B$ and $\Delta G/G$ have been measured to be comparable with 2-D calculations and the measured effective lengths are checked with a 3-D calculation.

INTRODUCTION

SRRC is building a 1.3 GeV synchrotron radiation electron storage ring with full energy injection. The design is a 6-superperiod Triple Bending Achromat (TBA) combined function lattice with circumference around 120 meter. Each achromat arc consists of three 1.22 m bending magnets with each magnet giving 20 degree bending. The bending magnets are curved with rectangular edges which produce a magnetic field of 1.24071Tesla and a 1.60444 T/m gradient strength at 1.3 GeVin the magnetic center.

Based on considerations of acceptable orbit distortions, the tolerances of the magnet has been derived. Analytic numerical 2-D calculation is difficult to interpret the real 3-D magnet problems. A prototype dipole magnet was constructed to evaluate the quality of the dipole magnets and obtain the valuable technical data. Typical criteria for good field are $\Delta B/B < 2*10^{-4}$ and $\Delta G/G < 3*10^{-3}$ within 24 mm width and 36 mm height. The center pole gap is 50 mm which permits a 2 mm clearance for vacuum chamber thermal insulation. The pole contour is designed to generate the dipole and quadrupole components and initial 2-D field calculations were done using the "MAGNET" program. The optimum solution was found by adding small shims at the end of the pole edge.The solution was also doubly checked using the "POISSON"program. The pole face tolerance of $\pm$ 0.02 mm was included in the computation.[1] It is found that the good field region is not affected by the mechanical error. The 1.5 mm sheets of AISI 1006 low carbon steel with good permeability and low coercivity were chosen for the laminations. The permeability was measured to have a deviation $\Delta\mu/\mu$ of about 5%. In order to ensure uniformity of the magnetic properties of the core, it is necessary to do a through shuffling of the steel sheets.

From our measurement results the absolute gradient corresponding to the nominal field is

1.22% too small at the magnetic center, but according to the effective length the gradient is shorter than the nominal dipole by 6%. That is to say, the pole faces are slightly tilted (edge focussing error) and bent (end sextupole field). This pole face tilt and bend can be compensated by different edge shims. After installing the 4-2-0 shims, the effective length of gradient increased to 0.64% longer than the nominal dipole. The results of $\Delta B/B$ and $\Delta G/G$ measurements are in agreement with the 2-D calculations.

MAGNET CONSTRUCTION

The maximum permissible error in the gradient of the dipole magnetic field is 0.025 mm over 30 mm in the slope of the pole profile. The mechanical tolerance of the individual sheets has to be set within $\pm$ 0.025 mm of the gap dimensions. The tight tolerances in the profile are determined by the punch dies. In order to minimize the effect of the difference in flatness and residual internal stresses in the steel sheets, the laminations were punched in a two stamping process. First the coil window and the gap was formed, then the exact gap profile was punched. The precision specified for the punch and die sets was also within $\pm$ 0.08 mm for the pole profile and reference surface[2]. In order to obtain a precise and clean cut, with minimum burr, the clearance between punch and die was within 0.025 mm. To measure the punch die sets and laminations, a measuring machine with sensitivity of 1 μm in three-coordinates was used to check the precision of the profile and the reference surface.

The iron core with a total length of 1170 mm consists of 8 blocks of laminations. Four stacking fixtures are used to assemble the cores. In order to obtain an accurate and smooth pole surface, stacking the gap between shims and reference surfaces of the lamination were designed as the principle stacking references. The lamination stacks were compressed by a bolt screw with a spring reactive force. To avoid welding distortions, the laminations were glued with the 3 M brand epoxy - the epoxy being as thin as possible to increase the packing factor of the core. The 8 block cores are held tightly together by four stainless steel bars and supported by a rigid steel girder. The straightness of the completed core is within 0.05 mm. The magnetic field will be excited by four pancake coils made of hollow water-cooled copper conductors, connected electrically in series. Each pancake has an individual water loop, the four water loops of the dipole magnet are connected in parallel. The current density is 4-5 A/mm^2 to compromise between the material and operating costs[3]. A winding tool was made to adequately obtain correctly wound coils and to maintain the cross-section of the conductor during the winding process. Leak test, 5 kv flash insulation tests and dimension measurements were performed to ensure the quality of coil.

EXPERIMENTAL RESULTS

For the purpose of checking the quality of this prototype dipole magnet, the Hall probe measurement system[4] has been used to map the magnet and analyse the results. After cycling up to 1000 Amps and then the magnet power is then reduced to the design current of 927 Amps. The measured performance at the center point compared to the design parameters is given in TABLE 1. The absolute gradient, corresponding to the nominal field is then 1.22% too small at the magnet center. By construction (rectangular magnet) the gradient is build along the lamination direction, however the gradient of interest for beam optics is the projection

TABLE 1. The dipole and quadrupole strengths at
the magnetic center.

	Measured	design	Δ%
Dipole [T]	$1.23957 \pm 2*10^{-5}$	1.24071	-0.09
Gradient [T/M]	$-1.58343 \pm .001$	-1.60444	-1.31

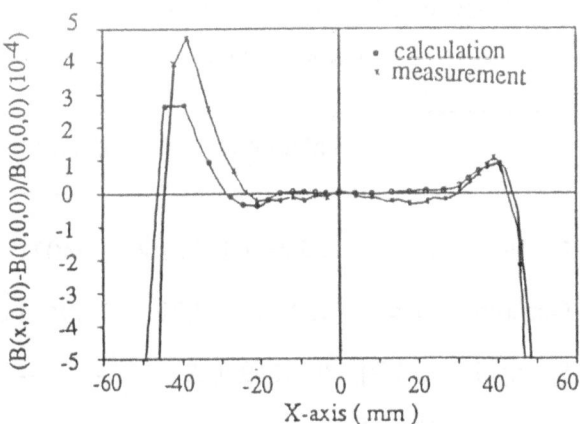

FIGURE1. The induced error from multipole of
measurement and 2-D calculation in y=s=0.

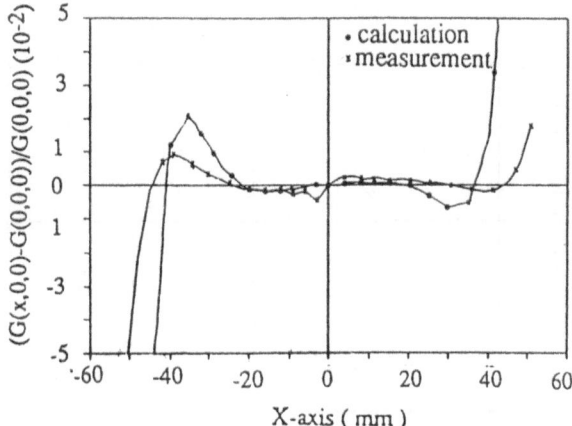

FIGURE2. The quadrupole error distribution of
measurement and 2-D calculation in y=s=0.

along the radial direction. That gradient value
increases as one goes along the reference trajectory,
from the magnet center-point towards the edge. For
a hard edge magnet having the nominal effective
length, the effective gradient would then be larger
by a factor α. For the present magnet with $\rho =$
3.49504 m and $\theta = 20^\circ$ one has

$$\alpha = 1.005$$

This 0.5% effect should then be subtracted from the
measured 1.22% to give 0.72%, if the quadrupole
effective length of the bending magnet corresponds
to the design one.

The remaining 0.72% error has little
importance at this stage since the integrated gradient
is the final quantity that must be compared with the
design. The quality of the pole profile has also been
checked by measuring the dipole field error along
the horizontal axis, at the magnet center, in the
midplane. FIGURE 1. and 2. show that the
agreement between measurements and 2-D
calculations is quite reasonable, and consequently in
that part of the magnet the overall multipole error is
within good tolerances knowing that the horizontal
beam stay clear region is ±12 mm. The field
expansion in the midplane:

$$B_y(x,0,0) = B_y(0,0,0) + \sum_{n=1} B_{ny}(0,0,0) x^n$$

with

$$B_{ny}(0,0,0) = \frac{1}{n!} \frac{\partial^n B_y(x,0,0)}{\partial x^n}$$

leads to the multipole strengths B_{ny} at the magnetic
center which are shown in TABLE 2. FIGURE 3.
shows the field error in the vertical direction. Near
the center region, the field deviation is very small
which is within good tolerances since the vertical
beam stay clear region is ±18 mm. The field
expansion along the vertical axis,

TABLE 2. The multipole strengths at the magnetic center.

		Measured	Design
Dipole	[T]	1.23957 ±2*10⁻⁵	1.24071
Quadrupole	[T/M]	-1.58343 ± .001	1.60444
Sextupole	[T/M²]	-0.29669 ± .2	----------
Octupole	[T/M³]	-0.52023 ± 8	----------
Decapole	[T/M⁴]	850.536 ± 800	----------

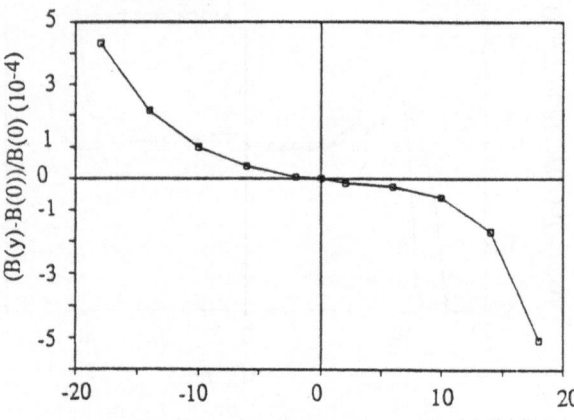

FIGURE3. Field homogeneity in the vertical direction at magnetic center.

$$B_y(0,y,0) = B_0(0,0,0) - B_{2y}(0,y,0)y^2 + B_{4y}(0,y,0)y^4 + \ldots$$
$$-SB_{1y}(0,y,0)y + SB_{3y}(0,y,0)y^3 + \ldots$$

where SB represents skew components which are given in TABLE 3.

TABLE 3. The measured values of skew components.

	Measured
Skew Quadrupole [T/M]	-0.000490 ±.001
Skew Octupole [T/M³]	-76.6 ± 8

The skew quadrupole, corresponding to the regular quadrupole is then 0.03% too small at the magnetic center. This means that the midplane rotation is only 0.3 mrad, if considered with respect to the S axis.

For the field integrals and effective length without shims, the effective length defined as:

$$L = \frac{\int B(x,0,s)ds}{B(x,0,0)}$$

permits location of the effective magnetic pole faces on each side of the magnet. The results, presented in FIGURE 4. show that :

a): The symmetry point of the magnet is shifted by 1.6 mm from the geometrical center. However, at the present stage the Hall probe center location is only known within ± 0.5 mm .

b): The pole faces are slightly tilted (edge focussing error) and bent (end sextupole field).

c): The left and right edges are slightly different. Note that the left edge contains the electrical connections.

These effects are also partly described in FIGURE 5. which shows the effective length as a function of x . At x=0, the effective lengths of dipole and quadrupole are shown in TABLE 4.

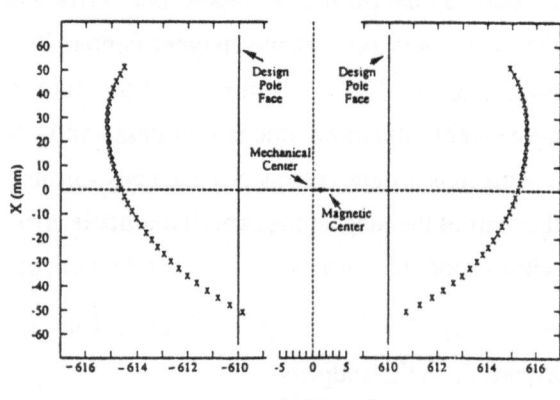

FIGURE 4. Magnetic center and effective pole faces deduced from the effective length distribution in the x direction.

TABLE 4. The effective lengths of dipole and quadrupole.

	measured [mm]	design [mm]
$L_{eff\ [dipole]}$	$1229.81 \pm .04$	1220
$L_{eff\ [quadrupole]}$	1170.59 ± 1	1220

The slope at $x = 0$ represents the quadrupole error, but essentially located at the edge the bent part shows the sextupole effect. This magnet behavior was checked by M.Harold using the 3-D code TOSCA which was shown in FIGURE 5. (SRRC Construction Workshop Proceedings, Feb. 1988).

Attempts to readjust the magnetic pole faces have been made using different end shim configurations. The shim locations and sizes are shown in FIGURE 6. The thickness of each shim is 1.5 mm and we have adopted the following notation:

3-0-0 shims means 3 at A and 0 elsewhere,

4-0-0 shims means 4 at A and 0 elsewhere,

4-2-0 shims means 4 at A, 2 at B and 0 elsewhere,

3-2-1 shims means 3 at A, 2 at B, 1 at C and 0 elsewhere.

As before, for each configuration field expansions are considered along the S trajectory and the multipole components are then integrated along the S trajectory. The results are shown in TABLE 5, which summarizes the performance of the different shim configurations. The brackets in this table indicate the integrated gradient that one would guess after readjusting $\int Bds$ to its design value. The field component values at the center have

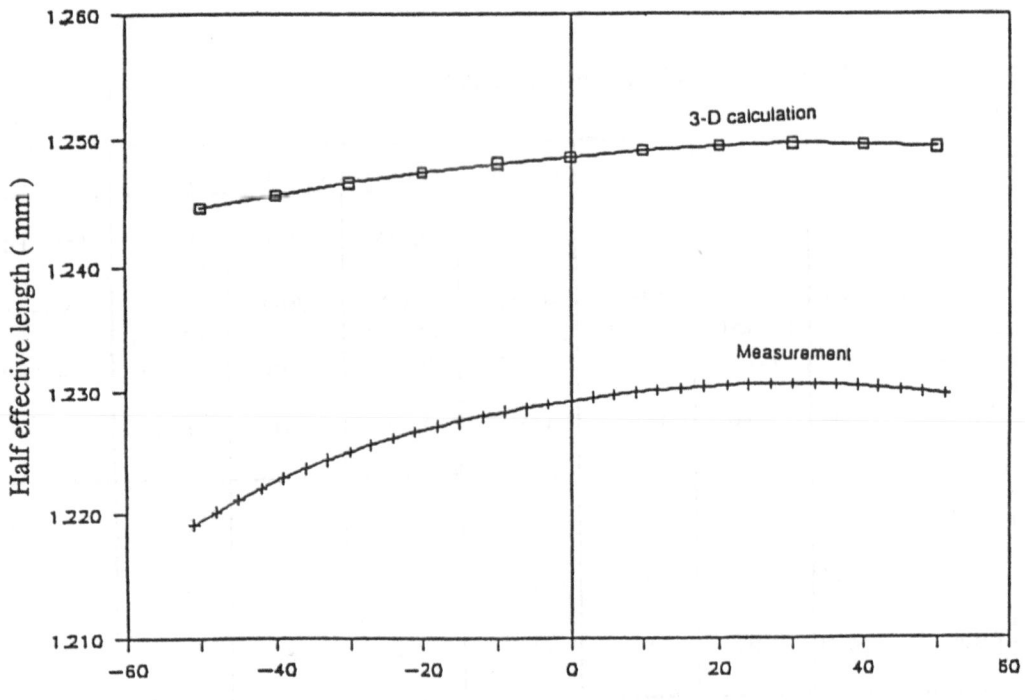

X-axis (mm)

FIGURE 5. The effective length of measurement and 3-D calculation in y=0.

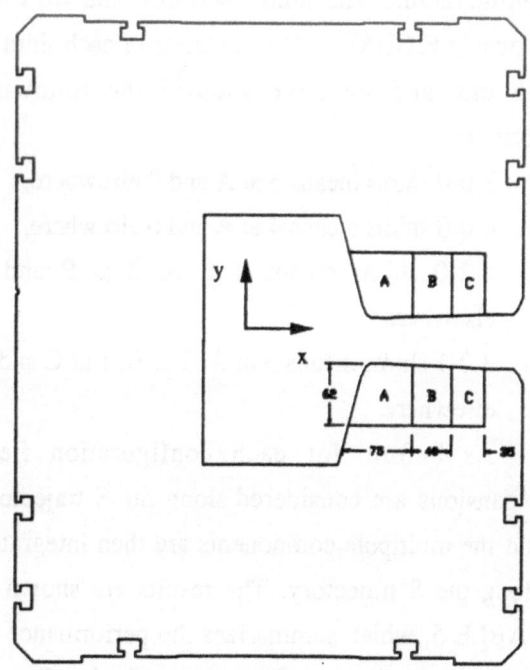

FIGURE6 Schemetic layout of shims.

not been reported in this table since they are not affected by the shims.

CONCLUSION AND DISCUSSION

For this rectangular magnet, the mechanical tolerance of each sheet has been set within +/- 0.025 mm. To avoid welding distortions, the laminations were glued by 3 M brand epoxy - the epoxy being as thin as possible to increase the packing factor.

From the measurement results, the quality of the pole profile is within the tolerance and the multipole strength at the magnetic center is within good tolerance, but at the edges of the magnet the sextupole strength exceeds the tolerance. The magnetic length is too long and 6 sheets of iron can be taken off to reduced the magnetic length to suit the design length. The pole faces are slightly tilted according to the effective length distribution , but the pole face tilt can be readjusted putting the 4-2-0 shims at the edges of the magnet. and the sextupole strength can be reduced.

Table 5. Comparision of Integated Strength for Different Shim Configurations.

			0 shim	3-0-0 shims	4-0-0 shims	3-2-0 shims	4-2-0 shims	3-2-1 shims	Specification
x,y,s coordinate	$\int B\,(0,0,s)\,ds$	[Tm]	1.5244	1.5252	1.5249	1.5265	1.5266	1.5268	1.5137 ± 0.0015
	$\int G\,(0,0,s)\,ds^{*}$	[T]	-1.8536 [-1.8409]	-1.9172 [-1.9027]	-1.9387 [-1.9245]	-1.9371 [-1.9209]	-1.9610 [-1.9443]	- 1.9098 [-1.8934]	-1.9574 ± 0.004
	$L_{bending}$	[m]	1.2298	1.2302	1.2303	1.2314	1.2316	1.2317	1.22 ± 0.001
	$L_{gradient}$	[m]	1.1706	1.2107	1.2249	1.2232	1.2394	1.2061	1.22
r,y,s coordinate	$\int S\,(0,0,s)\,ds$	[T/m]	- 1.56	- 0.15	0.37	1.56	- 1.30	- 1.52	0 ± 0.20
	$\int O\,(0,0,s)\,ds$	[T/m^2]	17.6	0.03	- 1.5	5.2	- 3.8	4.6	0 ± 8
	$\int D\,(0,0,s)\,ds$	[T/m^3]	- 806	- 855	- 1623	- 520	- 31	550	0 ± 314

* Including correcting factor 1.005.

REFERENCES

[1] J. Galayda, R. N. Heese, H. C. H. Hsieh, and H. Kapfer. " The NSLS magnet system " Proc. of the Particle Conf. (1979), San Francisco.

[2] H. C. H. Hsieh. BNL-NSLS. Private communication.

[3] D. Cornvet " The magnet system of the electron positron accumulator (EPA) ", Proc. of the European Particle Accelerator conference, (1988)

[4] C. S. Hwang, W. C. Chou, J. H. Huang, M. Y. Lin, Tzuchu Chang, P. K. Tseng. " High precision automatic magnetic field mapping system for the dipole magnet ", 11th International Conference on Magnet Technology, 1989.

DESIGN OF THE SEXTUPOLE MAGNET FOR AN 8 GEV STORAGE RING

J.Ohnishi and S.Motonaga
RIKEN-JAERI Synchrotron Radiation Facility Design Team
2-28-8 Honkomagome, Bunkyo-ku, Tokyo 113 Japan

ABSTRACT

A design study has been made to achieve a good field quality in the sextupole magnet for an 8 GeV storage ring for a highly brilliant synchrotron radiation source. The sextupole magnet is required to provide a horizontal or a vertical dipole field for the correction of closed orbit distortion in addition to a fundamental sextupole field. This paper describes the result of the field calculation and mechanical design of the magnet.

INTRODUCTION

An 8 GeV storage ring for a highly brilliant synchrotron radiation source has been designed at RIKEN [1]. The ring was designed with a Chasman-Green lattice to get emittance lower than 10^{-8} m-rad. Its major parameters are listed in Table 1.

The ring has a circumference of 1428.9 m, containing 48 unit cells. The layout of a cell of the storage ring is shown in Fig. 1. Total number is 96 dipoles, 480 quadrupoles, and 336 sextupole magnets. For closed orbit distortion (COD) correction, 192 steering magnets in addition to 384 correction elements are fitted to dipoles and sextupoles are installed in the ring. Guide lines for the required field quality of the dipole, quadrupole, and sextupole magnets are given in Table 2.

For the design of these magnets, we carried out numerical calculations of magnetic fields with a program code LINDA [2].

All magnets are constructed from a 0.5 mm thick silicon steel plate having insulation layers. The choice of the laminated structure was made because of

the requirement of producing a large number of units of an identical magnetic field quality.

In a low-emittance storage ring, very high chromaticity is induced by strong quadrupole magnets used to generate a high betatron tune and should be corrected by using strong sextupole magnets, which often lead to beam instability owing to their non-linearity. In order to avoid

TABLE 1.
Major Parameters of the Storage Ring

Energy	(GeV)	8
Current	(mA)	100
Circumference	(m)	1429
Number of cells		48
Tune	ν_x	51.22
	ν_z	19.16
Synchrotron tune	ν_s	0.00952
Momentum compaction	α	1.37×10^{-4}
Natural chromaticity	ζ_x	-119.35
	ζ_z	-40.34
Energy spread	σ_E/E	0.001
Natutral emittance	(πmrad)	5.3×10^{-9}
Harmonic number		2424
RF frequency	(MHz)	508.58

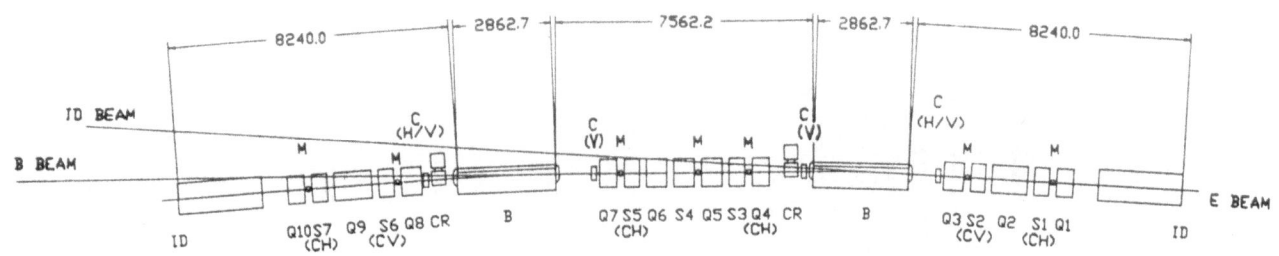

FIGURE 1. The arrangement of dipole, quadrupole, sextupole and correction magnets in the one cell of the storage ring

this instability, we added a few families of sextupoles to the ring; this method is called "harmonic correction". Three sextupole magnets located in dispersive regions are used to correct chromaticity and remaining four sextupoles to correct harmonics.

The arrangement of the sextupoles in a cell and their field strengths in a nominal operating mode are shown in Table 3. 'SF' and 'SD's stand for the chromaticity correction and 'S1' to 'S4' for the harmonic correction. 'SF' has an effective length of 0.6 m to keep the field strength less than 360 T/m^2. The other magnets have a effective length of 0.45 m.

All the sextupole magnets except an 'SF' family are capable of providing steering dipole fields for COD correction by adding auxiliary coils, that is effective to save space for steering magnets in the ring. The design is similar to that of APS and ESRF [3].

TABLE 2.
Required field quality of the dipole, quadrupole and sextupole magnets

Magnet	Dipole	quadrupole	Sextupole
Maximum field strength	0.61T	18T/m	$360T/m^2$
Gap distance or bore radius	65mm	45mm	55mm
Effective field length	2.863m	0.5m 0.6m 1.1m	0.4 0.6m
Field uniformity	$\Delta Bl/Bl$ $<5\times10^{-4}$ $x=\pm40mm$ $x=\pm35mm$ $y=\pm17mm$ $y=\pm15mm$	$\Delta Gl/Gl$ $<1\times10^{-3}$ $x=\pm35mm$ $y=\pm15mm$	$\Delta Gl/Gl$ $<3\times10^{-3}$

$\Delta Bl(x)/Bl(x), \Delta Gl(x)/Gl(x)$ indicate transversal variations in integrated field strength and integrated field gradient, respectively.

CONFIGURATION OF MAGNETS

Two types of magnets with different yoke structures have been designed: one is completely symmetric and the other

TABLE 3.
Family of sextupole magnets

	S1 (S1)	S2 (S2)	SD (S3)	SF (S4)	SD (S5)	S3 (S6)	S4 (S7)
B" (T/m^2)	210	-239	-259	347	-259	-178	286
L_{eff} (m)	0.45	0.45	0.45	0.60	0.45	0.45	0.45
Steering	H	V	H		H	V	H

H: Horizontal steering coils are incorporated
V: Vertical steering coils are incorporated

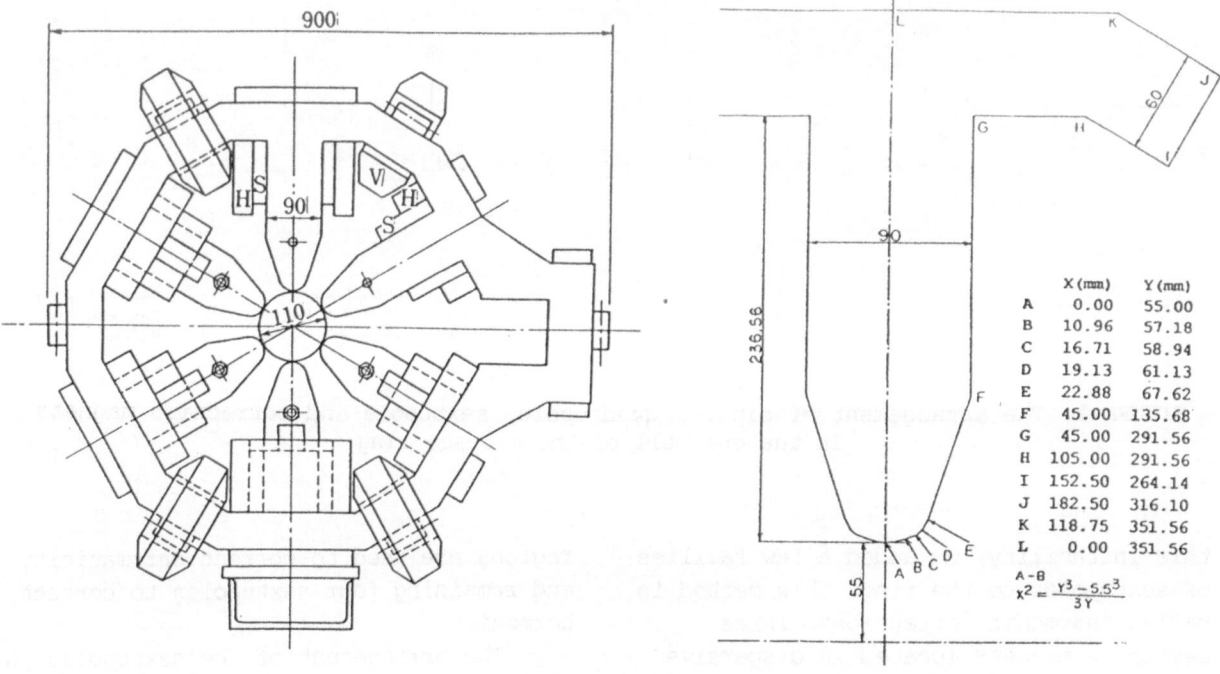

FIGURE 2. Side view of the sextupole magnet and the pole shape

asymmetric. Figure 2 shows the side view of a sextupole magnet, which has an asymmetric yoke structure on the left and the right-hand sides because an enlarged vacuum chamber equipped with a beam line for synchrotron radiation has to be inserted into yokes. Design parameters are listed in Table 4.

One magnet is divided into six parts, each having a pole. The bore diameter (110 mm) was decided from the size of a beam chamber and the uniform field region. The contour for all pole tips is the same, $r^3\cos 3\theta$ with radial shims. The shape of the pole is shown in Fig. 2. The widths of a pole and a yoke are designed to be thick enough so that the maximum strength of a sextupole field can be excited to 360 T/m^2 or more without correction fields or 300 T/m^2 with 0.06 T correction fields.

A sextupole field is generated with a main coil wound around a pole, as indicated as 'S' in Fig. 2.

FIELD QUALITY

The calculated field distribution of a sextupole magnet with symmetrical yokes is shown in Fig. 3. The Y-axis indicates $dB_y/dx/x$ on a median plane.

In the case of a magnet with asymmetrical yokes, field qualities will be inferior. The field distributions of a magnet with asymmetrical yokes of different thicknesses are shown in Fig. 4. The radial field distributions from center (x=0) toward two sides (+,−) do not agree. This disagreement results from a quadrupole component. The maximum value of

TABLE 4.
Designed Parameters of Sextupole Magnets

	S4	S1,S2,S3 S5,S6,S7
Number of magnets	48	288
Effective length (m)	0.6	0.45
Bore radius (mm)	55	55
Good field radius (mm)	35	35
Strength B", max. (T/m^2)	360	290
Number of turns per pole	9	9
Current (A)	950	750
Current density (A/mm^2)	6.1	4.8
Conductor size (mm)	10x16.5-ø3.5	
Voltage drop per magnet (V)	10.9	7.3
Power (kW)	10.4	5.5
Water circuits per magnet	6	6
Water flow (l/min)	7.4	3.9
Water pressure drop (bar)	3.6	1.0
Water temperature rise (°C)	20	20

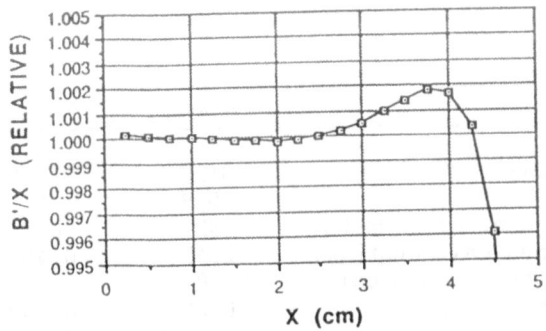

FIGURE 3. Calculated distribution of field gradient of the sextupole magnet with symmetrical yokes

the component is calculated to be 1.3×10^{-3} T/m, which is smaller than the gradient errors of a quadrupole magnet and will give no serious influence to a beam. On the other hand, field uniformity can be kept within 2×10^{-3} by choosing thickness. Sextupole field strength is intensified around both sides of a aperture as indicated in Fig. 4 because a integrated sextupole field along the beam direction is presumed to be smaller than in the central region due to the influence of 3-dimensional fields around the end of a magnet.

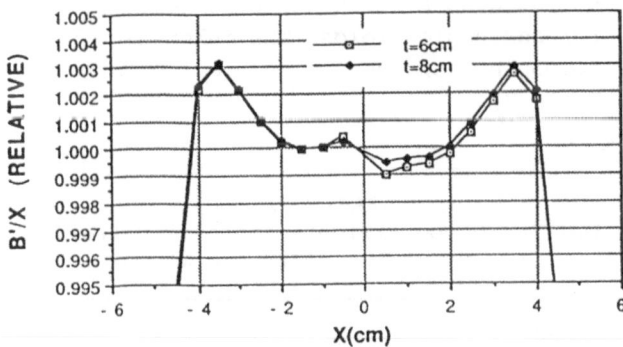

FIGURE 4. Calculated distribution of the field gradient of the sextupole magnet with asymmetrical yokes

COD CORRECTION SYSTEM

Closed orbit distortions are generated by field and location errors of magnets. Since large COD cause the reduction in a dynamic aperture, COD

should be corrected to be within an appropriate value, which is about 0.2 mm from beam dynamics. For the COD corrections, a large number of beam position monitors and steering magnets are placed over the ring as shown in Fig. 1.

Twelve steering elements per one cell are composed of four individual magnets (two of these magnets can provide horizontal and vertical dipole fields) and six steering elements added in sextupole magnets. Six of seven sextupoles in one cell have auxiliary coils and provide horizontal or vertical dipole fields in addition to fundamental sextupole fields.

The computer simulation of the COD corrections has been performed by RACETRACK [4]. The error conditions of magnets used in the simulation are listed in Table 5. The simulation results revealed that residual COD could be small enough, 0.08 mm for the horizontal direction and 0.14 mm for the vertical in

TABLE 5.
Error condition assumed in simulation of COD correction

BM	field error	(dB/B)	5×10^{-4}
	tilt error	(rad)	5×10^{-4}
QM	misalignment	(m)	2×10^{-4}
	gradient error	(dB'/B')	5×10^{-4}
	tilt error	(rad)	5×10^{-4}
SM	misalignment	(m)	2×10^{-4}
	tilt error	(rad)	5×10^{-4}
	gradient error	(dB''/B'')	5×10^{-4}

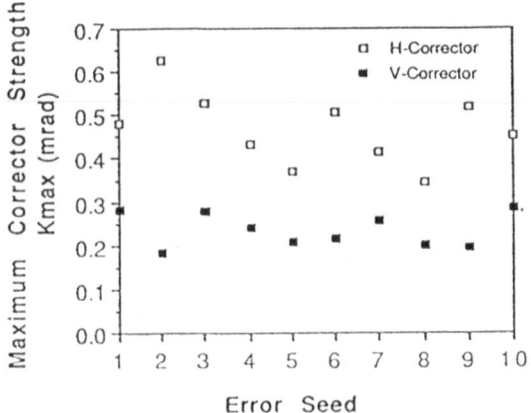

FIGURE 5. Maximum strength of steering fields estimated by COD simulation

root mean square values of ten simulated rings. We confirmed, therefore, that the arrangement and the number of beam monitors and steering magnets are sufficient to correct COD to be small enough.

Figure 5 shows the maximum strengths of the horizontal and the vertical correctors for ten simulated rings. This result indicates that all the correctors sufficiently have a capability of 1 mrad in the maximum kick angle.

CONFIGURATION OF STEERING ELEMENTS

The auxiliary coils in a sextupole magnet are indicated in Fig. 2 as "H" for horizontal steering and "V" for vertical steering. In real magnets, only one type of windings is installed.

The vertical dipole field for horizontal steering is provided by excitation of the respective upper and lower three poles with opposed polarities. Two coils around central poles and other four coils are excited separately with two power supplies. In contrast, the horizontal dipole field is provided by excitation of two left-hand side poles with one polarity and right-hand side ones with opposite polarity. For the excitation, four series coils which are wound on return yokes are used.

Table 6 shows the parameters of steering elements. The maximum field strength is 0.06 T, which gives a 1 mrad kick to a beam. Since the maximum current density through coils is 1.7 A/mm^2, the coils are cooled by air.

TABLE 6.
Parameters of Steering Windings

	horiz.	vert.
Family	S1,S3 S5,S7	S2,S6
Number of magnets	192	96
Physical length (m)	0.45	0.45
Kick angle (mrad)	1.0	1.0
Peak field (T)	0.059	0.059
Turns per pole	100,40	96
Current, max. (A)	42	40
Current density (A/mm^2)	1.7	1.7
Conductor size (mm^2)	5 x 5	5 x 5
Voltage per circuit (V)	9.6,7.7	14.7
Power per circuit (KW)	0.40,0.32	0.62

FIELD QUALITY OF SEXTUPOLE STEERING ELEMENTS

Figure 6 shows the calculated distributions of horizontal and vertical dipole fields by the steering elements incorporated in sextupole magnets. $B_x(y)$ on a X=0 plane is plotted for the horizontal fields and $B_y(x)$ on a Y=0 plane for the vertical. The distribution for the vertical was obtained by exciting the central poles 2.5 times more strongly than the other poles. These field qualities are not so good because the pole geometry is for sextupole fields. The multipole components of the fields are listed in Table 7.

TABLE 7.
Multipole coefficients in horiz. steering fields of the sextupole magnet

b_0	5.97	x 10^{-2}
b_2	-4.75	x 10^{+0}
b_4	4.77	x 10^{+3}
b_6	2.89	x 10^{+5}
b_8	-1.08	x 10^{+7}

$b_x = \Sigma b_n r^n \sin(n\theta)$

$b_y = \Sigma b_n r^n \cos(n\theta)$

$b_n [T/m^n]$

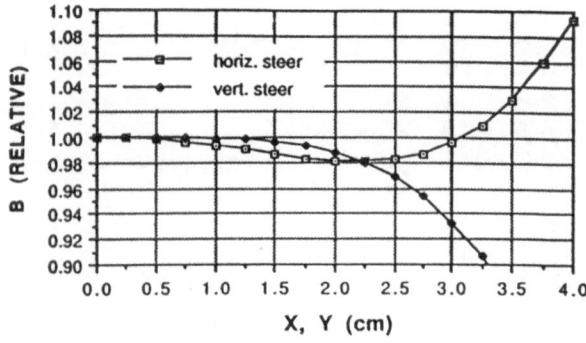

FIGURE 6. Calculated distribution of steering fields

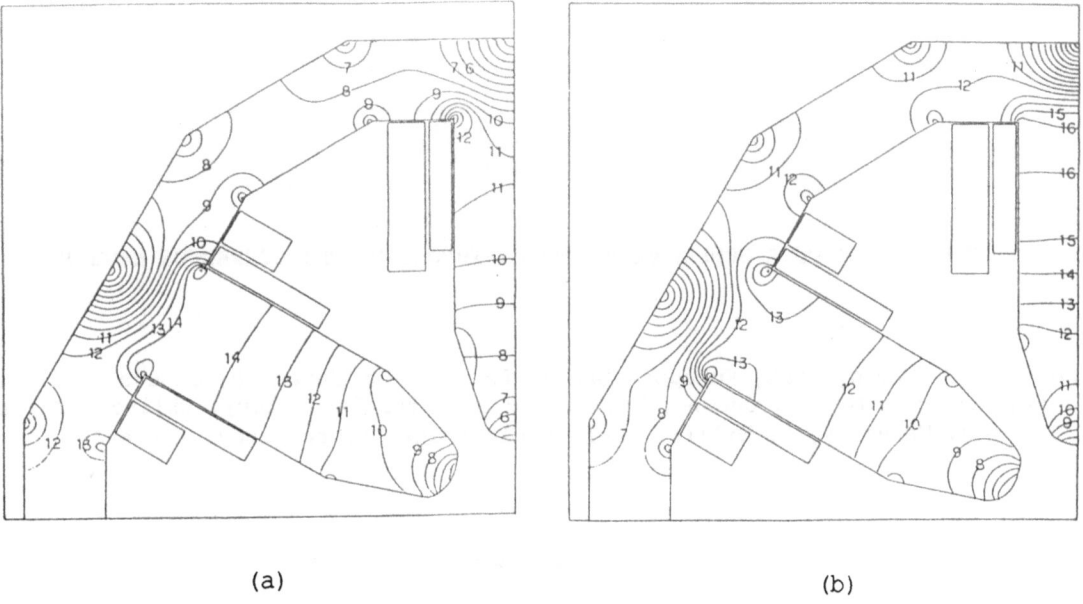

(a)   (b)

FIGURE 7. Distribution of flux density [kG] in iron yokes

SUPERIMPOSITION OF THE SEXTUPOLE AND THE STEERING FIELDS

Figure 7 shows the calculated distributions of the field strength in an iron yoke when both vertical dipole and sextupole fields are excited. In (a) and (b) of Fig. 7, a sextupole fields of 286 T/m^2 and a dipole fields of 0.06 T are excited, but the direction of the dipole fields is opposite. Since the field strengths around all pole tips are less than 10 [KG], no deviation from the fields distribution for a simple superimposition is seen. For (b), the field strength around the base of the central pole is slightly high and the exciting loss of 3% was found for generating steering fields on a sextupole field.

CONCLUSION

We have finished the field calculation and the basic design of a sextupole magnet and almost finished the detailed design of prototype model magnets. Measurement and improvement of the field of the magnets will be carried out in future.

REFERENCE

1. M.Hara, S.H.Be, R.Nagaoka, S.Sasaki, T.wada and H.Kamitubo, "Storage Ring Design for STA SR Project", Proceedings, IEEE Particle Accelerator conference, Chicago U.S.A, March 1989.

2. K.Endo and M.Kihara, "Manual of magnetostatic program, LINDA", KEK Accelerator-1,1972

3. W.F.Prag, K.M.Tompson and S.H.Kim, "Ring magnet for the synchrotron X-ray source at ANL", Proceedings, IEEE Particle Accelerator conference, pp.1486-1488, 1987

   "The Red Book Report B", ESRF

4. A. Wrulich, DESY Rep. 84-026,1984

   K.Tanaka, Private communication

DEVELOPMENT OF A SUPERCONDUCTING SEXTUPOLE-DIPOLE CORRECTOR MAGNET

A. Ijspeert, R. Perin, CERN, European Organ. for Nuclear Research, Geneva, Switzerland.
E. Baynham, P. Clee, R. Coombs, Rutherford Appleton Laboratory, Chilton, England.
J. Wheatley, D. Willis, Tesla Engineering, Storrington, England.

ABSTRACT

Each half cell of the proposed Large Hadron Collider (LHC) lattice [1] will be equipped with a 1.25 meter long superconducting corrector magnet which will combine a 4000 T/m² sextupole and a 1.5 T dipole. The correction magnets of the two rings will be mounted in pairs in the cryostat of the main quadrupoles. The sextupole coil is wound from a solid, NbTi based, superconducting wire. The dipole coil is wound from a preassembled ribbon containing 12 parallel wires. The two concentric coils are precompressed by shrink fitted aluminium rings. The yoke is simply a thick walled iron tube. The paper describes the magnet design, discusses the results of the field and stress calculations with emphasis on the superposition of the two types of field. It comments on the choice of the conductors and describes the developed fabrication techniques.

INTRODUCTION

The sextupole is designed to correct for chromaticity as well as for sextupolar errors from other machine elements like the main dipoles and will operate between − 4000 T/m² and + 4000 T/m². The dipole will provide a means for orbit corrections and operates between − 1.5 T and + 1.5 T.

Figure 1 shows a transverse cross-section of the paired magnets. Each pair will correct the horizontal orbit in one ring and the vertical orbit in the other ring and vice versa. The LHC will need at least 800 of these magnets and

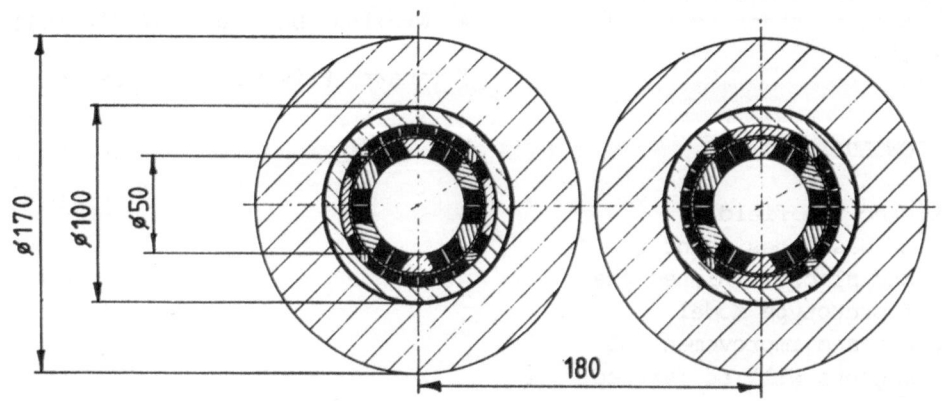

Vertical orbit correction − Horizontal orbit correction

FIGURE 1. Schematic cross section of two paired magnets.

therefore fabrication techniques have been chosen that are applicable to large series production. This project is being carried out in the frame of a cooperation agreement between CERN and Tesla Engineering with the active participation of Rutherford Appleton Laboratory.

GENERAL DESIGN

From the centre outwards the main components are :

the sextupole coils, the dipole coils, the aluminium alloy shrink fitted rings and the iron yoke. The sextupole coil is placed close to the centre in order to minimize the necessary excitation. Another reason is that the dipolar field does not create such intense fields in the inner sextupole as would be the case if their positions were reversed. The field quality aimed at is such that the effect of its higher

multipoles stays below about one tenth of that of the machine half cell at injection. The design parameters are given in Table 1.

The sextupoles are connected in series by means of superconducting lines in the machine cryostat. The dipoles are powered individually and their nominal current of 47 A is as small as possible to minimise the heat losses at the connections but large enough to keep quench voltages at an acceptable level. The wires are monolithic conductors consisting of NbTi filaments in a copper matrix and insulated with an enamel.

The overall diameter of the yoke has been limited to 170 mm to allow for 10 mm of air gap between the paired magnets. This will suppress the magnetic cross-talk between the two magnets which must operate independantly. The length was chosen to fit in the cryostat of the main quadrupoles.

Figure 2 shows a picture of a magnet model where the coils were made of copper wire.

TABLE 1

Design parameters
Values of the prototype given in brackets.

	Sextupole Coil			Dipole Coil	
Field (gradient)	4000	T/m^2		1.5	T
Peak field in coil	4.2	T		3.5	T
Current	458	A		47	A
Number of turns/coil	104			1200	
Ampere turns/coil	47.6	kA		56.4	kA
Stored energy/magnet	6.2	kJ		5	kJ
Inductance	0.059	H		4.6	H
Inner radius	25	mm		36	mm
Outer radius	34.6	mm		40.2	mm
Length straight sectn	1	m		1	m
Operating temp.	1.8	K		1.8	K
	Sextupole Conductor			**Dipole Conductor**	
Overall section	0.7x1.2		mm^2	Diam 0.35	mm
Metal section	0.5x1.0(0.58x1.08)		mm^2	Diam 0.3	mm
Filament size	~ 25	(64)	μm	~ 10 (21.5)	μm
Number of filaments	~ 340	(54)		~180 (36)	
Twist pitch	~ 15	(25)	mm	5 (15)	mm
Copper/NbTi ratio	2/1	(1.6/1)		4/1 (4.4/1)	
Insulation enamel	Polyimide	(PVA)		Polyimide (PVA)	
Breakdown voltage	2		kV	2	kV
Current density	915		A/mm^2	666	A/mm^2
Current dens. NbTi	2747	(2379)	A/mm^2	3332 (3596)	A/mm^2
% of short sample	59	(51)	%	65 (71)	%

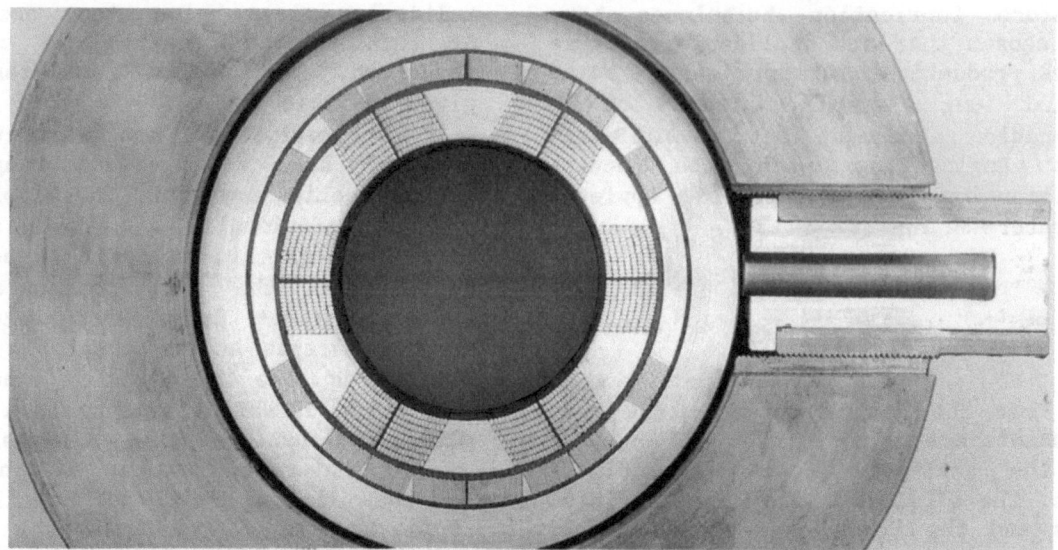

FIGURE 2. Cross section of a magnet model (coils of copper wire)

SEXTUPOLE COILS

Field calculations with POISSON [2] showed that the desired field quality could be obtained with coils consisting of one single block. The tangential height of the block has been optimised to suppress the 18-pole field component.

The iron yoke enhances the sextupole field only slightly; it accounts for ~ 5% of the total field. The loadline is perfectly linear up to the full excitation. Its slope is reduced by ~ 1 o/oo when the dipole is excited simultaneously. The latter causes additional sextupole errors and the overall effect is shown in Table 2.

The insulated conductor is wound around an insulated central island made of hard copper in 13 radial layers of 8 turns each. Each coil is vacuum impregnated. The required accuracy is +/- 0.02 mm on the position of the radial

sides. To obtain this, the central islands are precisely machined and the coils are preassembled on a mandrel, dowelled in position and the gaps between the coils are then filled with preimpregnated spacers.

The interconnections between the coils and the connections to the leads are made at one end by soldering the wire ends side by side in grooved copper pieces over a length of 40 mm. The specific resistance of such soldered joints has been measured to be a constant of 2.1 10^{-8} Ohm-cm at 4.2 K. The joint resistance is thus inversely proportional to the length of the joint. The total heat loss (3 joints) will be 3.3 mW at 4.2 K.

Quench calculations predict a local temperature rise of 136 K and a voltage peak of 290 Volts. The current decays with a time constant of 140 msec.

TABLE 2

Calculated multipole fields errors at nominal excitation of both magnets
At radius of 1 cm.(units of 10^{-6} Tesla meter)

Pole	Sextupole Coil	Dipole Coil	Sextupole + Dipole	Sextupole - Dipole	Specified Tolerance
2 -pole[*]	-69		-18'025	+2'483	150
6 -pole[*]		-188	-844	-3	700
10-pole	2	36	16	-36	100
14-pole	5	-4	-3	6	6
18-pole	-6	3	-1	-7	2

*) These errors can be corrected by adjusting the currents.

DIPOLE COILS

The desired field quality of the dipole field can only be obtained by subdividing the coil into at least two blocks. The iron yoke enhances the dipole field and contributes ~ 35 % to the total field. The load line is not perfectly linear; the field at full excitation is reduced by about 1 o/oo due to the iron saturation. A simultaneous excitation of the sextupole coil increases this effect to about 7 o/oo. Table 2 shows the field quality and the influence of the saturation effects in the yoke. The dipole error, a field reduction caused by the iron saturation when the sextupole is excited, appears to be far beyond the tolerance. However at injection energy, basis for the tolerance, the dipole is not expected to be powered to nominal field and at one tenth of its field, the dipole error is of the order of the tolerance value.

Winding a regular pattern with the extremely thin wire of 0.35 mm outer diameter appeared to be very difficult. Therefore the wire is preassembled to form a ribbon of 12 parallel wires which is then wound around a central island of insulated hard copper and connected at the end to form a series wound coil. Figure 3 shows a picture of the ribbon cross-section in the impregnated coil.

The whole coil is wound with one continuous ribbon, 100 layers overall, and the second block has been subdivided into 3 nearly rectangular sub-blocks separated by wedge shaped copper spacers in order to approximate a circular sector

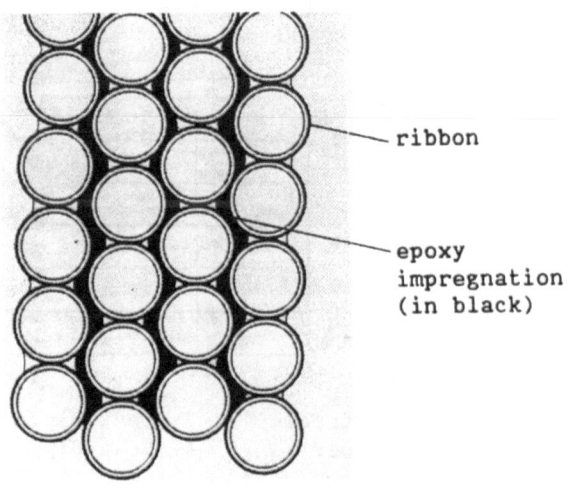

ribbon

epoxy impregnation (in black)

FIGURE 3 Detail of the coil cross section

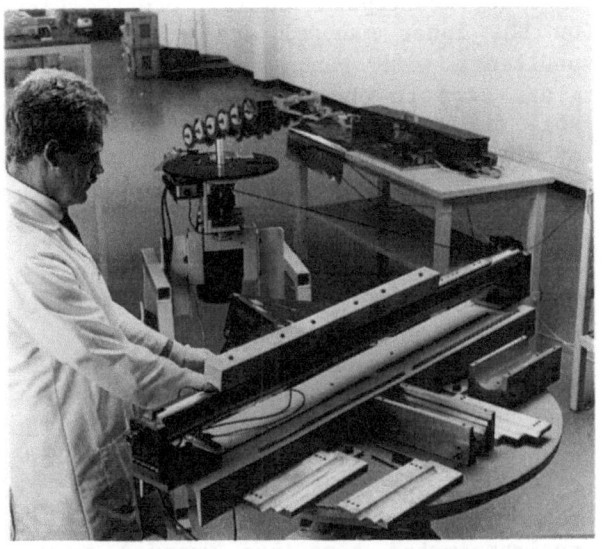

FIGURE 4 Winding operation.

geometry. Figure 4 shows the winding operation.

The interconnections in the coil and the connections between the coils and the leads are made at one end where the wires are soldered side by side in small copper tubes over a length of about 20 mm. The specific resistance over such joints has been measured to be $13 \cdot 10^{-8}$ Ohm-cm at 4.2 K. The total heat loss (25 joints) will be 3.6 mW at 4.2 K.

Quench calculations, assuming that all the energy is absorbed in the coil itself, showed that stabilisation with a copper content of 2:1 gives unacceptably high quench voltages and this ratio had to be increased to 4:1. The expected local temperature rise in case of a quench is 130 K and the voltage peak 950 Volts with a time constant of 180 msec for the current decay.

ASSEMBLY OF THE COILS

The assembly must be such as to guarantee the necessary mechanical precision, especially the +/- 0.02 mm on the radial positions of the coil faces. In addition, it must make the coils hold together as a single mechanical body to obtain a sufficient stiffness against radial bending of the coils due to the magnetic forces. This is obtained by a final vacuum impregnation. The assembly procedure goes as follows:

The sextupole coils, preassembled on the inner mandrel are wrapped at 7 positions with bands of preimpregnated glass tape to obtain a precise machineable outer diameter for the positioning of the dipole coil later on. Clamps are put on and the whole is cured in an oven.

The diameter of the 7 bands is measured and if necessary machined and one layer of preimpregnated tape is wound around. The space between the bands is filled by wrapping with glass tape.

The dipole coils are then put on the precise bands and the angles adjusted with jigs, one at each end. The spaces between the two coil halves are filled with preimpregnated spacers. As for the sextupole coils, bands of preimpregnated tape are wrapped around at 7 positions to obtain a precise machineable outer diameter, the coils are clamped and the whole is cured in an oven. The diameter of the bands is measured and machined if necessary and the spaces between the bands are filled by wrapping with glass tape.

The final impregnation is done in a mould composed of an especially designed retractable mandrel to fill up the bore and an outer mould located on the bands. Finally, the diameter is measured and machined if necessary and the aluminium rings are shrunk on.

CONTAINMENT OF THE E.M. FORCES

Due to the two independantly operated magnet coils, there are three basic types of magnetic forces: one caused by excitation of the sextupole coils only, one caused by excitation of the dipole coils only, and one which comes in addition from the interaction when the two coils are powered simultaneously. Figure 5 shows schematically these principal forces. The first two always have the same sign independantly of the polarity of the magnetic field but the third changes sign when the polarity of one of the two coils is inverted. It consists of strong bending forces in the sextupole coil whereas the additional forces in the dipole coil hardly contribute to the bending and therefore have not been shown in Figure 5. Thus there are four load cases to be considered: sextupole only, dipole only, sextupole + dipole, sextupole - dipole or vice versa. The bending moments are shown in Figure 6; the largest moments are due to the effect of the dipole field on the sextupole coil.

The stresses in the coils consist of the bending stresses together with

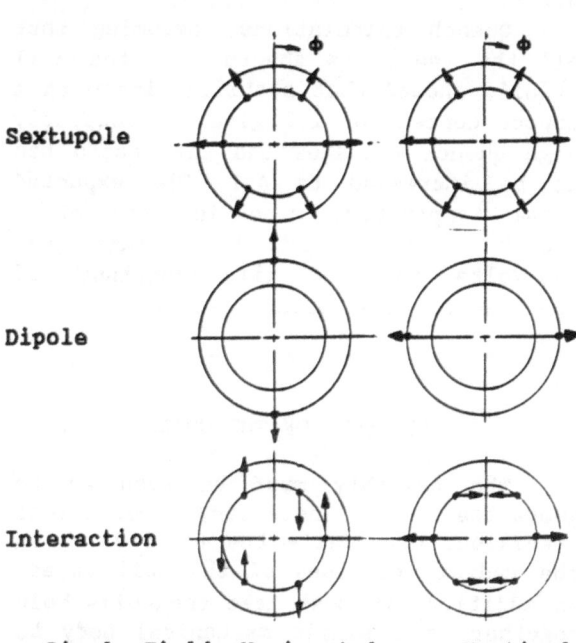

Dipole Field: Horizontal   Vertical

FIGURE 5. Principal bending forces
in coil assembly

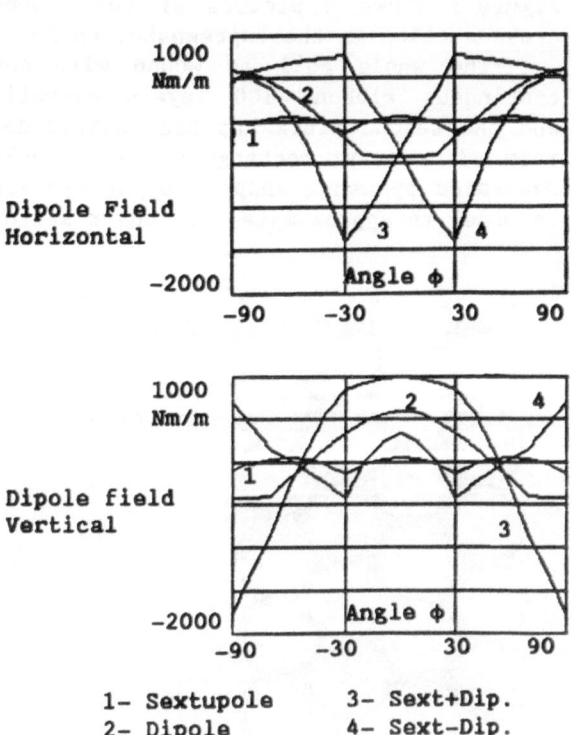

1- Sextupole   3- Sext+Dip.
2- Dipole   4- Sext-Dip.

FIGURE 6. Principal bending moments
in coil assembly

TABLE 3
Coil stresses (N/mm²)

		Min.	Max.
From Rings:	Shrink fit	-40	-15
	Therm.contract.	-10	+10
From Field:	Sext.+ Dipole	-13	+13
	Sext.- Dipole	-4	+11
Total*:	Sext+Dip+Rings	-58	-7
	Sext-Dip+Rings	-52	-8

*) The extreme stresses occur in different locations for each case and the stresses do not add up algebraically.

the local inner coil stresses. The stress distribution is complicated by the presence of materials of very different Youngs moduli ranging from the relatively soft insulation to the stiff copper. Finite element calculations have been made with the programme CASTOR [3] for the assumption that the coils and the shrink fitted rings hold together as one single body. The ranges of stresses in the coils are given in Table 3.

To avoid quenches caused by the tensile stresses, the 9 mm thick aluminium rings precompress the coils when shrink fitted on assembly and further compress the coils by contraction on cooldown from room temperature to 1.8 K. Table 3 shows the effect of the precompression. The shrink fit of 0.05 mm is just sufficient to keep the powered coils compressed.

The prestress calculations have been checked by strain gauge measurements made on the model shown in Figure 2.

YOKE

The yoke is a simple thick walled tube made of low carbon iron. The inner diameter has been fixed at 100 mm to keep the induction in the iron below 2 T. A calculation with POISSON, made to check the importance of the iron thickness, showed that an increase of the inner diameter (with a fixed outer diameter) although reducing the enhancement of the dipole field and increasing the saturation in the iron, does not notably increase the multipole components. However, the stray field

just outside the yoke changes from 0.016 T for the present inner diameter to 0.41 T for an extreme inner diameter of 160 mm

The coil assembly is centred in the yoke by a number of pins, stiff enough to counteract the force of 45 N per μm of radial offset that will pull the coil away from the centre in case of a slight off centre position. These pins are made of two different materials, aluminium and tungsten, to compensate for the difference in contraction between the coil and the yoke (Figure 2).

CONCLUSION

The development of the sextupole-dipole magnet is progressing satisfactorily. The complications caused by the presence of two types of coil in one magnet and by the extremely thin wire used for the dipole coil, have been solved in view of a future mass production. Critical points like the interconnections and the prestress have been checked by measurements on models. The prototype is expected to be powered by the end of this year.

ACKNOWLEDGEMENTS

The authors are pleased to acknowledge the contributions of R. Mauris (CERN) and M. Parr (Tesla) for the design, of D. Landgrebe (Tesla) for the development work and of B. Crépin (CERN) for the typing of the paper.

REFERENCES

1. Perin R., Progress report on the superconducting magnets for the Large Hadron Collider. IEEE Transactions on Magnetics, 1988, Vol. 24, Nr 2, pp 734-740.

2. Holsinger R. F. and Iselin C., The CERN-POISSON package (PIOSCR) user guide, Write up T604, CERN Geneva, August 1983.

3. Manuel d'utilisation du progiciel CASTOR PC2D, CETIM, Senlis France, 1989.

TOWARDS THE REALIZATION OF TWO 1.2 TESLA SUPERCONDUCTING SOLENOIDS FOR PARTICLE PHYSICS EXPERIMENTS,

P T M Clee and D E Baynham, Rutherford Appleton Laboratory, Chilton, Didcot, Oxfordshire, England.

ABSTRACT

Superconducting solenoids are required to produce a magnetic field of 1.2 Tesla in a room temperature volume of 145m^3 for the DELPHI particle physics experiment on LEP at CERN and for the H1 experiment on HERA at DESY.

The large solenoids which are in the region of 5-6m diameter and 5-7m long, were constructed with aluminium clad Nb-Ti conductor, wound on the inside of a liquid helium cooled support cylinder.

The coils are indirectly cooled with 2 phase liquid helium and suspended within an 80K gas cooled radiation shield inside a stainless steel vacuum vessel.

The coils have a stored energy > 100 MJ when powered at 5000A.

The paper covers aspects of the construction, installation and test of the two solenoids.

INTRODUCTION

The Rutherford Appleton Laboratory (RAL) has designed, built, installed and tested two superconducting solenoids providing very large volumes of magnetic field for High Energy Particle Physics experiments. The first was for DELPHI, one of four experiments operating on the Large Electron Positron (LEP) Collider at CERN, Geneva. The second was part of the UKs contribution to the H1 Experiment on the Hadron Electron Ring Accelerator (HERA) at Deutsches Elektronen Synchrotron (DESY), Hamburg.

The basic design concept used for DELPHI Solenoid (described in a previous paper)[1] was also adopted for H1. The overall parameters for the Solenoids are given in Table 1.

The DELPHI Solenoid has been fully installed and commissioned. H1 has passed the preliminary tests at RAL, been installed at DESY, and is at the stage of final commissioning and test.

TABLE 1
Parameters

		DELPHI	H1
Central Field	T	1.2	1.2
Peak Field	T	1.6	2.2
Current			
Main Coil	A	5kA	5.5kA
Trim Coil	A	±1kA	-
Ampere Turns	At	7.45x10^6	7.0x10^6
Run Up Time	Min	60	60
Run Down Time	Min	95	120
Inductance	H	8.75	7.8
Stored Energy	MJ	109.4	120
Discharge Volt.	V	750	750
Discharge Time	Sec	60	60
Inner Clear Dia	mm	5200	5205
Outer Clear Dia	mm	6200	6075
Radial Width	mm	500	435
Length Overall	mm	7400	5751
Weight Total	Tonnes	84	73

GENERAL DESIGN CONCEPTS

The overall design and development carried out for the DELPHI Solenoid produced a design and construction techniques which in practice proved to be sound and economic.

The main design concept adopted for the coils in both Solenoids is illustrated in Figure 1. The conductor, made up with Nb-Ti Rutherford cable and clad with aluminium for stabilisation, is wound on the inside of a support cylinder. Conductor joints within the coil were made by welding of parallel conductors over one complete turn of the coil (>15m); originally developed in Japan for CDF and Topaz Solenoids. Very low resistance joints were made - extrapolation from short joint tests indicate $<10^{-10}$ per joint.

Coils were fabricated in modular form of lengths up to 1.5m for both double and single layers, and were wound on a custom built machine using the 'inside winding' technique developed at RAL. This enabled a production line to be set up to manufacture all the modules and included - support cylinder cleaning - lining with glass/resin insulation - bonding and curing - electrical insulation test - conductor cleaning - winding of modules with continuous double half lap tape insulating and inline joint welding - bonding and curing - then final electrical test.

The earlier paper [1] indicated that 'B' stage glass/resin was to be used as the insulating and bonding medium. This material only carries a small amount of resin which is partially cured and requires large pressures between mating surfaces during the curing process to achieve high bond strengths.

Since that statement a similar but much improved system has been developed and used for both Solenoids. This consists of mixing solid and liquid epoxy resin to produce a thixotropic consistency. The glass cloth and tape, are vacuum impregnated with this mixture but unlike 'B' stage, the resin is not partially cured. The system enables a large amount of resin to be taken up by the glass. When finally vacuum bagged and/or pressure bonded with a 120°C cure, the final coils are indiscernible from a vacuum potted coil and very high insulation resistances are achieved.

H1 DEVIATIONS FROM DELPHI

There were many small changes to the DELPHI design and layout of components to meet the H1 requirements, but two major changes in concept were introduced. These were the reduction in the radial width of the Solenoid and the introduction of compression rather than tension supports for the coil within the vacuum vessel.

To meet the 435mm radial dimension, it was necessary to put the vacuum vessel stiffening rings inside the vessel and use the space between the rings to locate the circumferential radiation shields (see Figure 1). By designing the shields as simple aluminium sheets with He gas cooled pipes welded to the outer surface,

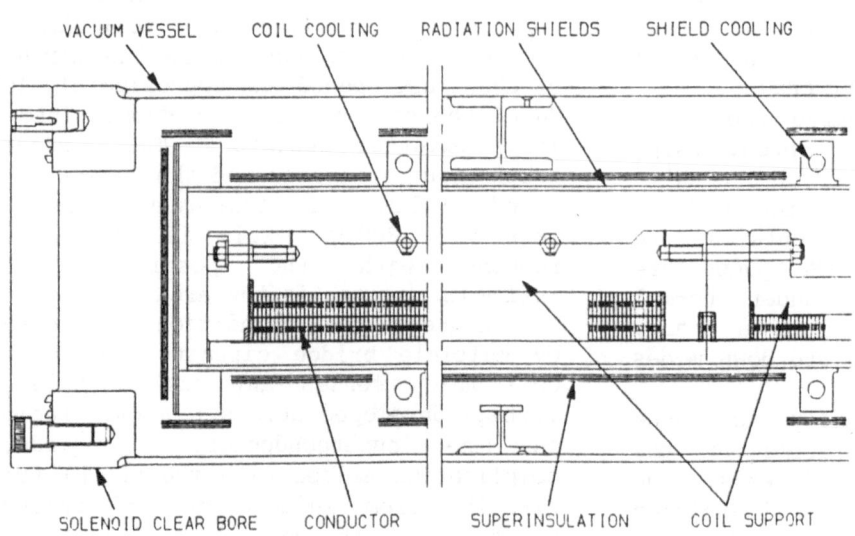

FIGURE 1 Cross Section of H1 Solenoid

it was possible to highly polish the surface facing the coil, thereby avoiding the need for any superinsulating blankets between the coil and shields.

The DELPHI coil was suspended at the ends from the vacuum vessel, on thin stainless steel rods and additional tie rods provided lateral stability. In the finished state this principle was perfectly acceptable, in practice however, during the assembly stage with the vacuum vessel end plates removed, the unsupported inner and outer shells moved a few millimetres. This made the setting up operation of centring the coil and then fitting the end plates very difficult.

For this reason, the support system for H1 was changed to two glass fibre support pillars located at each end on the horizontal axis. Linear bearings were built into the supports to accommodate the contraction of the coil during cooldown to 4.5K.

CRYOGENIC SYSTEMS

The cryogenic systems are required to cooldown and maintain the coil at 4.5K, the radiation shields at 80K and provide 4.5K gas and liquid to cool the current leads. Additional cooling is required during the current change in the coils to remove inductive heat generated in the support cylinder and a reserve of coolant is required for the coil and current leads when the coil is run down in the case of a power outage.

Cryogens for H1 are supplied from a large central refrigerator feeding a ring main, whereas for DELPHI a local refrigerator is provided by CERN.

In both cases a pumped two phase helium system has been adopted, using mechanical pumps developed by CERN[2] which are located in an intermediate dewar (ID) which also contains the reservoir of LHe. Two LHe pumps are installed, one operates under normal conditions with the second as a spare. During the current run up and down modes both pumps are operated.

The items most likely to cause problems during the lifetime of the Solenoid are the current leads, the superconducting/resistive transition joints, the electrically insulated joints in the current lead cooling circuits and the cryogenic interface connections. In the case of DELPHI these items are housed in special turrets located on top of the iron yoke at each end of the magnet and for H1 inside a valve/current lead box located on the floor outside the iron yoke.

The current lead system is based on a horizontal fin heat exchanger design developed by CERN[3]. For coils with stored energy > 100MJ, current lead reliability under fault conditions is of greater importance than minimising refrigeration heat loads at 4.5K. The fin type design allows extremely good safety in the event of gas flow failure with acceptable 4.5K heat loads for normal operation. Under test conditions lead temperatures did not exceed 350K when the current was run down from 5kA over 2 hours with zero gas flow.

The fin heat exchanger section between 4.5K and 300K is cooled with 4.5 gas taken from the top of the intermediate dewar. The base of the current lead is separately connected by a heat exchanger to the two phase circuit of the coil. In this way the base of the current lead is held at 4.5K by the two phase flow which absorbs the main heat load from 300K.

POWER SUPPLIES AND PROTECTION SYSTEM

The H1 Solenoid is powered from a 20 volt, 6kA power supply connected to the coil through high voltage, high power circuit breakers. DELPHI has a similar main supply but in addition there are two trim correction circuits powered by supplies that can operate up to $\pm$ 1kA.

Under quench conditions the circuit breakers are opened and approximately 100 MJ of the stored energy is dumped into a large external resistor. This process takes about 2 minutes and the resistor is sized to limit the temperature rise to less than 300°C. This is for safety reasons with the possibility of inflammable gases in the area.

A quench fault condition is detected by multiple bridge circuits across the coil which senses any out of balance voltage developed across a normal region in the superconductor. An added complication arises with the DELPHI trim circuits from induced out of balance voltages when the correction currents are changed. These voltages are compensated by a signal from mutual inductors located in the correction current busbar circuit.

For normal run down of the magnetic field the voltage and current decay are controlled by passive diode resistor assemblies. Due to the low inductance of the DELPHI trim coils the currents decay more rapidly than the main coil current. The voltage drop in the trim circuits is minimised to extend the decay time and avoid excessive eddy current heating in the coil support cylinder which could initiate a quench.

ASSEMBLY AND TRANSPORT

The available buildings at RAL were not ideal for constructing the Solenoids, therefore a horizontal assembly process was developed. Figure 2 shows the assembly area with the coil modules in the foreground and to the left, is the inner cylinder cantilevered from a large support frame. Successive cylinders, supported on a trolley, were slid over the cantilevered assembly.

On completion of assembly and tests, all connecting pipes and transfer lines were removed and additional supports to react high 'G' loads on the coil during

transit were fitted. The Solenoids were rolled out of the building on a trolley and transferred to a multi wheel (all steerable) transport trailer by a mobile crane, for transit by road and roll-on, roll-off ship to the destinations at CERN and DESY.

H1 SOLENOID TEST

Final assembly and pre-shipment test of the H1 Solenoid was made at RAL during March/April 1989. See Figure 3. The test programme was designed to demonstrate correct cryogenic operation of the coil, radiation shields and cryogenic systems and to establish correct operation of all electrical and protection systems at excitation currents up to 2kA. The upper limit of test current was set at 2kA since the test was executed without the iron yoke structure.

Cooldown and Cryogenic Tests

(i) <u>Cooldown of the Solenoid and Radiation Shields to 90K</u>. In this phase helium gas at 15bar was circulated using a compressor/heat exchanger unit built by DESY. Cooling was provided by heat exchange between the closed circuit helium loops and liquid nitrogen from a storage dewar.

During cooldown the temperature differential across the Solenoid was limited to 40K in order to keep thermal stresses within the coil to a minimum.

FIGURE 2   H1 Solenoid Assembly

FIGURE 3   H1 Test at RAL

With a helium mass flow rate of ~ 30 g/sec the cooldown of the 25 tonne coil to 90K was achieved in approximately 10 days.

The mass flow through the radiation shields was controlled in order to maintain a temperature differential between coil and shields <10K.

With the system cooled to 90K flow was directed to maintain the shields at 90K under the radiation heat load.

(ii) Cooldown and Operation of the Solenoid coil at 5K. The cooldown of the Solenoid coil to 5K was made using 5K helium gas supplied from a large mobile 30000 litre tanker. In this operation the mass flow through the coil was carefully controlled in order to achieve maximum enthalpy exchange with the gas. Cooldown of the coil required approximately 9000 liquid litres against a perfect efficiency requirement of approximately 7000 litres. With the Solenoid at <10K the helium supply was switched to transfer liquid to the dewar. Circulation of two phase liquid through the coil was established

Cryogenic tests were carried out to confirm heat loads to the coil and radiation shields:-
(a) radiation shields ~600 watts radiation and conduction.
(b) coil static heat load ~100 watts with radiation shields at 90K.

These measurements showed heat loads within the design limits.

Electrical Tests

Electrical tests on the Solenoid coil showed correct performance of joints and superconductor up to 2kA. In addition the electrical test programme was designed to commission the main power supply and protection system.

Correct operation of both slow dump and fast dump protection systems were demonstrated at currents up to 2kA.

Coil temperature rise under fast dump was in line with computed predictions.

DELPHI SOLENOID TEST AND COMMISSIONING

Although a large amount of stage and partial tests were carried out, it was not possible to fully test the solenoid until the final installation in the Experimental Hall.

Pump Down of the Vacuum System

It took 14 days to reach a pressure of 10^{-5} mbar and the overall leak rate was measured as 2×10^{-5} L mbar sec^{-1}.

Cooldown and Cryogenic Tests

The cold box, intermediate dewar (ID) and Solenoid were cooled as one at an average rate of 1.7K/hour from room temperature to 4.5K in 7 days.

Static heat load into the Solenoid, measured by temperature rise of the known mass was:-

	Measured	Calculate Max
Thermal Shield at 77K	900W	2500W
4.5K System	40W	87W

The dynamic heat leak of the coil, transfer lines and ID was found to be higher than expected, namely 147W at 4.5K. This was measured by recording the fall in LHe level in the ID with the refrigerator closed off. The reason for this high heat load is not fully understood at present. The overall heat load on the total system however, is well within the capacity of the refrigerator.

The current leads operated to specification and the total liquification load at 4.5K was 0.823 g/s. This is less than the design figure of 1.1 g/s because the final operating currents through the main leads are 4.1kA and 4.3kA instead of the nominal 5kA.

Powering and Protection Tests

Powering of the Solenoid was carried out in steps of 500A up to 5kA to enable all the equipment to be checked out and to ensure that the protection system could handle the extraction of 110MJ of energy. During this test the temperature rise of the external dump resistor was 100°C compared with the 300°C restriction and the temperature rise of the coil was 34K compared to the computed temperature of 42K.

In commissioning the protection system the response of the quench detector to current changes in the main windings was as expected, but the response to changes in the end trim windings was more complex. This was probably due to the effect of eddy current transients in the coil support cylinder, which made it difficult to maintain a null output from the detector during the first few seconds of the slow run down. This problem was solved by filtering the signal from the mutual inductors.

As mentioned earlier, problems could develop from the rapid decay of the trim currents creating eddy current heating in the coil support cylinder which could initiate a quench. In practice this was the case and the problem was overcome by by-passing the flywheel diodes in the trim power supplies with a shorting contactor during the slow run down, this doubled the decay time.

Magnetic Measurements

In measuring the magnetic field [4] it was found that the properties of the iron yoke were better than expected and the extra ampere turns at the coil ends over compensated for the field fall off. This was corrected by running the trim current in the end turns in the reverse direction to the main coil current. The field measurements also indicated that the Solenoid was offset axially relative to the iron yoke by 6mm. This was also corrected by adjusting the trim currents. The final trim current settings were -700A and -900A.

At these current settings, the field was measured over a cylindrical volume of 202cm radius and 560cm length and out to a radius of 252cm at the ends. The Bz

fields at three radii are shown in Figure 4 and are well within the required $\pm$ 0.01% over the volume of the TPC. Br was within specification and Bϕ was at most $\pm$ 1 Gauss within the volume of the TPC and $\pm$ 3 Gauss elsewhere. The stability of the field was 0.01% over a 12 hour period.

CONCLUSIONS

It is now clear that large solenoids can be designed and built to specification by careful use of suitable computation models[1]. In arriving at an optimised design all the requirements for space, materials, heat load, supports, ancilliary equipment etc, had to be carefully examined by several iterations to ensure a practical, safe and economic solution was obtained.

ACKNOWLEDGEMENTS

The RAL H1 and DELPHI Project Teams are grateful to CERN, DELPHI Collaboration, DESY and H1 Collaboration for the opportunity to be involved in these two interesting and challenging projects.

REFERENCES

1. Apsey R Q, Baynham D E, Clee P T M, Cragg D, Cunliffe N, Hopes R B, and Stovold R V, Design of a 5.5 Metre Diameter Superconducting Solenoid for the DELPHI Particle Physics Experiment at LEP. Applied Superconductivity Conference, 1984. Volume MAG-21, pp 490-3.

2. Morpurgo M, Design and Construction of a Pump for Liquid Helium, Cryogenics February 1977.

3. Gusewell D and Haebel E U, Current Leads for Refrigerator Cooled Superconducting Magnets, Proceedings ICEC 3, Berlin 187 (1970).

4. Evensen H, Fenyuk A, Stefanini G, Preliminary Report of the Field Measurements in the DELPHI Magnet. CERN internal report (CERN-EP 2-5-89).

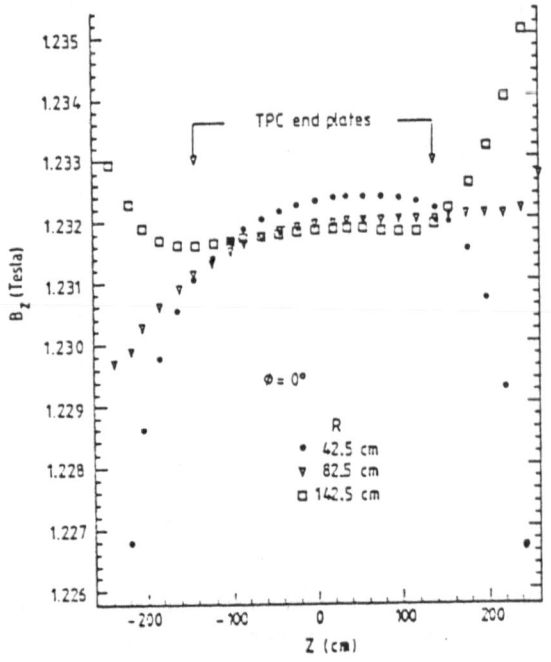

FIGURE 4   Bz Along Axis

MAIN RESULTS OF MAGNETIC MEASUREMENTS ON THE OAE MODIFICATION OF SSC2
(The second separated sector cyclotrons of GANIL)

D. Bibet, M. Barré, J. Blanchet, M. Duval, Cl. Eveillard, J. Fermé, A. Lemarié,
M. Legay, J.F. Libin
GANIL, BP 5027, 14021 CAEN Cedex - FRANCE

ABSTRACT

We describe magnetic measurements made to determine the perturbations induced by the injection magnets in the field pattern of a separated sector cyclotron. Results of a combination of shim plates and pole face winding to compensate for these pertubations are given. A method is developed to achieve nearly 10^{-4} accuracy for any radial field law required to accelerate an ion beam in the field strength range 0.65 T to 1.65 T. Some of the main characteristics of the magnetic measuring systems are also given.

INTRODUCTION

GANIL is a heavy ion accelerator composed of three cyclotrons CO2, SSC1, SSC2 connected in series. It has been running since January 1983 and has delivered about 28 000 hours of various beam from C at 95 Mev/u to Xe at 27 MeV/u.

During the last six months several parts of the machine complex have been modified to increase the energy range (by a factor 2) of heavy ion beams. This operation has been called "OAE"[1].

This paper deals with the magnetic field modifications of the second separated sector cyclotron (SSC2). The SSC2's magnet [2] consist of four 52 degrees sectors. Each sector has a pole area of 5 m^2, a 0.1 m air gap and weighs approximately 400 tons (fig n° 1).

Useful field strengths as a function of the magnetic rigidity of the accelerated ions are typically in the range 0.65 T - 1.65 T.

The main coils of the four sectors are supplied in series by a 1 MW power supply but each sector has an auxiliary

coil to equalize field level of the four sectors. A radial field law, - the isochronous field -, which fits relativistic accelerated ion mass variations is obtained by way of a pole edge profile and a set of pole face winding (trim coils).

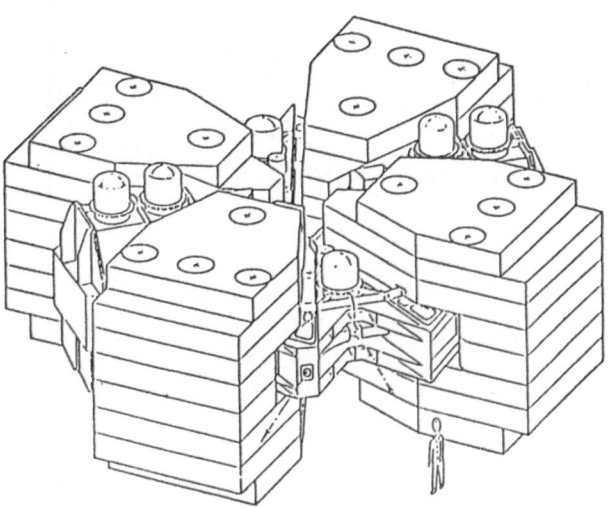

FIGURE 1 : isometric view of one of the GANIL separated sector cyclotrons sector magnets and vacuum chamber

In the central region of the SSC (fig n° 2) there are two window frame magnets (Mi1, Mi2) located between the noze of the sectors and two septum magnets (MSi3, MSi4) put inside the gap of sector A and B. These magnets are responsible for the beam injection.

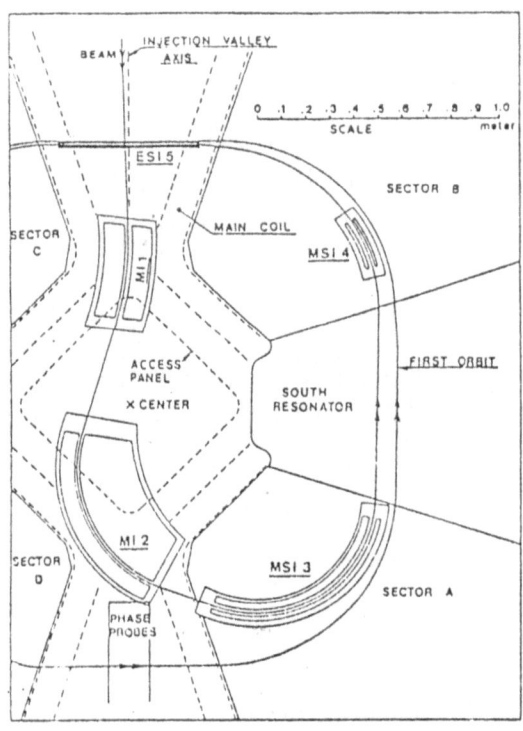

FIGURE 2 : beam injection system

With the OAE operation, the beam injection radius in SSC2 has been increased by a factor 1.4 and various components of the injection system were redesigned. Similarly the volume, position and the field strength of the injection magnets have been changed. One consequence of these changes is that field perturbations induced in the four sectors have to be redetermined and compensated for.

We sum up here results of magnetic measurements made to determine the new field pattern of SSC2 and how we recovered the field quality required to accelerate the heavy ion beams.

FIELD MEASURING SYSTEM

Details about the field measuring system are given in an internal re-

port [3]. Hence we recall only the main features here. There are two measuring devices. Both of them measure in cylindrical coordinates but do not have the same center.

Main system

A photograph of the main system is shown in fig n° 3.

FIGURE 3 : main measuring system

It is composed of a rotating ring rolling on a fixed ring (4.05 m diameter). A removeable fiber glass structure wich is attached to the rotating ring, support a hall assembly. The hall probes on this assembly are placed inside several copper blocks which are maintained at $37 \pm 0.1°$ C . The rotating ring is adjusted in concentricity on the machine center and the hall assembly is alligned on the mid plane level with an accuracy of 0.1 mm. The movable ring is moved by a step by step motor and the absolute azimuthal positions of the hall probes are red by an angular decoder. During the hall voltage measurements the rotating ring is locked in position by a pneumatic arm which is inserted in the peripherical grooves. 92 Siemens hall probes SBV601.S are used. They have been calibrated from 0.1 T to 1.8 T in a special magnet equipped with a NMR device.

Complementary measuring system

When the MSi3 septum magnet is installed in the gap of sector A, the inner part of the main hall assembly has to be turn away in order to avoid the septum. Since the MSi3 component is not concentric with the machine center and beam trajectories pass very close to it a special measuring apparatus had to be designed to get measurements in the useful area.

It consists of a carriage (which slides along a rail) pulled by an inox cable drawn from one end by a stepping motor and led on the other end an angular decoder. The hall holder which is fixed on the carriage is similar to that of the main apparatus but it only has 8 hall probes. The height of this measuring systems is only 39 mm because it must pass inside the gap of the compensating shim plate. These plates are attached to the MSi3 element.

To complete the data measurements 14 hall probes are also installed along the injection line. These probes are not temperature stabilized but their temperature are measured to correct for any increase of the hall voltage.

Finally, at the middle of each sector there is a NMR probe used to adjust exactly the field level.

Tests on the accuracy of the measurements have shown that they are better than 1 gauss (hall sensitivity is 4 μV/gauss, the hall voltage is measured by a SCHLUMBERGER 7075 voltmeter and noise and scanning effects are less than 2 μV).

Measurements are controlled by two BULL Micral 30 microcomputers. One is dedicated to the main apparatus, the other to the complementary system. Electronics are CAMAC modules and communications between command and high power are made by an opto coupler. Several interactive programs have been written to control the sequence of measurements and data acquisition. The operator is informed when a sequence error occurs and he can continuously view

the shape of the acquired field on a graphic display.

Mapping area

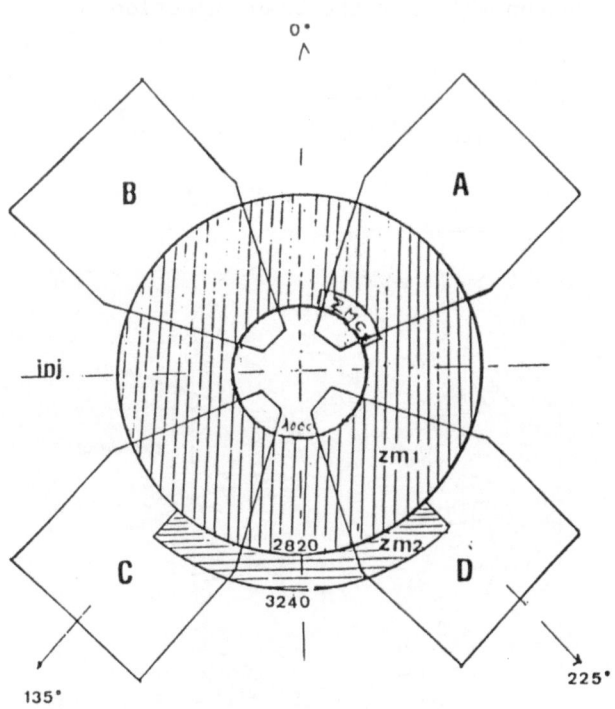

FIGURE 4 : mapping area

There are 3 measuring zones represented on fig n° 5. ZM1 and ZM2 are covered by the main apparatus, and ZMC by the complementary one. ZM1 radial extent is from 1 m to 2.82 m with 20 mm steps and the azimuthal extent is 360 degrees with 1 degree steps.

ZM2 is limited to only 90 degrees because of extraction elements. The main hall assembly is displaced to increase radius to 3.24 m. Two sets of hall probes are used for recovery measurements.

ZMC has a radial extent of 140 mm (20 mm steps) and an azimuthal extent of 120 degrees (with 1 degree steps) relative to the center of the complementary device.

ZMC and ZM1 have a large recovering area and a special code is used to transpose the measured points of ZMC in the ZM1 coordinate system.

DATA ANALYSIS AND RESULTS

In a separated sector cyclotron the radial field law for acceleration of ions is the same for the four sectors. In addition this law is made as symetrical as possible. However injection elements produce different magnitudes of perturbation in each sector. If we can cancel or more precisely compensate for these perturbations, then it is only necessary to supply the right current trim coils for the desired field law. Thus the measuring procedure will consist of the following :

1. The base field

This is the main magnet's field without any injection elements. In order to spare shutdown time of the machine, extraction elements were not removed. In fact nothing was changed in this region and the ZM2 measuring zone was used to substanciate this fact.

2. The perturbated field

It is the resulting field pattern after the injection elements are installed into the cyclotron.

3. The corrected perturbated field

It consists of determining the compensation necessary to come back on average to the base field of each 90 degree sector.

4. The corrected and isochronous field

It is the average field law $\left(\bar{B}(\bar{R})\right)$ suitable for any accelerated beam. Previous measurements made in 1986 with a prototype of the new MSi3 septum have shown that in order to have the necessary precision for interpolation of field data, we have to perform measurements for the 8 following field levels : 0.6 T, 1.0 T, 1.2 T, 1.4 T, 1.5 T, 1.6 T, 1.65 T, 1.7 T.

Base field results

Beam trajectories are circular inside the magnetic sectors, and almost straight lines between them. To calculate the average field on these trajectories, we integrate the measured field on the real beam path. The shim pole profile gives an average base field $\bar{B}$ (R) wich increases with the mean radius of trajectories, Υ – 1.05. This value of Υ is the middle radial evolution field law and it allows us to reduce the power in trim coils by $\approx$ 4 (Υ – 1.0 for heavy ions, Υ – 1.1 for light ions).

Perturbated field and corrected field results

As an example of perturbation fig 5 shows one existing at 1.6 T in sector A. Here we have made the difference between the perturbated field and the base field. On the left part we can see the fall-off field due to the Mi2 injection magnet which mainly picks up the fringing field of the sector. In the middle part of fig 5 we only see, in the accelerated region, the external effect of the septum magnet MSi3 which is composed of two iron-cobalt plates of 15.5 mm thickness. We can also see that its effect extends to almost the end of the sector. On the right part of fig 5 one sees the compensating shim effect. This shim is composed of two iron plates (1 mm thickness) located where the effect of the MSi3 septum is largest. The azimuthal and radial shape of this compensating shim have been experimentaly determined. It is a compromise between its local effect, which must be low enough so as not to affect the beam characteristics, and the reduction of the gradient $d\bar{B}/d\bar{R}$ of the average field perturbation. Because of the radial distance between the trim coil conductors

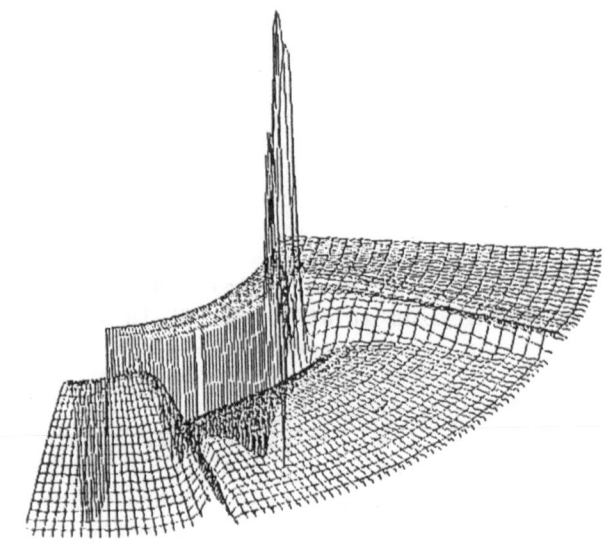

FIGURE 5 field perturbation pattern at 1.6 T in sector A

and the fact that the maximum current is limited to 300 A for the existing trim coils, reduction of the gradient perturbation was absolutely necessary. Unfortunately the resulting effect depends on the field level because permeability curves for the iron-colbat septum plates and those of the iron shim and pole sector are obviously very different.

Fig 6 shows an example of the resulting average perturbation obtained at 1.0 T, 1.4 T, 1.6 T and 1.65 T field levels. There is an overcompensation under 1.45 T but this is not a problem with trim coil current. In the other sectors,perturbations are lower than those of sector A. They are easily compensated for with trim coil currents.

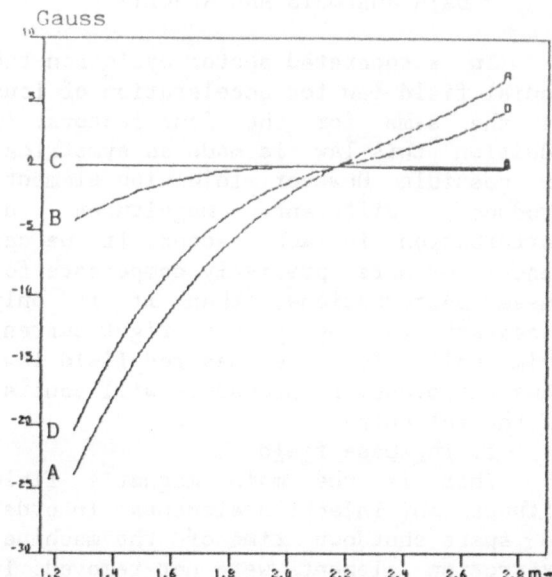

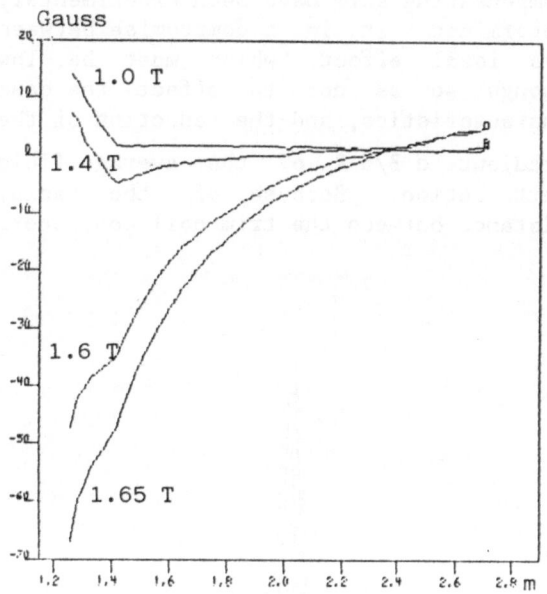

FIGURE 6 average field perturbation versus radius at different field level in sector A

Isochronous field results

In the original design when injection radius was 0.857 m, edge shims were added on one side of the poles of sector A and sector D to compensate for a part of the Mi2 effect. Currently with the 1.2 m injection radius these shims are no longer appropriate. Morover they modify the base field of these sectors . Nevertheless we have not

FIGURE 7 base field correction versus radius to equilize sectors A, B, D to the sector C

removed them because the base field of each sector are a little different due to iron dispersion in pole and yoke pieces. Indeed we decided to equalise as far as possible the base field of the four sectors before applying the isochronous law. To do this we chose sector C as a reference field and we used the independent trim coils available on other sectors to bring them to the base field of sector C.

We chose sector C for two reasons. Firstly, it is the least perturbated sector so its field pattern is the most constant. Secondly the base field of the other sectors are higher. The trim coils are used to decrease them. On the other hand the perturbation is greater in others sectors which means that there is a lack of field ; trim coils must increase the field. Since the trim coils used are the same for both effects the resulting current is lower. Fig 7 shows corrections we have to make at 1.6 T to equalize average base fields of sectors A, B and D to that of sector C. We can observe that it is the same order of magnitude as those of the injection perturbations. This means, that excess fields produced by edge shims, even if they are not well adapted, are still interesting.

Starting from the measured base field the ORBISO code was used to compute the trim coil contribution to the field law necessary to achieve the desired isochronous field to accelerate an ion. This field contribution depends on the sector. We calculate the difference between similar values for sector C. Then we obtain the necessary corrections as shown on fig 7. With regards to the isochronous correction for sector C, we compute, with the BOBO code, the current for the isochronous trim coils. It was also necessary to use the KB and KR functions. These functions give the correspondence between the average isochronous correction along the mean radius trajectories and axis isochronous correction along a sector axis where trim coil effects are given by the BOBO code.

An example of the axis isochronous corrections is shown in fig 8 for $^{12}C^{6+}$ ions at 95 MeV/u (dotted ligne, right scale).

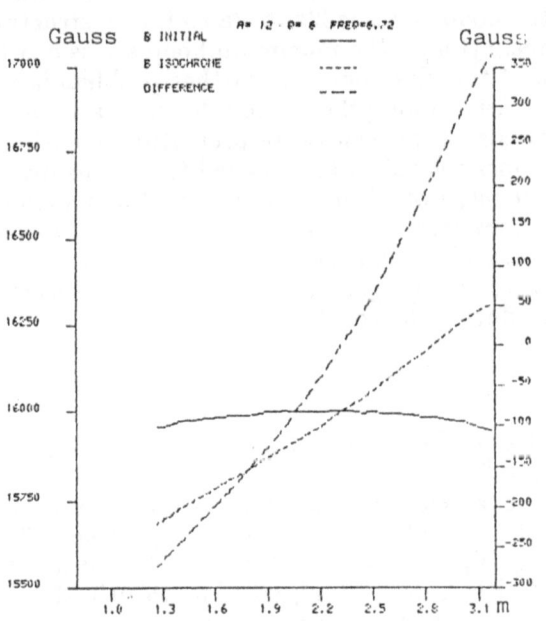

FIGURE 8 axis isochronous correction

The total current that one needs to apply in the independent trim coils will be the sum of values calculated to realize the previous three corrections (injection perturbation, equalization of sectors and isochronisation). However the trim coils supplied in series on the four sectors are only concerned with isochronous corrections.

After all corrections are applied, we measure the resulting field and fig 9 shows the residual difference from the desired field.

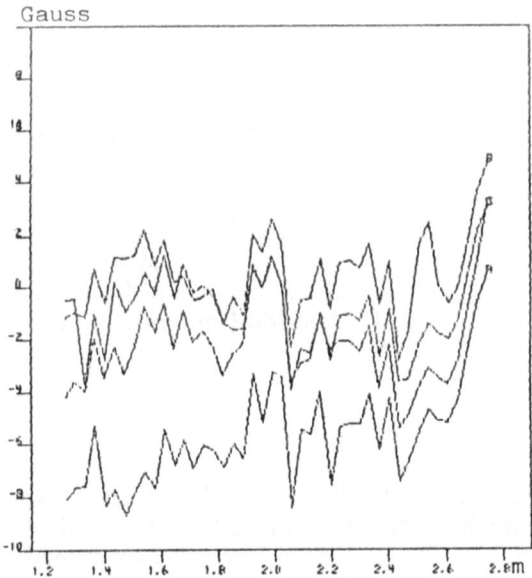

FIGURE 9 residual difference from the desired field

We should emphasise that the results presented here are the largest contributions to the corrections made by the trim coils. In spite of this the desired field is obtained on the average to nearly $\pm 10^{-4}$ ($\pm 2/16000$ Gauss). The small shift on sector A is related to field level and can be easily corrected with auxiliary coils.

REFERENCES

[1] J. Fermé, "Project OAE at GANIL", 11th International Conference on Cyclotrons and their Applications - Tokyo 1986

[2] J.M. Baze, D. Bibet, A. Dael, M. Gabrysiak, M. Ohayon, A. Riche, "The Ganil Magnet", 7th Int. Conf. on Magnet Technology - Bratislava 1977.

[3] M. Barré, Cl. Eveillard, A. Lemarié, J.F. Libin, "Bancs de mesures CSS2 et ligne d'injection OAE", GANIL internal report 89/14/Mag/03.

[4] D. Bibet and Al, "Mesures magnétiques sur les CSS, rapport n° 2, GANIL internal report 81R/022/TP/03.

THE UPGRADE OF MA.RI.S.A. FACILITY FROM 6.5 TO 10 TESLA DESIGN, CONSTRUCTION AND PRELIMINARY TEST

S.Angius*, P.Fabbricatore,R.Marabotto*, R.Musenich, R.Parodi
M.Perrella*, S.Pepe, R.Vaccarone
I.N.F.N., Sezione di Genova, via Dodecaneso 33, I-16146 Genova, Italy
(*)Ansaldo Componenti S.p.A., via N.Lorenzi 8, I-16121 Genova, Italy

ABSTRACT

The facility MA.RI.S.A. is based on a superconducting solenoid (MARISA I) with a maximum attanaible central field of 6.5 Tesla at 4.2 K and 1080 A supply current in a quite large bore: 500 mm I.D. The facility is upgraded by inserting into the basic solenoid a second one (MARISA II), that rises the field up to 8 Tesla at 4.2 K and 10 Tesla at 1.8 K in a bore 380 mm I.D. Mechanical stresses are the main problems of the inner solenoid; the hoop stress is 200 MPa so that a structure shrinking the solenoid is required. Using a stainless steel cylinder the maximum hoop stress can be reduced to 130 MPa without a significative reduction of the clear bore. A further problem is the interaction between the winding and the outer stainless steel cylinder; the friction due to relative axial motion, after the cool-down or under stress, of the two structures can cause premature quenching. To avoid this occurence, the banding structure of the inner solenoid was designed to be unisotropic i.e. composed by stainless steel and fiber glass thick strips, wound onto the solenoid and vacuum impregnated with epoxy resin. This structure behaves circumferentially like stainless steel but axially like the solenoid winding, so that no thermal stresses take place and, on charging the solenoid, the relative stress are reduced of a factor 2. In the paper these design guidelines are discussed together with the description of the technical solutions, adopted during the construction.

INTRODUCTION

The facility MA.RI.S.A.[1] is placed in the laboratories of Genova of I.N.F.N.. The facility was build-up in order to perform electrical measurements (critical current, magnetization, quench velocity,...) on Superconducting Cables for High Energy Physics applications. The magnetic field is generated by a solenoidal S/C magnet called MARISA I giving a central field of 6 Tesla in a bore of 500 mm at a supply current of 1000 A. Table I shows the main characteristics of the magnet. The maximum reached magnetic field is 6.42 T at 1070 A supply current. The samples under measurements are power supplied using a 10 KA Power Supply. These values of the magnetic field and sample supply currents allowed to perform some measurements on the S/C ca-

bles of the dipole magnets for HERA proton ring [2]
The accelerator machines of next generation (like LHC at CERN) requires High Field deflecting Dipoles (9-10 Tesla),so that facilities like MARISA must be upgraded up to 10 Tesla magnetic field. The upgrading follows two steps:
1) A insert S/C solenoid (MARISA II) is placed into the actual magnet and connected electrically in series, so that the magnetic field is raised to about 8T at 4.2 K.
2) Both the solenoids are cooled down to 1.8 K (using a superfluid subcooled refrigerator); at this temperature the Nb-Ti filaments, composing the winding of the magnets, shows critical fields of about 3 Tesla higher than at 4.2 K (at fixed current). This relevant property allows to increase the magnetic field up to 10 Tesla.

In the following sections our attention is mainly devoted to the insert solenoid MARISA II, the design problems of which are discussed in detail.

Next sections, after a brief description of the magnet geometry, it is shown the way this problem was solved. In Fig.1 an artistic picture of MARISA I+II is shown.

TABLE 1

MARISA I characteristics

ID(bore)	504.0	mm
OD	742.0	mm
Axial Lenght	622.9	mm
N.Layers	36	
N.turns	4176	
Wire dim.	3.2 x 5.2	mm^2
Weight	1.1	ton
Inductance	6.4	Henry
Magn. Field	6.0	Tesla
at Curr.	1000	A.
Stored Energy	3.2	MJoule

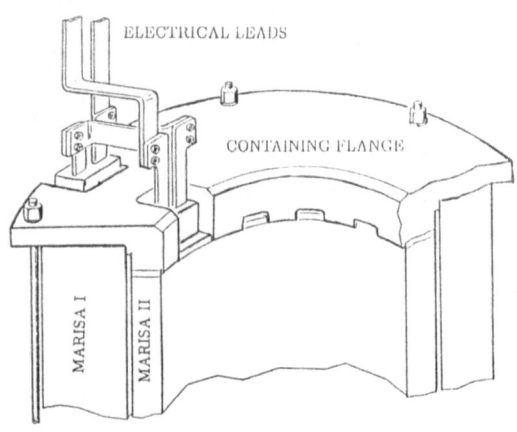

FIGURE 1: Artistic view of MARISA I+II

MAGNET STRUCTURE

The main characteristic of MARISA II is that (like the outer solenoid MARISA I) the winding is impregnated under vacuum with epoxy resin. MARISA I is mechanically self-supporting i.e. the magnetic circumferential and axial forces are supported by the winding, that behaves like an unisotropic but omogeneous medium.
When MARISA I was designed (during 1982), in order to have the minimum available space of the winding envelope we chose not to have a magnet cryogenically stabilized. We thought that a careful winding procedure, a good under vacuum impregnation and a suitable "recipe" of the resin would both forbid any (micro)movements of the turns and limit the energy release of cracking of the impregnant. Our assumption was drawn after the construction and testing of some prototypes performed in the ANSALDO laboratories [3]. When MARISA I was tested it reached 1.08% of nominal current before the first quenching proving that the good results obtained with small S/C solenoids (up to 30 Kg the big one) could be extended to bigger size magnet (at least 1 tons like MARISA I).
On designing MARISA II we decided to follow the same philosophy i.e. to construct a solenoid self supported with no mechanical restraining structures. Nevertheles this design concept found a strong difficulty in the large hoop stress of MARISA II winding due to the Lorenz force. It was clear since the begin of the design that a restraining structure was not avoidable.

GEOMETRICAL AND ELECTRICAL PARAMETERS

The above consideratons about the necessity to put a containing structure of the inner solenoid determinated the geometry of MARISA II. The internal diameter of MARISA I is 500 mm; leaving about 40 mm in diameter for the Liquid Helium access and the mechanical structure of MARISA II the Outer Diameter of MARISA II was put about 460 mm. After the choice of the conductor the geometry was defined as shown in Table 2.

TABLE 2

Inner solenoid parameters

ID(winding)	384.0	mm
OD(winding)	461.0	mm
Axial Lenght	600.0	mm
N.Layers	10	
N.turns	1580	
Wire dim.	2.6 x 3.8	mm^2
Weight	276.0	Kg
Inductance	0.525	Henry

The load lines with the working points of the MARISA I+II magnet are shown in Fig. 2.
The two relevant working points are: (I=920 A; B$_{central}$=8 Tesla) at 4.2 K and

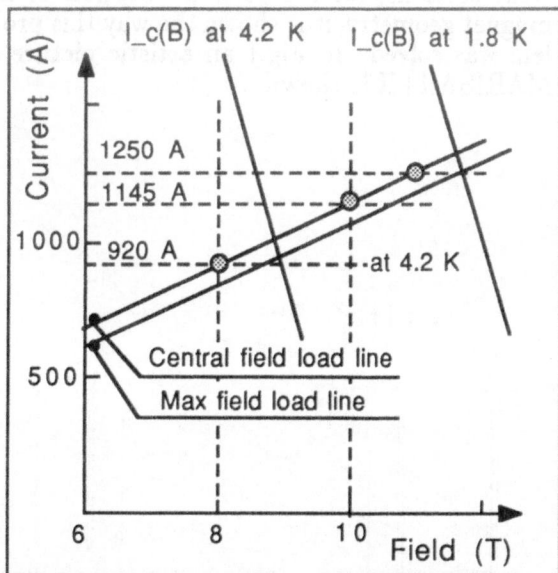

FIGURE 2:MARISA I+II load lines and
working point

TABLE 3

MARISA I+II Characteristics

Inductance	9.0	Henry
at 4.2 K		
Nominal Current	920	A
Central Field	8	Tesla
Max field at winding	8.6	Tesla
Stored Energy	3.8	MJoule
at 1.8 K		
Nominal Current	1145	A
Central Field	10.0	Tesla
Max field at winding	10.8	Tesla
Stored Energy	5.95	MJoule

(I=1145 A;$B_{central}$=10 Tesla) at 1.8 K. Really
the superconducting characteristic of the cable
would allow to obtain up to 10.9 Tesla at 1250
A. In spite of this limit should be considered an
ideal working point, we decided to perform the
mechanical design in order to be able to charge
the magnet up the ideal limit.

The conductor was chosen imposing the condition that the magnet should be stable against
disturbances induced by the transition of S/C
samples (the critical current of which is under measurement) placed very near the magnet
winding.

This requirement had two effects on the conductor:

1) It was composed by twisted sub-cables
(strands)

2) The strands were twisted onto a copper core
in order to rise the minimum energy release producing a quench.

The conductor was manufacted by Europa Metalli; the strands have a diameter 0.84 mm and
a ratio Cu/SC of about 1.8. The same strands
are used to compose the cables for HERA dipole
magnets. Table 3 shows the Electrical characteristics of MARISA I+II.

MECHANICAL DESIGN

Choice of the structure

The mechanical design has carried out in order to charge the magnet up to 1250 A. At

this current the maximum field at the winding (locate at the internal radius of MARISA
II) is 11.7 Tesla. The hoop stress results to be
200 MPa; this value is considered too high for
the winding. A suitable hoop stress value is
130 MPa. In order to reduce the hoop stress
at this value it is necessary to mechanically
shink the inner solenoid. A mechanical structure fitting our requirement could be stainless
steel cylinder 10 mm thick shinking the inner
solenoid.

In this case the structure of the inner would
be a solenoid vacuum impregnated with epoxy
resins, without internal mechanical structure,
srunk-fit into a stainless steel cylinder.

Nevertheless this technical solution is the cause
of several problems, due to relative axial movements between the inner solenoid and the mechanical structure. We point out two possible
situations:

1) Cooling down the magnet

The stainless steel cylinder contracts axially for
about 3060 μmm/mm, while the winding contraction is calculated to be 3360 μmm/mm.
The different contraction causes an axial relative force of about $1.5 \ 10^6$ N. This force is balanced by the friction between the cylinder and
the coil; nevertheless the friction does not forbid
micro-movements generating heat. During the
operation of the magnet this kind of mechanical
disturbance can cause a quench of the magnet.

2) Operating the magnet.

The radial magnetic field at the ends of the
inner solenoid causes an axial force inward directed of about $3.7 \ 10^6$ N; the force is applied
only to the solenoid but is balanced by the friction. We are again in the situation discussed
in item (1); releases of mechanical energy can
cause quenching.

In order to avoid friction problems between
solenoid and cylinder two solutions were analysed:

-I- Due to thermal contraction and magnetic force the solenoid would contract for about (300 +700) μ mm/mm respect to the cylinder. If the cylinder were compressed axially respect to the solenoid up to a deformation of 1000 μ mm/mm (equivalent to a force of $5.2\ 10^6$ N) the further relative movements would be compensated by the applied axial pre-stress

This solution shows the disantvantage of requiring complex and heavy mechanical structures (tick flanges, bolts,...) at the side of inner solenoid, so that we tried a different approach.

II- The chosen solution acts toward a modification of the -structure of the outer cylinder. We planned to wind a composite strip stainless steel-fiberglass onto the inner solenoid, already impregnated and machined; after winding a new impregnation is carried out. Regulating the amount of stainless steel and fiberglass in the strip it is possible to control the mechanical parameters in the axial direction. When the parameters are adjusted the thickness of the containing structure is calculated in order to have a maximum hoop stress of 130 MPa.

Such as kind of the mechanical structure behaves circumferentially like stainless steel and axially like the solenoid so that the most of problems would be solved:

Temperature - The thermal contraction does not give rise to any relative movements.

Magnetic force - The relative movements due to the Lorentz's force are forbidden by the gluing of the impregnant. Furthermore the Young modulus in axial direction of the outer containing structure is reduced respect to a stainless steel cylinder; this implies, as shown later, a reduction of the elastic energy at the interface and consequentely less possibility of relative micro-movements.

In next section the mechanical model used to calculate axial and circumferential stress is described.

Mechanical model
Axial direction

The solenoid and the mechanical structure are schematized like two slabs connected as shown in (Fig.3 (a)); let we call ΔX_s and ΔX_{ms} the two thickness, ΔY the deep and Z the length.

Applying an axial force F, i.e. a force normal to the surfaces $A_s = \Delta Y \Delta X_s$ and $A_{ms} = \Delta Y \Delta X_{ms}$, the solenoid contracts. The mechanical structure, being connected to the solenoid, contracts too, forbidding a complete deformation of the solenoid (Fig.3 (b)). Setting E_s and E_{ms} the Young moduli, the part of the force F supported by the solenoid is:

$$F_s = \frac{KF}{(1+K)} \quad (1.a)$$

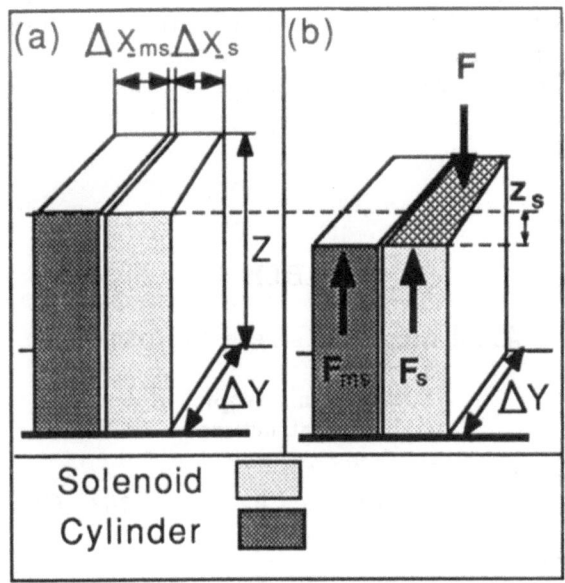

FIGURE 3 : Mechanical model for axial stress

being $K = \dfrac{\Delta X_s E_s}{\Delta X_{ms} E_{ms}}$. While the force supported by the mechanical structure is:

$$F_{ms} = \frac{F}{(1+K)} \quad (1.b)$$

Our aim was to obtain a mechanical structure with a very low axial elastic modulus so that it contracts following the solenoids but remains at low stress level. An interesting parameter is the elastic energy, stored in the mechanical structure of the magnet, that could be transformed to heat. Let we set $z_{s\ max}$ the maximum deformation of the solenoid if the whole force F would balanced only by the solenoid; furthermore let be z_s the axial deformation of outer mechanical structure and solenoid in the situation shown in Fig.3 (b). The energy that can be released as heat is:

$$H = \frac{1}{2\,Z}[A_s E_s (z_{s\ max} - z_s)^2 + A_{ms} E_{ms} z_s^2] \quad (2)$$

Rearranging Eq.2, the Energy H can be written as:

$$H = \frac{1}{2}\frac{F\ z_{s\ max}}{1+K} \quad (3)$$

Using Eqs. 1.a-1.b and 3 we calculated the axial forces and the energy H in two situations: Stainless steel cylinder 10 mm tick and unisotropic banding structure 14 mm tick. The results of this comparation are shown in Table 4. The thickness of the mechanical structure is adjusted in order to have a maximum hoop stress of 130 MPa. It is observed in Table 4 that the stored elastic energy at the interface is reduced of a factor 2 using an unisotropic restraining structure.

TABLE 4

Axial magnetic stress comparation

	S.Steel Cylinder	Unisotropic Structure	Solenoid	
Young mod. at 4.2 K	210	68	103	GPa
Δ L/L 300 → 4.2 K	3060	3360	3360	$\mu\frac{mm}{mm}$
$F_{m s-s}/F$	0.34	0.20	0.63-0.80	
H	30	15	-	Joule

Circumferential direction

The solenoid and the containing structure are schematized like a series of homogeneous cylindrical shells interacting mechanically between them. The solenoid is divided into 12 shelles (just the turns number); the outer structure into a single shell. A computer code was developed in order to calcultate the circumferential stress and strain. A correct determination of the magnetic stress at the solenoid requires the fully knowledge of the history of the magnet. We planned to wind the solenoid with a tension of 850 N, corresponding to a tensile stress of 60 MPa, onto a steel mandrel 20 mm in tickness. After winding the inner layers loss a part of the winding tension due to the deformation of the mandrel; the stresses at the layers range from 20 MPa (most innerlayer) to 60 MPa (most outer layer). After impregnation and mandrel removal the stresses fall down: the inner layer is now in compressive state (- 27 MPa) while the outer layer is in tension for about 23 MPa. The banding structure is wound onto the solenoid with a tension of about 100 MPa. After this winding the solenoid is compressed for about -21.5 MPa, while the tension of the banding structure ranges from 50 to 100 MPa. Cooling down the system, there is no significative variation of the stresses. Charging the magnet at 1250 A, the inner solenoid is put in tensile state for 130 MPa and the outer banding structure is put in tensile state too for about 250 MPa.

CONSTRUCTION

The solenoid MARISA II was constructed in ANSALDO COMPONENTI in Genoa during 1988 and earl months 1989. The manifacturing procedure followed without problems the design guidelines. Respect to the design there was some changes in the impregnation procedure. The solenoid was impregnated twice: first time only the solenoid winding was impregnated with epoxy resins reacted at 130 K. The second time the system solenoid-banding was impregnated. In the second impregnation a different epoxy was used in order to have a reaction temperature less that the glass-transition of the first epoxy resin. Following this procedure we avoided to loss the mechanical prestress of the solenoid and banding structure.

CONCLUSIONS

MARISA II is actually in mounting stage at INFN laboratory of Genoa. The first coold test of the MARISA I+II is planned for end 1989. The magnet will be operated at the maximum current at middle 1990. In the meantime the problems connected to the refrigeration down 2.17 K are to be solved. Against our first plans we think to use a lambda-plate refrigerator; the experiments performed on a prototype of superfluid refrigerator [4] gave us the indications for the design of a bigger refrigerator for MARISA.

REFERENCES

1. P.Fabbricatore, A.Parodi, R.Parodi, R.Vaccarone, MARISA a test facility for research in applied superconductivity. Proocedings ICEC 12 Butterworths, 1989, 879-882.
2. P.Fabbricatore, A.Matrone, A.Parodi, R.Parodi, C.Salvo, R.Vaccarone, A multiple sample holder for J_c measurements on HERA Cables. Proocedings ICEC 12 Butterworths, 1989, 903-907
3. P.Fabbricatore. Graduating thesis Universita' di Salerno A.A.1981-82
4. P.Fabbricatore, G.Gemme, M.Losasso, A.Parodi, R.Parodi, C.Salvo, R.Vaccarone, A Prototype refrigerator for subcooled superfluid helium at 1.8 K. Design and test. Proocedings ICEC 12 Butterworths, 1989, 286-289

DYNAMIC AND MAGNETIC BEHAVIOUR OF LARGE SUPERCONDUCTING COILS IN THE MAGNET OF THE MILAN HEAVY ION CYCLOTRON

E. Acerbi and L. Rossi

LASA - INFN and Universita' di Milano, Via Fratelli Cervi 201, 20090 Segrate (Mi) - Italy

ABSTRACT

The magnet of the Milan Superconducting Cyclotron was successfully excited at the beginning of this year. The maximum average field of 5 Tesla has been reached without trouble (no training and no short circuit). The superconducting coil, 2.4 meter in diameter, wound with a NbTi monolithic cable, has been assembled in two sections, independently excited, and installed in a complex iron circuit. Large forces and moments, strongly dependant on the current level, arise by the radial and axial decentering and tilting of the coil and by the iron yoke asymmetry. The main results of the first excitation of the coils and measurements of the forces and moments exerted on the coils are presented and compared with the calculated ones. Finally the effect of the coil decentering on the field quality is discussed.

INTRODUCTION

The Milan Superconducting Cyclotron [1] is a three sectors compact machine designed as a booster for the 15 MV Tandem of LNS,an Italian national lab for nuclear research in Catania (Sicily). It will be used to accelerate all kinds of ions with a maximum energy from 20 MeV/nucleon for the heaviest ions up to 100 MeV/nucleon for the fully stripped ions.

The magnet was excited at the beginning of the year (1989) and, after measurement of the most important parameters like inductance, stored energy, field stability and stress on the coils and on the cryostat, the field measurements have been carried out. Herein the main experimental results are presented and compared with the expected ones.

In fact the choice of superconducting coils has reduced drastically the magnet size and therefore the cost, but increased the technical problems. The iron circuit consist of two circular poles on which 3 spiral sectors are attached; the sectors and the circular poles are fully saturated in all the machine operating range. The return flux yoke has a lot of asymmetries and its magnetization varies from a low value up to near saturation according to the central field variation from 2 Tesla up to 5 Tesla.

Another peculiarity is the subdivision of the coils into two sections with the possibility of negative (with respect to the main field) excitation of one coil section. When this occurs the forces between iron and coils become more difficult to handle and the rest position of the coils with respect to the magnetic axis of the iron is changed too. This problem is quite important for the high field quality required by the beam acceleration. Furthermore, when one coil section is negatively excited, it becomes mechanically unstable. The system used to counteract the negative effects is described in the next section of this paper and its performance in the section on Coil and Cryostat Stress.

Another critical point is the stress on the median plane plate supporting the coils. Through this plate, which is part of the cryostat, big channels for beam injection and extraction have been drilled, so it undergoes severe stress. Data are presented and compared with the calculated ones.

Finally, the magnet has been mapped at different field levels; the results of the field map analysis con-

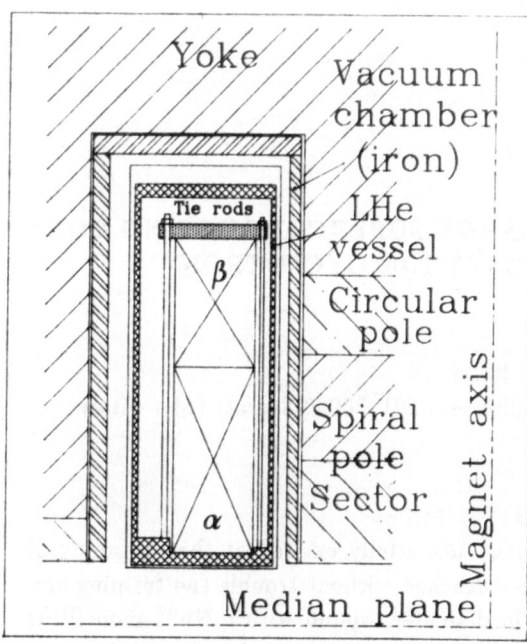

Figure 1: vertical section of the coils

cerning the coils behaviour are reported in the last section.

BRIEF DESCRIPTION OF THE COILS

The coils, built with the double pancake technique, look like split coils symmetric with respect to the median plane; in Figure 1 a vertical section of the upper coil is shown and in Table 1 the major parameters of the coils and of the superconducting cable are reported. The average maximum current density of the coils is 3500 A/cm2; their internal diameter is 2 m and the outer one is 2.4 m; the total height is 1.4 m and the gap, to allow the beam in and out, is 120 mm. Coil support is provided by a stainless steel ring plate where the horizontal links of the coil vessel are attached ; many holes are drilled in this plate, both for the beam trajectory and for different utilities like the beam extraction system and the probes for the beam diagnostic, making the plate itself weak with respect to the axial compression given by the coils. The same holes are drilled in the vacuum chamber of the cryostat and in the yoke with a strong azimuthal asymmetry. The effects on the field in the beam acceleration region are almost compensated (within 6-8%) by additional holes, but the compensation is not so effective in the coils region.

The helium vessel, where the superconducting coils are located, is made of stainless steel and is sur-rounded by a liquid nitrogen cooled thermal shield and by the vacuum chamber. This last is made of iron, so is part of the iron circuit. The vessel position is fixed by axial and radial titanium rods whose stress values are carefully monitored since they are the most precise indicator of a coil displacement. The radial tie rods are positioned every 120° and prestressed at 70 MPa, later increased to 130 MPa; while all the rods remain prestressed the radial rigidity is 55 kN/mm, against a calculated decentering force, arising from non-coaxiality between coils and iron magnet, of 10-15 kN/mm. The axial tie rods, also every 120°, have a rigidity of 65 kN/mm, whereas the cryostat weight is about 150 kN.

The coils are divided into two sections independently supplied, the α coil (the biggest one, 3/5 of the total Ampereturn, and the closest to the median plane) and the β coil. The current $I\alpha$ in the α coil can reach 1950 A, whereas in the β coil $I\beta$ ranges between -650 and 1800 A. When the β coil is positively excited, the total compression force is 15 MN. When it is negatively excited, the β coil is pushed away from the α (and from the median plane as well) by a 6 MN force; moreover the radial stress in the coils turns negative.

TABLE 1
Coils and Cable Parameters

inner radius (mm)	1000
outer radius (mm)	1165
total height (mm)	1480
midplane gap (mm)	120
α coil Ampereturn	3.8 10⁶
β coil Ampereturn	2.7 10⁶
maximum current (A)	2000
Inductance air core (H)	19.8
cable dimensions (mm²)	3.5x13
superconductor	NbTi
Cu/NbTi	20/1
NbTi Jc (kA/mm²) at 5 Tesla	1.25
Copper RRR (at B=0T)	160

To balance the destabilizing axial force a stainless steel plate on the top of the β coil keeps them with a 7 MN force ; 210 tie rods of Beryllium(2%)Copper bolted to the stainless steel midplane plate provide for the 7 MN. The choice of using a copper alloy has been done in order to minimize the load loss during the magnet cool-down at LHe temperature due to the differential thermal contraction between the tie rods and the coil stack. When both α and β coils are positively excited these 7 MN add to the axial

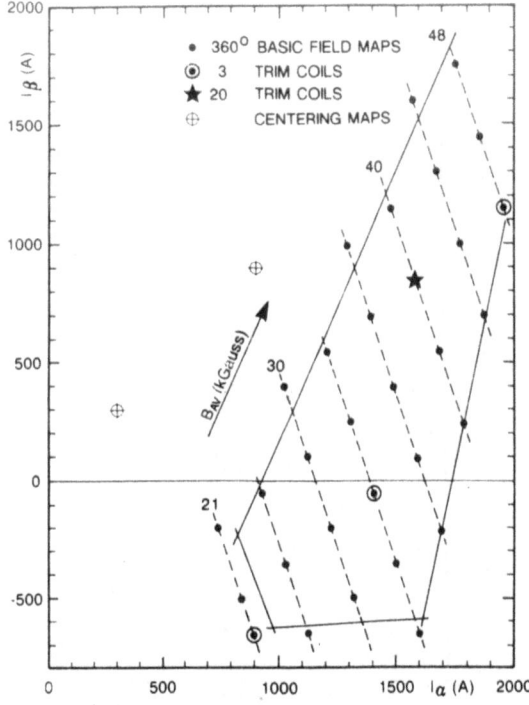

Figure 2: operating diagram of the cyclotron. The grid for the magnetic measurement and the flux contour lines (dashed lines) are indicated.

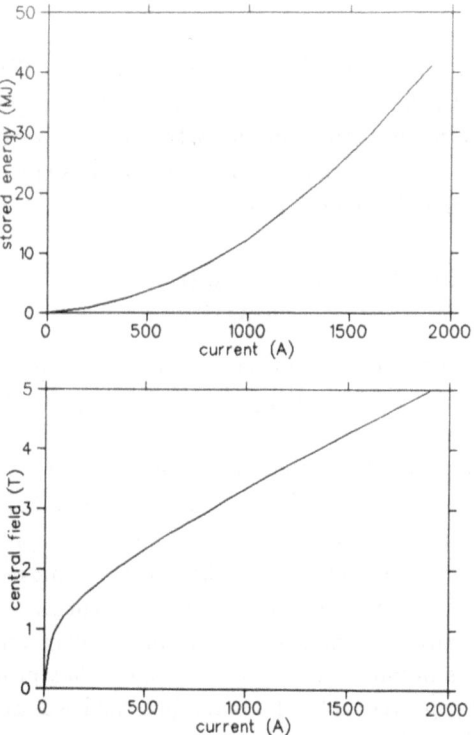

Figure 3: measured central field and stored energy vs current (along $I\alpha = I\beta$ line).

magnetic force, loading the stainless steel midplane plate whose stress in the weakest points, injection and extraction channels, approach the safety limit of 100 MPa.

The operating diagram on the $I\alpha, I\beta$ plane is shown in Figure 2, where the grid for the magnetic field measurements and the flux contour lines are shown. As can clearly seen, any field level can be achieved with different $I\alpha/I\beta$ values in order to get the suitable radial field profiles to accelerate any kind of ions,i.e., to fulfill the isochronism requirement for ions with different masses.

Because of the wide field range and of the relative variation of current of the two sections, the possible coil imperfections cannot be completely corrected with an adjustment of the coil position. Indeed the coils were assembled with precise magnetic measurements, and the Fourier harmonic content was minimized, especially the 1st and the 2nd harmonic [2], in order to keep low both the field imperfection and the forces arising from interaction between the coils and the iron magnet.

FIRST EXCITATION OF THE MAGNET

The superconducting magnet reached its maximum field value, 5 T at $I\alpha/I\beta$=1950/1750 Amps,

without any quench; the whole range of negative current in the β section was covered without indication of mechanical or electrical instability, too.

The behaviour of the Quench Detection System (QDS), comparing the voltage across double pancakes symmetric with respect to the median plane, was carefully investigated both in steady state and during the magnet charge. The noise was less than 5 mV in steady condition and 30 mV during the charge. Anyway the current ramp rate has been kept below 1 A/s in order to avoid spurious signals of the order of 70-100 mV, especially at low current, where the big inductance (200 H) due to the iron being not yet saturated can easily be a source of voltage noise.

The field and the stored energy are shown in Figure 3; it's noticeable that the measured maximum stored energy is 40 MJ, in good agreement with a simple analytical calculation and significantly less (20%) than the values predicted by a Poisson code used in our laboratory [3].

The reproducibility of the field level when the a current value is reached following different paths in the $I\alpha, I\beta$ plane has been found to be ±30 ppm (measured at 3.2 T); field stability has been measured to be ±20 ppm in a day operation.

At full excitation the attractive force on the poles is 15 MN; the measured reduction of midplane gap is .6 mm. The gap reduction achieves its total value when the current in the two coils is about 600 A; in fact the field is around 2.3 Tesla, so the iron poles are fully magnetized and the attraction between them does not increase anymore. This value is close to the calculated one of .5 mm.

INTERACTION BETWEEN COILS AND IRON MAGNET

Before their excitation the coils were centered to .1 mm radially (coaxiality with the iron poles) and .25 mm axially (coincidence between the median planes of the coils and of the iron). Above some current values we detected a sharp increase of the radial forces; analysing the data indicated by each link we found out which direction the coils had to be moved. The displacement was done without current in the coils and after the magnet was ramped again until the forces became once more too high. This procedure was repeated until the maximum field was achieved both in the positive and in the negative part of the operating diagram.

Eventually, after a total .15 mm of displacement, we found a position which allowed us to cover the entire operating diagram; in this position the forces on the coils were reasonable but still higher than expected. This position was kept as reference position for the field maps.

In Figure 4 the forces exerted on the coils on a current grid of the operative diagram are represented with arrows giving the intensity and the direction (positive $I\alpha$ axis is used to define the zero angle). When the coils are in the reference position, solid arrows, the force at the top diagram is about 13 kN; the force in the bottom part reaches 16 kN. The direction is almost opposite and this does not allow any force reduction over the whole operating region. Unfortunately in this position the coils generate a remarkable first harmonic component.

After the end of the field mapping, the coils were radially displaced by .5 mm and excited, in order to measure both the field imperfection and the force increment. The dashed arrows in Figure 4 represent the measured radial forces on the $I\alpha = I\beta$ loading line. The excitation was stopped at 1200/1200 (about 3.8 T) because the forces became so big that stress of the radial link approached its safety limit of 180 MPa. The measured force per unit displacement as a function of the field is shown in Figure 5. The values are much higher than the calculated ones based on a numerical code calculation of the field [4].

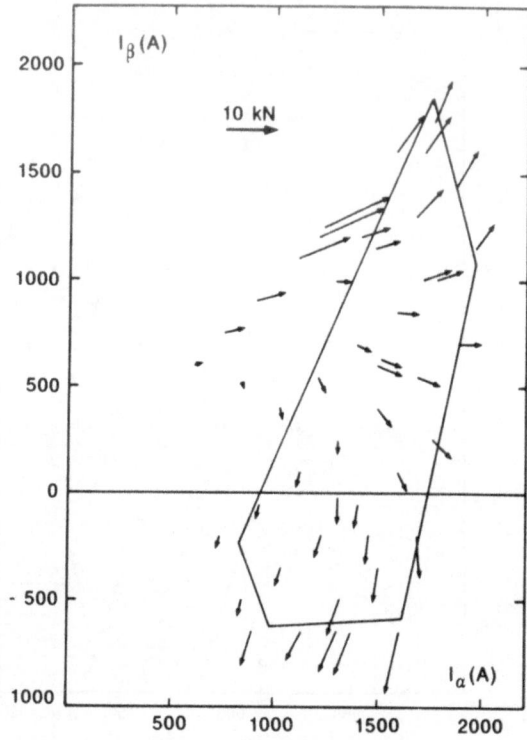

Figure 4: the arrows represent the force exerted on the coils at different current values.

A general comment on these problems is reported in the section on Field Measurement Analysis.

During the first excitation, the load on the axial links increased in a unbalanced way. The measured axial magnetic force was 22 kN on which a torque of 20 kN∗m was superimposed, with a corresponding axial shift of .3 mm and axis tilt of .2 mrad. Therefore the coils were lifted by .2 mm and tilted by .4 mrad. As a consequence the torque was virtually cancelled and the net axial force slightly decreased. Since the midplane error was within .2 mm at the maximum field no further reduction was attempted.

COILS AND CRYOSTAT STRESS

The hoop stress in the coils at maximum field is 80 MPa, with a corresponding maximum radius increase of about 1 mm [5]. As already reported in the coils description, the maximum total axial force toward the median plane is 22 MN, 15 MN given by the Lorentz force and 7 MN given by the prestress. This 22 MN load creates a strong coupling between the coils and the stainless steel plate supporting the coils. Because such a large external compression had not yet been attempted in this kind of magnets, any aspect was carefully studied, mainly to avoid the stick and slip regime which could be a source both of quench of the superconducting coils

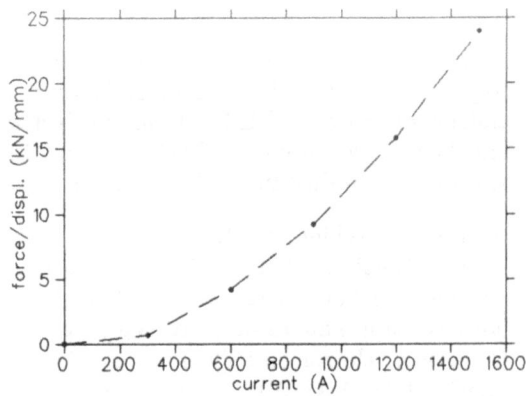

Figure 5: force between coils and iron magnet for a unit displacement of the coils axis.

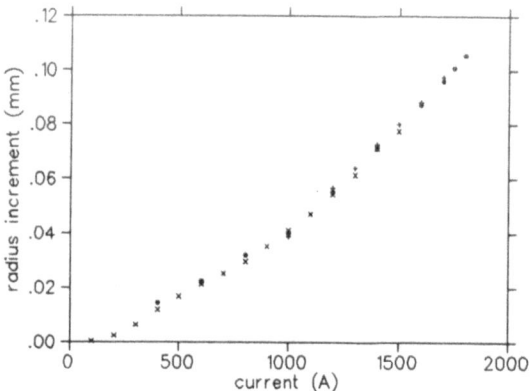

Figure 6: average increment of the midplane plate supporting the coils vs the current for the first three magnet excitations.

and of slow drift of the coil position with respect to the cryostat [7]. To allow the coils to slide without jumps, a thin teflon foil between the plate and the coil insulation had been inserted during the coil assembly.

The increase of the plate radius versus the magnet current was measured and results are plotted in Figure 6. The data are consistent with a static friction coefficient less than .1 [6], showing that under magnetic forces the radius of the coils grows smoothly and the teflon lubrication is effective even at low temperature. This was definitively confirmed also by the absence of any quench indication.

As far as the Be-Cu tie rods are concerned, their stress was measured by means of strain gages accurately calibrated to remove the influence of cryogenic temperature and of magnetic field from the signal. At the largest negative current in the β section their stress increases by 20 MPa only, starting from a value of 250 MPa, confirming the compression system works pretty well and allows us to reach negative current values higher than expected. In fact a current $I\beta = -750$ A was reached (the nominal maximum value being -650 A); even higher values will be tried in the next field mapping. In this way the operating diagram of the machine could be enlarged to include a final energy of 120 MeV/nucleon for He3 ions.

As regards the stresses on the stainless steel midplane plate where injection and extraction holes are drilled, they were measured by means of strain gages, too. Both in the extraction channel and the injection one the maximum value was about 80% of the calculated one, the highest stress level being 85 MPa in the injection. The small discrepancy between the calculations and the experimental values is probably due to underevaluation of the contribution of the coils in

supporting their own axial load.

FIELD MEASUREMENT ANALYSIS

An extensive field mapping on the magnet has been carried out and its main results concerning the field quality and the comparison with the calculation are reported elsewhere [8]. Therefore in this section we report some results obtained in a successive analysis and concerning mainly the magnet behaviour.

The coils, as already mentioned, were not positioned in the magnetic center. Indeed the first harmonic amplitude generated by the coils in the median plane at the extraction radius, r = 86 cm, ranges from .6 up to 1.6 mTesla, depending on the field level and on the ratio $I\alpha/I\beta$ used to reach that level. A detailed data analysis considering the whole set of maps indicated that the coils displacement from the magnetic center needed to produce such a 1st harmonic must be about 1.1 mm.

It also turned out that the radial profile of the 1st harmonic generated by a coils displacement is quite different from the air core calculation, even if the iron pole is well saturated for the whole operating diagram, from 2 up to 5 Tesla; full saturation was confirmed by the invariance of the flutter versus the field level. In Figure 7 both the experimental and the air core calculated 1st harmonic is plotted versus radius; the 1st harmonic grows linearly both versus current and displacement.

To the 1.1 mm basic displacement, an additional shift, variable according to the currents from -.28 mm to .23 mm, must be added; it is caused by the large forces the coils undergo within the operating diagram. Th 1st harmonic increment due to this additional shift is $\pm.3$ mTesla.

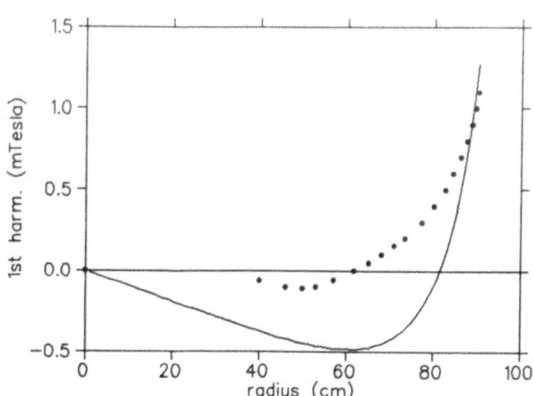

Figure 7: 1st harmonic given by 1 mm of radial displacement of the coils. Solid line represents air core calculation and dots the measured data.

The difference between the magnetic field center and the magnetic force center is almost due to lack of symmetry in the iron yoke holes; they produce a 1st harmonic component of 1.5 mT at the extraction radius (of which .5 mT will be eliminated in the next assembly of the iron vacuum chamber of the cryostat). This explains the almost quadratic increase vs current of the coils-iron magnet force per unit displacement shown in Figure 5. If the main force were between coils and pole, it would increase almost in linear way, being pole saturated over 300 A. Anyway a full understanding of this problem is not yet reached, especially at low field, below 3 T, where yoke contribution should be negligible but the forces are still higher than expected.

To overcome this problem some solutions have been envisaged: the most simple one is to increase the stiffness of the radial links, for instance by substituting titanium with a suitable steel alloy, but unfortunately this would mean an increase of the helium losses, too. Another solution is to replace the metallic links with links made of insulating material such as G11. Another possibility now under investigation is to improve the uniformity of the penetration into the yoke median plane to reduce the forces as well as its 1st harmonic contribution.

ACKNOWLEDGEMENTS

The authors thank Prof. G. Bellomo and Dr. P. Gmaj for their contribution in the field measurements analysis and Dr. F. Alessandria for the helpful discussions on the cryostat stresses.

REFERENCES

1. Acerbi, E. and Milan Cyclotron Group, Progress Report on the Milan Superconducting Cyclotron, presented at XII Int. Conf. on Cyclotron and Their Application, Berlin (FRG), May 1989, and to be published in the Proceedings.

2. Acerbi, E., Aghion, F., Baccaglioni, G., Mora, M. and Rossi, L., Magnetic Procedure for the Assembly of Superconducting Coils at the Milan Cyclotron Laboratory, Proceeding of X Int. Conf. on Cyclotron and Their Application, East Lansing (USA) April 1984, pp 75-78

3. Fabrici, E., private communication. Poisson code is POISCR the CERN-Poisson program package, unpublished CERN report.

4. Fabrici, E., Magnetic Forces on the Coils of the Superconducting Cyclotron at the University of Milan, Report n. INFN/TC-82/10, Physics Department of Milan University, Milan, July 1982.

5. Acerbi, E., Alessandria, F., Baccaglioni, G. and Rossi, L., Design of the Superconducting Coils of the Milan Cyclotron, Proceedings of the IX Int. Conf. on Cyclotron and Their Application, Caen (F), September 1981, pp.399-402.

6. Acerbi, E., Alessandria, F. and Rossi, L., Possible Movements of the Milan Superconducting Coils under the Influence of Mechanical Stresses, Proc. ofX Int. Conf. on Cyclotron and Their Application, East Lansing (USA), April 1984, pp.71-74.

7. Acerbi, E., Bellomo, G., Gmaj, P., Rossi, L. and Zhou, S., The Magnetic Field Measurements of the Milan Superconducting Cyclotron, presented at XII Int. Conf. on Cyclotron and Their Application, Berlin (FRG), May 1989, to be published in the proceedings.

ZEUS MAGNETS CONSTRUCTION STATUS REPORT

A. Bonito Oliva, F. Bordin, O. Dormicchi, G. Gaggero, M. Losasso, R. Penco, N. Valle
ANSALDO ABB COMPONENTI S.r.l, Genova, Italy
R. Bruzzese, M. Spadoni, N. Sacchetti
ENEA, Frascati, Italy
Q. Lin - WORLD LAB, Geneve, Switzerland

ABSTRACT

The construction progress status of the superconducting magnets for the ZEUS detector, commissioned by INFN Frascati and to be installed in the HERA e- p+ring (DESY, Hamburg), is reported. The first one is a double layer, two densities, aluminum stabilized coil 1849 mm in inner dia., 2487 mm in length and 32.6 mm thick, with a central field of 1.8 T and high particle transparency. The second one is a compensating magnet, wound by a copper stabilized Nb-Ti cable. Its coil has a central field of 5 T, inner dia. 370 mm, a length of 1200 mm and is inserted into a cold iron yoke. The main problems encountered during the large coil construction and the geometrical accuracy obtained are reported. Four splices among the high purity aluminum stabilized cable length were made. An outer support cylinder, 18 mm thick, was shrink fitted around the coil and then the temporary inner mandrel was removed. The distribution of mechanical stresses was measured in the different configurations. A large aluminum alloy vacuum chamber with high radiation transparency was built. The compensating coil is ready to be installed inside its stainless steel cryostat. The cryostat critical features are the high design pressure (20 bar) and the heavy cold mass.

INTRODUCTION

A thin solenoid and a compensating magnet for the ZEUS detector have been designed and manufactured by ANSALDO ABB COMPONENTI following the conceptual design carried out by INFN (1,2) that ordered the construction of the components. The commissioning includes also the transfer line, the current leads cryostat, power supplies, vacuum system and a complete computer control. The main features are reported in table 1. Before starting the construction, a model of the thin solenoid was built and tested, the results are reported in another paper at this conference (3).
After a liquid nitrogen test carried out in the factory, the thin solenoid has been installed inside the detector and is now under final test; the compensating magnet construction is almost terminated. The coil and iron yoke installation inside the cryostat is attended.

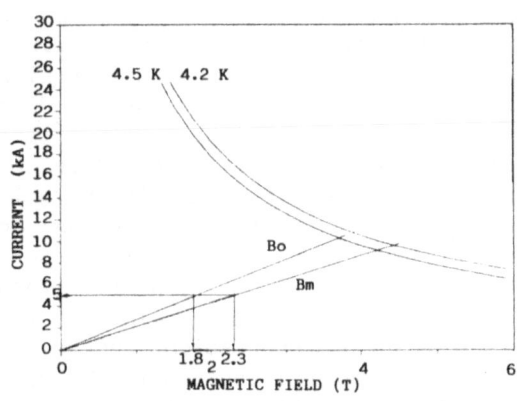

FIGURE 1 - Thin solenoid load line

- TABLE 1 -

- Thin solenoid -		- Compensating magnet -	
Type: double layer solenoid indirectly cooled by two phase helium		Type: cold iron bath cooled solenoid	
Coil inside diam.	1849 mm	Coil inside diam.	370 mm
Coil outside diam.	1914.22 mm	Coil outside diam.	481 mm
Coil length	2487 mm	Coil length	1200 mm
Turns	904	Turns	5083 in 18 layers
Central field	1.8 T	Central field	5 T
Rated current	5000 A	Rated current	1000 A
Conductor type: Rutherford (10 strands) with 99.996 % Al stabilizer		Conductor type: monolithic NbTi copper stabilized	
Al:Cu:NbTi ratio 14:1.1:1 (4.3 x 15) mm		Cable dim. (bare)	4 x 2.6 mm
18:1.1:1 (5.56 x 15)mm		Cu:NbTi ratio	6.5 : 1
Short sample critical current		Short sample critical current	
(2.3 Tesla, 4.5 K)	15000 A	(4.3 K, 5 Tesla)	1800 A
Stored magnetic energy	12.5 MJ	Stored magnetic energy	1.53 MJ
Dump resistor resistance	0.06 Ohm	Dump resistor resistance	0.4 Ohm
Radiation thickness	0.9		

Thin solenoid coil manufacturing

The coil was wound with a high purity aluminium stabilized cable supplied by EUROPA METALLI-Florence, and is inserted, without internal support, in an external aluminium alloy restraining cylinder. The external cylinder must support the magnetic hoop stresses during operation and provide the cooling of the coil by a cooling pipe welded on it. During the operation the coil experiences an axial magnetic force that reaches 400 tons on the middle section, so it must be axially precompressed by the cylinder, to avoid relative slipping during charging. The cylinder provides also a circumferential precompression of the coil, that in this way can withstand a magnetic pressure of 0.21 Kg/mm2 at the rated field, within the elastic limit. To provide this precompression the cylinder was shrink fitted on the coil with an interference of 0.86 mm in radius. The coil was manufactured winding the cable, under a constant tension of 1 Kg/mm2, on the aluminium winding form external side. In the winding line a taping machine insulated the cable by glass fabric tape 0.17 mm thick half overlapped. By an axial pressing we tried to reach a certain glass percentage (4) in the turn to turn insulation, to hold as a minimum the decreasing in precompression, due to the differential thermal contraction between the cylinder and the insulation. The reached percentage was 62%, resulting in an axial differential thermal contraction of approx. 100 μE. Every 25 turns the winding was axially pressed under a force of 40 tons by a ring with threaded rods. After winding, the first layer was permanently pressed with 80 tons; the same was done for the second layer. The coil was wound by 5.56 x 15 mm cable in the central part (1200 mm in length) and by 4.3 x 15 mm cable at each end (650 mm for each side), so with two current densities it is possible to have higher field uniformity inside the central tracking detector. The coil is made of five cable lengths and so there are two joints for each layer. A detailed R&D work was carried out for the conductor joints. The cables were welded for two meters on both narrow faces by TIG with high purity aluminium filler and then, at each joint edge, they were brazed by Sn(60)-Pb(40) for 300 mm. The measurements made by ENEA demonstrated that the brazed part of the joint allows to decrease the resistance from 2 x 10-9 Ohm (only welded part) to 5 x 10-10 Ohm, including magnetoresistance. After the winding completion, the coil was vacuum impregnated with epoxy resin and cured. Then the external surface was carefully machined, under a constant control of the insulation thickness, by a proximitor, to reach its specified value (fig. 2). It ranged from 0.7 up to 1.1 mm, while the dimensional accuracy obtained on the external diameter was ±0.02 mm.

FIGURE 2 - Coil turning

The ground insulation was tested up to 1000 V covering by aluminium foils the external surface of the coil.

The turn to turn insulation thickness ranged from 0.56 to 0.6 mm for the internal layer while for the external layer from 0.6 to 0.7. This is due to the higher friction in the second layer pressing.

No additional insulation was inserted between the two layers.

Finally the coil was axially compressed with a force of 150 tons by two flanges and 24 tie rods with Belleville springs. The external support cylinder was made by 25 mm thick 5083 aluminium alloy sheets. On the external surface a 20x20 mm 5754 aluminium alloy square pipe with a 12 mm in diameter hole was welded, to form the two cooling circuits in parallel, each 40 m long. Welds were staggered to avoid large cylinder deformations. The cylinder was heat treated and then machined on the internal surface to reach the requested shape and diameter. During the turning the cylinder was carefully hold in round shape by external rings but, due to the very low stiffness, at the end the interference with the coil ranged between 1.6 to 1.86 (1.73 mm average).

Shrink fitting
The cylinder installed on the fixture (fig. 3), was heated up to 140° C by

electrical heaters and the coil was suddenly lowered inside it with a clearance of 5 mm in diameter. After the return at room temperature, the two coil end aluminium rings were connected to the cylinder with 60 dowels for each side. The fixture for the coil axial pressing was removed and finally the inner mandrel was disassembled. The strains measured during the construction steps are reported in Table 2, while the axial forces distribution in each step is shown in fig. 4. The coil internal diameter measurements showed that the coil without inner support is cylindrical with a tolerance of +0.6 mm in diam.

Thin solenoid cryostat

The thin walled cryostat is formed by a 5083 aluminium alloy vacuum chamber with $\emptyset$ext. 2280 mm, $\emptyset$int. 1720 mm, overall length 2820 mm. The outer cylinder is 12 mm thick, the inner one 5 mm and the end flanges about 30 mm, so in radial direction the radiation thickness is 0.22 mm. Inside the detector the cryostat is supported by an aluminium cone bolted on the cryostat outer cylinder at one side and by three tie rods at the chimney side. Inside the warm bore the central tracking detector (weight 900 Kg) will be installed.

To avoid radiation damages, the vacuum

FIGURE 3 - Coil under pressing and behind the cylinder in the shrink fitting fixture

TABLE 2 - Strain measurement results
(average values $\mu\varepsilon$)

Support Cylinder		Mandrel		Configur.
axial	circum.	axial	circum.	
-	-	0	-60	1
-	-		-60	2
+307	+734	-60	-250	3
+325	+726			4
+249	+370	0	0	5

1) after winding and curing
2) after coil pressing
3) after shrink fitting
4) after tie rods disassembling
5) after mandrel removal

chamber is sealed by three metallic sealings made of indium wire pressed in a groove by a tooth.
In order to obtain leak tight sealings, a very precise machining of the coupling surfaces between flanges and cylinders was carried out. All the weldings were controlled by X-ray or ultrasonic methods and very low inclusions and porosities were detected. The total vacuum leak measured for the whole cryostat was 1.5 x 10-8 mbarl/sec under a Δ P = 1 bar. The thermal shield was made by Roll Bond panels of aluminium 99, 1.5 mm thick (fig. 5). In operation it is cooled by helium gas at 40 K and 15 bar. The superinsulation was made by both sides aluminized mylar foils and fiberglass net spacers. The restraining cylinder and the coil have a total weight of approx. 2 tons and are suspended by 16 radial tie rods, approx. 400 mm long and 8 mm in diameter (fig. 6). The five tons axial magnetic force acting on the coil at full current and due to the detector iron yoke is sustained by four axial tie rods 1300 mm in length and 8 mm diameter. At low currents the magnetic force (calculated value approx. 200 Kg) is directed in the opposite direction and counteracted by two axial tie rods of the same length. All the tie rods are made of Ti6Al4V extra low inclusion grade titanium alloy.
The magnet is connected to the HERA cryogenic line and to the electrical supply by a current leads cryostat placed outside the detector and an elbow shaped 5.5 m long transfer line. Inside the current leads cryostat there is a

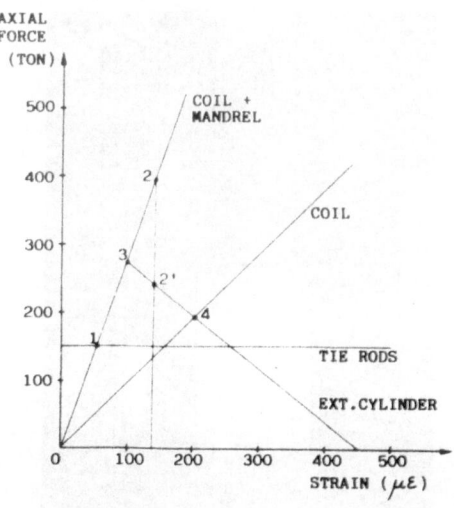

FIGURE 4 - Axial forces distribution
1) after coil pressing
2) after shrink fitting
3) after tie rods disassembling
4) after mandrel removal

lHe dewar for the 5000 A current leads cooling and the connections. pipes between the HERA line and the magnet cryogenic line. A R&D effort was made to build the electrical feed through (5000 A, 500 V, 4.2 K) from the current lead lHe vessel to the vacuum. The magnet cryogenic line has a diameter of 160 mm and contains: two phase lHe inlet and outlet pipes (4.5 K, 1 bar), helium gas (40 K, 15 bar) inlet and outlet pipes, two superconducting bus bars made with the same cable of the magnet and cooled by the lHe returned from the magnet. The cryogenic line acts also as vacuum pumping line between the magnet and the current leads cryostat where the vacuum pumps are connected.

FIGURE 5 - Cryostat external cylinder
and radiation shield

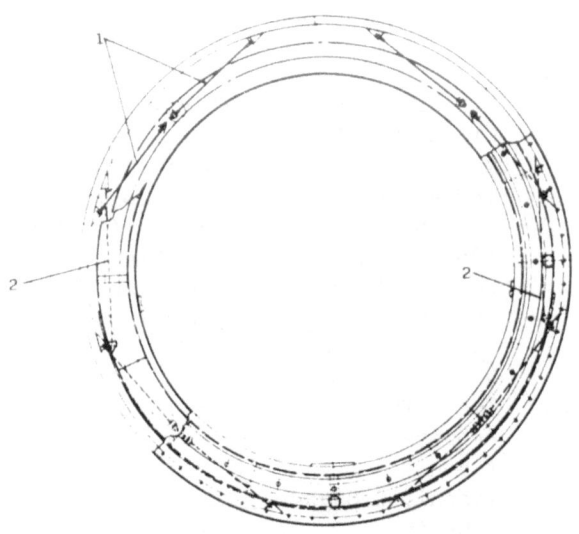

FIGURE 6 - Thin sol.radial tie rods (1),
thermal shield tie rods (2)

Compensating magnet

The coil has been wound with a monolithic copper stabilized cable (bare
dimensions 4 x 2.6 mm) insulated by the
supplier (Vacuumschmelze, Hanau) with
0.15 mm thick glass braid glued on the
conductor with a small amount of cured
epoxy resin. This insulation has good
mechanical resistance to abrasion so
that the occurrence of damages during
the coil manufacturing is low; it is
employed when the cable dimensions are
quite small. The layer to layer
insulation was made of glass fabric tape
that forms channels enhancing the coil
impregnation. The stainless steel
cryostat (fig. 8) is composed by a
cylindrical vacuum chamber with a tower
and a helium chamber designed to
withstand a maximum pressure of 20 bar

according to the HERA cryogenic plant
characteristics. The low carbon iron
yoke and the coil will be mounted around
the inner pipe of the cold chamber with
a small clearance and the differential
thermal contraction between the iron
yoke and the s.s. chamber will be
compensated by Inconel 718 springs. The
coil is in contact with the inner pipe
only at its ends, in the GFRP flanges
zones. The cold mass is about five tons
and is supported by four tie rods, 10 mm
diameter, and made of Ti6Al4V extra low
inclusion grade (yield strength 90
Kg/mm2 at room temperature). Another
four tie rods are installed symmetric-
ally with respect to the horizontal
axis, in order to withstand the magnetic
decentering forces, while three tie rods
for each side, 6 mm diameter, are
installed to withstand the axial
magnetic forces, that reach a maximum
calculated value of 2 tons. The cold
chamber has a neck with a stainless
steel bellow 0.8 mm thick, where the two
current leads have been installed. These
two components were tested up to a
pressure of 30 bars and a voltage of
1000 Volts ac.. The thermal shield was
made of copper sheet 2 mm thick with the
cooling pipe for the helium gas at 15
bar brazed around the external cylinder.
The superinsulation was made by both
sides aluminized mylar foils with GFRP
net as spacer. The cold test of the
magnet is expected for October 1989.

FIGURE 7 - Thin solenoid parts

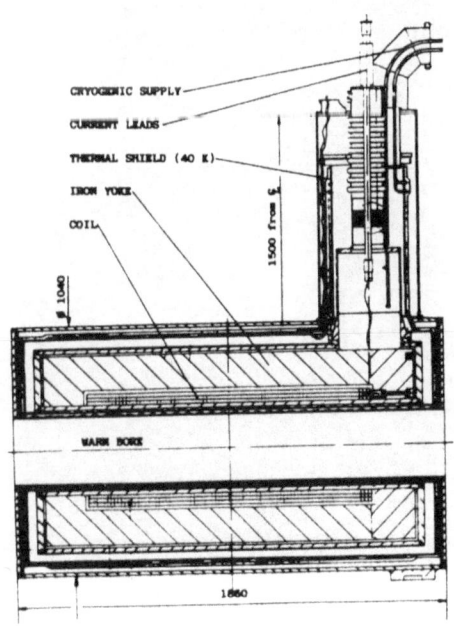

FIGURE 8 - Compensating magnet cryostat

234

REFERENCES

1. E.Acerbi, F.Alessandria, G.Bacca-glioni, E.Fabrici, L.Rossi (INFN/ Milano), R.Bruzzese, G.Pasotti, M.V.Ricci, N. Sacchetti,M.Spadoni (EURATOM-ENEA/Roma), O.Dormicchi, P.Fabbricatore, R.Penco (ANSALDO ABB COMPONENTI), Thin and Compensating Solenoids for ZEUS Detector. IEEE Transactions on Magnetics, March 1988, Vol.24, No.2, p. 1354
2. The ZEUS Detector Technical Proposal March 1986
3. A.Bonito-Oliva, G.Masullo, O.Dor-micchi, G.Gaggero, R.Penco, Quench behaviour of a thin solenoid. MT-11 1989, Int. Conf. on Magnet Technol.
4. J.Hamelin, Three-dimensional contraction and mechanical properties of glass-cloth-reinforced epoxy materials at cryogenic temperatures. A.C.E., Vol. 26, 1980, p. 295

FIGURE 9 - Thin solenoid assembly test

MAGNET DESIGN STUDIES FOR THE TRIUMF KAON FACTORY PROPOSAL

A.J. Otter, L. Ellstrom, C. Haddock, M. Harold*, P. Reeve, H. Sasaki†, P. Schwandt‡
TRIUMF, 4004 Wesbrook Mall, Vancouver, B.C. V6T 2A3

ABSTRACT

TRIUMF is engaged on a one year Project Definition Study of its KAON factory proposal. This proposal calls for a total of 1800 magnets to be installed in three storage rings, two synchrotrons, beam transfer lines and experimental facilities to increase the present beam from 200 μA at 500 MeV to 100 μA at 30 GeV. The paper discusses the current design status concentrating on the ac booster synchrotron ring magnets which will be driven by biased dc current modulated at 50 Hz. Methods of estimating the core losses for this excitation, the coil eddy current losses and design and fabrication features of prototypes will be presented.

INTRODUCTION

TRIUMF's KAON Factory was proposed in 1985 [1] and has been under continuous review and study since that time. The latest layout is shown in Fig. 1

it comprises transfer lines, three storage rings, two synchrotrons, a beam switch yard and experimental facilities. This complex will require over 1800 magnets of at least 100 different designs, (Table 1). The

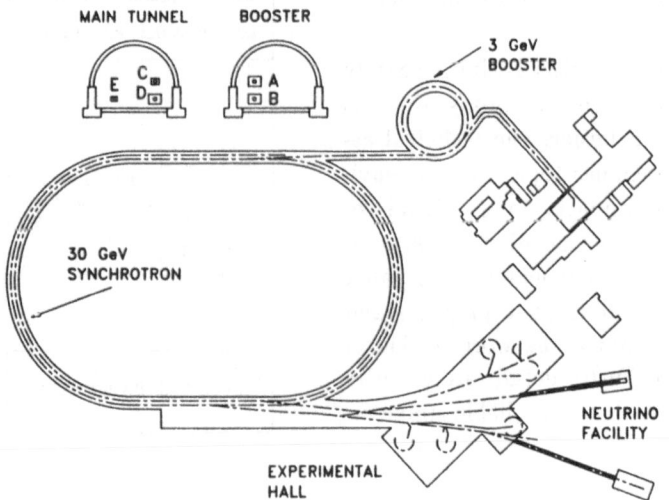

FIGURE 1 Proposed KAON factory layout

*Rutherford Appleton Laboratory, Chilton Didcot, Oxon. 0X11 OQX U.K.

†KEK, Oho-machi, Tsukuba-gun, Ibaraki-ken 305, Japan

‡Indiana University Cyclotron Facility, Bloomington, IN, U.S.A. 47405

TABLE 1 Estimated Magnet Quantities for KAON factory.

	Dipoles	Quads	Sextupoles	COD's	Total
450 MeV Transfer	11	47	—	6	64
'A' Ring	24	48	24	96	192
'B' Ring (AC)	24	48	24	96	192
'B–C' Transfer	5	19	—	6	30
'C' Ring	96	128	48	96	368
'D' Ring (AC)	96	128	48	96	368
'E' Ring	96	128	48	96	368
Switch Yard & Primary BL's	18	98	—	10	126
Secondary Beamlines	23	87	5	—	115
	393	731	197	502	1823

two synchrotrons are excited with a dual frequency, dc biassed excitation which will require ac magnets. The storage rings and experimental facilities magnets will have dc excitation. There are only two superconducting magnets, these will be in the secondary channels of the experimental facility. We do expect however that large superconducting magnets will be needed for the particle detectors of the experimental programs.

We are now completing a Project Definition Study (PDS) which ends in December of this year. This study is rather broad and besides the technical aspects of the project it includes topics such as environmental and economic impacts, international cooperation and the ability of Canadian industry to participate in building the components. This paper briefly describes the overall magnet program and some of the design highlights. The work on kicker magnets is not included. Furter details are given in internal design notes which are available upon request.

PDS PROGRAM

The project goals for the PDS program are set out in Fig. 2. A design workshop was held in October of last year [2] covering magnet and power supply technology. The major activity has been the design and procurement of full sized prototypes of the Booster dipole and quadrupole magnets. The dipole is shown in Fig. 3 and its parameters are given in Table 2. The detailed design of these magnets is discussed below.

We are making preliminary conceptual designs for all the magnets as their specifications are given to

MAGNETS

PROJECT DEFINITION YEAR

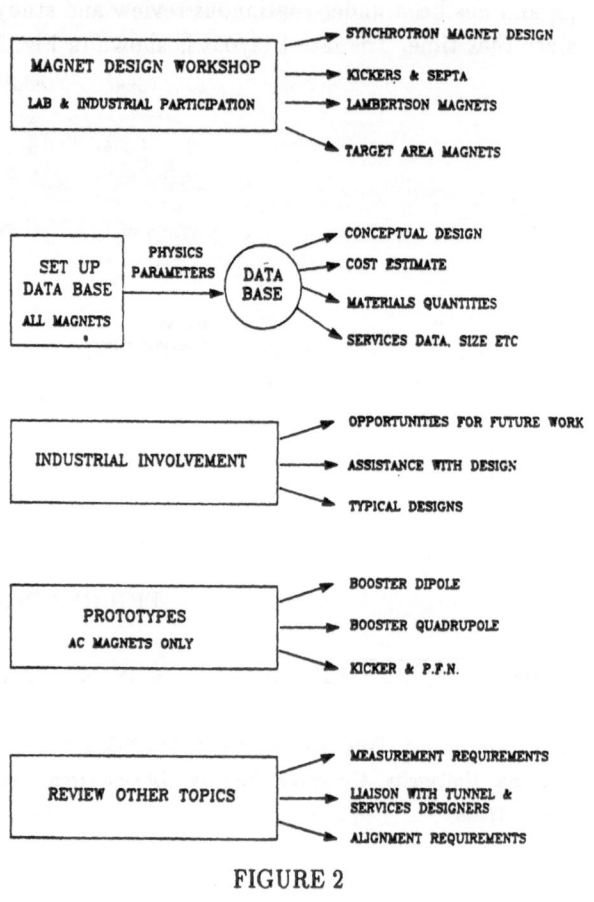

FIGURE 2

TABLE 2 Booster Synchrotron Input Parameters

Physics

	Dipole	Quadrupole	
Energy range	450 – 3000		MeV
Field rise frequency	33.3		Hz
Field fall frequency	100.0		Hz
Maximum pole tip field	1.05	0.70	T
Minimum pole tip field	0.277	0.175	T
Effective length	3.18	0.8	m
Pole gap or diameter	10.68	13.2	cm
Field uniformity	$B/B_0 \leq 1 \times 10^{-4}$	$C_n/C_2 < 2 \times 10^{-3}$	cm
Good field width	± 5.0	6.6	cm
Number required	25	48	cm

Power Supply

Maximum current	5000	1600	A
Maximum voltage/magnet(peak)	3	0.4	kV
Maximum inductance	5.75	1	mH
No. magnets/PS cell	5	24	
Maximum voltage/PS cell (peak)	15	10	kV

us by the beam opticians. This data is being stored in an "Oracle" database which will allow us to sum materials and give quick access to dimensional data and service requirements.

The design team comprises three full time professionals supplemented by technical staff. We have enjoyed the presence of visiting experts both from other laboratories and from industry for consultation and discussions on techniques and methods.

we wanted to gain experience with ac magnets which have not previously been needed at TRIUMF.

The magnets are driven in series by current wave forms similar to the one shown in Fig. 4 which is for the dipole. The waveshape rises at 33.3 Hz and falls at 100 Hz. Some aspects of the designs were dictated by the need to procure and build the prototypes within the PDS year, materials were ordered before the construction drawings were completed. A dipole and quadrupole magnet are being made and will be delivered by December of this year.

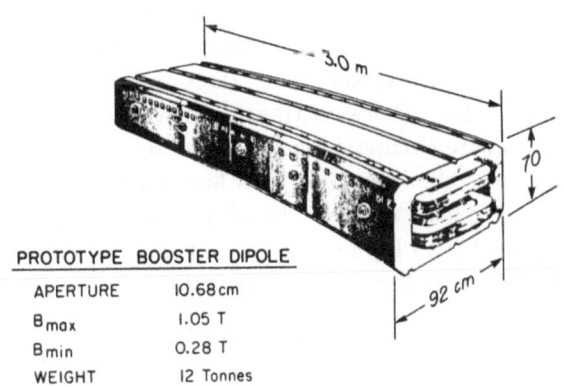

PROTOTYPE BOOSTER DIPOLE

APERTURE	10.68 cm
B_{max}	1.05 T
B_{min}	0.28 T
WEIGHT	12 Tonnes

FIGURE 3

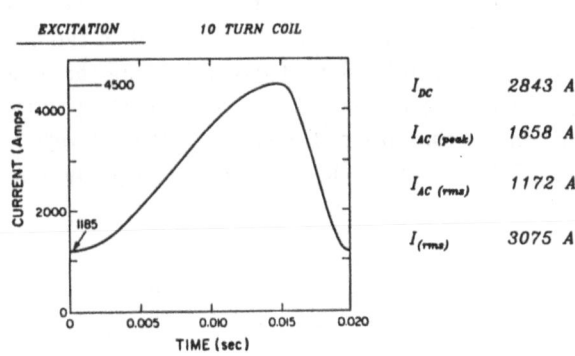

I_{DC}	2843 A
$I_{AC\ (peak)}$	1658 A
$I_{AC\ (rms)}$	1172 A
$I_{(rms)}$	3075 A

MEAN FREQUENCY $\bar{f} = 50$ Hz

FOR f^2 DEPENDENCIES $(\bar{f^2})^{1/2} = (f_1 f_2)^{1/2} = 57.73$ Hz

FIGURE 4 Dipole magnet excitation current

PROTOTYPE DESIGNS

It was decided that it would be prudent to concentrate our efforts on the Booster ring magnets because these are the most challenging technically and

DIPOLE STEEL DESIGN

The dipole magnet is 3.1 m effective length, with a 10.68 cm gap and field levels at injection and extraction of 0.277 and 1.049 T respectively. This converts to a dc field level of 0.663 T and an ac peak component of 0.386 T. At our magnet workshop [2] it was recommended that the field in the yokes not exceed 1.25 T for a 50 Hz magnet and this value was subsequently found to be the best value for minimizing the core losses.

It was decided at the beginning to build a curved magnet from one piece laminations. The single lamination reduces the amount of assembly labour and avoids uncertainties in the air gap due to the side joints. We also decided to use a non-oriented steel. Subsequent studies [3] showed that the lower losses of the grain oriented steel are not fully realized because the H magnet profile has significant regions where the flux is at 45 and 90° to the rolling direction. For our particular magnet the use of a thinner oriented steel would have resulted in a core loss reduction of only 1 kw compared to the M 17 steel 0.0185 in. (0.47 mm) thick chosen. The total magnet losses are estimated to be 58.5 kw [4].

In order to calculate the core losses for our excitation we separated them into the hysteresis and eddy current components. The hysteresis loss was then estimated at the peak flux density and adjusted for the fact that the steel only traverses about one third of the full hysteresis loop. The eddy current component of loss was calculated by estimating it from the ac flux component only and adjusting for the frequency. The methods of doing this are given in Ref. 5.

The magnet assembly will have the laminations parallel to each other and assembled on a curved bed. The laminations will be epoxied together and stiffened by stainless steel plates welded to the outside. There will be slots at intervals along the magnet to accommodate coil clamps and cooling arrays to remove core losses. The cross section is shown in Fig. 5. The pole profile and magnet end profile were

shaped [6] to avoid saturation in the steel which could cause excessive core losses and variation in the magnet properties between extraction and injection energies. The field uniformity is within the tolerances of 1×10^{-4} at both energies and the effective length is predicted to be constant to 1 part in 10^5.

COIL DESIGN

The physics specification of field level and volume coupled with the power supply requirement of a maximum allowable voltage determine the number of turns for the coil. Eddy currents due to the transverse field at the coil require that the conductor section be small so it is necessary to connect individual insulated conductors in parallel to form a single conductor. Our design limited the number of turns to 20 which requires a current of approximately 5000 A peak. We considered an array of conventional square hollow conductor and compared it with stranded hollow conductor of the type used at KEK [7], Fig. 6a. The square hollow conductor is arranged in a 1 × 12 array as shown in Fig. 6b.

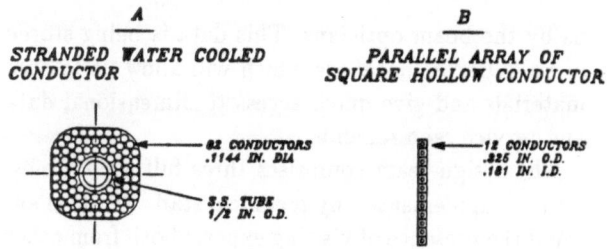

FIGURE 6 Conductor configurations

The prototype will be wound with the square hollow conductor which was chosen because of its better availability. Its eddy current losses are higher than the KEK type as shown in Table 3. The 1 × 12 array is chosen because transposition of conductors to eliminate circulating currents due to unequal flux linkages is eliminated for both horizontal and vertical components of flux. The eddy current losses are estimated from Eq. (1) for square hollow conductors which is correct as long as dimension a is less than

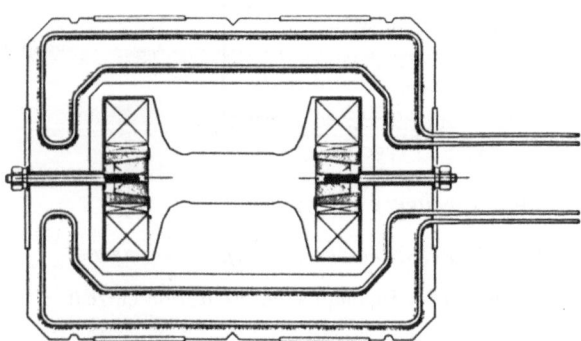

FIGURE 5 Prototype dipole cross section

TABLE 3 Comparison of dipole conductors

	A	B
	stranded conductor	1 × 12 square hollow array
dc resistive loss	42 kw	36.8
ac resistive loss	7.1 kw	6.3
ac eddy current loss ($B_{rms} = .08$ T)	1.8 kw	15.4
	50.9 kw	58.5 kw

the skin depth. We have verified this with PE2D calculations. A similar calculation for the stranded conductor which is larger overall than the skin depth shows that the outer turns shield the inner tube and the calculation assumption of a uniform field is incorrect, however the calculation is considered to be conservative.

$$\text{loss/meter} = \left[\frac{a^4}{24} - \frac{\pi d^4}{128}\right]\frac{B^2 w^2}{\rho} \qquad (1)$$

a = Conductor outer dimension
d = Conductor inner diameter (cooling hole)
ρ = material resistivity
B = Peak value of transverse sinusoidal field
      $B \cos wt$.

PROTOTYPE QUADRUPOLE

The quadrupole design [7] is based on an existing HERA quadrupole cross section because it was possible to procure laminations using existing tooling. the quarter cross section is shown in Fig. 7. Design principles were similar to those of the dipole. The coil is comprised of a bundle of narrow rectangular conductors indirectly cooled. Transposition of the conductors to eliminate circulating currents will be done at the coil terminations. The eddy current losses were studied using PE2D and the magnet is being modelled in three dimensions using TOSCA to compute the overall harmonic content.

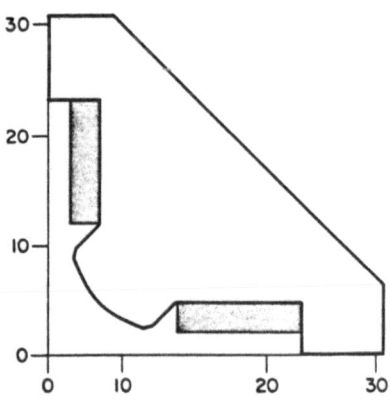

FIGURE 7 Prototype booster quad quarter section

DRIVER RING AND OTHER MAGNETS

Design of Driver ring magnets which operate at 6.6 and 20 Hz has been carried out to the conceptual stage only [8,9,10]. The pole and end profiles have been established but the main parameters are not yet fixed so the designs have not been finalized.

Preliminary data on all other magnets from size and cost estimating programs is being fed into a data base which allows us to quickly estimate power, space and service requirements. These are not considered to be final designs.

LOSS MEASUREMENT PROGRAMS

We have assembled an Epstein coil array and have started a series of core loss measurements with both sinusoidal and dual frequency excitation superimposed onto a dc bias field. These measurements will be compared with our estimating method given earlier. Some preliminary results are shown for an M22 steel (Fig. 8); from this data an estimate can be obtained for a waveform similar to ours. We have completed measurements on M4 steel samples and achieved agreement between estimating the loss from single frequency measurements and measuring them directly with biassed dual frequency excitation to within 6%. We are continuing the measurements on the prototype dipole steel so that we can compare the measured and estimated results.

PRELIMINARY CORE LOSS MEASUREMENTS

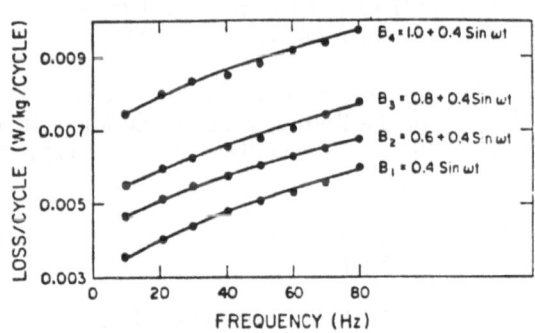

TOTAL LOSSES WITH VARYING DC BIAS M22 STEEL

FIGURE 8

Also we have installed square hollow conductors in a dipole magnet and measured the eddy current losses by cooling water flow and temperature rise. For 0.5 in. square hollow conductor the measured loss agreed within 5% of the calculated values at 100 and 33.3 Hz, with a field level of 600 G. A stranded cable sample with a copper cooling tube did not give such a good agreement but the field level was too low for an accurate determination. The tests on both types of conductors are continuing.

240

MAGNETIC MEASUREMENTS

Magnetic measurements on the prototype magnets
will be made with a variety of methods. The dipole
will be measured with both dc and ac excitation with
a Hall probe. In the ac mode about twenty measurements per pulse will be made using an HP3458A
multimeter connected to a 386 PC computer. The
quadrupole will be measured with a Morgan coil technique using a printed circuit board technique to measure the n=1,2,6 and 10 harmonics.

SUMMARY

These studies are continuing and will be extended
as time permits to other topics. We are starting to
look at radiation hard magnets for the target stations in the Experimental Halls. We have also been
involved with the RF Group's cavity tuner prototype
and choke transformers for the Power Supply Group.
A major effort will now be required to prepare new
cost, manpower and schedulle estimates for the final
PDS report.

REFERENCES

1. KAON factory proposal, TRIUMF (September 1985)

2. Proc. of the KAON PDS Magnet Design Work-shop, TRI-89-1, October 3-5, 1988.

3. Schwandt, P., Comparison of Realistic Core Losses in the Booster Ring Dipole Magnets for Grain Oriented and Ordinary Lamination Steels, TRI-DN-89-K31.

4. Otter, A.J., A Prototype for the Booster Dipole, TRI-DN-89-K18.

5. Schwandt, P., Estimation of Specific Core Losses in dc Biased Laminated Magnets, TRI-DN-89-K22.

6. Schwandt, P., An Improved Magnetic Design for the Booster Ring Dipoles, TRI-DN-89-K32.

7. Reeve, P.A., Preliminary Design on Prototype Booster Quadrupole, TRI-DN-89-K34.

8. Harold, M.R., A Preliminary design for the Driver Dipole, TRI-DN-89-K24.

9. Harold, M.R., Preliminary designs for the driver quadrupoles QF and QD, TRI-DN-89-K23.

10. Schwandt, P., Magnetic Design of the Driver Ring Dipole – A Second Look, TRI-DN-89-37.

DEVELOPMENT OF SUPERCONDUCTING BENDING MAGNETS FOR THE SR RING

T. Okazaki, Y. Hosoda, S. Isojima, T. Keishi, C. Suzawa, and T. Masuda
Osaka Research Laboratory, Sumitomo Electric Industries, Ltd.
1-1-3, Shimaya, Konohana-ku, Osaka, 554 Japan

ABSTRACT

The superconducting bending magnets with a bending radius of 0.5m and magnetic field of 4.0T have been developed. The bending angle is 90 degrees so that four magnets are installed in the synchrotron radiation ring whose size is 3m×5m. The main coils of the bending magnets are the iron-free curvature dipole coils and the conductors are 3-divided keystone type. The shim coils are epoxy-molded quadrupole coils installed on the beam duct directly. The design of these coils was accomplished using computer simulation. The designed electron beam energy is 615 MeV and the radiation power from the beam to the magnet is 1kW so that the absorber cooled by liquid nitrogen is set in the beam duct. Test results validated the design.

INTRODUCTION

Synchrotron Radiation (SR) for industrial applications is one of the key technologies for the next generation; however, the usual SR ring system is too large to be applied for industrial use. Several companies are developing compact rings. Sumitomo Electric Industries, Ltd. (SEI), began development of a compact SR ring system in 1984. Initially, Electrotechnical Laboratory and SEI constructed the NIJI-1 whose diameter is about 3.8m, as a part of a joint research project of the government and the private sector[1].

Subsequently, SEI has been developing a SR ring using four superconducting magnets(S.C. magnets). The S.C. magnet, with its strong magnetic field, presents a compact SR ring design, because the strong magnet field can produce a small particle bending radius.

In this paper, the general design of a SR ring with S.C. magnet and the excitation results of the magnet is reported.

GENERAL DESCRIPTION OF THE RING DESIGN

The electron beam energy is designed 615MeV with a stored beam current at 200mA and a peak wavelength of 5Å. Fig.1 shows the plan view of the SR ring and Fig.2 is the photograph of the superconducting magnet. It has four straight sections, two of them are 4.1m, and others are 2.1m. The designed bending radius is 0.5m, and circumference is 15.5m. It has eight quadrupole magnets which give a wide operating region, and in the longer straight section the lattice is set in achromatic for easy beam injection. Two sextupole magnets at this achromatic section are used to collect the end field effect of the bending magnets. A single wobbling magnet is in the long straight section to undulate the electron orbit. By using this magnet, exposure area is widened[2].

On June 21, 1989, this ring was constructed without using superconducting bending magnets, but with four normal conducting bending magnets for the beam

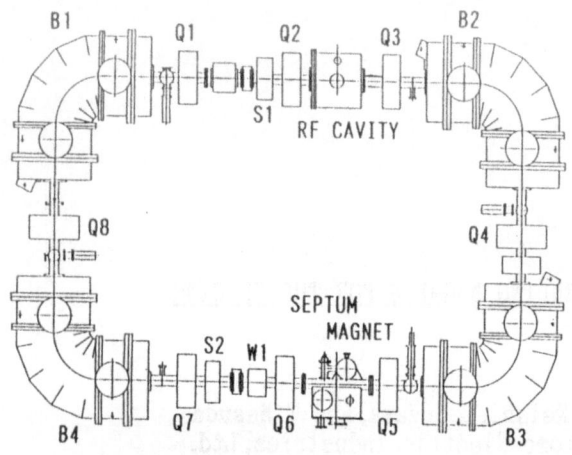

B1~B4   BENDING MAGNETS

Q1~Q8   QUADRUPOLE MAGNETS

S1,S2   SEXTUPOLE MAGNETS

W1   WOBBLING MAGNET

FIGURE 1. Plan view of SR ring.

FIGURE 2. Photograph of the superconducting magnet.

optics study and optimum injection
parameter search. After these experiments, the four normal conductivity magnets will be replaced by the superconducting bending magnets.

OUTLINE OF THE S.C. BENDING MAGNET

This magnet is made of two groups of
coils. One is curved dipole coil, and the
other is a curved quadrupole coil. Its
cross section is described in Fig.3.
Among the coil types for application to
the SR ring, a dipole type coil($\cos\theta$
winding) was chosen. The advantage of
this type is that the magnetic field distribution is the most homogeneous, but it
is difficult to get many S.R. beamports.
However, it is estimated that two beamports can be taken from the coil end.
This dipole and quadrupole coil can be
operated independently, so the magnetic
field gradient (field index) of this
bending magnet can vary with the quadrupole coil current. The magnetic field
produced by the dipole coil is 4.0T when
the main dipole coil current is 1774A and
the current density is about 270A/mm^2.
The field index can be varied from 0 to

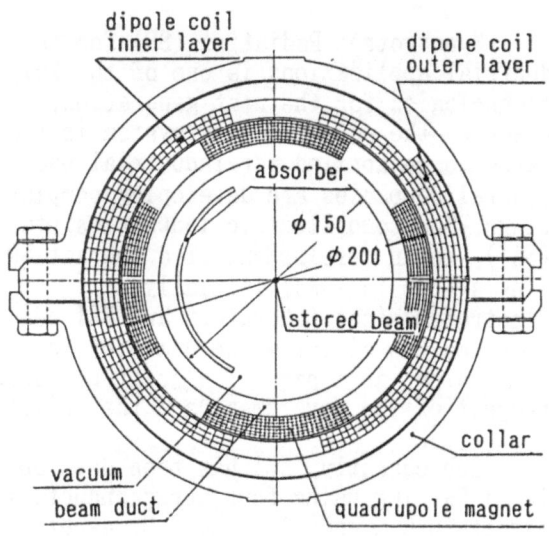

FIGURE 3. Cross section of the superconducting magnet.

0.5. The excitation rate is 0.34T/min.

No magnetic materials are used in this magnet to avoid magnetic saturation problems. Since the magnetic field distributions do not change with exciting current, its accuracy depends on only the power source's accuracy whose design value is 10^{-4}.

The structure of dipole coil is a double pancake coil, with an inner radius of 100mm and an outer radius of 123.5mm. The conductor type used in the dipole coil is a compacted strand cable whose dimensions are shown in Table 1. It is made of eight Nb-Ti strands and insulated by two kinds of Kapton thin tape (see Fig.4). The keystone angle is the average

TABLE 1.
Conductor characteristics

material	Nb-Ti
Cu/SC ratio	1.7 ± 0.1
strand diameter	0.865mm
number of strands	8
critical current	3.5kA at 6T
strand pitch	27mm

size of the outer and inner layers. By using this keystone strand cable, spacers are inserted in the coil ends, but aren't needed at the main body of the coil. The keystone figure cable is divided into three cables to reduce the current one-third. These 3 strands are wound in parallel and cured in a hot press. After forming this coil, these three cables are joined at the magnet end serially. This method has many advantages for making SR system compact. To make the same magnetic field, the volume of the power source, and current leads become smaller, and the consumption of liquid helium can be reduced.

The quadrupole coil set on the beam duct is molded by epoxy resin, and the dipole coils are fixed between the molded coils and the reinforcing collar which is fastened by bolts.

Its longitudinal length is about 2m. Since there is not adequate space for a vacuum pump near the magnet, a cold bore type similar to the cryo pump was adopted. The duct wall is cooled to liquid helium temperature. In order to avoid the increase of liquid helium consumption by SR power, there is a SR absorber inside of the beam duct. The absorber itself is cooled by liquid nitrogen, since it is also a kind of cryo pump.

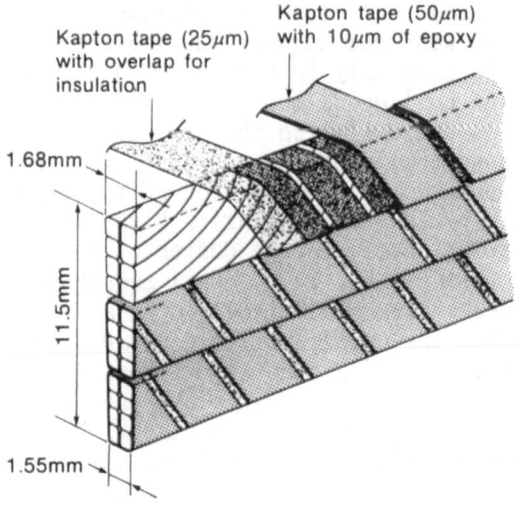

FIGURE 4. Structure of the 3-divided compacted strand cable.

DESIGN METHOD OF MAGNETS

Generally, in the case of this iron-free type magnet, its magnetic field leaks into a wide area and includes large amount of multipole components. These multipole components disturb electron beam accumulation, therefore, the design of the magnet coil ends is very important. The magnet is designed from the view-point of the multipole components reduction. At first, a large diameter of dipole coil was selected, thus greater than octa-pole components could be ignored. Quad and sextuple components are still large. To compensate for these multipole components, two unique designs were considered. One is to place the quadrupole coil in the dipole coils. The main purpose of this design is adjustment of the field index. The other is forming the sextupole component of inverse polarity at the coil end. This component compensated the original sextupole component in coil main body and coil ends.

Under these conditions, a good field uniformity of 5×10^{-4} more than 45mm at the coil center was obtained.

There are no precedents for SR ring construction using such a small bending radius and a iron-free magnet. The beam tracking calculations were performed to confirm the design results. For this kind of bending magnet, the linear isomagnetic treatment of beam optics and particle tracking is no longer adequate, because the end magnetic field of this type magnet leaks into a wide area, and the trajectory of the particles are changed by this field. Therefore a particle tracking computer code was developed. This code solves the 3-D motion equation by the Runge-Kutta method under a 3-D magnetic fields, so that their longitudinal distribution along the electron beam path was considered. To calculate the magnetic field of the magnet, we made a simulation model of dipole and quadrupole coils(see Fig.5 and Fig.6). Since this magnet is iron-free

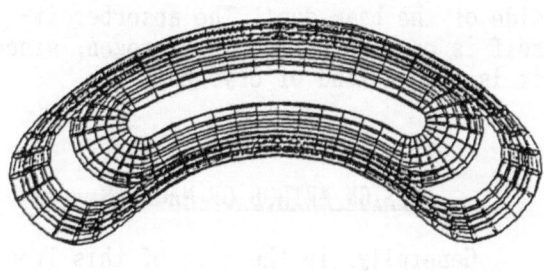

FIGURE 5. Simulation model of the dipole coil.

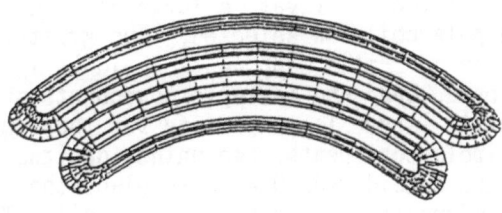

FIGURE 6. Simulation model of the quadrupole coil.

type, finite-element-method was not necessary. Therefore only the more important coil model was modeled. This program can treat multi-pole effects; however, one defect of using the Runge-Kutta method on beam tracking is the long solution time. Therefore, the dynamic aperture region cannot yet be precisely calculated. It is estimated horizontally about 70mm and vertically 15mm at the septum magnet.

The total ring design was done by the 2nd order beam tracking program. Results of the 2nd order simulation and 3-D real field tracking have good agreement when the initial emittance is small. The setting parameter of the current ratio between the dipole and quadrupole coil is calculated from the 3-D simulation code.

MAGNETIC FIELD MEASUREMENT

There are some difficulties when measuring the magnetic field of this magnet. One is the curvature of this magnet with a beam duct, and the other is the environment of the measuring point, such as at a liquid helium temperature and in a high vacuum. Therefore, it is difficult to introduce any moving object. Because of these conditions, the idea of introducing rotating coils in the beam duct was abandoned. As an alternative the NMR sensor was used, however the NMR sensor can be used only in a good field region. In the coil ends, the magnetic field varies radically, so the NMR sensor can't be used. Hall sensor was finally chosen to measure the distribution of the magnetic field along the axis. Five NMR sensors are placed in the center of the magnet (see Fig.7). Thirty Hall sensors are placed on two sensor holders so as to measure the cross section distribution of magnetic field at one time. These Hall sensor holders can move along the magnet axis on both sides of the magnet. Due to the variability of accuracy of Hall sensors, only thirty out of fifty could be used. All of them are measured inside a high uniform magnet with NMR sensor used to correct the Hall sensors. This measurement is performed twice (see Fig.8), and the most accurate one is placed at the center of the moving sensor holder.

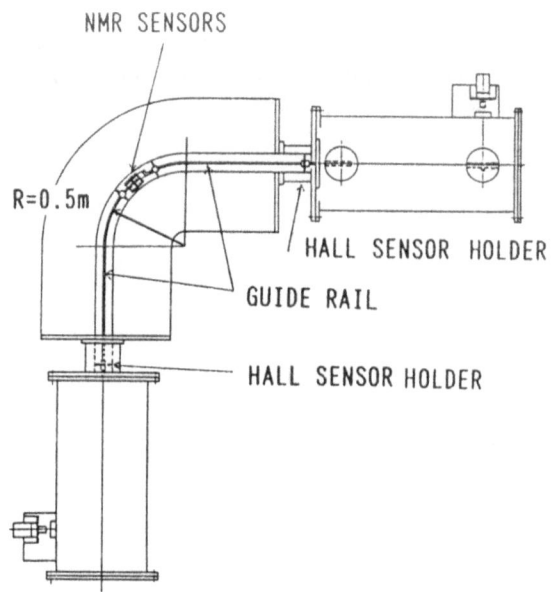

FIGURE 7. Plan view of the magnetic field measuring system.

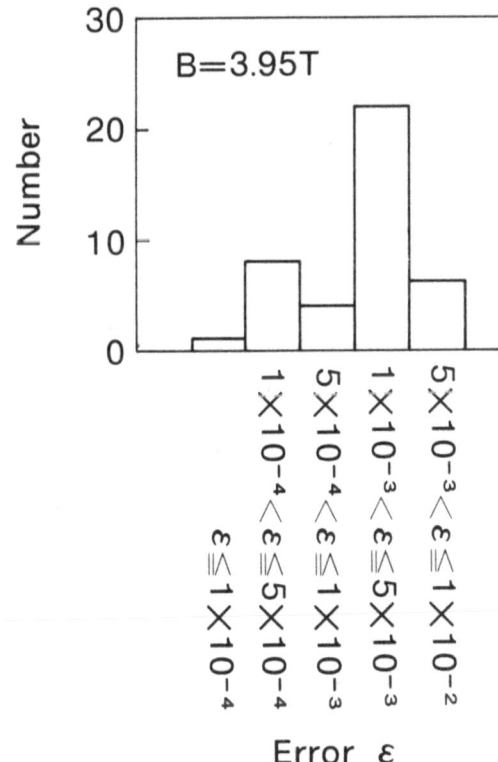

FIGURE 8. Accuracy of the Hall sensors $\varepsilon = |V1-V|/V$, where $V=(V1+V2)/2$, V1 and V2 mean first and second results on the same condition.

The magnet has been energized under a nominal current of 1322A. The results of the measurement dipole fields by using NMR sensors at the magnet center are displayed in Fig.9. In this figure, where solid lines represent design values, measured and design data are in close agreement. The variation of the dipole field along the electron path is shown in Fig.10. Again close agreement between measured and design values were obtained. In this experiment, the magnet quenched by the thermal disturbance due to the sensor holders, the magnetic field didn't reach the design value of 4.0T; however, in another series of experiments without sensor holders, the magnet reached the design field with ten times quenches.

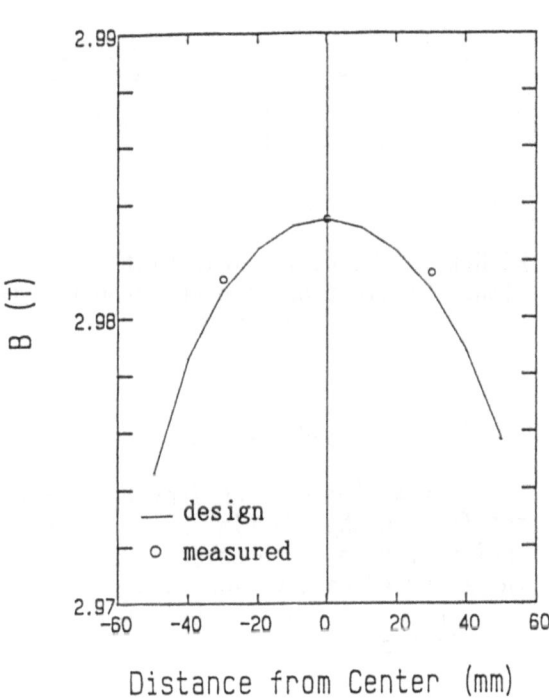

FIGURE 9. Magnetic field distribution at the magnet center measured by NMR sensors.

CRYOGENIC DATA

The initial cool-down of the magnet requires 500 liters of liquid nitrogen. Further cooling to 4.2K requires 350 liters of liquid helium. A filled helium vessel contains 280 liters. The helium boil-off rate is 10 liters per hour. This magnet has an SR absorber cooled by liquid nitrogen, and it is estimated that the helium boil-off rate doesn't increase much when electron storage is finished.

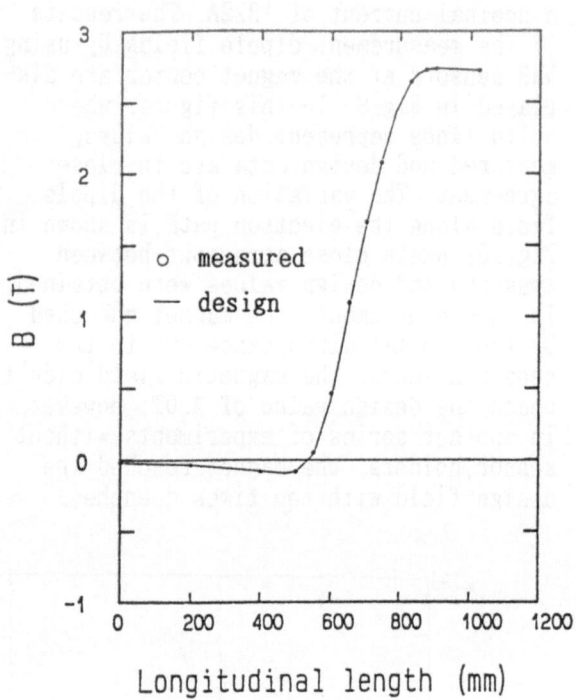

FIGURE 10. Magnetic field distribution along the electron path measured by Hall sensors.

CONCLUSIONS

A new type of iron-free bending magnet for a 615MeV SR ring with bending radius of 0.5m has been developed, and the measured data validated design.

These magnets will be set in the SR ring in September 1989.

ACKNOWLEDGMENTS

The authors wish to thank Dr.Tomimasu of the Electrotechnical Laboratory for his useful advise.

This project is entrusted by the Research Development Corporation of JAPAN.

REFERENCES

1. Takada, H., Tsutsui, Y., Tomimasu, T. and Sugiyama, S., High beam current storage at low energy for compact synchrotron radiation rings, _Rev. Sci. Instrum._, 1989, 1630-1634

2. Tomimasu, T., Noguchi, T., Sugiyama, S., Yamazaki, T., and Mikado. T., An electron undulating ring for VLSI lithography, _IEEE_ _Trans. Nucl. Sci._, NS-32, 1985, 3403

THREE DIMENSIONAL MAGNETIC FIELD ANALYSIS
FOR 5 m SUPERCONDUCTING DIPOLE MAGNET

H. Hirabayashi, T. Shintomi
K. Kondo*, M. Yamaguchi*, S. Itoh*, S. Murai*, M. Hirano*
KEK National Laboratory for High Energy Physics, Tsukuba,
* Toshiba Corporation, Keihin Product, Yokohama, Japan

ABSTRACT

A five-meter dipole magnet has been constructed and tested to verify electrical and magnetic performance. The magnetic field is analyzed with the three dimensional nonlinear magnetic field analysis program. The effectiveness of annular void at coil ends with iron yokes has been confirmed analytically as well as experimentally. The rated current of 6504A at the rated field of 6.6T was successfully achieved after several quenches.

INTRODUCTION

The accurate magnetic field analysis is required, in particular for the development of dipole magnets in view of superconducting magnet design and field quality. In connection with these, the impact of the winding, curing fixtures, end designs, yoke-collar-coil and such needs to be explored.

It is customary that the superconducting dipole magnets for the particle accelerators are surrounded with the iron yokes in order to achieve field enhancement. The characteristics of the superconducting magnets are governed by the field strength from the point of view of the critical current margin and the superconducting stability. Therefore, precise field analysis is required for the supercon-ducting magnets with the iron yokes, in

particular dipole magnets whose shape is not axisymmetric and which exhibit complicated electromagnetic performances, especially at the coil end regions. Saddle shaped coil end regions need to be tightly supported against the electromagnetic force and also thermal contraction in order to avoid conductor movements, however, it is not so easy to solve it, considering that configura-tion. Accordingly, it is a pertinent way to lower the magnetic field at this region. For this purpose, the iron yoke is voided annularly at the portion of coil ends. The three dimensional magnetic field analysis is performed for 5 m superconducting dipole magnet to evaluate the above effect. The magnetic field is analyzed with the three dimensional nonlinear magnetic field analysis program by integral equation method, and mesh generation programs for both the iron yokes and the saddle shaped coils. As a result of field analysis, it is made clear that the adoption of annular void is effective, reducing the peak field at the end regions. The above analysis is compared with the two diemensional nonlinear field analysis.

MAGNET CONSTRUCTION

The designed magnet is suitable for large colliding beam facility [1], with a central field of 6.6T, a winding bore of 40 mm, and a cold iron flux return yoke. A cross section of this magnet is shown in Figure 1, and main magnet parameter is in Table 1. The inner and

TABLE 1. Parameter summary

Magnetic field, dipole	6.60T
Dipole length	5.0 m
Excitation current	6504A nominal
Stored energy, 6.60T	330 kJ
Cold mass	2,040 kg
Inner coil	
Inner radius	2.00 cm
Outer radius	2.962 cm
Turns per pole	16
No. of wedges	3
Cu/SC area ratio	1.5/1
Short-sample limit	554 Amp/strand (5.0T, 4.2K)
Outer coil	
Inner radius	2.987 cm
Outer radius	3.993 cm
Turns per pole	20
No. of wedges	1
Cu/Sc area ratio	1.8/1
Short-sample limit	316 Amp/strand (5.0T, 4.2K)
Collar ID and OD	8.087, 11.09 cm
Collar material	YUS 170
Iron yoke ID and OD	11.14, 26.7 cm
Yoke material	low-carbon steel

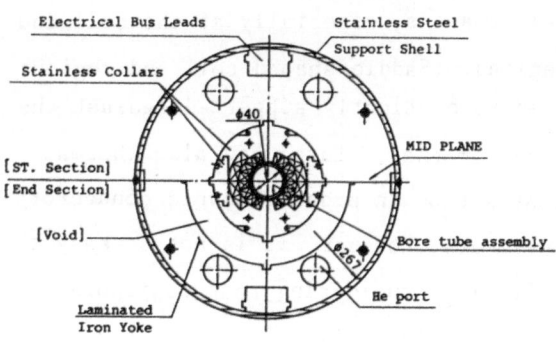

Figure 1. Cross-section of dipole magnet

outer "Rutherford type" compacted strand cables are insulated with YUPILEX polyimide tape by two layer wraps, the first layer of lap winding and the second layer of semicured epoxy coated gap winding. In order to secure mechanical rigidity of coil end, winding

is separated with several blocks at corresponding wedges in straight section and it is tightly compressed by machine made G10 spacers.

Figure 2 shows an axial cross section at the part of magnet end of 5 m model with annularly voided iron yoke.

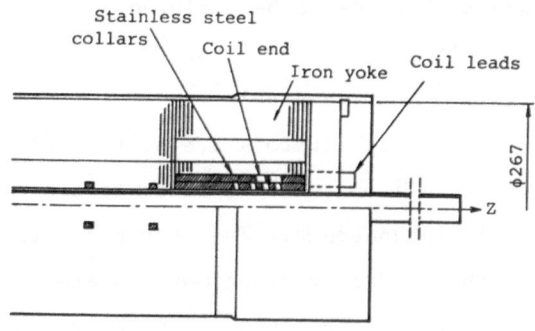

Figure 2. Axial cross-section at coil end

The void design is employed for the reduction of the field enhancement at the portion of saddle shaped coil end.

MAGNETIC FIELD ANALYSIS

METHOD

Three dimensional nonlinear magnetic field analysis program by integral equation method (3D by IEM) is applied to superconducting dipole magnet. In integral equation method [2,3], unknown quantity is located only in magnetic material, so it is comparatively easy to apply it to three dimensional open area problem. The permeability of magnetic material is nonlinear and therefore this problem is solved by iteration method. Magnetic field produced by superconducting coils is calculated with **Biot-Savart's law.**

Here, three kinds of dipole magnet, where the length of the iron yoke is different at coil ends as shown in Figure 3, are analyzed, i.e., iron yoke which is voided annularly at the portion of coil ends, full iron yoke which is 100 mm longer from the ends of coil straight section, surrounding the coil end completely, and short iron yoke which is 60 mm shorter from the ends of coil straight section.

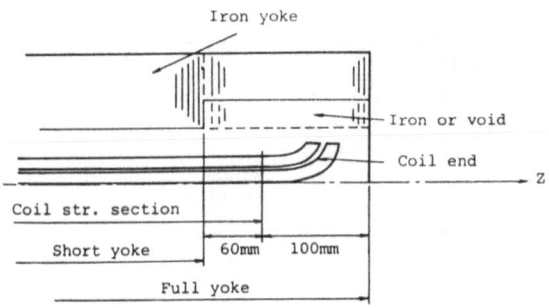

Figure 3. Three kinds of iron yoke

Only one-eighth of the space is analyzed, considering the symmetry of the dipole magnet. Figure 4 shows mesh division of iron yoke with annular void and superconducting coil. The iron yoke is expressed with the cluster of hexahedron elements, and the superconducting coil is expressed with the cluster of rectangular parallel piped elements.

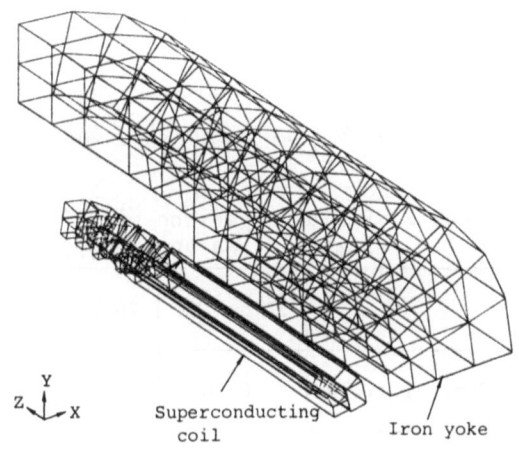

Figure 4. Mesh division

Supercomputer CRAY X-MP/22 was used for analysis and C.P.U. time was 6 minutes in the case of iron yoke with annular void for instance.

RESULT

The validity of three dimensional magnetic field analysis is first confirmed by comparing it with the two dimensional nonlinear analysis by finite element method (2D by FEM). The flux distribution by three dimensional analysis is shown in Figure 5. Table 2 shows the calculated field comparison of 3D by IEM with 2D by FEM. It is clear that the magnetic field at the center of dipole magnet is almost the same between two analyses.

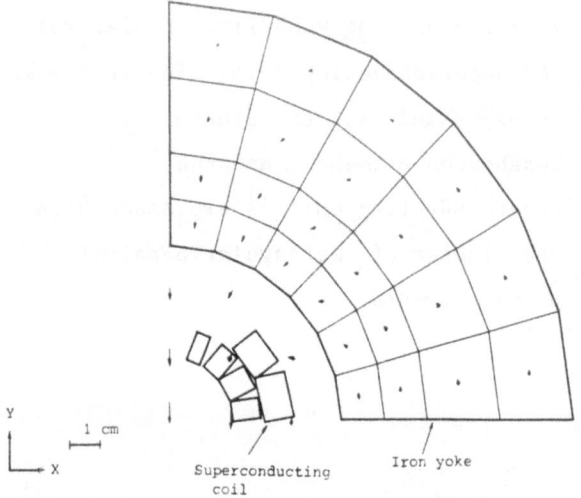

Figure 5. Flux distribution at the center
of Z axis (yoke with void)

TABLE 2. Field comparison with 3D
analysis and 2D analysis

	3D by IEM	2D by FEM
Central field	6.66	6.69

Unit; Tesla

Figure 6 shows the flux distributions along z axis for three kinds of iron yoke. From this figure, it is clear that the flux distributions are different among three kinds of yoke. At the ends of coil straight section, flux density in case of yoke with void is lower than flux density in case of full yoke. In case of short yoke, flux density is the lowest, therefore effective core length becomes shorter.

Table 3 shows maximum magnetic flux density of superconducting coil for three kinds of iron yoke. In full yoke, location of its field is at the coil end section, but in yoke with void, it is at the center of the coil straight section. As a result of field analysis, it is

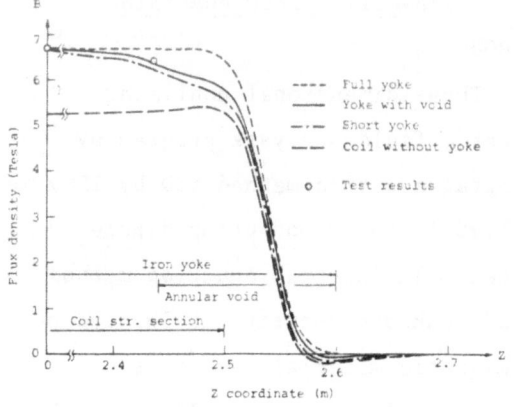

Figure 6. Magnetic field on Z axis

made clear that the adoption of annular void is effective for reducing the peak field at the end regions.

TEST RESULTS

The magnet was tested in a vertical cryostat with pool boiling condition. Figure 7 shows the 5 m model ready for test with iron yoke.

Figure 7. The 5 m model with iron yoke

TABLE 3. Maximum flux density of superconducting coil
for three kinds of iron yoke

		Full yoke	Yoke with void	Short Yoke
Coil str. section	center	6.94	6.93	6.85
	end	7.15	6.50	6.24
Coil end section		7.19	6.55	6.25

Unit; Tesla

Figure 8 shows the training behavior of the magnet [4,5].

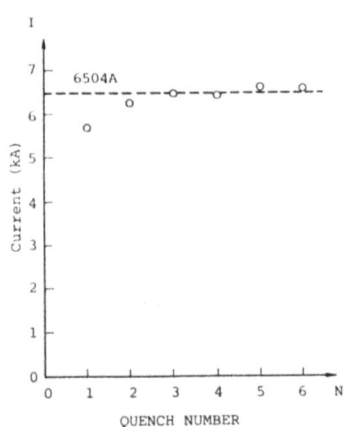

Figure 8. Training behavior of dipole magnet

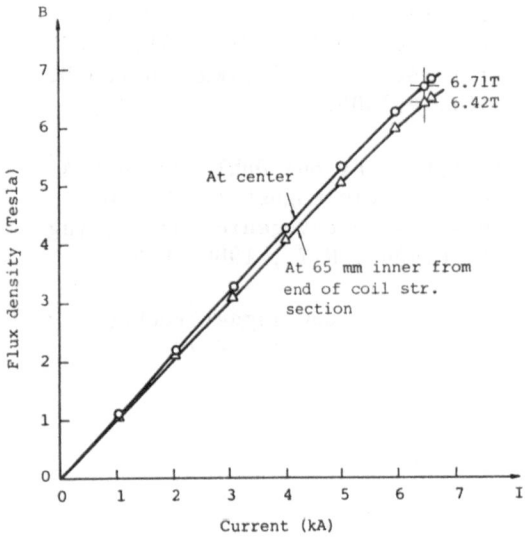

Figure 9. Saturation curve of central field

At the plateau field 6.7T, the experienced peak field reached about 95% of short sample data.

The accurate field measurement was performed with Hall effect elements in order to compare with analyses. Figure 9 shows magnetic field at center and end of magnet on Z axis. As a result of above test, it is clear that analysed results shown in Figure 6 agrees well with test results.

CONCLUSION

Three dimensional magnetic field analysis was performed for 5 m superconducting dipole magnet. The advantage of the magnet with annularly voided iron yoke was made clear analytically in comparison with full yoke and short yoke. The design field was relatively easily achieved, after several quenches. The technology developed for 5 m superconducting dipole magnet will be applicable to longer dipole magnets.

REFERENCES

1. SSC central design group, Conceptual design of the Superconducting Super Collider. SSC-SR-2020, Universities research association, Washington DC, 1986.

2. Newman, M.J., Trowbridge, C.W. and Turner, L.R., GFUN; An interactive program as an aid for magnet design. Proc. 4th Int. Conf. Magnet Technology, Brookhaven, 1972, P.617.

3. Gupta, R.C., Morgan, G.H. and Wanderer, P., A comparison of calculations and measurements of the magnetic characteristics of the SSC design d dipole. Proc. 1987 IEEE Part. Accel. Conf., Washington DC, 1987, P.1405.

4. Taylor, C.E. and Dahl, P., A 6.4 Tesla dipole magnet for the SSC. Advances in cryogenic engineering vol 31, Plenum, N.Y., 1986, P.25.

5. Taylor, C., SSC magnet technology. IEEE, 1988, MAG-24, P.726.

ANALYSIS OF THE PERFORMANCE OF THE EIGHT
SUPERCONDUCTING QUADRUPOLES FOR THE LEP LOW-BETA INSERTIONS

P.J. Ferry, D. Semal, ALSTHOM, Belfort, France
Ph. Lebrun, O. Pagano, S. Pichler, T.M. Taylor, T. Tortschanoff,
L. Walckiers, and L. Williams, CERN, Geneva, Switzerland

ABSTRACT

The LEP low-beta quadrupoles are iron-free superconducting magnets having a warm bore of 120 mm in diameter, an effective length of 2 m and a nominal gradient of 36 T m^{-1}. The coils form a two-block approximation of the cosine 2θ distribution, and are wound from monolithic multifilamentary NbTi conductor; mechanical prestress is obtained by means of aluminium shrinking rings. Following manufacture and appraisal of a prototype magnet, eight series units have been produced by industry and thoroughly tested and measured prior to installation into LEP. In this paper, a brief description of the manufacturing technique is given, and the results of tests and measurements are presented and compared with the performance expected from calculations and previous experience. The analysis reveals the tolerances on field quality which can reasonably be obtained with this type of construction.

INTRODUCTION

In LEP, the Large Electron Positron collider now operating at CERN, the luminosity, i.e. the rate of particle interactions at the crossing points, will be enhanced by using strongly focusing superconducting quadrupole magnets. The location of these magnets, which protrude into the spectrometer solenoids of the four LEP experiments, imposed severe constraints on their design[1]. A prototype and eight series quadrupoles were ordered from industry to CERN specifications. The design, test and performance of the prototype has been reported earlier[2]; here we report on the production and performance of the series units. These eight magnets, assembled in their horizontal cryostats, have now been delivered to CERN, and tested and measured prior to installation in the LEP collider.

MAGNET DESCRIPTION

The 2 m long ironless magnet is designed to provide a gradient of 36 T m^{-1} within a warm bore of 130 mm.

Coils are wound from monolithic multifilamentary NbTi superconductor and held together by a stainless steel and aluminium structure, calculated to give longitudinal rigidity and maintain compressive forces throughout in all working conditions[2]. Cooling is provided by a liquid helium bath contained in a close-fitting cryostat[3].

Coil

The coil design is based on that successfully developed for the ISR low-beta quadrupoles[4]. The cosine 2θ distribution is approximated by two conductor blocks, wound around a stainless steel centre post, and separated by copper wedges. The quadrants are epoxy-impregnated, bound together with glass-fibre tape and supported by an external structure.

The essential features of coil and support structure are shown in cross-section in Fig. 1. Dimensions given refer to unstressed components at room temperature.

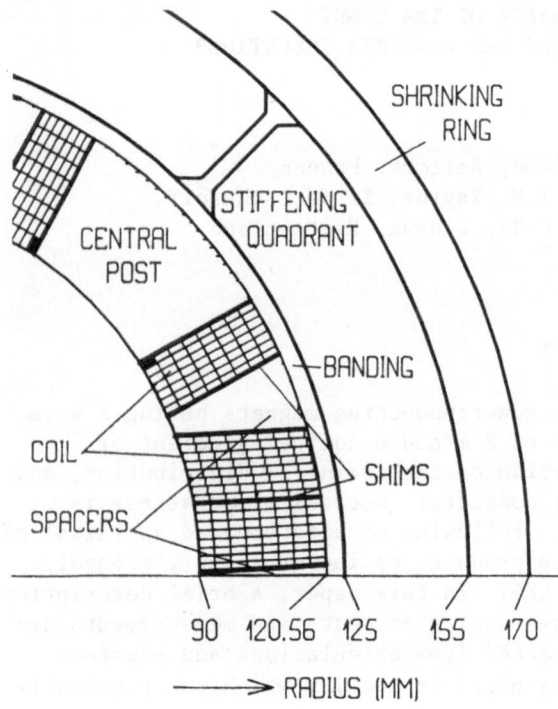

FIGURE 1: Coil and support structure

Structure

The pre-stress required to keep the coil in compression when the magnet is excited is provided by shrink-fitted aluminium alloy rings which apply external pressure to the coil via longitudinal stainless steel stiffening quadrants. Initial calculation of the required interference were performed using classical thick ring formulae. Resulting estimates were verified by a two-dimensional, plane-stress finite element analysis of the magnet structure, carried out using the program CASTEM. The coils were simulated by a homogeneous material having Young's modulus E_c in the ranges 3.5 to $6.5 \ 10^4$ N mm^{-2} cold, and half these values warm. Stresses and deformations for the structure were obtained for three load cases: at room temperature, after cooling to 4.2 K, and with magnetic loading (as calculated using the program POISSON) applied to the structure at 4.2 K. It was verified that separation of the innermost part of the face of the coil from the fixed centre post, the most likely cause of magnet quench, should occur at between 2350 A and 2650 A for E_c in the range considered, and compressive prestress in the cold coil of 70 N mm^{-2}. This value, and that of 35 N mm^{-2} at room temperature, were chosen as being the maximum admissible for safeguarding the integrity of the insulation. A fully trained magnet reaches 2250 to 2450 A, from which we deduce the equivalent Young's modulus of the coil blocks to be about $3.5 \ 10^4$ N mm^{-2} at 4.2 K. Displacements at assembly, cool-down, and with the nominal current of 1640 A are shown in Fig. 2.

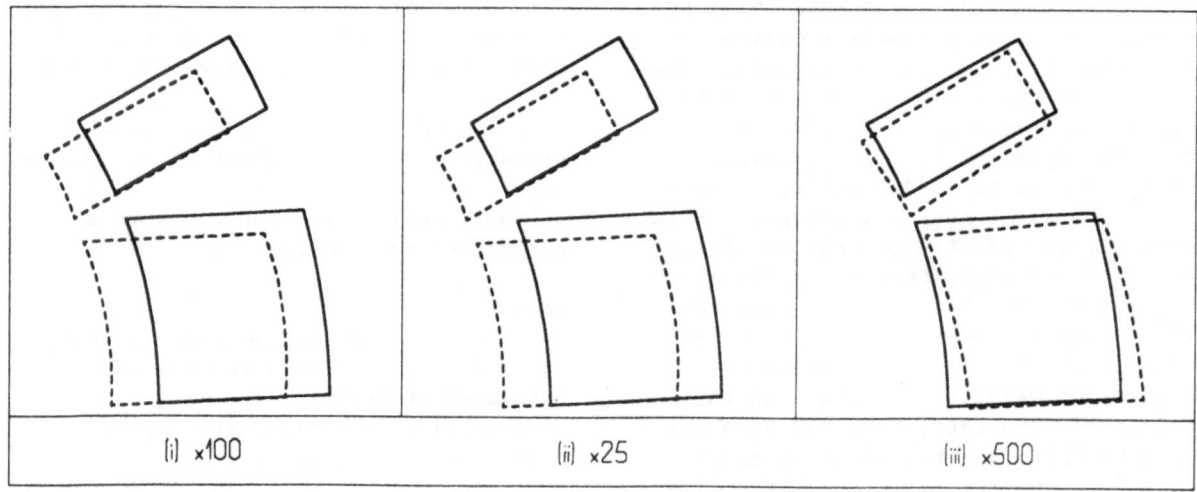

FIGURE 2: Magnified displacements at (i) assembly, (ii) cool-down and (iii) excitation

MANUFACTURE AND TEST

Conductor

The coils are wound from a solid composite wire of cross-section 1.8 mm x 3.6 mm, containing 2230 NbTi filaments of 37 μm diameter, and twisted with a pitch of 50 mm. This conductor is insulated with PVA enamel and polyimide tape; total insulation thickness is 0.1 mm. The wire was manufactured by IMI, U.K., and delivered in lengths of about 850 m or 1620 m, enough to wind one or two coils. The total length delivered was thus 31 km, and all batches were within specification. The critical current for 5 and 6 T fields parallel and perpendicular to the broad face, was measured on samples taken from the beginning and the end of each spool. The transverse mechanical dimensions were checked systematically on the finished wire and relevant insulation tests were performed. Short sample characteristics of the wire from the poorest batch used in each magnet are shown in Table 1.

TABLE 1
Short sample characteristics: critical current at 5 T and 4.2 K

	B $\perp$ to broad face	B $\parallel$ to broad face
MQC1	2610 A	3230 A
2	2720	3170
3	3080	3780
4	3130	3760
5	3080	3760
6	3100	3680
7	3170	3730
8	3590	3810

Coil Winding

In view of the similarity of their design, the techniques employed for winding the coils of these magnets were based on those developed for the manufacture for the ISR quadrupoles[5]. Besides the winding proper, this applies to the checks on interturn and ground insulation, and the assembly of the four coils. The dimensions of all components were checked before assembly, and the block positions were measured three times during winding: after the first block, half way through and at the end of the second block. These measurements were made with respect to regularly calibrated massive tooling ("heavy clamping"), which also served as the impregnation mould. Glass-fibre laminate shims were incorporated to correct the average block positions. Anticipated shrinkage during potting was taken into account. The measured azimuthal displacements of the coil blocks from their nominal positions were always less than 0.15 mm. With the exception of MQC5, however, all magnets were built up from quadrants with coil blocks displaced on the average towards the central post.

Assembly and Test

Once potted, and after careful machining of the mating surfaces, the coils were assembled on a mandrel, banded with epoxy-impregnated glass-fibre tape, and the outer diameter was machined to within 0.01 mm to match the inner diameter of the stiffening quadrants. The latter were then bolted onto the assembly to give about 15% of the final pre-stress. During this operation, the equivalent elastic modulus of the assembly was calculated using measurements of changes in the diameter, to give the interference fit necessary between stiffening quadrants and aluminium shrinking ring for obtaining the required pre-stress.

The shrinking rings were heated to 150°C to permit sliding over the horizontally placed assembly of coils and stiffening quadrants. Measurements were made of changes in inner and outer diameters before and after this operation; the derived value for the average elastic modulus of the warm coil assembly was found to be $3 \cdot 10^4$ N mm^{-2} ($\pm$ 20%).

After having been equipped with the necessary sensors to monitor temperatures, voltages, etc., the magnets were cooled to liquid helium temperature in a vertical test cryostat and subjected by the manufacturer (ALSTHOM, France), on his premises, to the specified operational tests. For this, the magnets were each powered to 2000 A and quenched five times; this was followed by a three hour endurance run at 1900 A. All magnets passed these tests without difficulty.

After testing, the magnets were assembled into their horizontal cryostats and delivered singly to CERN mounted in a shock-absorbing cradle.

Training

During all power tests, the potential across the coils was constantly monitored, and at the appearance of a voltage, a CERN-designed 16-channel fast transient recorder was triggered to record all relevant parameters in the time (programmable) leading up to and following a resistive transition (quench). Never more than 5 quenches were necessary to achieve the design test current of 2000 A; the lowest value of current at which a natural quench occured was 1640 A. For two magnets trained at CERN beyond 2000 A, up to 10 quenches were required to reach currents producing forces at the prestress limit. Once trained, the magnets do not require re-training after thermal cycling to room temperature; all eight magnets reached 1900 A without quench after installation in the LEP tunnel.

A summary of hot-spot temperature and peak voltage developed in the coil when no stored energy is extracted (the normal operating condition) is shown in Fig. 3. The inductance of the magnet is 0.23 H.

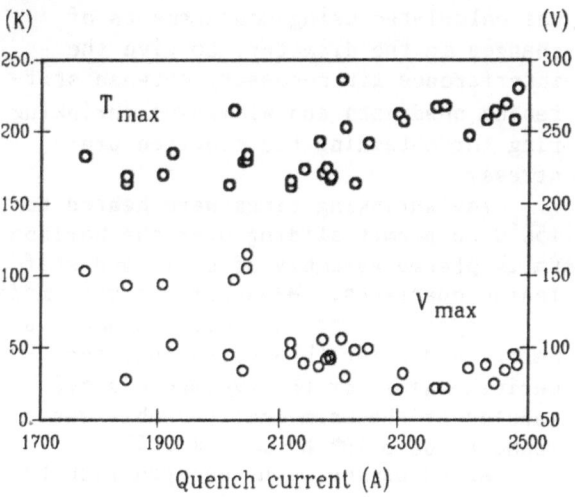

FIGURE 3: Hot-spot temperature, T, and peak voltage, V, at quench

MAGNETIC PERFORMANCE

Lessons from the Prototype

The results of the prototype measurements, in terms of relative errors on the integrated gradient at 50 mm radius, within which the field quality is required to be good, are given in ref.[2]. In addition to a dodecapole term of

- 0.4%, a skew sextupole term of - 0.1% and a skew decapole term of 0.1% are present. These errors were explained by a displacement of the coil blocks towards the central post and a combination of small random errors accentuated by the effect of the relative permeability of the stainless steel posts. The absolute values of the error terms were nevertheless somewhat larger than predicted from magnetic field calculations, including measured coil block displacements, thermal shrinkage and compression, and the measured permeability (1.009 - 1.013) of the stainless steel parts of the magnet. It was therefore decided to compensate the dodecapole component by the removal of one turn in the small coil block. The only price to pay for this was calculated to be a reduction in the main quadrupole gradient of less than half a percent and a small increase in the 20-pole component.

Gradient and Alignment

All measurements were made with a rotating coil, integrating output over angular increments as measured via an encoder. The main coil covered a length of 2.98 m; end-coils of 748 mm in length provided the information for alignment of the magnetic axis.

After cycling to nominal current, the transfer function of the magnets was measured to be 21.95 ± 0.05 T m^{-1} kA^{-1}, compared to 22.09 T m^{-1} kA^{-1} for the prototype magnet.

It was found necessary to correct the alignment of the assembled magnet with respect to the warm bore of the cryostat. This was done by measuring the magnet warm with an excitation of 7 A, and correcting the position via the suspension rods, taking into account calculated motion at cool-down. After cool-down, the measured axis was transferred to alignment targets mounted on the cryostat. Alignment was found to be preserved to within 0.02 mm after thermal cycling.

Field Quality in the Series Magnets

Table 2 summarizes the field quality measurements of all series magnets, given in terms of relative errors of the integrated gradient at 50 mm.

Table 3 shows the dodecapole component given in the same units, measured for the straight parts only (after correction for the actual position of the

TABLE 2

Measured integrated multipole relative gradient errors in 10^{-5} at 50 mm radius

	normal, n =							skew, n =						
	3	4	5	6	7	8	10	3	4	5	6	7	8	10
MQC 1	89	65	54	-246	16	8	-105	-183	26	66	-2	-33	8	0
2	-37	0	67	-277	12	1	-115	86	-46	16	8	4	-3	0
3	10	1	75	-259	10	1	-104	29	55	0	-8	-3	3	-1
4	-205	-201	-12	-274	-10	-29	-106	-139	42	46	8	-6	10	-1
5	9	-80	44	-115	12	11	-80	-83	15	5	14	-16	16	1
6	0	-24	-46	-277	-9	4	-103	-19	22	-17	17	12	-10	4
7	-131	-55	-102	-217	-30	-6	-95	5	60	5	-7	6	-5	-1
8	-43	-122	-9	-196	-7	-14	-93	54	-57	-18	-2	-4	-10	0

TABLE 3: Measured dodecapole gradient errors in the straight part

	normal, n = 6
Prototype	-174×10^{-5}
MQC 1	37
2	-42
3	34
4	2
5	169
6	-22
7	83
8	65

measuring coils). It can be seen that:
- The tendency of having measurable non-quadrupole components is confirmed. The integrated gradient errors at 50 mm due to sextupole and octupole components can be as high as 0.2%, and the decapole component 0.1%.
- The reduction of dodecapole expected from the introduction of the dummy turn is fully effective for MQC5, but less pronounced for the other magnets.
- The difference in dodecapole errors between straight part and integrals is up to 0.4% while computations predict a difference of less than 0.1%.

The first of the above points is an indication that the final coil block positions are not exactly that expected from the mechanical measurements performed on the single coils before assembly of the quadrants. The coils were, moreover, combined on the basis of these measurements in such a way as to minimize the appearance of unwanted terms. The non-quadrupole terms are an indication that asymmetries are present in the cold magnets of up to the equivalent of 0.4 mm block displacement. These effects could be explained by high values of relative permeability in the central coil posts and stiffening quadrants; the measured values of between 1.008 and 1.013 at 4.2 k explain less than 10% of the asymmetric terms, however.

The unexpectedly small change of dodecapole term observed between prototype and series magnets is explained partly by the measured mechanical geometry of the coil quadrants, all magnets except MQC5 being composed of "thin" coils, i.e. coils having the large coil block displaced towards the central post. Part of the integrated gradient errors can also be attributed to a slight difference between the geometry obtained in practice and that assumed in the computations, in the region of the transition from straight part to end. The straight parts have a relatively large effect on the error terms, and it was found that errors due to the ends are about twice that expected.

These considerations lead us to believe that a major contribution to the measured errors must originate in the ends of the magnets.

CONCLUSIONS

The magnets have been proved by the tests to be well constructed. In particular the support structure behaves very much as predicted from the calculations. The magnetic measurements indicate that the final block positions, and the transition from the straight part to the end, in the cold magnet may be somewhat less well-defined than that expected from the mechanical measurements made on individual coil quadrants.

The introduction of a dummy turn into the first small block has been shown to be a valid method for reducing the dodecapole error discovered on the prototype magnet.

The satisfactory performance of these magnets demonstrates the soundness of this design for a small series of individually powered units.

ACKNOWLEDGMENTS

Quality of fabrication is a vital ingredient to the success of a project such as this; the authors are therefore pleased to acknowledge the important contributions of H. Blessing (current feed-throughs), H. Depierre (instrumentation), D. Regin (mechanical follow-up), and G. Trinquart (design work), as well as the devoted effort of the production and test teams at ALSTHOM.

REFERENCES

(1) Taylor, T.M., Technological Aspects of the LEP low-beta Insertions, IEEE Trans. on Nucl. Sci., Vol. NC-32, No. 5, Part II, 1985, pp. 3704-06.

(2) Lebrun, Ph., Pichler, S., Taylor, T.M., Tortschanoff, T., and Walckiers, L., Design, Test and Performance of the Prototype Superconducting Quadrupole for the LEP low-beta Insertions. IEEE Trans. on Magnetics, Vol. 24, No. 2, 1988, pp. 1361-64.

(3) Blessing, H., Lebrun, Ph., Pichler, S., Taylor, T.M., and Trinquart, G., Design, Test and Performance of the Liquid Helium Cryostats for the LEP Superconducting Quadrupole Magnets. Proc. ICEC-12, Butterworth, 1988, pp. 112-16.

(4) Billan, J., Henrichsen, K.N., Laeger, H., Lebrun, Ph., Perin, R., Pichler, S., Pugin, P., Resegotti, L., Rohmig, P., Tortschanoff, T., Verdier, A., Walckiers, L., and Wolf, R., The Eight Superconducting Quadrupoles for the ISR High-Luminosity Insertion. Proc. XIth Int. Conf. on High Energy Accelerators, CERN, 1980, pp. 848-52.

(5) Billan, J., Perin, R., Resegotti, L., Tortschanoff, T., and Wolf, R., Construction of a Prototype Superconducting Quadrupole Magnet for a High-Luminosity Insertion at the CERN Intersecting Storage Rings. Yellow Report CERN 76-16, Geneva, 1976.

NUMERICAL ANALYSIS OF THE VOLTAGE CURRENT TRANSITION IN SUPERCONDUCTING CABLES

D. ter Avest and L.J.M. van de Klundert
University of Twente, Applied Superconductivity Centre, POB 217,
7500 AE ENSCHEDE, the Netherlands

ABSTRACT

A new method is presented to calculate the voltage current transition in a superconducting cable using any $E(J)$ relation for the superconductor in the cable. The method gives the possibility of comparing the voltage current transition in a superconducting wire with the transition in a cable composed of a number of these wires. Results are presented using a distribution function $g(J)$ to describe the relation $E(J)$. Differences in the transition of wires and cables will be shown as a function of a number of parameters.

INTRODUCTION

The properties of multifilament superconductors are usually determined by examining the voltage current transition. As a figure of merit the so-called 'n-value' is often attributed. This n-value is determined by locally fitting the relation $E \propto J^n$ to the measured voltage current graph. Another possibility to determine n is to use the relation

$$n(J) = \frac{J}{E} \frac{dE(J)}{dJ} . \qquad (1)$$

As the n-value appears not to be constant when considering several orders of magnitude of the voltage [1] it is still a poor quality standard. An alternative is to analyse the voltage current data in terms of a distribution function $g(J)$ [2]. $g(J)$ is a dimensionless function and is defined by

$$g(J) = \frac{J_m^2}{A} \frac{d^2 E(J)}{dJ^2} , \qquad (2)$$

or,

$$E(J) = A \int_{i=0}^{J/J_m} (J/J_m - i) \, g(i) di \qquad (3)$$

with J_m some maximum of J and A a normalization constant. The determination of the distribution function involves some numerical analysis to find the second derivative of E to J. The value of n can now be obtained by using:

$$n(J) = \frac{J}{J_m} \frac{\int g(i) \, di}{\int (J/J_m - i) g(i) di} \qquad (4)$$

where the integrals are taken from 0 to J/J_m.

Additional problems arise when analysing superconducting cables. In this report we will regard superconducting cables to be Rutherford type cables with a rectangular cross section. However, the described method is applicable to any geometry. Apart from side effects such as deformation of strands at the cable sides, the voltage current properties of cables will be determined by
1) the E(J) relation of the individual strands
2) the relative dimensions of the cable, i.e. the aspect ratio
3) the twist pitch of the strands in the cable
4) the direction of the applied field
5) the difference in conductivity

parallel to the superconductor and perpendicular to the superconductor

6) the insertion of a normal conducting core in the cable to reduce coupling losses.

To investigate these effects a numerical model was created and converted into a computer code. The important basis of the model is the implementation of an E(J) relation for the superconductor which has the form of equation (3). Due to twist of the superconductor it is necessary to define the direction parallel to the superconductor ($\parallel$) and the direction perpendicular to the superconductor ($\perp$). The relation for E(J) that we used is:

$$E_\parallel (J_\parallel) = \rho_\parallel J_m \int_0^{J_\parallel / J_m} (J_\parallel / J_m - i) g(i) di \qquad (5)$$

with ρ the electrical resistivity and J_m the critical current density as a function of the magnetic field $B_\perp$, defined by the Kim relation [3]. $B_\perp$ is composed of the applied field (in the n,t-plane) and the self field of the cable. $J_\parallel$ is the total current density in the direction of the superconductor:

$$J_\parallel = \eta J_s + E_\parallel / \rho_\parallel \qquad (6)$$

with J_s the superconducting current density and η the ratio of superconductor to the total volume in the cable. In order to examine the difference in n-value of superconductor and cable we used a distribution function that renders a constant n-value for the superconductor. The employed function is:

$$g(i) = \begin{array}{ll} (n-1)i^{n-2} & J_\parallel / J_m \leq 1 \\ 0 & J_\parallel / J_m > 1 \end{array} \qquad (7)$$

This function gives

$$E_\parallel = E_0 \; (J_\parallel / J_m)^n / n \qquad J_\parallel / J_m \leq 1$$

$$E_0 \; (J_\parallel / J_m - \frac{(n-1)}{n}) \cdot J_\parallel / J_m > 1 \qquad (8)$$

with $E_0 = \rho_\parallel J_m (B_\perp)$ and

$$dJ_s / dE_\parallel = 0 \qquad J_\parallel / J_m > 1 \qquad (9)$$

The n-value used in the calculations is 70

MODEL

In order to prevent sharp corners on the cable which would certainly lead to mathematical discontinuities, the cable is represented by a rectangular structure with a semicircle on each of the small sides as is represented in fig. 1. For convenience the (n,t,z) coordinate system is adopted in the cable. In this system n denotes the normal direction to the surface of the cable, t the tangential direction and z denotes the longitudinal direction. The direction of the superconductor is defined by the twist angle ψ as is shown in fig. 2. The current in the normal direction is always perpendicular to the direction of the superconductor and can be written as

$$J_n = \bar{\bar{\sigma}} \; \bar{E} \qquad (10)$$

in which $\bar{\bar{\sigma}}$ represents the conductivity tensor. We assume a conductivity $\sigma_\parallel$ parallel to the superconductor and a conductivity $\sigma_\perp$ perpendicular to the superconductor. Then, it follows in good approximation that

$$\sigma_\parallel = (1-\eta)\sigma_0 \qquad (11)$$
$$\sigma_\perp = (1-\eta)\sigma_0 / (1+\eta) \qquad (12)$$

with σ_0 the conductivity of the normal

FIGURE 1 Geometry of the cable as used in the calculations. The directions n and t are defined in the X-Y plane.

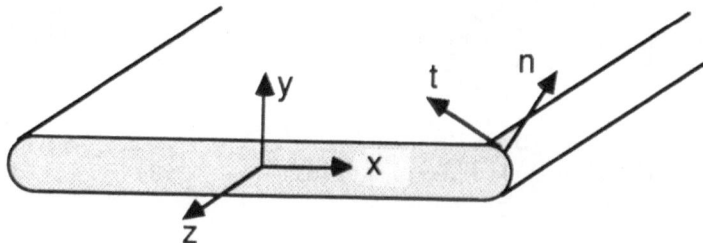

material. The superconducting current density Js can be written in the (n,t,z) coordinate system as

$$J_s = J_s \begin{bmatrix} 0 \\ \sin \psi \\ \cos \psi \end{bmatrix} \tag{13}$$

The components of the total current density are now

$$J_n = \sigma_n E_n \tag{14}$$
$$J_t = \eta J_s \sin \psi + \sigma_{tt} E_t + \sigma_{tz} E_z \tag{15}$$
$$J_z = \eta J_s \cos \psi + \sigma_{zt} E_t + \sigma_{zz} E_z. \tag{16}$$

Next we apply the Maxwell equations. In this problem we have z-invariance and the equations can be written as

$$(\nabla \times B) = \mu_0 J \tag{17}$$
$$(\nabla \times E) = 0 \tag{18}$$
$$\nabla . J = 0 \tag{19}$$
$$\nabla . B = 0 \tag{20}$$

The cable is now discretized using a so-called 'staggered grid'. In this grid, not all components of E and B share the same position in the grid, as is shown in fig. 3. The applied field is now fixed to some value. Starting with a given value of Ez, which is a constant over the cable cross section, an approximation of $J_\parallel$ is used to solve the components of E and B. The components of B, the self field of the cable, are used to calculate B⊥ which will affect Jm(B⊥). In an iterative method the values of Jm(B⊥) and $J_\parallel$ are adjusted until the transport current converges. This takes generally only two or three iterations.

Subsequently, the dependence of Ez on the transport current is investigated by varying two parameters. These are:

a) the angle α between applied field and the long axis of the cable in the n,t plane.

b) the twist pitch or angle ψ of the superconductor

Table 1 shows the relevant parameters in the calculation.

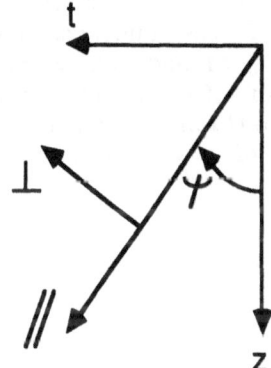

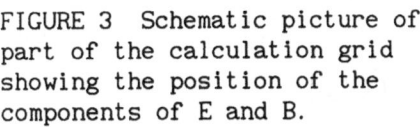

FIGURE 2 Definition of the twist angle ψ.

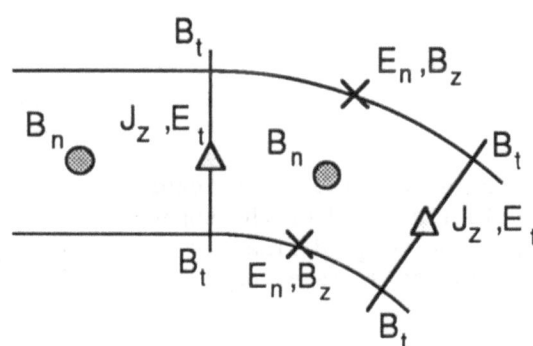

FIGURE 3 Schematic picture of part of the calculation grid showing the position of the components of E and B.

TABLE 1

Relevant parameters in the calculation.

applied field	0.5	[T]
dimensions of straight part of the cable	17x2	[mm]
radius semicircle	1.0	[mm]
η - ratio	0.3	
σ_0	$1.67 \cdot 10^8$	[Ωm^{-1}]
$J_m(B\perp=0)$	10^9	[A/m^2]
B_0 (in Kim relation)	2.0	[T]
twist pitch	10.0	[mm]

RESULTS

Fig. 4 shows n as a function of j^*:

$$j^* = \bar{J}_z/J_{\parallel} \, ,$$

which is a measure for the saturation in the cable. $\bar{J}_z$ is the average current density in the z-direction. Clearly, the n-value of the cable shows a decrease when j^* approaches 1, caused by the interaction of self field and applied field. The n-value of the superconductor is a hyperbole for $j^* > 1$ and can be obtained by substitution of (8) in (1). Decreasing n makes the transition broader because for a smaller value of n the derivative $dJ_s/dE_{\parallel}$

increases. For instance, an n-value of 20 shows that the n-value of the cable starts to decrease already for $j^* = 0.4$. The effect of the parameter α on the n-value is shown in fig. 5. Although there exists a dependence of the n-value of the cable on α, the difference for $\alpha = 0$ and 90 degrees is only 1 %. However, this difference can be biased by changing other parameters such as the applied field, n, or the twist pitch. Next, the twist pitch of the cable was changed in the range 50-5 cm, while the circumference of the cable is 4 cm. The result is viewed in fig. 6. A decrease of the twist pitch at a perticular value of the transport current naturally increases E_z and decreases the n-value due to additional saturation in the cable. Conversely, for a perticular value of E_z, or a voltage criterium, a decrease of the twist pitch decreases n. Finally, it is important to note that the solution $E_z(I)$ depends linearly on ρ as can readily be understood from eq. (8).

In the above section the effect of several parameters on the voltage current transition was described briefly. As the interaction between the parameters is complicated, care must be taken choosing the values of non-changing parameters. A next step in the development of this model will therefore be the implementation of a realistic distribution function and a measured $j_m(B\perp)$ relation.

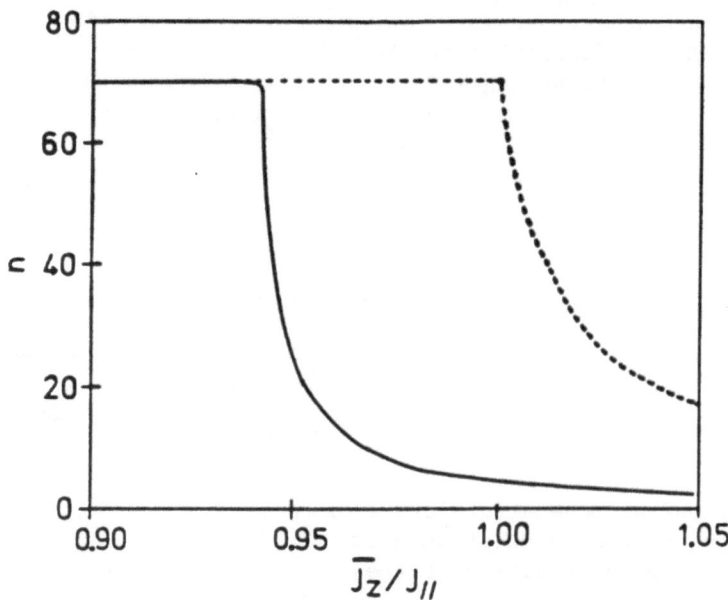

FIGURE 4 n-value of the cable (solid line) and of the super-conductor (dashed line) as a function of j^* (see text).

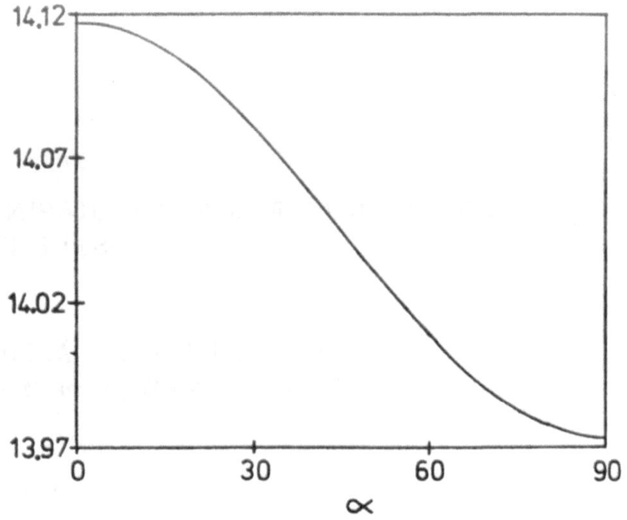

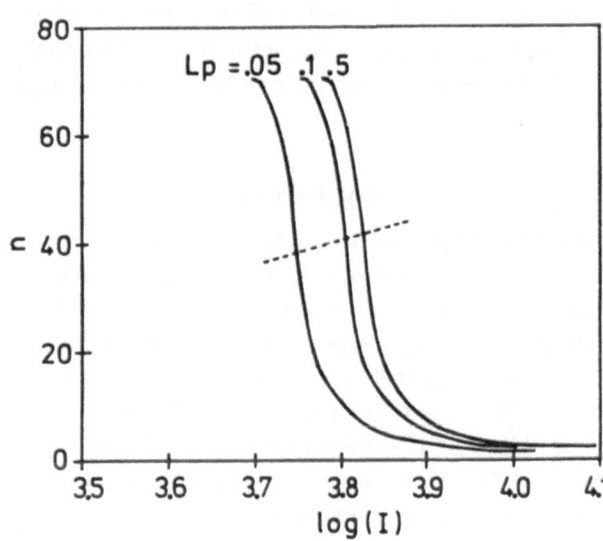

FIGURE 5 Dependence of n on
the angle α between applied
field and the long side of the
cable. E_z has some arbitrary
value.

FIGURE 6 Effect of changing
the twist pitch L_p on the
n-value of the cable. I is the
transport current. The dashed
line is a line of constant E_z.

CONCLUSIONS

A numerical method for the calculation
of the voltage current transition in
superconducting cables was presented. The
use of a distribution function for the
E(J) relation of the superconductor
allowed the comparison between the
transition in superconductor and cable
composed of this superconductor. It was
shown that while the superconductor was
given a constant n-value of 70, the
n-value of the cable decreases as the
saturation in the cable grows. For a
smaller value of n this effects becomes
more profound. Finally, changing the angle
between the applied field and the long
side of the cable showed a change in the
n-value of about 1 % when for n the value
of 20 is assumed.

REFERENCES

1. H. Boschman, H.H.J. ten Kate and L.J.M.
   van de Klundert, Critical current
   transition study on multifilamentary
   NbTi superconductors having a Cu, a
   CuNi or a mixed matrix, MAG 24, pp
   1141-1144, 1988.
2. D. ter Avest and L.J.M. van de Klundert
   On the J(E) relations in multifilament
   superconductors and composite wires.
   Internal report University of Twente,
   UT-KfK-1/26-6-89
3. Y.B. Kim, C.F. Hempstead and A.R.
   Strnad, Magnetization and critical
   supercurrents, Phys. Rev. 129 (1963),
   p. 528.

DESIGN OF AN AC MAGNETIC BIASING CIRCUIT FOR THE KAON FACTORY BOOSTER RF CAVITY.

C. Haddock, R.L. Poirier, A. Otter, T. Enegren and M. Zanolli*
TRIUMF, 4004 Wesbrook Mall, Vancouver, B.C., V6T 2A3

ABSTRACT

The resonant frequency of an rf cavity to be used in the booster ring of the TRIUMF KAON factory, is determined by the state of magnetization of a set of 6 rings of G810 aluminium doped yttrium-iron garnet ferrite. The design of an ac magnet which performs the required magnetization is presented. The results of the ac magnet code PE2D to predict eddy current induced power losses walls of the rf circuit, support structure and the G810 cooling jacket are presented.

INTRODUCTION

TRIUMF is presently completing the project definition stage of its proposed KAON factory. Acceleration of the beam in the booster ring will be accomplished using perpendicular biased yttrium-aluminum iron garnet type ferrite tuners [1] as opposed to the more conventional parallel biased Ni Zn ferrites [2]. The G810 ferrite is operated in saturation and offers higher magnetic Q's and lower power losses than conventional ferrites.

A dc prototype was constructed at LAMPF and is shown schematically in Fig. 1. A toroidal magnet surrounds six ferrite rings. The axis of the structure is the beam axis. Berylium oxide (BeO) cooling rings are placed between the ferrite. The BeO rings are in turn cooled at their outer radius by a copper cooling jacket.

An ac prototype based on the LAMPF dc tuner was designed at TRIUMF and is presently being constructed with an expected delivery date of October 1989. This paper describes some of the design considerations concerned with the ac version of the dc prototype.

*Permanent Address: CERN, PS Division, 1211 Geneva-23 Switzerland.

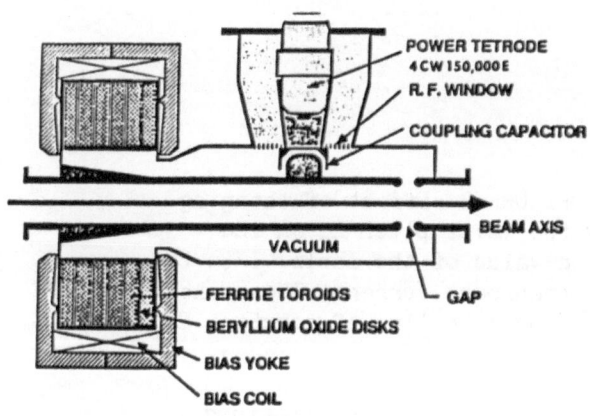

FIGURE 1. LAMPF DC Prototype Ferrite Biasing Circuit.

AC BIASING MAGNET

Since the magnet will operate under an alternating current waveform the circular yoke of the dc prototype was replaced by a laminated structure as shown in Fig. 2. The complete circular geometry was replaced with 12 rectangular laminated blocks. The blocks were then tapered by cutting and grinding to form the structure shown in the diagram. Grinding is performed in such a way that adjacent laminations are not shorted together. To save on the cost of dies, each lamination is made from three separate pieces.

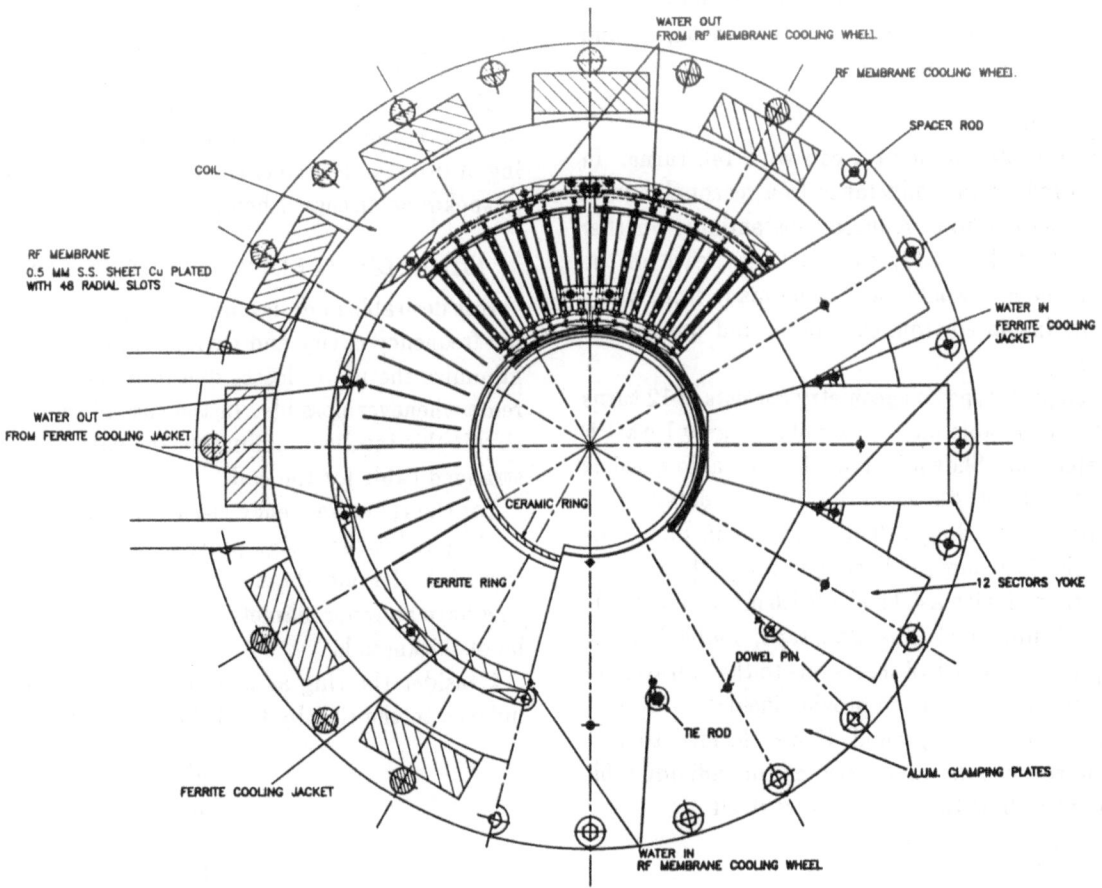

FIGURE 2. View of AC Biased Cavity Looking along Beam Direction.

This involves only two die types, one having the geometry of the pole and the other a simple rectangular shape which forms the return yoke.

The lamination thickness required to minimize eddy current loss in the yoke is given by evaluating the skin depth δ given by:

$$\delta = \sqrt{\frac{2}{\omega\sigma\mu}} \qquad (1)$$

Where:
ω = angular frequency
σ = electrical conductivity
$\mu = \mu_r\mu_0$ = permeability of the material.

The waveform shape which biases the ferrite rings is a dual frequency function with a dc offset and is shown in Fig. 3. The waveform rises at 50Hz in 15ms and resets in 5ms at 100Hz.

The skin depth evaluated for the 100 Hz component, using $\sigma = 1.02 \times 10^7$ $(\Omega m)^{-1}$ and $\mu_r = 1000$ is 0.498mm. The magnet yoke therefore is made using 0.0148 inch (0.37 mm) laminations of M6 steel.

The magnet laminations are glued together in blocks using an epoxy resin. This will provide some resistance to vibrations as well as providing the required inter lamination resistance.

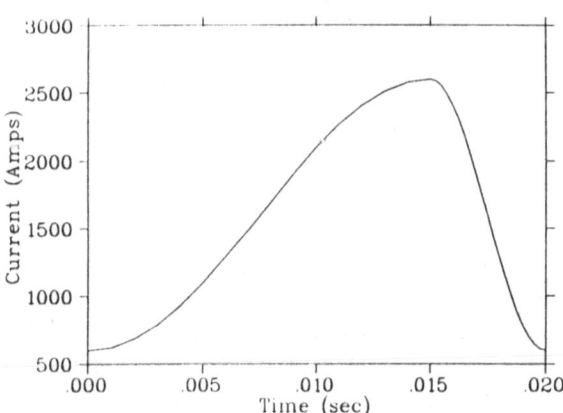

FIGURE 3. Dual Frequency Ferrite Biasing Waveform.

The magnet sectors thus formed will be held together by an aluminium clamping plate and a set of tie rods. The sectored design provides room for the entrance and exit of the stranded cable which requires a large bend radius and further provides easy access for the water jacket cooling lines. Furthermore this design will be useful for prototyping purposes which will likely involve dismantling the unit several times and for the installation of thermocouples and other diagnostic equipment.

Magnet Coil

The dc prototype used a coil with 140 turns. In order to reduce the inductance to a reasonable level this number had to be reduced substantially. Further in order to reduce eddy current losses, each of the turns of the coil would need to be made from many filaments insulated from each other and powered in parallel.

The ac prototype coil geometry consists of 12 turns of a stranded cable made available to us by LAMPF for evaluation. Shown in Fig. 4., the cable consists of 82 strands of #9 heavy formvar insulated magnet wire surrounding a 0.368 inch (9.35 mm) I.D, 0.499inch (12.67mm) O.D copper tube [3]. The cable is wrapped with a 4.0 inch (101.6 mm) wide, 0.005 inch (0.13 mm) thick, fiberglass tape, double lapped.

To prevent the risk of shorts due to the vibration of individual strands, and to provide insulation against the high voltages to ground, it was decided to vacuum impregnate the voids between the individual filaments as well as the coil structure itself.

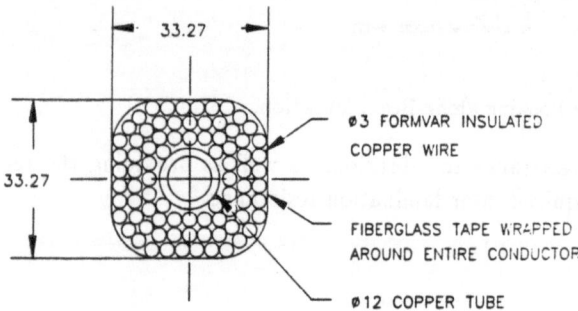

FIGURE 4. Cross Section of LAMPF Stranded Conductor.

RADIO FREQUENCY CAVITY

The radio frequency cavity surrounds the ferrite and BeO rings and extends radially inward toward the cone shaped center conductor as shown in Fig. 5. The cavity can be considered as consisting of four parts: the end walls which are two large disks, the outer conductor which is a large cylinder and the center conductor which is a tapered conical shape.

To reduce eddy current losses, the end walls and outer conductor of the cavity were made from 0.020 inch (0.5 mm) stainless steel "membrane". These walls are then plated with copper to a thickness of 0.0005 inch (0.0127 mm).

The center conductor is a machined stainless steel tube 0.078 inches (2.00 mm) thick and plated with copper on the outside to a thickness again of 0.0005

inches. The center conductor was made of thicker material than the rest of the circuit since it has to withstand vacuum loading.

The ferrite rings are held in place in the rf circuit using a ceramic ring and a support structure consisting of two stainless steel rings connected by an array of spokes as discussed below.

MINIMIZING EDDY CURRENT LOSSES

It is desirable in designing the magnetic circuit, radio frequency cavity and cavity support structure to minimize the power losses due to induced eddy currents whenever possible. In the case of the magnetic circuit this involved laminating the yoke and using a stranded cable for the coil.

For the radio frequency circuit some simple analyses help guide the design process. Consider the outer ring of the RF membrane support structure. Let the ring have an inner diameter r_i, outer diameter r_o and be of thickness h.

Consider the ring as a series of annuli. The emf induced in the ring by the field is given by:

$$e = -\frac{d\phi}{dt} \qquad (2)$$

$$= \pi r^2 \dot{B} \qquad (3)$$

Where r is the radial position of the annulus. The resistance dR of the strip is given by:

$$dR = \frac{2\pi r}{\sigma h dr} \qquad (4)$$

Where σ is the conductivity of the strip. The mean power due to eddy currents is:

$$dP_{mean} = \frac{1}{2}\frac{e^2}{R} \qquad (5)$$

$$= \frac{1}{4}\pi \dot{B}^2 h\sigma \int_{r_i}^{r_o} r^3 dr \qquad (6)$$

It can be shown that:

$$\dot{B} = \omega B_o$$

Upon integration:

$$P = \pi\frac{1}{16}(2\pi f B_o)^2 h\sigma[r_o^4 - r_i^4] \qquad (7)$$

If the ring is broken at opposite ends of a diameter the large eddy current buildup is avoided. The situation reduces to the eddy current losses in a long conductor of linear length $2\pi r_{mean}$. Consider the case in which a solid linear conductor is divided into elemental strips in the direction x, the thickness of the strips is again h.

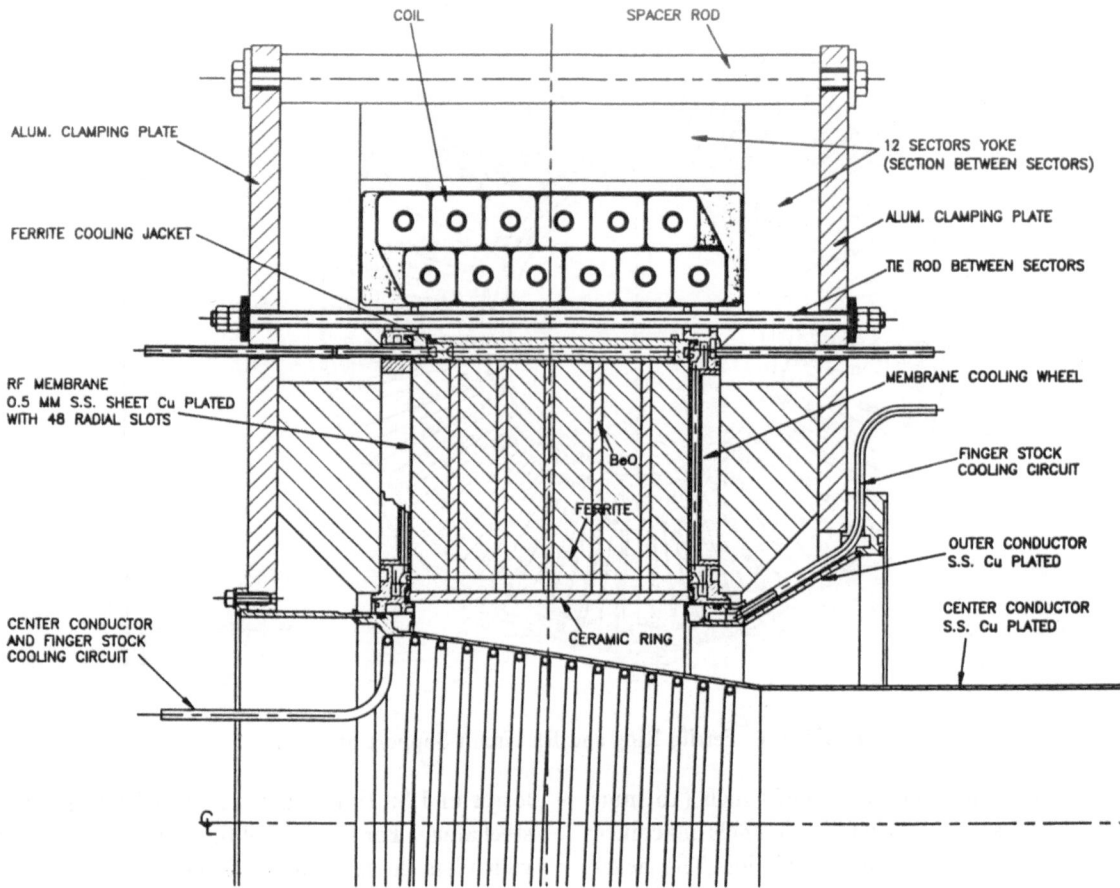

FIGURE 5. Cross Section of AC Biased Cavity.

The emf which drives the eddy currents is given by:

$$e = -\frac{d\phi}{dt} \qquad (8)$$

$$= A\dot{B} = 2xl\dot{B} \qquad (9)$$

The resistance dR of the strip is given by:

$$dR = 2\frac{l}{\sigma h dx} \qquad (10)$$

The mean power due to eddy currents is:

$$dP_{mean} = \frac{1}{2}\frac{e^2}{dR} \qquad (11)$$

$$= l\dot{B}^2 h\sigma \int_0^{w/2} x^2 dx \qquad (12)$$

$$P = \frac{l\dot{B}^2 h\sigma w^3}{24} \qquad (13)$$

Upon integration:

$$P = \frac{1}{6}\pi^2 w^3 l\sigma h f^2 B_o^2 \qquad (14)$$

Where w=width of the conductor=$(r_o - r_i)$.

The ratio of power losses with the break to those without is given by:

$$\frac{P_{break}}{P_{complete}} = \frac{2}{3}\frac{(r_o - r_i)^2}{(r_o^2 + r_i^2)} \qquad (15)$$

In the case of the membrane outer support ring the breaking of the complete ring results in a reduction of 900 times in eddy current losses. For the inner ring the reduction is a factor of 80 times. Consequently whenever a complete circular geometry can be broken a reduction in the eddy current losses will occur. This concept was extended to the large disc shaped rf membrane at the end of the rf cavity. Since the rf currents flow radially a large number of radial slots can be used to reduce eddy current losses with little disturbance to the rf currents. Consequently 48 radial slots are machined into the rf membrane. The inner and outer rings of the ferrite support structure were connected together via 48 slotted spokes as shown in Fig. 6., as a further extension of the basic methodology for reducing eddy currents.

The water jacket consists of a cylinder of copper

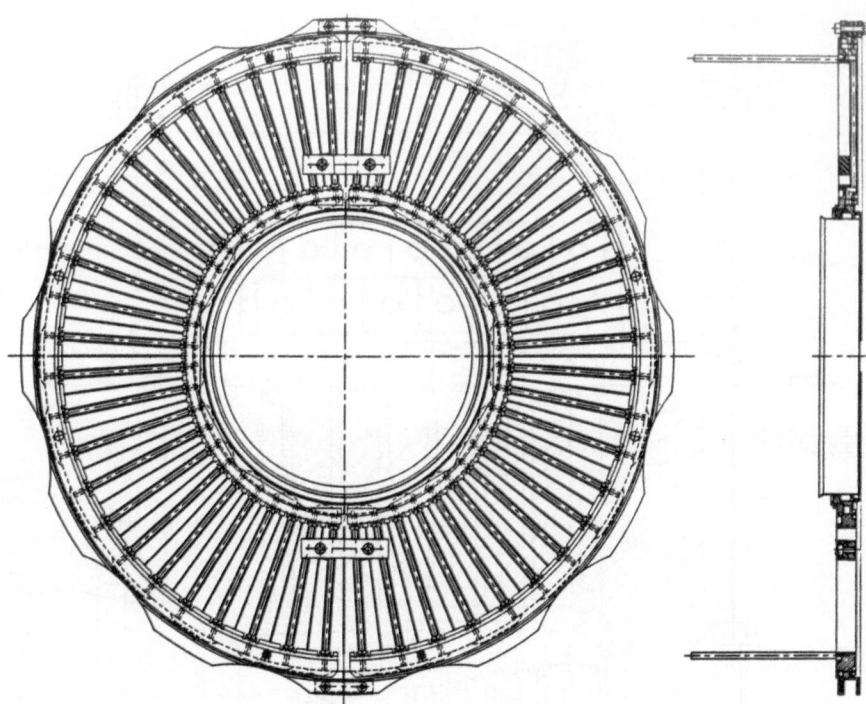

FIGURE 6. Ferrite Ring and RF circuit Support Structure.

6.25 inch (159 mm) long and 0.629 inch (16 mm) thick. It has a series of water channels machined longitudinally in the walls of the cylinder. The jacket has the potential to generate very large eddy current losses and several insulating breaks are required to minimize this effect.

MODELLING WITH PE2D

The essential elements of the radio frequency cavity were entered into the magnet code PE2D. The geometry is shown in Fig. 7.

The program has the capability to run magnetostatic, steady state (sinusoidal driving current), and transient driving current cases. Also shown in Fig. 7. is a flux plot when the code is run in the magnetostatic mode.

The field results from the magnetostatic case were used in analytical calculations shown above to verify the magnitude of the results from PE2D since it is imperative to include the break in the support structure and rf circuit into the PE2D geometry.

It was not practical using PE2D, to include 0.0005 inches of copper on the 0.5 mm stainless steel radio frequency circuit. If one take the relative thicknesses of the copper and stainless steel into consideration, and the relative magnitude of the conductivities, it is possible to show in an analytical calculation the effect of the two metals together [4]. In the case of that part of radio frequency circuit which has the form of a slotted disk, the result is an increase of eddy current

losses of 1.89 times that calculated for the stainless steel alone. For the part of the radio frequency circuit which has the form of a cylinder, there is a negligible addition to the eddy current losses due to copper plating.

The inner conical rf conductor was modelled as a circular cylinder of radius equal to the mean radius of the tube. The code was run in its transient driving waveform mode, using the waveform shown in Fig. 3. The eddy current induced power losses were evaluated at 10 intervals throughout the driving waveform cycle. The average of these losses for each part of the structure is presented in Table 1.

TABLE 1 Eddy current losses for ac geometry

Section	Eddy Current losses (W)
Inner Conductor	40
Outer Conductor	460
Disk Membrane	1890
Inner Support Ring	40
Outer Support Ring	20
BeO Cooling Jacket	1100
Total eddy current losses	3550

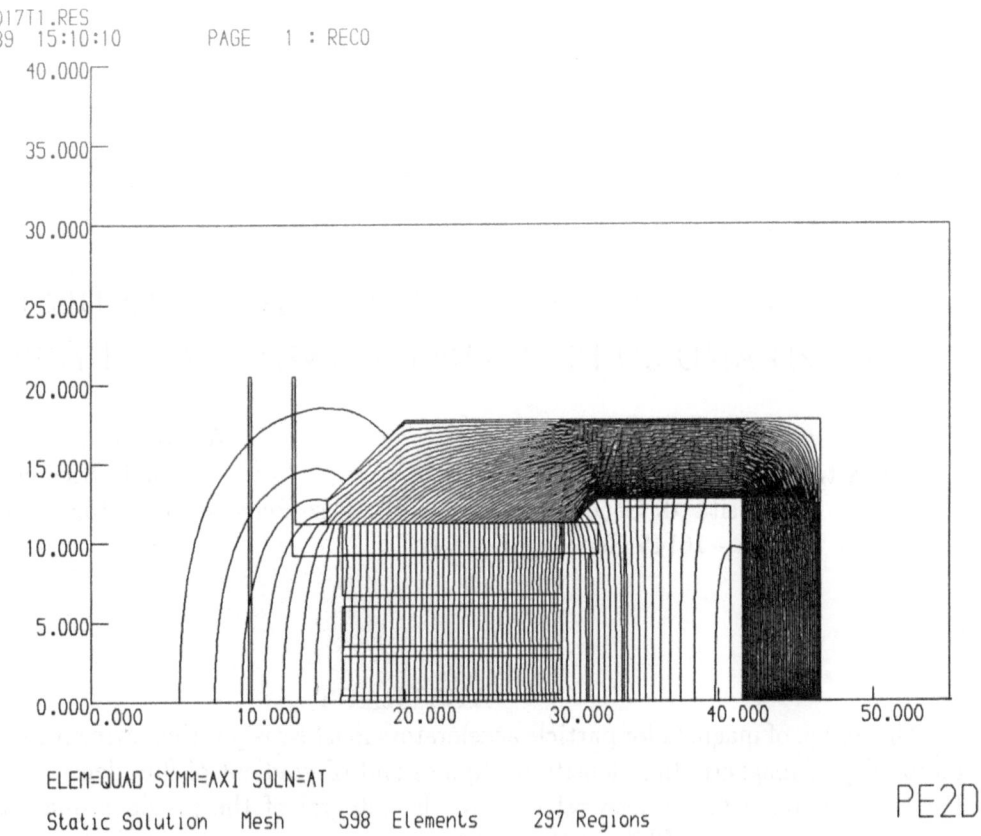

FIGURE 7. Cavity Geometry Modelled using PE2D.

SUMMARY

An ac version of a ferrite biased tuner has been designed at TRIUMF Assembly of the complete tuner is scheduled for October 1989. The eddy current losses in the support structure, radio frequency circuit and BeO cooling jacket have been evaluated using PE2D to be 3550 W total.

REFERENCES

1. Smythe, W.R., et al., "RF Cavities with Transversely Biased Ferrite Tuning", IEEE Particle Accelerator Conference, Washington D. C. p 2951 (1985).

2. "Status of RF Development Work on a ferrite tuned amplifier cavity for the TRIUMF to KAON factory" IEEE Particle Accelerator Conference Vancouver 1985 p2951

3. Otter, A.J., "LAMPF Style Stranded Magnet Cable" TRIUMF Internal report TRI-DN-88-K12 1988.

4. Haddock, C. "Eddy Current Analysis of the KAON RF Booster Cavity" TRIUMF Internal report 1989.

DIPOLE AND QUADRUPOLE MAGNETS DESIGN
WITH 2D AND 3D ELECTROMAGNETIC ANALYSIS CODES

P. Molfino, M. Repetto
Dipartimento di Ingegneria Elettrica
Universita' di Genova
Via Opera Pia 11a - 16145 Genova-GE, Italy

A. Matrone, P. Prati
Ansaldo Ricerche
C.so Perrone 25 - 16152 Genova-GE, Italy

ABSTRACT

The design of magnets for particle accelerators must satisfy rather stringent requirements on uniformity of magnetic flux density in dipoles and of gradient of flux density in quadrupoles. Further requirements are also relevant to the integral of the previous quantities along the longitudinal direction. While in the central part of the magnets satisfactory informations on magnetic field can be drawn from two dimensional magnetic analysis codes, the evaluation of fields in the end part of the device becomes more difficult due to the fringing effects, and would require a three dimensional analysis. In this paper different design strategies using approximate two dimensional analysis of the end part of the devices are presented, a comparison with a three dimensional analysis is performed and results are reviewed and discussed.

INTRODUCTION

The design of dipole and quadrupole magnets for particle accelerators requires the use of magnetic analysis codes in order to obtain a satisfactory estimate of the main magnetic specification. In dipole magnets design, two dimensional analysis codes are generally used to define the geometry of the cross section pole face in order to obtain the required uniformity of magnetic induction **B**. The design of quadrupole magnets requires also other magnetic computations, in order to define the pole profile satisfying the uniformity requirements of gradient of magnetic flux density required by the specifications. Usually, two dimensional codes available in the main research centers can provide the accuracy requested in the determination of the above quantities; however, the values obtained in this way are significant only in the central region of the device, where the two dimensional hypothesis is reasonably satisfied, while their accuracy in the end part of the magnet is much more doubtful, due to the real three dimensional behaviour of the field in that regions. The evaluation of fields in the end part of the magnet has to be taken into account when computing global quantities like the integral, along the longitudinal direction, of magnetic flux density in dipole magnets or of the radial gradient of field in quadrupole magnets. In fact, by neglecting

271

the end part contribution, a poor estimation of
the global quantities could be obtained requiring
adjustment of the magnet length in the prototyping phase. Since three dimensional analysis
codes are not up to now so readily available, several design strategies using only two dimensional
codes have been developed to take into account
end effects. Some of these strategies have been
used in the first design phase of the magnets
for the booster synchrotron of the storage ring
ELETTRA (Trieste, Italy) [1] and the results
obtained have been compared with the output of
a three dimensional code. The two dimensional
codes used to perform numerical computation
have been the finite element codes VF/PE2D [2]
and CEDEF [3], while three dimensional computations have been performed with the finite
element magnetostatic code VF/TOSCA [4].

DIPOLE DESIGN

As previously pointed out, the first step in
the design of the dipole magnets is generally the
determination of the pole shape providing the
required magnetic field uniformity in the center
of the gap; to this aim two dimensional computation are used to "synthetize" a pole shape providing the required field pattern. Once obtained
the optimal pole shape for the central region of
the device, the second design phase is relevant to
the determination of the pole end profile. This
has to be designed to obtain a field pattern in
the longitudinal direction giving rise to the required magnetic length and to a transversal behaviour of magnetic induction compatible with
that of the central part of the device, in order to
maintain undesired multipolar components below a given limit. Usually, the pole end profile
of dipole magnets is made up of a linear chamfer
to reduce manufacturing costs; the slope and the
length of the chamfer have to be determined by
means of numerical computations to satisfy the
requirements previously defined.

Two Dimensional Computation

The distribution of magnetic induction of the

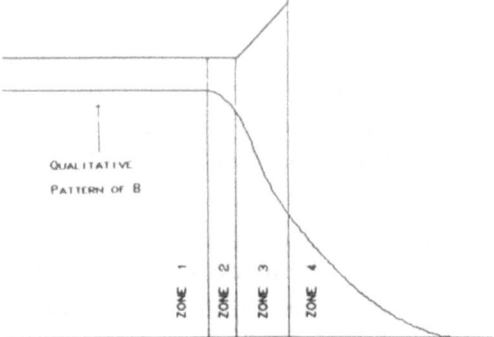

Figure 1: Subdivision of the integration domain
along the longitudinal profile of dipole magnet.

magnet in the chamfer zone along a line perpendicular to the magnet axis, can be approximated
by several cross sectional two dimensional analyses with increasing gap amplitudes. Furthermore, an approximate value of the longitudinal
distribution of the field along the magnet axis
can be obtained by a two dimensional analysis
in a longitudinal section of the device. In the
evaluation of the magnetic length of the dipole
magnet the field integral in the longitudinal direction along the dipole axis has to be computed while in the determination of the transversal pattern of the field the previous integral has
to be calculated at different values of the abscissa in the direction orthogonal to the axis.
In order to obtain the value of the above integrals, three different strategies have been proposed; each of these divides the integration domain in zones where the magnetic field is recovered from different computations. In the following the three methods are briefly described
whereas in Fig. 1 the different zones of the longitudinal domain used by method 2 are shown.

Method 1: with this method the longitudinal
where the integral has to be performed is divided in three zones. The first one spans the
central gap of the magnet up to the beginning
of the chamfer and the field in this zone is considered constant and equal to the rated value.
The second zone covers the segment of axis corresponding to the chamfer the field values are
computed by linear interpolation of some two
dimensional cross sectional analyses. The last
zone goes from the end of the magnet to the

value of longitudinal coordinate where the field reverses its sign. The value of this coordinate is obtained by a two dimensional analysis in the longitudinal section and the field behaviour is assumed linearly decreasing in this zone. Method 2: with respect to method 1 it adds a zone before the beginning of the chamfer. This zone begins where the field decreases from its rated value of more than 0.1 %, with the value of this coordinate obtained by a two dimensional computation in the longitudinal section. This method recovers field values in each zone by two dimensional cross sectional computation weighted on the two dimensional longitudinal slope [5]. Method 3: this method uses the field pattern in the cross section obtained by a two dimensional computation of the real geometry and multiplies this for a per unit function of the longitudinal coordinate obtained by a rough three dimensional analysis neglecting shims. All these methods have been used in the first design of the ELETTRA booster synchrotron dipole magnet with a gap of 40 mm, a rated value of flux density of 1.01 T and a field integral of 1165 [Tmm] corresponding to a magnetic length of 1150 [mm]; the results obtained in this way have been compared with a three dimensional computation of the structure including chamfer and shims. The results obtained for the magnetic length are shown in Table 1 where the integral value is also split in the contribution of the different zones. The results of the three dimensional analysis are given in the last line, headed TOSCA. In this table the physical length of the magnet has been computed by adjusting the length to match the rated design value of the global field integral. As it can be seen in the table, the zone affected by the greater error is the fourth one, giving rise to an overestimation of the physical length greater than 1 %. On the contrary, the field integral in the third zone, the one corresponding to the chamfer, is computed by method 1 and 2 in an efficient way. In fact, in this zone the values of field along the dipole axis obtained by two dimensional computations are in good agreement with those of TOSCA. It can be observed that method 3 estimates a physical length shorter than the magnetic one in reason of the fact that

TABLE 1

Comparison of integral of flux density in the longitudinal direction [Tmm] and computed physical length [mm] in dipole magnet.

zone method	1	2	3	4	total	Phys. Length
1	1121	-	31	9	1161	1158.0
2	1088	33	31	9	1161	1157.5
3	1073	30	34	24	1161	1140.0
TOSCA	1084	26	31	20	1161	1146.0

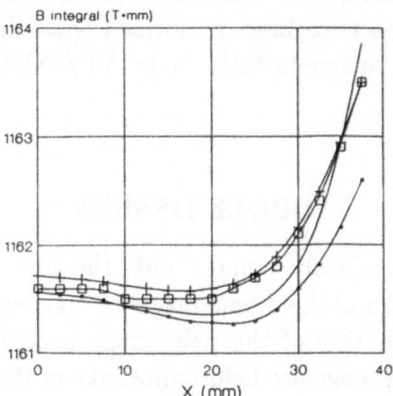

Figure 2: Integral of magnetic flux density along the longitudinal direction vs. the transversal coordinate, solid line TOSCA, □ method 1, + method 2, · method 3

it estimates better the contribution of the fourth zone. For the determination of the magneto-optical properties of the magnet, the previous field integral has to be evaluated at different values of the transversal coordinate. In Fig. 2 are shown the different pattern of the integral obtained by the three two dimensional different methods and by the three dimensional reference case. As it can be seen, the differences between values obtained by each of the three methods and the reference case are contained in a range of 0.1 % in the area extending up to about 40 mm from the axis giving then acceptable answers when used as input in a magneto-optical code.

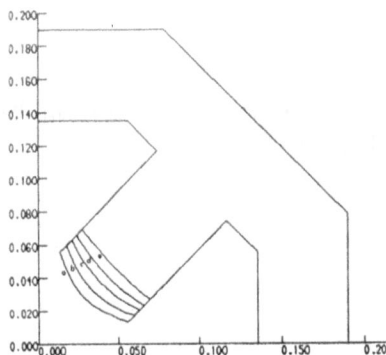

Figure 3: Stamping profiles: (a) used in the inner zone of the device, (b-e) used in the end part of the magnet.

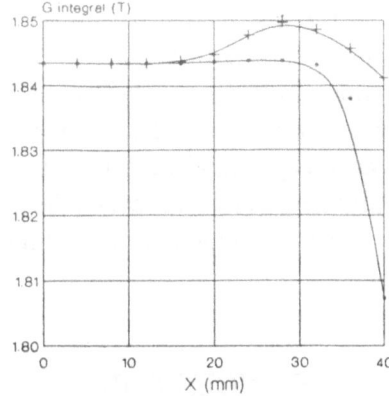

Figure 4: Integral of radial gradient of field along the longitudinal direction vs. the transversal coordinate. + TOSCA, · two dimensional computations

QUADRUPOLE DESIGN

Like in dipole magnets, the first step in the design of quadrupoles is the determination of the shape of the pole face in order to obtain the desired uniformity of the radial gradient of magnetic induction. For the quadrupole magnets of the ELETTRA booster synchrotron the pole shape adopted is the classical one made up of an hyperbolic profile in the central part of the pole, continued by a segment of straight line tangent to the profile in the outer region.

Two Dimensional Computation

The shape of the pole face satisfying the specifications has to be determined, for the rated value of inscribed radius, by the use of two dimensional analysis codes. Once determined the optimal profile in the central zone of the device, the end part shape has to be designed. Since a sharp cut could give unacceptable results in terms of multipole components, a smooth end is preferred. The chamfer of the end zone of the magnet is obtained with different stamping profiles. The end stampings have inscribed radius greater than the rated one, and their pole shape is made up of a circular segment. In Fig. 3 the central profile and four end profiles are shown.

In the quadrupole geometry the values of field and of its radial gradient are proportional to the inverse of the inscribed radius squared, so that, imposing a variation of the gradient in the lon-

gitudinal direction, it is possible to define the required longitudinal end profile. The effects of the increasing inscribed radius and of the change of the pole shape can be studied by means of two dimensional field computations in the transversal section; these computations can be combined together to estimate the three dimensional end effect. Since in quadrupoles it is not possible to make computations on the longitudinal section, there is no way to estimate the field out of the magnet by means of a two dimensional analysis. The results obtained by the approximate two dimensional approach have been compared with the ones of an accurate three dimensional analysis assumed as reference. The physical length computed with both methods differs of less than 0.5 %. This can be explained by the fact that the field out of the magnet drops off faster than in the dipole. The comparison of the two transversal behaviour of the integral of the radial gradient, as shown in Fig. 4, gives a difference contained in less than 0.5 % in all the zone of interest.

CPU TIME COMPARISON

A further comparison between two dimensional approximate analyses and three dimensional ones can be performed on the basis of the CPU time needed to obtain the required accuracy. The timing of the 2D methods have to be evaluated by summing the time of the single 2D runs, while the one of the 3D analysis is time of a single run. In both the cases under analysis 2D approximations needed about one half of the time requested by the 3D one with the required accuracy. As a matter of fact the 2D analysis provides a fast and rather cheap tool to the first design of complicate three dimensional geometry.

CONCLUSION

The analyses performed up to now show that, in a first design phase, the use of two dimensional codes gives enough accuracy to the designer with a computational burden that is small compared with that required by a three dimensional one, expecially when the output quantities required by the analysis are of integral nature like in the case under consideration. However the use of a real three dimensional analysis becomes necessary when design parameter are requested with an accuracy higher than 1 % or when it is necessary to verify local values of fields on the structure.

REFERENCES

1. Tazzari, S., Design study for the Trieste Synchrotron Light source, Ed. Lab. Naz. Frascati Report LNF-87/6, 1987

2. Armstrong, A.G.A.M. and Biddlecombe, C. S., The PE2D package for transient eddy current analysis, IEEE Trans. on Mag., 1982, 18, 411-5

3. Molinari, G., Albanese, R., Boglietti, A., Chiampi, M., Coco, S., DiNapoli, A., DelZoppo, R., Girdinio, P., Molfino, P., Repetto, M., Rubinacci, G., Santini, E., Savini, A., Tartaglia, M., A modular finite element package for research in electromagnetic analysis developed in a group of italian universities, Proceedings of BISEF, October 19-21, 1988 Beijing, China

4. Simkin, J. and Trowbridge, C.W., Three Dimensional nonlinear electromagnetic field computations using scalar potential. Proc. IEE, 1980, 6, 368-74.

5. Lieuvin, M. and Marks, N., Booster dipole design. European Synchrotron Radiation Facility/88-06

POLE SHAPING FOR IMPROVING THE DEVIATION OF THE EFFECTIVE LENGTH IN THIN QUADRUPOLE MAGNETS

S. Yamamoto, M. Morita, T. Matsuda and T. Yamada

Central Research Laboratory, Mitsubishi Electric Corporation
1-1 Tukaguchi-honmachi 8 Chome, Amagasaki, 661 JAPAN

ABSTRACT

Thin quadrupole magnets with or without shaping of pole ends had been studied experimentally and analytically. Multipole components in the median plane of the magnet had different signs compared with those in the pole ends. Deviation of the integrated field gradient was improved by the cut-off of the pole ends, because multipole components in the median region of the magnet were compensated by the components in the end region. Radial distributions of local field gradients in thin magnets were not improved by the pole end shaping. Experimental results were in agreement with the three dimensional analyses by TOSCA.

1. INTRODUCTION

Compact electron storage rings are useful as synchrotron radiation sources in X-ray lithography. Superconducting bending magnets are equiped to compact these rings [1][2]. Other components of the rings, for example, quadrupole magnet, skew-Q magnet, etc, are generally conventional type magnets. It is also effective to minimize the thickness of other components for compactness of the rings. In this paper, thin quadrupole magnet had been studied experimentally and analytically. Shapings of pole ends for improving the deviation of the integrated field gradient in thin quadrupole magnets had also been studied experimentally.

2. DESIGN AND FIELD ANALYSIS

2.1 Specification

Parameters of the thin quadrupole magnet are shown in Table 1.

TABLE 1
Parameters of the Thin Quadrupole.

Bore diameter	0.068 m
Core length	0.05 m
Pole width	0.08 m
Field gradient	3.5 T/m
Ampere turn	1.61 kA/pole
Core material	Pure iron

Two types of conventional quadrupole magnets, with a sharp cut-off [3] or no cut-off of pole ends, had been made respectively. The core length/bore diameter ratio of the magnet is 0.05m/0.068m=0.74. In order to enlarge the homogeneous region of the field gradient along x axis, the both sides of the hyperbola are extended to the straight line (0.01447m in length) parallel to the x axis as is illustrated in Fig.1. In the fabrication of the magnet, the end cut shape was approximated by three steps as is shown in Fig.2.

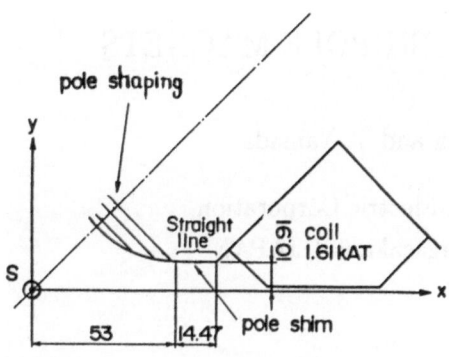

FIG.1 The pole shape of modified hyperbola. (in mm)

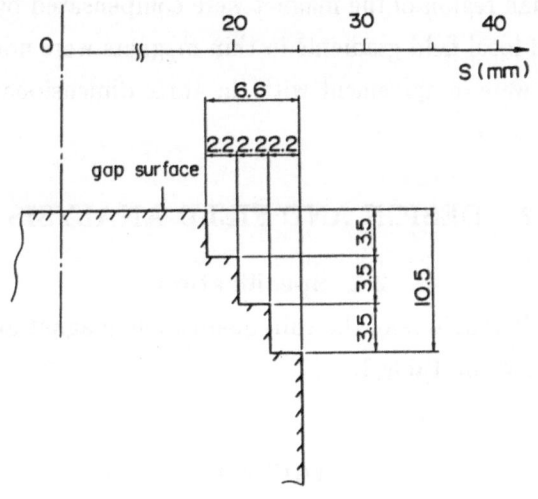

FIG.2 End cut shape of the quadrupole. The cut is approximated by three steps. The coordinate $s = 0$ is the median plane of the magnet.

2.2 Magnetic Field Analyses

Two dimensional computer code TRIM and three dimensional computer code TOSCA were used to study the possibility of the analysis of thin magnets [4]. In calculations, parameters with

no cut-off of the pole ends were used, because the influence of pole end cut is not analyzed by the 2D code. Fig.3 shows the finite-element mesh for one-sixteenth of the iron element of the problem. Radial distributions of the field gradient were calculated by TOSCA as is shown in Fig.4. The gradients are normalized by G_0 which is the gradient of the central region of the magnet.

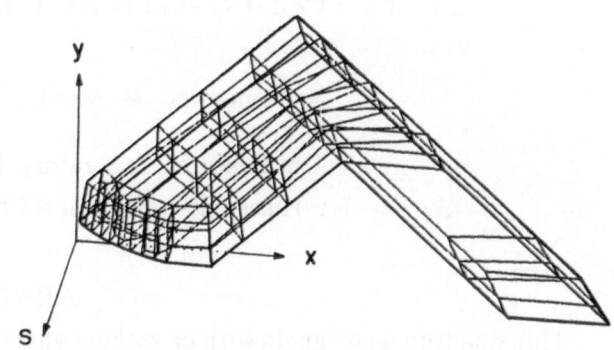

FIG.3 Finite-element mesh showing only iron element. No cut-off of the pole ends.

In the median plane, field gradient increases with radial position x. Two dimensional calculation was also shown in Fig.4 for comparison with 3D calculation. In the case of 2D calculation, radial distribution of the field gradient was homogeneous because of the hyperbola pole shaping.

The design value of the lineality of the gradient from the beam dynamics is $1 \times 10^{-3}/|x| \le 0.03$ m. On the basis of the 2D analysis, the design is reasonable for accelerator magnets. In the pole end region of the magnet, field gradient decreases with radial position x. Radial distributions by TOSCA were expanded to get the multipole components [5] as,

$$B = G_0(x + b_5 x^5 + b_9 x^9 + \cdots), \qquad (1)$$

where G_0(T/m) is the gradient at the center of the quadrupole, $x = s = 0$. Multipole coefficients b_5 and b_9 are 5th and 9th order of higher components generated by the quadrupole magnets. These coefficients are shown in Table 2. The sign of the each coefficient b_5, b_9 of the median plane is different from the sign of the pole end region. In the median plane, b_5 is negative and b_9 is positive. In the end region, b_5 is positive and b_9 is negative. The change of the sign is due to end effect of the magnets. Three dimensional analysis is essential for the calculation of thin magnets.

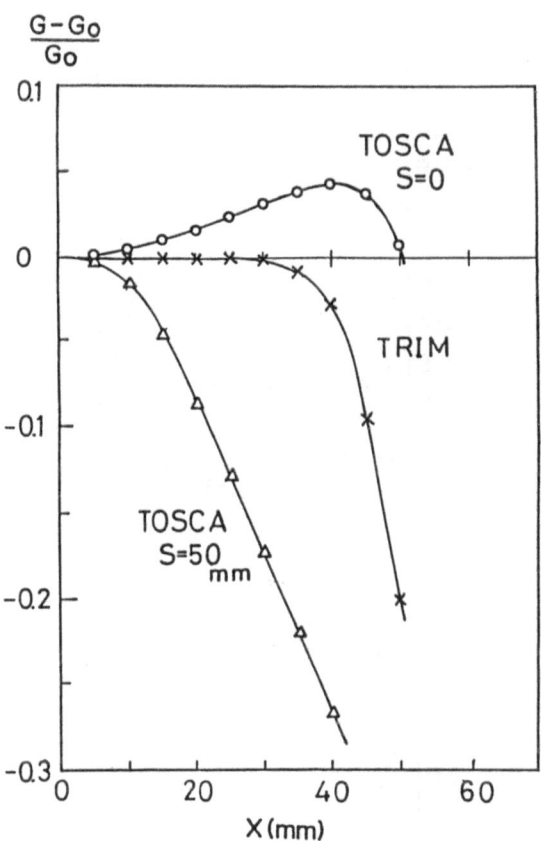

FIG.4 2D(TRIM) and 3D(TOSCA) calculations of the distribution of the field gradient $G(x)$, at the center($s = 0$) and the end region ($s = 50$ mm) of the magnet. No cut-off of the pole ends.

3. EXPERIMENTS

3.1 Measurement System

Magnetic fields were measured by moving search coils [6] (short coil, 0.0085m in length and long coil, 0.352m) and a thermally controlled Hall-probe. Moving distance of the search coil is 0.01 m. The accuracy of the measurement system is better than 1×10^{-3}. Measured data are stored in a host computer through the 12 bit A-D converter.

TABLE 2
Multi pole components calculated by TOSCA.
(with no cut-off of the pole ends)

	median plane ($s = 0$)	end region ($s = 0.05$m)
$b_5(\mathrm{m}^{-4})$	6.8×10^4	-3.4×10^5
$b_9(\mathrm{m}^{-8})$	-1.3×10^{10}	2.7×10^{11}
$b_{12}(\mathrm{m}^{-11})$	0	-1.3×10^{15}

3.2 Experimental Results

3.2.1 No Cut-off of the Pole Ends

Radial distribution of the field gradient is shown in Fig.5. In the median plane, field gradient increases with radial position x. In the pole end region of the magnet, field gradient decreases with radial position x. Three dimensional calculation is also shown in Fig.5. Experimental results are good agreement with theory. Distributions of the gradient along the magnet axis s was measured as is shown in Fig.6. The results were also agreement with the analysis.

3.2.2 With Cut-off of the Pole Ends

Magnetic field properties of the magnet with cut-off of pole ends was measured and compared with the magnet with no cut-off of the pole ends. Three dimensional calculation with our end cut-off model for estimating the effect of cut-off of the pole ends did not converge. So we estimate the effect from the experimental results. Radial distribution of the field gradient is shown in Fig.7. The distributions of the magnets with or without cut-off of the pole ends show the similar tendency. The field gradient with cut-off of the pole ends, however, had small coefficients of the multipole components. Radial distributions of the integrated field gradient GL using the long search coil was measured as is shown in Fig.8. GL is defined as $GL(x) = \int_{-\infty}^{\infty} G(x, s) ds$. The deviation of the distribution of the integrated field gradient was improved by the pole end shaping.

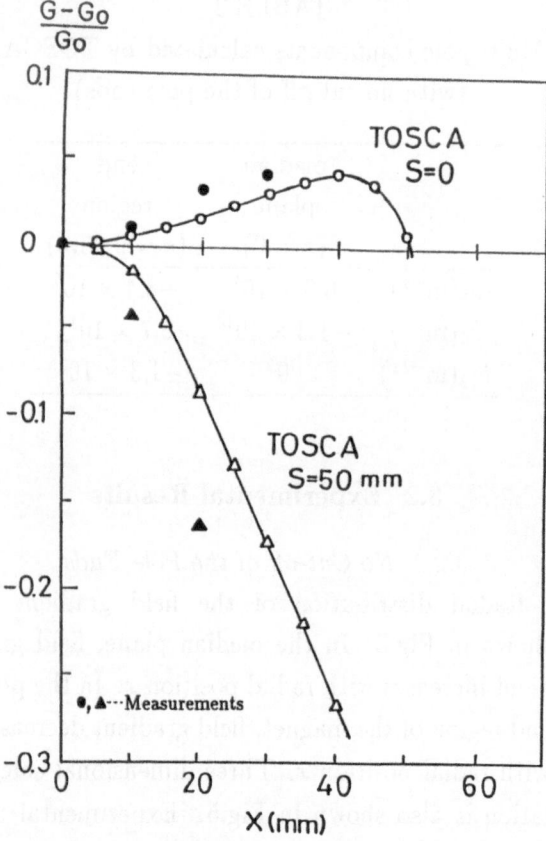

FIG.5 Experimental results of the distribution of the field gradient compared with 3D theoretical results. No cut-off of the pole ends.

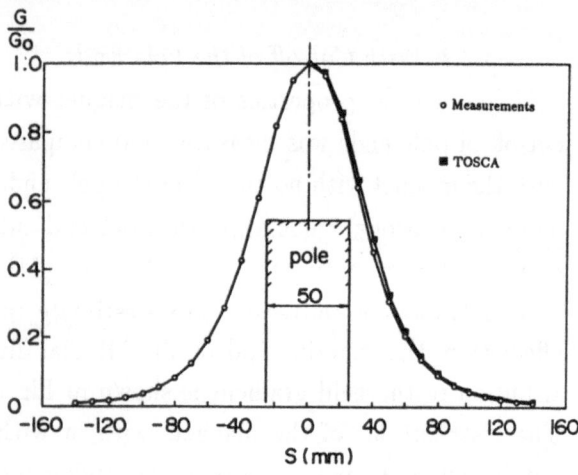

FIG.6 Theoretical and experimental results of the distribution of the field gradient along the magnet axis s. No cut-off of the pole ends.

The deviation of the integrated gradient with

sharp cut-off of the pole ends was $2.9 \times 10^{-3}/|x| \leq 0.03$ m. The deviation with no end cut was $6.1 \times 10^{-2}/|x| \leq 0.03$ m which was 21 times larger than that with end cut. The measured effective length on s axis L_{eff} which is defined as $L_{eff} = \int_{-\infty}^{\infty} G(s)ds/G(0)$ was 0.079m. It was 1.6 times longer than the core length.

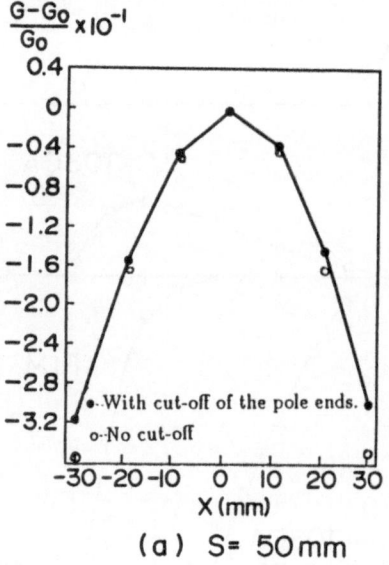

(a) S= 50mm

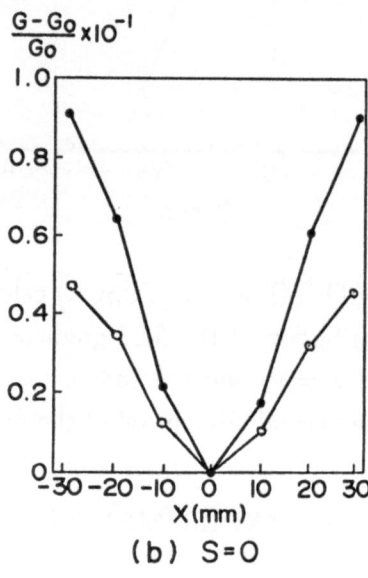

(b) S=0

FIG.7 Experimental results of the distribution of the field gradient, (a)$s = 50$mm,(b)$s = 0$, with cut-off of the pole ends, compared with no cut-off of the pole ends.

4. CONCLUSIONS

The magnetic field in the central region of the thin quadrupole magnet is influenced by the pole end region of the magnets.

Multipole components in the median plane, b_5 and b_9 had different signs compared with those in the pole ends.

Experimental results were in agreement with the three dimensional analyses by TOSCA.

Deviation of the integrated field gradient was improved by the cut-off of the pole ends, because of compensation effect of multipole components in each region.

In thin magnets, radial distributions of local field gradients were not improved by the pole end shaping.

Homogeneous magnetic field gradient calculated by 2D code cannot be realized in the case of thin magnets, for example, core length/bore diameter = 0.74.

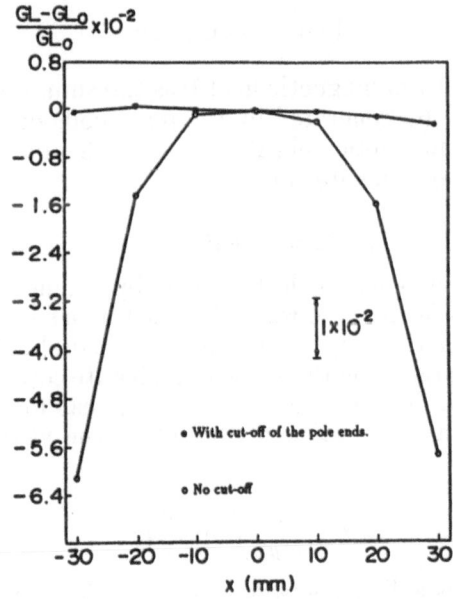

FIG.8 Experimental results of the radial distribution of the integrated field gradient $GL(x)$, with or without cut-off of the pole ends.

REFERENCES

[1] Jahnke, A., Marsing, H. and Meier, K., First superconducting prototype magnets for a compact synchrotron radiation source in operation. *IEEE Trans. on Magnetics*, **Vol.24**, No.2, March 1988, pp. 824-826.

[2] Krevet, B., Dustmann, C. and Flessner, H. -H., Superconducting magnets for compact synchrotrons. *IEEE Trans. on Magnetics*, **Vol.24**, No.2, March 1988, pp. 827-830.

[3] Kumada, M., Sasaki, H., Someya, H. and Sakai, I., Optimization on the end-shaping of a quadrupole magnet. *Nuclear Instruments and Methods*,211(1983), pp. 283-286.

[4] Early, R. A. and Cobb, J. K., TOSCA calculations and measurements for the SLAC SLC damping ring dipole magnet. *IEEE Trans. on Nuclear Science*, **Vol.NS-32**, No.5, October 1985, pp. 3654-3656.

[5] Parzen, G., Magnetic fields for transporting charged beams. BNL Report No.50536, 1977.

[6] Noda, A., Mutou, M., Hattori, T., Hirao, Y., Hori, T., Katayama, T. and Sasaki, H., Quadrupole magnet for TARN. INS-MUNA-23, Univ. of Tokyo, May 1980.

THREE DIMENSIONAL ANALYSIS OF A DEFLECTION YOKE WITH SURFACE CHARGE METHOD

Takafumi Nakagawa and Soichiro Okuda

Central Research Laboratory, Mitsubishi Electric Corporation
1-1 Tukaguchihonmachi 8 Chome, Amagasaki, 661 JAPAN

Abstract

Three dimensional magnetic field analysis program was developed with surface charge method for analyzing the magnetic field made by a deflection yoke (DY) with a slot core. The developed program was verified with the theoretical results. The magnetic flux density was estimated by the extrapolation with the inverse of square root of the number of elements. The calculated values were compared with the measured values of test DYs. The increased deflection sensitivity of a slot core was simulated in quantity. This program was proved to be efficient for designing a deflection yoke with a slot core.

I. Introduction

In a high performance cathode ray tube (CRT) system such as a high definition television system and high resolution color monitor display, a large screen size, high brightness and high resolution are required. In order to fulfill theses requirements, following items should be developed: (1) large deflection angle, (2)high accelerating voltage, (3)large diameter of CRT neck, and (4)high deflection frequency. These lead to the increase of deflection current and the rise in temperature of deflection yoke (DY). It is an efficient method to use a slot core for the increase of the deflection sensitivity. In this paper, the deflection sensitivity is defined as the proportion of the deflection angle to the coil current.

As the design of a core has been based on experiences of the designers and workers, it has taken much time and cost to obtain the design values of a slot core. The computer simulation is required to reduce time and cost. Since the core of a conventional DY is rotationally symmetric, its magnetic field can be calculated by a two dimensional model [1] [2]. As a slot core is not rotationally symmetric, a three dimensional field analysis is necessary. Although there are various three dimensional analyses [3] [4], the surface charge method was selected by the following reasons.

The finite difference method [3] is not effective for the analysis of the deflection field, since the three dimensional space which electron beams pass through should be divided into many elements. Although the boundary elements method with a current sheet approximation [4] is applicable to a conventional core, this method may not treat the coil distribution installed in the slots of a core. The surface charge method may not treat the coil distribution installed in the slots of a core. The surface charge method can calculate the wide area with a small number of variables and take into account the coil distribution in the slots. The developed program was verified with the theoretical values. The magnetic fields of a test DYs were simulated with this program and its results were compared with the measurements.

II. Field Calculation

The total magnetic field **H** is the sum of the coil field $\mathbf{H_c}$ generated by the main coils of DY and the magnetic field $\mathbf{H_m}$ made by the magnetization of a ferrite core.

$$\mathbf{H} = \mathbf{H}_c + \mathbf{H}_m \qquad (1)$$

The magnetic field $\mathbf{H_m}$ made by magnetization of a ferrite core was calculated by a surface charge method [5]. In this method virtual magnetic charges σ on the surface of a ferrite core are assumed. The magnetic field $\mathbf{H_m}$ is calculated by summing up the contribution of the virtual charges.

$$\mathbf{H}_m = -\frac{1}{4\pi\mu_0} \int_s grad\frac{\sigma}{r} dS \qquad (2)$$

where r is a distance from a point on the surface of ferrite core to a calculated point.

An integral equation is derived from the continuity of the normal component of the magnetic field on the boundary surface.

$$\mathbf{H} \cdot \mathbf{n}\,|_{out} = \mathbf{H}_c \cdot \mathbf{n}\,|_{in} + \mathbf{H}_m \cdot \mathbf{n}\,|_{in} \qquad (3)$$

where subscripts "out" and "in" denote the magnetic field in the outside and inside region of the ferromagnetic material, respectively. The second term in the righthand side is calculated by dividing the boundary surface into small surface elements. As the charge density σ and the susceptibility χ_s are assumed to be constant on the element and in the material, the following integral equation is obtained.

$$\frac{\sigma^{(i)}}{\chi_s\mu_0} = \mathbf{H}_c\cdot\mathbf{n}+(-\frac{\sigma^{(i)}}{2\mu_0})-\sum_{j\neq i}\frac{\sigma^{(j)}}{4\pi\mu_0}\int_{s^{(j)}} grad\frac{1}{r}dS\cdot\mathbf{n} \quad (4)$$

where the second and the third term of the right-hand side are the magnetic field produced by magnetic charge on a small surface element under consideration and on the outside surface elements, respectively. By solving this equation (4), the charge density distribution on the surface of ferrite core are obtained.

The coil field $\mathbf{H}_c$ was calculated by the following algorithm. The distribution of the coils is replaced by the current sheets. The Poisson equation is discretized by a finite difference method with the use of the scalar potential ϕ [1]. The current distributions i and the scalar potential ϕ are expanded in Fourier series with regard to the azimuthal angle θ as follows.

$$i(r,\theta,z) = i_1(r,z)\cos\theta+i_3(r,z)\cos 3\theta+i_5(r,z)\cos 5\theta+\cdots\cdots$$
$$(5)$$

$$\phi(r,\theta,z) = \phi_1(r,z)\sin\theta+\phi_3(r,z)\sin 3\theta+\phi_5(r,z)\sin 5\theta+\cdots$$
$$(6)$$

The scalar potential ϕ and current i are connected with the Ampere's law. These approximations allow us to use the scalar potential ϕ in the whole space. The final equation is a system of simultaneous linear equations. An iterative algorithm is used to solve the simultaneous linear equation. The magnetic field at an arbitrary location is calculated by differentiating the scalar potential ϕ.

III. Program

The flow chart of this program is shown in Fig.1. Coil position and coefficient of Fourier expansion is input in (a_1). The current distribution is expanded in Fourier series in advance. The magnetic scalar potential distribution is calculated for each harmonics (b_1) and the resulted potential data are stored (c_1). The coil field $\mathbf{H}_c$ is calculated by differentiating this resulted potential data. In the next step, the surface geometry of a ferrite core is input in (a_2). The surface is divided into small triangle elements. As the coil field $\mathbf{H}_c$ can be calculated at each mesh on the surface, the surface charge distributions can be obtained by solving eq.(4) and the results are stored (c_2). The integration is done theoretically. An U-L decompo-

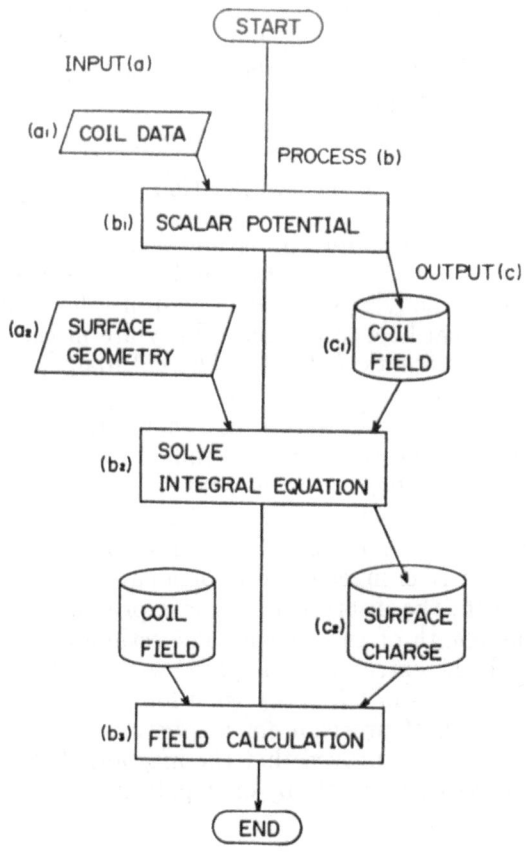

Fig.1 Flow chart of the developed program. (a), (b) and (c) show input, process and output, respectively

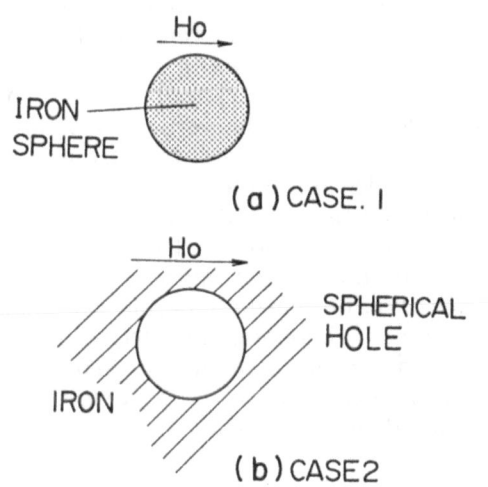

Fig.2 Schematic drawing of the theoretical model. (a) is an iron sphere in the uniform magnetic field (CASE 1). (b) is a spherical space in the iron in the uniform magnetic field (CASE 2).

sition is used to solve the simultaneous linear equation. Finally (b_3), the magnetic field in the whole space is calculated by eq.(1) with the resulted data of the scalar potential (c_1) and the surface charge (c_2).

IV. Verification

The developed program was verified with the theoretical models. Fig.2 shows the theoretical models, which were an iron sphere and a spherical hole in the iron. They are in the uniform magnetic field H_0. The calculated values in the center of the sphere were compared with the theoretical values. Fig.3 shows the discretizing error versus inverse of square root N, which is the number of the elements on the surface of sphere. This figure shows that the discretizing error is inversely proportional to square root of N. As N is inversely proportional to square of length, the discretizing error is proportional to the length of an element. The estimated values with the extrapolation agreed with the theoretical values in 0.5 % precision for an iron sphere and 0.05 % precision for a spherical hole in the iron. It was proved that the magnetic field was estimated precisely by extrapolation.

V. Application

V-1. Structure of a test model

The structure of a typical DY with a conventional ferrite core is shown in Fig.4. DY has a pair of saddle shaped horizontal coils, a pair of toroid shaped vertical coils and a conical ferrite core. The shapes of a conventional ferrite core and a slot core are shown in Fig.5. A slot core has some teeth and the space between the teeth formed slots. Since the vertical winding coils are installed in these slots, the distance of the magnetic poles of a slot core is shorter than the distance of a conventional core. As the magnetic resistance in the region of deflection field of a slot core becomes small, the stronger magnetic flux density is obtained even in the same excited coil current. Therefore, the current needed in the DY with a slot core will be smaller than a conventional core. As a result, a slot core will have high deflection sensitivity.

V-2. Measurement

We measured the magnetic field of a test DY with a slot core. The measurement was performed in the following conditions. The exciting current was at 5.0 A D.C in a saddle shaped horizontal coil. The horizontal coil was only excited. The terrestrial magnetism were canceled

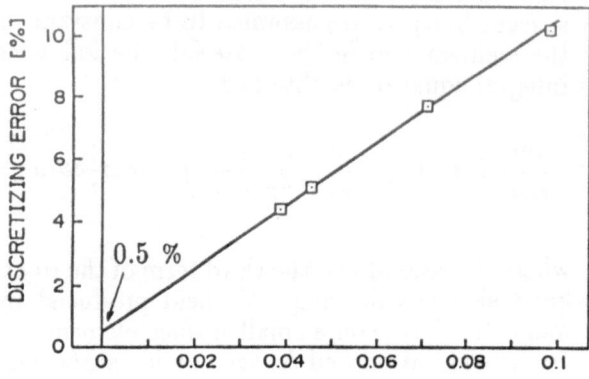

(a) CASE. I   MESH SIZE $1/\sqrt{N}$

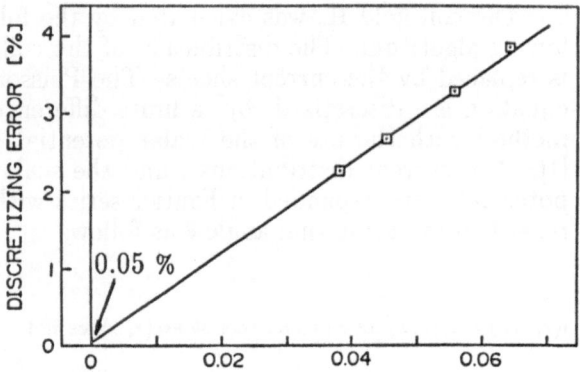

(b) CASE 2   MESH SIZE $1/\sqrt{N}$

Fig.3 The discretizing error versus the inverse of square root of N, N is the number of the elements on the boundary surface. (a) and (b) are the results for CASE 1 and CASE 2, respectively. Solid line is linear regression. Open squares show the calculated results.

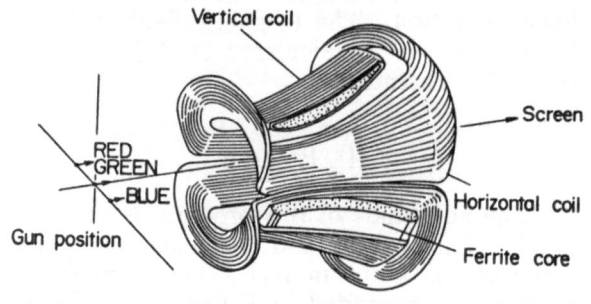

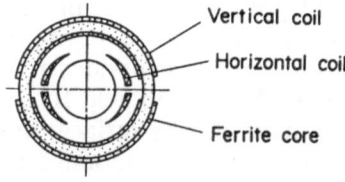

Fig.4 Schematic drawing of a typical DY

by Helmholtz coils. The measured values were obtained along the center axis with a gauss meter.

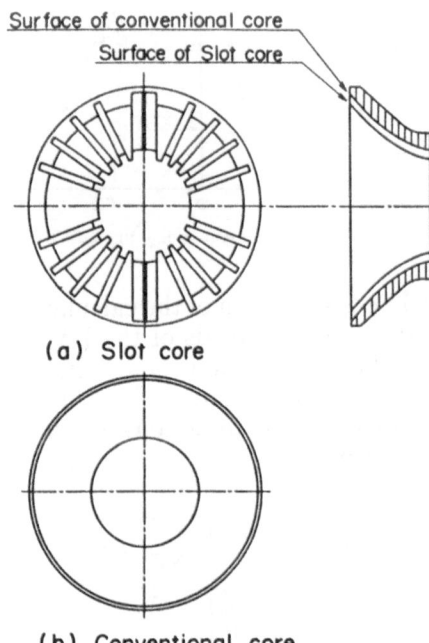

(a) Slot core

(b) Conventional core

F.g.5 Shape of a slot core (a) and a conventional core (b). The inside diameter of a slot core is shorter than a conventional core.

V-3. Results and Discussions

The calculation was performed under the same condition as the measurement conditions. The surface of a slot core was divided into small triangle elements as shown in Fig.6. CPU time was 14 hours in VAX-11/750 (0.6MIPS) to obtain the surface charge distribution in case of 980 unknown variables. Fig.7 shows the comparison between the calculation and the measurement in a test DY with a slot core. The calculated values were obtained by the extrapolation with the inverse of square root. Although the change of the calculated magnetic field agreed with the change of the measurements, the calculated values around the peak point were larger about 2 % in the maximum than measured values. But, approaching to the coil end, the calculated magnetic field became smaller. The possible reason of this discrepancy is that the coil field was not simulated exactly as shown in Fig.8. Since the structure of the coil ends (bending parts) is complicated, it is difficult to simulate the coil distributions. The exact simulation of the coil field is subject to the future study.

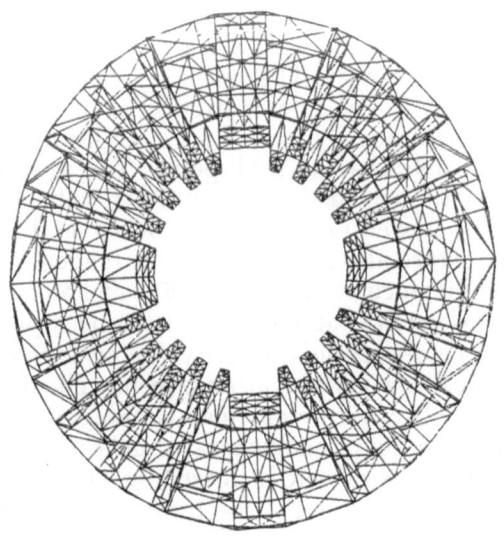

Fig.6 The division of a ferrite core with slots.

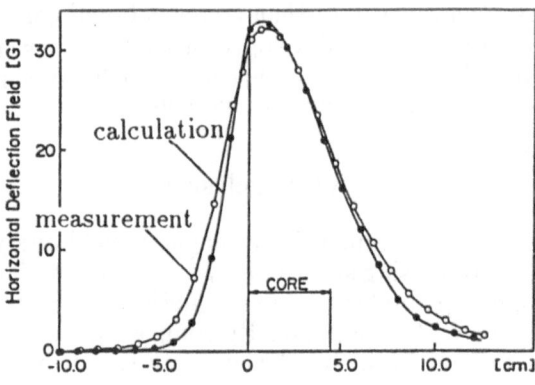

Fig.7 The distribution of the calculated and measured horizontal deflection magnetic field along the axis of DY with a slot core. Solid and open circles show the calculated values and the measured values, respectively.

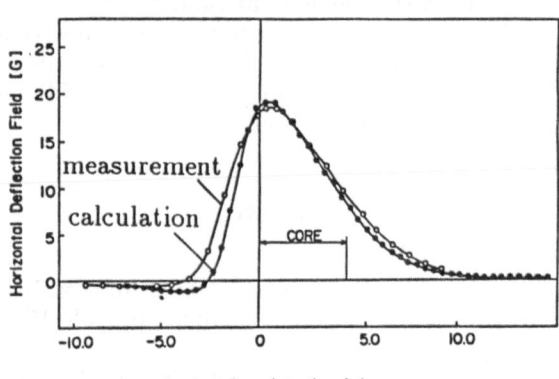

Fig.8 The distribution of the calculated and measured horizontal deflection magnetic field generated by the horizontal coil without a core. Solid and open circles show the calculated values and the measured values, respectively.

284

The calculated magnetic field distributions of DY with a slot and a conventional core are shown in Fig.9. The magnetic field density made by a slot core is stronger than the magnetic field density made by a conventional core about 6.9 % at the peak point. The measured magnetic field distributions of DY with a slot and a conventional core are shown in Fig.10. This figure also showed the same tendency and the difference between a slot core and a conventional core was about 6.7 % at the peak point.

The deflection sensitivity was compared for both calculation and measurement. The deflection angle could be obtained by integrating the deflection field. The integrated deflection fields of the slot core were increased by 7.0 % for the calculation and 7.4 % for the measurement of the conventional core. This result agreed with the comparison of the peak values. Therefore, this test DY with a slot core has 7% larger deflection sensitivity than a conventional core.

By using the developed program, it was simulated that a slot core has higher deflection sensitivity than a conventional core.

VI. Conclusions

A three dimensional magnetic field analysis program using surface charge method was developed for analyzing the magnetic field made by a DY with a slot core. By comparing the calculated magnetic flux density with the theoretical values and the measured values, the following conclusions were drawn.

(1) The program simulates the increased deflection sensitivity of a slot core in quantity.
(2) The magnetic field was calculated precisely by the extrapolation and the results agreed with the theoretical values.
(3) The developed computer program with surface charge method proved to be efficient for designing a slot core, which yields high deflection sensitivity.

Reference

[1] T.Nomura, *Transactions of the Institute of Electrical Engineers of Japan*, **Vol.91**, pp.155–163, 1971.
[2] Donald M.Fye, *J.Appl. Phys*, **50**, No.1, pp17, 1979.
[3] M.Watanabe, S.Shirai and M.Fukushima, *International Display Research Conference*, pp125–128, 1988.
[4] H.Sano, T.Takagi and M.Miki, *Transactions of the Institute of Electrical Engineers of Japan*, **Vol.108-A**, No.4, pp147–154, 1988.
[5] S.Nakamura, and M.Iwamoto, *Transactions of the Institute of Electrical Engineers of Japan*, **Vol.96**, pp.55–62, 1976.
[6] T.Nakagawa, S.Okuda, H.Nishino, M.Ogasa and T.Fujimura, *International Display Research Conference*, pp.23–26, 1988.

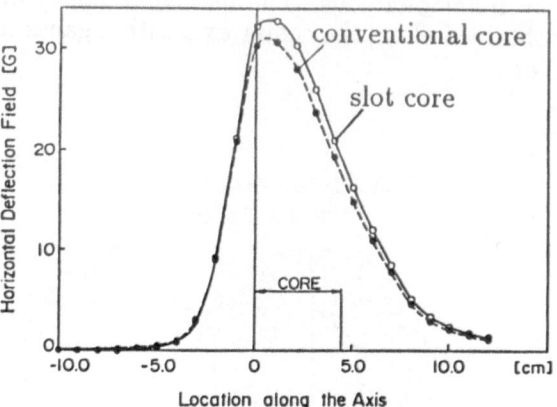

Fig.9 The distribution of the calculated horizontal deflection fields of the DY with a slot and a conventional core. Solid and dashed lines show the magnetic fields made by a slot core and by a conventional core, respectively.

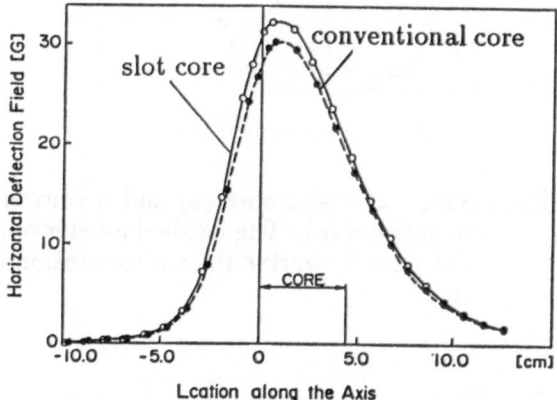

Fig.10 The distributions of the measured horizontal deflection fields of the DY with a slot and a conventional core. Solid and dashed lines show the magnetic field made by a slot core and by a conventional core, respectively.

QUENCH BEHAVIOUR OF A THIN SOLENOID MODEL

A. Bonito Oliva, G. Masullo, O. Dormicchi, G. Gaggero, R.Penco
ANSALDO ABB COMPONENTI Srl, via N. Lorenzi 8, I-16152 Genova, Italy

ABSTRACT

A thin superconducting solenoid of 0.76 m in internal diameter and 0.8 m in length has been manufactured and tested by Ansaldo under the ZEUS project contract with Istituto Nazionale di Fisica Nucleare (INFN). It is a model of the ZEUS detector magnet. The coil has two layers and it is wound with a Rutheford cable stabilized with high purity aluminium (bare dimensions 15 x 4.3 mm) manufactured by EUROPA METALLI,Florence. The coil has an external aluminium cylinder with helium circulating in an aluminium pipe welded around it. The quench propagation velocity and the minimum quench energy have been measured during the tests. The quench back behaviour has been analyzed. In this paper we present the measurements and the results obtained from the data analysis.

INTRODUCTION

Ansaldo designed and manufactured a thin superconducting solenoid to be installed in the ZEUS detector for HERA according to the conceptual design carried out by INFN, Milan (1,2,3). Previously, in order to check the feasibility of the project, Ansaldo designed, manufactured and tested a model of the coil.
A schematic representation of the model is shown in fig.1 and its characteristics are shown in table 1.

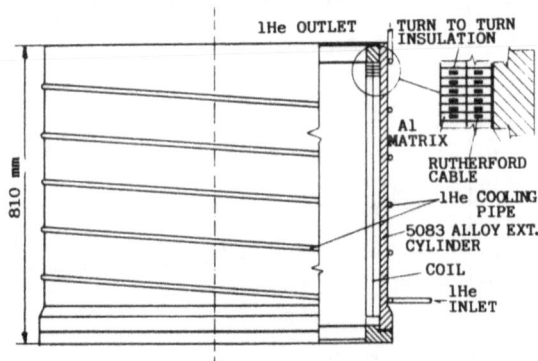

FIGURE 1 - Magnet and cooling pipe

TABLE 1 - Model characteristics

Geometrical characteristics:	
coil internal diameter	760 mm
coil external diameter	824 mm
coil length	677 mm
cylinder thickness	18 mm
number of turns/layers	140
number of layers	2
No. joints (int.layer)	1
turn to turn insul.thickness	0.5 mm
coil-cylinder insul.thickness	1 mm
Electrical characteristics:	
coil inductance	43 mH
cylinder inductance	0.6×10^{-6} H
mutual inductance	1.6×10^{-4} H
design current	5000 A
central field (5 kA)	1.8 T
dump resistance	3.6 mOhm
cylinder resistance (4.2 K)	3×10^{-3} mOhm
Cable characteristics:	
basic strand diam.	1.04 mm
Cu/NbTi ratio	1.1 : 1
filament diameter	40×10^{-3} mm
Al 99.996 stabilizer	
Rutherford cable (10 strands)	1.8x5 mm
bare conductor dim.	15 x 4.3 mm
critical current (2.3 T, 4.2 K)	15000 A
Al matrix res.(B=0 T,4.2K)	2×10^{-11} Ohm m

The cable is an aluminium stabilized Rutheford, manufactured by EM-LMI and tested by ENEA (4). The cable was insulated by glass fabric tape and the coil was impregnated with epoxy resin.
In order to support the magnetic hoop stresses and to have a better distribution of the quench transition, a 5083 aluminium alloy cylinder was shrink fitted around the coil. The cooling of the coil is ensured by two phase helium circulating in a pipe welded around the cylinder. A schematic view of the coil in the cryostat is shown in fig. 2.

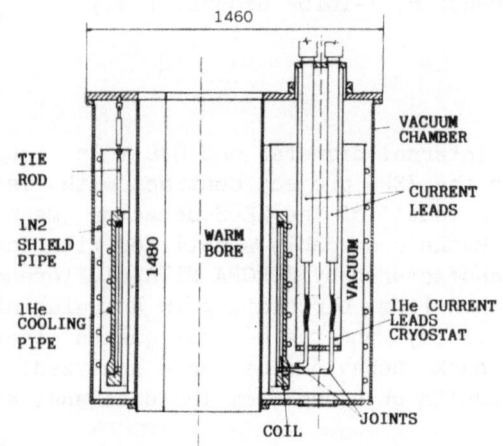

FIGURE 2 - Cryostat and magnet

The protection of the coil from the quench was ensured by an external dump resistor in parallel with the coil. Due to the bad behaviour of the junction between the two cable length in the winding, the maximum reached current was I=3200 A, whereas the design value was I=5000. The dissipated power on the junction was 2.5 W before the spontaneous quench. The coil mean temperature was 4.8 °K. In spite of the current limitation several measurements on minimum quench energy, quench propagation velocity and quench back were carried out. In this paper we describe the results obtained.

Minimum Quench Energy (MQE) measurements
To induce the transition in the coil, a very thin heater, 200 mm long and 5 mm wide, was stuck on the central turn internal face of the internal layer (fig.5). A capacitors bank was used to give the heat pulse. No differences in the MQE were noticed varying the heat pulse time constant from 6 msec to 12

msec. The results obtained for several current values are shown in fig.3.
The theoretical curves obtained by Wilson's theory (5) and Dresner's theory (6,7) are reported too. Note that we had low field on the cable (about 0.7 T) and so we had to fit the critical current data of ENEA (4) by Kim-Anderson's law (see fig.4). Applying Wilson's theory, we took into account also the different thermal conductivities in radial and axial direction of the winding section, being this difference not negligible for this cable geometry. In Dresner's theory the correction for the heat pulse spatial and temporal distribution was considered. The disturbance was considered as long as the heater and as wide as the cable narrow face; in the time duration the heat diffusion time through the insulation was included. Taking into account the non perfect applicability of these theories to our geometry, the accordance between calculated and experimental data seems to be reasonable.

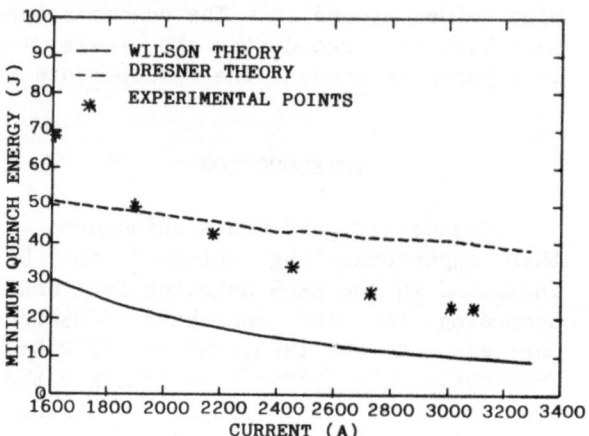

FIGURE 3 - M.Q.E. experimental and theoretical results

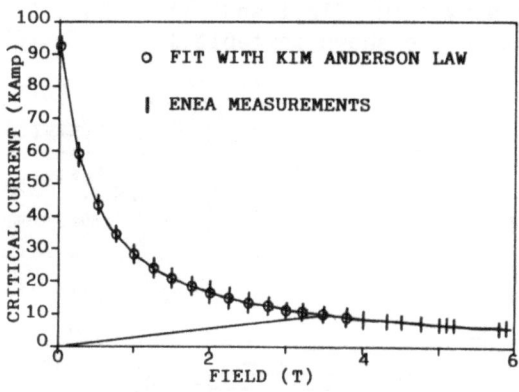

FIGURE 4 - Critical current at 4.8 K and load line

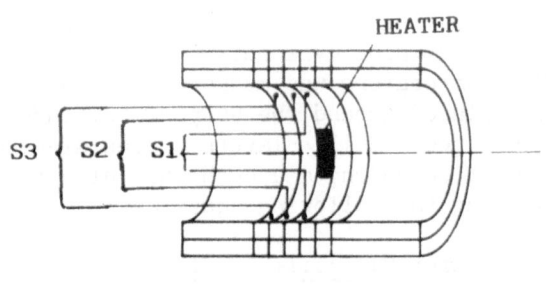

FIGURE 5 - Heater and transversal speed
voltage taps

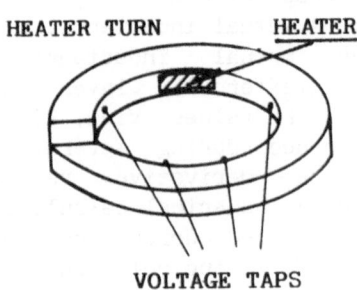

FIGURE 6 - Heater turn and longitudinal
speed voltage taps

Quench Propagation Measurements

We measured the longitudinal (along the cable) and axial (in the plane of the cable section, along the magnet axis) quench propagation velocity by a set of voltage taps distributed around the heater (fig.5,6). After each heat pulse we recorded the voltage evolution and the speed was calculated from the delay in voltage appearing between two adjacent voltage stations. The current was hold constant during the quench measurements inserting a 1 sec. delay in quench detector electronics. A typical voltage evolution obtained for a transversal velocity measurement is reported in fig.7.

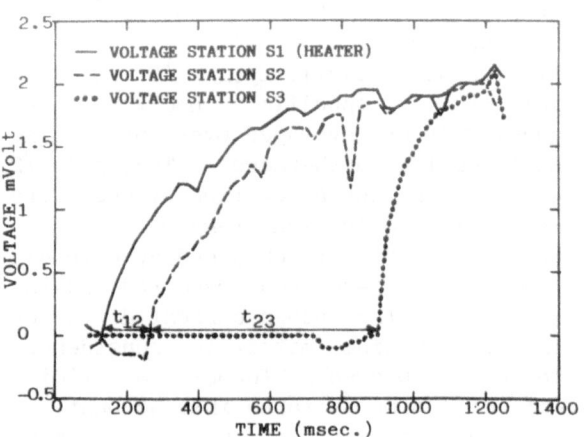

FIGURE 7 - Voltage evolution measured

As you can see the time t12 needed for the transverse propagation from the heater station (S1) to the adjacent one (S2) is smaller than the propagation time t23 between S2 and S3. That is due to the heater contribution to the S2 transition. In fact the measured time intervals t12 were the same for different current values. For this reason for the transversal velocity calculation we considered only t23 values. Regarding the longitudinal quench propagation velocity measurements, the voltage stations configuration is shown in fig.4. We observed an increase of the speed after a few tens of msec. from the starting, as noticed in other experiments (9). Also in these measurements, we excluded the heater station from our calculations to avoid taking into account the time delay for the normal zone formation after the heat pulse.

The results obtained for the longitudinal and transversal velocities are shown in fig. 8 while in table 2 the experimental and calculated transversal to longitudinal velocity ratio is reported too. The theoretical one was obtained by the following relation

$$\alpha = 0,7(Kal.a/(Kin.w))^{1/2}$$

where Kal, Kin are the thermal conductivities of aluminium and insulation at the critical temperature, a is the width of the cable (4.3 mm) and w the insulation thickness.

The theoretical longitudinal speed was calculated according to (10):

$$v_{adiab} = J/ C(\rho k/(Ts-To))^{1/2}$$

where k is the thermal conductivity, J is the current density, ρ is the electrical resistivity, C is the volumetric specific heat. These quantities are averaged over the winding unit cell. Furthermore $Ts=(Tg+Tc)/2$ where Tg is the generation temperature and Tc is the critical temperature.

Besides the results obtained by Devred's formula (8) have been reported. Using (10), an instantaneous current diffusion in the aluminium matrix is considered, while in Devred's formula a finite diffusion time is taken into account.

As you can see in the measurement current range the results obtained by (10) fit better the experimental curve, but Devred's curve seems to have the same shape.

That is probably due to the measurements current range, in fact for this kind of cables the effect of the finite diffusion time on the quench propagation velocity becomes appreciable at higher fraction of the critical current (9).

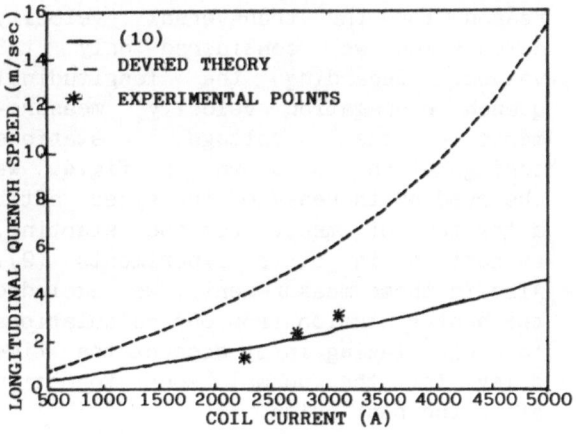

FIGURE 8 - Measured and calculated longitudinal speed

TABLE 2 - **Transversal velocity**

Current	Trans. velocity	Trans.to long. velocity ratio (10)	
(A)	(cm/sec)	EXP	THEOR.
1700	0.75	5.8	5.9
2400	1.1	5	6
3100	1.35	4.5	6

Quench Back Behaviour Analysis

In order to analyze the thermal quench back behaviour, the following quantities have been recorded during several quenches:

1. The cylinder temperature
2. The two coil layers resistances evolution
3. The induced current in the support aluminium cylinder.

The temperature was measured by several CGR placed around the support cylinder. The two coil layers resistances evolution during a quench was calculated solving the following equations:

$$Vint=(Lint+M)\frac{dIcoil}{dt}+Mic\frac{dIcyl}{dt}+Rint\ Icoil$$
$$Vext=(Lext+M)\frac{dIcoil}{dt}+Mec\frac{dIcyl}{dt}+Rext\ Icoil$$

where:
Vint, Rint, Lint (Vext, Rext, Lext) are voltage, resistance and inductance of the internal (external) layer, M is the two layers mutual inductance, Mic (Mec) is the mutual inductance between internal (external) layer and the cylinder. The values Vint, Vext, Icoil were recorded during several quenches while the derivative values were obtained by numerical calculation. The cylinder induced current was determined using the following relation

$$Icyl=(Baxial-Kcoil\ Icoil)/Kcyl$$

where Baxial is the total magnetic field in the coil centre, I is the current, Kcyl and Kcoil are two geometrical parameters calculated while Icoil and Baxial were measured during every quench by a transfoshunt and a hall probe.

The measurements were carried out charging the coil up to 3100 A and then opening the breaker and dumping the coil on the external resistor. The evolution of the internal layer resistance, during the spontaneous quench determined by the breaker opening, is shown in fig. 9, where is possible to see that the quench starts after about 4 sec. from the opening, almost simultaneously in the two layers.

Using Green's theory (11), a quench back time delay of approx. 2 sec. was calculated.

This discrepancy may be due to the non adiabatical heating of the cylinder with respect to the coil, contrarily to the theory hypothesis.

Finally we deduced for this case the average coil temperature evolution before the quench back starting using the following equation:

$$dEcoil = dEcyl - dHcyl$$

Where dEcoil is the energy dissipated during the time interval dt in the cylinder by the induced currents and dH cyl is the cylinder enthalpy variation calculated from the temperature measurements.
The results are shown in fig.10.
As you can see the calculated coil temperature reaches the critical temperature in approx. 4 sec. in good agreement with the experimental value.
The temperature curve A was calculated for adiabatical heating of the cylinder, the curve B for infinite thermal conductivity of coil and cylinder (isothermal condition).

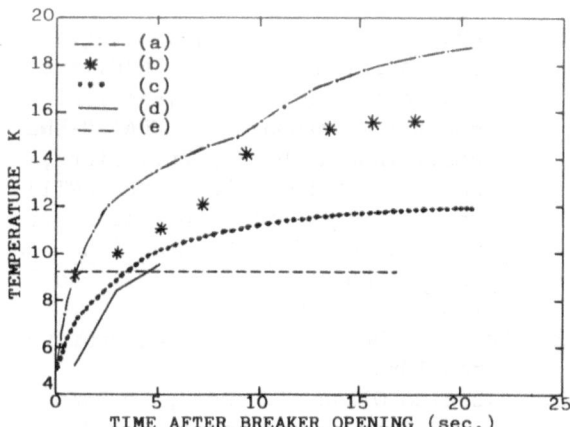

FIGURE 10 - Temperature evolution after breaker opening
(a) cylinder temperature (calculation A)
(b) experimental cylinder temperature
(c) cylinder and coil temperature (calculation B)
(d) coil temperature calculated from experimental data
(e) critical coil temperature

CONCLUSIONS

In summary, in spite of the current limitation, the measurements give us some interesting information:
- in spite of the winding anisotropy, the coil stability margin can be foreseen with sufficient accuracy using the existing theories
- in the quench speed calculation, a zero diffusion time of the current in the aluminium can be used at low current
- in order to foresee the quench back behavior of the coil the actual thermal exchange process between coil and cylinder should be considered.

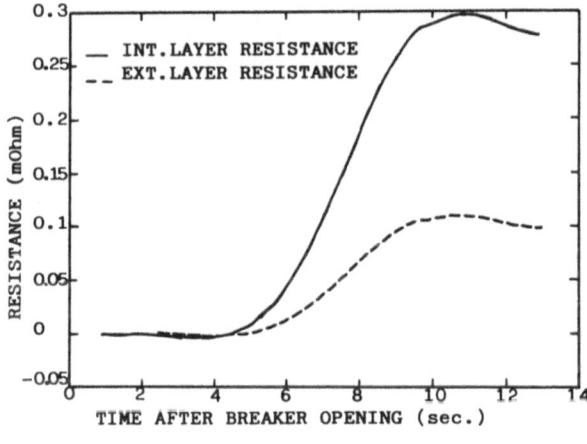

FIGURE 9 - Resistance evolution after breaker opening

REFERENCES

1. E.Acerbi,F.Alessandria,G.Baccaglio-ni,E.Fabrici,L.Rossi (INFN/Milano), R.Bruzzese,G.Pasotti,M.V.Ricci,N. Sacchetti,M.Spadoni (EURATOM-ENEA /Roma),O.Dormicchi,P.Fabbricatore, R.Penco (ANSALDO ABB COMPONENTI), Thin and Compensating Solenoids for ZEUS Detector. IEEE Transactions on Magnetics,March 1988, Vol.24, No.2, p.1354

2. The ZEUS Detector Technical Proposal March 1986

3. A.Bonito Oliva, F.Bordin, O.Dormic-chi, G.Gaggero, M.Losasso, R.Penco, N.Valle (ANSALDO ABB COMPONENTI), R.Bruzzese, M.Spadoni, N.Sacchetti (ENEA), Lin (WORLD LAB.), ZEUS magnet construction status report. 11° Int. Conf. on Magnet Technology, 1989

4. R.Bruzzese, S.Ceresara, G.Donati, S.Rossi, N.Sacchetti, M.Spadoni (EURATOM-ENEA), The aluminium stabilized Nb-Ti conductor for the ZEUS thin solenoid. IEEE Trans. on Magn., March 1988, Vol.25, No.2, p.1287

5. M.N.Wilson, Superconducting Magnets in Monographs on cryogenics Clarendon Press Oxford, 1983, p.79

6. L.Dresner, Quench energies of potted magnets. IEEE Trans. on Magn., 1985, Vol.MAG 21, No.2, p.392

7. L.Dresner, Quench energies of potted magnets, III. A.C.E. Vol. 31, 1985, p. 365

8. A.Devred, C. Meuris, Analitical solution for the propagation velocity of normal zones in large matrix stabilized superconductors. 9° Int. Conf. on Magnet Technology, 1985, p.577

9. M.Scherer, P.Turowsky, Investigation of the propagation velocity of a normal conducting zone in technical superconducting, Cryogenics, 1978, p.515

10. M.N.Wilson, Superconducting Magnets, in Monographs on cryogenics Clarendon Press Oxford, 1983, p. 204-207.

11. M.A.Green, Quench back in thin superconducting solenoid magnets, Cryogenics, 1984, p.3

HIGH PRECISION AUTOMATIC MAGNETIC FIELD MAPPING SYSTEM FOR THE DIPOLE MAGNET

C. S. Hwang, W. C. Chou, J. H. Huang, M. Y. Lin, Tzuchu Chang, P. K. Tseng*

Synchrotron Radiation Research Center, 8th Fl. No. 6, Roosevelt Road, Sec.1 Taipei, Taiwan, R.O.C.

*also Department of physics, National Taiwan University, Taipei, Taiwan, R.O.C.

ABSTRACT

A real time automatic measurement system has been developed and used to measure and analyse the Triple Bending Achromat (TBA) magnet at SRRC. An IBM/PC/AT has been used as the system controller due to its cost effectiveness with simple and expandable configuration of this Hall probe automatic magnetic field mapping system. The development concept, hardware and software systems as well as the precision are introduced and discussed in this paper.

INTRODUCTION

In recent years, the power of personal computers has grown very quickly and they are now relatively inexpensive. So the MMG (Magnetic Measurement Group) decided to use a PC to set up a simple, expandable, configurable and programmable automatic measurement system to monitor and control the whole status of our measurements.

The GPIB card can be used to connect many kinds of instrument to this system and save all the acquisition data in the data base for on-line/off-line analysis. This system can be divided into hardware and software, which offers:

(1) Connection of upto 15 instruments at the same time, for example, DMM(digital multimeter), NMR (Nuclear Magnetic Resonance), X-Y-Z table and many other instruments, which in corporate a GPIB interface.

(2) Presentation of results in both tabular and graphical formats.

(3) Monitoring of system status information at the control terminal.

The software package is written in the 'C' language. This software is divided into three major parts: (a) Device installation, (b) stepping motor driver, and (c) data acquisition. Each major part contains several subprograms to expand its capability.

The exact position of the three-dimension movable X-Y-Z table can be read from the optical linear scales. The resolution of these optical linear scales is 5 μm and the 5508A laser measurement instrument has been used to confirm this position accuracy. The precision of this automatic mapping system which includes Hall probe and X-Y-Z table is within ±30 ppm (the Hall probe accuracy is within ± 25 ppm).

The expansion capability of this system makes it useful in various applications. Besides dipole magnetic field measurement, it can also be used to survey the accuracy and stability of each instrument.

The position of the X-Y-Z table can be read from the optical linear scale signals and decoded simultaneously. We can correct the position error to within 20 μm in the whole distance at any time.

(3) Data acquisition:

This part is the kernel of this system. It can be divided into parameter setting, measurement process control, measurement type, monitor display and printer control. Moreover, it can deal with the results on-line.

The measuring time of the magnet is very long, so this program offers (a) different measurement interval spacing density, (b) different motion track paths (e.g., straight or curved line) to fit different analysis styles, (c) 1-D, 2-D, 3-D analysis software to analyse measurement results on-line to help the user sort out, analyse and perform diagnostics on the data of the magnetic field distribution quickly to confirm the quality of magnet.

X-Y-Z table For the field mapping, X-Y-Z table position and angle deviation will induce some error. The mapping result of the Hall probe will be influenced by the angle variation and while finding the integral field incorrect positions will induce integral field errors. Therefore, we had to identify the angle change and the position error of the

optical linear scales when the X-Y-Z table was moving. The accuracy measurements of the X-Y-Z table have been done to confirm the accuracy of this system . The range we have measured is :

(i) Distance measurement
(ii) Straightness measurement (vertical & horizontal)
(iii) Pitch measurement
(iv) Yaw measurement
(v) Whether the X-Y axis is orthogonal or not
(vi) Whether the table can be reset to zero or not at any time and at any point

The direction definitions for (i)-(iv) and measurement methods refer to reference [3], these measurement methods are almost the same but require changing the appropriate mirror. The measurement results are shown in TABLE 1. The y-axis position error is larger than for the other axes, but the error is only 0.1 mm in the whole movable distance of 0.6 m. The pitch error in the y-axis is also larger than for the other axes, but the angle change is within 20 seconds in the whole movable distance of 0.6 m.

(2) For testing whether the X-Y axis is orthogonal or not, the measurement method is to move the same distance along both X and Y axis at

TABLE 1. The accuracy of position and angle of X-Y-Z table   in the whole moving distance.

moving axis	distance error (mm)	vertical straightness error (mm)	horizontal straightness error (mm)	pitch error (arcsec)	yaw error (arcsec)
X	0.056	0.03	0.028	7.7	5.2
Y	0.1	0.04	0.037	20.6	3.4
Z	0.045	0.005	0.053	10	6

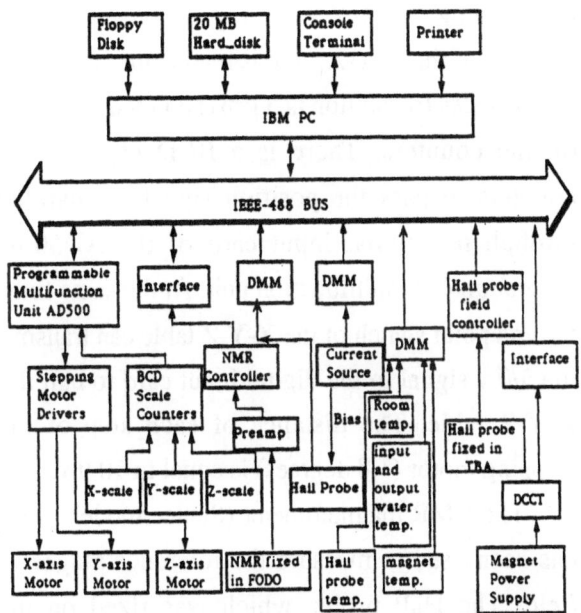

FIGURE 1. The hardware of dipole magnet measurement system.

example, from the keyboard interrupt signal) it performs the background operation which is an infinite loop. When an external interrupt signal temporarily changes states to the foreground operation (see FIGURE 2.) the whole I/O data must pass through the GPIB to be transmitted and through the I/O driver to be saved in the data-base. When the sampling-time of this system is reached, operation passes through the data base, I/O driver and GPIB to connect the peripheral instruments and transmit/receive data.

This system operation may be controlled interactively and the whole system is a multilevel tree configuration (see FIGURE 3.) which can be divided into three parts:

(1) Device installation:

This part can pre-establish the connection of certain kinds and numbers of instruments with some specified parameters, and can also simultaneously test whether the communication with the instrument controllers is normal or not.

(2) X-Y-Z table motor driver:

This part is used to control the stepping motor.

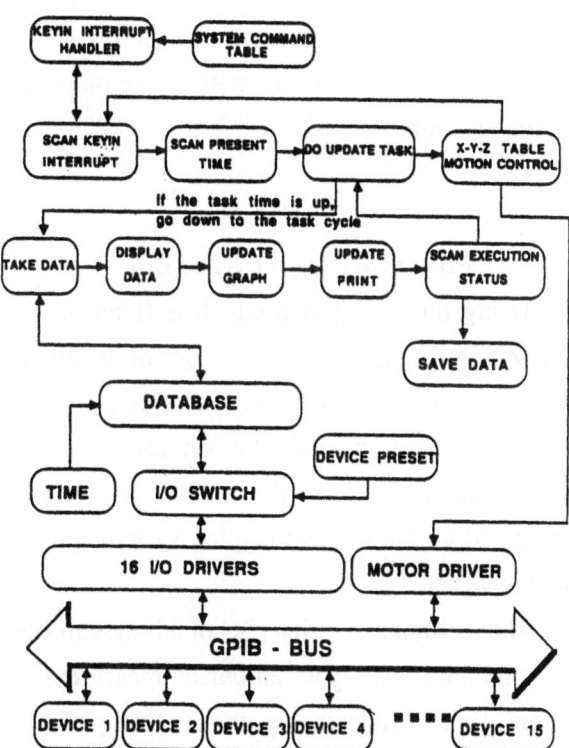

FIGURE 2. Hierarchical diagram of dipole magnet measurement system.

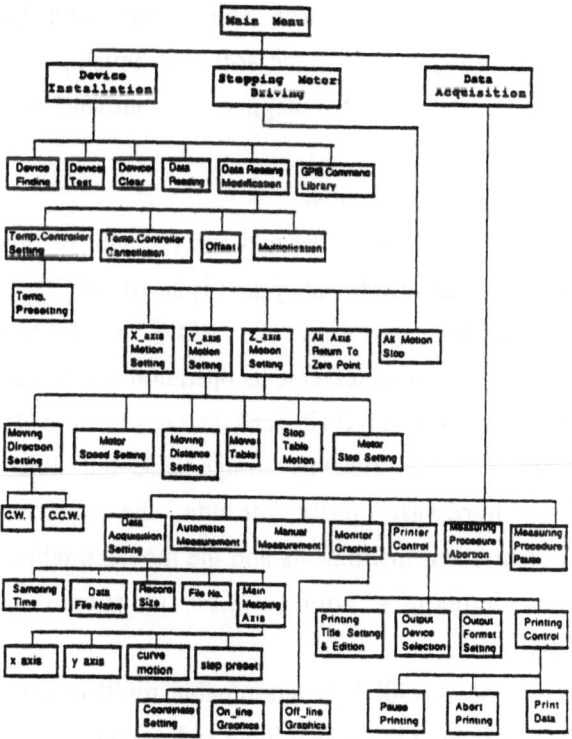

FIGURE 3. The software of dipole magnet measurement system.

That is to say, this system can apply the techniques of high accuracy automatic measurement methods to the variety of components within our mapping system and others.

MEASUREMENT SYSTEM

Using the Hall probe which is fixed on the X-Y-Z table to map the field strength of the dipole magnet requires a lot of time. Therefore, to save man-power and to avoid human errors in the measurement process, the SRRC MMG has developed an automatic mapping system for the dipole magnet.

The preliminary design goal of this system was to set up a multi-purpose automatic measurement system. From the beginning, the generality of this system was emphasized and it was built-up as an extensive set of modules. These modules can readily be used in other applications, by making trivial corrections in a short time, thereby significantly reducing the software development time for the multipole magnet. The hardware and software parts are:

1. Hardware description:

The main controller and analysis unit of this automatic measurement system is an IBM/PC/AT, which has installed the 8087 or 80287 microprocessor to increase its operation speed. The computer has a GPIB card to connect the 15 measurement instruments through IEEE-488 cables. The hardware parts include the various measurement instruments and the movable table , the elementary configuration is shown in FIGURE1.

The AD500A programmable multifunction unit[1] (which has digital input and stepping motor control cards) has been used to connect the computer for control of the stepping motor driver

and reading the X-Y-Z table position. The exact position of the X-Y-Z table is read from the optical linear scales (resolution is 5 μm) and then displayed on the counters. There is a BCD card on the counters to pass the position signals in parallel through the digital input card of the AD500A programmable multifunction unit. At the same time, the end point switch of the X-Y-Z table can transmit the 5/0 v signal to the digital input card to halt the X-Y-Z table. The distance of each step of the stepping motor (1.8 degrees) is equal to 10 μm.

The NMR magnetometer (Sentec model 1001) has been used to measure the reference magnetic field. The Hall probe, which was fixed on the X-Y-Z table and connected to the digital multimeter (Fluke 8505A), has been used to measure the field distribution of the dipole magnet. Additionally, significant temperature parameters are monitored (for example, ambient, Hall probe, magnet etc.). For measuring the temperatures, platinum resistance has been connect to HP model 3478A DMMs and the data sent through the IEEE-488 to be saved in the computer.

2. Software description:

This system is a real-time automatic measurement system. An operator can keep an eye on the system states from the monitor and modify the whole system at any time. The whole system program has been written in the 'C' language [2], because it has a good structural syntax and allows effective control of the low level computer system. This system has been divided into a few different modules and all these modules can be connected into an organic system by a global data base in order to develop, expand and debug the program in the future.

The operation of this system can be divided into foreground and background status. When this system has not received an external interrupt (for

the same time, then measure the total distance moved from the laser measurement instrument and compare this with the optical linear scales. If X-Y is orthogonal, then we should have the relation:

$$\sqrt{x^2+y^2}=D_{laser}$$

where x and y are read from the optical linear scales and D_{laser} is read from the laser distance measurement system. Although these two values are not identical, by transferring the relation to an angle, the angle of X to Y is $90°0'12.1''$. So the orthogonality of X and Y is very good.

(3) To test the zero point reset, the dial gauge is fixed and we gentle touch the X and Y axes individually, the table can be moved away and returned to the starting point repeatedly (according to the optical linear scale), the scale variation of the dial gauge can then be read and the maximum variation is about 8 μm. That is to say, the precision of zero point reset is within 8 μm.

MEASUREMENT RESULT

Provided that all the parameters have been set, the system can step automatically to measure the whole magnet and will not stop until the measurement is finished.

For mapping, our integrated Hall probe (the precision is within $\pm$ 25 ppm) was fixed on the X-Y-Z table to map the TBA prototype dipole magnet and obtain its effective length and multipole strength expansion; the effective length definition is

$$L_{eff}(x) = \int B(x,s)ds/B(x)$$

and the multipole strengths expansion in the midplane is

$$B_y(x,0,0) = B_y(0,0,0) + \sum_{n=1} B_{ny}(0,0,0)x^n$$

with

$$B_{ny}(0,0,0) = \frac{1}{n!}\frac{\partial^n B_y(x,0,0)}{\partial x^n}$$

We mapped at the y=0 plane three times at one week intervals to obtain the $L_{eff}(x)$. The results are shown graphically in FIGURE 4. The maximum error in these three effective lengths is $\pm$ 0.04 mm which is within $\pm$30 ppm. That is to say, the precision of this whole system is within $\pm$30 ppm. The result of multipole strengths expansion were shown in TABLE 2.

TABLE 2. The multipole strengths at the magnetic center.

		Measured
Dipole	[T]	$1.23957 \pm [2*10^{-5}]$
Quadrupole	[T/M]	$-1.58343 \pm [.001]$
Sextupole	[T/M^2]	$-0.29669 \pm [.2]$
Octupole	[T/M^3]	$-0.52023 \pm [8]$
Decapole	[T/M^4]	$850.536 \pm [800]$

The brackets in this table indicate the deviation of each pole strength according to this whole system precision simulations. That is to say the each pole strength precision value of this measurement system is indicated in the brackets.

DISCUSSION AND CONCLUSION

The automatic field mapping system not only saves the man-power required to perform measurements but also lets an untrained operator perform complex measurements easily within a

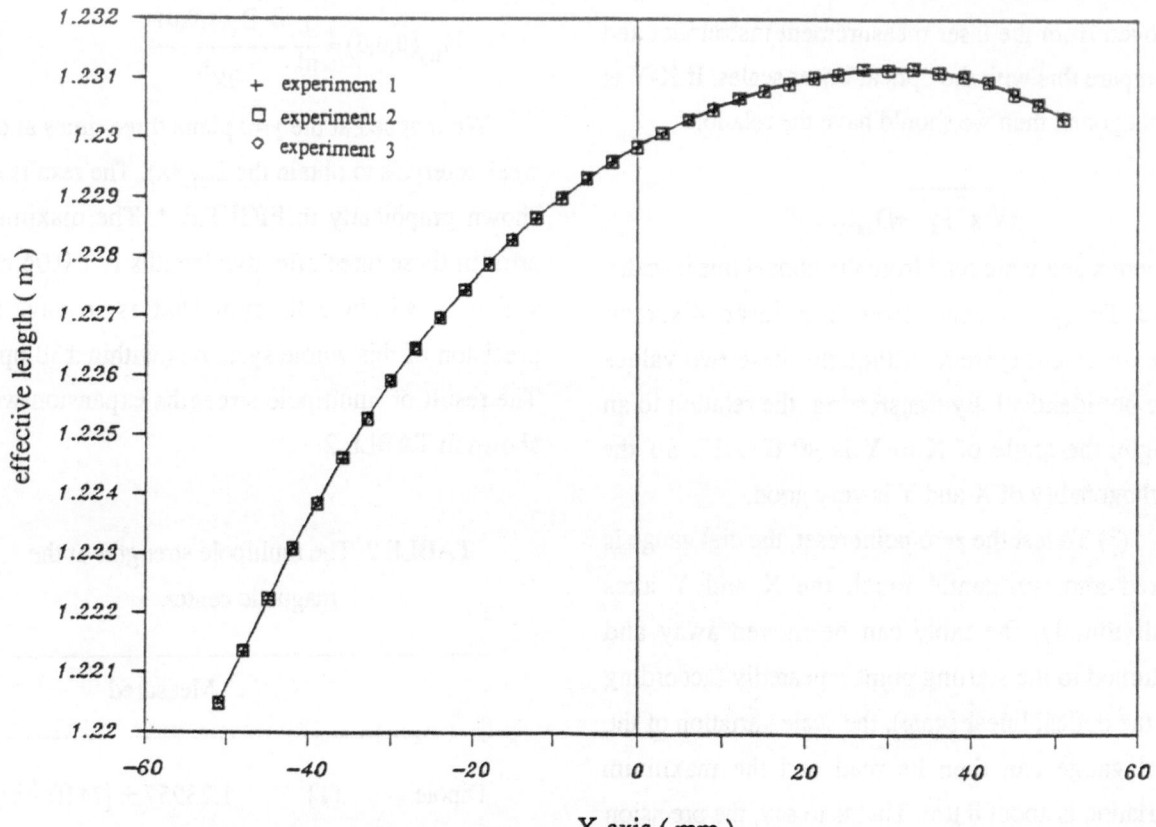

FIGURE 4. The three effective lengths distribution on three times along X-axis.

short time. To preserve system accuracy, we must take care of (1) the circuit connections, (2) the shielding, (3) the circuit vibration, (4) the software of the abnormal talk and listen of the IEEE-488 interface (which can induce unstable table movement), because these factors may induce some errors into this mapping system.

For the precision of this system at norminal field 1.24 Tesla is within ± 30 ppm. The precision values of each pole strength are indicated in the bracks of TABLE 2.

ACKNOWLEDGEMENT

The authors are indebted to professor G. J. Jan for his technical advice and Mr. In-Ho Lin in the Instrument Control group for his support of the real time servo-program. Thanks also to our laboratory colleagues.

REFERENCES

[1] AD500 programmable multifunction unit operation manual which is the IEEE-488 ATE professional of Advantage CO.,LTD.

[2] Programmer reference guide for the National Instruments GPIB DOS handle with lattice 'c' language interface.

[3] Service Manual 5528A laser measurement system and User's Guide of Hewlett-Packard Company,1982.

FIELD INHOMOGENEITY EFFECTS ON THE RELATION BETWEEN SHORT SAMPLE CRITICAL CURRENT AND THE QUENCH CURRENT OF HIGH FIELD DIPOLE MAGNETS

P.Fabbricatore, R.Musenich, R.Parodi
I.N.F.N.,Sez.di Genova, via Dodecaneso 33, I-16146, Genova, Italy

ABSTRACT

Superconducting cables, for High Energy Physics applications, can experience strong field inhomogeneity (1 Tesla) when critical current measurements are performed on short samples; this is due to the self field generated by the high current flowing through the cables (up to 10-15 KA). Recent studies showed that the critical magnetic field is not the maximum one at the cable but a lower field (the effective field). The effective field depends on both the maximum field and the cable geometry. A cable composing the winding of a high field dipole magnet experiences magnetic field inhomogeneity higher than the ones of the short sample case (up to 2.5-3 Tesla). In this paper we solved the problem to define both the critical magnetic field of a magnet, connected to the critical current Ic, and the quench current Iq. These critical parameters are related to the short sample values.

INTRODUCTION

This paper mainly deals with the problem to correctly define the Critical Magnetic Field B_c of a Superconducting (S/C) Cable carrying the Critical Current. We met first time such a kind of problem on measuring the critical current of S/C cables for HERA dipole magnet. There was no problem to define the critical current I_c i.e. the current flowing through the cable when a resistivity of $10^{-14} \Omega m$ is measured (the resistive criterium). The Critical Magnetic Field should be the Magnetic Field at the cable carrying the critical current.In spite of the applied magnetic field B_{ext} is space constant,the self field generated by the current flowing through the conductor added to the external field resulting in a strong field inhomogeneity. For HERA cable a current of 8000 A generates a field inhomogeneity of about 1 Tesla. The question was: "What is the Critical Field value?". Some people assume that the maximum field at the cable is the correct value, i.e. $B_c = B_{Peak}$ (at $I=I_c$)=$B_{c\ peak}$ (Peak Field assumption [1]).
We tried a different approach integrating the voltage along each strands composing the cable. This study [2] led to define an Effective Critical Magnetic Field $B_{c\ eff}$. The value of $B_{c\ eff}$ is lower than $B_{c\ peak}$ one, some times very close to the external applied field.

In High Energy Physic Application a cable composing the winding of a dipole magnet can experience field inhomogeneity stronger than a cable placed in a sample holder for critical current measurement, so that the problem of defining the Critical Magnetic Field again occurs. In this case the correct definition of B_c is connected to the critical working point of a magnet. Referrig to Fig.1,we suppose to wind a magnet using a Superconductor the characteristic of which is known (B_c vs. I_c curve). If the critical field is connected to peculiar field value at the conductor, for istance the Peak one, and the load line $B_{Peak}(I)$ is drawn, the intersection between the two lines (critical working point) gives information about the Critical Current of the magnet. Really the problem is slightly more complex because the only critical event observed in a magnet is the quench. For HERA cables we observe typical quench current values I_q 17% higher than the critical current one. The quench current, defined as the maximum current carried by the cable in the superconducting state, mainly depends on the thermal exchange between the

298

cable and the coolant.

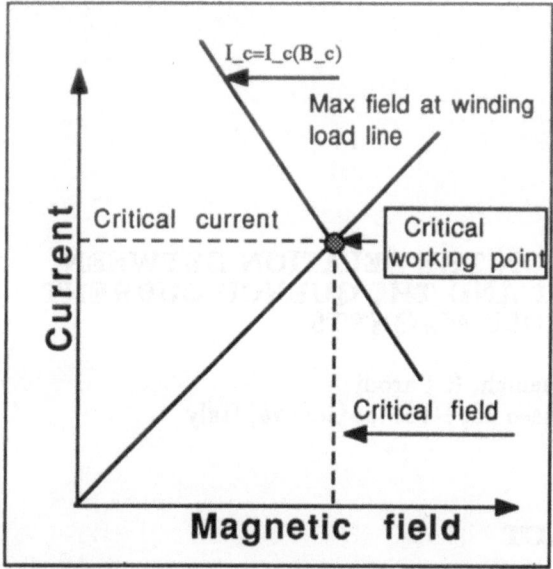

FIGURE 1: Determination of the critical
current of a superconducting magnet

Because the thermal exchange with the coolant
can be different for a cable placed in a winding
respect to a cable placed in a sample holder , the
ratio $i_q = I_q/I_c$ could be different too. Generally
we can expect to have the ideal working point
just few percent above the intersection point.
If the Critical Field is not the Peak one, this discussion must be fully revised for correctly setting the limits of a superconducting magnet.
In next sections we will show the criteria to find
the ideal working point (I_c, B_c) and the quench
current I_q of a magnet starting from the measured critical current on short sample.

THE EFFECTIVE FIELD

Short Sample Case
We start defining the Critical Current and the
Critical Field for a Superconducting cable.
The S/C cable is supposed to be composed by
several strands twisted with pitch l_p.
Let we introduce a orthogonal frame (x,y,z)
(Fig.2) so that the cross section of the cable
lies on the plane (x,y). A constant field $B_{ext}\hat{x}$
is applied normal to the cable . The field at
the conductor is given by two contributes: the
external field and the self field. The self field
is not constant moving on the cross section of
the cable; if we choose a close path on the cross
section of the cable it is possible to write the
self field variation moving along this path using
two quantities: The maximum self field ΔB_{max}

and a function of the position f(x,y) expressing
the field variation, being (x,y) the positions describing the choosen path.

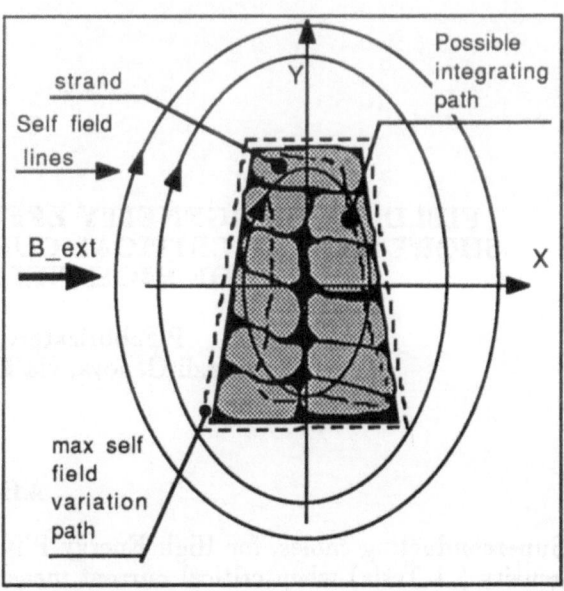

FIGURE 2: Field integration paths on a cable

We can write the component of the field normal
to the conductor in the region of the conductor:

$$B(I,x,y) = B_{ext} + \Delta B_{max}(I)f(x,y) \qquad (1)$$

Due to the twist, moving along each strand in
the direction of z axis, it is possible to find a
path with the same field values described by
Eq.1, i.e. the field variation along each strands
in a twist pitch is just the same of the field variations on every strands in the cross section of
the cable. We can define a function g(z) espress-
ing the variation of the maximum field moving
along the cable and write an equation equivalent
to Eq.1

$$B(I,z) = B_{ext} + \Delta B_{max}(I)g(z) \qquad (2)$$

where g(z) is a periodic function of z, being repeated every twist pitch. For a round cable we
can write:

$$g(z) = sin(\frac{2\pi}{l_p} z)$$

Choosing as path the outer border of the cable we obtain the situation characterized by the
maximum variation of the field; i.e. the most
significative situation.
Let we consider a single strand of the cable carrying a current I_s so that, setting N_s the number of strands, the current carried by the cable
is $I_{ca} = N_s I_s$.

The transition from the superconducting state to normal one is ruled by the relation:

$$V(I_s) = \frac{\rho_c}{A_s} \frac{I_s^{n+1}}{I_{cs}^n(B,T)} \quad (3)$$

where V is the voltage drop per unit lenght along the cable; ρ_c is the already mentioned critical resistivity; A_s is the cross section of the strands (S/C + matrix); $I_{cs}(B,T)$ is the critical current at a field B and temperature T; n is an integer called n-value. Due to the field inhomogenity the critical current I_{cs} is not constant moving along each strands. If the dependance of I_{cs} on the field were known, a spatial integration could be performed so that the voltage drop per unit lenght of the cable could be calculated. The behaviour of the critical current vs. the field for Nb Ti material is quite known (see Lubell's formulae [3]); in the field range 5-8 Tesla the field dependance of the critical current is quite linear:

$$I_{cs} = Constant \; [b(T) - aB_c].$$

By the knowledge of the constant a and b the voltage drop is calculated [2]:

$$V(I_s) = \frac{\rho_c}{A_s} \frac{I_s^{n+1}}{I_{cs}^n(B_{ext},T)} G \quad (4)$$

The function G is:

$$G = \frac{1}{2\pi} \int_0^{2\pi} \left(\frac{1}{1 - K \; g(z)}\right)^n \quad (5)$$

Where K is function of the maximum self field, the external field and the characteristic $B_c(I_c)$ of the cable:

$$K = \frac{a \; \Delta B_{max}}{b - a \; B_{ext}}$$

The function K depends on the self field and consequentely on the current: nevertheless in the transition region the variation of the self field is small respect to the field at the conductor and the value that K assumes at the critical current at the external field can be taken as constant value in the whole transition region.
From Eq.(4) it is possible to obtain the expected critical current of the cable, finding the current fulfilling the resistive criterium:

$$I_{c \; cable} = N_s \; I_{cs}(B_{ext})\left[\frac{2\pi}{G}\right]^{\frac{1}{n}} \quad (6)$$

We point out again that really G slightly depends on the current through K, in our approximation of G constant Eq.6 allows to calculate the critical current of the cable without using iterative methods. From the knowledge of the critical current at the external field

$N_s \; I_{cs}(B_{ext})$ of the expected critical current found in Eq.(6) and of the expected relative behaviour $I_{cs} = I_{cs}(B_c)$ we find:

$$B_{c \; cable} = \frac{1}{a_3}[1 - \left(\frac{2\pi}{G}\right)^{\frac{1}{n}}](a_2 - a_3 B_{ext}) \quad (7)$$

The critical field resulting from Eq.(7) is called Effective Critical Field $B_{c \; eff}$ meaning that it can be considered the field applied to the sample when the critical current is carried by the cable. We remark that there is no further meaning for this field i.e. it cannot be considered the effective field applied to the sample out of the critical conditions.

Winding Case
Now we consider the situation of a cable composing the winding of a magnet. In this case at a fixed current it is possible to write the field at the conductor just using an expression like Eq.2 but substituting the external field with a mean field at the conductor and the self field with a field oscillation around the mean value:

$$B(I,z) = B_{mean}(I) + \Delta B_{max}(I)g(z) \quad (8)$$

From this point of view the description is equivalent to the previous one; the main difference is that the function K cannot be considered constant. Consequentely the critical current of the magnet cannot be found by Eq.(6) but a simple numerical method must be applied. The method that we chose is:
(a) to fix the current of the magnet I_m
(b) to calculate the mean field at the conductor B_{mean}=Constant x I_m
(c) to calculate the maximum field oscillation $\Delta B_{max}(I_m)$
(d) to find the value of K and of G functions and to calculate the voltage drop using Eq.(4)
(e) to verify if the critical current is exceeded applying the resistive criterium, if that case occurs...
(f) ... to interpolate between the two last found values of voltage drop and the two last current values, so that the critical current is found.

The effective critical field can be obtained from Eq.(7) substituting G with $G(I_{c \; mag})$ and B_{ext} with $B_{mean}(I_{c \; mag})$.
A different approach can be tried, considering that the effective critical field can be considered a peculiar point of the function :

$$B_{eff} = \frac{1}{a_3}[1 - \left(\frac{2\pi}{G(I)}\right)^{\frac{1}{n}}](a_2 - a_3 B_{mean}(I)) \quad (9)$$

$B_{eff}=B_{c\ eff}$ at the critical current: $I=I_{c\ mag}$. We remark again that B_{eff} has no physical meaning at current values different than the critical one. The steps of this approach are:

(a) To follow the previous items (a),(b),(c)
(b) To calculate K and G functions so that $B_{eff}(I)$ is found using Eq.(9).
(c) Putting on a field vs. current (B-I) graph the measured short sample critical curve (determined using the effective field criterium) and the curve $B=B_{eff}(I)$, the intersection between the two curves is just the working critical point ($B_{c\ mag}$,$I_{c\ mag}$).

Fig.3 shows the modifications of the B-I graph when the Effective Critical Field is considered respect to the results of the Peak Field Criterium. The magnet critical current is quite higher than the limit determined by the Peak Feld Criterium; i.e. the expected critical current of a magnet like a high field dipole magnet is actually undervalued.

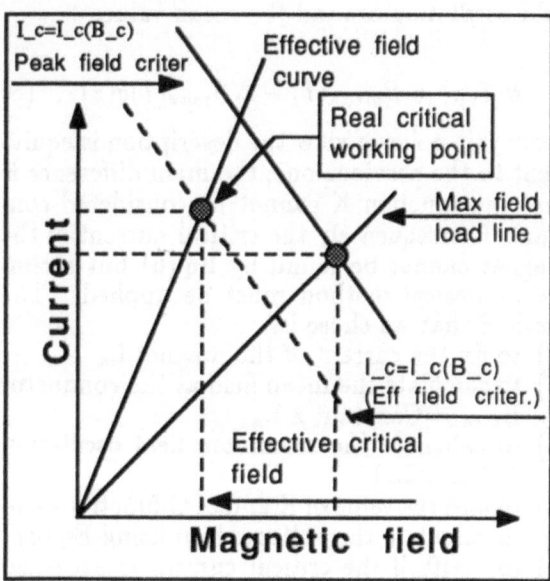

FIGURE 3:Critical current of a S/C magnet determined using both Peak Field and Effective Critical Field criteria

THE QUENCH CURRENT

The results reported in this section are widely discussed in a paper actually in preparation [4]. The quench current is determined by thermal considerations. The heat dissipated by Joule effect is balanced with the heat exchanged with the coolant. The ensemble of the equilibrium states determinates the quench current. The

formula we found connects the quench current with the critical current above defined:

$$i_q = \frac{I_{quench}}{I_c(B_c)} = \left[\frac{n^n}{(n+1)^{n+1}\alpha}\right]^{\frac{1}{n+2}} \quad (10)$$

the parameter α include the terms of dissipation, thermal exchange and effective field [4]. Using this model we calculeted with good accuracy the quench current of the S/C cables for HERA dipoles: foreseen $i_q=1.19$, measured $i_q=1.17$.

The cable of the winding of HERA dipoles can exchange much less heat with the coolant respect to the short sample, so that the relative quench current i_q is greatly reduced.

APPLICATION OF THE MODEL: HERA DIPOLE MAGNETS

The cable for HERA deflecting dipole magnets has a trapezoidal shape with dimensions 1.28/1.67 mm for the basis and 10 mm in height. It is composed by 24 strands of 0.84 mm in diameter able to carry about 450 A at 5.5 Tesla and 4.2 K. The strands are twisted with a pitch of about 100 mm. We take in consideration a given cable and suppose to wind a dipole magnet using that cable. We will show the differences in the determination of the ideal working point both following the standard method (using the Peak Field) and our method using the Effective Critical Field.

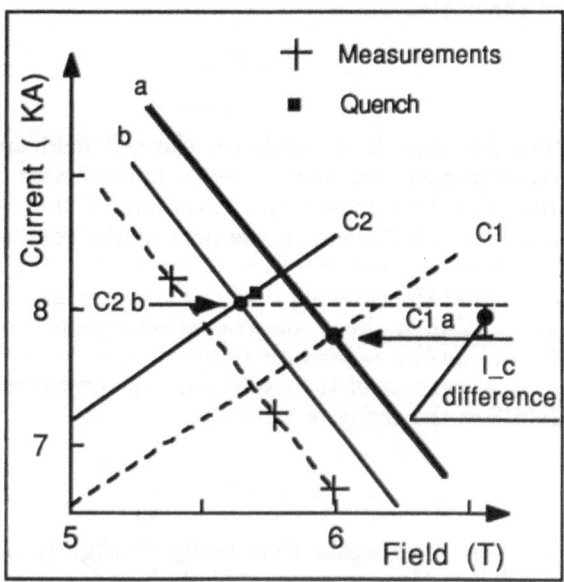

FIGURE 4: Determination of the critical current of HERA dipole magnets

The cable taken into consideration is called LMI199, manufactured by Europa Metalli (Lucca-I).

In Fig.4 a graph Field-Current is shown. The crosses designates the measured values as resulting from a measurement performed in the set-up of Genoa laboratory of INFN [5]; applying the self field correction we can draw the curve (a) using the Peak Field criterium or (b) using the Effective Critical Field criterium. The calculation of the critical current and field of item (b) is performed using Eq.(6) and (7); the function G is calculated using :

$$g(z) = sin^3\left(\frac{2\pi}{l_p} z\right)$$

and for the n-value the measured value n=17. The temperature is 4.6 K.

Next step is to find the working point of the magnet. This aim requires a map of the field at the conductor. Fig.5 shows the field lines of the HERA dipole magnet (only winding without iron yoke) at a cross section in the middle the magnet (z=0 being z the beam direction); the map was calculated using TOSCA code at current of 7000 A.

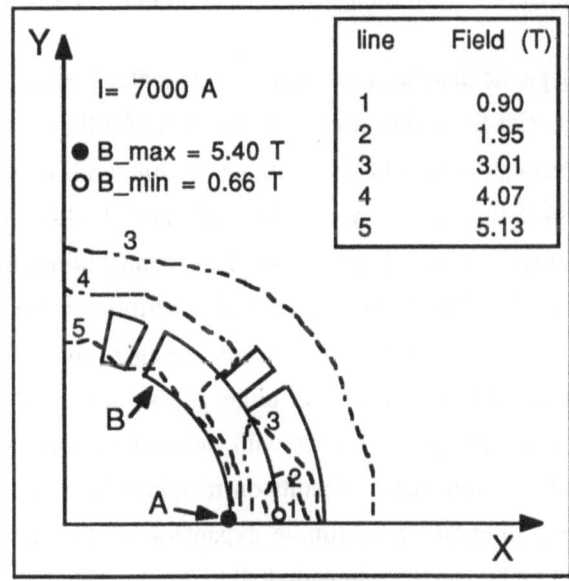

FIGURE 5:Magnetic Iso-field Lines at HERA dipole magnet calculated by using TOSCA

It is possible to observe that the cables experience different field situations. The cables placed in the zone A experience a strong inhomogeneity (from 0.9 to 5.40 Tesla); the cables in the zone B see a more homogeneous field (from 4.16 to 5.36). After some calculations we found the A zone to be the most critical. In Fig.5 the curve C1 represents the Peak field load line

while curve C2 indicates the curve B_{eff}. The intersection C1a is the standard working point, while C2b is the working point as results from our considerations. In that case the magnet has a critical current real margin of about 250 A higher than foreseen by the standard method (8050 A against 7800 A).

Furthermore if the quench current is calculated from Eq.10 considering the thermal exchange with the coolant, we find for this cable an ideal limit of about 8100 A.

CONCLUSIONS

The concept of Effective Critical Field not only contributes to correct the over-extimation of the critical current of S/C cable due to the Peak Field Correction but gives a basis to determinate the upper limit of a Superconducting Magnet. In the case of HERA cable the difference in determining the working point is quite small (250 A can be buried in the statistics), but we think that higher differences can be found for LHC or SSC cables due to higher field inhomogeneity.

ACKNOWLEDGEMENTS

We wish to thank Dr. P.Valente and Dr. P.Gagliardi (Ansaldo Componenti) for the informations about the magnetic field of HERA dipole magnets.

REFERENCES

1. M.Garber, A.K.Ghosh and W.B.Sampson, The effect of self field on the critical current determination of multifilamentary superconductors. IEEE Trans. Mag., 1989, Mag-25.
2. P.Fabbricatore, R.Musenich, R.Parodi, S.Pepe and R.Vaccarone. Self field effects in the critical current measurements of superconducting wires and cables. to be published on Cryogenics.
3. M.S.Lubell. IEEE Trans.Mag., 1983, Mag-19 No.3,754-757
4. P.Fabbricatore, R.Musenich, R.Parodi, R.Vaccarone, Effect of n-value and field inhomogeneity on the quench current of superconducting cables. In preparation.
5. P.Fabbricatore, A.Matrone,A.Parodi, R.Parodi, C.Salvo, R.Vaccarone, A multiple sample holder for J_c measurements on HERA Cables. Proocedings ICEC 12 Butterworths, 1989, 903-907

THREE-DIMENSIONAL COMPUTATION OF MAGNETIC FIELDS AND LORENTZ FORCES OF AN LHC DIPOLE MAGNET USING THE METHOD OF IMAGE CURRENTS

C. Daum

NIKHEF-H, Amsterdam, Netherlands

D. ter Avest

Applied Superconductivity Centre, University of Twente, Enschede, Netherlands

ABSTRACT

Magnetic fields and Lorentz forces of an LHC dipole magnet are calculated using the method of image currents to represent the effect of the iron shield. The calculation is performed for coils of finite length using a parametrization for coil heads of constant perimeter. A comparison with calculations based on POISSON and TOSCA is made.

INTRODUCTION

For the design of magnets, a detailed knowledge of fields and forces is needed as well in the straight sections of the coil as in the coil heads. The method of computation presented here is designed for structures with shell coils around a cylindrical aperture surrounded by a cylindrical iron yoke. The effect of the iron is taken into account using the method of image currents for a fixed value of the permeability, and, hence, the variation of the permeability in the iron is not taken into account. The fields and forces are calculated as the sum of the fields and forces of the strands out of which the conductors are composed. The strands are "ideal" strands, i.e. they are parallel to the axis of the conductor and thus do not follow the actual layout of the strands in a Rutherford cable. The current is concentrated in the centre of the strands. The magnetic field due to a single strand is calculated with the Biot-Savart law using delta functions for the radial and angular current distributions. The integrals in the Biot-Savart law can now be evaluated. Fields are always calculated as the sum of the contributions of the individual strands. A detailed description is used for constant perimeter coil heads. Lorentz forces are calculated at the centre of either the strands, or the conductors, or the blocks with the current concentrated at the centre. Results on magnetic fields and field integrals, a multipole expansion of the field integrals, and the magnetic length are presented. An extensive account of the method is given in Ref.[1].

AN LHC DIPOLE MAGNET

The shell coil of the prototype LHC magnet [2,3] consists of two layers. A quadrant of the coil is shown in Fig. 1, which also defines the coordinate system used. The inner layer has four blocks of conductors with four, four, three and two conductors, the outer layer has two blocks of conductors with seven and seventeen conductors. The conductors in the inner layer consist of two rows of thirteen "ideal" strands, those in the outer layer have two rows of twenty "ideal" strands. Both conductors have a slight keystone angle. Fig. 1 also shows a cross section of both conductors, which have a height of 17 mm, and have a width at top and bottom of 2.64 (1.81)mm and 2.18(1.44)mm for layer 1(2), respectively.

The LHC prototype magnet is a twin aperture magnet. The method of computation described in this paper is for a single aperture magnet with a cylindrically symmetric configuration. The shape of the "ideal" strands in the conductors is a race track with straight parts with individual length for each strand and with coil heads, which are half circles, if exposed in a flat plane, at the inner radius of the conductor, and half ellipses elsewhere for obtaining constant perimeter coil ends. The coordinate system is defined with the x-axis in the horizontal plane transverse to the symmetry axis of the magnet, the y-axis along the field direction and the z-axis along the symmetry axis.

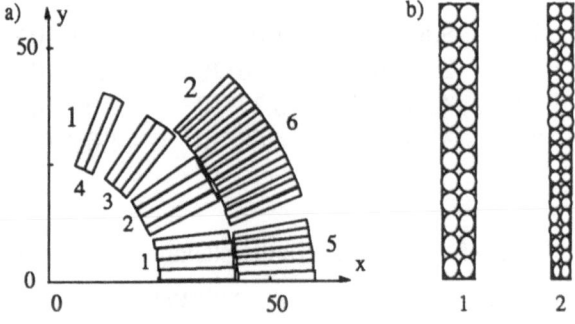

FIGURE 1   Layout a) of the conductors in the 6 current blocks of an LHC dipole magnet with dimensions in mm, and b) of the "ideal" strands in the conductors of layer 1 and 2.

COMPUTATION OF FIELDS, FIELD INTEGRALS AND FORCES

The basic integrals of the Biot-Savart law are expressed in a cylindrical coordinate system with the z-axis as symmetry axis of the cylinder. They can be evaluated analytically for the straight parts of each strand, if the current is taken to be concentrated at the centre of the "ideal" strands of Fig. 2. The integrals over the angular variables can be performed immediately using a delta function for the radial and azimuthal distributions at the position of the strand. The integration over z can be performed analytically for the exact length of the straight part of the strand. In the coil heads, the integration over the radial and angular variables can again be performed analytically. The integration over z is made using the Simpson rule, for which it is sufficient to make only 10 steps over a coil head.

The field at any field point is obtained by summation over the contributions of all strands1. In the same way, also field integrals are calculated.

The iron yoke has a radius r' in the central part of the magnet and jumps to a radius r'' in the outer part before the straight parts of the coils end. These radii are used for the calculation of the position of the image current of each strand and their contribution to the magnetic field as for the direct contributions of the straight parts of the strands. The following approximations are made. It is assumed that the iron yoke extends to infinity in the z-direction as well as in the radial direction outwards from radius r' and r'' separately. Images with respect to the plane transverse to the z-axis where the actual iron yoke changes from inner radius r' to r'' are neglected. The relative permeability of the iron is taken to be $\mu = \infty$.

A comparison[1] with three-dimensional field calculations using TOSCA[4] and with the multipoles of a two-dimensional calculation using POISSON[5] shows that these approximations can be made in the case that the radii r' and r'' are sufficiently large, and hence the contribution of the image currents to the field is small.

The Lorentz force on a point of a strand is the vector product of the strand current and the field at this point due to all other strands. It has been checked[1] that the force on a conductor, taking the current to be concentrated at the centre of the conductor, is within a few percent equal to the sum of the forces on the strands of the conductor. The forces on the conductors are used in the mechanical design of the magnet. In all cases, the field and field integrals are calculated from the individual strands.

RESULTS FOR AN LHC MAGNET

A program has been written for field and force computations with the method presented above. It has been applied to an LHC magnet of a nominal length of 1m with the conductor layout of Fig.1. The field and force configuration has been calculated for a central field $B(0,0,0) = 10T$ for which an excitation current $I = 14375A$ is needed. Then, the current in the strands of layer 1 and layer 2 are $I_{1s} = 552.9A$ and $I_{2s} = 359.4A$, respectively. The corresponding current densities are about $J_{1s} = 404A/mm^2$ and $J_{2s} = 558A/mm^2$, respectively. A graded current density is used for optimization of the current carrying capability of available conductors for the two coil layers. The iron yoke has an inner radius $r' = 0.100m$ between $z = -0.302m$ and $z = 0.302m$. Outside this range, it has an inner radius $r'' = 0.128m$. The contribution of the image currents in the iron yoke is less than 15% due to the large inner radius of the iron.

An important design criterion is the maximum field on the conductors. This occurs in layer 1 at the inner top edge of block 4 , and in layer 2 close to the inner top edge of block 6 (see Fig. 1). The maximum fields in layer 1 and layer 2 are $B_{1max} = 10.217T$, and $B_{2max} = 8.772T$ at $z_1 = 0m$, and $z_2 = 0.36m$, respec-

tively. In layer 2 the field at $z = 0$ equals 8.591T. These maxima should be compared with the properties of available superconductors at the required current densities. The magnetic length of this configuration is $L_{magn} = \int_{-\infty}^{\infty} B(0,0,z)\,dz / B(0,0,0) = 0.867m$.

The multipole expansion of the field integrals is

$$\int_{-\infty}^{\infty} B(x,0,z)dz = \left(\int_{-\infty}^{\infty} B(0,0,z)dz\right)\sum_{n=1}^{\infty} b_n\left(\frac{x}{x_0}\right)^{n-1} \quad (1).$$

Table 1 lists the coefficients b_n for a 1m LHC magnet with the configuration of Fig. 1.

TABLE 1

Multipole coefficients b_n of the series expansion (1) of the field integrals for a 1m LHC magnet; T, S and H are for Total field integral, and the contributions of the Straight part of the coil and the coil Heads, respectively, $x_0 = 20$ mm.

n	b_n (T)	b_n (S)	b_n (H)
1	$1.00*10^{+0}$	$8.60*10^{-1}$	$1.39*10^{-1}$
3	$9.17*10^{-3}$	$1.07*10^{-2}$	$-1.50*10^{-3}$
5	$-5.87*10^{-4}$	$4.18*10^{-4}$	$-1.01*10^{-3}$
7	$1.07*10^{-3}$	$1.05*10^{-3}$	$2.48*10^{-5}$
9	$6.40*10^{-4}$	$6.37*10^{-4}$	$3.75*10^{-6}$
11	$4.88*10^{-4}$	$4.91*10^{-4}$	$3.01*10^{-6}$
13	$-2.51*10^{-5}$	$1.41*10^{-5}$	$-3.92*10^{-5}$
15	$9.88*10^{-5}$	$1.02*10^{-4}$	$-3.16*10^{-6}$
17	$-2.16*10^{-4}$	$-1.96*10^{-4}$	$-2.07*10^{-5}$
19	$-4.34*10^{-5}$	$-6.37*10^{-5}$	$2.03*10^{-5}$
21	$3.60*10^{-5}$	$4.66*10^{-5}$	$-1.06*10^{-5}$

For a further detailed understanding, Fig. 2 shows the direct contributions of the coils to the field on the z-axis, and to the multipole coefficients

$$c_n = (a_n^2 + b_n^2)^{1/2} \quad (2)$$

of the field integrals, and those of their mirror images in the iron yoke for this magnet. Here, the coefficients

a_n are the skew multipole coefficients which vanish for the used dipole symmetry. We see that all multipole coefficients due to the mirror images for $n = 5$ and larger are negligibly small.

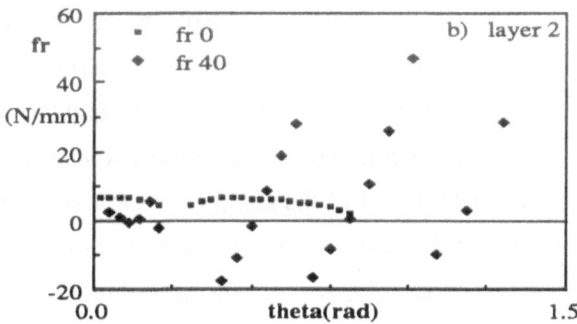

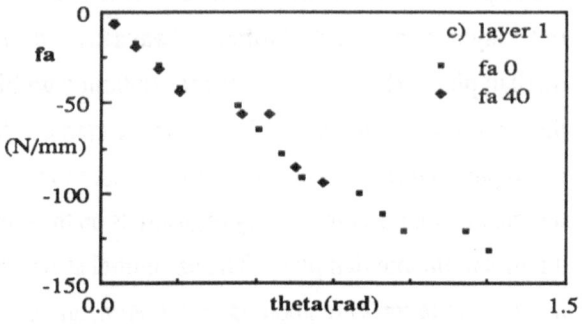

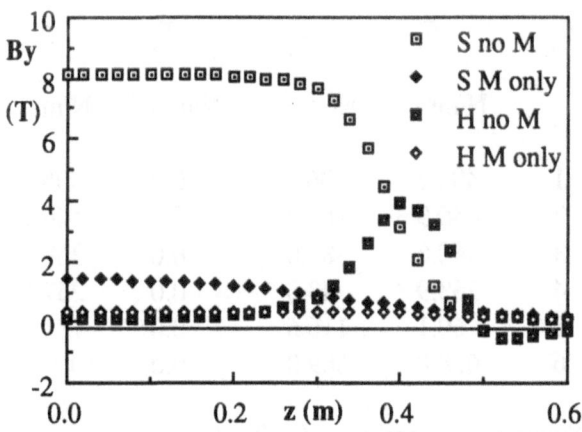

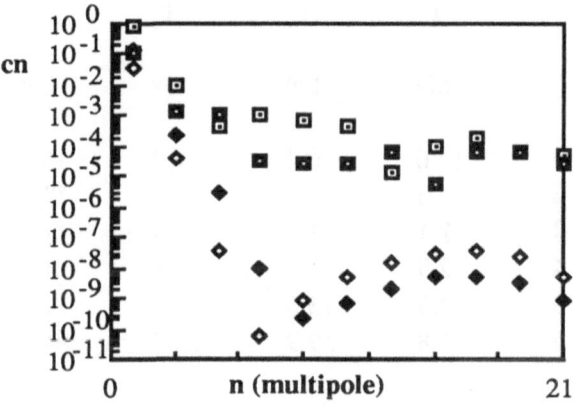

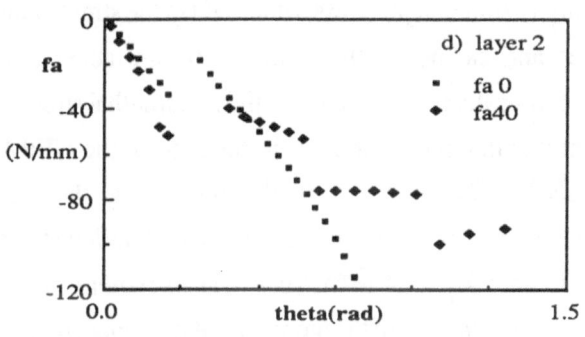

FIGURE 2 Field along the z-axis and multipole coefficients for direct contribution of the straight parts of the strands (S no M) and the coil heads (H no M), and for their mirror contribution (S M only and H M only).

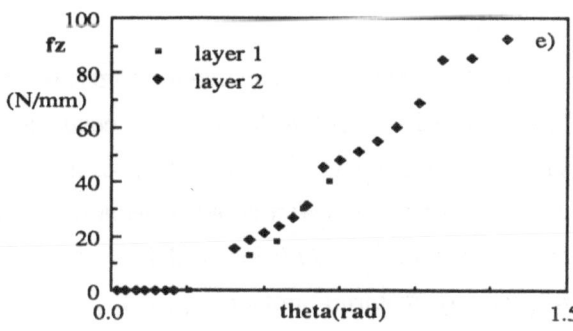

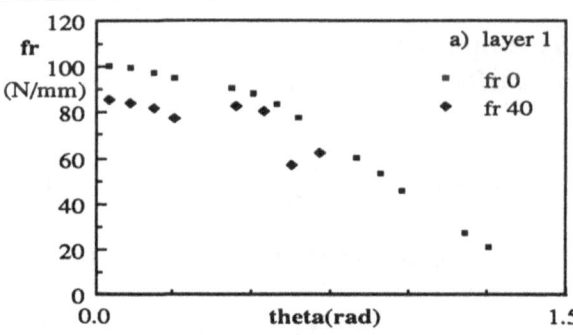

FIGURE 3 Radial (fr) component at z = 0 and 40 cm in a) layer 1, b) layer 2, azimuthal (fa) component at z = 0 and 40 cm in c) layer 1, d) layer 2, and e) longitudinal (fz) component of Lorentz force at z = = 40 cm in layer 1 and 2.

Fig. 3 shows the components of the Lorentz force as a function of the azimuthal angle θ of the conductors in the straight part of the coils at z = 0 and 40 cm, respectively. The radial force is always outward at z = 0 cm. The reversal of the radial forces at z = 40 cm within blocks in layer 2 needs attention in the mechanical design of a magnet. The overall radial force of the two layers is outward, which implies, however, a compression of the insulation between inner and outer layer. The azimuthal force is always towards the median plane. Hence, all forces are contained within the arch of the coil, and no forces are exerted on the beam pipe. The accumulated azimuthal force towards the median plane causes a very high stress on the conductors in the median plane. The azimuthal stress for layer 1 and layer 2 are about $\sigma_{1a} = 57$ MPa, and $\sigma_{2a} = 70$ MPa, respectively. The azimuthal forces require a high azimuthal prestress, which add to the stress values, and can not be chosen to be a comfortable factor of more than two higher than these azimuthal stresses at maximum excitation as in the case of the HERA magnets[6]. Also the radial force on e.g. the first conductor of block 1 causes a very high radial stress of about $\sigma_{1r} = 34$ MPa. These values require strong attention in view of the properties of the copper matrix of superconducting cables, in particular for Nb_3Sn conductors.

Table 2 shows the total Lorentz forces on the blocks from the sum of the forces on the conductors. The total longitudinal forces for an upper half coil head in the first layer and the second layer are outward and amount to $f_{1z} = 27.6$ kN, and $f_{2z} = 70.1$ kN, respectively, requiring strong end plates.

For the HERA dipole magnets[6] a study was made of the influence of the spacing between conductors in the coil heads on the maximum field in the conductor locally. The contribution of the sextupole component to the field integral was minimized locally at the coil head. For the LHC magnet, the sextupole contributions of the straight part of the strands and

TABLE 2

The Lorentz force on the block of conductors. N_b is the block number (see Fig. 1).

In median plane:

N_b	F_x	F_y	F_z	F_{tot}
	N/mm	N/mm	N/mm	N/mm
1	403.7	-36.8	0.0	405.3
2	440.9	-68.2	0.0	446.1
3	360.5	-81.0	0.0	369.5
4	253.9	-43.0	0.0	257.6
5	60.1	-116.6	0.0	131.2
6	698.8	-809.8	0.0	1069.7

Summed over coil head for $0 < \phi \le \pi/2$:

N_b	F_x	F_y	F_z	F_{tot}
	kN	kN	kN	kN
1	13.6	-0.3	6.5	15.1
2	14.3	0.4	10.5	17.7
3	8.1	-1.4	7.7	11.3
4	2.9	-0.4	2.9	4.1
5	4.5	-13.6	10.5	17.8
6	44.8	39.9	59.6	84.6

those of the coil heads have opposite sign for an open and a closed configuration of the blocks, as shown in Table 3. This permits the design of a coil head with a minimal sextupole component.

TABLE 3

Comparison of the sextupole component of the field integral for the open and a closed configuration of Fig. 4.

Configuration	$b_{3\,total}$	$b_{3\,straight}$	$b_{3\,head}$
open	$9.17*10^{-3}$	$1.07*10^{-2}$	$-1.50*10^{-3}$
closed	$-4.11*10^{-4}$	$1.09*10^{-3}$	$-1.50*10^{-3}$

with a three-dimensional TOSCA calculation yields $\mu_r = 18.4$.

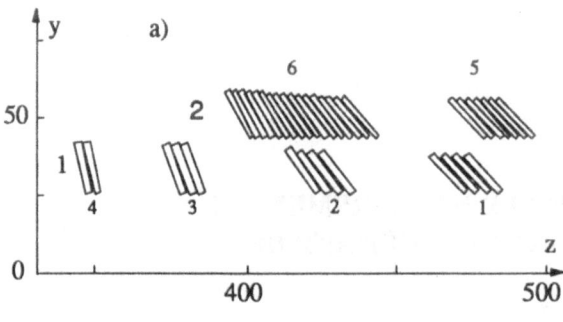

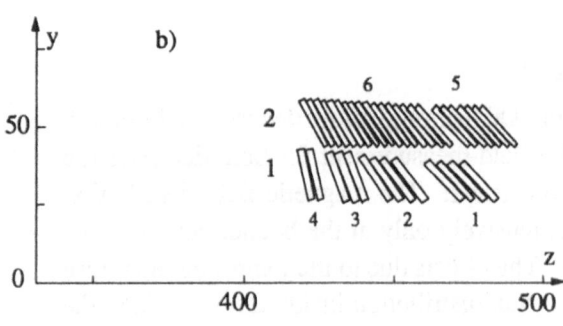

FIGURE 4   Layout of end of coil head for a) open, and b) closed configuration. Dimensions are given in mm.

TABLE 4

Comparison of the multipole expansion (1) of the field integrals. For a discussion, see text.

n	$b_{n\,Pois.}$	$b_{n\,1000}$	$b_{n1000re}$
1	$1.00*10^{+0}$	$1.00*10^{+0}$	$1.00*10^{+0}$
3	$6.21*10^{-4}$	$-6.56*10^{-5}$	$-1.06*10^{-4}$
5	$8.81*10^{-5}$	$-2.76*10^{-6}$	$2.11*10^{-5}$
7	$9.39*10^{-4}$	$9.51*10^{-4}$	$9.73*10^{-4}$
9	$6.71*10^{-4}$	$6.01*10^{-4}$	$6.05*10^{-4}$
11	$4.67*10^{-4}$	$5.53*10^{-4}$	$5.66*10^{-4}$
13	$-8.66*10^{-5}$	$-4.84*10^{-6}$	$-4.58*10^{-6}$
15	$9.68*10^{-5}$	$1.77*10^{-4}$	$1.80*10^{-4}$
17	$-2.80*10^{-4}$	$-2.37*10^{-4}$	$-2.42*10^{-4}$
19	$8.52*10^{-5}$	$-6.43*10^{-5}$	$-6.63*10^{-5}$
21	$4.03*10^{-5}$	$4.90*10^{-5}$	$5.02*10^{-5}$
I	14941A	14600A	14941A
μ_r	varying	∞	16.3

Table 4 shows the comparison[1] between the multipoles for a 1000m magnet using our method and a two-dimensional POISSON calculation for the configuration of Fig.1. Taking the same current for both cases, the direct contributions of the strands to $B(0,0,0)$ = 10T are the same. Scaling of the mirror contribution results in an average relative permeability $\mu_r = 16.3$ for our calculation, which is used to recalculated the coefficients for the 1000m magnet (1000mre). We observe the largest discrepancies for the sextupole and decapole coefficients which are particularly sensitive for the layout of the coil heads and are not described by the POISSON calculation. The POISSON calculation also shows that the configuration of Fig. 1 has not yet resulted in sufficiently small multipole coefficients for $n \geq 3$. A similar comparison[1]

REFERENCES

1. C. Daum and D. ter Avest, Three-dimensional computation of the magnetic fields and Lorentz forces of an LHC dipole magnet, NIKHEF-H 89/12 and LHC note No. 94.
2. The large Hadron Collider in the LEP Tunnel, ed. by G. Brianti and K. Hübner, CERN 87-05, 27 May 1987.
3. D. Leroy, R.Perin, G. de Rijk, W. Thomi, Design of a High Field Twin Aperture Superconducting Dipole Model, CERN SPS/87-32 (EMA), LHC note No 62.
4. TOSCA, Vector Fields, Oxford, UK.
5. POISSON, CERN program library, T604.
6. K.H. Mess, P. Schmüser, Superconducting Accelerator Magnets, DESY HERA 89-01, and in the Proceedings of the CERN Accelerator School on Superconducivity in Particle Accelerators, Ed. S. Turner, CERN 89-04, p.87.

THE MAGNETIC DESIGN AND FIELD MEASUREMENT OF FERMILAB COLLIDER DETECTORS: CDF AND D0

R. Yamada

Fermi National Accelerator Laboratory

P.O. Box 500, Batavia, Illinois 60510, U. S. A.

ABSTRACT

General magnetic characteristics of the CDF and D0 hadron collider detectors at Fermilab are described. The method and equipment for the field measurement for both detectors are described, and their field measurement data are presented. The magnetic field distribution inside the CDF solenoid magnet was measured extensively only at the boundaries, and the field values inside the volume were reconstructed. The effects due to the joints and the return conductor were measured and are discussed. The flux distribution inside the yokes and the fringing field of the D0 toroids were calculated and compared with measured data. A proposal to generate dipole magnetic field inside the D0 toroidal magnet is discussed.

INTRODUCTION

At Fermilab, there are two major hadron collider detectors: CDF (the Collider Detector at Fermilab) and D0; both are general-purpose collider detectors and weigh more than 4000 tons each.

The CDF is located in the B0 straight section of the Energy Doubler ring. It is composed of the central detector, the forward and backward detectors, and others.[1] This detector was built with the collaboration of U.S., Japan, and Italy and has been in operation since 1986 for high energy physics experiments.

The D0 detector is under construction at the D0 straight section, and is expected to be in full operation early in 1991.[2] The D0 detector has three major and two small toroidal magnets. The major toroids were completed and excited without two small toroids early this year (1989).

CDF

The central part of the CDF detector is constructed with a structurally complicated yoke, and excited with a cylindrical thin-walled superconducting solenoidal coil.

Design of central magnet structure

The central detector weighs about 2,200 metric tons and is shown in Fig. 1 together with its field mapping device. The magnetic circuit of this detector is composed of a 1.5 Tesla superconducting solenoid, the 2 in. thick iron plates used as pole pieces and also as part of the return yoke, and 8 in. thick iron slabs used in the end wall and return legs. The central hadron calorimeter is made of 1 in. thick iron plates which are partially magnetized unintentionally due to the proximity of the magnetic circuit.

TRIM calculation

A two-dimensional magnetostatic program, TRIM, was used to calculate the magnetic field distribution of the CDF magnet. Its accompanying program, FORGE, was applied to estimate forces on the yoke, coil, and calorimeters.[3] The magnetic structure of the CDF detector, which was used for

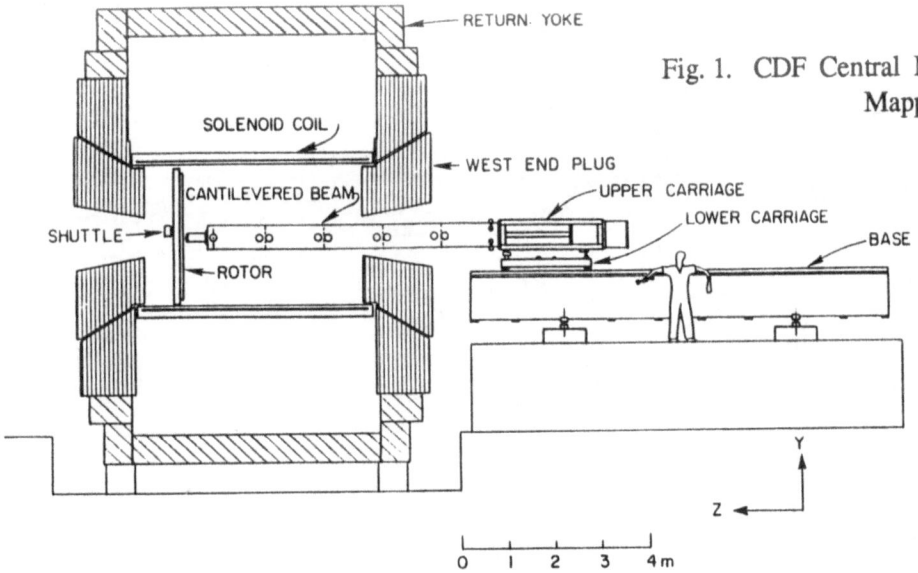

Fig. 1. CDF Central Magnet Structure and Field Mapping Device

the field calculation, is shown in Fig. 2. For reasons of simplicity and symmetry, its geometry was assumed completely axi-symmetric and only a quadrant was used. The resulting magnetic field and flux distribution is shown in Fig. 3.

At the excitation of 1.5 Tesla, each plug is pulled inward by a force of 540 metric tons. Therefore, an extensive study on the magnetic field distribution and its resulting magnetic force distribution was required. The stress and deflection of these components due to the magnetic force, as well as those due to gravitational force, were calculated using finite element analysis.[4]

Thin-walled superconducting solenoid

A thin-walled superconducting solenoid is used to magnetize the central magnet. This solenoid was designed by the collaboration of Fermilab, Tsukuba University, and Hitachi Ltd., and built at Hitachi Ltd.[5] This coil has been in operation since 1985, and it was run continuously for about a year from summer 1988 to spring 1989.

This solenoid has a 16-mm-thick outer aluminum bobbin, which works to restrain the coil conductors from moving and quenching. It also works as a one-turn secondary conductor with respect to the coil winding, when the coil quenches. Thus, the bobbin itself rapidly heats up uniformly and eventually heats the coil uniformly, leading to safe quenching, because it avoids dangerous local heating in the coil.

The superconductor is a monolithic conductor,

with the regular NbTi superconductor embedded in pure aluminum by the EFT method (extrusion with front tension), which was developed and manufactured at Hitachi Cable Ltd. The outside dimensions of the conductor are 3.89 mm x 20.0 mm. The ratio of NbTi/Cu/Al is 1/1/21.

The coil has 1164 turns and is made of 11 long conductors. Therefore, there are 10 joints along the length of the coil. The joints were made by welding pure aluminum of two adjacent turns completely around the circumference.

Automatic field mapping device for CDF

The general mechanical features of the mapping device are shown in Fig. 1. The major components are the base, the carriage, the long cantilevered beam, the rotor, and the shuttle. To eliminate eddy current effects, the beam, rotor, and shuttle are made of carbon-fiber composite. On the shuttle are mounted the sensing elements. The whole system is designed so that the sensing elements (search coils) can be placed with a positional accuracy of 0.75 mm in space coordinates and an angular accuracy of ±1 mrad.[6]

A search coil system with three components is used to measure the magnetic field in three dimensions. The output voltages of the search coils are connected to individual integrators and the resultant voltages, which are proportional to the field changes, are sent to an ADC in a CAMAC system. To monitor the drift of the integrators DVM's with a micro-volt range were used.

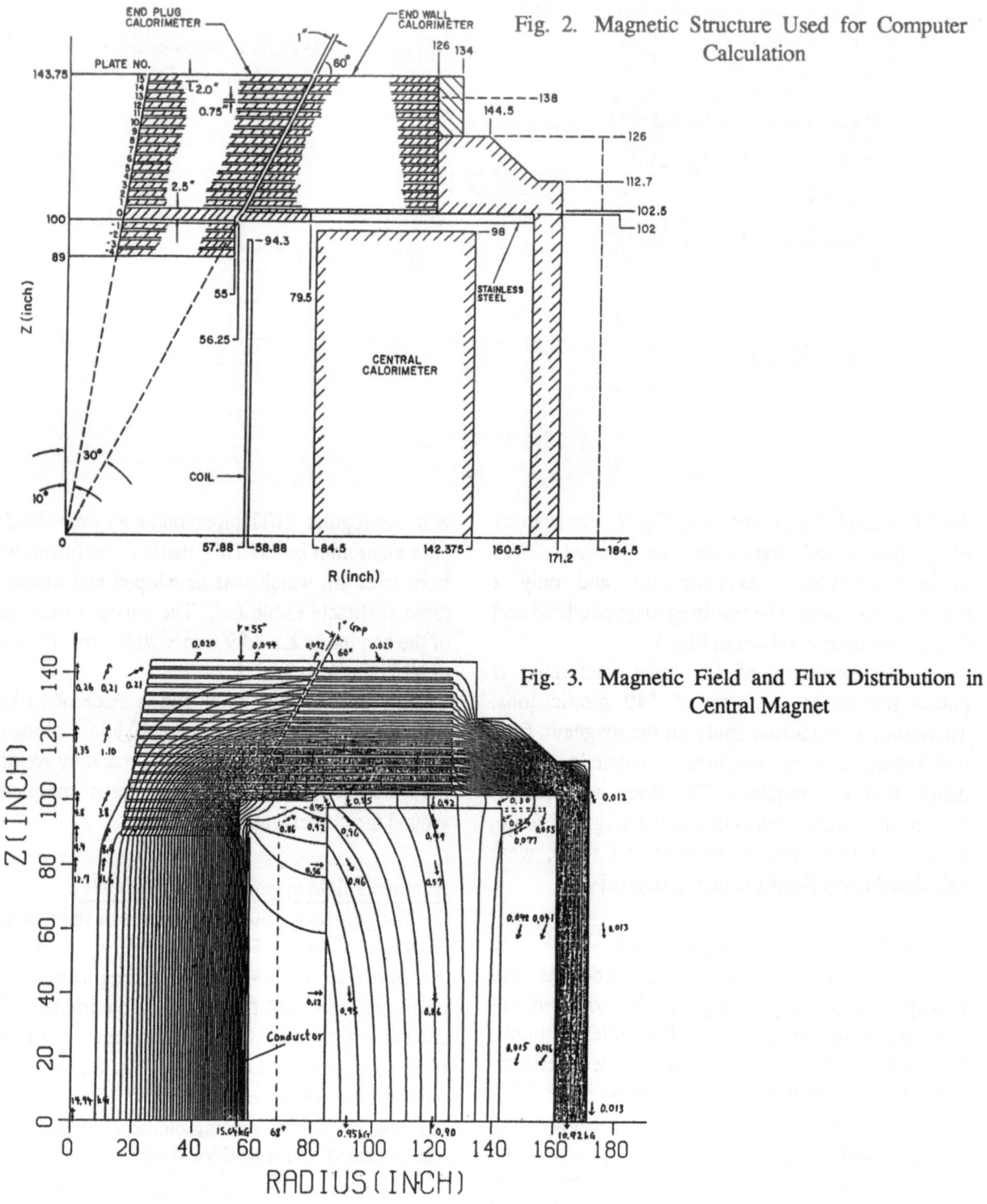

Fig. 2. Magnetic Structure Used for Computer Calculation

Fig. 3. Magnetic Field and Flux Distribution in Central Magnet

The sensitivity of the field measurement could be varied by changing the integrator time constants. In the central region, the axial fields (Bz) were measured to a precision of 0.5 Gauss. In the end-plug regions, Bz was measured with errors on the order of 14 Gauss. The radial component of the field was measured with a sensitivity of 0.5 Gauss in the central region and 4 Gauss in the endplug regions. The azimuthal component of the field was measured everywhere with a sensitivity of 0.7 Gauss.

In the central uniform field region, an NMR probe was used to get precise absolute field values with an accuracy less than 1 Gauss. In the central region, the magnitude of the magnetic field is essentially the same as the z component of the field, as the radial and azimuthal components are typically less than or on the order of 10^{-3} x Bz. During the measurement, a second NMR probe was mounted on the inside surface of the west end-plug and used to monitor the field at a fixed point for normalization purpose.

Data acquisition and manipulator motion control were handled by an IBM-PC interfaced to CAMAC via a Transiac Model 6002 crate controller. In addition, the Modulynx stepping motor controller was commanded by the PC through a CAMAC GPIB interface module. The IBM-PC was used only for the on-line data acquisition and system control. Raw data were transferred via an RS-232 serial link to a DEC VAX computer for further analysis and also for online graphical display of measurement data.

Field measurement data

For most of the measurements, the magnet was operated at its nominal current of 5000 amps corresponding to a field of 15 kG. In the inside volume of the solenoid most of the field measurement was done with the three-dimensional search coils and with a stationary NMR as a standard.

For complete field mapping, the field was measured on the boundary of a cylindrical surface and these data were then used to perform a fit based on Maxwell's equations. The fitting of the magnetic field data was performed using a procedure developed by Wind.[7] The fit coefficients were used to calculate values of the magnetic field inside the cylinder. These calculated values were then compared to measurements of the field within the cylindrical surface. The agreement was excellent. With this method the measurement time is reduced substantially because we have to measure only on the outer boundary surfaces.

The field distribution near the center of the magnet was scanned in detail with an NMR probe. The field distribution in the central region can be parameterized as follows:

$$B = 15083.86 - 7.46 \times 10^{-3} \times (z+25)^2 + 3.75 \times 10^{-3} \times r^2$$

where B is in Gauss and z are r are in cm. The B gives the magnitude of the field, but it is essentially equal to Bz. This shows the shifting of the mag-

netic center to the geometrical center by 25 cm, which may be due to the slight shifting of the coil, or slight non-uniform winding of the coil.

Non-uniformities in the field due to joints in the conductor and to the solenoid return lead were studied extensively. These joints cause local dips in the magnetic field strength near the joint. The effect of the joints is shown in Fig. 4, which was measured near the surface of the solenoid, at r=135 cm (14 cm away from the conductor) and on the axis. Similarly the return conductor, which runs perpendicularly to the winding conductors, causes local field disturbance near the surface of the solenoid.

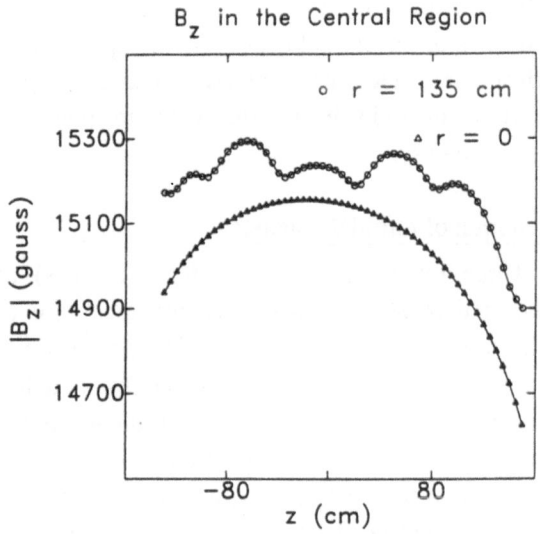

Fig. 4. Magnetic Field Distribution on Axis (r=0) and Near Coil (r=135cm)

D0 DETECTOR

A perspective view of the D0 detector is shown in Fig. 5. The overall dimensions of the D0 detector are 11.6 m wide, 12.9 m high, and 19.7 m long along the beam line. Its total weight is about 5000 metric tons. From the center, there is a set of the central tracking detectors, three uranium liquid argon calorimeters, and a set of toroidal magnets with proportional drift chambers.

Design of toroidal magnets

The muon detection system is designed to provide nearly complete coverage down to 5 degrees with momentum measurement through iron magnets. There are three large toroids: one central toroid (CF) and two end-wall toroids (EF's). In addition, there are two small toroids (SAMUS) being manufactured at Serpukhov Institute for High Energy Physics which will be mounted inside the opening of the EF's late this year. The thickness of the CF toroid is 108.6 cm and its total weight is 1973 metric tons. The thickness of each EF is 152.4 cm and each weighs 800 metric tons. Each SAMUS toroid weighs 32 metric tons. The magnetic flux distribution were calculated with the two dimensional program POISSON.[8]

The central toroid is made of three parts: one central fixed beam and two side-moving yokes. It is shown in Fig. 6, with a platform. It can be split apart during the installation of chambers. There is a 1/8th in. gap at the top and 3/32nd in. gaps at the both sides of the central beam. These gaps are filled with stainless steel plates and are provided to reduce the remnant field inside the yoke to make the separation easy.

Excitation of toroidal magnets

Three major toroids (CF and two EF's) are connected in series and excited together with the maximum current of 2500 amp. The power supply is provided with a solid state reversing switch to measure the magnetic flux density in the toroids. It takes about 10 minutes to go from positive maximum current to negative maximum current, because the coil current needs to drop below 2 amp to make sure all four SCR's are in the off state. The

SAMUS toroids will be excited with a separate power supply.

Magnetic flux distribution inside yoke

The total of 36 flux loops were used to measure the magnetic flux in the yokes of toroids. These loops are made of one-turn coils around the cross section of the yoke. By numerically integrating the output of these loops during toroid excitation, we can measure the total flux change through that loop. Absolute flux values are obtained by cycling the current from the positive maximum current to the negative maximum current and dividing the total flux change by two. The flux value in CF at zero current is about 1.7 kG and takes a long time to settle.

For data acquisition system, we used an existing Macintosh based system with a combination of a high accuracy scanner and a 6-1/2 digit DMM.[9] Due to the high accuracy but low speed mode of the system, we measured the output of each flux loop at a scanning rate of 8.3 sec/cycle. To correct its slow rate, each loop is provided with an R-C integrator at the end of an individual cable. With a 1 MΩ resistor and a 15 μF capacitor, its time constant is 15 sec.

The flux densities inside the CF and EF were measured 1.84 and 1.93 Tesla respectively at the straight part of the yokes. The flux density distribution diagonally across the corner is decreasing toward

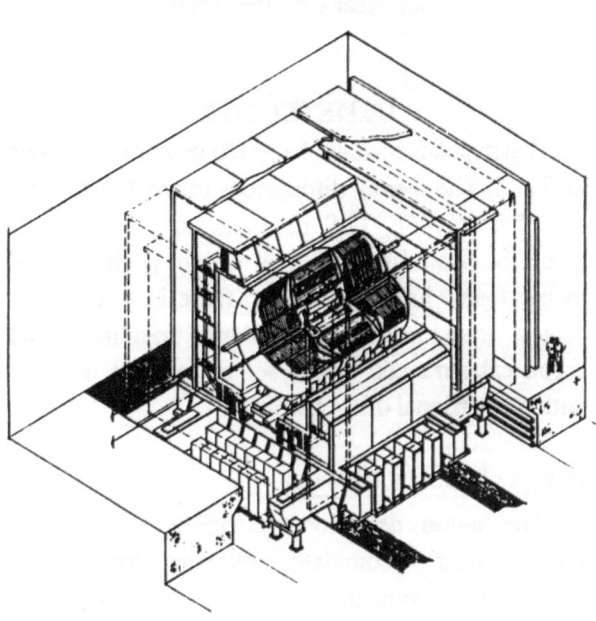

Fig. 5. Perspective View of the D0 Detector

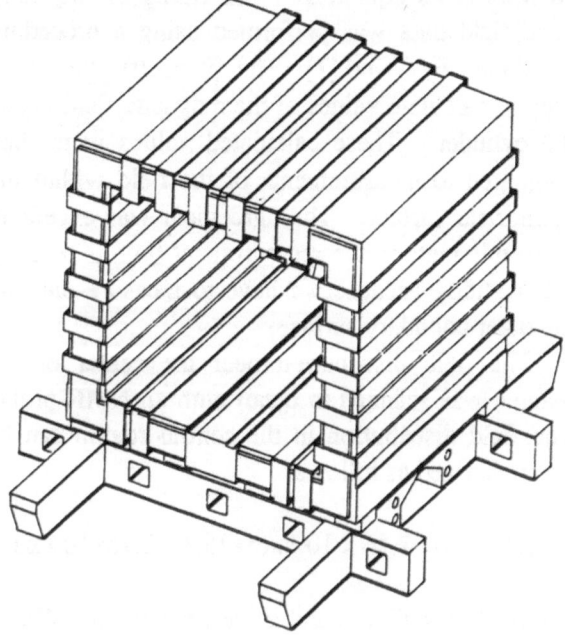

Fig. 6. CF Toroid Mounted on Platform

the outside as expected. The remnant flux density inside CF with zero current is 1.68 kG, with corresponding values in EF's are changing radially from 4 to 1 kG.

Fringing field inside CF

The fringing field inside the CF was measured with a Hall probe Gaussmeter. The magnetic field value is about 6 Gauss at the center and linearly increasing about 120 Gauss at the surface of the bottom beam at z=140 cm. These data show a similar pattern like the 2-dimensional computer calculation but differ somewhat due to the 3-dimensional structure of CF and EF.

Dipole mode operation of CF

If needed in the future, the polarity of the unit coils of CF can be rearranged to generate dipole field of about 400 Gauss inside the volume of CF with the maximum current of 2500 amp.

ACKNOWLEDGMENTS

The magnetic field fitting of the CDF detector was done by Dr. C. Newman-Holmes. The flux measurement of D0 toroids was carried out by Dr. H. Joestlein.

REFERENCES

1. Abe, F., et al., The CDF Detector: An Overview. Nucl. Instr. and Meth., 1988, **A271**, p. 387.

2. Design Report, the D0 Experiment at the Fermilab Antiproton-Proton Collider (1984).

   Brown, C., Eartly, D., Green, D., Haggerty, H., Hansen, S., Jostlein, H., Malamud, E., Martin, P., Oshima, N., Yamada, R., Marshall, T., Kunori, S., Fortner, M., Hedin, D., Kaplan, D., Kramer, T., Antipov, Yu., Baldin, B., Denisov, D., Glebov, V., and Mokhov, N., D0 Muon System with Proportional Drift Tube Chambers. Nucl. Instr. and Meth., 1989, **A279**, p. 331.

3. Yamada, R., Magnetic Field Calculation on CDF Detector (I). Fermilab TM-1162 and CDF Note No. 150, January 20, 1983.

4. Grimson, J., Kephart, R., Theriot, D., Wands, R., and Yamada, R., Magnetic Structure of CDF Central Detector. Proceedings of the 12th International Conference on High Energy Accelerators, p. 639, Fermi National Accelerator Laboratory, 1983.

5. Fast, R. W., Grimson, J., Kephart, R., Leung, E., Mapalo, L., Wands, R., Yamada, R., Minemura, H., Mori, S., Noguchi, M., Yoshizaki, R., and Kondo, K., Design Report for an Indirectly Cooled 3-m Diameter Superconducting Solenoid for the Fermilab Collider Detector Facility, Fermilab TM-1135, October 1982.

   Minemura, H., Mori, S., Noguchi, M., Yoshizaki, R., Kondo, K., Fast, R., Kephart, R., Wands, R., Yamada, R., Aihara, K., Asano, K., Kamishita, I., Kurita, I., Ogata, H., Saito, R., Suzuki, T., and Yamagiwa, T., Construction and Testing of a 3 m Diameter x 5 m Superconducting Solenoid for the Fermilab Collider Facility (CDF). Nucl. Instr. and Meth., 1985, **A238**, p. 18.

6. Yamada, R., Hawtree, J., Kaczar, K., Leverence, R., McGuire, K., Newman-Holmes, C., Schmidt, E. E., and Shallenberger, J., The CDF Field Mapping Device - "ROTOTRACK". Fermilab TM-1358 and Fermilab Internal Report CDF Note No. 345, August 1985.

   Yamada, R., Newman-Holmes, C., and Schmidt, E. E., Measurement of the Magnetic Field of the CDF Magnet. Fermilab TM-1369 and Fermilab Internal Report CDF Note No. 346. November 1985.

   Newman-Holmes, C., Schmidt, E. E., and Yamada, R., Measurement of the Magnetic Field of the CDF Magnet. Nucl. Instr. and Meth., 1989, **A274**, p. 443.

7. Wind, H., Evaluating A Magnetic Field Component From Boundary Observations Only. Nucl. Instr. and Meth., 1970, **84**, p. 117.

8. Yamada, R., Proposed Design of SAMUS Toroid and Its Magnetic Field Calculation. Femilab TM-1537 and Fermilab Internal Report D0 Note No. 717, June 9, 1988.

   Yamada, R., and Lee, P., Magnetic Field Calculation of CF Toroid. Fermilab Internal Report D0 Note No. 833, 1989.

9. Joestlein, H., D0 Toroid Field Measurements from April 19, 1989. Fermilab Internal Report D0 Note No. 846, 1989.

A FIVE POLE SUPERCONDUCTING VERTICAL WIGGLER

KAZUHITO OHMI and TATSUYA YAMAKAWA
National Laboratory for High Energy Physics, Ohho, Tsukuba, Ibaraki, 305 Japan.
TAKESHI DOI, HIDEO KAKUI and HIROSHI UKIKUSA
Ishikawajima-Harima Heavy Industries Co.,Ltd, Chiyoda, Tokyo, 100 Japan.

ABSTRACT

A superconducting vertical wiggler was installed into the KEK-PF ring in Aug. 1989. The wiggler has 5 magnetic poles and is operated in the permanent current mode with the field strength of 5 Tesla. We discuss the magnet and cryostat of the wiggler.

INTRODUCTION

A five pole superconducting vertical wiggler[1] has been constructed since Aug. 1987. A characteristics and schematic drawing of the wiggler are shown in Table 1 and Fig.1, respectively. The wiggler has 5 magnetic poles and is operated in the permanent current mode by using superconducting switches with the field strength of 5 Tesla.

The wiggler was installed into KEK PF ring at Aug. 1989 and will be used as a generator of a synchrotron radiation.

TABLE 1   Characteristics of the wiggler

Maximum field strength on the beam orbit	5 Tesla
Magnet gap	66 mm
Number of magnetic poles	5 poles
	arranged every 200mm
Rated exciting current	220 A at 5 Tesla
Superconducting wire	NbTi : Cu 1 : 1
	size 1.70 X 0.85 mm^2
	number of turn 2520
Liquid helium consumption in the permanent current mode 0.1 L/h	
Life time of the permanent current	3×10^4 hours
Inductance	1.31H/coil

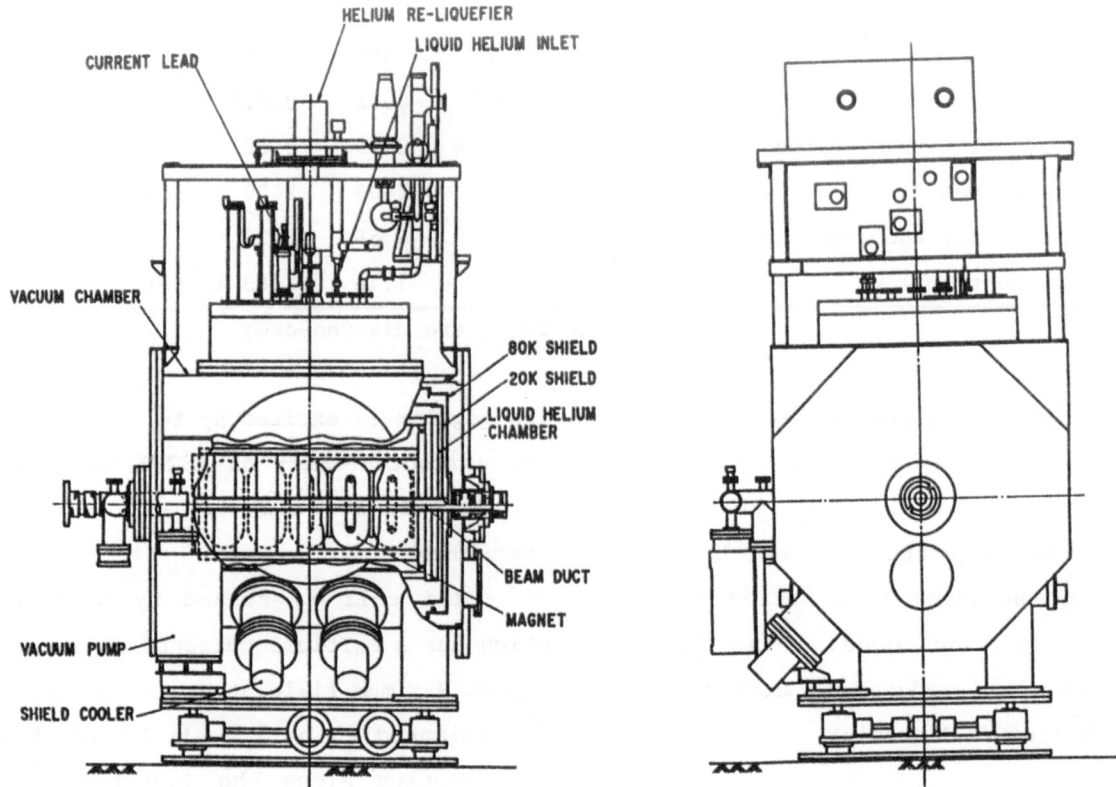

Fig.1 Side and front view of the superconducting wiggler.

CRYOSTAT

The wiggler cryostat includes two layer of thermal shields located between the vacuum chamber and helium vessel as shown in Fig.1. The outer (80K) shield is supported by SUS rods from the vacuum chamber and helium vessel is from the 80K shield, respectively. The inner (20K) shield is supported by G10 rods from 80K shield and performed a thermal anchor by mean of contact to the rods at the middle point between 80K shield and helium vessel. The thermal shields are cooled by two refrigerators. The refrigerator includes two heat stations, where the first station has a refrigeration capacity of 60W at 80K and the second station of 6W at 20K. The outer and inner shields are connected to the first and second station, respectively. The helium vessel is connected to a third refrigerator (2.5W at 4.2K) which re-liquefies the evaporated helium.

The designed heat load into each shield is shown in TABLE 2. Since most of the heat load comes from current leads, the current leads are pull up from the helium vessel in the permanent current mode. As a result, the total heat load into the helium vessel is reduced to 0.59W from 2.63W. When the exciting current is supplied, the current leads are connected to the magnet in the helium

TABLE 2　　Heat load

	80K	20K	4.2K
radiation	11.37	0.62	0.001
conductivity through the support rod	16.49	2.19	0.30
current lead		2.09	2.20 (0.16)*
helium inlet and instrument		2.56	0.13
total heat inflow		7.46	2.63 (0.59)

*　　(): current leads are disconnected

vessel and superconducting switches are warmed up by the heater of 2Wx4=8W.

The heat load into the helium vessel was estimated to be 1.4W by measuring the evaporated helium in the permanent current mode. The helium consumption was reduced to 0.04-0.25L/h by the helium re-liquefier.

DESIGN OF THE WIGGLER MAGNET

The wiggler magnet consists of 5 pairs of superconducting coils with iron poles as shown in Fig.2. Each coil is turned around the iron pole of 40mm in width and 260mm in height and is arranged every 200mm. A width of the beam duct is determined as 45 mm by taking into account of the beam size during the injection into the storage ring. The gap of magnet is chosen to be 66 mm, since a thermal shield of 80K is put between the magnet and the beam duct. The three inner pairs produce the maximum field strength of 5 Tesla at the center of gap and the two outer pairs are used to be canceled the field integral along the centerline of the magnet.

The 3 dimensional field calculation is done by using a program code JMAG[2].

The wiggler is excited up to 5 Tesla with the current density of 122 A/mm^2, where the cross section of the coil is 65mmx70mm.

The coil is turned by NbTi wire which has a critical current[3] as shown in Fig.3. Maximum field strength on the coil is estimated to be 7 Tesla by the field distribution along the vertical center line through the coil as shown in Fig.4. Since the superconducting wire contains NbTi and copper equally, the current density in the NbTi is 313A/mm^2. The maximum field strength by the current density is plotted in Fig.3 and is less than 80% of critical field.

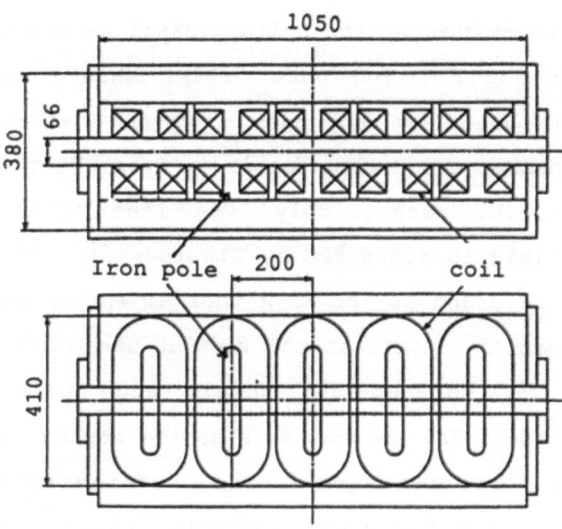

Fig.2 Magnet of superconducting wiggler.

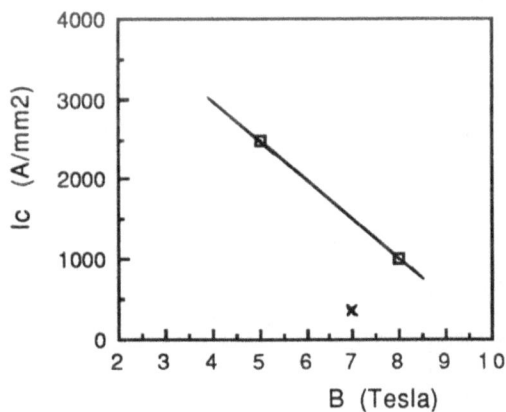

Fig.3 Critical current curve
x:operating point.

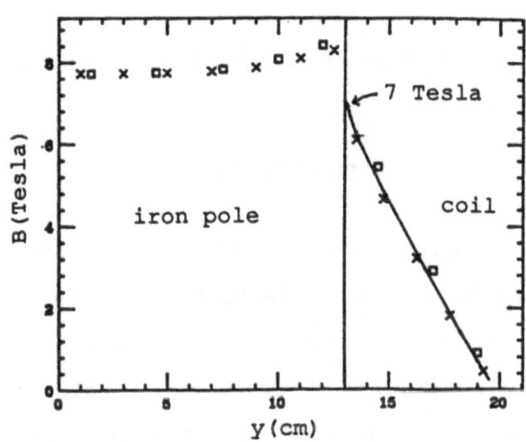

Fig.4 Maximum field strength in the coil.
Cross:single coil model with periodic
boundary condition. Square:5 pole model.

A voltage and heat induced in the
coil at quench is estimated by the
program code QAC[4] in the permanent
current mode and normal mode.   The

voltage and temperature are not serious
for the coils as shown in Table 3.

MEASUREMENT OF MAGNETIC FIELD

Some training was done to achieve the
rated current of 220 A. The progress of
the training is shown in Fig.5. The field
strength at the center of the wiggler
magnet was achieved 5.01 Tesla with an
exciting current of 223.6A.

The magnetic field along the
centerline of the wiggler magnet is
measured by using a calibrated hole
probe. The field distribution by the
measurement is shown in Fig.6 and is
compared with the calculation in the
figure.

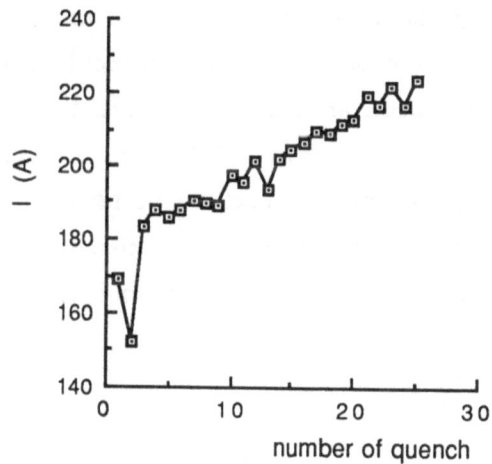

Fig.5 Training of the wiggler magnet.

TABLE 3   Specificity of quench.

Mode	Inner coil normal	Inner coil permanent	Outer coil normal	Outer coil permanent
Max. temp.(K)	92.4	77.7	37.2	42.9
Max. volt.(kV)	0.46	0.34	0.195	0.078
Released energy(kJ)	183	183	19.5	19.5

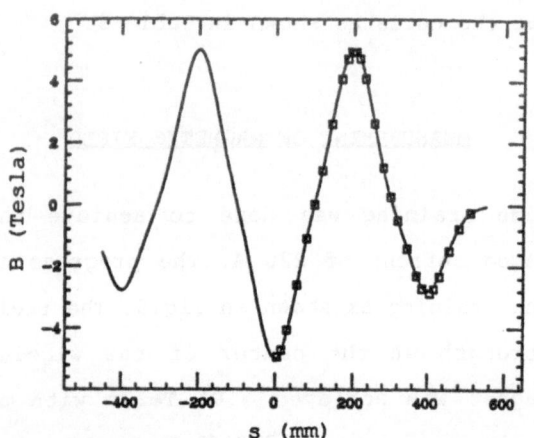

Fig.6 Magnetic field distribution along the centerline of the wiggler. Solid line and squared was given by the measurement and the calculation, respectively.

The current gradually decreases in the permanent current mode by the resistance of junctions between the coils. The life time was estimated to be 5.1×10^4 hours by the resistance of a test sample. The damping of the magnetic field was measured as shown in Fig.7. The mean life time of all coils is 3.01×10^4 hours. An orbit distortion of PF ring due to the current decrease is corrected by normal steering magnets. The current will

be restored every 300 hour, during which the decrease of the current is 1%.

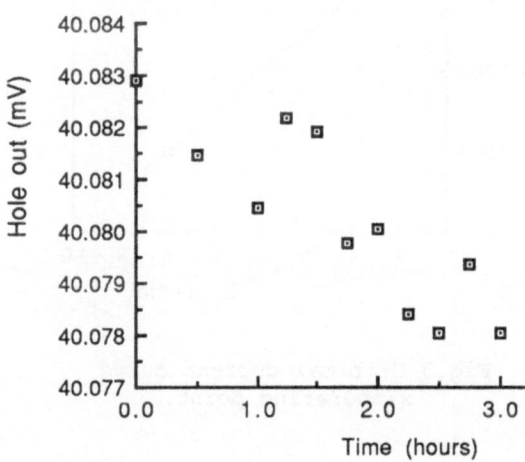

Fig.7 Decrease of the permanent current.

CONCLUSION

The wiggler could produce the field strength of 5.01 Tesla by the training of the coils and the helium consumption could be achieved the expected value of 0.1L/h. The decrease of the permanent current will have no influence on the operation of the PF ring. The wiggler will be in user operation since Oct. 1989.

REFERENCE

1. A three pole wiggler had been operated since 1983. The wiggler was replaced by the five pole wiggler.
    T.Yamakawa et al., NIM A246(1986)32-36.
2. developed by Japan Information Service.
3. date from Sumitomo Electric Industries.
4. Quench Analysis Code developed by IHI is based on Wilson's theory.
    M.Wilson, RHEL, Report RHEL/M151.

DESIGN OF SUPERCONDUCTING WIGGLER FOR A SYNCHROTRON RADIATION FACILITY

K.Aizawa, M.Sakiyama, M.Yokoyama, A.Iwata

Kawasaki Heavy Industries Ltd.
Kawasaki-cho, Akashi 673, Japan

and

I.Endo, T.Kasuga, M.Taniguchi, M.Tobiyama, M.Nomura

Hiroshima University
Higashisenda-machi, Naka-ku, Hiroshima 730, Japan

ABSTRACT

In a synchrotron radiation facility with an intermediate eletron beam energy of about 1.5GeV, hard X-ray is provided with the aid of a super-conducting wiggler which generates a magnetic field of more than 4T. The superconducting wiggler is installed into an electron storage ring, and its influence to life time of the electron beam is the most important problem. In this paper, designing method of 3-pole superconducting hori-zontal wiggler is discussed by taking account of bending angle, maximum excursion of the electron beam, sextupole component of wiggler field and superconducting characteristics. The typical design of wiggler coils and the whole wiggler system including refrigerator are also reported.

INTRODUCTION

Synchrotron radiation(SR) has many superior characteristics com-pared with other conventional light sources, and it has made rapid progress in science and industrial applications. With the extension of applications using SR, needs for SR are raised, espe-cially for generating X-ray range wavelength. In a multipurpose syn-chrotron radiation facilities with intermediate electron beam energy, the superconducting wiggler is a crucial component for obtaining the X-ray range wavelength, which must be designed more than 4T of magnetic field. As the wiggler is inserted into the electron storage ring, its influence to life time of the electron beam is the most important problem. Easy operation and free maintenance are also re-quired to the wiggler system.

Taking a case of the electron storage ring under planning at Hiroshima University(HiSOR:1.5GeV, 300mA), we describe a designing method and the optimum specifica-tions of a 3-pole superconducing wiggler in this paper.

DESIGNING CONDITION

HiSOR storage ring

TABLE 1 shows main parameters of HiSOR storage ring.[1]

TABLE 1. Parameters of the storage ring

Beam energy	1.5 GeV
Beam current	300 mA
Lattice function	Double forcusing achromat type (Chassman Green)
Dipole magnetic field	1.2 T
Injection system	Full energy injection from 1.5GeV electron cychrotron
Circuit length	100.17 m

FIGURE 1 exhibits structure of lattice and expected values of C.O.D.(Closed Orbit Distortion) with an empirical safety factor.

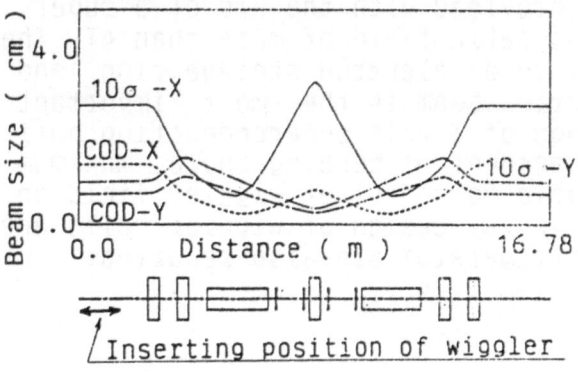

FIGURE 1 Structure of lattice and expected values of C.O.D.

The C.O.D. values and beam sizes are calculated under conditions shown in TABLE 2.

From FIG.1, maximum value of the beam width is about 24mm, so that height of beam duct needs more than 25mm. Considering heat insulational spaces and supporting structures against magnetic force, we assume the gap between upper and lower coil to be 56mm. Width of the beam duct is assumed to be 110mm in consideration of the beam width (10σ) and the value of C.O.D.-X.

Magnetic field strength of wiggler magnet

FIGURE 2 shows spectral brilliance of HiSOR. Requirement for the brilliance is more than 5×10^{11} photons/sec/mrad^2/mm^2 0.1% band width at photon energy of 35keV. Hence the necessary strength of magnetic field is more than 4.4T(see FIG.2).

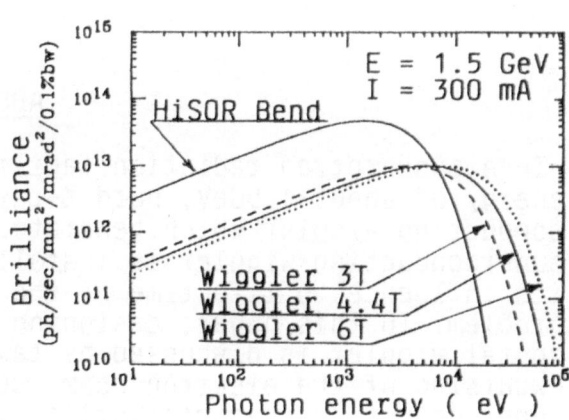

FIGURE 2 Spectral brilliance

Beam dynamics at the wiggler inserted into the storage ring

We executed the beam tracking in the case that the wiggler is inserted into the storage ring.

TABLE 2. Calculating conditions

The discrepancy of quadrupole magnetic field	$\Delta X = \Delta Y = 0.1$ mm
The magnetic field error of dipole magnet	0.1 %
The rotational error of dipole magnet	0.2 mrad
The permissible limit of the momentum gap	$\Delta P/P = 1$ %
X - Y coupling constant	10 %

and checked influences to the beam dynamics. As a result, we have found that beta functions are influenced by the magnetic field of wiggler, but they are able to be readjusted by the quadrupole magnets placed at both ends of the wiggler. Accordingly, we adopt the size of beam duct assumed above as one of the HiSOR's wiggler.

Influence to beam dynamics by sextupole component

As the magnetic field of wiggler is very high, its sextupole component of the field is expected to affect beam dynamics significantly. With various values of the sextupole component, we calculated dynamic aperture at injection point by beam tracking. Calculated result is shown in FIG.3.

mated to be about 25% of wire's guarantee.

TABLE 3 Specifications of wire

Wire	Formale covered monolith rectangular sectional wire
Sectional size of wire	0.6mm × 1.2mm
Copper ratio	0.9 ± 0.1
Filament diameter	20 μm
Twist pitch	25± 5 mm
R.R.R.	> 100

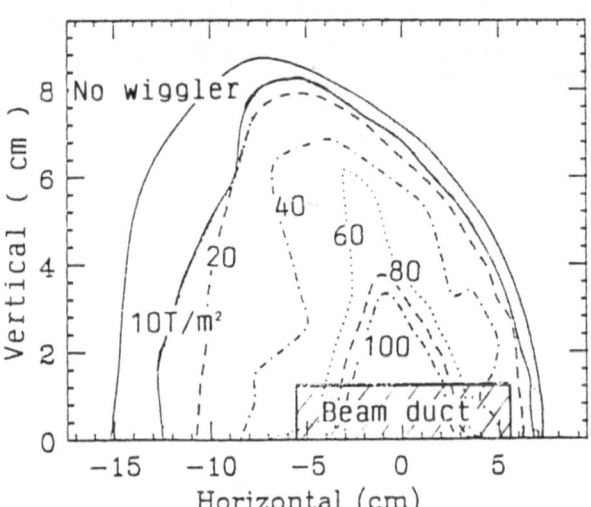

FIG.3 Dynamic aperture at injection point

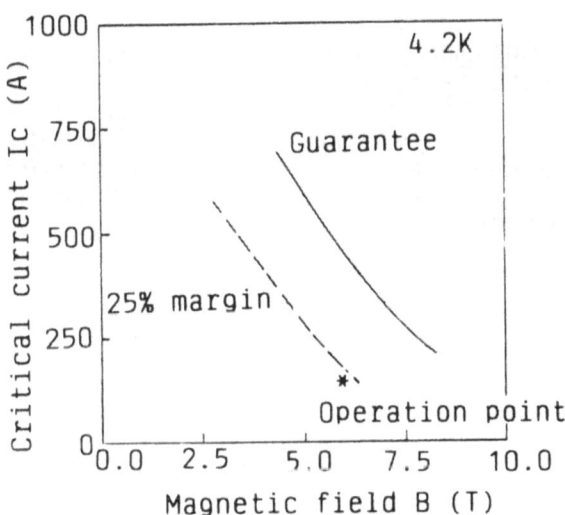

FIG.4 Ic-B characteristics of Nb-Ti wire

It is necessary that the dynamic aperture is much larger than the beam duct for stable rotation of electron beam in the storage ring. In our case of 110mm^W × 25mm^H beam duct, the sextupole component of wiggler field has to be less than 10T/m^2, from FIG.3.

Superconducting characteristics of wire

TABLE 3 represents a typical specifications of Nb-Ti wire. And FIG.4 shows Ic-B characteristics of this wire.

Because of packing factor and degradetion by winding, necessary superconducting margin is esti-

DESIGNING METHOD

TABLE 4 shows the conditions of designing a superconducting wiggler for HiSOR.

The 3-pole wiggler is composed of one pair of main coils and two pairs of auxiliary coils. Electron beam is bent by bending angles of ψ, -2ψ and ψ on passing through the first auxiliary coils, the main coils and the last auxiliary coils, respectively. We examine the coil parameters of the wiggler magnet under the above-mentioned designing conditions.

TABLE 4 Designing conditions

Size of beam duct	110mm^w × 25mm^H
Gap between upper and lower coil	56 mm
Muximum magnetic field	4.4 T
Sextupole magnetic component	< 10 T/m²
Superconducing characteristics of wire	75% of guarantee

Optimum positions of main and auxiliary coils

For calculating optimum positions, a simplified two dimensional model is adopted taking into account the symmetrization of magnetic field. (See FIG.5.)

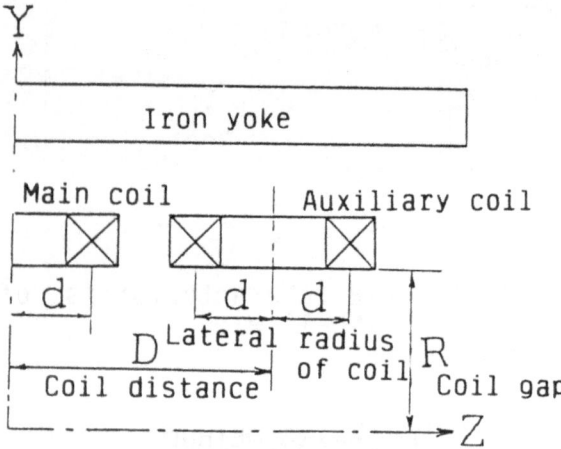

FIGURE 5 Two-dimensional calculation model

On the assumption that the iron yoke is magnetic charge which is distributed uniformly on a surface of belt-shaped plate, we can calculate the magnetic field distribution of wiggler by making use of analytical solutions. With this method, the calculation time of our program is shorter than other methods like F.E.M. So that, it is easy to calculate various input conditions. Compared this program with the program "POISSON", we confirm that this program has sufficient accuracy.

Following the calculated results, in FIG.6 is drawn the relation between D and d with the superconducting mergin as a parameter. From above results, we found that that it is possible to make coils of the wiggler for HiSOR with Nb-Ti wire which has 25% of supercon-ducting margin. The optimum relation between D and d exists due to mutual interfering between main and auxiliary coils.

Taking account of unequality of magnetic field, influence of multi-pole component of field and so on, the compact wiggler is better one. Therefore, we choose the relation of D=150mm and d=40mm as the optimum one. (* marks in FIG.7) At this condition, the sectional size of main coil is 44 mm× 44mm.

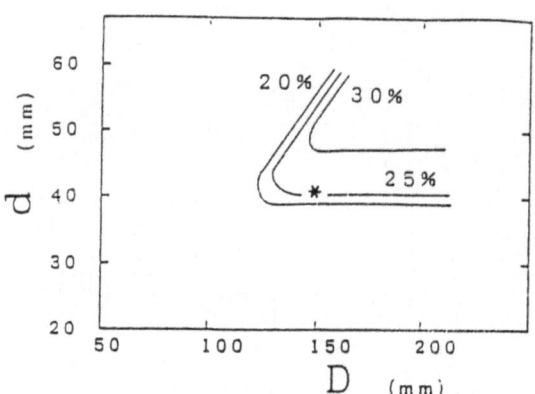

FIGURE 6 Relation between D and d

Longitudinal length of coil

For the purpose of appraising sextupole magnetic component on Z-axis of wiggler, we set up a simplified model assuming the race-track formed coil with finite area of section to rectangular single wired coil. FIG.7 shows a calculating model for sextupole component.

Strength of the sextupole component is calculated under the determined coil parameters. FIG.8 shows calculated distributions of sextupole component along Z-axis with various longitudinal length of coil. It follows that the longitudinal length of coil is necessary to be more than 150mm in order to satisfy the requirement about the sextupole component (within $10T/m^2$).

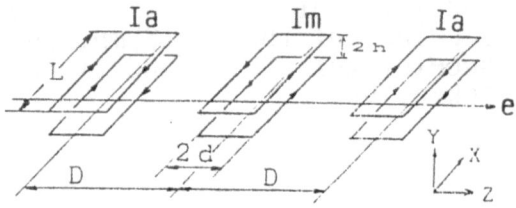

FIGURE 7 Calculating model for sextupole component

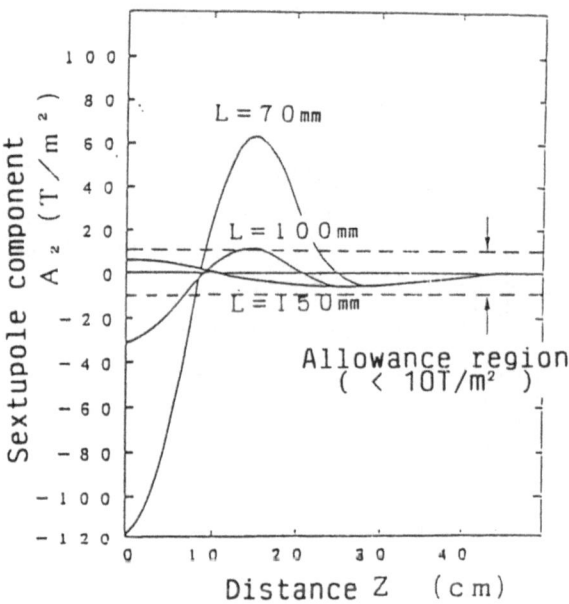

FIGURE 8 Strength of sextupole compnent

Specification of wiggler's magnet coils

We determined the specifications of wiggler magnet coils for HiSOR by two stages of magnetic field calculation under the above-mentioned designing conditions. TABLE 5 represents specifications of coils.

Whole system of the wiggler

FIGURE 9 shows organization and flow diagram of the wiggler's whole system. In this system, it can be continuously operated more than 6 months and maintenance personnels need not reside on site every time during the operation. This system is a closed one except cooling down and filling up inner vessel with liquid helium at first stage. Vaporized helium is automatically recondensed and stored. We are under consideration to magnetize the coils by electric current on permanent mode at the case of continuous operation.

Specification of organized equipments are shown in TABLE 6.

CONCLUSION

In this paper, we described a method of designing a wiggler which is a inserted light source for synchrotron radiation facilities, and showed specifications of trial designing of wiggler for HiSOR.

We will make a superconducting wiggler based on the specifications of this design. Measuring

TABLE 5 Specifications of coils

	Main coil	Auxiliary coil
Strength of magnetic field	4.4 T	2.2 T
Density of electron current	4.4×10^5 A Turn	2.2×10^5 A Turn
The gap between centers of upper and lower coils	100 mm	88 mm
Outer size of coil	124mm× 274m× 44mm	100mm× 250m× 32mm
Inner size of coil	36mm× 186m× 44mm	36mm× 186m× 32mm
Distance between centers of main and auxiliary coil	150 mm	

the distribution of its magnetic field, we will confirm the accuracy of calculation by simplified model, and justify the method of designing a wiggler. On the other hand, we are investigating also a wiggler having sector formed magnets,taking into account beam dynamics and its performances. We may report this sector type wiggler in future

REFERENCE

1. Endo,I.,Tobiyama,M. and Ohta, T., Conceptual Design of Hi-SOR, the Synchrotron Radiation Facility at Hiroshima. Proceedings of the 6th Symp. on Accelerator Sci. and tech., Tokyo,1981.

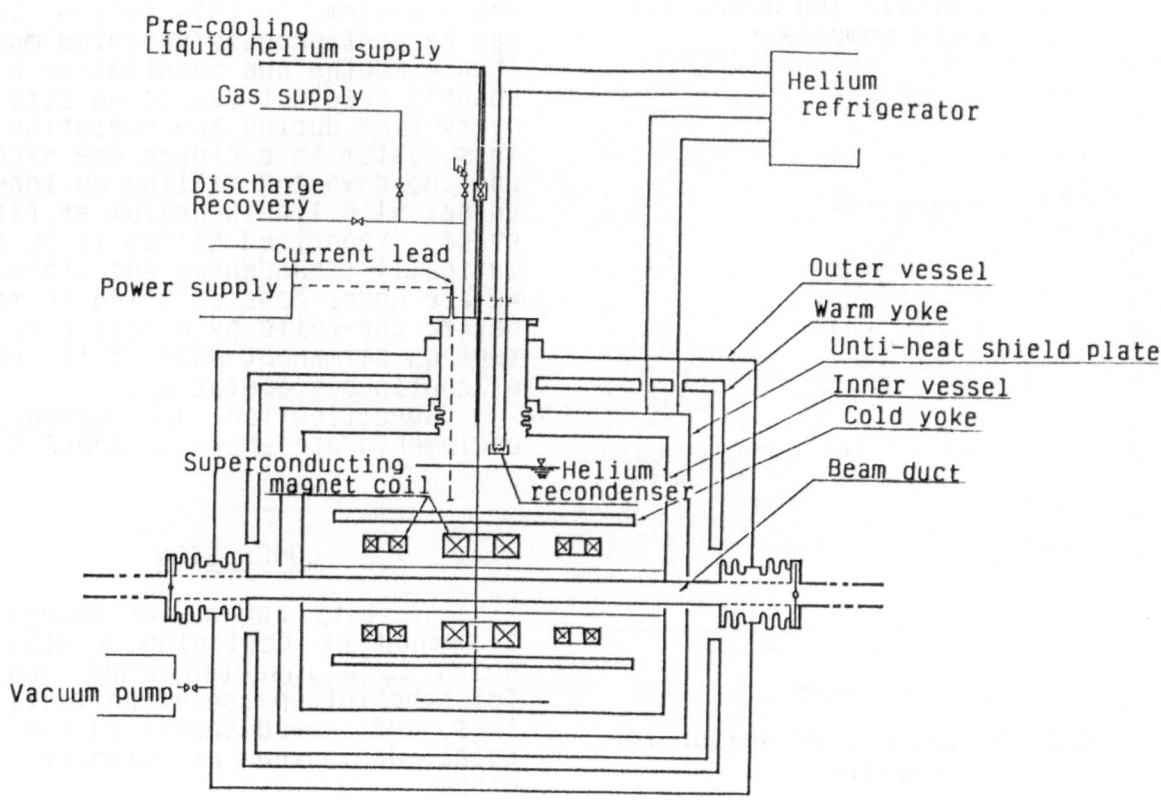

Fig.9 The whole system of the wiggler

TABLE 6 Specification of
organized equipments

Helium refrigerator	Stirling cycle with J.T. valve 10 W at 4.2K
Power supply for coils	1 set of DC 5V× 200A (Main) 1 set of DC 5V× 40A (Aux.) drift < ± 1× 10^{-4} nipple < ± 2× 10^{-4}
Vacuum pump	1 set of turbo-molecular pump 160 ℓ /s, 1× 10^{-8} Torr 1 set of ion pump 500 ℓ /s, 1× 10^{-11} Torr 1 set of rotary pump 180 ℓ /min, 5× 10^{-2} Torr

A SUPERCONDUCTING WIGGLER MAGNET FOR AN ELECTRON BEAM OF 120 MeV

Shigehiro OWAKI, Kazumune KATAGIRI, Shigehiro NISHIJIMA,
Tetsuya NISHIURA, Kazuya TAKAHATA, Kazutaka SEO and Toichi OKADA

I.S.I.R., Osaka University
Mihogaoka 8-1, Ibaraki, Osaka 567, Japan

ABSTRACT

According to a plan to generate synchrotron radiation in a VUV region using a 120 MeV electron beam, a superconducting wiggler magnet was designed considering minimization of the size and the cost. The wiggler magnet is a three pole vertical deflecting device and each dipole magnet forms racetrack shape with a wire of Nb-Ti alloy. The design aimed at formation of the highest central field of 5.0 T and both side peaks of 2.5 T on the electron beam axis. The magnet period is 190 mm and the deflection distance of the electron beam extended to 36 mm in spite of the design efforts to minimize their size because of the low energy of electron and the high field. The design concept and the result of its cooling down test are described here.

INTRODUCTION

In large scale synchrotron radiation (SR) facilities, SR as a high intensity X-ray source is generated from electrons or positrons accelerated to an energy of a few to several GeV, while SR generation by sharp bending low energy electrons is desirable in a small facility for factory use. Even if the energy of electron is nearly 100 MeV, the combination with a high field wiggler magnet can provide SR covering the region of VUV, e.g. a critical wavelength of SR reaches a few hundred angstrom, when an electron beam of the energy higher than 100 MeV and a superconducting (SC) wiggler magnet of the central field higher than a few T are combined.

In our Institute, a new linear accelerator is under construction and will provide an electron beam of 120 MeV (peak current 100 mA in a 1.0 s pulse) in this autumn. Basic concept of SC wiggler mag- nets for compact VUV SR facilities using the electron energy of this order was discussed, and trial manufacture of the magnets was performed. These results are described in this paper.

MAGNET DESIGN CONCEPT

The goal of highest central field of the wiggler magnet was determined to be 5 T, considering the state of the art of SC technology and the economy [1] - [4]. Therefore, as the critical wavelength λc of SR, 260 Å is expected by a equation [5],

$$\lambda c = 18.64/ E^2 B \text{ ------------ (1),}$$

where E and B is electron energy (GeV) and magnetic field (T) respectively. The peak light power of the SR is estimated to be 160 mW per single electron pulse of μs width by a equation [5],

$$P = 1.267 \times 10^3 E^2 <B^2> I L \text{ --- (2),}$$

where <B> is average field in the electron orbit, I is peak current and L is path length of electron pulse in the field.

For simplification and reliable operation the overall structure of the wiggler magnet was determined to consist of three pole vertically deflecting device packed in a pure iron yoke [6]-[8]. Each dipole magnet is wound in a racetrack shape with commercially available SC wire of Nb-Ti alloy and impregnated with epoxy resin.

The factors to determine the size of each magnet and the arrangement are as follows: 1) The highest central field of 5 T and the compensating field of -2.5 T at both sides are formed for an electron trajectory of single period to be realized. 2) Magnet size, of course, must be minimized. 3) Two power sources can supply DC current of maximum 200 A for main dipole and both end dipole magnets. 4) The pole gap should be about 60 mm because the inner size of beam duct requires 30 mm.

Wiggler field distributions along the electron beam axis were calculated by changing cross sectional size of each coil and the arrangement. Of course, the field path integral should be close to zero along the electron beam axis. Letting current density of SC wire of 3.2×10^8 A / m^2 (packing factor of 90 %), the designed field distribution was obtained under minimum size and stable configuration of the magnet. An electron trajectory with this field distribution was calculated and these results are shown in FIG. 1.

As shown in FIG. 1, a low energy electron beam is deflected so long (deflection distance of 36 mm), in spite of decrease of a distance of pole to pole (magnet period) as short as the minimum limit of practical sizes of the coils. This requires long uniform field along a long axis of the central racetrack magnet and resulted in the length of the magnet of 350 mm.

The designed specifications of the wiggler magnet with a SC wire are listed in TABLE 1. Three magnets were manufactured for the main pole magnet, and 5 magnets for the side ones. After SC

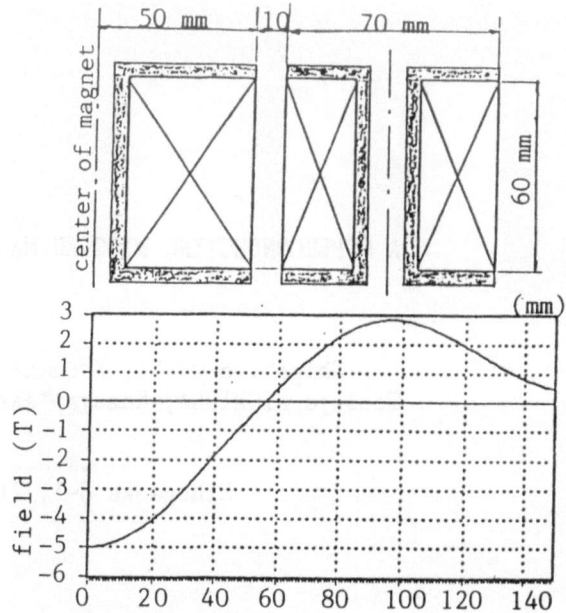

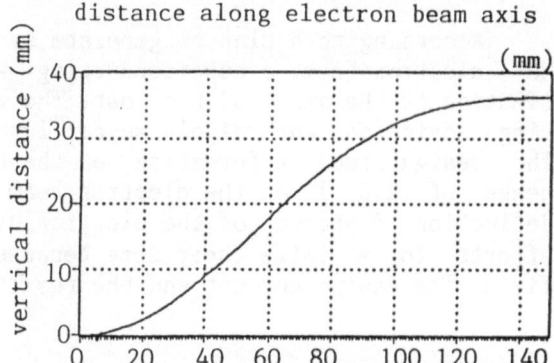

distance along electron beam axis

FIGURE 1. Arrangement of coils, expected field and electron trajectory with it.

characteristics of each magnet were individually tested, better ones were selected and assembled in a pure iron yoke. Training is performed both before and after the assembly.

The 2-dimensional field distribution of the wiggler magnet with an iron yoke and the electromagnetic force loaded there were calculated with a computer program. The results are shown in FIG. 2 and suggest a few problems, that is, the winding inside of the coils encounters a high field of nearly 7 T, and fringe fields may disturb the low energy electron beam. Moreover, we have to note the iron yoke is loaded with a large electromagnetic force at 4.2 K.

TABLE 1
Magnets Specification

specifications	main magnet	side magnet
inner diameter(mm)	20	20
outer diameter(mm)	100	70
height (mm)	60	60
length (mm)	350	270
cross section (mm^2)	2405	1496
number of turns	4321	2682
packing ratio (%)	90	90
average current density (A/m^2)	3.2×10^8	3.2×10^8
supply current (A)	179	179
central magnetic field(T)	5.0	2.5
wire length (m)	2978	1452
weight (kg)	13.4	6.5

wire specifications;

type	monolith, round wire
materials	Nb-Ti
insulator	formvar
conductor diameter(mm)	0.748 ± 0.01
total diameter (mm)	0.798 ± 0.01
filament diameter (μm)	30 ± 5
twist pitch (mm)	38 ± 5
Cu/SC ratio	1.0
typical critical current	10^{-11} (Ω cm) at 4.2 K ≥ 220 (A) / 8 (T)
residual resistance ratio	≥ 120

MANUFACTURE

The SC wire was wound on to a magnet bobbin of stainless steel (SUS 304 of 5 mm thick) with a winding tension of 44 N/mm^2. After that, each one was impregnated with epoxy resin in a vacuum. As for each magnet tightly packed from both sides by stainless plates, training tests were performed. Quench current of the magnets reaches nearly 85 A for main magnets and 120 A for side magnets after training of several excitations. The instability is supposed to be attributed to wire motion at the straight portion because the main and side magnets are different in the length of straight portion and the number of wire turns.

Two main and four side magnets were assembled in a yoke, which was shaped from pure iron blocks. This wiggler magnet shown in FIG. 3 is suspended from the top flange and the weight is partly supported at the bottom of the magnet vessel in a cryostat.

electron beam axis

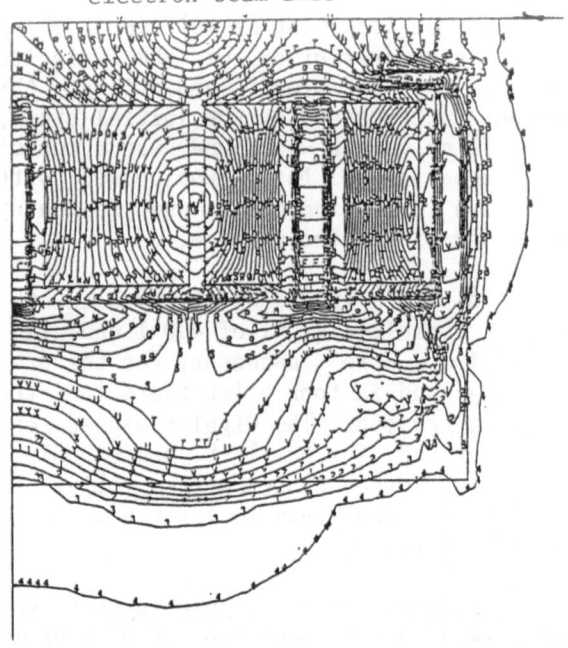

	contour and level (T)	
G ; 7.5965	O ; 5.0644	W ; 2.5322
H ; 7.2800	P ; 4.7478	x ; 2.2157
I ; 6.9635	Q ; 4.4313	Y ; 1.8991
J ; 6.5470	R ; 4.1148	z ; 1.5826
K ; 6.3305	S ; 3.7982	1 ; 1.2661
L ; 6.0139	T ; 3.4817	2 ; 0.9495
M ; 5.6974	U ; 3.1652	3 ; 0.6330
N ; 5.3809	V ; 2.8487	4 ; 0.3165

FIGURE 2. Calculated field distribution.

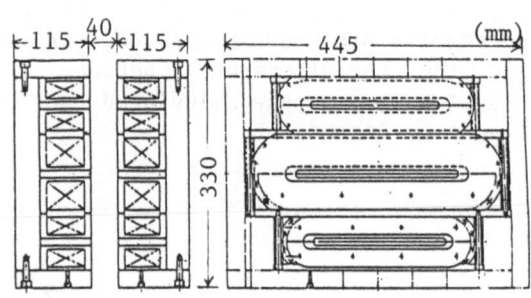

FIGURE 3. Configuration of wiggler magnet with an iron yoke.

A cryostat was designed and manufactured for the wiggler magnet. The detail will be described elsewhere.

EXPERIMENT

Vacuum leak test as well as cooling down characteristics of the wiggler system was performed, assembling the magnet in cryostat. At the initial stage, a small cold leak was found near a top of the magnet vessel but it was solved with addition of indium gasket and fastening pressure of the bolts there.

To store liquid He of 160 l in the magnet vessel, liq. N_2 of 400 l and liq. He of 150 l were necessary for the pre-cooling and it took a few hours for the liq. He transfer. The wiggler magnet can be operated for nearly 8 hours unless quenching of SC state happens. Liq. He evaporated less than 10 l per hour in a static state.

Training of the wiggler system was performed under connection of 6 magnets in series. From the result shown in FIG. 4, the increase of quench currents is not so much but training effect still continues. The present maximum current of 85 A, which produces the maximum field of 3 T, is about 55 % of the design value of 5 T.

The maximum field on the central axis in the beam duct was measured at each current by a magnetic flux meter of Hall element and the result is shown in FIG. 5. The field enhancement with the iron yoke is found from the result. Also the field distribution along the central axis in the beam duct were observed under the connection of 6 magnets in series and the results are shown in FIG. 6. It is found that the discrepancy between the measured value and the calculated one is not so significant.

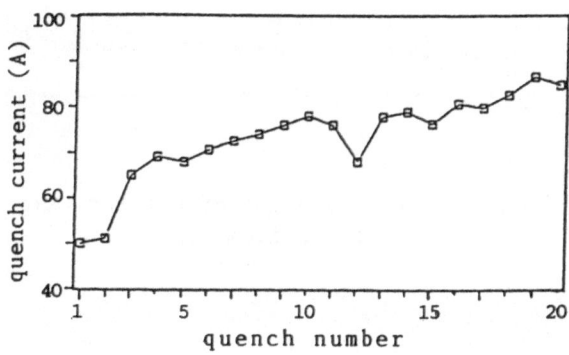

FIGURE 4.   Training of wiggler magnet.

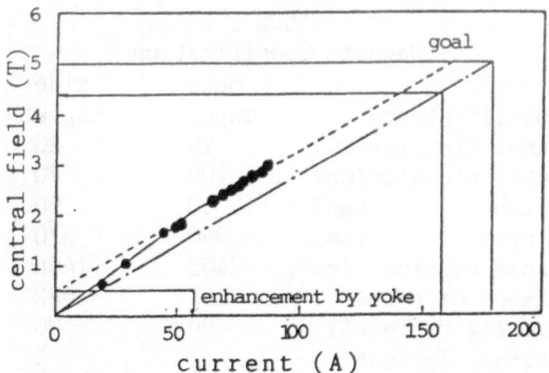

FIGURE 5. Observed **maximum** field on the central axis in beam duct vs. magnet current.

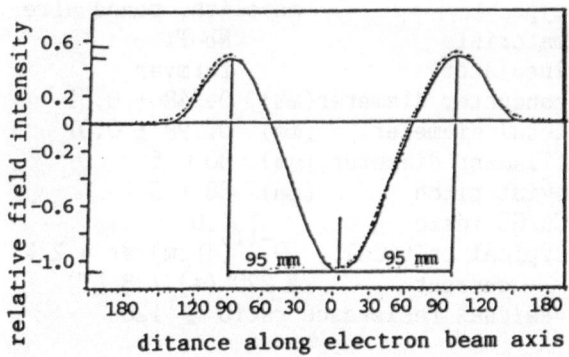

FIGURE 6. Observed field distribution along the central axis in beam duct.

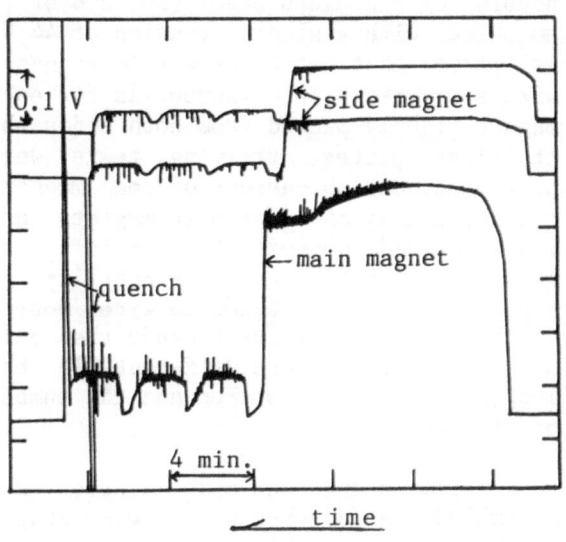

FIGURE 7. Voltage generation from magnets immediately before quench.

Voltage generation from each magnet was picked up and recorded during the training experiments. A typical example of the recording chart immediately before quenching is shown in FIG. 7. The voltage spikes are always generated from the main magnet and the induced pulses of opposite polarity are picked up from the side magnets. It is clear that the quenching originates from the main magnet.

The vertical uniformity of the magnets was observed at three peak fields on the electron beam axis and the results are shown in FIG. 8. The vertical uniformity for 40 mm long is maintained within a deviation of 0.5 % at the central and both side peaks. This results are very important for the case of electron beam of a long deflection distance.

Fringe fields around the cryostat were also measured and any value did not exceed a several mT. The fringe field seems to be too small to disturb the electron beam so much at the entrance and exit even if the absolute values increase to a few times with increase of the magnet current.

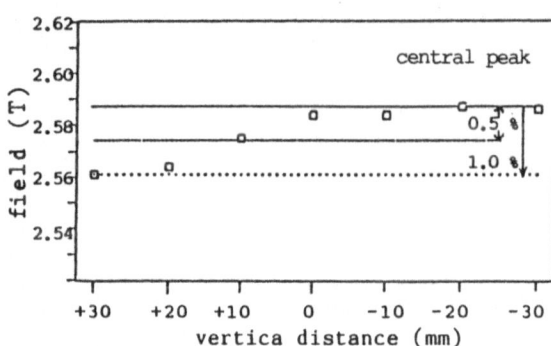

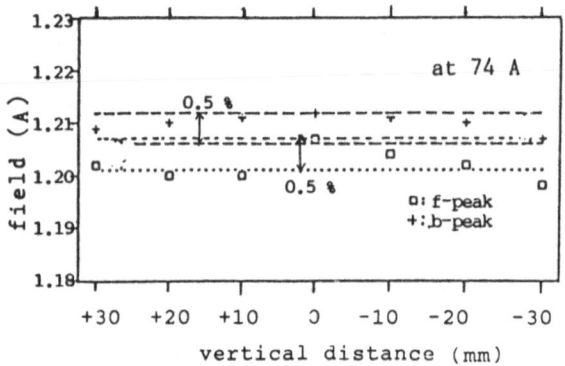

FIGURE 8. Vertical uniformity of magnets at three peak points.

DISCUSSION

A wiggler magnet system for a compact SR generator must be miniaturized and economical. However, it is found that a long uniform field along the deflecting direction is required, if SR in VUV region is desired with the combination of a electron beam of a low energy and a superconducting wiggler magnet of a high field. For the achievement of a uniform field of more than 40 mm long (maximum deflection distance of 36 mm), the length of racetrack magnet of 350 mm is not avoidable. This is one of the reasons why the wiggler magnet is not so compact.

The second reason depends on adopting an iron yoke, which cause a massive cryostat to support the weight of the wiggler system. This, however, met with so good results as field enhancement, field uniformity of long range, suppresses of fringe field and a mechanical support of the magnets.

A large amount of coolant is required to bring the wiggler magnet to a SC state but it can be operated for 8-10 hours, which fits to machine time of the electron accelerator.

The field distribution on the central axis in the beam duct is nearly similar to the calculated one. Because the experiment was performed under a series connection of 6 magnets, the calculated one would be established by adding one more power supply for both side magnets.

As for performance of the magnets, the training has not finished yet with quenching of 20 times. It is to be discussed, however, that the training procedure should be continued hereafter or the main magnet should be remodeled, because the quench current increases very slowly at present. The following two facts prove that the quenching originates from wire movement at the straight portion of the racetrack magnet. The first is the quench current of each magnet increases with increase of restriction pressure and points to prevent the coil expanding outside. The second, the long main magnet (350 mm) quenches at 85 A, while the side magnets of which shapes are similar except the length of 270 mm do at 120 A. Considering these facts, design of a dipole magnet including new idea is planned.

CONCLUSION

A three pole SC wiggler magnet of vertical deflection was completed for a compact VUV SR generator. The maximum field is 3 T at present and still increasing through training. The critical wavelength is expected to be 430 Å with an electron beam of 120 MeV and to become shorter in future.

A field distribution along the central axis in the electron beam duct was obtained with a good agreement with that calculated. The vertical uniformity of the field is obtained within 0.5 % in a wiggler plane.

The overall size of magnet system could not miniaturized so much. However, so many merits as a long vertical uniformity of the field, a small fringe field and a long machine time are obtained instead of the demerit.

These results would promise the wiggler magnet to cooperate easily with an electron linear accelerator for SR generation. The experiments will start immediately after the establishment of the accelerator.

ACKNOWLEDGMENT

The authors wish to thank Drs. A. Iwata, K. Haraguchi and S. Suehiro, Kawasaki Heavy Industry Co., for their help of the 2-D field calculation. And also they do Drs. K. Ikizawa, N. Takasu, T. Mori, N K K Co., for their help of the manufacturing and cooling down test of the system.

REFERENCES

1) Sampson, W., A model of super conducting wiggler with iron for the Brookheaven Ring. SSRP Rep. 1977,77/05 II51-53
2) Baynham, E., A superconducting wiggler for the Daresbury SRS. SSRP Rep., 1977 77/05, II54-60
3) Marks, N., Greaves, G.N., Poole, M.W., Suller, V.P. and Walker, R.P., Initial operation of a 5 T superconducting wiggler magnet in the SRS. Nucl. Instr. Methods 1983, 208, 97-103
4) Yonehara, H., Kasuga, T., Matsudo, O., Kinoshita, T., Hasumoto, M., Yamazaki, J., Kato, T. and Yamakawa, T., Undulator and wiggler of UVSOR. IEEE Trans Nucl. Sci. 1985, NS-32, 3412-3414
5) Winik, H. and Doniach, S., Synchrotron Radiation Research, Plenum Press, New York, 1980, pp.11-25
6) Yamakawa, T., Wiggler magnet. Teion Kougaku 1988, 23, 120-127 (in Japanese)
7) Jacquemin, J.P. and Perot, J., Design and test of a 5 T superconducting wiggler. IEEE Trans. Mag. 1988, 24, 1226-1229
8) Eriksson, J.T., Kettunen, L., Mikkonen R. and Sonderlund, L., Design and model test of a 7.5 T wiggler for the Max-Lab Synchrotron radiation facility IEEE Trans. Mag., 1989, 25, 1684-1687

SUCCESSFUL TEST OF A SUPERCONDUCTING WIGGLER FOR THE FRASCATI ELECTRON STORAGE RING ADONE

M. Barone, A. Cattoni, C. Sanelli,

INFN - Laboratori Nazionali di Frascati, P.O.Box 13, 00044 Frascati, Italy

Abstract

A superconducting wiggler magnet has been successfully tested at Ansaldo Componenti works. The design field of 6 Tesla has been reached after limited training. The field of 6.025 ± 0.003 Tesla has been measured by means of a flip coil connected to an integrating magnetometer. Field profile measurements were made at 5.0, 5.75 and 6.025 Tesla, during which time, the magnet showed high stability. The overall cryogenic losses have been measured and they are consistent with the computed ones.

Introduction

A superconducting wiggler, to be used as an insertion device in the existing storage ring Adone, has been designed at Laboratori Nazionali INFN, Frascati[1], and built by Ansaldo Componenti-Genova. The aim of this facility is to shift the "universal" spectral curve of the synchrotron radiation from the Adone bending magnets towards higher energy photons. At an electron beam energy of 1.5 GeV (B bending=1 Tesla) and for a circulating beam of 100 mA, the present photon flux at 1.5 KeV (critical energy) is about $2.4*10^{12}$ photons/s mrad/0.1% bw. (see Fig.1). The same flux can be obtained at 9 KeV from the 6 Tesla magnetic field of the wiggler.

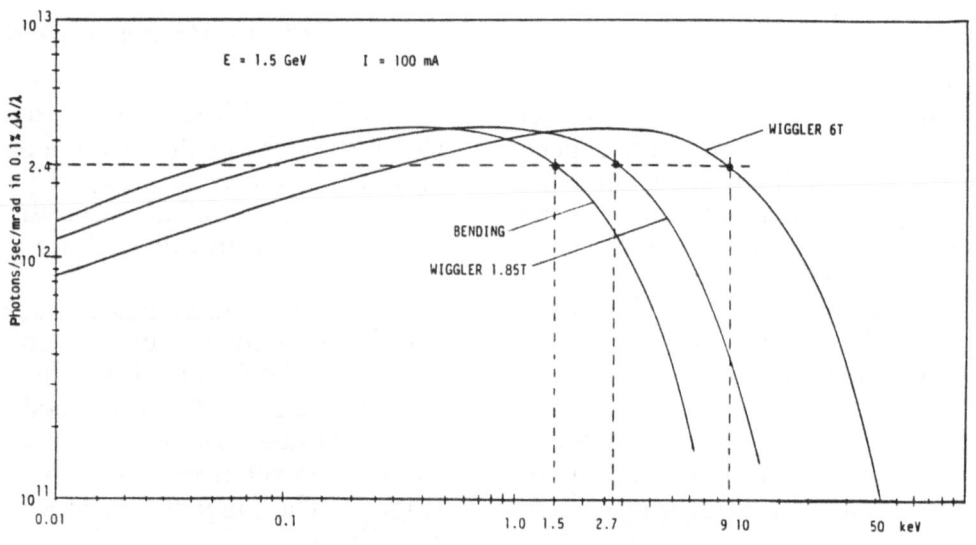

Fig. 1 - Universal spectral curve.

An important aspect is the structure of the synchrotron light source to be obtained, namely its horizontal and vertical phase space distributions (Fig. 2). The source created by a wiggler has a structure (in both planes) whose complexity depends on number of the "wiggles" (electron orbit oscillations in the wiggler).

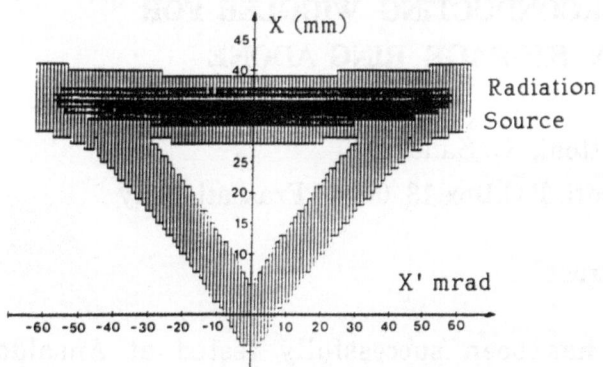

Fig. 2 - Phase space structure of the source.

To be compatible with the storage ring optics, the wiggler must also fulfill the following conditions:
1) $\int Bz\ ds \sim 0$ (the first integral of the vertical field component should vanish).
2) The orbit parameters, displacement and angle, at the entrance of the wiggler should be the same as those at the exit. This imposes a symmetry on the field distribution along the beam trajectory, with a symmetry axis orthogonal to the straight section, lying on the orbit plane and passing through the wiggler center.

To reach the goal of having a very simple light source structure and a compensated magnetic field integral, a peculiar magnetic field pattern along the beam trajectory has been adopted: a sharp vertical field peak (- 6 T, 14 cm f.w.h.m. Fig. 3) placed at the straight section center, compensated by two side tails (both decreasing from about + 1 T to zero in over 1 m of free straight section length).

The chosen field profile and symmetry, produce a single orbit bump in the horizontal plane, and the single "wiggle" gives rise to a single bright source spot in hori-

zontal and vertical phase space.

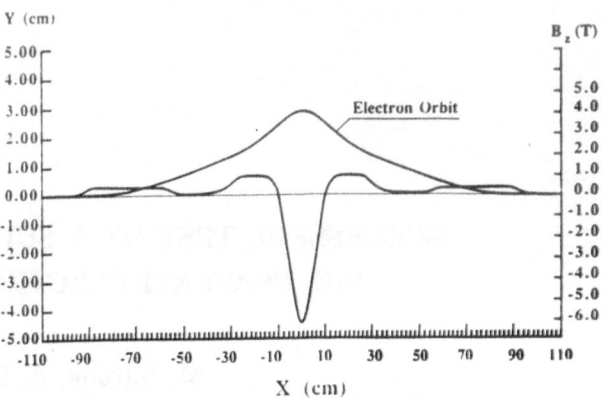

Fig.3 - Field profile and electron trajectory.

The magnetic structure consists of a superconducting 6 T dipole, placed in a warm bore cryostat, and two (warm) side dipoles for field compensation (Fig. 4).

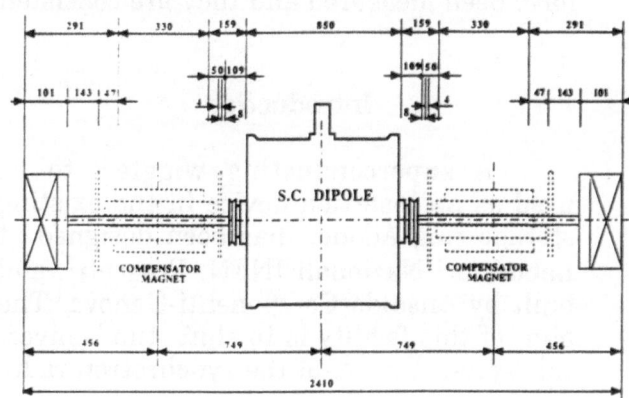

Fig.4 - Magnets lay-out.

The verification tests performed at the builder (Ansaldo Componenti)[2], concern the superconducting dipole only. A brief description of the procedures and results are given in this preliminary report.

The s.c. dipole has 2 race-track NbTi coils, separated by a central stainless steel plate in which an elliptic bore has been milled (on axis) to accommodate the vacuum chamber. The coils are kept together by two iron yokes weighing 356 Kg together (Fig. 5). The magnet gap is 6 cm, in order to allow a beam stay clear of 3.2 cm in the above mentioned vacuum chamber. The chamber wall is surrounded by a small annular gap, under vacuum, containing the radiation heat shield consisting of 5 NRC-2 superinsulation layers. The maximum design operating current is 360 A.

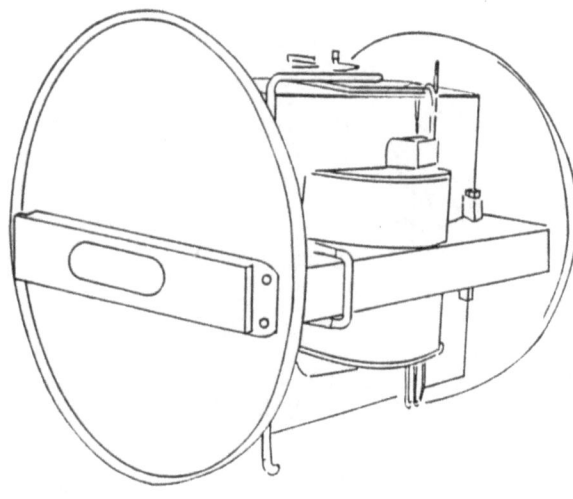

Fig. 5 - Cryomagnet's artistic view.

The computed static helium consumption of the cryostat is 5 l/h. The magnet is cooled with boiling helium, at 4.6°K and 1.4 atm., by a 1430S Koch liquefier/refrigerator connected to it through transfer lines.

The verification tests deal with cryogenics (cool down and cryostat operation) magnet excitation (training and magnetization curve), magnetic measurement (field profile at various excitation levels).

Cool Down

The on line refrigerator of the Ansaldo workshop, where the tests have been carried out (Fig. 6), was not available, and the cryostat (capacity: 70 liters of liquid helium) was therefore simply filled first with liquid N2 and then with liquid He using dewars. During the first cool down, a safe cooling speed was adopted, of the order of 2 K/hour. In the range from 300 to 80 K the maximum Δt across the magnet has been 50 K, measured through CLTS sensors. With the liquid helium at its maximum

level and after a thermalisation period, an average temperature of 50-60 K was measured at the radiation shields. The shields are cooled by boiling off the helium bath; the total evaporated volume (including the current leads cooling) was about 5000 Nl/hour. The insulation vacuum was $2 \cdot 10^{-6}$ Torr, maintained by a 100 l/hour diffusion pump.

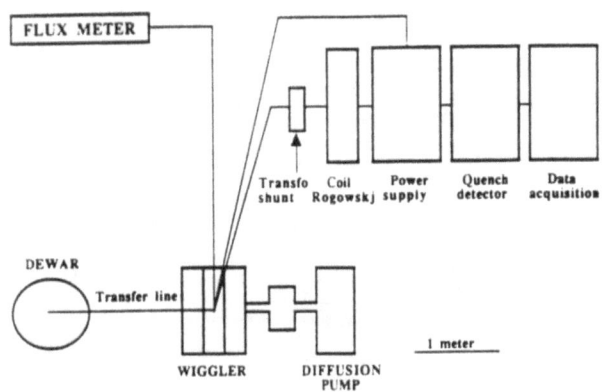

Fig. 6 - Wiggler testing area lay-out.

Magnet Excitation

The magnet mechanical structure is rather unusual if compared to that of a conventional superconducting dipole. This is due to the need of a large quasi-elliptic aperture (19x3.2 cm.) for the beam stay clear in Adone. In addition, as recalled above, the cryostat should have a warm bore, which means that the actual pole gap must accommodate also the bore vacuum insulation and this increases the transverse dimension of the magnetic yoke.

Fig. 7 shows the mechanical lay-out and it is clear that the large race track coils are supported by the magnet yoke only over a short length of the long side, and not at all at the curved heads. No reinforcing rings are allowed in order to avoid complicating the construction and enlarging the helium vessel transverse cross section. In spite of this mechanically unfavorable geometry, the magnet exceeded 6 Tesla after only few quenches all occurring above the 5.4 Tesla field level.

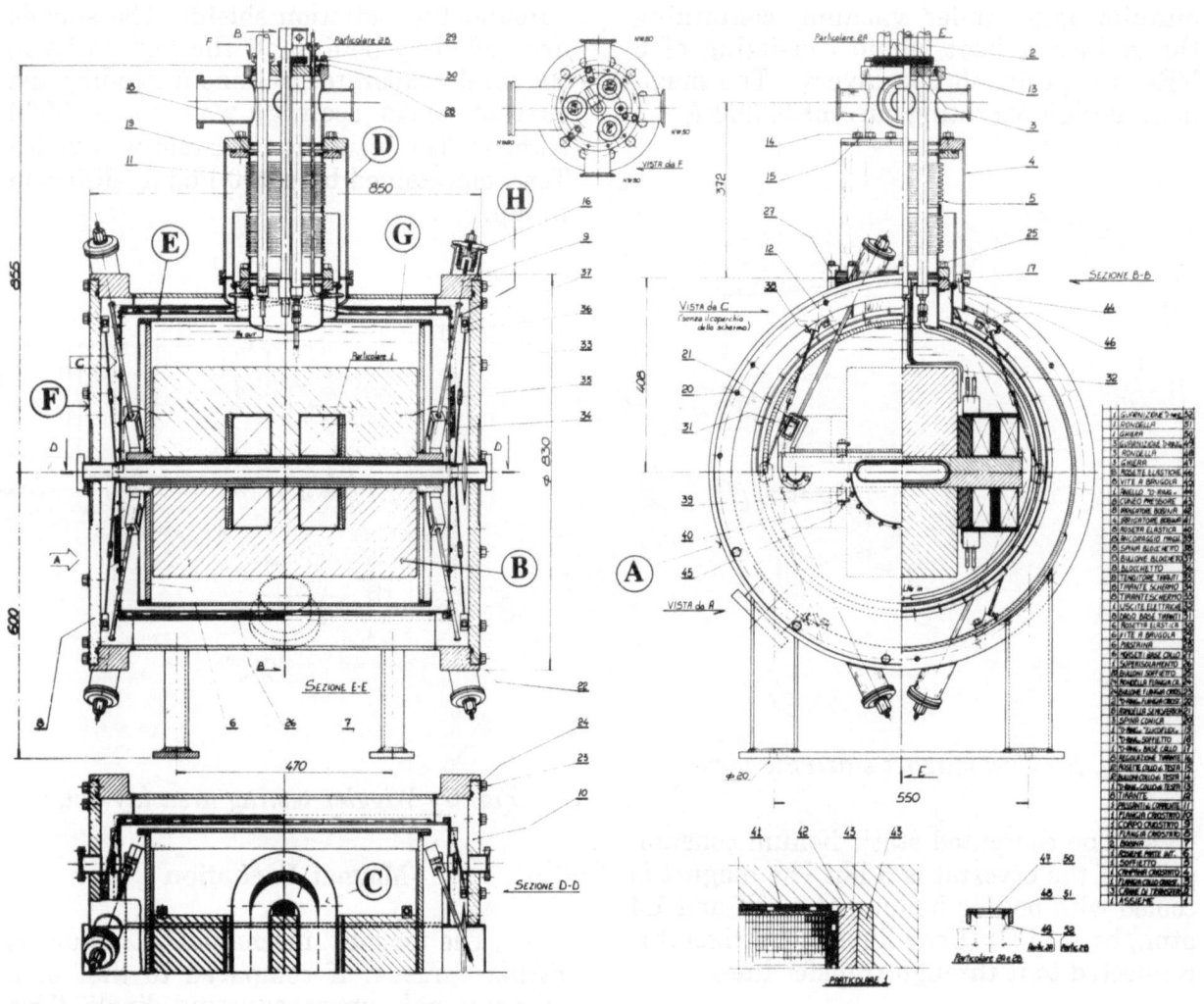

A: External vessel;   B: Magnetic yokes;   C: S.C. coils;   D: Service turret;
E: Helium vessel;   F: Flanges;   G: Radiation screen;   H: Suspension rods;

Fig. 7 - Transverse cross-section of the Cryomagnet.

Fig. 8 shows the magnet training. The quenches were due to mechanical settlement of the coil and yoke under electromagnetic forces and not to conductor current limits. This has been proven by increasing the bath temperature from 4.2 K to that corresponding to a bath pressure of 1.41 bar (about 4.6 K) and finding no difference in the current values. At the above mentioned bath pressure the blowout disk behavior also has been tested by inducing a quench at the maximum magnet energy. The disk broke at 1.7 bar according to the calibration value. Care was taken in finding the best ramping speed. A maximum ramp of 0.6 A/sec was adopted in the range from 0 to 290 A corresponding to a field of 5.560 Tesla; above this current value the ramp speed was reduced during training. The last quench occurred at 317.7 A corresponding to a field of 6.025 ± 0.003 Tesla. The maximum current level was attained by reducing the maximum ramp by a factor of ten. After reaching the 6 Tesla level no more quenches have been observed when repeating the ramp from zero to 6 Tesla. The magnet proved to be perfectly stable at every field level. The operating time was limited only by the helium availability. The best deenergizing ramp speed has been found to be 1.2 A/sec max. The current leads, during the magnet excitation, were cooled by a helium flux of 2500 Nl/hour.

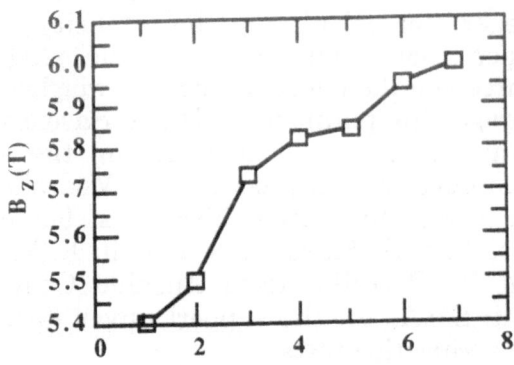

Fig. 8 - Quench.

Magnetic Field Profiles

The measurements have been made using an integrator magnetometer: a manually rotated small coil connected to an integrator made with very highly stabilized integrating circuits. The coil has a diameter of 5 mm and is 5 mm high. It has a measurement accuracy of 0.05% of full scale over the three ranges 200mT, 2T, 20T. The thermal drift, after two hours of running is less than 0.02 mT per minute. The instrument was built by Magnex, Abingdon, UK.

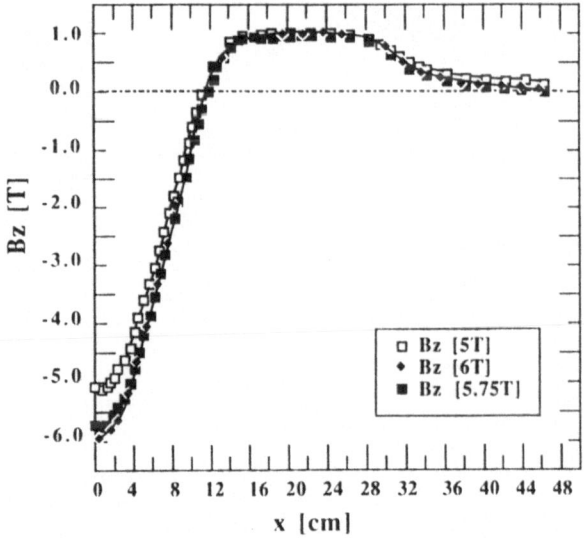

Fig. 9 - Wiggler field profile measurements.

The flip coil was held in position and moved in steps along the wiggler symmetry axes using a mechanical actuator with an accuracy of better than 1/10 mm.

Field profiles have been measured at 5.0, 5.75, 6.0 Tesla. Fig. 9 shows the results and Fig. 10 shows the comparison with the output of the tridimensional computer codes Tosca and Magnus. The experimental data are in good agreement with the code previsions.

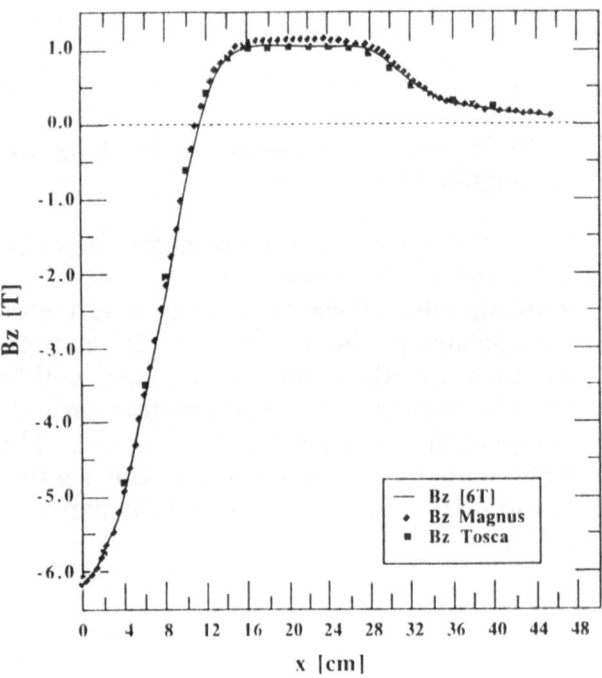

Fig. 10 - Code previsions and measurements.

Conclusions

The field profile obtained, exceeding 6 Tesla with high stability, is proof of the magnet reliability and confirms the design goal.

As mentioned above, the whole system consists of the superconducting dipole and two adjacent warm dipoles providing the field-integral compensation. The compensator magnets are "home made" and ready.

336

Magnetic measurements of the overall system are the most important task for the immediate future. In fact, to be compatible with the storage ring requirements, the field integral, along the beam trajectory (lying on the symmetry plane of the wiggler) should vanish at all currents. Furthermore, the higher order terms of the field, integrated along the straight section where the wiggler is installed, should be kept below well defined limits.

To fulfill these specifications two kind of magnetic measurements will be carried out:

- Point by point magnetic field map in the horizontal plane where the beam lies.
- Field integral measurements along the magnet axis.

Hall probes will be used for the point by point maps, while a three meter long rotating coil, whose sensitivity was tested by measuring the earth vertical magnetic field, (at Abingdon = 0.43 Gauss) will be used to measure the field integral and its integrated multipolar components. The field map measurements and the related equipment will be described in a paper to follow.

Aknowledgments

The authors wish to thank Dr. M. Preger for calculating the intensity, the phase space distribution of the photon source and the effects of the s. c. wiggler on storage ring parameters, Dr. A. Savoia for many useful discussions during the system development phase, Dr. G.Modestino, Dr. F. Sgamma from LNF; last but not least Drs. G. Masullo, S. Parodi, R.Penco and M. Perrella, from Ansaldo Componenti S.p.A., for the support given during the verification tests.

References

1. A. Aragona, M. Barone, A. Cattoni, U. Gambardella, G. Modestino, M. Perrella, M. Preger, C. Sanelli, A. Savoia, F. Sgamma: "Superconducting Wiggler for Adone: Design and Present Status" - to be submitted to Nuclear Instruments and Methods.

2. G. Masullo, S. Parodi: "Prove di vuoto, raffreddamento e di carica effettuate sul magnete wiggler INFN di Frascati" - Tech.Rep.Ansaldo PMA/PRMA/R89147.

FINAL DESIGN OF A 7.5 T SUPERCONDUCTING WIGGLER MAGNET

J.-T. Eriksson, L. Kettunen, R. Mikkonen, and L. Söderlund
Tampere University of Technology, Lab. of Electricity and Magnetism,
P.O. Box 527, SF-33101 Tampere, Finland

ABSTRACT

A 3-pole superconducting wiggler magnet is under construction. The magnet will provide a peak flux density in excess of 7.5 T. The wiggler will be placed in the MAX 550 MeV storage ring [1] and eventually later in the planned 1.5 GeV MAXII storage ring [2]. The field is produced by six NbTi/Cu racetrack shaped coils surrounding iron yokes. An adjustment current is supplied to the end coils to minimize the field integral over the whole field range. 3D field calculations were performed for the optimization of magnet geometry. The mechanical structure designed to counteract the magnetic forces is also discussed. The main objectives have been to keep the period length below 240 mm and the pole gap above 35 mm. The magnet parameters and expected performance will be presented. Integration of the magnet with the associated vertically orientated cryostat is also outlined.

Two model magnets have been built and tested. A linear magnet was built in order to illustrate the problems concerning the 3-pole concept. A one pole winding comprising iron provided information about magnet rigidity in the high field region and also verified the 3D calculations.

INTRODUCTION

The designed magnet is a three pole device in which the central pole produces a high flux density of 7.5 T, the outer poles provide the cancellation necessary to satisfy the beam dynamics of the storage ring. In the 550 MeV storage ring, the wiggler will produce soft X-rays with a critical wavelength of 8.2 Å and in the 1.5 GeV ring X-rays with a critical wavelength of 1.0 Å. The emitted power will be 0.91 kW and 6.8 kW respectively with a circulating electron current of 200 mA.

In order to minimize the focusing effect on the electron beam, the wiggler period length is limited to 240 mm. The field integral along the wiggler should remain below $2 \cdot 10^{-3}$ Tm. This can be attained with the adjustment current of 0-30 A. A sufficient field homogeneity over the beam aperture is attained by making the straight section of the coil long enough.

When optimizing the wiggler magnet regarding maximum field strength and small period length, the magnet gap height becomes an important parameter. The minimum gap was measured in the MAX ring using scrapers and the minimum vacuum chamber height was found to be 14 mm.

The cryostat has been designed so that no liquid nitrogen is required for shielding purposes. The traditional horizontal form has been abolished in favour of a vertical, top loading alternative, mainly because of the space limitations in the storage ring and the elongated heat conduction path to liquid helium.

The beam chamber is at room temperature and is separated from the coils by a superinsulated vacuum space.

The coils are made of NbTi/Cu super-conducting wire. The safety margin between the operating current and the conductor short sample value at the maximum field in the winding is about 50 %. The operating value of the magnet is naturally strongly dependent on the training procedure, winding technique and the choice of the impregnat. The magnet is cooled with pool boiling helium I.

MAGNET GEOMETRY

The main design principles of the 3-pole superconducting wiggler magnet have been a central field of 7.5 T, a periodicity around 240 mm, a total length of the system less than 900 mm and a accepted value of the field integral less than $2 \cdot 10^{-3}$ Tm. An iron core and a relatively small pole gap are needed in order to meet these specifications. Further, a rigid and compact construction is required due to the high magnetic field and the high current density. On the other hand, flexible connections between the magnet and the inner vessel of the cryostat are required due to thermal stresses.

The overall view of the magnet assembly and the longitudinal cross-section can be seen in figures 1, 2 and 5. Fig. 1 shows three parallel racetrack coils arranged perpendicular to the electron beam direction. The iron cores of the outer poles have the same dimensions as the central core but operate at half magnetic flux density at reverse polarity. Each winding dipole is made in modular form in order to simplify case handling and construction. The straight section of the middle cores and side cores are 190 mm and 175 mm respectively. The massive iron cores increase field homogeneity and, although highly saturated, a benefit in flux density. The core material and the cover plates are of stressannealed Armco iron.

The separation plate (stainless steel)between the upper and lower windings is made from two longitudinal parts by brazing. Bellows capable to withstand large axial movements form a flexible connection between the separation plate and corresponding inner vessel flanges. The poles are compressed in the vertical and longitudinal direction by vertical bolts and by end plates of stainless steel respectively.

The magnet is mounted to the support plate by four stainless steel rods. The three rods from the bottom primarily support the magnet during operation. The corrugated wall compensates the vertical movement of the assembly due to thermal shrinkage.

The main parameters of the 3-pole wiggler magnet are listed in table 1.

Table 1. Wiggler parameters and expected performance.

Main current	315 A
Adjustment current	16 A
Central field B_0	7.88 T
Ratio B_{max}/B_0	1.06
Integral value with 16 A	$0.09 \cdot 10^{-3}$ Tm
Beam deflection	0.02 m
Magnet energy	337 kJ
Total number of turns	13128
Pole gap	0.036 m
Period length	0.236 m
Vertical aperture of the vacuum chamber	0.015 m
Horizontal aperture of the vacuum chamber	0.074 m
Wiggler overall length	0.362 m
Height	0.356 m

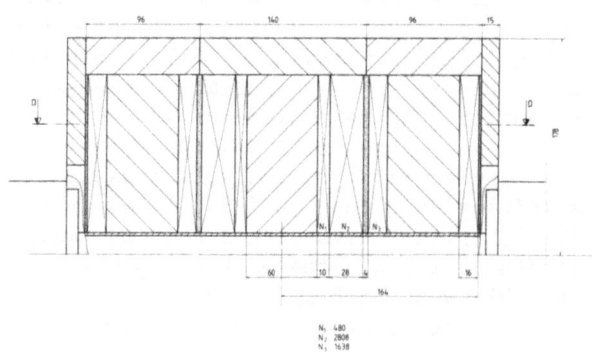

Fig. 1   Longitudinal cross-section of the 3-pole magnet.

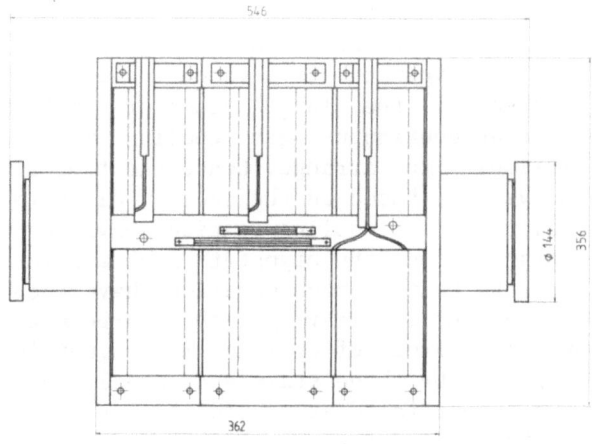

Fig. 2   Side view of the magnet.

The dimensions of the coils can be seen in figure 3. They are partly determined by the principle of minimizing B_{max}/B_o and the idea of getting the integral value close to zero with a small control current. Spacers 15 mm wide have been put between the end sections of the middle pole windings in order to lower the maximum flux density. With an adequate training and a proper impregnant the magnet should stand the operating current of 315 A at a maximum flux density of 8.4 T in the end section of the inner winding.

The coils are connected in series in order to obtain a minimum number of current leads. Every winding has an even number of layers so that the leads can be connected by the sides of the separation plate by soldering to the copper blocks.

SUPERCONDUCTING COILS

Flux density values above 8 T are close to the upper limit for common technological NbTi superconductors in magnets. However, Nb_3Sn was not chosen due to its brittleness. In order to simplify cryogenic arrangements and to reduce cooling costs superhelium is not used. A subdivision of the middle coils into two windings is necessary when using NbTi conductors. Conductors of ϕ 1.61 mm and ϕ 1.1 mm are for the low and high current density winding respectively. The side coils will be wound from the ϕ 1.1 mm conductor. The monohexagonal superconductors are manufactured by Alsthom Intermagnetics. Table 2 lists the measured parameters for both the conductors.

Table 2   Superconductor data

	Conductor 1		Conductor 2	
Dia (bare)	1.51 mm		1.02 mm	
Dia (ins)	1.61 mm		1.10 mm	
Dia (fil)	49.5 μm		34.6 μm	
Twist pitch	46 mm		30 mm	
Cu/NbTi	1.16		1.3	
RRR	121		110	
	5 T	2010 A	5 T	990 A
I_c(4,2 K,	7 T	1200 A	7 T	620 A
0.1 μV/cm)	8 T	800 A	8 T	420 A
	8.5 T	610 A	8.5 T	320 A
	9 T	430 A	9 T	220 A

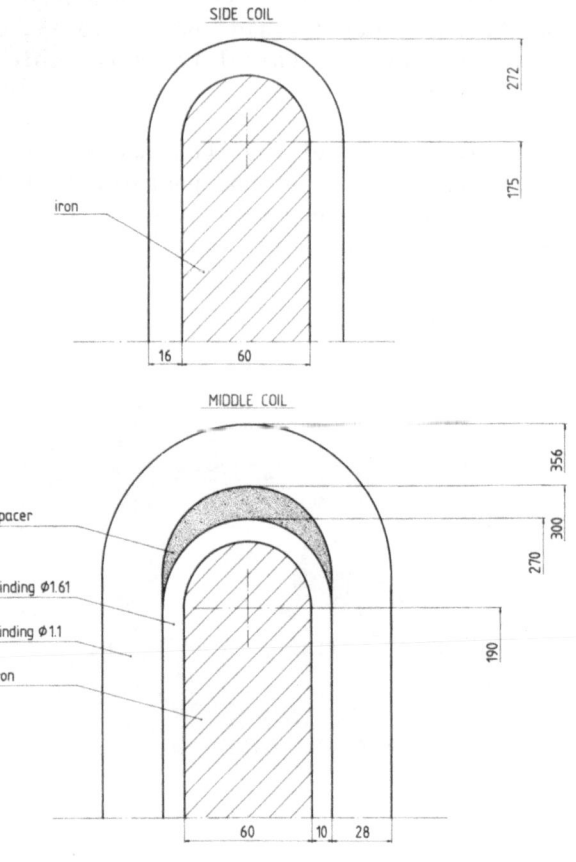

Fig. 3   Superconducting coils with iron cores.

FIELD CALCULATION

The magnetic field was designed in two parts. The preliminary cross-section of the magnet was optimized with 2D calculations. The final design was made with the aid of 3D static magnetic field computation (figure 4).

The 3D calculations were carried out with the program TOSCA [3], which uses the two scalar potential method. While the iron parts of the magnet are highly saturated and the main part of the field in infinity is caused by the coils, the reduced scalar potential was given for the exterior boundary as a homogeneous Neumann boundary condition. Also the field along the beam line was computed with the reduced scalar potential to minimize the numerical error.

The adjustment current was limited to 30 A. Thus the geometry was designed so that $\int B \cdot dl = 0$ with a 16 A adjustment, the central field being 7.8 T. It required several calculations to achieve this goal.

The track of the beam was computed with an option of the postprocessor of TOSCA. This feature made it easy to check that the beam dynamics will be satisfied.

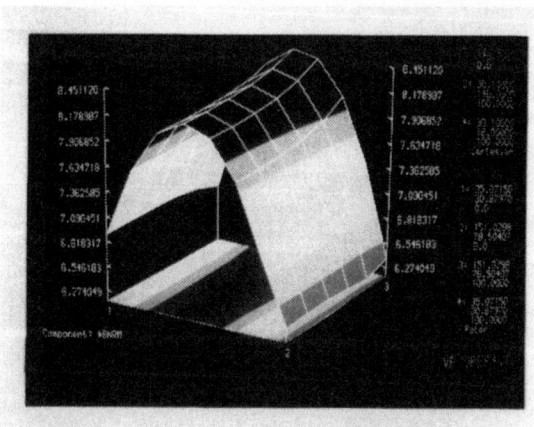

Fig. 4 Plot of the magnetic flux density.

CRYOGENIC ARRANGEMENTS

The vertical AISI 316 stainless steel cryostat is vapour shielded. Because of the large diameter of the cryostat continuous gas cooling may be difficult to arrange. Hence the thermal analysis is based on the idea of discrete cooling. Five cooling stations have been designed along the cryostat neck and for them the heat balance equations have been solved while leaving the in-between stretches free of coolant. The magnet is supported by three stainless steel rods. Five copper faced epoxy radiation baffles in the cryostat neck provide good emissivity and electrical insulation for the current leads. To place the wiggler to its right position, the magnet is supported by three adjustable supports between the convex bottom of the inner vessel and the magnet.

The upper part of the inner vessel wall is partly corrugated. The aluminium radiation shield is anchored to the inner vessel at the straight section of the wall. This makes the assembly easy although the optimal anchoring point would locate at the lower corrugations. Between the outer vessel and the radiation shield layers of multilayer superinsulation are situated, as well as between the radiation shield and the inner vessel. The benefits of the round corrugations are manyfold: (1) The heat conduction path can be increased by a factor of $\pi/2$. (2) The gas cooling is more effective. (3) The corrugated wall with a proper "spring constant" compensates the effect of the wall thermal shrinkage (essential for the beam tube assembly). (4) The wall thickness of the corrugations can be small. To prevent the possible excessive elongation of the inner vessel due to vacuum forces three epoxy rods support the inner vessel bottom through the vacuum space.

The theoretical overall helium evaporation is about 2.5 l/h of which the cryostat part is 45 %. The evaporation of the gas cooled current leads (two rated at 350 A, one at 30 A) is estimated 30 % of the total boil-off, the rest, 25 %, is due to the beam tube.

The wiggler will be used for synchrotron radiation research and accordingly continuously cooled down. This means that if the target boil-off 3 l/h is achieved, the cryostat will be filled in the morning and once refilled during a working day (the upper LHe level as indicated in the picture).

In the event of a quench one cannot expect any good quench propagation because each coil is thermally insulated from the others. One coil might then be loaded with the total stored energy [4]. Assuming adiabatic conditions the analytical theory of Wilson [5] gives a maximum quench temperature of about 110 K, under the assumption that the transition starts in the high field region. The magnet will be protected with small resistors or a thyristor protection circuit, in order to reduce rapidly the temperature and internal voltage.

The cryostat cross-section can be seen in Fig 5.

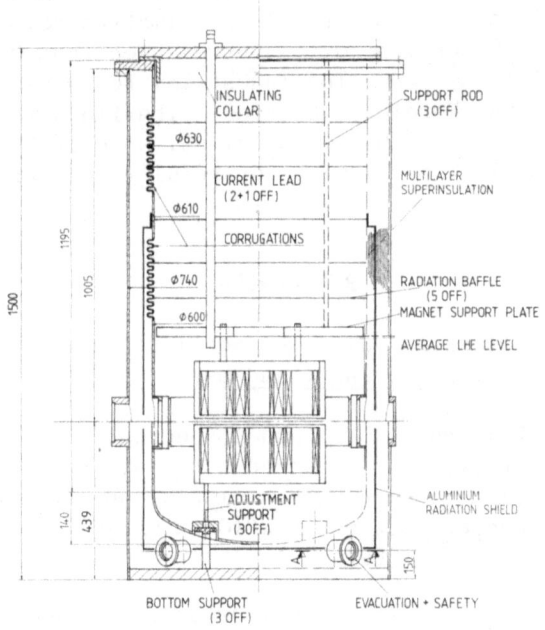

Fig. 5   The system cross-section.

MODEL WINDINGS

Two model magnets have been constructed in order to get information for the final wiggler design. The common features for the design of both magnets have been the same limitations: the amount and quality of the available superconductor, the weight of the magnet and the space in the cryostat.

The linear 3-pole model magnet was constructed to give information about the three pole concept and the winding technique and further to check magnetic field calculations. The magnet consisted of three parallel dipoles (figure 6). Each dipole was an assembly of racetrack coils with epoxy cores. The coil cores were 200 mm long and 29 mm wide. The total number of turns was 2300. The superconductor was rectangular, 0.85 mm × 1.38 mm, Cu/NbTi being 1/1.5. Because of a failure during the fabrication process the critical current values were quite low.

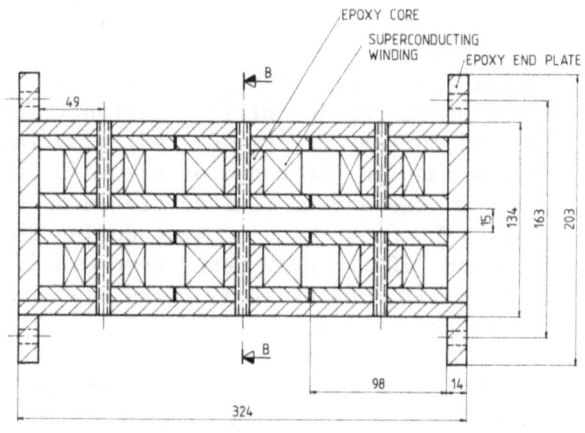

Fig. 6   Cross-sectional view of the linear 3-pole model magnet.

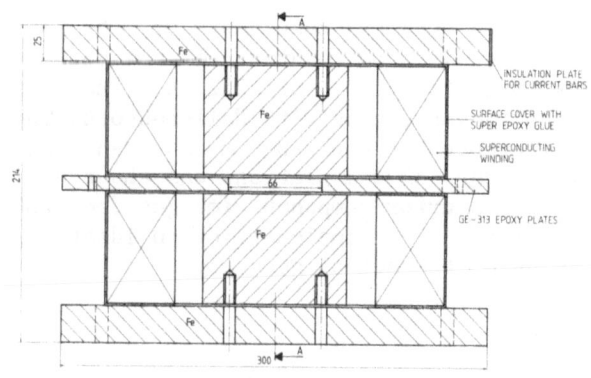

Fig. 7   Longitudinal cross-section of the one pole model magnet with iron core.

The second model magnet was aiming at getting experiences of the vacuum impregnated winding, its rigidity and training effect, the use of a winding machine and to check 3D field calculations [6]. The one pole magnet consisted of one dipole with iron cores (figure 7). The length and width of iron cores were 102 mm and 39 mm respectively. Both coils have 1381 turns. Round pieces were located in the end sections of the rectangular iron cores. The 12 mm wide gap between the coils was performed with epoxy plates. The rectangular, 1.0 mm × 2.12 mm, superconductor was used.

In the three pole magnet an operating current of 159 A (corresponding to a magnetic field of 1.75 T) was obtained without training. This is equal to 88 % of the short sample critical current. The critical current decreased radically for dI/dt values in the range of 4-10 A/s and remained thereafter quite stable up to 22 A/s. The operation losses during the 150 A and 100 A d.c. tests were 0.4 1/h and 0.26 1/h, respectively. The measured inductance was 160 mH.

The training effect was considerable in the one pole vacuum impregnated winding. During two cool downs the magnet was quenched several times. Due to training the critical values B_c and I_c increased about 40 %. This extreme change is partly explained by the incomplete epoxy impregnation.

The vertical component of the flux density in the center of the gap was measured to 6.2 T at 322 A. The flux density was probed with Hall-sensors. The difference between the measurements and the rough 3D calculations was about 4 %. The main error component imerges from the difficulty of implementing an identical image of the actual geometry.

CONCLUSIONS

A 3-pole, 7.5 T wiggler magnet has been designed. 3D field calculations indicate that the ratio B_{max} to B_0 is 1.06 and it is thus possible to utilize NbTi/Cu superconductors in pool boiling helium. Even though the field integral can be kept within acceptable limits, non-superconducting correction coils will be added to the storage ring. The LHe economy is expected to be kept below the target value by giving the cryostat a vertical orientation and utilizing a minimum number of current leads, i.e. two main leads and one lead for the adjustment current.

The magnet with its cryostat is scheduled to be delivered to the MAX-laboratory by the end of 1990.

ACKNOWLEDGEMENTS

The group at Tampere is grateful for the assistance they have received from people working at the MAX-lab at the University of Lund, especially Prof. Mikael Eriksson and Mr. Leif Thånell.

The wiggler project at Tampere University of Technology is sponsored by Asea Brown-Boveri, Vattenfall Ab, Aga Ab, Scanditronix Ab, University of Lund (Sweden), ABB Strömberg, Imatran Voima Oy, Outokumpu Oy, Huurre Cleanroom Oy (Finland) and The Industrial Foundation of the Nordic Countries. The financial support and the technical guidance of the steering committee is gratefully acknowledged.

REFERENCES

1. Eriksson, M., The accelerator system MAX. Nucl. Inst. and Meth. 196(1982).

2. Eriksson, M. and Lindgren, L.-J., MAXII, An advanced VUV synchrotron light source with a simple and compact magnet lattice. LUNFTOX(NTMX-7012) /1-19/ (1989)

3. Vector Fields ltd., Oxford, England.

4. Jacquemin, J.P. and Perot, J., Design and test of a 5 T superconducting wiggler. IEEE Trans. on Magn., Vol 24, No. 2, March 1988, pp. 1226-1229.

5. Wilson, M.N., Superconducting Magnets, Oxford University Press, New York 1983, pp. 200-231.

6. Mikkonen, R., Söderlund, L. and Kettunen, L., Performance of a non-linear One Pole Superconducting Wiggler Magnet. Tampere Univ of Tech, Lab of Electricity and Magnetism, Report 2, 1989, 32 p.

DETERMINATION OF LOADING ORDER OF PERMANENT MAGNET BLOCKS IN THE SOFT X-RAY UNDULATOR (SXU) AT THE NSLS

M. KITAMURA*, L. SOLOMON**, G. DECKER** and J. GALAYDA**

* *Hitachi Research Laboratory, Hitachi, Ltd.*
4026 Kuji, Hitachi, Ibaraki 319-12 JAPAN
** *National Synchrotron Light Source, Brookhaven National Laboratory*
Upton, New York 11973 U.S.A.

ABSTRACT

The optimal arrangement of the permanent magnet blocks in the SXU magnet was determined using the magnet transfer functions. Details of this method, some of the results of field measurements, and field calculations are discussed.

INTRODUCTION

A Soft X-ray Undulator (SXU) has been constructed for installation in the 2.5 GeV electron storage ring at the BNL-NSLS as part of the Phase II facility upgrade[1]. A cross sectional geometry of the

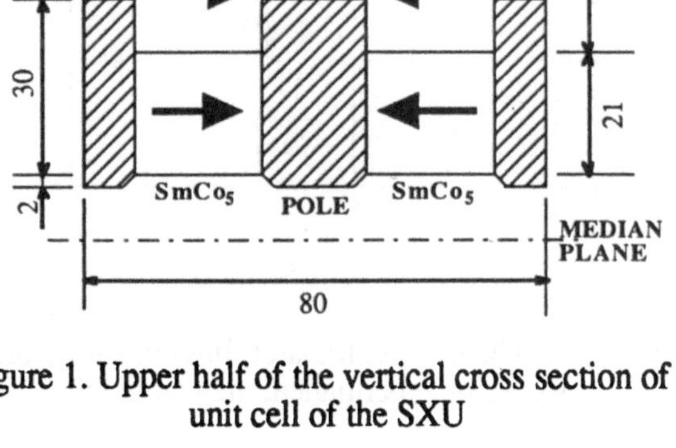

Figure 1. Upper half of the vertical cross section of a unit cell of the SXU

upper one half of a unit cell of the SXU is shown in figure 1. The SXU is a 3m long magnet with a period length of 80mm, a gap range of 31-100mm, 77 iron poles, a maximum on-axis field of 0.35T, with 8 blocks of $SmCo_5$ permanent magnet per period. The undulator parameter K, $(=0.934B(T)\lambda(cm))$ ranges from 2.6 to 0.15.

In order that the SXU should function properly, the following requirements on the undulator field must be satisfied. Namely:

i) The first integral of the undulator field in the median plane, with respect to the undulator center line, must vanish, so that the electrons emerge from the undulator at the same angle as at the entrance.

ii) The second integral of the undulator field also must vanish; then the horizontal displacement of the electrons becomes zero.

iii) The angular deviation of the electrons in the undulator, caused by error fields, most of which arise by error magnetizations in the permanent magnet blocks, must be less than a certain tolerance to avoid reduction of brightness in the undulator radiation. In the SXU magnet, the limit of transverse error fields theoretically is to be roughly $\approx 10^{-3}T$[2].

To accomplish these, we have studied a method

to determine optimal arrangements for loading the permanent magnet blocks in the undulator, using the magnet transfer function, which expresses the vertical magnetic field in the median plane of the undulator due to a individual magnet block.

This paper reports details of the calculation procedure for this method, properties of the measured transfer functions along with comparison with the 2-dimensional field calculations. Magnetic field measurements, after loading all the permanent magnet blocks into the two magnet halves of the SXU, are also presented.

TRANSFER FUNCTION

Let us introduce a coordinate system with the z-axis along the undulator center line, the y-axis orthogonal to the median plane, the x-axis both in the median plane and orthogonal to the undulator center line, and the origin on the center line of the magnet block.

As the maximum magnetic flux density in the iron pole is ≈ 1.3T, the non-linear effect by the magnetic saturation is very small. Thus we will superpose the magnetic fields due to a individual magnet block (the *transfer function*) to compute the undulator field and electron trajectory prior to the assembly of the SXU. Though the magnet block actually has three orthogonal components of the magnetization, i.e., the *major* component, M_z, and two *minor* components, M_x and M_y, we only consider the transfer functions on M_z and M_y. This is because the magnetic fields due to M_x were found to be $\approx 10\%$ of those due to M_y by the field measurement. The relative strength of the M_z and M_y of all permanent magnet blocks were also measured by a Helmholtz coil prior to the assembly of the SXU, and the standard deviation of the M_y/M_z was $\approx 2\%$.

Since the SXU has two magnet blocks per slot, four kinds of the transfer functions are needed. We denote them B_{hmaj}, B_{lmaj}, B_{hmin} and B_{lmin}, where the subscripts h and l indicate the permanent magnet block is *high* (closer to the center line) and *low* (farther from the center line), respectively, and the subscripts *maj* and *min* indicates the magnetization is major and minor, respectively.

To measure B_{hmaj} and B_{lmaj} a single magnet block with a representative major component was loaded into a magnet half with the major magnetization along the z-axis, and a steel plate was located, with a gap of 22mm, over the surface of the poles to impose midplane symmetry of the field distribution. Vertical magnetic fields on the surface of the steel plate were measured along the z-axis over a range from -0.3 to 0.3m by a Hall-effect Gaussmeter. To measure B_{hmin} and B_{lmin} the major component was oriented along the y-axis, and the other empty magnet half was placed, with a gap of 44mm, replacing the iron plate. The vertical magnetic fields in the midplane of the two magnet halves were measured over the whole length of the SXU.

The measured raw data, consisting of both fields due to the major and minor components, were numerically symmetrized to remove the unnecessary field component due to the minor one. In figure 2, the symmetrized transfer functions are plotted as a function of z. Figure 3 shows the horizontal electron trajectories denoted by x_{hmaj}, x_{lmaj}, x_{hmin} and x_{lmin}, due to the transfer functions shown in figure 2, given as the second integrals by:

$$x(z) = \frac{1}{(B\rho)_0} \int_0^z dz' \int^{z'} dz'' B(z'') \qquad (1)$$

where B represents the transfer functions, and $(B\rho)_0$, the magnetic rigidity, is 8.3391Tm for the 2.5GeV electron. The change of the electron trajectory due to the major transfer functions is quite different from those due to the minor ones, viz. the major

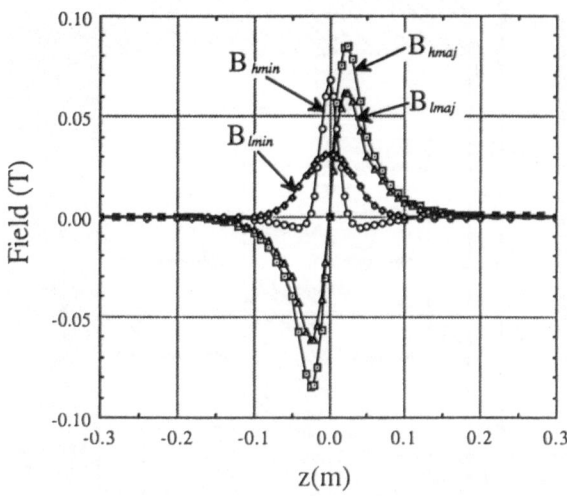

Figure 2. The magnet transfer functions measured for a single permanent magnet block

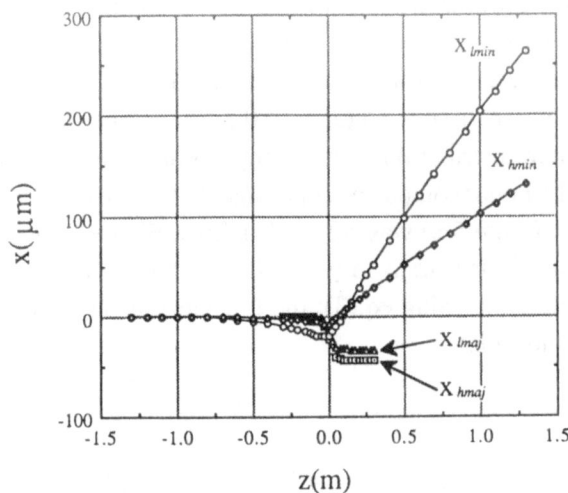

Figure 3. The second integrals of the magnet transfer functions shown in figure 2

magnet transfer functions cause only horizontal shifts, D_h and D_l ($D_h/D_l \approx 1.3$), while the minor magnet transfer functions cause deflections of the electron trajectory by angles, Θ_h and Θ_l ($\Theta_h/\Theta_l \approx 0.5$). The non-zero deflection angles, Θ_h and Θ_l, arise because B_{hmin} and B_{lmin} are even functions with respect to the origin, and the real magnetic fields differ somewhat from the 2-dimensional ones due to the side leakage flux caused by the finite horizontal width of the undulator. On the other hand, the gradient angles of the electron emerging from the undulator, due to B_{hmaj} and B_{lmaj} are unchanged being odd

functions with respect to the origin. To illustrate these features further, 2-dimensional magnetic flux lines of the transfer functions calculated by PANDIRA[3] in one fourth of the vertical cross sectional plane of the SXU are shown in figure 4. Obviously, the magnetic fields due to the minor transfer functions have longer tails than those due to the major ones. In particular, in the case of B_{lmin}, an almost constant field is produced in the gap between the back plates, while the tail for B_{hmin} becomes shorter due to the shielding of the magnetic field by the iron poles; Θ_l becomes larger than Θ_h because of its longer tail.

LOADING ORDER OF THE MAGNET BLOCKS

The procedure to determine the loading order of the permanent magnet blocks into the SXU consists of two computational steps.

Firstly, to determine combinations of four magnet blocks with the same slot number (two blocks for each magnet half), we create loading tables minimizing the following quantity:

$$\chi^2 = \sum_i \left\{ \left(<M_z>_i - <M_z> \right)^2 + \alpha \left(<M_y>_i \right)^2 \right\} \quad (2)$$

where i is the slot number, $<M_z>_i$ and $<M_y>_i$ are averages respectively of the measured major and minor

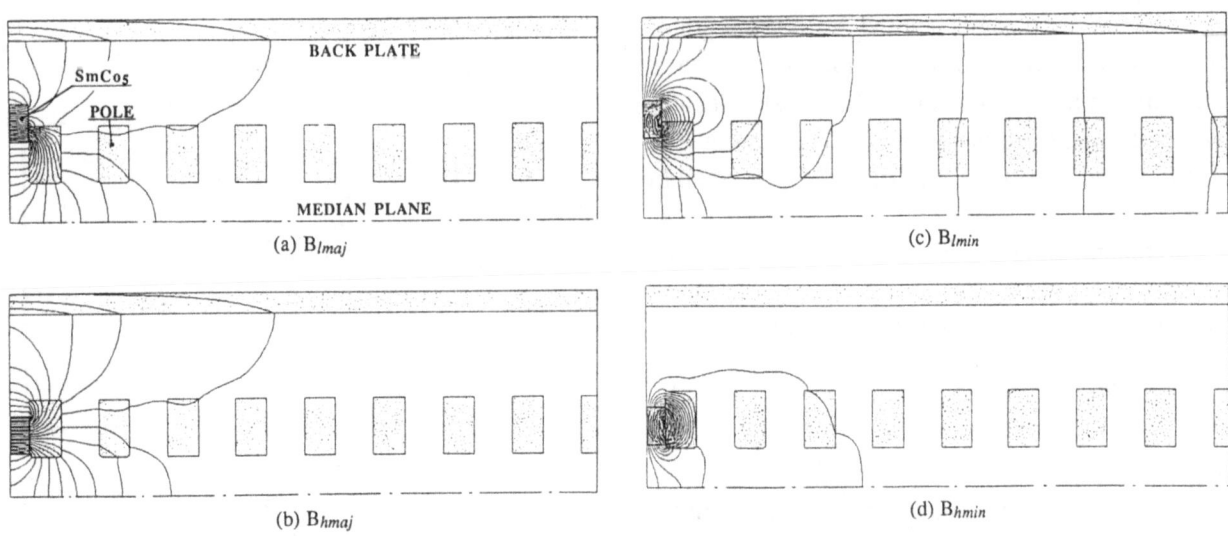

Figure 4. Magnetic flux distributions for the magnet transfer function, B_{hmaj}, B_{lmaj}, B_{hmin} and B_{lmin}, calculated by PANDIRA

components of the magnetization in the four magnet blocks in the ith slot of two magnet halves, $<M_z>$ is an average of $<M_z>_i$ over all the slots, and α is a weighting factor. In this way, several initial loading tables with α ranging from 0 to 5 were computed.

The second step renumbers the slots in the initial loading table, making the calculated electron trajectory as close to the ideal one as possible. Renumbering was performed in two ways:

i) choosing two slot numbers randomly and commuting them repeatedly until the difference between the calculated and ideal trajectories becomes less than a certain tolerance. Using the second integrals of the transfer functions shown in figure 3, the electron trajectory due to the undulator field was computed by:

$$x(z) = \frac{1}{M_0} \sum_i \{ <M_{z,h}>_i \; x_{hmaj}(z - \frac{i\lambda}{2})$$

$$+ <M_{z,l}>_i \; x_{lmaj}(z - \frac{i\lambda}{2})$$

$$+ <M_{y,h}>_i \; x_{hmin}(z - \frac{i\lambda}{2})$$

$$+ <M_{y,l}>_i \; x_{lmin}(z - \frac{i\lambda}{2}) \}$$

(3)

where $<M_{z,h}>_i$ and $<M_{z,l}>_i$ are averages of the measured major components of a pair of magnet blocks, respectively in the high and low positions of the ith slot of two magnet halves, $<M_{y,h}>_i$ and $<M_{y,l}>_i$ are the same averages taken for the measured minor components, and M_0 is the strength of the major component of the magnet block used to measure the transfer functions. The ideal trajectory was calculated by dropping the third and fourth term in eq.(3) and making $<M_{z,h}>_i$ and $<M_{z,l}>_i$ equal to $<M_z>$.

ii) renumbering the slots in the loading table considering the weighted mean strength of the minor component of the magnetization defined by:

$$\overline{M}_{y,i} = w_h <M_{y,h}>_i + w_l <M_{y,l}>_i$$

(4)

where $w_h = \Theta_h/(\Theta_h + \Theta_l)$ and $w_l = \Theta_l/(\Theta_h + \Theta_l)$.

Two plots of the weighted mean strengths of the minor components in the magnet blocks arrangements obtained by the renumbering methods i) and ii) are shown respectively in figure 5(a) and (b). In fig-

ure 5(a) the weighted mean minor components are distributed randomly. On the other hand in figure 5 (b) they are arranged in order. In both examples, the parameter α of eq.(2) was 0.5. Preferring the random distribution of the weighted mean strength of the minor component, the magnet block arrangement of figure 5(a) was selected as final. Figure 6 shows the electron trajectory calculated from the final loading table using eq.(3); the trajectory is, as desired, sinusoidal.

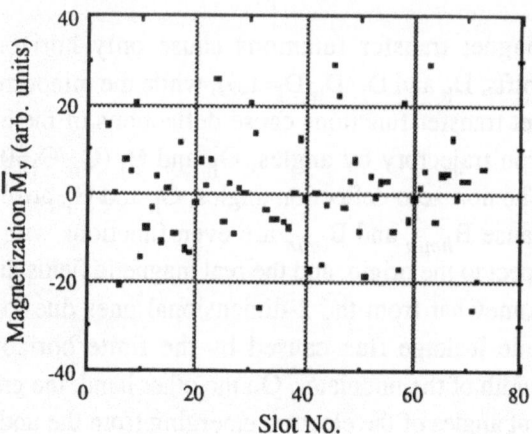

(a) renumbering method i)

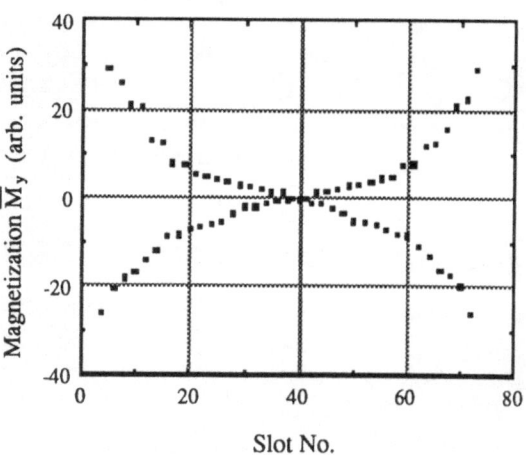

(b) renumbering method ii)

Figure 5. The weighted mean strengths of the minor components of the magnetization in each slot

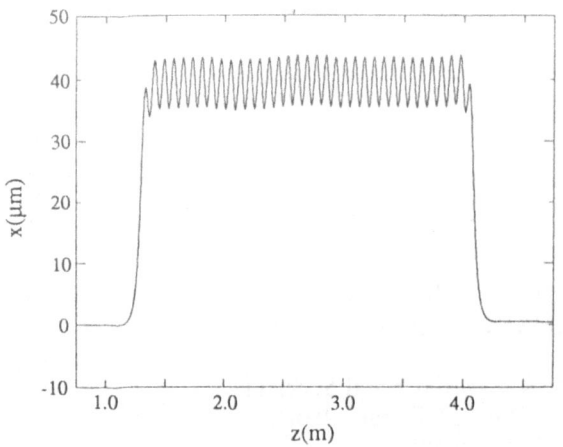

Figure 6. The electron trajectory calculated from the magnet block arrangement shown in figure 5(a)

FIELD MEASUREMENT

After loading the magnet blocks into the two magnet halves, vertical magnetic fields were measured 22mm above the surface of the iron pole, for each magnet half. The error fields were less than $\approx 10^{-3}$T. The second integrals of the measured fields, calculated by eq.(1), are shown in figure 7. The electron trajectories (a) and (b) correspond to the top and bottom halves, respectively, and the thin solid line was obtained by subtracting (a) from (b). The center line of the wiggling motion shown by the thin solid line is almost straight, as anticipated. Results of the field measurement performed with various gaps after the final assembly of the SXU were reported in ref. [4] and were also satisfactory.

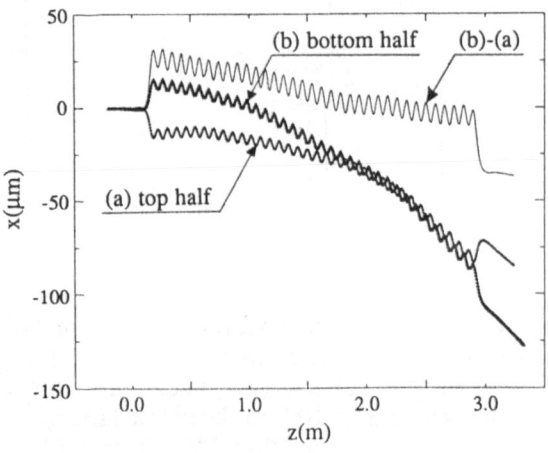

Figure 7. The second integrals of the measured magnetic fields. Measurements were performed for each magnet half.

CONCLUSION

A method to determine the loading order of permanent magnet blocks into undulators, based on the magnet transfer function has been studied and applied to the Soft X-ray Undulator(SXU) at the NSLS. The results of the magnetic field measurements indicate that an almost ideal electron orbit will be achieved.

ACKNOWLEDGEMENTS

The authors thank to NSLS staff who have helped in this work. Special thanks to G. Stenby for helping with the assembly and magnetic field measurements of the SXU.

REFERENCES

[1] H. Hsieh, S. Krinsky, A. Luccio, C. Pellegrini and A. van Steenbergen:"Wiggler, Undulator and Free Electron Laser Radiation Sources Development at the National Synchrotron Light Source," Nucl. Instr. Meth., Vol.208, 1983, pp.79-90.
[2] L. H. Yu and S. Krinsky:"Tolerances on Wiggler Magnets for NSLS Phase II," BNL 36211, 1985
[3] M.T. Menzel, H.K. Stokes:"User's Guide for The POISSON/SUPERFISH GROUP OF CODES," LA-UR-87-115, Los Alamos National Laboratory, January 1987.
[4] L. Solomon, G. Decker, J. Galayda and M. Kitamura:"Magnetic Measurements of Permanent Magnet Insertion Devices at the BNL-NSLS," in proceedings of the 1989 Particle Accelerator Conference

A MAGNET WITH A HIGHLY HOMOGENEOUS FIELD FOR A FROZEN SPIN POLARIZED TARGET OF PROTON AND DEUTERON

S. Isagawa, S. Ishimoto, K. Morimoto, A. Masaike* and S. Suzuki
National Laboratory for High Energy Physics, Ibaraki, Japan
E. Ban, JEOL Ltd., Tokyo, Japan

ABSTRACT

A 2.5 T polarizing field was prepared in a narrow gap between pole pieces which were placed near the return yoke on pole faces of a large aperture magnet called TELAS (Target Embedded Large Aperture Spectrometer). Through a careful design, proper choice of material and the so-called shimming technique, a homogeneity of the field better than 1 part in 10^4 could be obtained in pole gap of 95 mm over a useful target volume of 20 mm (height) x 20 mm (width) x 70 mm (length). A frozen spin polarized target system of proton and deuteron has been successfully operated, using this polarizing field in combination with TELAS spectrometer, to measure various spin dependent parameters of hadron-hadron reactions.

INTRODUCTION

The spin frozen target is a very useful technique for measuring spin dependent parameters of various nucleon-nucleon reactions. It is based on the fact that the nuclear spin lattice relaxation time is quite long at very low temperatures such as reached with a dilution refrigerator, even if the strength and the homogeneity of the magnetic field is reduced. After polarization one can therefore retract the target from a position with a high and very homogeneous field between heavy magnet poles and move it in a spectrometer to a position with a lower and less homogeneous field allowing a larger solid angle for observing scattered particles.

One of such kind of spin frozen polarized target was made at KEK [1], which enabled several studies of hadron-hadron scattering experiments at 12 GeV proton synchrotron. The magnetic field is given by a conventional C-shaped spectrometer, TELAS, a smaller version of multiparticle spectrometer [2]. Target material is polarized at about 0.25 ~ 0.3 K in a polarizing field of 2.5 T in a narrow gap of rectangular pieces which are attached on the pole faces of the spectrometer magnet. Then it is moved horizontally to be used in a beam at the center of the spectrometer where the high polarization is held at a temperature as low as 25 mK and in a magnetic field of 0.7 T. At the holding position the target is surrounded by scintillation counters, multi-wire proportional chambers and drift chambers.

In order to get high and uniform polarization and to measure the polarization with high accuracy, great homogeneity better than 1 part in 10^4 of the polarizing field is absolutely essential. Particularly

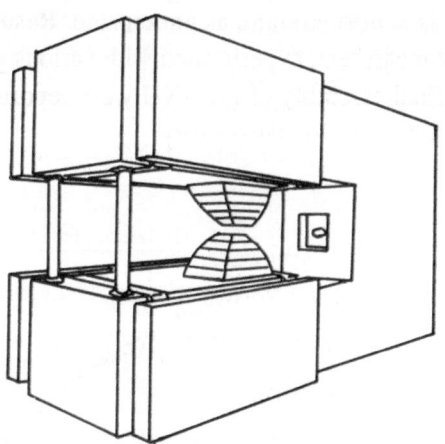

FIGURE 1   Perspective view of TELAS magnet. Characteristic parameters of TELAS are as follows.
   total weight: 128 tons (weight of coils: 8 tons);
   magnet aperture: 100 cm (height) x 115 cm
                          (width) x 150 cm (length);
   maximum current: 2500A;
   excitation: 8.5 x 10^5 A-turns;

*present address: Kyoto University, Kyoto, Japan

in case of deuteron, polarization achieved in the target depends very much upon the homogeneity of the magnetic field. This is because deuteron polarizations in the range 30 ~ 40 % are more directly proportional to inverse spin temperature than are proton polarizations near 90 %. Polarization is thus more sensitive to the inhomogeneity in the polarizing field[3].

DESIGN OF TELAS POLE PIECES

TELAS was designed and constructed at KEK for experiments with a polarized target to measure various spin dependent parameters of hadron-hadron reactions in the momentum range between 0.7 GeV/c and 2 GeV/c [2]. A perspective view of TELAS is shown in Fig. 1.

As a spectrometer it must have as large aperture and scattering angle as possible, provide as equal momentum resolution as possible for all directions and give as large field integral $B \cdot l$ as possible for the forward direction. On the other hand, as a polarizing magnet, it must be provided with a region of high magnetic field of 2.5 T with a good homogeneity. The useful volume of the region is, of course, the larger the better.

To meet these requirements TELAS consists of C-shaped magnet with a large gap (1 m) and a pair of small pole pieces placed near the return yoke on the pole faces of the magnet. The beam is incident to the spectrometer field through a hole in the return yoke. Although the magnet can be excited to the maximum current of 2500 A (excitation: 8.5×10^5 A-turns), the current was adjusted and fixed to 2010 A to get the analyzing field of 0.7 T at the center of the spectrometer, which is the optimum field for the trajectory reconstruction of scattered charged particles. On the other hand, the cryostat of the horizontal dilution refrigerator holding the polarized target was designed to have an outer diameter of 86.6 mm, which gave the lower limit of the air gap of the polarizing field. The starting conditions were thus to make a pair of pole pieces which can provide the highly homogeneous field of 2.5 T in the gap larger than 90 mm with the excitation current of 2010 A. Various shapes of pole including a conical column, a race-track mesa were examined, but the stack of rectangular pole bases and pole pieces was adopted as the final design due to its ease of machining, ease of shimming and its potential ability to elongate the homogeneous field region along the beam direction.

MAGNETIC FIELD MEASUREMENT

Overall measurement of the field was carried out using a calibrated Hall generator sufficiently long after the system was stabilized. The Hall element is a Siemens FC-33 made of InAs 3 x 6 mm^2 in size with a sensitivity of 37 mV/T. The voltage output was measured with a FLUKE 8375A

precisely kept at 20.000 mA. The Hall current was supplied from a DC precision current source (YEW2854) and its value was deduced from the voltage across a 1 ohm standard resistor which was monitored with a digital voltmeter (YEW2501).

The element was installed at the end of a search probe bar which could be accurately moved in three directions with a driving system firmly attached on the inner wall of the return yoke of the spectrometer. The accuracy to set the Hall probe at a position in the magnet is about 0.1 mm. The constant current source of the magnet is TRANSREX ISR2126-8 and its current stability is less than 1 part in 10^5. Calibration of the Hall probe was done by comparing the Hall voltage with the frequency of the wiggle appeared in the NMR signals of proton (^{1}H), lithium (^{7}Li) and deuteron (D) [4].

Although the temperature coefficient of the Hall probe is small (– 0.04 %/K between – 20 and + 65°C), a drift of the Hall voltage due to variation of the environmental temperature is not negligible. The drift in one sweep cycle along an axis is, however, very small as it takes only a very short time. When overall mapping of the field is needed, field profile for each measuring cycle can be compared based on the value of the common standard points and combined together without any inconsistency. A differential probe method was thus not used in this work.

CONSTRUCTION OF SMALL POLE PIECES

The starting point of the work producing the polarizing field is to take excitation curves for some combinations of gaps, pole tips and spacers. Figure 2 is an example of such excitation characteristics observed for rather final stage of the experiment.

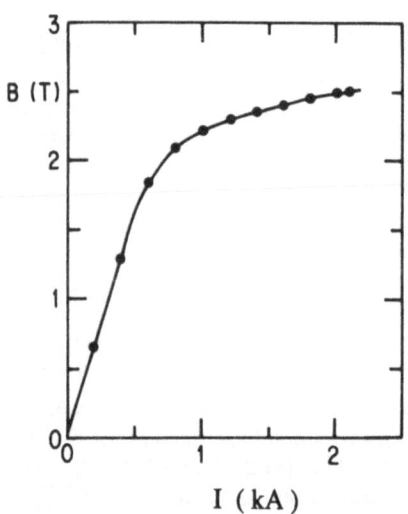

FIGURE 2   Excitation curve.

Dimensions of pole bases and pole pieces were determined very carefully. It is necessary that they have sufficient but minimum sectional area. It is because, on the one hand, serious magnetic saturation can not be allowed in any portion along the magnetic path to realize the necessary field intensity of 2.5 T in the air gap, on the other hand, much space for electronic counters is strongly required to detect particles scattered from the target as efficiently as possible.

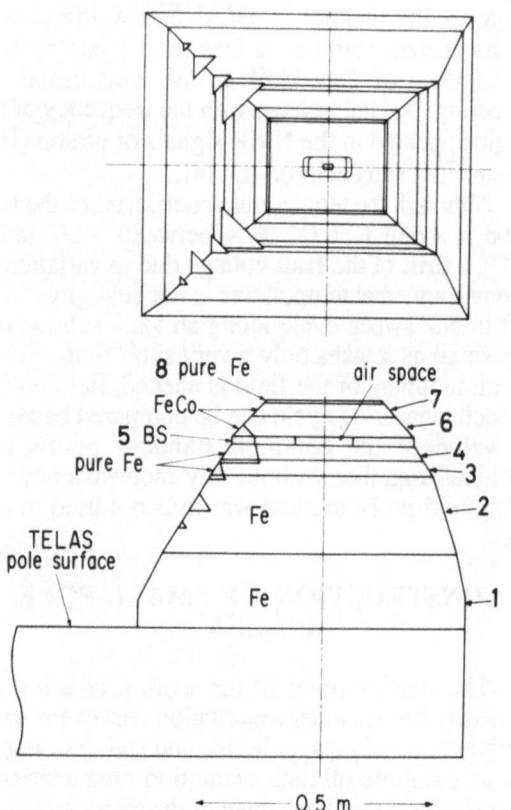

FIGURE 3   Final setup of the pole piece unit.

Figure 3 shows in detail the final setup of the pole piece unit. The construction is as follows. The 1st (1) and the 2nd (2) layer made of iron (S10C; carbon content 0.08 ~ 0.12 %) form the base stages which were designed fat enough to take in sufficient magnetic fluxes to give 2.5 T on the top, but were designed as slim as possible to give large scattering angles. The 3rd (3) and the 4th (4) layer are made of vacuum melted pure iron whose carbon content is below 0.002 %. The 5th layer (5) is two separate pieces of brass spacer ranged from 14.9 mm to 17.4 mm thick over the length of 380.5 mm in the beam direction. It is used to match the polarizing field of 2.5 T in the gap with analyzing field of 0.7 T in the spectrometer. The field strength in the gap was optimized by the thickness of the spacers. The ramp field caused by magnetic force and asymmetric structure of the magnet was also eliminated by adjusting the tilt angle of the spacers to 0.38°.

The 6th (6) and the 7th (7) layer of small pole pieces are made of cobalt iron alloy (Fe 49 %, Co 49 %, V 2 %) in order to sustain a high flux density with its higher saturation level (Bs = 2.36 T) and lower leakage factor. The final small linear inhomogeneity of the field was corrected by inserting thin liners made of copper and stainless steel (0.03 mm in thickness) into the space between the 6th and the 7th layer.

The top (the 8th, 8) layer is made of pure iron 6 mm thick which gives a pole face of 164 mm x 228 mm. It is Rose-type shim tip [5] having double rectangular ditches of 40 mm x 100 mm and 40 mm x 80 mm in the center of the pole face. The shim tip was used to compensate the higher order inhomogeneity caused by the finite dimensions of the pole piece 7.

SHIMMING A MAGNET

A higher order as well as a ramp inhomogeneity of the field was corrected step by step very carefully. The correction was, however, not straightforward due to irreproducibility of the field after partial dissassembling, for example, the upper parts above 6. Before fixing the setup, iterations totaled 45 times. They include the linear compensation by adjusting liners, shimming a pole tip by cut and try [4,6] and also, as a new trial, shimming a pole tip by adding small iron pieces and try. In order to compensate a dish-shaped field profile, a small thin iron slip 10 mm wide by 20 mm long with various thickness (0.1 – 1 mm) was placed like an island at the center of the shim top. They were first temporarily sticked on the ditch with double coated tape sticky on both sides, but were bonded semipermanently with epoxy adhesive after enough homogeneity was obtained. This method was very useful because the small piece can be easily changed and a large choice can be tried.

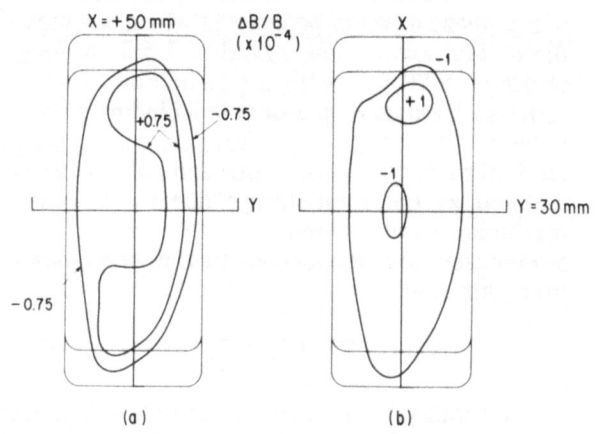

FIGURE 4   Contour maps showing a region covering (a) ± 0.75 x 10^-4 and (b) ± 1 x 10^-4 field homogeneity.

Bonding strength of double coated tape is sufficient for a temporary measurement. If the thickness of the tape (about 0.1 mm) cannot be disregarded, an instant bonding adhesive "Aron Alpha" can be used instead.

The first 7 iterations were performed first to optimize a combination of pole tips with brass spacers (4 independent discs at this stage), and then to adjust a tilt angle of the spacers and a ditch shape digged on the pole tip. For the 6 mm thick pole tip, depth of double ditch was determined to be 1 mm for the 1st step (40 mm x 100 mm) and 2 mm for the 2nd step (40 mm x 80 mm). The optimized edge angles of the pole tip are 24.5° and 25.0° along the long and short edges, respectively. The contour maps of this first approximation are illustrated in Fig. 4 a and b which show a region covering $\pm 0.75 \times 10^{-4}$ and $\pm 1 \times 10^{-4}$ field homogeneity on the median plane, respectively. A three dimensional view of the field profile at this stage is given in Fig. 5.

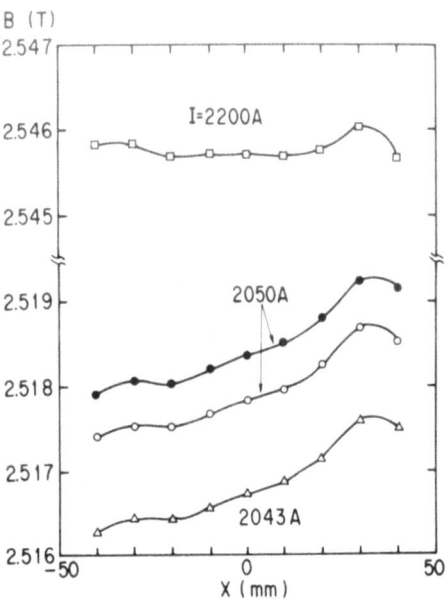

FIGURE 6   Hysteresis phenomena. Filled circles denote the field profile for 2050A after excitation up to 2200A.

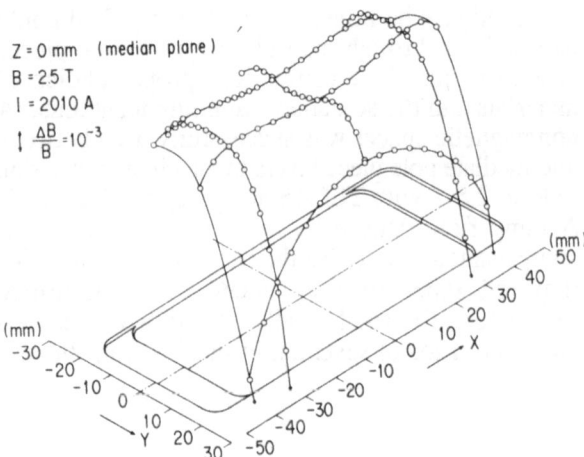

FIGURE 5   A three dimensional view of the field profile.

In the second series of a further 6 iterations, shimming was performed to compensate the dish-shaped field pattern by adding small pure iron slips 0.4 mm to 1.0 mm thick on the ditch. As the optimum thickness 1 mm was chosen, a new tip with an island machined to this size was made and used. The result was almost satisfactory, but as a whole setup was rather too temporary, a decision was made to fix the tilt angle by making wedge-shaped sturdy brass spacers (5 in Fig. 3).

During the last series of 32 iterations that were carried out after reassembling the poles above brass spacers, shimming was refined very well. One point is that the effect of hysteresis on the absolute value of the field and the linear inhomogeneity caused by varying excitation current were made clear as shown in Fig. 6 and Fig. 7, respectively. The second point is that at this stage the perturbation of the field is found to be almost linear and then the

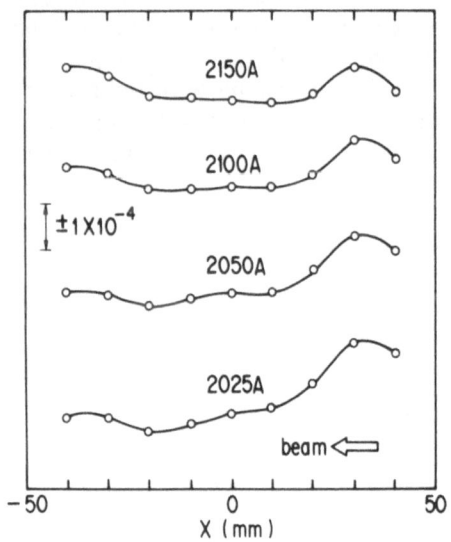

FIGURE 7   Dependence of the field profile along the beam direction on current. In this setup (not final) 2100A is optimum.

field pattern modified by island for instance can be estimated by linear approximation. Figure 8 shows the process of optimization performed on the thickness of island made of iron slips. In the lowest curve no islands are attached, while in the upper curve the field is overcompensated by use of iron slips of 1 mm in thickness. Just compensation is made by use of 0.4 mm thick slips which give the field pattern interpolated in a ratio 1:0.4.

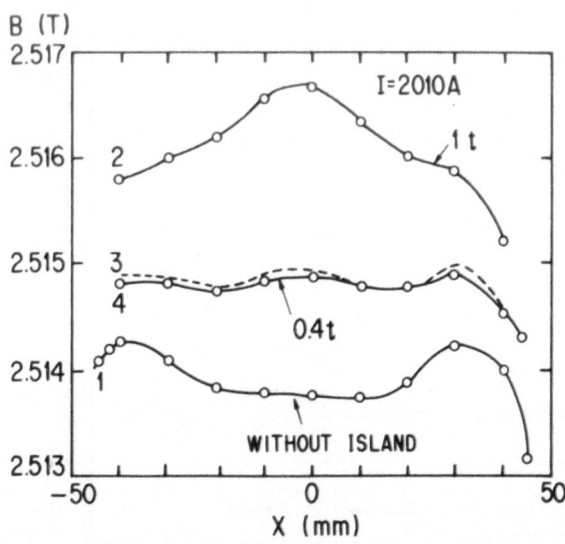

This is because the pole tip was installed to pole face of 7 by firmly tightening the 6 bolts. To eliminate this inhomogeneity last 3 iterations were made by changing the island thickness between 0.2 mm and 0.4 mm. Finally two small slips, 10 mm wide by 20 mm long by 0.1 mm and 0.2 mm thick, were stacked and bonded to the tip with an epoxy adhesive. The field profile obtained is shown in Fig. 9, spot measurements being made on a 5 mm grid. The field strength at the center of the polarizing gap was 2.506542 ± 0.0005 T for the coil current of 2010 A, which was ultimately checked by the frequency of the deuteron NMR signal. Contour maps of the field in the planes perpendicular to the field are shown in Fig. 10. Lastly, for the final setup, survey of the air gap between tips was made with a bar micrometer at twelve points on the pole tip surface, with a result that the gap was 95.66 ± 0.88 mm.

FIGURE 8 Effect of island-shaped shim. Curve 3 (dotted line) denotes the profile for 0.4 t island interpolated from curves 1 and 2. Curve 4 is the experimental results.

Further shimming continued using 0.03 mm stainless steel and 0.4 mm steel placed under the pole tips 8 at the outside edges to compensate the left-right asymmetry. Using this method, however, it should be noted that it produces its own difficulties, i.e. when a piece of foil spacer or liner is inserted between 7 and 8, it had induced another inhomogeneity caused by small bend of pole tip.

CONCLUSION

A highly homogeneous polarizing field could be obtained through careful design of pole pieces based on flux flow estimation, proper choice of materials and the so-called shimming technique. A nonmagnetic spacer was successfully used between intermediate pole piece layers to match the polarizing field of 2.5 T with TELAS analyzing field of 0.7 T. A ramp field caused by magnetic force etc. was compensated mainly by the tilt of the spacer. The required taper of the spacer was determined experimentally, final small correction was carried out by insertion of several thin stainless steel liners.

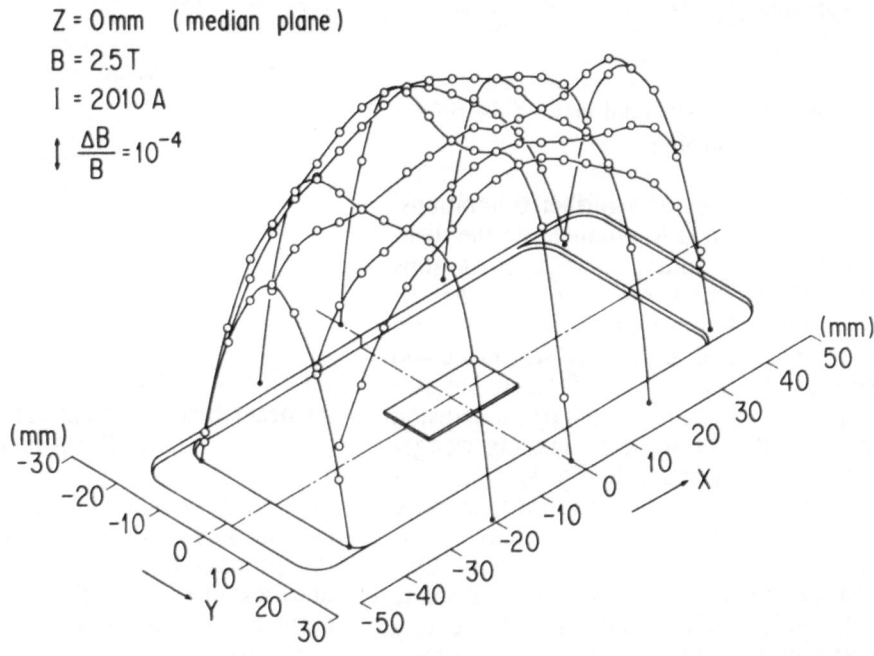

FIGURE 9 The field profile on the median plane in the gap with respect to the ditch on the shim tip.

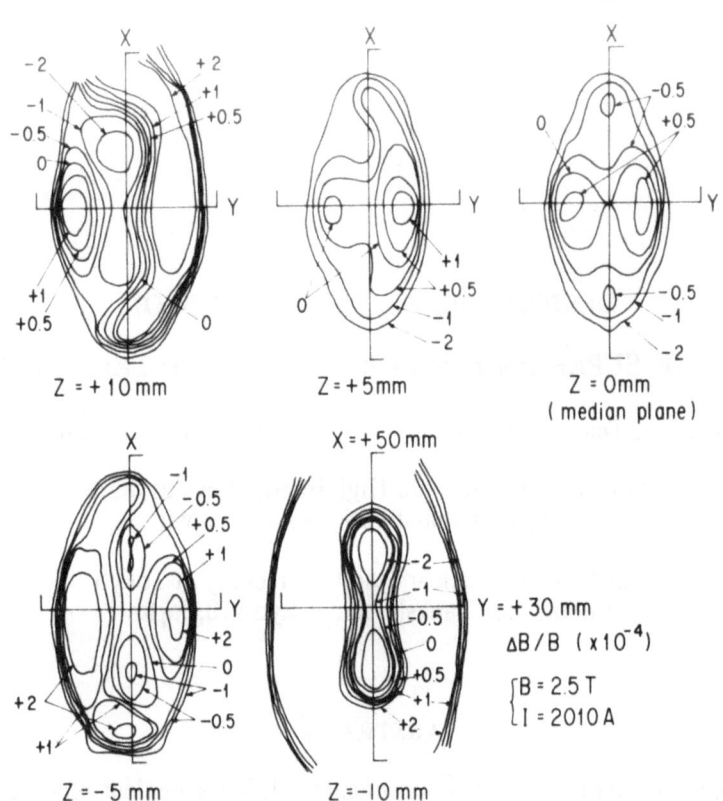

FIGURE 10 Final contour maps of the field in the planes perpendicular to the field.

The shim tip was used to correct the higher order inhomogeneity. A new method of "add and try" as well as "cut and try" worked well to compensate the dish-shaped inhomogeneity.

Required performance for spin polarized target was successfully achieved. The polarization of deuteron above 35 % was constantly obtained over the useful target volume of 20 mm (height) x 20 mm (width) x 70 mm (length). During the dynamic polarization process, if the microwave frequency (~ 70 GHz) for EPR was modulated, the maximum polarization could be made still higher as much as 4 – 5 %. The optimum FM peak frequency deviation was around 2.7 MHz, that is, the degree of modulation $\Delta f/f$ was $\pm$ 3.9 x 10^{-5}. This value is consistent with the result of field homogeneity in the target volume which was experimentally obtained in this work.

ACKNOWLEDGEMENTS

The authors express their sincere gratitude to Prof. S. Suwa and Prof. T. Nishikawa for their encouragement during the work. Thanks are due to Prof. H. Hirabayashi for his advice and valuable discussion. Their thanks also go to Prof. F. Takasaki, Prof Y. Watase and other members of KN experiment for their many discussions concerning the design and construction of TELAS. One of the authors (A. M.) is also much indebted to Dr. S.

Shimamoto for his advice and useful discussion.

REFERENCES

1. Ishimoto, S., Hiramatsu, S., Isagawa, S., Morimoto, K. and Masaike, A., A spin frozen polarized target of proton and deuteron, to be published in Jpn. J. Appl. Phys. KEK preprint 89-22, May 1989.
2. Nakajima, K., Ishimoto, S., Isagawa, S., Kabe, S., Kim, N., Kobayashi, S., Hirabayashi, H., De Lesquen, A., Masaike, A., Miyashita, S., Morimoto, K., Murakami, A., Ogawa, K., Sakuda, M., Suetake, M., Takasaki, F. and Watase, Y., The TELAS spectrometer, Nucl. Instr. and meth. 1982, 192, pp.175 – 191.
3. Niinikoski, T. O., Polarized target at CERN, Symposium on High-Energy Physics with Polarized Beams and Targets, Argonne, August 1976.
4. Ban, E., On the formation of homogeneous high field for polarized target experiment by the Rose shim (in Japanese), Trans. of IEE of Japan, 1985, A105, pp. 297 – 304.
5. Rose, M. E., Magnetic field corrections in the cyclotron, Phys. Rev., 1938, 53, pp.715 – 719.
6. Warner, G. P., Shimming a magnet to produce a high homogeneity field for muon spin rotation experiments, RL-81-083, November 1981.

DESIGN AND CONSTRUCTION OF
A LARGE SUPERCONDUCTING SPECTROMETER MAGNET

T. Shintomi, Y. Doi, O. Hashimoto*, T. Kitami*, Y. Makida and T. Nagae*

National Laboratory for High Energy Physics, KEK
Oho, Tsukuba-shi, Ibaraki 305, Japan

*Institute for Nuclear Study, University of Tokyo
Midori-cho, Tanashi-shi, Tokyo 118, Japan

ABSTRACT

The sector type superconducting spectrometer magnet, SKS, for nuclear physics experiments is under construction by Institute for Nuclear Study, University of Tokyo with collaboration from KEK. The central magnetic field is 3 tesla with the magnet gap of 50 cm and the stored energy is 11.2 MJ. Easy operation and maintenance have been taken into consideration in the design. It is intended to make the heat leak as small as 3 W at 4 K. Three dimensional magnetic field calculation and the stress analysis have been performed to design the coil and the supporting structures. The maximum operation current of 500 A is selected by calculating the maximum temperature of the coil at quench. A model coil has been tested to obtain the information on the stability and quench characteristics of the SKS magnet. A small refrigerator is to be used for thermal insulation at 80 and 20 K in combination with a medium size refrigerator at 4 K.

INTRODUCTION

There is a increasing demand in intermediate nuclear physics experiments for a large-acceptance and medium-resolution spectrometer in the momentum region around 1 GeV/c. Such a spectrometer will be a new tool to study nuclear responses in the unexplored energy region where new degrees of freedom in nuclear matter should be best investigated. In particular, studies of hypernuclei by pions in 1 GeV/c region will be intensively carried out by the spectrometer.[1]

A superconducting kaon spectrometer, SKS, for nuclear physics experiments is under construction at the Institute for Nuclear Study in collaboration with KEK[2]. It consists of a large sector type superconducting magnet which can produce up to 3 tesla in the gap of 50 cm. The estimated stored energy is 11.2 MJ for the maximum excitation.

In this paper the design and construction of the SKS magnet are described.

DESIGN AND CONSTRUCTION OF SKS MAGNET

Required Performances

The required performances of the SKS are listed in Table 1. The important features are medium momentum resolution of 0.1 % at 700 MeV/c and a large solid angle of 100 msr. From the requirements the magnet should generate magnetic field of 3 tesla with a magnet gap of 50 cm.

Table 1 Required performances of the SKS.

Momentum resolution	0.1 % at 700 MeV/c
Maximum momentum	1,500 MeV/c
Bending angle	100° at 700 MeV/c
Solid angle	100 msr
Central field	3 T
Magnet gap	50 cm

Another requirement is that the spectrometer should be movable around a target to measure angular distributions of reaction particles. It should have a flexible transfer line.

Parameters of SKS Magnet

From the required performances a sector type magnet is selected. Since the magnet is subjected to complicated electromagnetic forces due to the asymmetric shape, a H-type iron yoke is adopted for a simple coil configuration. To produce the magnetic field of 3 tesla, the necessary magnetic motive force is 2.1 MA turns. The magnet will be operated with a rated current of 500 A in a cryogenically stable mode. The estimated heat leak is very low around 3 W at 4.4 K. The parameters of the SKS are shown in Table 2.

Table 2 Parameters of the SKS magnet.

Central field	[T]	3
Field on conductor	[T]	4.5
Stored energy	[MJ]	11.2
Pole size	[m^2]	1.7x(1.2/2.3)
Magnet gap	[m]	0.5
Coil configuration		Sector
Coil cross section	[cm^2]	15x12
Conductor		NbTi/Cu
Cond. cross section	[cm^2]	0.3x0.2
Cond. Cu/SC ratio		10
Ampere turns	[MA]	2.1
Current	[A]	498
Current density	[A/mm^2]	58
Type of yoke		H
Total weight	[t]	~300
Heat leak at 4.4 K	[W]	3

Magnetic Field and Force Calculations

The magnetic field calculation has been carried out using two kinds of three dimensional computer codes to cross-check. One is based on a finite element method and the other on an integration one. From the calculated magnetic field, the electromagnetic forces were also estimated in two cases of the coil geometry, one is properly positioned and the other misaligned.

In case of the normal alignment, since the coil is symmetric in the x and z directions as shown in Fig. 1, each component of the forces balances for each symmetrical plane. On the contrary, force in the y direction is unbalanced and should be supported from the outside. The calculated forces acting on the quarter of the coil are 2.15 MN in the x direction, 110 kN in y and 1.98 MN in z.

The unbalanced forces were calculated for the coil position misaligned in the three directions. The magnetic forces due to the misalignment act on the coil as restituting ones in the x and y directions. Then the coil will be pulled back to the normal position. On the contrary, the force is unstable in the z directions. When the coil is misaligned by 10 mm in the z direction, the unbalanced force of 640 kN will act on the coil. As this is almost proportional to the displacement and the misalignment is less than 5 mm, the force will be one order of magnitude less than that for the normal alignment. Those magnetic forces are summarized in Table 3.

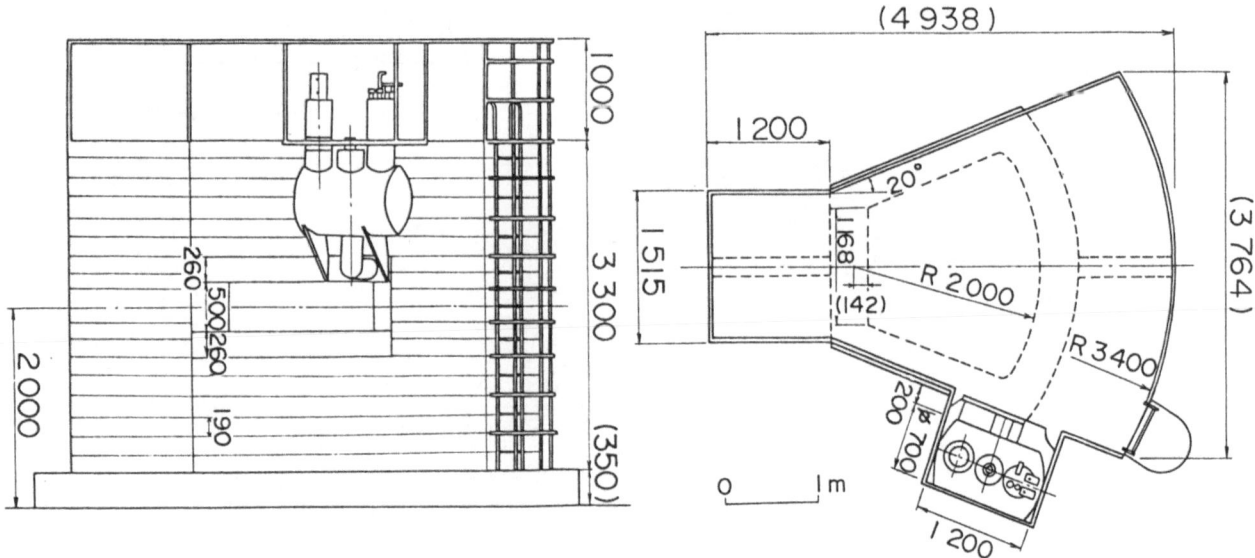

Fig. 1 Drawings of the SKS spectrometer magnet.

Table 3 Magnetic forces due to misalignment.

Δz +10 mm	160 kN	unstable
Δy +10 mm	150 kN	stable
Δy +10 mm	670 kN	stable
Δx +10 mm	220 kN	stable

The thermal stresses will be relatively large if the coil was supported from all directions. Therefore, the coil is supported at one end to be able to shrink freely using a supporting system devised to release the thermal stresses. The unstationary thermal stresses at cool-down will be made small by temperature difference less than 50 K in the coil and cooling slowly with a temperature gradient of 6 K/h.

Stress Analysis

The magnetic forces except for the asymmetric force in the y direction are supported by the rigid helium vessel. Using the finite element method, the stresses and displacements of the helium vessel were calculated under the condi-tion of magnetic forces and thermal contraction. The maximum stresses of 298 MPa will occur at the corner of the coil arc. For stainless steel these stresses are within an allowable value. The maximum displacement of 2.6 mm in the y direction occurs at the center of the arc. At the straight section of the coil, the displacements in the x and y directions are 2 mm, respectively.

Superconducting Coil

The superconducting coil of the SKS magnet consists of two parts, upper and lower coils, which are electrically connected in series. Each coil is set into upper and lower parts of a rigid helium vessel mechanically connected in order to support against electromagnetic forces in the z direction. Since the forces are mainly supported with the helium vessel to be cancelled for the symmetrical plane, the coil must be supported from the room temperature against only the asymmetrical forces in the y directions, the unbalanced forces by misalignment and gravity. GFRP cylindrical tubes are used for support as shown in Fig. 2.

The windings of coils are graded as shown in Table 4. The conductor is a monolithic type of NbTi/Cu of which Cu/SC ratio is more than 10. In general, a conductor with a large Cu/SC ratio and thick filaments does not have good properties in critical current density. To obtain a cable with high critical current density, the filament diameter should be thin to some extent. The cable is fabricated by burying a thin wire into a copper matrix.

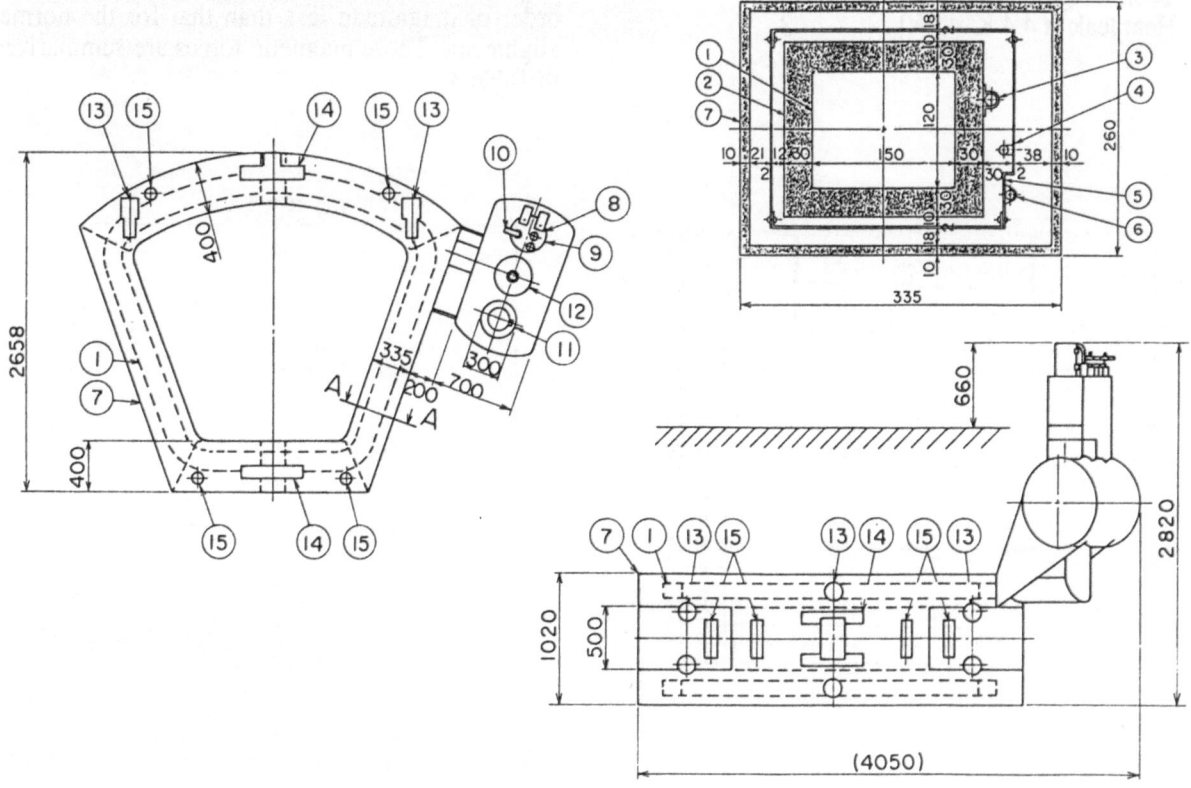

Fig. 2 Structures of the SKS superconducting coil. 1:coil, 2:He vessel, 3:precooling pipe, 4:20 K cooling pipe, 5:80 K shield, 6:80 K cooling pipe, 7:vacuum vessel, 8:power lead, 9:port, 10:safety valve, 11:80 K and 20 K ports, 12:extra port, 13 - 15:supports

Table 4 Parameters of the coil and the cable.

Coil	Inner	Outer
stored energy [MJ]	11.2	
ampere turns [MA]	2.1	
current [A]	498	
magnetic field B_0 [T]	3.0	
B_m [T]	4.5	3.0
no of turns/pole	990	1120
Cable		
size [mm^2]	3.5x1.9	3.3x1.8
filament size [mm]	27	24
no of filament	900	900
I_c at 4.2 K [A]	1380(4.5 T)	1540(3.0T)
insulation	25 μm Kapton half lap	

The load curve of the SKS magnet is shown in Fig. 3.

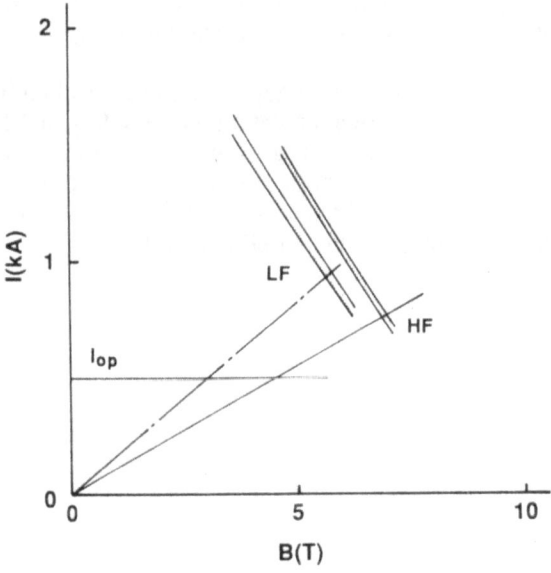

Fig. 3 Load curve of SKS coil.
LF: low field winding
HF: high field winding

Figure 4 shows the coil under winding and the finished coil with the helium vessel.

QUENCH TEST WITH MODEL COIL

Magnet stability and quench characteristics have been tested using a solenoidal model coil which was made to simulate the SKS magnet. The inner and outer diameters are 180 mm and 380 mm, respectively. The coil cross section is 100 mm in width and 120 mm in height, and almost the same

dimension of the SKS magnet. It has heaters for intentional quench test. The parameters of model coil are shown in comparison with the SKS magnet in Table 5.

At the operation below 282 A, the coil recovered from the normal state after removal of a heater pulse. In this condition, the heat flux was 0.213 W/cm^2 which corresponds to that obtained for a bare cable. At the operation of 300 A, and the heater power of 0.5 W with pulse duration of 80 ms, the coil quenched. The input heat corresponds to 1.5 J/cm^3. If the wire movement of the practical SKS magnet is 1 mm, the heat of 0.34 J/cm^3 will be generated. Therefore, the SKS magnet may be stable at the wire movement over 1 mm.

(a) coil under winding

(b) finished coil.

Fig. 4 Pictures of SKS coil.

Table 5 Parameters of the model magnet.

	Model coil	SKS coil	
I_c [A]	550	1,230	
I_{op} [A]	350	497.6	
E_{st} [MJ]	0.069	11.2	
B_0 [T]	3.35	3.0	
B_m [T]	4.35	4.5	
Cable size [mm²]	2.69x1.34	3.6x2.0	HF
		3.4x1.9	LF
Coil size [mm]	180 ID		
	280 OD		
Coil c.s. [mm²]	$120^h x 100^w$	$120^h x 150^w$	
No of layers	52	62	
No of turns/layer	43	33	HF
		35	LF
J_{av} [A/mm²]	65.2	58.6	
Inductance [H]	1.09	90.5	
E/s²R [A²s/cm⁴]	2.0x10⁹	1.5x10⁹	HF
		1.9x10⁹	LF
$\dot{Q}$/A [W/cm²]	0.375	0.13	HF
		0.30	LF

$$E/s^2R = \int I^2 dt$$

THERMAL HEAT LEAKS AND REFRIGERATION SYSTEM

Heat Leaks

As mentioned before, the SKS spectrometer magnet is intended to have low heat leak. To reduce the heat leaks, a thermal shield at 80 K and thermal anchors at 80 and 20 K are considered as shown in Fig. 2. A thermal shield at 20 K is omitted because of small contribution to reduce the heat leaks and also simple fabrication. The heat leaks to the cryostat are shown in Table 6.

Table 6 Heat leaks of the SKS magnet in W.

Temp	80 K	20 K	4 K
Power lead	--	--	0.6
Support	17.3	1.8	0.2
Gas conduction	--	--	<0.1
Service port	6.8	0.9	0.9
Radiation	19.0	--	1.0
Others	4.7	0.7	0.1
Total	47.8	3.4	2.9

Refrigeration System

The coil will be cooled by a refrigeration system of which flow chart is shown in Fig. 5.

A medium size refrigerator is to be installed for the initial cool-down and stationary cooling at 4 K. In combination with it, a mini-refrigerator, which can be operated without maintenance for a period more than 10,000 hours, is adopted for the thermal shields at temperatures of 80 and 20 K.

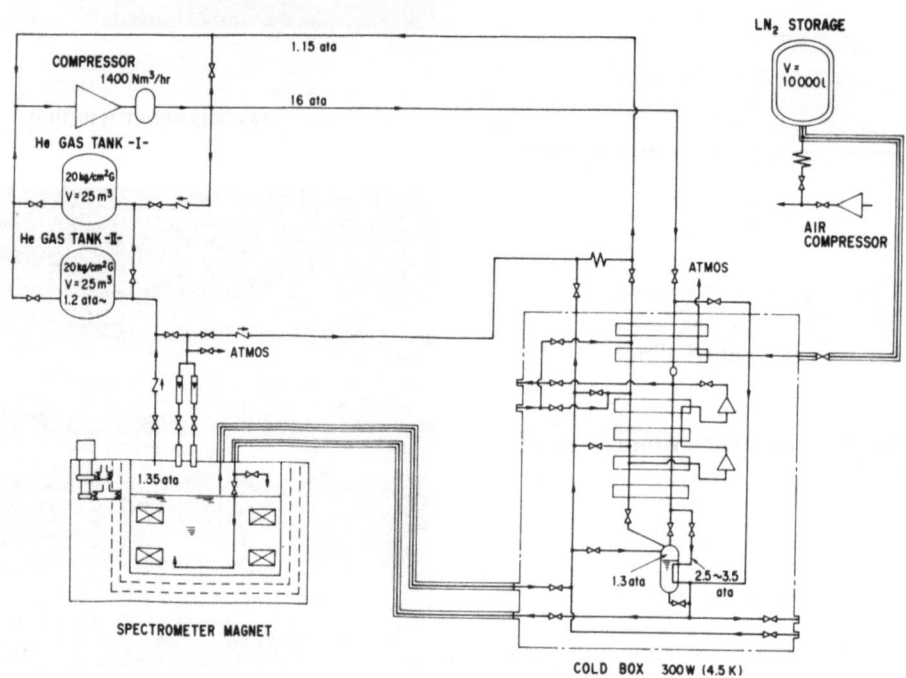

Fig. 5 The refrigeration system of the SKS magnet.

SUMMARY

A superconducting spectrometer magnet for nuclear physics experiments has been under construction and will be completed in the spring of 1990. The magnetic field of the sector type magnet is 3 tesla with the magnet gap of 50 cm. The stored energy is estimated to be 11.2 MJ. Easy operation and maintenance have been considered in addition to usual design concept. The magnetic field and forces were calculated with three dimensional computer codes. Using the results, stress analysis has been performed to determine the structures of the helium vessel and supports. The quench simulation was carried out to decide the operation current and to check the safety of the coil. As the result, the current of 500 A was selected. The stability has been tested using a model coil which simulate the SKS magnet. The heat leaks are estimated to be less than 3 W at 4 K. A small refrigerator is to be used for thermal insulations at 80 and 20 K in combination with a medium size refrigerator for 4 K.

REFERENCES

1. O. Hashimoto, et al., INS-report, To be published in Nuovo Cimento.

2. T. Shintomi, et al., Design of a Large Superconducting Spectrometer Magnet, IEEE Trans. on Magnetics, MAG- 25, 1667 (1989)

AN ULTRA-PRECISE STORAGE RING FOR THE MUON g - 2 MEASUREMENT*[1]

D. Brown, T. DeWinter, E. Hazen, C. Heisey, B. Kerosky, F. Krienen, D. Loomba, E. McIntyre, D. Magaud, W. Meng, J. Miller, L. Posnick, B. Roberts, D. Stassinopoulos, L. Sulak, W. Worstell - Boston University.
G. Bunce, H. Brown, B. Chertok, G. Cottingham, J. Cullen, G. Danby, B. DeVito, J. Jackson, M. May, J. Mills, C. Pai, A. Pendzick, I. Polk, A. Prodell, L. Snydstrup, R. Shutt, A. Soukas, K. Woodle - Brookhaven National Laboratory
K. Becker, M. Lubell - City College of New York.
T. Kinoshita, Y. Orlov - Cornell University. D. Winn - University of Fairfield.
K. Jungmann, H. Mundinger, G. zu Putlitz - University of Heidelberg
K. Endo, H. Hirabayashi, S. Kurokawa, T. Sato, A. Yamamoto - KEK. K. Ishida - Riken.
W. Lysenko - Los Alamos National Laboratory.
L. Barkov, B. Khazin, E. Kuraev, and Ya. Shatunov - Institute of Nuclear Physics, Novosibirsk, USSR. J.T. Reidy - University of Mississippi.
K. Nagamine, K. Nishiyama - University of Tokyo.
H. Ahn, J. Bailey, P. Cushman, S. Dhawan, A. Disco, F. Farley, V. Hughes, H. Venkataramania, - Yale University.

ABSTRACT

An ultra precise 3 GeV/c storage ring with a 14.5 kG super-ferric magnet is under construction at the Brookhaven AGS for the measurement of the muon anomalous magnetic moment to 0.35 PPM accuracy. This requires a magnetic field which is constant to $\approx$ 1 PPM
and is known sufficiently well that the magnetic field integral averaged over the muon
orbits can be calculated to 0.1 PPM. First the magnetic field will be statically
shimmed by various techniques. Pole face winding will be used for final small static
and dynamic corrections. Very elaborate NMR field monitoring techniques are required.
A "movable trolley" located inside the vacuum chamber and the electrostatic focusing
quadrupoles will measure the field throughout the muon storage volume. The trolley
"siding" is 180° from the injection point where no electric quadrupoles are located.
Injection can be interrupted so the trolley can circle the ring. Also $\approx$ 200 NMR probes
located outside the vacuum chamber monitor the field during physics running and control
the pole face windings. The very large ($\approx$ 15 m diameter) superconducting coils (SC)
are designed. Test winding will soon commence. Orders for the magnet steel can now be
placed.   R and D on various pulsed and SC dc injection methods is ongoing.

INTRODUCTION

A measurement of the muon g - 2
value, i.e. A μ = (g - 2)/ 2, to 0.35
PPM is a factor of 20 improvement over
present knowledge,[1] and is 20% of the
theoretical value. This accuracy might
indicate possible deviations from the
standard model. The method exploits the
spin motion of polarized muons in a
storage ring of uniform B.[2][3] The
spin precesses with angular frequency ω_s
greater than the cyclotron angular frequency ω_c by ω_a, the g - 2 precession

*Work performed under the auspices of the U.S. Department of Energy.

frequency $\omega_a = \omega_s - \omega_c = e/mc\ A\mu B$. Integral B and ω_a must be measured in the experiment. e/mc is known. To provide vertical focusing, electrostatic quadrupoles are distributed around most of the circumference. For a "magic" momentum of $P\mu = 3.09$ GeV/c, or $\gamma = 29.3$, the effect of the electric field on the precessional frequency drops out[2] requiring "only" producing and measuring a constant B field.

$$\omega_a = \omega_s - \omega_c = \frac{e}{mc} \left[A\mu B - \left(A\mu - \frac{1}{\gamma^2 - 1} \right) |B \times E| \right]$$

Figure 1 shows the layout of the experiment. Longitudinally polarized 3 GeV/c muons are captured in the ring from the decay of injected pions. Alternatively, muons can be directly injected, requiring a fast kicker. Calorimeters distributed around inside the ring detect electrons from muon decay. The electron emission direction is correlated with the muon spin direction, so the muon decay spectrum is modulated by ω_a. Table 1 summarizes this experiment and the previous (CERN) experiment. Table 2 lists parameters of the experiment, Table 3 expected errors. Note that the planned very small statistical error of 0.3 PPM requires knowledge of the field integral to 0.1 PPM.

TABLE 1

AGS Muon g - 2 Experiment

GENERAL APPROACH - Similar to CERN Experiment

IMPROVEMENTS

1. Primary Proton Beam Intensity (With AGS Booster x 100 to 200). Error, $\omega_a = 0.3$ PPM (CERN = 7.5 PPM)

2. Storage Ring Magnet: Field Homogeneity and Control
   Homogeneity ($\approx$ 1 PPM) [CERN $\approx$ 10 - 15 PPM]
   Control ($\approx$ 0.1 PPM) [CERN $\approx$ 0.5 PPM]

   Achieved by:

   a. Superconducting Coils
   b. Larger Magnet Gap
   c. Azimuthal Symmetry in Iron Construction
   d. Extensive Shimming Features
   e. NMR Feedback and Control

3. Magnetic Field Measurement (NMR) - Accuracy (0.1 PPM)

   Achieved by:

   a. NMR Trolley, Moveable Within Vacuum Chamber
   b. 200 Fixed NMR Probes Outside Vacuum Chamber
   c. Insertable NMR Probe for Absolute Calibration

4. Detector System - Increased Acceptance, Data Rate Capacity, and Time Measurement Accuracy

TABLE 2

Parameters of AGS Experiment

Pulsed Magnetic Inflector (or dc SC inflector)

   π Injection, μ from $\pi \to \mu$ decay
   μ Injection, need fast full aperture kicker

Muon Lifetime, τ (3.094 GeV/c)	64.4 μs
Orbital Frequency, f_c	6.70 MHz
g - 2 Precession Frequency, f_a	0.229 MHz
Orbit Radius	7112. mm
Central Magnetic Field	1.45 T
Magnet Gap	18. cm
Pole Width	56. cm
Storage Aperture Diameter	9. cm
Electric Quadrupole Field (Pulsed 1 ms)	38.7 kv

Superconductor 1.8 mm x 3.3 mm NbTi/Cu monolith 1:1 ratio, 1400 SC filaments of 50 μm.

Aluminum Extrusion 3.6 mm x 18 mm with SC Centered, Al/(NbTi + Cu) ratio 10, Al rrr > 2000

IC $\approx$ 10,000 Amps, Iop $\approx$ 5500 Amps (Bop $\approx$ 2 T)

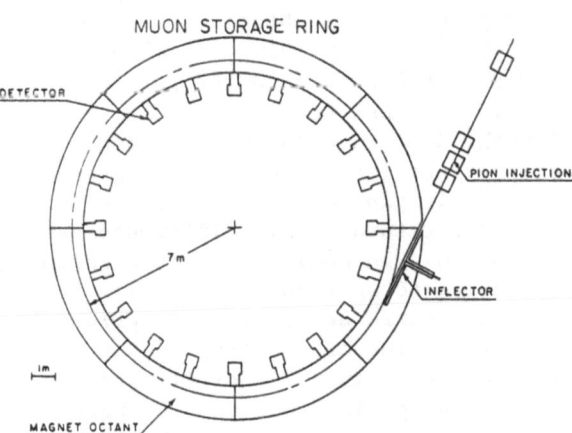

MUON STORAGE RING

DETECTOR

PION INJECTION

INFLECTOR

MAGNET OCTANT

FIGURE 1   General Arrangement of AGS
             g - 2 experiment.

Table 3

Counting Rates and Statistical Errors*

Storage Aperture Diameter	90 mm
Protons Per rf Bunch with Booster	4.2×10^{12}
Bunches Ejected per Ring Fill	2
Protons Per Fill	8.4×10^{12}
Pion $\Delta p/p$	$\pm 0.6\%$
Pions at Inflector Exit Per Fill	
(1.28×10^7 per 10^{12} protons)	1.07×10^8
Muons Stored Per Fill	
(at 134 PPM Capture Efficiency)	14×10^{13}
Electrons Counted Above 1.6 GeV Per Fill	
(20% of the Decays)	2.8×10^3
Fills Per AGS Cycle (1.4 s)	3
Fraction of AGS Protons Used	50%
Fills Per Hour	7714
Electron Counts Per Hour	22×10^6
Total Electron Counts N_e	2.8×10^{10}
Running Time for 0.3 PPM (1 std. dev)	1288 hours

*Pion Injection Case

	Systematic Errors	
Source	Comments	Error (PPM)
Magnetic Field	Includes Absolute Calibration of NMR probes and Averaging Over Space, Time, and Muon Distribution	0.1
Electric Field Correction	0.7 PPM Correction	0.03
Pitch Correction	0.4 PPM Correction	0.02
Particle Losses		0.05
Timing Error		0.01
	TOTAL (In Quadrature)	0.12

Figure 2 is a computer printout of the magnet cross section, showing the yoke, the pole, the SC coils, and the flux lines. The C magnet flux return provides access for the calorimeters. The pole base air gap decouples the aperture field shape from the properties of the yoke steel. Standard 1006 low (.06%) carbon steel, can be used for the yoke, the major Fe cost (640 tonnes). Very pure (10005)(.005%) iron pole pieces (the Nippon Steel Company) provide better permeability at high fields, and also expected greater chemical uniformity and freedom from voids. The pole base air gaps will be wedge shaped to correct the large gradient produced by the inside-outside radial asymmetry. The SC coils reduce greatly power supply field ripple and permit larger aperture gaps without the large power dissipation of copper coils. It should be possible to operate continuously for an entire muon run to minimize the effect of hysteresis and eddy currents on the magnetic field

shape. The SC coil pair at larger radius have a common cryostat. The smaller radius coils each have their own cryostat, located well above and below the horizontal mid plane (HMP) to provide access for shimming to the pole base air gaps. At this location the hoop stress on the inner coils is nulled: if located near the HMP a very strong negative hoop stress results.

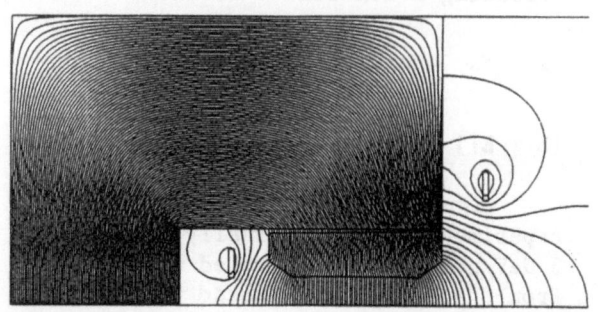

FIGURE 2 Computed Magnet Cross Section

The ring operates at a single field so extensive static shimming is planned i.e., iron and possibly other magnetizable materials. Improvement of up to 100 times standard magnet tolerances is expected from shimming. A model program is exploring this issue. Changes to the remote iron yoke will produce long "wavelength" changes in the aperture field. An extreme example is the penetrations near the HMP for injection apparatus, cryostat leads, etc. Extra iron will be added near these openings to try to restore the effective total reluctance. Iron shims can also be added or subtracted in the pole base air gaps. This can be used for all wavelengths down to ≈ 10 cm. The sensitivity to iron added is much greater in the air gap than on the surface of the yoke return. This difference can be exploited. In addition, static shimming by adding to or removing material from the pole surfaces provides in principle completely localized control, but with extreme sensitivity. For illustration, a 1 μm change in magnet gap changes the field by 5.6 PPM. Area can be exchanged for thickness. For example, a series of small spots covering 1% of the surface but 100 times as thick can be equivalent to the continuous layer. Techniques for accomplishing this are being explored in the model program.

Pole face windings employing economical printed circuits will make the final small corrections. Now simple correcting coils produce higher multipole errors, typically several percent of their desired fundamental. These errors should be less than 0.1 PPM to avoid complicating knowledge of the field, so static shimming is used to the maximum extent to minimize coil excitation. In addition, unless pole face currents are quite small, significant I^2R heating occurs. A 1 PPM dipole correction requires only 0.1 AT per pole.

THE STORAGE RING IRON AND COILS

Figure 3 shows the magnet cross section. The top and bottom blocks will be made of two plates of half thickness welded together. The HMP block will be two separate pieces. This provides a parting surface at the HMP for machining inflector and cryostat penetrations.

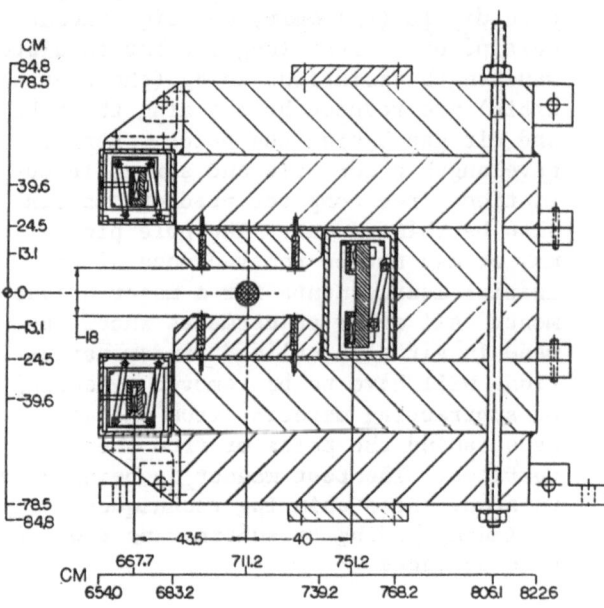

FIGURE 3 Magnet Cross Section

The outstanding construction feature of the SC coils is the very large diameter to cross section ratio. Very careful structural design is required. The Al stabilized conductor used is the same as for the Topaz detector solenoid.[3] The 24 turn single layer coils are wound on the inside of the mandrel under outward compression. Hinged supports accommodate the very large differential motions for 14 m diameter SC coils. The

forces (lbs/linear inch) are: outer coil, radially outward 207, vertically 88 away from HMP: inner coil, radially inward 3 and vertically 114 away from HMP. Supercritical helium passing through channels on the mandrels cool to 4.5 K. The cryostat vacuum vessel is anchored to the iron yoke. Eight mechanical stops define a circle concentric with the yoke, with diameter equal to the coil mandrel when cold and powered. The coil is located to ± 0.5 mm. The aperture field is iron dominated so tolerances on current locations are reasonable. Computations show for 1 mm vertical movement of the outer coil at ρ = 4.5 cm, a quadrupole of 1.1 PPM, sextupole = 0.2 PPM, others smaller. The dipole change is 56 PPM. Horizontal motion give comparable results. The inner coils are even less sensitive. Unless the coil location is reproducible and constant, field control will be more complicated.

Each pole is attracted to its yoke block by 800 lbs per linear inch. The pole support bolts and hole plugs will be carefully magnetically matched to minimize field errors. The pole edge shims will help smooth any residual effects (Fig. 3). The torque produced by attraction between poles is opposed by pretension in the large yoke tension bolts. All parting surfaces are flat and parallel to 38 μm. Critical radii are within 127 μm, as are the 30° sector mating surfaces. Although fabricated in 30° sectors, the ring will be as "continuous" as practical. The sectors fit tightly and the poles overlap the joints to minimize magnetic end effects. Figure 3 shows the iron yoke must be assembled around the double coil cryostat. Poles can be installed (or removed) after all SC coils are assembled. Electronic levels will continuously monitor relative vertical motion to high precision. The support pads will have freedom to accommodate small movement in the horizontal plane, so the ring formed by joining the 30° yoke sectors should maintain its circular integrity.

The magnet design used the computer code "POISSON". Extensive perturbation studies of small changes at the PPM level were made. For simplicity, two dimensional coordinates were normally used plus a simple transformation to the actual cylindrical coordinates. Some calculations using cylindrical coordinates confirmed the approximate method. Representative permeability data for 1006 steel was used in the yoke and 10005 steel in the pole pieces.

The air gap behind the poles and the detailed pole shape are used to adjust the field. Figure 4 shows the extreme uniformity computed over the required 4.5 cm radius ($\leq$ 1 PPM). The actual magnet will be much less uniform.

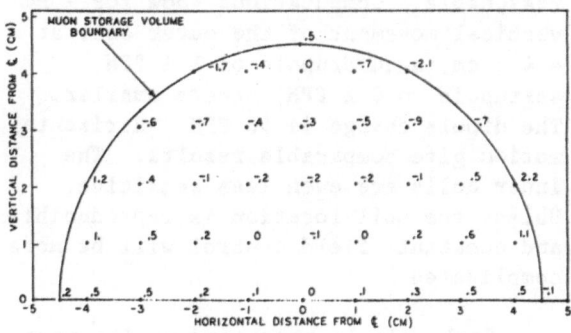

FIGURE 4   Computed Field Uniformity $\Delta B/B_0$ (PPM) in Storage Aperture.

The computer results can be used for perturbative correction following initial measurement. Table 4 shows computed 50 μm changes of the bumps at each pole edge and also the effect of changes in the air gap wedge angle. The wedge produces almost pure quadrupole whereas a quadrupole perturbation using edge bumps produces 41% octupole moment.

TABLE 4

Multipoles Computed at r = 4.5 cm for 3 Perturbations

Multipole Moment	I PPM	II PPM	III PPM
(Quad) 1	- .2	+ 9.2	+ 28.80
(Sext) 2	+ 7.5	0	+ 2.80
3	- 0.1	+ 3.8	0
4	+ 1.4	0	+ 0.10
5	0	+ 0.4	- 0.05
6	+ .1	0	- 0.05
7	0	0	0
8	0	0	- 0.05

Column 1 - All four pole edge bumps - 50 μm thicker.
Column 2 - Small ρ bumps 50 μm thinner, large ρ bumps 50 μm thicker.
Column 3 - Increase pole base air gap wedge by 27 millirads, 20 times fabrication tolerances.

Detailed design remains of the air gap behind the poles. Parallel machined surfaces are preferred over wedges. Preliminary design uses an Al plate with parallel surfaces and with $\approx$ 20 parallel grooves 2.5 cm wide spaced across the 56 cm plate, in the azimuthal direction. Iron shim stock inserted in the grooves stepwise approximates a wedge to correct the systematic gradient. On the other side of the Al plate, small radial grooves are used for "random" shimming, after measurements. Once the wedge design is complete, final adjustments to the pole shape will be computed using cylindrical coordinates and the final permeability data estimates.

PRECISION SHIMMING AND MEASUREMENT

The ring will probably start at $\Delta B/B_0 > 1 \times 10^{-4}$, (25 μm $\approx$ 100 PPM). Material variations and geometrical errors are always present. Computer calculations show yoke reluctance changes produce dipole (B_0) only, not significant multipoles. First long and intermediate wavelength variations in B_0 (the central orbit) are reduced by shimming the yoke and air gap behind the poles. Longer wavelength quadrupole and sextupole deviations from computer predictions can also be shimmed out. The pole pieces may or may not be removed once after initial measurements for a major adjustment. All other adjustments should be made in situo. Localized random variations will have to be removed by adding or subtracting material from the air gaps behind the poles or from the pole surfaces. The test magnet is being used to study these shimming techniques, including adding or subtracting from the pole surfaces.

Pole face windings provide the final adjustments. They are also necessary to adjust for environmental or cyclical induced field changes with time. The goal is to produce a field with $\Delta B/B_0 \leq 1$ PPM, including the pole face windings which will be controlled using the very elaborate NMR results. POISSON perturbation calculations will be a great aid in predicting the various correction processes.

NMR permits absolute measurement to 0.1 PPM. The trolley (Fig. 5) will be used for initial shimming and intermittently during actual data runs. A second long arm will be located at the opposite end of the trolley to that shown. With two probe matrices the restricted region at the injector can be traversed. The probes will probably be oriented in a polar fashion to give multipole fits rather than in the rectangular matrix.

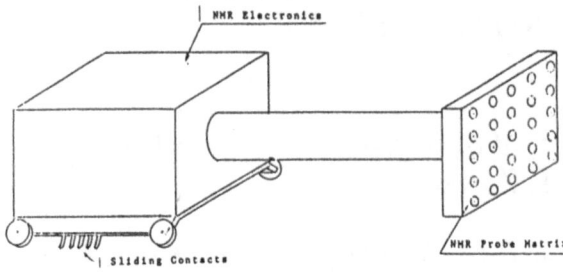

FIGURE 5 Schematic of NMR Trolley.

In addition to the 200 fixed probes outside the vacuum chamber (VC) and the moveable trolley, sets of plunging probes will be inserted into the VC. These absolute calibration probes will check the predictions of the two on-line systems. FM sweeping of the NMR line will be used. This requires no perturbing magnetic fields and uses small sample probes. The small 62 MHz rf transmitters around each probe are shielded. Each synthesizer is multiplexed to 16 NMR probes. NMR data will be collected from one probe at a time. A beam cross section measurement of 25 probes requires 2 seconds. The trolley then moves azimuthally 2 cm, stops, and repeats the process. In about three hours, a complete map with 50,000 points can be taken. Communication to and from the trolley is required and perhaps power for the electronics. Tests have been successful of sliding electrical contacts in vacuum. The trolley rides on tracks in the corners of the VC. The trolley azimuthal drive will either be mechanical or by a motor using the 14.5 kG field. An encoder will monitor the azimuthal location.

All 200 fixed probes will be sampled with a 1 second repetition time. The average NMR error frequency will be fed back to a precision current transformer to control the main power supply. In addition, the probe data will give "regional" information on dipole and multipole corrections to be performed by localized pole face windings.

An elegant alternate $\int B dl$ direct measurement is possible. For non-relativistic particles measuring the cyclotron frequency ω_a determines $\int B dl$ around the ring if their mass is sufficiently well known. At the AGS, p = 3 GeV/c Au^{+1} ions with β = 0.15 could be injected. If the incoming ion beam $\Delta p/p$ is measured to 4 x 10^{-4}, $\int B dl$ can be measured to 0.1 PPM.

COMMENT

The pulsed injector used at CERN[2] is difficult and costly to scale up to the desired AGS multipulse cycle. A DC inflector using SC coils is attractive. At least two versions are being pursued. Injector field compensating correction coils will probably be required to permit the 0.1 PPM orbital precision.

Direct muon injection is attractive but requires a fast kicker x 90° from the inflector. This is being studied.

REFERENCES

1. All the authors either work on this apparatus or will participate in the experiment. Only small portions of this work have been published. Extensive g - 2 Group Reports exist. These are available from the authors listed.

2. Bailey, J. et al., Nucl, Phys. B150, 1 (1979). 1 to 79.

3. Drumm, H. et al., Nucl. Instrum. Methods 158 (1979) 347 to 362.

4. Ikeda, M. et al., Proc. of ICEC-11, Berlin-West (1986) p. 675.

This paper was presented by G. Danby.

A SUPERCONDUCTING TOROIDAL MAGNET FOR CHARGED PARTICLE SPECTROSCOPY

J. Imazato, H. Tamura*, T. Ishikawa*, R.S. Hayano*, T. Yamazaki**,
A. Yamamoto and T. Kawaguchi***

National Laboratory for High Energy Physics (KEK), Tsukuba-shi, Ibaraki-ken, 305 Japan
*Department of Physics, University of Tokyo, Bunkyo-ku, Tokyo, 113 Japan
**Institute for Nuclear Study, University of Tokyo, Tanashi-shi, Tokyo, 188 Japan
**Mitsubishi Electric Co., Wadasaki-cho, Hyogo-ku, Kobe-shi, 652 Japan

ABSTRACT

A superconducting toroidal magnet has been designed and constructed for charged particle spectroscopy in nuclear and particle physics at KEK. It consists of 12 coils with iron cores, producing a maximum field of 1.8 T in 12 uniform gaps, and has a feature of indirect two-phase-helium cooling. In this paper we describe the basic design and the structure of the magnet and report the results of the first performance test.

1. INTRODUCTION

In nuclear and particle physics experiments, high momentum resolution and large acceptance are major requirements of a particle spectrometer. One of the best suited types of magnets satisfying these requirements for a center of mass system is a toroid. We designed and constructed a large-acceptance spectrometer based on a superconducting toroidal magnet with iron cores mostly for hypernuclear spectroscopy and rare decay studies using stopped K-mesons [1]. It can analyze emitted charged particles like pions in the momentum range from 100 to 300 MeV/c with a solid angle of 1.5 str. This solid angle is ten times larger than of spectrometers using a conventional dipole magnet. Fig.1 shows the function of the toroidal spectrometer with particle trajectories and

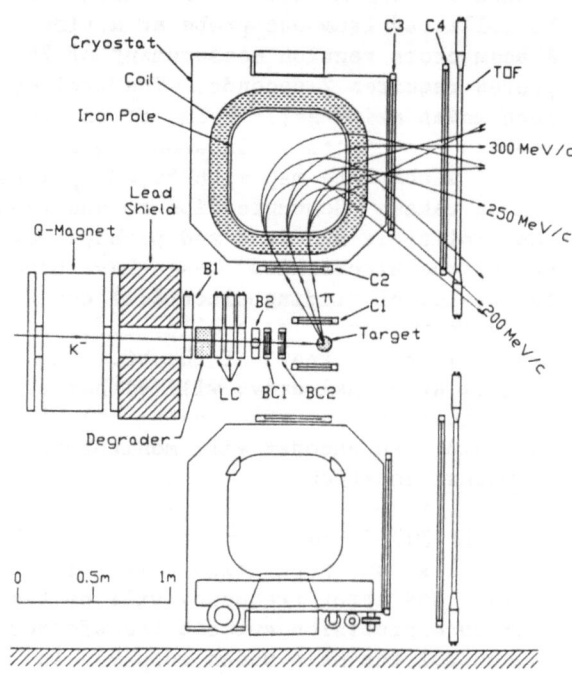

FIGURE 1 Function of the spectrometer

TABLE 1 Performance of the Spectrometer

Momentum range	100 to 300 MeV/c
Momentum resolution	0.2 % (FWHM)
Solid angle	1.5 str
Polarity	positive and negative

detector arrangement.

2. BASIC DESIGN

The magnet was designed to achieve an acceptance as large as possible with a bending power of about 2Tm, which is necessary for the required momentum resolution. Large acceptance means a wide momentum range and at the same time a large solid angle surrounding the source at the center of the magnet. The radial dependence of the field B~1/r of an air core toroidal magnet would be disadvantageous, because particles with larger momentum are bent less, resulting in narrower momemtum width for a given detector size. Therefore, we selected an iron-core type with uniform gaps. At the nominal maximum excitation of 1.8T the flat dipole field from the iron contributes 50 % to the median field and the toroidal field from the coils 50 %.

The magnet assembly comprises sectors, and one sector is composed of an iron core and a coil cryostat. The number of sectors was determined to be 12 by optimizing the solid angle acceptance with a trajectory tracking calculation. The field map was calculated with the three-dimensional field code TOSCA. The rectangular shape of the pole faces has the advantage that the trajectories have a quasi focal plane at the exit and that detector configuration is easy. By putting a source off-center as indicated in Fig. 1 we obtain even some vertical focusing due to the oblique entrance. We have also an option to set up detectors for both particle polarities.

A warm gap is a reasonable requirement for such a kind of spectrometer. Firstly, both the source and the detectors have to be outside the cryostat, and since the scattering from material spoils the resolution, we have to avoid to put cryostat windows into the flight path. Secondly a precise field measurement must be done in order to achieve high momentum resolution. Thus the 12 coils were contained in 12 separate cryostats. Futhermore, in order to attain a large solid angle each cryostat must be as thin as possible. Consequently warm iron and the so called "warm support" method were chosen; each coil is suspended in its cryostat without cancelling electromagnetic forces in the cold structure. The warm iron method also has advantages from the point of view of cooldown time, thermal shrinkage and fabrication.

Perfect symmetry of the whole magnet is important not only for electromagnetic force balance but also for the experiments. Each component was manufactured with sufficient precision, and structure analysis was performed to check the rigidity of the whole structure. Final assembly assured good alignment. Table 2 shows the main parameters of the magnet.

TABLE 2 Main parameters of the magnet

Maximum central field	1.83 T
Pole gap	20 cm
Pole face	82 x 76 cm^2
Outer diameter	3.8 m
Inner diameter	1.1 m
Number of sectors	12
Total weight	38 tons
Maximum field on wire	2.5 T
Operation current	1550 A
Ampere-turns / coil	369 kA
Average current density	133 A/mm^2
Inductance at 1550 A	1.8 H
Stored energy	2.2 MJ

FIGURE 2   View of the toroidal magnet

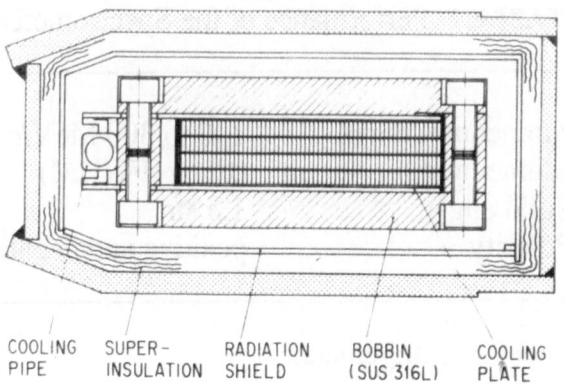

FIGURE 3   Cross section of the coil

3. MAGNET DESCRIPTION

Structure

The magnet has an outer diameter of 3.8 m, an inner field free region of 1.1 m in diameter and a weight of 38 tons. The twelve identical sectors are supported by a stainless steel outer ring of 15 cm thickness and 46 cm width. The inward attraction acting on each iron core is about 40 tons. In each gap two additional blocks are inserted at the innermost corners to guarantee the spacing and to increase the rigidity of the whole structure against this force. Each coil cryostat is fixed to the iron core of its sector.

The iron cores were made out of extra-thick electromagnetic steel with very low impurity content and large grain size to obtain good magnetic properties. The carbon content is less than 0.003 wt% and the total demagnetizing factor is about 0.01%.

Coil

The coils are rectangular and have a particularly thin cross section in order to realize a small cryostat (Fig.3). Flat NbTi-Cu monolith wire was wound in two double pancakes onto a bobbin made of SUS316L stainless steel of 15mm thickness. A high aspect ratio, 3.4, of the wire was selected because thereby tightness of the windings and high thermal conduction in the

axial direction are obtained. The main parameters of the wire are summarized in Table 3. The winding was impregnated with epoxi resin, and is indirectly cooled by two-phase helium through a one-turn of copper tube and copper plates. The operation current of 1550 A gives 369 kATurns of magnetizing force per coil, the average current density being 133 A/mm^2 which is

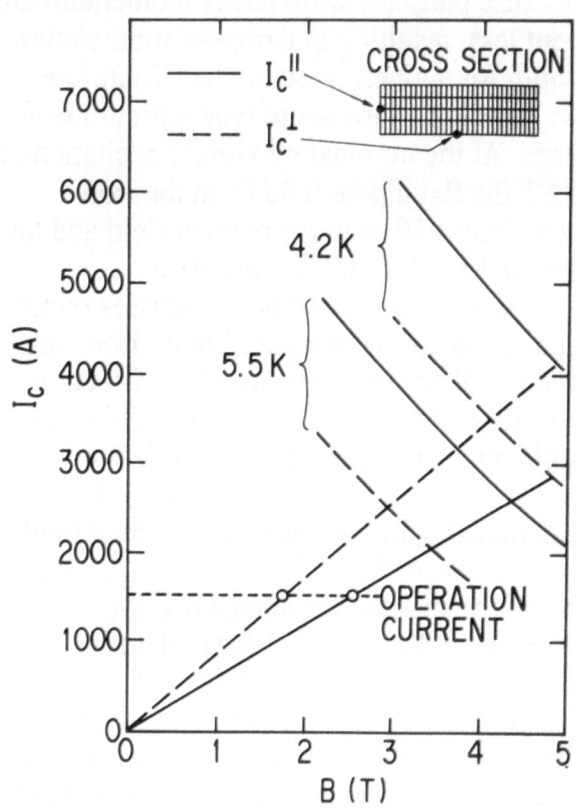

FIGURE 4   Critical current for short sample and load lines

relatively large value for this kind of indirectly cooled coil. The 12 coils are connected in series.

Since the aspect ratio is large, the critical current is degraded where the field is perpendicular to the wire plane. This is the case at both sides of the winding whereas the field is parallel at the top and the bottom part of the winding. Fig. 4 shows this critical current behaviour for a short sample together with the two corresponding load lines. Our rated current is designed to be about 60% of the critical value for 5.5 K which has enough margin compared with the actual temperature measured.

The hoop force along a coil is weekened to some extent due to the presence of core iron, but still amounts to 33 tons. Its distribution is illustrated in Fig.5, the residual inward attraction is 8.0 tons. The calculated maximum deformation of the coil is 75μm.

Cryostat

Each coil is suspended in its cryostat by means of 8 thermally insulated supports: two for radial , two for beam axis and four for azimuthal direction. The radial supports are hollow titanium alloy (Ti6Al4V) rods standing 4 ton tension each. The beam axis

TABLE 3 Parameters of wire

Material	NbTi/Cu monolith
Wire size	6.0 mm x 1.75 mm
Cu / Super ratio	3.5
Filament	35μ, 6cm pitch
Insulation	Kevlar fiber
Critical current (4.2 , 4 T)	
Field parallel	5000 A
Field normal	3600 A

supports made of 8mm$^\phi$ titanium alloy bars fix the coil in the neutral position. The azimuthal supports consist of a three-fold colum structure of G10 and stainless steel.

The coils are surrounded by 2mm thick copper radiation shielding which is cooled to 60 K. A layer between this plate and the vacuum jacket is super-insulation. Stainless steel of low permiability (SUS316) was used as vacuum wall. Fig. 6 shows the cryostat in perspective. Twelve cryostats are connected to one common vacuum system with ducts which also enable the transit of superconducting wires and cooling pipes to the neighbouring coils.

Cooling

The thermal and cooling chracteristics are summarized in Table 4. The calculated total heat leak into 4 K is 25 W, however, a smaller value, 15W, seems to be likely from the perfomance test. We aimed at a simple and reliable system and easy handling for

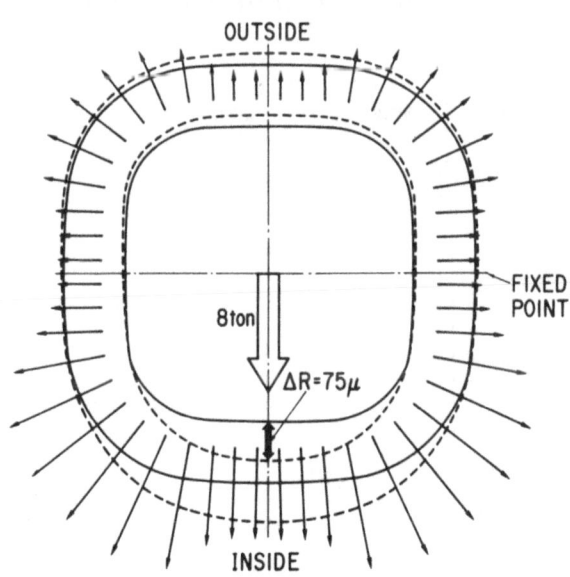

FIGURE 5 Hoop force distribution and coil deformation

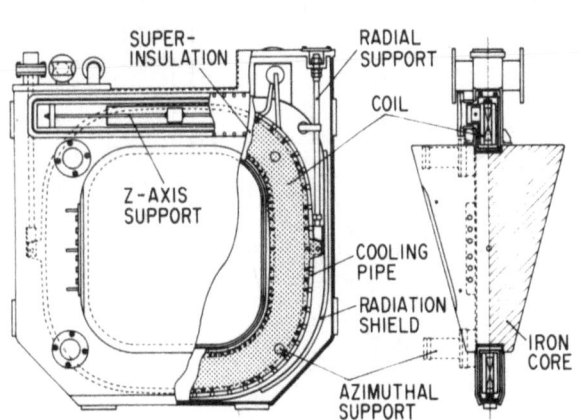

FIGURE 6 One sector structure of the magnet

TABLE 4 Thermal and cooling Parameters

Heat leak into 4.5 K	15 W
Heat leak into 80 K	250 W
Current lead cooling	10 l / hr
Cold mass	
4.5 K	2.7 ton
80 K	0.65 ton
Cooling method	forced flow
Coolant	two-phase He
Temperature	4.4 K
Mass flow rate	7 g/s
Cooling pipe	12 mm$^\phi$ Cu
Rad. shield cooling	60 K GHe
Precooling time	2 ~ 3 days
Warm-up to 150 K	6 days

cooling. The consequence is the adoption of forced flow of two-phase helium (4.4K) for coil cooling and cold GHe for shield cooling instead of using LN$_2$. Both coolants will be supplied from a stand-alone refrigerator TCF50. Fig.7 shows the flow diagram; 12 coils are connected in series, and the return of the two-phase flow fills a 20l liquid vessel located in a service port, and cools the current leads.

The design value of the mass flow rate of two-phase helium is ṁ = 7g/s and the pressure drop along a 60m long 12mm$^\phi$ pipe is calculated to be less than 0.15 kg/cm^2. The gas quality of the return flow is small. Current lead cooling requires 0.3g/s of evaporating gas. Precooling from room temperature proceeds automatically providing temperature-controlled helium gas.

Protection

Since the coils are not perfectly stabilized a reliable quench protection system is necessary. Fig.8 shows the concept of quench detection and circuit breaking. Two quench detectors are picking up the voltage balance from different points. If any of them detects a quench, both DC circuit breakers open and the stored energy is damped by the protection resister R_p1 with a time constant of about 5sec. If these circuit breakers open, coils become nomal conducting due to AC loss effects, although the normal-zone propagation is confined to one coil only before circuit breaking. The temperature rise calculated is 30 K, thus immediate restart of cooling and operation is possible. For other less important faults only one circuit breaker, DCB1, opens, maintaining

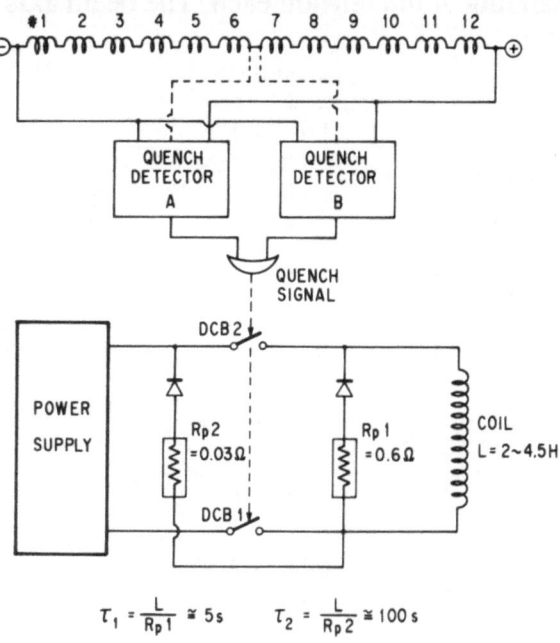

FIGURE 8 Schematic diagram of quench detection and circuit breaking

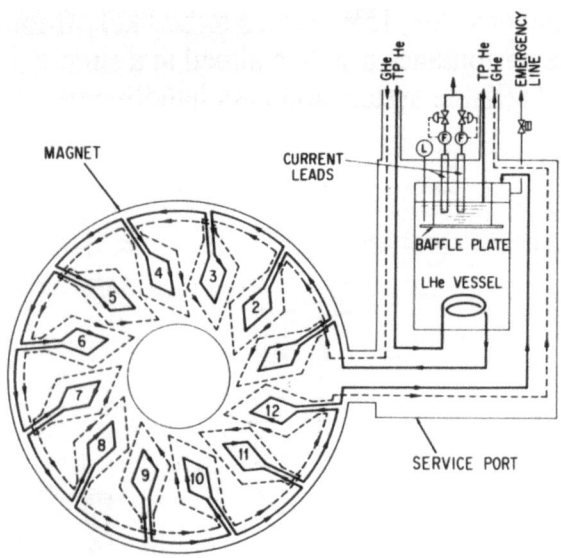

FIGURE 7 Schematic diagram of cooling

the current through R_p2. In this case the energy is damped with a long time constant of about 100 sec and no quench is induced.

The coolant in the magnet is discharged by opening a recovery line valves at the entrance and the exit of the magnet. The pressure rise in the middle of the cooling pipe was calculated be less than 50 kg/cm^2 and the discharge time 3sec.

4. PERFORMANCE

The first test has been performed at the factory immediately after assembly . The evacuatation took about 72 hours to reach 1.0×10^{-3} Torr where cooling was started. A 150l /hr He refrigerator supplied temperature-controlled gas of almost the rated mass flow during precooling. The supply temperature was adjusted such that the temperature difference between the first coil and the coolant did not exceed 30 K and between the first and the last coil 80 K. Fig .9 shows the cooldown curve. It took about 70 hours to cooldown the mass of 2.7 ton of the coils and of 0.65 ton of the radiation shiedings.

At steady cooling with a mass flow rate of 5.9 g/s the temperature in the coils was measured to be nearly 4.5 K. The pressure drop was about 0.1 kg/cm^2 and no instability of the two-phase-helium flow was

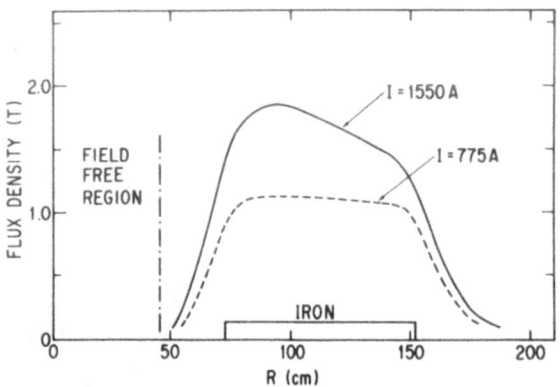

FIGURE 10 Median plane field distribution

observed. The thermal stresses in coil bobbins and supporting rods were measured but no abnormal effect was detected. The total heat input could be estimated from the coil temperature rise and the rate at which the LHe level decreased in the vessel just after stopping the coolant supply.

Excitation of the magnet was performed clearing several check processes at lower cuurent. The rated current of 1550 A could be attained without a quench and the maximum median plane field of 1.83T was confirmed (Fig.10).

Because no natural quench happened, the characteristics of current, voltage, coil temperature and gas pressure at a quench were measured by cutting off the circuit artificially. The measured temperature rise of 30 K and the energy recovery rate of 75% agreed with theoretical predictions. Normal functioning of the quench detectors and the safety processes were also checked by artificially inducing a quench by stopping the helium supply.

ACKNOWLEDGEMENTS
The authors would like to thank Prof. K. Nagamine and Prof. H. Hirabayashi for the helpful discussions, and Prof. K. Nakai for stimulating this project. We also indebted to Mr.T. Haruyama and other staff of the KEK cryogenic group for discussions in designing a cooling system.

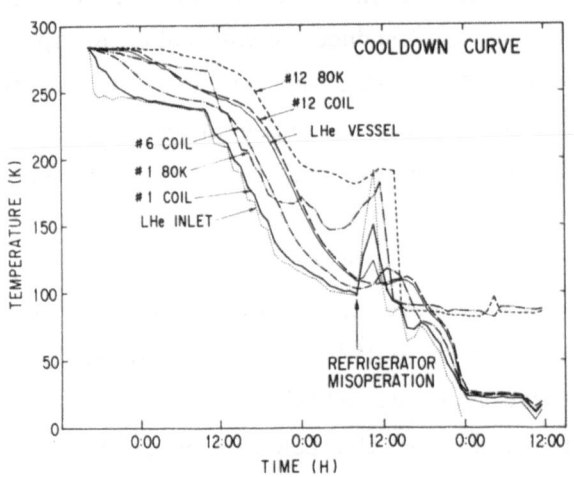

FIGURE 9 Cooldown curve

REFERENCES
1) T. Yamazaki et al., Nuovo Cimento 102A (1989) 695

CORRECTIVE COILS FOR THE
ELECTRON COOLER AT ESR (GSI)

Th. Odenweller

GSI Darmstadt, P.O. Box 110552, D-6100 Darmstadt, FRG

ABSTRACT

The Experimental Storage Ring ESR, presently under construction at GSI, is designed for accumulation, storage und cooling of heavy ion beams up to uranium with energies up to 500 MeV/amu either directly out of the heavy ion synchrotron SIS or via a stripper target, where electron-free or electron-poor ions are produced. The SIS/ESR-facility will also be capable of delivering beams of unstable nuclei produced by fragmentation of fission of heavy projectiles. After preselection in a large acceptance Z/A-separator the secondary beams can be prepared for experiments by stochastic precooling, intensity accumulation and final electron cooling. For electron cooling a high degree of homogeneity is required in the cooler solenoid. Therefore corrective coils are needed for the electron cooler.

INTRODUCTION

At GSI (Gesellschaft für Schwerionenforschung, i.e. center of heavy ion research) heavy ions up to uranium have been accelerated in the UNILAC (universal linear accelerator) [1] since 1975. The maximum specific projectile energy of 10 MeV/amu was upgraded to 20 MeV/amu in 1982. Plans to extend the accelerator facility by the addition of the synchrotron SIS (Schwerionensynchrotron, i.e. heavy ion synchrotron) and the storage and cooler ring ESR (Experimentier-Speicherring, i.e. experimental storage ring) were approved in 1985 [2], [3]. A schematic overview of the GSI with the SIS/ESR facility is given in figure 1.

With the SIS/ESR facility, which is presently under construction at GSI, heavy ion research will be extended to relativistic energies.

The combination of these two rings will allow to produce completely stripped heavy ion beams up to U^{92+} with the highest possible phase space densities, achieved by various beam cooling techniques. In addition SIS/ESR will provide beams of radioactive nuclei in the energy range of several MeV/amu up to 1-2 GeV/amu, again cooled to the highest possible phase space densities.

The beams of the ESR may be used either circulating with high currents or extracted with a great variety of time

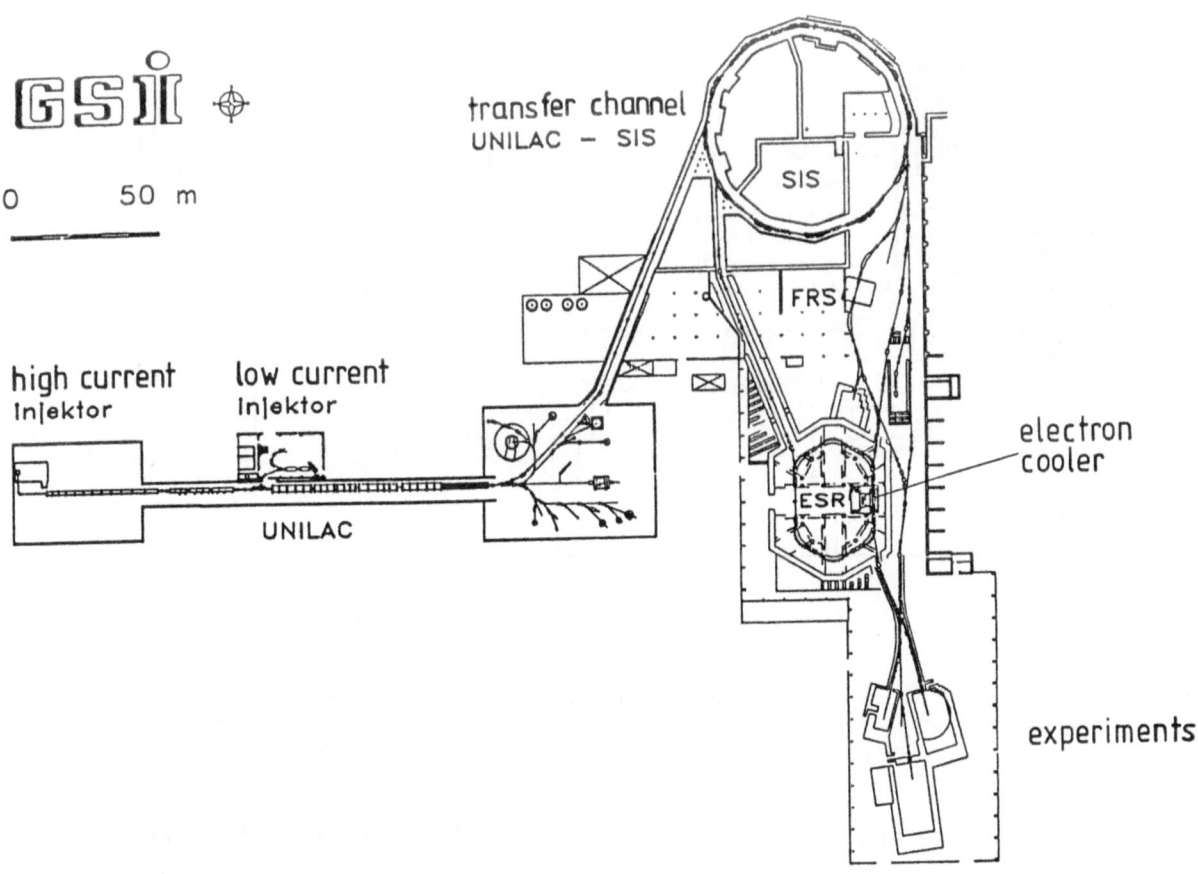

Figure 1: UNILAC, SIS, ESR

structures and intensities. They may also be reinjected into SIS for further acceleration or deceleration. There will be a large experimental area with several experiments set up on beams from both SIS and ESR.

Main features of the ESR design are the maximum bending power of 10 Tm, large transverse and momentum acceptances, stochastic precooling and electron cooling of the beam. For beams with low phase space density stochastic pre-cooling will be used. For cooling to very high phase space density, electron cooling is fore-

seen in the electron cooler solenoid. The electron cooler consists of an electron gun, an electron collector, two 90° toroid systems and a solenoid of 2.5 m length, in which the beam cooling by electrons will proceed (see figure 2). A "cool" electron beam of 5-10 A will be produced by the electron gun. The electron beam will be adjusted in such a way, that it circulates collinearly along the heavy ion beam inside the electron cooler solenoid on a radius of 25 mm at the corresponding average velocity. The maximum electron beam energy of 310 keV cor-

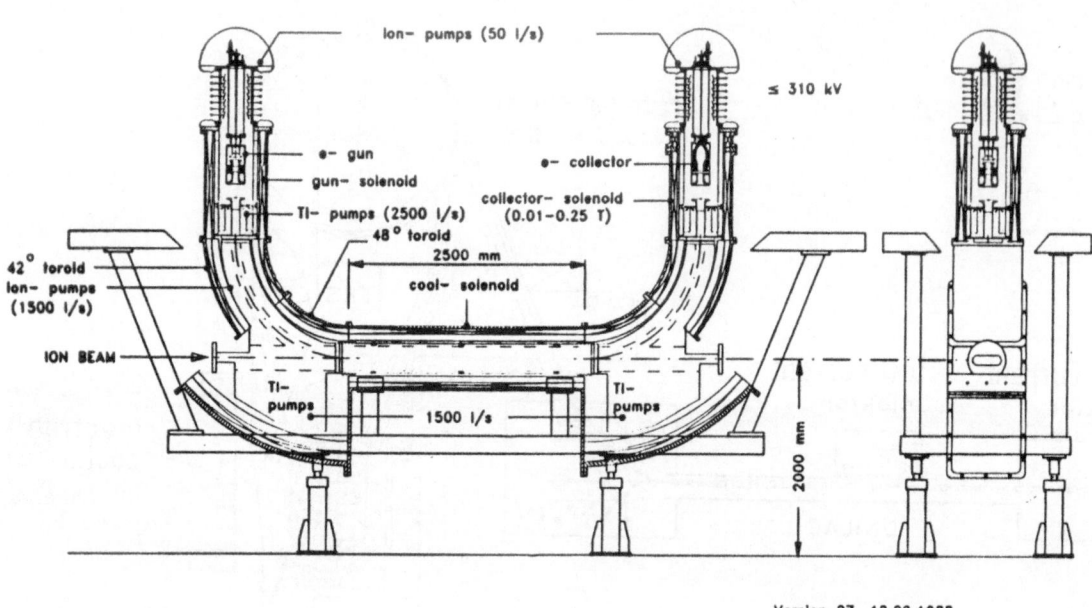

Figure 2: Electron Cooler

responds to the maximum specific energy of the U^{92+}-ions in the ESR.

With an electron beam current density of up to 1 A/cm^2 and ion beams of initially $\Delta p/p = 0.1$ % and $4\ \pi$ mm mrad, cooling times of 30 ms for U^{92+} at 500 MeV/amu are expected. Heavy ion beams with emittances as small as 0.1 π mm mrad and momentum spreads of less than 10^{-5} may be produced.

METHODS

General

Two different kinds of corrective coils are needed for the electron cooler:
1.: "steerer coils" in the gun- and collector-solenoids, the cooler-solenoid and both 90-degrees toroids and 2.: coils for "fine correction" only in the cooler-solenoid.

The basic shape of both kinds of coils will be similar. It will be the shape of a coil, which is wound on the surface of a cylinder. The steerer coils will always be used in pairs as shown in figure 3, whereas the fine-correction coils may also be used in sets of three or four, corresponding to the sextupole and octupole correction respectively.

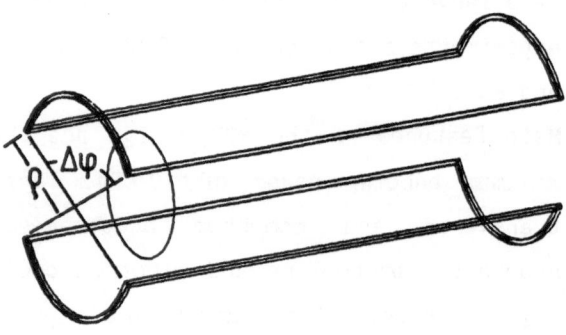

Figure 3: Optimized dipole coils

The magnetic induction-field B produced by such a pair of coils can be described in terms of a Fourier-analysis of the radial component B_ρ on a cylinder between the pair of coils, which in this case is equivalent to the multipole expansion. If the current in this pair of coils flows in the same sense in each coil, the pair of coils will produce mainly a magnetic dipole-moment, if the direction of the current is reversed in one coil, the pair of coils will produce a magnetic quadrupole-moment in the space between the coils. The multipole expansion of the coil system shown in figure 3 will in general exhibit not only one but a lot of multipole components and this is valid if the coils are operated as a dipole as well as if they are operated as a quadrupole.

If the coils in figure 3 are operated as a dipole, then the first nonzero multipole moment will be the Fourier component of the order 1 and the only nonzero higher orders will be the odd orders 3,5,7 etc. because of the symmetry of the configuration.

Correspondingly, if the coils are operated as a quadrupole, then the first nonzero multipole moment will be the Fourier component of the order 2 and the only nonzero higher orders will be the even orders 4,6,8 etc. Thus, in the dipole case the sextupole will be the next nonzero multipole.

The quotient of the sextupole and the dipole coefficient depends on the azimutal angle $\Delta\phi$ which is covered by each of the symmetric coils. Making use of the TOSCA-routine installed on the IBM-3090

at GSI [4] and [5] it was found, that the sextupole to dipole ratio varies from positive to negative values when $\Delta\phi$ is changed from 160° to 20°.

Thus it can be seen, that there must be a certain value for $\Delta\phi$ which will correspond to a sextupole to dipole ratio of zero. This angle was found to be $\Delta\phi_{opt}$ = 120° for the dipole with help of the TOSCA-routine.

When each coil of the dipole pair covers this azimutal angle, the pair of coils will produce a nonzero dipole but a zero sextupole moment. In this way it is possible to obtain a pair of optimized dipole coils.

In the same way, optimized azimutal angles can be found for the higher multipoles especially the quadrupole, sextupole and octupole. The criterion for optimisation is always, that the next higher nonzero multipole moment must be brought to zero by the choice of the azimutal angle $\Delta\phi$.

In this way, making use of the TOSCA-routine, the following list of optimized azimutal angles can be found for the first multipole moments:

Number of Multipole	1 (Dipole)	2 (Quad.)	3 (Sext.)	4 (Oct.)
Number of Coils	2	2	3	4
$\Delta\phi_{opt}$	120°	90°	60°	45°

Steerer Coils

Steerer coils are needed in the gun, collector and cooler solenoids as well as in both 90°-toroids.

The steerer coils in the gun, collector and cooler solenoid are needed to compensate for any possible nonzero inclination between the geometrical solenoid-axis and the magnetic field-lines. To achieve this compensation the steerer coils must be arranged in such a way to superpose a small transversal field component to the longitudinal field so that the vector sum of the two field components can be brought into coincidence with the geometric axis.

The steerer coils in the 90°-toroids are needed to prevent the circulating electrons from moving out of the deflection plane and to guide them on the desired circle's arc.

In all cases except the 90°-toroids two pairs of steerer coils, which are turned around 90° relative to each other are needed, because the required transversal component may have any direction in a plane perpendicular to the axis. When the azimutal angle of optimized dipole coils is used for the steerer coils, they will have four overlapping sections with an overlap of 30° each.

Fine-Correction Coils

The fine-correction coils in the cooler solenoid are planned to compensate for inhomogeneities of the magnetic field near the solenoid axis, where the beam-cooling by electrons will proceed. The intention is, to reduce the field inhomogeneities to less than 10^{-4} in a cylinder of 50 mm diameter around the axis of the cooler solenoid.

As a basis for this correction the inhomogeneity of the magnetic field must be measured. This will be done by a measuring set-up with cylinder symmetry, which yields the radial component of the magnetic induction field B_ρ on a series of circles inside the solenoid around the solenoid axis. These measurements are carried out by a measuring set-up, which consists of an aluminum cylinder of 190 mm external diameter and 4560 mm length carrying a guide rail with a sliding carriage and a temperature stabilized hall probe on it.

For the measuring procedure this aluminum cylinder, which can be turned mechanically more than 360° around the azimutal coordinate ϕ, is inserted into the cooler solenoid. The radial component on a circle, B_ρ, obtained in this way, is submitted to a Fourier analysis, to obtain the multipole moments in a series of planes perpendicular to the solenoid axis. The Fourier analysis is done according to equation 1.

$$B_\rho(\phi) = c_o + c_1 \cdot \cos 1(\phi - \alpha_1) + c_2 \cdot \cos 2(\phi - \alpha_2) + c_3 \cdot \cos 3(\phi - \alpha_3) \ldots \qquad (1)$$

In this way, the multipole intensities c_i and the corresponding azimutal angles α_i are obtained. With these values together with the optimized azimutal angles $\Delta\phi_{opt}$ covered by the corresponding set of multipole coils all free parameters of the fine correction coils can be determined. These free parameters are: α_i, $\Delta\phi_{opt}$ and the local number of turns n. The local number of turns n is calculated from the coefficients c_i by comparison with the corresponding coefficients obtained from the TOSCA program. The number of turns is called "local", because it may surely be different from one plane to the other.

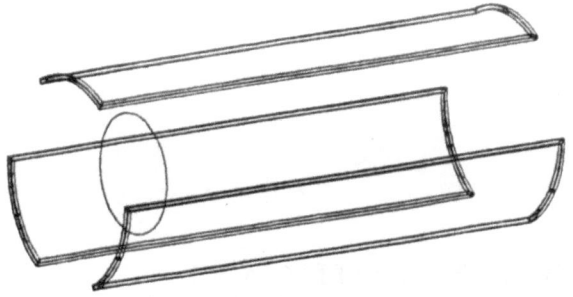

Figure 4: Optimized sextupole coils

The fine corrections will be executed by mounting coils in the order of the successive multipoles. For every multipole, the corresponding set of corrective coils (e.g. fig. 3, fig. 4 etc.) will be mounted, but, as the parameters α_i and n may vary over the length of the solenoid, the coils may be distorted, and not every turn will extend over the whole length of the solenoid.

In this correction procedure a so-called "local approximation" has been made. This means, that it is assumed, that a turn of a corrective coil in one plane has no effect on $B_\rho(\phi)$ in the neighbouring plane. This can surely be only an approximation, and this is why the correction procedure described will not yield a perfect result after the first application. Therefore the correction procedure will have to be applied several times, but it is hoped, that this iteration will converge rapidly.

RESULTS

The measuring program at GSI is delayed by several months. The measurements are about to start in August/September 1989.

REFERENCES

[1] J. Glatz; Linear Accelerator Conference; Stanford (1986)

[2] K. Blasche, D. Böhne, B. Franzke, H. Prange; IEEE Trans.Nucl.Sci. NS-3 (5) (1985) p. 2657

[3] P. Kienle; The SIS/ESR Project of GSI; GSI-Report; GSI 85-16 (1985)

[4] C.C. Sahm et al.; TOSCA-PE2D-Programs at GSI; GSI Scientific Report; Darmstadt, 1986, p. 331

[5] J. Simkin, C.W. Trowbridge; Three-dimensional nonlinear electro-magnetic field computations, using scalar potentials; IEE PROC., Vol. 127, Pt. B, No. 6, 1980, p. 368

A METHOD FOR THE HIGH ENERGY DENSITY SMES
—— SUPERCONDUCTING MAGNETIC ENERGY STORAGE

Y. Mitani and Y. Murakami
Osaka University
2-1 Yamada-oka, Suita
Osaka, 565 JAPAN

ABSTRACT

The energy density of superconducting magnetic energy storage (SMES), 10^7 [J/m^3] for the average magnetic field 5T is rather small compared with that of batteries which are estimated as 10^8 [J/m^3]. This paper describes a method for the high density SMES on supposition of the use of novel superconductors whose critical current and magnetic field are far more larger than the conventional ones. We propose an integration of solenoids of long axial lengths with small diameters. The calculation results show that the energy density can be increased up to 4 or 5 x 10^7 [J/m^3] with the magnetic field of 20 [T] and the mechanical stress is kept in the allowable limits even in this high density of magnetic field.

INTRODUCTION

Superconducting magnets have been investigated in various areas of science and engineering: the magnets in a pulsed accelerator, the pulsed poloidal coils of a tokamak, the energy storage used to even out the load variations in electrical power demands or to stabilize power systems, the magnetic levitation system, the magnetic resonance imaging and so on [1].

When we design a superconducting magnet, the electrical properties of the material, that is, the critical current density (J_c) and the critical magnetic field (H_c) at the operating temperature, should be primarily taken into account. In addition to this, the design of magnet with large energy capacity may be restricted by the mechanical stress applied to the super-

conductors as well. For example, in the case of the SMES (superconducting magnetic energy storage) with huge energy storage which is considered to be the substitution for the pumped storage hydroelectric power plant, it is very important how to support their enormous electromagnetic forces. In contrast with this, for the design of smaller sized magnet whose cable is made of the conventional superconducting materials like NbTi, the mechanical stress may not come into question because the critical current and magnetic field are not so large.

In the future, however, we can expect the advent of novel superconductors whose J_c and H_c are far more larger than the conventional ones. In this case, the restricting condition is the mechanical stress due to the large magnetic field and the large current of magnet. Therefore, a new concept for the design of SMES is required in order to make the most of the high qualities of the superconductors.

The energy density of 5 [T] field for SMES is approximately 10^7 [J/m^3] which is comparable with flywheels and rather less than that of batteries which are estimated as 10^8 [J/m^3]. In this paper a method of design for the high energy density and compact sized SMES, is described. We propose an integration of solenoids of long axial lengths with small diameters which are closely assembled such that the directions of axial fields of solenoid is reversed each other. The circumferential hoop stress of solenoid is decreased by small diameter of solenoid, since the stress is proportional to the diameter of coil. Furthermore, the leakage field is absorbed due to the ampere turns by the adjacent coils. The calculation results show that the energy density has increased up to 4 or 5 x 10^7 [J/m^3] with 20 [T] and the mechanical stress is kept in the allowable limits even in this high density of magnetic field.

CONFIGURATION OF MAGNET

In this paper we propose the configuration of SMES shown in FIGURE 1.

It contains $N \times N$ of solenoids with long axial lengths and small diameters located in a cubic space l^3 [m^3]. They are closely assembled orderly such that the directions of axial field is reversed each other in order to cancel the leakage of magnetic field. Small diameter of solenoid is for decreasing the hoop stress. A solid cylinder with the thickness of s [m] is located outside of the winding to support the stress and to keep the insulation. Here, the self

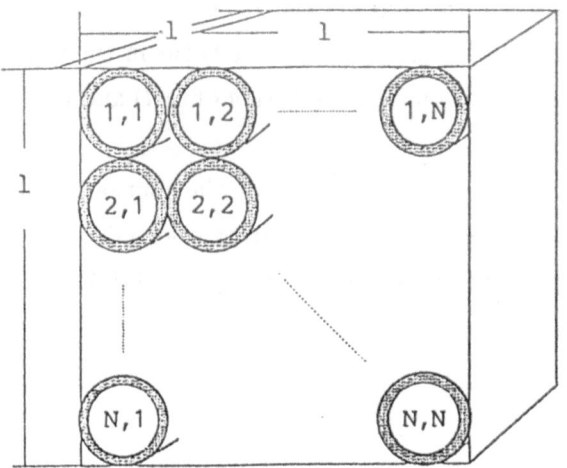

(a) whole configuration

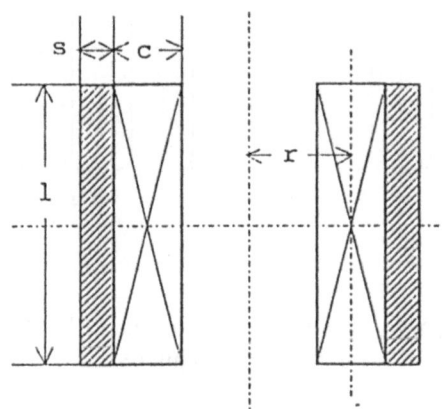

(b) one of solenoids

FIGURE 1   Configuration of the SMES.

and mutual inductances of each coil are represented by functions of the magnet shape to calculate the capacities of total energy storage.

The self inductance L [H] of each coil is expressed by

$$L = \mu_0 N_t^2 r^2 (K - k)/l,$$

where $\mu_0 (= 4\pi \times 10^{-7}$ [N/A^2]) is the magnetic permeability of free space and N_t is the number of turns. The quantity K is the constant in Nagaoka's formula for a solenoid and is expressed as a function of $(2r/l)$. The quantity k takes into account the decrease of inductance due to the separation of the turns in the radial direction which is a function of (c/l) and $(c/2r)$ [2].

For the calculation of the mutual inductance, we consider a pair of solenoid coils located in parallel each other (see FIGURE 2).

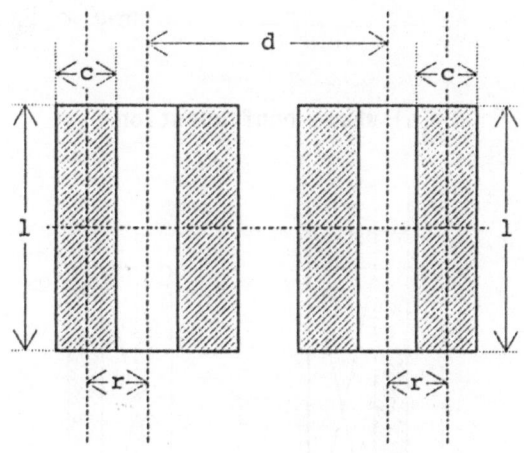

FIGURE 2   Model for the calculation of mutual inductances.

The mutual inductance M' [H] arranged as a series involving zonal harmonics is

$$M' = 0.1 \times 10^{-6}\pi^2(rN_t/l)^2$$
$$(Z_1/R_1 - Z_2/R_2 - Z_3/R_3 + Z_4/R_4),$$

in which the quantities $Z_m (m = 1, 2, 3, 4)$ are given by

$$Z_m = t_2^2 - (1/4)(r/R_m)^2(2t_4t_2)P_2(u_m)$$
$$+ (1/8)(r/R_m)^4(2t_6t_2 + 3t_4^2)P_4(u_m)$$
$$- (5/64)(r/R_m)^6(2t_8t_2$$
$$+ 12t_6t_4)P_6(u_m)$$
$$+ (7/128)(r/R_m)^8(2t_{10}t_2 + 20t_8t_4$$
$$+ 20t_6^2)P_8(u_m) - \cdots .$$

The four radii vectors R_m are given by

$$R_1 = R_4 = (l^2 + d^2)^{1/2},$$
$$R_2 = R_3 = d,$$

and the zonal harmonics $P_{2n}(u_m)$ are for argument

$$u_1 = -l/(l^2 + d^2)^{1/2} = -u_4,$$
$$u_2 = u_3 = 0.$$

The factors t_n are given by

$$t_n = 1 + \{n(n-1)/3!\}(c/2r)^2$$
$$+ \{n(n-1)(n-2)(n-3)/5!\}(c/2r)^4$$
$$+ \cdots + [n(n-1)\cdots\{n-(n-1)\}$$
$$/(n+1)!](c/2r)^n.$$

The convergence of the series Z_m is sufficient for most purpose as long as all the distances R_m are greater than $2r$ [2].

Total mutual inductance for the i-th solenoid M_i [H] is to be obtained as a superposition of the calculated values of M' [H] for every combinations.

Consequently, the total stored energy E [J] is represented by

$$E = \sum_{i=1}^{N^2}(L - M_i)(jcl)^2/2$$

where j [A/m^2] is the overall current density.

CALCULATION OF ENERGY DENSITY

It has been evaluated how high energy density can be obtained by the SMES with proposed configuration for different values of $N(N^2$ is the total number of solenoids) and for different thicknesses of superconducting winding c, where the cubic space is full-filled with the solenoids. The concise space with 1 [m] × 1 [m]

× 1 [m] cubic volume is supposed as the size of SMES.

Here the maximum hoop stress applied to the magnets is constrained by determining the overall current density not to exceed the constrained stress, where the hoop stress is calculated by using the finite element method after calculating the distribution of magnetic field by applying the Biot-Savart law to each subdivided element.

TABLE 1 shows conditions for calculation. The constrained maximum stress 3.0×10^8 [N/m^2] is beyond the yield strength of annealed copper but within that of hardened copper [1].

TABLE 1
Conditions for calculation

Young's modulus	1.0×10^{11} [N/m^2]
Poison's ratio	0.3
thickness of the outside cylinder s	0.01 [m]
the maximum hoop stress	3.0×10^8 [N/m^2]

For conciseness, Young's modulus and Poison's ratio are assumed to be uniform over the winding including the outside structure.

The results are shown in FIGURE 3 and TABLE 2.

These results show that subdivision into an adequate number of solenoids makes the total energy capacity maximum. Besides, the greater thickness of windings makes the energy storage capacity larger. The reason may be understood as follows.

The energy is stored as static magnetic field which exists mostly inside space of the solenoids and somewhat inside of the superconducting windings. When a large coil is located

TABLE 2
Results of calculation

(a) The case that c=0.01 [m]

N	r[m]	E[MJ]	j[x10^{9}A/m^2]	B$_{max}$[T]
1	0.485	9.55	0.546	6.8
2	0.235	17.2	0.675	7.6
4	0.110	28.5	0.902	10
6	0.068	36.4	1.09	13
8	0.048	41.0	1.26	15
10	0.035	42.6	1.41	17
12	0.027	41.7	1.54	18
14	0.021	38.9	1.78	20
16	0.016	34.5	1.78	21
18	0.013	29.4	1.90	22
20	0.010	24.2	2.03	24

(b) The case that c=0.02 [m]

N	r[m]	E[MJ]	j[x10^{9}A/m^2]	B$_{max}$[T]
1	0.480	13.7	0.332	7.4
2	0.230	24.3	0.413	9.3
4	0.105	37.6	0.553	13
6	0.063	44.4	0.671	16
8	0.043	45.6	0.773	19
10	0.030	42.6	0.866	21
12	0.022	36.8	0.958	23

(c) The case that c=0.04 [m]

N	r[m]	E[MJ]	j[x10^{9}A/m^2]	B$_{max}$[T]
1	0.470	21.5	0.214	9.0
2	0.220	35.8	0.266	12
4	0.095	47.9	0.357	17
6	0.053	47.3	0.436	21
8	0.033	40.3	0.521	25

in the cube, therefore, the ununiformity of magnetic field inside of the solenoid decrease the average overall energy density. When a great number of solenoids with small diameter are located in the cube, the outside cylindrical structures for stress support and for insulation occupy the great part of the cube and decrease the average energy density.

It is possible to get the high energy density SMES of 40 or 50 [MJ/m^3] which keeps

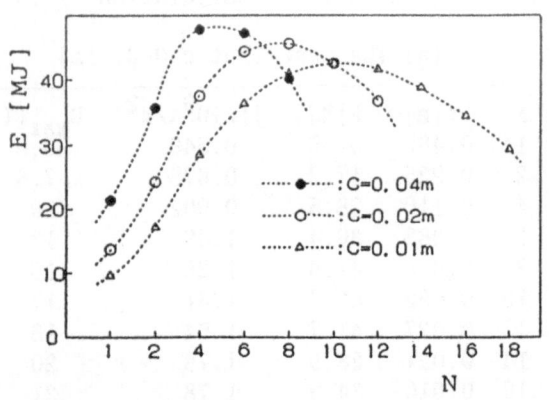

FIGURE 3   Energy storage capacity of SMES for different numbers of solenoids and for different thicknesses of superconducting winding.

the hoop stress in the limit of 3.0×10^8 [N/m^2].

On the other hand, the current density of $0.5 \sim 1.5 \times 10^9$ [A/m^2] at the maximum magnetic field of $17 \sim 20$ [T] is necessary for the realization of the proposed high energy density SMES.

CALCULATION OF FLUX LEAKAGE

The effectiveness of decreasing the flux leakage due to the ampere turns by the adjacent coils is evaluated.

With picking up one coil and a pair of coils, the intensity of magnetic flux produced by them has been calculated, respectively.

FIGURE 4 shows the contour maps of the flux intensity; (a) is the case when the only one coil is located and (b) is the case when a pair of coils are assembled where the coil with $c = 0.02$ [m] and $r = 0.043$ [m] is used for these evaluations.

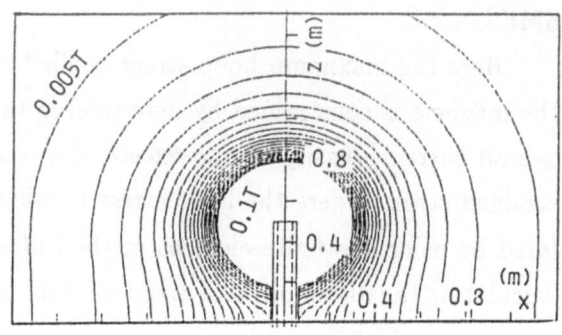

(a) the case when the only one coil is located

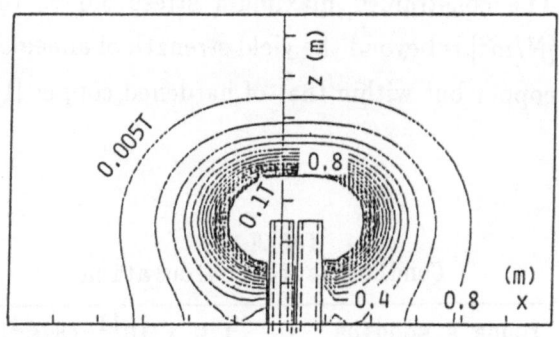

(b) the case when a pair of coils are located

FIGURE 4   Contour maps of the intensity of magnetic field.

It can be said that the adjacent coil decreases the flux leakage each other effectively as expected.

CONCLUSIONS

(1) A method for the high energy density Superconducting Magnetic Energy Storage has been proposed.

(2) The proposed configuration is an integration of solenoids of long axial lengths with small diameters.

(3) A high energy density SMES of 50 $[MJ/m^3]$ has been obtained with the current density of 1×10^9 $[A/m^2]$ and the magnetic field of 20 [T] in which the stress is kept within 3.0×10^8 $[N/m^2]$.

(4) The effectiveness of decreasing the flux leakage due to the ampere turns by the adjacent coils has been demonstrated numerically.

ACKNOWLEDGEMENT

This work was financially supported by Grant-in-Aid for Scientific Research of the Ministry of Education, Science and Culture, Japan.

REFERENCES

1. Wilson, M.N., Superconducting Magnets, Clarendon Press Oxford, New York, 1983.

2. Grover, F.W., Calculation of Mutual Inductance and Self Inductance, Dover Publications, Inc., New York, 1946, pp.228-234.

AN OPTIMAL DESIGN OF SOLENOIDAL SUPERCONDUCTING COIL USED FOR ELECTRICAL POWER SYSTEM STABILIZATION

Y. Mitani and Y. Murakami
Osaka University
2-1 Yamada-oka, Suita
Osaka, 565 JAPAN

ABSTRACT

This paper describes a systematic method to design an optimal solenoidal superconducting coil for the use as an electrical power system stabilizer. The proposed method derives an optimal shape of coil so as to minimize the volume of superconductor under constraints of the current density and the maximum magnetic field. In order to solve the problem systematically the stored energy and the maximum magnetic field are represented by analytic functions of the current density and the shape of coil. Several numerical results derive the optimal shapes of several hundred MJ magnets. The finite element method evaluates the mechanical stress applied to the superconductors numerically.

INTRODUCTION

Much attention has been recently paid to superconducting magnetic energy storage (SMES) in electrical power system application, especially power system stabilization [1-4]. The SMES with a proper power converter is capable of controlling active power and reactive power simultaneously for electrical power systems. This controllability makes it possible to stabilize power system oscillations as well as voltage fluctuations significantly. In this paper a systematic method to design an optimal solenoidal superconducting coil for the use as a power system stabilizer, is proposed.

The SMES for the electrical power system use must have an enough energy capacity to charge and discharge the specified power. In the case of the large scale SMES used to even out the load variations, the necessary energy capacity has been estimated to be 10^{10} through 10^{13} J and the mechanical stress which is applied to the superconductors due to the large magnetic field is the major restricting condition for the magnet design. On the other hands in the case of the SMES for the power system stabilizations, the necessary capacity has been estimated to be 10^7 or 10^8 J, which is rather small. Therefore, the restricting conditions for the magnet design would be the critical current density and the critical magnetic field.

There are lots of applications of small or medium sized superconducting magnet; magnetic resonance imaging, magnetic levitation, accelator magnet in high energy physics and superconducting generator are the examples. They utilize the intense magnetic field inside of the superconducting magnet. However, the

SMES in question is specified to have enough energy storage rather than the intensity of magnetic field.

The proposed design method in this paper derives an optimal shape of magnet to minimize the volume of superconductor under constraints of the current density and the maximum magnetic field. In order to solve the problem systematically the stored energy and the maximum magnetic field are represented by analytic functions of the current density and the shape of magnet. The numerical results show that the optimal volume is proportional to $E^{0.6}$, where E is the stored energy. For further informations, the mechanical stress applied to the superconductors is evaluated by the finite element method

NUMERICAL MODEL
OF SOLENOIDAL COIL

FIGURE 1 shows a model configuration of solenoidal coil which has a circular cylindrical winding of rectangular cross-section.

The SMES which is used to stabilize electrical power systems must have an enough energy capacity to charge and discharge the specified power. In addition to this the current density and the maximum magnetic field must be below the critical ones respectively. In order to solve the problem systematically it may be desired that the stored energy and the maximum magnetic field are represented by analytic functions of the current density j [A/m^2] and the shape of magnet r [m], l [m], c [m].

There are lots of approximated expressions of the inductance L [H] by analytic functions of the shape of magnet. Here, we use the following equations:

$$L = 21.0 \times 10^{-7} N_t^2 r^{1.75}/(l+c)^{0.75}$$

$$(l+c \geq r),$$

$$L = 21.0 \times 10^{-7} N_t^2 r^{1.50}/(l+c)^{0.50}$$

$$(l+c < r), \qquad (1)$$

where N_t is the number of turns. The stored energy E [J] equals $LI^2/2$ and the current I [A] is represented by

$$I = jlc/N_t$$

where j [A/m^2] is the average overall current density. Therefore,

$$E = 10.5 \times 10^{-7} l^2 c^2 r^{1.75}/(l+c)^{0.75} j^2$$

$$(l+c \geq r),$$

$$E = 10.5 \times 10^{-7} l^2 c^2 r^{1.50}/(l+c)^{0.50} j^2$$

$$(l+c < r). \qquad (2)$$

The field B_0 [T] at the central point may easily be calculated by integrating the contributions from individual circular current filaments, to find

$$B_0 = (\mu_0/2)jl \ln[\{\alpha + (\alpha^2 + \beta^2)^{1/2}\}/$$

$$\{1 + (1 + \beta^2)^{1/2}\}] \qquad (3)$$

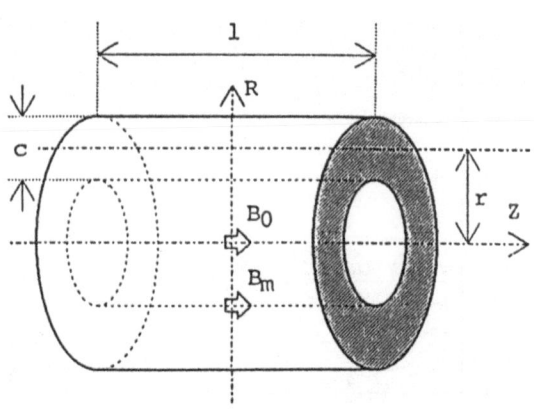

FIGURE 1   A model configuration of solenoidal winding.

where

$$\alpha = (r + c/2)/(r - c/2)$$
$$\beta = (l/2)/(r - c/2)$$

On the other hand, the maximum field B_m on the winding occurs at the center of inner surface indicated in FIGURE 1 and it is this field rather than B_0 that determines the critical current density in the winding. B_m is also represented as a function of the shape of magnet, however, B_m cannot be found analytically [5]. Here, B_m may be represented approximately by series functions as follows.

The variations of field along the axis of a solenoid (Z- axis) $B_Z(Z, 0)$ and along the radial direction (R-axis) $B_Z(0, R)$ may be represented by

$$B_Z(Z, 0) = B_0\{1 + E_2(Z/a)^2 + E_4(Z/a)^4$$

$$+ E_6(Z/a)^6 + \cdots\} \qquad (4)$$

$$B_Z(0, R) = B_0\{1 - (1/2)E_2(R/a)^2$$

$$+ (3/8)E_4(R/a)^4$$

$$- (5/16)E_6(R/a)^6 + \cdots\} \qquad (5)$$

$$a = r - c/2$$

where the error coefficients are given by

$$E_{2n} = \{1/B_0(2n)!\}$$

$$\{d^{2n}B_Z(Z, 0)/dZ^{2n}\}\,|_{Z=0} \qquad (6)$$

which may be found by evaluating the (2n)th derivative of equation

$$B_Z(Z, 0) = (1/2)ja\mu_0$$

$$[\beta_1 \ln[\{\alpha + (\alpha^2 + \beta_1^2)^{1/2}\}$$

$$/\{1 + (1 + \beta_1^2)^{1/2}\}]$$

$$+ \beta_2 \ln[\{\alpha + (\alpha^2 + \beta_2^2)^{1/2}\}$$

$$/\{1 + (1 + \beta_2^2)^{1/2}\}]] \qquad (7)$$

$$\beta_1 = (l/2 - Z)/a, \beta_2 = (l/2 + Z)/a$$

at $Z = 0$ [5]. Consequently we can derive $B_m = B_Z(0, a)$ to write the following series expansions.

$$B_m = B_0[1 + \mu_0 j(r - c/2) \cdot$$

$$\lim_{N \to \infty} \sum_{n=1}^{N} \{(C_{2n} + D_{2n})P_{2n}\}/B_0] \qquad (8)$$

where

$$C_k = (1/k!) \sum_{m=1}^{k-1} \{K_{(k,m)}\alpha^{2k-2m+1}$$

$$b^{-k+2m-1}\}(\alpha^2 + \beta^2)^{(-2k+1)/k}$$

$$D_k = -(1/k!) \sum_{m=1}^{k-1} \{K_{(k,m)}\beta^{-k+2m-1}$$

$$(1 + \beta^2)^{(-2k+1)/k}\}$$

$$K_{(k,m)} = -(k - 2m)K_{(k-1,m)}$$

$$- (3k - 2m - 1)K_{(k-1,m-1)},$$

$$K_{(2,1)} = -1, K_{(k,k)} = 0,$$

$$P_k = -\{(k - 1)/k\}P_{k-2}, P_0 = 1.$$

The convergence of the series in eq. (8) has been investigated for the various iteration number N. FIGURE 2 shows the result where N is set to 5. It agrees with that of ref. [5] very well. Thus, B_m has been defined by the analytical function of eq. (8) with setting N to be 5.

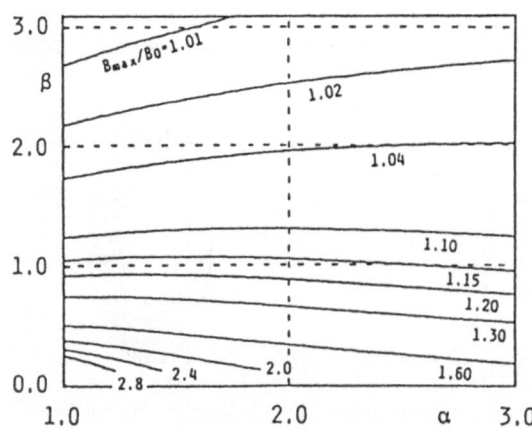

FIGURE 2 Ratio of maximum to central field, B_{max}/B_0.

DESIGN OF OPTIMAL
SOLENOIDAL COIL

The winding volume of superconductor V [m^3] is represented by

$$V = 2Nrlc. \tag{9}$$

The best choice of r, l and c would be that which minimizes winding volume and thus the quantity of superconductor required.

Here, if the stored energy E, the current density j and the maximum field B_m are given in eqs. (2), (3) and (8), an optimal solution of r, l and c which minimize the volume V may be obtained.

We can use the following algorithm for seeking a solution.

Let us assume that an adequate set of r, l and c which satisfies the eqs. (2), (3) and (8) for given E and B_{max} is determined. Then resetting the value of volume V to kV ($k < 1$, but nearly equals to 1), a new set of r, l and c, which gives smaller volume, may be obtained in eqs. (2), (3), (8) and (9). The iteration of this routine will reach the optimal solution. The Newton-Raphson method is available for solving these equations numerically.

In this method, the determination of the initial values of r, l and c is a key to the solution. Here, we consider the specific shape of solenoid with long axial length, in which B_m is approximately equal to B_0, that is

$$B_m = B_0 = \mu_0 cj. \tag{10}$$

Appropriately setting l to be sufficiently longer than r, the other initial values c and r are determined in eqs. (10) and (2), respectively for the given values of B_m and E.

NUMERICAL STUDIES

It has been estimated numerically that the necessary energy capacity of SMES for the stabilizing control of real scale power system of 1,000 [MVA] class is several hundred MJ [4]. Here we design the optimal shapes of SMES with 200 through 1,000 [MJ] capacities. We prepared two sets of constraints for the design of SMES as follows

case 1: $j = 1 \times 10^7$ [A/m^2], $B_{max} \leq 10$ [T]
case 2: $j = 3 \times 10^7$ [A/m^2], $B_{max} \leq 8$ [T].

The calculated results of the optimal design are shown in TABLE 1.

TABLE 1
Optimal design of SMES

(a) Case 1

energy [MJ]		200	400	600	800	1000
r	[m]	1.96	2.28	2.42	2.56	2.70
l	[m]	1.06	1.25	1.26	1.33	1.37
c	[m]	0.94	1.04	1.24	1.31	1.39
α		1.63	1.59	1.69	1.69	1.69
β		0.36	0.36	0.35	0.35	0.34
B_{max}/B_0		1.70	1.71	1.69	1.69	1.70
B_{max} [T]		5.3	6.0	6.7	7.1	7.4
volume [m^3]		12.2	18.5	23.6	28.1	32.1

(b) Case 2

energy [MJ]		200	400	600	800	1000
r	[m]	1.70	2.23	2.59	2.87	3.12
l	[m]	.561	.613	.640	.643	.677
c	[m]	.568	.613	.649	.697	.701
α		1.40	1.32	1.29	1.28	1.25
β		.199	.159	.141	.127	.122
B_{max}/B_0		2.27	2.53	2.66	2.72	2.80
B_{max} [T]		8.0	8.0	8.0	8.0	8.0
volume [m^3]		3.40	5.25	6.77	8.10	9.31

These results show that the relations between the optimal volume V_0 [m³] and the stored energy E [MJ] are approximately represented by

case 1: $V_0 = 0.52E^{0.6}$,
case 2: $V_0 = 0.15E^{0.6}$.

The superconducting magnet with several hundred MJ may be realized by the volume within several ten m³. It is interesting that the size factors α and β are almost constant independent of the energy storage capacity in each case.

Next, the hoop stress applied to the magnet has been calculated by using the finite element method, where the magnet has been subdivided into small fractions like FIGURE 3. Young's modulus and Poison's ratio were set to 1.0×10^{11} [N/m²] and 0.3, respectively, where they were assumed to be uniform over the winding.

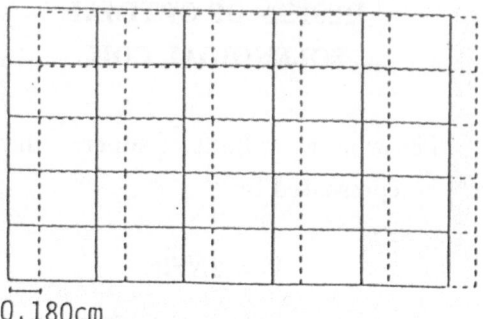

0.180cm

(a) displacement (40 times)

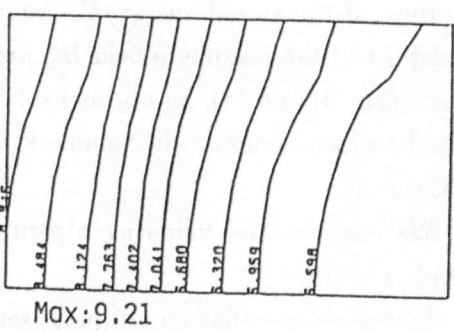

Max:9.21

(b) contour map of the hoop stress ($\times 10^7$ N/m²)

FIGURE 4   Calculated results of the force applied to the magnet (E=400 MJ in the case 1).

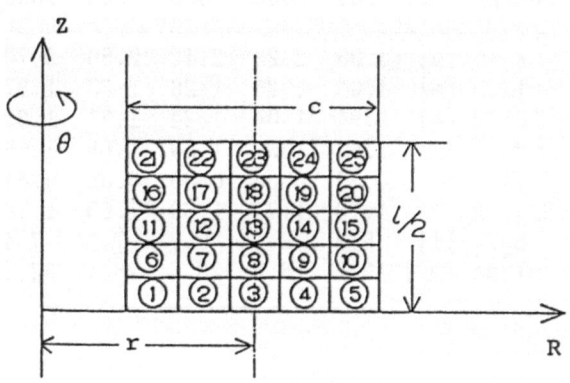

FIGURE 3   Subdivision of magnet for the application of finite element method.

An example of the calculation is shown in FIGURE 4, which corresponds to the case 1 with the energy capacity of 400 [MJ].

FIGURE 5 shows the maximum circumferential hoop stress for different stored energy

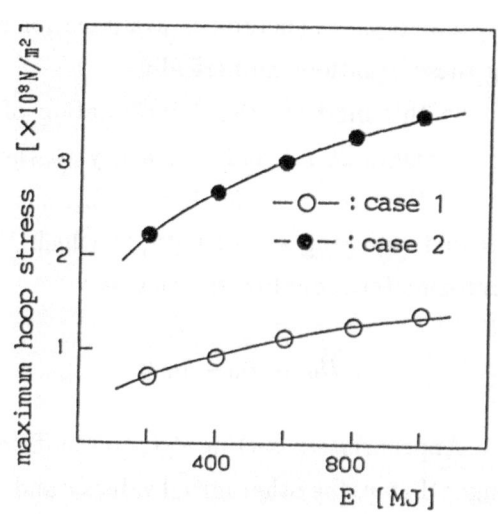

FIGURE 5   The maximum circumferential hoop stress for different stored energy capacities.

capacities. The maximum hoop stress of the magnets in case 2 is nearly three times as large as that in case 1. It can be observed that the maximum hoop stress is inversely proportional to the volume of superconductor.

CONCLUSIONS

(1) A systematic method to design an optimal solenoidal superconducting coil for the use as an electrical power system stabilizer, has been proposed.

(2) The stored energy and the maximum magnetic field has been represented by analytic functions of the current density and the shape of coil.

(3) It has been demonstrated by several numerical results that the superconducting magnet with several hundred MJ can be realized by the volume within several ten m^3.

(4) The mechanical stress applied to the coil has been evaluated by the finite element method. The results show that the hoop stress should be taken into account for the magnet with high current density.

ACKNOWLEDGEMENT

This work was financially supported by Grant-in-Aid for Scientific Research of the Ministry of Education, Science and Culture, Japan.

REFERENCES

1. Shintomi, T., Masuda, M., Ishikawa, T., Akita, S., Tanaka, T. and Kaminosono, H., Experimental study of power system stabilization by superconducting magnetic energy storage, IEEE Trans. on Magnetics, 1983, **19**, 350-353.

2. Boenig, H.J. and Hauer, J.F., Commissioning tests of the Bonneville Power Administration 30 MJ superconducting magnetic energy storage unit, IEEE Trans. on Power Apparatus and Systems, 1985, **104**, 302-308.

3. Nitta, T., Shirai, Y. and Okada, T., Power charging and discharging characteristics of SMES connected to artificial transmission line, IEEE Trans. on Magnetics, 1985, **21**, 1111-1114.

4. Mitani, Y., Tsuji, K. and Murakami, Y., Application of superconducting magnetic energy storage to improve power system dynamic performance, IEEE Trans. on Power Systems, 1988, **4**, 1418-1424.

5. Wilson, M.N., Superconducting Magnets, Clarendon Press Oxford, New York, 1983, 20-67.

DESIRED CAPACITY AND CONTROL STRATEGY OF SMES
FOR LOAD CHANGE COMPENSATION AID OF POWER SYSTEM

J. Hasegawa   and   E. Tanaka
Hokkaido University, N13 W8, Sapporo, 060, Japan

ABSTRACT

A superconducting magnetic energy storage system (SMES), which is characterized by a very high efficiency and a very quick response time, is expected much in power systems not only as a energy storage system but also as a power system stabilizer or as a load change compensator.   In this paper, some guidelines to determine fundamental specifications and a control strategy of SMES for load change compensation aid are discussed.   Dynamic performances of systems are simulated by assuming a various time constant and MW capacity of SMES under several system configurations and restrictions. As a result, a required MW and MWh capacity of SMES for this aid becomes very clear. The optimal control strategy for load change compensation with SMES are also proposed based on the simulation results.

INTRODUCTION

A number of energy storage systems are being developed for load leveling aid of power systems. A superconducting magnetic energy storage system (SMES), which is characterized by a very high efficiency and a very quick response, is one of the most promising systems and expected much in a power system not only as a energy storage system but also as a power system stabilizer or as a load change compensator [1].

In a power system load frequency control, a possible shortage in a total amount of load following capability of power plants has been a serious problem. In addition, outputs of thermal plants cannot be changed so quickly because of the constraints on response rate related to thermo-mechanical dynamics. On the other hand, a response time of SMES is less than 100 ms, which is much faster than those of conventional power plants.

And also there is no constraint on a response rate in SMES. Therefore, it is desirable to use the SMES as well as the power plants for a load change compensation. In practice, however, there are some reasonable MW and MWh capacities and some desirable operating patterns of SMES for this aid.

In this paper, authors propose some guidelines in order to determine fundamental specifications and a control strategy of SMES for this aid. In the concrete, the control effects of SMES are examined by a frequency response and a step response technique against a load change which must be compensated by LFC. The time constant and the MW capacity of SMES are parameterized. And a number of load changes are simulated under several system configurations and restrictions. The optimal control strategy for a load change compensation with the SMES are proposed based on simulation results.

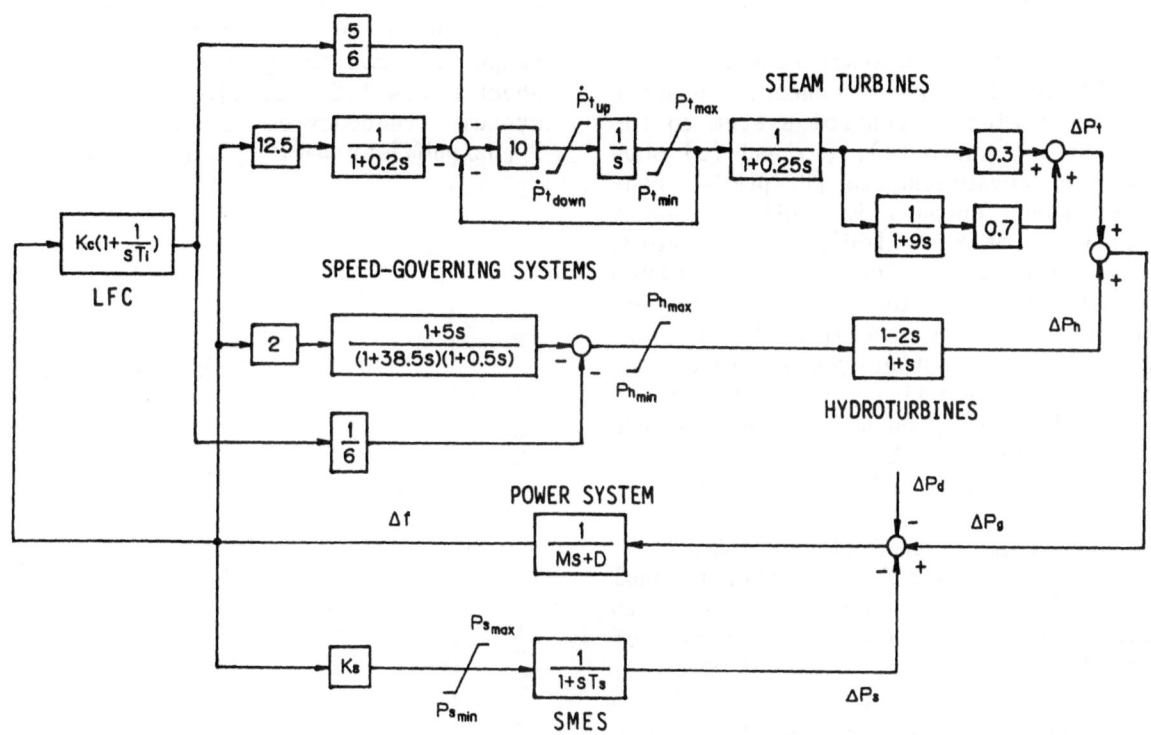

FIGURE 1   Power system model.

POWER SYSTEM MODEL

A system model used for simulation is shown in FIGURE 1 [2].

A power system dynamics between a load change and a frequency deviation is modeled in FIGURE 1 by a first order lag element having the following data:
- Total rated area capacity;
  $P* = 10000$ MW = 1 puMW,
- Nominal system frequency;
  $f* = 60$ Hz = 1 puHz,
- Inertia constant; $M = 15$ puMWs/puHz,
- Load-frequency sensitivity constant;
  $D = 3$ puMW/puHz.

A relative ratio in the generating capacity of thermal, hydro and nuclear power plants is assumed to 5:1:4. The thermal and the hydro power plants are regulated to follow load changes by governor-free operation and LFC, and are summarized to the typical plant models. However, the nuclear power plants are excluded from consideration because they are usually operated at fixed outputs.

The restrictions on power plants are assumed as follows:
- Upper and lower limits;
  Thermal; Ptmax = 0.05 puMW,
          Ptmin = -0.45 puMW,

Hydro; Phmax = 0.01 puMW,
        Phmin = -0.09 puMW,
- Changing rate limits for thermal;
  Electro-hydraulic speed-governor;
      Ptup = 0.05 puMW/s,
      Ptdown = -0.05 puMW/s,
  Mechanical-hydraulic one;
      Ptup = 0.05 puMW/s,
      Ptdown = -0.5 puMW/s.

Parameters of a proportional and integral controller for LFC shown in FIGURE 1 are determined by the ultimate sensitivity method:
- Proportional gain; Kc = 21.78,
- Integral time;   Ti = 3.82 .

The SMES with a proportional controller is modeled by a first order lag element:
- Time constant;   Ts = 0.1 s,
- Proportional gain; Ks = 20 or 40,
- MW capacity;   variable,
- MWh capacity;   no restriction.

The load change considered here is unpredictable one with a magnitude of about 3% of the total load and with a period of below ten and several minutes, which is a typical one for the LFC.

RESULTS AND DISCUSSIONS

Frequency Response Characteristics

FIGURE 2 shows frequency response characteristics of control system to the load change, where (a), (b) and (c) show frequency deviations in [Hz/puMW], generator power outputs in [dB] and power outputs of SMES in [dB], respectively. These characteristics can be derived directly from a closed loop transfer function W(jω) to the load change, as neglecting the output power limits and the changing rate limits of generators.

In FIGURE 2, each lines denote the characteristics of the followings:
- Broken lines; LFC without SMES,
- Solid lines; SMES(I),
- Dashed-and-dotted lines; SMES(II).

In a case of SMES(I), the proportional gain of SMES is set to an equal level to a conventional LFC, Ks=20. In a case of SMES(II), Ks is set to twice of SMES(I), Ks=40, besides the integral time of LFC is set to 1.0, about 26% of that of a conventional value.

In the case of LFC without SMES, we can observe a clear resonant peak in the power output of generators at a specific range of load change frequencies; i.e., about ω=0.4-1.5 [rad/s]. And the power system frequency deviation reaches to 9 [Hz/puMW] at ω=0.9-1.0 [rad/s].

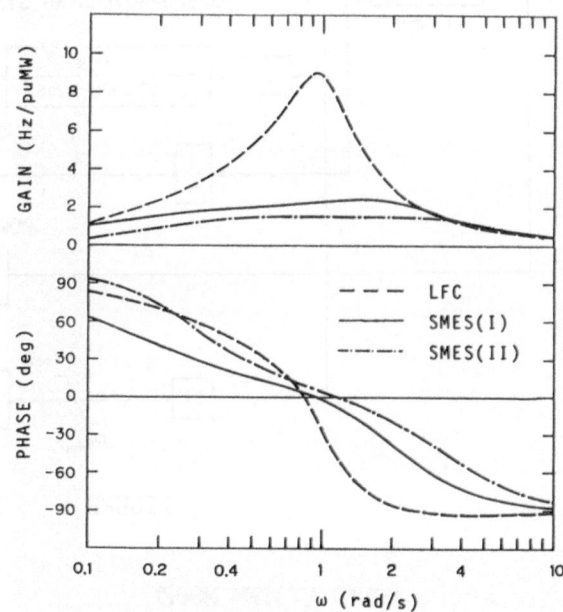

(a) System frequency.

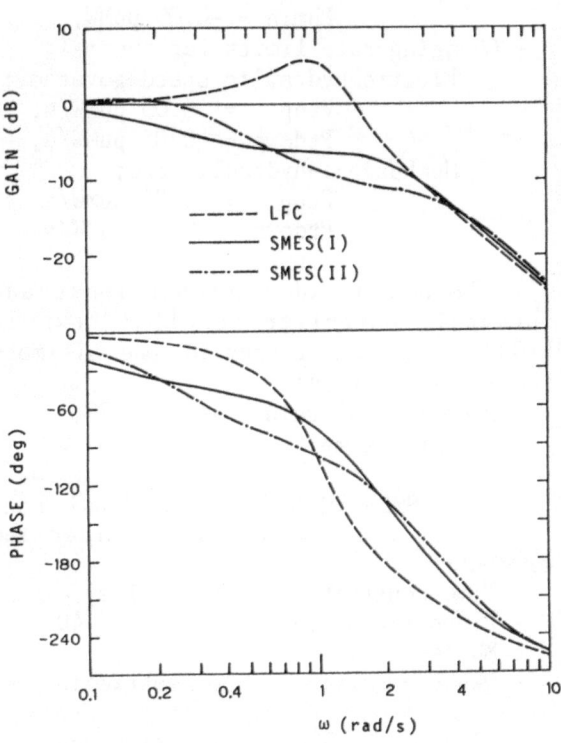

(b) Generator power output.

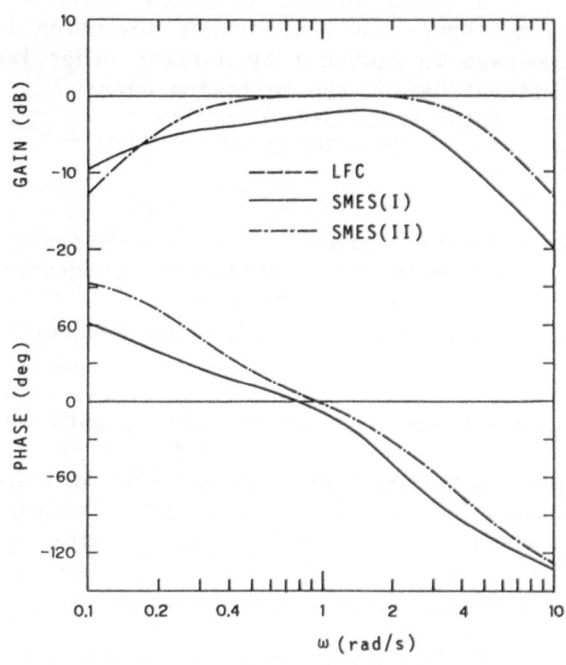

(c) Power output of SMES.

FIGURE 2   Bode diagram --- Frequency response.

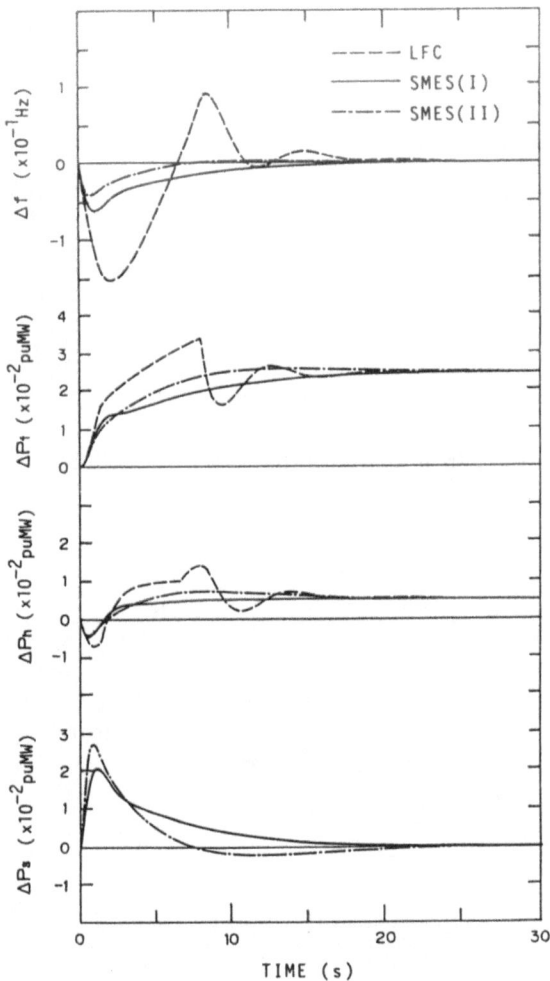

FIGURE 3   Step response.

By using SMES, such a resonant peak disappears completely and the maximum frequency deviation decreases greatly. This means that the SMES is absorbing effectively this frequency range of load changes. Thus a gain of the generator power output decreases clearly in this range; i.e., 10 [dB] or more of decrease at almost ω=1.0 [rad/s]. This tendency becomes more remarkable as increasing the proportional control gain of SMES.

However, if the gain of SMES would become larger, then the SMES should bear heavy burden because it would absorb even a long cycle load change as well as one within the range mentioned above. To avoid such a heavy burden of SMES, it may be desirable to share in load change compensation duties between generators and the SMES, such that the long cycle load change is absorbed by generators and a relatively short cycle one by the

SMES. Such a coordination is possible by shortening the integral time of LFC.

A setting of SMES(II) is a result of such a coordination. We can see that the gain of SMES decreases in this case at low frequency region than SMES(I), instead of an increase in the gain of generator. In the case of SMES(II), the gains of generator and SMES are crossing each other at ω=0.33 [rad/s], and the generator plays dominant role at a low frequency region of load change and the SMES at a high frequency region.

From FIGURE 2, we can see too that a required MW capacity of SMES is almost the same magnitude of the maximum load change considered.

Step Response Characteristics

FIGURE 3 shows a simulation result of a response against a step load change of 0.03 puMW. A settling time, which is assumed in this paper as a required time for a frequency deviation to settle down within ±0.01Hz, is affected much by restrictions of power plants, especially by the changing rate limits for thermal power plants, if there is no SMES. The settling time goes to 16.3 seconds as shown in FIGURE 3 by a broken line when the changing rate limits are set to (-0.5, +0.05), while it requires about 1 minute, more than 3.6 times of 16.3 [s], for relatively more severe rate limits of (-0.05, +0.05).

On the contrary, by using the SMES, the settling time goes to less than 10 seconds even for the severe rate limits of (-0.05, +0.05). And also the step response itself becomes very good by the SMES; no overshoot response is observed in the generator output, the maximum system frequency deviation decreases to 1/2 - 1/3 of that without SMES.

TABLE 1 summarizes main results shown in FIGURE 3.

TABLE 1   Results of step response.

		SMES(I)	SMES(II)	LFC
$|\Delta f|$max	[Hz]	0.063	0.041	0.153
Settling time	[s]	10.0	4.3	16.3
$|\Delta Pt|$max	[puMW]	0.025	0.0259	0.0339
$|\Delta Ph|$max	[puMW]	0.0054	0.0073	0.014
$|\Delta Ps|$max	[puMW]	0.0208	0.027	----
Energy	[puWh]	29.9	21.4	----

Effects of MW Capacity of SMES

FIGURE 4 shows the settling time for a step load change against the MW capacity of SMES. The SMES(I) is less affected by the MW capacity restriction than the SMES(II). We can see from this FIGURE that SMES may work effectively even if the MW capacity decreases to 1/6 of the maximum load change for SMES(I), and 1/2 for the SMES(II), as far as the settling time. However, it is worth to note here that the lower the MW capacity is, the larger the maximum frequency deviation is. The best selection of MW capacity of SMES is still the same level of the maximum load change considered.

Effect of Time Constant of SMES

FIGURE 5 shows an effect of the time constant of SMES to the maximum frequency deviation and FIGURE 6 shows similarly one to the maximum power output of SMES. From these figures, we can see that the SMES(I) is less sensitive than the SMES (II) to the time constant change of SMES.

Effect of Power System Parameter Change

In order to examine a robustness of the control system, dynamic performances are simulated by changing the inertia constant M and the load-frequency sensitivity constant D within a range of 50% to 200% of standard values of them. FIGUREs 7 and 8 show relative changes in

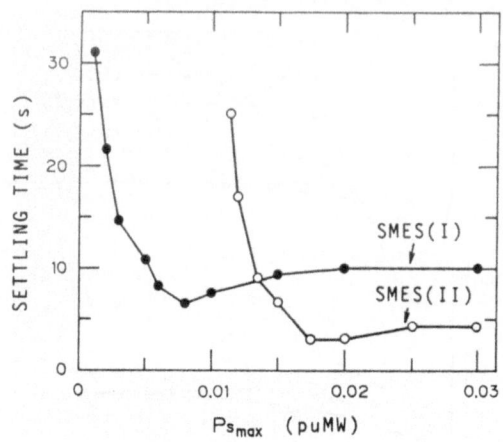

FIGURE 4　Settling time vs. SMES MW.

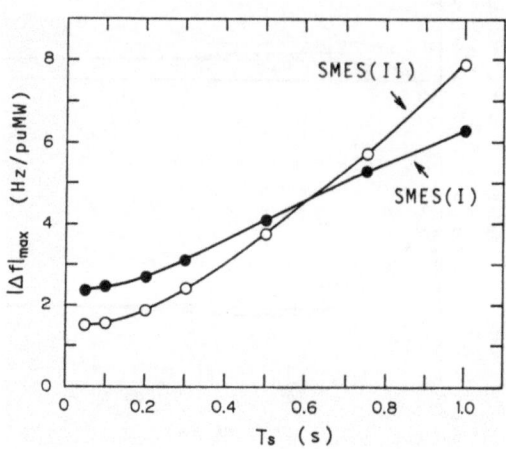

FIGURE 5　Ts vs. frequency deviation.

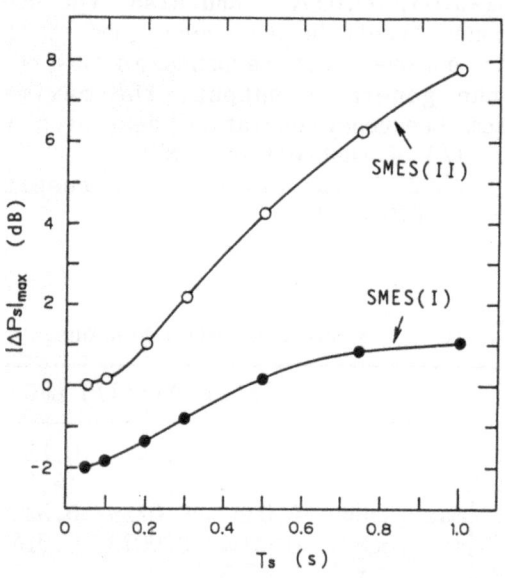

FIGURE 6　Ts versus SMES output.

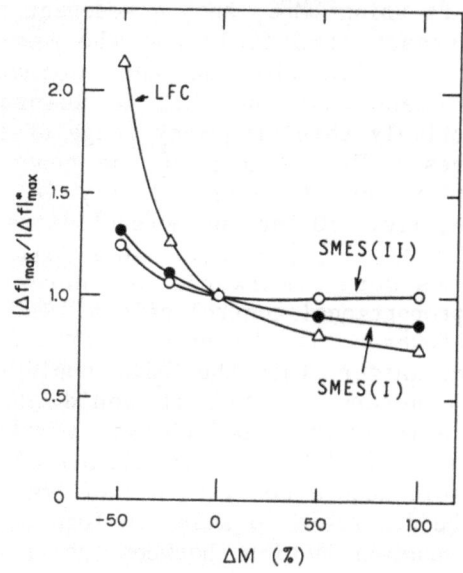

FIGURE 7　Effect of inertia constant.

the maximum system frequency deviations in each cases. Although the frequency deviation is much affected by decreasing M in the case without SMES, the effects of system parameter changes are sufficiently controlled by using SMES.

SMES is very effective to make the control system be robust against a power system state change.

Effects of The Restriction in Capacity of The Regulating Power Plant

The effects when the capacity of regulating power plants are restricted are examined by step response analyses. Results are summarized in TABLE 2. The upper limit of generator output is set to the same level with the load change. Here, we can see that the frequency deviation and the required MW capacity of SMES are seldom affected by this level of restriction, though the settling time tends to be prolonged and a required MWh capacity of SMES tends to be large. Again, the SMES(I) is less affected than SMES(II).

TABLE 2   Effects of total regulating capacity restrictions.

(a) Thermal 0.03 ; Hydro 0.0

		SMES(I)	SMES(II)
$|\Delta f|$max	[Hz]	0.056	0.037
Settling time	[s]	16.7	10.8
$|\Delta Ps|$max	[puMW]	0.0186	0.024
Energy	[puWh]	50.9	55.1

(b) Thermal 0.025 ; Hydro 0.005

		SMES(I)	SMES(II)
$|\Delta f|$max	[Hz]	0.064	0.041
Settling time	[s]	14.9	9.0
$|\Delta Ps|$max	[puMW]	0.0212	0.027
Energy	[puWh]	45.9	49.2

(c) Thermal 0.02 ; Hydro 0.01

		SMES(I)	SMES(II)
$|\Delta f|$max	[Hz]	0.066	0.041
Settling time	[s]	17.5	7.5
$|\Delta Ps|$max	[puMW]	0.0219	0.0271
Energy	[puWh]	56.5	47.7

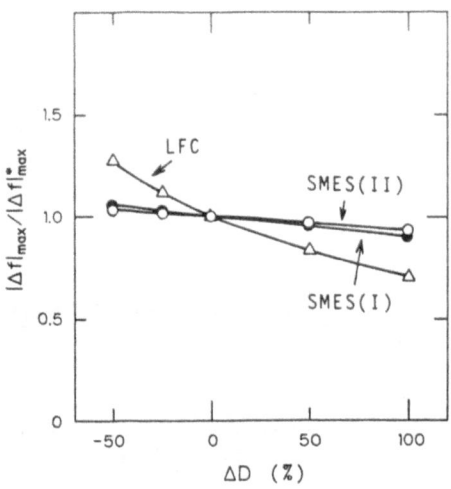

FIGURE 8   Effect of constant D.

CONCLUSIONS

(1) Applying a proportional control to the SMES, it is shown that a required MW capacity of SMES is almost the same level with the maximum amounts of load change, and a required MWh capacity of SMES is less than 100 puWh; i.e., about 30 puWh in standard case.
(2) The SMES is especially effective to the load change with a cycle range of few seconds to few minutes. Thus it is preferable to share the duties between generators and the SMES. Generators may respond to long cycle change and the SMES to a cycle range stated above.
(3) The robustness of the control system is greatly enhanced by using the SMES.
(4) A setting of SMES(II) is preferable if operating conditions are free from the any restrictions of generators. Otherwise, a setting of SMES(I) may be better.

REFERENCES

[1] Peterson, H.A., Mohan, N. and Boom, R.W., Superconductive energy storage inductor-converter units for power systems. IEEE Trans., 1975, **PAS-94**, pp. 1337-46.
[2] IEEE Committee Report, Dynamic models for steam and hydro turbines in power system studies. IEEE Trans., 1973, **PAS-92**, pp. 1904-15.

STUDY OF SMES SYSTEM WITH DC INTERTIE FOR POWER LINE STABILIZATION

Hidehiko OKADA   Tadao EZAKI   Kokichi OGAWA   Hiromi KOBA
Oita University, 700 Dannoharu, Oita 870-11, Japan
Fujio IRIE
Kyushu Engineering Department, Kinki University, Kashi-no-mori, Iizuka 820, Japan
Masakatu TAKEO   Seiki SATO
Kyushu University, 6-10-1 Hakozaki, Fukuoka 812, Japan
Kanichi TERAZONO
Kyushu Electric Power Co.Inc., 47-1-2 Shiobaru, Fukuoka 815, Japan
Masahiro TAKAMATU
Kobe Steel, LTD., 4 Iwayanaka, Nada-ku, Kobe 657, Japan
Noriko KAWAKAMI and Masaru HIRANO
Toshiba Co. 1 Toshiba, Fuchu 183, Japan

ABSTRACT

A new superconducting magnet energy storage (SMES) system proposed by one of authors for power transmission line stabilization is studied by an experimental model system and computer simulation. It has a SMES in a dc intertie section connected to ac transmission lines via convertors. The distinctive characteristics of this system is that independent stabilization can be made for each ac line and also that the power flow through the dc section can be controlled. For verifying these characteristics we developed a model system consisted of a superconducting pulse magnet (100kJ), pulse width modulation (PWM) convertors, simulated power line system with a generator (10kVA) and a short-circuiting device to give power line fault.

INTRODUCTION

The application of SMES to stabilization of power transmission lines is being studied (1)-(4), but all of these other than ours have a SMES connected as a branch of a power line. We have proposed a new system having a SMES in the section of dc intertie in ac lines(5)-(6).

This system has higher freedom of control which enables to stabilize each part of ac line independently as well as to control power flow from one side to other side as shown in Fig.1. A peculiar characteristic is to disconnect P and Q change and to cut disturbance propagation from one side to the other.

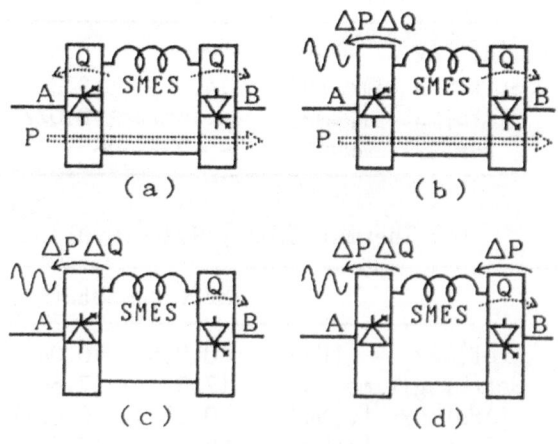

FIGURE 1   Operating modes of SMES with dc intertie.

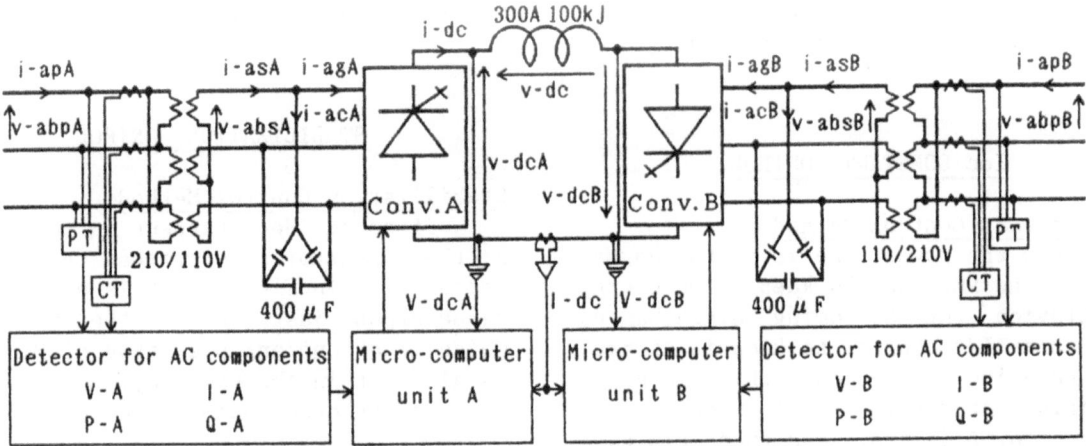

FIGURE 2   Schematic diagram of SMES system with dc intertie in ac lines.

In our previous papers (5) and (6), experiments just showing the principle of our system were made. However full characteristics were not made clear in them due to the lack of the convertor capacity. In this paper full characteristics have been obtained by building up the convertor capacity twice and by installing a short-circuiting device.

SMES SYSTEM AND PQ CONTROL

SMES System

To show the above mentioned characteristics, an experimental SMES system is constructed which consists of a superconducting magnet and two GTO convertors as shown in Fig.2.

A superconducting magnet of 100kJ has pulse rate of 100A/sec. For making the magnet compact, the highest magnet field is fixed in 7T.

A convertor consists of the Graetz bridge with GTO's (300A, 2500V) and is controlled by the pulse width modulation of 6 pulses using a computer. Condensers are connected to ac lines to absorb switching surges from GTO's, which causes additional reactive power of about 5kVar to the reference value Qr of a convertor.

PQ Control

Pulse width M and phase angle α to a convertor are determined by given active-power Pin and reactive power Qin.

$$M = [(Pin^2+Qin^2)/(VdcIdc)^2]^{1/2}$$
$$\alpha = \cos^{-1}[Pin^2/(Pin^2+Qin^2)]^{1/2}$$

where Vdc is the mean dc output voltage at M=1 and α=0, and Idc is the dc current,

that is, the magnet current.

The flow chart to get M and α from the reference values of Pr and Qr in the computer is shown in Fig.3. If $Pr^2+Qr^2 > (VdcIdc)^2$, Pr and Qr are limited to Pin and Qin within VdcIdc by limiters.

At every 60° calculated M and α are given to a gate logic (UGL). Then Qbias is added to correct the reactive power of the condenser 400μF in convertor ac lines, and Pbias is zero.

If energy flow from A and B is not zero, the magnetic current increases or

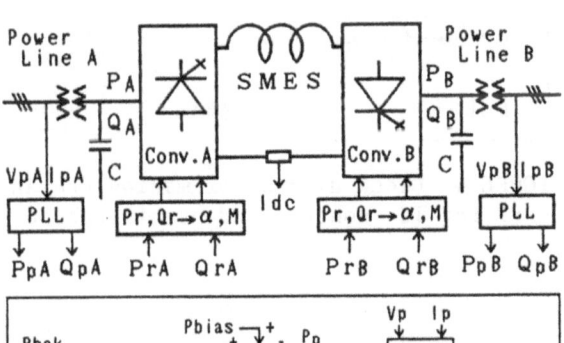

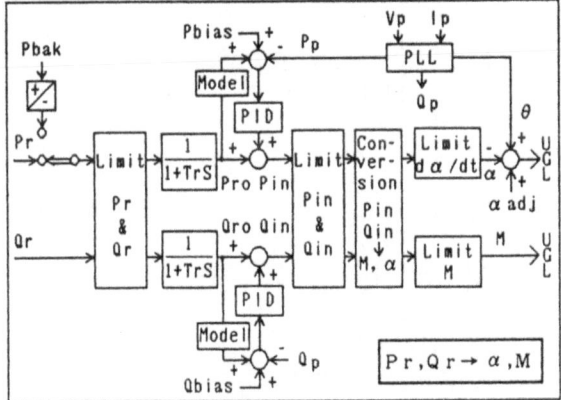

FIGURE 3   Block diagram of active and reactive power control by PWM.

decreases. At the max. or min. value of Idc, the Pbak mode cuts Pr and turns Idc back to the original value Idorg.

Idc CONSTANT CONTROL

Idc Constant Control Method

When the power line is in the normal state, the SMES should keep some energy to provide for a coming disturbance. Therefore Idc control to keep it at prescribed value is needed, which is shown in Fig.4. At first, PintA is entered to a convertor A as PrA, and then PrB to a convertor B is made through thick lines. The PrB consists of the opposite sign of PrA and the deviation of Idc from the reference value Idcr.

Experimental Results

Fig.5 shows the controlled Idc when the assigned power is changed. In this

control mode, Idc is well kept around the reference value of 70A at each operation.

STABILIZATION CONTROL

Stabilization Control Method

If an accident happens on power line A, the convertor B operates through the dotted line in Fig.4. PrB is set to the value PfpoB just before the accident, which is a constant operation of P. QrB is also set in the same way.

On the other hand, the convertor A

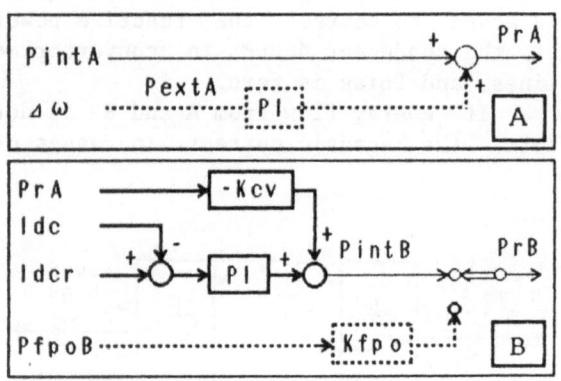

FIGURE 4 Block diagram of making PrA and PrB for Idc control and stabilization.

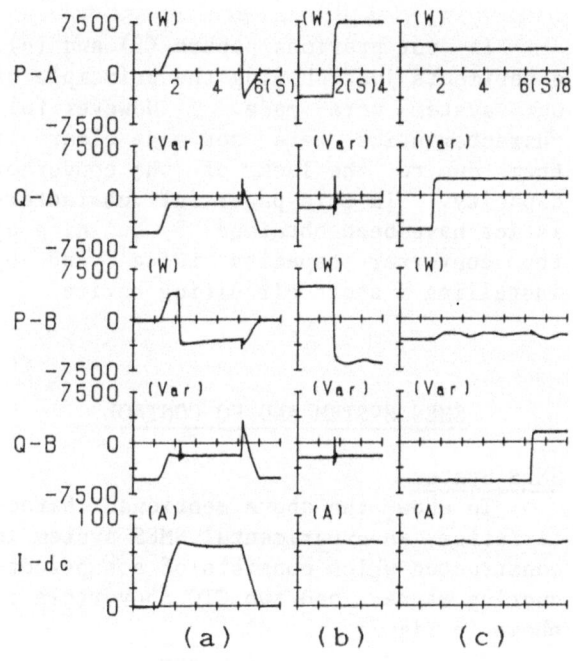

FIGURE 5 Operations of Idc control mode.
(a) start and stop
(b) change P-A
(c) change Q-A and Q-B

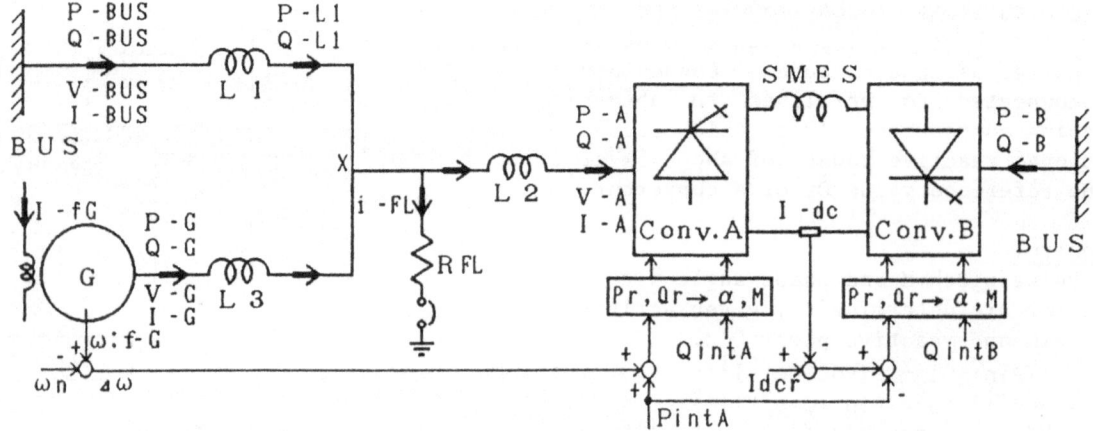

FIGURE 6 Schematic diagram of an experiment for stabilization.

must operate to stabilize power line A. For this stabilization control, only the deviation $\Delta\omega(=\omega-\omega_n)$ of angular velocity ω from the rated value ω_n of the generator is now feedback as PextA.

Experimental Circuit

A schematic diagram of this experiment is shown in Fig.6 where a generator is connected through a transmission line to the convertor A.

A generator (10kVA, 200V, 60Hz) is driven by a dc motor coupled with a flywheel to adjust GD^2, and has AVR control. A dc motor is driven by the three phase fully controlled bridge rectifier with TG feedback speed control.

A short-circuiting device consists of Triac's for ON and a mechanical breaker for OFF, and is able to make a short-circuit for a few cycles.

Experimental Results

As shown in Fig.7(b), the stabilization mode starts by detecting a deviation of V-A at the constant Idc mode. That is, the convertor A outputs the adjusted value of P-A to suppress the deviation of generator revolution fG, while the convertor B is operated in constant P and Q

mode. Fig.7(a) shows the case of no stabilization control for comparison.

POWER FLOW CONTROL

The power flow control was confirmed experimentally by Fig.8. Fig.9 shows each waveform when P-A, that is active power of ac line 2, decreases gradually. Active power P-L1G of ac line 1 increases as decreasing of P-A.

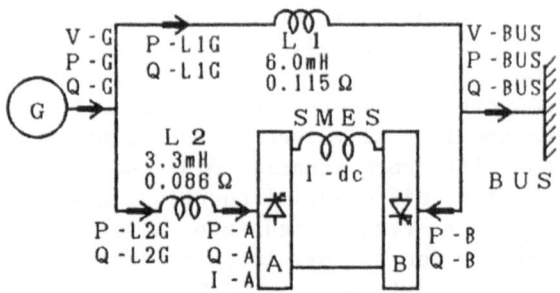

FIGURE 8 Schematic diagram of an experiment for power flow control.

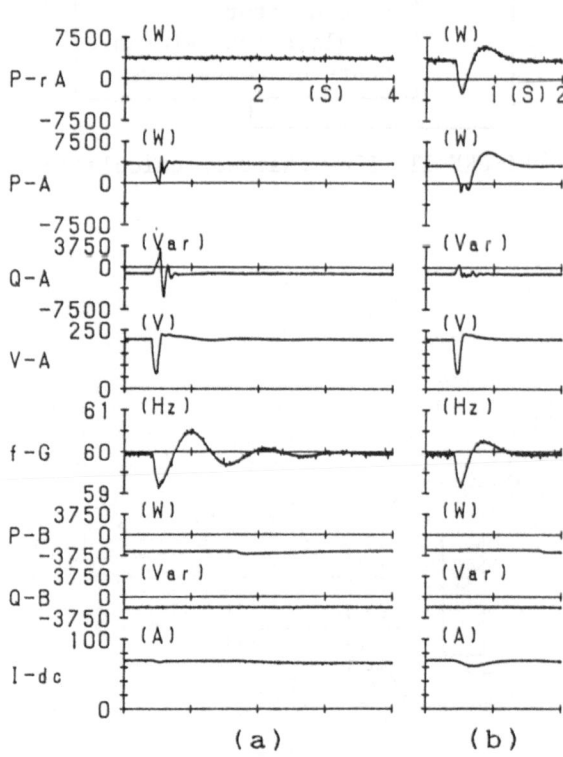

FIGURE 7 Stabilization for 5-cycle short
(a) no stabilization control
(b) by stabilization control

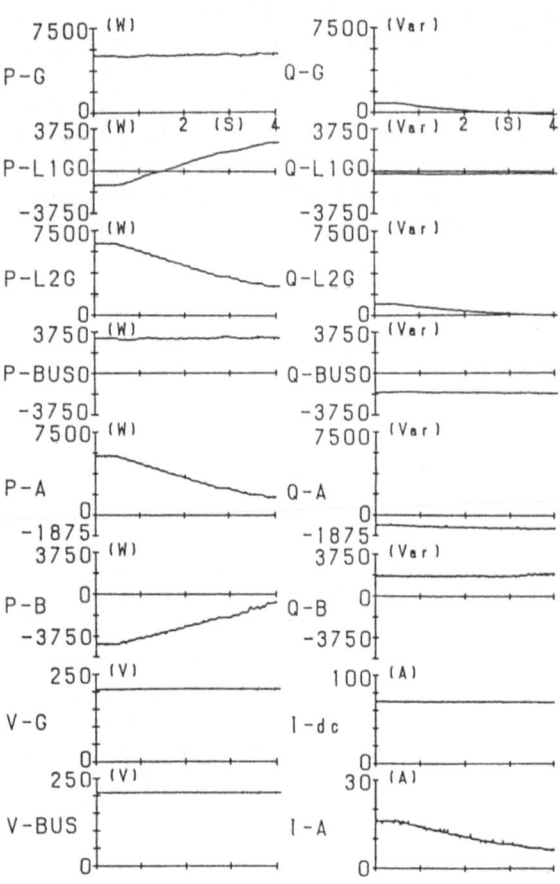

FIGURE 9 Waveforms at decreasing P-A.

COMPUTER SIMULATION

Analysis Method

Computer simulation is carried out for the experimental model system to study the complex power line network in the future. An equivalent circuit used for the analysis is shown in Fig.10. The generator is represented by the Park's equation. The iron loss and saturation in the transformers are neglected. The convertors are assumed to consist of ideal switches, in which switching losses are ignored.

At first, initial power flow is calculated. All state variables are decided by the given values of active and reactive powers of the generator and convertors at both sides. These are used as the initial values of the solutions of the difference equation discussed later.

The differential equation given by Fig.10 is transformed into a difference form, which is formulated as follows,

$$[A]\frac{[x_{t+dt}]-[x_t]}{dt} = [B][x_{t+dt}]+[C][u_{t+dt}]$$

where dt is sampling period. $[x_{t+dt}]$ and $[x_t]$ are state vectors at t+dt and t respectively. $[u_{t+dt}]$ consists of bus voltages of side A and B, and the field voltage of generator. The solutions are calculated at every 2° and integrated over previous 2 cycles. The effective values of voltages and currents are decided at every 60° and give the active and reactive powers at each node. These

are used in calculations of the velocity of the generator and the operation of of convertors.

If an accident happens, slip s of the generator is calculated from the following equation.

M ds/dt + D s = P - Pm ,

where M is the inertia constant, D the damping coefficient, P the output of the generator and Pm the output of the DC motor for drive.

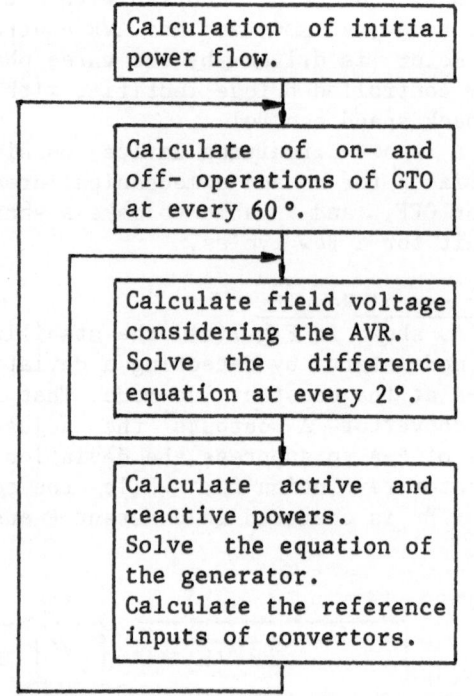

FIGURE 11   Flow chart of calculation.

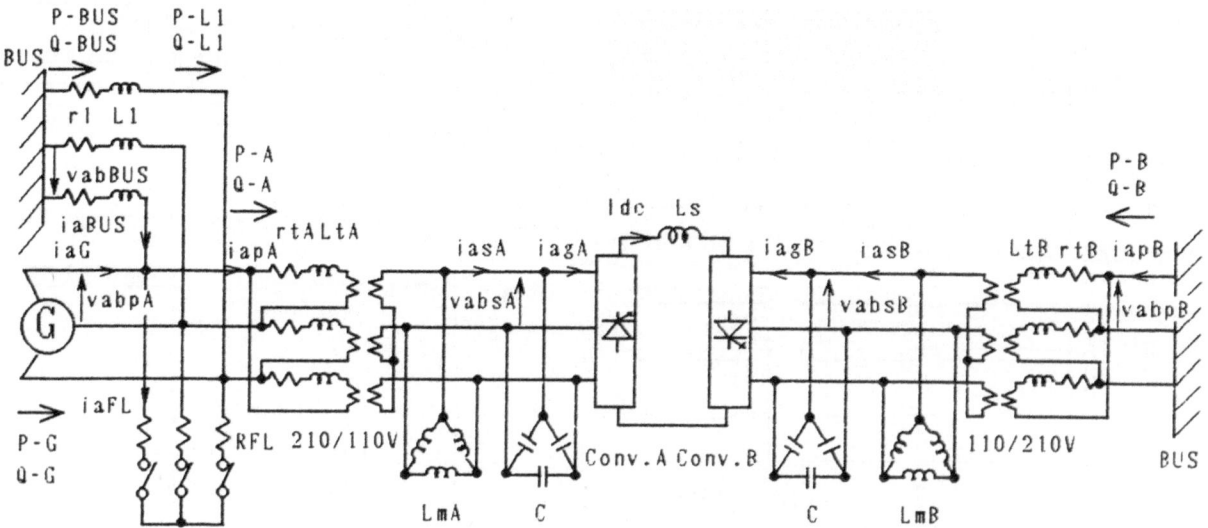

FIGURE 10   An equivalent circuit of model power system.

The description of active and reactive powers control process simulates the experimental system. At normal operation, the convertor at side A operates with PrA and QrA which are given in the initial power flow. The convertor at side B operates to keep the magnet current the reference value Idcr. When an accident happens, the deviation of the angular velocity is added to PrA. PrB in convertor B maintains the value just before the accident. The pulse width M and the phase α of the convertors are calculated with these reference inputs and replaced at every 60°.

The AVR is represented by the first order system. The calculation procedure is shown in Fig.11.

Simulation Results

The numerical solution for the stabilization control is shown in Fig.12. It corresponds to the experimental results shown in Fig.7. In calculation, the magnetic saturation of the generator is neglected. Numerical results are seen to be in good agreement with experimental ones.

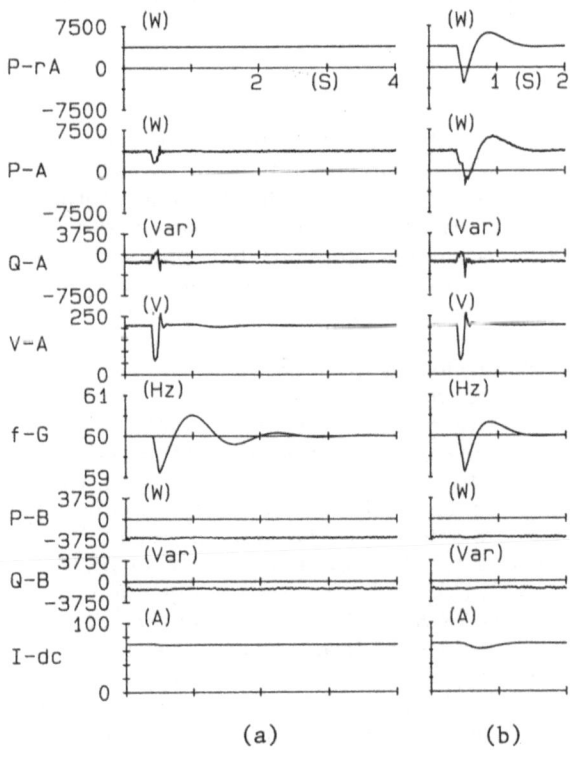

FIGURE 12 Simulation results of
   stabilization
   (a) no stabilization control
   (b) by stabilization control

CONCLUSION

The characteristics of SMES system with dc intertie in ac power lines were presented.

At normal condition, Idc control is able to keep the magnet energy in constant while transmitting the assigned power from one side to the other.

At accident mode, the stabilization control suppresses the deviation of the generator revolution for the accident line side, while a constant P and Q operation keeps the normal operation for the normal line side.

The power flow control in parallel lines was shown as further application of this system.

These operations were confirmed by experiments and computer simulation.

REFERENCES

(1) Rogers, J.D., Boenig, H.J., J.C.Bronson, Colyer, D.B., Hassenzahl, W.V., Turner, R.D. and Shermer, R.I., 30-MJ SMES Unit for Stabilizing an Electric Transmission System. IEEE Trans. on Magnetics, 1979, MAG-15, 820-823.

(2) Nitta, T. and Okada, T., Control and Characteristics of SMES Connected to Power Systems. Proc.of US-Japan Workshop on SMES, Madison, USA, 1981, 64-78.

(3) Mitani, Y., Tsuji, K., Ise, T. and Murakami, Y., Fundamental Study on Electrical Power System Stabilization Using Superconducting Magnet Energy Storage. Proc. of MT-9, 1985, 353-356

(4) Ishikawa, T., Akita, S. and Taniguchi H., Power System Stabilization by SMES Using Current-fed PWM Power Conditioner. PESC'88 Record, 1988, 334-341.

(5) Irie, F., Workshop on SMES as satellite meeting of ICEC-10, 1984.

(6) Okada, H., Ezaki, T., Ogawa, K., Koba H., Takeo, M., Funaki, K., Sato, S., Irie, F., Chikaba, J., Terazono, K., Takamatu, M., Kawakami, N. and Hirano M., Experimental Study of SMES with DC Intertie for Power Line Stabilization. PESC'88 Record, 1988, 326-333.

ON THE MULTIPLE-RING COIL SYSTEM FOR SMES

Tadao EZAKI
Oita University, 700 Dannoharu, Oita 870-11, Japan

Fujio IRIE
Kyushu Engineering Department, Kinki University, Iizuka 820, Japan

ABSTRUCT

A new coil system named "multiple-ring coil system" for the intermediate SMES is proposed. This system consists of concentric multiple low aspect ratio coils in which the direction of current is opposit to those of adjacent coils. Characteristic features of this system such as the reduction of hoop force acting on the coil conductor, total force on them and leakage magnetic field are discussed in comparison with a single solenoid system and a toroidal coil system. A conceptual design for an intermediate scale SMES of 100MWh shows that this system has advantages of the small hoop stress and the small leakage magnetic field to the single solenoid system and an advantage of the small total force to the toroidal coil system.

INTRODUCTION

As for the application of the superconducting magnetic energy storage system [SMES], two kinds of applications have been mostly considered, that is, the applications for the diurnal storage of electric power of a very large scale SMES (~10[GWh])[1-2] and for the stabilization of electric power line of a small scale one (~1[GJ])[3-4]. But the large scale SMES might be unsuitable for Japan, considering following conditions for a location of SMES: It is rather hard to find such site for the large scale one which requires a rigid bedrock to support huge magnetic forces and a large uninhabited area to avoid effects of a leakage magnetic field. The power line system in Japan has complicated network structure with distributed heavy power load area. Therefore numbers of intermediate scale SMES's (10~100[MWh]) distributed near urban area may be effective in Japan. These SMES's can play as rolls of both the peak load shaving for local demands and the stabilization of power line system as well as controlling of the reactive power.

The cost of this intermediate SMES is relatively higher than that of the large scale SMES which is to be discussed in comparison with hydro-pumped storage. But the same criterion in cost as the large scale one can not be applied to the intermediate scale one, because this SMES has an additional merit which is the improvement of the total effeciency of transmission system by the stabilization of the power line system.

The intermediate scale SMES settled in the urban area should be designed with regard to following points. At first, the magetic force acting on the coil conductor itself should be as possible as small, because a rigid bedrock is hardly available at urban area. This is also important in case of using very brittle high temperature superconductor or high field compound type superconductor. Secondary, the plottage which includes area need for the decrement of leakage magnetic field should be small because of higher cost of ground in such area.

As a conventional low aspect ratio solenoid coil system has a large leakage field and relatively large hoop force on the coil conductor, this coil system might be not suitable for the purpose discussed above. Contrasively, a toroidal coil has only little leakage fields, however the total magnetic force acting on the coil is very large and it requires much quantity of superconductors.

We propose here a new coil system named "multiple-ring coil system", which consists of many concentric coils with low aspect ratios. With this configuration, a inner space of the outer most coil can be utilized effectively to store magnetic energy, because the magnetic energy of a low aspect ratio coil is concentrated largely near the coil conductor. Therefore more efficient space utilization is realized than by single coil system. In this system, we can expect the reduction of the magnetic force acting on one coil, because magnetic energy is divided to many coils. In addition to this, we can also expect the reduction of leakage magnetic field if currents in adjacent coils have differnt direction to each other, because the magnetic field near the system coil cancell each other.

In the following, characteristic features of this system such as the reduction of hoop force acting on the coil conductor, total force on them and leakage magentic field are discussed in comparison with a single solenoid system and a toroidal coil system.

OPTIMUM CONFIGURATION OF THE SYSTEM

A schematic figure of the multiple-ring coil system which consists of N coils and has an outer radius R is shown in Fig.1. Circular cross section is assumed for coils for simplicity. As all of the ring coils carry same current density, the radius of each coil is proportional to the square root of its ampere turns. An aspect ratio $\beta = a/R$ is used to designate a form of this system, where a is the maximum radius of cross section among those of the ring coils.

In this system, the stored enegy of the system W, the hoop force acting on the ℓth ring coil F_{ℓ} and the magnetic field at the ℓth coil B_{ℓ} are

$$W = \mu_0 I^2 R w, \qquad (1)$$

$$F_{\ell} = \mu_0 I^2 f_{\ell}, \qquad (2)$$

$$B_{\ell} = \mu_0 (I/R) b_{\ell}, \qquad (3)$$

where I is the muximum ampere turns among those of ring coils, μ_0 the permeability of vacuum. Expression of w, f_{ℓ} and b_{ℓ} having the meaning of shape factors which depend on the configuration of the

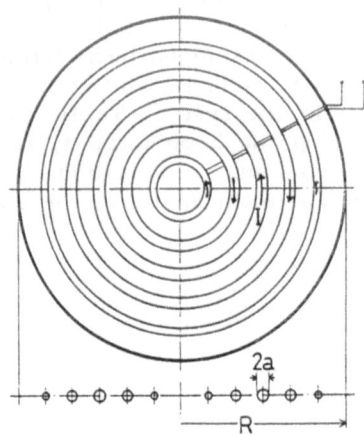

FIGURE 1 A schematic figure of multiple-ring coil system.

system such as aspect ratio β, a number of coils N , normalized ampere turns of each coils $[i_1, i_2, ..., i]$, normalized radii of each coils $[r_1, r_2, ..., r_N]$, and are obtained as

$$w = \frac{1}{2} \sum_{l=1}^{N} i_l \left\{ r_l \left(\log \frac{8r_l}{\sqrt{i_l}\beta r_p} - \frac{7}{4} \right) \right.$$
$$\left. + \sum_{\substack{m=1 \\ m \neq l}}^{N} i_m (r_l + r_m) \left[\left(1 - \frac{k_{lm}^2}{2} \right) K(k_{lm}) - E(k_{lm}) \right] \right\} \qquad (4)$$

$$f_l = i_l \left\{ \frac{1}{2} i_l \left(\log \frac{8r_l}{\sqrt{i_l}\beta r_p} - \frac{3}{4} \right) \right.$$
$$\left. + \sum_{\substack{m=1 \\ m \neq l}}^{N} i_m \left[\frac{K(k_{lm})}{1 + r_m/r_l} + \frac{E(k_{lm})}{1 - r_m/r_l} \right] \right\} \qquad (5)$$

$$b_l = \frac{1}{4\pi r_l} \left\{ \left(\log \frac{8r_l}{\sqrt{i_l}\beta r_p} - \frac{3}{4} + \frac{2r_l}{\sqrt{i_l}\beta r_p} \right) \right.$$
$$\left. + 2 \sum_{\substack{m=1 \\ m \neq l}}^{N} i_m \left[\frac{K(k_{lm})}{1 + r_m/r_l} - \frac{E(k_{lm})}{1 - r_m/r_l} \right] \right\} \qquad (6)$$

where K and E are complete elliptic integrals of the first and the second kinds, respectively and k_{lm} is defined as

$$k_{lm} = 2\sqrt{r_l r_m}/(r_l + r_m) \qquad (7)$$

and r_p is the normalized radius of the coil with maximum ampere turns. In this derivations diameters of coil cross sections are assumed to be very small compared with distances between adjacent coils.

As one of the purpose of this multiple-ring system is to reduce the hoop force on one coil and the magnetic field on the superconductor in the system should be less than the critical value, we have to discuss the characteristic of the system with the maximum hoop force F_h and the maximum field B_m among those of all of the coils in the system. These are given by replacing the hoop force f_l in eq.(2) with its maximum value f and b_l in eq.(3) with its maximum value b, which are

$$F_h = \mu_0 I^2 f, \qquad (8)$$

$$B_m = \mu_0 (I/R) b, \qquad (9)$$

where f and b are

$$f = \max(f_1, f_2, ..., f_N), \qquad (10)$$

$$b = \max(b_1, b_2, ..., b_N). \qquad (11)$$

Now we optimize the configuration of the system in terms of fixed values of the stored energy W and the maximum fiel B_m. Equations of the maximum hoop force F_h and the system radius R are derived for this condition as,

$$F_h = f/(bw)^{2/3} (B_m W/\mu_0)^{2/3} \qquad (12)$$

$$R = (b^2/w)^{1/3} (\mu_0 W/B_m^2)^{1/3} \qquad (13)$$

And, also the maximum ampere turns I and current density J are given as

$$I = (bw)^{-1/3} (WB_m/\mu_0^2)^{1/3} \qquad (14)$$

$$J = (w/b^5)^{1/3} \beta^{-2}/\pi (Bm^5/W\mu_0^4)^{1/3} \qquad (15)$$

The maximum value of the hoop force should vary with a distribution of ampere turns at each coils. To get the condition for minimizing the maximum hoop force, a distribution of ampere turns of coils is optimized numerically. An example of distributions of ampere turns and the hoop force in each coil are shown in Fig. 2 (a) and (b).

Decrease of the muximum hoop force with the number of coils N is seen in Fig.3. The hoop force is rapidly decreased with an increasing of N in a region N is smaller than 4, but in the region of larger N it has the minimum value for large value of β or it saturates. This shows that too large numbers of coils is not effective for the reduction of the hoop force, and it only increases of superconductor quantity. The optimum value of N might be determined taking account of conductor cost, support material cost and construction cost, however we tentatively chose here the optimum value of N as the 90 % of a reduction of the hoop force from the value with one ring coil system to the saturated value or the minimum value is attained.

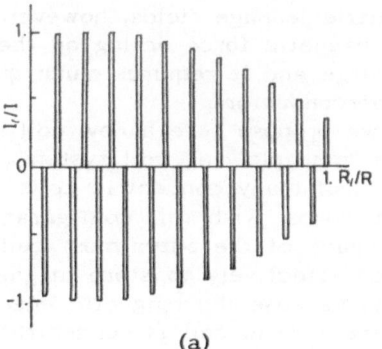

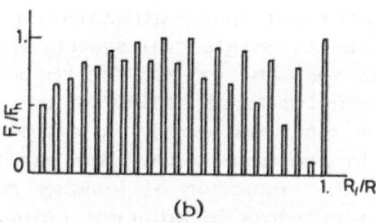

FIGURE 2 (a) Optimized distibution of ampere-turns of coils, (b) Optimized hoop forces of coils.

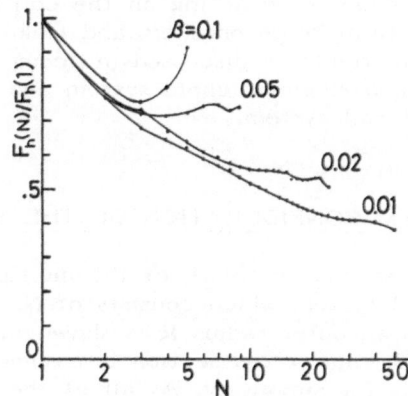

FIGURE 3   Normalized hoop forces vs coil numbers.

RESULT AND DISCUSSIONS

The Maximum Hoop Force

To show the effectiveness of multiple-ring system for the reduction of hoop force, the normalized maximum hoop forces in multiple-ring system and in the single solenoid system are plotted to the normalized system radius in Fig.4. Both of them decrease with radius, but the reduction in multiple-ring system is more rapid. Then the hoop force in multiple-ring system is getting smaller than it in the single solenoid with the increase of radius. From this result, it can be said that the multiple-ring system is effective to reduce the hoop force on one coil for the system with a large radius of certain extent.

Total Force

A quantity of supporting structure, which is one of principal factors of the system cost, is proportional to the total force F_t acting on coils in the system. The total forces in the multiple-ring system are plotted to the system radius in Fig. 5. For comparison, the single solenoid system and the toroidal system are also plotted there. Rugged characteristics of the multiple-ring system in this figure and also in Fig. 6 are due to discrete changes of optimum numbers of coils in the system.

This figure shows that the total force of the multiple-ring system is larger than that of the solenoid system and that it is much less than that of the toroidal coil system. The less total force of the multiple-ring system than that of the toroidal coil system is one of the advantage of this system.

Quantity of Superconductors

Quantity of superconductor Q which is the product of current capacity and length of the superconductor is also related to system cost. Those of the three systems are shown in Fig. 6. Both in the single solenoid system and the toroidal coil system, there are minimum of the quantity of superconductor at about $\beta=0.36$ and $\beta=0.64$, respectively, and that of the toroidal

coil system is 2~3 times larger than the single solenoid system.

The multiple-ring system need the conductor quantity 2~5 times larger than

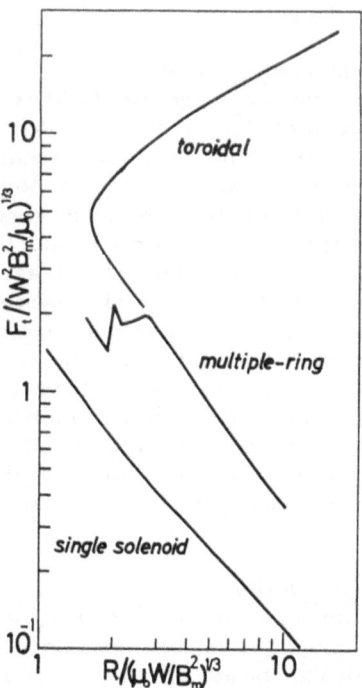

FIGURE 5 Maximum radius dependences of total force of multiple-ring system, that of single solenoid system and that of toroidal system.

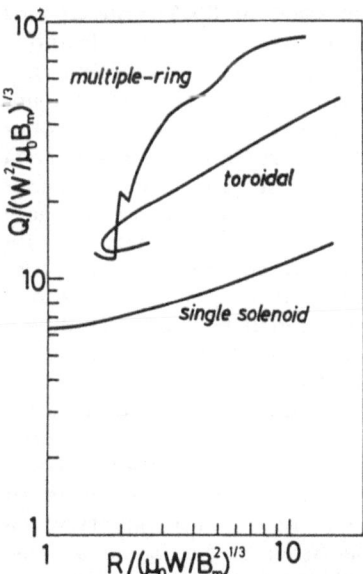

FIGURE 6 Muximum radius dependences of superconductor quantity of multiple-ring system, that of single solenoid system and toroidal system.

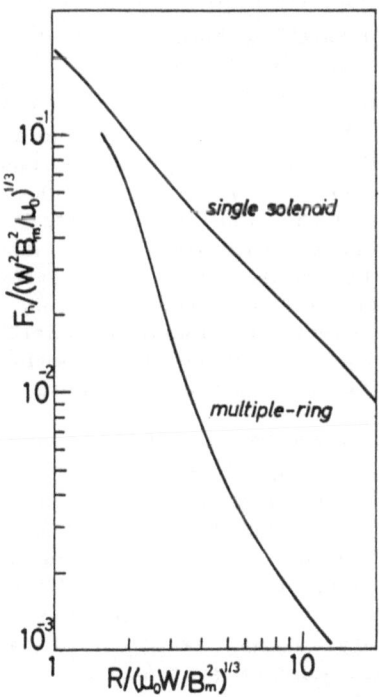

FIGIUER 4 Maximum radius dependences of maximum hoop force of multiple-ring system and that of single solenoid system.

the single solenoid system. The much quantity of superconductor in the multiple= ring system comes from the many numbers of coils. If we reduce the coil numbers with allowing a little increase of the hoop force, the quantity of superconductor could be decreased.

Leakage Magnetic Field

The leakage magentic field is the most important factor for the SMES near the urban area. The toroidal system is the most excellent system from the point of the leakage field, because it is very small except only near coil conductors. Examples of distributions of leakage magnetic field of the multipl-ring system with β=0.01 and N=22 and the single solenoid system with β=0.05 are shown in Fig.7 (a) and (b), respectively. In these figures the leakage field of the multiple= ring system is about 1/10 of the single solenoid system, and a similar result is obtained for the system with a large value of N in certain extent.

Example of Design

To demonstrate a feasibility of the multiple-ring system, an example of conceptual design based on the result mentioned above is made for the intermediate scale of SMES, which can be installed in the underground of a substation near urban area. As mentioned above, the quantity of superconductor for the condition of the 90% reduction of the hoop force is so much. Then two designs are made for the case of 90% reduction of hoop force (case A) and the case of 80% reduction of it (case B). Conditions for the design are: Stored energy 360GJ or 100MWh, maximum magnetic field 8T on conductors, and leakage magnetic field on the ground 50G, 100G or 100G. As for the system radius, a large radius is suitable for the reduction of the hoop force, but it brings an increase of the current density, as well as the plottage of the system. A value of the system radius is chosen here about 100m. The current densities corresponding this condition are $3\times10^7 A/m^2$ for case A and $2\times10^7 A/m^2$ for case B.

For the comparison, designs of other two systems are also made. As for the system radius, two specifications are chosen here, one is the value which give the minimum superconductor quantity (case C, E) and the other is the same as the case A (case D, F). And the hoop stress on the superconductor are obtained for the current density of $3\times10^7 A/m^2$. Obtained design parameters are shown in Table 1.

Comparing the two designs of the multiple-ring system, the characteristic

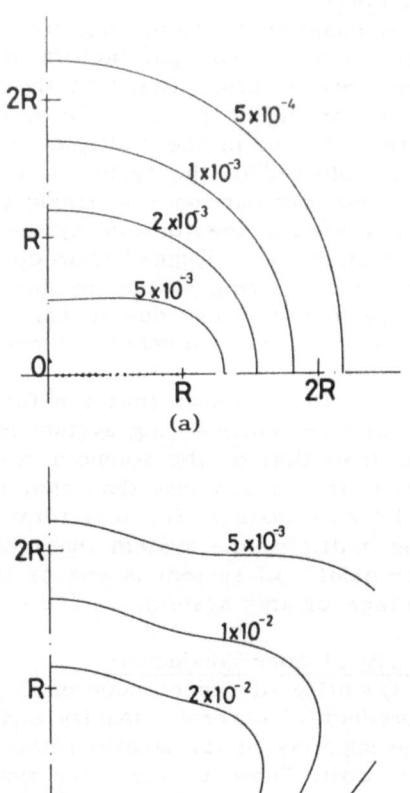

FIGURE 7 Distributions of normalized leakage field; (a) multiple-ring system N=22, β=0.01, (b) single solenoid system β=0.048

features of the design for case A with N=22 are the small hoop force and the small leakage field, but those for case B with N=10 are the small superconductor quantity and the small total field. While the optimum conditin for the coil number N should be decided with considerations of cost factors such as superconductor, supporting structure and magnetic shielding system, etc., it seems that the optimum condition for N exists around those numbers.

One of the distingushed characteristics of the multiple-ring system is that it has the much less hoop stress on the coil conductor than that of the single solenoid system. Only a simple reinforcement is enough to hold this small stress of 320MPa, but a firm reinforcement might be necessary to hold the stress of 753MPa in the sigle solenoid system.

The decicive advantage of this system over the single solenoid system is the smallness of the leakage magnetic field. To reduce the leakage magnetic field less than 100G, the depth of the multiple- ring

TABLE 1. Design parameters of the multiple ring, the single solenoid and the toroidal coil system

		Multiple-ring		Single solenoid		Toroidal	
		case A	case B	case C	case D	case E	case F
Aspect ratio		0.01	0.012	0.36	0.048	0.60	0.076
Outer diameter of system	[m]	113	117	37.2	113	71.8	113
Hight of system	[m]	1.13	1.48	13.5	5.36	27.0	7.98
Number of coils		22	10	1	1	--	--
Quantity of conductor	[GAm]	97	58	15	18	30	49
Current density of conductor	[MA/m^2]	30	22	30	30	30	30
Hoop stress on superconductor	[MPa]	320	264	1088	753	806	444
Hoop force on one coil	[GN]	0.32	0.45	1.28	4.56	--	--
Total force	[GN]	33.7	22.9	28.7	8.0	60.1	180.7
Underground depth of system [m] Max. field on the ground 50[G]		118	148	287	317	~0	~0
100[G]		86.7	110	200	225	~0	~0
200[G]		59.3	68.4	144	162	~0	~0

Stored enrgy W=360[GJ](100[MWh]), Maximum magnetic field B_m=8[T]
Maximum leakage magnetic field on the ground 50, 100, 200[G]
case A, B: 90%, 80% reduction of hoop force, respectively
case C, E: Minimum conductor quantity, case D, F: Same system radius as case A

system for case A is only 87m, but the single solenoid system with a low aspect ratio requires depth of 225m which might be too deep to construct the system.

The toroidal system has the very small leakage field, but the total force of it is very large and a large volume of space to house the system is necessary. On the other hand, the multiple-ring system needs the relatively large superconductor quantity. Though estimation of total costs including support material cost and ground treatment cost have not been made at this moment, the multuple-ring system seems to be feasible.

CONCLUSION

In this paper, we propose a new coil system named multiple-ring coil system, which consists of many concentric coils with low aspect ratio. Characteristic features of this system such as the reduction of hoop force acting on the coil conductor, total force on them and leakage magnetic field are discussed in comparison with a single solenoid system and a toroidal coil system, and conceptial designs for the intermediate scale SMES of 100MWh are also made. Obtain results shows an effectiveness of the multiple-ring coil system for the intermediate scale SMES in urban area.

REFERENCES

(1) Boom, R.W. and Peterson, H.A., Superconductive energy storage for large system. IEEE Trans. on Magnetics, 1972, MAG-8, 701-703.

(2) Hassenzaahl, W.V., Will superconductive magnetic energy be used on electric utility system?. IEEE Trans. on Magnetics, 1975, MAG-11, 482-488

(3) Nitta, T. and Okada, T., Control and characteristics of SMES connected to power system. Proc. US-Japan Workshop on SMES, Madison, USA, 1981, 64-78

(4) Mitani, Y., Tsuji, K., Ise, T. and Murakami, Y., Fundamental study on electrical power system stabilization using superconducting magnet energy storage, Proc. MT-9, 1985, 353-356

DESIGN AND TESTS OF THE SUPERCONDUCTING MAGNET
FOR ENERGY STORAGE

T. Tominaga, O. Takashiba, H. Fujita

Chubu Electric Power Company, Inc.,

20-1, Aza Kitasekiyama, Odaka-cho, Midori-ku, Nagoya-shi, Aichi-ken, 459, Japan

and

K. Asano

Hitachi Ltd., Hitachi-shi, Ibaraki-ken, 317, Japan

ABSTRACT

A superconducting magnet with a 400 mm coil inner diameter and a stored energy of 1MJ has been built and tested for use of superconducting magnetic energy storage (SMES). The conductor, a compacted strand cable made up of 13 multifilament wires, is a NbTi superconductor containing a copper and a copper-nickel alloy. The coil consisting of 32 double pancake coils with open cooling channels in the horizontal and vertical directions in order to remove the AC loss of the coil are bound with fiberglass reinforced plastic. The magnet is provided for the purpose of investigation of an electric utility transmission stabilization system and various experiments with use of SMES.

This paper describes the design and testing results of this magnet.

INTRODUCTION

Superconducting Magnetic Energy Storage (SMES) has many possible applications such as in an electric utility power stabilization system, leveling of load fluctuation, a substitute for a pumped storage plant etc.

In order to investigate those application, 1MJ SMES system, which consists of pulsed superconducting magnet, GTO thyristor converter main circuit system and control system, was constructed and various kinds of system stabilization tests are currently underway with fluctuating load using SMES connected to simulated power transmission lines.[1]

The superconducting magnet is a pulsed solenoid coil and it is provided with a room temperature space in the center to allow high magnetic field tests. The stored energy of this coil is 1MJ and the central magnetic field is 4.2T at a rated current of 1000A.

The coil was assembled with the vertical cryostat and successfully tested.

The outline and some details on the magnet construction and testing are given in the followings.

DESCRIPTION OF THE MAGNET

General design

A vertical cross section of the magnet, which consists of a superconducting coil and a cryostat, is shown in Fig. 1. The superconducting coil consists of 32 double pancake coils with open channels in the horizontal and vertical directions and its inner and outer diameter are 400 mm and 697 mm, respectively.

The cryostat is made of 304 stainless steel. The whole size of the cryostat is designed to be 200 mm in inner diameter, 1m in outer diameter and 2 m in height.

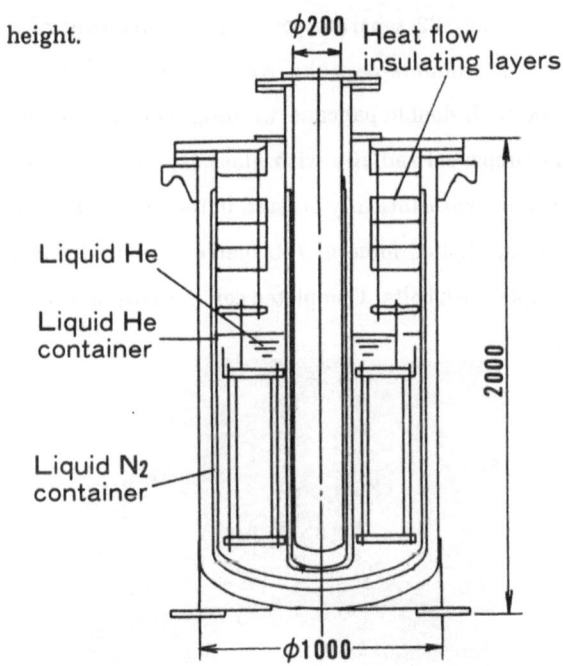

FIGURE 1.   Vertical cross section of the superconducting magnet

The cryostat includes a liquid helium reservoir, which contains the superconducting coil, to maintain liquid helium temperature. The height is determined to decrease the heat load to be permissible.

The main parameters of the magnet are listed in Table 1.

TABLE 1

Magnet parameters

Coil configuration		Solenoidal pancake
Current		1000 A
Inductance		2 H
Stored energy		1 MJ
Central magnetic field		4.2 T
Max. field on the conductor		4.5 T
Coil	inner diameter	400 mm
	outer diameter	697 mm
	height	667 mm
Cryostat		
	Material	304 stainless steel
	Inner diameter	200 mm
	Outer diameter	1000 mm
	Height	2000 mm

Conductor

The superconductor is designed with the purpose in mind of improving both stability and reduced AC losses. The conductor is Rutherford type and is composed of 13 strands of 1.18 mm diameter, each strand contains $7\mu m$ NbTi filament containing copper and a copper-nickel alloy. The cross-sectional view of the conductor is shown in Fig. 2.

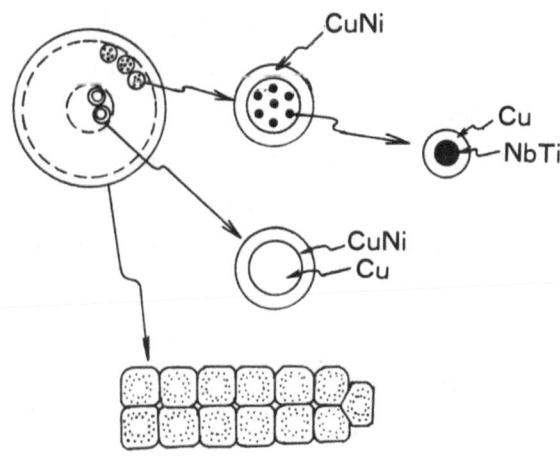

FIGURE 2.   Cross-sectional view of the superconductor

To keep the stability the ratio of the copper to NbTi is determined as 3.6 and the residual resistivity ratio is determined as 80. On the other hand to reduce the AC loss, copper is devided finely by copper-nickel. The cable compaction factor is near 0.9.

The main parameters of the conductor are listed in Table 2.

TABLE 2.
Conductor parameters

Cable height	8.45 mm
Cable width	2.1 mm
Number of strands	13
Strand diameter	1.18 mm
Number of filaments	78000
Filament diameter	7 μm
Cu/CuNi/NbTi ratio	3.6/1.3/1.0
RRR of stabilizer copper	80
Cable twist pitch	94
Critical current in cable (at 4.2 K)	4550 A at 5T

Magnet production

This magnet is planned to perform various extensive researches besides energy storage. the principal object of its design is to develop a slightly high field (~4T) and large warm bore (~200 mm) in addition to the required stored energy (1MJ). Basic parameters of the magnet are determined through the above requirements.

The important features for manufacturing the coil are: improvement of cooling efficiency; increase of the stiffness of the coil winding; decrease of the AC loss.

For the improvement of cooling efficiency, double-pancake winding coils with cooling channels in the horizontal and vertical directions were adopted. For the increase of the stiffness of the coil, the spacer structure that vertical spacers were fitted in horizontal spacers as shown in Fig. 3 was adopted and the coils were wound with high tension. For decrease of the AC loss, FRP as a material for the coil

structual objects such as bobbins, coil flanges and binding, was adopted.

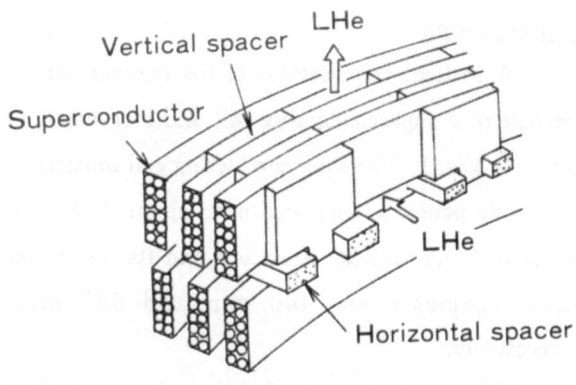

FIGURE 3. Concept of the winding structure

Thus, 32 separate coils, each consisting of a two-layer double-pancake winding, were stacked. After each double-pancake winding were connected and supported radially with glass tape binding, the coil was hydrostatically pressed in the axial direction with an applied force of 150 metric tons and finaly clamped with bolts. Completed coil is shown in Fig. 4.

FIGURE 4. Completed coil

Completed magnet is shown in Fig. 5.

FIGURE 5. Completed magnet

TESTING RESULTS

After completion of leakage checking, the magnet cooled with liquid nitrogen from the room temperature to 77K within 12 hours. At this temperature, nitrogen replaced with liquid helium. It took 5 hours to cool down the coil in the cryostat. For the excitation test, GTO thyristor converter and DC circuit breaker were connected to the magnet. The quench protection resistor was 250 mΩ and the threshold level of the quench detection circuit was set for 0.5V for at least 500 msec. Main circuit is shown in Fig. 6.

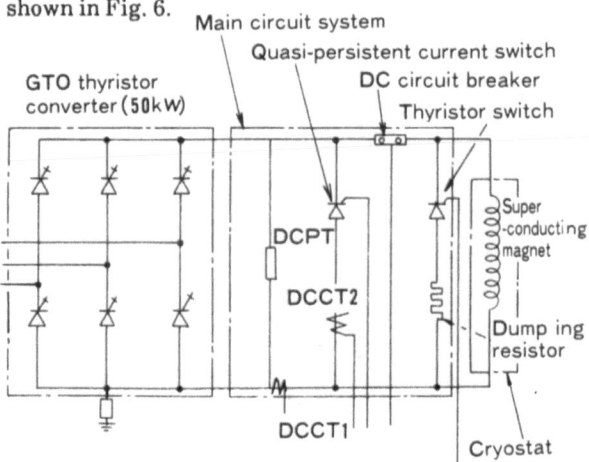

FIGURE 6. Main circuit

Excitation tests were carried out by raising the flat top current gradually with the charging rate of 25 A/sec (0.11 T/S). The flat top currents tested were 100, 200, 500, 870 and 1000 A. For each current two tests were made. In the first test the coil was discharged with a rate of 25 A/S after charged up. In the second dump test the power supply was disconnected by opening the switches after the coil was charged up. Neither quench nor abnormal response was detected through the excitation tests. In the fast dump test the coil current was measured to decrease consistently with a circuit time constant for the coil of 8.5 sec, so the coil did not have any normal region. In addition, no imbalance of voltages among the voltage taps was observed. This implies that a series of dump tests caused no detectable normal section in the coil.

The rated current corresponds to the load ratio of 60% as shown in Fig. 7.

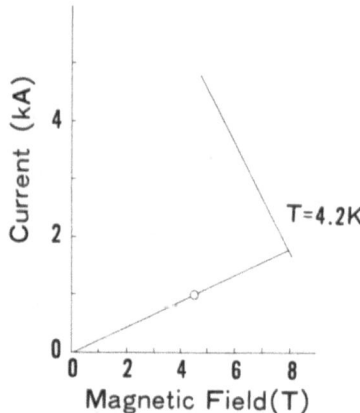

FIGURE 7. Load line of the magnet and its performance

Rapid changes of the magnetic field induces eddy currents in the conductor and in the stainless steel walls of the helium vessel. The heat generated by these eddy currents cause the pressure in the helium vessel to increase and the liquid helium level to lower. The pressure rise was only 0.01 kg/cm^2 and the amount of liquid helium loss was about 2 liters after fast discharge from 1000 ampere excitations.

Permanent current mode tests with the quasi-persistent current switch was carried out under the condition of the initial current of 300 A with no trouble.

Furthermore continuous pulsed-charging tests were carried out at low current. Excitation pattern is shown in Fig. 8.

Increases of the liquid helium evaporation were observed caused by increases of the AC loss during excitation, but those were sufficiently lower than design values and permissible.

Besides excitation test, heat load to 4.2K was measured to be about 3W for the steady state operation and also permissible.

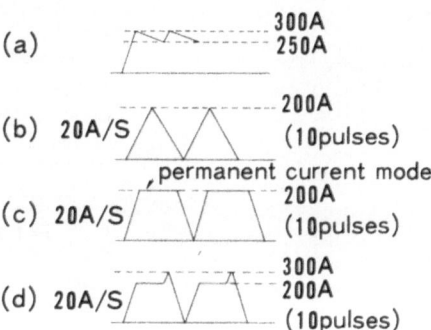

FIGURE 8. (a) shows the permanent current mode test pattern (b), (c) and (d) show the continuous pulsed-charging test pattern

CONCLUSIONS

A 1MJ superconducting magnet for use of SMES was successfully constructed and tested.

The magnet was charged up to rated current of 1000 A with rated ramp rate of 25 A/s with no quench. Furthermore, it was confirmed that the superconducting magnet was free from trouble through permanent current mode tests and continuous pulsed-charging tests.

The SMES system has been installed in the simulated transmission lines, and is presently being used for checking various utility system stabilization effects.

ACKNOWLEDMENT

The authors would like to thank the other members of Electric Power Engineering Research Section of Chubu Electric Power Company, Inc. for their enthusiastic support.

REFERENCES

1. Tominaga, T et al., Power control experiments using a PWM GTO thyristor converter in a 1MJ Superconducting Magnetic Energy Storage System. The 24th IAS Annual Meeting, 1989, to be published.

25 kV SUPERCONDUCTING FAULT CURRENT LIMITER

T. Verhaege - J.P. Tavergnier - A. Février
Laboratoires de Marcoussis, Centre de Recherches de la C.G.E.
Route de Nozay - 91460 MARCOUSSIS (France)
Y. Laumond, M. Bekhaled
GEC ALSTHOM-DEA, 3, avenue des Trois Chênes - 90018 BELFORT (France)
M. Collet, V.D. Pham
GEC ALSTHOM-DAT, 130, rue Léon Blum - 69611 - VILLEURBANNE (France)

ABSTRACT

Our technological progress in the field of superconductivity over the last ten years made possible the manufacture of industrial lengths of conductors, consisting of NbTi ultra-fine filaments, embedded in a Cu-30 wt % Ni matrix ; 50 Hz losses are greatly reduced, and the electrical resistance beyond the critical current is very large. Such conductors offer numerous new perspectives, through which the design of electrotechnical machines could be reconsidered. This paper describes the main features of a 50 Hz single-phase fault current limiter, constructed in our laboratories, which present some electrical and cryogenic properties : S-N transition limiting the fault current to a value a few percent above the threshold current within a few µs, high voltage insulation capabilities, moderate cryogenic losses during steady state and in transient conditions

INTRODUCTION

The continually growing need for electrical energy implies an increase in short-circuit power.

In order to reduce the destructive effects of high fault currents, circuit-breakers with even faster interrupting times have appeared ; nowadays most HV (high-voltage) and EHV (extra-high-voltage) circuit-breakers can eliminate a fault in 33 milliseconds.

At HV, all the AC current circuit-breakers require current zero crossing in order to interrupt the fault.

Breaking in one cycle is then the limit for HV circuit-breakers of the conventional type.

Breaking or limiting the fault current before its first zero crossing however, is the domain of current limiting devices.

The recent progress of NbTi superconducting wires for use at 50-60 Hz makes this new application possible and attractive, as demonstrated by the experiment described below.

The use of high T_c materials could also lead to efficient solutions.

PRINCIPLE

The "limiter effect" developped by a superconducting cable appears when the current density J exceeds the critical current density $J_c (B, T)$, due to a network overload or short circuit. Current crosses non-superconducting zones, and creates an electrical field $E = \rho \left[J - J_c (B, T) \right]$

Transient Joule losses $P = E.J = \rho J \left[J - J_c(B, T) \right]$ are very high, due to the high values of current densities ($\cong 10^9$ A/m^2) and electrical resistivities ($\cong 6.10^{-7}$ $\Omega.$m for NbTi, 4.10^{-7} $\Omega.$m for Cu-30 wt % Ni). As the conductor's enthalpy is very small at low temperatures, T reaches the critical value T_c (B) extremely rapidly and the model $P = \rho J^2$ applies.

The subsequent temperature rise decelerates, due to the form of the enthalpy curve H (T), to thermal diffusion, and eventually to a reduction of J, imposed by the developed resistance. It is however necessary to break the residual current rapidly, in order to limit the temperature at an acceptable level.

The superconducting fault current limiter consists essentially of an AC high performance superconducting cable, in series with the load (Figure 1). It is wound so as to minimize the self inductance, and causes very low active and reactive losses under normal operation. The cryogenic losses are mainly from the cryostat and the current leads.

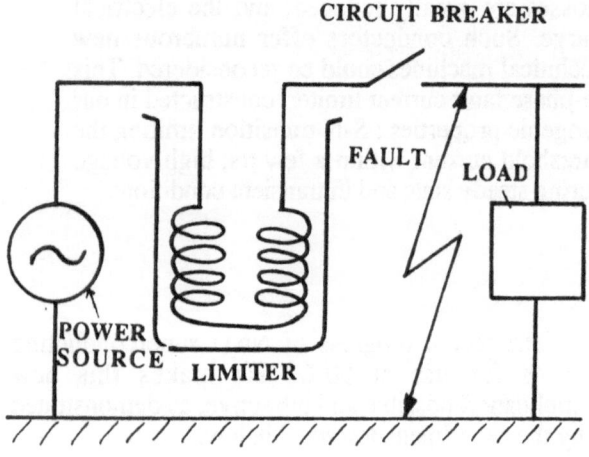

CIRCUIT BREAKER

FAULT   LOAD

POWER SOURCE   LIMITER

FIGURE 1. Line diagram

The rated current is one half or one third of the conductors critical current.

Under short-circuit operation, the current rises at a very high speed, only limited by the network inductance. It reaches a threshold value I_t after a few microseconds, inducing a conductor's bulk transition. The phenomenon does not rely on quench propagation, because incomplete transition is inconsistent with a current exceeding the threshold value.

The current is immediately limited by the conductor's resistance in the normal state (figure 2). Its peak value approaches :

$$I_{max} \cong \sup (I_t, V_{peak}/R_{max})$$

I_t = threshold current (near the DC-critical current)
V_{peak} = peak value of the network's voltage
R_{max} = limiter total resistance in the normal state.

The conductor length is determined so as to obtain $V_{peak}/R_{max} \leq I_t$, and then $I_{max} \leq I_t$.

A HV circuit-breaker will interrupt the residual current in the next zero crossing.

CONDUCTOR CHARACTERISTICS

The limiter specifications imply high current transport capability, a high resistivity in the non-superconducting state, stability and low losses in 50-60 Hz conditions.

To this end we used wires made of ultra-fine NbTi filaments, in a Cu-30 wt % Ni matrix. The spectacular features of these products in AC conditions are presented in [1].

The particular conductor used for that demonstration is made of a Cu Ni core, and six wires of 0.2 mm diameter, each containing 96756 NbTi filaments of 0.4 µm diameter. It is entirely free of copper, and has a normal resistance of 1.7 $\Omega/$m.

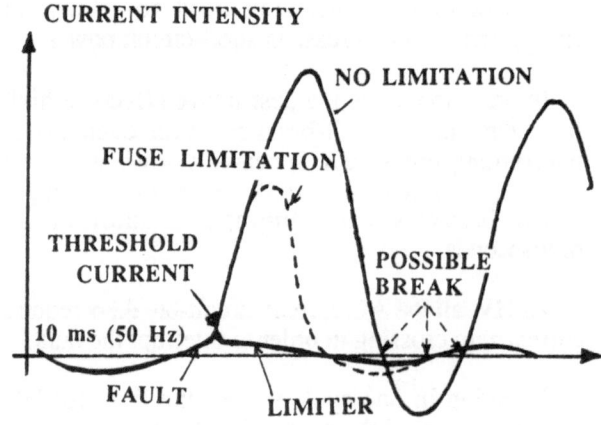

CURRENT INTENSITY

NO LIMITATION

FUSE LIMITATION

THRESHOLD CURRENT

POSSIBLE BREAK

10 ms (50 Hz)

FAULT   LIMITER

FIGURE 2 - Typical current evolution

Its threshold current at 50 Hz is about 330 A, whereas its critical DC current is 600 A. The reduction is explained by a stability defect at low magnetic fields. The problem has been investigated in [2], which concludes to optimum wire diameter and NbTi ratios, function of the magnetic field applied. The rules have been applied to the definition of the last generation of wires and cables produced by GEC ALSTHOM, which present both better stability and very low losses in AC conditions.

25 kV LIMITER DESIGN

A demonstration model of one phase of a 25 kV-three-phase superconducting limiter has been built and tested by ALSTHOM and the Laboratoires de Marcoussis.

The limiter's active part is made of 200 m of the above-mentioned conductor, wound in two layers on epoxy mandrels. Their directions of rotation are inversed, reducing the self-inductance to 1.0 mH. The coil has a height of 450 mm and an outer diameter of 270 mm.

It is immersed in a helium bath at atmospheric pressure.

Current leads cooled by vaporised helium are each constituted of 24 copper wires in a tube of Ø 50 mm.

The cryostat is conventional, with a height of 2080 mm and an internal diameter of 360 mm.

During dielectric tests the 25 kV superconducting limiter withstands a peak voltage of 51 kV to ground and across terminals.

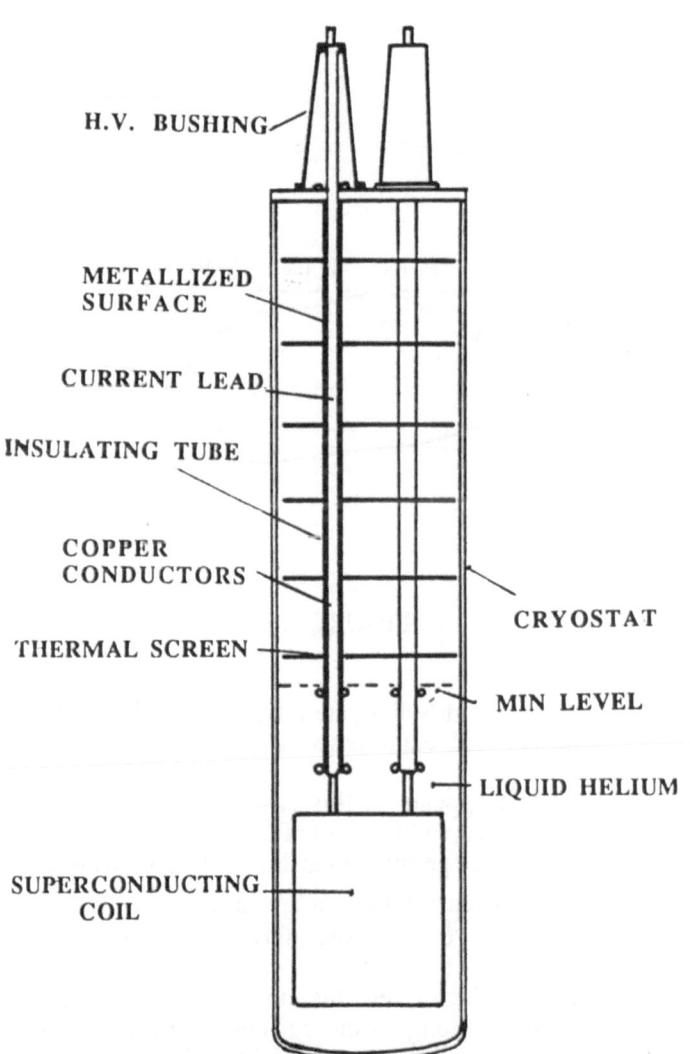

FIGURE 3. The fault current limiter

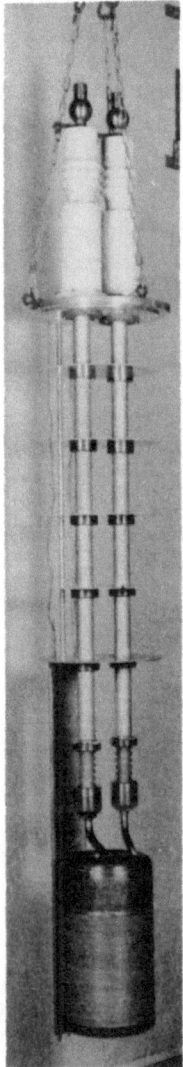

FIGURE 4. A view of the fault current limiter

It implies sufficient spacings in boiling helium, which presents poor dielectric properties. Hypercritical helium is a better dielectric medium ; it would be useful for higher voltages and nominal ratings. The layer end-turns need a particular care : the field concentration is reduced by metallic rings, which develop eddy currents ; they have to be redefined, to limit the associated losses, which were excessive in that case.

The limiter's characteristics are summarized in Table 1.

We did not record accurate helium consumption measurements ; theoretical rated values are about 0.3 l/h for the conductor and 0.6 l/h for current leads and 4 l/h for the cryostat. The relative part due to the cryostat would be less important for higher rating devices.

The consumption due to the current leads is proportional to the rated current ; its relative part decreases for higher voltages.

Helium consumption in the case of short-circuit and limitation is of the order of a few liters.

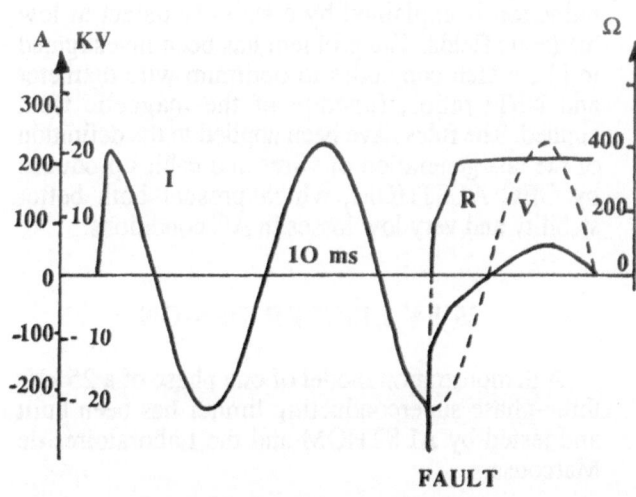

FIGURE 5. RUN n° 22

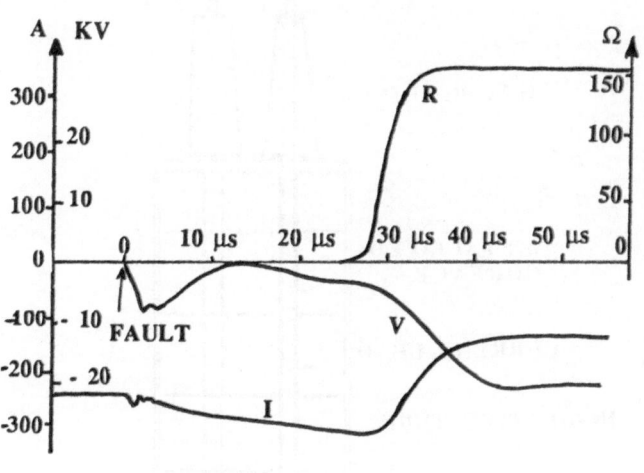

FIGURE 6. Detail

Frequency	50 Hz
Rated voltage	25 kV
Test voltage	51 kV
Rated current	200 A r.m.s.
Threshold current	330 A
Maximum resistance	340 Ω
Theoretical short circuit current	up to 15000 A r.m.s.
Short circuit current limitation	350 A

TABLE 1 - Limiter's characteristics

EXPERIMENTAL RESULTS

The 25 kV superconducting fault current limiter has been submitted to 24 runs, in different conditions of test voltage and prospective fault currents, theoretical short-circuit current (determined by the network's inductance and voltage).

Fig. 5 and 6 illustrate a typical run (RUN 22) ; with two different time scales :

- a fault appears in this case near the top of an alternance, for a current of 240 A ;
- after some perturbations due to capacitive effects, the current ramps at a speed of $\cong 2.6$ A/μs, in accordance with the network's voltage and inductance ;
- the conductor remains superconducting during 25 μs, as long as the current is lower than the threshold value ; an appreciable voltage is induced by the limiter's inductance ;

- when the current reaches the threshold value ($\cong 330$ A), a large resistance appears in a few microseconds ; it is coherent with a bulk transition of one of the two layers, initiated simultaneously in numerous points ;
- the developped resistance is sufficient to reduce the current below the threshold value, and the second layer does not quench immediately ;
- the normal phase extends progressively in the second layer, which is completely normal after a few milliseconds, as indicated by the smooth variation of the resistance in figure 5.

In the run illustrated by figure 7 (RUN 13), the fault appears at a low value of the current, just before its zero passage. The quench begins only two milliseconds later, as the current reaches the threshold value. Just one of the two layers quenches, creating a sufficient current limitation.

Normal phase propagation in the second layer does not exist, or is very slow. The abrupt resistance variation indicates that the transition is initiated simultaneously in numerous points, for a given current value. It is only possible when the conductor and winding characteristics are very homogeneous.

The limitation of the bulk transition to one of the two conductor's layers can be attributed to small differences in the temperature and in the magnetic field : the first layer quenches and reduces immediately the current, which never reaches the second layer's threshold value.

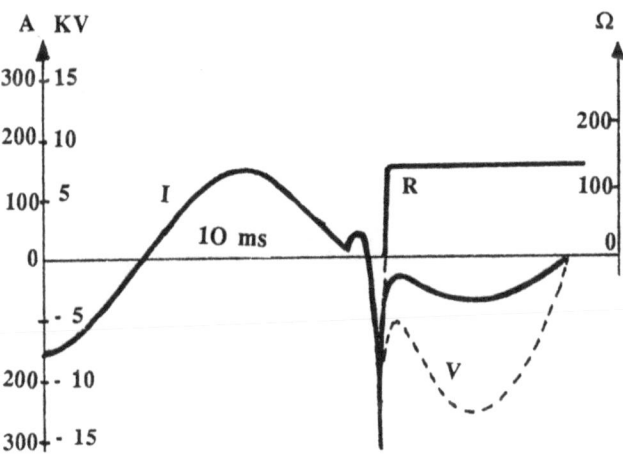

FIGURE 7. RUN n° 13

Such properties could be particularly useful in pulsed power supplies [3]. Energy dissipated in helium during a fault is given by

$$Q = \int R.I^2 \, dt = 9.75 \text{ kJ} \quad \text{for Run 22, or 4 kJ for}$$

Run 13. It corresponds to the vaporization of respectively 3.73 and 1.53 liters of helium.

The 24 runs confirmed an absolute current limitation.

In all the cases and at every moment, $I(t) < 350$ A.

Recovery of the superconducting state was generally completed after a few seconds.

Overvoltages do not appear if the limiter maximum resistance is correctly defined, that is to say comparable to the rated load impedance ; it was adjusted here by a normal resistance, mounted in parallel with the superconducting winding.

The problem of quench detection and circuit-breaking was not investigated here, as circuit-breaking was simply programmed in advance. Rapid detection is possible using the central CuNi wire as a detector [4].

HIGH T_c FAULT CURRENT LIMITER

High T_c materials could make the use of limiting devices more simple and economical, due to the possibility of use in liquid nitrogen ($\cong 77$ K), reducing considerably the cryogenic equipment and running costs.

Their normal resistivity is about ten times greater that of than of NbTi - CuNi wires. Critical current densities are not sufficient at the present time ; 100 A/mm^2 at 0.3 T would be a good goal for that function.

The cryogenic cost of losses is reduced by a factor $\cong 50$. It seems nevertheless unrealistic to hope for ultra-fine filaments made from ceramic powders ; bulk conductors like monofilaments, bars or tubes, are conceivable, with acceptable losses. A first experiment made on an YBaCuO tube of Ø 9/11 mm gave 105 A r.m.s. at 50 Hz, with losses of 10 kW/m^3. That value is relatively high due to high intra-grain j_c, compared with relatively low inter-grain j_c. The high T_c AC limiter's emergence is therefore conditioned to the future improvement on inter-grain j_c.

418

CONCLUSION

The superconducting fault current limiter can fulfill a function without equivalence in classical materials : it guarantees an absolute current limitation at a designed value, moderately above the rated current.

Its feasibility has been demonstrated here, although some technical details were not completely investigated.

Adequate NbTi conductors are now available for the limiter's function; High T_c materials are in progress, and could present economical advantages.

ACKNOWLEDGMENTS

The authors wish to thank A. Coquelin, C. Cottevieille, J.P. Lucas, M. Quemener and J. Jacquelin for their contributions to the design and high voltage testing of this current limiter.

REFERENCES

[1] J.Y. Georges, T. Verhaege, A. Février, Y. Laumond, A. Lacaze and P. Bonnet. Industrial performances of superconducting cables for 50-60 Hz, This conference.[2]
A. Guéraud, J.P. Tavergnier, A. Février, Y. Laumond, A. Lacaze, B. Dalle and A. Ansart. Thermo-electromagnetic stability of ultrafine multifilamentary superconducting cables for industrial frequency use. This conference.

[3] J.R. Argouarc'h, R. Mailfert, F. Moisson, Pulsed power supplies for electromagnetic launchers : comparative evaluation of system weight and applications of superconductivity. 1 st European Symposium on E M L Technology, Delft, September 13-14, 1988.

[4] A. Février, J.P. Tavergnier, H. Nithart, M. Kiblaire, J.L. Duchateau. High sensitive quench detection method using an integrated test wire. IEEE Transaction on Magnetics. Vol. MAG.17 n° 5, September 1981.

SWITCH OF MAGNETIC FIELD BY ROTATING SUPERCONDUCTING SHIELD

K.Takahata, S.Nishijima and T.Okada

ISIR Osaka University, Ibaraki, Osaka, Japan

ABSTRACT

The switch of magnetic field up to 1T has been developed using tubular superconducting magnetic shield. The switch is operated by rotating the shield in the transverse field. The shield was constructed by winding the NbTi multifilamentary composite wires and impregnating entirely with the Wood's metal. The wires were wound spirally at the angles of +45 and -45 degrees with respect to the axis. the uniform transverse field was applied using a superconducting split magnet. The switching ability was examined with rotating the shield around the axis. The experimental results indicate that the field can be erased and regenerated by rotating the shield because the shielding ability shows anisotropy. The present technique will be applicable to change the field made by the superconducting magnet under the persistent-mode operation.

INTRODUCTION

Several tubular magnetic shields for transverse field have been fabricated using a type II superconductors such as NbTi or Nb_3Sn so far [1-4]. These shields were mainly used to shield or trap the large and complex transverse multipole magnetic fields with high accuracy. It is, however, required to put smaller sections of superconducting tubes or strips together to form a large shield. Sufficient overlaps in the connected regions are also necessary in order to suppress field penetration from the connected area [1,4]. It was reported that even the winding of strips could shield high field without the penetration if the overlaps were sufficient [3].

We have proposed a new type of shield without overlaps [5]. Shielding current flows through a superconducting fusible alloy among the superconductor in this shield. In fabricating a long shield, this method must be suitable. In previous reports [6], we tried to fabricate small tubular shields and the achievement of the shielding capability in longitudinal and transverse fields was up to 2T. Commercial NbTi multifilamentary wire was chosen as a superconductor in order to prove that overlapped regions did not need as in the conventional shield. Under the present state, the field which can be shielded has been limited to 2T because the critical field and current density of the fusible alloy were fairly low compared with those of NbTi. The shielding current mainly flowed through the wires and only the neighboring wires were coupled electrically by sharing current through the fusible alloy.

Although the wires were spirally wound into the form of a conventional solenoid for longitudinal field, it was suitable to wind them obliquely at angels of +45 and -45 degrees with respect to the axis in the shield for transverse field. When the transverse field is

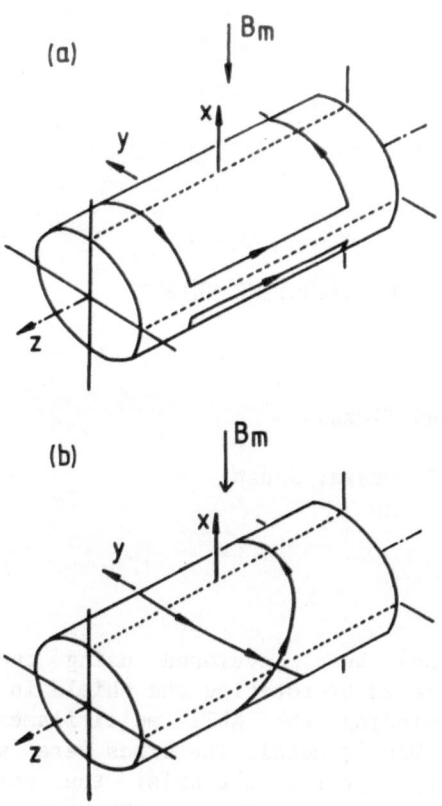

Fig. 1 Schematic illustration of shielding current paths in the conventional superconducting tube (a) and the shield investigated in this study (b).

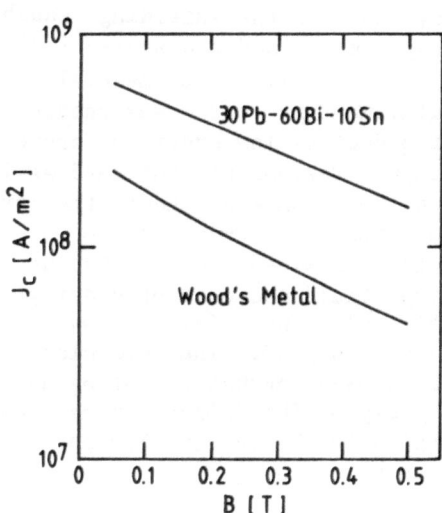

Fig. 2 Jc vs. B curves of 30-Pb-60Bi-10Sn fusible alloy and the Wood's metal.

applied to the shield, the current paths follow ellipses around the cylinder surface as illustrated in Fig.1(b). The conventional shields provide a different configuration of current path which follow the saddle-shapes as shown in Fig.1(a). It was suggested that even the performance of the discontinuous superconductors were able to be analyzed using the saddle-shaped current model [4].

In this paper, the behavior of the shield with the superconducting wire and fusible alloy is discussed and the switching device of the magnetic field is also proposed. The effects of field direction on the shielding ability were examined rotating the shield around the axis and anisotropy of the shielding ability was clarified. Applicability of the rotating shield to the switching device of the magnetic field was investigated.

FABRICATION OF SHIELDS AND EXPERIMENTALS

The superconducting wires used for test shields were commercial multifilamentary NbTi composites with copper matrix. They had the diameter of 0.35mm, the filament diameter of 0.03mm, and copper/super ratio of 1.3. Lower copper ratio may be suitable to reduce the current decay due to the normal component because the shielding currents must be shared between the wires. The copper, however, plays an important role which is to ensure the dynamic stability. We, here, chose the commercial wire for the shields considering the both situation.

The fusible alloy for connection must have superconductivity. It is noticeable that Pb-Bi and Pb-Bi-Sn alloys have properly high critical current density, Jc and critical field, Bc2. The Jc and Bc2 of these alloys have been measured changing the composition by means of the magnetization measurement and the four-terminal method. The 30Pb-60Bi-10Sn has highest Jc in the field up to 0.5T as a result of measuring the magnetization. Commercial fusible alloy, Wood's metal (25Pb-50Bi-12.5Sn-12.5Cd), was chosen for the test shield because it was easy to obtain. Figure 2 shows Jc vs. B curves of 30Pb-60Bi-10Sn and Wood's

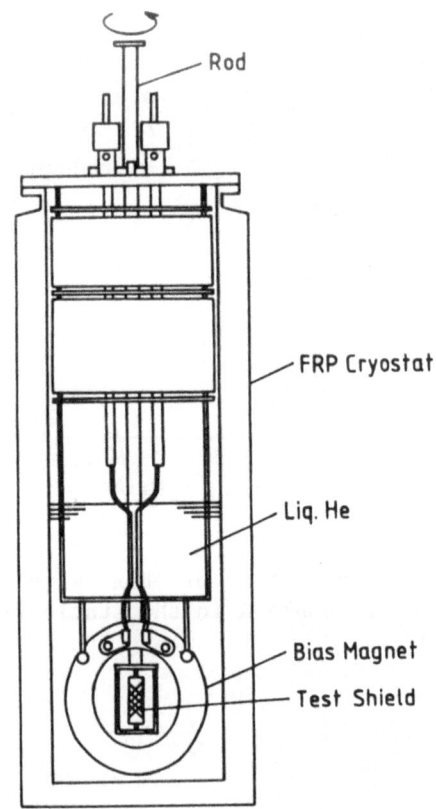

Fig. 3 Experimental arrangement.

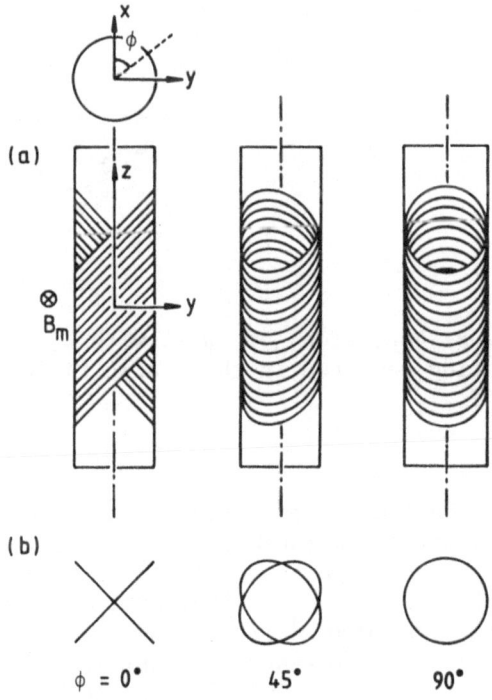

Fig. 4 Appearances of the shield as rotating. A pair of superconducting rings extracted from the shield is shown in (b).

alloy. The Jc of 30Pb-60Bi-10Sn is 3.5 times as high as Wood's metal. At the practical application, it is desirable to use the fusible alloy which has higher Jc. Wood's metal had Bc2 of 1.5T and Tc of 9K.

The wires without insulator were wounded spirally at angles +45 and -45 degrees with respect to the axis around the bakelite bobbin of which outer diameter and length were 17mm and 50mm, respectively. Two coils were made with two and four layers (called sample A and sample B, respectively). Each layer had 65 turns. The coils were impregnated the Wood's metal. Final outer diameters of sample A and B were 19mm and 21mm, respectively.

The uniform transverse field was supplied by a superconducting split pair oval shape magnet of which clear bore size was 24 x 70mm. The shields could be rotated around the axis by the rod led to the flange. The experimental arrangement is shown in Fig. 3.

Figure 4 shows schematic illustrations of the appearances of the shield from the field direction (defined as x direction). The appearance changes as shown in Fig.4(a) as rotating the shield from the rotation angle φ=0 to 90 degrees (φ is defined as shown in the figure). The shield can be considered to be the gathering of superconducting rings. Figure 4(b) shows a pair of rings extracted from the gathering. When φ=0, no flux is linked by the rings. It is expected that the shield has no shielding ability in this case. As increasing the angle up to 90 degrees, the magnetic flux across the ring is increased. In previous report[6], the shielding ability was measured using another shield when φ=90 degrees and high shielding efficiency was obtained.

The anisotropy of shielding ability was examined by two procedures. In procedure I, the angle was first fixed and the external field was increased until the shield quench occurred. In procedure II, the external field was first turned on and the shielding was rotated. The Hall probe was set at the center of the shield and only the x-component of the magnetic field, Bsx was measured. The y-component of the magnetic field may be generated by the shield

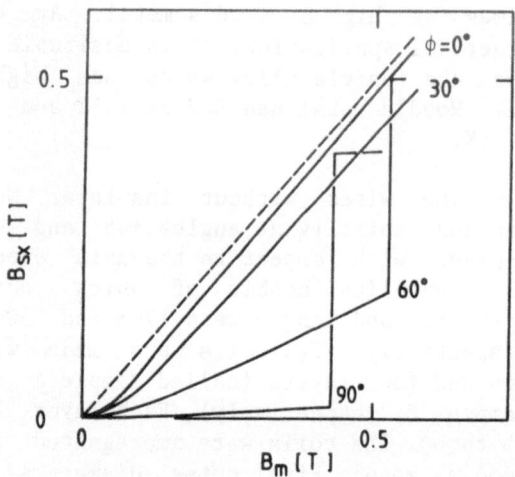

Fig. 5 Relationships between the external field, Bm and the measured field, Bsx as applying the field to sample A with 2 layers.

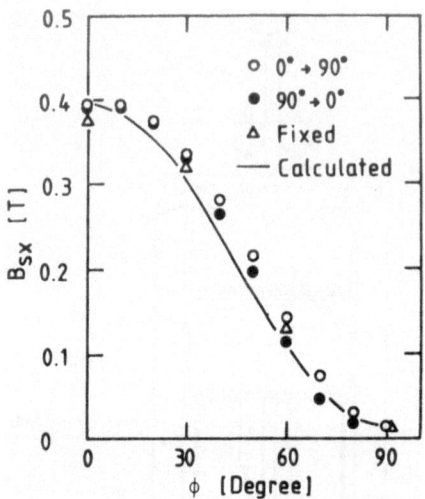

Fig. 7 Variation of Bsx with φ when rotating sample A in the static field of 0.40T.

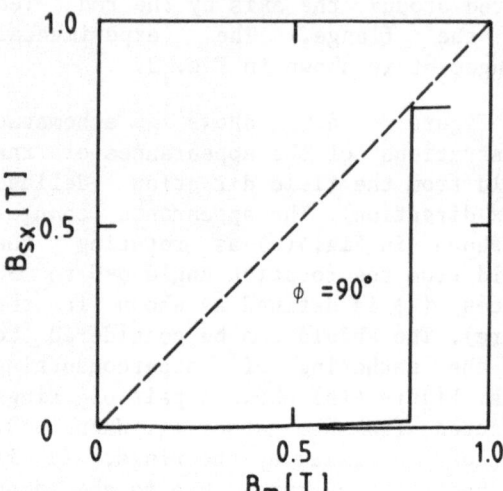

Fig. 6 Relationship between the Bm and Bsx as applied the field to sample B with 4 layers.

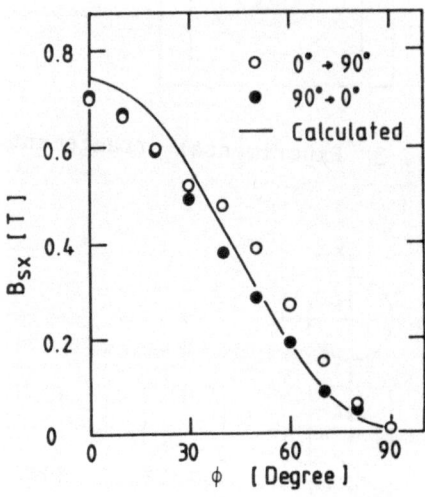

Fig. 8 Variation of Bsx with φ when rotating sample B in the static field of 0.74T.

except φ=0 and 90 degrees. The y-component was, here, neglected because the situations were important as φ=0 and 90 degrees.

RESULTS AND DISCUSSION

Figure 5 and 6 show the relationships between the external field, Bm and the measured field, Bsx when the rotation angle was fixed and the field

was applied to sample A and B respectively (procedure I). The broken line shows the measured field without the shield. The sweep rate of the external field was 6×10^{-3} T/sec. When φ=0, the shielding efficiency of sample A was rather small. The efficiency was increased with increasing φ. When φ=90 degrees, the efficiency was 96% just before the quench at the field of 0.4T. The anisotropy of shielding ability was clarified as expected. Sample B kept the shielding ability until the external field reached 0.79T. The efficiency was 97% just before the quench.

Figure 7 and 8 show the variation of Bsx when the shields, were rotated in the static field (procedure II). The fields were measured at intervals of 10 degrees. It took about one minute to rotate the shield for 10 degrees. The external fields were set just below the quench fields in the previous experiments, 0.40T for sample A and 0.74T for sample B. The quench did not occur during rotation under the fields. The shields were first rotated from $\phi=0$ to 90 degrees and returned to its original position. Open and close circles in the figures indicate the measured field when rotating from $\phi=0$ to 90 degrees and returning, respectively. The measured fields at 0.40T in procedure I are also plotted in Fig. 7 as triangular markings. The anisotropy of the shield ability agreed with that in procedure I. Note that the magnetic field inside the shield could be erased and regenerated under the constant external field by rotation of the shield.

It is found from Fig. 4(b) that magnetic flux across a ring is proportional to $\sin\phi$. The shielding current through the ring makes the field to the normal direction. This field is simply given by $\eta Bm\sin\phi$ where η is the shielding efficiency as $\phi=90$ degrees. The x-component is thus $\eta Bm\sin^2\phi$. Consequently we have

$$Bsx=Bm(1-\eta\sin^2\phi). \qquad (1)$$

The calculated values were also represented in Fig. 7 and 8. The calculated values agreed well with the experimental ones. The results demonstrated that the configuration of the shielding current could be expressed as Fig. 1(b) instead of Fig. 1(a).

To examine the possibility of fast erasing and regeneration, ϕ was changed trapezoidally. Figure 9 shows time variations of ϕ and Bsx when sample A was rotated from $\phi=0$ to 90 degrees and returned for 3 sec under the field of 0.31 T. The field could successfully be erased for 3 sec. It took, however, approximately 30 sec to regenerate the original field unexpectedly. The results indicated that several wires were coupled electrically by normal currents through

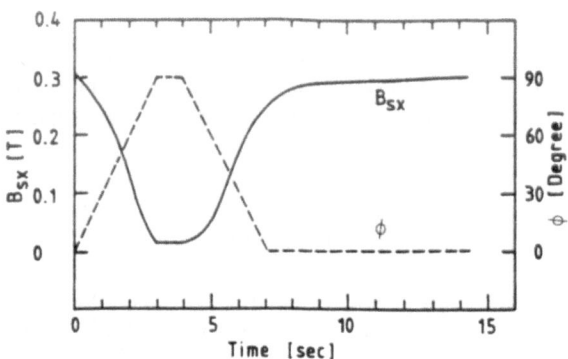

Fig. 9 Time variation of ϕ and Bsx as rotating sample A tapezoidally.

the Wood's metal and saddle-shaped currents were macroscopically produced when the rotation rate was fast. The current might decay slowly because the resistivity of the Wood's metal was relatively low. The coupling provides enhancement of the shielding ability. When erasing the field, response is almost independent of the coupling. To cut the coupling, the resistivity of the fusible alloy must be increased under the normal state.

CONCLUSION

The tubular shields which have anisotropy of shielding ability were fabricated using the superconducting wire and fusible alloy. The NbTi multifilamentary wires were wound spirally at the angles of +45 and -45 degrees with respect to the axis and the coils were impregnated entirely with the Wood's metal. The anisotropy was demonstrated by rotating the shields in the transverse field. This feature can be applied to the magnetic field switch which erases and regenerates the field inside the shield under the constant field. In our experiment, the field of about 0.7T could be erased and regenerated by rotating the shield. If using the magnet with a persistent-current switch, it will be hard to control the magnetic field. The magnetic field switch using the shield will, therefore, be suitable to change the field in the magnet under the persistent-mode operation for such as HGMS, accelerator and so on.

ACKNOWLEDGMENT

This work is partly supported by Grant in Aid for Scientific Research No. 0105002, Ministry of Education in Japan.

REFERENCES

1. Martin, F., Lorant, S. T. St, and Toner, W. T., Nucl. Instr. Meth. 103, 1972 503-514.
2. Rabinowitz, M., IEEE Trans. on Magn., Vol. MAG-11, 1975 548-550.
3. Rabinowitz, M., Arrowsmith, H. W., and Dahlgren, S. D., Appl. Phys. Lett., Vol.30, 1977 607-609.
4. Frankel, D. J., IEEE Trans. on Magn., Vol. MAG-15, 1979 1349-1353.
5. Okada, T., Takahata, K., Nishijima, S., Nakagawa, S., and Yoshiwa, M., IEEE Trans. on Magn., Vol. 24, 1988 895-898.
6. Takahata, K., Nishijima, S., Ohgami, M., Okada, T., Nakagawa, S., and Yoshiwa, M., IEEE Trans. on Magn., Vol. 25, 1989 1889-1892.

DEVELOPMENT OF THE THERMAL SUPERCONDUCTING SWITCH AND RESEARCH ON ITS CLOSED PROCESS

Y.B. Lin, Y.L. Chen, L.Z. Lin, S. Han
Institute of Electrical Engineering, Academia Sinica
P.O. Box 2703, Beijing 100080, China

ABSTRACT

A series of thermal superconducting switches for the superconducting magnet systems in the persistent mode have been developed at the Institute of Electrical Engineering, Academia Sinica. The maximum current is up to 300 A (at 1T). The multifilament NbTi wire and cable with CuNi matrix and the multifilament Nb_3Sn wire with bronze matrix are used for the superconducting switches respectively. Two types of NbTi superconducting switch have been made and operated in different superconducting magnet systems successfully. The researches on the closed process of the superconducting switch and some problems in the process have been engaged. The development of the thermal superconducting switch and the experimental results have been presented in this paper.

INTRODUCTION

The superconducting magnet systems with the superconducting switch operating in the persistent mode have many advantages. First, the magnetic field with very high stability can be attained. Second, the DC source with a suitable stability can be substituted for that with very high stability, so the investment in the DC source will be reduced. Third, the source current is not necessary any longer for the superconducting magnet system during the persistent operation, so the consume of liquid helium and the operational cost can be minimized. Last, the noise immunity of the superconducting magnet system will be better, thus the reliability of the system will be increased. Therefore they have been found wide application.

To advance the application of the superconducting magnet system operating in the persistent mode, since 1978 a series of thermal superconducting switches have been developed at the Institute of Electrical Engineering, Academia Sinica. The multifilament NbTi wire with CuNi matrix and the multifilament Nb_3Sn wire with broze matrix have been used for the superconducting switches. Two types of NbTi switch of the maximum currents up to 300 A (at 1 T) have been operated successfully, one for a superconducting magnet system for large format electro–camera reseach and another for the high field superconducting magnet systems. The design and the experimental results of the thermal superconducting switch have been described. The researches on the closed process of the superconducting switch and some key problems in the process have been made.

DESIGN OF SUPERCONDUCTING SWITCH

Requirements for the SC Switch

The sc switch for the sc magnet system operating in the persistent mode had to fulfil the following requirements.

1. The necessary maximum current for the magnet can be caught.
2. It is possesed of the requisite level of the withstand voltage for system.
3. The sc switch resistance in the normal state should be large enough, so that the energy to be lost in the switch will be less during the energization and the energy transfer of the sc magnet.
4. The sc switch can be opened as fast as possible in case of the quench in the sc magnet.
5. The sc switch will not overheat when it transits to the normal state spontaneously during the persistent operation.

Construction and Material of the SC Switch

The developed thermal sc switch was wound spirally onto a coil former. It consists of superconductor and heater which is close to the superconductor. The sc switch must be wound noninductively, so the magnetic energy will not be stored in it and the fast response can be obtained. The heater was wound bifilarly and was arranged uniformly. Silicon grease with better thermal conductivity at low temperature was applied between the heater and the superconductor. These would minimize the delay time to thermal trigger for opening the switch. The sc switch can be opened within a few milliseconds and a good repeatability was obtained.

The twisted multifilament NbTi wire in Cu(70)Ni(30) matrix was selected as the conductor for the low current sc switches. The superconducting cable consisting of 7 NbTi–CuNi strands was used for the high current sc switches. They have a very high power density of the order of $10^{13} Wm^{-3}$ and a good performance in stabilization. The NbTi–CuNi conductor is most useful for the construction of the sc switch at the present state of art [1] [2]

The parameters of the superconductors for NbTi switches are listed in Table 1. Moreover, the multifilament Nb_3Sn–bronze conductor of 0.5 mm in diameter for sc switch has also been developed. The filament diameter is $6\mu m$, number of filaments is 250 and $\rho(300K) = 18\mu\Omega cm$.

TABLE 1
The Parameters of the Superconductors for the NbTi Superconducting Switches

		NbTi wire	NbTi cable
Diameter	mm	0.3	0.92
Fil. diameter	μm	15	23
Number of fil.		200	85×7
Twist pitch	mm	10	7.5
Matrix		Cu(70)Ni(30)	
Matrix / NbTi	ratio	1 : 1	1 : 1
I_c (1.3T)	A	61	> 400
$\rho(10K)$	$\mu\Omega cm$	35	35
Insulation		Polyester lacquer	Fibre–glass fabrics

The heater of manganin wire insulated with polyester lacquer was used for the switches. Capacitor papers were inserted between the insulated superconductor and heater. The withstand voltage of the sc switch has been reached more than 1 kV.

Determination of the Superconductor Length for Switch

During the persistent operation of sc magnet system the severest condition for sc switch occurs on the spontaneous transition of the superconductor from superconducting to normal state without quenching of the sc magnet. In this case, the stored energy in the magnet will discharge to the resistance of the sc switch conductor R_s and the dump resistance R_p. The equivalent circuit of the sc magnet system when the sc switch turns into normal state is shown in Fig.1. Part of the energy from the magnet will lose in the sc switch conductor which is in the normal state and the conductor temperature will rise from 4.2 K to T_m. The process is considered to be adiabatic process because the time constant of circuit is generally small. Assuming R_s is constant, the following equation should be tenable

$$\frac{R_p}{R_s + R_p} \cdot \frac{LI_o^2}{2} = Al_s \int_{4.2K}^{T_m} C_v dT \qquad (1)$$

Where L is the inductance of the magnet, I_o is the maximum working current of the

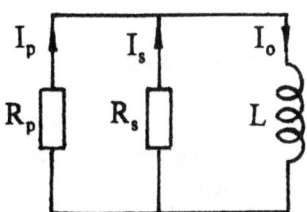

FIGURE 1   The equivalent circuit of the sc magnet system when the sc switch turns into the normal state

system. A and l_s is the cross section and length of the switch conductor respectively, C_v is the specific heat of constant volume for the switch conductor.

Although there are no ready–made data on C_v for the used NbTi–CuNi wire, we can calculate the increment of internal energy $\triangle U_i = U_i(T_m) - U_i(4.2K)$ for all metallic components in the conductor by means of their Debye characteristic temperature θ_D. The increment of internal energy for the conductor is $\triangle U = U(T_m) - U(4.2K)$, because the expression

$$\frac{U(T) - U_0}{T} = \frac{1}{T}\int_0^T C_v dT = f\left(\frac{\theta_D}{T}\right)$$

is commonly used to all metals and the data are ready–made [3]. Thus the equation (1) can be written as

$$\frac{R_P}{R_s + R_P} \cdot \frac{LI_o^2}{2} = Al_s \triangle U \qquad (2)$$

Using the relation $R_s = \rho_s l_s / A$ the conductor length for switch can be got from equation (2)

$$l_s = \left[\left(\frac{R_P A}{2\rho_s}\right)^2 + \frac{R_P LI_o^2}{2\rho_s \triangle U}\right]^{\frac{1}{2}} - \frac{R_P A}{2\rho_s} \qquad (3)$$

where ρ_s is the average resistivity of the switch conductor.

In the design of the sc switch, taking $T_m = 200K$ the calculated value of $\triangle U$ is $304 \, J / cm^3$.

In addition, the efficiency of the energy transfer for the system after quench of the sc magnet should be considered. The conductor length must be extended if the efficiency of the energy transfer is on the low side in

accordance with the calculated value of l_s from equation (3). However, the instabilities of the sc switch operating at steady currents have been observed [1]. The instabilities depend on the cable length for sc switch and are not observed for a cable length up to 12m. To assure the stability of the sc switch in the persistent operation the conductor length for sc switch should not be extended too long.

Design of the Heater

The heater must provide the sc switch with the energy which is necessary for the fast transition to normal state. The minimum energy per volume unit for warming up the sc switch conductor from 4.2K to T_c is given by

$$Q_o = \int_{4.2K}^{T_c} C_v dT \qquad (4)$$

The maximum heating energy of heater is taken $Q_{max} > 10Q_0$ to open the sc switch fast. The calculated value of $Q_0 = 45 \, mJ / cm^3$ for the used NbTi–CuNi wire.

EXPERIMENTAL RESULTS

A series of thermal sc switches have been developed. The main parameters of the typical sc switches are listed in Table 2.

A low current sc switch has been operated in the sc magnet system for large format electro–camera research in the persistent mode [4]. The experimental results prove that the sc switch can be operated safely and reliably in the system. Owing to the high stability of the field due to persistent operation and the high homogeneity of the magnet, very good experimental results have been obtained. The microphotographs obtained with the camera indicate that it has a resolution better than $1.5\mu m$ over the entire field of 100mm diameter.

The high current sc switches have been operated in the high field sc magnet systems in the persistent mode. The switches have very good stability during the petsistent operation at the current up to 300 A. The closed experiments between the high current sc switch and magnet have been engaged using the sc switch No.1. The sc magnet is a 12T Nb$_3$Sn high field magnet which consists

TABLE 2
The Main Parameters of the Typical Superconducting Switches

Types		Low current switch	High current switches		Nb₃Sn switch
Purpose		Sc magnet system for large electro—camera research	12T sc magnet system (Switch No.1)	14T sc magnet system (Switch No.2)	Test
Inner diameter	mm	11	58	262	40
Outer diameter	mm	12	65	275	43
Height	mm	26	27	51.5	110
Conductor material		NbTi—CuNi wire	7 cabled NbTi—CuNi wires		Nb₃Sn—bronze
Conductor length	m	2.7	6.6	80	5.4
Normal resistance	Ω	15	5.3	65	4.5
Max. Closed current	A	61(at 1.3T)	300(at 0.8T)	300(at 1T)	> 120(at 500Gs)
R(300K) of heater	Ω	350	105	316	900
I_{min} of heater	mA	75	120	300	160

of two inserting Nb₃Sn solenoids of 31 mm inner diameter, 150mm outer diameter and 149.5mm in the height. The magnet was energized to 350 A, then the switch was transited to sc state turning off the switch heater. When the source current was decreased to 45A (current in the switch is 305A), the switch quenched without quenching of the magnet. In this case the losing energy in the switch is 40% higher than the design value, but the switch is in safety. The experiments have shown that the high current sc switch has a high overload capacity.

The experiments on the sc switch No.1 have shown that the maximum closed current of the sc switch depends on the rate of decrease of the source current when the closed current is near to the critical current. The higher the maximum closed current of the sc switch, the lower the rate of decrease of the source current. For example, the maximum closed current is more than 300A at the decreasing rate of 0.7 A / s. The maximum closed current of 293 A and 266A are corresponding to the rate of 2.8A / s and 3.3A / s respectively.

The Nb₃Sn sc switch has been operated in the persistent mode with a small Nb₃Sn magnet of $I_c = 260A$. A DC source of 120 A was used for the experiments. The experiments have shown that Nb₃Sn switch can carry 120 A without quenching at zero—magnetic field and close more than 112 A of current at 500 Gs successfully. The measured relation curve of the minimum control current I_{min} of heater to the closed current I_0 for the Nb₃Sn switch is shown in Fig.2. The minimum control current of the switch heater decreases slowly with the increase of the closed current of the switch. The experiments have shown that the

multifilament Nb₃Sn—bronze wire can also be used for the sc switch.

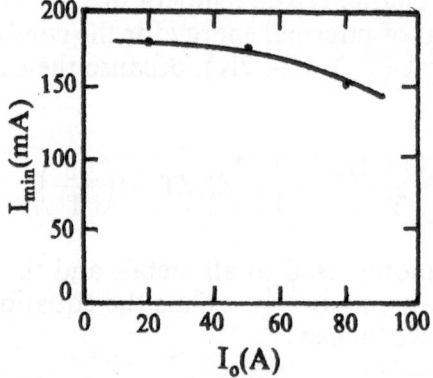

FIGURE 2   The relation of the minimum control current of heater with the closed current for Nb₃Sn switch

THE RESEARCHES ON THE CLOSED PROCESS

The circuit diagram of the sc magnet system operating in the persistent mode is shown in Fig.3. L is the inductance of the sc magnet, R_s is the resistance of the sc switch and $R_s = 0$ when the switch is in the sc state. R_p is dump resistance, E is voltage of the DC source and R is the charging resistor.

During the energization of the sc magnet the voltage V_s across the magnet is remained constant by means of adjusting the charging resistor, so the current in the sc switch which was turned into the normal state is V_s / R_s and the current in the magnet increases linearly with times. The resistance of the charging resistor equals R when

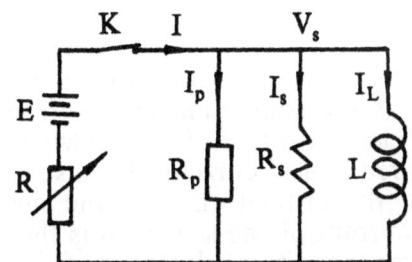

FIGURE 3 The circuit diagram of the sc magnet system operating in the persistent mode

the adjusting is stoped. Assuming R_s is constant and neglecting alternating component of the DC source, the differential equations can be written as follows

$$\left. \begin{aligned} L\frac{dI_L}{dt} &= I_s R_s \\ L\frac{dI_L}{dt} + IR &= E \\ I_s R_s &= I_p R_p \\ I &= I_s + I_p + I_L \end{aligned} \right\} \qquad (5)$$

From equations (5) the expressions for the currents in the magnet and sc switch can be obtained

$$I_L = \frac{E}{R}(1 - e^{-t/\tau}) \qquad (6)$$

$$I_s = \frac{R_p E}{RR_p + R_p R_s + R_s R} e^{-t/\tau} \qquad (7)$$

where $\tau = L(R^{-1} + R_s^{-1} + R_p^{-1})$.

The equations (6) and (7) show that the magnet current I_L increases with times exponentially and the switch current I_s decreases with times exponentially after the adjusting of the charging resistor was stoped. When $I_L = E/R$ and both I_s and V_s approach to zero, the energization process of the magnet is finished. However, the DC source, in fact, has the alternating component. Assuming the ripple factor of the DC source is k, the voltage across the switch is an alternating voltage after energization. If the voltage is expressed as rms, then

$$V_s = \frac{R_s R_p}{RR_p + R_p R_s + R_s R} kE.$$

Therefore we find the solutions of the currents I_L and I_s if $\omega L \gg R_s R_p / (R_s + R_p)$:

$$I_L = E/R \qquad (8)$$

$$I_s = \frac{R_p}{RR_p + R_p R_s + R_s R} kE \qquad (9)$$

The magnet will be closed by sc switch only if k is low enough. In this case, the heat generation due to the alternating current in the switch is lower than the heat dissipation when the switch heater is turned off, so the switch temperature will be able to drop below T_c and the switch will turn into sc state. We have observed that the sc switch can not be closed due to a poor quality of DC source (k = 10.9%). The difficult problem was solved when the DC source was filtered by capacitor bank in parallel and a inductor in series. Therefore the DC source to meet the closed requirement for sc switch must be possessed of a low enough k. The researches into the effect of quality of the DC source on the closed capacity of the sc switch have shown that the maximum closed current of the sc switch I_o is proportional to $1/k$ [5]. The experiments on the low current sc switch have shown that the quantitative relation between k and I_o is

$$k \leqslant 0.86 I_o^{-1} \qquad (10)$$

The equation (10) is applicable to the sc switches having the same construction as the developed one.

When the sc switch is turned into the sc state, $R_s = 0$, hence $I_s = kE/R$ and the source current $I = (1+k)E/R$. In order to catch to the persistent current the source current must be decreased to zero. During decreasing the source current the voltage $V_s = 0$, so the magnet current is constant

$$I_L = E/R \qquad (11)$$

Assuming that I is decreased uniformly from $(1+k)E/R$ to 0 within time t_s, the source current will change with times t as follows

$$I(t) = (1+k)(1-t/t_s)E/R \qquad (12)$$

The equivalent circuit of the sc magnet system describing the closed process of the sc switch is shown in Fig.4. The current equation can be written as

$$I_s(t) = I(t) - I_L \qquad (13)$$

From equations (11), (12) and (13) we can obtain

$$I_s(t) = K(1-t/t_s)E/R - (t/t_s)E/R \qquad (14)$$

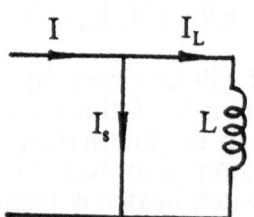

FIGURE 4 The equivalent circuit of the sc magnet system describing the closed process of the sc switch

The equation (14) shows that as the source current decreases with times, the alternating current in the sc switch decreases too, but the derect current or the closed current in the switch increases oppositely. When $t = t_s$, the source current is turned off and the switch current is

$$I_s = -E / R = -I_L \qquad (15)$$

Hence the closed loop between the sc switch and magnet is formed.

The closed current of the sc switch increases as the source current decreases during the closing of the sc switch, so it is unsuitable to decrease the source current at an excessive rate. When the desired closed current is near to the critical current of the sc switch, the fast decrease of the source current will cause the sc switch to transit to normal state, thus the closed loop cannot be formed .

CONCLUSIONS

The developed thermal sc switches with sc magnet systems have operated in the persistent mode safely and reliably. It proves that the design and manufacture of the sc switch are successful.

The multifilament NbTi–CuNi wire and cable are most useful for the construction of the sc switch. The multifilament Nb$_3$Sn–bronze wire can also be used for the sc switch.

The maximum closed current of the sc switch is proportional to the reciprocal of the ripple factor of the DC source. One of the key problems for closing the sc switch is the DC source having a low enough ripple factor.

The closed current of the sc switch increases as the source current decreases during the closed process of the sc switch. When the desired closed current of the sc switch is near to the critical current, the maximum closed current of the sc switch is the higher, the lower the rate of decrease of the source current. Therefore it is another key problem for closing the sc switch to decrease the source current slowly.

ACKNOWLEDGEMENTS

The authors wish to thank Wang Shuyuan, Luo Kunlun, Li Xiaoshan, Song Naihao, Cha Delong, Xu Li, Li Kewen, Lin Gunying, Yi Changlian and Lei Yuanzhong for their participation of partial experiments. The authors also wish to thank Prof. Zhang Yong for his support and help.

REFERENCES

[1] Ulbricht, A., A resistive superconducting power switch with a switching power of 40MW at 47kV. Cryogenics, 1979, 19(10), 591–602.

[2] Grawatsch, K., Köfler, H., Komarek, P., Kornmann, H. and Ulbricht, A., Investigations for the development of superconducting power switches. IEEE Trans. on Mags., 1975, MAG–11(2), 586–9.

[3] White, G.K., Experimental Techniques in Low–Temperature Physics, Clarendon Press, Oxford, 1979, pp.275–81.

[4] Zhang, Y., Gao, G.Z., Lin, Y.B., Wang, S.Y., Lin, M.H., Qiao, X.F., Zhao, H.L. and Du, Y.H., A superconducting magnet system for large format electro–camera research. IEEE Trans. on Mags., 1981, MAG–17(5), 1994–6.

[5] Lin, Y.B., Wang, S.Y. and Zhang, Y., The effect of the quality of direct current source on the closed capacity of the superconducting switch. Acta Physica Temperaturae Humilis Sinica, 1980, 2(4), 323–30.

THEORY AND VERIFICATION TESTS OF THE Nb-Ti RIBBON THERMALLY CONTROLLED SWITCH FOR 50 Hz APPLICATIONS

I.Hlásnik, [+]N.V.Markovsky, [+]O.A.Shevchenko, J.Kokavec, Basic Laboratory for Technical Applications of Superconductivity, Electro-Physical Research Center, Slovak Academy of Sciences, Dúbravska cesta 9, 842 39 Bratislava, Czechoslovakia, [+]Institute of Electrodynamics, Academy of Sciences, Ukrainian SSR, Pr. Pobedy 56, 25268 Kiyev 57, USSR

ABSTRACT

Theoretical analysis of the influence of the Nb-Ti ribbon thickness, of the bath and working temperatures on the recovery time constant and the losses of a thermally controlled superconducting switch (TCSS) is presented. It reveals the possibility to increase the repetition frequency of such switch in the range of the industrial 50-60 Hz frequency. Successful verification tests of a thermally controlled switch made from 20 um thick, 10 mm wide and 11 cm long Nb-Ti ribbon in a half-wave rectifier mode at 50 Hz have been performed.

INTRODUCTION

In the last decade superconducting rectifiers for powering the DC high current superconducting magnets have been intensively developed /1/ - /5/. They allow to reduce substantially the current leads cross-section and by this the heat leak into the cryostat as well as the price of the current supply /4/.

The most powerful rectifiers built so far were thermally controlled models operating at frequencies $f \leq 5$ Hz /5/. The power of superconducting transformer is proportional to the product of the frequency and of its volume. As the modern low loss multifilamentary conductors allow to built superconducting transformers working at $f > 50$ Hz /6/ the development of superconducting switches working at these frequencies would reduce the transformer volume by more than one order of magnitude. In order to achieve a repetition rate higher than 10 Hz only thin films / 7/ or ribbons /2/ are practical.

The aim of this work is to

demonstrate theoretically as well as experimentally the possibility to build from the Nb-Ti ribbon thermally controlled superconducting switches working at f 50 Hz.

THEORETICAL CONSIDERATIONS

The repetition frequency as well as the efficiency of a TCSS made from Nb-Ti ribbon are determined mainly by its recovery time $_{rec}$, as well as by the activation energy E_{act} and loss energy in the off state E_N.

Recovery Time Calculation

In accordance with /5/ we calculate $_{rec}$ as the time necessary to cool the parts of the conductor that have the highest temperature T_N during the off state to the temperature T_{sc} during the on state at which the critical current of the conductor is higher or equal to the working current I_L. We shall assume a one dimensional heat flow in the direction perpendicular to the ribbon surface. If we neglect the temperature drop in the superconductor in comparision with that in the helium boundary layer then the equation for the thermal equilibrium has the form

$$- A\, C_s\, \frac{\partial T}{\partial t} = P\, \dot{q}(T) \qquad (1)$$

where A is the cross-section and P the cooled perimeter of the ribbon. T the superconductor temperature and $\dot{q}(T)$ is the heat flux density from the superconductor surface into the helium. Integrating the eq. (1) we obtain

$$\tau_{rec} = \frac{A}{P} \int_{T_N}^{T_{sc}} \frac{-C_s(T)\, dt}{\dot{q}(T)} =$$

$$= \frac{A}{P}\left[\tau^+(T_{No},T_N) - \tau^+(T_{No},T_{sc})\right] (2)$$

where reduced recovery time $\tau^+(T_{No},T)$ is given by the expression (3)

$$\tau^+(T_{No},T) = \int_{T_{No}}^{T} \frac{-C_s(T)}{\dot{q}(T)} \quad T \qquad (3)$$

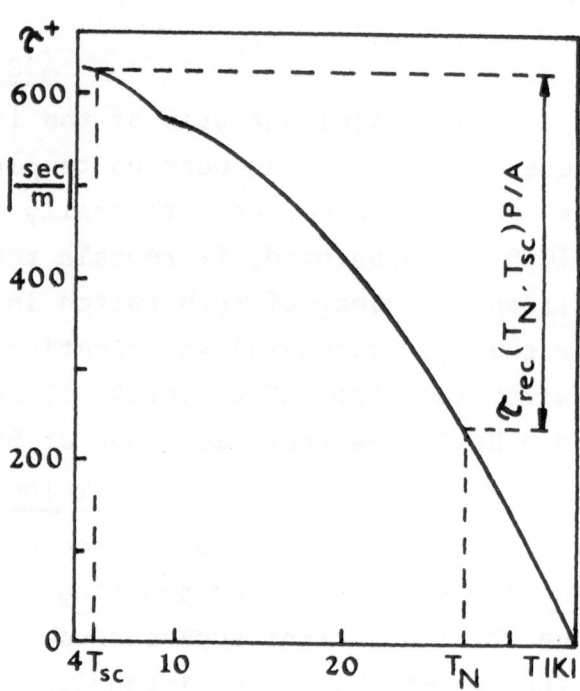

FIGURE 1   Reduced recovery time $\tau^+(T_{No},T)$ for $T_{No}=34$ K

Fig.1 represents the function $\tau^+(T_{No},T)$ for $T_{No}=34$ K calculated by numerical integration with $C_s(T)$ and $\dot{q}(T)$ taken from /5/ and /8/ respectively . For A/P=10 μm, $T_N=10$ K, $T_{sc}=4.5$ K $\tau_{rec}=0.6$ msec.

Activation Energy E_{act}

E_{act} equals to the sum of the superconductor enthalpy increase $\Delta H_s(T_{cs}, T_N)$ given as

$$\Delta H_s = A L \int_{T_{sc}}^{T_N} C_s \, dt \qquad (4)$$

where L is the length of the Nb-Ti ribbon and of the latent heat of evaporation of the helium diffusion layer Q_{tr} necessary to change the nucleate pool boiling into the film boiling during the activation heat pulse /8/,/9/

$$Q_{tr} = \left\{ \mathcal{T}_{tr}^{1/2} = 900 \, \mathcal{T}_{tr}^{1/2} = h_{tr} \right. \qquad (5)$$

$\mathcal{T}_{tr}$ being the heat pulse duration.

Loss Energy E_N in the Off State

E_N consists of the loss energy in the heater E_{hN} and of that in the superconductor element (SE) E_{SEN}, i.e.

$$E_N = \int_0^{\mathcal{T}_N} (U_{hN}^2/R_h + U_{SEN} I_{SEN}) dt \qquad (6)$$

where U_{hN} is the voltage across the heater with resistance R_h and U_{SEN} and I_{SEN} is the voltage and the current applied to SE. E_{NSE} depends on the temperature distribution along SE.

When $T > T_c$ in all SE $I_{SEN} = U_{SEN}/R_{No}$ with $R_{No} = \rho_n L/A$ being the resistance of SE in the normal state. In this case

$$E_N > LP \, \dot{q} \, (T_c) = E_{No} \qquad (6a)$$

When SE consists of normal and superconducting regions i.e. $R_N < R_{No}$, I_{SEN} equals to minimum propagation current I_{mpr} /10/-/12/. For superconductors with Stekly parameter $\alpha \gg 1$ I_{mpr} is given as

$$I_{mpr} = \left[\frac{(2 \div 8) \, A \, P \, \dot{q}(T = T_c)}{\rho_N} \right]^{1/2}$$

$$= I_c \left[\frac{2 \div 8}{\alpha} \right]^{1/2} \qquad (7)$$

The factor $(2 \div 8)$ before $\dot{q}(T = T_c)$ means that at I_{mpr} the heat flux from the normal zones of the SE is $2 \div 8$ times higher than $\dot{q}(T_c)$, i.e. the superconductor temperature is also nearly $2 \div 8$ times higher than T_c. In this case $R_N/R_{No} < 1$ and $E_N \leq E_{No}$.

EXPERIMENTAL and RESULTS

For the verification tests in the half wave rectifier mode (see Fig.2) a TCSS made in IED from 20 μm thick, 10 mm wide and 110 mm long bare Nb-Ti ribbon with $R_{No} = 0.29\Omega$ and $I_{cDc} > 800$ A. It

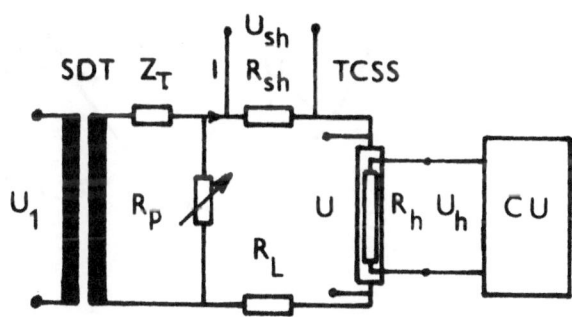

FIGURE 2 The electrical circuit used for verification tests

was thermally controlled by a resistive heater ($R_h = 5.65\Omega$) sticked on one side of the Nb-Ti ribbon, so that the cooled perimeter $P \doteq 1$ cm. The heater was powered by a control unit (CU) giving variable pulse voltage magnitudes, du-

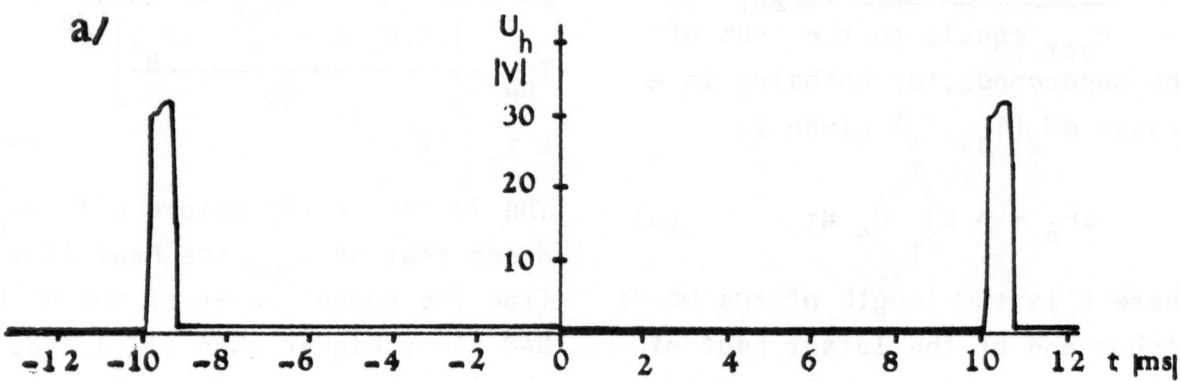

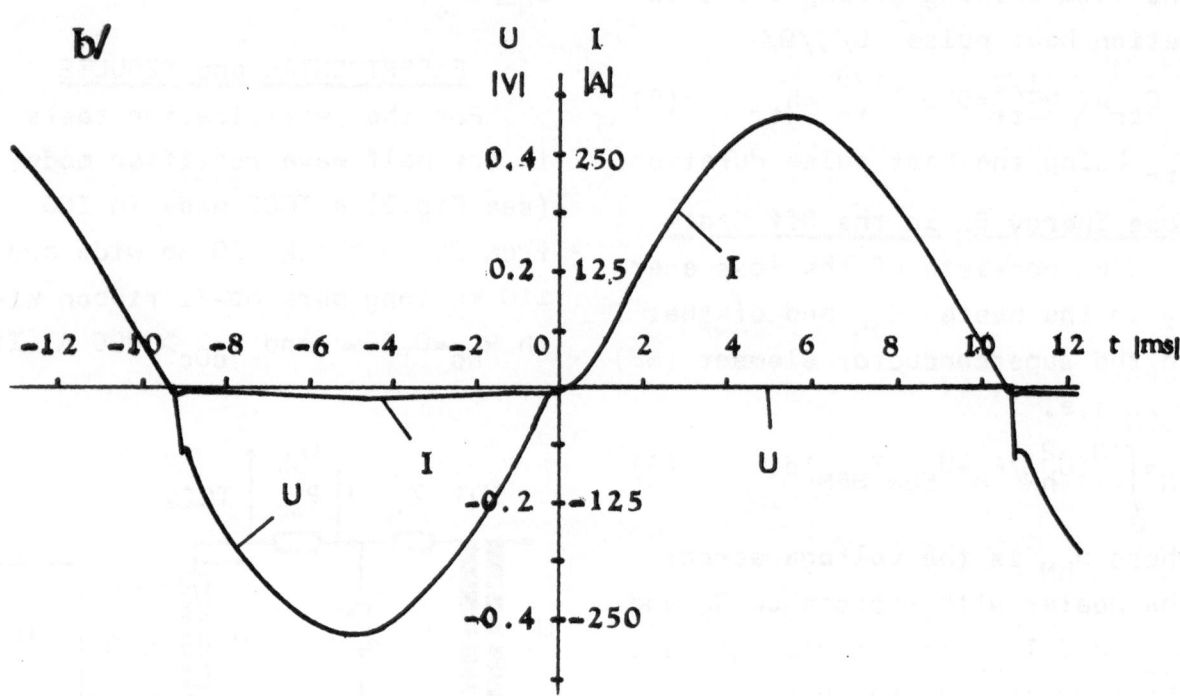

FIGURE 3   Traces a/ of voltage on the heater, b/ of voltage and current of the superconducting element

ration and triggering times. The
measuring scheme allowed to set
the maximum current I_{max} in the on
mode, as well as the maximum voltage U_{max} in the off mode of the
TCSS by changing the parameters
of the heat pulses, the input voltage U_1 on the step-down transformer (SDT) as well as the resistance of the parallel resistor
$R_p.R_{sh}\approx 0.4$ mΩ. The maximum obtained amplitudes U_{max} and I_{max}
were 0.57 V and 306 A respectively. The traces of the voltage on
the heater, on the superconducting element and of the current
in SE registrated for a typical
run are shown on the Fig.3. From

it we obtain the activation heat pulse energy $E_{act} = 90$ mJ compaired to 10 mJ calculated from ex. (4) and (5).

The time of the transition into the resistive state determined from the duration of the voltage step on the SE is 75-150 µsec.

From our measurements τ_{rec} cannot be determined exactly. However the experimental results give $\tau_{rec} < 5$msec (this is the time in which I reaches I_{max} value in the on state). Moreover we see that I_{SEN} at low voltage ($U_{SEN} < 0.15 \div 0.3$ V) is practically constant and equal to 6.3-6.9 A compaired to $I_{mpr} = 4.5 - 9$ A calculated from ex. (7). At the same time $R_N/R_{No} \leq 0.15$. This means that SE in the off state was in the minimum propagation current mode.

The loss energy in the off state determined from Fig. 3 $E_N = 27$ mJ equals to E_{No} calculated from ex. (6a) using $\dot{q}(T)$ from /8/.

The load power during the on mode for an ideal power supply ($U_L = U_N$) would be $P_L = U_{max} I_{max}/2 = 87$ W i.e. $E_L = 870$ mJ. Then for the TCSS efficiency /5/ we obtain $\eta = 1 - (E_{act} + E_N)/E_L \doteq 87$ %.

CONCLUSIONS

The theoretical analysis and experimental results have shown that NbTi commercially available ribbon with the thickness of the order of 10 micrometers can be successfully used for TCSS wor-

king in the half wave rectifier mode at sinusoidal input voltage of the industrial frequency $50 \div 60$ Hz.

However further work is necessary in more detailed study of the influence of different design, technology and regime parameters mainly on the recovery time τ_{rec} as well as on the activation energy E_{act} and the loss energy in the off state E_N of a TCSS for superconducting magnet power supply. Such parameters are the thickness of the superconducting ribbon eventually that of the insulation layer covering it, the heating and cooling regimes time dependence of the working voltage and current as well as of the heat power magnitude generated in the heater and in the superconductor element. This informations are necessary for the optimisation of the design, technology and regime of such TCSS.

Acknolegement

The autors are indebted to prof. R.G.Mints for valuable discussion on the minimum propagation current mode of the TCSS.

REFERENCES

1 Ten Kate,H.H.J., Bunk,P.B., Britton,R.B., Van de Klundert, L.J.M., High current and high power superconducting rectifiers, Cryogenics 1981, 21, 291

2 Markowski,N.V., Pan,V.M., Scha-
stlivyi,G.G., Flis,V.S., Shev-
chenko,O.A., Applications prospects of high power cryotrons
in electrical machine systems
for excitation of cryoturboge-
nerators, Superconductivity in
Technology, Proceedings of the
2nd All Union Conference on Technical Applications of Superconductivity, Leningrad, Sept.
26-28, 1983, p. 86

3 ten Kate, H.H.J., Superconducting rectifiers, Thesis, University of Twente, April 1984

4 Sikkenga,J., ten Kate, H.H.J.,
A full scale superconducting rectifier for powering a MRI-mag-
net, IEEE Trans. on Magn. 1989,
Vol. MAG-25, 1771

5 Mulder, G.B.J., Increasing the
operation frequency of superconducting rectifiers, Thesis,
University of Twente, March
1988

6 Hlásnik,I., Prospects of multifilamentary superconductor AC
50 Hz applications, J.Phys.,
1984, 45, C1-459

7 Gray,K.E., Kampwirth,R.T., Design for a repetitive superconducting opening switch, Cryogenics 1984, 24, 21

8 Wilson,M.N., Superconducting
Magnets, Clarendon Press, Oxford, 1983

9 Schmidt,C., App.Phys.Lett.,
1978, 32, 827

10 Keilin,V.E., Klimenko,E.Yu.,
Kremlev,M.S., Samoilov,N.B.,
Stability criteria for curren-
ts in combined conductors,Proc.
of the International Conference
Les Champs Magnétiques Intenses,
C.N.R.S. Paris, 1967, p. 231

11 Dresner,L., Analytic solution
for the propagation velocity
in superconducting composites,
IEEE Trans. on Magn. 1979, Vol.
MAG-15, 328

12 Gurievich, A., Mints,R., Rakh-
manov,A., Physics of the composite superconductors, Moscow,
Nauka 1987, in Russian

BASIC TEST OF A 3-PHASE SUPERCONDUCTING FAULT CURRENT LIMITING REACTOR

H.Kado, T.Ishigohka

Seikei University
Tokyo, Japan

ABSTRACT

A novel 3-phase superconducting fault current limiting reactor (SCFCLR) for power system is presented. It has 3-phase superconducting windings on single iron core. Small experimental devices are fabricated and tested. Through some experiments, the fundamental behavior of the limiter is confirmed. In the experiment, two SCFCLRs are inserted in the sending and the receiving ends of the model power-transmission line. In a normal operation, the SCFCLR exhibits very small impedance for balanced three-phase current. In the case of single-line-to-ground fault, the fault current is limited to very small value by the large zero-phase-sequence reactance of the SCFCLR. In this case, the superconducting windings do not quench. In the case of two-phase or three-phase short circuit, the superconducting winding of the SCFCLR quenches, and the short circuit current is limited by the normal conducting resistance of the winding. It is experimentally confirmed that the SCFCLR can limit the fault current in all fault conditions.

INTRODUCTION

With the growth of electric power system, the short circuit current is increasing year by year. The problem of damage at the fault point and induced noise to communication lines are getting serious. So, various types of fault current limiters have been studied for the protection of power system. Because fault current limiter has to carry full load current continuously, the resistance should be very small in normal operating condition. In this point, the SCFCLR would be very desirable. The authors propose a new type SCFCLR with 3-phase superconducting windings on single iron core. Small experimental devices are fabricated and some experiments for single-line-to-ground and three-phase short circuit are carried out. The fundamental idea for current limiting behavior is confirmed in each case.

PRINCIPLE

The SCFCLR has three superconducting windings with the same numbers of turns wound on single iron core. In a normal operating condition, it exhibits very small impedance because the sum of the three phase current is zero. On the other hand, in single-line-to-ground fault, the reactance becomes very large because the balance of the three-phase current is lost. Therefore, the fault current is limited by the reactance of the SCFCLR. In this case, the winding does not quench. Because more than 90% of the power system faults are the single-line-to-ground fault, the SCFCLR can limit most of the fault current without quench. However, for two-line-to-ground, and three-phase short circuit fault, the reactance of the SCFCLR does not increase. Therefore, in this case, the SCFCLR quenches at the instance when the fault current exceeds the critical value of the superconducting winding. After the quench, the fault current is limited by the normal conducting resistance of the superconducting winding. As a result, the SCFCLR can limit the short circuit current in all fault conditions. This method does not need any overcurrent detection sensors or switching elements. Therefore, it would be a highly reliable fault current limiter.

Figure 1 shows the fundamental principle of the SCFCLR. Let us consider that a single-line-to-ground fault has occurred in phase-U. It is assumed that the superconducting reactor is ideal, and that induced voltages of each winding are equal. Equations for the currents are;

$$\dot{i}_{us} + \dot{i}_v + \dot{i}_w = 0 \quad \ldots (1)$$

$$\dot{i}_{ur} + \dot{i}_v + \dot{i}_w = 0 \quad \ldots (2)$$

where,

$\dot{i}_{us}$: current of phase-U in sending side

$\dot{i}_{ur}$: current of phase-U in receiving side

$\dot{i}_v$: current of phase-V

$\dot{i}_w$: current of phase-W

Voltage equations are;

$$\dot{e}_u + \dot{v}_s = 0 \quad \ldots (3)$$

$$\dot{e}_v + \dot{v}_s + \dot{v}_r = \dot{Z}\dot{i}_v \quad \ldots (4)$$

$$\dot{e}_w + \dot{v}_s + \dot{v}_r = \dot{Z}\dot{i}_w \quad \ldots (5)$$

$$\dot{v}_r = \dot{Z}\dot{i}_{ur} \quad \ldots (6)$$

where,

$\dot{e}_u$: phase-U generator voltage

$\dot{e}_v$: phase-V generator voltage

$\dot{e}_w$: phase-W generator voltage

$\dot{v}_s$: reactor voltage of sending side

$\dot{v}_r$: reactor voltage of receiving side

$\dot{Z}$: impedance of load

Solving equations (1)-(6), we get;

$$\dot{i}_{us} = \dot{i}_{ur} = \dot{e}_u / \dot{Z} \quad \ldots (7)$$

$$\dot{i}_v = \dot{e}_v / \dot{Z} \quad \ldots (8)$$

$$\dot{i}_w = \dot{e}_w / \dot{Z} \quad \ldots (9)$$

$$\dot{i}_e = 0 \quad \ldots (10)$$

where,

$\dot{i}_e$: ground-fault current

So, the each phase current does not change after the fault, and the ground-fault current becomes zero. However, this is the result for the ideal case. In a practical case, the effect of small excitation current will appear.

As for the voltages;

$$\dot{v}_s = -\dot{e}_u = -\dot{v}_r \quad \ldots (11)$$

On the other hand, the line-to-ground voltage at the fault point of the sound phase, for example phase-v, $\dot{v}_{vf}$;

$$\dot{v}_{vf} = \dot{e}_v + \dot{v}_s \quad \ldots (12)$$

Substituting equation (11) to equation (12), we get;

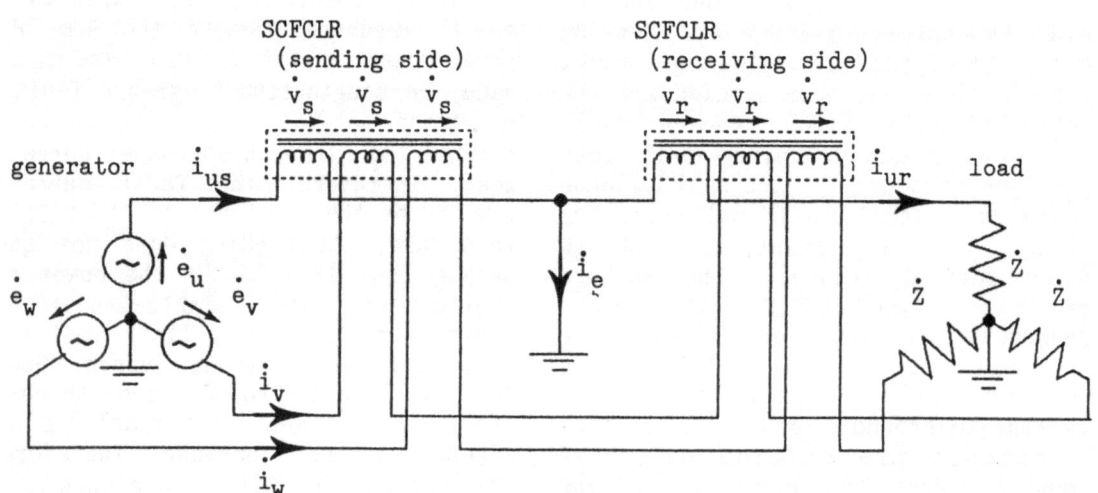

Figure 1   Fundamental Principle of SCFCLR

$$\left|\dot{v}_{vf}\right| = \left|\dot{e}_v + \dot{v}_s\right| = \left|\dot{e}_v - \dot{e}_u\right|$$

$$= \sqrt{3}\left|\dot{e}_v\right| \quad \ldots(13)$$

Therefore, in the fault condition the line-to-ground voltage of the sound phase becomes √3 times larger than that in a normal condition. However, this voltage increase does not appear both at the sending and the receiving ends.

STRUCTURE OF SCFCLR

The structure of the SCFCLR is shown in Figure 2. The SCFCLR has three superconducting windings with the same number of turns wound on single iron core. The connections of windings are made

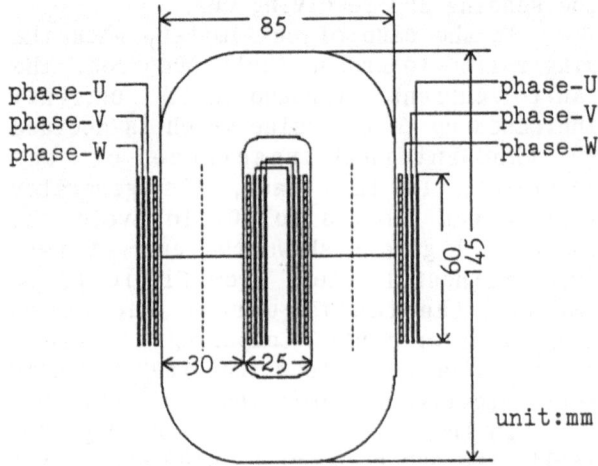

Figure 2 Structure of SCFCLR

taking into consideration the balance of the leakage reactances of each phase winding. The rating of the SCFCLR is 200V, 10A. The specification is shown in Table 1.

Table 1 Specification of SCFCLR

superconducting wire		
diameter	:	0.103mm
twist pitch	:	0.98mm
filament diameter	:	0.444 μm
filament no.	:	15367
matrix NbTi/CuNi	:	1/2.5
Ic	:	24.0A at 1T
		12.6A at 2T
iron core	:	ZT100 0.1t
winding		
No. of turns/phase	:	400x2
length	:	60mm
diameter(avg.)	:	35mm

THE EXPERIMENTAL CIRCUIT

The experimental circuit is shown in Figure 3. A generator and a three-phase load are connected to a model transmission line. Two SCFCLRs are inserted in the sending and the receiving ends. In this circuit, two switches are used. The SW1 is for the single-line-to-ground fault, and the SW2 is for three-phase short circuit. Each fault is assumed to occur when these switches are activated.

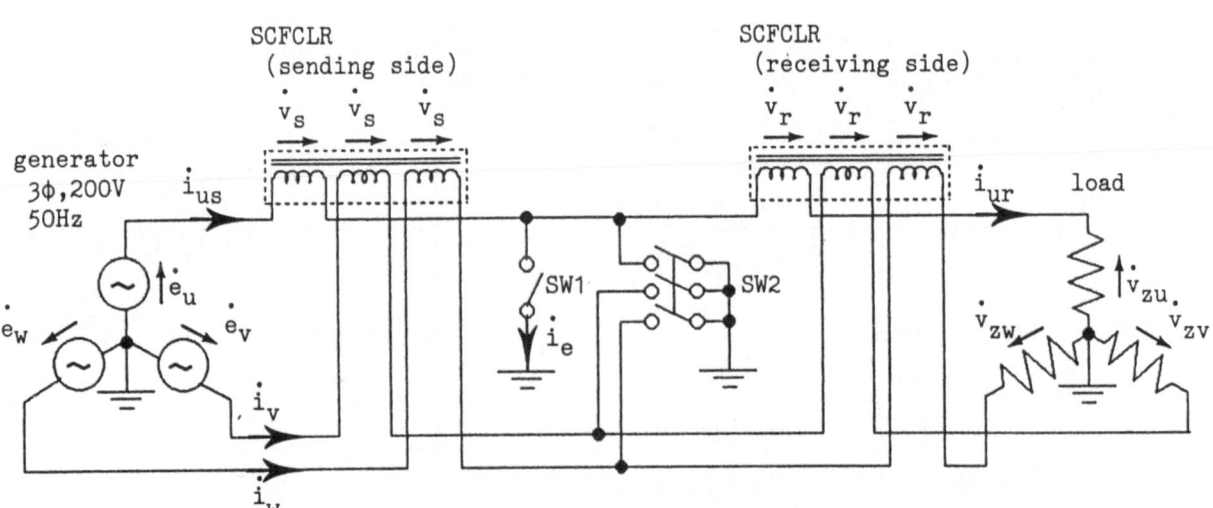

Figure 3   Experimental Circuit

RESULT AND DISCUSSION

Single-line-to-ground fault

First, we operated the circuit shown in Figure 3 at the normal state condition of 200V, 10A. In this condition, three-phase current is balanced, so the reactance drop of the SCFCLR is zero.

Then, in this circuit, the SW1 is closed, and a single-line-to-ground fault is produced. Figure 4 (a), (b) and (c) show the fault phase current of sending and receiving side, and the ground-fault current, respectively. And, Figure 4 (d), (e), (f), (g) and (h) show the reactor voltages of each side, the supply voltage, the load voltage and the line-to-ground voltage at the fault point. As can be seen in Figure 4, at the instance of the fault, the reactor voltage increases instantly. And, the ground-fault current is almost perfectly limited by the large zero-phase-sequence reactance of the SCFCLR. This fault current limiting process does not need any detection or switching circuits. On the other hand, we can recognize that the line-to-ground voltage of the sound phase at the fault point increases by factor of 3 after the fault. However, as can be seen in Figure 4 (f) and (g), this voltage increase does not appear both at the sending and receiving end.

In the case of no limiter, when the single-line-to-ground fault occurred, the fault current (ground-fault current) increases up to the value which is decided by the internal reactance of the generator. In this case, the generator voltage was limited to 50V to avoid the damage. Figure 5 shows the current wave forms without limiter. From Figure 5, we can see the fault current increases greatly after fault. Comparing Figure 4 and 5, it becomes clear that the SCFCLR is very effective to limit the fault current.

In the case of single-line-to-ground fault, the superconducting windings do not quench. From a practical point of view, it is very important that it can limit the fault current without quench. More than 90% of the power system short circuit faults being the single-line-to-ground fault, this ability of the SCFCLR is very attractive.

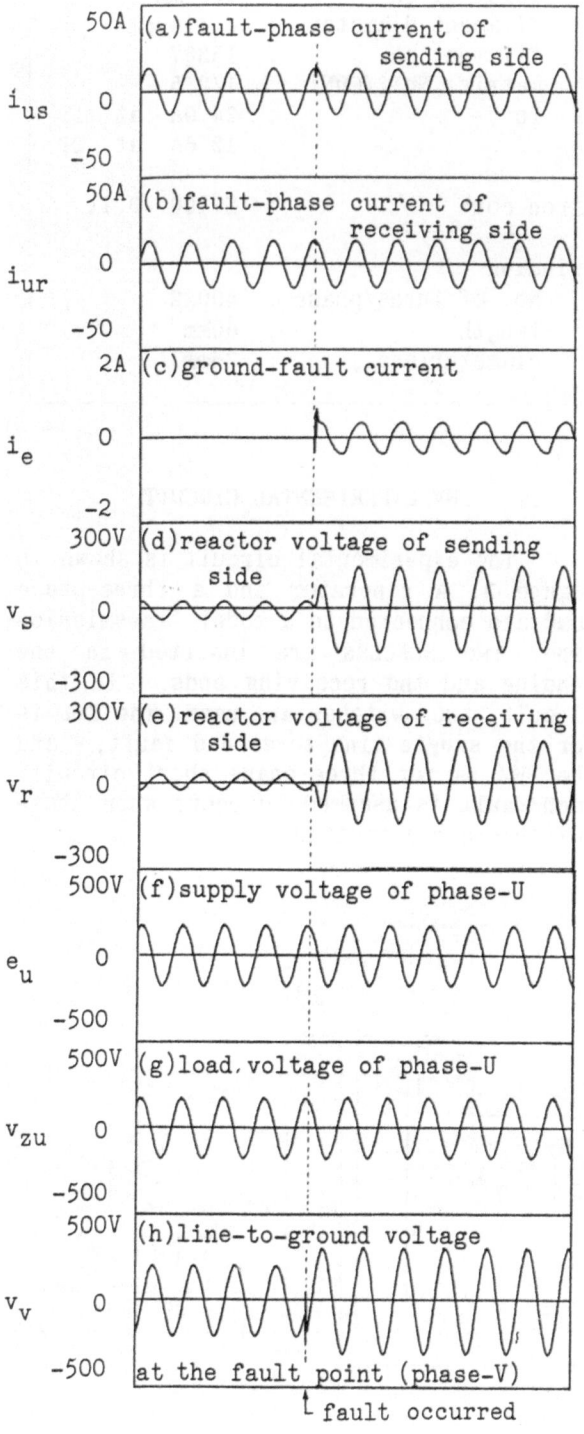

Figure 4 Single-line-to-ground Fault

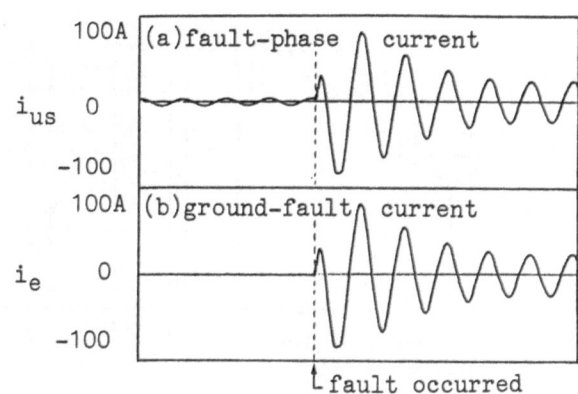

Figure 5 Single-line-to-ground Fault without SCFCLR

Figure 6 shows the results of simulation for the single-line-to-ground fault. This is a numerical analysis by differential equations for the three-phase circuit. The experimental result shows good agreement with the analytical result.

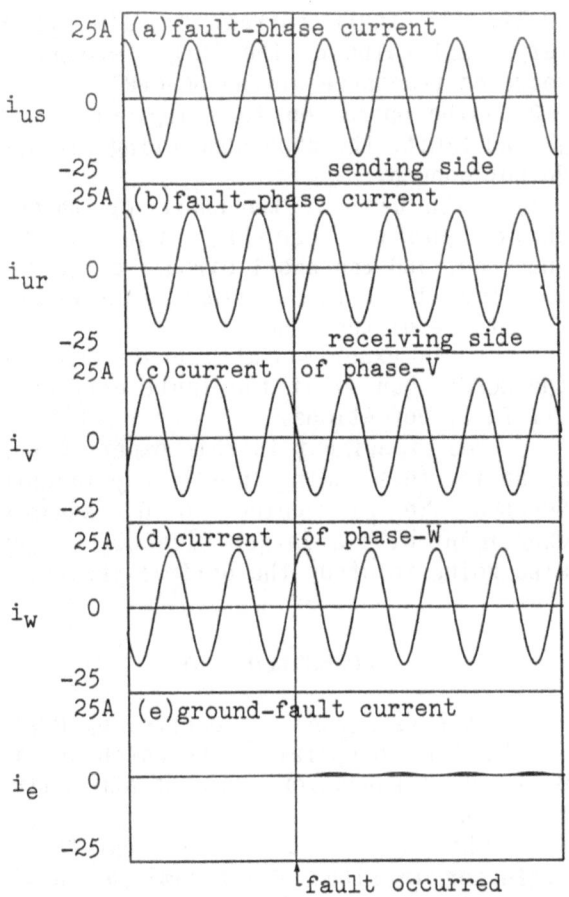

Figure 6   Result of Simulation for
Single-line-to-ground Fault

Three-phase short circuit fault

The initial operating condition is 200V, 8A in the circuit shown in Figure 3. In this state, the SW2 is closed suddenly, and three-phase short circuit fault is produced. Figure 7 (a), (b), (c) and (d) show each phase currents of sending side, and ground-fault current, respectively. And, Figure 7 (e), (f) and (g) show reactor voltages. When the fault occurs, the superconducting windings quench at that instance. So, the short circuit fault current is limited by the normal conducting resistance. In the reactor windings, the voltage by normal conducting resistance appears.

In the case of no limiter, when the three-phase short circuit fault occurs,

the fault current (short circuit current) increases to the value which is decided by the internal reactance of the generator. Also, in this case the generator voltage was limited to 50V to avoid the damage. Figure 8 shows the current wave forms in the case without the limiter. When the fault occurs, we can see the increased fault current. From these experiment, the current limiting ability of the SCFCLR becomes clear.

In the same way, also for the two-line-to-ground fault, the SCFCLR limits the fault current by the normal resistance of the superconducting windings after quench.   So, in these cases, the circuit

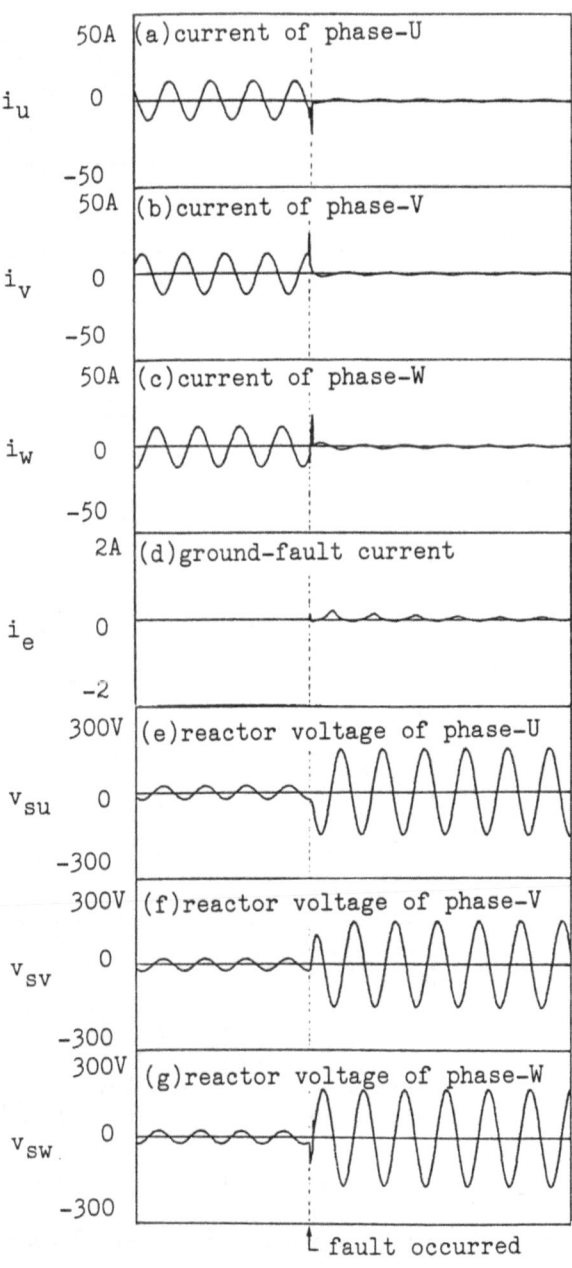

Figure 7   Three-phase Short Circuit Fault

must be broken off without delay to suppress excessive heat loss in the superconducting windings. That would be essential to get fast recovery to superconductivity for the fast reclose.

Figure 9 shows the simulation result of the current at the three-phase short circuit fault. The analytical result agrees with the experimental one very well.

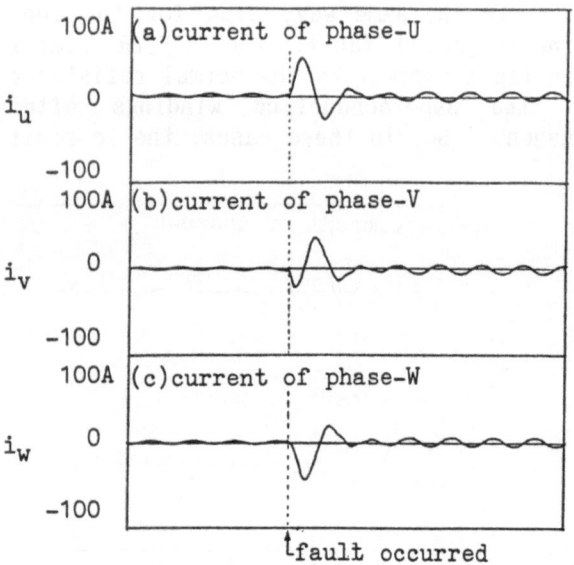

Figure 8 Three-phase Short Circuit Fault without SCFCLR

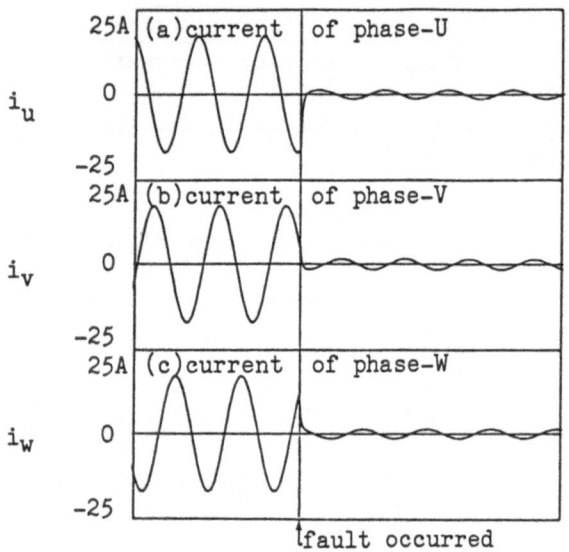

Figure 9 Result of Simulation for Three-phase Short Circuit Fault

CONCLUSION

We have proposed a novel 3-phase superconducting fault current limiting reactor (SCFCLR). And small experimental devices are fabricated, and some experiments are carried out. The fundamental behavior of this reactor is confirmed. The experimental results are summed up as follows;

1)for a balanced 3-phase current, the SCFCLR exhibits very small impedance,

2)in the case of single-line-to-ground fault, the fault current is limited to very small value by the large zero-phase-sequence reactance of the SCFCLR,

3)in the operation for single-line-to-ground fault, the superconducting windings do not quench,

4)in the case of two-phase or three-phase short circuit, the SCFCLR quenches, and the short circuit current is limited by the normal conducting resistance of the winding.

It is experimentally confirmed that the SCFCLR can limit the fault current in all fault conditions.

The experiment in this paper is only a basic test using small experimental SCFCLRs. We are going to do advanced experiment with a larger size SCFCLR, and also going to study the optimal design.

ACKNOWLEDGMENT

This research is supported by Grant-in-Aid for Co-operative Research of the Ministry of Education, Science and Culture of Japan.

The authors are very grateful to Professor Emeritus N.Mizukami of Seikei University for many important suggestions, and to Mr.Y.Seki and Mr.H.Yachi for their devoted support throughout the experiments.

REFERENCES

[1]T.Ishigohka, "SUPPRESSION OF SINGLE-LINE-TO-GROUND FAULT CURRENT BY INSERTION OF THREE-WINDING REACTORS TO A TRANSMISSION SYSTEM", Canadian Communications and Energy Conference, Montreal, Oct. 1982

[2]T.Ishigohka, Y.Seki, K.Sakaniwa, "SUPERCONDUCTING FAULT CURRENT LIMITER USING THREE-WINDING REACTOR", 1988 Annual Meeting of IEEJ, No.819

[3]H.Kado, T.Ishigohka, "FABRICATION AND BASIC TEST OF 3-WINDING SUPERCONDUCTING FAULT CURRENT LIMITING REACTOR", IEEJ paper SA-88-43, Tokyo, Dec. 1988

Nb$_3$Sn PERSISTENT CURRENT SWITCH

M. Urata, H. Maeda, M. Tanaka, S. Murase, *Y. Oda, *S. Nakamura,
*E. Suzuki, **S. Kageyama and **S. Kabashima.

Toshiba R & D Center, 4-1, Ukishima, Kawasaki, 210, Japan.
*Showa Electric Wire and Cable Co.,LTD., 2-1-1, Odasakae, Kawasaki, 210.
**Tokyo Institute of Technology, 4259, Nagatsuda, Midoriku, Yokohama, 227.

ABSTRACT

A Nb$_3$Sn persistent current switch (PCS) has been developed in the Toshiba R & D Center. The Nb tube processed Nb$_3$Sn conductor with Cu-10wt%Ni matrix has been developed for the PCS. The quench currents were as high as 1100 A at 1.5 T and 380 A at 6 T, with the "off" resistance of 35 Ω. The switch suffered from degradation at low fields. They were caused by flux jumpings. The winding temperature rise due to self field AC loss and resulting quench behavior were also studied in this paper.

INTRODUCTION

Persistent current switch (PCS) is one of the key components of persistent current system such as MRI or magnetic levitation vehicle. By using high resistivity Cu-Ni alloy matrix, however, PCS involves inevitable instability problem; i.e., PCS suffers from unexpected premature quenches. This is especially the case at low fields such as 1- 2T.

Nb$_3$Sn PCS is a promising device to improve stability, because of its larger temperature margin. Moreover, Nb$_3$Sn PCS is required to develop Nb3Sn persistent current system such as high field NMR [1].

This paper describes (a) the fabrication process of the Nb$_3$Sn conductor and (b) the fabrication process and test results for the Nb$_3$Sn PCS. Origins of in-stability will be discussed, based on magnetization measurement and numerical calculations.

FABRICATION OF CONDUCTOR AND PCS

The Nb tube processed Nb$_3$Sn conductor has been developed for the PCS at Showa Electric Wire and Cable Co.,LTD. The parameters for the conductor are listed in Table 1. By using Cu-10w%Ni alloy as the matrix material, reduction performances are better than those for copper-matrix conductor. It results in a smaller outer diameter of Nb tube, 43 μm, than those for the copper-matrix conductor, 70-80 μm. Best heat treatment condition was at 700 ° C for several days . Cross-sectional view for the conductor after heat treatment is shown in Fig. 1.

Table 1. Nb₃Sn PCS Conductor Parameters.

Outer Diameter	0.87 mm
Nb-Tube Number	258
Nb-Tube Diameter	43 μm
Tin content in Cu Tube	25 %
Matrix to Super Ratio	1.02
Matrix Material	Cu-10w%Ni
Insulation Thickness	120 μm

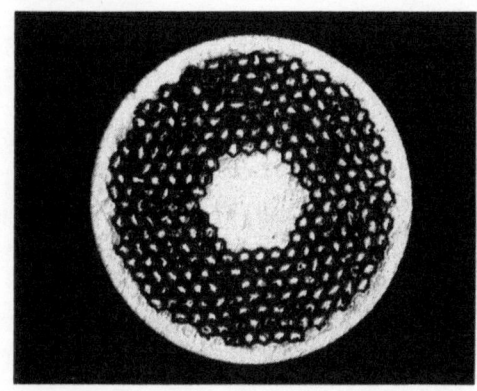

Fig. 1. Cross sectional view of the Nb₃Sn conductor.

A Nb₃Sn PCS was made through a wind and react process, with inner diameter of 18 mm, outer diameter of 106 mm and the coil length of 74 mm. First, the Nb₃Sn conductor was wound non-inductively around a 1 mm thick stainless steel bore tube. Finally, the winding was heat treated and impregnated with epoxy resin.

Two sets of heater were wound between layers. To reduce heater power for keeping the " off" resistance, the winding was thermally insulated from liquid helium; i.e., The 10 mm coil bore was filled with epoxy-glass, while the winding is surrounded by a 5 mm thick fiber-reinforced plastics blanket.

EXPERIMENTAL PROCEDURE

The PCS was mounted inside a NbTi magnet, producing the field in the range from 0 T to 7 T. A search coil was wound around the PCS to detect any flux change in the winding. An acoustic emission (AE) sensor was attached to the top flange to detect mechanical disturbances. The PCS voltage, search coil signal, coil current and the AE signal were recorded in a multi-channel data recorder, which were played back with multi-channel transient digital memories.

EXPERIMENTAL RESULTS

PCS Performance

Resistance for the Nb₃Sn PCS at fully resistive state was 35 Ω. Heater power required to keep full resistance was 22 W, about 10 times as high as NbTi PCS with the same dimension. Recovery time from full resistance to superconducting state (not to 4.2 K) was 100 sec.

The PCS quench current (open circle) is shown in Fig. 2 together with a short sample quench current (solid line). The results are summarized as follows:
(i) The PCS quench current is as high as 70 % of the short sample critical current at above 1.5 T.
(ii) On the contrary, both the PCS and short sample quench current show enormous *degradation* at below 1.5 T.

Thus, it is shown that the Nb₃Sn PCS has sufficiently good stability at high fields as expected beforehand, while it is unstable at low fields. The degradation might be related to flux jumping as mentioned as follows.

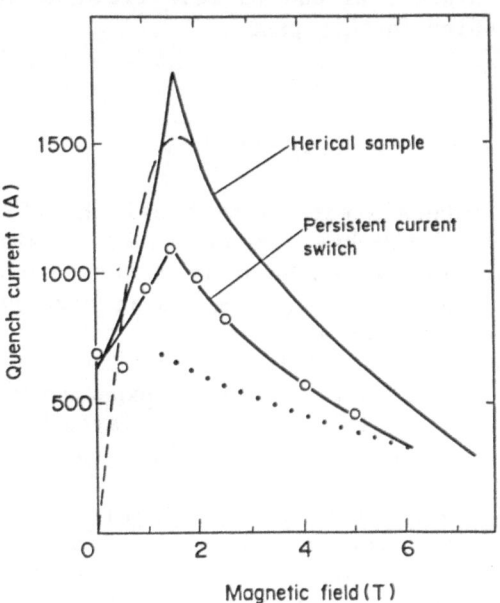

FIGURE 2 Quench current for the PCS: Solid line; the observed short sample quench current: Open circle; the quench current for the PCS: Dashed and dotted lines; critical lines for flux jumping.

Flux Jumpings

Voltage spikes were frequently observed during ramping up the current or during changing the magnetic field. Occurrence of spikes depended on the history of current sweep and magnetic field. Figure 3 shows such examples:

(a) The magnetization was cleared out by "heating" the PCS for a few minutes, at first, then the PCS current was ramped up at 0 T. Only a few voltage spikes were observed until its quench at 712 A (see Fig. 3 (a)).

(b) The applied field was increased to 1.5 T and decreased to 0 T. Then the PCS current was ramped up. During ramping up the current, many voltage spikes, larger than 4 mV, were observed (see Fig. 3 (b)).

These results suggest that the voltage spikes are due to magnetic instability i.e., flux jumping. Transient signals for voltage, current, AE and search coil voltage at the quench at 0.5 T are shown in Fig.4. The normal voltage was preceded by a negative voltage spike. A kink was simultaneously observed in the search coil voltage. If the spike is due to mechanical disturbance, AE burst will be observed simultaneously. As no AE burst appeared, the quench is not of the mechanical origin.[2]

These voltage spikes did not emerge at fields above 1.5 T

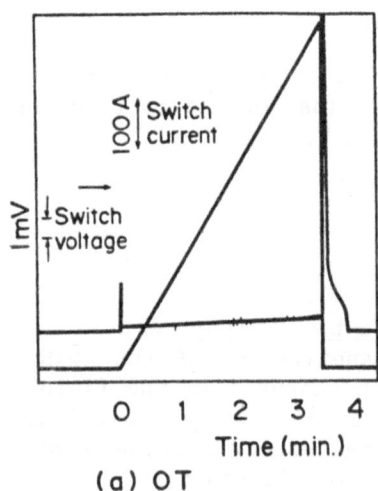

(a) OT

(quenched at 712A)

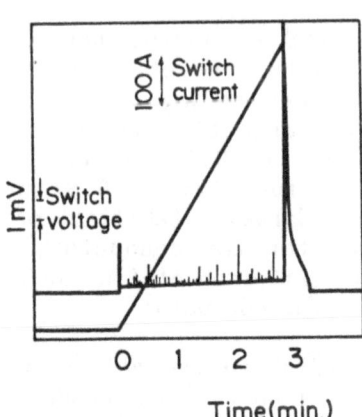

Time(min.)

(b) OT → 1.5T → OT

(quenched at 576 A)

FIGURE 3   Voltage spikes recorded during ramping the PCS current. After the magnetic field was sweep to 1.5 T, many numbers of flux jumps were observed.

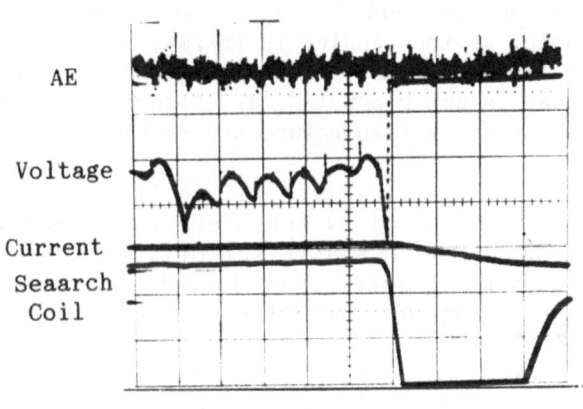

(4 msec./div.)

FIGURE 4   Transient signals observed at a PCS quench at 0.5 T. A voltage spike is caused by flux jumping.

Effect of Current Sweep Rate

The quench current depended on the current sweep rate, especially at high fields above 1.5 T. The example is shown by the open circle in Fig. 5; the applied field is 3 T. Quench current for 200 A/sec., 600 A, is smaller than that, 730 A, for 3.3 A/sec.

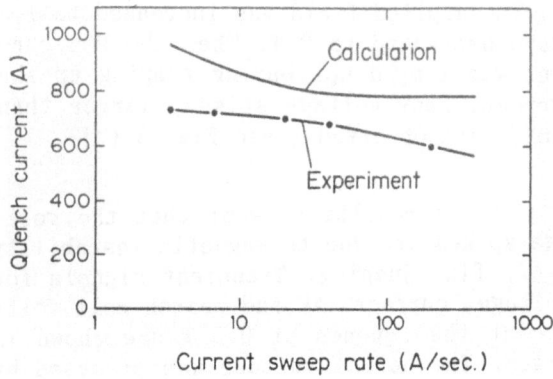

FIGURE 5 Quenching current dependence on the current sweep rate at 3 T. The solid line was obtained based on the curve in Fig. 6.

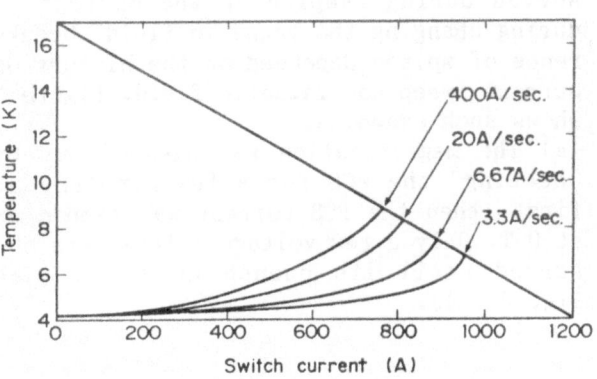

FIGURE 6 Temperature rise due to self field AC loss during raising the PCS current at 3 T.

DISCUSSION

Current Sweep Rate Dependence

The PCS is wound bifilarly; moreover, the applied magnetic field is kept constant. Thus, heat generated during ramping up the current is due to self field AC loss. As the winding is thermally isolated with the thick insulator cylinder, the heat is not conducted out efficiently thus the winding temperature are raised.

Temperature rise within the winding was calculated by computer. The program assumes the PCS as a set of multi-layer infinite cylinders. It includes (i) heat generation by the self-field A.C. loss, [3][4] (ii) heat conduction through the winding body, and (iii) heat transfer to liquid helium at both the inner and outer surface. Details of calculation will be published elsewhere[5].

Maximum temperature within the winding at 3 T is shown in Fig. 6. A straight line in Fig. 6 gives the critical current[6]. The winding temperature gradually increases with current: The more quickly the PCS current is ramped up, the hotter the winding becomes as heat is not effectively conducted to liquid helium.

The intersection between the winding temperature and the critical current gives the maximum attainable quenching current for the PCS. The quenching current decreases with current sweep rate as shown by a solid line in Fig. 5; it qualitatively agrees with the sweep rate dependence of the observed quenching current indicated by the closed circle.

Magnetization Measurement

Magnetization of the Nb_3Sn conductor has been measured in the field range from -1.0 T to +1.0 T(Fig. 7). Magnetization fluctuates during ramping down at 0.36 T-0.26 T. The fluctuations are due to flux jumping. Magnetization measured for a thinner diameter conductor, e.g. 0.501 mm, did not show these fluctuations.

The PCS conductor has a Cu-10w%Ni matrix. As the resistivity of Cu-10w%Ni is 1000 times as high as the copper at 4.2K, the magnetic diffusivity for the CuNi matrix is 0.12 m^2/sec., while the thermal diffusivity is $6.9*10^{-4}m^2$/sec. Therefore, dynamic stabilization effect is neglected for the conductor, i.e., the adiabatic criterion of flux jumping is applied for the conductor.

According to the adiabatic stability criterion[7] on the slab conductor, critical half width, a, for flux jumping is given as follows:

$$\frac{\mu_0 J_c{}^2 a^2}{\gamma C(\theta_c - \theta_0)} = \frac{3}{1 + 3 i^2} \quad (= \beta) \quad (1)$$

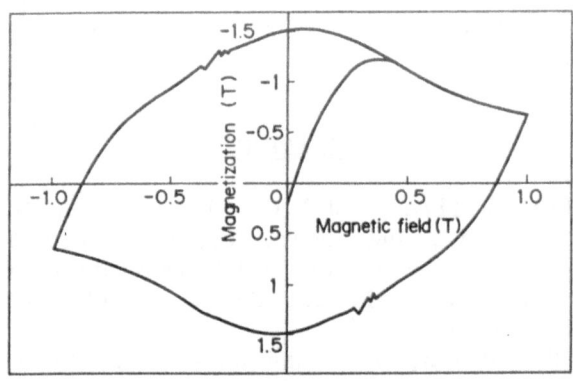

FIGURE 7   Magnetization for the 0.87 mm diameter PCS conductor.

where γ is a density, C is the heat capacity, θ_0 is a bath temperature, and i= transport current/ critical current. The critical half width in case of magnetization measurement(i.e., $i = 0$) is 5.1 μm at 0.5 T, if J_c of $3.6*10^{10}$ A/m^2 is assumed for the Nb$_3$Sn layer.

The Nb$_3$Sn layer thickness for 0.87 mm diameter conductor is about 4.5 μm, while that for the 0.501 mm diameter conductor is 2.5 μm. Thus, it is probable that the 0.501 mm conductor is almost free from flux jumping, although 0.87 mm conductor may suffer from them in the magnetization experiment, as shown in Fig. 7.

Degradation origins: Short sample

Short sample quench current densities are compared between three kinds of conductors(see Fig. 8): with diameter of 0.87mm, 0.587mm, and 0.501mm, respectively. Low field degradation, observed in the 0.87 mm conductor, is removed as the conductor diameter is reduced.

Decay time for self field effect is 0.23 sec. in the critical current measurement. The current distributes uniformly within the conductor during critical current measurement.

Assuming a Nb$_3$Sn layer thickness, a, as a thin slab with a thickness, $2a$, Eq. (1) gives a critical transport current for flux jumping. The result is shown by a dashed line in Fig. 2, which coincides with the experiment (solid line). Thus, it is shown that the degradation at low fields is due to flux jumpings within Nb$_3$Sn filaments.

Degradation Origins: PCS

Diffusion time of the self field effect in case of the PCS is so long as 56 hours. Thus the filaments are fully coupled during current ramping; the problem in this case is occurrence of self-field flux jumping. Dotted line in Fig. 2 shows the critical line for self field flux jumping[7]. Though the current seems to agree with the experiment at high fields, it cannot explain the positive slope at low fields.

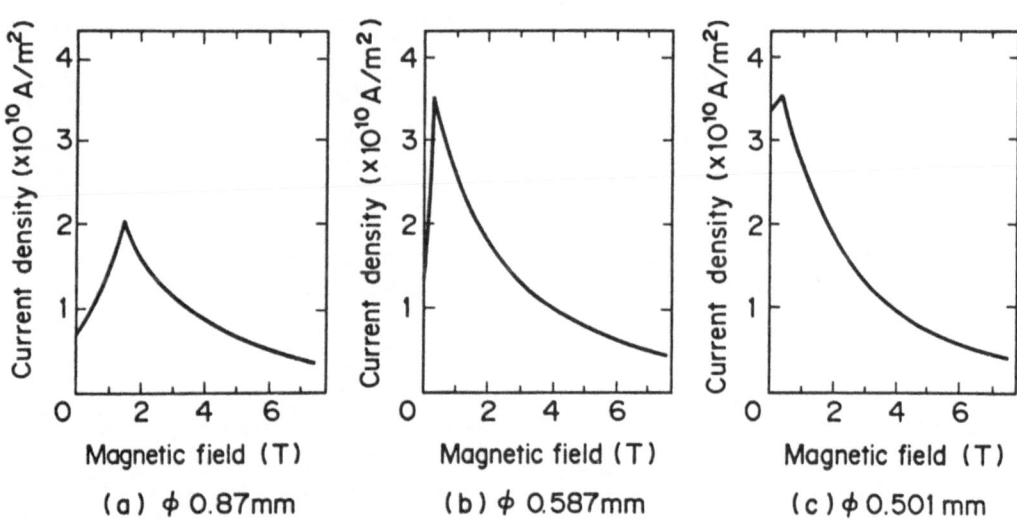

FIGURE 8   Short sample quench current densities for three kinds of Nb$_3$Sn PCS conductors. Each conductor was reduced from the same rod.

The degraded behavior for the PCS may be explained as follows; During charging the PCS, the conductor consists of (i) a critical state outer region and (ii) a unoccupied inner region. The critical state region penetrates with a transport current. Fluxes of the critical state region cannot move individually; however, fluxes around a leading front of current penetration can move individually, which might result in flux jumping as the short sample.

These flux jumpings for the PCS might be eliminated, if the PCS is wound by thinner diameter conductor. Stability for the PCS may be sufficiently improved: Such a kind of PCS is being fabricated in our laboratory.

CONCLUSIONS

The Nb_3Sn persistent current switch (PCS) with 35 Ω normal resistance has been developed. The PCS attained the sufficiently large current at high fields; however, degradation is observed for short sample quench current and PCS quench current, at low fields. The degradations are due to flux jumpings. Stability will be improved, if a thinner diameter conductor is wound for the PCS, as occurrence of flux jumping is suppressed.

As the PCS is thermally insulated, heat due to the self-field A.C. loss cannot be conducted away effectively. The heat, thus generated, increases the winding temperature, resulting in quench current decrease with the current sweep rate.

REFERENCE

(1) Williams, J.E.C., Pourrahimi, S., Iwasa, Y., and Neuringer, L.J., 600 MHZ spectrometer magnet, IEEE Transactions on Magnetics, Vol. 25, No. 2, March 1989, 1767-1770.
(2) Urata, M., and Maeda, H., Stabilization of Superconducting Dry Solenoids, IEEE Transactions on magnetics, Vol. 25, No. 2, March 1989, 1528-1531.
(3) Duchateau, J.L., Turck, B., Krempasky L., and Polak, M., The self-field effect in twisted superconducting composites, Cryogenics, 1976, 97-102.
(4) Ogasawara, T., Yasukochi, K., Nose, S. and Sekizawa, H., Effective resistance of current-carrying superconducting wire in oscillating magnetic fields 1: Single core composite conductor, Cryogenics, 1976, 33.
(5) Maeda, H., Kageyama, S., and Kabashima, S., Stabilization of NbTi persistent current switch, to be published.
(6) Spencer, C.R., Sanger P.A. and Young, M., Temperature and magnetic field dependence of superconducting critical current densities of multifilamentary Nb_3Sn and NbTi wire, IEEE Transactions on Magnetics, Vol. MAG-15, No. 1, JE, 1979, 76-79.
(7) Wilson, M.N., Superconducting Magnets, Clarendon Press Oxford, 1983.

TEST ON SUPERCONDUCTING AC FAULT CURRENT LIMITER

Daisuke Ito,   Eriko S. Yoneda,

Energy Science and Technology Lab.
Toshiba R&D Center
Kawasaki,   Japan

Tsutomu Fujioka

Power and Fusion Technology Development Department
Toshiba Corporation, Tokyo, Japan

Kazuyuki Tsurunaga

Fuchu Works
Toshiba Corporation, Tokyo, Japan

ABSTRACT

The authors have developed a new superconducting fault current limiter whose impedance during normal operation is very small. During fault conditions, the limiter behaves as a superconducting reactor. The limiter consists of a superconducting limiting coil and a superconducting trigger coil. The former coil has a larger critical current than the latter coil. These coils are wound non-inductively on coaxial cylindrical formers and are connected in parallel to each other. They are wound with AC superconductor having ultra-fine NbTi filaments. The limiter has a very little impedance, because both coils are wound non-inductively and in superconducting state during normal operation. On the other hand, in the case of fault conditions, the trigger coil quenches at a critical current. After the trigger coil quenching, the limiter becomes a superconducting reactor, because non-inductiveness is broken by trigger coil current. The fault current is, therefore, limited by the superconducting limiting coil to a certain value determined by the coil inductance. In experiments, the authors have succeeded in limiting a fault current level to 200A, with a limiter whose terminal voltage under limiting conditions was 54V.

INTRODUCTION

In the 1980s, several superconducting fault current limiters were proposed[1-5]. Among their designs, the NEI, Toshiba, and Marcoussis limiters were actually developed and tested[1,5,6]. In these limiters, Toshiba and Marcoussis limiters have non-inductive windings wound with ultra-fine filament superconductors.

The superconducting fault current limiters are not only compact in size, very low impedance, and low AC losses during non faulted operation, but have submilli-second response time under faulty conditions. In addition, in the limiting condition, an arbitrary limiter impedance can be selected and liquid helium consumption can be small for the Toshiba limiter, because the fault current is impeded, not by resistive but by inductive impedance by a superconducting reactor.

A small model of such a limiter was developed to demonstrate the concept. In this article, experimental results concerning the model are described.

STRUCTURE

A diagram of the limiter structure is shown in Fig. 1. The limiter consists essentially of two layer coils. Individual coils are wound in opposite winding directions and the parallel connection of the coil results in double layer non-inductive windings.

Both coils have nearly equal inductance, as indicated in Table 1, and the apparent total inductance for the limiter is 23µH. The outer layer coil and the inner layer coil are labeled limiting coil and trigger coil, respectively, in the following. The coils are wound with different critical current AC superconductors.

The trigger coil is wound with a superconducting strand, whose cross-section and parameters are shown in Fig.2 and Table 2. Its field dependence of critical current is shown in Fig.3.

TABLE 1 Limiting coil and trigger coil parameters.

	R.T.	77K	4.2K
Resistance [Ω]			
Limiting coil	15.5	6.5	0
Trigger coil	619	441	0
Inductance[µH] (at 120 Hz)			
Limiting coil	767		768
Trigger coil			731
Total inductance [µH]			23

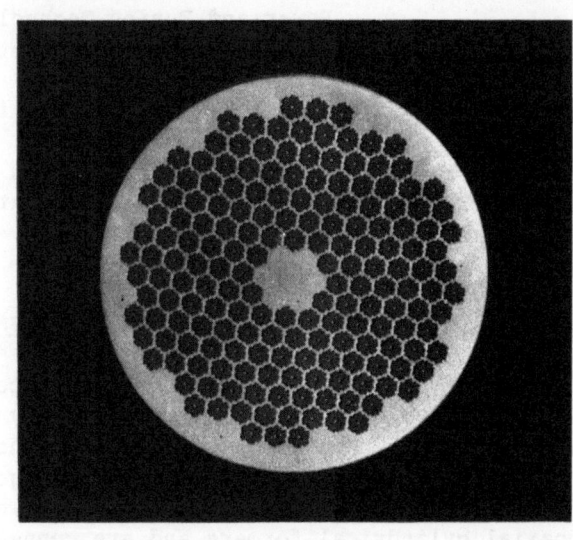

FIGURE 2 Trigger coil superconductor cross-section.

TABLE 2 Trigger coil conductor parameters.

Wire diameter	0.103mm
Filament number	9720
Filament diameter	0.44µm
Filament twist pitch	1.13mm
Matritix ratio NbTi:Cu:CuNi	1:0.8:3.7
Strand twist pitch	3.7mm

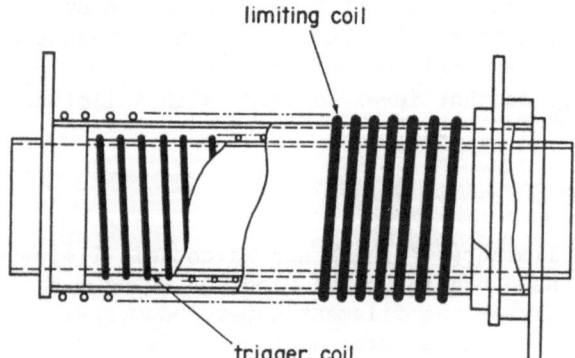

FIGURE 1 Current limiter diagram.

The coil has 78mm diameter, 300mm length and 200 turn winding.

The limiting coil is wound with 42 strand cable, whose conductor parameters are listed in Table 3. This conductor, whose critical current near zero field is twenty times larger than the trigger coil strand, is the same kind of cable as that used to wind the 500kVA coil[7]. The conductor can transport the maximum fault current I_{fmax} = 400 A without resistance. The coil has 80mm diameter, 300mm length and 200 turn winding.

There is a thin fiber reinforced plastic layer between coils in order to insulate them thermally and electrically from each other.

Both conductors were manufactured by Showa Electric Wire & Cable Co.,Ltd.

OPERATION PRINCIPLE

Since both coils are wound non-inductively with low AC loss superconductors, the limiter impedance in superconducting state is very small. This limiter is connected in series with a power network system.

During normal operation, both coils carry nearly equal currents and generate opposite directional field. These fields cancel each other. Therefore, the limiter has very small impedance, due to residual inductance and approximately zero resistance. At a 20 ampere rated current, voltage drop across the limiter, due to the 23 µH residual inductance listed in Table 1, is 0.14 volts.

Under fault conditions, the trigger coil quenches when the fault current exceeds the critical current. Since normal zone propagation velocity is several km per second[6], the trigger coil current falls instantaneously, i.e. within a submilli-second, to a very small value, defined by the conductor matrix resistivity. At that very instant, the' limiter has an inductance whose value is equal to that of the limiting coil, which is still kept in superconducting state. Therefore, the fault current limiter can limit the fault current to a current value determined by the limiting coil inductance L_{lim} as a reactor.

EXPERIMENT AND RESULTS

The limiter CLE, whose specifications are shown in Table 4, was tested in the circuit shown in Fig.4. The object was to show that a device working on the principle described in the previous section is feasible and to demonstrate its performance feature.

TABLE 3 Limiting coil conductor parameters.

Wire diameter	(mm)	0.112
Number of filaments		14478
Filament diameter	(µm)	0.49
Filament twist pitch	(mm)	0.98
Matrix ratio NbTi : Cu : CuNi		1 : 0.1 : 2.5
Strand twist pitch	(mm)	3.05
Subcable twist pitch	(mm)	6.7

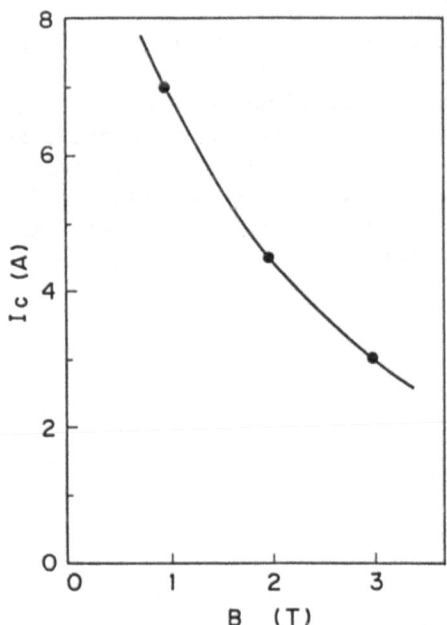

FIGERE 3 Critical current for trigger coil superconductor.

The CLE was in series with a switch S_1, load impedance Z=9.8 ohm and simulated line impedance Z_1=0.1 ohm. In order to simulate a fault current condition, Z was short-circuited with switch S_2, after S_1 was connected to a 100 volt power supply.

Transient phenomena during the fault condition, recorded with digital memory oscilloscopes, are shown in Figs.5 a),b),c) and d).

At the moment of the short circuit, terminal voltage e_0, for the CLE shown in Fig.5 a), dropped from 100 volts to 55 volts according to large internal impedance in induction voltage regulator as a power supply.

Before S_2 closing, terminal voltages for limiting coil e_1 and trigger coil e_2 are approximately zero, as shown in Figs. 5 c) and d), because the limiter's residual inductance is very small.

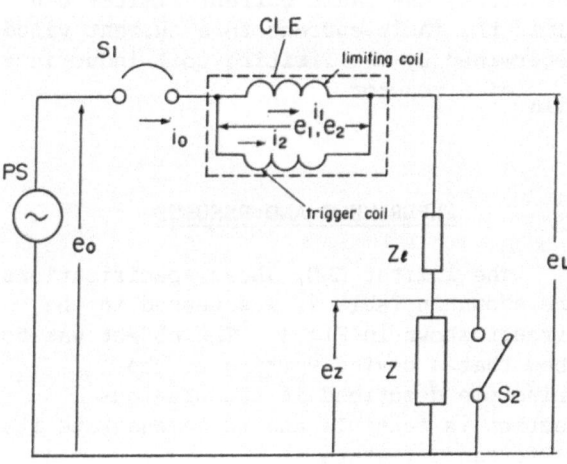

FIGERE 4 Circuit used for limiter test.

TABLE 4. Limiter specification.

Rated voltage	100	V
Frequency	50	Hz
Normal rated current	20	A
Fault current level I_f	56	A
Maximum fault current I_{fmax}	400 A ~ 80 A	
Fault impedance L_{lim}	0.75	mH
Normal voltage drop across device	0.14	V

At the moment S_2 closes, trigger coil current i_2 increases steeply from 5 amps to a 28 amp peak current. That is followed by rapid decreasing to 0.7 amps during time period T_{on} within a sub-millisecond. In order to study the trigger coil current variation more precisely, magnified time scale results are shown in Fig.6. These results indicate that the T_{on} value is 0.5 msec.

After i_2 rapid decreasing, the i_2 to e_2 phase angle becomes approximately zero. The result indicates that the winding becomes resistive. Apparent trigger coil resistance Rt is 78.6 ohms, i.e. Rt = e_0/i_2 = 55/0.7. Based on the Rt value and the 77 K resistance value, 441 ohms, as shown in Table 1, normal zone length after trigger coil quenting can be estimated to be more than 9 m. Normal zone propagating velocity v_p is , therefore, roughly estimated as approximately 18,000 to 50,000 m/sec. This value is extremely large, compared with author's obtained result[7].

The difference might be due to high quench voltage between the trigger coil and large current density as shown in the following.   The 28 amp peak current corresponds to critical current Ic_2 for the strand near zero field. Critical current density, 1.9×10^{10} A/m^2, is very large still the A.C. superconductor has very thin filaments.

Limiting coil current i_1 must be nearly equal to i_2 before trigger coil quenching. Fault current level I_f, over which the CLE behaves as a reactor, is 56 amps. This value can be controllable to an arbitrary value, by varying the NbTi amount in the superconducting wire for the trigger coil.

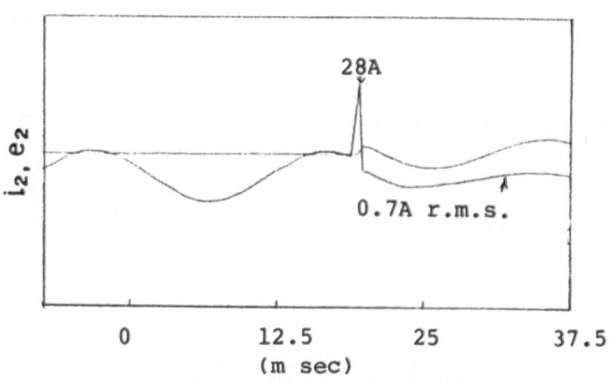

Figure 6 Trigger coil current and voltage variation.

453

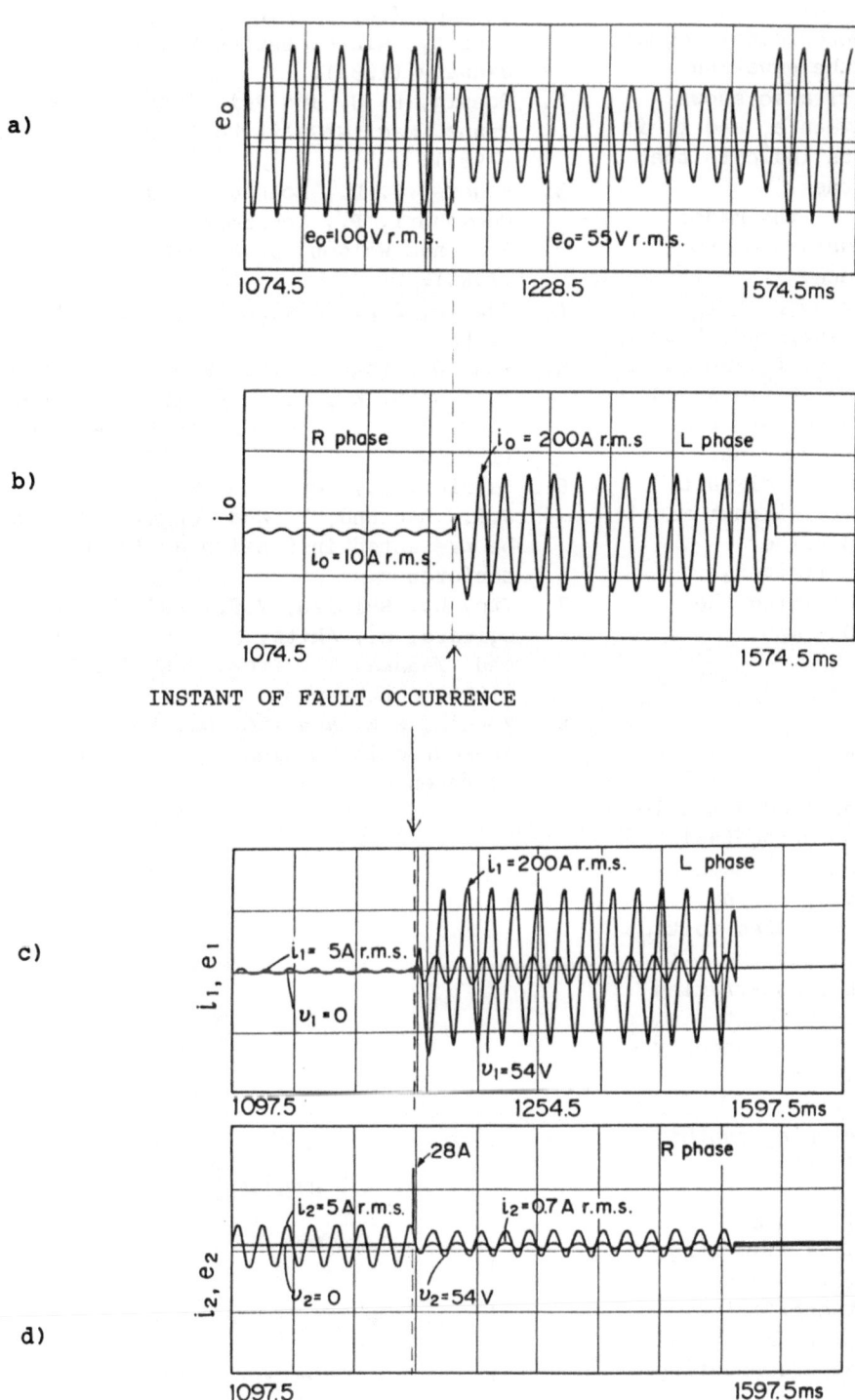

FIGURE 5 Transient behavior of limiter at 50 Hz operation. a) Terminal voltage e_0.
b) Total current i_0. c) Limiting coil voltage e_1 and current i_1. d) Trigger coil
voltage e_2 and current i_2.

As shown in Fig. 5 c), limiting coil current i_1 increases from 5 amps to 200 amps after S_2 closing. Terminal voltage e_1 also increases from 0 volt to 55 volts. This value is approximately equal to that calculated with the equation $e_1 = i_1 \times L_{lim}$. Figure 5 c) also shows that the phase angle between e_1 and i_1 is 90 degrees. These results indicate that the limiting coil is in the superconducting state, i.e. the major coil impedance mainly consists of no resistance, only inductance.

In Fig.5 b), limiter current i_0, i.e. total i_1 and i_2 current, variation is shown. After closing S_2, i_0 increases from 10 amps, i.e. the sum of i_1=5A and i_2=5A, to 200 amps, which is the same value as the limiting coil current.

If the CLE is not in the circuit, fault current after a short circuit might be increased to a 550 amp value, determined by e_0/Z_1. The limiter, therefore, succeeded in limiting the fault current value to 200 amps.

CONCLUSION

The authors have developed a single phase model of a superconducting fault current limiter, whose rated voltage is 100 volts and whose rated operating current is 56 amperes. The rated voltage test could not take place, but the authors have succeeded in demonstrating that the limiter can limit a fault current to 200 amps inductively under the 55 volt terminal voltage, with sub-milisecond operation time.

The authors are developing another kind of superconducting limiter, consisting of a double layer non-inductive trigger coil and a superconducting reactor ,i.e. limiting coil, connected in pararell as shown in Fig. 7. The limiter characteristics, including the trigger coil recovery, with a switch from normal to superconducting state, are being tested.

ACKNOWLEDGEMENT

The authors thank Mr. Y. Sugiyama of Utility Power System Engineering Div. of Toshiba Corporation and Mr. T. Kuriyama of R & D Center of Toshiba Corporation for their helpful discussions.

REFERENCES

1. Raji, B. P. , Parton, K.C. and Batram, T.C. , IEEE Power Engineering Society, Wintermeeting New Year January,(1982).
2. Boenig, H. J. and Paice, D. A., IEEE Trans. Mag. MAG-19, No.3,1051 (1983).
3. Rogers, J. D., Boenig, H. J., Chowdhuri, P., Schermer, R. I, Wollan J.J. and Weldon, D. M., ibid.,1054 (1983).
4. Sabrie, J.L., Alsthom Review, No.5 (1986)
5. Ito, D., Yoneda E.S., Fujioka, T. and Tsurunaga, K., to be published in Adv. Cryo. Eng. , Proc. CEC in Los Angeles (1989).
6. Fevrier, A., Verhaege, T., Tavergnier, J.P., Laumond, Y. and Bekhaled M., to be presented in Session JC-12 in this conference.
7 Ito, D., Shimizu, Y.S., Fujioka, T., Ogiwara, H., Akita, S., Ishikawa T. and Tanaka, T., Proc. ICEC-12, 719 (1988) Southampton.
8 Yoneda, S.E. and Ito, D., to be presented in Session LB-05 in this conference.

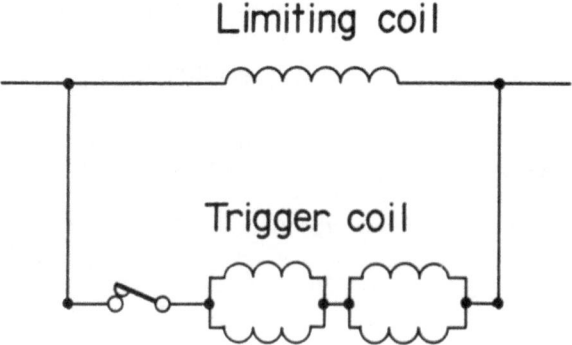

FIGURE 7. 2nd phase model fault current limiter.

OFF-ON CHARACTERISTICS OF MAGNETICALLY CONTROLLED SUPERCONDUCTING SWITCH

T. Nitta*, M. Tada*, T. Okada* and S. Isojima**
*Department of Elect. Eng., Kyoto University, Kyoto, 606, Japan
**Sumitomo Electric Industries, Ltd., Osaka, 554, Japan.

ABSTRACT

This paper describes structure of an experimental switch, the experimental results and the theoretical approach for the off-on characteristics of the switch. The experimental magnetically controlled superconducting switch, which is composed of superconducting gate wire (NbZr/CuNi) and superconducting field coil (NbTi/CuNi), has been made. Some experiments on the switching characteristics of the switch have been carried out. The experimental results show special characteristics of switching: 1)at failure of switching-on, the gate current at zero magnetic induction is independent of the initial current and is constant and 2)the gate current at the off-on transient state, that is, from the normal state to the superconducting state of the superconducting gate wire, does not depend on the voltage across the gate wire, but depends uniquely on the magnetic field around the gate wire. The gate current can be explained to be a minimum normal-zone propagation current given by the equal-area theorem on cryogenic stabilization.

INTRODUCTION

Superconducting switches by use of transition between superconducting state and normal state of superconductor have been examined. Two type of the switches, that is, thermally one and magnetically controlled one have been considered. In this paper, we examine characteristics of magnetical switches. We have many papers on them and their applications presented mainly by Twente University.[1] However, the switching characteristics may not be discussed in detail. This paper describes experimental results and some considerations on switching characteristics, especially, off-on characteristics of magnetically controlled switch.

MAGNETICALLY CONTROLLED SUPERCONDUCTING SWITCH

A magnetically operated super-conducting switch was made. A gate wire which is of superconducting wire has two states of on and off corresponding to transition of superconducting state and normal one, respectively. The items required for the gate wire are 1) large resistance in the normal state, 2)low critical magnetic field Bc2, 3)high current density, and 4) high critical temperature. After surveying super-conductor fitted to the above items, Nb-1%Zr which is easy to get was selected. The specifications of the gate wire is shown in Table 1. The gate wire, the length of which is 5 m, is on the inner coil fixed by epoxy resin. The inductance is $2.6*10^{-6}$ H.

The field coil which generates magnetic flux is requested to have properties of low loss at change magnetic flux, low self inductance and homogeneous magnetic flux generation. The double solenoid was chosen for the configuration. Tables 2 and 3 show the specifications of wire and configurations of the field coil, respectively. The inductance of the inner and the outer coils are calculated to be 1.8 mH and 10.0 mH, respectively. The total inductance of the two coil is calculated to be 7.1 mH. The magnetic induction in the switch volume for the field current of 1 A is calculated to be 15.04 mT.

EXPERIMENT FOR SWITCHING CHARACTERISTICS

The following two experiments were

Table 1 Specification of gate wire

material superconductor	Nb-1%Zr
matrix	70Cu30Ni
matrix/superconductor	1.5:1
No. of filaments	1369
diameter of wire (with insulation)	0.355 mm
diameter of filament	0.006 mm
resistivity along wire	$3.16*10^{-8}$ ohm*m
critical current	100 A
critical magnetic induction	0.74 T
critical temperature	about 9.3 K

Table 2 Specification of wire for field coil

material superconductor	Nb-50%Ti
matrix	70Cu30Ni
matrix/superconductor	1.08:1
No. of filaments	1369
diameter of wire (with insulation)	0.66 mm
diameter of filament	$11.3*10^{-6}$ m
twist pitch	11.5 mm
critical current (B=1.5 T,T=4.2 K)	100 A
critical magnetic induction	about 10 T
critical temperature	about 10 K

Table 3 Specification of field coil

inner coil
inner diameter	11.8 mm
outer diameter	23.5 mm
length	60.0 mm
No. of turns	704

outer coil
inner diameter	30.0 mm
outer diameter	43.1 mm
length	60.0 mm
No. of turns	796
switch volume	16.4 cm2

carried out. 1) Experiments, where the gate current is constant. 2)Experiment for switching characteristics.

-Constant Gate Current Experiment

The experimental results for the constant gate current are shown in Figs. 1 and 2, which show change of the gate-resistance at turn-off and turn-on due to the field current, respectively. The results lead to
1) the off-resistance is 2.1 ohm. It agrees with that of the short-sample test, and
2) gate resistance exhibits hysteresis with varing field current at turn-on and turn-off.

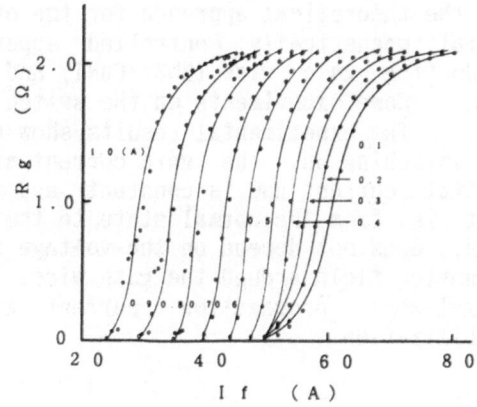

Fig.1 Resistance of gate wire at turn-on of constant gate current experiments (when field current decreases)

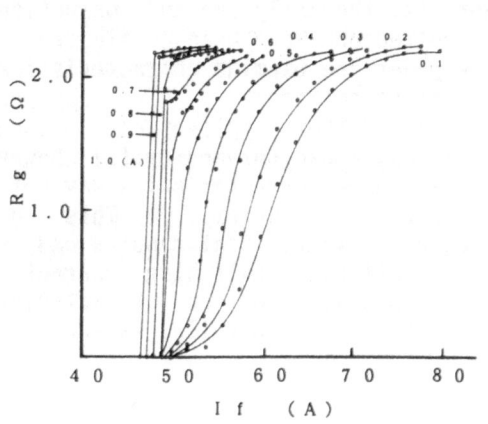

Fig.2 Resistance of gate wire at turn-off of constant gate current experiments(when field current increases)

-Transient Switching Characteristics

The experimental circuit is shown in Fig.3. The resistor RL of the load is paralleled to the switch. The current source is paralleled, too. The circuit is equivalent to one, where the load resistor and the voltage source are connected in series to the switch. The resistor r is the resistance for the current lead and current sensor. The resistance of the load resistor RL must be much lower than the off-resistance of the switch(2.1 ohm). It is selected to be 0.1 ohm.

In the experiments, at first, the gate, where an initial current flows, is in on-state. By increasing field current as shown in Fig. 4, the gate current decreases and the gate turns off. After that, by decreasing the field current, the gate current increases and the gate turns on in some cases. In other cases, the gate does not turn off, and the gate current for the field current of 0 A is less than the initial current. The variation of the field current is 3.33 A/sec.

The results are shown in Fig.4. They lead us to:
1) The field current for turn-off, which is independent of the initial gate currents, is about 45 to 48 A.
2) When the initial gate current is less than 1.7 A, turn-off is possible.
3) When turn-off is failed, the final gate current, that is, the gate current when the field current becomes zero, again, is about 1.5 A. It is independent of the initial gate current.
4) In the process of turn-off, when the field current is less than 50 A, the gate current is a constant value for the field current and independent of the initial current.

The other experiments, where resistance of the load resistor was changed or the variations of the field current was changed, were carried out. The characteristics of gate currents at turn-on are the same that those of the above mentioned experiments. The experiments for switching in steady states, that is, the variation of the field current is very small, were carried out. The results during turn-on is shown in Fig.5. The gate current is independent of the initial current and depends uniquely on

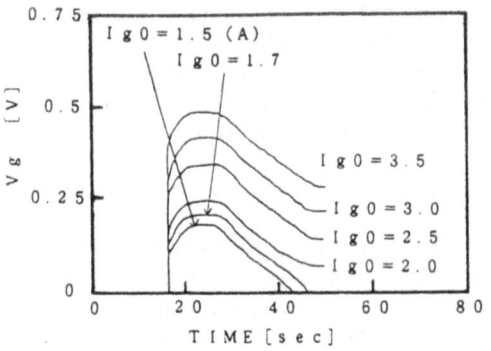

(a) Gate voltage

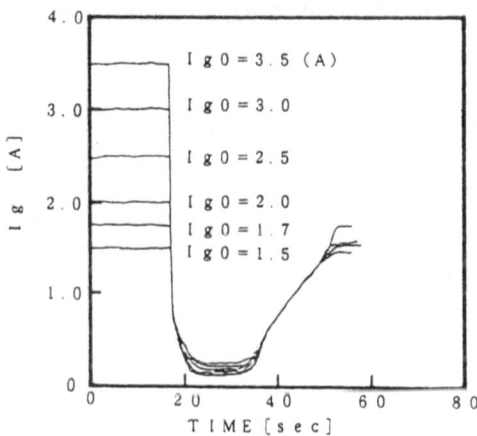

(b) Gate current

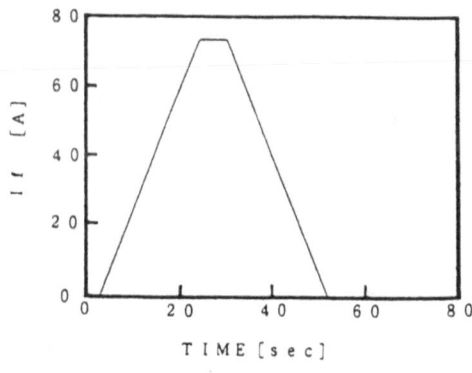

(c) Field current

Fig.4 Switching transient characteristics

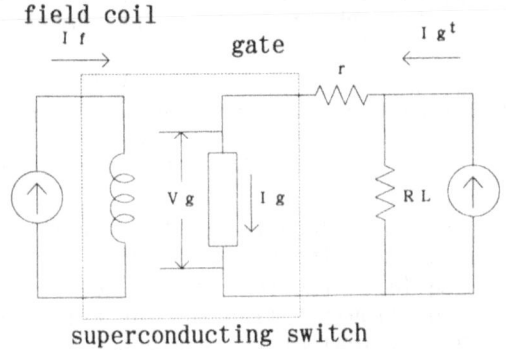

Fig.3 Circuit for switching characteristics

the filed current, that is, magnetic induction around the gate wire. Figure 5 shows that for example, in case of Ig0=1.0 A, the gate current Ig becomes 1.0 A when the field current If becomes 30 A, and turns on.

THEORETICAL APPROACH TO TURN-ON CHARACTERISTICS

Theoretical approach to turn-on characteristics of the switch, especially the constant current characteristics (see the conclusion 3)and 4) in the previous section), is tried.

-Critical Characteristic of Gate Wire

The critical characteristics of the gate wire, that is, the relation among the temperature, the current and magnetic induction at the transition of normal state and superconducting one is derived.

The relation between critical magnetic induction Bc and critical temperature Tc is approximated experimentary by

$$B_c(T) = B_{c\theta} \left\{ 1 - (T/T_{c\theta})^2 \right\} \qquad (1)$$

Bc0: critical magnetic induction at I=0 and T=0.
Tc0: critical temperature at B=0, I=0.

The relation between critical current Ic and critical temperature Tc is approximated with the equation

$$I_c(T) = I_{c\theta}(1 - T/T_{c\theta}) \qquad (2)$$

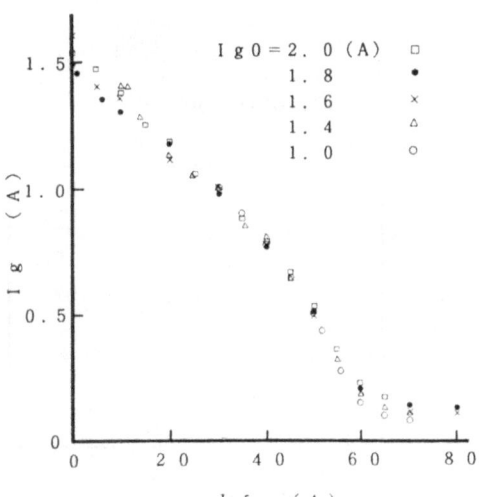

Fig.5 Switching characteristics at turn-on in steady state.

Ic0: critical current at B=0 and T=0.

The relation between the critical current and the critical magnetic induction Bc at temperature T is approximately given by

$$I_c(T, B) = I_c(T, 0) \left\{ 1 - B/B_c(T, 0) \right\} \qquad (3)$$

From the above three equations, we obtain the approximate equation which expresses the relation among the critical current, temperature and magnetic induction as

$$I_c(T, B) = I_{c\theta} \left\{ 1 - \frac{T}{T_{c\theta}} - \frac{B}{B_{c\theta}(1 + T/T_{c\theta})} \right\} \qquad (4)$$

The values of Ic0, Tc0 and Bc0 are obtained by the specification of the gate wire. The critical temperature Tc0 is given by Tc0=9.3 K, which is the critical temperature for Nb. (The critical temperature for Nb-1%Zr could not be investigated.) The critical magnetic induction at I=1.0 A and T=4.2 K is estimated to be 0.74 T by the experiments for constant gate current characteristics. Then we obtain Bc0=0.917 T. The experiments lead us to the critical current Ic=100 A at T=4.2 K and B=0.4 T. Then we obtain Ic0=403 A.

-Transient State at Turn-on

The transient state at turn on is considered by use of thermal equilibrium state between the gate wire and liquid helium, the cooling characteristics of liquid helium, and the critical characteristics of the gate wire. The following two states may be estimated:
1) The whole gate wire is in the resistive state. The resistance is changed by the magnetic induction around the gate wire.
2) The part of the gate wire is in the normal state and the other part is in the superconducting state. The resistive area is changed by the magnetic induction.

Consider which state is valid. The heat flux q of liquid helium is given by

$$q = 0.12 + 0.022 \Delta T \quad (10^4 W/m^2) \qquad (5)$$

From the above equation, the experimental results for the constant gate current, and Eq.(4), the temperature difference between liquid helium and the gate wire at the critical state is given by 2.9 K. The heat flux of liquid helium is given to be 1800 W/m^2.

The heat produced in the gate wire is 0.05 W just before turn-on since the gate current is 1.0 A and the resistance is 0.05 ohm. (see the results of the experiment of the constant gate current experiments.) Consider the heat per unit surface. By corresponding to the above two states, the surface S for cooling are, respectively:

1) S=P*L, P: cooling perimeter, L: the length of the gate wire.

2) S=P*L/2.1, since the resistive zone (normal zone) is considered to be changed in proportional to the resistance.

The numerical results for both cases are

1) 20 W/m^2 and

2) 840 W/m^2,

respectively. Then in comparison with the heat flux of liquid helium of 1800 W/m2, Case 2) may be valid. The heat produced is less than the heat flux of liquid helium. It is because the heat conduction along the the gate wire is neglected in the above consideration. It is concluded that in turn-on process, part of the gate wire is in normal state and that the part become smaller for the smaller of magnetic induction.

-Minimum Propagation Current

The experimental results show that we have cases where the gate wire cannot be completely turned on. The final current in the cases is about 1.5 A which is independent of the initial current. From the consideration of the previous section, the normal zone exists partially. Therefore, by use of the equal area theorem, the above phenomena can be considered.

-The Equal Area Theorem

The condition for equilibrium is give by

$$\int_{T_b}^{T_N} K(T)\{W(T)-G(T)\} \, dT = 0 \qquad (6)$$

where
K(T):thermal conductivity along the gate wire.[W/m/K]
W(T): heat transfer of liquid helium. [W/m^3]
G(T): heat generation. [W/m^3]
TN: highest temperature
Tb:temperature of liquid helium.

When the current that is satisfied with the above equation flows, the normal zone cannot grow nor shrink. Then the current is called the minimum propagation current. The current 1.5 A of the experimental results, which is the final current at failure of turn-on, is considered to be the minimum propagation current.

-Final Current at Failure of Turn-on

The heat generation is given by

$$\begin{cases} G(T)=0 & (W/m^3) \quad (T<T_c) \\ \quad\quad = \rho\,(I/A)^2 & (T \geq T_c) \end{cases} \qquad (7)$$

where A:sectional area, ρ is the resistivity of the gate wire in the normal state at 4.2 K.

From Eq.(4), the critical temperature is given by a function of the gate current as

$$T=9.3 \times \left(1 - \frac{I}{403}\right) \qquad (8)$$

The heat transfer in case where the gate wire is directly in contact with liquid helium is given by

$$W(T) = \frac{q \times P}{A} \qquad (W/m^2) \qquad (9)$$

where P denotes a cooling perimeter and q is heat flux.

In our case, however, the gate wire is covered by epoxy resin. The heat transfer of epoxy resin must be taken in account. Then we have

$$w(T)=\int_{T_b}^{T} \frac{P}{dA} \lambda_0 T^a \, dT \qquad (10)$$

$$= \frac{\lambda_0 P}{(a+1)dA} \{T^{(a+1)}-T_b^{(a+1)}\}$$

where d is the width of epoxy resin, and $\lambda_0 T^a$ is a function which expresses heat transfer of epoxy resin.[2]

Assumed that heat conduction along the gate wire is given by

$$K(T)=\lambda_{w0}T^K \qquad (11)$$

we have a solution of Eq.6,

$$\frac{\lambda_0 P}{(a+1)dA}\left(\frac{T_N^{(a+k+2)}-T_b^{(a+k+2)}}{a+K+2} - \frac{T_b^{(a+1)}}{K+1}\right.$$

$$\left. \times \{T_N^{(k+1)}-T_b^{(k+1)}\} \right)$$

$$= \frac{\rho\,(I/A)^2}{K+1}\{T_N^{(k+1)}-T_c^{(k+1)}\}$$

$$(12)$$

If thermal conductivity K(T) is independent of the temperature T, we have

$$\frac{\lambda_\emptyset P}{(a+1)dA}\left(\frac{T_N^{(a+2)}-T_b^{(a+2)}}{a+2}-T_b^{(a+1)}(T_N-T_b)\right)$$

$$=\rho\,(I/A)^2(T_N-T_C) \qquad (13)$$

From $G(TN)=W(TN)$, we have

$$T_N=\left(T_b^{(a+1)}+\frac{\rho\,I^2(a+1)d}{\lambda_\emptyset PA}\right)^{1/(a+1)} \qquad (14)$$

From the above equation and the property of the gate wire shown in Tables 1 and 4, we obtain the minimum propagation current at zero magnetic induction,

$$Imin=1.66\ A \qquad (15)$$

where the cooling perimeter is assumed to be half of the perimeter. The constant for the epoxy resin is the same that those in reference [2]. The width d is assumed to be a value in Table 4. The minimum propagation current depends on the width d and the cooling perimeter. The calculated value of Imin is close to the experimental value of 1.5 A. Then the final current, if the gate current cannot be turned on, is considered to be the minimum propagation current.

-Relation Between Gate Current and Field Current During Turn on Process

During the turn-on process, the phenomenon that the gate current depends on only the field current can be explained by the above theory. By using Eq.4 in place of Eq.8, the current during turn-on process can be calculated. The calculated value is shown in Fig.6 with the experimental values of the switching in the steady state. The calculated values are close to the experimental one when the field current less than 40 A. When the field current is larger than 40 A, the assumption of the cool end, may not be valid.

Table 4 Values for calculation

ρ mean value of resistance of gate wire along wire	$3.2*10^{-8}$	ohm*m
A sectional area	$7.6*10^{-8}$	m2
P cooling perimeter	$4.9*10^{-4}$	m
a, $\lambda_\emptyset$ constant for heat conduction of epoxy resin	1.8 $5.1*10^{-3}$	
d width of epoxy resin	$8.0*10^{-4}$	

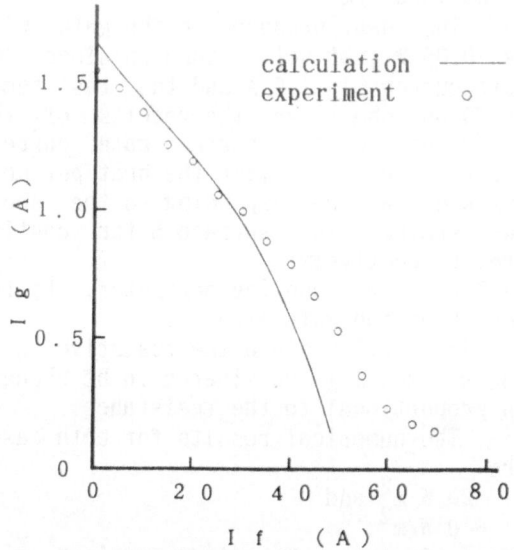

Fig.6 Minimum propagation current (calculation) with comparison to experimental results of switching characteristics in steady state.(Ig0=2.0 A and RL=0.1 Ohm).

CONCLUSION

On. a magnetically controlled superconducting switch, some experiments and some considerations are performed. Especially, the turn-on process is considered. The current during turn-on process is concluded to be the minimum propagation current. Therefore, in order to have a high quality superconducting switch, both of the minimum propagation current and the resistance of the gate wire must be larger.

References

[1] For example, G.B.J. Mulder and et.al.,"A FAST OPERATING MAGNETICLLY CONTROLLED SWITCH FOR 1 kA", IEEE Trans. on MAG, Vol.21, No.2, 1985.
[2] G.B.J. Mulder and et.al.,"EXPERIMENTAL RESULTS OF THERMALLY CONTROLLED SUPERCONDUCTING SWITCHES FOR HIGH FREQUENCY OPERATION", IEEE Trans. on MAG, Vol.24, No.2, 1988.
[3] M.N. Wilson, "Superconducting Magnets", Clarendon Press, Oxford, 1983.

This work is supported in part by the Ministry of Education, Science and Culture of Japan.

OPERATION OF A NEW-TYPE RECTIFIER FLUXPUMP
WITH SATURABLE CORE TRANSFORMER

K. Funayama, T. Isono, M. Suzuki, M. Sato and T. Anayama

Dept. of Elect. Eng., Tohoku University, Sendai 980, Japan.

ABSTRACT

A new-type transformer-rectifier fluxpump has been constructed and tested. This fluxpump consists of a power and control unit, a superconducting transformer with a saturable iron core, two superconducting switches and a load magnet. High efficiency and constant increasing (decreasing) rate of load current have been accomplished by using a saturable core transformer. The basic principle of this fluxpump is given and the sequence operation of charge, hold and discharge are presented. On the basis of experimental results, the loss and efficiency of this fluxpump are discussed.

INTRODUCTION

The transformer rectifier fluxpump is known as the power supply apparatus of charging superconducting magnets with a small ac current.[1,2,3,4,5] In this apparatus, the input ac current with a small amplitude is converted to a high dc current flowing in a superconducting magnet by a transformer and two switches. The ratio of the secondary current to the primary one of transformer ranges from ~10 to ~100 and the diameter of the input power lead can be appreciably reduced. This yields a much reduction in the thermal loss which takes place at the power leads, and makes the capacity of the power supply apparatus very low.

Recently, we have made a new-type rectifier fluxpump [4] where the superconducting transformer with a saturable core is substituted for the frequently used air-core one. This fluxpump

offers a high pumping current and maintains a constant increasing (or decreasing) rate of load current.

In this report, components of the fluxpump such as a power and control unit, a transformer and two superconducting switches are presented, and the method of operating the fluxpump and experimental results are presented. Further, the total loss and energy efficiency involved in charging the load magnet up to its allowable current are examined.

BASIC PRINCIPLE OF THE PRESENT FLUXPUMP

The circuit of a fullwave transformer-rectifier fluxpump with a saturable core transformer, which has been tested in this study is given in Fig.1. Here, L is a superconducting magnet (load), SW is superconducting switch and ST is a superconducting transformer with a saturable core. The operation of the fullwave transformer-rectifier fluxpumps is principally divided into two regions, namely a pumping region and a commutation region. The two regions run during half-cycle and yield an increase in load current, I_L.

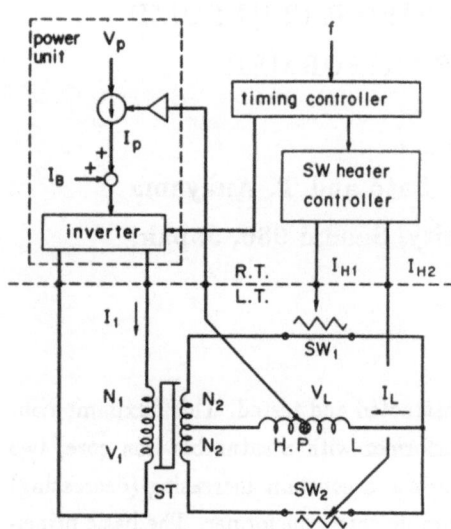

Fig. 1 Circuit of a present fullwave transformer rectifier fluxpump.

The current increase during each half-cycle, ΔI_L, is given by

$$\Delta I_L = \frac{\Delta \Phi_L}{L} , \qquad (1)$$

where L is the self inductance of the load magnet, and $\Delta \Phi_L$ is an increase of the magnetic flux, Φ_L, in the load magnet. Meanwhile, in transformer-rectifier fluxpumps where the load current from one loop to the other loop is inductively transferred.

The $\Delta \Phi_L$ is equal to a change in the flux linkage of the secondary winding of the superconducting transformer and is written by

$$\Delta \Phi_L = 2N_2\phi_m - 2\ell_S I_L , \qquad (2)$$

where, ϕ_m is the maximum flux through the windings of the transformer, N_2 the turns of the secondary winding and ℓ_S the self-inductance of the secondary winding. Here, $2N_2\phi_m$ is usually written [1,2,3] by $2MI_P$ using the mutual-inductance, M, of the transformer and the primary current, I_P. This term is caused by reversing the current of the primary winding $2\ell_S I_S$ is induced by an inductive commutation operation.

The present fluxpump which has a superconducting transformer with an iron core indicating a nearly rectangle loop operates at non-saturation state of the iron core for a pumping region and works at saturation states for a commutation region. Accordingly, ℓ_S becomes very small ($2N_2\phi_m \gg 2\ell_S I_L$) and eq. (2) follows that

$$\Delta \Phi_L = 2N_2\phi_m = 2N_2\phi_S , \qquad (3)$$

where ϕ_S is the saturation flux of the core. eq. (3) indicates that $\Delta \Phi_L$ is constant and independent of I_L.

The charging voltage, v_L, across the load magnet is governed by $v_L = d\Phi_L/dt$. Therefore, the mean voltage, V_L, during each half-cycle is expressed by

$$V_L = \frac{2}{T} \int_0^{\frac{T}{2}} v_L dt = \frac{2\Delta \Phi_L}{T} , \qquad (4)$$

where T is the time required for one cycle.

If the operating frequency $f(= 1/T)$ is introduced, eq. (4) follows that

$$V_L = 2\Delta \Phi_L f = 4N_2\phi_S f . \qquad (5)$$

When the pumping operation is repeated n times to increase the load current from $I_L = 0$, the achievable value of I_L is given using eqs. (1) and (5) by

$$I_L = \frac{\Delta \Phi_L}{L} n = \frac{2N_2\phi_S}{L} n . \qquad (6)$$

Further, using the time, $t(= n/2f)$, required for n pumping operations, eq. (6) is rewritten by

$$I_L = \frac{V_L}{L} t = \frac{4N_2\phi_S f}{L} t . \qquad (6')$$

Load power, P_L, and energy stored in the magnet, W_L, are given by

$$P_L = V_L I_L = \frac{(4N_2 \phi_S f)^2}{L} t , \qquad (7)$$

and

$$W_L = \frac{L I_L^2}{2} = \frac{8(N_2 \phi_S f)^2}{L} t^2 . \qquad (8)$$

As shown in eqs. (6') and (7), the conspicuous feature of this fluxpump is to offer a linear increase in I_L and P_L with t.

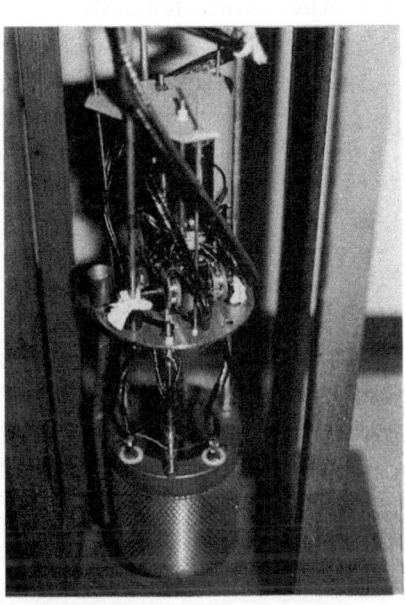

Fig. 2 Photograph of the fluxpump and the load magnet.

COMPONENTS OF THE PRESENT FLUXPUMP

The present fluxpump is made of a power unit, a superconducting transformer (ST) with an iron core, two superconducting switches (SW) and a load magnet (L). Their dimensions and characteristics are summarized in Table 1. Fig.2 shows the photograph of the present fluxpump and the load magnet.

A power unit which supplies a primary current I_1 to the transformer consists of an inverter and a dc constant voltage power source with current limiter. The Hall voltage corresponding to I_L is fed back and the I_1 is electrically controlled to values which satisfies the following equation:

$$I_1 = \pm(I_P + I_B) , \qquad (9)$$

where $I_P = (N_2/N_1)I_L$ and $I_B (<< I_p)$ is a small current capable of exciting the iron core of the transformer. N_1 and N_2 are the turns of the primary and secondary windings, respectively. During discharging of the load magnet, this power unit can acts as a load.

A high saturation magnetization, low loss and rectangular hysteresis loop at 4.2 K are required for the magnetic core of the transformer. In this study, from the results of magnetic measurements for several ferromagnetic materials,

Table 1 Dimensions and characteristics of components of the fluxpump.

superconducting saturable transformer : ST	core : effective cross sectional area S = $4.19 \times 10^{-4} m^2$, magnetic circuit length =0.179 m, saturation flux density Bs = ~2.0 Wb/m2 winding : turns ratio $N_1 : N_2 = 6000 : 180 (= 100/3)$
superconducting magnet ; load : L	size : o.d.: 125 mm, i.d.: 85 mm, l : 137 mm self-inductance L = 1.135 H stored energy = 5.68kJ (at 100 A)
superconducting switch : SW	conductor : Ic= 256 A, Rn =17 Ω, non-inductive winding, thermally switch, heater input per one oparation = 60 mJ
Hall probe : H.P.	F.W.BELL,INC. Hall GENERATOR BHA-921 Hall voltage V_H=4.34 mV/T (at 4.2 K)

the U-shaped cut-core of grain oriented Fe-Si was sellected as a magnetic core. The primary winding was wound from a NbTi FM wire (NbTi : Cu : CuNi =1 : 0.3 : 2.88) of 0.153 mm in diameter. For the secondary winding, a bundle conductor consisting of 7 FM wires was used. Both open-circuit and short-circuit tests on this transformer at 1 Hz yield a capacity of 22 VA and a efficiency of 99.96 %.

Each superconducting switch was made of a NbTi FM wire with CuNi matrix (diameter = 0.311 mm, length = 3.7 m, Ic = 256 A at 1 T) indicating high normal-state resistivity. The FM wire was non-inductivly wound on FRP bobbins (outer diameter = 30 mm, length = 10 mm). A manganin heater wire (diameter = 0.3 mm, length = 0.95 m) was inserted between layers of the FM wire. The heating power required for driving switches to a resistive state is 60 mJ. The resistance at the opration R_S is $\sim 3.5\ \Omega$ by the primary voltage V_1 of 10 V.

The used load magnet is made of a NbTi FM wire. Its self-inductance is 1.135 H and the stored energy is 5.86 kJ at the allowable current, 100 A.

OPERATION FOR ENERGYING THE LOAD MAGNET

Fig. 3 shows waveforms of the primary current I_1, the primary voltage V_1 and the load voltage V_L obtained in charging the load magnet at the operating frequency of 0.4 Hz. The commutation and pumping operations of the fluxpump during half- cycle are found in this figure.

The commutating and pumping operations are performed in the following manner. After one pumping operation in the upper loop of the secondary circuit (see Fig. 1), SW2 is closed. Then, the iron core of the transformer is in one saturation magnetization state and the current commutation to the lower loop occurs by reversing I_1. As shown in Fig. 3, the impulse voltage

(at time A) for V_1 indicates a commutating operation. Next pumping operation is simply caused by opening SW1. By this operation on SW1, the magnetization of the iron core moves from one saturation state to the other and the resultant flux change provides an incease (or decrease) in the load current.

The pumping action continues during period BC (see Fig. 3) and V_1 keep a constant. Subsequently, next current commutation to the upper loop is performed at time A', when the iron core is in the other saturation state.

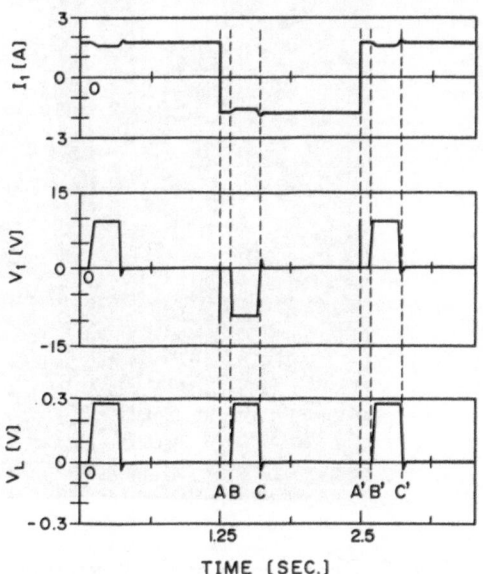

Fig. 3 Primary current I_1, primary voltage V_1 and load voltage V_L of the fluxpump during the load magnet.

A typical result on the sequence operation of charge, hold and discharge is given in Fig.4. The load current, I_L, was increased up to $\sim$100 A at a constant charging rate, as expected from eq. (6'). The operating frequency, f is 0.2 Hz and the amplitude of V_1 is 10 V. The amplitude of I_1 is adjusted according to eq. (9), namely $I_1 = 0.03I_L + I_B$ [A]. Here, the magnitude of I_B is 0.05 A, and its positive (negative) sign yields a charge (discharge) mode. A hold mode was achieved by stopping the operation on switches.

The maximum operating frequency is approximately 1 Hz and limited by the recovery time of switches from a resistive state.

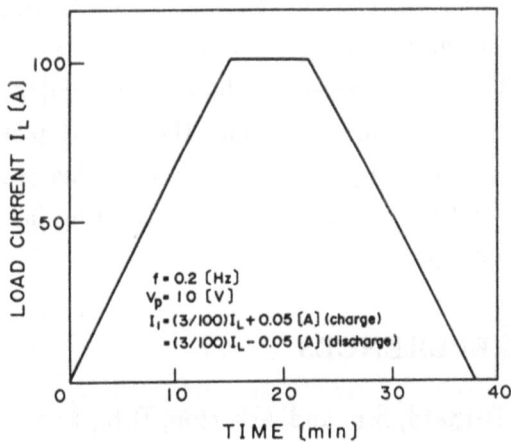

Fig. 4 Variation of the load current I_L during the sequence operation of charge, hold and discharge.

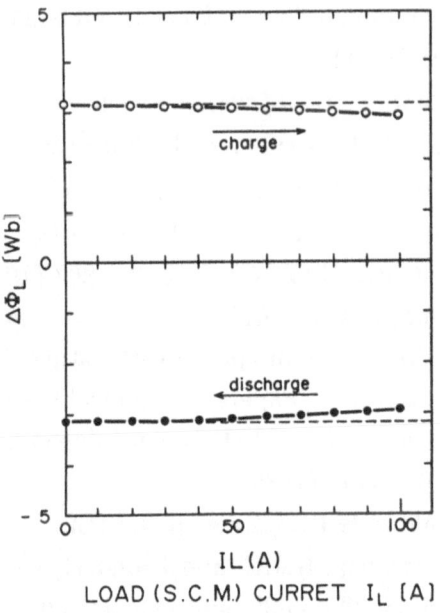

Fig. 5 Dependence of the pumping flux $\Delta\Phi_L$ on the load current I_L. The broken lines indicate a pumping flux estimated from the saturation magnetic flux density B_S of the iron core.

Fig. 5 shows relations between $\Delta\Phi_L$ and I_L for this operation. The value of Φ_L was measured with a Hall prove. The $\Delta\Phi_L$ keep a constant in the entire range of I_L and is in good accord with the value, $(2N_2 B_S S =) 0.316$ Wb, estimated from the saturation magnetization of iron core (see Table 1).

LOSSES AND EFFICIENCY

Energy efficiency η_E of fluxpump is defined as

$$\eta_E = \frac{W_L}{W_I + W_{SH}} \times 100[\%] , \qquad (10)$$

where W_L is the energy stored in the load magnet, W_I the input energy introduced through the transformer, W_{SH} the energy required for driving thermally switches.

The input and output energies involved in charging the load magnet from $I_L = 0$ to 100 A have been electrically measured under the same operation conditions as used in Fig. 3. The driving energy for switches has been also obtained by measuring the power input $(= VI)$ to the heaters. The results obtained are as follows :the number of pumping operations is 377, $W_L = 5684$ J and $W_I = 5734$ J. Since the heating energy required for opening a switch during each pumping operation is $\sim$60 mJ, 377 pumping operations yield $\sim$23 J for W_{SH}. From the results of these energies, the value of η_E becomes 98.7 %.

Thermal loss W_{LOSS} which occurs in the fluxpump is defined as

$$W_{LOSS} = W_I - W_L + W_{SH} . \qquad (11)$$

The $W_{LOSS} = \sim$73 J is obtained by substituting $W_I = 5734$ J, $W_L = 5684$ J and $W_{SH} = 23$ J for eq. (11).

In order to classify these losses, the loss of each component, namely an iron core, windings and so on, has been independently checked. Losses E_{LOSS} obtained on various components during half-cycle and the estimated entire loss ΣE_{LOSS} during the course of charging for $I_L = 0$ A to 100 A, are summarized in Table 2. The total sum of ΣE_{LOSS}, is estimated ~65 J and very close to $W_{LOSS} = 73$ J obtained by measuring directly W_I and W_L. On the other hand, from Table 2, it turns out that the most of W_{LOSS} occurs at superconducting switches.

These results imply that the sum of all losses (E_{LOSS}) during every half-cycle is approximately constant and less depends on I_L.

Table 2 Losses per half sycle E_{LOSS} and ΣE_{LOSS} of component parts of the present fluxpump. Operating parameters are $I_L =$ from 0 A to 100 A, f=0.2 Hz, $I_1 =$ from 0.05 A to 3.05 A, $V_P =$10 V and the number of pumping operations, $n = 377$, and $\Sigma E_{LOSS} = \Sigma_{n=1}^{377} E_{LOSS}$.

item	E_{LOSS} (mJ)	ΣE_{LOSS} (J)
iron loss	~7.5	~3
winding loss	0~ 5.5[*1]	~0.7
commutation loss	~ 0	~ 0
SW leakage loss	~ 100	~38
SW heater input	60	23
joint ohmic loss	0 ~ 2.5[*2]	~0.3
total	168~175	~ 65

*1 : results of short circuit tests of the transformer by the sine wave currents of the frequency of 1Hz.

*1,*2 : winding loss and joint ohmic loss depend on the load current I_L. Accordingly, dependences of E_{LOSS} on I_L were measured and on the basis of the results ΣE_{LOSS} were estimated.

SUMMARY

A new-type rectifier fluxpump with a saturable core transformer has been constructed. This fluxpump operates at non saturation state of the magnetic core of the transformer for flux pumping and works at saturation states for current commutation. From the results on charge, hold and discharge, it becomes evident that the amount of the pumping flux is strongly connected with the saturation magnetization of the the magnetic core and the constant charging (or discharging) rate is achievable.

The sum of losses per half-cycle is independent of load current I_L and the output power $P_L(= V_L I_L)$ may increase with I_L. This property will be very useful for constructing large-scale fluxpumps.

REFERENCES

1   Bernard, S.P. and Atherton, D.L., Performance analysis of transformer rectifier fluxpumps. Rev. Sci. Instrum., **48**, 1245, (1977).........

2   Klundert, L.J.M.van de and Kaet, H.H.J. ten, Fully superconducting rectifiers and fluxpumps, Part 1: Realized methods for pumping flux. CRYOGENICS **21**, 206 (1981).

3   Klundert, L.J.M.van de and Kaet, H.H.J. ten, On fully superconducting rectifiers and fluxpumps. review. Part 2 : Commutation modes, characteristics and switches. CRYOGENICS **21**, 267 (1981).

4   Funayama, K., Akiyama, K. and Anayama, T., New-type fluxpump with saturable transformer. Papers of technical meeting on magnetics of IEE Japan (Japanese), MAG-86-8, (1986).

5   Mulder, G.B.J., Kaet, H.H.J.ten, Krooshoop, H.J.G. and Klundert, L.J.M. van de, Thermally and magnetically controlled superconducting rectifiers. ASC-88 (1988).

DESIGN, CONSTRUCTION AND TESTING OF A 50 kA SUPERCONDUCTING TRANSFORMER

G. Pasztor, E. Aebli, B. Jakob, P. Ming, E. Siegrist, P. Weymuth
Paul Scherrer Institute, 5232 Villigen, Switzerland

Abstract

In order to enable tests on future full size NET conductors a superconducting transformer was designed and constructed at PSI. The transformer unit which will be used in the SULTAN Test Facility is specified for currents up to 50 kA. The main advantage of such a device where high currents are induced in a superconductor test loop is that large and power consuming current leads are eliminated.

In this paper the design and construction of the transformer components and cryostat are described in detail. Information concerning future operation of this device in the SULTAN-III Test Facility is also given. Some results of the performed test are presented.

INTRODUCTION

To fulfil the special requirements imposed by the NET fusion program, the magnet system of the SULTAN Test Facility is being modified by replacing the NbTi middle coil by a pair of Nb_3Sn split coils and by dividing the outer 6T coil into two halves. The resulted splitted magnet system SULTAN III, allows radial access of samples to the high field region and will provide the test capability of full size NET conductors and of 2m diameter - small pancakes [1].

The facility will be equipped with a sample unit allowing short sample critical current measurements in transversal magnetic fields up to 11.5T. The unit is designed to permit sample insertion and removal without warming up the coils.

To supply the sample current a superconducting transformer will be used instead of direct feeding of samples from an external power supply. This solution eliminates the problem of large heat losses produced by the current leads and does not need large and expensive power supplies.

This paper describes the design principles and fabrication of a 50kA superconducting transformer for use in the SULTAN-III Test Facility for short sample measurements of high current superconductors and presents the results of transformer test.

DESIGN AND FABRICATION

The induced current method to feed superconducting loops is particularly advantageous at high currents as required for testing candidate conductors for fusion applications. The present transformer was designed to produce 50 kA which is a high enough value compared to the current carrying capacity of the most conductors proposed for the NET TF coil [2].

Basically the transformer consists of two concentric superconducting coils with the primary inside the secondary. While energizing the primary coil a current will be induced in the secondary turns.

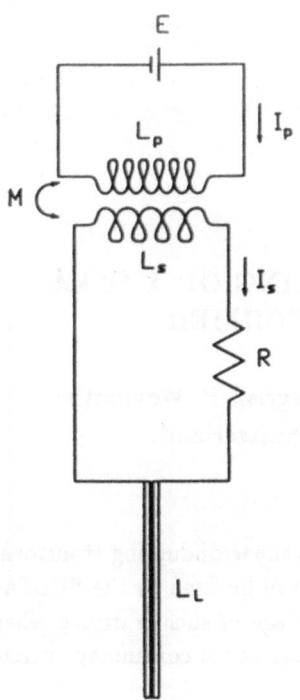

Figure 1 - Transformer equivalent circuit

Figure 1 shows the equivalent circuit of the transformer. The governing equations for this circuit are:

$$L_p \dot{I}_p + M \dot{I}_s + R_p I_p = E \qquad (1)$$

$$(L_s + L_L) \dot{I}_s + M \dot{I}_p + R_s I_s = 0 \qquad (2)$$

where L_p and L_s are the self-inductances of the primary and secondary respectively, L_L is the load inductance of the secondary, R_p and R_s are the resistances of the primary and secondary respectively, M is the mutual inductance between primary and secondary, E is the charging voltage of the primary and I_p and I_s are the primary and secondary currents respectively.

Assuming that

$$I_p(t) = \begin{cases} kt & 0 \le t \le t_o \\ kt_o & t \ge t_o \end{cases}$$

equation (2) gives during charging

$$(L_s + L_L) \dot{I}_s + R_s I_s = -Mk \qquad (3)$$

with the general solution

$$I_s = x e^{\alpha t} - \frac{Mk}{R_s} \qquad (4)$$

where

$$\alpha = -\frac{R_s}{L_s + L_L}$$

and x is any real number.

Taking into account the initial conditions $I_s = 0$ for $t = 0$, x can be determined and (4) becomes

$$I_s = -\frac{Mk}{R_s}(1 - e^{\frac{R_s}{L_s + L_L} \cdot t}) \qquad (5)$$

In the case that the time constant of the secondary $\tau = \frac{L_s + L_L}{R_s}$ is much larger than the charging time t_o of primary, (5) gives

$$I_s = -\frac{M}{L_s + L_L} \cdot I_p$$

$$where \quad M = K\sqrt{L_p L_s}$$

The currents of primary and secondary can be related by

$$\frac{I_s}{I_p} = K\sqrt{\frac{L_p}{L_s}} \frac{1}{1 + \frac{L_L}{L_s}} \qquad (6)$$

Equation (6) says that in order to reach a high secondary current for a given primary current, a high turns ratio and coefficient of coupling are necessary [3]. On the other hand, the current transformation ratio I_s/I_p will have a maximum value at zero inductance of the secondary load i.e. of the test samples. This implies the use of a noninductive circuit. As a consequence the sample unit will consist of two parallel straight conductor pieces tightly clamped together and short circuited by joints at the lower end. At the upper end, the test conductors will be connected to the secondary winding.

The induced current in the secondary circuit decays with a time constant $\tau = L/R$ where L is the total inductance of the loaded secondary and R the sum of the joint resistances in the circuit. On the other part, the measuring time for critical current or quench energy is in the range of 10 to 100 seconds. In order to achieve a time constant which is long compared to the measuring time, the sum of the joint resistances must have a value lower than $10^{-8}\Omega$.

Based on these design principles, a transformer which supplies 50kA by charging the primary with 200 A has been designed at PSI and subsequently manufactured at ABB's Oerlikon works. The transformer system consists of a layer-wound indirectly cooled, superconducting primary solenoid and ten superconducting secondary turns wound on the outside of the primary coil. The conductor used in the primary is a 0.97 diameter NbTi superconducting wire with a 1.8÷1 copper - to - superconductor

ratio. 30 turns of copper cooling tubes surround the primary winding laterally and at the inner diameter.

The secondary winding is made of two parallel electrically insulated NbTi forced flow cabled conductors based on the conductor used in the 6T SULTAN outer coil. The current leads are brought out from the secondary coil by winding them onto the grooved circumference of a G11 cylinder mounted at the lower end of the winding. In order to minimize the mechanical disturbances during operation, the secondary turns and current leads were impregnated with resin in one step. The end of the leads are soldered together by pairs into a cop-

TABLE I
Characteristics of the SULTAN Superconducting Transformer

Primary

Configuration:	indirectly cooled solenoid
Number of turns:	3240
Winding dimensions:	i.D. 255 mm
	o.D. 297 mm
	Length 163 mm
Inductance L_p:	2.46 H
Max. operating current:	200 A
Peak conductor field:	3.5 T
Superconductor:	insulated MF NbTi, 0.97 diameter
	Cu: NbTi = 1.8
	I_c (4.2K,5T) = 589 A

Secondary

Configuration:	forced flow cooled solenoid
Number of turns:	2×10
Winding dimensions:	i.D. 308 mm
	o.D. 372 mm
	Length 192 mm
Inductance L_s:	13 μH
Mutual inductance M:	6.6 mH
Coefficient of coupling:	0.77
Desired circuit resistance:	$< 10^{-8}\Omega$
Max. operating current:	50 kA
Peak conductor field:	3.6 T
Superconductor:	NbTi tube conductor, $2 \times (16 \times 16$ mm^2)
	Cu: NbTi ≈ 7
	I_c (4.2K, 5T) = 2×24 kA

per block which is subsequently clamped to the test sample current terminations. Table I presents the main characteristics of the transformer.

COOLING

The transformer coils are cooled by forced flow of supercritical helium. The coils and the cryostat are both mounted on the same flange (see figure 2). All subcomponents of this compact unit are cooled in series having heat exchangers in between to cool the incoming helium back to 4.5K. A schematic flow scheme is shown in figure 3. The massflow in the cooling circuit can be regulated with a JT valve located at the end of the cooling chain. The liquid helium level in the cryostat which contains the heat exchangers is kept constant by a heater.

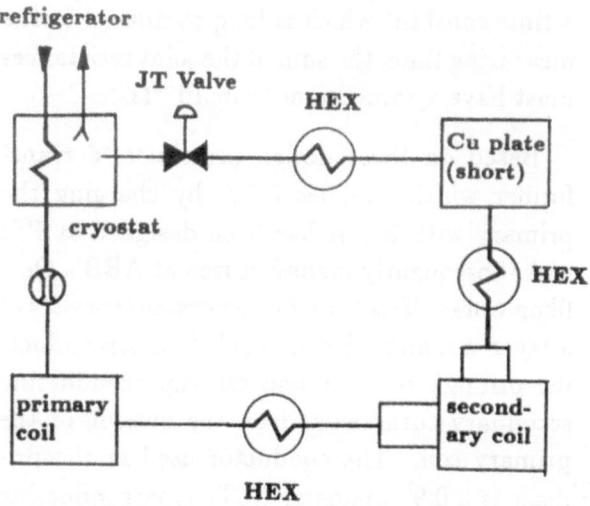

Figure 3 - Schematic flow scheme

INSTRUMENTATION

To measure the temperature at different locations of the cooling circuit CLTS- and CGRT sensors are installed.

The massflow measurement is based on a differential pressure method, whereby an orifice is mounted in the cross sectional area of the conduit. In this way a pressure drop results between inlet and throttled cross section behind the orifice.

Voltage taps with protective resistors were installed on the primary coil current connections and at the middle of the winding. In this way the coil is divided for quench protection into two equal parts. The energy stored in the coil needs to be removed in case of a quench in about 1s to prevent excessive temperature increase.

The current in the secondary circuit is determined by a Rogowski coil [4]. This is a small toroidal coil which surrounds the secondary conductor and measures its self field. The current is obtained by integrating the voltage induced in the coil.

Rogowski coils were mounted on both secondary conductors. Previously they were calibrated at room temperature by replacing the

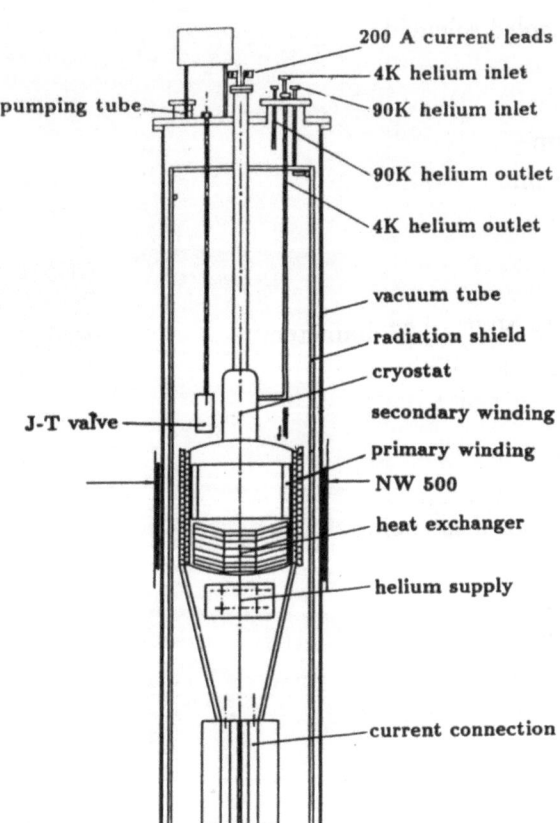

Figure 2 - Top of sample holder with cryostat and superconducting transformer

secondary conductor by a copper bar of same dimensions and feeding it with a known current. By charging the copper conductor with 1 kA, the integrated voltage signal could be determined for both Rogowski coils to be 459 μVsec and 456 μVsec respectively.

Additionally to the Rogowski coils, the conductors were fitted with Hall-probes to measure their self field and consequently their current. The probes can be used for long period recordings like determination of the time constant of the secondary circuit.

Included in the secondary circuit is a heater mounted on the conductor pair which allows to cancel residual currents before starting a new measurement.

TESTS

The test objectives were mainly to evaluate the transformation ratio and the current carrying capability of the secondary coil. For this purpose the current terminations of the secondary were short circuited with a copper bar. Besides the usual room temperature sensor- and quench detection circuit - checks, the tests were performed in four major steps:

Cooling

In order to test the cooling characteristics of the system, pressure drop measurements were performed. The massflow was varied between 4 g/sec and 16 g/sec. The lowest massflow is given by the power dissipation of the system. The upper value was determined by the cooling power limitation of the refrigeration system.

The whole transformer unit could be stably operated at any massflow within the mentioned range. A slight change of massflow was observed when energizing the transformer too fast due to the fact that the whole system has to absorb additional heating power dissipated in the joints of the secondary circuit. In order to avoid this effect the energizing time of the transformer has to be matched to the time constant of the heater which regulates the helium massflow in the transformer circuit.

Charging and Discharging of the Transformer

As a second step of the test programme the transformer unit was charged and discharged for several times to determine the transformation ratio between primary and secondary. For this measurement the signals of the two Rogowski coils, were used. The transformation ratio was determined to be 210.

The system was energized with various current ramps in the range of 100 - 200 A/s in the secondary. During these runs the calibration of the Hall-probes was also done.

Joint Resistance in the Secondary Circuit

In a third series of measurements the resistance of the secondary circuit was determined by measuring the current in the secondary circuit as a function of time keeping the current in the primary coil constant.
It is easily seen that

$$R_s = \frac{L_s}{\tau}$$

where L_L can be neglected being certainly small in this experiment.

The time constant τ was evaluated graphically. The current in the secondary decays exponentially. The time constant turned out to be 62 minutes yielding a joint resistance in the secondary circuit of $3.5 \cdot 10^{-9} \Omega$.

Bipolar Power Supply

In the SULTAN-III Facility the transformer unit will be operated by a bipolar power supply. As no suitable power supply was available during the test period we simulated such an operation by a simple switch (see figure 4).

With such an array the operating range of the transformer unit can be practically doubled, e.g. driving the primary coil from -200A to +200A. Of course, this does not mean that one can have twice the specified current in the secondary circuit because of other limiting parameters. The big advantage of this operating mode is that the current in the secondary coil can be kept constant for a rather long time.

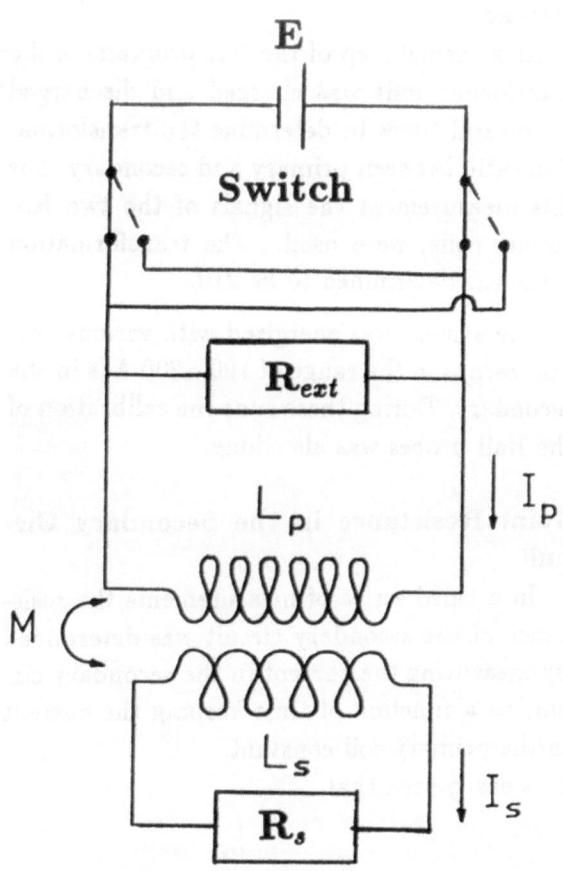

Figure 4 - Electrical circuit with switch

In our experiment we were first loading the primary up to -150A and were waiting for several hours until the current in the secondary decayed to about 3kA. Subsequently the secondary was energized again by sweeping the primary from -150A to +70A using the mentioned switch at the moment when the current of the primary coil was zero. As last step, the current ramp in the primary circuit was changed to a value which allows the compensation of the current decay in the secondary coil. During this experiment a current of 50kA in the secondary could be kept constant during 25 minutes.

CONCLUSIONS

From the test results it can be concluded that this device fulfills the specifications. The test has proven that especially the needs of the NET-Team are clearly covered by the following testing capabilities

- Max. transport current : 50 kA
- Max. background field : 11.0 T
- Max. cond. self field : ~ 0.5 T
- Range of current
  ramp rate : 100-200 A/s
- Range of conductor temp. : 4.5-6.5 K*
- Range of massflow rate : 3-10 g/sec
- Conductor inlet pressure : 4-10 bar

⋆ at 6.5K one has to work at reduced massflow rate i.e. ≈ 1.5 g/sec

ACKNOWLEDGEMENTS

Thanks are due to Drs. E. Balsamo, ENEA, C. Schmidt, KfK, and H.H.J. ten Kate, THT, for helpful discussions during the design phase.

REFERENCES

1. Della Corte, A., Pasotti, G., Ricci, M., Sacchetti, N., Spadoni, M., Dal Mut, G., Spigo, G., Veardo, G., Roeterdink, J.A., Elen, J.D., Gijze, A.C., Franken, W.M.P., Aebli, E., Horvath, I., Jakob, B., Marinucci, C., Ming, P., Pasztor, G., Vécsey, G., Weymuth, P., " The SULTAN-III Project", Proc. 15th Symposium on Fusion Technology, Utrecht 1988 in:" Fusion Technology 1988, North Holland, Vol. 2, pp 1476

2. Poehlchen, R., Bottura, L., Katheder, H., Malavasi, G., Minervini, J., Mitchell, N., Ricci, M., Salpietro, E., Rauch, J., Benz, H., Dal Mut, G., Perella, M., Di Meglio, A., Nebuloni, I., "Design Status of the NET Toroidal coils", ibid. pp 1576

3. Voelker F., and Acker, R.C., "Experimental superconducting transformer for current step-up", Particle Accelerators, Vol. 1, 209, (1970).

4. Schmidt, C., "Critical current, stability and AC loss measurement of EURATOM LCT conductor", IEEE Trans. Magn., MAG-19, 707, (1983)

QUENCH RECOVERY TEST OF 3-WINDING SUPERCONDUCTING TRANSFORMER WITH AN AUXILIARY WINDING ONLY IN LOW VOLTAGE SIDE

H.Kamijo*, T.Ishigohka*, N.Mizukami*, M.Yamamoto**
* Seikei University, Musashino 180 JAPAN
**Takushoku University, Hachioji 193 JAPAN

ABSTRACT

Superconducting transformers (SCT) have to be protected against any power system faults. So far, to protect a SCT, it is proposed to introduce auxiliary windings both into the low and the high voltage sides of a SCT. While, the authors proposed to introduce the auxiliary winding only into the low voltage side. The authors fabricated a small experimental SCT of this type, and carried out some transient experiments simulating actual power system faults. Through the experiments, it was confirmed that 1)the SCT can operate continuously at any power system faults, 2)the SCT has the effect of the current limiting ability, 3)the low voltage main winding should be isolated from the circuit after quench, 4)the direct current component appears when the low voltage main winding is reclosed, 5)the heat dissipation during quench is reduced by the presence of the auxiliary winding.

INTRODUCTION

The superconducting transformer (SCT) has advantages as follows; 1)high efficiency, 2)small size, 3)light weight. However, it has a problem of quench of superconducting windings at short-circuit faults in a transmission line. The SCT has to operate continuously through any power system faults. However, it is impractical to design a SCT not to quench at any power system short-circuit faults because of the excessive winding loss in normal operation. To solve this problem, the Westinghouse group proposed to introduce auxiliary windings both into the high and the low voltage sides of the SCT[1]. In this case, when the main windings quenched, the load current is carried by the two auxiliary windings. So the transformer can operate continuously. This idea was already confirmed by several researchers[2][3]. However, it would be rather complicated to install two auxiliary windings. Particularly, the preparation of the auxiliary winding in the high voltage side is highly troublesome because of the insulation requirement. Therefore, the authors proposed the new idea to introduce the auxiliary windings only into the low voltage side[4][5]. By designing properly, the possible maximum fault current can be suppressed considerably by the high leakage reactance of the auxiliary winding. Besides, in this case, the heat dissipation due to the quench is reduced greatly because the quench occurs only in the low voltage side. The authors fabricated a small experimental SCT(2kVA) of 3-winding structure with an auxiliary winding only in low voltage side, carried out some transient experiments simulating those in a actual power system. In this paper, the results of these experiments are shown.

PRINCIPLE OF OPERATION

The newly proposed SCT has a 3-winding structure with an auxiliary winding only in the low voltage side. At power system fault, we design that only the low voltage main winding (L.M.) will quench. While, the high voltage winding (H.M.) and the low voltage auxiliary winding (L.A.) maintain superconducting state. So, we have to design that the critical current capacities of H.M. and L.A. windings have enough margin to the possible maximum short-circuit current calculated by parameters of power system including the high leakage reactance of the auxiliary winding. In our design, the current capacities of the H.M. and L.A. windings are 1.5 pu. In the normal operation, the power is transported between the L.M. and the H.M. windings. When the L.M. winding quenches at a sudden short-circuit fault, the power is transported between the L.A. and the H.M. windings. In this way, the SCT can operate continuously.

DESIGN AND CONSTRUCTION

An iron core made of grain oriented magnetic steel sheets with a thickness of 0.1mm (ZT-100 0.1^t) was prepared.

As for the superconducting wire, we used 7-strand cables for all the windings for the convenience of the procurement. The characteristics of superconducting wire are shown in Table 1. The current capacity of the L.M. winding has to be made smaller than the L.A. winding. So, only five element wires in 7-strand cable were used in the L.M. winding. In the other windings, all the element wires were used.

Table 1. Characteristics of
superconducting wire.

Conductor
Structure	:	7-strand cable
Diameter	:	0.370-0.380mm
Twist pitch	:	3.05mm

Characteristics of the element wire
Diameter	:	0.103mm
Twist pitch	:	0.98mm
Filament diameter	:	0.444μm
Number of filaments	:	15367
Matrix ratio	:	NbTi/CuNi=1/2.5
Critical current	:	24A(at 1T)

In the 3-windings transformer shown in Figure 1(a), the leakage reactances between each windings are expressed as follows;

$X_{12}=2\pi fc\{\delta_{12}+(\Delta_1+\Delta_2)/3\}$
$X_{23}=2\pi fc\{\delta_{23}+(\Delta_2+\Delta_3)/3\}$
$X_{31}=2\pi fc\{\delta_{12}+\Delta_2+\delta_{23}+(\Delta_3+\Delta_1)/3\}$
where; $c=\mu_0(w^2 m/h)$, $\mu_0=4\pi\times10^{-7}$,
w:number of turns, m:mean length / turn, h:height of coil.

From these equations, equivalent leakage reactance of each winding in Figure 1(b) are determined as follows;

$X_1=2\pi fc(\delta_{12}+\Delta_1/3+\Delta_2/2)$
$X_2=2\pi fc(-\Delta_2/2)$
$X_3=2\pi fc(\delta_{23}+\Delta_3/3+\Delta_2/2)$

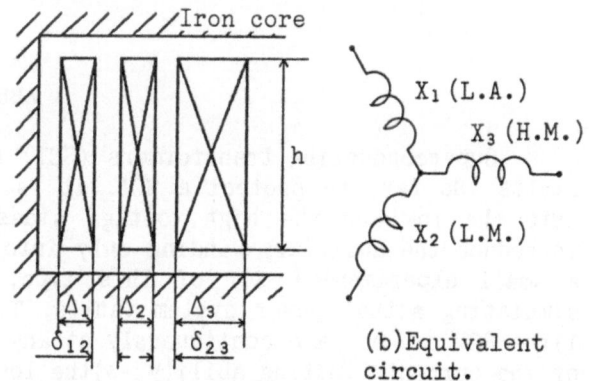

(a)Winding arrangement.
Figure 1. Leakage reactance of 3-winding transformer.

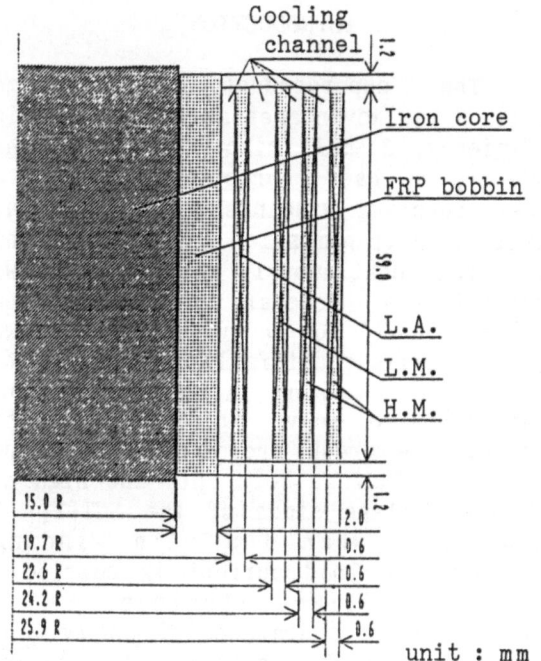

unit : mm

Figure 2. Structure and dimensions of experimental SCT.
　　L.M. : Low voltage main winding.
　　L.A. : Low voltage auxiliary winding.
　　H.M. : High voltage winding.

In the design, the L.M. winding is arranged in the center between the L.A. and H.M. windings. The H.M. winding is located in the outside considering the insulation. And, the L.A. winding is located in the inside. All the windings are wound to a single core leg. Figure 2 shows the structure of the SCT.

CHARACTERISTICS OF 3-WINDING SCT

The characteristics of the 3-winding SCT are shown in Table 2, and the leakage reactances are shown in Table 3. The capacity of the SCT is 2kVA. The magnitude of the critical current of each winding is measured by the quench test. The result is shown as follows;

    L.M. : 40.0 A (peak)
    L.A. : 45.0 A (peak)
    H.M. : 32.5 A (peak)

From these results, the rated current in the low voltage side is determined to be 25.0 A(rms). This value has a margin of about 10%.

Distributions of transport current in the normal operation between the L.M. and the L.A. windings are determined by equivalent leakage reactance of each winding. In the normal operation, the current flowing into the L.A. is 4 %. This value agrees well with theoretical one.

Table 2. Rating of experimental SCT.

Capacity	:	2kVA
Phase	:	Single
Frequency	:	50Hz
Number of turns	:	300/600
Voltage	:	80/160V
Current	:	25.0/12.5A
Exciting current	:	1.7%
Impedance	:	5.6%
No-load loss	:	9.0W

Table 3. Leakage reactances of experimental SCT.

L.M. - H.M.	:	0.179Ω
L.A. - H.M.	:	0.377Ω
L.M. - L.A.	:	0.182Ω
(Equivalent leakage reactances)		
L.M.	:	0.008Ω
L.A.	:	0.190Ω
H.M.	:	0.187Ω

EXPERIMENTAL CIRCUIT

Considering the use of the SCT in a power system, the experiment simulating that in power system was carried out. When a sudden short-circuit fault occurred in an actual power system, the circuit-breaker clears the fault in 3cycles, and it recloses in about 0.5 sec (high-speed reclosing). Figure 3 shows the circuit for this experiment. The SCR.1 and SCR.2 are the switches for simulating sudden short-circuit fault, 3-cycle clearing, and high-speed reclosing. A magnetic switch is used for the L.M. circuit to isolate it from the main circuit. All the switches are controlled by personal computer as shown in Figure 4. Both the main clearing time:T1, and the supplementary clearing time:T2 for the low voltage main winding can be adjusted freely.

Two kinds of experiments are carried out shown as follows.

Case-A:L.M. winding is not isolated.
  1)sudden short-circuit fault
  2)3-cycle clearing
  3)fault   removal
  4)high-speed reclosing
Case-B:L.M. winding is isolated.
  1)sudden short-circuit fault
  2)3-cycle clearing
  3)fault removal
  4)the L.M. winding is isolated
  5)high-speed reclosing
  6)the L.M. winding is reclosed

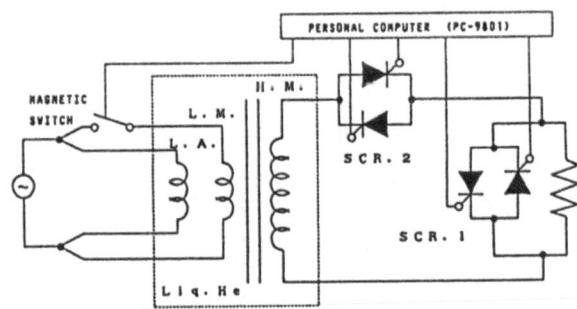

Figure 3. Experiment circuit for SCT.

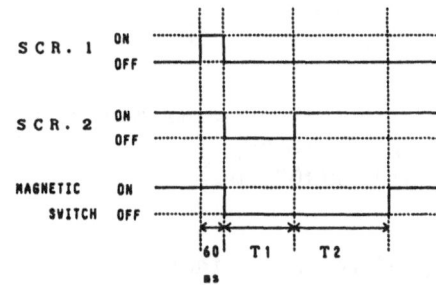

Figure 4. Timing chart of each switches.

EXPERIMENTAL RESULT

Sudden short-circuit faults

At sudden short-circuit faults, the magnitude of the current in the L.M. and the H.M. windings abruptly increase. Then, only the L.M. winding quenches by excess current, but the H.M. winding does not quench. After the quench of the L.M. winding, the short-circuit current in the low voltage side is shifted from the L.M. winding to the L.A. one. So, short-circuit current can be transported between the L.A. and the H.M. windings. At this time, short-circuit current is limited.

After 3cycles, the fault is cleared. At this time, there are two cases depending on the length of cleared time. In one case in which the cleared time is long enough, the quenched L.M. winding can recover superconductivity by the time of reclosing. Therefore, at the time when the power system is reclosed, the SCT can be operated in a normal operating condition by the L.M. and the H.M. windings. In the other case in which the cleared time is not long enough, the L.M. winding cannot recover its superconductivity during the clear time as shown in Figure 5. In this case, the power can be transported between the L.A. and the H.M. windings. However, it can't be expected that the L.M. winding recovers superconductivity spontaneously after the reclosing. This is because of the continuous heat dissipation in the winding when the full system voltage is applied to it.

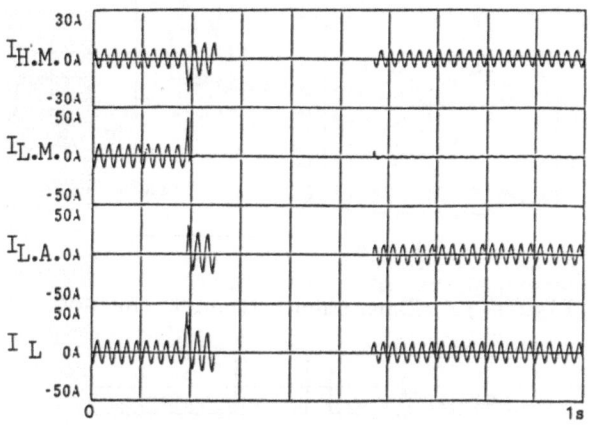

Figure 5. Result of experiment Case-A.
T1:300ms, Lord:2.5Ω,
Primary voltage:6V(rms).

Recovery of the L.M. winding

In the later case, the L.M. winding that has quenched by excess current has to recover superconductivity by all means. So, the L.M. winding should be isolated from the main circuit by additional switch. After the L.M. winding recovered superconductivity, it is reconnected to the main circuit. This experimental results are shown in Figure 6. When the L.M. winding is reclosed after recovery of superconductivity, the load current shifts from the L.A. to the L.M. winding. Thus, the SCT can return to the normal operation. In this case, the DC components appears on each current of the low voltage windings. The magnitude of this DC component depends on the phase-angle at the instance of the reclosing. Two typical cases are shown in Figure 6.

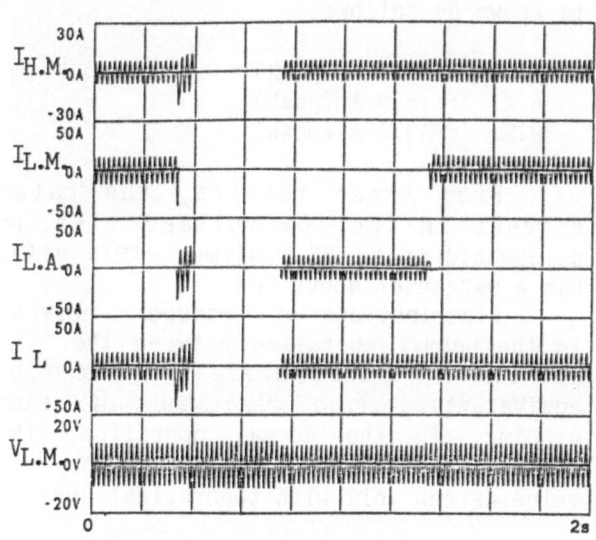

(a)The case of small DC component.

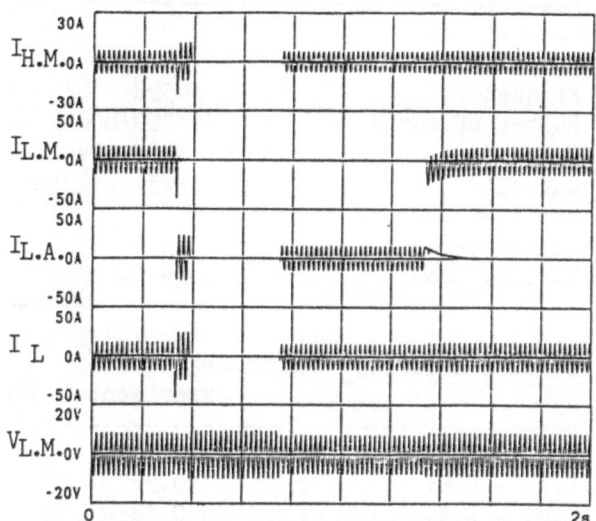

(b)The case of large DC component.
Figure 6. Results of experiment Case-B.
T1:350ms, T2:600ms, Load:2.5Ω,
Primary voltage:7V(rms).

DISCUSSION

The effect of current limiting

Similarly 4-windings SCTs[3], the 3-winding SCT also has a current limiting ability for sudden short-circuit faults. The leakage reactance between the L.A. and the H.M. winding is about twice as large as that between the L.M. and the H.M. winding in the experimental SCT as shown in Table 2. Therefore, the short-circuit current is suppressed by this high leakage reactance when it shifted from the L.M. to the L.A. winding.

For a typical example, the authors discuss about Figure 6(a). The current in low voltage side increases up to about 41.6 A (peak) at the instance of sudden short-circuit fault. However, after the L.M. quenched, the short-circuit current shifts to the L.A. winding. Then, the short-circuit current is suppressed to about 33.3 A (peak). And, it decays to about 25.0 A (peak) just before the fault clearing. The short-circuit current is smaller than the critical currents of the L.A. and the H.M. windings. If the L.M. winding did not quench, the short-circuit current would increase up to 55.3 A (peak) except for the DC component. So, our experimental SCT has an ability to suppress the short-circuit current to about 50 - 70 %.

The effect of current limiting can be adjusted by regulating the leakage reactance. And, the leakage current which flows into the auxiliary winding in the normal operation can be also adjusted by selecting the appropriate values of the leakage reactances.

Phenomena at reclosing of L.M. winding

From the experimental result, the following phenomena at the reclosing of L.M. winding are confirmed .
1) the load current increases.
2) the DC component appears to the each current of the low voltage windings.

The reason for item 1) is because the L.M. winding has a smaller leakage reactance than that of the L.A. winding.

The phenomenon of item 2) would become a problem at the reclosing of L.M. winding. The magnitude of the DC component depends on the timing at which the magnetic switch is reclosed.

In the case shown in Figure 6(b), the current of the L.M. winding becomes as follow;

normal operation : 13.8 A (peak)
at reclosing : 24.3 A (peak)

So, the current of the L.M. winding increases up to about 1.78 times as much as that in a normal operation. Theoretically, this factor is calculated to be about 2 in the maximum case.

When this peak value exceeds the critical value, the L.M. winding which recovered superconductivity quenches again at reclosing. An example of this phenomenon is shown in Figure 7.

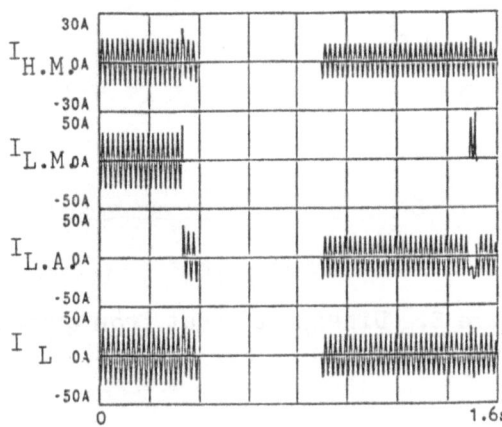

Figure 7. The case in which L.M. winding quenches again at reclosing.
T1:500ms, T2:600ms, Load:1.0Ω,
Primary voltage:7V(rms).

The DC component flows in both windings of the low voltage side. The direction of those two currents are opposite to each other. The DC component does not appear in the high voltage side.

It was found that the magnitude of the DC component is determined by the timing of reclosing. And, the DC component is equal to the instantaneous value of the L.A. winding at reclosing. To prevent the quench by this DC component, some countermeasures must be considered as a phase-angle-controlled reclosing, etc..

Recovery of superconductivity of L.M. winding

As shown in Figure 6, the quenched L.M. winding can recover superconductivity by isolating it from the circuit for certain time. The SCT recovers normal operating state by reclosing the L.M. winding after this.

While, it is reasonable to consider that the recovering time is related to the heat dissipation in the winding during quench. Then, the heat dissipated in a duration of 3-cycle at the quenched L.M.

winding is shown in Figure 8. In Figure 8, we can see that the heat dissipation of the SCT with the auxiliary winding is smaller than that of the SCT without it. The reason would be that as the resistance of the quenched L.M. winding increases, the current of the L.M. can transfer to the L.A. winding.

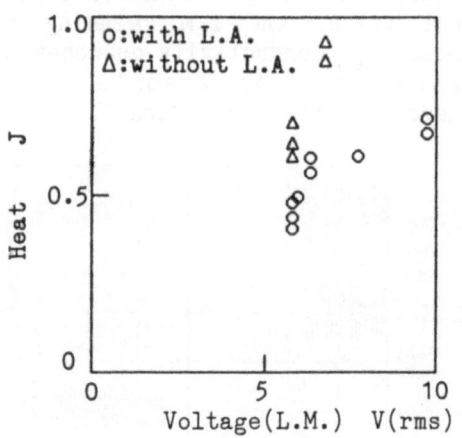

Figure 8. Dissipated heat from quenched L.M. winding.

In our experimental SCT, the recovering time becomes about 500 ms for the heat dissipation of about 0.5 J. However, this value would not have a universality. For a SCT with larger capacity, the recovering time may be longer. And, the L.M. winding may not recover superconductivity during the cleared time of power system. Therefore, the quenched L.M. winding should be isolated from the circuit. On the other hand, it can be considered that the recovering time of the quenched winding is considerably small. The main reasons are; 1) the stored energy of a transformer is much smaller than a conventional superconducting magnet, and 2) when the auxiliary winding is introduced, the heat dissipation during quench is considerably small as shown in Figure 8.

CONCLUSIONS

Through the experiments on the 3-winding-structure SCT with an auxiliary winding only in low voltage side, it becomes clear that;
1) the SCT of this type can operate continuously for any power system faults,
2) the SCT of this type has an ability of current limiting by the high leakage

reactance between the L.A. and the H.M. windings,
3) the L.M. winding should be isolated from the circuit after its quench. And, it should be reconnected to the circuit after the L.M. winding recovered superconducting state. Thus, the SCT can recover the normal operation.
4) the DC component appears to the low voltage windings as the L.M. is reclosed. The maximum current can be about twice as much as that in normal operation. Then, there is a possibility that the L.M. winding quenches again.
5) the heat dissipation in the quenched L.M. winding can be suppressed considerably by the presence of the auxiliary winding. Therefore, the recovering time of the quenched L.M. winding to superconducting state would be considerably small.

ACKNOWLEDGMENT

This research is supported by Grant-in-Aid for Co-operative Research of the Ministry of Education, Science and Culture of Japan. The authors are very grateful to Mr.Y.Kushiro for his devoted support throughout the experiments.

REFERENCES

[1] H.Riemersma, M.L.Barton, D.C.Litz, P.W.Eckels, J.H.Murphy, J.F.Roach : "Application of Superconducting Technology to Power Transformers", IEEE Transactions on Power Apparatus and Systems, Vol. PAS-100, No.7 July 1981. pp.3398-3405.
[2] F.van Overbeeke, K.Oordt, LJ.M.van de Klundert : "Design and Operation of a Protection System for Transformers with superconducting windings", Cryogenics Vol.25, December 1985. pp.687-694.
[3] K.Funaki, M.Iwakuma, M.Takeo, K.Yamafuji : "Preliminary Test and Quench Analysis of a 72kVA Superconducting Four-Winding Power Transformer", ICEC 12, 1988.
[4] M.Yamamoto, T.Ishigohka, T.Shimohka, N.Mizukami, M.Yamaguchi : "Preliminary Study on AC Superconducting Machines", IEEE Transactions on magnetics, Vol.24, No.2, March 1988. pp.1473-1476.
[5] T.Ishigohka, T.Shimooka, H.Kamijo, N.Mizukami, M.Yamamoto : "Superconducting Power Transformer with an Auxiliary Winding only in Low Voltage Side and its Behavior at Line Faults", T.IEE Japan, Vol.108-D, No.11, 1988. pp.984-989.

ON THE INDUCTIVE METHOD FOR MAXIMUM CURRENT TESTING OF SUPERCONDUCTING CABLES[*]

G.B.J. Mulder, H.H.J. ten Kate, H.J.G. Krooshoop and L.J.M. van de Klundert,
Applied Superconductivity Centre, University of Twente,
P.O.B. 217,   7500 AE Enschede,   The Netherlands.

ABSTRACT

In order to test superconducting cables at high currents it is convenient to generate the required transport current inductively, i.e. by means of a superconducting transformer. The paper gives a survey of the devices in different laboratories that apply this technique to test cables above 20 kA. An existing test facility at the University of Twente, suitable for 50 to 200 kA, is treated in more detail. Specific aspects of such a facility are discussed, for example the design of the transformer, the methods to measure the current in the superconducting secondary circuit and the fabrication of joints with a sufficiently low electrical resistance.

INTRODUCTION

As a result of various large scale projects for fusion and magnetic energy storage there is an increasing interest in the experimental investigation of superconducting cables having large operating currents, often exceeding 20 kA. When developing facilities to test such cables, a decision should be made on how the cable is supplied with the transport current.

The straightforward and conventional method is to connect the cable under test directly to a low-ripple high-current power supply by means of two vapor cooled current leads going in the cryostat. This method is rather expensive, mainly due to the power supply and the costs of helium liquefaction. An optimized pair of current leads dissipates about 2 W/kA at the operating current, i.e. 300 litres/hour in the case of a 100 kA test facility.

A more economical alternative is to generate the current through the sample inductively. In that case the sample is connected to the secondary turn(s) of a superconducting transformer having a high current amplification factor. The power supply and current leads now have to match the primary current of the transformer, which is relatively low, typically several tens of amperes.

The induced current method was successfully used by several laboratories [1-15]. In our laboratory it has been applied in several cases to test cables up to 100 kA and measure their maximum current, quench sensitivity, V-I characteristics, a.c. losses, quench propagation velocity and the quality of electrical joints. The experience at our university with this technique is based on 10 years of research in the field of superconducting rectifiers for powering superconducting coils.

---

[*] These investigations in the programme of the Foundation for Fundamental Research on Matter (FOM) have been supported in part by the Netherlands Technology Foundation (STW).

PRINCIPLES AND DESIGN

Figure 1 shows the basic scheme of the test set-up. The transformer is usually of the air-core type. It has a large number of turns on the primary coil compared to the secondary. Therefore, the amplification factor is large so the primary coil can be fed with a moderate current. In general, the secondary circuit consists of:
* a permanent section L_1 that couples well with the primary coil,
* a removable section L_2 made of a short sample of the cable under test,
* a section that can be heated above T_c and therefore acts as a switch,
* one or more low-resistance electrical joints connecting the components.

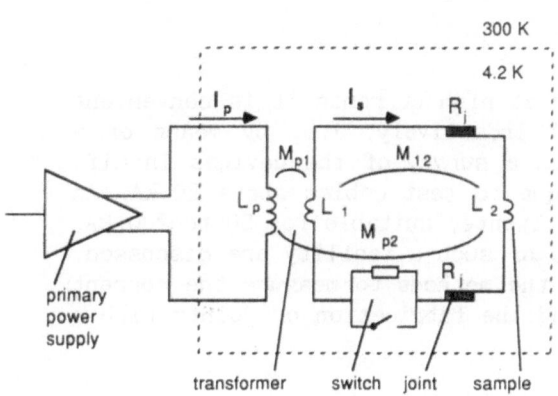

FIGURE 1. *Scheme of the test set-up.*

It is possible to treat the secondary circuit as an LR-circuit having a self inductance $L_s = L_1 + L_2 + 2M_{12}$, a mutual inductance with the primary coil $M_{ps} = M_{p1} + M_{p2}$, and a time-dependent resistance $R_s(t)$. The resistance R_s represents the total dissipation in the secondary, i.e. the sum of:
* ohmic dissipation in the joints,
* current sharing voltage over the cable,
* transport current losses caused by ramping the field and/or the current.

The procedure to generate a sweep of the current in sample is as follows. The thermal switch is opened so that I_s decays within a few milliseconds and I_p is adjusted to a certain level. Then, the switch is closed and the actual measurement starts. The primary current is swept up or down, thereby generating a large current in the secondary circuit given by Kirchhoff's law

$$M_{ps}\frac{dI_p}{dt} = L_s\frac{dI_s}{dt} + I_sR_s(t). \qquad (1)$$

If the decay time $L_s/R_s(t)$ is much longer than the duration of the measurement, the secondary current varies linearly with the primary current. The amplification factor is then M_{ps}/L_s. On the other hand, if the decay time and the measuring time are of the same order of magnitude, the relation between I_p and I_s is non-linear. In that case a prescribed shape of I_s can be generated by measuring I_s and operating the transformer in the feedback mode.

The highest current that can theoretically be induced in the secondary circuit is $2I_p^{max}M_{ps}/L_s$, when applying the full primary current sweep from $-I_p^{max}$ to $+I_p^{max}$. Usually, however, the maximum secondary current is limited by the quench current of the sample.

Transformer Design

An important parameter of the transformer is the necessary magnetic flux that the primary coil should generate in the secondary circuit in order to perform the measurement. Integration of (1) yields

$$M_{ps}(I_p(t)-I_p(0)) = L_sI_s(t) + \int_o^t I_sR_s(t)dt,$$

where I_s was taken zero at t=0. This equation clearly shows that the flux change $M_{ps}\Delta I_p$ is used partly to induce I_s and partly to sustain the resistive voltage in the secondary circuit during the measurement. Suppose the desired measuring time is t_m, the expected average resistive voltage V_s and the expected quench current of the sample I_s^{max}. In that case, the transformer should satisfy the following condition,

$$2M_{ps}I_p^{max} > L_sI_s^{max} + V_st_m . \qquad (3)$$

This implies that the required size of the transformer depends on the type of experiment that will be performed. For example, when measuring V-I characteristics, the current sharing voltage of the sample adds up to the voltage over the joints and V_st_m may well become the dominant term in the right-hand side of equation (3).

It should be noted that the transformer is essentially an a.c. device. The occurring a.c. losses in the primary and secondary conductors represent a heat load that has to be considered during the design. If the transformer will be used at high current rates, it may be necessary to improve its cooling and/or to apply low loss a.c. superconductors.

Measurement of the Secondary Current

A consequence of the inductive method is that the current has to be measured in the closed superconducting secondary circuit. Most laboratories use one of the following methods and obtain an accuracy of a few per cents :

* Measure the primary current change and determine I_s as $\Delta I_p M_{ps}/L_s$. This method is only suitable if the duration of the measurement is much shorter than the decay time of the secondary circuit. In addition, an accurate calculation of M_{ps} and especially of L_s is in most cases rather difficult.

* Integrate the voltage of a calibrated Rogowski coil placed around the secondary conductor. Due to inevitable drift of the integrator, this method is also inaccurate for long measuring times.

* Measure the field produced by I_s using a Hall probe or other magnetic sensor. The sensor should by preference be placed in the permanent section of the secondary circuit and have an orientation that minimizes the sensitivity to disturbing magnetic stray fields of the primary coil and the magnet. It is also possible to shield undesired field components by means of a superconducting shield or to compensate them using a set of balanced Hall probes.

Ten Kate [16] describes a far more elegant and accurate method based on a self-integrating short-circuited superconducting Rogowski coil. His current sensing device is insensitive to external fields, easy to calibrate and gives an accuracy in the order of 0.1 %. It is actually applied in one of our test facilities [13].

The Joints

The preferred technique for making the high-current electrical connections in the secondary circuit is that of soldered lap joints. This method is applied in most of the laboratories, but the achieved joint resistances R_j vary by three orders of magnitude between about 50 pΩ and 50 nΩ, see Table 1. A few guidelines should be observed to obtain good quality joints:

- It is important to use a low-resistive solder, but the solder should not become superconducting in order to avoid flux jumps or quenches in the joint. According to our experience, Sn3%Ag solder is well suited for applications at 4.2 K.
- Soldering must be done sufficiently fast

otherwise the Cu matrix of the conductor will dissolve in the solder.
- A "clasping hands" type of joint is in principle better than a "praying hands" joint, see figure 2. In the latter type, due to eddy currents under a.c. conditions, only a part of the joint length is effectively used for current transfer between the conductors.
- The joint can be improved by soldering extra superconducting wires parallel to it.
- R_j is a decreasing function of the contact area between both conductors. Therefore, the joint should be as long as possible.

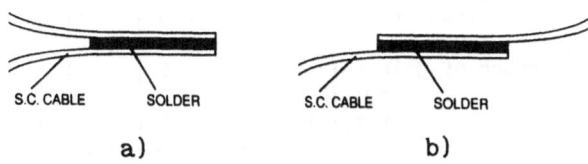

a)   b)

FIGURE 2. *Two types of joints: a) praying hands, b) clasping hands.*

It should be kept in mind that R_j is by no means constant. Due to magneto-resistivity and current distribution effects in the joint, R_j is an increasing function of the applied magnetic field as well as the current. Therefore, integration of eq. (1) is in general only possible if the term $I_s R_s(t)$ is small.

Protection of the Sample

When a quench occurs in the sample, the current will decay quickly because of the small self-inductance of the secondary circuit. The stored secondary energy W_s is also small. Furthermore, only a part of W_s will be dissipated in the secondary during a quench, the rest is coupled out to the primary coil or is dissipated in metallic constructional parts of the apparatus due to eddy currents. As a result, the energy dissipated in the sample itself is usually insufficient to cause a burn-out. This passive "self-protecting" behaviour is an advantage of the induced current method.

AN OVERVIEW

The first devices in which superconducting transformers were applied in order to test cables in the kA-range date back to about 1970 [1,2]. In 1977, Kullmann [3] briefly described a test set-up, with a

TABLE 1

Survey of laboratories that apply the inductive method for currents above 20 kA.

Laboratory	KFK	KFK	GD	PSI	TAC	CERN	University of Twente			
Reference	[6]	[7]	[8]	[9]	[10]	[11]	[12]	[13]	[14]	[15]
Layout, see Fig.3	3.A	3.E	3.C	≃ 3.D	3.G	3.F	3.B	3.H	3.A	3.D
Appl. field [T]	4.9	1.8	7.5	SULTAN	5	9.3	1	7.5	5	13$
Year of operation	1982	1988	1987	1989*	1989*	1989*	1985	1986	1988	1989
Primary turns		538	700	5400	200	3018	7555	7555	4026	3600
L_p [H]		≃ 0.1	0.17	7	0.024	2.58	10.1	10.1	7.1	1.0
I_p^{max} [A]		300	120	200	12000	300	75	75	200	90
Secondary turns	1	1	1	10	2	18	1	1	1	1
L_s [μH]	2.1		1.27	30	3.1	101	0.30	0.9	0.46	0.33
I_s, design [kA]		60	25	50	300	40	100	40	200	50
I_s, sample# [kA]	70	40	0.8	--	--	--	40	25	95	40
M_{ps} [mH]			0.23	11	0.24	14.9	1.3	1.4	1.07	0.17
$2M_{ps}I_p^{max}$ [Vs]			0.06	4.4	5.8	8.8	0.20	0.21	0.43	0.03
Amplification		≃ 200	180	360	77	147	4300	1600	2300	520
Number of joints	1	1	3	2	2	3	1	2	1	2
R_j [nΩ]	0.13	2	≃ 50	--	0.05*	--	1-4	0.5	0.05	0.2
τ_s [s]	16000	500	≃ 10	--	30000*	--	75-300	1000	10000	800

\# Maximum achieved value, limited by the sample.
$ 15 T in october 1989.
* Expected values.

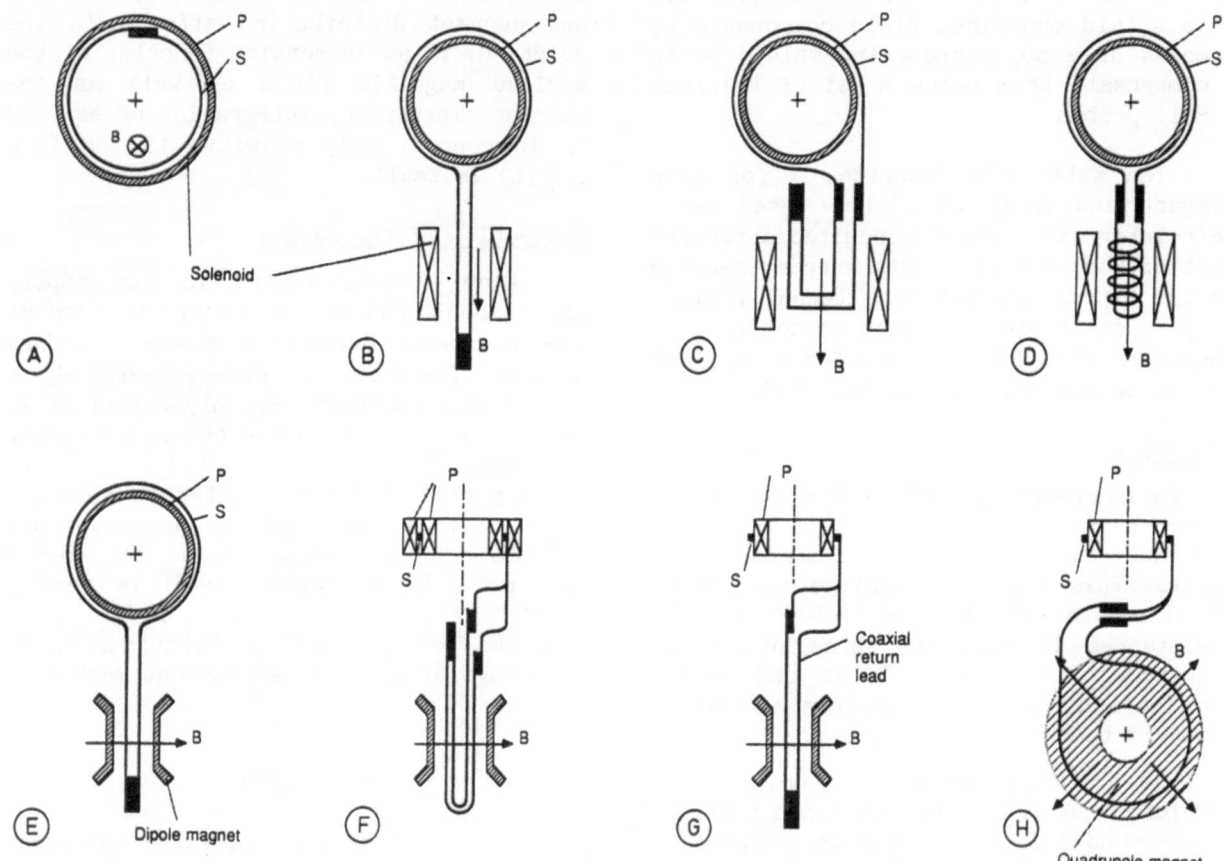

FIGURE 3. *Schematic layout of various test facilities showing the primary coil (P), the secondary coil (S), the joints, and the magnet providing the background field.*

transformer intended to measure the super-conducting characteristics of generator cables up to 40 kA. Table 1 gives a survey of other devices that were designed for cable tests at 20 kA or more. Their lay-out is schematically shown in figure 3.

Transformers were used at KFK to test toroidal and poloidal fusion conductors [6,7]. At General Dynamics a laboratory was put into use for measuring conductors up to 25 kA [7]. Devices at PSI [9] and TAC [10] have been constructed but are at present not yet operational. The most impressive of them is the apparatus of TAC in which a 200 kA SMES conductor will be tested in the fall of 1989. Finally, we have to mention the design of a 40 kA transformer at CERN [11] which will be applied for measurements on the dipole conductors for LHC.

Test Facilities at Twente

Over the years, testing short samples by induction of the transport current has become a rather standard technique in our laboratory. We use this method as soon as the required current level exceeds 1 kA. For example, the quench current and quench sensitivity of a high-current switch con-ductor was measured using the the tennis-racket geometry shown in figure 3.B. In this apparatus, currents of over 40 kA were induced in the sample [12]. Another example is a device [15] in which several cables for the LHC project at CERN were tested concerning their critical currents, up to 40 kA at magnetic fields up to 13 T.

One test facility has been developed especially to perform loss measurements on superconducting cables [13]. The sample can be exposed to very general conditions: a combination of d.c. and a.c. currents up to 40 kA, a magnetic background field of up to 7 T and a superimposed a.c. field of about 0.2 T. In this apparatus, the losses of several conductors were successfully measured, including a prototype conductor for TORE SUPRA, a Siemens generator cable and a prototype cable for the LHC dipoles.

A 200 kA Device

A special type of device is shown in figure 3.A. Here, the primary coil gener-ates the transport current as well as the applied field. In other words, the current in the sample is induced by slightly changing the magnetic background field.

Prior to each measurement, the field is set to a desired level while the switch is open so that I_s remains zero. Thus, it is possible to change I_s and B independently although they are generated with the same coil. This approach was used in a test device designed for currents up to 200 kA [14]. Due to the symmetrical geometry, see figure 3.A, the enormous Lorentz forces acting on the sample can be controlled easily. In this arrangement, a relatively long length of the test cable is exposed to the external field that has an orien-tation perpendicular to the cable, which represents the realistic case. A problem is that the joint section experiences the same magnetic field as the rest of the sample, therefore it can easily become the critical part of the secondary circuit. This problem has to be solved by enhancing the critical current of the joint, either by taking it out of the magnetic field or by reducing its temperature. The latter solution was applied in our apparatus when testing a cable with a critical current of 95 kA. As illustrated in figure 4, we reduced the temperature of the joint to 3.1 K in order to measure the correct quench current between 4 and 5 K.

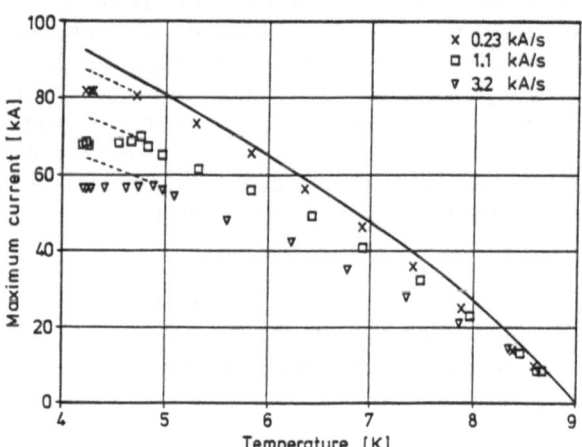

FIGURE 4. *Quench current of a cable [14], measured as a function of temperature, at several values of dI/dt. The dashed curves correspond with measurements where the joint was cooled to 3.1 K, in the other points the joint was at 4.2 K.*

CONVENTIONAL VERSUS INDUCTIVE METHOD

It is interesting to discuss the in-ductive method in terms of advantages and disadvantages compared to the conventional method.

The most obvious advantage of the inductive method is that very high currents can be generated with relatively low costs. For example, in one facility [14] two cables were tested up to ≃ 100 kA with a helium consumption of 1 to 2 litres per hour of operation. With current leads the same experiments would have required about 300 litres per hour. A power supply for 100 A, 10 V was used to feed the primary. A further advantage is that the current ripple is low, especially if the secondary circuit is in the persistent mode. The method is very well suited for measurement of joint resistances. Instead of measuring the voltage over the joint (usually in the μV-range) it is possible to determine the joint resistance from the measured decay time during persistent mode. Finally, the "self-protecting" behaviour of the sample is important.

A shortcoming of the inductive method is the difficulty of accurately measuring the current in the sample. Secondly, the energy that can be coupled from the primary into the secondary is limited to roughly the stored energy of the primary coil. This may exclude certain experiments such as the measurement of propagation velocities over long lengths of sample, the study of V-I characteristics at large current sharing voltages and tests of large inductance samples. In these cases the required size of the transformer increases and the costs can grow to the level of the conventional solution.

It can be concluded that the inductive method is a worthwhile alternative for the conventional way of short sample testing, especially at very high current levels.

REFERENCES

1. Gillani, N.V., Britton, R.B., Critical currents of superconductors in low fields. Review of Scientific Instruments, 1969, 40, 949-51.
2. Purcell, J.R., DesPortes, H., Short sample testing of very high current superconductors. Review of Scientific Instruments, 1973, 44, 295-6.
3. Kullmann, D., Intichar, L., Investigation of superconductors for large turgenerators. Proc. MT-6, Alfa, Bratislava, 1977, 189-93.
4. Shirshov, L.S., Enderlein, G., Apparatus for critical current measurement of high current superconductors. Cryogenics, 1985, 25, 527-9.
5. Albrecht, C., Marsing, H., Neumüller, H.W., Electrical joints for the European LCT coil, Proc. MT-8, Journal de Physique, 1984, 45 Colloque C1, 607-10.
6. Schmidt, C., Stability tests on the LCT conductor. Cryogenics, 1984, 24, 653-6.
7. Schmidt, C., Stability of poloidal field coil conductors: test facility and subcable results. In Proc. ICEC-12, Butterworths, Guildford, 1988, 794-8.
8. Leung, E.M.W., Arrendale, H.G., Bailey, R.E. and Michels, P.H., Short sample critical current measurements using a superconducting transformer. In Advances in Cryogenic Engineering, vol. 33, Plenum Press, New York, 1988, 219-26.
9. Pasztor, G., PSI Switzerland, private communications.
10. Colvin, J., et al., SMES conductor test program. IEEE Transactions on Magnetics 1989, 25, 1586-8.
11. Gao, Z., Knezovic, A., Proposed device for measuring the joint resistance and critical current of superconducting cables short sample up to 40 kA current. CERN internal note, EMA 88/8.
12. Ten Kate, H.H.J., Pijper, H., Nijhuis, A., van de Klundert, L.J.M., Maximum current and quench sensitivity test of 40 kA multistrand NbTi/CuNi conductor. In Proc. MT-9, SIN, Zurich, 1985, 584-7.
13. Ten Kate, H.H.J., Uyttewaal, W., ten Haken, B. and van de Klundert, L.J.M., The Twente high-current conductor test facility, first results on critical current and propagation in two cables. In Advances in Cryogenic Engineering, vol. 33, ed. R.W. Fast, Plenum Press, New York, 1988, 211-18.
14. Mulder, G.B.J., Krooshoop, H.J.G, Nijhuis, A., ten Kate, H.H.J. and van de Klundert, L.J.M., A convenient method for testing high-current superconducting cables. Presented at CEC/ICMC-89, Los Angeles, July 1989.
15. Ten Kate, H.H.J., ten Haken, B., Wessel, S., Eikelboom, J.A., Hornsveld, E.M., Critical current measurement of prototype cables for the CERN LHC up to 50 kA between 7 and 13 tesla using a superconducting transformer circuit. Pres. at MT-11, Tsukuba, August 1989.
16. Ten Kate, H.H.J., Nederpelt, W., Juffermans, P., van Overbeeke, F. and van de Klundert, L.J.M., A new type of superconducting direct current meter for 25 kA. In Advances in Cryogenic Engineering, vol. 31, ed. R.W. Fast, Plenum Press, New York, 1986, 1309-12.

AC LOSS CHARACTERISTICS
OF SUPERCONDUCTING POWER TRANSMISSION CABLE

Kazuaki ARAI, Naotake NATORI, Noboru HIGUCHI
Electrotechnical Laboratory
1-1-4 Umezono, Tsukuba-shi, Ibaraki 305, JAPAN

and Tsutomu HOSHINO
Saga University
1 Honjo,Saga-shi,Saga,840 JAPAN

ABSTRACT

Characteristics of 10m-long Nb_3Sn superconducting power transmission cables are described. Cables are designed to be suitable for power of 1 to 3 GW and are assembled using Nb_3Sn superconducting tapes.

Ac losses, which are of major concern in superconducting cables, is greatly dependent on the design of the cable. And it is very important to assemble cables to avoid the influence of the thermal contraction. Several superconducting cables have been assembled and tested so far[1]. In this paper, two cables among them are picked up, then their ac losses and assembling method are discussed. One of them(cable"N") is assembled by the following method: The electrical insulation of the cable is composed of many pieces of tyvek paper, each of which is helically wound with an appropriate lay angle to reduce axial thermal contraction.

Ac losses of this cable were reduced to about 1/5 to 1/10 of that of the other cable(cable"I"). To investigate the causes of the higher ac losses in cable"I", it was disassembled after the ac loss measurement. Then some degraded Nb_3Sn tapes with traces of edges of the insulator tapes and wires for measurement were found. In order to estimate the degradation of tapes, their critical currents are measured. There are some tapes whose critical currents reach less than half that the original tape.

INTRODUCTION

Power demand in the biggest cities in Japan is showing rapid increase recently, 5-6% per year, because of the high concentration of economic activities in those areas. In such big cities, it can be said that applying superconducting cables is almost inevitable in the near future. The reason is that they are able to transmit a large block of power through densely populated areas where big overhead lines can not be acceptable.

Results of economic evaluations performed so far indicate that superconducting transmission lines are not economically feasible, unless the capacities are over 5GVA or so.

As a result, it can be related that superconducting power transmission lines are superior candidates for power corridors, technically, but they need efforts to bring down the economical breaking point to the reasonable value, in order to bring the superiority into full operation. Therefore, the

A_1:Lay angle of shielding conductor tape.
A_2:Lay angle of transport conductor tape.
d_1:Diameter of shielding conductor.
d_2:Diameter of transport conductor.

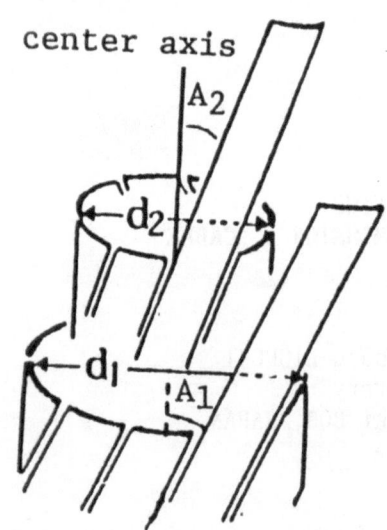

FIGURE 1. Lay angles and
diameters of the cable.

objective of our research is assigned on the fundamentals of superconducting cables suitable for the capacity of around 1-3GVA class.

DESIGN OF THE CABLES

Ac loss characteristics are of the major significance with regard to choosing the material of conductors. And cables assembled with tape conductors are more advantageous than other types of cables with regard to flexibility. As a result, Nb_3Sn superconducting tapes have been adopted for the cables.

If cables with oxide superconductors are manufactured in the future, Nb_3Sn superconducting tapes could be replaced with them without difficulty because both have similar mechanical characteristics, e.g. brittleness.

Cables are composed of two layers of conductors, one for the transport of current, the other for the shielding current to reduce eddy current losses in the cooling channels[2].

The conductor tapes and insulation in the cable are to be helically wound at appropriate lay angles. For the conductor tapes, the most appropriate lay angles are defined by the following

expression(See FIGURE 1).

$$\tan A_2/d_2 = \tan A_1/d_1 \qquad (1)$$

If conductors are wound according this condition, the magnetic flux density at the center of the cable becomes zero so that the eddy current losses inside the transport conductors can be eliminated. On the other hand, there exists a condition to be fulfilled for both the conductors and the insulators shown below.

$$C_t/\sin^2 A = \text{constant}, \qquad (2)$$

where C_t is the coefficient of thermal contraction and A is the lay angle.
This expression indicates the condition to avoid damages which may be caused by the unbalanced thermal contraction inside the cable.

But those two conditions are not consistent each other. In case the first one is satisfied, the second one can not be fulfilled throughout the entire cable. Therefore, to achieve a compromise them, the following way of assembly is applied. For the transport conductors and the shielding conductors, the lay angles are determined to satisfy condition (1). The lay angles of the insulator tapes are gradually changed on the compromising condition. That is, the lay angles are chosen to satisfy condition (2) for both the transport conductor and its adjacent layers of insulator tapes. It is the same for the shielding conductor and its adjacent layers of insulator tapes. The lay angles are gradually changed between those of the above two portions. Cable"N" is the only one of this type of cables designed so far.

Cable"I" is almost the same as cable"N", but there are some differences. Dimensions of these cables are shown in TABLE1. The most important point is the difference of the pitch -which is related with the lay angle and the diameter of the cable - of the insulator tapes. While the pitches are gradually changed from 64 to 112mm for able"N", those for cable"I" are almost constant(about 34mm). And SUS corrugated tube is used for cable"N" while SUS rigid tube is used for cable"I". FIGURES 2(a) and 2(b) show a model made from a piece cut from cable"I" and cable"N", respectively.

TABLE 1. Dimensions of cable"I" and cable"N" (unit:mm)

item	material		outer diam.		pitch	
	I	N	I	N	I	N
former	SUS **rigid** tube	SUS **corrugated** tube	27.2	27.0	-	-
tranport conductor	Nb3Sn tape		30.0	30.6	257	250
insulator	tyvek		43.7	45.0	**33.7-34.8**	**64-112**
shielding conductor	Nb3Sn tape		46.6	45.5	257	250

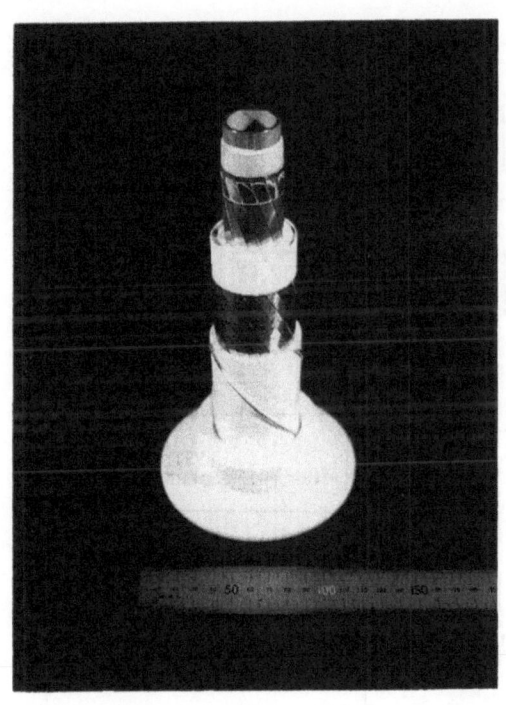

FIGURE 2(a). Cable"I".

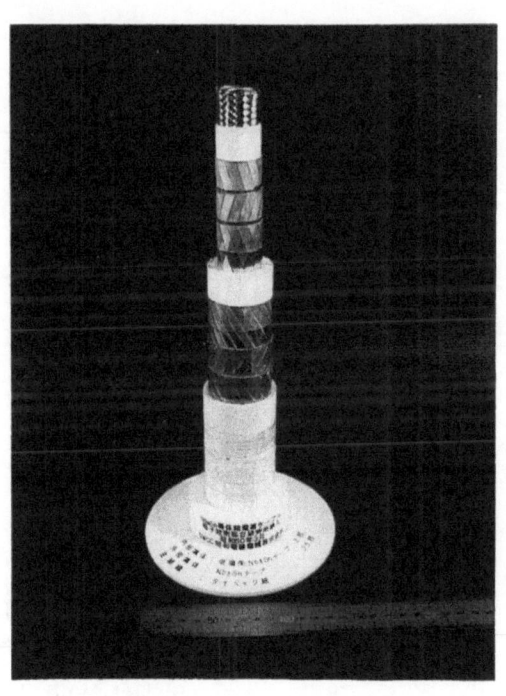

FIGURE 2(b). Cable"N".

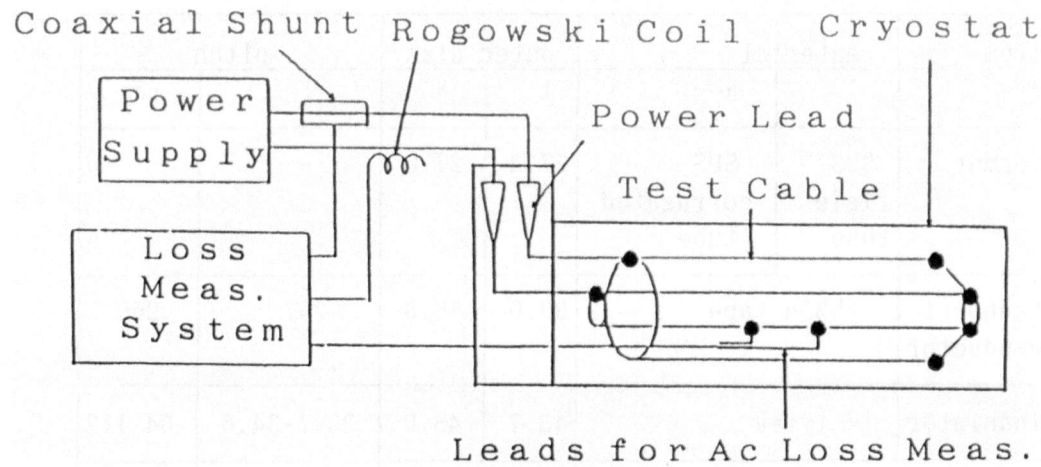

FIGURE 3. Equipment for current tests.

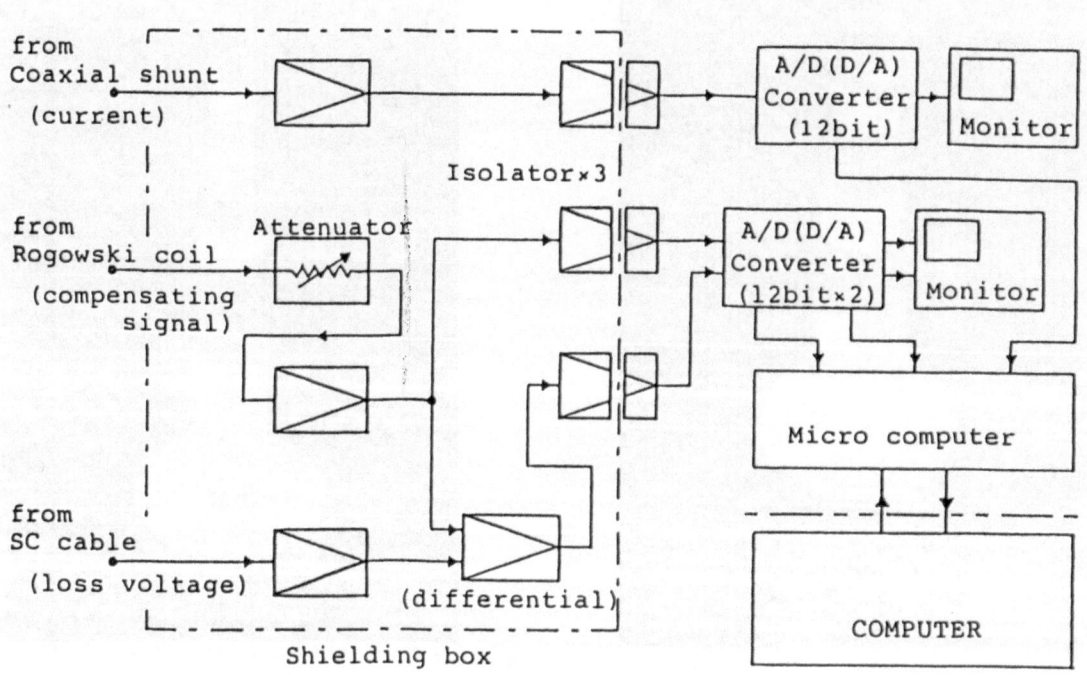

FIGURE 4. Ac loss measurement system.

EXPERIMENTS

The two cables described in the previous section were tested. FIGURES 3 and 4 show schematic diagrams of the equipment comprising the loss measurement system. The conditions of experiment were as follows: (a) The cables were short circuited at one end. (b) The power supply was a single phase transformer with rating 15V and 15kA. (c) The cables were cooled with liquid helium.

The results of the ac loss measurement are plotted in FIGURE 5. The ac losses of cable"N" were reduced to about 1/5 to

1/10 of those of cable"I" for maximum surface current in the range from 150 to 600 A/cm. The difference in the ac losses may not be attributed to the difference of the their SUS tubes because there is little difference between the thermal contractions of a radial directions in the tubes. It can be said that the reduction in the ac losses is due to the assembly of the insulator tapes in cable"N".

DISASSEMBLING OF CABLE"I"

After the experiment, cable"I" was disassembled and some degraded Nb_3Sn tapes were found. To investigate the cause of the degradations in them, their critical currents were measured. They are shown in FIGURE 6. FIGURE 6 shows the degree of degradation of a tape with traces of the edges of insulator tapes(tyvek) and of a tape with traces of wires for measurement, and it also shows the ac losses of the original Nb_3Sn tape. The degradation due to the wires for

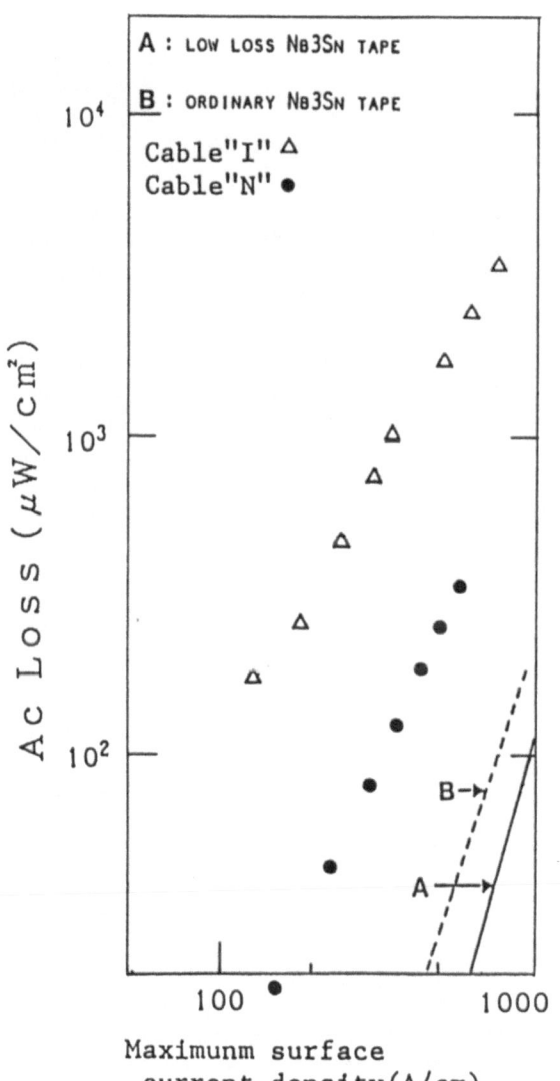

FIGURE 5. Ac loss characteristics of cables.

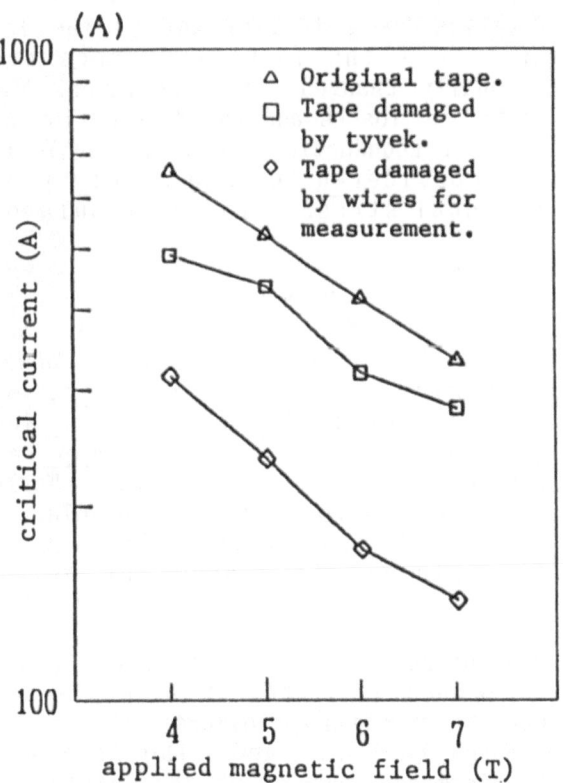

FIGURE 6. Critical current characteristics of degraded tapes and original tape.

490

measurement is more severe and the critical current is 41% to 48% of that of the original tape. On the other hand, the critical current due to the insulator tapes(tyvek) is 74% to 84% of that of the original tape. But the degradation caused by the wires for measurement influenced only little the test of the cable because it was observed as local damage. The damage caused by the insulator tapes makes their ac losses higher, and it is supposed to be a main reason of higher ac losses.

Cable"N" is not yet disassembled since it is scheduled for further experimentation. But it can be said that the conductors in cable"N" were not be damaged by the wires for measurement because wires such as the ones used in cable"I" were not used in cable"N".

CONCLUSIONS

It can be concluded from tests of cable"N" and cable"I" that ac losses in superconducting cables employing Nb_3Sn can be reduced to about 1/5 to 1/10 by adopting the method by which the lay angles of the insulator tapes are gradually changed layer by layer. The higher ac losses and the degradation of the superconducting tapes in cable"I" are considered to come mainly from residual stresses in the insulator tapes.

REFERENCES
1) Higuchi,N., Natori,N.,Arai,K. and Hoshino,T., Nb_3Sn superconducting power transmission cable. In Proc. of International Symposium on New developments in applied superconductivity, ed. Y.Murakami, World Scientific, Osaka, 1988, pp. 668-673.
2)Sutton,J. and Ward,D.A.,"Design of flexible coaxial cores for as sc cables",Cryogenics,1977,17,495-500

Some of references not quoted here about superconducting power transmission cables are shown as follows:
3)Forsyth,E.B. and Thomas,R.A., "Performance summary of the Brookhaven superconducting power transmission system",Cryogenics,1986,26,599-614
4)Itoh,N., Natori,N. and Higuchi,N. "Superconducting power cables and their current tests",Transactions of JIEE of Japan (B).,1981,101,443-450

SECTIONALIZING AND OPTIMIZATION OF TOROIDAL SUPERCONDUCTIVE WINDINGS

V.V.Andrianov,*A.Y.Archangelsky,**V.V.Zheltov,*S.I.Kopylov,**M.B.Parizh*

* Institute for High Temperatures, Academy of Sciences of the USSR, Moscow, USSR

* * Moscow Power Engineering Institute, USSR

ABSTRACT

This paper deals with the problem of optimizing sectionalized toroidal magnetic systems (TMS) by the critical performance of the superconductors, by mechanical stresses, and by deformation. Sectionalization of windings provides a more efficient use of the superconductor through an equal distribution of current. The most regular criteria of efficiency are the ratio of current to its critical value, the mechanical stresses, and the deformation of winding sections or any combination of these factors.

INTRODUCTION

High current TMS are used in fusion units and in inductive energy storage devices. The major advantages of TMS are the high homogeneity of the field in the direction of the flux lines and the practical absence of leakage fields. The latter is especially important for large inductive energy storage devices. In practice, TMS do not made as continuous windings, but are assembled out of a finite number of separate elements made in the shape of circular rings [1]. In that case, leakage fields turn out to be several orders smaller than in the equivalent windings of the solenoid type.

A major disadvantage of the usual TMS is the higher cost of superconductor per unit of stored energy than in solenoid windings. This performance may be improved a lot by the sectionalization of the TMS winding. Sectionalization allows a more efficient use of the superconductor by providing a distribution of current in the sections that results to equal loading by any criterion. Usually the following criteria serve as determinants: ratio of current to its critical value, mechanical stresses, deformation of the system sections or any combination of these factors.

For solenoid windings, optimization by sectionalization is developed in detail [2,3]. Sectionalization of TMS has not been examined practically (though a detailed research of their optimization by selection of winding shape and size has been undertaken [4]).

METHOD OF CALCULATION

The magnetic field flux density in TMS (Figure 1) may be represented as:

$$B(r,\varphi) = B_m(r)(R-r)/(R+r\cos\varphi). \quad (1)$$

Here $B_m(r) \cdot B(r,\pi)$ is the maximum value of the flux density on a coil of radius r, which is reached at $\varphi = \pi$, and can be calculated if a distribution of the average current density in the winding $j_\kappa(r)$ is present:

$$B_m(r) = \frac{M_0(R-r_2)}{R-r} \int_{r_1}^{r_2} j_\kappa(r) dr. \quad (2)$$

Ignoring leakage fields, TMS stored energy may be also expressed through B_m :

$$E = E_1 + E_2 \quad,$$

$$E_1 = \frac{2\pi^2}{M_0} \int_{r_1}^{r_2} \frac{B_m^2(R-r)^2 r dr}{\sqrt{R^2-r^2}} \quad, \quad (3)$$

$$E_2 = \frac{2\pi^2}{M_0}(R-r_1)^2 B_m^2(r_1) \int_0^{r_2} \frac{r dr}{\sqrt{R^2-r^2}} \quad,$$

where E_1 is the energy stored within the winding, and E_2 is the energy of the torus' internal space.

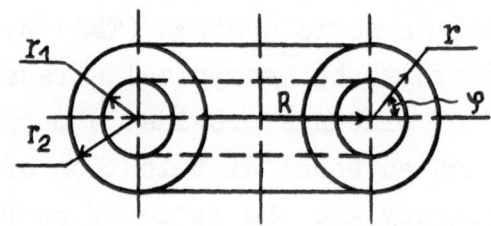

FIGURE 1. Cross-section of the TMS.

To design a sectionalized system it is necessary to calculate a distribution $j_\kappa(r)$ that meets a chosen optimizing function f. The latter depends on the maximum flux density of each section and its radial size:

$$j_\kappa(r) = f(B_m, r) \quad. \quad (4)$$

Optimization is most effective within an infinitely large number of sections, when (4) can be met for each section of the winding. In that case the following equation can be written down on the basis of (2) and (4):

$$B_m(r) = \frac{\mu_0(R-r_2)}{R-r} \int_{r_1}^{r_2} f(B_m,r)\,dr \quad . \quad (5)$$

Differentiating (5) with respect to r :

$$\frac{dB_m}{dr} - \frac{B_m}{R-r} + \frac{\mu_0(R-r_2)}{R-r} f(B_m,r) = 0 . \quad (6)$$

Solution of (6) with the obvious boundary condition $B_m(r_2) = 0$ allows to find a dependence $B_m(r)$. Then, according to (4) an optimal dependence $j_k(r)$ may be determined.

The above analysis is true for TMS with an infinitely large number of sections when the current density continuously changes over the section of the winding.

In case of a finite number of sections, the flux density B_m can be regarded as a sum of the flux densities of all the sections which are external with respect to the point under examination. Taking into account that $j_k = const$ within the boundaries of each section and that it must meet an optimization condition in the most loaded cross-section of the section, at $r = r_{on}$ (usually it is its internal or external surface) we shall obtain the following recursive formula for the calculation of the distributin of B_m in an arbitrary section n :

$$B_m(r_n) = B_m(r_{n-1}) + \frac{\mu_0(R-r_2)}{R-r}(r_{n-1} - r) f(B_m, r_{0n}), \quad (7)$$

where $r_{n-1} > r > r_n$. When using (7), calculation should be done in succession beginning from the outer section.

Correlation (7) allows to determine an optimum distribution of currents when the sizes of the sections are given. The choice of geometry of the sections (parameters r_{n-1} and r_n) depends upon the method of power supply. In case of a parallel connection of the sections, the necessary distribution of current must be provided by the ratio of the effective inductances of sections [1 - 3] .

If each section is power-supplied from a separate source (independent excitation), the sizes of the sections represent additionally variable values and are chosen from the condition of maximum stored energy through an iterative procedure.

RESULTS ANALYSIS

Actual calculations were shown for an optimization function f , providing for:
1. optimization by critical current ($f = A/B_m$, which ensures

fulfilment of condition $j_\kappa B$ =const over the entire winding);

2. optimization by mechanical characteristics. Among them the following ones are of interest: the maximum electrodynamic force on the coil (in calculation per unit of volume) P_m , mechanical stress on the conductor σ_m , radial deformation of the winding coil $\Delta \rho_m$. All these values can be expressed through B_m (values σ_m and $\Delta \rho_m$ are obtained on the assumption that forces are not transmitted from coil to coil) [1] :

$$P_m = B_m j_\kappa \quad ,$$

$$\sigma_m = \frac{\lambda_m B_m j_\kappa r\, r_2}{2\,(r_2 - r_1)\,\lambda} \quad ,$$

$$\Delta \rho_m = \sigma_m r / E \quad .$$

Here: λ and λ_m are the relative coefficients of the winding, filled with superconductor and material that take up the load accordingly, and E is Young's moduls.

In non-sectionalized TMS, owing to the dependence of B_m on r_1 , the functions σ_m and $\Delta \rho_m$ have maxima at $r = r_2/2$ and $r = 2 r_2/3$ or on the torus' internal space. Therefore, if a non-sectionalized magnetic system is limited by these values, the choice for the most dangerous

section of the winding will depend on its relative thickness. At $r_1/r_2 > 0{,}5$ a maximum stress is reached on the internal surface of the winding, otherwise $\sigma_{max} = \sigma_m(r_2/2)$ By analogy the maximum value of $\Delta \rho_m$ takes place at $r = r_1$ in the winding with $r_1/r_2 > 2/3$, otherwise $\Delta \rho_{max} = \Delta \rho_m (2 r_2/3)$.

As parameter $\mathcal{E}$, that determins the energy efficiency of sectionalized TMS, we take the ratio of its stored energy E to the energy E_s of the part of a non-sectionalized infinitely long solenoid winding of length equal to $2\pi R$, with the same material characteristics and values for r_1 and r_2 as TMS. Besides the characteristics of the sectionalization and the type of the optimizing function, this value depends only on the relative size of the windings (variable within the limits from 0 to 1):

$$\alpha = \alpha_s = r_1/r_2 \; , \; \beta = r_2/R \; .$$

This allows to do a sufficiently compact analysis.

The curves shown in Figure 2 determine a maximum efficiency of the sectionalization, corresponding to the limit of an infinitely large number of section $N \to \infty$. In this limit there is no difference between the cases of parallel and independent excitation, therefore $\mathcal{E}$ depends only on the geometry and the type of the optimizing function. For each function the maximum value is reached on the curves $\beta =0$, corresponding to the limit range of the solenoid winding. Reduction of $\mathcal{C}$ by an increase in the parameter

β is connected with two effects: reduction of the energy of the non-sectionalized torus in comparison with an equivalent solenoid winding, and reduction of the efficiency of the TMS sectionalization in comparison with the efficiency of the solenoid sectionalization. Influence of the first is shown on the insertion in Figure 2, which examines the case of a non-sectionalized torus (N=1). The second effect is explained by the dependence of the field in the TMS on the corner coordinate φ , which makes it impossible to satisfy the optimizing function over the entire length of the coil. We have to remark that at $\beta < 0,4$ a sectionalized TMS of any thickness still remains more efficient than a non-sectionalized solenoid winding (the corresponding dependencies lie above the straight line $\mathcal{C} =1$). From the point of view of the type of optimizing function, the maximum effect of the sectionalization is reached for $f = A/B_m$, where A =const. This type of function also allows for a simultaneous optimization by the critical performance of the superconductors and by the electrodynamic stresses, provided

equation $A = P_m$ is satisfied.

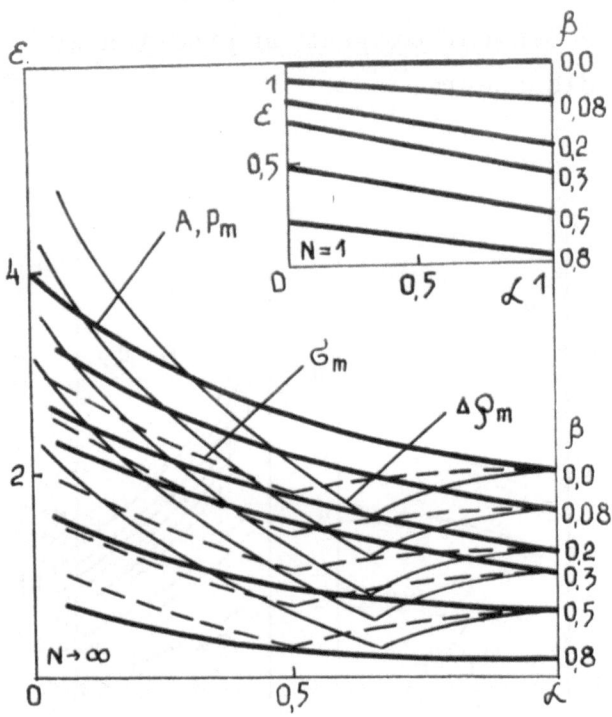

FIGURE 2. $\mathcal{C}$ vs relative sizes α and β .

Calculations for a finite number of sections are shown on Figure 3. As a parameter that determins the efficiency of sectionalization, $\mathcal{C}_T$, is chosen the ratio of the sectionalized and non-sectionalized TMS of the same size. Introducting this new parameter $\mathcal{C}_T$ allows not to vary the radius of the axis, because $\mathcal{C}_T$ practically does not depend on R . Corresponding changes do not exceed 15 % with R varying from r_2 to ∞ . Ratios shown in the analysis of Figure 3 remain valid, with the sectionalization efficiency turning out close to the highest possible extreme even

for a number of sections N > 8-12.
For N > 5-7, the difference between the cases of parallel and independent excitation practically
disappears.

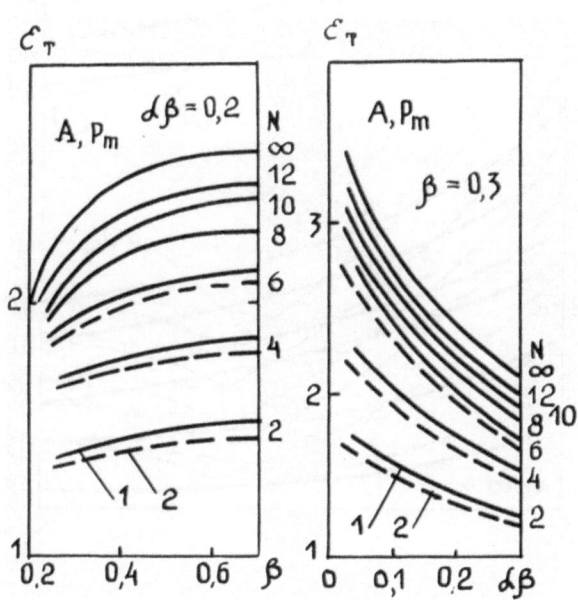

FIGURE 3. $\mathcal{C}_T$ vs α and β .
(1 - independent and
2 - parallel connection)

CONCLUSIONS

This work covers an analysis
of the limit and of the practically attainable efficiency of TMS
sectionalization, and the results
show that the real efficiency of
sectionalization is close to the
theoretically attainable one for
a number of sections N > 8-12.
The possibility of a simultaneous
optimization of TMS by the critical performance of the superconductors and by mechanical stres-

ses has been shown. A range of
sizes has been estimated in which
a sectionalized TMS stored a lar-
der energy than an equivalent non-
sectionalized solenoid winding.

REFERENCES

1. Foner, S., and Schwartz, B.B.,
Superconducting Machines and
Devices, Mir Publishers, Moscow,
1977, pp. 194-201.

2. Andrianov, V.V., and Kopylov,
S.I., The parameters of superconducting coil with parallelly connected sections. Electricity,
1983, 12, 43-46.

3. Andrianov, V.V., Parizh, M.B.
and Kopylov, S.I., Superconducting windings with parallel connected sections. IEEE Trans. on
Magn.,1983, 19, 1105-08

4. Leites, L.V., Electromagnetic
calculations of transformer and
reactor parameters,Energy,
Moscow, 1981, pp. 17-31.

THE INFLUENCE OF A REAL MAGNET ON THE PERFORMANCE OF COLD MHD DEVICE

P. Del Vecchio*, G. Pasotti** and G. M. Veca*
* Electrical Energy Dept. of University "La Sapienza"
Via Eudossiana n. 18 - 00184 Roma (Italy)
** ENEA - Fusion Dept. of Frascati Labs, Via E. Fermi,
Frascati (Italy)

ABSTRACT

With reference to a device for pumping conducting fluids the behavior of this device subjected to a magnetic field produced by a real magnet was compared with that obtained utilizing an hypothesized map of the magnetic field. Through the use of the calculation code GFUN it has been seen that the magnet which most closely approximates the ideal shape is the saddle coil type. The features of the device were computed through a calculation code which schematizes the fluid with an electrical network that has lumped parameters: this was done taking into account the armature reaction and the magnetic field produced by the current of the leads which feed the electrodes.

INTRODUCTION

MHD-type devices (see Fig. 1) have not been widely applied to practical use due, above all, to the difficulty in generating, without high Joule losses, sufficiently intense magnetic fields in large volumes. This difficulty was worsened by the necessity of having in large volumes even very high magnetic fields (> 3T) in order to have elevated mechanical power transmitted to the fluid with low velocity of the fluid itself (containment of hydraulic losses). The practical possibility of obtaining high magnetic fields on large air-gaps through the use of superconducting magnets, has sparked new interest in the study of MHD devices. Some years ago the authors began a feasibility study of a device that would pump sodium in the secondary circuit of fast nuclear breeder reactors [1-4]. Recently, however, this program was shelved due both to the change in the National Energy Plan in Italy and to very low efficiency as a result of Jou-

le losses concentrated in the duct walls. Nevertheless, the authors have continued their analysis of the feasibility of a device for pumping conducting fluids in general.

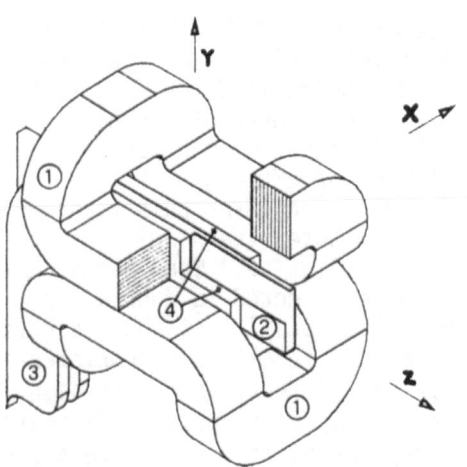

Fig. 1 - Schematic drawing of an electromagnetic dc pump: (1) magnet, (2) duct, (3) current leads and (4) electrodes

In addition to the work covered in this paper a study on the possibility of using these machines both as marine thrusters and as direct conversion of electrical energy devices is under study. In a previous paper [4] the authors have carried out the best magnetic field profile and the duct geometry in electromagnetic pump and its performance. Among different feasible magnets, the authors have taken into account the saddle coil type magnet which most closely approximates the ideal shape and optimizes pump performance. When magnetic field profile was obtained, a hypothesized profile and the real curve has been implemented: according to computations previously made, this hypothesized profile would improve pump performance. The aim of this paper is to compare performance of the pump subjected to the magnetic field of the magnet actually feasible by techniques commonly used with the one of the same pump subjected to the magnetic field obtained by a hypothetical magnet which would produce the hypothesized magnetic profile in order to quantify the depletion of pump performance. The goal of this work is to determine the usefulness of technological and project efforts capable of introducing those changes which would enable the real magnet to be close to the hypothesized curve.

top of the duct near the lateral wall, still as a function of the length of the device. Curve C is the hypothetical abstract profile of these curves. In fact, the ideal profile of the magnetic flux density along the median plane of the channel would be constituted by a flat top (entire length of the electrodes) declining to zero according to a predetermined law [4] at the two extremes.

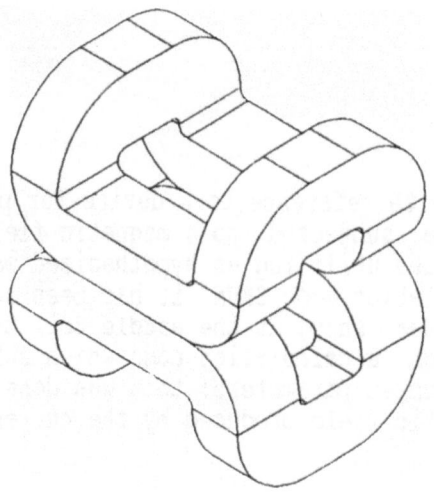

Fig. 2 - Dipole magnet saddle coil type

THE MAGNET

The magnetic field outside the pump is a dipole; it must be as uniform as possible in the area included between the electrodes which carry the electrical current to the fluid. Since the volume occupied by the duct of the pump under analysis is a few tenths of a cubic decimeter, and the magnetic field produced is of 5T, the forces of attraction between the coils are very strong. For this reason the authors have chosen the saddle coil shape as shown in fig. 2. The heads of the two coils were selected in such a way as to obtain the profile of the magnetic field that cancels itself in the briefest space possible; in fact, as can be seen in fig. 3, curves A and B approximate quite well the trapezoidal line C. Curve A indicates the profile of the magnetic field on the axis of the duct as a function of the length of the device. Curve B indicates the profile of the magnetic field on the

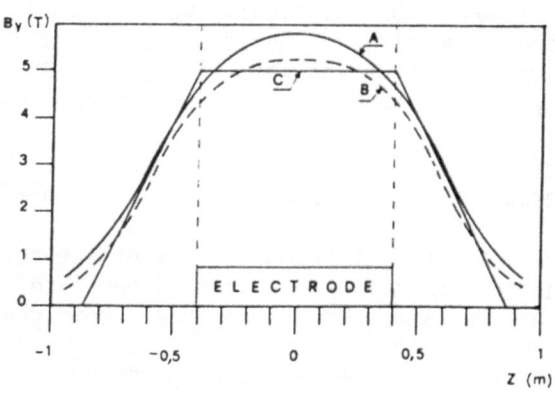

Fig. 3 - Magnetic field produced by the magnet along the z axis: curve A = on the channel axis, curve B = on the corner, curve C = hypothetical profile

TABLE I
Principal parameters of the Magnet

Internal width	X = 0.40 m
Length of the rettilinear tract	Z = 0.80 m
Section of a coil	(0.35 x 0.30) m²
Max magnetic field on the axis	5.8 T
Max magnetic field on the coils	6.8 T
Average current density	J = 3 10⁷ A/m²

The saddle coil configuration has an advantage over others such as the race track: it has a low relationship between the maximum magnetic field in the super-conductor and the magnetic field in the useful region: this allows for sufficiently high current density in the superconductor wire. One should note that a significant space is left between the duct and the magnet permitting the lodging of the dewar but the design of the latter is not described in this paper.

METHOD OF ANALYSIS OF THE MAGNETIC FIELD

The magnetic fields which involve the device are all in the air. To compute the contribution from the two superconducting coils and from the bars leading the current to the electrodes, the GFUN code was used [5]; while for the computation of the magnetic field arising from the currents in the conducting fluid the authors used their own quasi tri-dimensional calculation code which outlines the behavior of the electrical currents in the fluid in motion through an electrical network with lumped parameters [1-3]. The curves A and B of Fig. 3 were then computed with the GFUN code. In Fig. 4 is shown the profile on the axis of the pump of the magnetic flux density produced by the current in the leads as a function of the length of the device (see Fig. 5). The direction of the current was chosen specifically to create a magnetic field with the same polarization as the magnet: in this way, the force that is generated on each bar for the interaction with the field of the magnet (F/l = 500 kN/m) is directed in such

a way as to distance the bars and is easily containable with the proper external tying system. If the two magnetic systems (magnet and bars) produced opposing fields, the two bars would tend to crush the channel and it would be more difficult to contain the forces. Theoretically, it would have been easier to feed the electrodes with the bars positioned perpendicularly to the electrodes to thus avoid producing a magnetic field on the fluid: this would have complicated the construction of the dewar without producing any appreciable gains: the value of the magnetic field generated by the bars in fact is only 4 percent of that of the principal field.

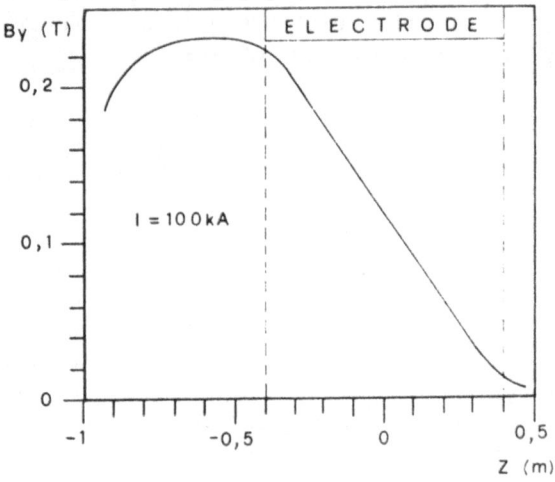

Fig. 4 - Magnetic field produced by the current leads on the axis of the MHD channel

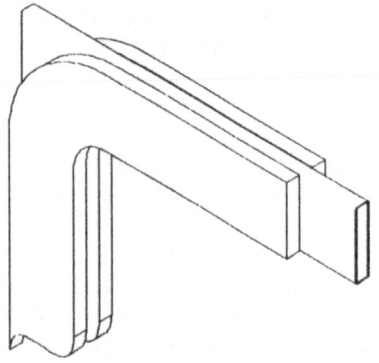

Fig. 5 - Schematic drawing of the leads of the electrodes

The diagram in Fig. 4 was plotted with a device feeding current equal to I=100kA. One notes that the magnetic field remains fairly constant (B_y = 0.22T) along the entire tract in which the bars run parallel to the channel before reaching the electrodes; it then lowers linearly in the tract included between the electrodes.

NUMERICAL RESULTS

Figure 6 shows the results from the computation of the map of the magnetic field created by the current which passes through the conducting fluid for the same value of the current with which Fig. 4 was constructed (I = 100 kA): the curve furnishes the profile of the magnetic flux density on the reference axis as a function of the length of the device.

TABLE II
Data used for the MHD device

Resistivity of the fluid	$7.2 \cdot 10^{-4}$ Ohm cm
Cross section of the duct	300 cm²
Shape of the duct	rectangular
Length of the active part	0.80 m

With reference to the values of the parameters indicated in Tables I and II it was useful to compare the performance of the device that has a real magnet with that of the same device using an hypothesized magnetic field. Figure 7 compares the curves in terms of efficiency as a function of flow rate, figure 8 compares the curves in terms of efficiency as a function of mechanical power and finally, figure 9 compares the curves in terms of head as a function of flow rate. In all the diagrams "RE" and "ID" refer to computations made with a real or an ideal magnetic field, respectively. One notes that the projected magnet approximates very closely the features obtained with a distribution of a hypothetical magnetic field.

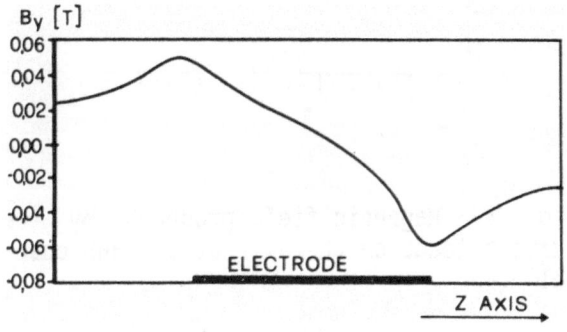

Fig. 6 - Magnetic field produced by the current in the fluid on the axis of the pump

The calculation was performed made with the quasi tri-dimensional code mentioned before. Combining the results of the magnetic fields of the bars, the coils and the fluid, and inserting the results into the model developed by the authors the features of the pumping device were calculated with a rapidly converging iterative process. The main characteristics of the pumping device are indicated in Table II.

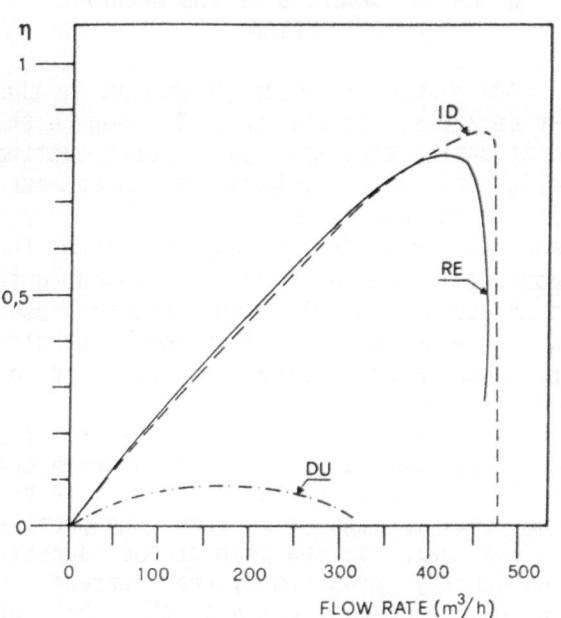

Fig. 7 - Efficiency's progression as a function of flow rate: ID=hypothesized magnetic field, RE=real magnetic field, DU=real magnetic field with dissipative effects in the duct

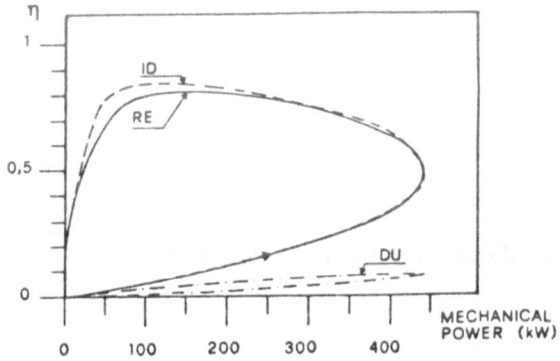

Fig. 8 - Efficiency's progression as a function of mechanical power: ID=hypothesized magnetic field, RE=real magnetic field, DU=real magnetic field with dissipative effects in the duct

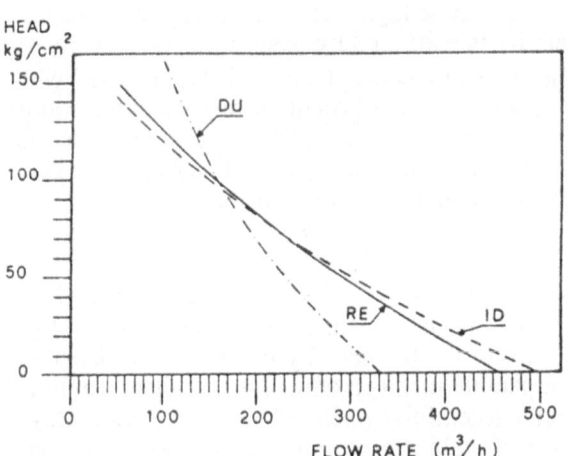

Fig. 9 - Head's progression as a function of flow rate: ID=hypothesized magnetic field, RE=real magnetic field, DU=real magnetic field with dissipative effects in the duct

The performance illustrated by curves in figures 7, 8 and 9 refers to a pumping device in which it may be possible to insulate the electrodes from the duct and the walls of the duct from the fluid. In in-stances where this would not be possible, as for example for sodium, the performance of the pump deteriorates to a very low level. This is demonstrated by the dotted line curves indicated by letters DU, computed by considering a dissipative effect of the electrical currents in a stainless steel duct of only 1 mm.

ACKNOWLEDGMENTS

The authors wish to thank Enzo di Ferdinando of ENEA for his enthusiastic collaboration on the computer graphics. This work was financed in part by ENEA and by MPI (research contract 60/88).

REFERENCES

[1] Del Vecchio P. and Veca G.M. "Eddy currents and field computation in a super-conducting electromagnetic pump"; IEEE Transactions on Magnetics, Vol. Mag-19, No. 5, September 1983

[2] Del vecchio P. and Veca G.M. "Eddy current computation in a superconducting electromagnetic pump"; IEEE Transaction on Magnetics, Vol Mag-19, No. 6, November 1983

[3] Del Vecchio P., Geri A. and Veca G.M. "Superconducting magnets for electromagnetic d.c. pumps"; IEEE Transaction on Magnetics, Vol Mag-21, No. 2, March 1985

[4] Del Vecchio P., Geri A. and Veca G.M. "Superconducting MHD devices: their performance according to components geometry"; Paper presented at Applied Superconductivity Conference on 22th of august 1988, in S. Francisco,CA

[5] Newmann, M.J. et al. "GFUN: an interactive program as an aid to magnet design"; Proc. 4th Int. Conf. on Magnet Technology, Brookhaven 1972.

MHD MAGNET DESIGN FOR THE J. E. CORETTE PLANT RETROFIT

H. Gurol, General Dynamics Space Systems Division, P.O. Box 85990, MZ-92-8260
San Diego, California, 92138, U.S.A.

ABSTRACT

The purpose of this paper is to discuss the conceptual design of the superconducting magnet system for the coal-fired MHD retrofit of the Corette plant. The basic magnet requirements are driven by increased MHD system performance, and a high degree of reliability. High magnetic fields increase the net channel output from MHD systems: the peak channel field used is 4.5 T. These high magnetic fields require the use of superconducting coils to reduce operating costs. Reliability is enhanced by using a current density of about 3,000 A/cm^2, resulting in a high stability margin. The conductor concept chosen is NbTi cooled with pool-boiling 4.2 K helium. An efficient superstructure is used to carry the Lorentz forces when the coil is operating. Wherever possible, off-the-shelf components are used for power supplies, refrigerators, etc. In terms of overall size and stored energy this coil is similar to other large magnet systems that have been built and operated successfully, and hence will not require significant technology development.

INTRODUCTION

The J. E. Corette coal-fired power plant located in Montana, U.S.A., is one of two candidates for the retrofit of a magnetohydrodynamic (MHD) generator. General Dynamics Space Systems Division has been a member of the MDC (MHD Development Corporation) team to develop conceptual designs of critical components The purpose of the retrofit is to serve as a test bed that will lend confidence in the scale-up to even larger MHD units.

The basic magnet requirements were driven by increased MHD system performance, and a high degree of reliability. Reliability of the superconducting magnet was considered to be extremely important for this retrofit. Hence, a rather conservative approach was taken, whenever the benefits of a more aggressive design were not overwhelming. High magnetic fields increase the net channel output from MHD systems; typically 6 Tesla peak for subsonic channel flow and 4.5 Tesla for supersonic flow. These high magnetic fields require the use of superconducting coils. High reliability is obtained by using current densities comparable to superconducting magnets that have been successfully operated for years, with a high level of stability margin. In addition, wherever possible, off-the-shelf components are used for power supplies, refrigerators, etc. We have relied in some cases on a previous MHD magnet designed by General Dynamics called the "CASK MHD Magnet" [1]. Since this design study was performed at a high level of detail, many of its structural, and conductor characteristics were carried over. Also, we used the detailed cost estimates of that report, and scaled them to the MHD Retrofit magnet.

COIL PARAMETERS

The peak channel magnetic field required was 4.5 Tesla; the active channel length was 10 meters. The coils were positioned in a manner consistent

Supported by the MHD Development Corporation, contract number MDC-1002.

with the channel space requirement, and requirements for thermal isolation of the coil windings. The major coil parameters are given in Table 1.

Table 1   Major Coil Parameters

Coil length = 11 m
Coil width (without superstructure) = 3.6 m
Stored energy = 0.75 GJ
Peak channel field = 4.5 T
Conductor current = 3800 amps
Superconductor = NbTi
Ampere turns = 14.6 MA
Coil pack dimensions = 30x30 cm^2 and
   40x40 cm^2
Coil pack current density = 3220 Amps/cm^2
Cooling = LHe-I at 4.2 K, pool boiling
Coil configuration = nested circular saddle
Maximum field on conductor = 8 T
Heat load into He system = 120 Watts
Overall length = 13 m
Overall width = 8.3 m
Overall height = 10 m

The overall coil dimensions in Table 1 include the superstructure, as well as the vacuum vessel. The height includes the vacuum vessel support and the stack for the conductor leads. Figure 1 is a cross-section of the coil, showing the major components of the magnet cold and warm masses. The superstructure was designed to carry the Lorentz forces in an efficient manner. It is designed, following the CASK magnet design [1] to use support structure for the coil where it is needed, and to reduce the amount of structure where the loads fall off. Since the required magnetic field taper in the channel is small, cylindrical helium and vacuum vessels were used.

MAGNETIC FIELD CHARACTERISTICS

The magnetic field calculations were performed using the code EFFI ("Electromagnetic Field, Force, and Inductance"). The coil configuration used for the magnetic field calculations is shown in Figure 2. The coil nesting was adjusted to give the required field drop-off rate at the channel ends. Ninety (90) degree end-turns were used for overall coil compactness, but can also be placed at an angle if a slower field drop-off rate is desired. The magnitude of the magnetic field as a function of position along the channel center line is also shown in Figure 2. At positions outside the main channel (z < -1 m, and z > 11 m) the magnetic field reverses, hence the two small "bumps" in the curve near those positions. Even though the field drops off rapidly outside the coil ends, its value may still be large enough to be of concern for components near the channel inlet and outlet. Another important consideration is the presence of structural iron near the coil ends, which can undergo magnetization resulting in significantly enhanced loads. For this reason we also calculated the magnetic field near the coil ends. These magnetic field values must be taken into consideration in properly designing the structural and electrical aspects of the combustor. The maximum field on the coil is about 8 T, and occurs in the coil ends as the coil pack goes through the 90 degree turn.

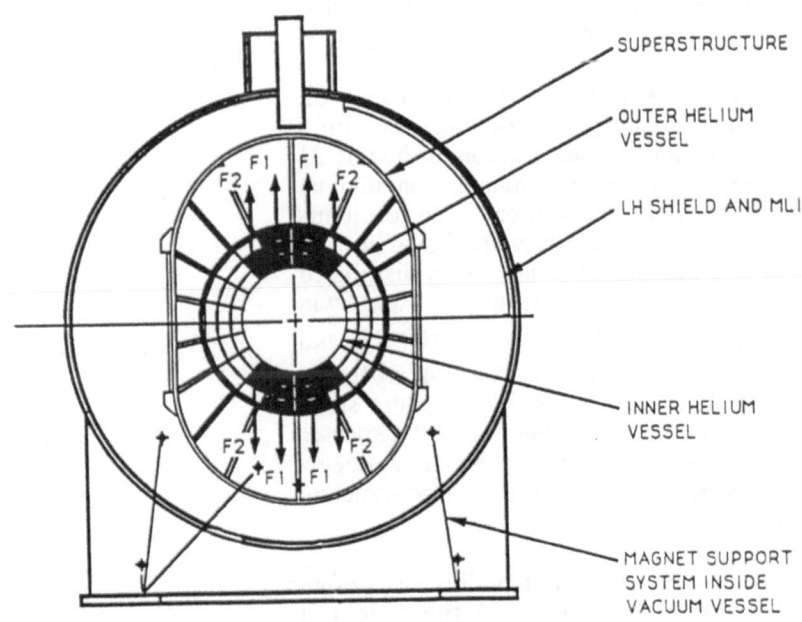

Figure 1   The superstructure, coil pack, and helium vessel are shown at a typical cross-section.

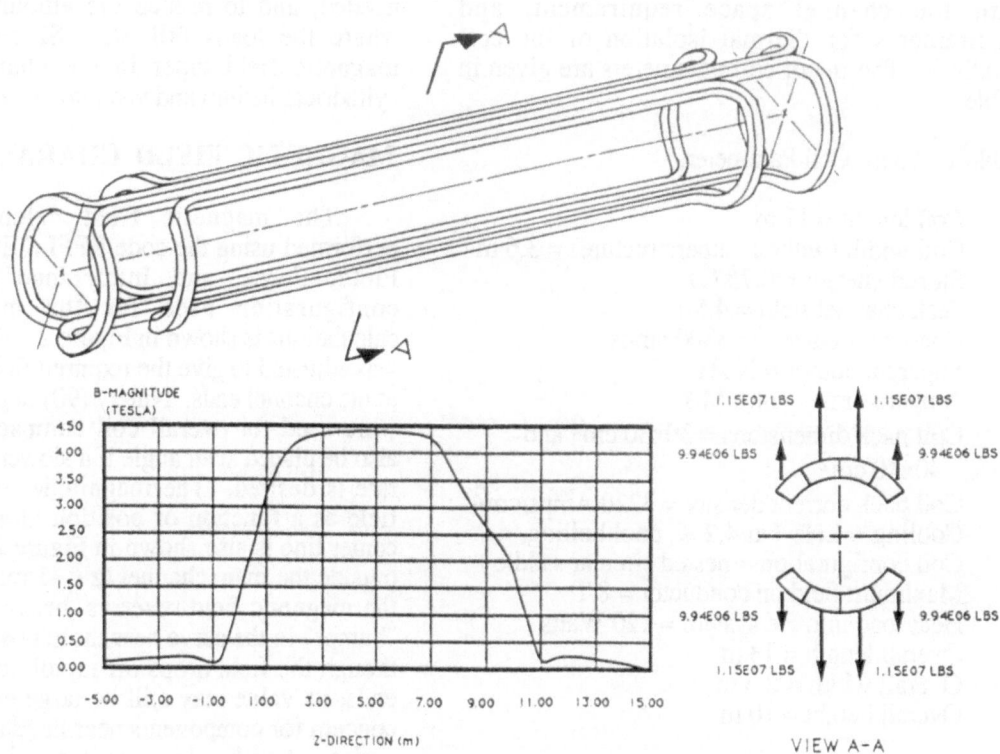

Figure 2 The magnetic field requirements determine the induced Lorentz loads.

STRUCTURAL CONSIDERATIONS

Radial, non-axisymmetric Lorentz forces are generated within the coil during the operation of the magnet. The vertical component of these forces, distributed along the horizontal portions of the winding, tend to pry the coil apart. The vertical force resultants (i.e., the integral of the distributed forces) are shown in Figure 2.

A structure is required to restrain the coil windings and in so doing, provide a load path for these magnetic forces. This is accomplished by 23 ring stiffeners equally spaced along the longitudinal axis of the magnet. The vertical force resultants applied to a typical ring stiffener are shown in Figure 1. The Lorentz forces are evenly distributed into the 23 ring stiffeners. The forces F_1 and F_2 in each of the 23 ring stiffeners are 5×10^5 lbs. and 4.32×10^5 lbs., respectively. The ring stiffeners are to be constructed of 304 LN stainless steel. The required beam depths and flange thicknesses are sized for the developed bending moments around the ring. Figure 3 shows the superstructure and its method of asembly.

This must be viewed as a preliminary design which will be optimized later. A finite element analysis will be required for verification of the final configuration.

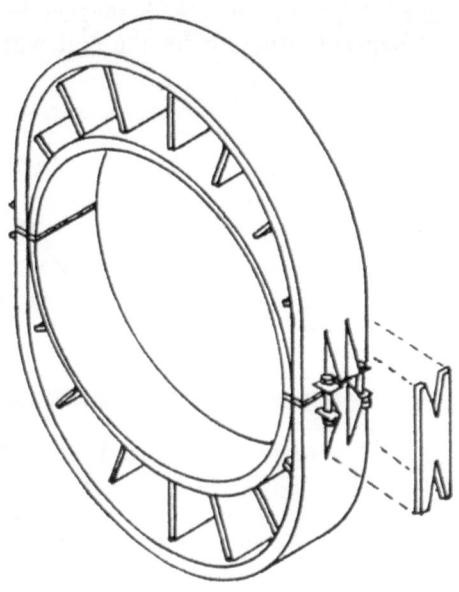

Figure 3 The superstructure assembly is sized to restrain the induced Lorentz loads in an efficient manner.

CONDUCTOR DESIGN

The conductor design chosen for the MHD retrofit magnet is shown schematically in Figure 4. It is a monolithic, pool-boiling, He-I cooled conductor. The conductor consists of the superconducting region which contains NbTi superconducting strands in a copper matrix in a ratio of about 2.5:1. This region is surrounded by high purity copper stabilizer. Its purpose is to carry all of the current, in case a normalization event occurs in the conductor. The overall copper to superconductor ratio is about 20:1. This leads to an unconditionally cryostable magnet according to the Stekley criteria, which states that to be completely cryostable the parameter α, defined below, must be less than unity [2]. That is:

$$\alpha = \rho \, I_c^2 \, / \, [hAfP(T_c-T_b)] \; < \; 1,$$

where ρ is the electrical resistivity of the copper, I_c is the critical conductor current, f is the fraction of copper area, P is the wetted perimeter, A is the conductor cross-sectional area, and T_c, T_b are the critical and bath temperatures, respectively. The critical current for this conductor is about 5800 A. With an operating current of 3800 A, the ratio of operating to critical current is a very safe 66%. In addition, the cryostability parameter α is less than unity (about 0.95); hence, the conductor is unconditionally cryostable.

We have considered the possibility of using an Internally Cooled Cable-in-conduit Conductor (ICCS). One of the main potential advantages of this concept is structural. Basically, the superconducting strands are placed in a stainless steel conduit, which takes the place of an outer He vessel. The conduit serves to carry some of the magnetic loads imposed on the coils. Also, a separate He vessel is not required. However, the length of the MHD coils (10 m) means that they are basically a very long, hence flexible structure. This means that the conduit of an ICCS conductor will not be capable of carrying the magnetic forces, and an external support structure will still be needed. Therefore, many of the advantages of an ICCS conductor are not nearly as pronounced as they are for example for a solenoidal geometry. For this reason, we have, at this conceptual level, decided to use a pool-boiling concept. However, the magnet community is gaining considerable experience with ICCS conductors; one of the biggest applications being a 129 m diameter energy storage coil (SMES) being designed by Bechtel and General Dynamics. Therefore, if later studies show worthwhile advantages, there is sufficient

technical expertise to adapt the concept for use on MHD coils.

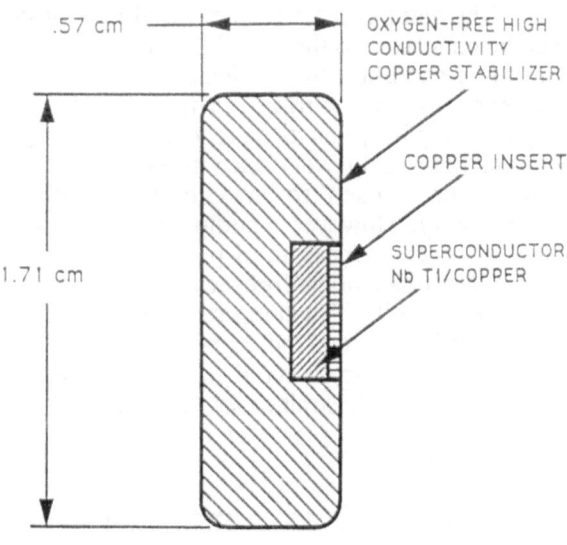

Figure 4   The conductor was chosen to be unconditionally cryostable.

POWER SUPPLIES

Power supplies for a magnet system are basically divided between the power needed to charge the coil, and the power supply requirements for cryogenic refrigeration.

The power supply requirements for charging a magnet depend on the inductance of the coil, and how fast it needs to be charged up to full current. From magnetics calculations, the stored energy is found to be $E = 0.75$ GJ. The chosen conductor operating current is 3800 Amps. This is a reasonable current, in line with magnets built to date. The resulting inductance L is 104 Henries. The required power supply charge voltage can then be determined, given the charge time. Given a nominal one hour charge time the required power supply voltage is approximately 110 volts. The power supply rating is then 417 kW. Assuming about an 80% efficiency leads to a power rating of about 0.5 MW for a 1 hour charge time.

The refrigeration power is based on the heat leak rate into the 4.2 K He-I. The major heat leak occurs at the vapor-cooled leads. The estimated total heat leak rate into the MHD Retrofit magnet is estimated to be about 120 W. Using a refrigeration efficiency of 600 W/W for He-I, leads to refrigeration power requirement of about 72 kW.

COIL PROTECTION

The dump resistor is an external resistor that can be connected to the coil if the instrumentation detects a condition that indicates that the coil is undergoing a normalization, and that superconductivity cannot be recovered. This is an extremely unlikely event since the coil parameters have been chosen to provide unconditional cryostability. However, a dump resistor is always used as an extra measure of protection to ensure that there exist no unforeseen events that can damage the coil. The dump resistor resistance is determined by the coil inductance and the required energy dump time to avoid damage to the coil. This needs to be determined for this coil.

FACILITY REQUIREMENTS

The space requirements for the coil and its subsystems have been determined based on similar magnets built or designed to date, or by inquiring with subsystem vendors. The major components sized include: coil, cryogenic storage systems, refrigeration equipment, power supplies, and dump resistor (based on a one minute dump). All components sized have been arranged to fit in the space provided by the general facility plan.

COST AND SCHEDULE

The cost for the MHD Retrofit magnet system was estimated by scaling from the CASK magnet study performed by General Dynamics in 1980 [1]. The first step in estimating the cost, was to estimate the masses for the major magnet components. The masses of the different components of the coil were evaluated first by calculating the amount of conductor required to produce the field along the channel. Then the forces generated in the conductor due to the interaction of the current in the coil with the generated magnetic field were evaluated. This allows an approximate structural mass to be evaluated. The cold mass of the coil is 3.65×10^5 kg. The total coil weight is 5.43×10^5 kg. This gives a value of stored energy per unit mass of about 1.4 Joules/gm. This figure is in line with other large superconducting magnets that have been built.

The cost for the retrofit magnet was estimated by scaling from the CASK cost estimate which was in 1980 dollars; use was made of the Department of Energy inflation tables through 1986, and 4% yearly inflation after 1986. The scaling used was of the form:

Retrofit cost = (Labor for coil ends + fee + contingency) + ($37/kg) x (Total coil weight).

This resulted in a total cost of about $50 M in 1989 dollars.

The expected schedule for manufacturing and site installation of the MHD retrofit magnet is expected to span 5 years. This schedule is compatible with other large superconducting magnets that General Dynamics has built. A good example are the MFTF magnets built for LLNL.

CONCLUSIONS

A conceptual MHD magnet has been designed for the J. E. Corette plant retrofit. The basic design philosopy was to reduce risk as much as possible by using a conservative approach to choosing the conductor (monolithic NbTi), coolant (pool-boiling He-I, and structure (1.4J/gm). The expected construction time for this magnet is about 5 years.

REFERENCES

1. "CASK Commercial Demonstration Plant MHD Conceptual Design Program", General Dynamics Report Number CASK-GDC-031, prepared for Massachusetts Institute of Technology Francis Bitter National Laboratory, December 1979.

2. "Superconducting Magnets", by M. N. Wilson, Clarendon Press Oxford, 1983, Chapter 6.

HIGH FIELD SUPERCONDUCTING MAGNET FOR A CLOSED CYCLE DISK MHD GENERATOR

T.OKAMURA, Y.OKUNO* and S.SHIODA**

Tokyo Institute of Technology, Oh-okayama 2-12-1, Meguro-ku, Tokyo, 152 Japan
* Saga University, Honjo-machi, Saga, 840 Japan
** Tokyo Institute of Technology, Nagatsuta 4259, Midori-ku, Yokohama, 227 Japan

ABSTRACT

Advantages of implementing high magnetic field up to 10 T for a closed cycle disk MHD generator (thermal input of 1000 MW) are discussed, and technical feasibility of its superconducting magnet is studied. The present study suggests a high performance of the generator with high magnetic field. A possible construction of the magnet, in which a special structure is arrenged to sustain a large electromagnetic force induced between the coils, is shown.

NOMENCLATURE

B : magnetic field strength
Bo : magnetic field strength at the center of the disk
E_r : Hall electric field
e : electron charge
J_r : Hall current density
m_e : electron mass
n_{gas} : concentration of working gas atom
P : static pressure
β : Hall parameter
β_{eff} : effective Hall parameter
ν_{ej} : collision frequency between electron and jth particle
ν_{e-gas} : collision frequency between electron and working gas atom
σ_{eff} : effective electrical conductivity
$\langle \ \rangle$: spatial average

INTRODUCTION

An utilization of a high magnetic field is examined with an aim of achieving a better performance of closed cycle disk type supersonic MHD generators. An attention is paid to MHD generators with a high enthalpy extraction percentage of 40 %, in which a ratio of pressures at the entrance and the exit of the MHD generators is large. In this case, a high pressure of a working gas is required at the entrance, otherwise a pressure in downstream components such as a compressor, heat exchanger, boiler, etc, is undesirably low.

In the present work, firstly, advantages of the high magnetic field for the generator are clarified, and secondly, a technical feasibility of a high field superconducting magnet coupled with the generator is discussed. As a first step toward these problems, conceptual design studies on the generator and the magnet are conducted for the case under typical conditions of closed cycle MHD generators.

The magnet is considered to be a Helmholtz type to obtain a good uniformity of magnetic field in the MHD channel. In this magnet, two circular coils face each other with a space left in between to accomodate the generator. In this case, a special care has to be taken to support the separated coils against a large electromagnetic attractive body force induced between them, while keeping heat loads to the coils as low as possible. For this purpose, a new supporting structure is proposed. And its feasibility and cryostatic stability of a superconductor are examined.

BASIC PARAMETERS FOR DESIGNING MHD GENERATORS AND SUPERCONDUCTING MAGNETS

A typical configuration of the disk type supersonic MHD generator to be dealt with in the present paper is shown in Fig.1. Both sides of the disk (nozzle & MHD channel & diffuser) are attached to hot ducts through which a working gas enters the nozzle and then flows along the radial direction of the disk.

Basic parameters for designing the present MHD channel are listed in TABLE 1. As a working gas, potassium seeded argon or helium is considered, respectively. The thermal input to the MHD generator is assumed to be 1000 MW, which is a typical value for commercial generators. Electron temperature in the MHD region is kept to be 6000K, which makes the seed fully

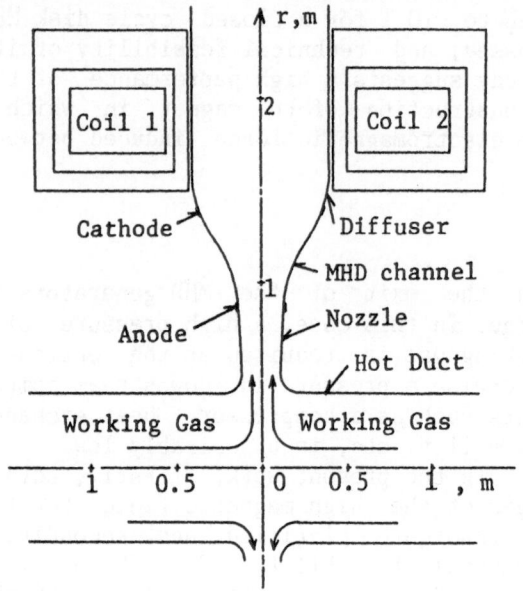

FIGURE 1. A typical configuration of the disk type MHD generator and the magnet.

TABLE 1
Basic parameters adopted for designing an MHD channel

Working gas	Ar+K	&	He+K
Thermal input, MW		1000	
Gas temperature, K		2200	
Electron temperature, K		6000	
Wall temperature, K		600	
<Hall parameter>		20	
Inlet Mach number	2.1		1.7
Outlet Mach number		1.2	
Seed fraction	3×10^{-5}		2×10^{-5}

ionized. The spatial average of Hall parameter in the MHD channer1 is specified to

be 20 and Mach number for the flow at the inlet of the MHD channel is 2.1 and 1.7 for the argon and helium cases, respectively, aiming at a high performance of the generator. The Mach number at the outlet of the MHD channel is close to unity, leading to better diffuser efficiency. The seed fractions are assumed to be 3×10^{-5} and 2×10^{-5} for the argon and helium cases, respectively, from the view point of the forrowing fact. That is, as the seed fraction increases the output power density of the generator becomes higher and the stored energy of the magnet becomes lower, on the contrary, as the seed fraction decreases, the stagnant gas pressure at the outlet of the generator becomes higher and the Hall current density becomes lower. The static gas pressure in the MHD channel varries in order to achieve a constant value of the Hall parameter by changing the magnetic field strength.

Since a generator geometry and a magnetic field distribution in the MHD channel are coupled with each other, calculations for the generator and the coils have to be made simultameously. The shapes of the coil cross section are assumed to be rectangular, and the average current density in the coils is fixed to be 100 A/mm^2 on the basis of the current technology level. The inner radius of the coil and the distance between the coils are assumed to be (cathode radius) + 30 cm and (MHD channel hight at cathode location) + 30 cm, respectively, taking into account of an adiabatic space and a supporting structure. The magnetic field distribution in the MHD channel is calculated for the selected coil shape, and the performance of the generator under the magnetic field distribution is calculated on the basis of one-dimensional MHD equations under the condition of a fully ionized seed. The detailed procedure for the numerical calculations is described in Reference [1].

CHARACTERISTICS OF GNERATORS AND MAGNETS

The enthalpy extraction and the isentropic efficiency as functions of Bo are shown in Fig.2. It can be seen that large values of about 40 % and 80 % are obtained, respectively, and that these values have little dependence on Bo. This is because the performance of the disk type MHD generator is primarily determined by the value of the Hall parameter.

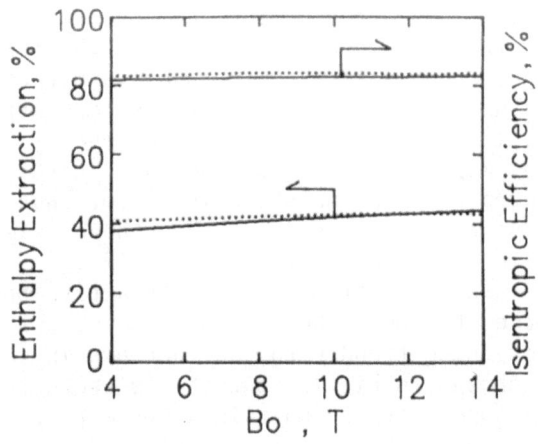

FIGURE 2. The enthalpy extraction and the isentropic efficiency as a function of Bo; —— Ar, ••••• He.

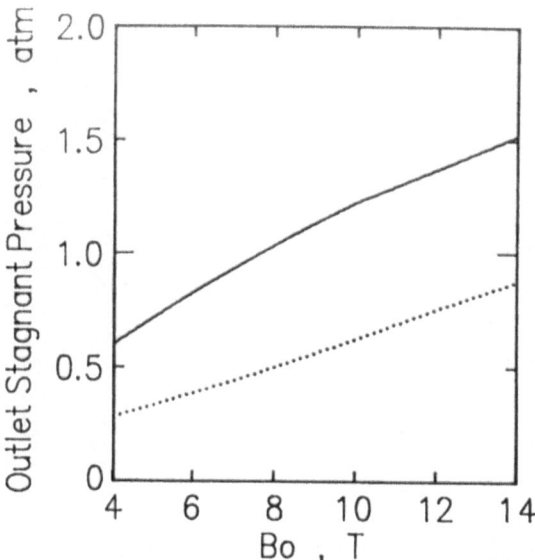

FIGURE 4. The stagnant gas pressure at the outlet of the MHD channel as a function of Bo; —— Ar, ••••• He.

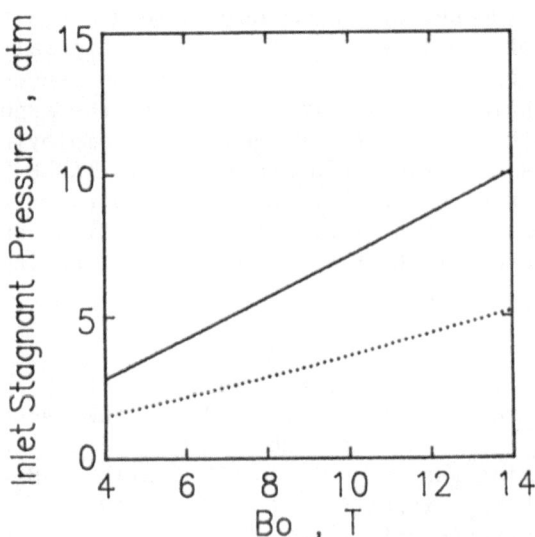

FIGURE 3. The stagnant gas pressure at the inlet of the MHD channel as a function of Bo; —— Ar, ••••• He.

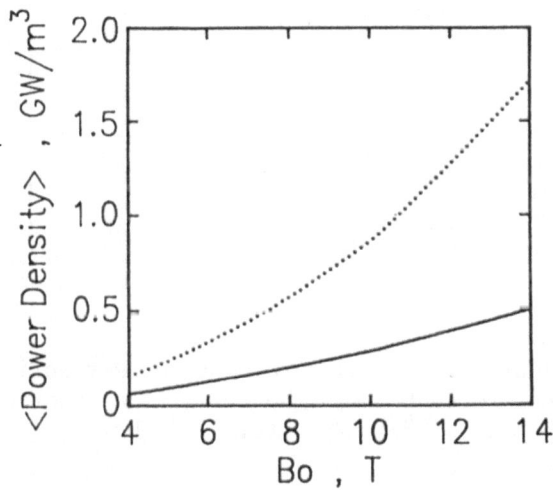

FIGURE 5. The spatial average of output power density of the MHD generator as a function of Bo; —— Ar, ••••• He.

The Hall parameter is written as

$$\beta = \frac{eB}{m_e \nu_{ej}} \propto \frac{B}{\nu_{ej}} \quad .$$

Since the following relation holds for the present case;

$$\nu_{ej} \simeq \nu_{e-gas} \propto n_{gas} \propto P \quad ,$$

we have

$$\beta \propto B/P \quad ,$$

which indicates that when higher magnetic field is applied the static pressure in the MHD channel can be increased keeping a constant value of the Hall parameter. Fig.3 and Fig.4 show relations between the stagnant gas pressure at the MHD channel

inlet and outlet and Bo, respectively. It can be seen that a ratio of the outlet pressure and inlet pressure is about 1/6. The outlet pressures in the generator becomes considerably low. For instance, when Bo is 4 T the outlet pressures in the argon and helium-driven generator are 0.6 atm and 0.3 atm, respectively, which are considerably low.

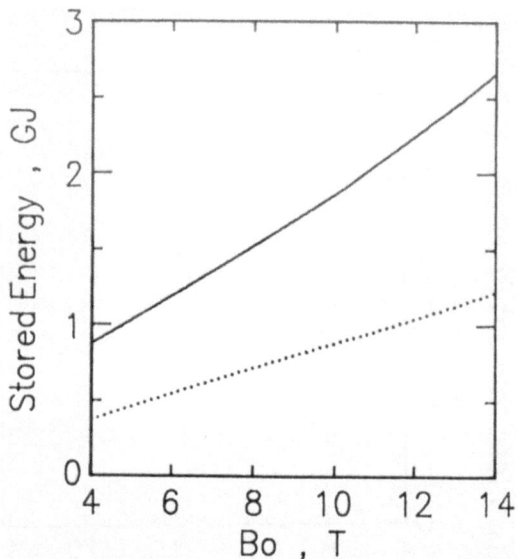

FIGURE 6. The stored energy of the magnet as a function of Bo; —— Ar, ••••• He.

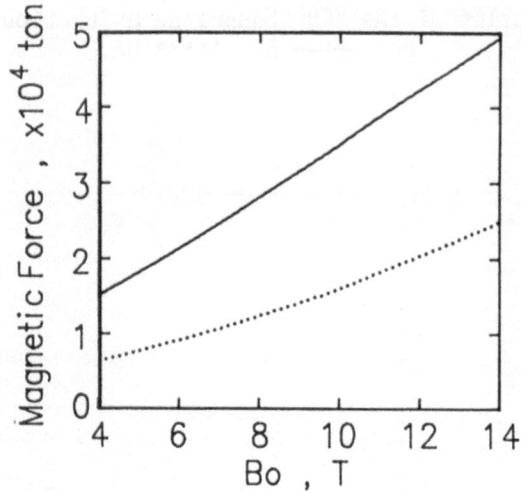

FIGURE 7. The electromagnetic attractive body force between the coils as a function of Bo; —— Ar, ••••• He.

The other advantage of the high magnetic field, which corresponds to the high pressure, is that the MHD channel can become compact. This leads to a relation that the output power density tends to remarkably increase with an increase in Bo, as shown in Fig.5. An important fact, which can be seen from this figure, is that the output power density increases larger than five times while Bo increases from 4 T to 12 T. This fact is caused by the decrease of the MHD channel size as Bo increases.

The inner radius of the coil and the distance between the coils of the super-conducting magnet can become smaller as Bo increases, corresponding to the decrease of the MHD channel size. On the other hand, the coil length and width (outer radius - inner radius) increase for larger Bo. In consequence, the stored energy of the magnet tends to increase with an increase in Bo, as shown in Fig.6, but it does not increase more than three times while Bo increases from 4 T to 12 T.

The electromagnetic attractive body force between the coils also shows an increasing trend with an increase in Bo, as shown in Fig.7. But it is shown that the force can be sustained by a proposed structure, which will be explained in the next section.

TYPICAL GENERATOR AND MAGNET

We select a magnetic field strength Bo of 10 T, which is a reasonable value taking into a consideration of the present technical level. Parameters of the argon and helium-driven MHD generators employing this magnetic field are listed in TABLE 2. The size of the helium-driven generator is smaller than that of argon-driven one because of the higher velosity of helium.

TABLE 2
Parameters of typical MHD generators for the magnetic field strength at the center of the disk of 10 T

Working gas	Ar+K	He+K
Enthalpy extraction, %	42	43
Isentropic efficiency, %	82	84
Seed fraction	3×10^{-5}	2×10^{-5}
Hall current, kA	66	38
J_r at anode location, A/cm^2	3.6	4.2
J_r at cathode location, A/cm^2	1.1	1.0
Hall voltage, kV	6.4	11
$\langle E_r \rangle$, V/cm	136	399
$\langle$Power density$\rangle$, MW/m^3	289	851
$\langle \beta_{eff} \rangle$	20	20
$\langle \sigma_{eff} \rangle$, S/m	56	20
Inlet Mach number	2.1	1.7
Inlet stagnant pressure, atm	7.1	3.6
Outlet stagnant pressure, atm	1.2	0.6
Heat loss in the MHD channel, kW	1547	1534
Pressure loss in the MHD channel, atm	0.11	0.04
Anode radius, cm	88	59
Cathode radius, cm	135	87
MHD channel height at cathode location, cm	68	67
Hot duct radius, cm	49	39

The output power density of the helium-driven generator amounts to about 1 GW/m^3 and the Hall field increases to about 400 V/cm. In this case, prevention of the Hall voltage breakdown in the MHD channel is a key point. However, it should be noted that the higher power density or the higher Hall field for higher magnetic field does not necessarily lead to a Hall voltage breakdown along the MHD channel because the gas static pressure increases for higher magnetic field. For the case of argon-driven generator, the output power density is about 0.3 GW/m^3 and the Hall field is 140 V/cm.

In the forrowing section, we discuss about an assemly of the magnet coupled with the argon-driven generator. The maximum magnetic field and the electromagnetic force of this magnet are larger than those of the magnet for the helium-driven generator. Parameters of the magnet are listed in TABLE 3. The coils are assumed to be cooled by superfluid helium (He II) and a Nb$_3$Sn conductor is partially employed at the location of high field taking account of the maximum magnetic field in the coil of 20.2 T.

TABLE 3
Parameters of a typical manget for an argon-driven MHD generator

Coil inner radius, cm	165
Coil outer radius, cm	199
Coil length, cm	53
Distance between the coils, cm	98
Single coil weight, ton	19
Average current density, A/mm^2	100
Rated current, kA	3
Megnetic field at anode, T	10.3
Magnetic field at cathode, T	9.0
Maximum magnetic field in coil, T	20.2
Electromagnetic attractive body force, ton	35008
Self inductance, H	202
Mutual inductance, H	4.3
Stored enegy, GJ	1.9

A large electromagnetic attractive body force between the coils of about 35000 tons is induced in this magnet. We propose that the two coils are separated against this force by eight supporting rods made of stainless steel which are connected with a He II vessel. Each rod, having a diameter of 36 cm and length of 1 m, penetrates through an insulated space into the diffuser part of the generator channel, as shown in Fig.8. It is note-worthy that the coils are arranged outside of the cathode. The rods serve to support the coils against the attractive body force and the He II vessel help them to withstand circumferenctial tension, since the coils themselves cannot sustain the hoop force. The He II vessel is enveloped by a He I vessel providing a vacuum space for thermal insulation. A He II reservoir, the upper part of the He II vessel, is connected with the He I reservoir by a narrow communication channel.

When the magnet is charged, a compressive force of 4400 tons, which is about 1/160th of the buckling strength of the supporting rod, is induced in each rod. A strain along the symmetry axis of the rod is calculated to be 0.23 %. The junctions between the rod and the He II vessel have the maximum possibility of getting damaged by tention induced by the magnetic field. This tension is estimated to have a value of 1800 kg/cm^2 which is about 70 % of the allowable stress of the stainless steel. Hence, there is no possibility of suffering from any damage.

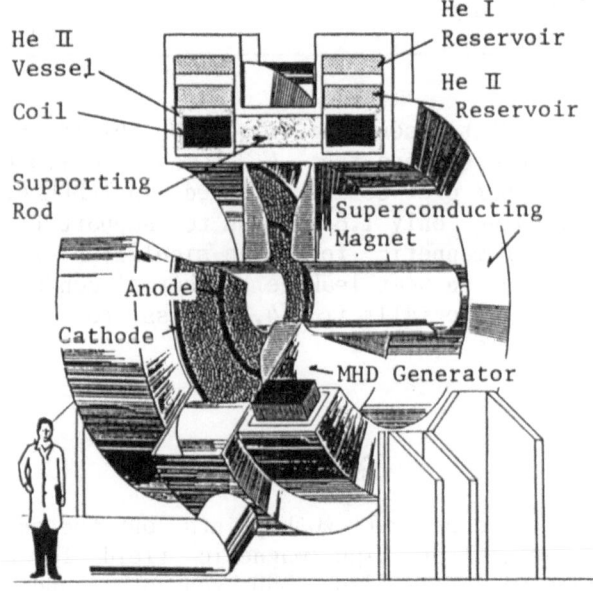

FIGURE 8. A configuration of the typical disk type MHD generator and the magnet for the magnetic field strength at the disk center of 10 T.

At the same time, a strain develops in the coils in tangential direction by the hoop force. It is calculated that the maximum strain that develops in the coils along this direction is about 0.88 % when the wall thickness of the He II vessel is 10 cm. Though no experimental result about the strain dependence on the critical current in Nb$_3$Sn conductors at 1.8 K and 20 T is obtained, the critical current density in the present case is cosidered to be about 110 A/mm^2 on the bases of results about the critical current densities for some Nb$_3$Sn conductors as a function of magnetic field at 1.8 K and 4.2 K [2] and the starin dependence on the critical current at 4.2 K [3],[4],[5]. Therefore, the present strain can be considered not to have a prominent effect on the stability of the superconductor.

TABLE 4
Heat loads in the magnet

From 80 K to 4.2 K	
Supporting structure for He I vessel	171 W
Current leads	60 W
Radiation from LN$_2$ shield	5 W
Ducts	4 W
Other parts	10 W
Total	250 W
From 4.2 K to 1.8 K	
Supporting structure for He II vessel	15 W
Current leads	5 W
Communication channel	1 W
Other parts	1 W
Total	22 W

Heat loads from liquid N$_2$ shield to 4.2 K mass and from 4.2 K mass to 1.8 K mass per one coil are listed in TABLE 4. By using only 1.8 K mass to support the electromagnetic force, a significant reduction in heat leakage by thermal conduction, especially from 4.2 K mass to 1.8 K mass, is assured.

CONCLUSIONS

The present work points out the advantages of high magnetic field for a closed cycle disk MHD generator and a possible construction of the superconducting magnet coupled with the generator.

A configuration of the magnet is clarified. The coils are cooled by He II employing partially a Nb$_3$Sn conductor at the location of high magnetic field. They are supported against the large electromagnetic force by the proposed structure which consists of a He II vessel and eight rods.

The output power density of the MHD generator increases larger than five times while *Bo* increases from 4 T to 12 T, but the stored energy of the magnet does not increase more than three times.

High output power densities of 0.3 GW/m^3 and 1 GW/m^3 are expected to be obtained for the argon and helium-driven generator, respectively, when *Bo* of 10 T is applied, though experimental studies on the Hall voltage breakdown in the MHD channel and a durability of the electrodes under these high density conditions are required.

REFERENCES

1. Y. Okuno, S. Kabashima, H. Yamasaki, N. Harada and S. Shioda, Comparative studies of the performance of closed cycle disk MHD generators using argon, helium and argon-helium mixture, Energy Convers. Mgmt., 1985, Vol. 25, No. 3, pp. 345-353.

2. M. Suenaga, K. Tsuchiya, N. Higuchi and K. Tachikawa, Superconducting critical-current densities of commercial multi-filamentary Nb$_3$Sn(Ti) wires made by the bronze process, Cryogenics, 1985, Vol. 25, No. 3, pp. 123-128.

3. J.W. Ekin, Strain scaling low for flux pinning in practical superconductors. Part 1: Basic relationship and application to Nb$_3$Sn conductors, Cryogenics, 1980, Vol. 20, No. 11, pp. 611-624.

4. M. Hong, D.M. Maher, M.B. Ellington, F. Hellman, T.H. Geballe, J.W. Ekin and J.T. Holthuis, Further investigations of the solid-liquid reaction and high-field critical current density in liquid-infiltrated Nb-Sn superconductors, IEEE Trans. on Mag. 1985, Vol. MAG-21, No. 2, pp. 771-774.

5. T. Okada, M. Fukumoto, K. Yasohama and K. Yasukouchi, Strain effects in cabled and braided "In Situ" formed Nb$_3$Sn conductors, IEEE Trans. on Mag. 1985, Vol. MAG-21, No. 2, pp. 775-778.

SUPERCONDUCTING MAGNETS FOR DISK-SHAPED MHD GENERATORS REVIEW AND PREDESIGN.

H.G. Knoopers, H.H.J. ten Kate, L.J.M. van de Klundert,
P. Massee*, H.A.L.M. de Graaf*, W.J.M. Balemans*.
Applied Superconductivity Centre, University of Twente, P.O.Box 217,
7500 AE Enschede, The Netherlands and
(*) Eindhoven University of Technology, Eindhoven, The Netherlands.

ABSTRACT

In this paper possible magnet systems that can be used in combination with disk-shaped MHD generators are reviewed. The main coil configurations are the single solenoidal magnet and the split-pair magnet. Advantages and disadvantages of these magnet systems are discussed. After an enumeration of a few basic design considerations, a predesign of a split-pair magnet for a compact open disk demonstration unit with a thermal input power of 10 MW is presented. Furthermore, the use of Nb_3Sn technology is envisaged to obtain a compact magnet with 9 T in the centre of the MHD channel. The preliminary design configuration consists of a NbTi and a Nb_3Sn section. The distribution of winding volume among both sections is based upon minimum conductor costs.

INTRODUCTION

Magnetohydrodynamic (MHD) energy conversion in combination with a conventional steam cycle can improve the efficiency of the energy production of a coal fired electricity power plant considerably. The main efforts in MHD-research concerned the linear channels both in open and closed cycle operating mode. However, there are indications that disk-shaped MHD generators from the view-point of efficiency alone can have a performance comparable to linear generators.

An MHD system study has recently been started in a collaboration between the Group Electrical Energy Systems at the Eindhoven University of Technology and the Applied Superconductivity Centre of the University of Twente. The total plant efficiency of linear and disk generators in open and closed cycle applications has been investigated [1]. Furthermore a conceptual design of a small scale open cycle disk MHD demonstration generator will be completed.

Due to its geometry, the disk generator has several advantages over the linear generator, as for example a more simple electrode geometry, a more compact generator configuration and lower construction costs. To a large extent, reduction of costs can be ascribed to the less complicated construction of the superconducting magnet system. These promising advantages were the main motivation to study further the steady state combustion driven MHD generator in disk configuration operating at high magnetic field.

MAGNET CONFIGURATIONS

The number of realistic magnetic configurations that can be used in disk-shaped MHD generators for large and small scale application are limited. The main coil constructions found in literature are solenoids and split-pair magnets, each having its own advantages and disadvantages.

Solenoids

One of the main arguments to use a single coil magnet is optimum accessibility of the MHD channel for cleaning and for maintenance of the channel, electrodes and insulations. Solenoids placed between combustor and generator channel as shown in figure 1 is a practical solution [2-5].

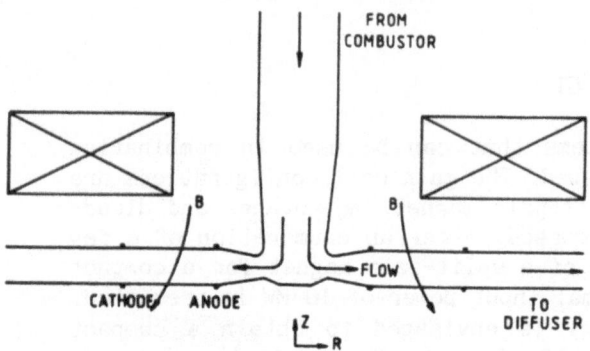

FIGURE 1. Schematic representation of an MHD channel with a single coil magnet.

In an MHD disk generator study, Retallick [2] considered four slightly different open cycle generators working at a field of 7 T. In order to enhance the field uniformity the coil is split into parts with respectively a low and a high current density. A nearly constant B_z-component between cathode and anode can be created. Marston [3], in his conceptual space-based MHD-generator design, also uses a radially increasing current density by subdivision of the magnet in four sections. Optimum use of the magnetic field is obtained by placing an MHD channel at both sides of the magnet. A reduction of weight is here the main argument to use a single coil magnet. This reduction is caused mainly by the simple support structure of the single coil magnet.

A disadvantage of the concept using a single solenoid is a large radial magnetic field component in the MHD channel. This field component introduces undesirable nonuniformities in the plasma velocity in

the generator. To overcome this, adjustment of the channel configuration in a way that the total field vector is perpendicular to the plasma flow was suggested [2]. However, the influence of a curved shape of the MHD channels on the plasma flow and on the performance of the generator has not yet been investigated thoroughly. Another solution to eliminate the radial field component is to place the MHD channel in the plane of symmetry of the single coil [4]. A large increase of coil volume and field components in the diffuser section however, does not make this design a practical one.

Split-pair Magnet

When using a split-pair magnet [4-6], the MHD-channel is placed in the gap between the coils, so the plasma will flow in radial direction as shown in figure 2. A main advantage of the split-pair magnet compared to the single coil magnet is the small radial field components in the MHD channel. Also the gradient in the magnetic field component B_z perpendicular to the plasma flow, is very small.

A considerably less conductor volume is necessary to produce the same field level in the active MHD volume of the single solenoid which is also an attendant advantage of the split-pair magnet.

The main problem of applying this coil configuration in an MHD disk generator is the containment structure that has to handle the attracting electromagnetic

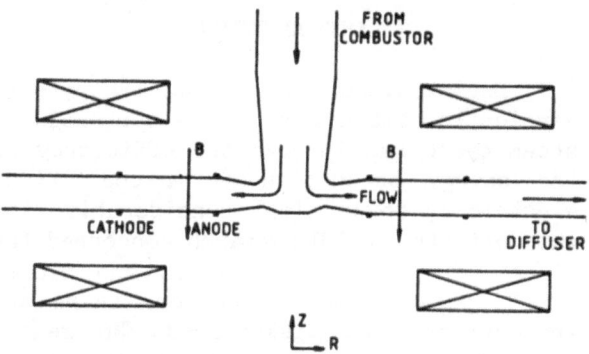

FIGURE 2. Schematic representation of an MHD channel with a split-pair magnet

force between the coils. This structure must penetrate the vacuum of the cryostat and the disk channel. Therefore the introduction of obstructions in the plasma flow are inevitable. As a consequence, these force containing structures are complicated and maintenance of the channel becomes more difficult. Besides this, the penetrating struts going through the vacuum chamber of the cryostat enhances the heat load of the cooling system considerably.

Another solution that was suggested [7] is a pair of concentric coils placed between two MHD disk channels. The radial field component of such a coil systems is small in comparison to a single coil magnet. However, because the current direction in both coils has to be opposite, the coils repel each other and a mechanically unstable coil configuration has been created.

COIL DESIGN

In our study we selected the open cycle MHD disk generator working at a magnetic field of 9 tesla and a 10 MW thermal input. A preliminary design of a demonstration generator will be completed. The main design parameters of this generator are collected in table 1.

TABLE 1
Design parameters of the 10 MW_{th} disk generator.

Combustor temperature	2870	[K]
Stagnation temp. at gen. inlet	2800	[K]
Inlet Mach number	1.80	[-]
Inlet radial Mach number	1.44	[-]
Inlet combustor pressure	5	[bar]
Inlet swirl	0.75	[-]
Inlet height	0.0354	[m]
Anode radius	0.073	[m]
Cathode radius	0.473	[m]
Outlet height	0.0091	[m]
Outlet pressure *	1.1	[bar]
Field at anode	9.0	[T]
Field at cathode	6.4	[T]
Enthalpy extraction	8.1	[%]

---

* down stream of diffusers

The obtained parameters were deduced by solving the radial- and angular component of momentum equation, the energy equation and continuity equation in the one dimensional approximation. Under the restrictions of a minimum allowable pressure at the exhaust of the diffusers of 1.1 bar and a maximum electric field in the MHD channel of 12 kV/m, optimum enthalpy extraction was obtained at a linear decrease of the radial Mach number.

Basic Design Considerations

- Due to the mentioned advantages the split-pair magnet was selected as the most suitable coil configuration. The design parameters of the magnet are determined for the major part by the dimensions of the channel, insulation - and cryostat wall thickness. Therefore the minimum inner radius and distance between the coils are 0.3 and 0.4 m respectively.

- A magnetic field of 9 T was selected to gain experience with this more difficult type of MHD coils and to obtain a relatively compact unit. To attain the desired field strength in the active area of the MHD channel, a Nb_3Sn superconductor will be used. The magnet will be split into two sections, a background field coil of NbTi and a high field (insert) coil of Nb_3Sn that will enhanced the field in the MHD channel up to 9 T.

- Forced flow cooling is preferred above bath cooling in order to get a compact rigid coil construction that can handle the high stresses we are dealing with. Attendant advantages of this type of cooling is a smaller change on shorted turns due to the possibility of better insulation of the conductors. Also less coolant is necessary in comparison with cooling by immersion.

- Application of the react- and- wind method for Nb_3Sn coil fabrication is preferable because this concept prevents degradation of properties of the composite conductor constituents. Furthermore, problems with insulation of the composite conductor are not to be expected.

- By using forced flow cooling, it is advantageous to apply the double pancake winding technique. With this winding method the coolant temperature can be arranged in such a way that it will increase with decreasing field strength [8]. A second advantage is that the coil will be build from relative small modules that can be handled easily.

Coil Optimization

Coil dimensions were calculated by minimizing a function containing design field in the MHD channel and maximum allowable field $Bmax_{NbTi}$ at the inner windings of the outer coil section. If we assume a fourfold difference between the costs of Nb_3Sn and NbTi composite conductor per unit volume, then the variation of total conductor costs as a function of coil thickness can be calculated as shown in figure 3. Characteristics of three different adjustments of the outer

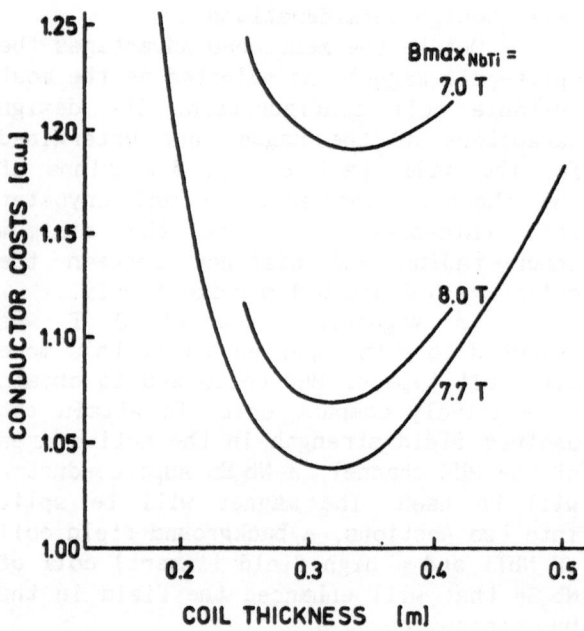

FIGURE 3. Conductor costs versus coil thickness.

coil are given. A minimum will be obtained at a coil thickness around 0.32 m. Here a linear relationship between current density and temperature is assumed. Although $Bmax_{NbTi}$ has no considerable effect on the optimum coil thickness, minimum conductor costs will change dramatically as indicated in figure 4. Here the effect of $Bmax_{NbTi}$ on conductor costs is shown. A sharp minimum appears at a value of 7.7 T at an overall current density in this coil section of 3.7×10^7 A/m^2. The current density in the inner coil is assumed to be 3.2×10^7 A/m^2 which is imposed by the maximum allowable hoop stress in the conductor.

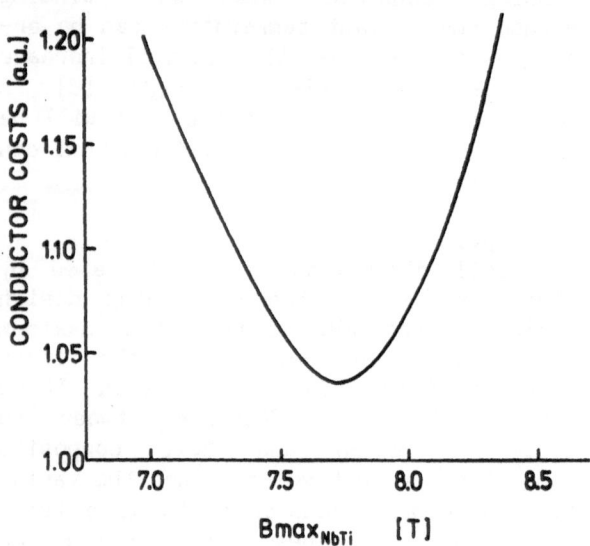

FIGURE 4. Conductor costs versus maximum field at outer coil winding.

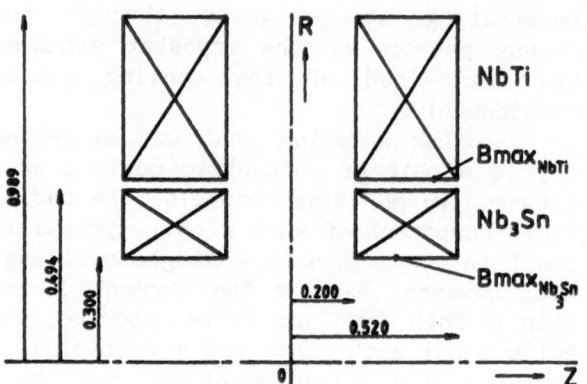

FIGURE 5. Optimum dimensions of both sections of the split-pair magnet.

Coil dimensions are shown in figure 5 and supplementary parameters defined at an operating current level of 6 kA in both sections of the split-pair magnet, are listed in table 2 .

TABLE 2
Parameters of the magnet system for a 10 MW$_{th}$ MHD disk generator.

Section		Nb$_3$Sn	NbTi
Winding inner diameter	[m]	0.300	0.524
Winding outer diameter	[m]	0.494	0.989
Coil thickness	[m]	0.320	0.320
Number of turns	[-]	312	864
Mean current density	[A/mm^2]	32	37
Maximum field	[T]	11.1	7.7
Operating current	[kA]	6.	6.
Self inductance magnet	[H]	1.481	
Mutual inductance coils	[H]	0.401	
Stored energy	[MJ]	67.8	

Field Distribution

The NbTi-section of the split-pair magnet with current densities and dimensions, found in the optimization process described above, generate a field of 6.4 T in the centre of the MHD channel. The Nb$_3$Sn insert coil will enhance this field to the desired 9.0 T. The decay in radial direction of the axial field component Bz is shown in figure 6. At the anode of the disk generator this field is still 9.0 T, but will decrease to 6.4 T at the position of the cathode. The maximum radial field component that will occur in this part of the MHD channel is limited to 0.2 T.

Stress

The current carrying capacity of multifilamentary Nb$_3$Sn, in contrast to NbTi, will be affected appreciably by

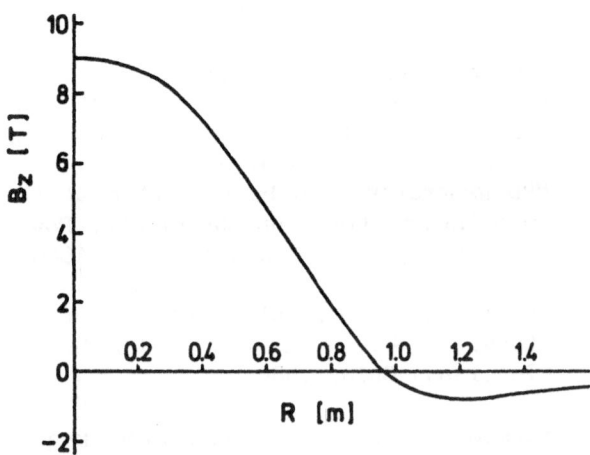

FIGURE 6. Field distribution as a function of radius in the centre of the MHD channel

strain. Uniaxial strain in the super-conducting windings will mainly be caused by pretension of the conductor during coil fabrication, by radial expansion and mutual attraction of the coil parts due to Lorentz forces and by internal forces. The latter are introduced by fabrication of the conductor and by differences in thermal contraction of the composite materials after cooldown of the magnet to liquid-helium temperature. This internal stress is compressive for the superconducting filaments which usually results in a prestrain in the range 0.1 - 0.3 % [9].

The peak field in the Nb_3Sn-section of our magnet is 11.1 T. This field occurs near the centre of the inner windings and decay to 6.9 T at the outer edge of this section. These turns will experience a Lorentz force that will be balanced by a tensile force in tangential direction of $F = I \cdot B \cdot R$. From the calculated fields and dimensions of the Nb_3Sn-section it becomes clear, that the force acting on the windings decreases just a little from the inside to the outside of the coil section, so the inner turns will be supported by the outer turns. The tangential stress caused by the radial Lorentz force was calculated using a layer model [10]. The composite conductor is modeled by layers alternately carrying current and non-current. The distribution of available space among the constituents of the composite conductor is dictated by cryogenic stability and maximum allowable stress in the reinforcement material. A cable thickness (current carrying layer) of 1.8 mm and a stabilizer /reinforcement layer (noncurrent carrying layer) of 5.7 mm was assumed. When using the following material properties

-$E_{stainless\ steel}$ = 210 GPa,
-$E_{cold\ worked\ copper}$ = 154 GPa,
-$E_{Nb_3Sn\ multifilamentary\ cond.}$ = 60 GPa,
-Poisson ratio = 0.33,

a stress level in the Nb_3Sn cable of 110 MPa was found. The calculated stress in the additional copper stabilizer is 260 MPa which is roughly 2/3 of the tensile yield strength of the cold-worked copper. The stress in the stainless steel reinforcement layer is 360 MPa. The maximum strain due to the hoop stress is limited to 0.17 %.

The total attracting body force between the coil parts of the split-pair magnet that have to be absorbed by the force containment structure is 28 MN. For global computation of the bending out of its plane, the body force is assumed to act at a distance R = 0.72 m on the girder of the containment structure. Solving the Saint-Venant's equations for a curved beam [11] the displacement out of its initial plane of the girder can be found. This displacement can introduce an additional stress in the conductor. Strain as function of the thickness of the stainless steel girder and the number of supporting rods N is shown in figure 7. A rectangular profile of the girder has been assumed.

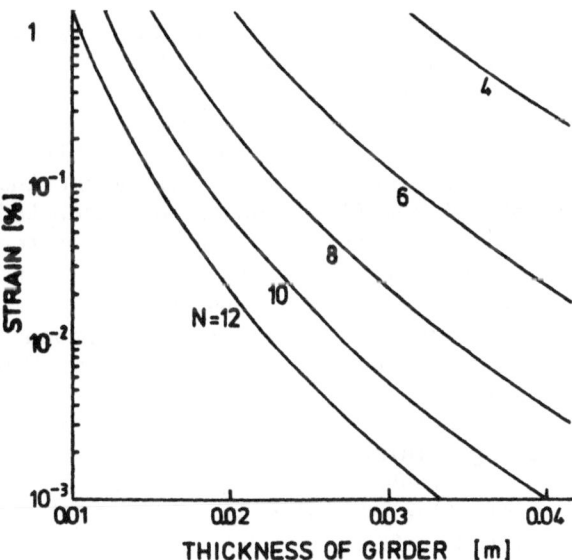

FIGURE 7. Strain introduced by bending of the girder due to attractive body force as a function of the girder thickness.

The total uniaxial strain in the conductor when the magnet is energized is a combination of the mentioned effects. We aspire to a situation in which the sum of these strain effects is neglectable because, at a field of 11 T, an uniaxial intrinsic strain of 0.2 % reduces the maximum current density of Nb_3Sn multi-filamentary conductors with 10 % maximum. Besides this, an increase of strain with 0.1 % will cause an additional degradation in the current carrying capacity of roughly 10 % . Due to filament fraction the irreversible strain point above which permanent degradation occurs is reached between 0.4 and 0.7 % [9].

Due to the adopted react-and-wind technique, bending strain in the Nb_3Sn multifilamentary conductors will be introduced. To limit this strain contribution, the Nb_3Sn-layer in the composite conductor must be located at the neutral axis of the conductor and its thickness must be as small as possible. Therefore cabled conductors in Rutherford geometry are usually applied. Maximum bending strain ε_b will appear at the inner windings and can be given by the relation $\varepsilon_b = h / (2 R_{in})$, where h is the cable thickness and R_{in} the inner radius of the coil. With the assumed cable thickness the maximum critical current degradation due to the bending strain, that we have to accept, is 5 % .

CONCLUSIONS

Although the single coil geometry for a disk type MHD channel has several important benefits, the split-pair magnet geometry was selected for the open cycle demonstration disk generator under study. This selection was based mainly on the almost optimum field conditions generated by such a coil system. A better performance of the generator can thus be expected. Also a considerably less conductor volume is necessary to produce the same magnetic field in the MHD channel.

Optimum coil dimensions were found assuming minimum conductor costs.

The predesign of the 10 MW_{th}, 9 T demonstration MHD disk generator will be completed after which the practical realization of an MHD demonstration unit is pursued.

ACKNOWLEDGEMENT

These investigations in the programme of the Foundation for Fundamental Research on Matter FOM have been supported in part by the Netherlands Technology Foundation STW, Utrecht, The Netherlands.

REFERENCES

1. Massee, P., de Graaf, H.A.L.M., Balemans, W.J.M., Knoopers, H.G. and ten Kate, H.H.J., System studies of closed and open cycle disk and linear MHD generators. To be presented at the 10th Intern. Conf. on MHD Electr. Power Generation, Tiruchirapalli, dec. 1989.

2. Retallick, F.D., Disk MHD generator study. NASA report CR-159872, Cleveland, Ohio, 1980.

3. Marston, P.G., Superconducting magnet system for a space-based 100 MW MHD disk generator. IEEE MAG-24, (1988), 885-886.

4. Okamura, T., Kabashima, S. and Shioda, S., Superconducting magnet for a full scale helium-driven disk MHD generator. 9th Intern. Conf. on MHD Electr. Power Generation, Tsukuba, (1986), 125-134.

5. Nakamura, T., Lear, W.E. and Eustis, R.H., Feasibility study of the inflow disk generator for open-cycle MHD power generation. 19th Symp. on Eng. Aspects of MHD, Tullahoma, June 1981, pp.3.1.1-3.1.14.

6. Okamura, T., Kabashima, S., Shioda, S and Sanada, Y., Superconducting magnet for a disc generator of the FUJI-1 MHD facility. Cryogenics, 1985, 25, 483-491

7. Louis, J.F., Disk generators. AIAA Journal, vol 6, no 9, (1968), 1674-1678

8. Roeterdink, J.A., Klok, J., Elen, J.D. and Franken, W.M.P., Application of A-15 conductors in NET TF-coils. Proc. MT-9, Zurich (Sept. 1985), 622-625.

9. Ekin, J.W., Mechanical properties and strain effects in superconductors. In superconducting Materials Science: Metallurgy Fabrication and Application (Foner, S. and Schwartz, B.B). Plenum Press, New York, 455-510.

10. Liedl, J., Gauster, W.F., Haslacher, H. and Grossinger, R., Calculation of the mechanical stresses in a high field magnet by means of a layer model. IEEE MAG-17, (1981) 3256-3258.

11. Volterra, E. and Gaines, J.H., Advanced strength of materials. Englewood Cliffs, N.J. :Prentice Hall, 1971.

FIRST RESULTS OF THE START-UP PHASE OF THE 2 MVA SUPERCONDUCTING GENERATOR SMG

ao. Univ. Prof. Dr. H.Köfler, Dipl. Ing. F.Ramsauer,
Anstalt für Tieftemperaturforschung
Joanneum Research, A-8010 Graz Steyrerg.19, Styria, Austria
Dipl. Ing. H. Fillunger ,
ELIN UNION AG., Vienna, Austria

ABSTRACT

The superconducting synchronous generator SMG - part of the development programme for superconductor application of ELIN Austria - is now ready for acceptance tests for use in a public powerstation. The paper will report the final assembly and some of the start up tests of the machine done in the laboratory. The generator was manufactured in Anstalt für Tieftemperaturforschung. The instrumentation of the rotor for monitoring stationary and transient phenomena in the machine is discussed. The influence of the measuring systems chosen on the running performance is evaluated critically. Furthermore particulars are given on the preliminary examinations with respect to the refrigeration system for operation of the generator in the public grid. The concept of the refrigeration system is outlined.

INTRODUCTION

Development of synchronous machines with superconducting excitation is on the way in several countries [1,2] of the world. In Austria this development task is tackled in a different way than elsewhere. The Austrian program on superconducting devices does not aim at demonstration that superconducting generators are technically possible. This has been demonstrated in several places all over the world. The main objective of the report is the detection of problem areas, their investigat and demonstration of solutions for this problems with respect to reliable operation of such a machine in the public grid. It is convenient having a synchronous generator small enough to allow quick changes in the construction on one hand and big enough to show all problems likely to occur in power generators of this type.
Just to recall the steps of the austrian program, the intended time schedule is shown in figure 1.

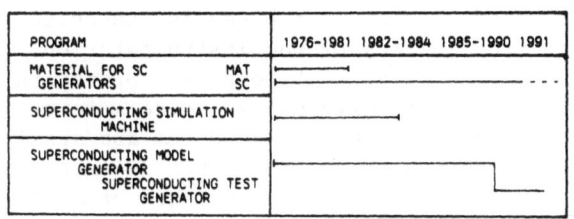

PROGRAM	1976-1981 1982-1984 1985-1990 1991
MATERIAL FOR SC GENERATORS	MAT SC
SUPERCONDUCTING SIMULATION MACHINE	
SUPERCONDUCTING MODEL GENERATOR SUPERCONDUCTING TEST GENERATOR	

Figure 1   Time schedule

It is shown that starting with a cryogenic machine having only some superconducting windings for demonstration of proper temperature management in a rotor the project proceeds to a fully superconducting excited generator using part of the available cross section for the exciterwinding and the final machine SMG now under test. Beside this, it was recognized as necessary during the tests of the secondstage machine, a small super-

conducting generator for continous operation in the grid was built (20 kVA nominal rating). This machine allows the convenient test of computer codes for operation and supervision of superconducting
excited generators. The computer codes
are part of the development program for
reliable operation of superconducting generators. They are not the only measure
for reliability. The international test
specifications for synchronous machines
even do not cover the quality assessment
tests done by producers of conventional
generators. This fact holds even more for
superconducting synchronous turbogenerators with their drastical changes in the
construction.

One major change in quality tests has to
be done with respect to the superconducting winding, which have to be tested
with the critical current specified by
magnetic field, operating temperature and
superconductor material.

Therefore an area of problems is connected to tests of winding modules which have been reported at MT 10 [3]. Further
information on this subject can be found
in this paper.

Another very different task compared to
conventional test procedures are the vacuum tests of structural elements of the
rotor and of the completed rotor structure. Thermal insulation of the low temperature parts of the rotor in long term
operation depends critically on this
tightness, especially when the rotor is
built as a permanent sealed vacuum system. The big advantage of this concept is
that no sliding vacuum seals and no pump
line for continous pumping is necessary.
A third circle of problems is connected
to the quality of the instrumentation.
So far no results of long term operation
of sensors in cryogenics inside a rotor
spinning with 3000 or 3600 rpm are available. Therefore the SMG rotor is instrumented heavily. The performance of different sensors and signal transfer systems
in the rotating frame will be reported in
the future, but the logistic ideas for
the measuring inside the SMG rotor is described here.

Last area of interest is management of
the cooling fluid for the generator. Due
to the connection with the generator its
performance is controlled by the generator itself. The stationary and transient
operation of the synchronous generator
impose great difficulties on the refrigerator. A possible solution of such a refrigeration system is shown.

TEST SYSTEMS AND METHODS

Test of reliable operation of superconducting devices is time consuming. This
was shown also in a preceeding report [3]
where tests of winding modules have been
reported. Testing of the modules was
continued by a test of a setup of all
flat modules of the exciter winding in a
special rig. This rig was a simple
copy of the central rotor structure on
which 10 flat winding modules forming
about 60 % of the total magnetic excitation were mounted. Due to reasons of
accessibility the retaining structure was
formed by several short tubes. The contacts between the coils, however, had
to be prepared in a preliminary way, because final assembly on the rotor needed
the individual modules seperated. The
total rig was immersed in liquid helium
and tested as usual. In a second test all
winding modules are fixed on the central
rotor structure of the SMG rotor. Contacts between the 14 coils now are constructed in the final version and can also be fixed properly in place. Figures 2,
3 and 4 show the subsequent growing of
the exciter winding. One can recognize
that superconductors are embedded in an U
shaped copper tube squeezed at the contact section. The soldered joint then is
bandaged with glass tape and impregnated
with epoxy. The resulting well insulated
connection between the coils is fixed
in special mounting pieces which match
into the individual places of the contact location. The total exciter winding
is enclosed in a shrink tube, which is
welded helium tightly with the side flanges of the central rotor. At this stage
the exciter winding was ready for the integral coil test. The central rotor was
fitted to a special transferhead at the
supply end of the rotor and then mounted
with a special flange vertically inside
a cryostat. This cryostat formed the

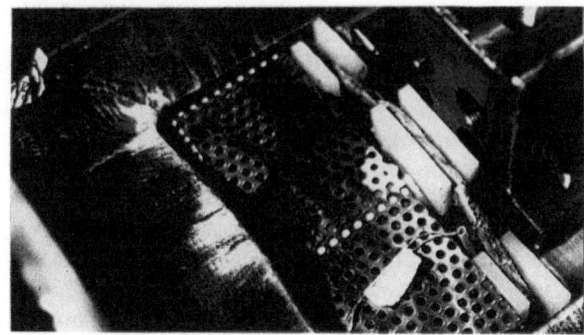

Figure 2: partially assembled rotor

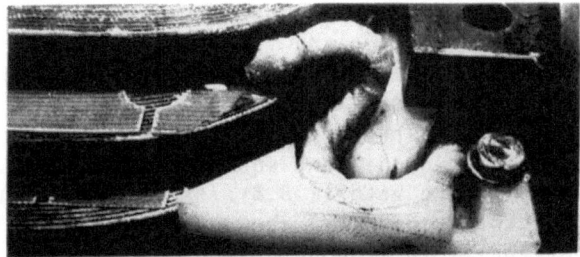

Figure 3: view on winding contact

Figure 4: assembled rotor

thermal insulating vacuum. Helium was transfered into the central rotor along the same path as in use during normal rotating operation. Due to upright test position the heat exchanger at the driven end of the rotor is flooded by liquid helium although in normal operation this component will only be cooled by cold helium gas. This has to be accounted when reading the results of the test.

After these tests in the cryostat the rotor was completed with radiation shield, acting also as cold damper, superinsulation and its vacuum system. Mechanical elements such as driven shaft, balancing rings, sliprings for signal transfer and for the exciter current and helium transfer equipment were mounted.Therefore just a few words on the testing of components for tightness. During the manufacturing process an escorting quality test of tightness of material and weldings is necessary. In fact this is one of the procedures which consume most of the test time during manufacturing. The completed structure then with great probability will also pass the integral leak test. In case of this stationary test the thermal insulating vacuum around the central rotor after pump down to 10^{-5} mbar was monitored by a mass spectrometer. First warm helium gas was filled into the central rotor to see tightness against warm gas either at ambient and at enhanced helium pressure. Then the system was cooled to the temperature of boiling nitrogen and the test for tightness repeated.

Long term observation of the vacuum, eliminating the influence of released gas from the surfaces, was done. The cool down to temperature of liquid helium improved the insulating vacuum due to the cryopumping considerably. At this temperature level another long term test was done to reveal possible leaks against helium due to cracks which are opened by thermal contraction.

Tests of the completed rotor in the first stage concentrate on the performance of the fieldwinding. These tests crucially depend on the thermal stability of the central rotor if ultimate values of current are to be achieved. Reliable operation calls also for knowledge about the performance of the rotor winding during transient thermal state. Due to the rather big thermal time constants it is difficult to adjust the desired thermal state. In the rotor of SMG, however, an intermediate state can be achieved and measured with great accuracy. A quick cool down of the central rotor allows to reach superconductivity of the winding, while the mechanical structure remains between 6 to 7 Kelvin inside the central rotor. The radiaton shield, serving as a cold damper stays at 140 to 150 Kelvin. In this thermal state the rotor was excited several times at various rotational speeds. Induced quenches give information on the electrical insulation performance of the exciter winding. The voltage level can be adjusted by properly chosen dump resistors. Excitations of the field winding are also used to adjust the quench protection in the rotating frame. This leads to the problem area of measurement inside the rotor. The SMG rotor is instrumented with 84 sensors at different locations. The measuring system is split into a so called safety circle and a control circle and demands 182 transferpaths. This amount is managed by 40 channels of slipring transducers and by rotating electronics which offer the rest of the channels. The direct pathes from the rotor to the stationary ambient are reduced to 24 channels. The other channels are supply or transfer channels of the rotating electronic equipment. The 24 direct channels are used for safety inspection of the spinning rotor. The safety circle includes potentials of the exciterwinding and sensorwindings for detecting of quenches, temperatures of the central rotor at the periphery and the inner surface, temperatures of current lead, cold damper and torque tube and a superconducting level detector. Some of these sensors deliver redundant signals, by different measuring procedures. So even when one measuring principle fails the rotor supervision is possible by the remaining signals. The electronics are based on transfer of 4 to 20 mA, have different amplifiers and for some sensors logical elements which can reduce the power consumption in standby operation. Level detection inside the rotor runs over different systems. Level of liquid helium is measured by the superconducting detector already mentioned, by a detector built with carbon resistors [4] and by different temperature sensors sitting on extreme positions inside the rotor. In connection with flow measurement at the entry and the exit of the rotor the liquid helium management is controlled. Naturally the flow of coolant in the rotor is also governed by the status of operation of the generator. So in a careful analysis all possible problems of operating a generator during cool down, stationary operation with normal regulation and in case of faults and quench have been checked. The outcoming is a specification for a refrigeration system allowing reliable operation of the generator.

RESULTS

The results of the tests can only show a small spectrum of work. As indicated previously thermal insulation vacuum is to be tested first. The vacuum volume of SMG amounts to 0.05 m³. This volume at the beginning of one year of operation should have the value of 10^{-5} mbar in warm condition and at the end of same year 10^{-4} mbar again at ambient temperature. In this rotor therefore the tolerable leakrate is as low as 1.4×10^{-10} mbar.l/s .
The integral leak test of the parts filled either with liquid or gaseous helium was done at the lowest possible measuring range and showed leakrates smaller than 2.10^{-10} mbar.l/s. This was the lowest detectable leakrate of the measuring device. Leakrates of welded seams to ambient are less dangerous because cold faces of the central rotor will pump a lot of air and other gases freezing out at liquid helium temperature.
With respect to the performance of the exciter winding some results at enhanced temperature will be shown. It was said, that the rotor has a rather long timecon-

stant in cooldown. The intermediate temperature state already mentioned can be classified best by the temperature of the end of the normal conducting current lead. This point has a temperature of 7 Kelvin at conditions equal to the cooldown of the rotor (0,5 g/s LHe and 200 A current). Figure 5 shows the load line of the field winding in the machine together with the critical current limit at 7 K . At this point a handicap of the measurments turned out. The temperatures taken inside the rotor are dedicated to govern the flow of helium. They give only very limited information from the temperature field of the field winding during transient cool down or warm up conditions especially at temperatures lower than 20 Kelvin. Anyhow the experiments showed that the rotor winding after a quench returns to superconductivity very quickly. So temperature monitoring is one of the problem areas that has to be investigated more intensively. In a

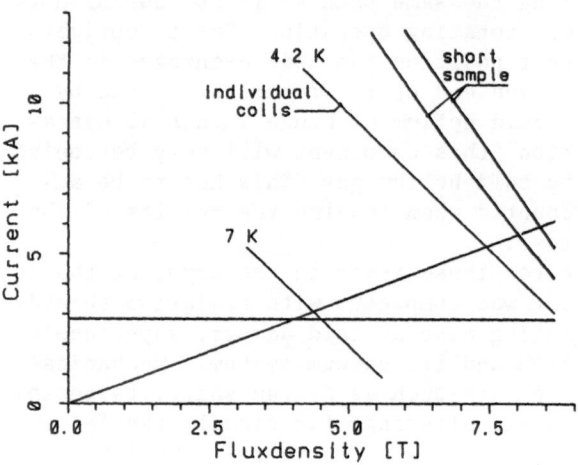

Figure 5 Load line of rotor

commercial generator only few temperatures will be taken and the question is, at which points inside the machine shall one measure these temperatures. The temperature sensors should give information on the coolant flow as well as on the temperature distribution in the central rotor.
In SMG the induced quench was also used for voltage test of the winding and the current lead. The discharge resistor was chosen with an appropriate value to have voltages up to 200 Volt. No evidence of insulation breakdown could be found.
In table 1 the measuring system is outlined. The subdivision in a safety and in a supervision circle is indicated. The total measuring system is built under the

assumption, that the data gained, can be cross checked by different sensing principles at any location inside the rotor.

TABLE

location in the rotor	thermoelement	carbon resistor	pressure gage	PT 100	strain gage	voltage tap lead	voltage tap Wdg.	hall probe	search coil BS	search coil WS	search coil SC	level detector CR
1		1										
2	2	2+2*	1	1*			13+3*	2	4	2*	1*	1
3	18+1*	9	2	1*		9						
4	3+2*			1	2							
5	3											
Σ=84	24+3*	12+2*	3	1+2*	2	9	13+3*	2	4	2*	1*	1

1..LHe supply, 2..field wdg.,
3.. lead, 4..torque tube,
5..damper
* safety circle

For instance ac fields penetrating the dampershields can be measured either by their integral influence on the exciter current, by hall probes and by search coils different in size. This redundant measuring hopefully will give results of equal meaning. Further enhancement of security of measurement is gained by special tests of the sensors. The superconducting level sensor for instance must measure level in pressurized liquid helium at enhanced magnetic fluxdensity. Results of a dummy test of this sensor are shown in figure 6. At pressure above 2 bar the superconducting level indicator fails. So in the rotating frame the level indicators with carbon resistors has to give all informations necessary.

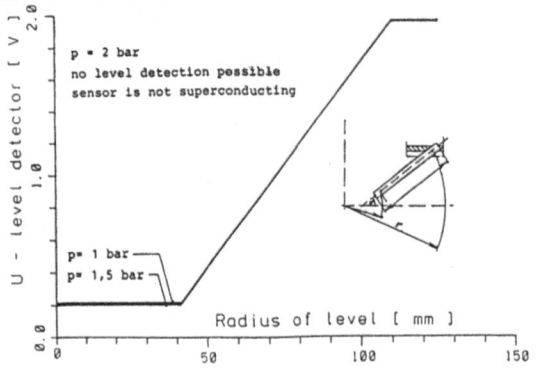

Figure 6 Level signals at different p

The refrigeration system has to provide the proper ammount of liquid helium to replenish the evaporized coolant. When the returning gas is cold enough to use the enthalpy for cooling purpose in the refrigerator, it should be possible to pass the gas through an adequate heatexchanger. The system also has to manage transient electrical states of the generator which will cause enhanced cooling flow and finally a quench at current overflow. The sketch of a system capable of all these challenges is shown in figure 7.

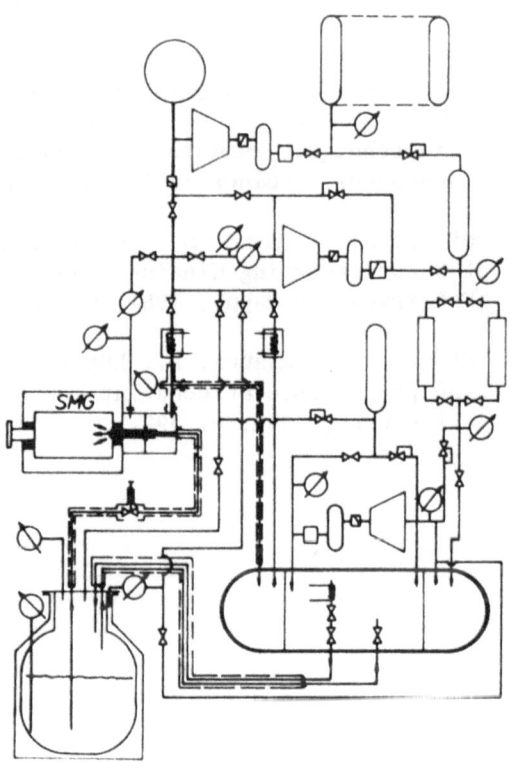

Figure 7 Scheme of refrigeration

CONCLUSIONS

Only few results of the start up phase of SMG could be presented within the limited space of this paper. Therefore in some concluding remarks should be emphasized what work has to be done in future to make best use of SMG. The test bed of the machine in Anstalt für Tieftemperaturforschung cannot master full load tests of the machine nor long term tests at operating temperature of the machine. Latter is due to other occupations in research which do not allow blocking of the staff with this task alone. The way one can get experience for long term operation runs over a stepwise increase of operating time. This opens the opportunity to adapt by analysis of results the control schemes and the supervision of the generator. In the near future the generator will be driven to full speed. The monitoring will first make use of all available informations and will then be reduced to find the minimum of measurement inside the rotor. In addition the supply scheme will be simulated by an existing refrigerator. At the end of all this work should stand the knowledge how to run reliable superconducting generators with NbTi or Nb_3Sn superconductors.

ACKNOWLEDGEMENTS

The authors want to mention the substantial support of this work by their colleagues, by ELIN and by Forschungsförderungsfond für die gewerbliche Wirtschaft.

REFERENCES

[1] Lambrecht D.,Status of development of superconducting AC generators. IEEE Trans. on Magn., 1981, MAG-17,No. 5.

[2] Kominosono, H., Iwamoto, M., Fujino, H., Maki, N., Development of superconducting synchronous machines in Japan. CIGRE Report 11-02, Paris, 1982.

[3] Köfler, H., Hubmann, M., Ramsauer, F., Fillunger, H., Design, Fabrication and Test of Superconducting Winding Modules for a 2 MVA Synchronous Generator. IEEE Trans. on Magn., 1981, MAG-24, No. 2.

[4] Köfler, H., Ramsauer, F., LHE level detector for superconducting generators, operating performance at constant current or constant voltage supply. Proc. ICEC 11, Berlin-West, 1986, p. 592.

DESIGN CONCEPTS AND EXPERIMENTAL RESULTS OF SUPERCONDUCTOR FOR FIELD WINDINGS OF 70MW CLASS SUPERCONDUCTING GENERATOR

A. Ueda, S. Maeda, T. Hirao, K. Mio, Y. Nagata, T. Yamada,
M.Morita and H. Yoshimura
Mitsubishi Electric Corporation, Japan

ABSTRACT

AC losses and critical currents of three types of superconductors were measured to evaluate the characteristics of the superconductors which are designed to meet the requirements for the field windings of a 70MW class superconducting generator with low response excitation system. The superconductors used for the test are compacted stranded cables composed of Cu-CuNi-NbTi strands, and have the same cable size. The cross sectional structure of the strands and matrix ratios for tested cables are different. This paper presents the results of the experiment and design concepts of the superconductor for the 70MW class generator.

INTRODUCTION

The 70MW class machine, planned to be built in the early 1990's, will be supplied with a low response excitation system that has a peak voltage of several ten times the rated voltage. When designing the superconducting field windings, consideration must be given to abrupt increases in field current and magnetic field due to short-circuits in the power system and the following changes in current and field due to the response of the excitation system.[1]

This means that the superconducting cable is required not to quench due to the induced AC losses. The critical value of the superconducting cable is also required to be sufficiently larger than the maximum value of the current and magnetic field developed during the fault.

In order to determine the structure of the superconducting cable which will meet these requirements, three kinds of superconducting cables were selected and evaluated for the critical currents and the AC losses, before the 70MW class superconducting generator could be built.

CABLE DESIGN

Cable Specifications

Field current and magnetic field of the 70MW class low response type superconducting generator for normal operations and fault conditions are as follows:

o Rated operating current and magnetic
  field   : 3000 A, 4.6T.
o The maximum current and magnetic
  field at the fault : 3600 A, 5.4T
o The maximum change rate of magnetic
  field at the fault: 5T/sec

To meet the above conditions, the field winding of superconducting generator would have the following specifications: Current density per outer cross section of cable is $140A/mm^2$ for rated operation ($170A/mm^2$ during the fault). The critical current is determined so that the Tc margin is over 1.5K at the fault, from the view point of the stability against quench and mechanical strength.

In order to reduce the AC losses, compacted stranded cable, each strand consisting of Cu-CuNi-NbTi, is used for the machine. And in order to restrict the temperature rise of the superconducting cable caused by the AC losses, the cables are directly cooled with liquid helium.

Two kinds of matrix ratio were selected to compare their stability against quenching. One of them has high Ic margin (Cu/CuNi/NbTi=1/1/1), the other contains more copper (2/1/1).

Cable Design

The superconducting cable for the generator were designed according to the specification described above.

Fifteen strands per cable were adopted in order to reduce the AC losses and to have sufficient mechanical strength. The strand diameter is 1.24mm so that the current density requirement is meet. Strand insulation was provided to restrict the coupling current between the strands. The thickness of the strand insulation (PVF) is 10μm, which is the minimum value needed to insulate each strand.

The temperature rise of the winding was calculated for the fault and the cross sectional structure of strand is examined. The temperature rise of the cable induced by the AC loss, which occurrs in the superconducting cable at a sudden short circuit in the power system, is shown in figure 1. The superconducting cable having the ratio of 2/1/1 (Cu/CuNi/NbTi) was analyzed using a transient thermal analysis program. The superconducting cable was modeled as having a rectangular cross section, and a thermal conductivity which is the average of Cu, CuNi, and NbTi. One-forth of the out side of the superconducting cable was assumed to be cooled directly with liquid helium. The temperature rise should be made sufficiently small compared to the 1.5K margin. Restricting the temperature rise to under 0.1K will reduce the AC loss to under 25 kW/m³.

The construction of the superconducting cable which has an AC losses under 25 kW/m³ was examined. First, the hysteresis loss was examined.

The hysteresis loss Ph is shown as follow.

$$Ph = \frac{2}{3\pi} \, Jc \cdot df \cdot \dot{B} \cdot \lambda \, (W/m^3)$$

Where Jc: critical current density
df: filament diameter
$\dot{B}$: change rate of field
λ : fraction of NbTi

Assuming Jc = 2500 A/mm², filament diameter = 9μm and λ = 0.25, the hysteresis loss is 6 kW/m³. Since this value is smaller than 25 kW/m³, ultra fine multi-filament, which have diameteres of only a few microns, are not required. The filament diameter was determined to be 9μm.

The eddy current loss is:

$$Pc = \frac{\tau}{\mu_o} \cdot \dot{B}^2 \qquad (W/m^3)$$

Where τ : loss time constant calculated from the coupling losses and eddy current loss.

Assuming the eddy current loss is 19 kW/m³, the loss time constant τ is 0.96 ms. This level of the loss time constant is achievable with the mixed matrix strand, which consists of Cu-CuNi-NbTi and has 1.24mm diameter.

Three kinds of fundamental structure of the superconducting cable are determined, as shown in figure 2. Details of the configurations are given in table 1. The ratio of Cu/CuNi/ NbTi is 2/1/1 for conductor A and B, and 1/1/1 for conductor C. The difference between conductors A and B is the positioning of the copper stabilizer. Conductor A has the copper stabilizer located in the center and along the circumference of the

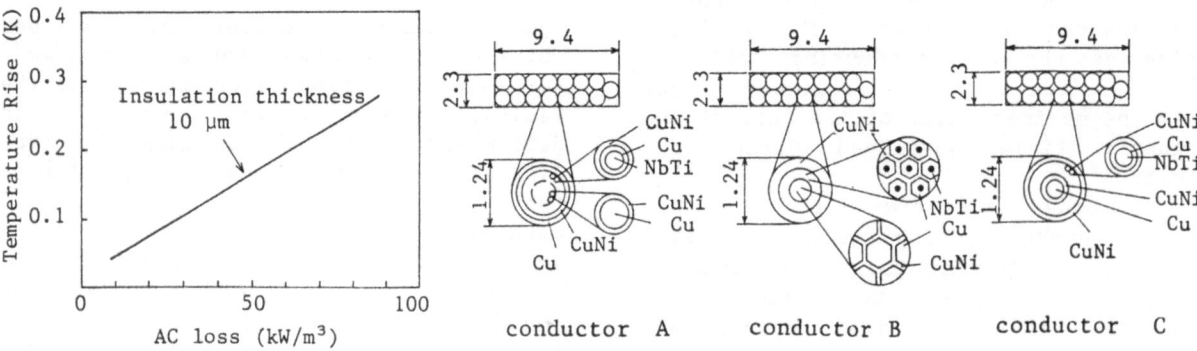

Fig.1 Temperature rise by AC loss

Fig.2 Cross sectional Structure of the Cable

strand. Conductor B has the copper stabilizer located in the center.

Critical Current

The critical current of the cables and strand untwisted from the cables are shown in figure 3. Figures 3(a) and 3(b) show the results of measurement for the cable A and B (Cu/CuNi/NbTi=2/1/1) respectively. Figure 3(c) shows the results of measurement for the cable C (Cu/CuNi/NbTi=1/1/1). The values represented by solid lines are products of the number of strands (15) and the critical currents of the untwisted strand. The criterion for determining the critical current of the strand is :

$$\rho = 10^{-13} \Omega m.$$

Points o and ● show the quench current and critical current of the cable respectively. The symbol o indicates that the cable quenched before the flux flow state was observed. The symbol ● indicates that the flux flow state was observed, and that the resistivity was approximately $5 \times 10^{-14} \Omega \cdot m$. The numbers shown in the figures represent the measurement sequence. For this critical current measurement, in which large current flowed in the cable, the self-excited field was not negligible. This self-excited field has been taken into account in the figure. Training effect was observed in conductors A and B, and their quenching point almost reached the critical current estimated from untwisted strand data. On the other hand, in conductor C, quench occurred at lower current than the critical value of the untwisted strand, and the flux flow state was not observed. The training made small increase of quench current, i.e., the original critical current of the

TABLE 1
Superconducting Cable

		Conductor A	Conductor B	Conductor C
Strand				
Diameter	mm	1.24	1.24	1.24
Filament diameter	μm	9	9	9
Twisted pitch	mm	12	12	12
Cu/CuNi/NbTi ratio		2/1/1	2/1/1	1/1/1
Cable				
Number of Strands		15	15	15
Conductor cross section	mm	2.3x9.4	2.3x9.4	2.3x9.4
Twisted pitch	mm	95	95	95
Insulation		PVF,10μm	PVF,10μm	PVF,10μm

EXPERIMENTAL METHOD

Critical Current Measurement

The critical current of the cables was measured by changing the field of superconducting magnet, in which the cable was disposed. Since the current of the cable is approximately 10 kA, large electromagnetic force acts between the cable and superconducting magnet. For the purpose of reducing the movement of the cable by electromagnetic force, the cable was wound to form a noninductive coil. As a results, electromagnetic forces acting between cable and super-conducting magnet was reduced.

AC Loss Measurement

The cable wound on a bobbin, which had a 28mm diameter and 100mm length, was located in the magnet. The AC losses of the cable was generated by a half cycle sinusoidal magnetic field, and measured by the magnetization measurement method.

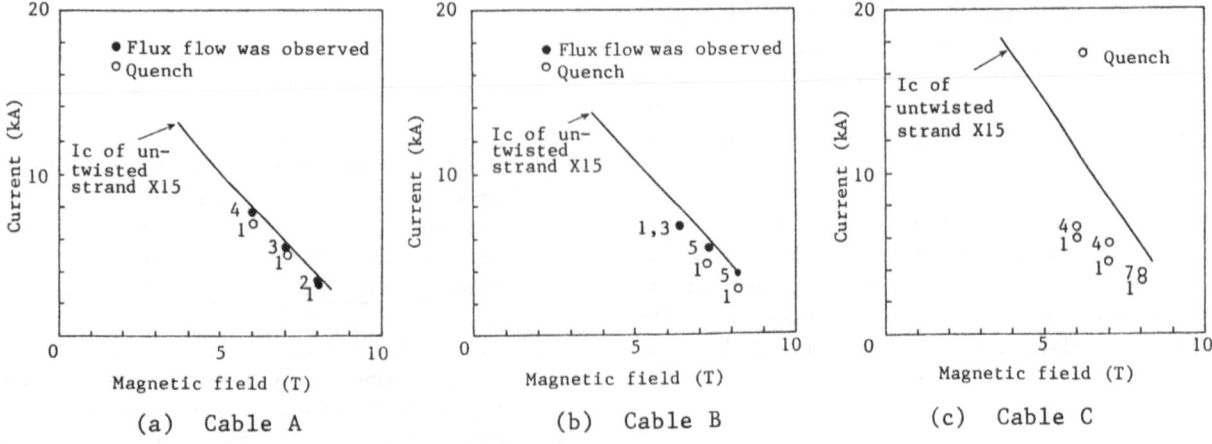

(a) Cable A　　　(b) Cable B　　　(c) Cable C

Fig.3　Critical Current of Cables

cable could not be attained.

The difference between the first-quenched current of the cable and the critical current of the untwisted strand exposed to the same magnetic field, is considered as a index of the cables stability. Since the first quenched current of the cable with Cu/CuNi/NbTi as 2/1/1 was close to the critical current of the untwisted strand, it is concluded that the application of this strand to the coil should create no problems. The application of the cable with Cu/CuNi/NbTi as 1/1/1, is considered less straight forward, since its critical current was significantly smaller than the critical current of the untwisted strand, and the training had little effect. The reasons for their difficulties are as follows. As the spring back of the superconductor was large when the coil was manufactured, the winding could not be tightened to have enough strength to restrict the wire movement. As the conductor contained smaller amount of copper, the cables stability was insufficient.

AC Loss

AC loss measurements were carried out for three kinds of superconducting cables. The loss time constant was calculated from the difference between the measured AC loss and the calculated hysteresis loss. The loss time constant and AC losses at 4.6T and 5 T/S are shown in table 2. The AC losses contain the hysteresis loss. The AC losses of each superconductor were approximately 20 kW/m^3. The loss time constant of the conductors are calculated using Turck's equation [2], and this result shows good correspondance with the measured result.

The calculated results of the loss components are shown in table 3. The cable A has the greatest loss at copper sheath, the cable B and C have the greatest loss at copper core located in the center.

Since the AC losses are approximately 20 kW/m^3, the temperature rise of superconductor at the fault is estimated to be under 0.1 K, and it is concluded that its application for the field winding should create no problems.

TABLE 2
Measured AC loss of Superconductors

	Cable A	Cable B	Cable C
loss time constant (msec)	0.79	0.68	0.62
AC loss (kW/m^3)	20.3	19.4	20.4

TABLE 3
Calculated values of loss component

	Cable A	Cable B	Cable C
Total losses (mS)	0.79	0.63	0.57
Copper core (mS)	0.10	0.39	0.36
Filament bundle (mS)	0.13	0.23	0.21
CuNi barrier (mS)	0.23	-	-
Copper sheath (mS)	0.32	-	-

CONCLUSIONS

The superconducting cables for the field winding of the low response type superconducting generator, were examined. Three kinds of samples were manufactured and the critical currents and AC losses were measured. It is concluded that the superconductor which has the ratio of Cu/CuNi/NbTi as 2/1/1 can be applied for the field winding of the low response type superconducting generator.

ACKNOWLEDGEMENT

This work was performed as a part of "R & D on Superconducting technology for Electric Power Apparatuses" under the Moonlight Project of Agency of Industrial Science and Technology. MITI, being consigned by New Energy and Industrial Technology Development Organization (NEDO).

REFERENCE

1. A. Ueda, T. Hirao, H. Hatanaka, M. Morita, Experimental results of field winding and concepts of rotor component development of superconducting generator. IEEE Trans. on Magnetics, VOL MAG-23, No.5, 1987, 3548

2. B. Turck, Coupling losses in various outer normal layers surrounding the filament bundles of a superconducting composite. Journal of Applied Physics, VOL50, 1979, 5397

300 MWA SUPERCONDUCTING GENERATOR AND FIRST
EXPERIMENTS OF ITS COOLING-DOWN

G.M.Khutoretsky, I.F.Filippov, V.D.Varschawsky,
Yu.L.Rybin, S.G.Stefanovich
Electrosila Corp., Leningrad, USSR

ABSTRACT

Within the scope of the national programs on the commercial superconductivity applications, effort is being continued at the Electrosila Corporation in developing the production prototype of a 300 MW superconducting generator (1). The work includes further theoretical and experimental studies on mathematical and physical modelling and testing of a full-scale prototype at the Manufacturer's test stand and, possibly, at a commercial station in the course of trial operation.

INTRODUCTION

It should be noted that the current difficulties in financing big research projects will perhaps affekt the schedule of this work. The main parameters of the superconducting generator developed should be as follows:

power, KW	300000
power factor	0,9
voltage, KV	20
speed of rotation, rpm	3000
frequency, Hz	50
efficiency, %	99
inductive reactances:	
Xd	0,362
Xd'	0,282
Xd"	0,15
stator mass, t	132
rotor mass, t	25
overal dimensions, m	
length	11.47
width	4.46
height	4.9

STATOR

The stator casing provided with two gas coolers affords either air or hydrogen filling, the gas circulation being performed by a single centrifugal fan mounted on the rotor from the turbine side. The main bearings are outborne ones of uprise-type, with cylindrical inserts; the tail bearing, from the side of the helium inlet, is a segmental shoe.

The active zone of the stator, a structure close to a conventional design, is made toothless which makes it possible to use the active zone to best of advantage and decrease permeance for the leakage fluk. With the view of cutting losses, magnetic shunts are installed before the press rings. The stator winding is secured in the active zone in the slots of the monolith frame made of glass-reinforced plastic rigidly fastened to a laminated ferromagnetic core. The end parts of the winding are secured with the aid of brackets and insulating rings. To cut additional losses, use is made, in the stator winding are secured with the aid of brackets and insulating rings. To cut additional losses, use is made, in the stator winding bars, of multiwire conductors consisting of insulated copper elementary conductors 1.18 mm dia, and also of multistage patterns of transposing and twisting of the elementary conductors.

The bars of stator winding are cooled by the distilate circulating through square-section Cu-Ni pipes arranged between the strands, the pipes being transposed in the slot and end sections.

ROTOR

The rotor with a superconducting field windings is substantially a rotating cryostat. A distinctive feature of the rotor is that it contains of cylindrical shell made of titanium alloys displaying good mechanical and thermophysical properties both at room and cryogenic temperatures.

The superconducting field winding, of saddle shape, is installed in radial slots of the rotor frame. Slot wedges and a cylindrical bandage are used to secure the winding against the centrifugal forces. Seven coils are installed on each pole. To equalize the magnetic field and reduce the maximum induction in the supercondector zone, the internal coil at each pole is decreased in height. Each coil consists of six sections and is divided by jackets to form radial cooling channels. A section is made of insulated superconductor of niobium-titanium alloy in a copper matrix, the superconductor filling ratio being about 0.35. The critical current of the shorter specimen of this conductor in a magnetic field of 5 T at a temperature of 4.2 K amounts to about 2000 A. Electrical intersection and intercoil connections of superconductors in the winding are made on the upper turns easily accessed. The resistance of these conductions made in overlap in a copper pipe by magnetic-discharge welding amounts to the order of 10^{-8} Ohms in a magnetic field of 5 T at a temperature of 4.2 K. Liquiefied helium at a temperature of about 4.2 K is supplied into the rotor with the aid of a coaxially arranged vacuum-insulated pipe from a cryogenic support system.

The helium temperature at the rotor outlet should be in the range of 200-300 K, and the pressure, close to the atmospheric one.

Provided in the zone of passing of the helium from the stationary central pipe to the rotating helium vapours are partly removed, the central pipe being cooled by vapours. Arranged along the way of helium going to the cryostatting zone in a separator wherein the helium liquid and vapour phases separates.

Thence, the vapour is supplied for cooling of the current lead and, partially, of the torque extensions from the side of the current lead, and the liquid, having passed the stabilizing vats, enters the two distribution chutes.

The liquid helium is further supplied to the periphery of the superconducting field winding through the radial channels arrenged in the zone of the major channel. The superconducting field winding is cooled by helium circulation under the action of the termosiphon effect in the channels formed with the aid of glass-cloth-base laminate gaskets between the frame and sections of the superconducting field winding. Helium boils intensively at the interface liquid-vapour found in the central space under the slot sections and in the spaces under the frontal sections of the superconducting field winding. The vapour formed is fed, from the central space and from the end sections from the turbine side, for cooling of the torque tube from the turbine side and further, through the central vacuum-insulated pipe, is wihtdrawn from the rotor. From the end sections of the turbine side, the vapour, mixing with the vapour from the separator, is supplied for cooling of the current lead and the torque tube from the side of the latter and further is also withdrawn from the rotor through a special coupler to the pipelines into the cryogenic support system. The cryostatting zone is insulated from heat influxes from the outside by a vacuum space between the frame and the external sections of the rotor, and from the ends, by the helium vapour cooled thermoinsulating plugs made of glass-reinforced plastic. Disposed between the outside cylinder with damping system and the superconducting field winding frame is a radiation screen at an intermediate temperature of 80-100 K protecting the cryostatting zone from radiation and serving as an additional damper of magnetic fields having a rather low rotational speed as regards the rotor. Both the warm and cold damping screens are substantially copper cylinders placed upon solid and rigid titanium cylinders taking up the centrifugal forces and electrodynamical loads and transmitting, to the rotor shaft, the torques brought about while in transient modes of operation.

By the present time, the wound stator of toothless type and main assemblies of the rotor have been made and passed operation-wise technological tests. Technology has been worked out and modules made of the saddle-shaped superconducting field winding, and the modules have been installed, fastened with wedges and baked into the rotor frame. Technology has been worked out and a two-layer radiation and a three-layer electromagnetic screens of the rotor have been made with the use of cooling-down the internal cylinders with liquefied nitrogen and induction heating of the external cylinders. The vacuum surfaces of the screens have been polished.

TESTS

A program of complex tests of the generator has been developed, the program including several stages and providing for the determination of main parameters and characteristics of the generator, conditions and modes of operation of the cryogenic support system, excitation and protection systems. Based on the complex test results, the design of particular subassemblies of the generator and its support system will be finished, programs and methods of testing of the generator and its systems will be specified and supplemented.

One of the first and perhaps most important stages has been the one connected with testing the superconducting field winding on the rotor frame at the temperature of liquified helium in a purposely developed static vertical cryostat having an internal diameter of 1200 mm and a depth of 8 m.

In the process of cooling-down, studies have been conducted on the regimes and patterns of supplying gaseous and liquified helium into this cryostat . The total coolingdown time amounted to about 280 hours.

In the second test cycle, to reduce the helium consumption and prevent excessive helium pressure rise in the cryostat during the quench, a special container of stainless steel was installed in the static vertical cryostat, the container accomodating the rotor frame with the superconducting field winding

embraced by a technological bandage.

After 17 trainings, a maximum current was obtained in the superconducting field winding amounting to 1300 A, that is, about 70% of the critical current of the shorter superconductor specimen.

Prior to the final assembling of the rotor, static balancing of the rotor frame was carried out on parallels. Only a slight imbalance of 400 g was found. In assemblying the rotor, vacuum-tight joints were applied with testing these without fail.

A first stage of mechanical tests of the shafting was then carried out with the use of a run up-and-balancing machine. As a result of fine balancing of the rotor, a satisfactory vibration condition of the rotor was obtained (about 2 mm/s on the bearings). Afterwards, tests were performed for overspeed of the rotor. Before bringing the rotor into the stator, the main space of the rotor was evacuated to 10^{-4} torr, the helium inlet (previously evacuated to 10^{-5}) was installed into the rotor, and evacuation was being performed for 18 days of the main vacuum space of the rotor with a continuous induction heating of the rotor shells to 380–390 K. In this, a vacuum of 10^{-5} torr was obtained in the main space.

With the rotor brought into the stator and mounted on the two main bearings, a preliminary fine balancing of the rotor was performed. The installation of a third auxiliary bearing in the zone of the slip-rings and a final fine balancing at the nominal rotational speed of 3000 rpm resulted in a satisfactory vibration condition of the main bearings. The shaft end vibration amounted to 100 microns (double amplitude).

Operating cycles were carried out with the rotor running at nitrogen temperatures. The rotor shaft vibration condition, upon multiple changes of the temperature from 300 to 77 K, remained practically unchanged.

The oil seals of the effluent gas chamber performed satisfactorily in all regimes, they gave no gas leakage, the oil temperature was normal. Oil leakage into the sump chamber were slight (less than 1.0 l/min). Thanks to the adequately performing drain system, no oil traces were found in the effluent gas system.

"Fresh gas" leakages were observed into the effluent gas chamber through the "fresh gas" seals at the shaft end.

In the experiments, the liquified nitrogen delivery varied from 5 g/s to 20 g/s (forced delivery). For the most time the flow-rate amounted to about 7 g/s. With the forced nitrogen delivery, frosting of the effluent gas chamber and overcooling of the torque tubes were observed.

With the nitrogen delivery level of 5-11 g/s, no frosting occurred. The total operating time in the nitrogen coolingdown experiments was about 500 hours.

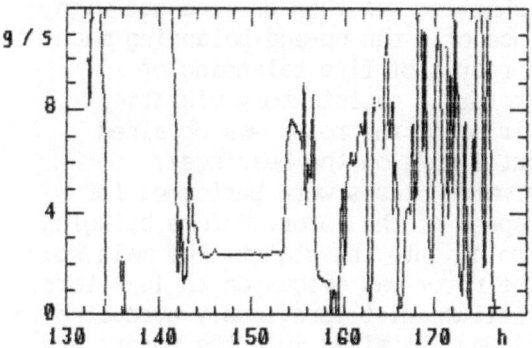

Fig.1.Coolant flow-rate dependence

Illustrated in Fig.1 there is one of the test stages which was being carried out for about 200 hours. In this experiment, 12 starts were performed and 7 tons of nitrogen consumed. The rotor was kept at a temperature of 77 to 80 K for about 120 hours.

In evaporation the nitrogen flow-rate 2.44 g/s and practically independent on the rotor rotational speed. The heat influx to the cryostatting zone amounted to 480 W.

A most characteristic measure of cooling-down of the cryostatting zone is the change of the field winding resistance.At a room temperature the field resistance was 84 ohms, and at the nitrogen temperature, about 11.6 Ohms.

The cooling-down process is shown in Fig.2 from which one can readily see that the nitrogen temperature was obtained practically within 50 hours after the beginning of intensive cooling-down. The evaporation experiment was being carried out between the 144 and 154 hours of operation. Shown in Fig.1 there is a steady-state flow-rate during this period of time.

The abrupt change in nitrogen flow-rates and temperatures corresponds to the change in the rotor rotational speed. This is attributed to the expander effect (isentropy compression or expansion).

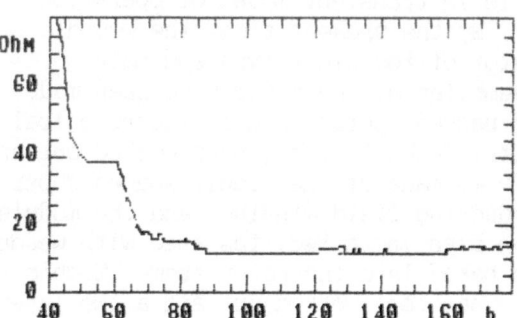

Fig.2.Time dependence of field winding resistance in cooling down

The temperature-sensitive elements and magnetic field frensducers of various types, 100 in number, are installed on the rotor. These were installed in the torque tubes, in the zone the helium inlet, current leads, winding and liquid bath.

Measurement data characteristic of local temperatures in the zone of the end parts of the winding and in the rotors "large tooth" through channels in which nitrogen was supplied to the field winding are shown in Fig.3.

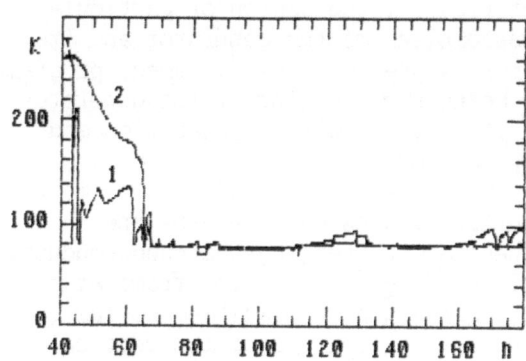

Fig.3. Time dependence of temperature in the zone of field winding in cooling-down, 1 - large tooth, 2 - end part of winding

Because of design features, the "right hand" torque tube had a lower temperature in the warm zone, and, with increasing nitrogen flow-rate, the whole tube overcooled rapidly.

The work is now under way preparatory to the next stage of testing the generator with cooling-down the rotor with liquied helium.

REFERENCE

1. Fomin, B.I., Kurilovich, L.V., Khutoretsky, G.M., Filippov, I.F., Rybin, Yu.L., Varschavsky, V.D. and Tjurin, Yu.G., Main stages of manufacturing a 300 MW superconducting generator.Cryogenics, 1987,27,May, 243-8.

534

ELECTRICAL POWER SYSTEM STABILIZATION UTILIZING ENERGY IN FILED WINDING OF SUPERCONDUCTING GENERATOR

Y. Mitani, Y. Kowada, K. Tsuji and Y. Murakami
Osaka University
2-1 Yamada-oka, Suita
Osaka, 565 JAPAN

ABSTRACT

This paper proposes a new control concept for superconducting generator to improve power system dynamic performances more significantly. The stored energy in the field winding of the superconducting generator is used to stabilize the power oscillation in electrical power systems. The proposed controller makes it possible to regulate the generator terminal voltage almost constant as well as to stabilize the power oscillation significantly. Some numerical studies demonstrate the effectiveness of the power system stabilizing control. Specifications of field winding necessary for the stabilization of model power system are evaluated from the numerical results.

INTRODUCTION

The advantages of a synchronous generator with superconducting field winding can be stated as: 1. Reduction in losses, 2. Improved system performance, 3. High-Voltage armature feasibility, 4. Small size and weight, and so on. Some feasibility studies have been carried out, and currently some model machines for demonstration are under construction or tested [1-5].

As stated above, the power system stability may be effectively improved just by replacing a conventional generator by a superconducting generator. In this paper a new control concept for superconducting generator which is capable of improving power system dynamic performances more significantly, is proposed.

The superconducting field winding which produces strong magnetic flux has another fea-

ture as an energy storage. On the other hand, superconducting magnetic energy storage (SMES) is well-known as a stabilizer of power system oscillations. Therefore the application of stored energy in superconducting generator to the power system stabilization may be feasible. However, it would become a subject of discussion that the system voltage might fluctuate due to the varying current of field winding.

In this paper it is proved qualitatively that the stabilization of power swing by using the energy of superconducting generator has the possibility of suppressing the voltage fluctuation of power system as well. A power system stabilizing control scheme can be devised according to this characteristic.

Some numerical studies demonstrate the significance of the improved system performances by the proposed control scheme. Furthermore, specifications of field windings necessary for the stabilization of model power system are evaluated from the results. In the case of 1,120 MVA generator with 10 kA field current, the maximum DC voltage used for the control is about 3,500 V, which corresponds to the evaluated values for the superconducting generator with a quick response excitation.

EVALUATION OF ENERGY IN FIELD WINDING OF SUPERCONDUCTING GENERATOR

To evaluate the possibility of applying the energy in field winding of superconducting generator to the stabilization of power oscillation, the energy capacity of field winding was estimated. TABLE 1 shows an example of parameters of field winding circuit for a 1,120 MVA superconducting generator. Voltage V_f is used for the steady state power supply and V_f' is for the quick response excitation. The capacity of power conversion was estimated for the quick response excitation.

Next, the capacities of SMES for the electrical power system stabilization were also estimated from the literature [6,7]. TABLE 2 shows the results. The capacities surveyed here are the amount of energy transfer between the SMES and the power system, and the magnitude of power conversion during the power system stabilization. The values in TABLE 2 were converted in terms of the stabilization of 1,120 MVA power system which corresponds to the capacity of the superconducting generator in TABLE 1, where the assumed fault was a three phase short circuit lasting for 1 cycle.

In the case of the stabilizing control of 1,120 MVA power system by SMES, the amount of 20 or 30 MJ energy and 60 or 70 MVA power conversion are used. Comparing these values

with that in TABLE 1, it can be said that the field winding of superconducting generator has nearly the same amount of capacities as the SMES for the power system stabilization.

TABLE 1
Machine constants of
field winding circuit

open circuit time constant	T_{do}'=1,900 s
current	I_f=10,000 A
inductance	L=0.94 H
voltage for steady state	V_f=5 V
voltage for quick response	V_f'=5,000 V
stored energy	$E=LI^2/2$=47 MJ
power conversion for quick response	$S=I_f V_f'$=50 MVA

TABLE 2
The capacities of SMES used for the
1,120 MVA power system stabilization

system	energy	power
one machine to infinite bus system	19 MJ	75 MVA
6 machine interconnected power system	30 MJ	60 MVA
experiment in artificial power system	19 MJ	75 MVA

RELATION BETWEEN POWER SYSTEM STABILIZATION AND CONTROL OF SYSTEM VOLTAGE

It has been made clear that the field winding energy has a sufficient capacity to be used for the power system stabilization. Here the influence of the varying current in the field winding on the voltage of power system, is to be discussed.

To illustrate the characteristics of power system with a superconducting generator, a simple system configuration shown in FIGURE 1 is assumed. The system consists of a long distance bulk power transmission line, a superconducting generator and a power converter to exchange power between the field winding circuit and the power system.

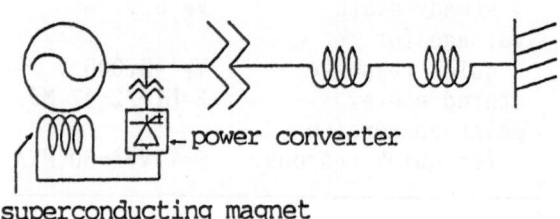

power converter

superconducting magnet

FIGURE 1   A model power system with a superconducting generator.

Let the control scheme of the power converter for the power system stabilization be

$$\Delta P_{SM} = -K_1 \Delta \omega \qquad (1)$$

where P_{SM} [W] is the active power controlled by the power converter and ω [rad/s] is the angular velocity of generator. Δ denotes deviation from operating point. This control scheme is the same as that by SMES.

Now, we assume that some kind of disturbance occurs in the power system. After the disturbance the rotor of generator may begin to oscillate caused by the imbalance between the mechanical input power and the electrical output power. The system voltage may also fluctuate following the oscillation of rotor angle. The system characteristics of these oscillations are schematized in FIGURE 2 where the damping of oscillation is neglected for conciseness.

Then the energy and the current in the field winding may change proportionally to the integral of power absorption following the power system stabilizing control scheme of equation (1). These relations are also drawn in FIGURE 2.

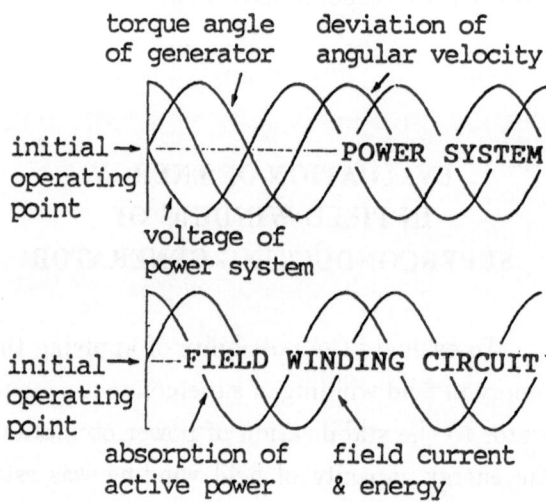

torque angle of generator   deviation of angular velocity

initial operating point

POWER SYSTEM

voltage of power system

initial operating point

FIELD WINDING CIRCUIT

absorption of active power   field current & energy

FIGURE 2   Characteristics of power system oscillation and field winding circuit when the power system stabilizing control is applied.

FIGURE 2 shows that the field current becomes larger (or smaller) when the system voltage is to fall (or to rise).

Thus, it can be concluded that the power system stabilizing control scheme represented by equation (1) has a possibility to suppress the voltage fluctuation at the same time.

SIMULATION STUDIES FOR POWER SYSTEM STABILIZATION

In order to verify the effectiveness of the proposed control scheme, digital simulations have been carried out. The model power system shown in FIGURE 3, which consists of four 1,120 MVA superconducting generators with the proposed control scheme and a 200 km double circuit transmission line connected to the infinite bus, was used for the analysis. Machine constants of the superconducting generator are shown in TABLE 3. Each generator is modeled as a third-order system with torque equations and response of the field winding circuit.

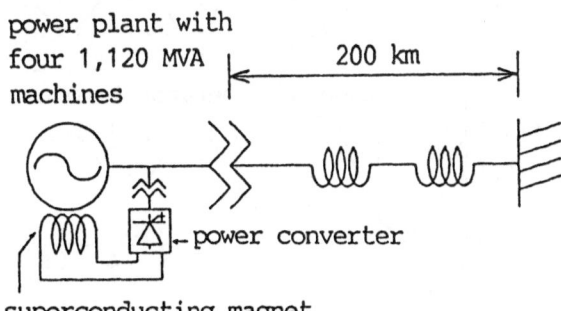

FIGURE 3   A power system model with superconducting power plant connected to the infinite bus through long distance transmission lines.

TABLE 3
Machine constants of
superconducting generator

capacity   1,120 MVA
power   1,000 MW
power factor 0.9
 q-axis synchronous reactance x_q=0.3 pu
 d-axis synchronous reactance x_d=0.3 pu
 d-axis transient reactance x_d'=0.2 pu
 inertia constant M=6 s
 damping coefficient D=3 pu

Active power of the converter was controlled according to the equation (1) where the reactive power was set to be always zero for conciseness.

The power system stability and the fluctuation of system voltage were evaluated for different values of control gain K_1 where the criteria were defined by

$$J = \int_0^5 (\Delta\omega)^2 dt \qquad (2)$$

and the maximum magnitude of voltage fluctuation ΔV_t after the fault, respectively. The results are shown in FIGUREs 4 and 5, where several kinds of three phase short circuit faults in TABLE 4 were assumed as system disturbances.

The power system becomes more stable as the gain K_1 becomes larger. This effectiveness is the same that by SMES. The voltage fluctuation becomes smaller first when the gain K_1 be-

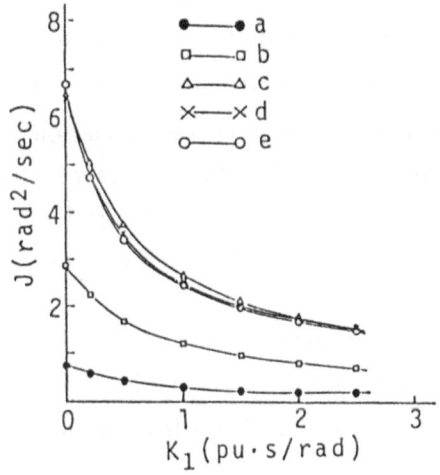

FIGURE 4   Power system stability for different values of control gain K_1.

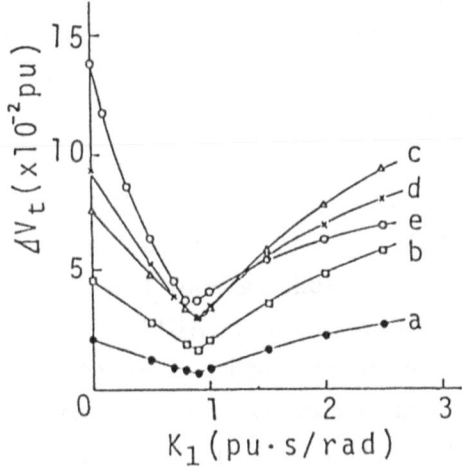

FIGURE 5   Voltage fluctuation for different values of control gain K_1.

TABLE 4

The conditions of three phase short circuit applied to the power system

	fault duration	system condition after the fault
a	1 cycle	pre-fault condition
b	2 cycles	pre-fault condition
c	3 cycles	pre-fault condition
d	3 cycles	1 cct (20km) was cut off
e	3 cycles	1 cct (40km) was cut off

comes larger since the deviation of field current due to the absorption of active power following the power system stabilizing control suppresses the voltage fluctuation. However the enlarged K_1 beyond the certain value makes the voltage fluctuation larger since the deviation of field current swings the voltage in reversed phase. Therefore, the value of K_1 should be carefully tuned for the design of power system stabilizing controller.

Consequently ΔV_t becomes smallest when the gain K_1 equals about 0.9 [pu·s/rad]. These tendencies are little dependent on the fault conditions.

The system behaviors after the fault e are shown in FIGURE 6 (a) (with the constant excitation control) and (b) (with the proposed control).

The results show that the power system stabilizing control makes it possible to damp out the power swing effectively and to suppress the fluctuation of the system voltage significantly at the same time. From the simulated waveforms, the system capacities used for the power system stabilizing control are estimated. The results are shown in TABLE 5.

The values are smaller compared with the power and energy capacities of SMES in TABLE 2. These capacities correspond to the estimated values for the superconducting generator with quick response excitation (see TABLE

1), however, the proposed control scheme is significantly effective compared with the quick response excitation.

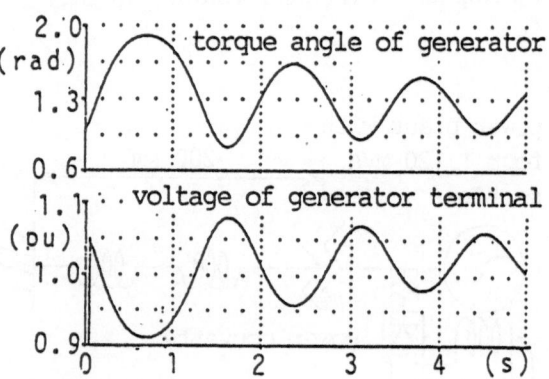

(a) with constant excitation control

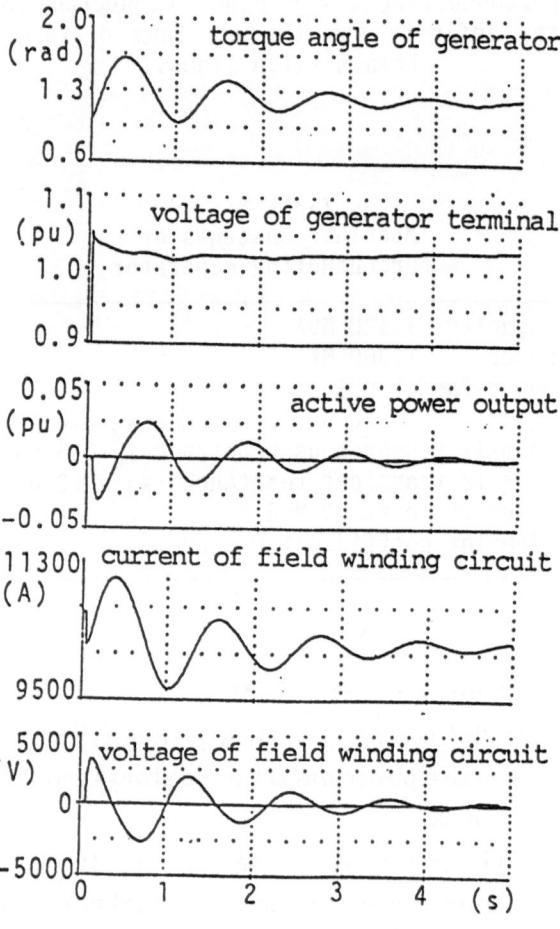

(b) with the proposed excitation control

FIGURE 6　Results of simulation.

TABLE 5
System capacities used for the
stabilizing control

the maximum active power	: 35 MW
the exchanged energy	: 14 MJ
the maximum voltage of the field winding	: 3,400 V

CONCLUSIONS

(1) A new control concept for superconducting generator to improve power system performances significantly, has been proposed.

(2) It has been demonstrated that the field winding circuit of superconducting generator has a sufficient energy capacity to stabilize power system oscillations.

(3) The proposed control scheme makes it possible to suppress the fluctuation of system voltage as well as to improve the damping of power swing.

(4) It has been evaluated that the capacities of field winding circuit used for the proposed stabilizing control correspond to the values for the superconducting generator with quick response excitation.

ACKNOWLEDGEMENT
This work was financially supported by Grant-in-Aid for Scientific Research of the Ministry of Education, Science and Culture, Japan.

REFERENCES

1. Introhar, L. and Lambrecht, D., Technical overview of the German program to develop superconducting AC generators. IEEE Trans. on Magnetics, 1983, 19, 536-540.

2. Sabrie, J. L. and Goyer, J., Technical overview of the French program. ibid, 1983, 19, 529-532.

3. Smith, J. L., Overview of the development of superconducting synchronous generators. ibid, 1983, 19, 522-528.

4. Fujino, N., Technical overview of Japanese superconducting generator development program. ibid, 1983, 19, 533-535

5. Tomiyama, S. and Aiyama, Y., The superconducting generator development program in Japan. ibid, 1987, 23, 3533-3535.

6. Mitani, Y., Tsuji, K. and Murakami, Y., Application of superconducting magnet energy storage to improve power system dynamic performances. IEEE Trans. on Power Systems, 1988, 3, 1418-1424.

7. Mitani, Y., Tsuji, K. and Murakami, Y., Application of superconducting magnet energy storage to power system stabilizing control. Technology Reports of the Osaka Univ., October 1986, pp.305-315.

FINITE ELEMENT MODELLING OF ELECTROMAGNETIC SHIELDS IN SUPERCONDUCTING AC GENERATOR

Songyop Hahn, Dongchul Han, Hyochul Sin and Byoungsuk Lee
Seoul National University, Seoul, Korea
Gueesoo Cha
Soonchunhyang University, Onyang, Korea
Sungchin Hahn
Donga University, Pusan, Korea
Youngjin Won
KEPCO Research Center, Daejun, Korea

ABSTRACT

Shield system of double-shielded Superconducting AC Generator(SCG) prevents the flux variation into the superconducting field winding. It is essential to analyze the role of shield system during the transient state for the design of shield system. This paper describes the finite element analysis of shield system. Transient characteristics of the shield system during three phase short and line-to-line short have been shown. Currents of field and armature winding have been also calculated. Most of ac component of flux variation has been blocked by the damper which is closer to the armature winding than the electro-thermal shield. In electro-thermal shield there exists only dc component of flux variation. More stress is exerted on damper in line-to-line short than three phase short.

1. INTRODUCTION

Double-shielded single-rotor Superconducting AC Generator(SCG) is the most popular type among various SCG's reported yet. Outer and inner shield system are called damper and electro-thermal shield, respectively.[1]

In this arrangement, damper is designed to maximize damping and to bear the majority of short-circuit force. Electro-thermal shield is introduced to screen the low frequency flux variation and to prevent heat penetration into superconducting field winding.

The shield system must ensure that heat losses by the flux variation in the superconducting field winding should not drive the superconductor into normal state. Therefore exact analysis is required for the design of shield system.

For analysis of shield system during the transient state, equivalent circuit modelling based on Park's equation and frequency domain analysis has been used. But results obtained by the above methods give only global effects of the shield system.[2]-[5]

This paper presents finite elemet analysis of the shield system. Finite element analysis not only gives more accurate results with the effect of harmonics but also provides local behaviour in any point. Transient behaviours of 2 GVA SCG, such as, three phase short and line-to-line short, are calculated in this paper. In each case, induced current of the shield system and variation of line current and field current are shown.

2. FINITE ELEMENT MODELLING

At any region of SCG consisting of conductor and magnetic material, Maxwell's equations are applicable.

$$\nabla \times \vec{H} = \vec{J}_o + \vec{J}_e \qquad (1)$$

$$\nabla \times \vec{E} = -\frac{\partial \vec{B}}{\partial t} \qquad (2)$$

$$\nabla \cdot \vec{B} = 0 \qquad (3)$$

By the following relations,

$$\vec{B} = \mu \vec{H} \qquad \vec{J}_e = \sigma \vec{E}$$

and vector identity, governing equation results,

$$-\nabla^2 \vec{A} = \mu \left(\vec{J}_o - \sigma \frac{\partial \vec{A}}{\partial t} \right) \qquad (4)$$

Source currents are initiated by the EMF, so that circuit euqation can be expressed as follows.

$$\frac{d\phi_j}{dt} + (r_j + R_j)I_j + L_j \frac{dI_j}{dt} = V_j, \qquad (5)$$

$$j = f, a, b, c$$

where r : internal resistance of armature

R, L : external inductance

Applying the Galerkin's method and discretizing by the first order triangular element, Eq.4 and Eq.5 become,

$$\begin{bmatrix} A & B \\ C & D \end{bmatrix} \begin{bmatrix} E \\ F \end{bmatrix} = \begin{bmatrix} G \\ H \end{bmatrix} \qquad (6)$$

where, A : stiffness matrix
B : conversion matrix I to J
C : conversion matrix ψ to A
D : impedance matrix
E : $A_1, A_2, \ldots A_{n-1}, A_n$
F : I_f, I_a, I_b, I_c
G : forcing matrix
H : V_f, V_a, V_b, V_c

Time derivatives in Eq.4 and Eq.5 are discretized by backward difference method.

3. SCG MODEL

To demonstrate the effect of shielding system during the transient state, 2 GVA SCG with double-shielded single-rotor is chosen. Rated voltage and current of the model SCG are 30000 [V] and 38490 [A], respectively. Cross section and dimension of SCG are given in Fig.1 and Table 1, where subscript i and o means inner and outer radius.

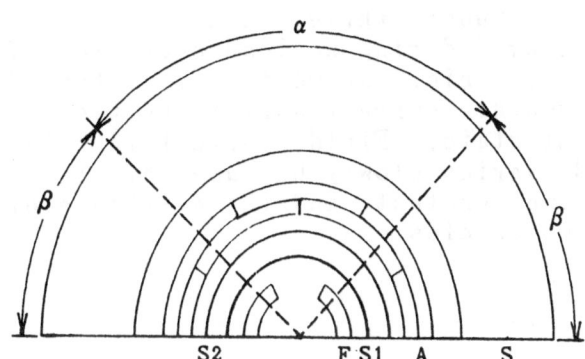

FIGURE 1. Cross section of SCG

F : field winding
S1 : electro-thermal shield
S2 : damper
A : armature winding
S : environment shield

TABLE 1. Dimension of SCG [m]

$r_{fi} = 0.277$	$r_{fo} = 0.377$
$r_{Ei} = 0.411$	$r_{Eo} = 0.5$
$r_{Di} = 0.55$	$r_{Do} = 0.65$
$r_{ai} = 0.75$	$r_{ao} = 0.95$
$r_{si} = 1.15$	$r_{so} = 1.8$

Major specifications of SCG are given in Table 2, where T, σ, D and E stand for time constant, conductivity, damper and electro-thermal shield.

TABLE 2. Specifications of SCG

$V = 30000\,[V]$ $I = 38490\,[A]$

$X_d = 0.55\,[p.u.]$ $X_{d'} = 0.37\,[p.u.]$

$X_{d''} = 0.29\,[p.u.]$ $X_{d'''} = 0.16\,[p.u.]$

$T_D = 0.27\,[S]$ $T_E = 1.6\,[S]$

$\sigma_D = 5\times10^7\,[\mho/m]$ $\sigma_E = 5\times10^8\,[\mho/m]$

2 Dimensional analysis is done under the assumption of Z direction current only. In analysis region, mesh is generated with 5 division. In this case the numbers of total node and element are 409 and 756.

4. THREE PHASE SHORT

Sudden three phase short is assumed during no load operation of SCG. Fig.2 shows the variaton of armature current during the transient state. Field current in Fig. 3 varies slowly because of large time constant and the existence of two shields.

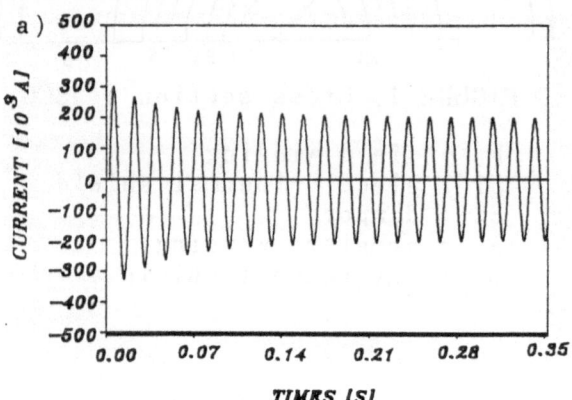

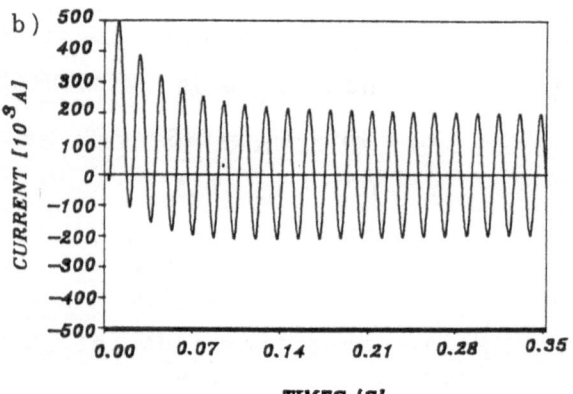

FIGURE 2. Armature current
a) A phase   b) B phase

Fig.4 and Fig.5 show the average induced current density in damper and electro-themal shield in q axis direction ($\alpha = 90°$ in Fig. 1) and d axis direction ($\beta = 45°$ in Fig.1).

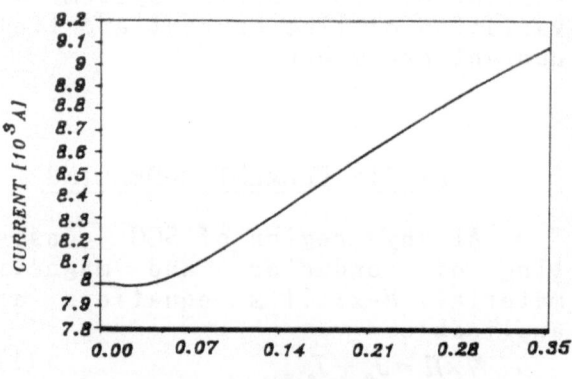

FIGURE 3. Field current

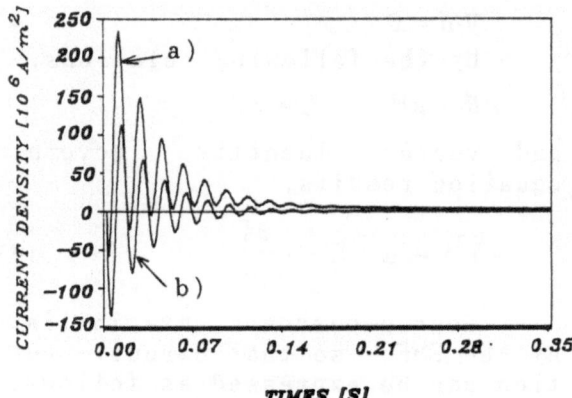

FIGURE 4. Induced current in damper
a) q axis direction
b) d axis direction

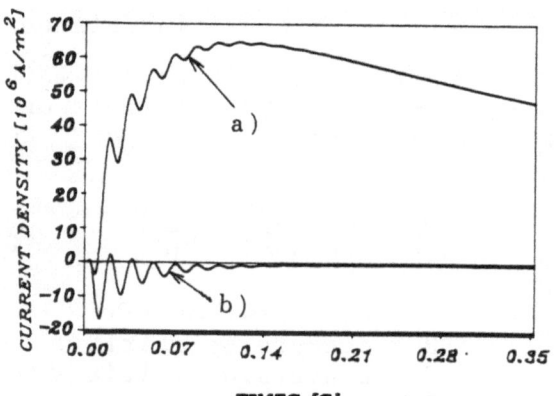

FIGURE 5. Induced current in
electro-thermal shield
a) q axis direction
b) d axis direction

Fig.4 and Fig.5 show that current is also induced around d axis but its magnitude is about 1/2 1/5 of q axis direction and it vanishes faster than that of q axis direction.

Damper is closer to the armature winding than the electro-thermal shield, so that magnitude of induced current in damper is greater than that of electro-thermal shield. Futhermore ac component of flux variation due to armature current is screened by the damper. Therefore little ac component of induced current is induced in elector-thermal shield.

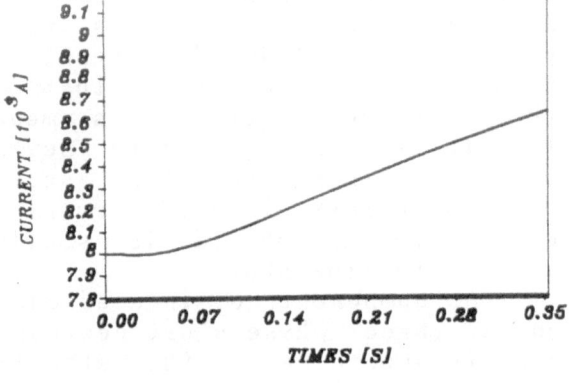

FIGURE 7. Field current

5. LINE-TO-LINE SHORT

Line-to-line short occurs during no load operation of SCG as the previous case. A phase is assumed to be opened to emphasize the effect of B and C phase current. Armature current of B phase is shown in Fig.6. Increase of field current in Fig.7 is about 1/2 compared with that of three phase short.

Armature currents of B and C phase are the same in amplitude but differs 180° in phase. Therefore resultant MMF is not revolving but pulsating. Futhermore frequency of pulsating MMF is twice than that of armature current.

Fig.8 and Fig.9 show the induced current of damper and electro-thermal shield whose fre-

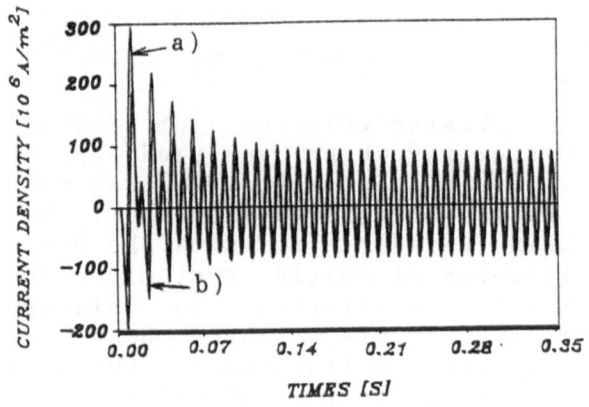

FIGURE 8. Induced current in damper
   a) q axis direction
   b) d axis direction

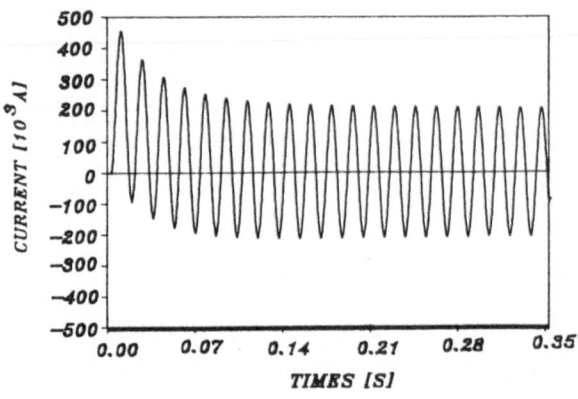

FIGURE 6. Armature current, B phase

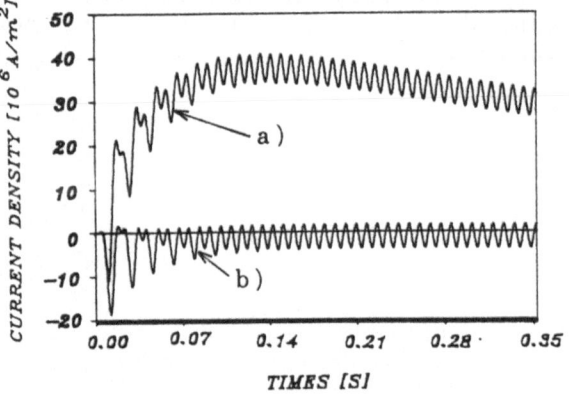

FIGURE 9. Induced current
   in electro-thermal shield
   a) q axis direction
   b) d axis direction

quency are twice as much as line current. Meaning of q and d axis direction is the same as three phase short. Induced current in damper is greater than that of three phase short. This means line-to-line short is more severe to damper than three phase short. On the contrary, induced current of electro-thermal shield is smaller at line-to-line short.

AC component of induced current at three phase short decreases rapidly according to the stabilizing of armature current but it does not decrease in two phase short. That is due to the continuous pulsation of armature MMF.

6. CONCLUSIONS

Characteristics of shield system have been analyzed by the finite element method. Three phase short and line-to-line short have been considered to show the effectiveness of shield system. Larger stress is exerted on electro-thermal shield at three phase short than line-to-line short. But it is the line-to-line short to exert larger stress on damper. Induced current decreases rapidly in three phase short, but it continues in line-to-line short.

Results of this paper give more precise behaviour of SCG than equivalent circuit analysis.[6] Ettects of harmonics in armature and field winding can be completely considered in finite element analysis.

ACKNOWLEDGEMENT

The authors would like to thank Korea Electric Power Co. Research Center for their support.

REFERENCES

1. N.Maki et. al., "Development of Superconducting AC Generator", Trans. on MAG, Vol.MAG-24, No.2, 1988, pp.792-795
2. T.J.E.Miller et.al.,"Penetration of Transient Magnetic Fields through Conducting Cylindrical Structure with Particular Reference to Superconducting A.C. Machines",Proc. of IEE, Vol.123, No.5, 1976, pp.437-443
3. T. Bratoljic, H.Fursich and H.W. Lorenzen, "Transient Small Perturbation Behaviour of Superconducting Turbogenerator",IEEE Trans. on PAS, Vol.PAS-96, No.4, 1977, pp.1418-1429
4. T. Bratoljic, "Negative Sequence and Eddy Current Losses in the Rotor of a Superconducting Turbo generator", 6'th Int'l Conf. MT Bratislava,Aug/Sep, 1977,pp.206-211
5. T.Okada, T.Nitta and T.Shintani, "Transient Performance of the 20 KVA Superconducting Synchronous Generator in Power System", Trans. on MAG, Vol.MAG-23, No.2, 1987, pp.1340-1343
6. T.okada et.al., " The Basic test on the 20 KVA Superconducting Synchronous Generator ", Trans. on MAG, Vol.MAG-19, No.3, 1983, pp. 1043-1046

CALCULATION OF MAGNETIC FIELD CAUSED BY ARMATURE WINDING OF SUPERCONDUCTING GENERATOR

Kiyoshi Yamaguchi, Naoki Maki, *Hitachi Research Laboratory, Hitachi Ltd.*

Kuji-chou Hitachi-shi Ibaraki-ken, Japan

ABSTRACT

A superconducting generator has an air-gap armature winding which is different from a conventional generator's. The superconducting generator(SCG) also has a cylindrical iron magnetic shield. The straight section of the winding is just inside the shield and for the end section the winding is outside of the shield. Particularly, in the end section of the winding, the magnetic field has a three dimensional distribution. To calculate accurate synchronous reactance and leakage reactance, magnetic field and magnetic flux-turn must be calculated precisely. We calculated the whole magnetic flux distribution by TOSCA[1],[2]. The reactance of the strait section of the winding calculated by TOSCA's results is almost the same as the reactance by a two dimensional analytic method. But the leakage reactance is not the same by the two methods.

INTRODUCTION

Superconducting generators have rotors made of non-ferromagnetic substances and have no iron teeth in the armature winding. Therefore, the magnetic field generated by a superconducting generator's armature winding has a quite different distribution from a conventional generator which has ferromagnetic rotor and iron teeth. In case of SCG, the only ferromagnetic substance is the magnetic shield which also acts as a magnetic yoke for the magnetic flux of the field winding. Because of these factors SCG has about one-fifth synchronous reactance of conventional generator. Smaller synchronous reactance gives greater stability to the power transmission line system.

It is essential to make a precise calculation of synchronous reactance. The reactance can be calculated from self-flux linkage of the armature winding, and the linkage is worked out by the self-flux density distribution of the winding.

MODELING

A superconducting generator has a schematic structure showed in Fig.1. The rotor and vacuum space form a concentric cylinder cryostat. The outermost cylinder is a damper which is the shell of the vacuum space and acts electromagnetic shield, too. Innermost section, the winding support shaft, supports the superconducting field winding. A radiation shield between the winding support shaft and the damper intercepts radiant heat from the damper. Liquid helium is fed to the rotor through a helium transfer coupling. Gaseous helium, vaporized in the helium tank by the heat, is collected through helium transfer coupling. The field winding is continuously excited by the current through collectors and leads.

Fig.2 shows a model of the armature winding and iron core. The figure also shows the dimensions of the winding and the iron core of a 70MW class SCG. The winding is made with a straight section in the iron core and end section out of the core. Particularly, the shape of the conductor of armature winding end section has a geodetic line on a cone surface, as shown in Fig.3, which is the shortest distance of two points. The line is straight if the cone surface is rolled out. Armature windings have two layers of conductor and in the end section the conductors cross over to the opposite side as much as 180 degrees. In practise, armature winding conductors cross over less than 180 degrees to improve the current wave form. Fig.4 shows a bird eye view of the armature winding. The end section is represented by a chain of three conductors.

The armature winding has 9 pairs of conductors per phase. To simplify, three neigh-

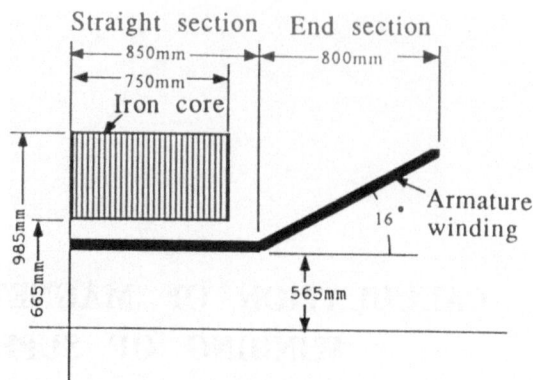

Fig.2 Dimension of armature winding

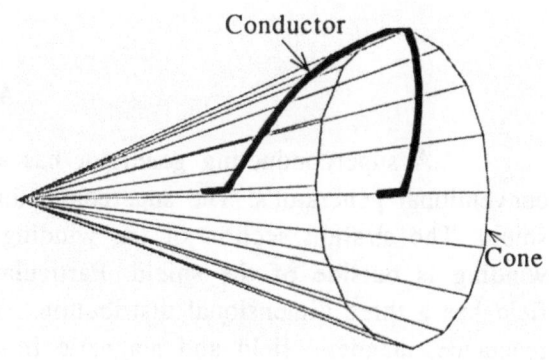

Fig.3 End section shape of armature winding

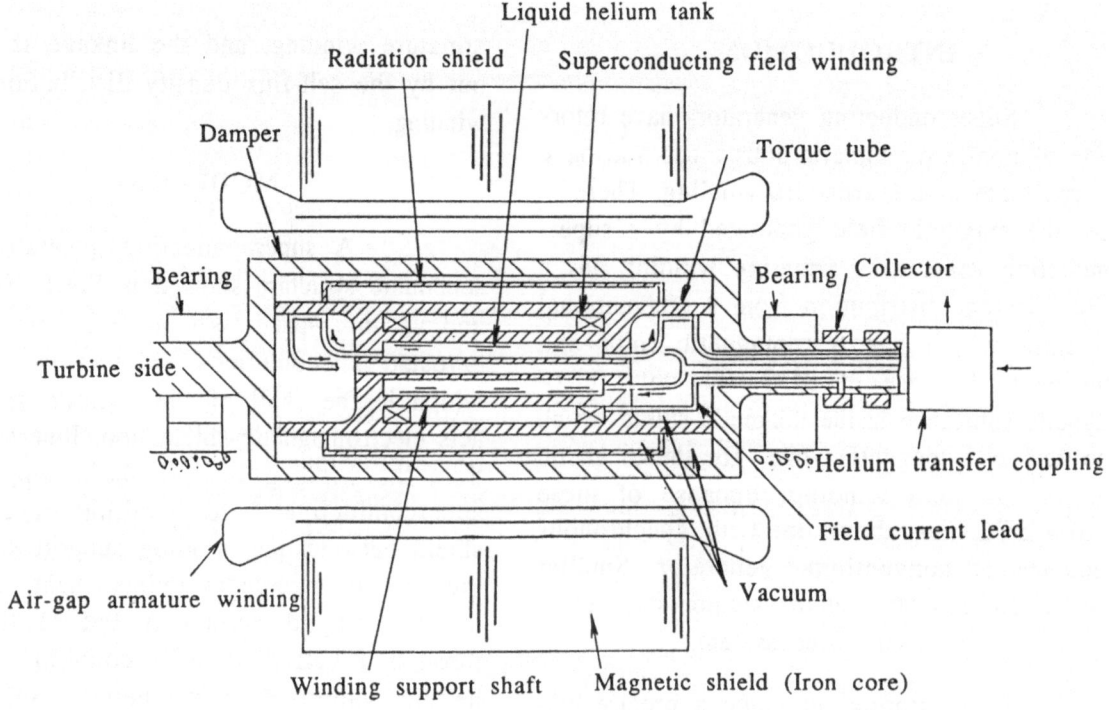

Fig.1 Schematic structure of a superconducting generator

boring conductors are represented by one conductor, Fig.5. To improve the current wave form of the armature winding, three conductors in inner layer, represented by one character, project into neighbor phase.

ANALYSIS

Actual conductor's current amplitude is rated at 6770 A. The armature winding is three phase with 120 degree shifted phase angle. It is assumed that U-phase current is at its peak. Z axis has the same direction as the axis of rotation. X and Y axis is shown in Fig.5.

Fig.6 shows linkage flux distribution along Z axis at the armature winding. The ordinate is flux density in Tesla and the abscissa is the Z axis position measured from the body center of the generator. Fig.7 shows linkage flux density distribution on diameter of U-phase winding at body center. The abscissa is radial position measured from the Z axis. Fig.8 shows the linkage flux density along the tangential direction at the armature winding of the straight section. The abscissa is the angle from the center of the U-phase winding axis.

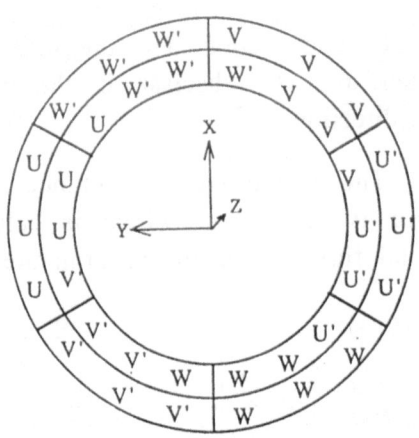

Fig. 5 Phase distribution of armature winding

(one character represent three conductor)

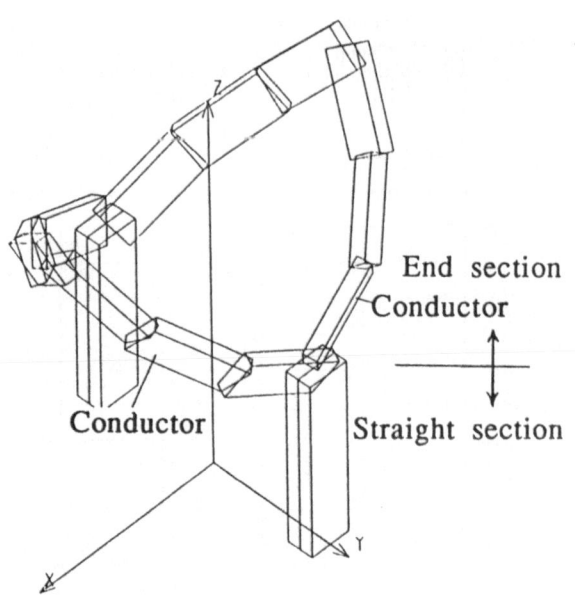

Fig.4 Bird's eye view of armature winding

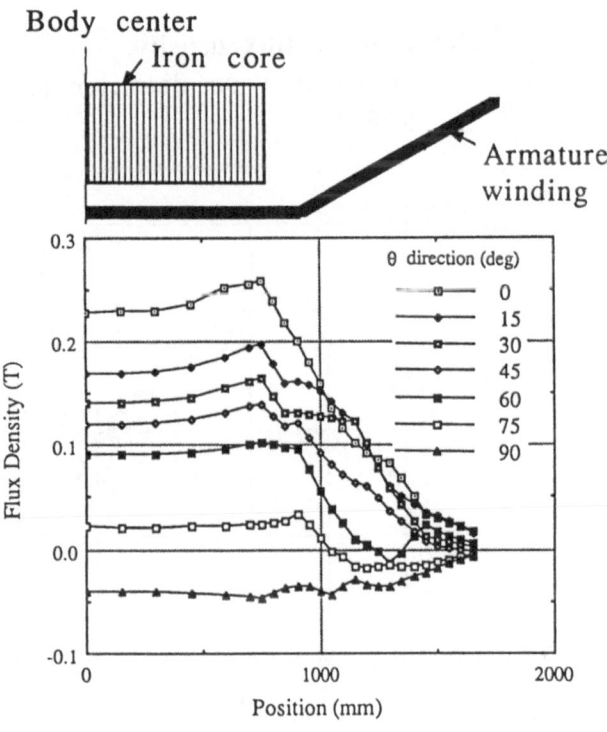

Fig.6 Linkage flux density distribution at armature winding

Fig.9 shows the linkage flux density distribution at the winding in the end section. Fig.10 shows flux linkage distribution along the Z axis in the end section. The ordinate is flux in Weber and the abscissa is the position along Z direction.

The synchronous reactance of the generator is 0.398 p.u. which is calculated from total flux linkage of the armature wind-ing. 0.261 p.u. is the reactance of the straight section of the winding and 0.137 p.u. is the leakage reactance of the end section.

DISCUSSION

As indicated in Fig.6, the flux density has a relatively uniform distribution in the

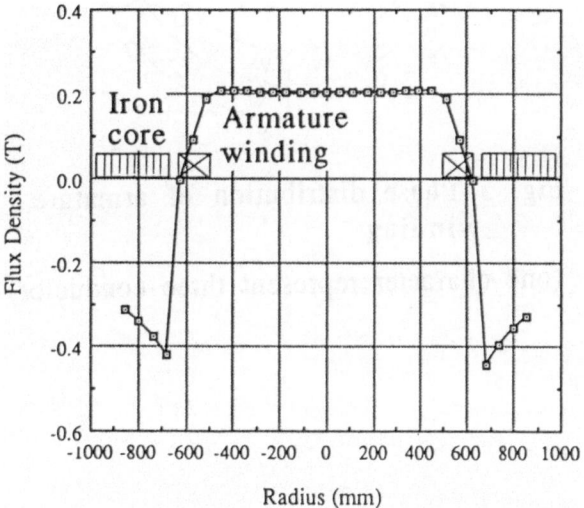

Fig.7 Linkage flux density distribution on diameter (U-phase)

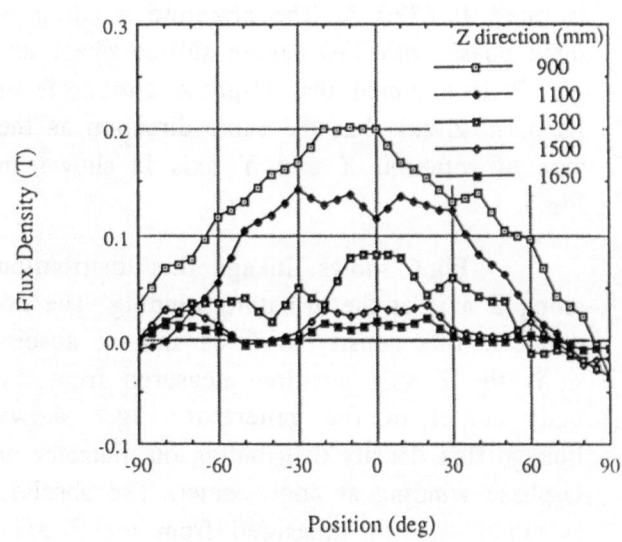

Fig.9 Linkage flux density distribution at armature winding (End section)

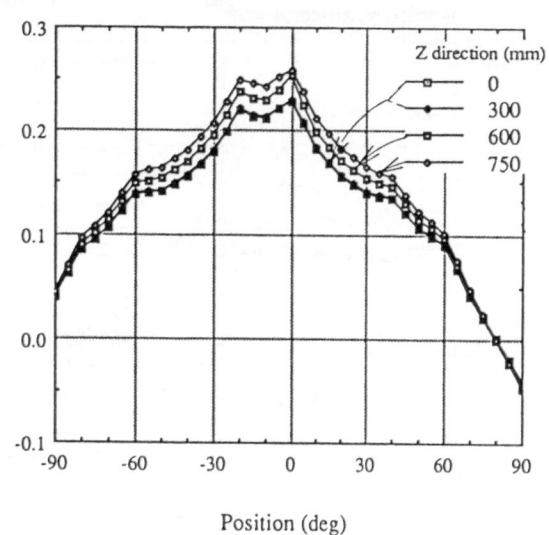

Fig.8 Linkage flux density distribution at armature winding (Straight section)

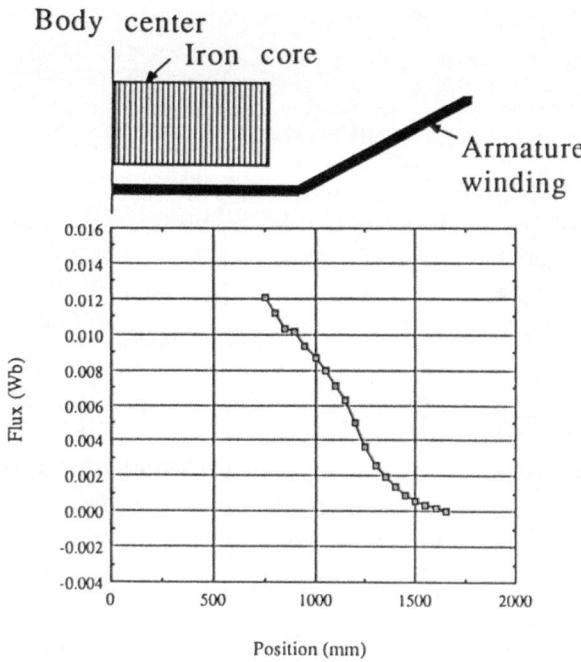

Fig.10 Flux density distribution at end section

straight section. In the end section, flux density descends linearly to the end. Fig.7 shows that the areas surrounded by the curve and 0 Tesla line is equal in its plus and minus part. Fig.8 shows that the armature winding current creates a sinusoidal distribution of flux density in the iron core. Fig.9 shows that the winding creates an almost sinusoidal distribution of flux density out of the iron core. But, there are many ripples because the conductors of the winding are complicated in the end section.

The synchronous reactance calculated by two-dimensional analytic method is 0.35 p.u. made up of 0.255 p.u. of reactance in the straight section and 0.095 p.u. of leakage reactance. The reactance of the straight section is almost the same by the two methods. But the leakage reactance is not the same. In two-dimensional treatment, the leakage reactance is calculated with equivalent length that is well established for a conventional generator. It is not clear which method is preferred. The flux density, used in the leakage reactance calculation, is the value in the armature winding. As the flux density vector near the end section of the winding changes rapidly, the flux density in the winding of the end section may not be exact. Because the rotor of SCG is non-ferromagnetic substance, the flux density distribution is not same as that of a conventional generator's in the end section. So the equivalent length of the end section for SCG must be computed from more accurate field calculation.

CONCLUSION

Superconducting generators have no ferromagnetic substance in the rotor and iron teeth in the armature winding. Thus the magnetic field which is generated by a superconducting generator's armature winding has a quite different distribution from a conventional generator which has ferromagnetic rotor and iron teeth.

A flux density distribution created by armature winding of 70MW class superconducting generator is calculated by TOSCA. And the synchronous reactance of the generator is calculated from the flux density distribution.

ACKNOWLEDGEMENT

This work was performed as a part of "R&D on Superconducting Technology for Electric Power Apparatuse" under the Moonlight Project of Agency of Industrial Science and Technology, MITI, beeing consighned by New Energy and Industrial Technology Development Organization (NEDO).

REFERENCES

[1] TOSCA Reference Manual, VECTER FIELDS LTD, 24 , BANKSIDE, KINGTON, OXFORD OX5 1JE, ENGLAND, 1988

[2] J. SIMKIN AND C.W.TROWBRIDGE, "THREE DIMENSIONAL NONLINEAR ELECTROMAGNETIC FIELD COMPUTATIONS USING SCALAR POTENTIALS", PROCEEDINGS OF THE IEE, VOL127, No.6 1980

SPECIFICATIONS AND SELECTING PROCEDURES ON SUPERCONDUCTING FIELD WINDINGS OF SUPERCONDUCTING GENERATORS

K.Uyeda,S.Meguro,T.Saitoh,K.Takahashi,H.Hatakeyama
Engineering Research Association for Superconductive Generation Equipment and Materials
(Super-GM),Umeda UN Bldg.,5-14-10 Nishitenma,Kita-ku Osaka City,530 Japan

ABSTRACT

R&D program on the supercoducting generator at Super-GM includes three types of NbTi superconducting field windings as the main supercoducting component. This paper describes the target specifications of NbTi supercoductors used for field windings of model rotors and the selecting procedures of optimum superconductor from two or three promising ones as one of the most important R&D subjects executed by both of generator and supercoductor development groups under the promotion of Super-GM. Repetitive checks and reviews are scheduled based on the research progress until the final choice of the design in order to utilize the most up-to-date results in the model rotors to be constructed between FY 1991∿1993.

1. INTORODUCTION

The superconducting generators are able to enhance power output by improving the energy density as the consequence of generating a high magnetic field using superconductors in rotating field windings. The superconducting generators have many advantages over the coventional machines such as lower operation cost due to decrease of loss, lower manufacturing cost due to reduction in size and weight, and higher stability of power system due to drop of reactance. There are many research problems yet to be solved for practical application of the generators, and among them, stabilization of the rotating superconducting magnet which constitutes the field winding is most important and is a basis in the design and manufacture of generators.[1]

This report describes the concept adopted for designing the superconductors of the field windings and the testing methods used for evaluating the developed superconductors.

2. OPERATIONAL CONDITIONS OF GENERATOR AND REQUIREMENTS FOR FIELD WINDING

In designing a field winding of a generator, the conditions under which the field current varies must be clarified. According to the recommendation of CIGRE WG11-05, operational conditions of generator in power system have to be examined as shown in TABLE 1. In the column of "Quench-free", "Yes" means that as a result of examining the stability of the superconducting field winding against the disturbance concerned, it has been concluded that the generator should be designed not to be quenched, and "No" means that the generator may be quenched if inevitable. The values of field current variation shown are approximate at the present level of design, since the variation greatly depends on the design of the generator dampers. From the table, it can be seen that the most severe disturbance condition for the field winding requiring the quench-free condition is power system fault (removal of fault point by primary relay protection). FIG.1 shows the variation of field current at the time of power system fault.[2]

The slow response excitation type superconducting generator is intended to gently change the field current for the purposes of controlling the generator voltage variation at the time of system

TABLE 1. Operational Conditions for Superconducting Generator
Related to the Field Winding design

Disturbances		Variations in field current		Quench-free
		Frequency (Hz)	Quantity (%)	
Normal Operation	Control of system voltage	≥ 0.1	≤1.0	Yes
	Unbalanced load	50~100	≤1.0	Yes
	Pararelling	0.25~2.0	≤0.5	Yes
Emergency Operation	Load shedding	0.25~2.0	≤0.5	Yes
	System fault	50~100/0.25~2.0	≤0.5/a few	Yes
	Short circuit at the terminal	50~100	a few	No
	Forced pararelling	0.25~2.0	a few	No

fault and controlling the system voltage in normal condition. The quick response excitation type superconducting generator is intended to positively control excitation for the purpose of enhancing power system stability. In terms of variation frequency of field current, the slow response machine corresponds to about 1 Hz, and the quick response machine, about 10 Hz.

Based on these quantities of variation, the requirements on the performance of the superconductors for field windings can be derived.

3. DESIGN OF SUPERCONDUCTORS

3.1 Design Procedure

The operational conditions of a

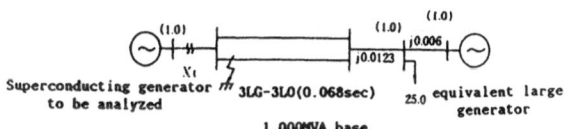

(a) One-machine to equivalent large generator system model

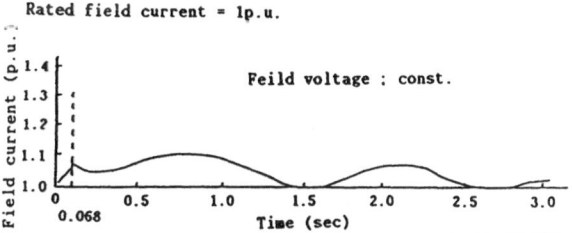

In case of slow response excitation machine

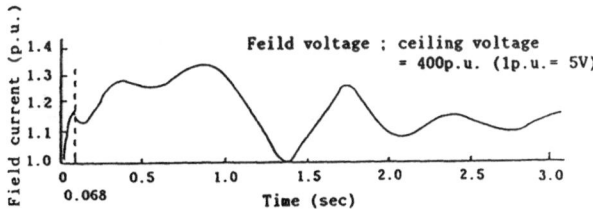

In case of quick response excitation machine

(b) Results of analysis

FIG.1 Example of field current oscillation after power system fault

superconducting generator impose severe requirements on the performance of the superconductor for the field winding. For reasonable design of superconductors satisfying the requirements of performance, manufacturers in charge of developing generators and manufacturers in charge of developing conductors carried out joint work under the promotion by Super-GM. The design work is shown in FIG.2 as a flow chart. The field windings developed by Super-GM include two types; slow response excitation type and quick response excitation type. The conductors for the respective field windings were developed by several manufacturers. Therefore, for the same requirements of performance, several conductors are designed and developed. This joint development scheme aims at developing more advanced superconductors under positive competition.

However, since a manufacturer who will finally be in charge of manufacturing the conductor for the 70 MW class model machine is to be selected, objective and fair management is inevitably required. For this reason, the exchange of information between manufacturers in charge of developing generators and the manufacturers in charge of developing conductors was always carried out in a field open to all the manufacturers, and Super-GM carefully arranged to set targets of design and development of superconductors aiming at generators with higher performance. Furthermore, in the respective stages of design, experts of universities and national research institutes outside Super-GM were asked to extend cooperation in evaluating the results of design.

3.2 Features of Superconductor Design
(1)Superconductors for Slow Response Machine

Of the 70MW class field windings,two were for slow response excitation type. For one of them, design was carried out

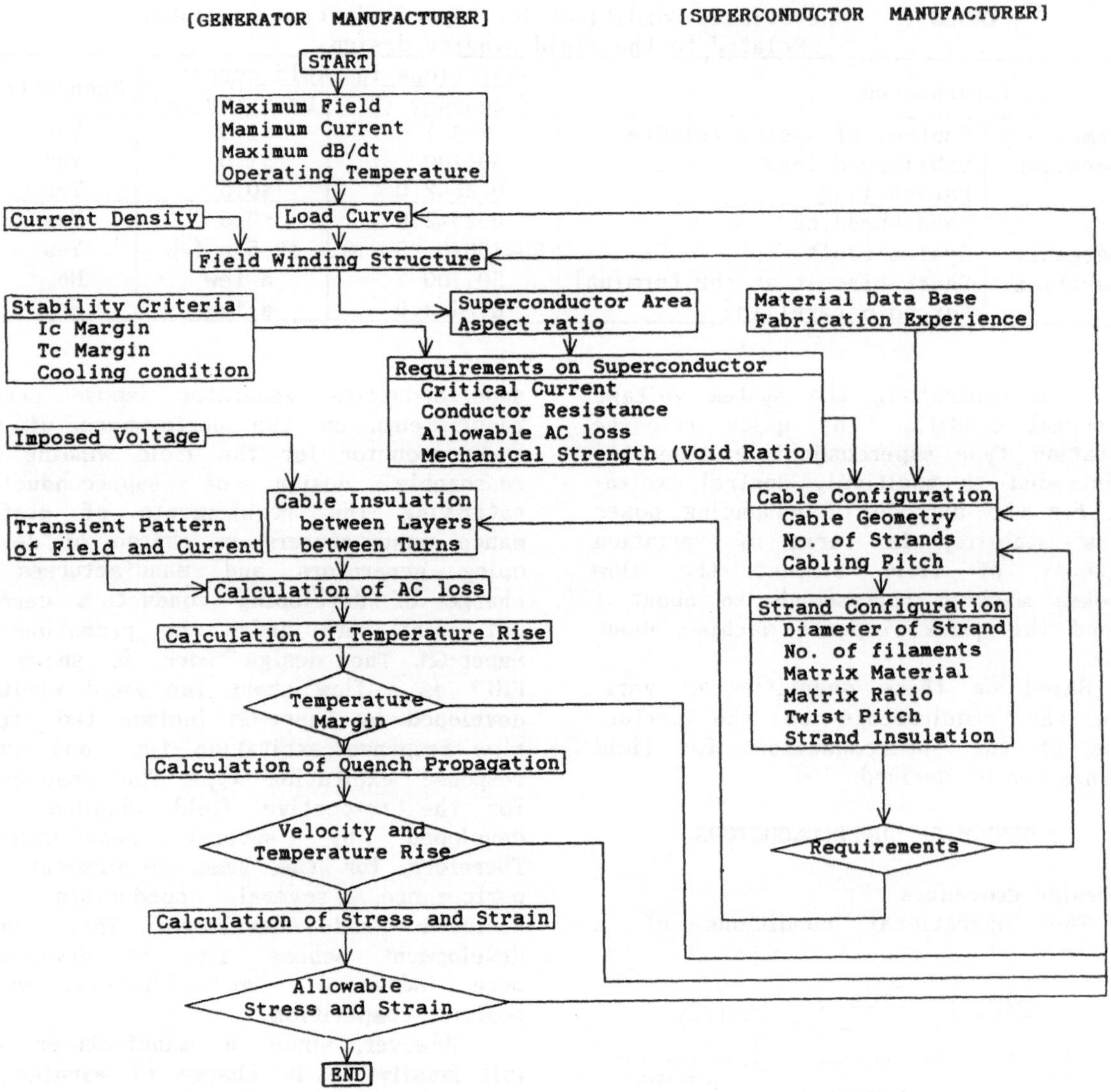

[GENERATOR MANUFACTURER]
[SUPERCONDUCTOR MANUFACTURER]
START
Maximum Field
Mamimum Current
Maximum dB/dt
Operating Temperature
Current Density
Load Curve
Field Winding Structure
Stability Criteria
Ic Margin
Tc Margin
Cooling condition
Superconductor Area
Aspect ratio
Material Data Base
Fabrication Experience
Requirements on Superconductor
Critical Current
Conductor Resistance
Allowable AC loss
Mechanical Strength (Void Ratio)
Imposed Voltage
Cable Insulation
between Layers
between Turns
Transient Pattern
of Field and Current
Cable Configuration
Cable Geometry
No. of Strands
Cabling Pitch
Calculation of AC loss
Strand Configuration
Diameter of Strand
No. of filaments
Matrix Material
Matrix Ratio
Twist Pitch
Strand Insulation
Calculation of Temperature Rise
Temperature Margin
Calculation of Quench Propagation
Velocity and Temperature Rise
Requirements
Calculation of Stress and Strain
Allowable Stress and Strain
END

FIG.2 Design procedures of superconductor for field windings

Jc (NbTi 5T)
(A/mm²)
Slow response type A
Slow response type B
Quick response type
RRR
Stabilizer
Cu Ratio
Matrix
Strand Insulation
Filament Diameter
Twist Pitch
Cabling
Rigidity
High Jc
Stabilization
AC Loss Reduction
Cu+Al
Cu
Cu+CuNi
None
Insulated
Single
Double
(μm)

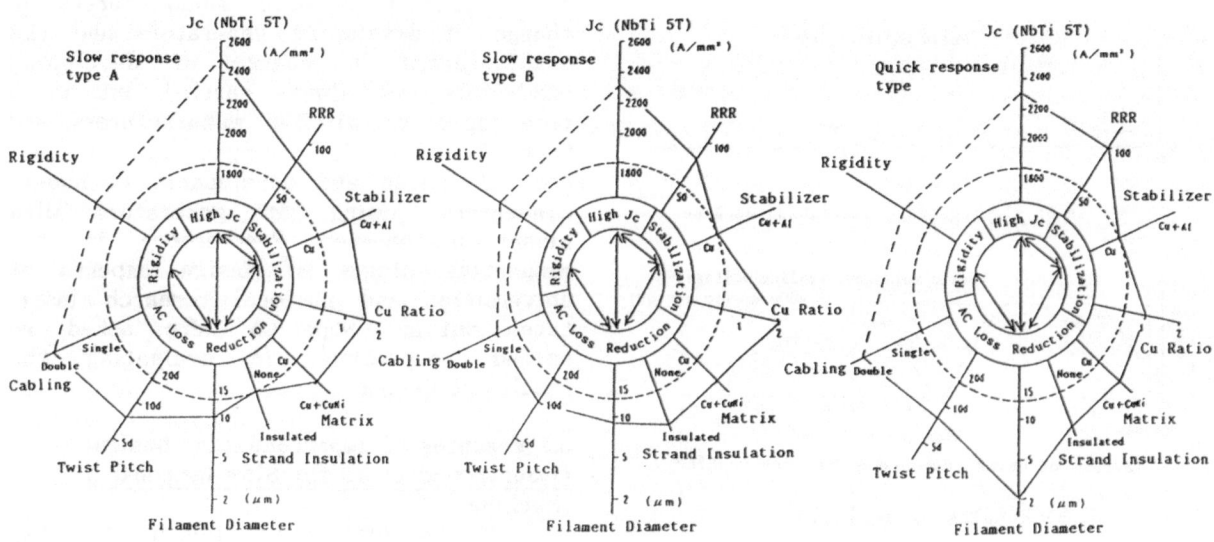

FIG.3 R&D Target for Characteristics of Superconductors

mainly to improve the stability of super-conductor, by enhancing the performance of stabilizing material and elaborating the conductor structure, to decrease the conductor resistance. For the other one, design was carried out mainly to enhance the current density, for enhancing the current density of the conductor from both viewpoints of NbTi material proper-ties and conductor structure, also considering long-term durability.

(2) Superconductor for quick response machine

Design was carried out mainly to decrease AC loss in view of severe exci-tation control conditions, by elaborating the strand structure and the cable struc-ture, and also to secure high current density and high rigidity.

3.3 Target Specifications of Superconductors

All the superconductors for the 70MW class model machine have a critical current of about 10kA at 5T 4.2K. FIG.3 is a diagram showing the target values of the designs. For the designs, four objec-tives of higher current density, higher stability, lower loss and higher rigidity were set, and design target levels were set in reference to the present industrial production level (shown by dotted circles). Arrows in the center indicate the trade-off relation, such as between high current density and reduced AC loss. Therefore target level is most advanced when both levels at opposite position are high at the same time.

Super-GM is using these diagram in order to evaluate the level of R&D. The rigidity of the conductor cannot yet be expressed quantitatively in relation with mechanical stability, and is expressed only by a dotted line. Considering the clarification of the quantitative rela-tion as R&D, Super-GM is promoting the research.

4. METHODS FOR TESTING SUPERCONDUCTORS

The superconductor used for a field winding is designed and manufactured, and wound into a saddle shaped coil, in order to sufficiently exhibit its performance as a generator. The verification of per-formance as a conductor is to be finally carried out in the tests of the 70MW class model machine, but on the way to the final test, the performance as a con-ductor alone, partial model, etc. is being verified in the respective steps of development. It is important to decide

the test items and methods required for feeding back these results to generator design and conductor design. Especially considering that the optimum conductor is selected by analyzing the results of studies concurrently conducted in many research institutions, it is necessary to establish test methods which allow analy-sis under the same standard. Since no such test standard is established at pre-sent, test methods were examined under the present project, for the purpose of establishing common test methods.

4.1 Test Items

R&D subjects of field windings for generators and test items required for verifying them are listed in TABLE 2. According to the progress in the deve-lopment of generators and conductors, standardization should be intensified cumulatively. The respective test items can be considered in two categories; (1)properties which must be considered together with the environment of use in the generator, and (2)properties which can be evaluated independently of the environment of use. These are shown in TABLE 3 with the former as complex properties and the latter as simple properties.

Presently, standardization of test methods under the project is being pro-moted mainly for the simple properties.

4.2 Test Methods and Data Analysis Methods

For standardizing test methods, it is necessary to prepare test codes which clarify (1)pretreatment condition of sample (2)form of sample (3)test condi-tion, (4)test method, and (5)method of analysing test results. Especially for the method of analyzing test results, what properties should be evaluated as a conductor for a generator must be suffi-ciently considered. For example, with regard to the critical current, which of Jc value of NbTi, overall Jc value of strand or overall Jc value of conductor should be adopted as an important pro-perty value must be examined, and as for AC loss, the frequency dependency and changing field amplitude dependency should be analyzed in relation with the structures of strand and cable.

5. PERFORMANCE TEST OF SUPERCONDUCTORS

The performance tests required for development of superconducting field windings were respectively allocated to manufacturers in charge of developing conductors, manufacturers in charge of developing generators and neutral laboratories as shown in TABLE 4, in oder to avoid unnecessary double efforts and to secure compatibility of data. Of course, performane test will be carried out according to the common test methods described in Section 4.

TABLE 2. R&D Items and Test Items for Superconducting Field Windings

R&D ITEMS	TEST ITEMS
Higher Critical Current Density	·Critical current (Ic) ·Temperature dependency of Ic ·Variation of Ic relating to cabling ·Variation of Ic under strain ·Current distribution in strand and cable
Higher Stability	·Resistivity (ρ) of stabilizing metal ·ρ of superconducting composite ·Temperature dependency of ρ (RRR) ·Effect of magnetic field on ρ ·Heat transfar coefficient to LHe ·Heat capacity of a composite ·Heat conductivity of a composite ·Minimum Propagating Current ·Minimum Propagating Zone ·Minimum Recovery Current ·Minimum Propagating Energy ·Quench current of model coil ·Quench Propagation Velocity ·Simulation of wire movement ·Temperature (Ic) margin ·Effect of impregration
Reduction in AC losses	·Total loss measurment (thermal and electrical) ·Dependency on frequency of changing field ·Dependency on amplitude of changing field ·Evaluation of transverse resistivity (ρ) ·Effect of trnsport current ·Effect of stress on strand coupling ·Proximity effect ·Effect of biasing DC field ·Temperature dependency of loss
Higher Mechanical Strength	·Stress-strain property (tension and compression) ·Bending property ·Fatigue property ·Evaluation of rigidity

5.1 Test at Manufacturers in Charge of Developing Conductors

The manufacturers in charge of developing conductors will carry out tests of strand and stranded wire levels as their particular efforts, and will cover all simple properties and some complex properties with special attention paid to the basic properties of materials.

5.2 Test at Manufacturers in Charge of Developing Generators

The manufacturers in charge of developing generators will carry out tests at final conductor and partial model levels as their particular efforts, for the purposes of confirming the requirements on the superconductor and obtaining characteristic data as static and rotating magnets at partial model level in order to verify the magnet design.

5.3 Common Tests

For the items which do not coincide or show large dispersion in test results among the tests conducted by respective manufacturers, common tests using the same test bed must be carried out to assure the compatibility in test data.

Furthermore, basic data concerning long-term reliability and endurance against severe load in actual service conditions of conductors and coils for generators which are surmised to difficult to obtain by the individual test bed of respective manufacturers will be obtained and fed back for the analysis of the data obtained by the respective manufacturers.

The common tests will be carried out

TABLE 3. Test Items Being Standardized in Super-GM

Classifications		Test Items
Simple property	Geometrical properties	Dimensions
	Electrical properties	Resistivity(RRR)
		Strand critical current
		Conductor critical current
		AC losses
	Mechanical properties	Tensile properties
		Compressive properties
Complex property		Resistivity under strain
		Critical current under strain
		Stability
		Rigidity

TABLE 4. Role Distribution in Superconductor Tests

Test Apparatuses	Organization in Charge	Test Samples	Test Items	
Individual	Manufacturer in charge of superconductor development	·Strand ·Stranded wire	·Critical current ·Resistivity ·AC losses ·Mechanical strength	
	Manufacturer in charge of generator development	·Final conductor	·Critical current ·Resistivity ·AC losses ·Mechanical behavior	
		·Partial model	·Windability ·Quench current ·Stability against transient operation	
Common	Cooperative operation	·Strand ·Stranded wire ·Final conductor ·Small coil ·Winding model	Static	·Critical current under the real operational conditions ·AC losses under the real operational conditions ·Determination of SC properties after the repetition of excitations and quenches ·Determination of insulation properties after the repetition of excitations and quenches
			Dynamic	·Stability of rotating magnet ·Large scale dynamic test of model windings

under the cooperation of the respective manufacturers and neutral research institutes.

5.4 Comprehensive Selection

The data obtained in Item 5.1 to 5.3 will be objectively compiled, and examined in comparison with the required specifications. At the same time, comprehensive examination will be made, considering the items which cannot be covered by test codes, such as the windability, quality control, cost, manufacturing experience, etc. in the respective manufacturers in charge of developing generators.

These results of work will be reported and discussed in an open field of Super-GM including outside experts for the selection of the best conductor, as done in the conductor design stage.

6. CONCLUSION

For the development of superconducting field windings by Super-GM aiming at the practical application of superconducting generators, a joint research system is organized, in which manufacturers in charge of developing generators and manufacturers in charge of developing superconductors work together under positive competition. In the promotion of such a project, it is inevitable and also effective to design jointly, to standardize test methods and to specifically allocate test responsibilities.

7. ACKNOWLEGEMENTS

This work was performed as a apart of "R&D on Superconducting Technology for Electric Power Apparatuses" under the Moonlight Project of Agency of Industrial Science and Technology, MITI, being consigned by New Energy and Industrial Technology Development Organization (NEDO).

The authors are much indebted to the members of committee, subcommittee and working group at Super-GM and wish to thank them for their contributions and cooperative works.

REFERENCES

[1]"Feasibility study on superconducting machinery and materials technology to electrical power generation", prepared by Technova Inc. for Moonlight Project of AIST, MITI, March, 1986 and 1987
[2]"Feasibility study on superconducting machinery and materials technology to electrical power generation", prepared by Super-GM for Moonlight Project of AIST, MITI, March, 1988

CURRENT SITUATION OF R&D ON SUPERCONDUCTING GENERATORS CARRIED OUT BY ENGINEERING RESEARCH ASSOCIATION FOR SUPERCONDUCTIVE GENERATION EQUIPMENT AND MATERIALS

T.Tanaka, S.Hirose, M.Tanaka, T.Kitajima and M.Sunada,
Engineering Research Association for Superconductive Generation Equipment and Materials
(Super-GM), Umeda UN Bldg.,5-14-10 Nishitenma, Kita-Ku Osaka City, 530 Japan

ABSTRACT

The application of superconducting technologies to electric power apparatus is very important from the viewpoint of promotion of energy saving, and the technological development attracts keen attention at home and abroad. In Japan, for the purpose of establishing technologies of superconducting generators , the R&D project of Super-GM started from FY 1988 for a scheduled period of eight years.

This paper describes the outline of superconducting generator and its R&D program of Super-GM and the results of FY 1988 R&D on superconducting generators carried out by Super-GM.

INTRODUCTION

Super-GM was established in September, 1987 in order to proceed the research and development (R&D) on applications of superconducting technology to electric power apparatus, based on the result of feasibility studies for superconducting generator carried out in FY 1985 and FY 1986 under the Moonlight Project of the Agency of Industrial Science and Technology, the Ministry of International Trade and Industry. Super-GM R&D project sponsored by the New Energy and Industrial Technology Development Organization (NEDO) under the Moonlight Project started from FY 1988 for a scheduled period of eight years. During the first four years, the development of elemental technologies and partial models is conducted to establish high reliable design philosophy and method, and in the latter four years, 70MW class model machines with two kinds of excitations, a slow response and a quick response, will be manufactured and tested as a series of R&D extending to 200MW class pilot generator for practical use in the future. Conceptual designs of 200MW class pilot generators were carried out in FY 1987 and a R&D with some of elemental technologies was carried out in FY 1988.

OUTLINE OF SUPERCONDUCTING GENERATOR AND ITS R&D PROGRAM

(1) Outline of superconducting generator structure

Fig.1 shows the schematic structure of superconducting generators, which are under development in this project. The field windings of superconducting generator using NbTi conductors generate high magnetic field, as a result, higher energy density of the generator is achieved in comparison with conventional ones. The rotor has multi-cylindrical structure, which includes vacuum thermal insulation layer for preventing external heat penetration, and dampers for shielding external flux invasion to field winding. The superconducting magnet fixed to the rotor has such features not observed in the conventional stationary magnet, that liquid helium is subject to

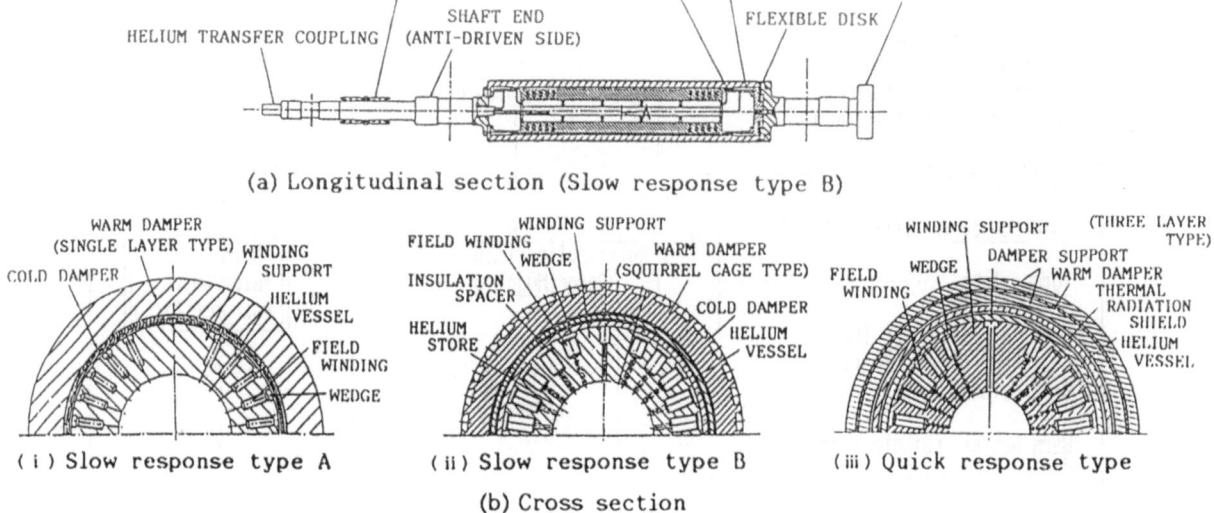

(a) Longitudinal section (Slow response type B)

(i) Slow response type A   (ii) Slow response type B   (iii) Quick response type

(b) Cross section

FIG.1 Schematic structure of superconducting rotor

high centrifugal force and that helium transfer coupling (HTC) for feeding and discharging liquid helium from a shaft end is required. As cooling loads for rotor, transient loads include AC loss of superconductor due to field magnetic flux and current variation, and eddy current loss of cryogenic structural materials, and steady state loads include Joule loss of current leads, heat conduction from torque tube, and heat entering from outside due to the convection and radiation in the thermal insulation layer.

Because of high magnetic flux from field winding, stator has non-magnetic teeth and so-called air gap winding, and that is quite different from conventional ones.

(2) Overview of R&D project and outline of elemental technologies and partial models

In the design of superconducting generator rotor, there are many alternatives, and various methods are researched and developed in respective countries. It is necessary in this project to judge which method is most advantageous for practical application. For this purpose, generator manufacturers who participate in Super-GM share the efforts under a research organization to complement each other.

R&D step for superconducting generator is shown in Fig.2. In the first four years of eight-year period, R&D with elemental technologies and partial models will be conducted to search the most

suitable method. In the latter four years, model machines will be built and tested in order to verify those developed technologies. The outline of these elemental technologies and partial models are described below.

a)Superconductor test ; The superconductivity, mechanical properties and insulation capabilities of various superconductors for the slow response excitation and the quick response excitation are measured. Feasibility and suitability as conductors for field windings will be examined by these data.

b)Field winding model ; The superconductivity after formation of field winding is identified, and the method for manufacturing winding will be established. Furthermore, the cooling and supporting structure will be examined.

c)Multi-cylindrical rotor model ; The technologies for manufacturing and assembling rotors with multi-cylindrical structure will be established, and vibration characteristics and so on will be examined.

d)Rotating cooling model ; The heat transfer characteristics of liquid helium under high centrifugal force is identified, and the rotor cooling structure will be examined.

e)Helium transfer coupling (HTC) model ; The sealing characteristics of magnetic fluid seal and so on are investigated, and the long-term reliability of HTC in various operating conditions will be verified.

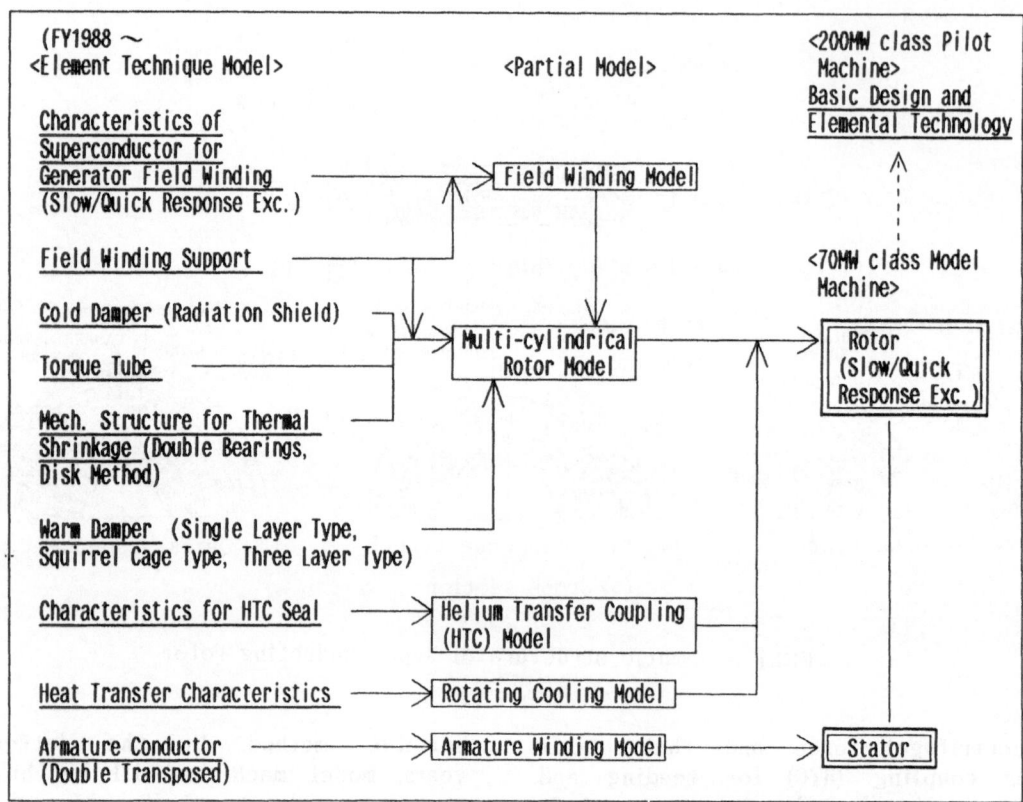

FIG.2 R&D step for superconducting generator

f)Air gap armature winding model ; The electric and mechanical characteristics of a double transposed conductor aiming at loss reduction are studied, and the structure for supporting the straight and end portions of winding and the method for cooling the core for magnetic shield will be examined.

RESULTS OF FY 1988 R&D ON SUPERCONDUCTING GENERATORS

(1) Basic designs of 70MW class model machines

Based on the conceptual designs of the 200MW class pilot machines, R&D program for element techniques of 70MW class model machines was worked out, and the study of basic specifications was carried out as the first basic designs of model machines. Table 1 shows main design parameters of the model machines, in comparison with 200MW class pilot machines.

Moreover, basic designs of the rotor and stator of the model machines were made by the basic specifications, and the main structure and electric characteristics were clarified.

For the rotor, the magnetic field and electromagnetic force of field windings, and electromagnetic stresses at respective portions of the rotor were analyzed. For the stator, the structure of air gap armature winding was studied, and the temperature distribution at cross section of stator coil was calculated. Furthermore, the electromagnetic forces at the straight and end portions of armature winding, the stress at the teeth portion and the temperature distribution of magnetic shield were calculated.

(2) Superconductor tests

The structure of the superconductor used for the field windings of the model machines was studied, and development specifications were set for both slow and quick response excitation machines. Table 2 shows development specifications. In the slow response excitation machines, short strands for candidate conductors were manufactured using the present engineering, and critical current characteristics and AC loss characteristics were measured to confirm the adequacy of design specifications. Furthermore, using the strands, partial short conductors were manufactured. Moreover, the methods of insulating, cooling and supporting superconductors, and winding workability were examined.

TABLE 1 Ratings and parameters of 70 MW class model machine
compared with 200 MW class pilot machine

		70 MW class model machine			200 MW class pilot machine		
		Slow response type		Quick re-	Slow response type		Quick re-
		A	B	sponse type	A	B	sponse type
1. Ratings							
Capacity	(MVA)	83	83	73	223	223	223
Voltage	(kV)	10	10	10	20	20	16
Current	(A)	4,790	4,792	4,215	6,438	6,438	8,047
Power factor		0.9	0.9	0.9	0.9	0.9	0.9
Pole number		2	2	2	2	2	2
Speed	(rpm)	3,600	3,600	3,600	3,600	3,600	3,600
2. Electric parameters							
Reactances Xd	(p.u.)	0.35	0.35	0.45	0.40	0.30	0.40
Xd'	(p.u.)	0.25	0.26	0.36	0.29	0.21	0.29
Xd"	(p.u.)	0.21	0.19	0.25	0.23	0.14	0.18
Open circuit const. Tdo'	(sec)	260	260	250	560	650	410
Armature const. Ta	(sec)	0.11	0.11	0.14	0.28	0.10	0.05
Field winding Inductance Lf	(H)	0.46	0.43	0.39	1.0	1.1	0.45
3. Field winding (at rated output)							
Field voltage Vf	(V)	5	5	5	5	5	5
Field current If	(A)	3,000	3,000	3,200	2,800	3,000	4,500
Maximum magnetic flux density Bfmax	(T)	4.0	4.6	4.2	4.0	5.2	4.5

TABLE 2 Specification of superconductors for 70 MW class model machine

ITEM	SPECIFICATIONS		
	Slow response excitation		Quick response excitation
	A	B	
Superconductor Configuration	NbTi Rutherford type cable with two stage cabling	NbTi Rutherford type cable	NbTi Rutherford type cable with two stage cabling
Section area	3.6 × 6 mm	3 × 7.3 mm	≦ 29 mm2
Aspect ratio	1.7	(2.4)	≅ 3
Critical current	≧4.8kA at 4.2K and 6.6T ≧8.5kA at 4.2K and 4.0T	≧ 6.3kA at 4.2K and 7T ≧10.5kA at 4.2K and 5T	≧ 8.5kA at 4.2K and 7.5T ≧12.8kA at 4.2K and 6.0T
Copper ratio	—	2/1/1 (Cu/CuNi/NbTi)	≧ 1
R.R.R.	≦5μΩ/m at≅10K and 0T	≧80 at 10K and 0T	≧ 100 at ≅10K and 0T
AC loss	≦30kW/m3	≦25kW/m3	≦10kW/m3 (B=4.2→6T, Ḃ=10T/s)
Strand wire	—	With insulation	With insulation

Five kinds of one slot model coils were designed and manufactured by simulating the structure in the slot of the field winding. A superconductor testing was carried out to measure the quenching limit by adding DC and variable magnetic field. Fig.3 shows the configuration of testing model coil, and Fig.4 shows the test result of the short sample conductor for example.

The conductors for quick response excitation machine are subject to the most severe conditions in current density and loss. For conductor test, using conductors available now, the critical current charcteristics of strands and primarily stranded wires and the deterioration characteristics after mechanical history were measured. As a result, it was confirmed that they were little deteriorated in the critical current after mechanical history. Furthermore, the magnetization characteristics of strands and primarily stranded wires were measured in both cases of the strands with and without insulation to clarify the relation between strand insulation and AC loss.

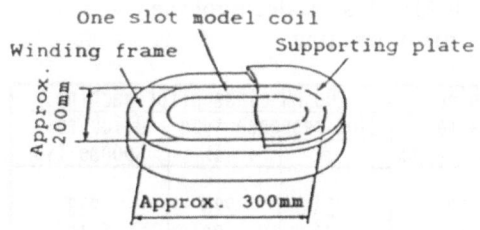

FIG.3 Configuration of one slot model coil

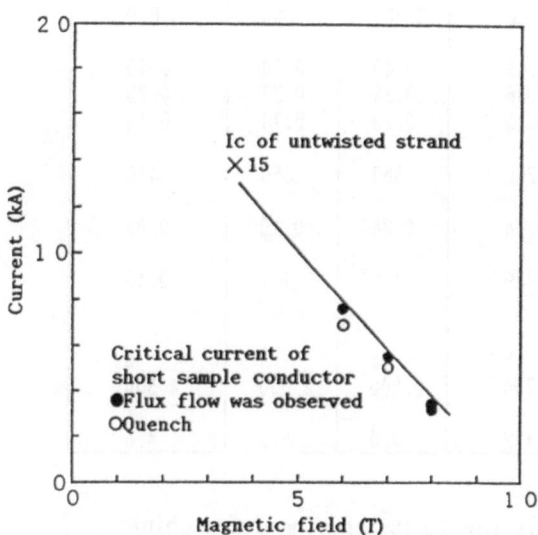

FIG.4 Result of short sample conductor test (for example)

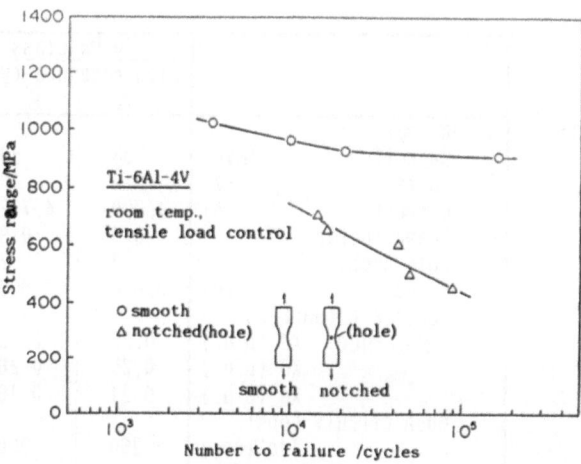

FIG.5 Result of low cycle fatigue test (Ti-6Al-4V Titanium alloy)

(3) Multi-cylindrical rotor model

As to the slow response excitation type, the basic properties of the materials were examined for the field winding support, thermal contraction joint and cold damper, and candidate materials were selected. For each material, test pieces were taken from the base metal and the welded portion of the field winding support, and mechanical property tests of tension, fracture toughness and low cycle fatigue were carried out to obtain basic data for design of multi-cylindrical rotor. For example, Fig.5 shows the low cycle fatigue property of titanium alloy candidate for structural material of the thermal contraction joint.

As to the quick response excitation type, thin-walled cylinder models were manufactured using high strength stainless steel (A286) and nickel based alloy (Inconel 718) as main structural materials of rotor to establish thin-walled ring forging techniques for the respective materials. Furthermore, test pieces were taken from the rings,

and tested to measure mechanical properties (yield strength, tensile strength) and physical property (electric resistance). It was found that all the trially manufactured rings satisfied the specified values. The appropriate welding method (welding material and welding conditions) for the nickel based alloy was clarified. As a result of welding procedure tests (TIG welding), it was confirmed that the weld metals and heat affected zones were free from weld cracking.

(4) Warm damper

As to the warm damper, single layer damper, squirrel cage type damper and three layer damper have been developed as shown in Fig.1.

For the single layer damper, many test sample alloys based on Monel K500 (Ni-30%Cu-3%Al alloy) were manufactured to find optimum metal composition and heat treatment conditions. Furthermore, small rings were manufactured, and it was confirmed that the alloy had good hot forgeability on an industrial scale.

For the squirrel cage type damper, mechanical property tests of candidate materials were carried out to obtain various basic data. Smaller-scale model with double damper structure of squirrel cage type damper and cold damper was manufactured to test and measure the magnetic shield characteristics and the electromagnetic force acting on the damper at the time of short-circuit, for confirming the applicability as a damper for superconducting generators.

For the three layer damper, chromium copper as the damper material, and A286 as the damper support material were selected. The manufacturing methods (explosion bonding and shrink fitting) were studied.

(5) Cooling characteristics

A heat transfer characteristic model for testing the heat transfer characteristics of liquid helium under centrifugal force and the cooling characteristics of radiation shield was planned, and the basic design for test blocks simulating and evaluating heat transfer characteristics of field winding portion was made. The model allows the measurement of the heat transfer rates of liquid helium under high centrifugal force over 1000G considering cooling paths by insulation spacers for field windings.

(6) Helium transfer coupling

The functions of various seals as important components of helium transfer coupling (HTC) were studied and a seal characteristic model was designed and manufactured. Furthermore, an HTC general verification model as large as the HTC of the model machines was planned and partially designed, and specifications of the testing equipment necessary for HTC model tests were decided.

(7) Stator coil

For research of stator coil, conductor structure, insulation method and cooling method were studied. The stator coil has a structure of double transposed small wires, and the arrangement of rectangular cooling pipe was studied. It was clarified that the insulation method by vacuum pressurized impregnation with epoxy resin is adaptive. Furthermore, short sample of stator coil was manufactured, and configuration of coil without earth insulation is shown in Fig.6.

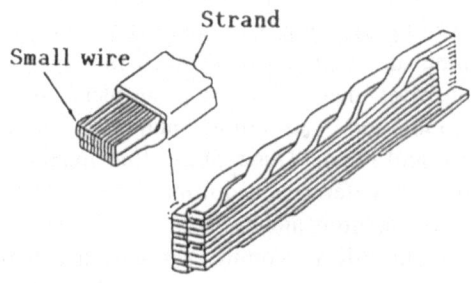

Strand

Small wire

FIG.6 Configuration of stator coil

CONCLUSION

The results obtained in FY 1988 are summarized as follows;

1) Basic designs of 70MW class model machines with two kinds of excitation type, a slow response and a quick response, were carried out in which design specifications were clarified.

2) R&D of elemental technologies such as superconductor tests, testing for candidate materials of multi-cylindrical rotors and warm dampers, studies for cooling characteristics, studies for HTC sealing and testing and manufacturing for stator conductor were conducted with some models, and basic data were obtained for 70MW class model machines.

ACKNOWLEDGEMENT

This work was performed as a part of "R&D on Superconducting technology for Electric Power Apparatuses" under the Moonlight Project of the Agency of Industrial Science and Technology, the Ministry of International Trade and Industry, being consigned by the New Energy and Industrial Technology Development Organization (NEDO). The authors are much indebted to the members of superconducting generator subcommittee and other people concerned for their assistance and cooperation.

REFERENCES

1. "Feasibility study on superconducting machinery and materials technology to electrical power generation", prepared by Technova Inc. for Moonlight Project of AIST, MITI, March, 1986

2. "Feasibility study on superconducting machinery and materials technology to electrical power generation", prepared by Technova Inc. for Moonlight Project of AIST, MITI, March, 1987

3. "Feasibility study on superconducting machinery and materials technology to electrical power generation", prepared by Super-GM for Moonlight Project of AIST, MITI, March, 1988

DESIGN AND MANUFACTURE OF SUPERCONDUCTING GENERATOR WITH HIGH-RESPONSE EXCITATION

T. Okada,   T. Nitta,   S. Hayashi,   K. Saikawa,   M. Tari,   M. Kumagai
Kyoto University   The Kansai Electric Power Co., Inc.   Toshiba Corporation

ABSTRACT

Superconducting generators have numerous advantages, one being their capability of improving the stability of power systems because they enable the selection of lower synchronous impedances. This advantage can be increased by adoption of high-response excitation.

In this joint study initiated in 1986, Kyoto University, the Kansai Electric Power Co., Inc., and Toshiba Corporation recently designed and manufactured, as a world's first project of its kind, a 100-kVA high-response-excitation superconducting generator, and successfully conducted an extensive series of tests including rapid changes of excitation currents. This experimental generator also realized the concept of utility-use superconducting generator in the future, and will contribute much to research and development of superconducting power applied technology in line with the "Moonlight Project" being directed by the Japanese Government.

INTRODUCTION

Research and development is being promoted worldwide on superconducting generators—generators using superconducting field winding that can achieve high efficiency, light weight, small size, low impedance, large unit size, and so on. All these are superconducting generators with constant excitation. Superconducting generators that appear to fulfill the most hopes for the future will be high-response-excitation ones that enable rapid changes in field voltage, resulting in improvement of transient stability. Studies along these lines are also included in research and development under the title of Superconducting Power Applied Technology within the "Moonlight Project" started in 1987.

Prior to that time, in 1986, Kyoto University, the Kansai Electric Power Co., Inc., and Toshiba Corporation jointly began research and development on the world's first high-response-excitation superconducting generator.

During designing, manufacturing, and testing of the 100-kVA "High-Response-Excitation Superconducting Generator" (Hesper 1), we solved a multitude of engineering problems, and through factory tests,

we evaluated basic properties of the generator and the total systems.

This paper describes the results of our study.

NECESSITY OF EXPERIMENTAL HIGH-RESPONSE-EXCITATION SUPERCONDUCTING GENERATOR TEST SYSTEM

One of the potential merits of superconducting generators is improvement of the permissible limit of transmission capacity, and some system analyses have proved that such extension can be even further improved by using high-response excitation. However, designing and manufacturing such new-type generators involves new engineering problems related to superconductors, structural materials, insulation, cooling, and so on. This requires that maximum care be taken in determining optimal principal specifications for generators and exciters.

From this viewpoint, we started a feasibility study in 1986, and completed it in 1988, including designing and manufacturing the experimental

TABLE 1
Test program of Hesper 1

1986	1987	1988	1989	1990	1991	1992
Preparatory study and planning	Principal specification setting					
	Analytical calculation and characteristics evaluation					
		Test/evaluation of basic characteristics				
	Design and manufacture		Improvement of generator characteristics			
	Development of key components				Evaluation of power-system stability	
	Study on high-response-excitation system					Long-term reliability test
	Feasibility study on advanced superconducting generator					

generator.

After installing Hesper 1 in Kyoto University, we are planning to study the improvement of power-system stability by connecting it to a model power system, and to conduct long-term reliability tests of the superconducting generator.

The entire plan of the test program is shown in TABLE 1.

DESIGN AND MANUFACTURE

Basic Design

Major specifications of the experimental generator are listed in TABLE 2. For excitation-system ceiling voltage, 150 p.u. (300 V) was chosen so that the magnetic flux density change during high-response excitation would be identical to that for 1000-MW superconducting generators. Prior to manufacturing, we analyzed the magnetic flux, losses,

TABLE 2
Specifications of experimental generator Hesper 1

Capacity	(kVA)	100
Voltage	(V)	220
Current	(A)	263
Power factor (lagging)		0.9
No. of poles		4
Rotating speed	(rpm)	1800
Excitation system		Thyristor excitation
Ceiling voltage	(V)	300
Driving motor		DC motor
Driving system		Thyristor Leonard

temperature, vibration and stress under various operation conditions, and checked the system for soundness during operations. A typical analysis of three-dimensional magnetic flux distribution along the field winding is shown in FIG. 1.

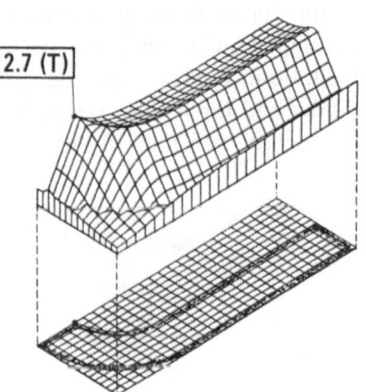

2.7 (T)

FIGURE 1 Three-dimensional flux density distribution along the field winding

The field coils are resin-impregnated racetrack type, for which a monolithic NbTi conductor with a three-layered cross section (shown in FIG. 2) is used to reduce ac losses caused by high-response excitation and to ensure stability.

For the room-temperature damper, chrome copper is used to adjust resistivity so that the damper will serve as a magnetic flux shield against a negative-phase and asynchronous magnetic field, but will not act as a magnetic flux shield during high-response excitation (FIG. 3).

For the armature winding, we use an air-gap winding mounted on a supporting insulator. For the conductor, rectangular-shaped Litz wire is employed to reduce eddy current losses.

Conductor diameter	(mm)	1.2
Filament diameter	(μm)	2.4
No. of filaments		47 000
NbTi/Cu/CuNi		1/1.5/2.5
Twist pitch	(mm)	15
Rated current	(A)	140

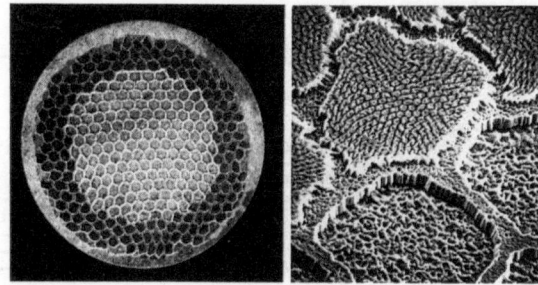

FIGURE 2 Specifications and micrographs of
superconducting conductor

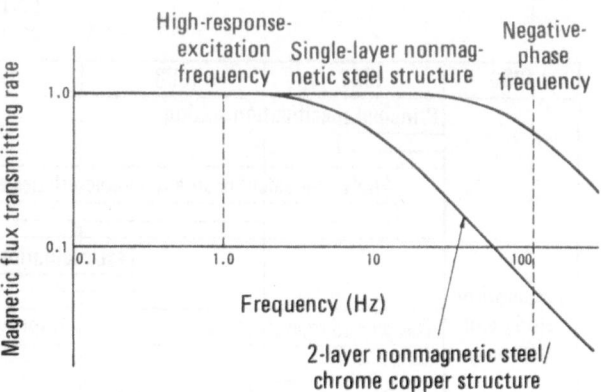

FIGURE 3 Room-temperature damper flux-shielding
characteristics

Structural Design

Structure of the main portion of the generator is shown in FIG. 4 and TABLE 3. Structural design of this generator incorporates advanced design for the purpose of studying engineering problems regarding large-capacity superconducting generators with high-response excitation. For example, as to the coil-support cylinder and the helium vessel, Inconel 718, having high strength and high electric resistance, is used to reduce eddy current losses and to endure large-fluctuation electromagnetic force during high-response excitation. Also, for the radiation shield, high-electric-resistance stainless steel is used by considering shielding characteristics during high-response excitation, and a helium gas forced-cooling system is adopted because of low thermal conductivity of this material.

Development of Key Components

We engaged in advance development of key components and also verification of the manufacturing stage. Among them, we paid much attention to developing the field winding and conducted extensive

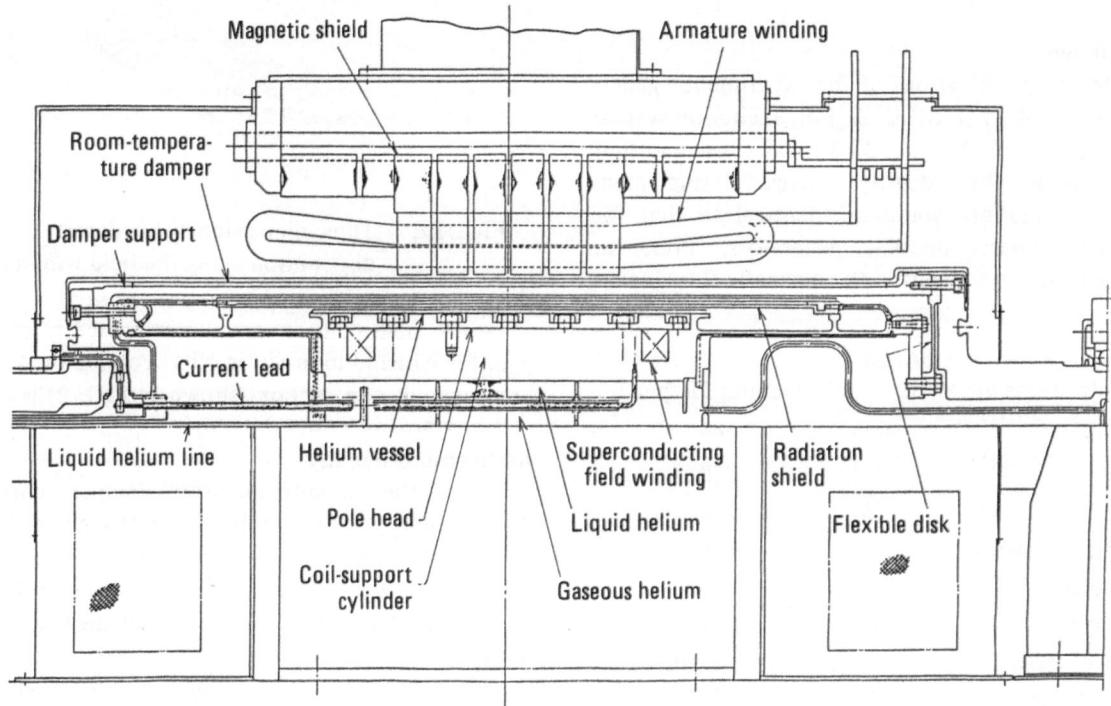

FIGURE 4 Structure of main portion of generator
(Cross section of upper half and appearance of lower half)

TABLE 3
Comparison between Hesper 1 and
high-response-excitation, large-capacity generator

	Large-capacity generator	Hesper 1
Superconducting conductor	NbTi/Cu/CuNi stranded	NbTi/Cu/CuNi monolithic
Field winding	Saddle shaped	Racetrack Conductor fixing method and cooling structure can be reflected in large-capacity generator.
Coil-support cylinder	Inconel 718	
Helium vessel	Inconel 718	
Radiation shield	High-electric-resistance material Helium gas forced-cooling	
Room-temperature damper	Proper electric-resistance material	Chrome copper
Damper support	A286	
Support structure of cold rotor	Flexible disk (multilayer)	Flexible disk (single layer)
HTC	Magnetic fluid sealing employed	
Armature winding	Air-gap winding	

FIGURE 6 Excitation test using cold-temperature
rotor assembly

tests as shown in FIG. 5. FIGURE 6 shows the excitation test of an assembled rotor at standstill.

The helium transfer coupling (HTC) providing liquid helium to the rotating part must be highly reliable. We checked its required functions and soundness considering the operational condition of large-capacity generators to be used in the future (FIG. 7). Also, we developed and tested an automatic pressure regulator for stably feeding liquid helium to the cold rotor, as well as the protection and control system for continuously exhausting gas from the adiabatic space.

FIGURE 7 Main part of helium transfer coupling
evaluation system

Rotor Structure Materials

Considering future large generators, for rotor structural materials we use Inconel 718 and high-strength austenite steel A286 (TABLE 3). The main rotor components are shaped as hollowed cylinders; thus, for the helium vessel and the damper support, pipe forging was adopted, and solid forging was employed on the coil-support cylinder (FIG. 8).

A good deal of knowledge was obtained as to the ingot making, forging, welding quality, and heat treatment of these materials.

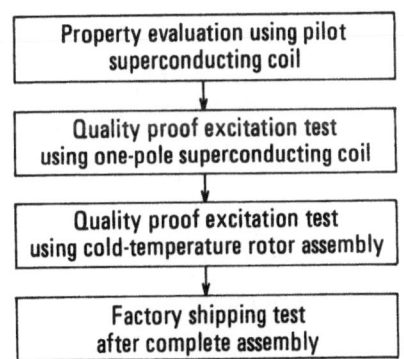

```
┌─────────────────────────────────┐
│ Property evaluation using pilot │
│    superconducting coil         │
└─────────────────────────────────┘
              ↓
┌─────────────────────────────────┐
│  Quality proof excitation test  │
│ using one-pole superconducting coil │
└─────────────────────────────────┘
              ↓
┌─────────────────────────────────┐
│  Quality proof excitation test  │
│ using cold-temperature rotor assembly │
└─────────────────────────────────┘
              ↓
┌─────────────────────────────────┐
│     Factory shipping test       │
│   after complete assembly       │
└─────────────────────────────────┘
```

FIGURE 5 Schematic diagram of field winding
developing procedure

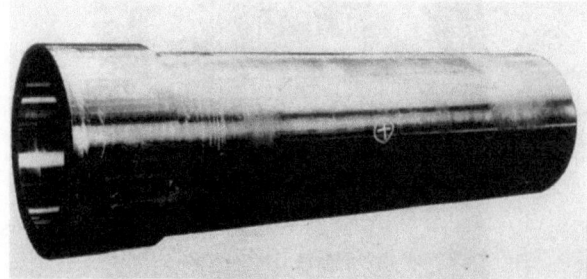

FIGURE 8 Coil-support cylinder made of Inconel 718

Measurement

For protective, monitoring and experimental purposes, sixty-six sensors are installed in the generator (25 on the rotor, 29 on stationary portion, and 12 on the HTC). Measurement items and measuring devices for the rotor are listed in TABLE 4. Signals from the rotor are transmitted through a 40-way slip ring.

TABLE 4
Measurement on rotating parts

Item	Object		Sensor	Qty.
Temperature	Helium	Cold rotor	Carbon resistance thermometer	4
	Structural component	RT rotor, torque tube, radiation shield	Thermocouple	10
Magnetic flux	Cold rotor, RT rotor		Hall generator, search coil	8
Liquid level	Hole in coil-support cylinder		Carbon resistance thermometer	1
Quenching detection	Field winding		Voltage tap	1
Rotating speed, phase	Surface of rotor shaft		Optical sensor	1
			Total	25

Excitation Control and Protection/Monitoring Control

The excitation system is also an important key technology of a superconducting generator. We conducted extensive studies on the control system and its configuration. The present exciter for Hesper 1 is designed to evaluate the basic characteristics of this high-response-excitation superconducting generator, which is confined to manual field current regulating functions. A functional block diagram of the excitation system is shown in FIG. 9. A minor loop is used to maintain field voltage control at a constant level, and a field current control loop is provided outside the minor loop to facilitate regulation. Excitation-system ceiling voltage can be changed between 300 V (structural material loss evaluation) and 125 V (power-system stability evaluation), and an external signal input terminal is provided to conduct high-response-excitation tests as required.

We also studied the concepts of protection and monitoring of the superconducting generator. The specific monitoring items are quenching detection, liquid helium level in the rotor, rotor internal pressure, and vacuum pressure of the adiabatic space. As for the protective interlocking of Hesper 1, we adopted the following: opening of generator and field circuit breaker, cessation of helium feed, isolation of adiabatic space from the exhausting system, intermediate rotating speed hold, and other items.

FACTORY TESTS

We have conducted an extensive series of tests that included cooling down, generator characteristics, and checking on protective and monitoring control system.

Referring to temperature distribution in the rotor, we established the effective procedure of

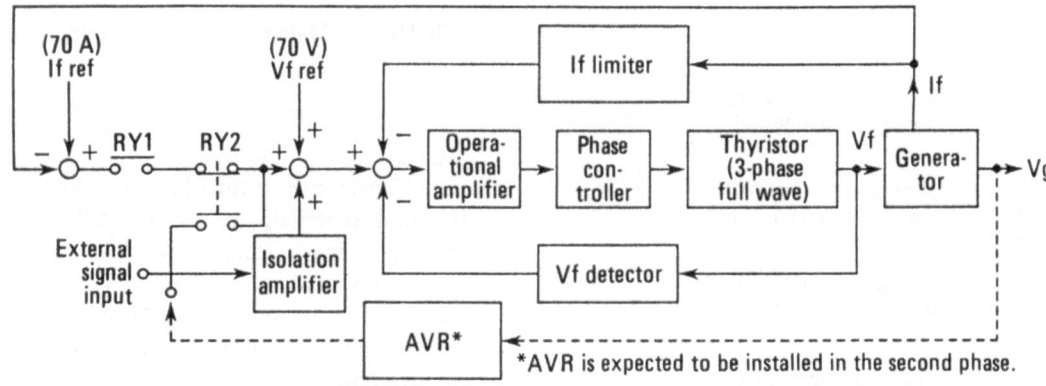

*AVR is expected to be installed in the second phase.

RY1: For making If control loop live/dead, RY2: For switching manual (MEC) and automatic (AVR)

FIGURE 9 Functional block diagram of excitation control system

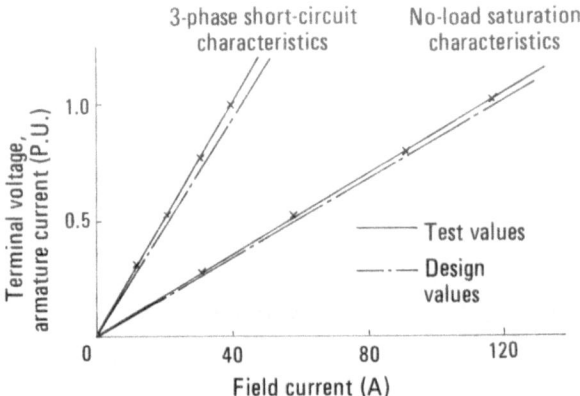

FIGURE 10 No-load saturation and three-phase short-circuit characteristics

TABLE 5
Generator constants of Hesper 1

Item			Measured	Design
Synchronous reactance	xd	(p.u.)	0.34	0.36
Transient reactance	xd'	(p.u.)	0.33	0.30
Subtransient reactance	xd''	(p.u.)	0.20	0.16

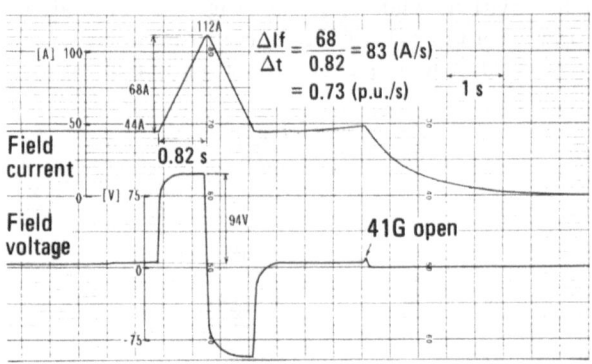

FIGURE 11 Results of field forcing test

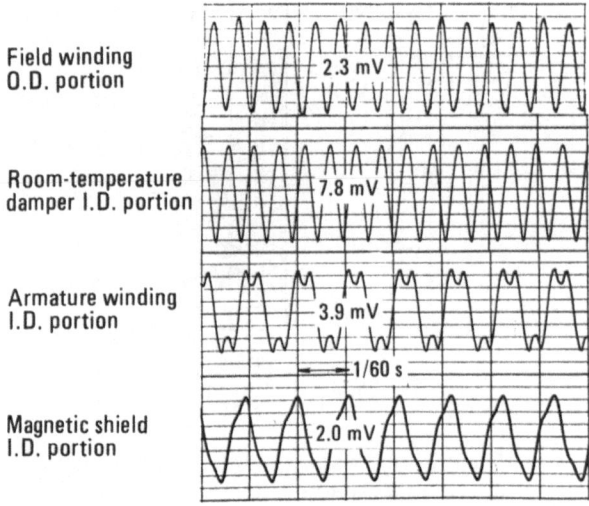

FIGURE 12 Search coil output voltages at short-circuit test

cooling down. We tested no-load saturation and three-phase short-circuit characteristics (FIG. 10) and load characteristics, and measured the generator constants (TABLE 5), while conducting a field forcing test (FIG. 11). A great deal of data were successfully obtained from sensors in the generator. Typical outputs at a short-circuit test are shown in FIG. 12. A dynamic characteristics test was conducted on the exciter to check its response. An interlock test was also conducted on the protective and monitoring panel. FIGURE 13 shows Hesper 1 being tested.

FIGURE 13 Hesper 1 under factory test

CONCLUSION

The results of studies and factory tests on Hesper 1, the world's first high-response-excitation superconducting generator, have been outlined herein.

Hesper 1 has proved feasibility of the design concept of such a new-type generator.

The experimental generator has shown good agreement between the design and the test values, as well as successful operations of all measuring sensors on the rotating parts and other parts, proving that Hesper 1 functioned well as a test generator.

The wealth of knowledge accumulated from designing, manufacturing and testing Hesper 1 is expected to largely contribute to future development of superconducting generators.

We are planning to conduct tests on Hesper 1 already installed in Kyoto University, according to the comprehensive test program as described herein. These tests will provide much useful data.

REFERENCE

1. S. Hayashi, T. Kaito, T. Okada, T. Nitta, M. Tari, and K. Ito, Designing and engineering problems on high-response-excitation superconducting generators (in Japanese). Papers of Technical Meeting on Rotating Machinery, The IEE of Japan, RM-88-13.

FULLY SUPERCONDUCTING AC GENERATOR WITH BRUSHLESS EXCITATION SYSTEM

Muta,I., Tsukiji,H., Tsutsui,Y., Hoshino,T., Mukai,E. and Furukawa,T.
Department of Electrical Engineering, Saga University
1 Honjyo-machi, Saga-shi, 840 Japan

ABSTRACT

A study on efficient excitation system must be important to enhance the potential of superconducting AC generators further. After the first successful demonstration in a feasibility of magnetic-flux pump based brushless excitation system suitable for superconducting generators in 1983 [1], we have been constructing a testing machine able to generate actual electric power. Concurrently with it, we have been studying a fully superconducting generator with both of armature and field coils made of superconductors.

The paper presents the experimental machine system and its testing results about pumping-up performances of magnetic-flux pump and electrical characteristics when operated as fully superconducting brushless generator with the capacity of less than 20 kW.

INTRODUCTION

A common scheme of superconducting AC generators developed so far over the world is of rotating field type in which the use of electric power leads, collector rings, brushes and so on are generally required for supplying excitation power to superconducting field coil from external source. So such a collector ring system may be accompanied with some problems in cooling efficiency and maintenance caused by heat leak, ohmic loss, mechanical contact wearing-out and so forth. To cope with this situation, the newly-developed generator introduces, expelling the collector ring system, a so-called "magnetic-flux pump" system capable of both indirectly feeding persistent current to the superconducting field coil and concurrently adjusting its magnitude. After the proposal of such a new scheme by Mawardi [2] in 1977, we had constructed the first scanty experimental facility to demonstrate the possibility of the novel concept as a first phase and then successfully confirmed it [1],[3]~[5].

Our newly-developed superconducting generator this time, in addition, consists

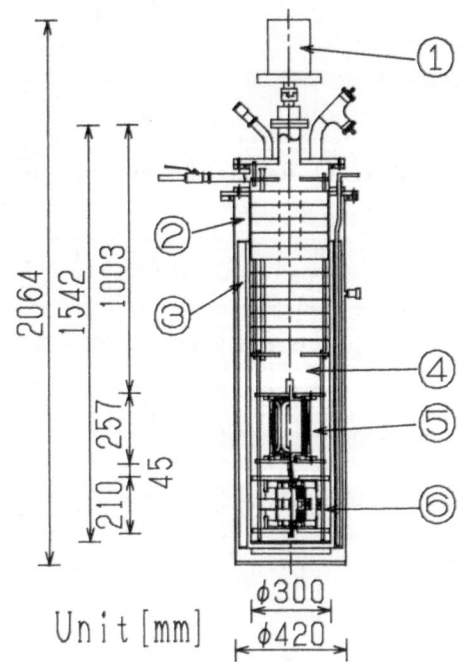

Unit [mm]

① Driving motor　④ LHe cryostat
② Vacuum space　⑤ Generator
③ LN₂ jacket　⑥ Flux pump

FIGURE 1 General assembly of test facility

of superconducting armature coils with very low AC loss. Thus we have developed a so-called fully superconducting brushless generator capable of generating electric power of less than 20 kW.

The paper describes experimental machine system, performances of magnetic-flux pump, electrical characteristics when operated as fully superconducting brush-less generator, and calculated characteristics.

EXPERIMENTAL MACHINE SYSTEM

FIGURE 1 shows the developed fully superconducting brushless AC generator assembled in a cryostat made for general use. To minimize the cost of the study, all members of vertical type of the machine are wholly immersed in liquid helium coolant.

Fully Superconducting Generator

In FIGURE 2(a) is illustrated the schematic diagram of the laboratory-made generator, the rotor of which is composed of 4-pole saddle-shaped superconducting field coil with straight section of 150 mm and outermost diameter of about 110 mm, while the armature is composed of 3-phase, 12 spiral-pancaked superconducting coils fixed to FRP frame as shown in FIGURE 2(b). Both of stator and rotor are not backed with magnetic cores, that is, fully air-cored.

The superconducting field coil can be excited indirectly through the magnetic-flux pump mentioned below without mechanical brush system, and forms a persistent current circuit with the magnetic-flux pump.

Major specifications of the generator are provided in TABLE 1. The field coil is wound by single superconductor developed for DC use, while the armature coil is wound by single fine multi-filamentary superconductor with very low AC loss for AC use and is wired in series of each of 4-pole windings at the present experiment.

Magnetic-Flux Pump

A magnetic-flux pump is such a device as superconducting DC homopolar generator capable of accumulatively producing large DC current and so changing magnetic flux confined in a closed superconductive electric circuit. A type of rotating flux spots has been known as one of the most practical and promising version. On end-

less or segmented thin cylindrical strips made of superconductive material e.g. Nb are produced moving normal-conductive regions by one or more magnets with the same polarity. A superconducting load coil is connected through many parallel super-

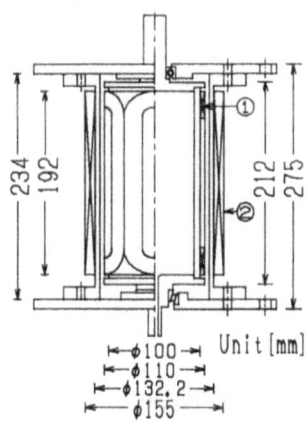

① Field winding ② Armature winding

(a) Whole assembly

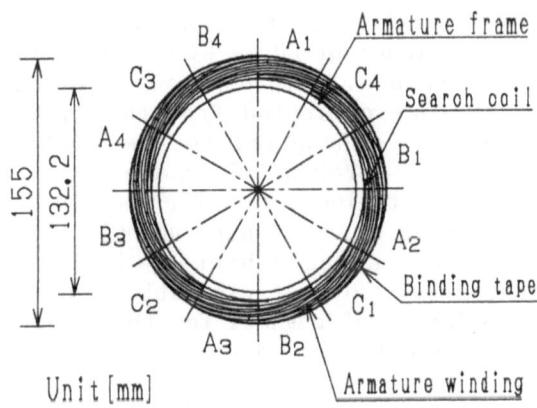

(b) Armature winding
FIGURE 2 Cross section of generator

TABLE 1
Parameters of fully superconducting generator

magnetic flux pump	excitation coil : 8-poles, 183 turn/pole NbTi wire of φ0.535mm Nb cylinder : 20μm thick, 60mm wide 100mm dia.
field winding	4-poles, saddle-shaped, NbTi wire(φ0.385mm, φ0.375mm), No. of series turns 1420, Resistance 336[Ω](at R.T.), Self-inductance 67.5mH
armature winding	NbTi wire for AC applications of 0.209mm dia. (dia. of filament 0.93μm, No. of filaments 15367, Composite ratio 1/0/2.3 twist pitch 0.98mm) Spiral pancake shaped coil, No. of turn 154/phase/pole, Mutual-inductance 22.78mH(phase), 40.69mH(lime-to-lime)

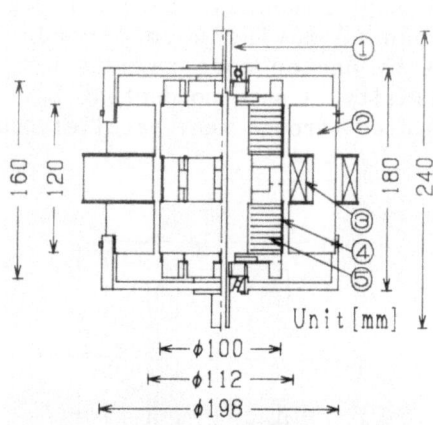

① Rotor shaft   ④ Nb sheet
② Magnet core   ⑤ Back core
③ Exciting coils

FIGURE 3 Magnetic flux pump

conducting wires bonded at both edges of each strip.

FIGURE 3 shows a cross-section of developed flux pump. The system comprises two flux pump components connected in series, each of them having rotatable Nb strip cylinder of 60 mm wide, 20 μm thick and 10 cm dia. as presented in TABLE 1. The current collected from the Nb strip cylinders is supplied to the generator field coil through NbTi superconducting wires. Excitation system for the flux pump is composed of stationary unipolar superconducting magnets backed with a soft steel core. Both of excitation intensity and polarity of magnets for the flux pump can adjust the field current, holding a persistent current circuit without any transition-to-superconductivity switching devices.

EXPERIMENTAL RESULTS

For the present, the capacity of prime mover is 2.9 kW. The experiment has been intermittently done, limiting the top revolving rotor speed to 1500 rpm, because of low mechanical strength of the whole system and poor cooling method.

Performances of Magnetic-Flux Pump

FIGURE 4 shows a diagram of pumping-up process of the generator field current in case of the flux pump excitation current of I_{ex} = 13.8 A and the revolving speed of 400 rpm, calibrated from output of hall generator element. From this figure it should be noted that the pumping-up current is almost linearly increasing with elapsed time over a range of low self pumped-up field current, but gradually getting a saturation over a range of its high level. The maximum value of pumped-up current or the quenching current in this field system has been experimentally recognized to be about 70 A.

The pumping-up rate or an increase of the field current against revolving rotor speed N [rpm] and the flux pump excitation intensity I_{ex} [A] are plotted in FIGUREs 5 and 6, respectively. From these figures, it can be clearly observed that the pumping-up rate increases approximately in proportion to the revolving speed N and has an optimal value for the excitation intensity I_{ex}. The latter phenomenon can be explained well by a magnetic Reynolds number dependent on normal-conductive spot area which is spreading wider over superconductive strips with the excitation intensity. The scattering in the measured pumping-up rates appears as found in the figures, it should be commented, because results for different values of pumped-up

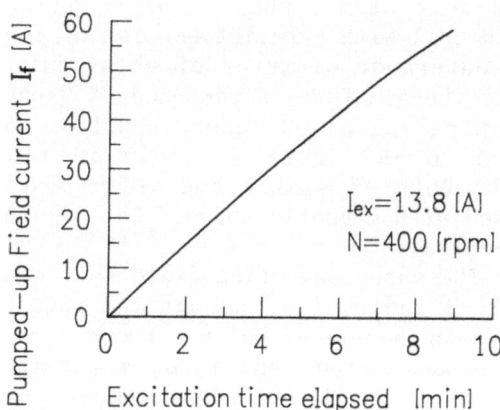

FIGURE 4 Example of flux pumping behavior

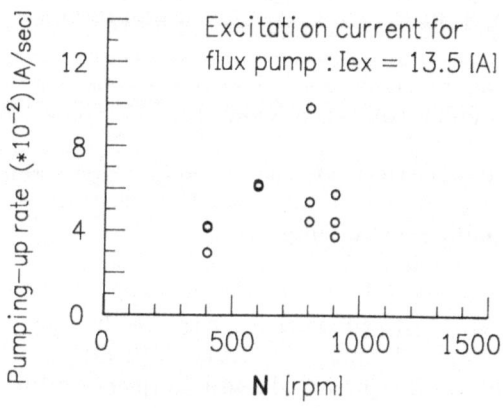

FIGURE 5 Pumping rate of field current vs. revolving field velocity

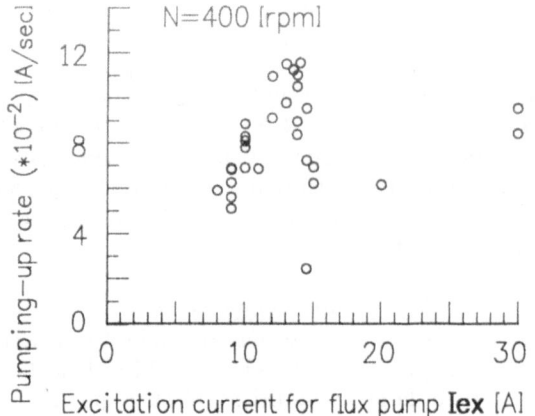

FIGURE 6 Pumping rate of field current vs. flux pump excitation

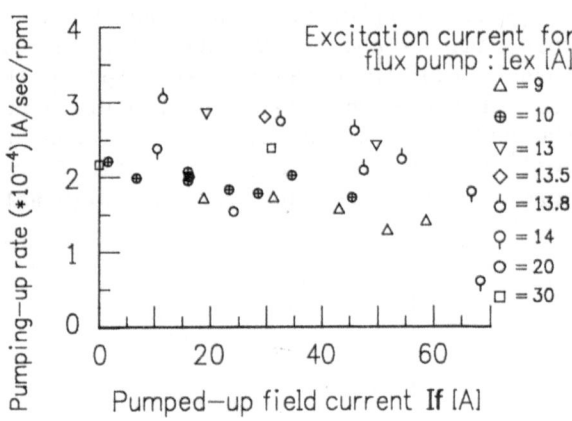

FIGURE 7 Pumping rate of field current vs. pumped-up field current

field current are plotted in the same figure.

FIGURE 7 illustrates that the pumping-up rate normalized by revolving rotor speed tends to decrease with the increase of self pumped-up field current I_f [A]. Also, the scattering appears in the figures because of the same reason mentioned above.

Characteristics of Generator

No-load and load characteristics

Up to the present, the test has been run, by limiting the revolving rotor speed to 1500 rpm. No-load terminal voltages of the brushless generator for a different pumped-up field current and a rotor speed are shown in FIGURE 8, in comparison with calculated values based on the mutual-inductance between field and armature coils. Experimental and calculated values agree well.

FIGURE 9 gives load characteristic

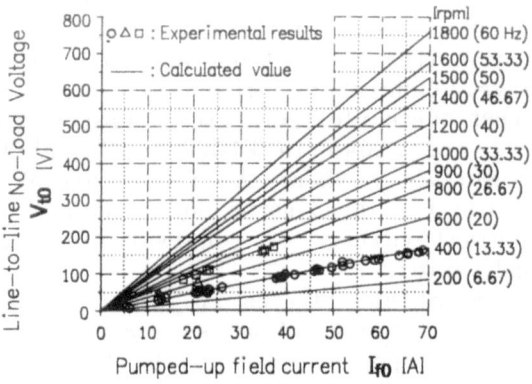

FIGURE 8 No-load characteristics

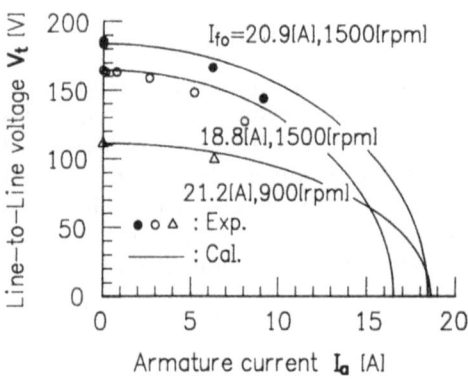

FIGURE 9 Load characteristics

curves or terminal voltage vs. load current for the case of a persistent field current I_{fo} = 20.9 and 18.8 A at 1500 rpm and I_{fo} = 21.2 A at 900 rpm, in comparison with calculated values. Since the armature reactance is designed to be quite large for small machine, load curves substantially droop. At present, the maximum output power of 2.3 kW at unity power factor has been obtained in case of I_{fo} = 20.9 A and N = 1500 rpm.

FIGURE 10 shows output characteristics or output power vs. load current in the same situation as in FIGURE 9. FIGUREs 9 and 10 are also quite in agreement with calculated results.

Predicted load characteristics

Although the experiment has not been done yet completely, some other characteristics of the generator will be calculated in a sense of prediction here, based on machine constants. Besides the preceding FIGUREs 8, 9 and 10, the calculated results are shown in FIGUREs 11, 12 and 13.

FIGURE 11 presents the change of the persistent field current caused by armature reaction, implicitly increasing with load, in the same situation as in FIGUREs 9 and 10. The curve showing the field current required to maintain a constant terminal voltage V_t is known as a compounding curve. Generator compounding curve and output curve plotted against armature current are shown in FIGUREs 12 and 13, respectively, where the field current at no-load I_{fo} is adjusted so as to hold the terminal voltage V_t constant while the load varies.

From both figures it is recognizable that the maximum output power available in the developed machine may be limited by the stability of armature superconducting coil because the armature coil of critical current of less than 30 A_{rms} turns to quenching before the field coil of critical current of about 70 A_{dc} turns to quenching. Consequently, though a stable output power depends on the magnitude of no-load excitation I_{fo} and induced voltage V_{to}, it will be about 10 kW in case of I_{fo} = 28.3 A. If, assuming that the critical current of armature and field coils are 30 A_{rms} and 70 A_{dc}, respectively, we can

design a machine so that both coils turn to quenching simultaneously, I_{fo} at no-load should be set 64 A_{dc} to yield no-load terminal voltage V_t = 561 V_{rms} and then the available maximum output power may be around 24.5 kW at N = 1500 rpm or 29.4 kW at 1800 rpm. Needless to say, because of weak mechanical factors such a large output power cannot be gained in this laboratory-made machine.

AC loss in armature coil

AC loss in the superconducting armature coil as presented in TABLE 1 will be estimated here. Assuming that a peak value of magnetic field applied to the armature coil is 1 Tesla and a critical current in peak is around 53 A, hysteresis and coupling losses will be about 0.9 Watt and 1.0 Watt, respectively and thus total AC loss will be about 1.9 Watt. However, AC loss of 0.19 Watt can be estimated in the case of three-phase shorted armature since the magnetic flux density on the armature coil is about less than 0.3 Tesla. Therefore, AC loss must be less than 0.19 Watt

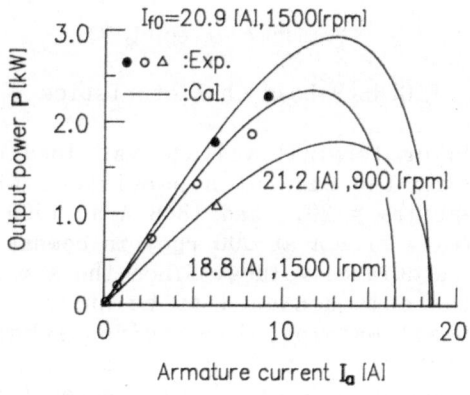

FIGURE 10 Output characteristics

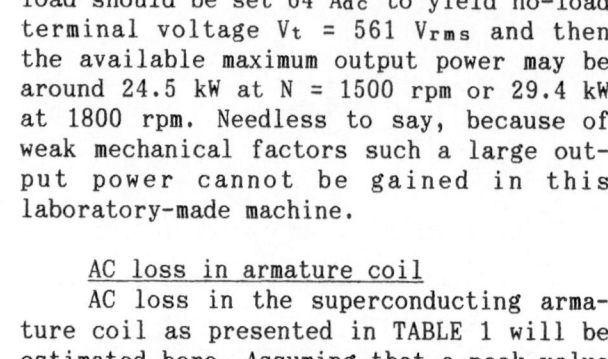

FIGURE 12 Compounding curves (V_t : const.)

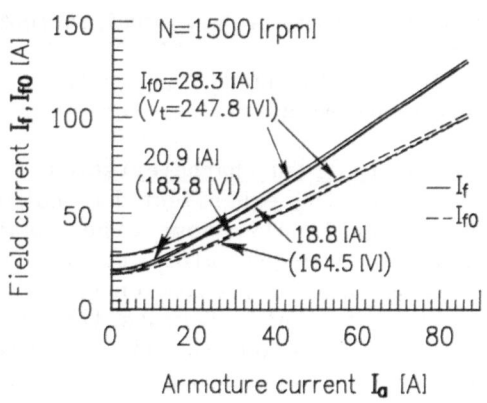

FIGURE 11 Change of the persistent field current

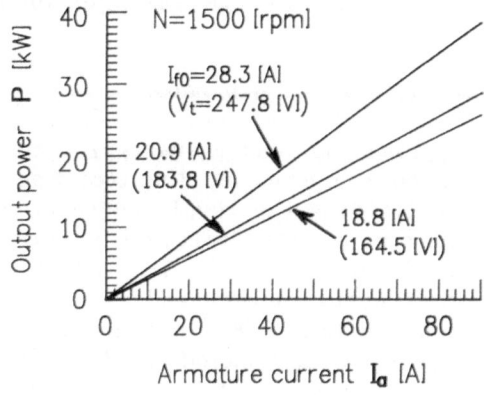

FIGURE 13 Output characteristics
(V_t : const.)

even under any on-load.

CONCLUSIONS

First, the following items have been revealed with respect to performances of the magnetic-flux pump set-up in the generator system:

1) Pumping-up rate tends to decrease with increase of self pumped-up current in the field circuit.

2) Pumping-up rate increases in proportion to traveling speed of normal-conductive spots on Nb strips as far as a certain point.

3) There is an optimum excitation intensity of magnets for magnetic-flux pump, so that the maximum pumping-up rate of field current is achieved.

4) Critical current of superconductive field circuit system of the generator might be about 70 A.

Second, experimental characteristics with respect to the generator have been obtained as follows:

1) Experimental results have been gained as was expected and calculated characteristics agree well with them.

2) Output terminal voltage of spiral-pancaked armature coil is of pretty sinusoidal waveform. Calculated waveforms and magnitudes during no-load, on-load and transients also agree well with experimental results.

3) Output power has been gained of 2.3 kW, nearly corresponding to the capacity of the prime mover.

4) Unsymmetrical quenching owing to thermal rather than electromagnetic factor happens to be observed among three-phase armature coils, but without quenching in the field circuit system. Such an unsymmetrical quenching in fully superconducting generators must be naturally one of crucial problems in practical machine design.

Finally, expected characteristics calculated from measured machine constants are as follows:

1) When the terminal voltage can be maintained a no-load voltage V_{to} of 247.9 V induced at $I_{fo} = 28.3$ A and N = 1500 rpm by adjusting the field current, an available maximum output power will be about 10 kW, taking account of critical current of less than 30 A_{rms} in armature superconductors.

2) When maximum values of field and armature currents are simultaneously lim-ited to its respective critical value, the initial field current at no-load I_{fo} can be set of 64 A_{dc} and then the available maximum output power will be about 24.5 kW at 1500 rpm or 29.4 kW at 1800 rpm.

3) AC loss of the armature coil can be estimated to be less than 0.19 W under on-load.

ACKNOWLEDGMENTS

The research was performed in part under seven-year Grant-in-Aid for Energy Research Project of the Ministry of Educational Science and Culture, Japan, and has been supported in part by Isahaya Densi Co., Ltd., Isahaya, Nagasaki. The authors would like to thank Professor Okada, T. in Kyoto University, Director in superconducting generator group of Energy Research Project. And, they are very grateful to Mr. Izaki, H.; President of Isahaya Densi Co., Ltd. for his financial support.

REFERENCES

1. Muta,I., Preliminary experiments for possibility of brushless superconducting synchronous generator excited by magnetic flux pump. Proceedings of International Conference on Electrical Machines, Lausanne, Swiss., Sept. 1984, 1118-1121.

2. Mawardi,O.K., Brushless superconducting alternators. IEEE Trans. on Magnetics, MAG-13, 1977, 780-783.

3. Muta,I., Tanimizu,H. and Mukai,E., Feasibility test of brushless superconducting synchronous generator excited by magnetic flux pump. Proceedings of Shanghai International Conference of Cryogenic Engineering '85, Shanghai, China, June 1985, 139-142.

4. Muta,I., Tsukiji,H., Mukai,E. and Furukawa,T., Preliminary experiment for brushless superconducting synchronous generator with magnetic flux pump. Electric Energy Conference '87, Adelaide, Australia, Oct. 1987, 301-305.

5. Muta,I., Tsukiji,H., Sasano,T., Tsutsui, Y. and Furukawa,T., Experiment of burshless fully superconducting generator with magnetic flux pump. In Progress in High Temperature Superconductivity - vol.15, ed. Y.Murakami, World Scientific, London, 1989, pp.645-650.

A STUDY OF APPLICATION OF HIGH Tc SUPERCONDUCTORS TO SUPERCONDUCTING GENERATORS

N.Maki and K.Yamaguchi   Y.Yagi
Hitachi Research Laboratory   Hitachi Works
Hitachi Ltd., Hitachi-shi, Ibaraki-ken, Japan

Abstract

Conceptual designs of 1000MW superconducting generators, in which field windings are cooled by helium (~4.2K), nitrogen (~77K), air (~300K), were carried out. Structural features and performance of these generators were clarified by comparing design results of superconducting and conventional generators.

Technical problems of high Tc superconducting generators, e.g. improvement of filament current density, winding method and stabilization, are then discussed.

Introduction

A superconducting AC generator consists of a non-magnetic cryogenic rotor containing a super conducting field winding and an airgap winding stator. The superconducting generator (SCG) has advantages of improved power system stability and higher efficiency compared with a conventional generator. Therefore, development work, through different approaches, is being carried out in several countries[1]-[5].

A ceramic oxide superconductor, which has a high critical temperature beyond liquid nitrogen boiling point (77K) at the pressure of 1 barr, has been discovered recently[6],[7]. Various research activities to improve its critical temperature (Tc) and critical current density (Jc) are continuing.

When the high Tc superconductors are used for the SCG, it is expected that the rotor structure and cooling system are simplified, and stability of the superconductor is improved due to the increase in temperature margin and specific heat. Therefore, SCG reliability can be enhanced.

When developing the high Tc SCG, many problems, e.g. improvements in H-Jc characteristics of high Tc superconducting wire and the winding method, must be solved. In order to determine the feasibility of the high Tc SCG, a conceptual design study was carried out and technical problems in its development were examined.

Design

When designing a SCG, the following basic conditions for the SCG must be met.

1) A 1000MW, 50Hz, 2 Pole generator with a power factor of 0.9 should be used.

2) Terminal voltage is 26kV considering possible application of present isolation techniques.

3) Synchronous reactance is set as 0.3 p.u. from the viewpoint of power system stability.

4) Current density, which is defined by the ratio of the rated field current to a rectangular section circumscribed by the field conductor, is 60-150 A/mm^2.

5) Coolants are liquid helium (~4.2K), liquid nitrogen (~77K) and air (~300K).

Structure of SCG

Using these basic conditions, a design of the SCG was made and a cross section of the helium cooled generator and a schematic section of the rotor are shown in Figs.1 and 2, respectively. The rotor consists of the warm damper (~330K), cold damper (50-100K) and cryogenic winding support shaft (~4.2K). The spaces between them are evacuated to give thermal isolation.

To reduce the heat input through the torque tube to the cryogenic winding support shaft and to withstand the large torsional stress at the same time, a thin mechanical torque tube is fixed between the cryogenic winding shaft and room temperature shaft.

A helium transfer coupling is used to transfer liquid helium into the rotor and exhaust helium gas through the rotor outlet. In order to accommodate the differemce in thermal contraction between the cold torque tube and the warm damper, a double bearing system is adopted.

Saddle-shaped superconducting field coils are put in the slots of the winding support shaft and fixed by metallic wedges. Supercritical helium flows toward the rotor center through the radial passage in the slots between the insulation spacers and superconductors. Armature copper coils are put in the slots hollowed out of the insulation structures

Table 1 Specifications of 1000MW generator (Liquid helium cooled)

Rated		Field Winding	
Capacity	1120 MVA	Inner dia.	0.59 m
Voltage	26000 V	Outer dia.	0.75 m
Current	24870 A	MMF	1.83 MA/pole
Pawer factor	0.9	Currrent Density	150A/mm^2
Speed	3000 rpm	Max. flux density	4.6 T
Generator		Stored energy	24 MJ
Rotor outer dia.	1.12 m	Armature winding	
Stator outer dia.	3.8 m	Inner dia.	1.23 m
Rotor weight	60 ton	Outer dia.	1.62 m
Stator weight	390 ton	MMF	2330 A/cm
Xd	0.30 p.u.	Current density	8.5 A/mm^2
Xd'	0.20 p.u.	Inner dia. of core	1.69 m
Xd"	0.13 p.u.	Length of core	6.6 m

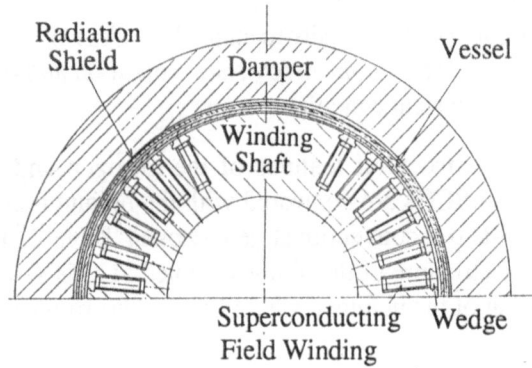

Fig.2 Schematic section of the superconducting rotor

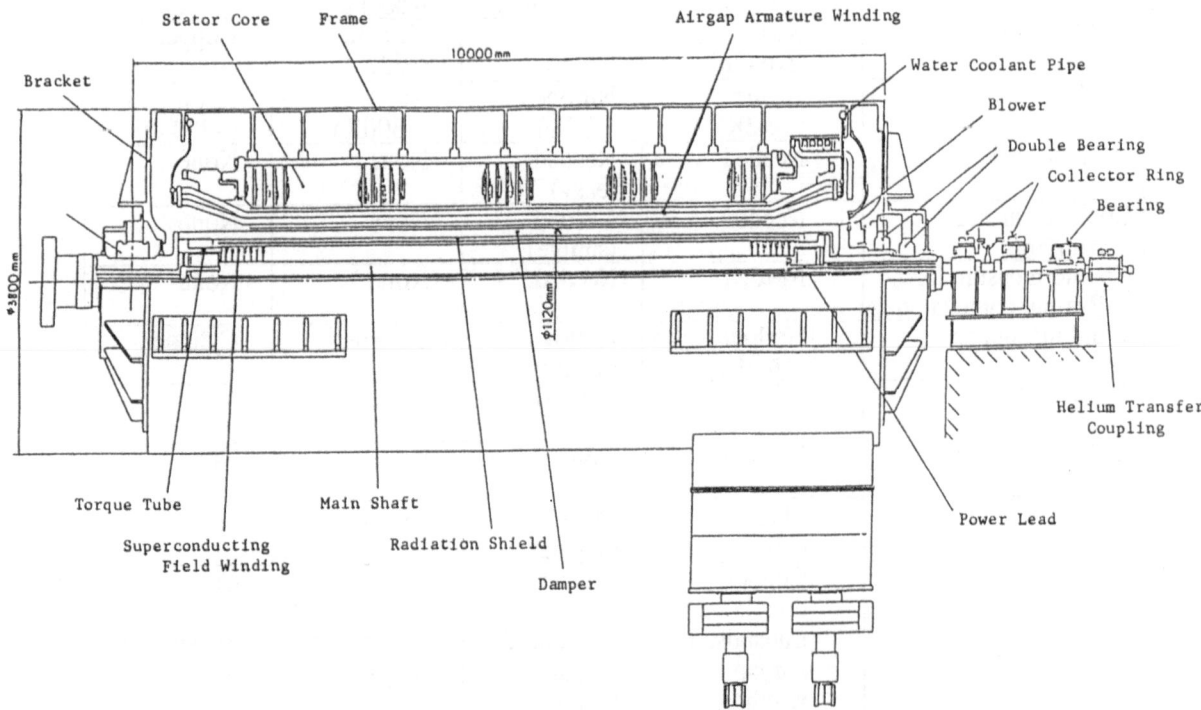

Fig. 1 Cross-section of 1000MW superconducting generator

inside the cylindrical magnetic shield (stator core), which forms the airgap winding.

Major specifications of the 1000MW generator are listed in Table 1. Ratios of transient or sub-transient reactances to synchronous reactance are about 0.7 and 0.4, based on the position of the field winding and damper in relation to the armature winding.

The thicknesses of the field winding and armature winding are decided from their current density and packing factor. Current density of the field conductor is considerably higher, and current density of armature conductor is almost the same as that of the conventional generator. The operating point, which shows the ratio of the rated field current and flux density to attainable values of short wires, is 0.55 in the steady state condition and 0.72 in the transient condition.

Structural features of the nitrogen and air cooled SCGs are compared with the helium cooled SCG and a conventional generator in Table 2. The basic structure of the nitrogen cooled SCG is similar to that of the helium cooled SCG, but the former has

no radiation shield, easier vacuum sealing and a simpler structure for coolant transfer coupling. Two useful cooling system are considered. One is an open system in which coolant is consumed, and the other is a closed system in which coolant is recycled by using a liquefier. Slot structure for the field coils of the nitrogen cooled SCG is almost the same as that of the helium cooled SCG. However, the pressure of nitrogen in the rotor generated by centrifugal force is higher due to the density difference between nitrogen and helium, and therefore the coolant vessel in the rotor must be mechanically strengthened for the former.

The basic structure of the air cooled SCG is rather similar to that of the conventional generator, because the coolant transfer coupling, vacuum seal system, countermeasure for thermal contraction, thick cylindrical damper, radiation shield and torque tube are all unnecessary, and a strip shaped damper is inserted on the field coil in the rotor slot, just the same as in the conventional generator.

As for the stator, the basic structures of the nitrogen and air cooled SCGs are almost the same as

Table 2 Structural features of superconducting and conventional generator

Components	Superconducting generator			Conventional generator
	Cryogenic	Low temp.	Room temp.	
Field conductor	NbTi (Nb$_3$Sn)	Ceramic oxide	Ceramic oxide	Copper
Coolant	Helium (~4.2K)	Nitrogen (~77K)	Air (~300K)	Hydrogen (~300K)
Vacuum seal	Needed	Needed (easy)	None	None
Coolant transfer coupling	Needed	Needed (simple)	None	None
Countermeasure for thermal contraction	Needed	Needed	None	None
Winding support	Hollow cylinder	Hollow cylinder	Cylinder	Cylinder
Torque tube	Needed	Needed	None	None
Radiation shield	Needed	None	None	None
Damper	Thick cylinder	Thick cylinder	Al strip	Al strip
Armature	Copper double transposed Airgap winding Indirect water cooling	Copper double transposed Airgap winding Indirect water cooling	Copper double transposed Airgap winding Indirect water cooling	Copper single transposed Stator slot Direct water cooling

Table 3 Comparison of 1000MW SCG parameters

Coolant / Field current density	(A) helium 150A/mm^2	(B) nitrogen 150A/mm^2	(C) nitrogen 75A/mm^2	(D) air 150A/mm^2	(E) air 150A/mm^2	(F) air 60A/mm^2
Generator parameters						
synchronous reactance (p.u.)	0.30	0.30	0.30	0.30	0.30	0.30
transient reactance (p.u.)	0.20	0.20	0.20	0.19	0.12	0.12
sub-transient reactance (p.u.)	0.13	0.13	0.12	0.17	0.10	0.10
open circuit time const. (s)	1600	1600	1600	1500	900	2500
open circuit sub-transient time const. (s)	0.08	0.08	0.12	0.02	0.02	0.02
Size & weight						
rotor diameter (m)	1.12	1.12	1.30	1.12	1.12	1.12
stator diameter (m)	3.8	3.8	4.0	4.5	3.8	3.8
bearing span (m)	10.0	10.0	10.0	6.8	10.0	10.0
rotor weight (ton)	60	58	85	40	45	65
generator weight (ton)	450	448	500	430	425	445
Performance						
thermal load (w)	90	1250	1300	---	---	---
coolant feed rate (l/h)	120	30	30	---	---	---
maximum flux density (T)	4.6	4.6	4.4	4.5	2.5	2.9
generator efficiency (%)	99.54	99.57	99.54	99.56	99.54	99.54

that of the helium cooled SCG, because iron stator teeth are not used considering the high airgap flux density, and double transposed copper bars are used to reduce the eddy current loss in the bars due to a strong alternating magnetic field.

Features of high Tc SCG

Design parameters of the 1000MW generator are given in Table 3. The nitrogen cooled SCG has the following features compared with the helium cooled SCG.

(1) Generator parameters are almost the same.

(2) Size and weight of the generator increases as field conductor current density decreases.

(3) Since the outer diameter of the rotor is limited to below 1.3 m due to mechanical stress, current density of the superconductors must be more than 75 A/mm^2 at 4.6 T.

(4) Since the coolant feed rate is decreased by one-forth, the volume and cost of the coolant decreases considerably.

(5) Generator efficiency is increased by 0-0.3%.

Next, features of the air cooled SCG are compared as follows with the helium cooled SCG.

(1) A large diameter, short shaft generator (D type) and a small diameter, long shaft generator (E,F type) can be designed, due to the free selection of the field winding thickness.

(2) Generator parameters of the D type are almost the same that of the A type. However, xd', xd" and Bm of E and F types decrease to 0.6, 0.8 and 0.6 times that of the the A type. Rotor weights of the D and E types decrease to approximately 0.7 times that of A type due to lack of a large damper.

(3) Generator efficiency increases by 0-0.3% considering no loss for liquefire use or increase in mechanical loss.

(4) Current density of superconductors must be more than 60 A/mm^2 at 2.9T.

Technical problem

Technical problems associated with the high Tc SCG are shown in Table 4 and discussed below.

(1) Field current density of the SCG must be more than 80 A/mm^2 at 5T, which means that the filament current density of the superconducting wires is more than 1000 A/mm^2 at 5T. At present, the maximum filament current density of superconducting wires is about 200 A/mm^2 at 0T and decreases below one-tenth that at the magnetic field of 5T. Therefore, filament current density of the superconducting wires must be improved more than 50 times that of present high Tc wires. Furthermore, high Tc superconductors are required to maintain good

Table 4 Technical problems of high Tc SCG

Items	Contents
Superconducting wire	
i) H-Jc characteristics	Conductor current density 80 A/mm^2 at 5T
ii) Stabilization	Multi-filament, stabilized material
Low copper loss	Optimum structure
iii) Manufacturing	Lengths of several hundred
long wire	meters
Field winding	
i) Winding method	Mechanical properties (torsion, tensile, compression)
ii) Insulation, support	Conductor insulation, winding support structure
iii) Connection	Low resistive loss, high mechanical strength
Rotor	
i) Cooling of field	Coolant feed system
winding	Cooling of current leads
ii) Structure member	Material, optimum structure
iii) Protection system	Detection and countermeasures for quenching

performance levels under any circumstances for a long time.

(2) High Tc superconductor have high temperature margins for coil quenching because of their large thermal capacity. However, when coil quenching occurs, the superconductors are easily burned or damaged because of their low conductivity. Therefore, high Tc superconductors must be stabilized with high conductivity materials and a good cooling system. In order to wind superconducting coils, long wires of more than several hundred meters are required.

(3) Becausef high Tc superconductors are more brittle than compound superconductors, winding works is very difficult and therefore, a wind and react (W&R) method is preferred. An insulation system suitable for transient high voltage and a winding support system to withstand electromagnetic force and thermal stress must be developed. Furthermore, a connecting method for superconductors, which have lower resistive loss and higher mechanical strength, must be developed.

(4) Regarding the cryogenic rotor, a rotor design and a vacuum seal system suitable for simpler rotor operation under high temperature conditions, must be developed. Materials, which are used for the rotor, must be changeable due to the higher operating temperature. For the nitrogen cooled SCG, a self-pumping effect for feeding coolant is generated

in the same way as in the helium cooled SCG, and the centrifugal force becomes large due to the higher density of the coolant. Therefore, flow resistance must be reduced. For the air cooled SCG, a coolant flow system containing inlet and outlet structures, must be developed. Furthermore, a protection system including detection and countermeasures for coil quenching must be developed.

Conclusions

Conceptual designs of 1000MW SCGs were carried out. The superconducting field windings are cooled by helium (~4.2K), nitrogen (~77K) and air (~300K). These design results were compared for the three SCGs and a conventional generator.

Basic structures of the nitrogen and air cooled SCGs are similar to that of the helium cooled SCG and conventional generator, respectively. In the case of the nitrogen cooled SCG, cost of coolant decrease considerably because coolant feed rate decreases to one-fourth. For the air cooled SCG, rotor weight decreases approximately 0.7 times that of the helium cooled SCG because a large damper is unnecessary.

The greatest problem to face is improvement of the filament current density of the high Tc superconducting wires to more than 1000 A/mm^2 at 5T. Stabilization, manufacturing long wire, winding work and connection techniques development must also be undertaken. Rotor fabrication, and vacuum seal, cooling and protection systems also must be developed.

At present , 100-300MW helium cooled SCGs are being developing in West Germany, the Soviet Union and Japan and their realization are good. High Tc SCGs have significant advantages of high reliability and easy maintenance, and they will have a great impact on many industries when they become practical.

Acknowledgements

The authors would like to thank Dr. H. Ogata and Mr. M. Ooi of Hitachi Ltd. for their support.

References

1. D. Lambecht, "Superconducting turbogenerators: status and trends" Cryogenics Vol.25, pp.619-627, November 1985.

2. J.L. Smith, "Development of superconducting synchronous machines in the USA." IEEE Transaction on Magnetics, Vol. MAG-23, No.5, pp.3529-3532, September 1987.

3. B.I. Fomin, et al,"Main stages of manufacturing a 300MW superconducting generator." Cryogenics, Vol.27, pp.243-248, May 1987.

4. A. Fevrier, "Latest news about superconducting AC machines," IEEE Transaction on Magnetics, Vol. MAG-24, No.2, pp.787-791, March 1988.

5. K. Uyeda, "Research on superconducting generator and materials in Japan." Proceedings of the American Power Conference, April 1988.

6. T. Matsumoto, et al, "Ag sheathed high Tc superconducting tape with Jc=10^4 A/cm^2." ISTEC Workshop on Superconductivity, pp.111-114, February 1989.

7. K. Sato, et al, "BiPbSrCaCuO superconducting wires with high critical current density," ISTEC Workshop on Superconductivity, pp.119-122, February 1989.

TEST RESULTS OF MOTORS WITH A SUPERCONDUCTING ROTOR

M. Takahashi, K. Arai, M. Satoh, M. Watanabe,
N. Takahashi and R. Takahashi
Hitachi Research Laboratory, Hitachi Ltd.,
Ibaraki, Japan

ABSTRACT

Experimental studies of two-pole, three-phase superconducting motors using Y-Ba -Cu-O (high-Tc) and Nb$_3$Sn rotors were carried out. The former high-Tc rotor shapes were cylindrical, cylindrical with axial slits, or reluctance types. The rotor outer diameters were 38mm and their lengths were 30-32mm. Each high-Tc rotor started to rotate by itself and kept a speed with a large slip at superconducting conditions in liquid nitrogen. For example, the cylindrical rotor rotated at 46 rpm when the motor was excited by a 5 Hz sinusoidial alternating current of 1.5 VA input.

The cylindrical Nb$_3$Sn rotor was confirmed as able to rotate not only at asynchronous speed, but also synchronous speed. Further, once the rotor reached the synchronous speed, it kept a constant speed even if the stator voltage was decreased.

These results indicated that the superconducting motor shoud be able to be driven by the AC magnetizing loss of the superconductor.

INTRODUCTION

Studies for a superconducting motor have been made since the early 1960's.[1] With the discovery of high-Tc superconductors, the possibility of major practical applications has become more likely. One example is a motor which utilizes the repulsive force caused by the Meissner effect.[2] We have been developing new motors using a ceramic superconductor rotor of the Y-Ba -Cu -O system and an inter metallic compound rotor of Nb$_3$Sn in order to investigate the possibility of a superconducting AC motor.

In this paper, we report on the motor rotating speed and the starting torque as obtained experimentally, and attempt to clarify features of the torque generating mechanism. These rotors are assembled with a two-pole, three -phase conventional stator which is submerged in coolant. A synchronous rotation is induced in the rotor by the application of a three -phase rotating magnetic field. A probable mechanism for this rotation is discussed on the basis of the flux penetration behavior and AC magnetizing loss of the superconductor.

STRUCTURE OF THE SUPERCONDUCTING ROTOR

An exterior view of the motor is shown in Fig. 1, and its main specifications are listed in Table 1. Three rotor structures using ceramic superconductor of a Y-Ba-Cu-O system were proposed and manufactured as shown in Fig. 2. The rotor shapes were completely cylindrical, cylindrical with axially slits, or reluctance types. The reluctance type has eighteen superconductor bars and its rotor slots are filled with FRP (fiber reinforced plastics).

The starting materials for the high -Tc superconductor were prepared from stoichiometric quantities of yttrium, barium and copper oxides calcined at 900 ℃ in air. After grinding, the powder was pressed into the required shape, sintered at 950 ℃, and annealed for several hours at 300 ℃. The transition to zero resistivity was completed at 89 K. The inter metallic compound of Nb_3Sn rotor was also manufactured for experiments. The cylindrical rotor was prepared from stoichiometric quantities of Nb and Sn powder. The powder was pressed and

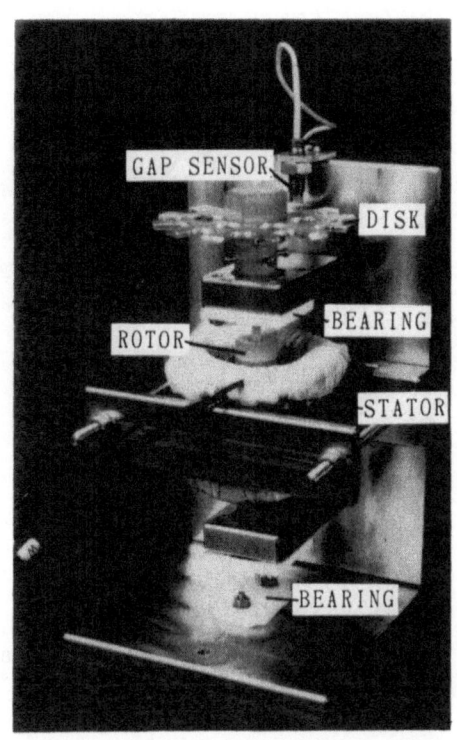

Fig. 1. View of Superconducting Motor.

sintered for 5 hours at 1000 ℃ under 10^{-6} Torr. The critical temperature Tc was 17.5 K as determined by resistivity measurements. The stator size for the Nb_3Sn cylindrical rotor was smaller than that for the high-Tc rotors which corresponded to this rotor size.

The rotors were fabricated with a shaft and inner ring as one body. All

TABLE 1. Major Specifications of Motors

1. Rotor	Y-Ba-Cu-O			Nb_3Sn
Type	Cylindrical	Cylindrical with Axial Slits	Reluctance	Cylindrical
Outer Dia. (mm)	38	38	38	25.5
Superconductor Length (mm) Thickness Width	30.5 2.75 –	30 2.75 5.5	32 10 2	25.6 10.25 –
No. of Bars	–	16	18	–
2. Stator	2-Pole, 3-Phase Conventional Type			
Outer Dia. of Core(mm)	65			60
Inner Dia. of Core(mm)	39			29
Length of Core (mm)	30			29
No. of Slots	18			18
Coil Turns/Phase	51			51

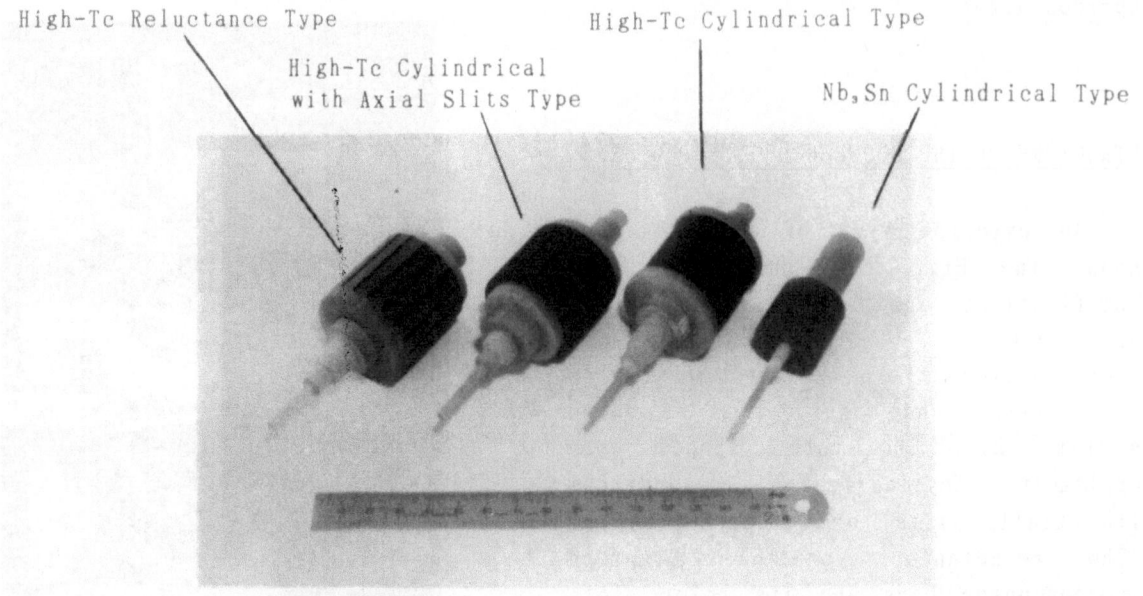

High-Tc Reluctance Type

High-Tc Cylindrical
with Axial Slits Type

High-Tc Cylindrical Type

Nb,Sn Cylindrical Type

Fig. 2. Photograph of Trial Manufactured Rotors.

rotor parts were made of FRP except for the superconductor at the rotor surface.

As a result, it was easy to clarify the rotating phenomena because torque action by the magnetic field was restricted to within the superconductor.

EXPERIMENTAL PROCEDURE

As shown in Fig. 3, each rotor is vertically inserted into the conventional two-pole, three-phase stator, and supported not only by a thrust load of the pivot bearing, but also by a radial load of the teflon plate to reduce the mechanical friction loss.

Each rotor, with the conventional stator, was set into an adiabatic dewar made of glass. This allowed easy observation. Liquid nitrogen was added to above the high-Tc rotor, while the motor with the Nb,Sn rotor was separately submerged in liquid helium. The motor input power for the stator was supplied by a three-phase 120kVA sinusoidial VVVF inverter.

The starting torque was measured with a spring balancer by winding ordinary sewing thread around the top of the rotor shaft. The rotating speed was estimated with an electro-magnetic gap sensor by counting the pulses corresponding to the number of blades on a rotating aluminum disk. Other electrical quantities were measured by a digital multi-meter.

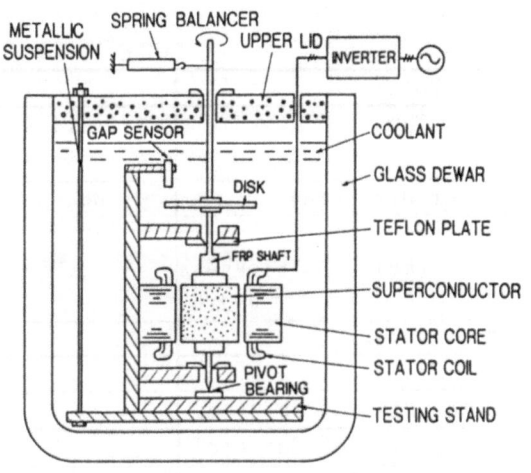

Fig. 3. Schematic Diagram of Experimental Equipment.

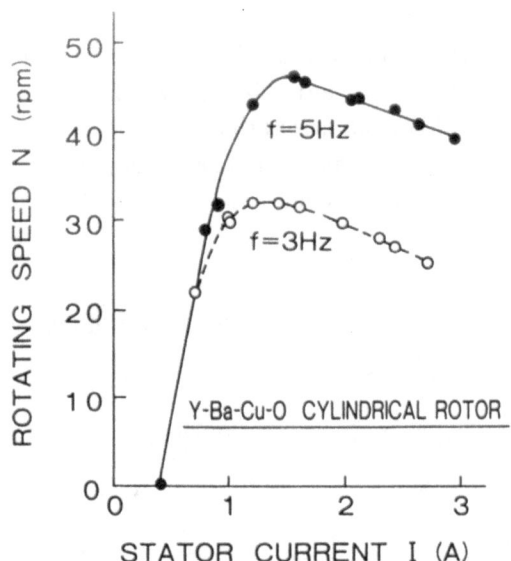

Fig. 4. Rotating Speed versus Stator
Current.

TEST RESULTS

Rotating Speed of the High-Tc Motor

In tests, not every rotor rotated at room temperature no matter what the stator exciting power. On the other hand, when the rotors were dipped in liquid nitrogen and kept at superconducting conditions, all of them started to rotate. As one example, Figure 4 shows the rotating speed dependence on the stator armature current for the cylindrical rotor with the frequency as a parameters. The rotor began to rotate at a current of 0.4 A and reached a maximum speed of 46 rpm at a 5 Hz sinusoidial alternating current of 1.5 VA input. But, the viscous drag of liquid nitrogen reduced the rotating speed when the current was increased.

The rotating speed had the same tendency for the other rotors, the maximum speeds were 28 rpm and 20 rpm for the slit cylindrical rotor and reluctance types at currents of 4.8 A and 5 Hz, respectively. As mentioned above, all the rotors could start to rotate by themselves and keep a speed

with a large slip for synchronous speed.

The rotational effect for the reluctance rotor which was not able to create a pole return current at the rotor end was assumed as follows. The motor utilized the force caused by an induced current loss which occurred at the boundary between superconducting grains or magnetic hysteresis which appeared below Tc and disappeared above that.

Measurement of Magnetic Flux

The applied alternating magnetic flux was estimated with one-turn search coils which were arranged at a pole pitch interval on the inner surfaces of the stator core and the superconductor. Figure 5 plots measured values which give the magnetic flux versus stator current at a locked rotor. No difference in flux density between 77 K and room temperature was observed because the critical flux density for the superconductor was above 40 mT. Besides, below 40 mT, the flux density at 77K was less than that at room temperature. This verified that the flux was reflected by the current at

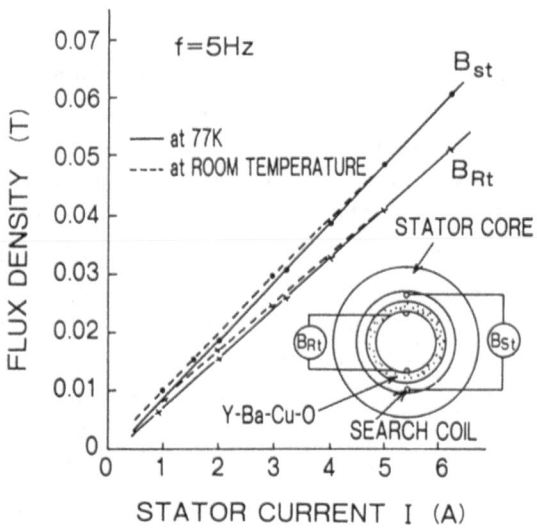

Fig. 5. Flux Penetration on Applied AC
Magnetic Field.

the intergranular region and superconducting grains.

From these results, the phenomena at currents 1-3A in Figure 4 happened when most of the flux lines passed through the grains.

Motor Fundamental Characteristics with the Nb₃Sn Rotor

A similar motor test was done with the cylindrical Nb₃Sn rotor. The motor was immersed in liquid helium after it was established experimentally that this rotor did not rotate at room temperature. No load and lock tests were carried out to evaluate the motor characteristics.

No Load Test

The relations between the applied motor voltage and rotating speed with the frequency as a parameter are indicated in Figure 6. The exciting current frequency was limited to less than 10 Hz taking into account the mechanical strength of the rotor and the glass dewar. The rotor was confirmed at as able to rotate smoothly not only at asynchronous speed, but also synchronous speed. Further, the following interesting behaviour was observed with the transition of the rotating speed.

At first, when the stator voltage was small, the flux did not penetrate into the rotor because of the persistent current caused by the Meissner effect. Therefore, no rotor torque would be induced because of no electric phase difference between the flux and persistent current. Next, if the stator current was increased more, the magnetic field intensity was above the lower critical field Hc_1. Then the flux began to penetrate into the rotor, consequently, hysteresis and eddy currents of the superconductor

contributed to torque generation, and eventually the motor started to rotate. Further, magnetic flux penetration and a pinning effect facilitated the start of synchronous rotation.

Once the rotor reached the synchronous speed Ns, it kept its

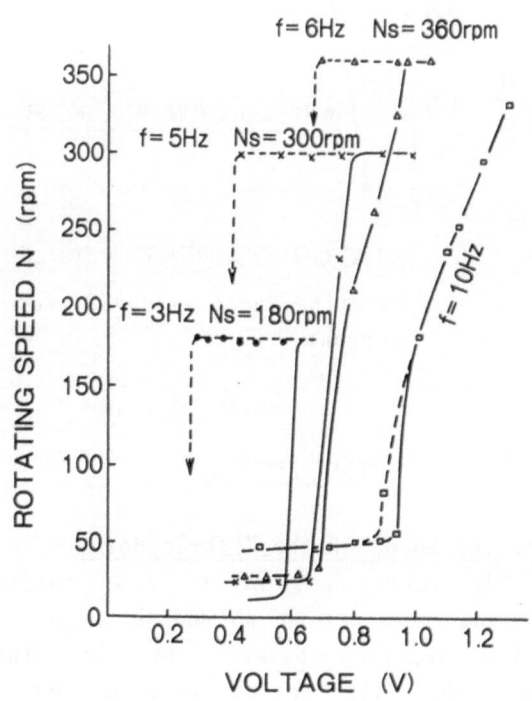

Fig. 6. Rotating Speed at No Load Test.

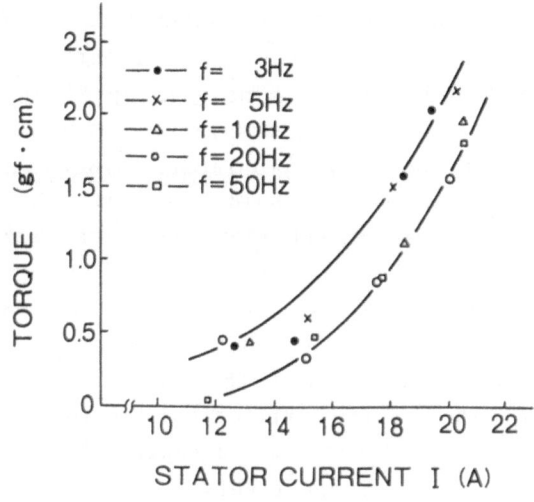

Fig. 7. Starting Torque at Lock Test.

constant speed as a synchronous machine even if the applied motor voltage was decreased. This behavior was similar to that of over-exciting characteristics for a conventional hysteresis motor and it occurred in every test case.

Lock Test

The measured torque is shown in Figure 7. The starting torque in the lock test was approximately proportional to the square of the exciting current and independant of its frequency. The maximum torque was obtained of 2.2 gf cm at 5 Hz.

CONCLUSION

Two-pole, three-phase motors were developed using Y-Ba-Cu-O ceramic (high -Tc) superconducting rotors, and a Nb₃Sn rotor. The fundamental characteristics were investigated.

The cylindrical high-Tc rotor rotated at a maximum speed of 46 rpm under a 300 rpm (5 Hz) rotating stator magnetic field. The Nb₃Sn rotor was found as able to rotate by itself and reach a synchronous speed. It generated a starting torque of 2.2 gf cm at 5 Hz. The mechanism of the rotation for these motors was regarded as depending on the AC magnetization loss of the superconductor.

These motors had a much smaller torque than in the present industrial motors, so that further improvements in rotor construction and characteristics of the superconductor are needed before commercial applications are possible.

REFERENCES

1. J.D. Jones and P.W. Matthews, Three-Phase Superconducting Motor, The Review of Scientific Instruments, vol.35, no.5, pp.630-633, May 1964.

2. A.Takeoka, A.Ishikawa, M.Suzuki, K. Niki and Y.Kuwano, Meissner Motor Using High-Tc Ceramic Superconductors, IEEE Transactions on Magnetics, vol.25 no.2, pp.2511-2514, March 1989.

FUNDAMENTAL DESIGN FOR MEISSNER MOTOR

A. Ishikawa, A. Takeoka, Y. Kishi
Functional Materials Development Center, Sanyo Electric Co., Ltd.,
Moriguchi, Osaka 570, Japan

ABSTRACT

We have developed a new superconducting motor, called the "Meissner Motor", using high-Tc ceramic superconductors. The output of this motor depends on the repulsive force caused by the Meissner effect and the heat flow rate from superconductors to liquid nitrogen. The motor output has been estimated by computer simulations in view of the Meissner effect and the heat flow rate. The maximum output of the motor with a radius of 30 mm is simulated at about 0.03 W. It is a fairly low output compared with commercial motors. Superconductors with a higher Meissner effect are required to improve the motor output.

INTRODUCTION

While a number of materials and manufacturing methods have been investigated for high-Tc ceramic superconductors, we have developed a new superconducting motor using ceramic superconductors in the Y-Ba-Cu-O system. This motor, called the "Meissner Motor", utilizes the repulsive force caused by the Meissner effect. We have not seen any prior investigations on a motor based on this idea except the precedent-setting research into thermomagnetic motors [1]. In the present paper, we will design a calculation model to estimate the motor output by computer simulations, and also we will show the influence of the cooling ability on the motor output.

SIMULATION METHOD

Motor Structure

The structure of the experimental Meissner Motor is shown in *Fig.1*. Superconductors were prepared by pressing and presintering a coprecipitated compound of YBa$_2$Cu$_3$O$_{7-x}$ at 960 °C for 3 hours and sintering at 980 °C for 1 hour.

Cooled down in liquid nitrogen, the superconducting plates near the magnet were repelled by the magnetic field due to the Meissner effect and this provided the motor with torque. The rotating plates were heated in turn at the top of the wheel and returned to a normal state. In this state, they could easily move into the magnetic field.

This motor rotates at a maximum speed of 40 rpm and generates a maximum torque of about 0.6 gf-cm. Though these characteristics were considered to be limited by low magnetization and still cooling ability [2], the influence of a superconducting shape on the motor output and the maximum output of the motor have not yet been investigated. These factors must be estimated to further develop the Meissner Motor.

Simulation Model

Figure 2 shows a calculation model,

where a is the width, b the length, t the thickness, r the average orbital radius of the superconductor and θ the position. The intersection of the orbit and the permanent magnet's axis is considered as the origin of the position. It is assumed that the superconductor begins to cool in liquid nitrogen at the origin.

Models of magnetic dipole (instead of two permanent magnets) supply a magnetic field. The repulsive force F between the magnets and the superconductor is proportional to the magnetization and the gradient magnetic field, and is defined as

$$F = V \cdot I \cdot \frac{dH}{dz} \qquad [\text{gf}], \quad (1)$$

where V is the volume of the superconductor, I the magnetization and H the magnetic field. The apparent magnetization is assumed to be a fixed value here because we confirmed experimentally that its variation was much less than the

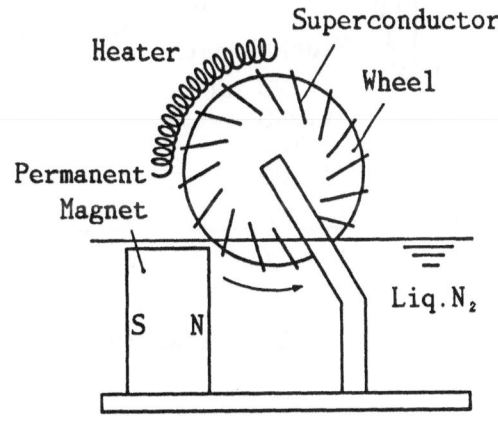

Fig.1 Schematic diagram of the Meissner Motor.

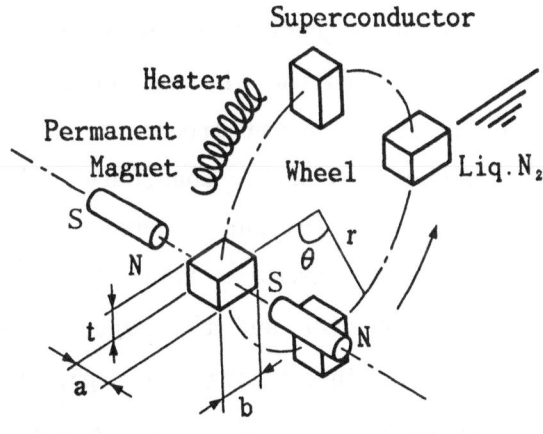

Fig.2 Configuration of the calculation model for the repulsive force distribution of the Meissner Motor.

TABLE I Numerical values used for calculation

Maximum magnetic field	$Hmax$	4000	Oe
Magnetization (at 77 K)	I	0.25	emu/cm³
Orbital radius	r	30	mm
Space between the magnets	$2d$	$a + 5$	mm
Thermal conductivity of the superconductor	λ	3.00	J/msK
Critical temperature of the superconductor	Tc	93	K
Temperature conductivity of the superconductor	h	2.73×10^{-6}	m²/s
Heat flow density (liquid nitrogen, $\Delta T = 7$ K)	q	16	W/cm²

variation of the gradient magnetic field. Numerical values used for the calculation are listed in *Table* I.

Repulsive Force Caused by the Meissner Effect

In order to calculate the repulsive force caused by the Meissner effect, we first considered a small segment of the superconductor, ignoring any variation of the gradient magnetic field within that segment. In the magnetic field, electric current flows around the segment as the Meissner effect is caused. Then, magnetization is generated in proportion to the electric current. As the electric current around the whole superconductor can be considered the total of the current around small segments, the repulsive force between the superconductor and the magnets is calculated by integrating the repulsive force between its small segment and the magnets. The small segment $dxdydz$ has the magnetic moment dI as shown in the following equation,

$$dI = k \cdot \mu_0 \cdot i \cdot dxdydz \quad [\text{emu/mm}^3], \quad (2)$$

where k is a proportional constant, μ_0 space permeability, and i the electric current. The repulsive force dF of the small segment is obtained by the following equation,

$$dF = dI \frac{dH}{dz} \quad [\text{gf/mm}^3], \quad (3)$$

The repulsive force F of the supercon-

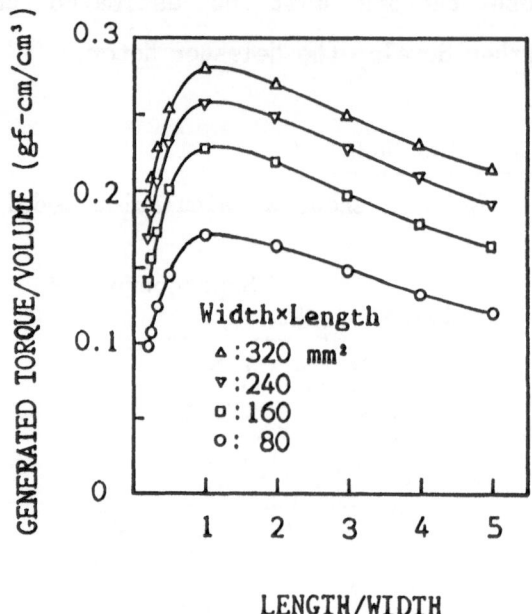

Fig.3 Influence of the ratio of width and length on generated torque with the product of them as a parameter.

ductor is obtained by integrating equation (3).

RESULTS AND DISCUSSION

Generated Torque

The generated torque Ts applied with one superconductor is obtained by integrating its repulsive force, which is averaged per one revolution, and multiplied by orbit r. *Figure 3* shows the influence of the ratio of the width and the length on generated torque Ts per volume, with the product of the width and the length as a parameter. When the ratio is about 1 to 1, the generated torque was at its maximum. The generated torque was determined by detailed calculations when the ratio at the maximum generated torque was 1 to 1.2. This ratio will be used in the following calculations.

Figure 4 shows the influence of thickness on the generated torque per volume, with the width as a parameter. The results show that two 10 mm-thickness superconductors provide the motor with higher torque than one 20 mm-thickness superconductor. We found that superconductors with less than 3 mm thickness are preferable for development of a motor with higher torque. As the thinner superconductor tends to cool down quicker, the thickness of about 0.5 mm will be used in the following calculations.

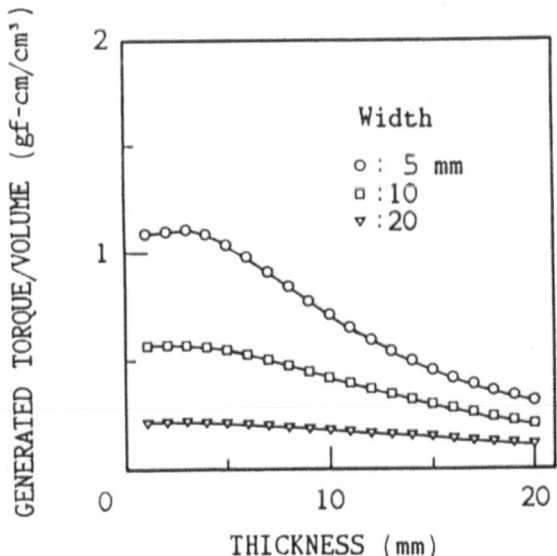

Fig.4 Influence of the thickness on the generated torque per volume with the width as a parameter. The ratio of width and length is 1:1.2.

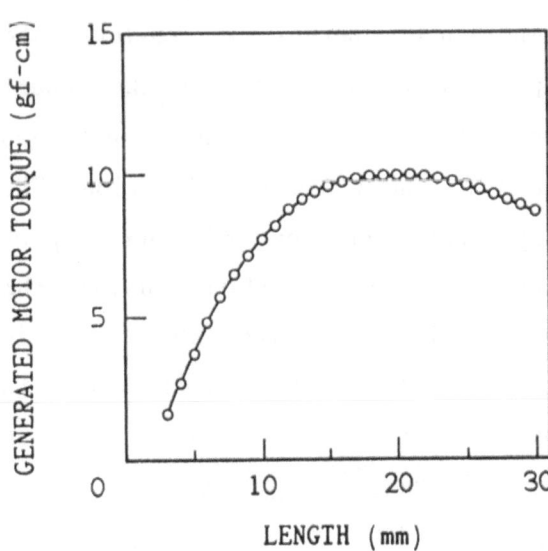

Fig.5 Generated motor torque dependence of the length. The ratio is 1:1.2, thickness is 0.5 mm, the number of the superconductors determined according to the length.

Generated motor torque Tm is obtained by multiplying the torque Ts and the total number n of the superconductors. Though the number is determined by thickness and length, the former is already determined and the latter is determined in the following. *Figure 5* shows the calculation results concerning the influence of length on the generated motor torque. We found that the maximum generated torque is about 10 gf-cm, when the length is 21 mm, the width is 17.5 mm and the number is 210.

Motor Output

The above calculations showed the generated torque under the condition that the temperature of the superconductor was fixed (= 77 K). The motor output was calculated by taking account of temperature changes as follows. We suppose that the heat from the superconductor to the liquid nitrogen flows in the thickness direction. The temperature changes of the superconductor are calculated using the Heisler diagram and the values shown in *Table* I. As a result, temperature T is approximated as a function of cooling time t in the following equation,

$$T = \frac{4.03 \times 10^4}{(t + 0.0537)^{3.62}} + 77 \ [K]. \quad (4)$$

The relationship of the cooling time t, the position θ and the rotating speed N is given by,

$$t = \frac{60}{2\pi} \cdot \frac{\theta}{N} \qquad [s]. \quad (5)$$

The magnetization is 0.25 emu/cm^3 at 77 K, and is 0 at 93 K, as listed in *Table* I. Supposing that it varies linearly between the two temperatures, the repulsive force F while rotating is,

$$F = s \cdot F_0 \qquad [gf], \quad (6)$$

where F_0 is the repulsive force at 77 K and s is the ratio of the magnetization at 77 K and magnetization at between 93 K and 77 K, and changes according to the temperature. The force F is obtained from equation (4), (5) and (6), and the force decreases remarkably as rotating speed increases as shown in *Fig.6*.

Friction on bearings and viscos drag caused by liquid nitrogen are considered as inner loads. Motor torque Tr is obtained by subtracting torque

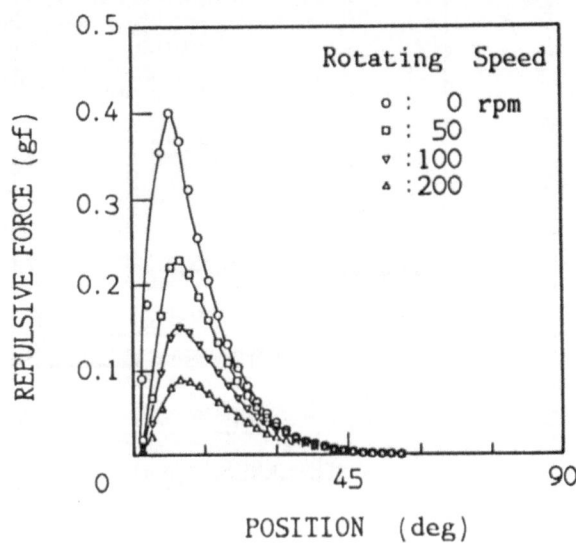

Fig.6 Repulsive force at 0, 50, 100 and 200 rpm. The superconductor is 21×17.5×0.5 mm^3.

caused by these loads from generated motor torque Tm. And the motor output is estimated by multiplying the torque Tr and the rotating speed N. The torque and the output of the motor is shown in *Fig.7*. The torque decreases monotonously as the rotating speed increases, and reaches 0 at 130 rpm. The maximum output is estimated to be 0.027 W at 57 rpm. It is a fairly low output compared with commercial motor, and superconductors with a higher Meissner effect are required to improve the motor output.

The above output showed calculation results when considering the heat flow rate, that is, when the superconductors were naturally cooled down in liquid nitrogen. By ignoring the heat flow rate, that is, assuming that the temperature of the superconductor changes quickly from 93 K to 77 K, the maximum output improves to 0.15 W in the calculation when $s = 1$ in equation (6). Thus it is also important to develop a new motor structure to increase the cooling rate.

CONCLUSION

We have estimated the output of a Meissner Motor by computer simulations. The maximum output of the motor with a radius of 30 mm is simulated at about 0.03 W, which is a fairly low output compared with commercial motors. By ignoring the heat flow rate, the maximum output improves to 0.15 W in the calculation. Superconductors with a higher Meissner effect and a new structure with a faster cooling rate are required in order to improve the motor output.

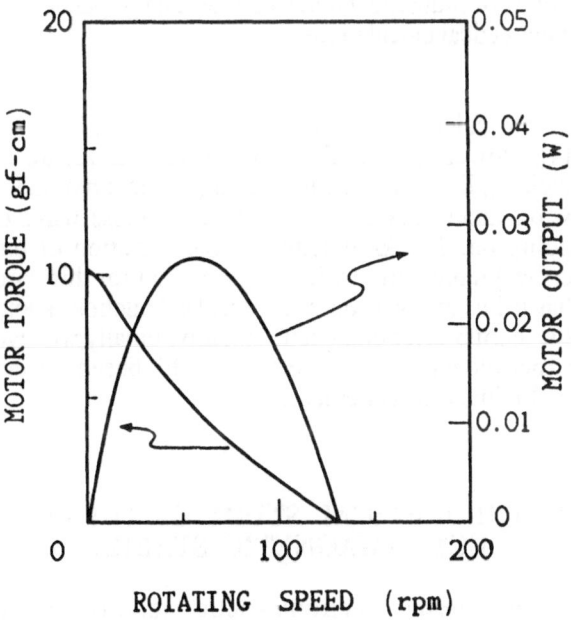

*Fig.*7 Estimated torque and output of the Meissner Motor.

REFERENCES

1. K. Murakami and M. Nemoto , Some Experiments and Considerations on the Behavior of Thermomagnetic Motors. IEEE Trans. on Magn., 1972, MAG 8, 387
2. A. Takeoka, A. Ishikawa, M. Suzuki, Y. Kishi and Y. Kuwano, Fundamental Characteristics of a New Superconducting Motor Called the "Meissner Motor". Jpn. J. Appl. Phys., 1988, Vol.27, No.10, 1875.

THERMO-ELECTROMAGNETIC STABILITY OF ULTRAFINE MULTIFILAMENTARY SUPERCONDUCTING CABLES FOR INDUSTRIAL FREQUENCY USE

A. Guéraud, J.P. Tavergnier, A. Février
Laboratoires de Marcoussis, Centre de Recherches de la CGE, route de Nozay
91460 MARCOUSSIS (France)
Y. Laumond, A. Lacaze
GEC Alsthom - DEA, 3, avenue des Trois Chênes - 90018 BELFORT (France)
B. Dalle, A. Ansart
EDF - DER, 1, avenue du Général de Gaulle - 92141 CLAMART (France).

ABSTRACT

The development from 1983 onwards of Nb-Ti ultrafine multifilamentary wires for 50-60 Hz applications has opened up many interesting perspectives in electrotechnology. The AC losses have been greatly reduced due to decreasing the filament diameter to values of 0.1 to 0.2 µm and by using a highly resistive CuNi matrix with a 30% nickel content. The low thermal conductivity and specific heat of this CuNi matrix and the very high critical current densities induce a very acute problem of thermoelectromagnetic stability. As the electromagnetic diffusivity is much higher than the thermal diffusivity, the stability is governed by an adiabatic criterion. The last experiments about various coils and different assembly configurations of industrial cables give us a large panel of results which permit us to check the validity of the theoretical calculations.

I - INTRODUCTION

The development from 1983 onwards of Nb-Ti ultrafine filamentary and economically viable wires for 50-60 Hz applications has opened up many interesting perspectives in electrotechnology [1]. The AC losses have been greatly reduced due to decreasing the proximity effect, with filament diameters of value from 0.1 to 0.2 µm and by using a highly resistive CuNi matrix with a 30 % nickel content [2].

The thermo-electromagnetic stability of wires to be used in 50-60 Hz applications is a very important problem because of the following reasons :

- The CuNi matrix has a very low thermal conductivity
- Critical current densities are very high in low applied magnetic fields due to the ultra-fine nature of the filaments [3].

The calculations and experiments made last year [4] allowed us to define several wires in regard to their applications. The industrial production of these wires and of several cables made of these wires is going on. The knowledge of the variation of the critical temperature with the induction for this wire has led us to revise our theoretical calculations about the thermo-electromagnetic stability. Finally we give experimental and theoretical results based on the preliminary measurements.

II. THEORETICAL STUDY OF THERMO-ELECTROMAGNETIC STABILITY.

The thermo-electromagnetic stability of the individual filaments at industrial frequencies is an important problem because of the low values of magnetic induction generally used in applications. We typically defined three values from 0.1 to 1 T :

the first value 0.1 T or self induction, corresponds to the fault current limiters [5], the second one 0.5 T is the typical induction of a transformer, the last one, about 1 T, is the value relative to the stator of an alternator. These low values give rise to very high critical current densities, J_c, and thus to large self field effects so that the current distribution within the wire becomes very inhomogeneous. The consequence of this is that the external layers of filaments will tend to reach their critical current before the inner layers transport a significant proportion of the current. This effect is more marked when k, the coupling coefficient between the magnetic induction and the current is low. The detrimental aspect of this effect, from an electromagnetic point of view, is that several saturated layers behave as a massive layer of the same thickness [6].

One of the main characteristics of CuNi is a magnetic diffusivity much higher than the thermal diffusivity. This leads to an adiabatic stability criterion [7].

Referring to the adiabatic stability model of a semi-infinite sheet, the criterion that concerns a multifilamentary wire can be written for stability :

$$N_s \leq N_{sM} = \left(\frac{1}{\delta F}\right)\left[\frac{12}{\pi \mu_0} \frac{C}{J_c \left|\frac{dJ_c}{dT}\right|}\right]^{1/2}$$

where : N_s is the number of saturated layers,
    N_{sM} the maximal number of saturated layers before instabilities can occur
    δF the filament diameter,
    C   the specific heat of the superconducting material,
    J_c the critical current density,
    dJ_c/dT the variation of J_c with respect to temperature.

This expression shows that the electromagnetic stability is improved when the magnetic induction and the specific heat increase and when the temperature gradient of the critical current. Moreover, a wire will be completely stable if N_{sM} is higher than the number of layers of filaments in the wire.

To calculate the number of saturated layers (N_{sM}) we need to know the variation of the specific heat and of the critical current temperature with the induction.

The critical current density is specific to each wire, which is not the case of the specific heat. Figures 1 and 2 show the variation of this parameter and of the critical temperature.

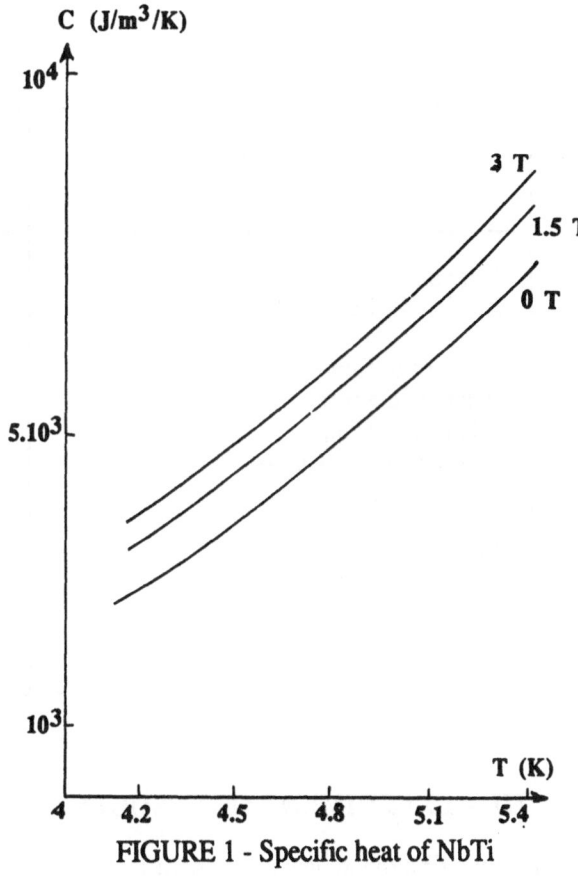

FIGURE 1 - Specific heat of NbTi

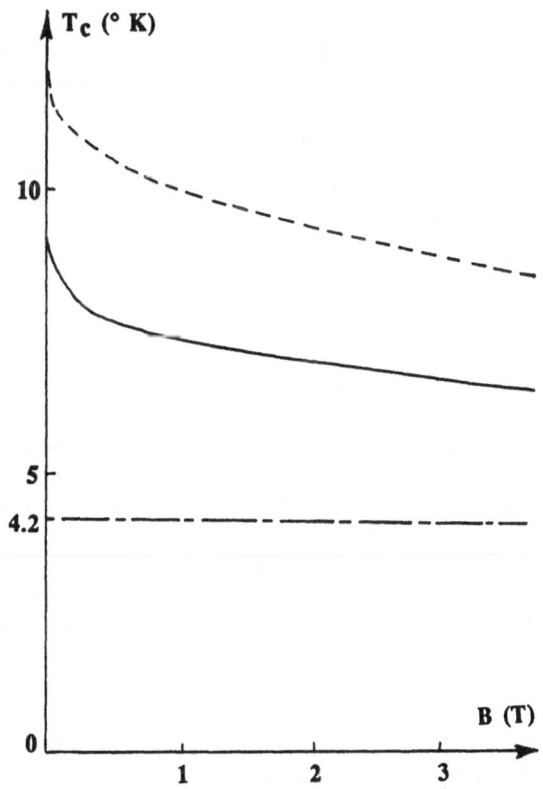

FIGURE 2 - Variation of the critical temperature with the induction for filaments having a diameter of :
* 5 μm (---)
* 0,13 μm (——)

Conductor	AC 87/4 A	AC 88/S1
Wire diameter	0.247 mm	0.3 mm
Insulated wire diameter	0.29 mm	0.36 mm
Twist pich	0.96 mm	1.5 mm
Number of filaments	299 880	920 304
Filament diameter	0.2 μm	0.136 μm
Filament spacing	0.15 μm	0.123 μm
Filamentary zone diam. :		
outer	232 μm	277 μm
inner	48 μm	40 μm
Components		
superconductor	23.1 %	19.1 %
Cu-30 wt % Ni	74 %	79.6 %
Cu	2.9 %	1.3 %

III - EXPERIMENTAL RESULTS

Since the first theoretical and experimental results obtained last year, we performed all the measurements on a wire (AC 87/4 A) and a cable made of six wires twisted around a central CuNi massive filament (6+1). We are now obtaining the first experimental results on a new collection of wires called AC 88. All the characteristics of these wires are presented in Table 1.

a) Results concerning the wire AC 87/4 A.

In Figure 3, we present, in bias field, the comparison between theoretical and experimental results for the short wire sample. We have represented the experimental results by a bar limited by the highest and the lowest values obtained for each induction. From a theoretical point of view, we observe a good stability for an induction higher than 0.9 T. We notice that the theoretical results are closed to the highest experimental values. If we only retain the best experimental results we see the good agreement between theory and experiments. The dispersion of the results is due to the fact that it is impossible to increase slowly and regularly the current intensity in the short wire sample. Then some aleatories phenomena appear and show the sensibility of wires in regard to magnetic disturbances.

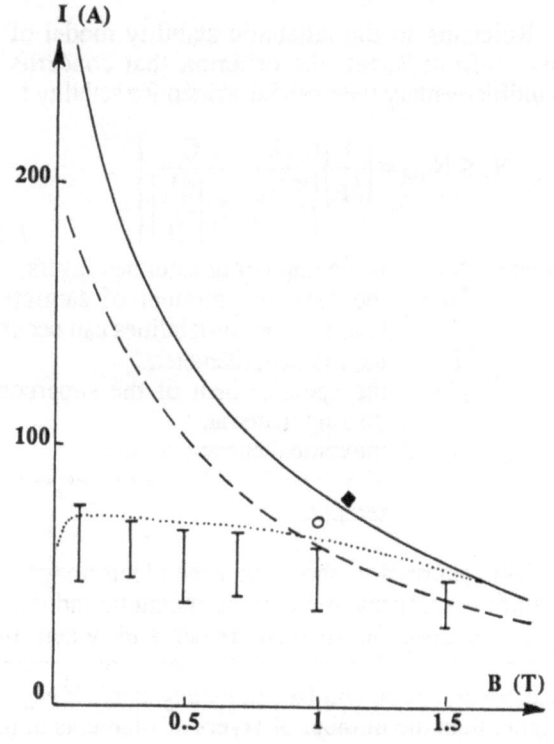

FIGURE 3 - Critical current of the wire AC 87/4 A at 4.2 K

———	DC critical current 1 μV/cm
-------	DC critical current 0.1 μV/cm
.....	AC theoretical values
I	AC Experimental values
♦, o	DC and AC quench current of a coil

We have also calculated and measured the stability in a coil. The main difference of behaviour between a short wire sample and a coil is the current distribution between the layers of filaments. The calculated results show clearly that the current progressively penetrates the wire from the surface, saturating successive layers of filaments. The current is shared between more layers of filaments as the value of B increases ; this is because the time varying transverse field component acting on the wire forces the current to penetrate towards the center of the wire in order to ensure a flux-balance between the filaments of different layers. Thus, the presence of the time varying transverse field leads to more of the layers carrying currents which are a fraction of their critical currents. This effect improves the stability of the wire as the total transport current is now higher before the instability limit is reached in the wire. We can obviously see that in figure 3, where the AC quench current of a coil is higher than the one of the short wire sample, at the same induction. The small difference between the AC and DC quench currents is due to a little increase of the temperature of the winding.

Concerning the wire AC 87/4 A, we also made calculations and experiments for a (6+1) cable. Experimental measurements of the AC and DC critical currents were obtained as a function of the field-current coupling coefficient, k, where $B = kI$ by progressively reducing the number of layers of windings from 12 to 2. The major difficulty during the experimentation was the mechanical instability. The cable is an assembly of wires and the coils were not impregnated, so that little movements are possible and sufficient for the winding to quench. So we need to remark that the DC theoretical performances were not reached because of an interminable training. For the 6 layers coil, we observe after 41 quenches that, the divergence between theory and experiment was small (74 A for 78 A). Consequently, the electromagnetic stability was unavailable.

We finally made measurements on an epoxy-impregnated coil. The problem of the mechanical stability disappeared but the heat transfer was an obstacle to keep the bath temperature (4.2 K) in the windings. Moreover, this wire had a relatively high level of losses so that the rise of temperature in the winding is important (about 1K) and leads to a decrease of the critical current density. The quench current value calculated at 5.4 K (250 A) corresponds to the experimental result of this coil (252 A). We can deduce from this, that the coil was magnetically stable at this temperature.

In conclusion for this wire, we can say that the theoretical approach of electro-magnetic stability is correct. When the current is coupled with the induction, the wire is stable at 4.2 K for magnetic induction higher than 0.5 T. If we are able to maintain the temperature close to 4.2 K, which is possible by using some helium circulation channels, the AC quench current is very close to the DC value.

The main problem to solve is the lack of mechanical stability. Impregnation can be a solution. Due to the bad coefficient of heat transfer of the epoxy, we must have good cooling conditions of the winding (channels, ...).

b) Results concerning the wire AC 88/S1

All the characteristics of this wire are given in Table 1. The definition of this wire resulted from all the calculations and experiments on the previous wires. The losses are as low as the best performances obtained until now. The diameter and the number of filaments of the wire have been computed in relation with the inductions corresponding to several possible applications in order to have an electromagnetically stable, wire.

The calculation and the results of the first experiments about losses are described in another paper [8] presented at this conference The measured dependency of the D.C. critical current as a function of magnetic field is shown in Figure 4. We obtained the DC critical current for three values of the temperature, 4.2-4.8 and 5.4 K. Verifying that the temperature increases linearly when the critical current decreases, we obtained the critical temperature in relation with the induction. Until now, the computation of the stability used some values measured on filaments with a diameter of about 5 µm. But the filaments now have a diameter of about 0.1 µm. We have reported in the Figure 2 the previous and the new values and we can see that the difference is large. It means that all our calculations of the electromagnetic stability at 4.2 K were optimistic.

The safety margin is reduced by more than 50 % at 1.5 T. Then, it seems difficult with a good value of the overal critical transport current density to improve the stability by increasing the temperature beyond 5.4 K in all the winding.

We have also presented in Figure 4, the experimental measurements of the AC critical currents obtained first with the short wire sample in bias field, then with a coil. We notice that :

- On the contrary of the other wire shown in this paper, we can see that the theoretical results are close to the lowest experimental values

- A better penetration of the current in the successive layers of filaments improves the stability of the wire, which we clearly observed last year with another conductor [4]

- The comparison with theoretical calculations obtained with the new values of the critical temperature shows that the experimental trend follows the theoretical prediction .

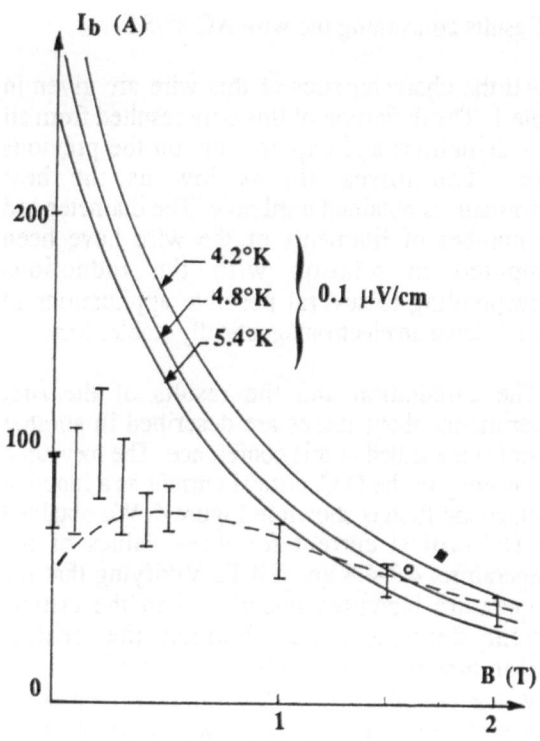

FIGURE 4 - DC critical current at 0.1 μV/cm as a function of induction and temperature.

- AC critical current at 4.2 K for :

. a short wire sample in bias field

     ---   Theoretical values

     I   Experiment values

. a coil (8 layers) :

     ♦ DC quench current (65)
     O AC quench current (58,5)

CONCLUSIONS

On the one hand, measurements on short wire samples and wires are working well. On the other hand, the experiments on cable about thermo-electromagnetic stability are linked to the mechanical stability of the conductor and to the possibility to have a low increase of the temperature in the critical point of the winding : in effect, we saw that a small increase of the temperature in the case of ultrafine multifilamentary wires could lead to a severe deterioration of performances.

Last year measurements and this year's first results on the AC 88 wires allow us to have a good prediction of the electromagnetic stability behaviour of these industrial frequency wires, in particular with the new values of the critical temperature varying with the induction. More work must be done on the mechanical stability of wires and cables and are in progress now.

REFERENCES

[1] P. Dubots, A. Février, J.C. Renard, J.C. Goyer and Hoang Gi Ky - Behaviour of multifilaments under A.C. magnetic fields. Journal de Physique. sup. 1 (1984) pp. 467-470.

[2] J.R. Cave, A. Février, T. Verhaege, A. Lacaze, Y. Laumond - Reduction of AC losses in ultra-fine multifilamentary NbTi wires. Applied Superconductivity Conference. San Francisco (1988).

[3] P. Dubots, A. Février and Hoang Gia Ky - Specific properties of NbTi ultra-fine filaments wires. International Symposium on flux pinning and electromagnetic properties in superconductors (1985), Fukuoka, Japan.

[4] A. Février, A. Guéraud, J.P. Tavergnier - Thermo-electromagnetic stability of ultra-fine multifilamentary superconducting wires for 50-60 Hz uses. Applied Superconductivy Conference, San Francisco (1988).

[5] A. Février, T. Verhaege, J.P. Tavergnier, Y. Laumond, M. Bekhaled, M. Collet, V.D. Pham - 25 kV superconducting fault current limiter. This conference.

[6]   A. Février - Losses in a twisted
       multifilamentary superconducting composite
       submitted to any space and time variations of
       the electromagnetic surrounding. Cryogenics
       23, p. 185 (1983).

[7]   J.L. Duchateau - ·Les instabilités
       magnétiques dans les composites
       supraconducteurs multifilamentaires. Thèse
       Orsay (1975).

[8]   J.Y. Georges, A. Février, Y. Laumond,
       A. Lacaze and P. Bonnet. - Performances of
       industrial superconducting wires for 250-
       60 Hz applications. This conference.

PERFORMANCES OF INDUSTRIAL SUPERCONDUCTING WIRES FOR 50-60 Hz APPLICATIONS*

J.Y. Georges, A. Février
Laboratoires de Marcoussis, Centre de Recherches de la CGE, route de Nozay
91460 MARCOUSSIS (France)
Y. Laumond, A. Lacaze, P. Bonnet
GEC Alsthom - DEA, 3, avenue des Trois Chênes - 90018 BELFORT (France)

ABSTRACT

Multifilamentary superconducting wires with a greatly reduced level of losses have been produced in lengths of several tens of kilometers. In spite of the reduction of the filament diameter, proximity effects are avoided and we make the best possible use of the reversible motion of the flux lines, so that the hysteretic losses are lower. The reduction of losses due to induced currents can be obtained by choosing a twist pitch as short as possible. These concepts lead us to realize, first at a small scale, then at an industrial scale, conductors comprising filaments of Nb-Ti with a diameter of 0.1 to 0.2 μm and a highly resistive CuNi matrix. The twist pitch of these filaments is four times the diameter of the conductor. Some coils have been tested at the frequency of 50 Hz and show very reduced losses.

INTRODUCTION

For multifilamentary superconducting wires to be industrially competitive in AC applications, the AC losses must be lower than that of conventional conductors, such as copper, taking into account refrigeration efficiencies (1 Watt at 4.2 K is equivalent to 500-1000 Watts at room temperature). Based on a full treatment of loss calculations, the loss contributions occurring in a superconducting wire can be expressed approximately as : [1]

$$P = \alpha \, dF J_{cov} \left(1 + \left(J_{ov}/J_{cov}\right)^2\right)\left|\frac{dB_t}{dt}\right| + \beta \frac{P^2}{\rho_t}\left|\frac{dB_t}{d_t}\right|^2$$

where α and β are wire dependent constants, dF is the filament diameter, J_{ov} the overall current density, J_{cov} the overall critical current density, p the

*This work was partly funded by Electricité de France

filament twist pitch, ρ_t the transverse resistivity. The first term represents the hysteretic losses in fully penetrated filaments using the Bean model, and the second term represents the eddy current coupling loss.

Thus, it can be seen that a lowering of the losses will be obtained by reducing the filament diameter, decreasing the twist pitch and increasing the matrix resistivity. These considerations have led to the development by GEC ALSTHOM and the Laboratoires de Marcoussis, since 1983, of long lengths of sub-micron multifilamentary wire. In attempts to further reduce the losses by reducing the filaments diameter it was observed that proximity coupling of closely spaced filaments led to an increase in losses. [2]

In order to better define the possible applications to industrial power systems, it is absolutely necessary to investigate in a quite precise way the losses in submicronics filaments. For this preliminary study on hysteretic losses, we propose to approach them by the determination of an effective filament diameter, dF*, in the BEAN model calculation dF* = F (wire, $\vec{B}$) .dF where F will be a phenomenological law.

ULTRA-FINE FILAMENTS : CHARACTERISTIC AND THEORETICAL APPROACH OF F (wire, $\vec{B}$) [3, 4, 5]

Proximity effects

Proximity coupling becomes important when the distance between filaments becomes comparable to the coherence length of the normal metal, ξ_n. An estimate of the value of ξ_n for Cu-30 % Ni is thus

$\cong 15$ nm at 4.2 K.

The consequence of proximity effects on losses is an increase in the effective filament diameter and a decrease in the transverse resistivity.

Reduction of these effects can be obtained by increasing the filament spacing or by reducing ξ_n, either by increasing the matrix resistivity or by the addition of magnetic impurities such as manganese.

When this effect becomes insignificant, losses decrease faster than filament diameters. This very favorable effect is attributed to reversible flux-line motion.

Many studies of proximity effects have been made, and a general feature is an exponential dependence on both normal layer thickness and field.

Consequently we approximate these effects by :

$$F_1 (wire, \vec{B}) = Z_1 Z_2 \left(1 - \exp - \left(\frac{\alpha}{B}\right)\right)$$

where terms Z_1 and Z_2 represents these effects in the first and second stacks

$$Z_1 = 1 + \left(NCP1 \left(1 + D_1/dF\right) - 1\right) \exp - \left(\frac{\left(D_1 - 2\xi_n\right)^2}{2\xi_n D_1}\right)$$

$$Z_2 = 1 + \left(NCP2 \left(1 + \frac{D_2}{NCP1*dF}\right) - 1\right) \exp - \left(\frac{\left(D_2 - 2\xi_n\right)^2}{2\xi_n D_2}\right)$$

Where NCP_1 and NCP_2 respectively represent the layer numbers of the ranges 1 and 2.

D_1 = distance between filaments
D_2 = distance between stacks of filaments
α = constant linked with the superconducting composite and with the magnetic history of the superconducting material.

It can be noticed that for a high value of ξ_n, the coupling phenomenon becomes important $(Z_1 \sim NCP1 (1 + D_1/dF))$; in return, for ξ_n close to zero, there is no longer a proximity effect $(F_1 = 1)$. The term $\left(1 - \exp - \left(\frac{\alpha}{B}\right)\right)$ keeps also the limit conditions on the magnetic field so that the proximity effects are favoured at a low field and destroyed at a high field.

The proximity effects are very closely linked to the interfilamentary distance D_1 and to the distance between the stacks D_2. For important distances between filaments, in the order of ten times ξ_n, these proximity pinning effects no longer exist. On the contrary, for values of $3.\xi_n$, the proximity effects become important.

Reversible flux-line motion.

We can simply explain in a qualitative way the interaction of vortex lattice with the pinning sites.

Two pinning interactions exist : on the one hand a surface pinning interaction created by the transition between the normal zone and the superconducting zone ; on the other hand a "volumic" pinning interaction created by the inner metallurgical structural defects.

For fine filaments, surface pinning interaction will be strengthened with regard to the volumic pinning interaction. In view of a positive field variation, there is a vortex creation from the relation $\frac{dn}{dt} = \frac{1}{\phi_0} \frac{dB}{dt}$ this vortex creation form can be attributed to a thrust force. In the intersection normal-superconductor zone, the vortex movement is slow and quasi-reversible, that means the majority of the potential energy supplied during the penetration phasis will be recovered during the depenetration. In the volumic pinning interaction zone, the movements are abrupt and accompanied by damped vibratory movements. In practice, it is desirable to favour the surface pinning interaction. We search for a better reversible flux-line motion.

with regard to AC losses, in producing the most regular ultra-fine filaments.

The surface pinning is dependent on the filament apparent diameter, dF, and on the magnetic field, with an exponential law.

We defined the filament effective diameter by $dF^* = dF + 2\,\xi_n$ because there is an extension of the superconducting zone in the normal zone on a depth ξ_n.

We approximate this effect by :

$$F_2\,(\text{wire}, \vec{B}) = 1 - \exp - \left(\beta\,F_1\,dF^*\right)^2 \sqrt{B}$$

where β is a constant linked with the superconducting composite and with the magnetic history of the superconducting material.

The function F_2 respects the limit conditions :

- tends to 1 for high magnetic field
- tends to 0 for low filament diameter.

$F\,(\text{wire}, \vec{B})$ function

Taking into account the increase of hysteretic losses due to the filament effective diameter dF^* and to the fact that the critical current density is higher at the filament surface we take in this preliminary study for the function $F\,(\text{wire}, \vec{B})$

$$F\,(\text{wire}, \vec{B}) = (1 + A)\,F_1\,F_2\left(1 + 2\,\xi_n/dF\right)$$

The term A lower than one, will be chosen equal to O for large values of dF and for filament diameters lower than twice the penetration length, λ. Near $dF = 4\,\lambda$, we take

$$A \cong 1 - \frac{2\,\lambda}{dF}.$$

sample	Filament number	Fil. diam wire diam.	% superc.	% Cu	NCP1	NCP2	D_1	D_2
AC 87/3 A	242892	875.10^{-6}	18.6	2	17	10	0.99 dF	4.2 dF
AC 87/3 B	965320	437.10^{-6}	18.5	2,6	17	16	0.88 dF	4.2 dF
AC 88/S1	920304	$455\ 10^{-6}$	19.1	1.3	16	17	0.9 dF	3.18 dF

Table 1 - Characteristics of wires AC 87/3A and 87/3B and AC 88/S1

EXPERIMENTAL DETERMINATION OF F (wire, $\vec{B}$)

Different values of F are obtained by the ratio of the critical current density obtained by magnetic measurement, J_{cm}, over the transport critical current density, J_c : $F\,(\text{wire}, \vec{B}) = J_{cm}/J_c = \dfrac{dF^*}{dF}$

- Critical current measurements from the electric method.[6]

This method is based on the measurement of the conductor characteristic (V.I) at a given temperature T, with a known excitation field H.

From the critical current measurement, $I_c\,(H,T)$, and taking into account the self field effect, we defined a model which takes into account :

- the average local magnetic induction transverse to the filament.
- the form of the critical current density chosen at a low field : KIM model.

From deconvolution of this critical current $I_c\,(H, T)$ we obtain the transport critical current density normalized to the filament section on the superconducting zone. Btrc is a value for which the self field becomes prevalent compared to the external magnetic field. For field values lower than Btrc, a good critical current density approximation is given by the KIM model (Figure 1), such as l

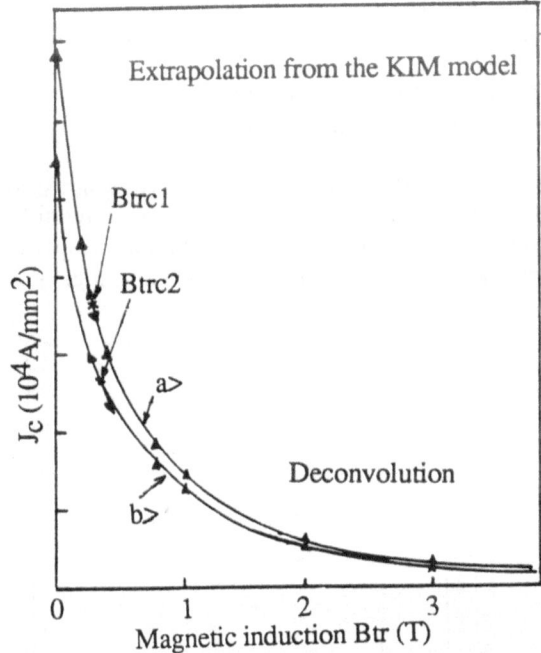

FIGURE 1 - Critical current density calculation. For a non-twisted wire at 4.2 K. The first part gives J_c by deconvolution of the critical current I_c (B,T), for high field values until the field limit equals Btrc. In the second part, we applied the KIM model for the magnetic field lower than Btrc.

a>Df = 105 nm for AC87/3A serial
b>Df = 105 nm for AC87/3B serial

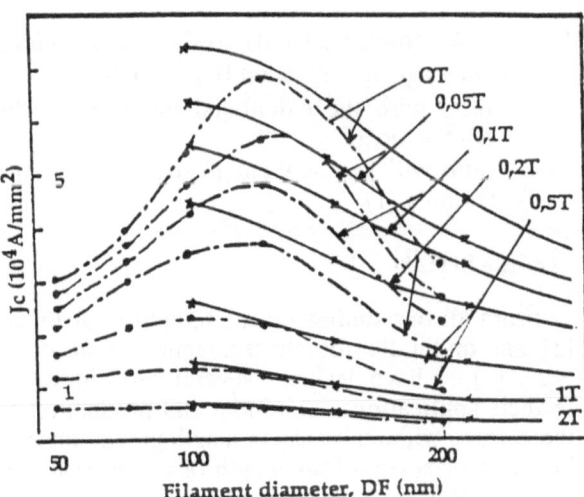

FIGURE 2 - Critical current density as a function of the filament diameters given at different values of the local magnetic field to the filament. The continuous lines agree with the serial AC 87/3A, the discontinuous lines with the AC 87/3B.

$$J_c = \frac{\alpha}{B + B_o} \qquad O \leq B \leq Btrc$$

where α and B_o are material constants.

We calculate, from the above-mentioned expression, the critical current density J_c as a function of the local magnetic induction for the AC 87/3 A and 3 B series, showed in Figure 2. Characteristics of these wires are given in Table 1.

There is a maximum in J_c for filament diameters of about 100 nm for the AC 87/3 A and of about 130 nm for the 3 B series.

- Critical current density measurement from the magnetization method.[7]

This well known method allows to eliminate the self field. When the variation of magnetic induction with time is slow enough, the magnetization due to eddy currents through the matrix becomes insignificant and the observed behaviour is due to the superconducting filaments.

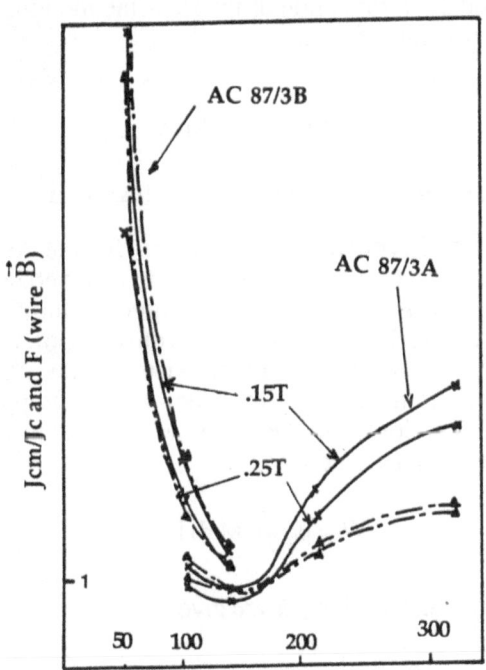

FIGURE 3 - Theoretical and experimental results for F (wire, $\vec{B}$) with B = 0.15 T and 0.25 T.

—— Experimental curves with J_c magn/J_c

---- Theoretical curves with the function F (wire, $\vec{B}$)

The resolution of the Maxwell-Bean equation $\overrightarrow{\text{curl}}\,\vec{B} = \mu_o \vec{J}_c$ leads to : $M\uparrow - M\downarrow \cong \alpha.dF.J_c\,(\vec{B})$ for a transverse magnetic induction.

- Results

Figure 3 shows theoretical and experimental results for the function F (wire, $\vec{B}$) at two magnetic induction values (0.15 T, 0.25 T) different values of the filaments diameter of the wire AC 87/3A (324 nm, 210 nm, 131 nm, 105 nm) and AC 87/3B (131 nm, 105 nm, 65 nm, 52 nm).

These different values of F (wire, $\vec{B}$), give us useful informations on discrepancies between observed behaviour and Bean model predictions with ultra-fine filaments. For a ratio inferior to one, there is a prevalence of the flux line motions. When the ratio is higher than 1., we are in presence of proximity effects. It is difficult to separate these two effects which always exist but which take more importance for low values of the field. Those first theoretical and experimental agreements confirm the physical interpretation of the flux-line motions and the proximity effects.

AC LOSSES

- Hysteretic losses with no transport current.

At different filament diameters, for the AC 87 series, we obtain the losses from magnetization curves for the maximum magnetic induction $B_{max} = 0.34$ T. The samples consisted of many short (4 mm) lengths of untwisted wires placed transverse to the low frequency (0.4 Hz) applied field. The losses in the wires were obtained from cycled magnetization loops :

$$W = \mu_o \oint M\,dH$$

The calculated losses are given by :

$$W = \int F\,(\text{wire}, \vec{B})\,.\,P\,(\text{Bean})\,.\,dt \qquad (1)$$

The F (wire, $\vec{B}$) function uses the parameters values :
$\alpha = 0.13$, $2\xi_n = 2.16\ 10^{-8}$ m, $\beta = 5.5\ 10^6.$

In Figure 4 for the wires AC 87/3A and 3B series we present :

- the ratio of the experimental losses by magnetization over the calculated losses from the BEAN model
- the ratio of the experimental losses over the new calculated losses by the formula (1).

These results confirm that we have progressed in the right direction so far, and we are now able to calculate hysteretic losses within about 10 per cent for a ratio of filament diameter over spacings of 1.

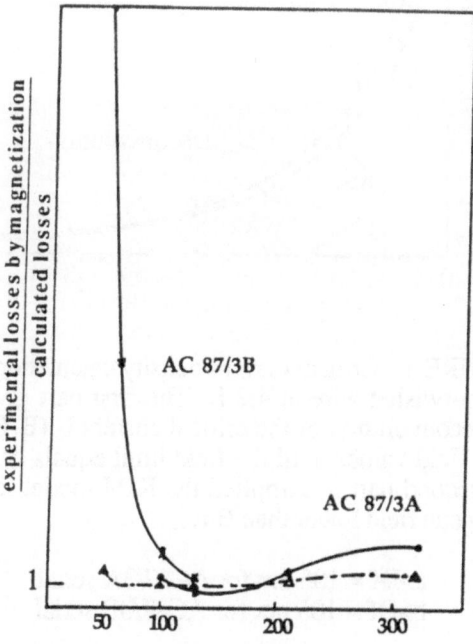

FIGURE 4 - Theoretical and experimental ratio for a maximum magnetic induction $B_{max} = 0.34$ T.
— ratio with the calculated losses from the BEAN model
---- ratio with the new calculated losses by the formula (1).

- Losses in coils

From all our studies on the reduction of losses [3], and on the thermo-electromagnetic stability of wires [8] we have designed several wires in regard to their applications. The industrial production of these wires without breaks in length greater than 10 kms and of several kms length of cables made of these wires is going on.

In Figure 5, we present :
. the experimental values of losses in a small coil (inner diameter = 2.5 cm, outer diameter = 3.6 cm, length = 9 cm, K = B/I = 2.55 10^{-2} T/A, winding : 200 m of AC 88/S1 wire)
. the theoretical values of losses obtained by the classical BEAN model calculation .

We can observe that the theoretical values are high than the experimental ones. The lower the induction, the higher the relative difference. This confirms that proximity effects are now negligible. At 1 T we got 34 kW/m^3 with an overal current density of 550 A/mm^2 and an overal critical current density of 1650 A/mm^2. These values obtained on an industrial scale manufactured wire is close to those obtained last year on a laboratory scale.

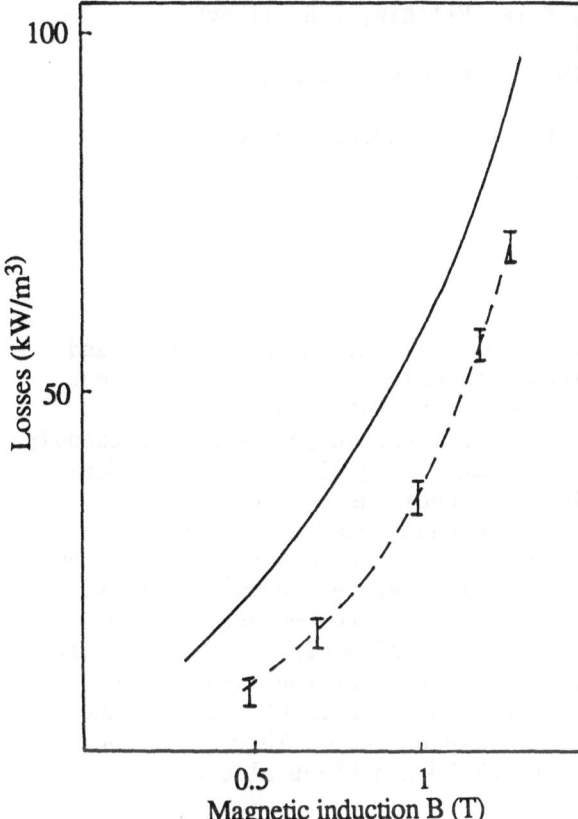

FIGURE 5 - Losses of 8 layers coil made with AC 88/S1

—— Theoretical values

I Experimental values

CONCLUSION

These results show that the discrepancies between calculated losses and observed values for filament diameters below ~ 400 nm is interpreted as due to proximity effects and to flux line-filament size effects. Our phenomenological calculation agree well for hysteretic losses in magnetization measurement. In winding conditions the wire is submitted to transport currents and to an axial component of the magnetic induction, consequently we must take into account the distribution of the currents between the layers of superconducting filaments. Now that we have successfully completed the first step, our future work will be to take into account all of these effects on an industrial scale. Nevertheless our new industrial wires (operating at a third of J_{cov}) present losses at least one order of magnitude lower than for copper in typical applications : stator at 1 T ; superconductor ~ 2 10^{-5} W/Am, copper (15A/mm^2) - 2 10^{-4} W/Am (using a refrigeration factor of 500).

AKNOWLEDGEMENT

We thank C. Agnoux, J.P. Tavergnier, F. Saint-Saens and the Group at GEC ALSTHOM for heir technical support.

BIBLIOGRAPHY

[1] A. Février - Losses in a twisted multifilamentary superconductive composite submitted to any space and time variations of the electromagnetic surrounding. Cryogenics, 23 p. 185 (1983).

[2] P. Dubots, A. Février, J.C. Renard, J.P. Tavergnier, J. Goyer and H.G. Ky - NbTi wires with ultra-fine filaments for 50-60 Hz use : influence of the filament diameter upon losses. IEEE Trans-Mag 21, p. 177 (1985).

[3] J.R. Cave, A. Février, T. Verhaege - Reduction of AC losses in ultra-fine multifilamentary NbTi wires. IEEE Trans-Mag 25 n° 2, p. 1945 (1989).

[4] S. Takacs and A.M. Campbell - Hysteresis losses in superconductors with very fine filaments. Supercon. Sci. Technol. 1 p. 53, (1988).

[5] A.K. Ghosh and W.B. Sampson, "Anomalous low field magnetization in fine filament NbTi composites. IEEE Trans-Mag.24 n° 2, p. 1145 (1988).

[6] A. Février, J.C. Renard - Critical current density in multifilamentary composites. Advance in Cryogenic Engineering 24, p. 363 (1977).

[7] M.N. Wilson - Superconducting magnets. Clarendon Press Oxford (1983).

[8] A. Guéraud, J.P. Tavergnier, A. Février, Y. Laumond, A. Lacaze, B. Dalle, A. Ansart. Thermo-electromagnetic stability of ultrafine multifilamentary superconducting cables for industrial frequency use. This Conference.

INFLUENCE OF THERMOMECHANICAL WORKING SCHEDULES ON CRITICAL CURRENT DENSITY AND STRUCTURE OF SUPERCONDUCTOR Nb-Ti BASE ALLOYS

G.K. Zelenskiy, A.V. Arsent'ev, A.P. Golub', E.A. Golub', V.E. Klepatsky,
V.V. Medkov, E.V. Nikulenkov, A.D. Nikulin, V.Ya. Fil'kin, V.S. Titov*,
P.P. Pashkov*, V.A. Vasil'ev**, A.I. Nikulin**
A.A. Bochvar's All Union Scientific Research Institute of Inorganic
Materials, Moscow, USSR.
*All Union Scientific Research Institute of Electromechanics, Moscow, USSR.
**Institute of High Energy Physics, Serpuhov, USSR.

ABSTRACT

Results are presented of staged determination of critical current density and structure of superconductor HT-50 and HT-55 wires unannealed and subjected to four intermediate anneals at 400°C. The wires produced without intermediate anneals have a single phase structure. With an increase of deformation jc grows monotonously, the size of β-matrix subgrains is reduced. The maximum jc equal to 0.48×10^5 A/cm^2 in the field of 5T is reached in the HT-55 alloy. In specimens subjected to intermediate anneals the first anneal leads to a more than order of magnitude increase of jc which is related to the α-phase precipition along the boundaries in the HT-50 alloy and also within the subgrains in the HT-55 alloy. The subsequent deformation cycles effect an increase of jc, a decrease of the sizes of the α-phase and the β-matrix subgrains while intermediate anneals have an opposite effect. A non-uniform precipitation of the α-phase is noted across the filaments. At the final stages of deformation a drastic increase of jc to 3.2×10^5 A/cm^2 and to 3.9×10^5 A/cm^2 in the field of 5T is observed in HT-50 and HT-55 alloys, respectively. There is a common dependence of jc on the β-matrix subgrain size for HT-50 and HT-55 alloys.

INTRODUCTION

Advances in the superconductor wire manufacturing processes and, in particular, the introduction of a niobium diffusion barrier between copper and niobium-titanum alloy lead to a considerable increase of the current carrying capacity [1,2]. Progress in process developments made it possible to extend the temperature range, to increase the time and number of intermediate anneals in the superconductor production [3,4]. The correlation between the structural states and the wire critical current densities (jc) indicates the importance of the β-matrix subgrain sizes, the amount and morphology of α-precipitates [5-8]. However, for the rare excepition in a majority of papers no information is available on structural and jc changes during wire manufacture and this does not allow a purposeful achievement of the required structural state, corresponding to a high current carrying capacity of a material [9]. This work is an attempt in this direction.

SUPERCONDUCTOR WIRE INVESTIGATIONS

Using the traditional methods of extrusion and cold work two assemblages were fabricated from superconductor alloys HT-50 (Nb-48 mass. % Ti) and HT-55 (Nb-55 mass. % Ti) in a copper matrix. The assemblages comprised several tens of strands, each containig 55 superconductor filaments. The filaments are surrounded with a niobium diffusion barrier that prevents the interaction between the Nb-Ti alloy and copper during manufacture [1, 10]. To make staged measurements of the critical current and perform structural investigations the strands were removed from an extruded bar or at different stages of a wire manufacture.

The investigations were carried on using specimens received with four intermediate heat treatments (IHT) at 400^0 C and of a 24 h duration and without anneals. The Nb-Ti filamens without copper and the Nb barrier were investigated in the X-ray diffractometer ДРОН-3 at the continuous motion of a counter using a carbon monochromator at the inlet (Cr K_α) or reflected (Cu K_α, Cu K_β) beams and at the filtered (Zr) radiation, Mo K_α without a mono - chromator. During the analysis the intensities of the diffraction maxima $(100)\alpha$, $(110)\alpha$ and $(200)\beta$ measured from which the amount of the phases present could be estimated. The fine structure of the Nb-Ti filaments was studied by the method of transmission electron microscopy

TABLE 1.

Characteristics of HT-50 (numerator) and HT-55 (denominator) wire spesimen

Condition	β-subgrain size, nm	$Jc \times 10^4$ A/cm², 5T.	Intensity		
			$(100)\alpha$	$(110)\alpha$	$(200)\beta$
Cold worked prior (CWP) to 1-st intermediate anneal (IA)	165/173	0.2/0.1	0/0	0/0	45/65
After intermediate anneal (AIA) N 1	190/208	1.9/7.8*	-/6	-/5	-/63
CWP to 2-nd IA	119/137	3.7/14*	0/0	1.7/3	41/61
AIA N 2	131/140	3.0/8.8*	1/1.5	1.4/1	39/46
CWP to 3-d IA	80/79	7.8/11	0/0	2.5/5	45/45
AIA N 3	95/108	5.8/7.3	1/0.4	1/5	31/40
CWP to 4-th IA	78/53	8.4/14	0/0	3/6	48/63
AIA N 4	108/89	4.6/5.8	0/0.5	3/6	48/41
Final drawing strain:					
$(\mathcal{E}=2 \ln(D_{initial}/D_{final})$					
$\mathcal{E}=0.585$	67/80	-/8.5	-/0	-/4	-/76
$\mathcal{E}=1.300$	56/53	12/13	0/0	2.5/7	67/72
$\mathcal{E}=1.970$	49/46	16/17	0/0	2.2/5	59/55
$\mathcal{E}=2.630$		22/27			
$\mathcal{E}=3.040$		29/36			
$\mathcal{E}=3.800$	39/36	32/39	0/0	2/2	61/55

Note: *The α-phase is along the boundaries and within of the subgrain of the β -matrix

using light and dark field (in the α-phase reflections) images and microdiffraction [11]. The arithmetic mean values of the β-matrix subgrain width were determined by the method of secants normal to the drawing axis in 30-125 points of measurements. Jc was measured at 4.2 K in a longitudinal electric field Eo=1 μV/cm. The critical current density was determined using only Nb-Ti alloy.

Table 1 lists the results of the investigations into the HT-50 and HT-55 superconductor alloy speciments. The extruded NbTi filaments reveal a structure of slightly elongated (1:d=3:1) differently oriented blocks of β-grains. At a considerable scatter in their cross-sectional sizes (100-1000 nm) the anerage values are 500 nm and 680 nm for HT-50 and HT-55 alloys, respectincly. After drawing (strain ε =1.02) the β-grain size is reduced by a factor of three, the deformtion texture is revealed with the [110] direction along the drawing axis, the values of jc are at level of 2×10^3 A/cm^2. With an increase in the deformation the β-matrix subgrain width is

monotomensly decreased and jc reaches 0.5×10^5 A/cm^2 (Fig.1). No α-phase precipitates were identified.

As compared to the drawing state the first IHT effects a more than an order of magnitude growth of jc. The subgrins became wider, in the HT-55 alloy the α-particles precipitate essentially along the boundaries, but are also noted within the subgrain, in the HT-50 alloy they are arranged along the subgrain boudaries only. The presence of the α-phase is evidenced by the microdiffractions and diffraction maxima (100)α and (110)α on X-ray patterns (Fig.2, table 1). If the relation between the intensity of (200)β to that of (100)α or (110)α can describe the volume content of the α-phase, that the amount of the α-phase is more in the HT-55 than in the HT-50 (Table 1).

At the second stage of the thermo-mechanical working (TMW) the cold deformation by drawing reduces the cross-sectional size of subbands and imparts an elongated shape to the α-particles, jc is increased. In this case the X-ray patterns show of the (110)α-phase. IHT of a drawing wire having a corresponding diameter lead to a reduction of jc. The mode of structual changes identified at the subsequent stages of TMW is generally retaind. A substantial increase in the α-phase content is observed after the third and fourth cycles of TMW. Following a 24h anneal at 400^0 C of the (100) α-phase is recorded in the HT-50 and HT-55 alloys up to the third and up to and including the fourth anneal, respenctively. It is likely that on particular TMW all the α-phase has not yet fully precipitated from the HT-55 alloy. An increase of the average cross-sectional size of the subgrains is less significant after the third and fourth anneals, as some subbands decorated with the α-particles retain their dimensions. The jc are higher in the HT-55 speciments than in the HT-50 ones in similiar states. The most drastic change in the jc values depending on the amount of deformation was noted at the final deformation stage after the fourth heat treatment (Fig.1). The subgrain size is reduced, the α-particles become finer, the density of their precipition grows; jc=3.2×10^5 A/cm^2 and jc=3.9×10^5 A/cm^2 were received for the alloys HT-50 and HT-55, respectively, in the field of 5T (Fig.2).

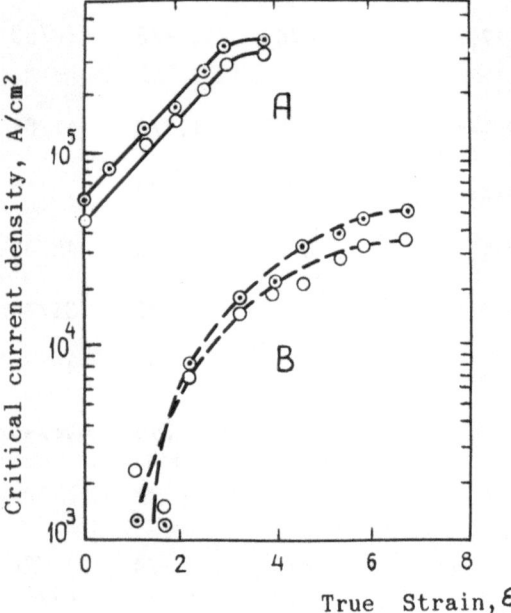

Figure 1. Influence of deformation on jc of HT-50(O) and HT-55($\odot$) alloys subjected to intermediate anneals (A) and unannealed (B). The deformations were calculated for A after the last anneal and for B after extrusition.

607

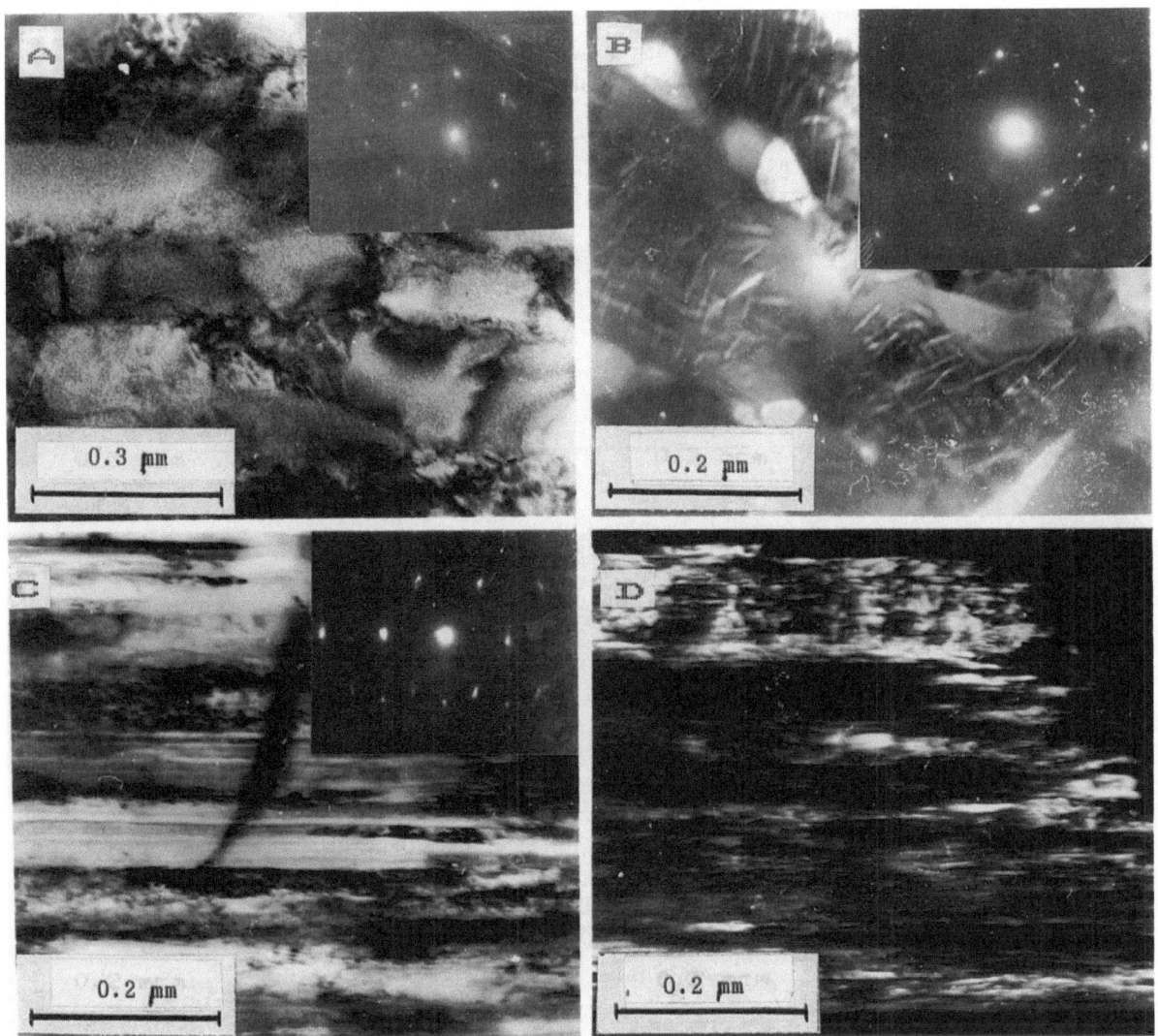

Figure 2. Structure of HT-50(A) and HT-55(B) filaments after the first anneal;
HT-55(C,D) after the final strain, $\varepsilon = 3.8$
A)longitudinal B)cross-sectional C)longitudinal D)cross-sectional, dark field
in α-phase reflection

Metallographic investigations of HT-50 and HT-55 filaments revealed concentric zones of different etching characteristics at different stages of TMW. At the last stages of TMW (first in the HT-55 alloy and then in the HT-50 alloy) and at the final deformation stage the layers having different etching characteristics disappeared (Fig.3). To find out the cause of the difference in the etching characteristics X-ray investigations were performed at four sources of X-ray radiation. The filaments stripped of the copper and Nb barrier were 50 μm diameter. Judging from the intensity relations the data indicate a

non-uniform α-phase distribution across the filaments (Table 2).

Independent of the alloy composition all the experimental data on the speciments both subjected not subjected to intermediate anneals are plotted in Fig.4. As can be seen, the corresponding lines can be drawn through the points belonging to the drawed unannealed specimens and through the main portion of the points related to the specimens subjected to IHT. The points related to the HT-55 alloy specimens where the α-phase precipitated both along the boundaries and within the β-matrix

608

TABLE 2.

X-ray diffrction analysis date on HT-55 wire specimen after fourth intermediate anneal at tenfold adsorption of X-rays.

Type of radiation	Integrated intensity		Intensity relation	Penetration depth for
	$(200)\beta$	$(110)\alpha$	$(110)\alpha/(200)\beta$	$(110)\alpha$, mcm
Cr Kα	785	165	0.21	2.96
Cu Kα	820	135	0.16	5.54
Cu Kβ	410	45	0.11	7.32
Mo Kα	370	80	0.22	22.9

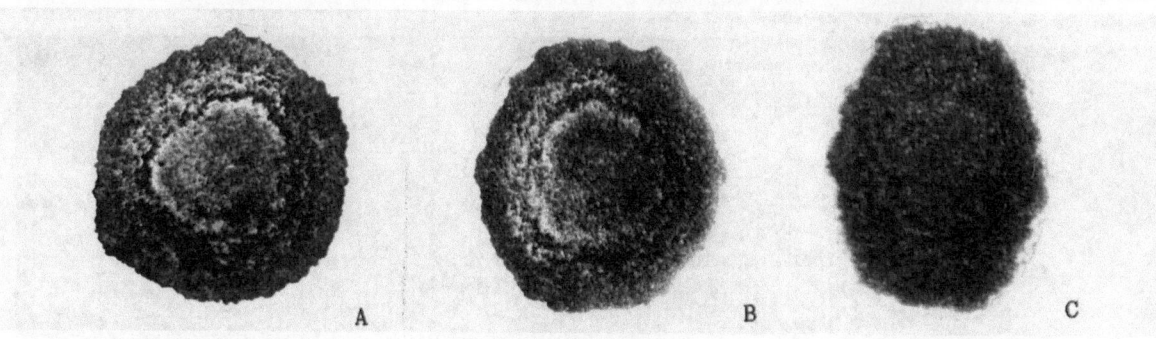

Figure 3. Microstructure of HT-50 filaments prior to first(A), third(B) anneals and after forth anneal at deformation stage(C).

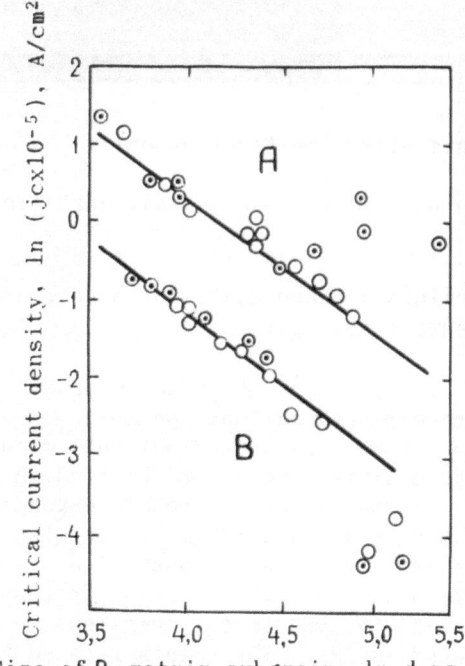

Figure 4. Dependence of jc on size of -matrix subgrain for HT-50(O) and HT-55(⊙) alloys subjected to intermediate anneals(A) and unannealed(B).

subgrains fall out of this regularity:

$$\ln (jc\times10^{-5})=5.96+1.54xp-1.78 \ln d,$$

where :
jc is a critical current density, A/cm^2.
d is a width of β-matrix subgrains, nm.
p=1 for specimens subjected to
    intermediate anneals
P=0 for unanneal specimens.

The data presented allow the discrimination between the contribution made by the subgrain size and the α-precipitates to the jc values. Thus, the results of this work indicate that a fine cellular substructure favours a higher probability of the α-phase precipitation from a supersaturated solid solution and the preferred sites of these precipitates are subgrain boundaries. To produce high jc wires even if there are changes in the phase composition across the wire the TMW schedules are required that would induce a fine cellular substructure of the β-matrix with fine disperse α-particles along the subgrains.

CONCLUSIONS

The comprehensive investigations into the changes of the structure and critical current density at different stages of HT-50 and HT-55 superconductor wire manifacture made it possible to establish that the decisive part in achieving high jc is played by the β-matrix fine cellurar structure decorated with fine disprse α-particles which is reached as a result of multisycle TMW. Thus for wire specimens having superconductor filaments 4-6 μm dia and subcells of 35-40 nm the critical current densities of 3.2×10^5 and 3.9×10^5 A/cm^2. (the field of 5T) were attained for the HT-50 and HT-55 alloys, respectively. It is possible that further developments at the TMW and progressive deformation methods that effect a finer cellular substructure of the β-matrix will lead to higher jc [2].

REFERENCES

1. G.K. Zelenskiy, V.A. Vasil'ev, L.D. Bogdanova, A.P. Golub', E.V. Nikulenkov, HT-50 composite superconductors for UNK magnets ,Report at the 24 International Conference of Countries-member of CMA on Physics and Techniques of Low Temperatures, 17-20.09.1985, Berlin, DDR.

2. E. Gregory, Recent advantages in commercial multyfilamentary Nb-Ti wires in the United States, in: "Cryogenic Materials '88", vol 1 Superconductors, eds. R.P.Reed, Z.S. Xing, E.W. Collings, ICMC, Boulder, Colorado(1988), pp. 361-371.

3. Li Cheng-ren, We Xiao-zu, Zhon Nong, Nb-Ti superconducting composite with high critical current density, in: "IEEE Transactions on Magnetics", vol.Mag 19, N3, (1983), pp. 284-287.

4. D.C. Larbalestier, P.J. Lee, Li Chengren, W.H. Warnes, New developments in Nb-Ti superconductors, in: "Proceed of Workshop on Superconducting Magnets and Cryogenics", ed. P.F. Dahl, BNL-52006, May 12-16, 1986, (1986), pp. 45-50.

5. A.D. McInturff, Metallurgy of the NbTi superconductors, in: "Metallurgy of Superconductor Materials", eds. T. Luhman, D. Dow-Hughes, Academic Press, New York (1979), pp. 67-96.

6. G.K. Zelenskiy, A.P. Golub', A.D. Nikulin, V.Ya. Fil'kin, V.P. Kosenko, V.L. Mette, E.V. Nikulenkov, L.V. Potanina, Composite superconductors for UNK magnets, Report CB-3, ICMC, june 7-10, 1988, Shenyang, P.R.China.

7. Wang Keguang, Zhang Tingle, Wu Xiaozu, Li Chengren, Zhou Nong, The pinning force and critical current NbTi superconducting wire with plate-like Nb-Ti precipitates, in "Advance Gryognetic Materials", vol.32, N4(1986), pp. 903-909.

8. E.W. Collings, The physical mettalurgy of titanium alloys, ASM, (1984). G.K. Zelinskiy, A.V., Arsent'ev, A.P. Golub', V.E. Klepatsky, E.V.Nikulenkov, A.D. Nikulin, V.Ja.Fil'kin, V.S. Titov, P.P. Pashkov, V.A. Vasil'ev, A.I. Nikulin, Influence of thermomechanical working schedules on structure and properties of HT-50 and HT-55 superconductor alloys. The paper to be presented at the CEC/ICMC, Los Angeles, USA, 24-28 July, 1989.

10. V.Ja. Fil'kin, V.F. Gogulya, V.P. Kosenko, E.V. Nikulenkov, A.D. Nikulin, P.J. Slabodchikov, G.K. Zelinskiy, K.P. Myznikov, A.I. Nikulin, V.A. Vasil'ev, Composite superconductors for UNK magnets, in: "Processing of Workshop on Superconducting Magnets and Cryogenics", ed, P.F. Dahl, BNL-52006, May 12-16, 1986,(1986), pp. 56-59.

11. V.S. Titov, G.N. Vlasov, Foils prepered method for electromicroscopic in investigations of superconducting wire. Proceedings of the All Union Scientific Research Institute of Electromechanics, vol.40, Moscow,(1974), p. 133-136.

DEVELOPMENT OF KA-CLASS SUPERCONDUCTING CABLES FOR AC USE (I)-DESIGN AND FABRICATION

T. Hamajima, M. Shimada, M. Ono, M. Yamaguchi, D. Itoh,
and T. Fujioka
TOSHIBA CORPORATION, Yokohama Japan
K. Funaki, K. Tasaki, M. Iwakuma, M. Takeo and K. Yamafuji
Kyushu University
T. Kumano and E. Suzuki
Showa Electric Wire & Cable Co., Ltd., Kawasaki, Japan

ABSTRACT

A large current AC superconductors were fabricated and tested. A strand with more than 20000 NbTi filaments was extruded down to 0.153 mm in diameter to produce filaments of 0.57 μm in diameter. The mixed matrix composed of Cu and CuNi was employed. A final conductor was made of 12 2nd-level sub-cables (6x7 strands) with a mandrel and was compressed on its four sides like a Rutherford type one. Four kinds of conductor were fabricated; one had an FRP mandrel, and the other three had a stainless steel mandrel with various compression rate. They were wound on FRP bobbins to form coils. Three of four coils were successfully operated up to about DC 5 kA.

INTRODUCTION

A considerable progress in the low AC loss superconductor fabrication at the operation of commercial frequency with a better understanding of electro-magnetic behavior of the conductor in alternating field allows us to develope a superconducting machine for AC use(1). However, the superconductor developed so far for AC use can carry the current of only some hundreds amperes(2). Therefore, it is required for the application to AC machines that the current of the superconductor should be more than some thousands amperes.

We have been developing kA-class AC superconductors applying to the secondary winding of a superconducting transformer. The design and fabrication of the superconductors as well as coils are mainly described. The DC operation results of the coils which were wound with the final conductors having various compression rate are described. The AC operation results are reported at a companion paper(3).

SUPERCONDUCTOR

Strand

A double stacked NbTi strand with more than 20000 filaments was processed to 0.153 mm in diameter to reduce the AC loss, and 0.57 µm in diameter of the NbTi filament was finally obtained. The cross section of the strand is shown in Fig. 1, in which the inner part of the strand is composed of Cu surrounded with CuNi, and the outer part is composed of NbTi and

CuNi matrix. The ratios of Cu and CuNi to NbTi are 0.16 and 2.16, respectively. The copper used here plays an important role to protect the conductor from excessive temperature rise at the occurrence of quench. A reliable quench detection system allows us to discharge the current in about 0.1 sec, and therefore, the amount of the copper is adequate for the 3 kA operation. The main characteristics of the conductor are listed in Table 1. The twist pitch of the filament is 1.3 mm which reduces the coupling loss to the level of the hysteresis loss at the condition of a commercial frequency.

Sub-cable

The 1st-level sub-cable was assembled with 7 strands and the 2nd-level sub-cable was wound with 6 1st-level sub-cables around a stainless steel coated with an insulator.

Final conductor

The final cable conductor was made of 12 2nd-level sub-cables wound on a mandrel and was finally compressed, like Rutherford type one, as shown in Fig. 2. Two kinds of material were used as the mandrel, namely, stainless steel and FRP. Three final conductors with the stainless steel mandrel were fabricated with the parameter of the compression rate which was defined as follows,

$$C_r = 1 - d_f/d_n, \qquad (1)$$

where, d_f was a compressed 2nd-level sub-cable diameter and d_n was nominal diameter of the 2nd-level sub-cable.

TABLE 1 Main parameters of the strand

Strand diameter	0.153 mm
Filament diameter	0.57 μm
Number of filaments	21336
Twist pitch	1.3 mm
NbTi:Cu:CuNi ratio	1:0.16:2.16

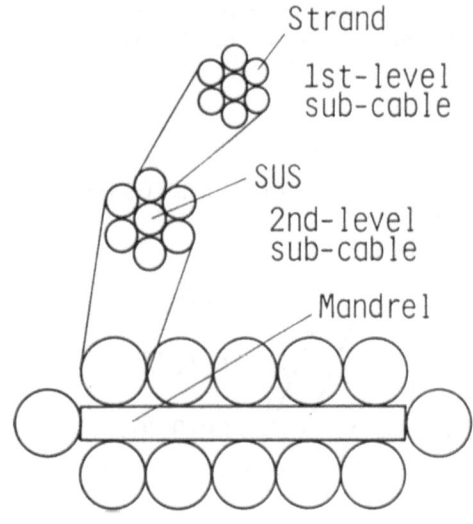

Fig.2 Construction of the final conductor

TABLE 2 Main parameters of the final conductor

Number of 2nd sub-cables	12
Mandrel	SUS or FRP
Compression rate	21.4% (FRP)
	27.5, 36.2, 49.6% (SUS)

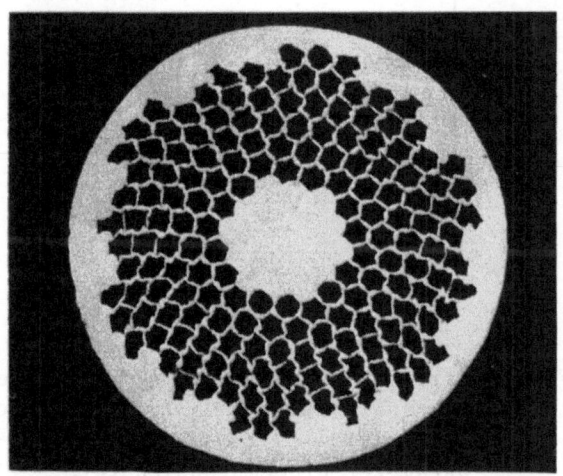

Fig.1 Cross section of the strand

CRITICAL CURRENT

Sub-cable

We investigated the characteristics of the strand through the fabrication process from a strand to the final conductor. Fig. 3 shows the critical current of the 2nd-level sub-cable, which is six times that of 1st-level sub-cable. Any change of the critical current density was not observed up to the 2nd sub-cable, because the strand of the 2nd sub-cable had not experienced forced compression.

The deterioration of the critical current density of the final conductor was of particular concern, because it was forcedly compressed to form rigidly. Therefore, we studied a relation between the critical current and the compression rate of the 2nd sub-cable as shown in Fig. 3, prior to fabrication of final conductor. In this case, the 2nd sub-cable was compressed from four directions by Turks head roller in order to form square cross section. The critical current decreases with large compression rate. This comes from that the NbTi filament begins to break. It was observed that the strand in the 2nd sub-cable began to break down with more than 22% compression rate.

Final conductor

We measured the 2nd sub-cable which was untwisted from the fabricated final conductor, and the results are shown in Fig. 4. Comparing with the previous results (Fig. 3), the critical current of the final conductor decreases moderately with the compression rate. Although the final conductor was compressed in four directions, the 2nd sub-cable was subject to simple compression only between two neighboring sides, and deformed freely towards the other directions. Therefore, the fracture of the filaments in the final conductor was not much.

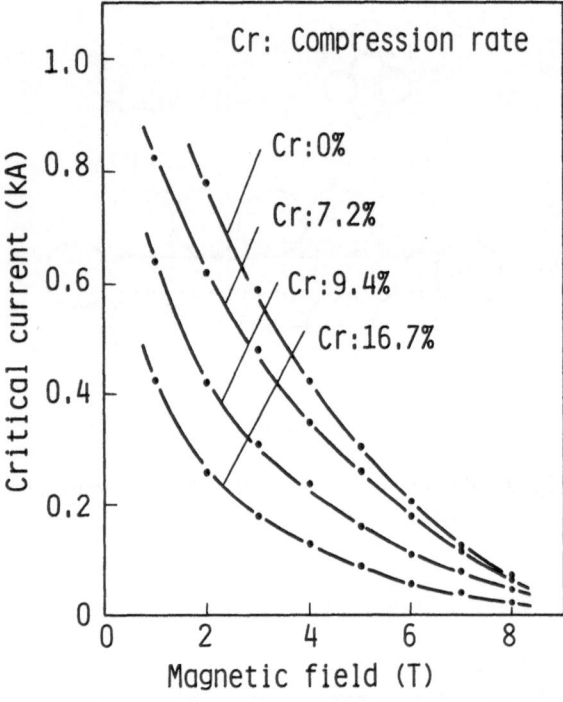

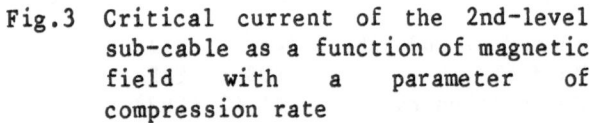

Fig.3 Critical current of the 2nd-level sub-cable as a function of magnetic field with a parameter of compression rate

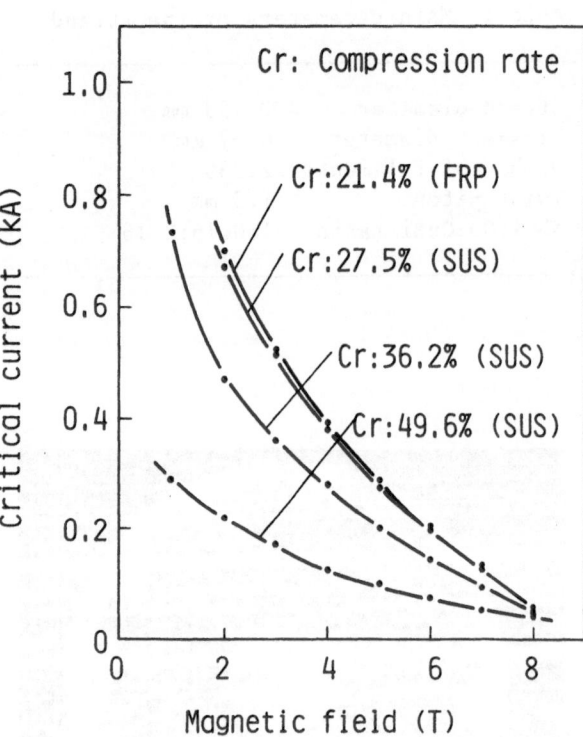

Fig.4 Critical current of the 2nd-level sub-cable which was untwisted from the final conductor with a parameter of compression rate

COIL

We made four coils; one with an FRP mandrel, and the other three coils with stainless steel mandrel. The final conductors used for the three coils were compressed in various degree, namely, 27.5, 36.2, and 49.6%, in order to investigate the relation between coil performance and compression rate.

The final conductor was wound on a bobbin in a form of one layer and 10 turns. The bobbin was made of FRP to reduce the eddy current loss, and was machined to provide axial cooling channels and spiral grooves in which the conductor was tightly fixed, as shown in Fig. 5. The width of the groove was matched to that of the final conductor in order to prevent conductor motion from causing a quench.

TABLE 3 Main characteristics of the coil

Outer diameter	94 mm
Height	142 mm
Number of turns	10
Max. field @ 5kA	0.38 T
Inductance	5.0 H

Fig.5 Coil wound on an FRP bobbin

DC TEST RESULTS

The coil tests were carried out by using a DC power supply up to about 5 kA. The results are shown in Fig. 6. The coil wound with the conductor with FRP mandrel experienced training to reach 5 kA. The training phenomena of the other three coils with stainless steel mandrel were not observed explicitly up to 5 kA charging. However, one coil wound with the final conductor subject to the most heavy compression showed a small normal voltage at 1.0 kA which gradually grew and finally let the coil quench at 2.4 kA.

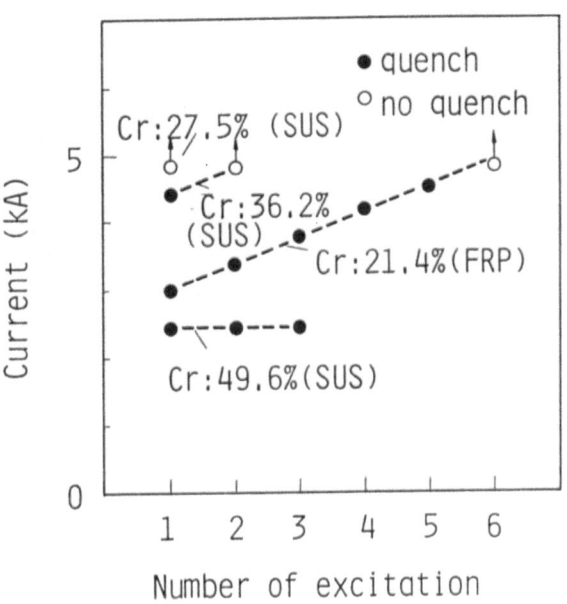

Fig.6 Test results of four coils

CONCLUSIONS

We fabricated kA-class AC superconductor in a form of Rutherford type conductor. Three coils wound with the final conductors were successfully operated up to about DC 5 kA, but the one coil with the final conductor compressed heavily quenched at 2.4 kA. It is found from the results that the coil wound with the final conductor compressed loosely has better performance. This encourages us to develope a large current secondary winding of a superconducting transformer.

REFERENCES

(1) P. Dubots, A. Fevrie, J.C. Renard, J. Goyer and Hoang Gia Ky, Behaviour of multifilamentary Nb-Ti conductors with very fine filaments under A.C. magnetic field. Journal de Physique 48 (1948) C1-467.

(2) D. Itoh, E. Shimizu, T. Fujioka, H. Ogiwara, S. Akita, T. Ishikawa, and T. Tanaka, Development of 500 kVA A.C. 50 Hz superconducting coil. Proc. of ICEC 12 (1988) 724.

(3) K. Funaki, K. Tasaki, M. Iwakuma, M. Takeo, K. Yamafuji, M. Ono, M. Shimada, T. Hamajima, M. Yamaguchi, D. Itoh, T. Fujioka, T. Kumano, and E. Suzuki, Development of kA-class superconducting cables for AC use (II), in this conference.

DEVELOPMENT OF kA-CLASS SUPERCONDUCTING CABLES FOR AC USE (II) ELECTROMAGNETIC PROPERTIES

K. Funaki, K. Tasaki, M. Iwakuma, M. Takeo, K. Yamafuji,
T. Hamajima,* M. Shimada,* M. Ono,* M. Yamaguchi,* D. Itoh,*
T. Fujioka,* T. Kumano** and H. Suzuki**
Faculty of Engineering, Kyushu University, Fukuoka, Japan
*TOSHIBA CORPORATION, Yokohama, Japan
**Showa Electric Wire & Cable Co. Ltd., Kawasaki, Japan

ABSTRACT

Large-current-capacity superconducting cables for AC use have been successfully fabricated and tested with regard to AC losses and quench behaviors. The final cables are compacted flat conductors made by triply twisting superconducting composites with NbTi filaments of 0.57 μm in diameter.
A superconducting transformer was constructed for 60Hz operation of the large-current-capacity cables which were installed as the secondary winding. The maximum current of the secondary winding attained to 2650 A(rms) and 3290 A(rms) for the test cables with FRP and stainless steel mandrels, respectively.
AC losses of the cables were evaluated by collecting helium vapor generated only by the secondary winding.
These experiments were safely performed by a reliable method of quench detection using pickup coils mounted inside and along the windings.

INTRODUCTION

Since superconducting composites with very fine filaments for AC use were developed,[1] various types of coils[2],[3] and test transformers[4],[5] were constructed by bundled cables of the composites to study fundamental properties and technological problems of the AC superconducting cables. So far most of the cables were 100A-class made by doubly twisting the composites. It is important to increase the current capacity of the cables up to the level of kA-class for the next step in the development of the AC superconducting cables.

We designed and fabricated triply twisted cables of the superconducting composites with NbTi filaments of 0.57 μm in diameter. The cables were located in a test transformer as the secondary winding and AC losses and quench processes of the cables were evaluated experimentally.

EXPERIMENTS

Test Cables

Final cables are compacted flat conductors which are composed of triply twisted superconducting composites with 21,336 NbTi filaments of 0.57μm in diameter. The diameter of the composites is 0.153 mm and the ratios of matrix materials, Cu-10%Ni and Cu, to NbTi are 2.16 and 0.16, respectively. The first-level sub-cable was assembled with 7 composites, and the second-level sub-cable with 6 first-level sub-calbles was wound around a stainless steel wire coated with Formver. The final cable was made of 12 second= level sub-cables with a mandrel of FRP or stainless steel. Specifications of the test cables are listed in Table 1. One= layer coils of the final cables were made for test operations. The previous paper [6] describes the design of the coil and DC operation results.

Transformer

A supersonducting transformer was constructed for 60Hz operation of the test coils. Each test coil was located as the secondary winding in the transformer and shorted by the same test cable. The primary winding was equivalent to the first-level sub-cable of the test cables. Characteristics of the primary and secondary coils are listed in Table 2. Inductances of the transformer were evaluated by numerical calculation. The ratio of the secondary current to the primary was 44.8. It was confirmed that the numerical results of the primary self-inductance and leakage inductance agree with measured ones within a error of 2%. The secondary

current was obtained from the measured primary current with the current ratio. Figure 1 shows the outline view of the transformer. The outermost coil is the secondary winding wound by the test cable. The coil formers were made of FRP to reduce the eddy current loss.

AC-loss measurement

The test cables were composed of bare superconducting composites. Additional AC loss is generated by electromagnetic coupling current between the composites in an alternating magnetic field.[7] AC losses at the levels of the elementary composites and the final cables were evaluated by measuring DC magnetization of the first= level sub-cable and collecting helium vapor generated only by the secondary winding in the transformer, respectively. The helium vapor was gathered by a jacket which covered three fourths of the secondary winding as shown in Fig.2. Helium vapor from the primary winding and the current terminals of the secondary winding was not guided into the jacket. The amount of helium gas gathered by the jacket was measured by a flow meter with the thermal senser of semiconductor in a room-temperature space. The relation between the amount of helium gas and AC loss was calibrated by a manganese heater mounted on the inside wall of the jacket.

Table 1 Characteristics of test cables

Strand	
daimeter	0.153 mm
diameter of filaments	0.57 μm
number of filaments	21,336
twist pitch	1.3 mm
Cu-10%Ni:Cu:NbTi	2.16:0.16:1
Final cable	
structure	7 × 6 × 12
size of cross-section	3.56×9.18 mm^2
mandrel	FRP or
	stainless steel

Table 2 Specifications of transformer

	primary coil	secondary coil
self-inductance	22.8 mH	5.04μH
mutual-inductance		225.7 μH
inner diamerter	61.2 mm	94.0 mm
outer diameter	72.1 mm	101.1 mm
height	142 mm	142 mm
number of turns	972	10
number of layers	4	1
conductor	first-level sub-cable	final cable

Fig. 1 Test superconducting transformer.

Fig. 2 Jacket for collection of helium vapor.

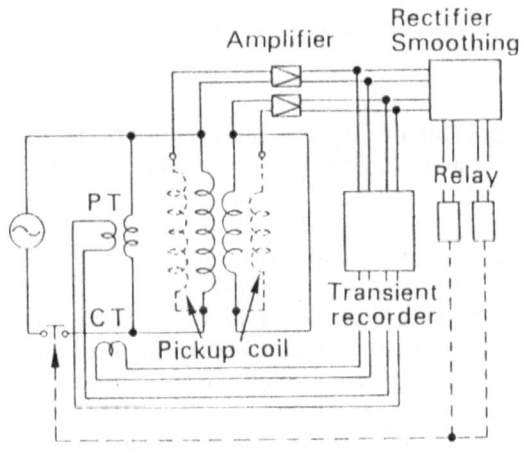

Fig. 3 Experimental circuit.

Quench detection

In AC superconducting coils, the amount of the stabilizer is restricted to diminish the coupling-current loss, and the density of heat generation after quenching is much higher than that in DC and pulsed coils. The conventional method of quench protection with an external dump resistor cannot be used because high voltage is supplied between both the terminals of the AC coil. Therefore, the protection of the AC superconducting coil is to be assured by reliable quench detection and interruption of power line without time lag after the detection.

The initiation of quenching is usually detected by measuring resistance of the normal region generated. Reliability of the quench detection depends on separation of the resistive voltage from the inductive one between the terminals of the coil. In the test transformer, potential taps wound inside the primary winding and along the secondary one were used as the pickup coil for the resistive signals, which was proposed to measure AC losses and instabilities due to flux jump in supeconducting wires.[8] This method is also effective to the quench detection for the usual superconducting coil and especially for the AC coil with very high terminal voltage. The quench detection system for AC operation of the test trasformer is illustrated in Fig. 3. The signals from the pickup coils were directly regulated by amplifiers, full-wave rectifiers and smoothing circuits. The pickup coils were electrically isolated from the outside measurement system by the amplifiers. When the signals exceed a threshold level, the relaies actuate the circuit breaker. The signals after amplification were stored by a transient recorder with the primary voltage and current.

RESULTS & DISCUSSION

AC operation

The secondary coil of the large-current-capacity cable was excited by supplying AC voltage to the primary coil in the transformer. The signals from the pickup coils indicated which of the primary and secondary coils was quenched as shown in the following section. For the cable with stainless steel mandrel, the training of the quench current was scarcely observed. The results of the AC operation were represented in Fig. 4. Here critical current was evaluated from that of the second= level sub-cable untwisted.[6] In this case, the transformer was steadly operated for 5 minutes at the secondary current of 3290A(rms). For the cable with FRP mandrel, on the other hand, the secondary current more than 2650 A(rms) was not attained because of quenching of the primary coil.

AC losses

The total loss of the cable at the AC operation is composed of the internal loss of the multifilamentary composites and additional ones due to multi-strand struc-

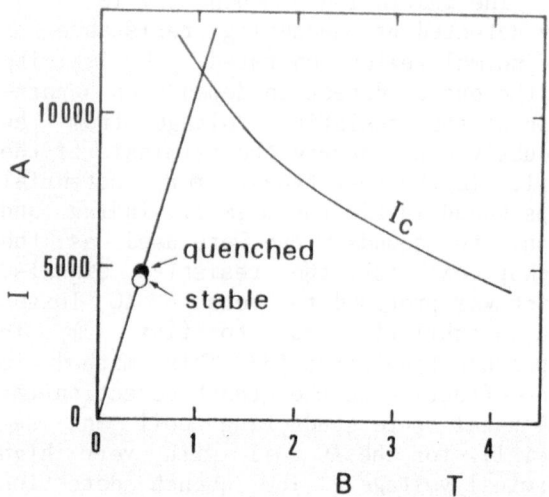

Fig. 4 Load line and critical current for the cable with stainless steel mandrel.

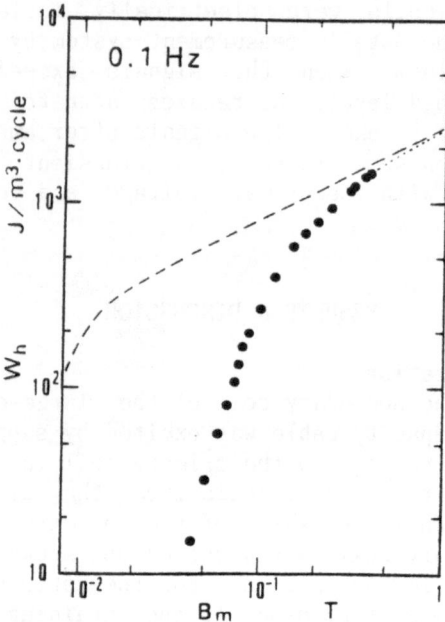

Fig. 5 Measured hysteresis loss of the first-level sub-cable. The dashed curve shows a theoretical result calculated by the 'critical state model'.

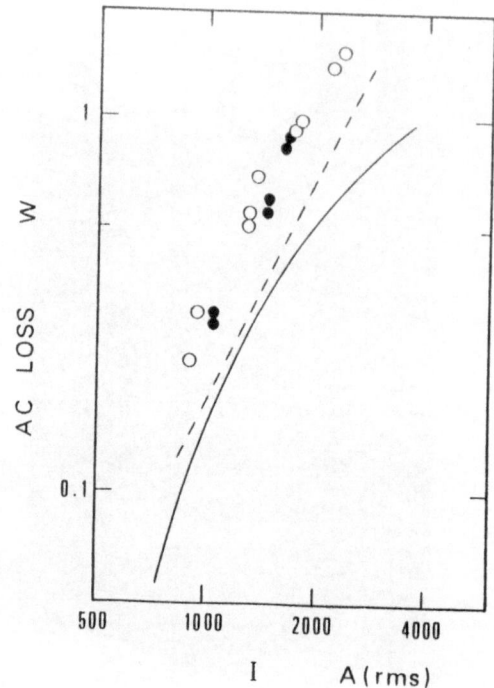

Fig. 6 Total AC loss of the final cables. The solid line shows the internal loss of the composites. The dashed line is the additional loss.

the region of lower amplitude of magnetic field may originate from the effect of reversible motion of the fluxoid lattice in very fine filaments of sub-micron in diameter.[10] Another component of the internal loss, the coupling-current loss among the filaments, can be negligible in the present electromagnetic condition.

Total AC loss measured by collecting helium vapor is represented in Fig. 6. The white and black circles show the AC losses in the cables with stainless steal and FRP mandrels, respectively. The AC loss scarcely depends on the material of the mandrel. For comparison, the internal loss of the composites in the coil with spatial variation of magnetic field was numerically calculated using the experimental data in Fig. 5. The result is shown by the solid line in Fig. 6, where the coupling-current loss in the composites was neglected. The difference between these two results is the additional loss which is shown by the dashed line. The additional loss is comparable to the internal loss in the region of current lower than 2000 A, while that becomes major component of the total loss with increase in the amplitude of the current.

The additional loss may be attributed

ture of the cable.

The main component of the internal loss of the composites, the hysteresis loss of the NbTi filaments, was evaluated by measuring DC magnetization of the first-level sub-cable. The result is shown in Fig. 5, where the corresponding theoretical prediction by the 'critical state model' is indicated by the dashed line.[9] Remarkable deviation of the theoretical results from the measured one in

to the coupling current between the composites at each level of the sub-cable and/or the mechanical dissipation due to frictional motion of the windings etc.. Further consideraton is neccessary to depress the additional loss.

Quench analysis

Wave forms of the primary current and the signals of quench detection around the initiation of quenching are illustrated in Fig. 7. The amplitude of the signal from the secondary pickup coil abruptly increased at the point A which is at the peak of the current. The trapezoidal wave form of the signal after the point A oreginates from overflow of the amplifier. The change in the wave form at the point A implies that the quenching of the transformer was initiated in the secondary winding. The primary current was interrupted by the quench detection system at the point B . The time interval between A and B was 53 msec(3.2 cycles). Before the point A , the amplitude of the background signal from the secondary pickup

coil is about 50 times larger than that from the primary. Because the secondary pickup coil is mounted along the winding and the main component of the signal is out-phase with the current. The signal from the primary, on the other hand, is almost composed of the in-phase component which corresponds to a part of the hysteresis loss in the NbTi filaments. This means that the primary pickup coil gives more sensitive signal of quench detection than the secondary. In any case, the resistive signal from the secondary is sufficiently larger than the background level and hence the interruption of the current is compeleted in about 3 cycles after the initiation of quenching.

Increase in temperature of the secondary winding after transition to normal state can be approximately estimated by

$$\int_{T_b}^{T} \frac{C(T)}{\rho(T)} \, dT = \int_{0}^{t} J^2(t) \, dt , \qquad (1)$$

where C is heat capacity uper unit volume of the cable, ρ average resistivity, J overall current density and T_b bath temperature. The temperature of the test cables at the interruption of the current was calculated for the parameters of interruption time and copper ratio (percent-

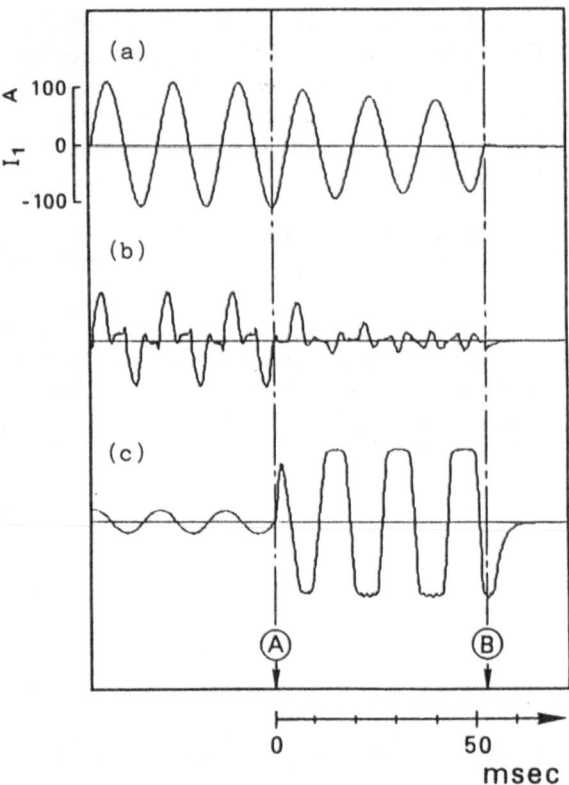

Fig. 7 (a) The primary current (I=76.6A). (b),(c) Signals of quench detection from the primary and secondary pickup coils, respectively.

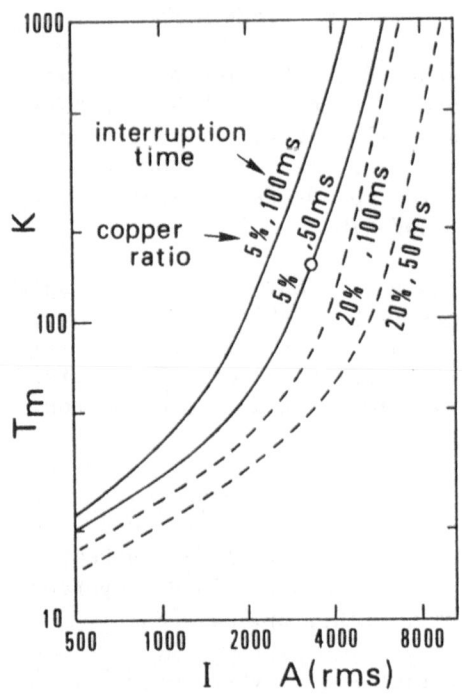

Fig. 8 Increase in temperature of the test cable after quenching.

age of copper in the composite). The dependence of the maximum temperture on the AC current is shown in Fig. 8. In the case of quenching of the test cable with stainless steel mandrel at 3380 A(rms), the temperature rose up to about 150 K as marked by a white circle in Fig. 8. The maximum temperature decreases with the increase in copper ratio, but increases with the extension of interruption time. In the case of 100 msec interruption time and 20% copper ratio, for example, the maximum current is 4300 A (rms) for maximum temperature of 150 K. Further investigation is neccessary to optimize the cable structure under the viewpoints of AC loss, stability and quench protection.

CONCLUSION

kA-class superconducting cables for AC use were fabricated by triply twisting superconducting multifilamentary composites. The stable maximum current attained to 3290A(rms) for the cable with a stainless steal mandrel. The AC losses of the cables were measured by collecting helium vapor. Additional loss due to multistrand structure was equivalant to the internal loss of the composites. Operational conditions of the AC cable was also discussed in relation with the quench detection.

REFERENCES

1. Dubots,P., Fevrier,A., Renard,J.C., Goyer,J. and Ky,H.G., Behaviour of multifilamentary Nb-Ti conductors with very fine filaments under a.c. magnetic fields. **J. de Phys.**, 1984, **45**, pp. C1-467-470.

2. Itoh,D., Shimizu,E., Fujioka,T., Ogiwara,H., Akita,S., Ishikawa,T. and Tanaka,T., Development of 500 kVA A.C. superconducting coil. **Proc. of 12th Int. Cryo. Eng. Conf.**, Southampton, 1988, pp.719-723.

3. Tsukamoto,O., Yamamoto,M., Ishigohka,T., Tanaka,Y., Takao,T. and Torii,S., Development of large-current capacity epoxy-impregnated 50 Hz superconducting coil. **ibid.**, pp.724-728.

4. Février,A., Tavergnier,J.P., Laumond, Y. and Bekhaled,M., Preliminary tests on a superconducting transformer. **IEEE Trans. Magn.**, 1988, **24**, pp.1477-1480.

5. Funaki,K., Iwakuma,M., Takeo,M. and Yamafuji,K., Preiminary test and quench analysis of a 72 kVA superconducting four-winding power transformer. **Pro. of 12th Int. Cryog. Eng. Conf.**, Southampton, 1988, pp.729-733.

6. Hamajima,T., Shimada,M., Ono,M., Yamaguchi,M., Itoh,D., Fujioka,T., Funaki,K., Tasaki,K., Iwakuma,M., Takeo,M., Yamafuji,K., Kumano,T. and Suzuki,E., Development of kA-class superconducting cables for AC use (I) Design and fabrication. to be presented in this conference.

7. Sumiyoshi,F., Nagaichi,H., Ichimaru,O. and Yamafuji,K., Increase of coupling-current losses in supersonducting cables due to first-stage cabling. **Cryogenics**, 1986, **26**, pp.39-44.

8. Ezaki,T., Irie,F., Klipping,G., Lüders, K., Matsushita,T., Ruppert,U. and Takeo,M., System for coil simulation measurements of losses and instabilities in superconducting wires. **Cryogenics**, 1979, pp.731-735.

9. Noda,M., Funaki,K. and Yamafuji,K., Hysteresis loss in a superconducting rod with a round cross section exposed to a transverse AC magnetic field. **Mem. Fac. Eng. Kyushu Univ.**, 1986, **46**, pp. 63-76.

10. Sumiyoshi,F., Matsuyama,M., Noda,M., Matsushita,T., Funaki,K., Iwakuma,M. and Yamafuji,K., Anomalous magnetic behavior due to reversible fluxoid motion in superconducting multifilamentary wires with very fine filaments. **Jan. J. Appl. Phys.**, 1986, **25**, pp.L148-L150.

SUPERCONDUCTORS AND RARE EARTH MAGNETS : AN EXCITING COMBINATION FOR ELECTRICAL MOTORS

P. Tixador*, Y. Brunet *, P. Brissonneau**, M. Ivanes**

* CNRS-CRTBT/LEG , B.P. 166 X - 38042 Grenoble Cedex, FRANCE.

** L.E.G., ENSIEG , B.P. 46 - 38402 St-Martin-d'Hères Cedex, FRANCE.

ABSTRACT

Due to the technological progress on ultra-fine NbTi filaments associated to a high resitive matrix, the use of superconductors (S.C.) at industrial frequencies (50/60 Hz) is now possible. 50 Hz losses in the best a.c. S.C. wires do not exceed 20 kW/m^3 for 1 T peak value magnetic induction. The first applications of these wires have been the S.C. power transformer and the synchronous machine with S.C. coils for both field and armature windings. First studies show that the magnetic induction in such machines remains relatively low so that rare earth permanent magnets (P.M.) with a remanence greater than 1 T (Nd-Fe-B, Sm-Co) are good candidates to create the excitation field. The combination of these two very performant materials opens up new possibilities for large torque motors. Better magnetic properties of the P.M. (intrinsic coercive force above all) at low temperature and high current density at the armature are two of the key improvments led by this new type of S.C./ P.M. machines. A small amount of soft magnetic materials in such motors may even improve their specific power. First design of a small scale machine is presented.

Notation list

E	: no-load voltage, V_{rms}/phase,
I	: nominal armature current, A_{rms}/phase,
K	: current sheet density [1] ; $K = 2\dfrac{N_s k_d}{\pi r_0} I$
Γ	: torque,
w	: angular frequency, $w = 2\pi f$ (f:frequency),
p	: number of pole pairs,
N_s	: total number of turns in series for a phase,
k_d	: fundamental armature winding factor,
k_f, k_f'	: fundamental magnet shape factor,
r_1, r_2, r_0, r_s:	radius (see Fig.3),
L	: active length,
B_r^0	: no-load fundamental radial induction at r_0,
J	: armature current density, A_{rms}/mm^2,
X_d, x_d	: synchronous reactances (Ω and p.u.),
J_r	: remanence, T.

INTRODUCTION

It is now possible to use superconducting (S.C.) wires at industrial frequencies (50/60 Hz) [2] as their losses amount to about 20 kW/m^3 for 2T peak to peak induction variations at 50 Hz and 4.2K [3]. Considering that the high current densities in superconductors reduce the volume of conductor (at this field level the "overall" current density may reach 1500 A/mm^2) and the cost of refrigeration, these losses are of the same order of magnitude than the copper losses at normal temperature [3]. These performances are achieved with submicronic filaments (0.1-0.2 μm) embeded in a high resistive CuNi matrix (ρ = 3.9 10^{-7} Ωm at 5 K) associated to a small pitch length for the wire. To preserve the thermo-electromagnetic stability of the wire at low 50 Hz fields [4] the wire diameter is small, typically 0.25 mm if stability at 1T is required. For a large rated current conductor several wires must be associated to get the wanted current. Such conductors can be used to realize a.c. windings [5]. One of these new applications is the fully superconducting synchronous machine [6] : both field and armature windings are wounded with superconducting wires. A small scale machine is now under construction in our laboratory [7]. These machines are characterized by great reactances to get a high power density, by the suppression of the electromagnetic shields at the inductor and by a relative simple combined cryogenic design [8]. Nevertheless a rotating horizontal helium vessel is needed. As the magnetic field at the armature is relatively low (<1T) to have low losses and high current densities, a superconducting armature may be associated to rare earth magnets for the rotating inductor of a synchronous motor.

The advantages are :

S.C. armature :
- high current density,
- static cryogeny,
- air-cored light structure,

magnet inductor :
- "cold" operation (< 250 K),
- better magnetic properties.

The aim of this report is the description of such a machine designed for a power of 500 kW, 1000 rev/min, 50 Hz.

GENERAL STRUCTURES

Cryogenic constraints make disk type solutions not very attractive. Two cylindric structures are possible : classical (inner rotating magnets) and inverse. For the classical solution, the coupling is lower but the rotor cryogenic design is more simple and the magnetic circuit of the inductor is reduced. An interesting solution for smale-scale machines with a single source of cooling is an overhanging rotor inside the armature cryostat.

FIELD MAGNET

Magnetic Properties

Taking advantage of the cryogenic (5 K) environment of the S.C. armature, it is easy to cool the magnets at low temperature to improve their magnetic properties. The remanence follows a Curie-type law and increases slightly when the temperature decreases. This increase is larger for NdFeB compounds (0.11%/K) than for SmCo as their Curie temperature is lower (310°C compared to 800-1000°C). Higher are the increases of the anisotropy field H_a, of the intrinsic coercive field H_C (fig. 1) and then of the critical demagnetization field $H_k^{0.9}$

($H_k^{0.9}$ is the field for which the magnetization reachs 90 % of the remanence [9]). For NdFeB coupounds $H_k^{0.9}$ is very close to H_c as the hysteresis loop is practically a rectangle. The magnetic stability is better at low temperature [11].

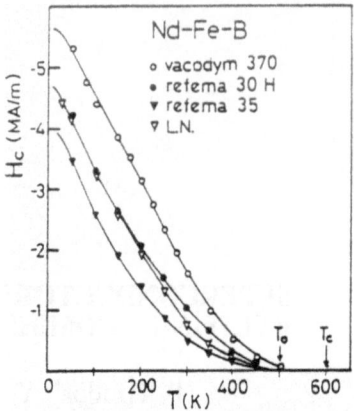

FIGURE 1 : variation of H_c with the temperature for different NdFeB sintered magnets [10].

Nevertheless some care must be taken with NdFeB compounds as their anisotropy is no more uniaxial when the temperature becomes lower than the temperature of spin reorientation transition at about 135 K [12]. The performances of the magnet decrease in the working zone of the magnetization curve as shown by fig. 2 at 100K [10].

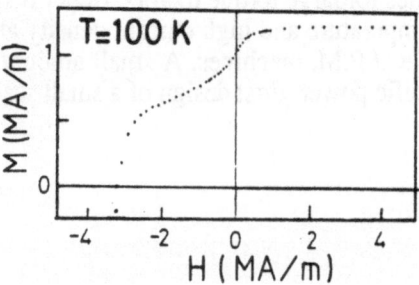

FIGURE 2 : hysteresis loop of a NdFeB magnet at 100 K

The addition of some elements (Tb,Dy) may lessen the minimal operating temperature of NdFeB [13].

Magnet Field Calculations

To have a high excitating field, both surface magnets radialy magnetized and buried magnets are mounted on a magnetic yoke (fig.3). An external magnetic yoke closes the flux.

To get the expression of the induction the surface magnets have been represented by fictitious surface charges at the outer surface of the magnets and fictitious volume charges inside the magnets. The buried magnets, supposed to be tangentially magnetized, have been represented by two distributions of current sheet density at r_1 and r_2 and a volume current inside the magnets.

The relative quantity of the two types of magnets (characterized by the angle ß) may be optimized in function of the ratio r_1/r_2 and the pole pair number p.

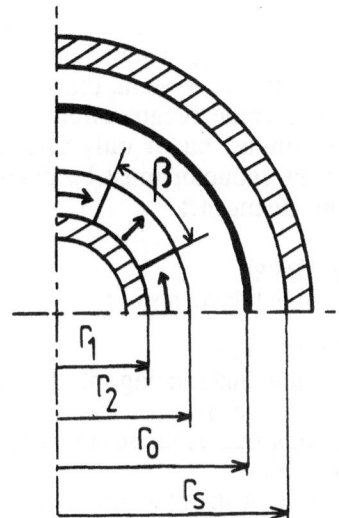

FIGURE 3 : magnet disposal of the studied machine.

The fundamental radial component of the induction at the radius r_0 ($r_2 < r_0 < r_s$) is (p>1) :

$$B_r^0 = \frac{J_r}{2} \left(\frac{r_2}{r_0}\right)^{p+1} \frac{1+\left(\frac{r_0}{r_s}\right)^{2p}}{1-\left(\frac{r_1}{r_s}\right)^{2p}} \ p \left[\frac{1-\left(\frac{r_1}{r_2}\right)^{p+1}}{p+1} \{k_f + k_f'\} \right.$$

$$\left. + \frac{\left(\frac{r_1}{r_2}\right)^{p+1} - \left(\frac{r_1}{r_2}\right)^{2p}}{p-1} \{k_f - k_f'\} \right] \qquad (1)$$

$$k_f = \frac{4}{\pi} \sin\frac{\beta}{2} \qquad\qquad k_f' = \frac{4}{\pi}\cos\frac{\beta}{2}$$

ARMATURE

To avoid losses under induction variations (50 Hz or rotating) at 5 K neither magnetic nor metallic materials are used for the helium vessel. A 800/1 refrigeration ratio is used to take into acount the refrigeration process between 5 K and 300 K.

The superconducting windings are made with thin layers of S.C. wires as the current densities are high.

ELECTROMAGNETIC DESIGN

Hypothesis
- classical cylindric type synchronous machine,
- buried and surface magnets are used,
- inner and outer magnetic yokes,
- 2D calculations,
- time and space harmonics are neglected,
- armature is located at the mean radius r_0.

Basic Relations for Design
E may be deduced from Lentz's law :

$$E = \sqrt{2} \ N_s k_d \ B_r^0 \frac{Lw}{p} \ r_0 \qquad (2)$$

$$\Gamma_{max} = \frac{3EI}{\frac{w}{p}} = \frac{3\sqrt{2}}{2} B_r^0 K \ \pi r_0^2 L \qquad (3)$$

From the expression of L_d [1] it comes :

$$x_d = L_d w \frac{I}{E} \qquad (\underline{definition})$$

$$x_d = \frac{3\sqrt{2}\mu_0}{2\pi} \frac{N_s k_d I}{B_r^0 r_0} \frac{\left(1+\left(\frac{r_1}{r_0}\right)^{2p}\right)\left(1+\left(\frac{r_0}{r_s}\right)^{2p}\right)}{1-\left(\frac{r_1}{r_s}\right)^{2p}} \qquad (4)$$

The torque may be expressed with x_d and B_r^0 :

$$\Gamma_{max} = \frac{2}{\mu_0} x_d B_r^{0^2} \pi r_0^2 L \frac{1-\left(\frac{r_1}{r_s}\right)^{2p}}{\left(1+\left(\frac{r_0}{r_s}\right)^{2p}\right)\left(1+\left(\frac{r_1}{r_0}\right)^{2p}\right)} \qquad (5)$$

The cross section of the conductors of one phase is noted S_a and the "active" volume V_a. As $I = S_a J$ and $V_a = 6N_s S_a L$ the maximum torque is then expressed :

$$\Gamma_{max} = \frac{\sqrt{2}}{2} k_d B_r^0 J V_a r_0 \qquad (6)$$

Superconducting Armature Advantages
S.C. wires have much higher current densities (100 to 500 A/mm^2 for practical use) than copper wires. The armature current sheet density K is thus higher as it is possible to wound more series turns N_s; the volumic density of the torque is then better (equation (3)) for a S.C. armature than for a classical one. Nevertheless the induction created by permanent magnets is smaller in a superconducting machine as the cryogenic insulation increases the magnetic path in air (air-cored machines) and this effect is more critical for smale scale machine as the cryogenic distances are quite independant from the electrical power.

Due to the increase of the current sheet density, the per unit reactances are higher than in classical machine. As for fully superconducting machines [6] equation (5) shows that the volumic density of the torque may be increased for a given excitation induction only for high x_d.

For high values of x_d (1 to 3 p.u.) the power factor is small. The power supply must be a voltage supply as for current supply commutation times would be prohibitive. On the other hand short-circuit torques are small and the mechanical overdimensioning needed to support this transient is greatly reduced.

Geometric Ratios

From expressions (1) and (5) it may be deduced that for a given x_d the ratios r_2/r_0 and r_0/r_s must be as near as possible from unity. They are imposed by thermal conditions (thickness of the helium and vacuum vessels, minimum thermal isolating distances, use of coreless windings) through the following relations :

$$r_0 = r_2 + d_1 \quad (7) \qquad r_S = r_2 + d_2 \quad (8)$$

where d_1 and d_2 are the minimal "cryogenic" distances which depend little on the electrical power; the radius r_2 is choosen as the relevant variable.

Armature Conductor Design

The armature is designed to cope with a severe electrical fault without quenching. This no-quench criterion [14] is very severe for the wire but makes the machine very reliable. The current density under nominal operation is much less than the critical current density of the superconductor under nominal conditions. The losses are higher than for a machine designed only for nominal operation as the "active" volume is greater.

One of the most severe transient for the superconducting conductor is a short-circuit between two phases and the neutral when the machine runs as a synchronous capacitor absorbing reactive power. The load equation of the conductor is :

$$I_{max} = 0.8 \, S_a \, J_c \, (B_{max}) \qquad (9)$$

where J_c (B) is the short sample critical current density characteristic of the wire, I_{max} and B_{max} are the highest fault quantities appearing during the transient.

The coefficient 0.8 gives a good margin of safety to avoid any normal transition.

The armature conductor volume, i.e. the "active volume", influences directly the torque as shown by equation (6). Nevertheless it is limited by the accepted level of losses and by the possibilities of heat exchanges between the windings and the cooling fluid. For a given radius r_0 and N_c full layers of conductors, the maximun quantity of superconductor is given by :

$$N_s \, D = N_c \, k_s \, \alpha \, r_0 \qquad (10)$$

k_s : coefficient of stack; α : electrical angle of the coils and D the equivalent diameter of the conductor defined by $\pi D^2/4 = S_a$.

An interesting value for α is 72° as it suppresses the first armature harmonic.

Magnets

The reliability criterion must apply for the permanent magnets too : the demagnetization field at the magnets must be always lower than the critical field H_k. A demagnetization of the magnets is much more penalizing than a quench of the superconducting windings : with a fast responce and well designed protection system, the windings may work again after some recuperation time (in the region of the minute) but if only one magnet is demagnetized the inductor must be dismantled to magnetize again the magnet.

Parameters for Design

If we assign the values of :

- maximum torque Γ_{max},
- frequency and nominal rotating speed (f and p),
- ratio of magnet radius r_1/r_2,
- N_c full layers of armature phase conductor,
- no-load voltage E,

all the parameters of the machine may be deduced from the external radius of the magnets r_2 as the radius r_0 and r_s are expressed by (7) and (8). The numerical resolution of (9) gives the value of x_d then the values of L and N_s through expressions (5) and (4) respectively. From N_s equation (10) it is possible to know D, the equivalent diameter.

The value of E is not very important for the electromagnetic design, it interferes mainly on the design of the coils : if E is small few turns are wounded with a "large" conductor as the current is important, on the contrary for a given torque a large value of E increases N_s (then N_c) and decreases D.

Armature Losses.

The S.C. conductor is subjected to an elliptic rotating induction. As the losses are only well known for an uniaxial varying field, we assume that the losses for the armature are those given by the highest component (i.e. radial) of the field varying at f. We assume a linear axial variation of the magnetic field in the coil ends to take into account the losses in that area. The losses are calculated when the motor works at maximal torque and they are multiplied by the refrigeration ratio ζ to be corrected for temperature.

Weight and overall Volume of the Machine

The weight is mainly determined by the magnetic cores (armature and inductor) and the magnets as non metallic and non magnetic materials (fiber-glass or carbon-glass) are used for the cryostat. The volume of the machine is the volume of the armature cryostat V_T; outside radius: r_e and total length: L_t. L_t and r_e are defined by cryogenic constraints :

$$L_t = L_b + L_c \qquad r_e = r_s + e_c + d_3$$

where L_b is the total length of the coils (active part and coil ends), e_c the thickness of the armature magnetic core, L_c and d_3 minimum cryogenic distances.

Design of a 500 kW, 1000 rev/min, 50 hz Machine.

The following curves are plotted for :

E = 150 V, 2 full layers of conductor with $\alpha = 72°$ and $k_s = 0.7$, a remanence of 1.4 T (NdFeB at 150 K) and $\beta = 125°$, $r_1/r_2 = 0.7$.
$d_1 = 40$, $d_2 = 70$, $d_3 = 25$, $L_c = 300$ (figures in mm).

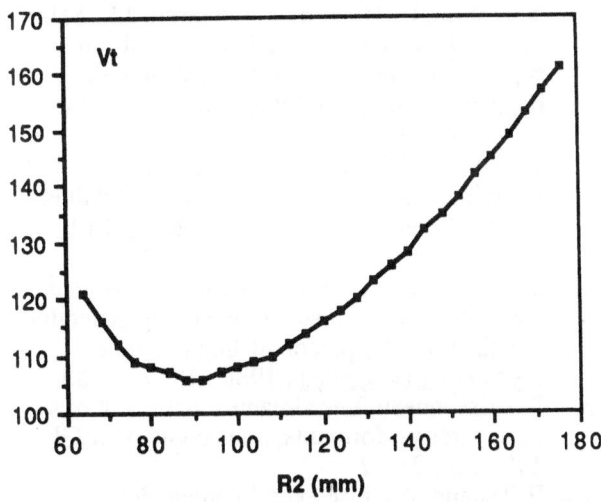

FIGURE 4a : overall volume $V_T = \pi\, r_e^2\, l_t$ (10^{-3} m³)

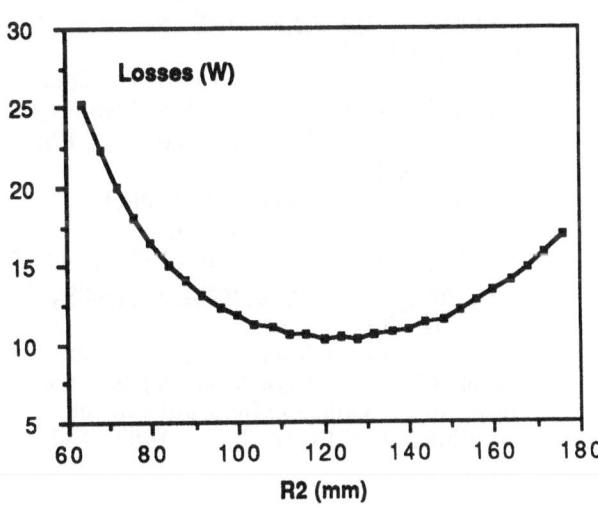

FIGURE 4b : losses of the S.C. armature (W at 5K)

For small radius r_2 the coupling between armature and magnets is weak and the no-load induction B_0^r is low. To reach the designed torque, x_d (fig.4c) L and the overall volume (fig.4a) are high. The operating field $B_0^r\sqrt{1+x_d^2}$ and the amount of superconductor are high leading to high loss values (Fig.4b).

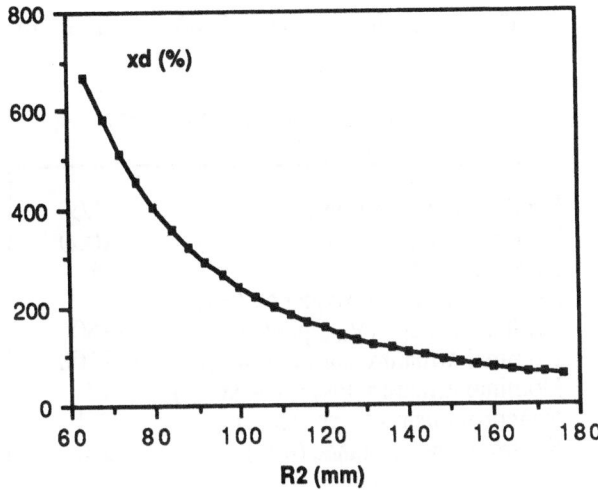

FIGURE 4c : synchronous reactance (%)

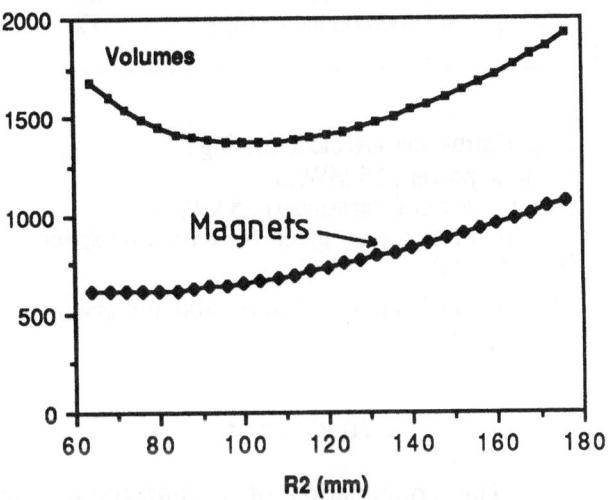

FIGURE 4d : volume of magnet and of magnetic cores (10^{-5} m³)

For large radius r_2, due to a better coupling, the no-load induction increases : L and x_d (Fig.4c) decrease but as the total length is then imposed by the cryogenic distance L_c the overall volume increases (Fig.4a). As the coupling is better the fault quantities increase and the armature conductor section is important to insure the no-quench criterion. Associated with a higher field, it explains the increase of the losses (fig.4b).

For our design, from the curves the value of 100 mm is optimun for r_2. For this choice, the volumic and specific powers are maximum with the lowest value of x_d for a minimum overall volume.

For only one full layer of conductor V_T increases by 50%. For three full layers V_T decreases only by 16% whereas the losses are too high (35 W; 300% increase).

The parameters of the 500 kW machine are reported in the table.

TABLE: 500kW machine parameters

Maximum torque (Nm)	4800
Nominal rotation speed (rev/min)	1000
Number of pole	**6**
Maximum supply frequency (Hz)	50
No-load voltage (V_{rms}/phase)	150
Nominal armature current (A_{rms}/phase)	1100
Nominal armature losses (W at 5K)	12
Synchronous reactance (p.u.)	2.4
Subtransient reactance (p.u.)	2.0
Overall diameter (mm)	430
Overall length	730
Weight of magnet (kg)	50
Weight of superconductor (kg)	17

The performances are the following :
- volumic power : 14 MW/m^3
- specific power (assessment) : 5 kW/kg
(these two figures are given for a rotation speed of 3000 rev/min).
- electrical efficiency : 98 % (ζ=800 and cryogenic losses out of account).

CONCLUSION

The combination of a superconducting armature and rare-earth magnets seems to be interesting for volumic and specific power. Such machines are characterized by a high per unit synchronous reactance. This first design based on a no-quench criterion (maximum reliability) shows the possible oportunities and the construction of a small scale machine (20 kW) will check the real possibilities of this new configuration.

REFERENCES

1. A. Hughes, T.J.E. Miller , Analysis of fields and inductances in air-cored and iron-cored synchronous machines. , Proc. IEE , 1977 , **124** , 121-126.

2. P. Dubois, A. Février, J.C. Renard, J.C. Goyer, H.G. Ky, Behaviour of multifilamentary NbTi conductors with very fine filaments under AC magnetic fields, J. Physique , 1984 , **45** , 467-471

3. J.R. Cave, A. Février, T. Verhaege, A. Lacaze, Y. Laumond, Reduction of AC losses in ultra fine multifilamentary NbTi wires, IEEE Trans. on Mag. , 1989 , **25**, 1945-1948

4. A. Février, A. Gueraud, J.P. Tavergnier, Y. Laumond, A. Lacaze, Thermo-electromagnetic stability of ultra-fine multifilamentary superconducting wires for 50-60 Hertz use , IEEE Trans. on Mag. , 1989 , **25**, 1496-1499

5. A. Février, Y. Laumond, Prospective uses of superconductors for 50/60 Hertz applications , In ICEC 10 proceedings , Butterworths , 1986 , pp. 139-152

6. Y. Brunet, P. Tixador, T. Lecomte, J.L. Sabrié, First conclusions on the advantages of full superconducting synchronous machines , Electric Machines and Power System , 1986 , **11** , 511-521

7. Y. Brunet, P. Tixador , P. Vedrine, Experimental results of an experiemental three-phase AC superconducting armature , IEEE Trans. on Mag. , 1989 , **25**, 1811-1814

8. Y. Brunet, P. Tixador, H. Nithart, Cryogenic conceptions for full superconducting generators : realization of superconducting armature cryostat , Cryogenics , 1988 , **28**, 751-755

9. P. Brissonneau, Les aimants à base de terres-rares : trés performants, mais couteux , RGE , 1987 , **3** , 21-27

10. P. Tenaud, Analyse expérimentale des mécanismes de coercivité dans les aimants NdFeB frittés , Thesis , 1988 , University Joseph Fourier Grenoble 1

11. D. Givord, A. Lienard, P. Tenaud, T. Vadieu, Magnetic viscosity in NdFeB sintered magnets , J. of Magnetism and Magnetic Materials , 1987 , **67** , L281-L285

12. D. Givord, H.S. Li, R. Perrier de la Bâthie , Magnetic properties of $Y_2Fe_{14}B$ and $Nd_2Fe_{14}B$ single cristals , Solid State Commun. , 1984 , **51** , 857-860

13. H. Fujii, W.E. Wallace, E.B. Boltich, Concerning magnetic characteristics of $(R_{2-x}R_x')Fe_{12}Co_2B$ (R=Pr and R'= Tb and Dy) , J. of Magnetism and Magnetic Materials , 1986 , **61** , 251-256

14. Y. Brunet, J.L. Sabrié, P. Tixador, Electrical design of air-cored synchronous generators using superconducting field and armature windings , IEE Proceedings , 1987 , **134** , 47-52

45 T, STEADY STATE

R.J. Weggel, M.J. Leupold, J.E.C. Williams, and Y. Iwasa

Francis Bitter National Magnet Laboratory*

Massachusetts Institute of Technology, Cambridge MA 02139

ABSTRACT

We discuss methods and results of calculations for hybrid magnet systems for extremely intense continuous magnetic fields. We focus on a system consisting of a 20 MW radially-cooled insert magnet of 33 mm bore and 406 mm outer diameter, operating within the bore of a 15 T superconducting magnet. We predict such a system to be capable of 45.5 T, given typical cooling-system parameters and the structural efficiency which we expect of our "monohelix" form of construction.

For generality we examine the effect of varying each of seven input parameters. Most significant is background field, for which each increment of one tesla changes the total field by 0.75 T. Also crucial is structural efficiency: conventional Bitter construction is scarcely adequate for this field. Significant, too, are power consumption and insert volume. Less important are available water pressure and flow, and permissible temperature rise.

INTRODUCTION

In anticipation of a possible request for a competitive proposal from the National Science Foundation for a higher field magnet laboratory, the Francis Bitter National Magnet Laboratory of M.I.T. is proposing the construction of a 45 T hybrid system with an insert magnet consuming approximately 20 MW.

A new plant must allow continued operation of some 24 magnets which are matched to the present power supply and cooling system. We are free, however, to consider an additional cooling system with higher water pressure, as well as to connect modular power supplies in series for the sake of keeping current within bounds. Part of the exercise is to define the basic parameters of plant enhancements.

HYBRID MAGNET SYSTEM

A hybrid system consists of a compact high performance water-cooled magnet (insert) operating in the room temperature bore of a large superconducting magnet housed within a cryostat. With respect to the plant the insert can be characterized by its electrical and hydraulic impedances, which have to be matched to the power supply and coolant circulating system. The superconducting magnet should be as high in field and as large in bore as one can afford, and tolerant of inductive interactions with the insert magnet. The cryostat should be efficient, compact and robust. Insert burnouts, which must be accepted as inevitable, dominate the cryostat design because of the tremendous fault forces which can occur between the magnets.

---

* Supported by the National Science Foundation.

WATER-COOLED MAGNET DESIGN

All of our water-cooled magnets, of which inserts for hybrid systems are an important subset, employ either the Bitter plate construction or the "monohelix" variant thereof.[1,2] Variations in current density, j, with axial coordinate, z, can be achieved by adjusting turn thicknesses. Similarly, turn composition can be adjusted to achieve electrical or mechanical property grading. The variation of current density with radial coordinate, r, within any turn automatically is inversely proportional to the product of radius and resistivity. Magnets consisting of two concentric coils have a two step variation of $j(r)$ and material properties. Fields are computed from the current density throughout the winding volume.

A water-cooled magnet should be reliable and efficient, consuming no more power than necessary. Our design procedure strives for reliability by limiting the peak temperature anywhere in the magnet, and matching the local strength of the windings everywhere to the local stresses. Our procedure maximizes efficiency by allocating power among all regions of the magnet in an "optimum" way. At this optimum all regions, except those constrained by peak temperature limitations, have the same incremental efficiency, measured in T/MW, as defined by one of two equations.[3] One applies for regions which are not stress limited, the other for regions which are. In the latter, the conductor resistivity depends not merely on its temperature, but also on the strength which it is required to have. Temperature depends upon power density, thermal conductivity, heat flux, and cooling water temperature and velocity. To a first approximation, stress depends on the product of three factors: current density, radius, and ambient magnetic field.

We partition a magnet into many axial zones. In each we tentatively choose a conductor of approximately appropriate strength and conductivity, and adjust the turn thickness until the local strength of the structure matches the local stress. In order to match the local stress arbitrarily closely, the conductor will typically be a composite of two conductors, one weaker, the other stronger than required, and combined according to the rule of mixtures. If the resulting peak temperature is higher than permissible, we substitute a conductor of higher conductivity but lower strength, increasing the turn thickness to once again match strength with stress.

Then we compare the incremental efficiencies of all zones. In every zone whose temperature is less than allowed, and whose efficiency is above average, we replace the conductor with one of higher strength but lower conductivity. In zones with lower than average efficiency we increment the conductivity, if possible. If not (because the zone already employs pure copper), we usually increase the turn thickness, even though the turn thereby becomes stronger than necessary. But if this results in excessive efficiency as now computed from the equation applicable when incremental efficiency no longer depends on stress, the turn thickness is left unchanged.

For purposes of analysis and design we use an empirical linear relationship between conductivity c, measured in % I.A.C.S., and strength s:

$$c = 100 \qquad (s < 0.2S) \qquad (1a)$$

$$c = 125\left(1 - \frac{s}{S}\right) \quad (s > 0.2S) \qquad (1b)$$

For heavily cold worked conductors in the form of thin sheet such as used in Bitter magnets, S typically is about 1725 MPa (250 ksi). For billets of the sort needed for monohelices, S is only 80% to 90% of this value; this percentage we define as the "material efficiency factor" of the conductor.

To estimate the strength of a structure, not merely that of the materials with which it is built, the materials' strength must be scaled down by a "structural efficiency factor" due to the presence of slits and other discontinuities and fatigue effects; for Bitter magnets S must be reduced to about 45% of the value appropriate for the conductors themselves.

We then define a "stress factor" to quantify the strength of coils. This stress factor is the product of the structural efficiency and material efficiency factors. For example, a construction with a structural efficiency factor of 75% and a material efficiency factor of 80% has a stress factor of 60%.

HYBRID SYSTEM ANALYSIS

Base Case

There are so many parameters to be considered in the design of a hybrid magnet that it is not feasible to consider all possible combinations. Instead, one can focus on a "base case" and consider variations from this design in which only one, or at most two, parameters have been varied simultaneously. These base case parameters may be more or less arbitrary, provided that they enable one to predict the increment in total field resulting from an increment in any parameter.

The base case assumes a 33 mm bore insert consuming 20 MW. This bore size is standard at the FBNML for magnets of the highest performance. The power is postulated to be delivered at 40 kA and 500 V, from four modular units each delivering 20 kA at 250 V. Such a supply, a generous doubling of our present capability, retains compatibility with present operations.

The insert consists of two radially-cooled coils, electrically and hydraulically in series and each with two cooling plates per turn, as in our present hybrid systems. The inner coil is of 156 mm outer diameter, as in our present 8-MW hybrids; its length has been increased 25% to 254 mm, to accept more power. The outer coil is of 406 mm length and o.d., the diameter of the largest plates in any of our magnets. Both dimensions are about 20% larger than in present hybrids, again to accept more power. Each coil is partitioned axially into eight zones: four equal length zones on either side of the midplane.

Water flow is 100 liters per second, which is nearly optimum, given the other inlet parameters, and the water pressure drop across the stack is 15 atmospheres. The peak temperature rise anywhere in the magnet is limited to 100° C above the inlet temperature of 20°C.

We assume a stress factor of 60%; i.e., in Eq. (1), S has the value 1035 MPa (150 ksi). From experience with Bitter coils we can make a good estimate of the structural efficiency of the design; that is, the maximum permissible stress factor consistent with acceptable longevity. Our experience with monohelices is as yet not sufficient for us to be able to make such well calibrated estimates.

We estimate the stresses as the mean of the unsupported hoop stresses at the inner and outer radii of each zone. Each hoop stress $s(r)$ is simply

$$s(r) = B(r)j(r)r, \qquad (2)$$

where $B(r)$ and $j(r)$ are the magnetic induction and current density at radius r. $j(r)$ is the peak current density at that radius in the coil, averaged over the minimum conductor thickness at any circumferential position. That is, it is the overall current density, divided by the fraction of coil which is unbroken conductor, not insulation, cooling passages, or slits. In turns of the midplane zone of the inner coil of the base case, conductor plates constitute 92% of the axial thickness; with the removal of conductor for cooling passages the effective thickness is reduced to 84% of this. In Bitter coils and partial monohelices, some fraction (9% for Bitter coils) is lost to slits. Experience shows that these slits also greatly reduce the structural efficiency of the coil. Bitter coil are reliable only for a stress factor of about 45% or so. The par-

tial monohelix construction appears to have a structural efficiency factor which is significantly higher, between 55% and 60%. The complete monohelix, which is completely devoid of slits, should have a structural efficiency factor which is considerably better still, perhaps 80% or so.

Given the above input parameters, the background field required to generate a total field of 45 T turns out to be about 15 T. This, then, is our base case from which our parameter study proceeds.

Parameter Study

To determine the most profitable avenues for improving the performance of the base case design, we have varied, over a range of plus and minus 25% to 40%, each of five input parameters: background field, insert size, water pressure, flow, and allowable temperature rise. For greater thoroughness we have also varied simultaneously the two most important insert parameters, power and stress factor.

RESULTS AND DISCUSSION

The results of our parameter analysis are summarized in Table 1 and Figures 1 through 4. Table 1 is an abbreviated listing of our computations of the total field of systems differing from the base case by one of its seven input parameters. The full table constitutes the basis for Figs. 1 through 3.

Figure 1 plots the dependence of total field on the parameters: background field, insert power, insert size, and water pressure drop. We note that background field is extremely significant; power and size are important; and water pressure much less so.

Figure 2 plots the dependence of total field on the remaining parameters: insert strength, water flow, and peak temperature limit. Here we note that strength is crucial, whereas neither flow nor the temperature limit is so important. In an apparent paradox it would seem that there is such a thing as too much cooling flow. To achieve such flow without increasing water pressure requires enlarging the cooling passsages, thereby removing conductor better retained for strength and conductance.

The relative importance of the various parameters is shown in the bar chart of Fig. 3, which shows the rate at which total field changes per unit percentage change in each parameter. We see that background field is by far the most influential parameter. All other parameters must be incremented by at least 6% to achieve a 1% (0.45 T) improvement in total field.

Figure 4 plots the results of the only bivariable parameter analysis. Here we see the fields resulting from combinations of insert power and strength, each ranging over a factor of two. From the graph we can also see, by scanning horizontally and noting the intercepts, the strong dependence of required power on insert strength. For example, to generate 45 T with a system with a stress factor of 80% requires only 14.5 MW. Were the stress factor only 56%, 20 MW would be necessary.

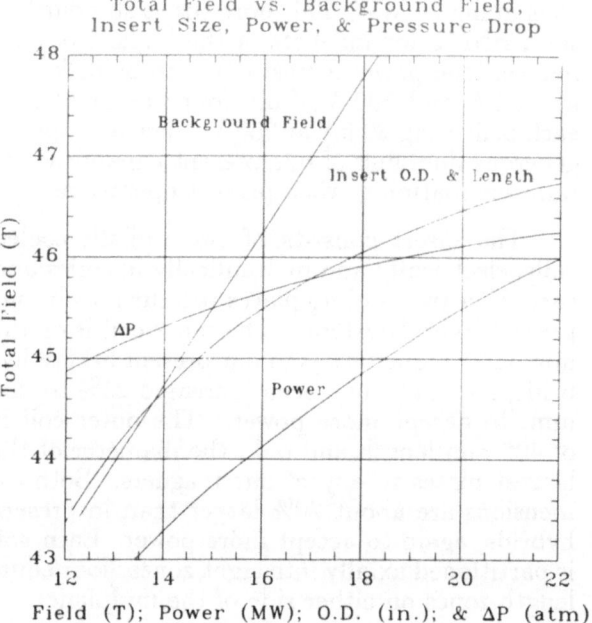

Fig. 1 Total field of 33-mm bore hybrid systems as four of seven input parameters are varied, one at a time, from their base values: background field=15 T; insert power=20 MW; o.d.(=length) of outer coil=406 mm (16 in.); water flow=100 liters/s; water pressure drop through magnet=15 atm; peak temperature rise=100 °C (above supply temperature of 20 °C); insert strength "stress factor" (defined in text)=60 %.

Table 1 Total Field of Hybrid System *vs.* Input Parameters

Param.*	Zone Code†	Inner Coil (MW);(T)	Field (T)	T/MW
Base‡	4433;4333	6.71;16.66	45.50	0.295
12T	4433;4331	6.65;17.00	43.21	0.330
14	4433;4332	6.69;16.77	44.73	0.305
16	4433;4333	6.70;16.54	46.26	0.288
18	4433;4333	6.70;16.31	47.78	0.275
20	4433;4333	6.70;16.08	49.31	0.262
45%	4433;4333	6.51;15.29	43.32	0.243
50	4433;4333	6.58;15.80	44.15	0.262
70	4433;4331	6.75;17.32	46.51	0.331
80	4432;4331	6.93;17.93	47.37	0.356
8MW	4311;3311	4.14;15.28	39.24	0.893
12	4331;3321	5.18;16.08	42.18	0.576
16	4433;3331	6.07;16.49	44.12	0.415
24	4433;4433	7.92;16.95	46.40	0.190
12in.	4443;4443	8.36;17.41	43.38	0.165
14	4433;4433	7.16;16.93	44.67	0.251
18	4433;4331	6.42;16.46	46.05	0.332
20	4433;4331	6.25;16.32	46.47	0.360
80°C	4443;4433	6.54;15.94	44.59	0.205
120	4433;3332	7.44;17.37	45.96	0.334
140	4333;3332	8.01;17.88	46.25	0.351
12atm	4433;4333	6.35;16.08	44.87	0.275
18.0	4433;4332	6.97;17.06	45.91	0.312
80ℓ/s	4433;4433	7.13;17.04	45.67	0.263
120	4433;4333	6.41;16.14	44.90	0.286

* The parameters in column 1 are consecutively, 1) background field, in T; 2) stress factor (defined in text), in %; 3) insert power, in MW; 4) o.d. (=length) of outer coil, in inches; 5) peak temperature rise, in °C; 6) water pressure, in atm; and 7) water flow, in liters/s.

† four zones, numbered from midplane, in each coil.
    1: copper, efficiency limited;
    2: copper, stress limited;
    3: stress and efficiency limited;
    4: temperature limited.

‡ Defined in caption of Fig. 1.

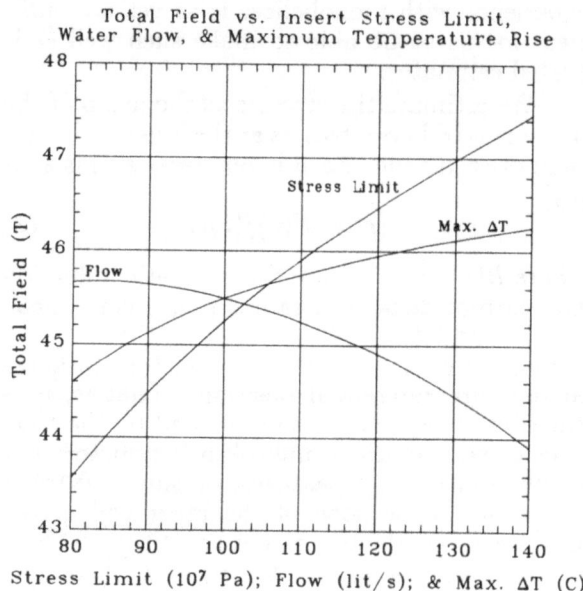

Fig. 2 Total field of 33-mm bore hybrid systems with one-at-a-time variations of the three input parameters not varied in Fig. 1. Base values of all parameters are as in Fig. 1.

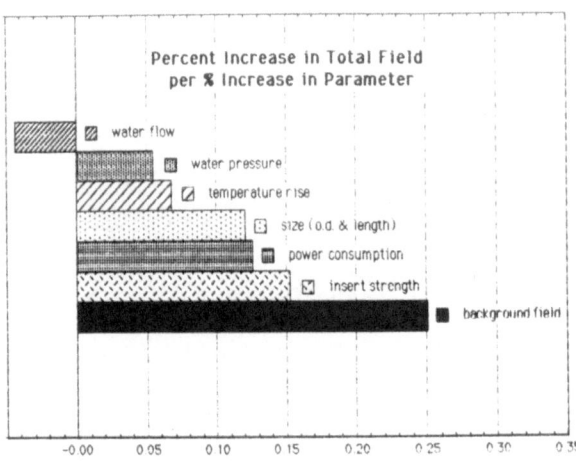

Fig. 3 Fractional variation in total field of 33-mm bore hybrid systems with one-at-a-time variations in each of seven input parameters. Base values of all parameters are as in Figs. 1 and 2.

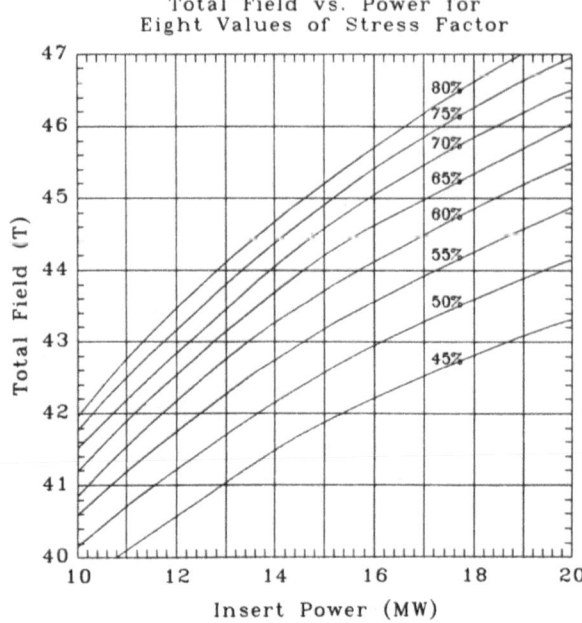

Fig. 4 Total field of 33-mm bore hybrid systems *vs.* power consumption and strength of insert, as quantified by its "stress factor" (defined in text). All other input parameters are as in Figs. 1 through 3.

SUPERCONDUCTING MAGNET

The degree of stability, ranging from fully cryostable to adiabatic, is an important design consideration for superconducting magnets. In a hybrid magnet system the superconducting magnet can be almost adiabatic. There has to be enough cooling to maintain superconductivity in the face of dissipations arising from splice joints and sweep induced phenomena, but a quench is acceptable in a catastrophe such as an insert burnout, which in itself would necessitate a shutdown. We coined the term quasi-adiabatic to describe such a design.[4]

Structural efficiency, best achieved through the use of intrinsic conductor strength rather than reinforcing members, requires that superconductors have stabilizer material which is either heavily cold-worked copper or one of its higher-conductivity alloys. The quasi-adiabatic design concept, if properly executed, enhances structural efficiency by bringing otherwise unused conductor strength into play. In a stabilized winding containing voids for cooling by liquid helium, radial stiffness is inherently low, and, consequently, radial Lorentz forces are resisted solely by local hoop stresses. But by removing voids and thereby increasing the radial modulus, otherwise overstressed conductor near the inside radius can be reinforced by understressed conductor further out. The magnet can be smaller, because the conductor can be sized for a lower peak stress level.[4] Table 2, giving vital statistics for a cryostable coil and a QAD coil, shows the advantages which the QAD concept confers.

Table 2

Cryostable and QAD Magnets for Base Case

Parameter	Cryostable	QAD
central field (T)	15	15
peak field (T)	15.25	15.64
operating current (A)	2,875	2,940
stored energy (MJ)	123.0	50.7
overall conductor j (A/cm^2)	2,497	4,623
heat flux (W/cm^2)	2.0*	8
inside diameter (mm)	520	520
outside diameter (mm)	1748	1151
length (m)	1.55	1.10

* conditional stability.

Superconducting performance of both Nb_3-Sn and NbTi is improved by lowering temperature, and there are additional cooling advantages with subcooled, superfluid liquid helium. The elimination of the need for convection is especially beneficial in regard to QAD. Surface heat fluxes can be higher. For instance, with helium at 4.2 K unconditional cryostability allows only ~ 1 W/cm^2, and conditional cryostability with cold-end recovery, ~ 2 W/cm^2. Cooling of conductor by sub-cooled, superfluid helium in short passages allows heat fluxes as high as ~ 8 W/cm^2 for a short time. This is the basis of the QAD stability.

Winding pack details are known in principle at this time if not in detail. There is a lingering apprehension concerning the use of Nb_3Sn on account of its unforgiving strain properties, and as long as it persists, we look for a hedge. In the case of Hybrid III it took the form of confining the NbTi and Nb_3Sn to separate coils. Alternatively, both materials can be combined in pancake coils, since one can replace those pancakes which are found to contain defective or damaged conductor.

COMMENTS

While our analyses are generally applicable to the design of hybrid systems, they were conducted in the context of magnet technology as practiced by us. In particular is the radially cooled Bitter concept with certain inherent possibilities with respect to how current and strength can be distributed. Other construction styles, such as the polyhelix, place different possibilities and constraints before the designer.

REFERENCES

[1] R.J. Weggel, J.E.C. Williams, M.J. Leupold, E.S. Bobrov, and J.A. Dalessandro, The Radially-cooled continuous helix: Design study and fabrication of test coil, *9th Int'l. Conf. on Magnet Tech.*, 439 (1985).

[2] R.J. Weggel, Monohelix research at the MIT Magnet Lab., *IEEE Trans. on Mag.* **24**, 1384 (1988).

[3] R.J. Weggel, Optimum partition of power to maximize field of magnet system, *J. Phys. (Paris)* **25**, C41 (1984).

[4] M.J. Leupold, R.J. Weggel and Y. Iwasa, Hybrid III system, *IEEE Trans. Magn.* **MAG-24(2)**, 1070 (1988).

A DESIGN OF 50 T HYBRID MAGNET
FOR QUASI-STATIONARY OPERATION

Y. Nakagawa, G. Kido, S. Miura, A. Hoshi, K. Watanabe and Y. Muto
Institute for Materials Research, Tohoku University, Sendai 980, Japan

ABSTRACT

The Tohoku hybrid magnet has produced the field of 31.1 T, where the superconducting and water-cooled coils generate 11.5 T and 19.6 T, respectively. It is possible to raise the field to 40 T by a simple reinforcement. It seems, however, that the continuous 50 T field is difficult to be produced. Thus we are planning to construct a 50 T quasi-continuous magnet with a duration of 1 s. The Joule heat of a resistive coil is partially removed during the field generation. The repetition frequency is much higher than that of a conventional pulsed magnet. The required energy is much reduced if a 10 T superconducting coil is placed outside of a 40 T resistive coil.

INTRODUCTION

The High Field Laboratory for Superconducting Materials, Institute for Materials Research, Tohoku University has three hybrid magnets in operation [1]. The largest one, named HM-1, has produced the field of 31.1 T in a room temperature bore of 32 mm. The continuous field of around 30 T has also been produced at MIT [2], Grenoble [3] and Nijmegen [4], where a project to enhance the field to 35 T or 40 T is being pushed forward. In Japan, the National Research Institute for Metals at Tsukuba will complete the 40 T hybrid magnet in a few years [5].

More intense fields are difficult to be produced continuously. Thus the future project of our laboratory is directed to the construction of a quasi-continuous magnet producing the field more than 50 T. The 40 T magnet with a duration of 0.1 s is in operation at Amsterdam University; the field will be enhanced to 60 T in near future [6].

There are two different approaches to the problem on the quasi-continuous magnet: an elongation of the duration of the pulsed magnet and an extension of the hybrid magnet to higher fields. The present paper deals with the latter after a general discussion on various limitations in generating the high magnetic field.

LIMITATIONS
IN PRODUCING HIGH FIELDS

Mechanical Limitation

We examine a cylindrical field coil with

an infinite length. The magnetic flux density B varies with both the current density j and the radius r, as given by

$$dB/dr = -\mu_0\lambda j, \qquad (1)$$

where λ is a packing fraction of the conductor in the coil. The Lorentz force exerted on the conductor gives rise to the azimuthal stress:

$$\sigma = jBr. \qquad (2)$$

The coil is assumed to consist of thin conductor layers supported independently to eliminate the stress increase due to the radial strain.

If the current density is uniform ($j = j_0$) in the coil with an inner radius a_1 and outer radius a_2, the stress σ has a maximum σ_m at $r = a_2/2$, provided that $a_1 < a_2/2$. The field inside the coil is given by

$$\begin{aligned} B_0 &= \mu_0\lambda j_0(a_2 - a_1) \\ &= 2(\mu_0\lambda\sigma_m)^{1/2}(1 - a_1/a_2). \end{aligned} \qquad (3)$$

Higher fields can be produced by the uniform-stress coil ($\sigma = \sigma_m$). Combining eqs.(1) and (2), we obtain

$$B_0 = (2\mu_0\lambda\sigma_m \ln a_2/a_1)^{1/2}, \qquad (4)$$

where we assume that λ is constant in contrast to a real magnet with non-uniform current density. Equations (3) and (4) are plotted in Figure 1 using typical values of σ_m to be compared with the yield strength of some conductor alloys: 36 kg/mm^2 for cold-worked pure copper and Cu-0.17% Ag alloy, 55 kg/mm^2 for Cu-0.4% Al$_2$O$_3$ and 100 kg/mm^2 for Cu-18% Nb. Here we have tentatively assumed that $\lambda = 0.8$.

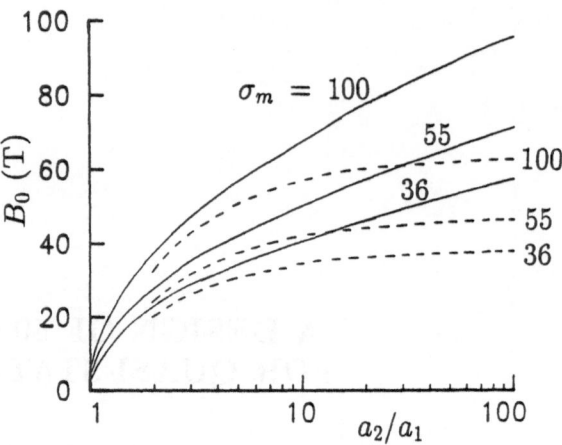

FIGURE 1. Maximum field produced by uniform-current-density coil (- - -) and uniform-stress coil (——) with outer radius a_2 and inner radius a_1. Conductor yield strength is σ_m in kg/mm^2. The packing factor $\lambda = 0.8$.

When the dimension ratio a_2/a_1 is large, the uniform-stress coil is much more advantageous than the uniform-current-density coil. It should be emphasized that the mechanical limitation is equally important for continuous and pulsed fields. The same principle should be applied also to the hybrid magnet; the superconducting coil should withstand the same stress as the resistive coil in an ideal design.

Thermal Limitation

In a continuous resistive magnet the Joule heat should be removed stationarily. Usually the coil is cooled by water flowing through channels in the conductor. The contact area S per unit volume should be large enough to satisfy the following relation:

$$\kappa\Delta TS = \rho j^2, \qquad (5)$$

where ΔT is the temperature difference between the conductor and water, ρ the electrical resistivity and κ the heat transfer coefficient depending on the flow velocity of water. If the coil is cooled by cryogenic fluids such as liquid nitrogen and liquid neon, the power consumption is much reduced because of the decrease in the resistivity. The circulation of the cryogenic fluids, however, requires elaborate techniques. Such a magnet producing continuous 30 T field was constructed at NASA in U.S.A. [7].

The cryogenic magnet is very useful for the pulse operation where the Joule heat is not removed during the field generation. The heat balance during the time interval dt is expressed as

$$\rho j^2 \, dt = C \, dT, \qquad (6)$$

where C is the specific heat per unit volume of the conductor. The resistivity of copper alloys may be approximated by

$$\rho = k\rho_0 + \rho_r, \qquad (7)$$

where ρ_0 is the resistivity of annealed pure copper, being proportional to T^5 or T at low or high temperature, respectively; k is a numerical factor not so different from unity; ρ_r is the temperature-independent residual resistivity. The specific heat C, on the other hand, is insensitive to alloying.

We derive from eq.(6) that

$$\tau j_0^2 = \int_{T_0}^{T_1} C/\rho \, dT, \qquad (8)$$

where T_0 and T_1 are, respectively, the initial and final temperatures of the conductor, which carries the constant current density j_0 during the time interval τ. Figure 2 shows typical examples of C/ρ vs. T curves. The case (a) corresponds to annealed pure copper, the case (b) to Cu-Ag alloy, the case (c) to Cu-Al$_2$O$_3$, and the case (d) to Cu-Al$_2$O$_3$ in high magnetic fields. We have measured the magnetoresistance of this material at 4.2 K which amounts to about 0.10 $\mu\Omega$cm at 23 T.

The integral of eq.(8) is evaluated from Figure 2. Since C/ρ is very small below 20 K, the Amsterdam magnet [6] has been cooled by liquid neon rather than liquid helium. In most cases, it may be sufficient to choose that $T_0 = 77$ K, i.e. the coil is cooled by liquid nitrogen. The final temperature T_1 should be lower than a permissible value for the insulator of the coil. Here we choose that $T_1 = 373$ K and $T_0 = 77$ K. From eq.(8) with $\tau = 1$ s we obtain $j_0 = 2.6 \times 10^8$ A/m^2 for the case (d). If this value is inserted in eq.(3) for the uniform-current-density coil with $a_2 = 0.2$ m and $\lambda = 0.8$, we obtain $B_0 = 52$ T. It should be noticed that the uniform-stress coil is less advantageous than the uniform-current-density coil because of inhomogeneous heating of the conductor. This fault may be somewhat reduced if the conductor is exposed to the cooling fluid. The partial cooling by fluid during the field generation enables us also to increase the repetition frequency of the pulsed field.

We also deal with a quasi-continuous coil cooled by 10°C water, as described in a later

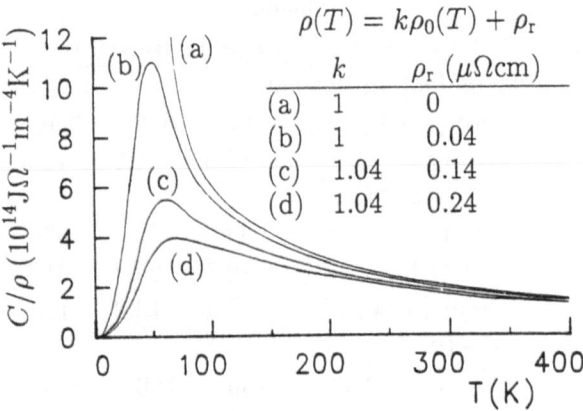

FIGURE 2. C/ρ vs. T curves. C: specific heat per volume, ρ: electrical resistivity, T: temperature.

section. If we take $T_0 = 283$ K in the above calculation, we obtain the field of only 23 T. To obtain the higher field the coil should be cooled by flowing water during the field generation.

Energy Limitation

The resistive magnet for continuous field is energized by a DC source of several MW. The electric power W is proportional to the square of the field B_0, if both the dimension and temperature of the coil are kept constant.

As can be seen in eq.(3), higher fields can be produced by larger coils. As a result, the increase of W is much more than that proportional to B_0^2. A use of the hybrid magnet enables us to save the power considerably.

For the pulsed coil, the high field production is limited by the total energy Q rather than the power W. Instead of a continuous supply of MW power, we can use a storage of MJ energy such as a fly-wheel, a chemical condenser bank and a

large superconducting coil. This energy is consumed by the coil which is heated from T_0 to T_1. The heat balance is given by

$$Q = V \int_{T_0}^{T_1} C \, dT, \qquad (9)$$

where V is the volume of the conductor. We can also save the energy for the pulsed magnet by using the hybrid magnet. The superconducting coil, however, should withstand the large induced voltage due to the rapid change of the field produced by the resistive coil.

DESIGN OF QUASI-CONTINUOUS MAGNET

Modification of Present Hybrid Magnet

Table 1 lists the characteristics of our hybrid magnet HM-1a [1,8] together with those of virtual magnets HM-40 and HM-50 which may produce 40 T and 50 T, respectively. The size of these magnets and the ratio of the fields of superconducting and water-cooled coils are assumed to be the same. It is also assumed that both the electric power consumed by the water-cooled coil and the electromagnetic stress exerted on the conductor are proportional to the square of the field. The stress of each helical coil is nearly constant, satisfying the optimal condition.

The characteristics of HM-40 seem to be reasonable: the superconducting coil producing 14.8 T can be realized by the present technique, the electric power of 12.2 MW is tolerable and the high-conductivity alloy with the strength of 40.4 kg/mm^2 may be available. It should be noticed, however,

TABLE 1

Modification of HM-1a to produce higher fields

	HM-1a	HM-40	HM-50
Superconducting magnet			
bore (mm)	360	360	360
field (T)	11.5	14.8	18.5
Water-cooled magnet			
bore (mm)	32	32	32
field (T)	19.6	25.2	31.5
power (MW)	7.4	12.2	19.1
conductor			
outer radius (mm)	150	150	150
inner radius (mm)	19	19	19
half height (mm)	150	150	150
electromagnetic stress (kg/mm^2)	24.4	40.4	63.1
Total field (T)	31.1	40	50

that the water cooling is rather difficult. The flow rate of cooling water in HM-1a is 350 m³/h and the delivery water pressure is 20 kg/cm², giving a reasonable temperature rise of water. In order to keep the water temperature of HM-40 similar to that of HM-1a, the flow rate should be 579 m³/h with the delivery pressure of 55 kg/cm². The pressure may be lowered by widening the cooling channel which may result in smaller packing factor λ. Moreover, the temperature difference ΔT in eq.(5) should be lowered by increasing the contact area S. This means that the number of helical coils should be increased with the fixed value of the outermost radius of the polyhelix coil. Thus the practical design of 40 T hybrid magnet becomes more elaborate although it does not seem impossible.

The HM-50 in Table 1, on the other hand, is lacking in reality. Even if the 18.5 T superconducting magnet could be realized, the continuous 50 T magnet would not be constructed because of the difficulty in cooling the resistive magnet. Moreover, the stress of 63.1 kg/mm² is too high for usual high-conductivity alloys. The stress can be reduced if the outer radius of the water-cooled magnet is enlarged. This alteration, however, results in the power consumption more than 19.1 MW. Also it required a larger bore of the 18.5 T superconducting magnet. It is unlikely that such a superconducting magnet is realized at present.

Partial Cooling of Quasi-Continuous Coil

Thermal limitation of the pulsed coil may be relaxed by the forced cooling during the pulse. The full cooling has been experienced in designing a high-power

water-cooled magnet for stationary operation. The partial cooling is meaningful if the energy removed by the cooling is comparable with the integrated heat capacity of the coil given by eq.(9).

Figure 3 shows an example of the field and temperature of a quasi-continuous magnet energized by a large condenser bank and cooled by circulating water. The coil has an inner radius of 19 mm, an outer radius of 200 mm, a half length of 200 mm, a conductor weight of 377 kg, an electrical resistance of 1.15 mΩ at 20°C with a temperature coefficient of 2.75×10^{-3} and an inductance of 207 μH. The heat transfer to water corresponding to κS in eq.(5) is estimated to be 893 kW/K. These characteristics have been determined by our experience of the water-cooled magnet with the flow rate of 350 m³/h and the initial temperature of 10°C. The condenser bank has a capacity of 494 F and an initial voltage of 360 V. The calculated maximum field is slightly more than 40 T, and the maximum temperature of the conductor is about

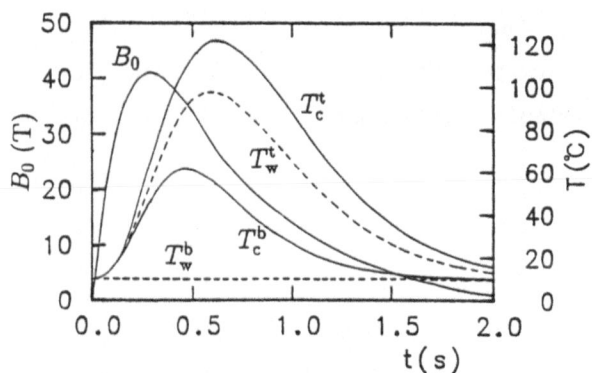

FIGURE 3. Time dependence of magnetic field B_0, coil temperatures T_c and water temperatures T_w. Superscripts t and b stand for the top and bottom of the coil, respectively (see text).

120°C. If this magnet is placed inside of a superconducting magnet of 10 T, the center field will amount to 50 T. The maximum stress, however, exceeds 60 kg/mm^2 not permissible for a usual conductor. The condenser bank mentioned above has an energy of 32 MJ. The energy storage using a large fly-wheel may be less expensive than that of the condenser bank.

We are also planning to use a cryogenic insert coil for the quasi-continuous hybrid magnet. This may enable us to save the energy source. We may possibly use the present 8 MW power DC supply for the cryogenic insert coil. We are making some preparatory experiments on the heat transfer to flowing liquid nitrogen. The final design of our quasi-continuous 50 T hybrid magnet will be completed afterwards.

ACKNOWLEDGEMENTS

The authors are greatly indebted to Messrs. T. Fujioka, H. Takano and S. Hanai of Toshiba Corporation for valuable discussion, especially for cooperation in calculating the result shown in Figure 3.

Thanks are also due to Professor K. Noto of Iwate University for the cooperation in the establishment of our High Field Laboratory.

REFERENCES

1. Nakagawa, Y., Noto, K., Hoshi, A., Miura, S., Watanabe, K., Kido, G. and Muto, Y., Present status of hybrid magnets at Tohoku University, Proc. 9th Int. Conf. on Magnet Technology, SIN, Zurich, 1985, pp. 424-7.

2. Leupold, M.J., Hale, J.R., Iwasa, Y., Rubin, L.G. and Weggel, R.J., 30 tesla hybrid magnet facility at the Francis Bitter National Magnet Laboratory, Massachusetts Institute of Technology, IEEE Trans. Mag., 1981, MAG-17, 1779-82.

3. Schneider-Muntau, H.-J., The generation of high d.c. magnetic fields, IEEE Trans. Mag., 1988, 24, 1041-4.

4. Perenboom, J.A.A.J. and van Hulst, K., The Nijmegen High Field Magnet Laboratory, Physica B, 1989, 155, 74-7.

5. Inoue, K., Takeuchi, T., Kiyoshi, T., Ito, K., Wada, H., Maeda, H., Nii, K., Fujioka, T., Sumiyoshi, Y., Hanai, S., Hamajima, T. and Maeda, H., Primary design of 40 tesla class hybrid magnet system, Proceedings of this conference.

6. Gersdorf, R., Roeland, L.W. and Mattens W.C.M., Design of magnet coils for semicontinuous magnetic fields up to 60 T, Physica B, 1989, 155, 10-7.

7. Prok, G.M., Swanson, M.C. and Brown G.V., Design and prototype fabrication of a 30-tesla cryogenic magnet, Adv. Cryog. Eng., 1978, 23, 170-7.

8. Nakagawa, Y., Miura, S., Hoshi, A., Ito, S., Nakajima, K., Takano, H., Inoue, K. and Hanai, S., Design and performance of water-cooled magnets for hybrid magnets, Sci. Rep. RITU, 1986, 23, 251-9.

THE CONSTRUCTION OF PULSED FIELD SOLENOIDS IN THE OXFORD - LEUVEN COLLABORATION

H. Jones[*], F. Herlach[**] and J. A. Lee[†]

[*]University of Oxford, Clarendon Laboratory, Oxford OX1 3PU, UK
[**]Katholieke Universiteit Leuven, B3030 Belgium
[†]AEA Technology, Harwell Laboratory, Didcot, Oxon OX11 0RA, UK

ABSTRACT

At MT10 we reported our success in exceeding the 50 Tesla barrier using a simple conductor / coil configuration. In this paper we report the progress of the collaboration over the last two years. During this time we have explored the use of alternative conductors and coil construction techniques. We describe our experiences with the use of various high strength conductors, insulation, coil winding and impregnation technology. We discuss the problems we have identified and our attempts to overcome them. We describe the relevance of this programme to a much larger European collaborative venture now in progress and the installation of a new pulsed field facility at the Clarendon Laboratory.

INTRODUCTION

Having demonstrated that it is possible to generate 50T or more using simple technology [1], the last two years have seen considerable activity directed towards the setting up of programmes which will build on this work and that of others. Two major relevant projects have been funded. The first is to install and develop a pulsed field facility at the Clarendon Laboratory Oxford, this is funded by the SERC. The second is an award under the CEC SCIENCE programme to develop high field pulsed conductor. This latter programme involves 5 European magnet laboratories including Oxford and Leuven.

In the last two years involvement in the organization of the new initiatives has precluded much effort in pushing the field level higher but we have managed to develop our magnet construction techniques and we feel that some of the lessons learned are worth reporting here. In the following sections we address Conductors and Coils, Wire Insulation, Winding Techniques and Impregnation.

CONDUCTORS AND COILS

Steel / copper

The 52T coil we have reported [1] used a square section 316L stainless steel clad copper wire. The ratio of steel to copper was 0.8:1. Since then we have been using rectangular section conductor of nominal size 2.5 x 2.0mm. In general the stainless steel: copper ratio is lower ~ 0.6:1. This adjustment to the reinforcement ratio was made to get some in-service experimental information on the lower limit of reinforcement required to produce "50T" conductor. This is important as we need to maximise the conductivity of the wire and this implies minimizing reinforcement. A further change in the latest wire is use of type 304L stainless steel as its tensile properties are marginally better at 77K. Once again the wire was produced for us by our colleagues at AEA Technology Harwell Laboratory.

This lower reinforcement wire has been used to build 4 coils at Oxford. All four were wrapped with a thick shell of Kevlar fibre as described by Jones et al [1]. The first two of these used the 316L clad wire and were wet wound (see Impregnation). They both had

18mm bores with an alpha (outer radius / inner radius) of 3.5. They achieved 40T and survived but the inductance change was ~ 2% at this field so their operating field was derated to ~38T. They have been in use at Leuven for some 18 months now.

The second two coils were wound from the 304 clad wire, had 20 mm bores (alpha = 3.14) and were vacuum impregnated. They also incorporated layer end pieces described under Winding Techniques and shown in fig. 1. These coils achieved 39T before the inductance had changed 0.5%. As they were urgently needed for an important series of experiments about to begin at Leuven, they were not tested to higher fields. These coils compare favourably with the first two described which showed 0.5% inductance change as low as 35T. Furthermore stress models such as that of Herlach et al [2], indicate that the maximum achievable field is a strong function of alpha. As the alpha of the 20mm bore coils is ~ 10% lower than that of the 18mm versions, a lower stable field would normally be expected. No comparative data exist yet for the strengths of the two conductors used but the main inference is that the 304L reinforcement gives a significant improvement.

Copper - Niobium

We were fortunate to obtain, courtesy of Dr. G. Boebinger of AT & T Bell Labs, a length of the Cu-Nb conductor produced by Supercon inc., USA. This was similar to that used by Foner to generate 68T [3]. Unfortunately the two coils produced in Oxford to our usual pattern failed due to voltage breakdown. However one of them achieved over 40T with no discernible shift in inductance and this is encouraging for future attempts with this type of wire.

Conductors under development

At Harwell, fabrication trials on conductors which can be heat treated after winding to impart high strength, are under way. One is a CuBe clad copper, as proposed by Gersdorf and Mattens [4] the other is similar but clad with a Cu-Ni-Si alloy. As both require precipitation hardening at temperatures in excess of 300C, we are considering using glass woven insulation. This process is also being developed at Harwell.

CONDUCTOR INSULATION

In reference [1] we describe our use of Kapton ribbon to insulate the stainless steel/copper conductor. We still find this the most convenient insulation to apply in-house to the development conductors which we are evaluating.

In general we have found Kapton successful when used with the steel clad wires. The only precaution we take is to ensure that the wire surface is chemically clean and degreased before the Kapton is wrapped on and heat sealed. Adhesion appears to be excellent and the only reservation we have about this method is that the insulation appears to be rather weak in shear.

We encountered significant problems when we tried to insulate our first length of the Supercon wire this way. It became clear during our initial attempts to wind, that the adhesion of the Kapton was indifferent in places. The surface of this wire is a skin of pure copper. This led us to speculate that the problem was associated with oxidation of the surface during the heat treatment of the Kapton (~275C). Tests revealed that thorough mechanical cleaning and degreasing of the surface of the wire prior to wrapping were insufficient to effect a significant improvement. We decided that the surface had to be rendered inert immediately after cleaning. We found that the application of a copper primer, Redux 108 (TM Ciba-Geigy), which is used at Oxford in constructing water cooled magnets [5], was effective in significantly improving the adhesion. Subsequently we carried out further trials and concluded that in addition to the primer, heat treatment in an argon atmosphere further improved the bond.

WINDING TECHNIQUES

The Layer End Problem

In principle these small coils are easy and quick to wind. Typically they have only ~ 200 turns. If only copper is used, then few difficulties are encountered provided reasonable care is taken. As the conductor strength increases however, some problems become more acute. These are usually associated with the stiffness of these stronger conductors and the small winding diameters of the inner layers, typically ~ 10-20mm. Also, although rectangular section conductor is much more tractable in winding than square section, it can give greater problems at the layer ends where the wire climbs onto the next layer - the "cross-over point". Essentially the gap in the lower layer is relatively wide as the wire starts to climb to the upper layer. Local forces are such that the wire cants towards the end cheek causing severe shear stress on the wire insulation at this point. Naturally precautions are taken to ensure that the glass interleaving layer is in place at this point, but, as discussed in the section on impregnation, when using wet winding techniques it can be difficult to see whether this is fully the case. Also at the edge of the glass cloth, fraying takes place, and where a secure layer is most needed, it is least likely.

It is of course possible to reinforce locally the wire insulation at this point. A number of different techniques are used for this. One example is binding the relevant length of the wire with glass fibres before proceeding with the cross-over. Most of these methods are time consuming and tedious. More seriously, they locally thicken the conductor and, in a multilayer coil, the end stacking faults can accumulate so that the whole coil is unevenly wound.

Those coils we have produced which have failed to reach the theoretical field calculated from conductor strength, have done so because of voltage breakdown. Post mortems carried out on these failed coils indicate that the problem is invariably at the layer ends. This was particularly obvious on the two coils constructed from Supercon Cu Nb wire. It is possible that the difficulties encountered in adhering the Kapton insulation to this wire may have exacerbated the problem.

At Oxford the technique now in use to try to overcome these difficulties has been developed by Rob Harris and Martin Whitworth. Specially shaped pieces are prefabricated from Tufnol (T.M) 6F/45, a cloth/epoxy composite. One type acts as a ramp to bring the wire smoothly up on to the next layer so that the cross-over point is supported evenly from below. The other type acts as a filler for the wedge shaped void which occurs at the beginning of each layer. A set of these pieces, sufficient to wind a coil, can be machined from a single disc of Tufnol prior to beginning winding. Examples are shown in Fig. 1.

Layer – by Layer Reinforcement

From experimental results and theoretical calculations, e.g. Herlach et al [2], we have come to the conclusion that the mechanical strength of a coil ought to reside in the conductor itself; outer reinforcement can help but is of limited value as we confirmed in the MT10 paper [1]. If outer reinforcement is used, it appears to be most efficient if the coil is prestressed and completely enclosed in a hard shell. This is not easy to achieve reproducibly

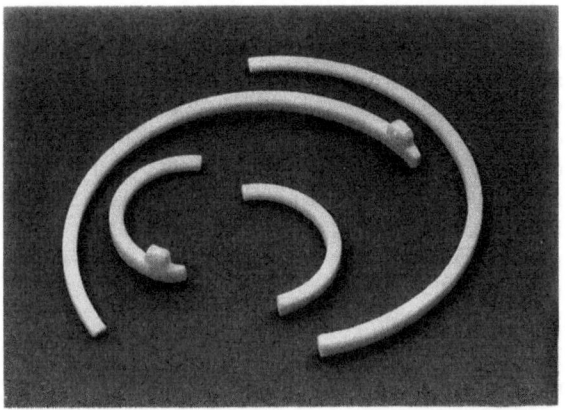

Fig.1. Layer end pieces. Left; two "ramp" pieces which facilitate the cross-over. The small keys lock into a groove in the end cheek in order to prevent movement as the wire climbs to the next layer therefore there are right and left hand versions. Right; two layer end fillers.

in practice. Therefore apart from developing ultra-strong wires, we have considered the use of high-strength fibre as insulating material between the layers of the coil. This reduces the filling factor but there is the additional advantage that the interlayer insulation is reinforced. Very strong fibres have become available recently. In practice the actual strengths of some composites is less than the values quoted in the literature. For pulsed field coils, the strength at 77K is very important. This is usually not known and needs to be determined experimentally. In this regard Kevlar (T.M. Du Pont) so far appears to be superior.

At Leuven a coil winding machine has been constructed which allows winding a layer of fibres between each layer of wires. The thickness of each fibre layer is precisely adjusted by microprocessor control. This enables us to wind an optimised coil where the strain will be the same in each layer. It was found that wet winding is not suitable for this procedure therefore these coils will have to be vacuum impregnated (see section entitled Impregnation). The first coils have been wound and are awaiting impregnation at Oxford.

IMPREGNATION

These types of pulsed coils are subject to great stresses and high voltages ($\sim$ 5 to 10 kv). For these reasons alone it is essential to ensure that all the interstices of the windings are impregnated with some medium which obviates voids.

Whilst it is not possible to eliminate completely mechanical movement, the monolithic structure which results from impregnation means that the whole coil moves, whereas the existence of voids can allow excessive local movement, often of a percussive nature and this can have a seriously adverse effect on the conductor insulation. Further, the existence of a void, particularly in regions of high potential difference (e.g. at alternate layer ends), can allow flashover when the trapped air becomes ionized. In practice the impregnation medium is invariably an epoxy resin.

Methods of Impregnation

There are basically two ways to impregnate a coil, wet winding and vacuum impregnation. With wet winding, the impregnant is brushed or trowelled on during the winding process. This technique has two major advantages; (a) it is quick (b) it allows the use of a filled resin e.g. Stycast 2850 FT (T.M. Emerson and Cuming, UK). This is an epoxy resin with an Al_2O_3 filler. With the correct catalyst, furthermore, a room temperature cure is possible and thus less restrictions are placed on the wire insulation used.

At Leuven this method has been used exclusively to date for the copper coils produced there over a number of years [2] and the process has been found satisfactory.

At Oxford, the experiences using wet winding for constructing the present series of high strength conductor coils, have been mixed. The relatively intractable nature of these development conductors, i.e. mainly their stiffness, means that the winding process is a little more difficult. The cross-overs for example have to be performed with great care. Although the "ramp" and "end fill" pieces now in use greatly facilitate this operation, it is still necessary to be able to inspect visually the winding at all times. It has been found that the combination of wet, opaque Stycast and the interlayer woven glass cloth used [1] can obscure undesirable winding faults particularly at the layer ends. A further difficulty with wet winding, at least with a cold cure resin, is the relatively short lifetime of the resin system. If a problem develops with the winding process it can happen that the resin "goes off" whilst the problem is being rectified. This can result in hardened spots of resin on the last layer wound, even if attempts to clean it are made. The result is stacking faults and pressure points both of which can cause problems in use. Also it is questionable that complete freedom from voids can be guaranteed given that even distribution of

this high viscosity resin relies, in the end, on the windings squeezing the resin into the interstices. Another problem is that some of the conductor and coil types we are investigating completely rule out wet winding because they require post-winding heat treatment at relatively high temperatures.

In the section on winding techniques we describe the layer-by-layer reinforcement method being developed at Leuven. This method does not lend itself easily to wet winding. For these reasons we are turning our attention once more to vacuum impregnation which, if done properly, is undoubtedly the best method. At Oxford this has always been the preferred method. Techniques have been developed for the vacuum impregnation of superconducting coils over a number of years.

The construction of small pulsed coils is in many ways analogous to that of small superconducting windings. Just one example is the similarity between "wind and react" Nb_3Sn coils (post-winding heat treatment of $\sim 700\,C$) and the proposed pulsed coils which will use CuBe or Cu-Ni-Si based conductors. Both of these types, of necessity, will require vacuum impregnation. As both types of coil are operated at cryogenic temperatures, large thicknesses of the unfilled resin used are undesirable because of the cracking which results. For this reason, at Oxford, we developed the "spit roast" curing technique [6]. Basically this involves flooding a pre-warmed and evacuated coil with de-aerated, long working life, resin which is then warmed to $\sim 40C$. This temperature is the optimum as there are two conflicting requirements; lowest possible viscosity, which decreases with increasing temperature, and long usable life which also reduces with temperature. The pressure is then increased from -1 bar to $+5$ bar and the coil is left to soak for anything from several hours to several days, depending on the thickness and density of the windings. The coil is then withdrawn from its bath of resin whilst still wet and the surplus is drained off, the close windings retain their resin inside by capillary action. The coils are then revolved slowly on their axes under infra-red heaters so that the resin gels without any preferential long term migration of the resin in the windings. We have produced many fully impregnated superconducting windings in this way. The method does not translate well to these pulsed coils. The principal reason is that the more open nature of the coil ends allows dry areas in these vulnerable regions and some of our early failures we attribute to voltage breakdown at the coil ends.

TABLE 1

Characteristics	RESIN SYSTEM		
System Components (parts by wt.)	CY 1300 (100) HY 906 (80) DY 063 (≦ 2)	CY 1300 (*) (100) HY 932 (*) (26) D 2000 (†) (37)	Stycast 2850 FT (100) (Blue) Catalyst 24 LV (7)
Manufacturer	All Ciba-Geigy	(*) Ciba-Geigy (†) Texaco	Emerson & Cuming
Viscosity 20C (Poise) 25C 40C 70C	- 15 - 1	55 - 11 -	150 - - -
Usable life (At stated temperature)	≦ 3 weeks (40C) ≦ 1 week (60C)	72h (40C) 15h (60C)	~ 1h (25C)
Cure tempe- rature	180C	100C	Room temperature
Advantages	Long life Very low viscosity	Moderate cure temp. Low viscosity Crack resistant at 77K	Filled resin Room temperature cure Fast process Crack resistant
Disadvantages	Hot cure Slow process	Moderate usable life not suitable for deep or dense windings	Wet wind only Short usable life

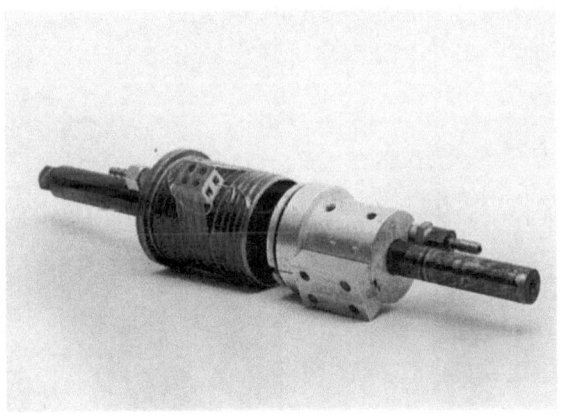

Fig.2. Winding mandrel and impregnation jig. The outer cylinder, on which is wound a heater coil, is shown partly withdrawn to reveal the space occupied by the windings. Note the resin ingress and egress nozzles at either end.

The method we are now using for the pulsed coils comprises winding on a collapsible mandrel on which is a thin centre tube fabricated from Kapton, Fluon FEP (T.M. DuPont) coated tape. This forms a resin seal. After winding, an outer cylindrical jacket is fitted and sealed to the end cheeks of the mandrel. Each end cheek is fitted with a resin nozzle, one ingress the other egress. A view of this apparatus (the jig) is shown in fig. (2). The jig is mounted vertically in a vacuum/pressure vessel and evacuated. The outer jacket carries a heater which can raise the jig to a temperature appropriate to the stage of the impregnation reached. Warm de-aerated resin is introduced via the ingress nozzle (in the lower end cheek) and allowed to rise slowly through the coil exiting through the egress nozzle. This nozzle is fitted with a header funnel which is allowed to fill to ~ 2/3 its volume. The resin supply is valved off and the vessel is overpressured to 5 bar. The resin is then gelled and cured in this state. The various resin systems used are summarised in Table 1.

CONCLUSIONS

We are developing a technology for producing extended life pulsed coils, energised by a capacitor bank of ≲ 200 kJ stored energy, for fields ~ 40 - 50 T over pulse lengths ~ 10 ms. The significant feature of our work is that the technology is maintained at as simple a level as possible. This is important for a multi-purpose experimental facility, such as exists at Leuven and is being installed at Oxford, as the requirement for coils may not always be dominated by the maximum possible field but flexibility in bore sizes and pulse lengths to suit individual experiments. Consequently the coils need to be capable of being produced quickly and cheaply. The coils need to be reliable over long periods but at the same time they can be regarded as consumables with a finite life.

In this paper we have shown that using a judicious mix of predictably behaved conductor, a simple in-house wire insulation, careful winding methods and impregnation techniques, it is possible to fulfil these requirements up to ~ 50 T. It seems however that to achieve fields significantly in excess of this routinely, a more sophisticated conductor and coil technology will be necessary at least for small, low alpha coils. It is for just this objective that the 5 European institution collaboration has been formed.

ACKNOWLEDGEMENTS

The original support for this work was provided by the Paul Instrument Fund of the Royal Society (UK) to whom we are grateful. Present support is from the Science and Engineering Research Council and the Commission of the European Communities.

We would like to thank many individuals on both sides of the English Channel for their enthusiastic support and assistance in this collaboration. In particular Martin Whitworth, Jim Sherratt and Rob Harris of the Clarendon Laboratory; Maarten van der Burgt, Luc Van Bockstal and Guido Heremans of K.U. Leuven. We acknowledge the skilled work of the technical staff at Harwell who produced the conductors. It is a pleasure to thank the other participants in the European Collaboration at Amsterdam, Grenoble and Toulouse for many helpful discussions.

REFERENCES

1. Jones, H., Herlach, F., Lee, J.A., Whitworth, H.M., Day, A.G., Jeffery, D.J., Dew-Hughes, D., Sherratt, G., 50 Tesla pulsed magnets using a copper conductor externally reinforced with stainless steel. IEEE Trans. Mag., 1988, 24, 1055-1058.

2. Herlach, F., de Vos, G., Witters, J., Stresses in coils for strong pulsed magnetic fields. J. de Physique, 1984, 45, supplement C1, 915-921.

3. Foner, S., Bobrov, E., Renaud, C., Gregory, E., Wong, J., 68.4T pulsed magnet fabricated with a wire-wound metal-matrix Cu/Nb microcomposite. IEEE Trans. Mag., 1988, 24, 1059-1062.

4. Gersdorf, R., and Mattens, W.C.M., Feasibility study for a non-destructive 100 ms/60 Tesla magnet coil. Proc. 9th international magnet technology conference. Eds. Marinucci, C. and Weymuth, P., Published by S.I.N. 1986, 447-449.

5. Hudson, P.A., Jones, H., Whitworth, H.M., A continuous winding / bonding technique for the manufacture of multi-layer helical magnet windings. J. de Physique, 1984, 45, supplement C1, 55-58.

6. Hudson, P.A., and Jones, H., High field facilities and services of the Clarendon Laboratory, Oxford. IEEE Trans. Mag., 1981, 17, 2242-2245.

GENERATION OF LONG FLAT-TOP PULSE FIELDS FOR SOLID STATE PHYSICS

Noboru Miura, Shojiro Takeyama and Katsuyoshi Watanabe*

Institute for Solid State Physics, University of Tokyo
Roppongi, Minato-ku, Tokyo 106, Japan

Flat-top pulse magnetic fields up to 41 T were generated by using a pulse-forming-network type condenser bank of 112 kJ. The duration of the top flat part was as long as 1.5 ms, so that the pulsed fields were conveniently employed for solid state physics experiments such as magneto-optical or transport measurements, where other parameters were swept at constant high magnetic fields.

I. INTRODUCTION

Generation of steady high magnetic fields by water-cooled magnets or hybrid magnets has been limited to about 30 T so far. The reason of the limitation is partly the difficulty of large power supply and partly the large electromagnetic force exerted on the magnets[1]. Much higher magnetic fields can be conveniently generated by pulse magnets. Although the duration of the field is usually of the order of 1-100 ms or even shorter depending on the maximum field, high magnetic fields up to 40-70 T are obtained with smaller scale facilities, by using wire wound coils. Higher magnetic fields in the megagauss range are also generated with a duration of several microseconds by various methods[2].

For solid state experiments, we often need to fix a magnetic field constant for sometime during which we want to sweep other parameters at a constant field. In such a case, we can perform the measure-

ments not only in steady fields but also in pulsed magnetic fields, if the duration of the pulsed field is sufficiently long. Usually, pulsed fields have a nearly sinusoidal form, because we discharge a current from a condenser bank to a solenoid coil. We can regard the region in the vicinity of the maximum of the field as constant and complete the measurement within this time interval[3]. When we need a constant field with an accuracy of 1 % for 1 ms, for example, we need 15 ms as a half period of a sinusoidal cycle. To generate such a long pulse field, we need a large condenser bank and a large magnet.

By generating a nearly rectangular pulse field, we can elongate the top flat part retaining the total pulse width. Attempts to generate nearly rectangular pulse fields were first made by Misu et al. in 1969[4]. Later, Dworschak et al. obtained a flat-top field with 1 ms duration up to 44 T[5]. Since they employed a relatively small condenser bank with a total storing energy of 17 kJ, the maximum field was obtained in a small bore diameter of 7 mm.

In the present paper, we report a generation of flat-top pulse fields with a

---

*Permanent address: Department of Physics, Yamanashi University, Takeda, Kofu 400 Japan

duration of 1.5 ms in a relatively large bore diameter of 20 mm. The long duration and large bore diameter are very convenient for the use in various solid state experiments such as magneto-optical spectroscopy and magneto-transport measurements. By employing a condenser bank with a storing energy of 112 kJ, a maximum field up to 41 T was obtained in the bore diameter of 20 mm. The use of strong wire which was reinforced by Nb-Ti filaments embedded in copper matrix for the solenoid allowed many shots at the maximum field. Some examples of the application to solid state experiments are also presented.

II. DESIGN OF THE CONDENSER BANK

We show in Fig. 1 a circuit diagram of the condenser bank to generate flat-top pulse fields. When the switches S_1 - S_4 are all open, the circuit composes a pulse-forming-network (PFN). When we turn on the ignitron switch G_1, after charging the capacitors C_1 - C_5 to the same voltage V_0, the current is supplied to the load inductance L_1 from the capacitors in different phases owing to the inductances L_2 - L_5. A method using the Fourier transformation to choose the parameters L_2 - L_5 and C_1 - C_5 for obtaining a flat-top pulse is described in [5]. With an infinite number of circuit elements we can determine the parameters to make any waveform of the discharge current in the load coil. When the number of capacitances and inductances is limited to a small number as in the present system (5 in this case), the discharge current deviates from an ideal shape showing ripples on the flat-top part. The final determination of the parameters has to be made by a computer simulation for obtaining the best flat-top shape.

The current I_i in the i-th circuit element is given by

$$V_i - I_i R_i - \dot{I}_i L_i = V_{i-1}, \quad (i = 2 - 5)$$

$$V_1 - I_1(R_1 + R_c) - \dot{I}_1 L_1 = 0, \quad (1)$$

where V_i is the voltage across the i-th condenser C_i and given by

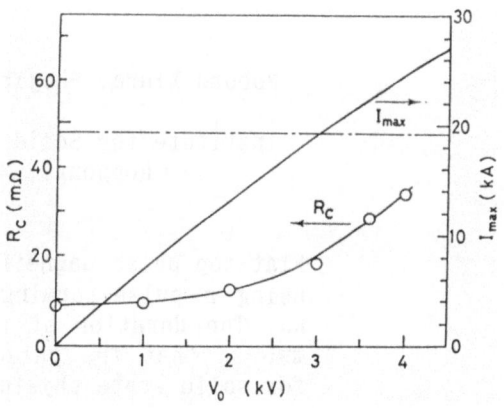

Fig. 2   Variation of the coil resistance R_c due to the Joule heating of the wire after the pulse as a function of the charge voltage of the condenser bank V_0. The solid line is calculated with (3). The open circles represent the experimental data for the coil described in the next section. The horizontal dot-broken line shows the coil resistance at room temperature. The maximum current I_{max} is also shown against V_0.

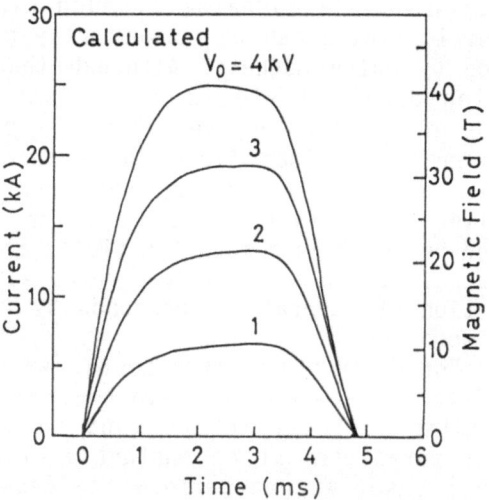

Fig. 3   Calculated profile of the current and magnetic field at various charge voltages

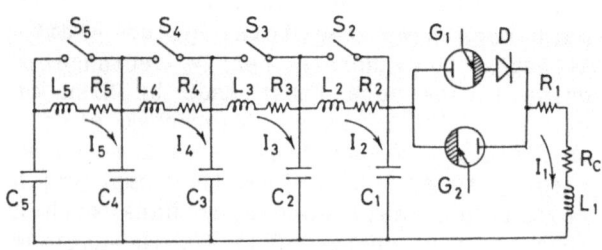

Fig. 1   Circuit diagram of a condenser bank for flat-top pulsed magnetic fields

$$V_i = V_0 + 1/C_i \int (I_{i+1} - I_i)dt,$$
$$(i = 1 - 4)$$
$$V_5 = V_0 - 1/C_5 \int I_5 \, dt. \qquad (2)$$

We restrict the size of the condenser bank to be of order 100 kJ, voltage to be 4 kV, the maximum field to be 40 - 50 T in a bore of 15 - 20 mm, the load inductance to be about 100 - 130 µH, and number of circuit elements to be 5. Finally, the following parameters were determined: $L_1 =$ 100-130 µH, $L_2 = 37.5$ µH, $L_3 = 33.6$ µH, $L_4 = 28.8$ µH, $L_5 = 39.6$ µH, $C_1 = 3.0$ mF, $C_2 = 2.5$ mF, $C_3 = 2.0$ mF, $C_4 = 2.0$ mF, $C_5 = 4.5$ mF. $V_0 = 4.0$ kV (the maximum voltage).

With these parameters, a computer simulation was done to obtain a theoretical profile of the current waveform. The resistance $R_2 - R_4$ in each circuit element was neglected since it has only a small effect in the final results. However, the resistance of the load coil R_c and that of the output circuit including the diode R_1 are large and have a significant effect to the current. R_c varies with time because of the Joule heating by the large current depending on the charge voltage. It was assumed that the resistance increase ΔR_c during a time interval Δt is expressed by the power $P = R_c I_1^2$ dissipated in the wire as

$$\Delta R_c = a(P \cdot \Delta t) + b(P \cdot \Delta t)^2. \qquad (3)$$

The constants a and b were determined in such a way that the value given by (3) agreed with the actual experiment soon after the pulse. R_1 was estimated to be 18 mΩ. In Fig. 2, the final resistance after the shot thus determined is illustrated as a function of the charge voltage V_0, together with the experimental data. The agreement between the calculation and experiment is satisfactory.

The calculated profiles of the current and the field for the coil described in the next section are shown in Fig. 3. It was designed in such a way that for low charge voltages, the top flat part had a small positive gradient. With increasing charge voltage, the top becomes flatter because of the gradual resistivity increase by the Joule heating during the pulse. At around $V_0 = 3.6$ kV, the best flatness of the top is obtained, and at larger V_0, the gradient of the top becomes slightly negative.

The inductances $L_2 - L_5$ are solenoid coils with a diameter of about 300 mm. They are wound with multi-wire copper

cables whose cross-section is 60 mm^2. When these are short-circuited by switches S_1-S_4, the condenser bank can be used to generate ordinary sinusoidal pulse fields.

Another unique point of this condenser bank is that it has two ignitron switches (G_1 and G_2) connected inversely parallel to each other. For obtaining a rectangular pulse, only G_1 is triggered. A semiconductor diode is put in the circuit, so that the current is never reversed even if the ignitron G_1 looses the rectifying function after a large current flow. On the other hand, when the switches S_1-S_4 are closed, the current can be reversed by triggering G_2 after some delay from G_1, to obtain a full period of a nearly sinusoidal form.

III. CONSTRUCTION OF MAGNETS

To generate high magnetic fields, it is essential to construct magnets which have a sufficient mechanical strength. We employed copper wires reinforced by Nb-Ti filaments inside as a material for the coil winding. They are in fact commercial superconducting wires, and have a yield strength as high as 70 kg/mm^2. We use the wire in the normal state at 77 K, just utilizing the mechanical strength. The wire has a rectangular cross-section 4 x 2 mm^2 with insulating layers of capton tapes and glass tapes on the surface. Figure 4 illustrates the design of the magnet.

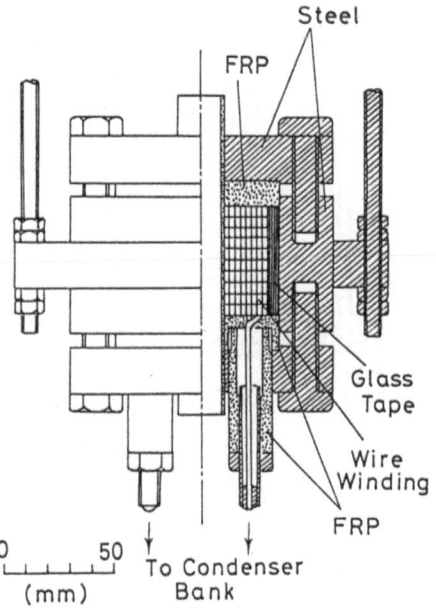

Fig. 4 Design of the magnet

The wire was wound on a bobbine made of glass epoxy (FRP) with a tension of about 20 kg. Further, we wound glass tapes firmly on the wire winding. After winding these, epoxy was impregnated and solidified at high pressure of 150 kg/cm^2 at about 100 °C. The high pressure was applied to the coils sealed in rubber tubes filled with epoxy, by using a piston cylinder with oil as shown in Fig. 5. The oil was heated by a heater installed inside the cylinder. After the epoxy was solidified, the coil was shaped by a lathe and fit into a steel cylinder for further reinforcement. The magnet has an inductance of 130 µH. The resistance is 8.89 mΩ at 77 K and 48.14 mΩ at 300 K.

The coil was mounted in a liquid nitrogen bath to be operated at 77 K. An insert for liquid helium temperature was set inside the magnet bore to cool samples below 4.2 K.

IV. RESULTS AND EXAMPLES OF APPLICATION

Figure 6 shows typical waveforms of the generated pulse fields at the charge voltage V_0 of 1, 2, 3 and 4 kV. The observed pulse shapes are in good agreement with the calculated curve shown in Fig. 3.

At B = 36.2 T (V_0 = 3.6 kV), the field variation at the top is 0.62% within 1 ms and 2.1% within 1.5 ms. For all the V_0 range above 2 kV, the field variation is less than 1.7% during the time interval of 1 ms and less than 3.8% in 1.5 ms. Thus the flat-top field is very useful to sweep other parameters in the measurement at a constant field. In order to obtain an equivalent top flat part with a sinusoidal pulse form, we have to generate a long pulse field by using a large magnet which requires a long cooling time after each shot. The present magnet is fairly small, so that the cooling time is as short as 20 min. even at the highest charge voltage. The energy stored in the load coil L_1 is $1/2L_1I^2$. It turned out to be as large as 35 % of the total energy stored in the entire condenser bank $1/2CV_0^2$ at V_0 = 4kV.

Finally, we show two examples of the application of the field to solid state experiments. Figure 7 shows the data of the resonant magneto-tunneling in n-type GaAs-AlGaAs double barrier resonant tunneling devices. The barrier layers had a thickness of 56 Å and the well width was 600 Å. The magnetic field was applied perpendicularly to the current through the sample. The bias voltage to the sample was applied as a sinusoidal voltage with a frequency of 1 kHz. Both the voltage and current signals were recorded by a digital memory. The voltage v.s. current curve at constant magnetic fields was obtained by reading the data within the duration of

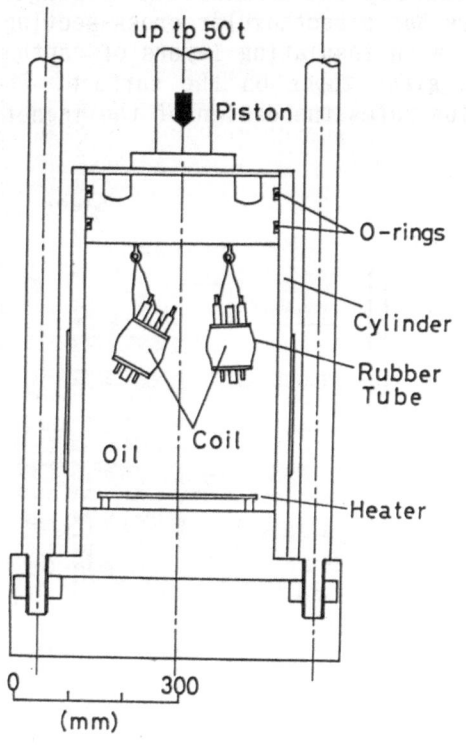

Fig. 5 High pressure apparatus to solidify the impregnated epoxy under high pressure at high temperature

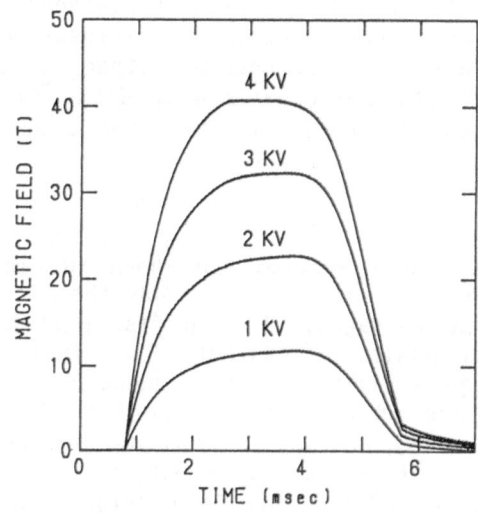

Fig. 6 Experimentally observed waveforms of the magnetic field generated at various values of V_0

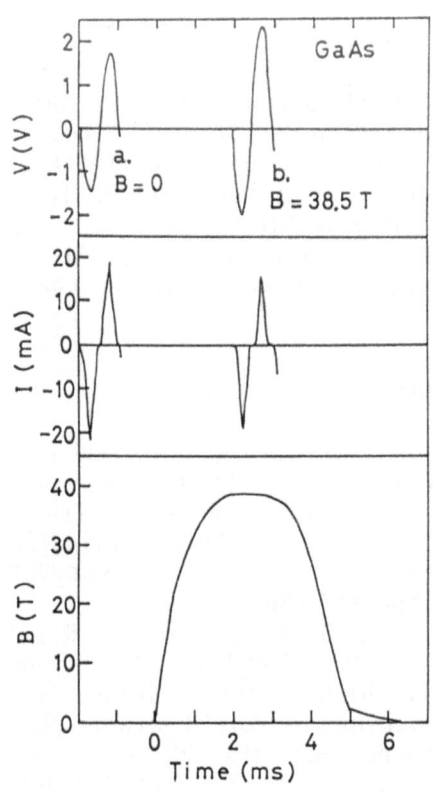

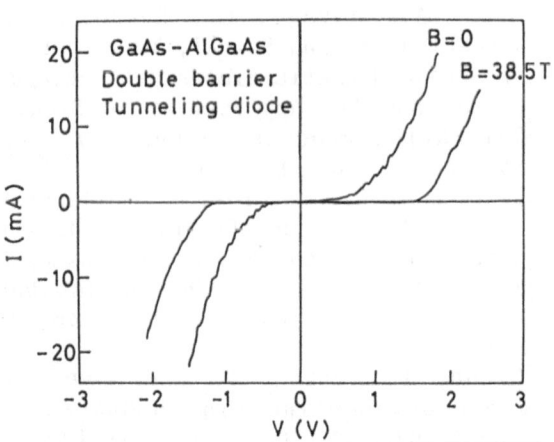

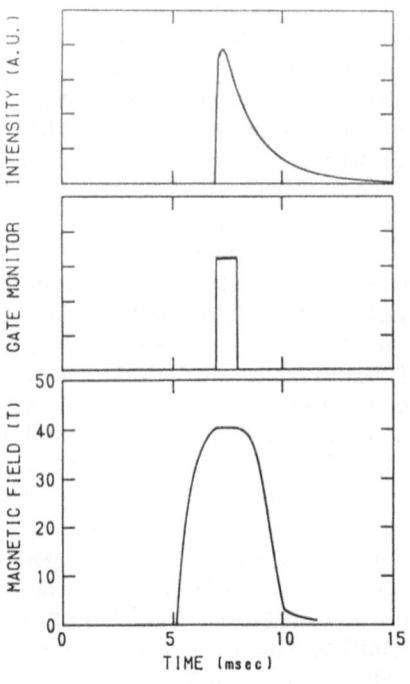

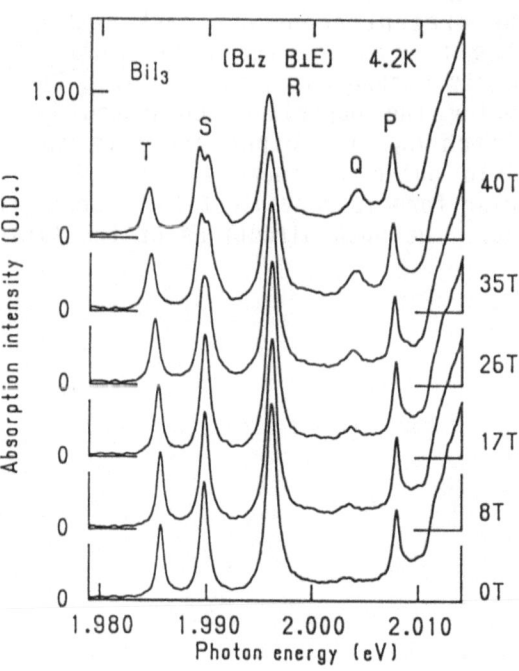

Fig. 7 Experimental traces for the voltage-current characteristics in an n-type
GaAs-AlGaAs double barrier tunneling structure. The temperature was 4.2 K.
(Upper)Waveforms of the ac voltage applied
to the sample V, the corresponding current
I and the magnetic field. The ac voltage
was applied continuously, but we show only
the periods which were read out for obtaining the V-I curve at the constant top
field (b) and zero field (a).
(Lower) V-I curve at B = 0 and B = 38.5 T.

Fig. 8 Data of the magneto-absorption
spectra of excitons in BiI_3 measured using
an optical multi-channel analyzer (OMA).
(Upper)The pulse profiles of the intensity
of the light source (flash lamp), the gate
pulse of the OMA and the magnetic field.
(Lower) Magneto-absorption spectra at
various magnetic fields. P, Q, R, S, T
denote the exciton lines localized at
stacking faults of the crystal.

the flat-top. The zero field curve was obtained by reading the data before the field pulse. It is evident that the threshold voltage for the current is significantly increased by a high magnetic field. In addition, the fine oscillatory structure observed at B = 0 due to the rather wide well width is greatly modified by the high magnetic field which bends the electron orbit in the well[6].

The second example is the measurement of magneto-optical spectra. Figure 8 shows the magneto-absorption spectra in BiI_3 at various magnetic fields which are applied perpendicularly to the crystal c-axis (z-direction). The gate of an OMA (Optical Multi-channel Analyzer) was opened during 1 ms on the flat-top part of the field, as shown in the upper figure. It is well known that in BiI_3, there exist very sharp absorption lines P, Q, R, S, T which are associated with stacking faults of the crystal. As shown in the lower figure, these stacking fault exciton lines exhibit considerable shifts and splittings in high magnetic fields[7].

As described in the previous section, the present condenser bank can be used also to produce the full cycle of the nearly sinusoidal magnetic field by triggering the oppositely connected ignitron switch G_2. The experimentally observed field pulse is shown in Fig. 9. Such a pulse-form is convenient for a measurement where the both directions of the field is essential, such as the Hall effect or magnetic hysteresis measurements.

ACKNOWLEDGMENT

The condenser bank used in this work was constructed by Shizuki Electric Co. Inc. The authors thank the engineers who contributed to the construction for their effort and valuable discussion. They are also indebted to T. Osada for the contribution in the experiment on tunneling devices.

REFERENCES

1. See for example, Miura, N. and Herlach, F., Pulsed and Ultrastrong Magnetic Fields, in Strong and Ultra-strong Magnetic Fields and Their Applications , ed. by F. Herlach, Springer-Verlag, 1985, pp. 247-350.
2. Miura, N., Goto, T., Nakao, K., Takeyama, S and Sakakibara, T., Production of Ultra-high Magnetic Fields and Their Application to Solid State Physics, Physica B, 1989, **152**, 23-32.
3. Miura, N., Iwasa, Y. and Itakura, T., Shubnikov-de Haas Effect in Si MOS-FET in High Magnetic Fields up to 37 T, J. Phys. Soc. Jpn., 1982, **51**, 1228-1235.
4. Misu, A., Aoyagi, K., Kuwabara, G., Kinoshita, K. and Suzuki, T., Production of Strong Magnetic Fields of Rectangular Pulse, and Its Application to Study of Relaxation Process in Ruby, J. Phys. Soc. Jpn., 1969, **8**, 57-64.
5. Dworschak, G., Haberey, F., Hildebrand, P., Kneller, E. and Schreiber, D., Production of Pulsed Magnetic Fields with a Flat Pulse Top of 440kOe and 1ms Duration, Rev. Sci. Instrum., 1974, **45**, 243-249.
6. Osada, T., Miura, N. and Eaves, L., Magneto-tunneling in n-GaAs/AlGaAs Double Barrier Tunneling Structures in High Magnetic Fields Parallel to the Barrier Planes, to be published, 1989.
7. Watanabe, K., Takeyama, S., Komatsu, T., Miura, N. and Kaifu, Y., Magneto-optical Study of Excitons Localized around Two-Dimensional Defects of BiI_3 in Pulsed High Magnetic Fields up to 47 T, to be published in Proc. Int Conf. the Application of High Mgnetic Fields in Semiconductor Physics, Wurzburg, 1988, ed. by G. Landwehr, Springer-Verlag, 1989.

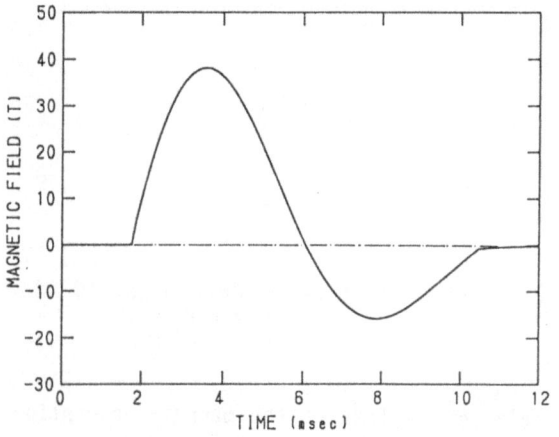

Fig. 9 The waveform of the magnetic field when the switches S_1 - S_5 were closed and the ignitron switch G_2 was triggered 6.5 ms after G_1. V_0 was 3.6 kV. The maximum field was 36.3 T in the positive side and 17.5 T in the negative side.

PRIMARY DESIGN OF 40 TESLA CLASS HYBRID MAGNET SYSTEM

K. Inoue, T. Takeuchi, T. Kiyoshi, K. Itoh, H. Wada, H. Maeda, K. Nii
National Research Institute for Metals
1-2-1, Sengen, Tsukuba-Shi, Ibaraki 305, Japan
T. Fujioka, Y. Sumiyoshi, S. Hanai, T. Hamajima and H. Maeda
Toshiba Co. Ltd.
1-6-1, Uchisaiwaicho, Chiyoda-Ku, Tokyo 100, Japan

ABSTRACT

The construction program of 40 T-class hybrid magnet is proceeding on the base of a 5-year plan starting from 1988 fiscal year. At the 1st stage we carried out a R & D study on the candidate materials for magnet conductors, such as Nb-tube processed $(Nb,Ti)_3Sn$ multifilamentary conductors and $Cu-Al_2O_3$ alloys. The primary design of a 40 T class hybrid magnet with associated facilities were also worked out. The design assumes generation of a back-up field of 15 T (B_{max} of 15.7 T) at 4.2 K with a superconducting magnet with a room temperature bore of 400 mm. A water-cooled magnet is designed as a polihelix type one to generate incremental fields up to 25 T in a clear bore of 30 mm with power consumption of about 14 MW.

INTRODUCTION

The recent discovery of high T_c oxide superconductors has increased extremely the requirement of developing ultimately high field magnet. Corresponding to the discovery of high T_c superconductors, the Science and Technology Agency, Government of Japan, started the Multi-Core Research Project on Superconductivity (MCRPS) at 1988 fiscal year. Based on this project, we are constructing such extremely high-field facilities as 80 T-class long-pulsed magnet, 20 T-class large-bore superconducting magnet and 40 T-class hybrid magnet at the National Research Institute for Metals (NRIM). When these facilities will be completed, a high-field research center will be established in the NRIM. The construction of 40 T-class hybrid magnet is the major subject in the MCRPS.

The hybrid magnet is well known to be able to generate the highest steady field; about 31 T is the world's highest steady field at present [1-3]. However, the field strength of 31 T is not enough to study high T_c superconductors. On the other hand, the construction of 35-45 T hybrid magnet has become possible due to the recent development of high-field superconducting magnet.

The construction program of 40 T-class hybrid magnet is proceeding on the base of a 5-year plan starting from 1988 fiscal year. At the 1st stage we carried out a R & D study on the candidate materials for magnet conductors; the high-field superconductors and the high strength/high conductivity Cu-alloys. The primary design of a 40 T class hybrid magnet with associated facilities were also worked out. In this paper, we describe the results of the 1st stage developments.

R & D STUDIES ON CONDUCTOR MATERIALS OF 40 T CLASS HYBRID MAGNET

As the candidate superconductors in high fields, we studied on the $(Nb,Ti)_3Sn$ multifilamentary wire made by the bronze process, the bronze/external Sn process and the Nb tube process, and Nb_3Al multifilamentary wire made by the Nb/Al-Mg composite process. The best results, overall J_o's of 500 A/mm² without Cu at 4.2 K and 16 T, were obtained for the $(Nb,Ti)_3Sn$ multifilamentary monolith of 1 mm × 1 mm, made by the Nb tube process [4]. The effects of aspect ratio and bending strain were also studied on the Nb-tube processed $(Nb,Ti)_3Sn$ multifilamentary monolith. The monoliths with aspect ratio of 2 showed overall J_c(16 T, 4.2 K)'s of about 380 A/mm² without Cu at bending strain of 0.4 %. Overall J_c(15.7 T, 4.2 K) of 250 A/mm² in $(Nb,Ti)_3Sn$ monolith is used as a condition for designing the superconducting magnet in 40 T class hybrid magnet system. The detailed results are reported in this conference

[4].

As the candidate Cu alloys with both high-strength and high-conductivity, we studied on the $Cu-Al_2O_3$ and Cu-Nb alloys. The $Cu-Al_2O_3$ alloys, including 0.33 wt% Al and cold-worked 40 % in the area reduction ratio, exhibit the 0.2 % yield strength of 48 kgf/mm² and the electric conductivity of 81.9 % IACS, which satisfy the efficiency required for the Cu alloy conductors used in the water-cooled magnet in 40 T class hybrid magnet system. However the processes for fabricating the large scale ingots of $Cu-Al_2O_3$ alloys and for cold-working them should be developed for constructing the water-cooled magnet. On the other hand the Cu-Nb alloy should be heavily cold-worked for obtaining high yield strength. The

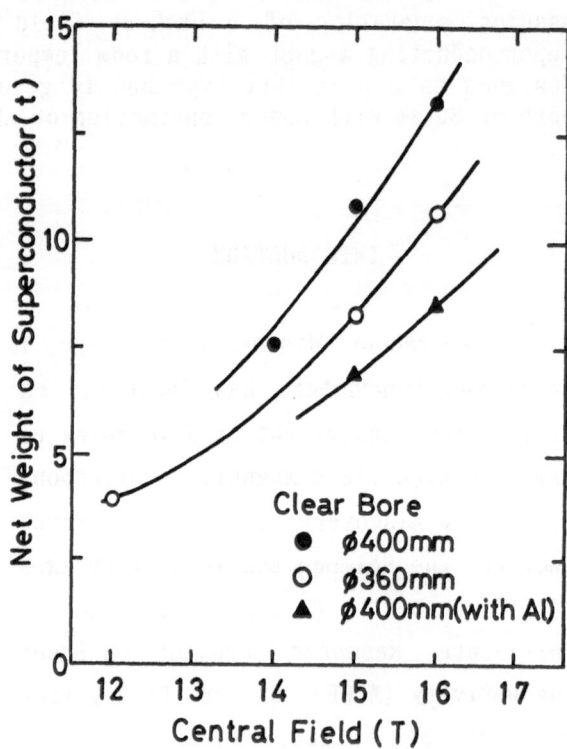

FIGURE 1 Calculated net weight of superconductor required for constructing large high-field magnets as a parameter of central field. Its dependence on room-temperature clear bore and effect of Al-stabilizer are also calculated.

production of large scale Cu-Nb alloy cold-worked heavily seems to be more difficult than that of Cu-Al$_2$O$_3$ alloy.

DESIGN OF 40 T CLASS HYBRID MAGNET

In order to determine the field allotment between the superconducting magnet and the water-cooled magnet in 40 T-class hybrid magnet system, we calculated the net weight of superconductor required for constructing large high-field magnet as a parameter of central magnetic field, as shown in Fig. 1. Basic assumptions are as follows; operation temperature of 4.2 K, cryogenically-stabilized magnet composed of the superconductors with Cu or Cu + Al stabilizer, cooling channel of three-dimensional structure, and room-temperature clear bore of 360 mm or 400 mm. The net weight

of superconductor, i.e. the construction cost of superconducting magnet, increases rapidly with increase of central field and clear bore. The superior stabilization with embedding pure Al in Cu-housing can reduce remarkably the net weight and the stored energy of the superconducting magnet.

Considering both the construction cost and the degree of technical difficulty, the main parameters of superconducting magnet were designed as shown in table 1. The design assumes generation of a back-up field of 15 T (B_{max} of 15.7 T) at 4.2 K with a superconducting magnet consisting of 50 double pancakes windings of 470 mm i.d.; each pancake is wound with 4 different cryogenically-stabilized superconductors (G1, G2, G3, and G4), where the grading depends on field distribution. The designed parameters of superconductors are shown in table 2. Both the G4 and the G3 conductors are Nb-

TABLE 1   Main parameters of the superconducting magnet design for 40 T-class hybrid magnet system.

central field		15.0 T
operation temperature		4.2 K
number of grading		4
grading position	G1: i.d., o.d.	470,0 mm, 631.2 mm
	G2: i.d., o.d.	631.2 mm, 859.2 mm
	G3: i.d., o.d.	859.2 mm, 1041.6 mm
	G4: i.d., o.d.	1041.6 mm, 1236.6 mm
coil length		910 mm
winding		pancake
number of pancakes		100
number of turns		G1: 13, G2: 19
		G3: 19, G4: 25
magnetomotive force		15.2 MAT
inductance		24.5 H
stored energy		49 MJ

TABLE 2   Main parameters of the superconductor design for 40 T-class hybrid magnet system.

		G1	G2	G3	G4
superconductor		$(Nb,Ti)_3Sn$	$(Nb,Ti)_3Sn$	Nb-Ti	Nb-Ti
structure		monolith + Cu-housing + Al	monolith + Cu-housing	monolith	monolith
stabilizer		1/2 H OFHC + Cu-clad Al	1/2 H OFHC	1/2 H OFHC	1/2 H OFHC
dimen-sion (mm)	cond.	8.5×5.6	8.5×5.4	8.5×4.2	8.5×3.3
	supercon.	4.8×2.4	2.9×2	8.5×4.2	8.5×3.3
Cu ratio (Al)		3.7, 0.2	8.3	8.5	13
maximum field		15.74 T	12.6 T	8.2 T	3.6 T
critical current		3000 A	3000 A	3000 A	3000 A
operation current		2000 A	2000 A	2000 A	2000 A
resist.($\times 10^{-10}\Omega$m)		6.2	7.5	5.6	5.0
Joule heat flux		0.35 W/cm²	0.4 W/cm²	0.41 W/cm²	0.48 W/cm²

Ti multifilamentary monoliths. Both the G2 and the G1 conductors are composed of $(Nb,Ti)_3Sn$ multifilamentary monoliths. For the G2 conductor the monolith is embedded into the Cu-housing. The design of G1 conductor includes the use of tape-shaped pure Al embedded in the Cu-housing in order to accomplish sufficient stability, especially in the highest magnetic field.

The stresses induced in the superconducting magnet are designed to be reduced by setting 3 non-bonding places into each pancake, where the stress transfer between layers is obstructed. The maximum circumferential stress, σ_θ, of 19 kgf/mm² and the maximum axial compressive stress, σ_z, of 13.9 kgf/mm² are induced in the G1 and the G4 conductors, respectively. At every place in the superconducting magnet the summation,

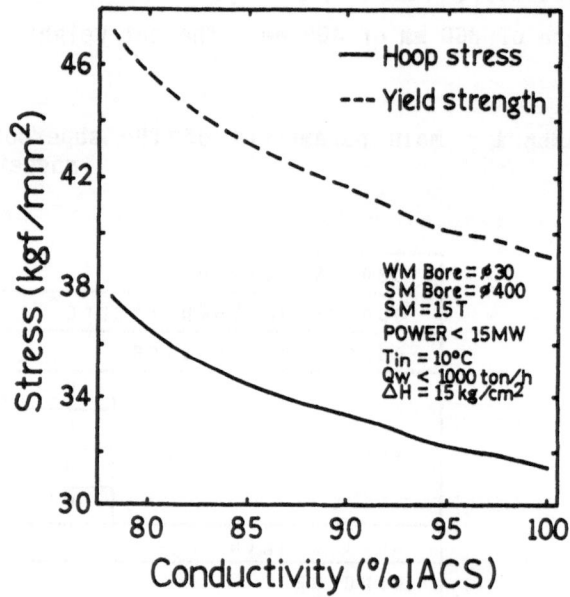

FIGURE. 2   Calculated hoop stress and designable yield strength vs electric conductivity curves for the conductor materials used in the water-cooled magnet of 40 T class hybrid magnet system. The efficiency of the conductor materials should be above the dashed line.

σ_θ + σ_z, does not exceed 27 kgf/ mm^2, which is lower than the 0.2 % yield strengthes of G1 to G4 conductors.

The development on the high-strength/ high-conductivity Cu alloys is one of the most important targets in this research project. Fig. 2 shows the calculated hoop stress and the designable yield strength, as a parameter of electric conductivity, required for the conductor materials used in the water-cooled magnet of 40 T-class hybrid magnet. High strength and high conductivity above the dashed line in Fig. 2 are required for the conductor materials. These values are attainable but need the development efforts. The following conditions were assumed for this calculation; clear bore of water-cooled magnet is 30 mm, room temperature bore of superconducting mag- net is 400 mm, back-up central field of

superconducting magnet is 15 T, power consumption at water-cooled magnet is lower than 15 MW, temperature of cooling water at inlet is 10 ℃, mass flow rate of cooling water is lower than 700 ton/ hr, head loss of cooling water is 15 kgf/ cm^2.

A water-cooled magnet is designed as a polihelix type (Fig. 3), composed of 15 layers to generate incremental fields up to 25 T in the clear bore of 30 mm. The electric conductivity and the thermal conductivity of the conductor material are assumed to be 89 % IACS and 353 W/m · K, respectively. The maximum power consumption in the magnet is assumed to be about 14 MW. When the hybrid magnet system generates its highest field of 40 T, the calculated hoop-stresses induced

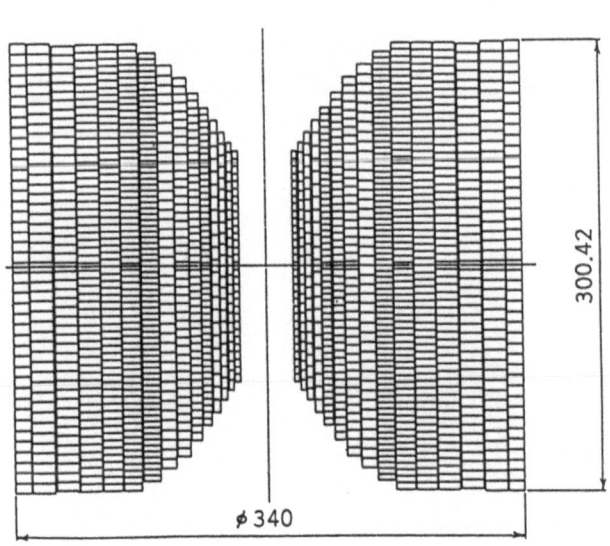

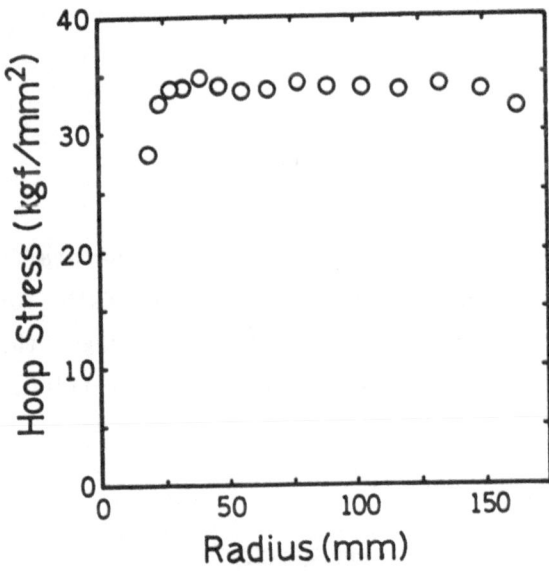

FIGURE 3   A design of 25 T water-cooled polihelix magnet used in a 15 T supercon- ducting magnet with 400 mm ϕ room-tem- perature bore.

FIGURE 4   Calculated hoop stress induced at the equatorial plane of the 25 T wa- ter-cooled polihelix magnet as a parame- ter of the helix radius.

in the water-cooled magnet are lower than 35 kgf/mm², as shown in Fig. 4. The calculated temperature rise of cooling water is lower than 25 °C.

We also designed the water-cooling system composed of turbo-refrigerator (15 MW cooling power), cooling towers, water tanks, and many water-pumps, which is similar to the system adopted at Tohoku University. The 15 MW d.c. power supply system (430 V × 35 kA) is also designed of being composed of main transformer (6.6 kV/410V, 20.4 MVA), thyristor converter (d.c. 430 V, d.c. 17.5 kA), passive filter, active filter, and phase controller. In the case of power breakdown, an energy of about 5 MJ stored in the chemical condenser of 60 F is instantaneously released into the magnet to retard its discharge rate; this method should effectively suppress the damage to the water-cooled magnet and the quench of the superconducting magnet.

ACKNOWLEDGMENTS

The authors wish to thank the staffs of Iten-suishin-shitsu in the NRIM for their valuable support to design the buildings for the high field facilities.

REFERENCES

1. Schneider-Muntau, H.J. and Vallier, J.C. The Grenoble hybrid magnet. IEEE Trans. Magn., 1988, MAG-24, 1067-1069.

2. Noto, K., Watanabe, K., Hoshi, A., Muto, Y., Nagamura, J., Osaki, O., Sumiyoshi, Y., Hamajima, T., Satow, T. and Murai, T. Design and performance of superconducting magnets for hybrid magnets. Sci. Rep. RITU, 1986, 33, 238-250: Nakagawa, Y., Miura, S., Hoshi, A., Ito, S., Nakajima, K., Tanaka, H., Inoue, K. and Hanai S. Design and performance of water-cooled magnets for hybrid magnets. Sci. Rep. RITU 1986, 33, 251-259.

3. Leupold, M.J., Weggel, R. and Iwasa, Y. Design and operation of 25.4 and 30.1 Tesla hybrid magnet systems. Proc. 6th Intl. Conf. Magn. Technology (MT-6), (ALFA, Bratislava, Czechoslovakia, 1978) pp. 401-405.

4. Inoue, K., Takeuchi, T., Itoh, K., Murase, S., Shiraki, H., Nakayama, S., Fujioka, T., Sumiyoshi, Y. and Hamajima, T. High field superconducting properties of 16 T class $(Nb,Ti)_3Sn$ conductor by the tube method. to be published in Proc. MT 11 (Tsukuba, 1989).

REMOVAL OF DISSOLVED OXYGEN IN WATER
FOR HIGH-POWER RESISTIVE MAGNET

A. HOSHI, S. MIURA and Y. NAKAGAWA
Institute for Materials Research, Tohoku University, Sendai 980, Japan

ABSTRACT

When a water-cooled magnet is burnt out, an abnormal resistance decrease of a coil is observed. The resistance decrease may be due to thermal and mechanical deterioration of insulator plates. It was also observed that the destroyed Bitter disks and insulator plates were contaminated by copper oxide. The quantity of the copper oxide at a lower part of Bitter coils was larger than that at an upper part. Usually the burn-out occurs at the lower part of the coil although the temperature is higher at the upper part since the cooling water flows from the bottom to the top of the coil. Thus it is thought that the copper oxide is created from the copper plate and dissolved oxygen in deionized cooling water. The dissolved oxygen can be removed by bubbling of pure nitrogen gas. It was found that the amount of copper oxide was much reduced when we used the cooling water processed by the nitrogen bubbling.

INTRODUCTION

In our old facilities of the high magnetic field [1], the Bitter coils were cooled by deionized water of high resistivity more than 10^6 ohm cm which was stored in a large reservoir of 200 m^3 capacity. Because of the insufficient cooling capacity of the secondary cooling system the primary cooling water temperature became higher during the experiments and the major cause of burn-out of the coil was a thermal catastrophe. Therefore, a new cooling system [2] was selected on the occasion of the establishment of High Field Laboratory for Superconducting Materials (HFLSM).

The new cooling system has two turbo-refrigerators which can chill the deionized water directly and keep the water temperature between 6 °C and 15 °C. The cooled deionized water is stored in a large reservoir of 160 m^3 capacity. Then the water is supplied to the water-cooled magnets by four circulation pumps which pressurized it up to 20 kg/cm^2. The refrigerators are cooled by city water flowing through two huge cooling towers of so-called open type. The cooling capacity of primary and secondary cooling systems are fully satisfied to remove the Joule losses of the magnets. At the maximum flow rate (350 m^3/h) and the rated electrical power (8 MW), the temperature difference between inlet and outlet water does not exceed 30 °C. The new cooling system has a good efficiency and capacity to allow the stable operation of magnets and the life time of the magnet becomes longer. Nevertheless, another problem still remains as follows.

If a water-cooled magnet is burnt out during a operation of a hybrid magnet, a superconducting magnet combined with the water-cooled magnet is also destroyed. In order to prevent the expensive superconducting magnet from the failure, the electric resistance of the water-cooled magnet is monitored during operation of magnet [3] so that the water-cooled magnet is renewed immediately when an abnormal resistance decrease is observed. The resistance decrease may be due to the deterioration of insulator plates. This deterioration of insulation is thought to be caused by the deposition of copper oxide on the surface of the insulator plate; the copper oxide is created by the dissolved oxygen in the cooling water. The present paper describes the relation between the copper oxide and the dissolved oxygen in the cooling water.

EXPERIMENTAL METHOD AND RESULTS

It is an important problem that an apparatus with cooling water such as a boiler at a steam-power plant is affected by the amount of dissolved oxygen in the water. In spite of using deionized cooling water, a trouble is caused frequently by oxidation of a pipe within a boiler. The oxidation depends on the amount of dissolved oxygen in the water. Then 60 % hydrazine hydrate ($NH_2NH_2 \cdot H_2O$) solutions are mixed in the water to decrease the dissolved oxygen.

Nevertheless, the mixing of the hydrazine hydrate solutions can not be used in the cooling system of magnets because of an increase of electric conductivity of the cooling water. Therefore a method of bubbling nitrogen gas into water is the best to remove the dissolved oxygen in a cooling system of water-cooled magnets.

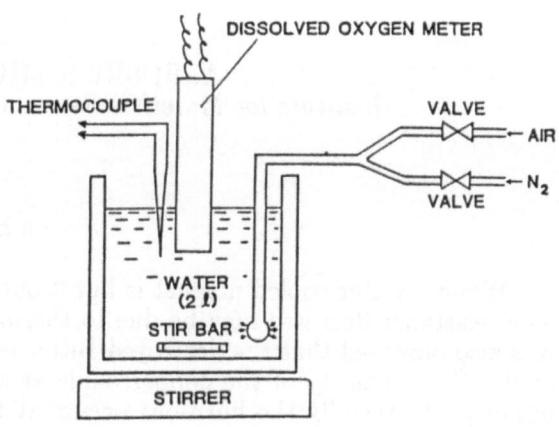

FIGURE 1. Experimental apparatus for bubbling air or nitrogen gas into water.

Before obtaining the decreasing curves of the content of dissolved oxygen with bubbling nitrogen gas, the saturated dissolved oxygen solution was prepared by bubbling air into deionized water. The apparatus for bubbling is shown in Figure 1. The nitrogen gas flow rates were changed by the diameter of bubbling holes of the pipe and the pressure of nitrogen gas. The amount of oxygen in water was measured by a dissolved oxygen meter with an electrode of galvanic cell type (Central Kagaku Co., Ltd. of Japan, Model 100-M). Figure 2 shows the relation of the amount of dissolved oxygen versus the bubbling time in the case of air bubbling. In this case, the diameter of air holes is 0.7 mm and the gas pressure is 130 kPa. It is shown that the amount of dissolved oxygen is nearly saturated after three hours.

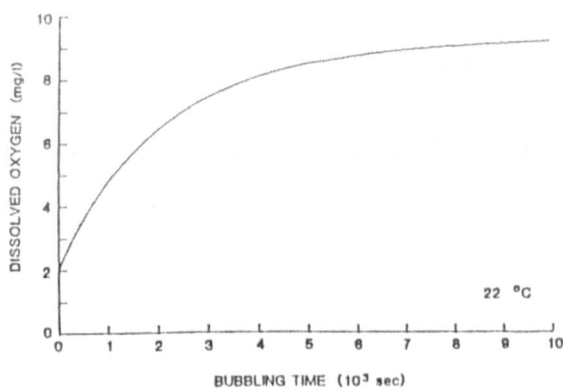

FIGURE 2. Amount of dissolved oxygen in water versus time for air bubbling with bubbling holes of 0.7 mm in diameter and pressure of 130 kPa.

Figure 3 shows an example of decreasing curve of dissolved oxygen in water versus time for nitrogen bubbling with bubbling holes of 0.7 mm in diameter and the pressure of 130 kPa at 22 °C. The upper curve is the logarithmic value of dissolved oxygen versus time. The straight line shows that the decrement of dissolved oxygen is expressed by the simple logarithm; the time constant of the decrement is given by the gradient of the straight line.

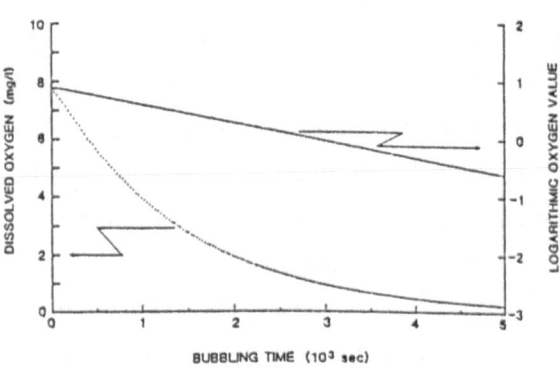

FIGURE 3. Dissolved oxygen concentration in water (lower curve) and the logarithmic oxygen value versus time. A diameter of bubbling hole is 0.7 mm and bubbling pressure is 130 kPa. Water temperature is 22 °C.

The experiments for decreasing dissolved oxygen were made using 0.7 mm and 0.5 mm diameter holes. Figure 4 shows the plots of the time constants against the flow rate of bubbling nitrogen gas. The curves indicate that the smaller diameter of bubbling hole is better than the larger one for the efficiency of decreasing dissolved oxygen.

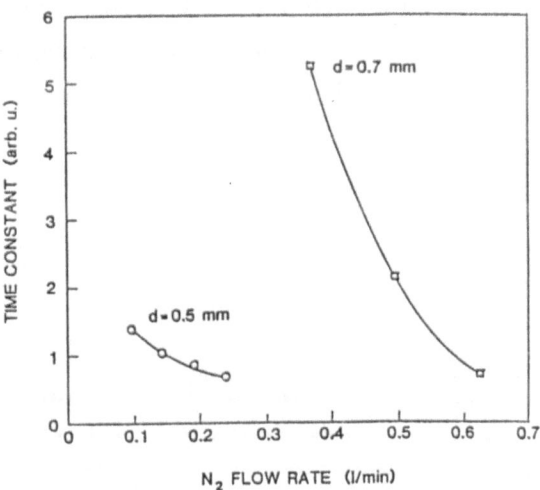

FIGURE 4. Time constants of decreasing oxygen versus bubbling gas flow. d is a diameter of bubbling hole.

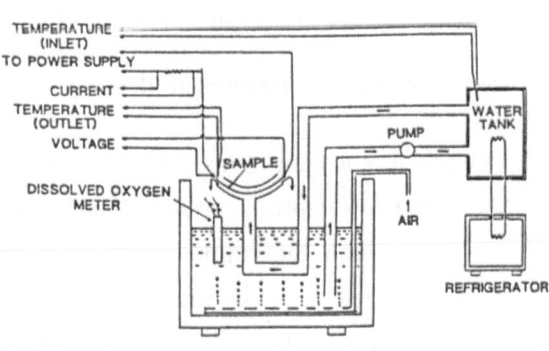

FIGURE 5. Apparatus to create copper oxide. Cooling water was chilled by a refrigerator and kept 10±0.1 °C.

In order to examine the creation of copper oxide, a copper strip was heated in circulating deionized water, as shown in Figure 5. The dimension of copper strip is $0.1 \times 1 \times 90$ mm^3. The copper strip was heated by Joule heat. The electric current through the copper strip had different values from 10 to 50 A. The deionized water flew through the glass pipe in which the copper strip was placed. The inner diameter of the glass pipe was 4.0 mm. The flow rate of water was 0.024 m^3/h and the inlet temperature was held at 10 ± 0.1 °C. Air bubbling was made during the experiment so that the amount of dissolved oxygen was constant.

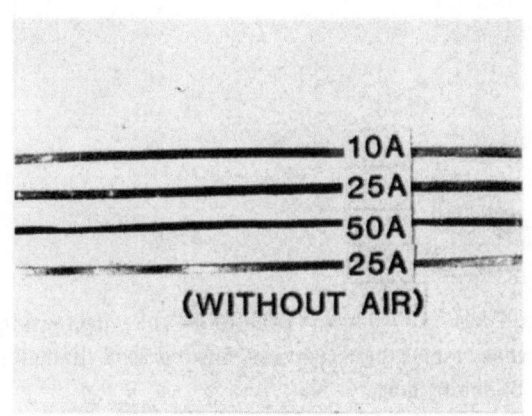

FIGURE 6. Photograph of copper strip contaminated by copper oxide. From the top to bottom, I is 10, 25, 50 A respectively. Lowest is the case with nitrogen bubbling at 25 A.

Figure 6 is a photograph of copper strips oxidized by the dissolved oxygen. The values of current through the copper strips were 10 A, 25 A and 50 A from the top to the bottom in the picture respectively. The black parts of copper strips were the copper oxide created by the dissolved oxygen. It is clearly observed that the amount of copper oxide increases with an increase current through the strip. For comparison, the strip at 25 A without dissolved oxygen was also shown at the lowest in the same picture.

Figure 7 is the experimental result of the oxidizing process. It shows the resistance of copper strip and the concentration of dissolved oxygen. The resistance of copper strip slightly increased with the times. It was the effect of oxidation of copper by the dissolved oxygen.

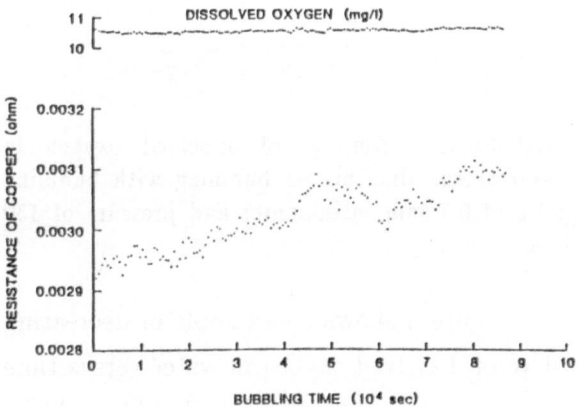

FIGURE 7. Dissolved oxygen concentration and resistance of copper strip versus time for I=50A.

APPLICATION TO WORKING SYSTEM

From the fact that the copper strip without dissolved oxygen was not contaminated by copper oxide, it is thought that the nitrogen bubbling method is useful to prevent the oxidation of copper. Then we examined the nitrogen bubbling to the cooling system of water-cooled magnets. A large amount of pure nitrogen gas is needed for bubbling the working system. Fortunately, we can use the nitrogen gas evaporated from a liquid nitrogen reservoir. The amount of the naturally evaporated nitrogen gas is about 2.1 m^3/h. The piping work of nitrogen bubbling was carried out in the deionized water tanks, as shown in Figures 8 and 9.

Figure 9 is the working drawing diagram of the piping in the reservoirs. The pipe is bored with a diameter of 0.5 mm at intervals of 100mm. The amount of bubbling nitrogen gas is 1.25 m³/h. Copper wires of 0.8 mm diameter were set in a strainer on outlet side of ion exchanger during 20 days. The copper wires were only slightly contaminated by copper oxide. If nitrogen bubbling was not done, copper wires were deep-black by contamination of copper oxide. It was confirmed that a rate of contamination of copper oxide was much reduced by nitrogen bubbling.

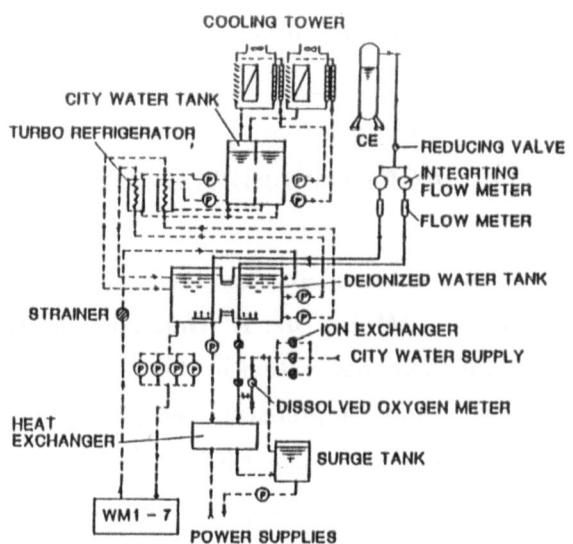

FIGURE 8. Cooling system for water-cooled magnet in HFLSM. Piping work of nitrogen bubbling is added on the upper right hand. CE is a cold evaporator of liquid nitrogen (liquid nitrogen reservoir).

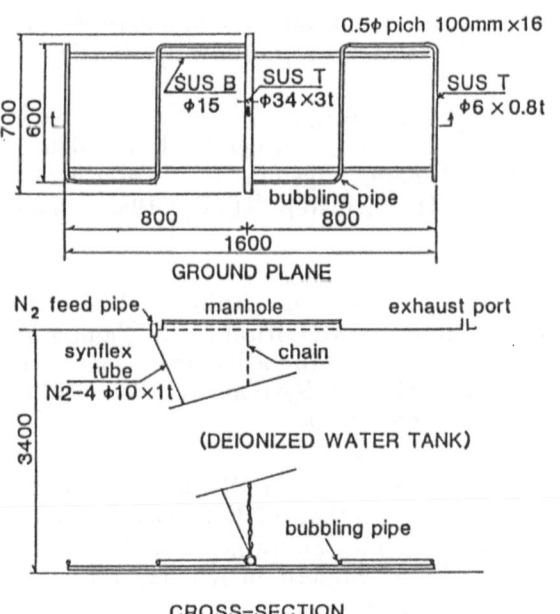

FIGURE 9. Working drawing diagram of piping in deionized water tank.

Figure 8 is a block diagram of the cooling system of water-cooled magnets.

DISCUSSIONS

When a water-cooled magnet was burn-out, it was observed that the destroyed Bitter disks and insulator plates were contaminated by copper oxide. The quantity of the copper oxide at a lower part of the Bitter coil was larger than that at an upper part. Usually the burn-out occurs at the lower part of the coil although the temperature is higher at the upper part since the cooling water flows from the bottom to the top of the coil.

The amount of saturated dissolved oxygen in water, $C_{T,P}$, is expressed by

$$C_{T,P} = A - BT + CT^2 - DT^3, \qquad (1)$$

where A, B, C and D are numerical constants determined by empirical methods and T the temperature of water. For example, A, B, C and D are 14.452, 0.3855, 0.00680 and 0.0000572 respectively [4]. According to the relation (1), the amount of saturated dissolved oxygen is the smaller at the higher temperature. When the cooling water is heated by Joule loss dissipated in the magnet, the amount of dissolved oxygen is decreased and a part of the excess oxygen will be combined with the copper of magnet. The quantity of excess oxygen is larger when the water temperature is lower. Thus it is thought that the copper oxide is created by the dissolved oxygen in cooling water on the Bitter disks.

The Joule loss of the copper strip was estimated from the electric current, I, and the across voltage of the strip, V. The temperature rise of the copper strip was estimated from the electric resistivity and its thermal coefficient, and the dimension of the strip. The estimated temperature rising of strip was 47 - 48 °C.

A quantitative measurement of deposited copper oxide was not done. All excess oxygen in water can be estimated from equation (1). The amount of copper oxide was estimated from resistance change of copper strip. Nevertheless, the ratio of the excess oxygen to transfer the copper oxide was about 0.04 % and the ratio was too large.

CONCLUSIONS

The oxidization of conductor by dissolved oxygen in water introduces a failure of water-cooled magnet. The dissolved oxygen is removed by bubbling nitrogen gas into the water. Now our water-cooled magnets are processed by nitrogen gas bubbling. It is expected that the oxidization of magnets is suppressed remarkably, and that the life time of the magnet becomes much longer.

ACKNOWLEDGMENTS

The authors express their thanks to Professor Y. Muto, Drs. K. Watanabe and G. Kido for their cooperation in this work and to Mr. M. Kudo for preparing the photograph. The authors also acknowledge Professor K. Noto of Iwate University for useful advice. They also thank to members in HFLSM for many assistances.

REFERENCES

1. Maeda. S., High Field Research at Tohoku University, In High Magnetic Fields, Proc. Int. Conf. on High Magnetic Fields, MIT, Massachusetts, 1961, pp. 406-11.

2. Hoshi A., Nakagawa Y., Miura S., Kudo M., Sai K. and Ishikawa Y., Power Source and Cooling System for Water-cooled Magnets, Sci. Rep. RITU, 1986, **A33**, 271-80.

3. Miura S., Hoshi A., Noto K., Watanabe K., Shimizu M., Hirata Y., Yamao A. and Kato T., Operation and Control of Hybrid Magnet System, Sci. Rep. RITU, 1986, **A33**, 281-8.

4. Tanaka Y., Uchida H. and Saito H., Preparing Method of Reference Standard Solutions for Calibrating Dissolved Oxygen Meter, Bulletin of NRLM, 1981, **30**, 22-31. (in Japanese)

A 20 T, 130 mm BORE HYBRID MAGNET

P. Rub, M. Ohl*, H.-J. Schneider-Muntau*, J.C. Vallier

Service National des Champs Intenses, CNRS, BP 166X
F-38042, Grenoble Cedex, FRANCE.
* Max-Planck-Institut für Festkörperforschung
Hochfeld-Magnetlabor, BP 166X, F-38042, Grenoble Cedex, FRANCE

ABSTRACT

The studies of significantly long of superconductors with high H_{c2} and high critical current densities in high magnetic fields require very strong fields in a large volume.

In order to perform such experiments, we have built a new resistive insert for our 10 MW hybrid system, giving a field of 20 T in a room temperature bore of 130 mm diameter.

The design and the test results of this magnet are reported.

INTRODUCTION

All magnetic field laboratories are interested in generating the highest magnetic fields possible, very often at the expense of the available experimental volume.

The Grenoble HFML builds also magnets with large volumes, mostly used for the characterization of superconductors in significant dimensions.

The production of very high magnetic fields with s.c. materials leads us to the necessity to dispose of a big bore and high field magnet.

The reason is that for realization of high field s.c. magnets the wire properties have to be known precisely and tests have to be perfomed with small coils wound from these materials.

We are describing here the magnet system which has been built and succesfully tested recently in our laboratory in Grenoble, and which generates a total magnetic field of 20 T (9 from the s.c. part and 11 from the resistive one) in a room temperature bore of 130 mm diameter, with a power consumption of less than 9 MW.

S.C. PART

As superconducting part the 420 mm room temperature bore, Nb-Ti magnet, built in Grenoble for our 30 T, Φ 50 mm hybrid system, is used.

This magnet is described in detail in previous papers [1], [2], and has been designed to work continuously for long periods.

The main characteristics of the coils are given in Table 1.

According to the cooling efficiency available in our 1.8 K device, we can calculate that the electrical dissipation of the magnet is smaller than 50 mW, which is completely negligible compared with the total cryogenic losses at 1.8 K : 4 W.

With this magnet, we have taken a lot of care concerning the transient losses induced into the Nb-Ti material during a fast switching-off of the current in the 10 MW internal resistive magnet.

TABLE 1

PARAMETER	INNER COIL	OUTER COIL
Inner diameter (mm)	500	742
Outer diameter (mm)	667.2	1087
Height (mm)	855	852
Conductor size (mm x mm)	5.6 x 2.8	5.0 x 2.2
Material	Nb-Ti	Nb-Ti
Overall Current density (A/cm^2)	3970	5230
Number of filaments	10000	830
Diameter of filaments (mm)	0.022	0.042

In the last 3 years, the cryogenic installation has been operated at 1.8 K for more than 10000 hours without any major trouble.

During the last months, we have experienced a maximum continuous operation of 3400 hours, deliberately stopped for the refrigerator maintenance.

The s.c. magnet is able to produce 11.0 T on the axis (11.6 T on the s.c. material). At this level of magnetic field, the stored energy is 23 MJ for an operating current of 860 A.

In the case of the 30 T, 50 mm hybrid magnet, the superfluid bath temperature increased only by 19 mK for a slow switching-off of the current in the 10 MW magnet, when the superconducting part generates 10 T.

To obtain a similar behaviour with the 20 T, 130 mm bore magnet, we have voluntarily limited the field given by the superconducting part to 9.0 T, thus the total field available will be 20 T.

RESISTIVE PART

The resistive insert is a classical Bitter type consisting of 2 concentrical stacks, electrically in series and hydraulically in parallel.

TABLE 2

PARAMETER	INNER COIL	OUTER COIL
Disk diam. mm	140 x 255	260 x 358
Disk thick. mm	4.4	4.4
Number of disks	74	74
Current kA	26.5	26.5
Voltage V	110	216
Power MW	2.92	5.72
Field T	6.07	5.06
Cooling m^3/h	140	220

With a current of 26.5 kA, the total voltage is 326 V and the corresponding power is 8.64 MW when a field of 11.13 T is generated.

Both the positive and negative current connections are on the bottom of the magnet, which is easily realized by connecting the even number of coils in series.

The cooling flows are distributed proportionaly to the powers and the same pressure drop is obtained for the 2 stacks by using 2 different patterns of cooling holes:
- inner disk : 450 holes of 2.4 mm Φ
- outer disk : 540 holes of 2.7 mm Φ

The inlet of the water is at the bottom of the housing and the return by channels on the outside diameter. It is impossible to measure directly the flow in each coil, but the average temperatures of stacks are in good accordance with the calculations.

The preliminary design of this magnet was presented last year [3]. The main characteristics of the coils are listed in Table 2, and a cut view of the whole resistive part is shown in Figure 1.

Figure 1

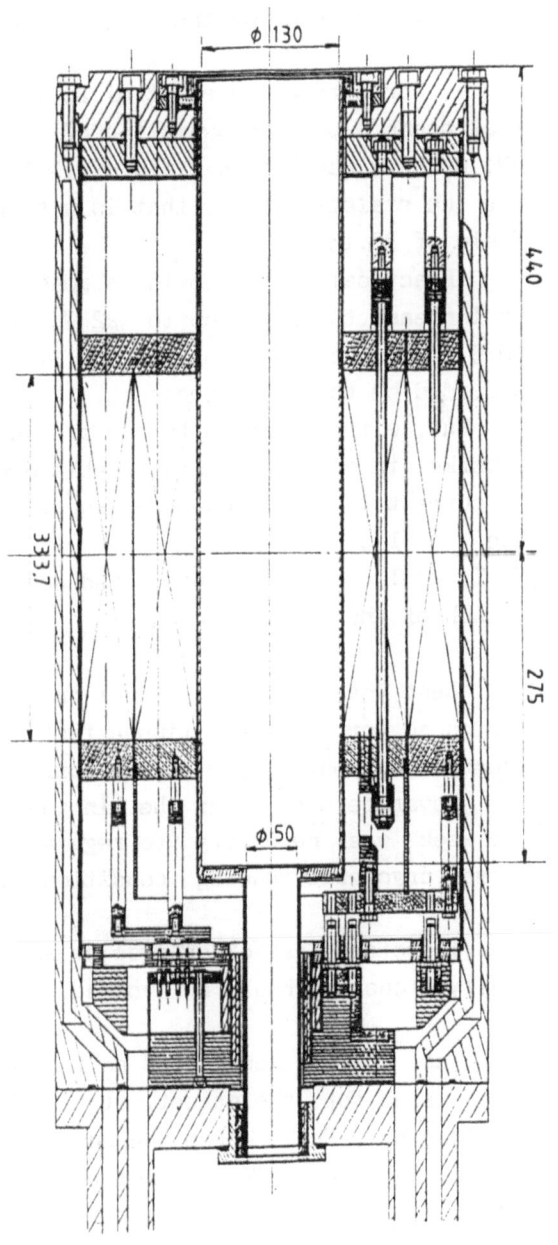

HYBRID SYSTEM

At first, we have performed tests and field mapping of the resistive part alone, and we have found that the distance between the magnetic centers of the two parts was less than 1 mm.

The corresponding magnetic forces, due to this little shift, being small and compatible with the calculated mechanical limits, we have energized the resistive magnet in a background field of 9.0 T generated by the s.c. magnet.

For a total current of 26.5 kA, that means a power consumption of 8.64 MW, we have reached a field of 11.13 T for the resistive part, that is a total field of 20.13 T.

Practically, it should be possible to increase the field up to 22 T, with 10.5 T for s.c. contribution and 11.5 T provided by the resistive part.

But, with reference to our previous remarks, in case of a catastrophic drop of current in the 10 MW power supplies, the level of transient losses in the Nb-Ti would be sufficient to induce the quench of the s.c. magnet.

Last year, the s.c. magnet has experienced for the first time a fast discharge, without damage to the coil.

Nevertheless, after the incident, one week was necessary to regain the normal cryogenic running conditions at 1.8 K.

Therefore fast discharges or especially a quench should be avoided.

Thus, 20 T is considered as the nominal value for the magnet. It will be obtained by 9 Tesla from the superconducting part and 11 Tesla from the resistive one.

Figure 2

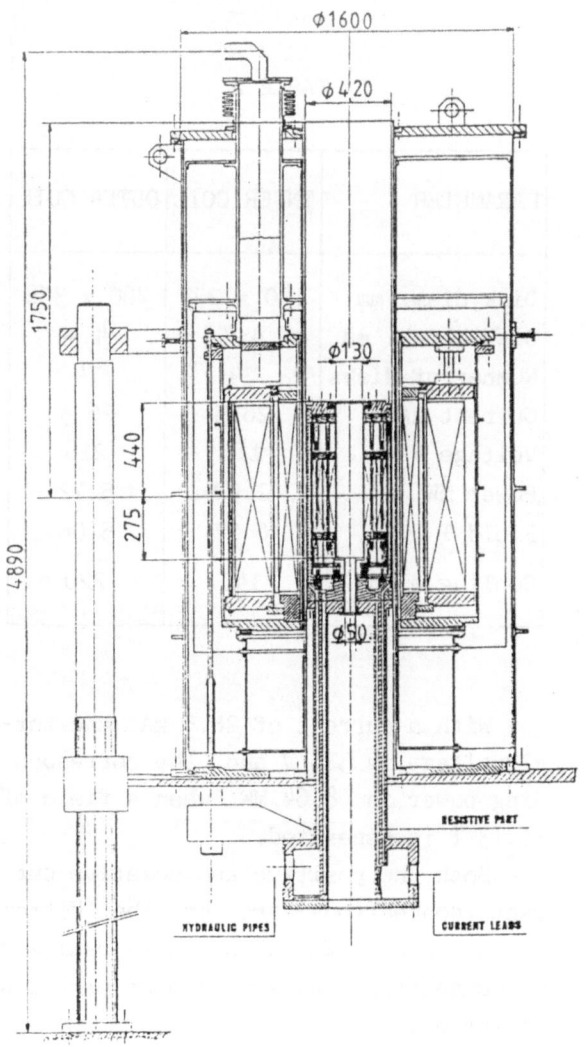

HYBRID MAGNET 20 TESLA ⌀130mm

Exceptionally, and for a limited number of experiments, it should be possible to reach 21 or 22 Tesla.

The Figure 2 above, presents the hybrid system.

CONCLUSION

By combining the field produced by the superconducting part of our 31 T, 50 mm hybrid system with the field of a new resistive Bitter magnet, we obtain 20 T in 130 mm, for a total power consumption of less than 9 MW.

The new system has mainly been built for testing superconducting materials, in significant lengths and under stress conditions.

Such studies are of major importance for the design of large bore, high homogeneity, superconducting magnets in the range of 20 Tesla.

For the first time, the 4 power supplies were operated in parallel, giving us valuable information for the design of the future magnets, in the connexion with the new 20 MW facility.

REFERENCES

[1] M.P.I. für Festkörperforschung, HML and CNRS, SNCI, A 30 Tesla hybrid magnet with 5 cm bore : a french-german project. IEEE Trans. on Mag. 1981, Vol. 17, 5, 1783-85.

[2] H.-J. Schneider-Muntau, J.C. Vallier The Grenoble hybrid magnet. IEEE Trans. on Mag., 1988, Vol 2, 1067-69.

[3] M. Ohl, P. Rub, H.-J. Schneider-Muntau and J.C. Vallier. Project of a 20 Tesla magnet in a bore of 130 mm for studies of significant lengths of high HC2 superconductor materials. Proceed. of European Workshop on high Tc superconductors and potential applications. Genova, 1987, 453-454.

THE NIJMEGEN HIGH-FIELD MAGNET LABORATORY: EXPERIENCE WITH THE NEW 30-T HYBRID MAGNET SYSTEM

K. van Hulst, H. van Luong, J.A.A.J. Perenboom, J. Rook and J. Singleton

High-Field Magnet Laboratory, University of Nijmegen,
Toernooiveld, NL-6525 ED Nijmegen, The Netherlands

ABSTRACT

The Nijmegen high-field facility is configured around a 6-MW electrical power supply and the associated water cooling system. Five magnet stations are available: three water-cooled magnets, with a maximum field of 20 T, and two hybrid magnet systems: the latest hybrid, Nijmegen-II, generates fields up to 30.4 T in a 32-mm bore. The cryogenic characteristics of this 30-T hybrid magnet have been much improved by modifications to the support of the coil tank.

The University of Nijmegen has been operating a facility to generate high static magnetic fields since 1976: the Nijmegen High-Field Magnet Laboratory (NHFML) is housed on two floors of the physics wing at the Faculty of Science. The basic infrastructure of the facility consists of five magnet stations, an electrical power supply that presently delivers 6 MW, and a closed-cycle water cooling system.

The main effort of the in-house scientists is in magneto-optical and magneto-transport studies of III-V and II-VI semiconductor materials and devices. Current research programmes also include the study of conventional and high-T_c superconducting materials, heavy-Fermion systems and quasi-one-dimensional conductors, and magnetic separation. The NHFML is strongly committed to international collaborative research programmes and it operates as a user-oriented facility.

1. THE INSTALLATION

The dc current to energise the high-field magnets is provided by two 3-MW power supplies (300 V × 10 kA), which are usually operated in parallel. Each unit consists of a transformer (with induction control), diode rectifiers and a passive LC-filter. Further regulation is provided by a transistor bank of 2000 germanium transistors in parallel. An hydraulic system will set the induction control and so realise coarse control of the current. The voltage across the transistor bank is used for fine control, and may be varied rapidly for the purpose of accurately regulated magnetic-field sweeps or magnetic-field modulation. To control the field, two remote-control panels are available to the experimenter. The first one, which has been in use for several years, is manually operated. Its reference is an 18-bits DAC, which may be programmed for constant

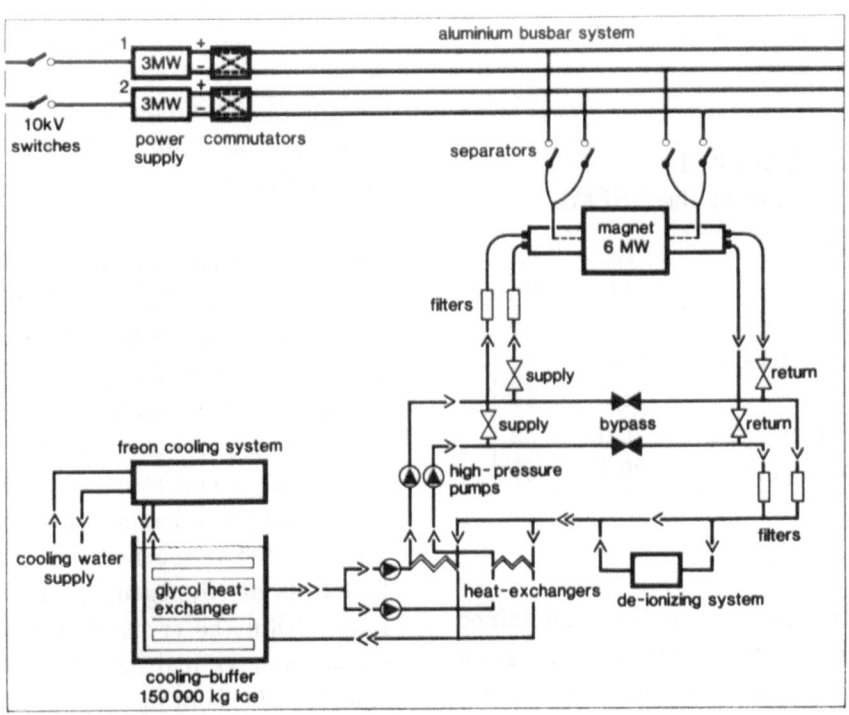

FIGURE 1. Schematic representation of the installation of the NHFML, showing a magnet station, the 6-MW electrical power supply, and the closed-cycle water cooling system with 18-MWh cold buffer.

field levels or for sweeps between any two levels. The maximum sweep rate is 200 A/s. The second reference unit, which has a 16-bits DAC, became available recently. It can be controlled via its IEEE-interface by a personal computer. Push buttons, overruling the bus input data, have been provided for safety reasons. This unit is used especially for experiments with a routine character, where the desired field levels can be programmed beforehand. It allows field levels to be set with efficient speed and without any overshoot, which is an important feature, e.g. with certain magnetisation experiments.

The cooling of the resistive magnets and of the electrical power supplies is achieved with a closed-cycle water cooling circuit. Figure 1 shows the general layout of the installation, indicating a magnet station, the 6-MW power supply, and the cooling system. A maximum flow of 400 m^3/h of deionized water is pressed through the magnet-coils at a pressure up to 23 bar. The cooling water is kept cool through its contact with a mass of 150 tons of ice in a 450 m^3 basin, frozen using a compressor-driven refrig-

erator cooled with ground water. This mass of ice and cold water represents a cold buffer of 18 MWh.

The NHFML operates (with certain restrictions on the average energy consumption) on work-days starting after 13:00 h, except during four weeks in December; the nominal power of 6.3 MW is available without any restrictions on workdays from 23:00 h to 8:00 h, and at weekends around the clock. On a typical day two shifts of nominally four hours will be scheduled, although regularly very long sessions lasting until the early morning will take place.

2. THE HIGH-FIELD MAGNETS

The NHFML has five magnet stations operational. Operation started in 1976 with several 15-T magnets. In recent years, demand has been growing for the higher field levels, and with a new magnet, to be installed later this year, the configuration will be as listed in Table 1.

TABLE 1
Magnet stations at the NHFML.

Magnet system		B_{max} [T]	Bore [mm]
1.	Hybrid magnet N-I	25.0	53
2.	Duplex Bitter coil	20.0	32
3.	Bitter coil	15.2	60
4.	Duplex Bitter coil	20.0	32
5.	Hybrid magnet N-II	30.4	32

The magnets are being run and maintained by the laboratory staff. There is no program in Nijmegen to develop and construct magnets however. The hybrid magnet systems Nijmegen-I and Nijmegen-II have been built under a collaborative contract at M.I.T. in Cambridge Ma. Nijmegen-I has been in operation since 1978 and will now be outfitted with a 6.0-MW 53-mm bore insert coil. The second, more advanced hybrid magnet system has been in operation since early 1987.

3. NIJMEGEN-II

The design of the 30-T magnet, the features of the subcooled superfluid helium cryostat and the first testing at M.I.T. and at Nijmegen have been reported earlier [1-5].

During performance tests [4] we found that the helium boil-off became excessive when the coils were energised. In the original cryostat design, the cold mass was suspended from the cryostat top cover by means of six steel cables attached to the bottom end of the coil tank. Centricity of the suspension points at both ends and adjustment of the cables to equal lengths would take account of proper alignment and of centricity of the field axis with the water-cooled insert coils. Fiberglass straps at the bottom end

of the coil tank were used to secure its radial and rotational position. These straps together with the weight of the cold mass would suffice to withstand the interaction forces between insert and superconducting coil under normal operating conditions.

Six heavy members, mounted on the bottom cover of the cryostat and adjustable from the outside, would act as "stops" so as to cope with excessive forces (e.g. those occurring during an insert failure). They would limit the vertical movement of the cold mass to 1 mm in either direction and so protect the suspension structure from being overstressed. The stops, when centered, could travel approximately 1 mm in either direction before touching the cold mass. During first tests at Nijmegen, the position of the coil tank within the cryostat was found to be less stable than we had hoped: at moderately high combined fields of insert-coils and superconductive magnet, we had to readjust the setting of these stops to prevent, or at least reduce, thermal shorts at the higher fieldstrengths. Shortly after the magnet system was made available for physics research, the situation worsened to the point where 1.8-K operation was no longer possible. By itself this was enough of a reason to dismantle the cryostat, but problems with degrading contacts to the gas-cooled current leads at the cold end and unsatisfactory efficiency of the helium transfer made the decision easier.

M.J. Leupold, one of the original designers of this hybrid magnet system, proposed to replace the six stops by as many fiberglass struts, each strut consisting of a number of 3-mm thick straps, to rigidly support the cold mass from the bottom cover of the vacuum tank. The struts, while consisting of relatively thin straps, would combine stiffness in one sideways direction with flexibility in the other. After we had convinced ourselves of the centricity and accurate positioning of the existing mounting holes in the bottom structure of the coil tank and in the bottom cover, we opted for a high precision design, which would avoid the necessity of final readjustments. The design was such, that no modifications to the major cryostat parts had to be made. The struts are 180-mm long dog-bone-shaped members. Their ends are clamped to

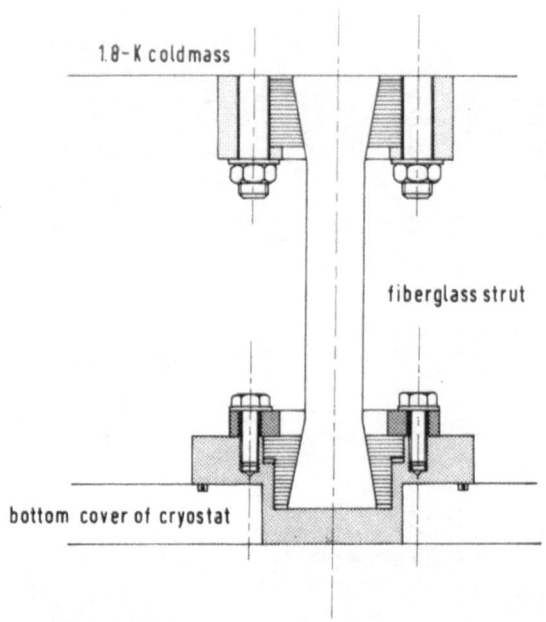

FIGURE 2. Diagram of the dog-bone-shaped fiber-glass struts (a stack of ten 3-mm thick elements) supporting the 1.8-K cold mass of the coil tank from the room-temperature bottom cover of the cryostat.

the bottom cover and the bottom of the coil tank (see Figure 2). Each strut consists of a stack of ten 25-mm wide and 3-mm thick elements. The struts are mounted so, that they are flexible in the radial direction and thermal contraction of the cold mass will cause them to become slightly bent into an S-shape, without introducing any significant stress, while radial and rotational forces on the cold mass will be met by the full stiffness of part or all of the struts, respectively. The struts are strong enough to withstand axial as well as rotational forces under all anticipated conditions. The maximum radial forces that might occur, however, are close to the limit and we therefore added six radial straps for additional support in that direction. Figure 3 is a photograph of the bottom cover of the cryostat with the six fiberglass struts mounted and ligned out; at this stage, the support structures for the six radial straps were put in place. The spring constant in the radial direction of the six struts together was estimated to be about 3×10^4 N/mm. Measurements indicate that it is actually about 25% higher.

To keep the heat input into the 1.8-K space as low as possible, the struts and the radial straps have been heat-sinked to the radiation shield at mid-length. The extra heat-load on the radiation shield amounts to about 5 W, which is almost negligible with respect to the cryogenerator's cooling power at 80 K. The estimated heat input into the coil tank through the new suspension parts will be approximately equal to that of the parts that have been removed.

In solving the problem of heat dissipation in the current contacts, mentioned earlier, we were facing some serious limitations. Certain parts could only have been replaced by cutting major welds of the helium reservoir and we deemed it wise to avoid this. The male contacts at the cold end of the gas-cooled leads were replaced with fixed contacts, and the expansion of the leads is now accomodated with a sliding seal at the top of the chimney. The access for helium transfer was modified in such a way, that during the filling of the coil space the liquid is forced to the bottom end of the coil, with the benefit of more efficient precooling of the coil mass with the cold helium gas. For helium transfer during operation of the system, needed to maintain a sufficiently high helium level in the reservoir, the transfer syphon can be lifted somewhat resulting in significantly less boil-off losses because the bottom end of the syphon is no longer immersed in the liquid, as was the case previously.

The tests performed after the modifications had been completed showed results well in agreement with our expectations. The standing loss of the cryostat, with no current in the superconductive coil, had not changed. From the data we have collected so far, it can be concluded, that coil losses, if any, are small. At 1500 A at 4.2-K operation, they may be 1 W or less. At 1950 A at 1.8-K operation, with the water-cooled insert swept up and down in the course of the experiments, the total heat load on the 1.8-K space is about 8 W. If the insert is swept more or less continuously, this figure is somewhat higher; with no field variations in the insert coils, it might be as low as 7 W, indicating that coil losses can not be more than about 2 W. These data should be considered with some reservation, however, as the settling times of the system are very long.

FIGURE 3. Photograph of the bottom cover of the cryostat: the six fiberglass struts are all clamped to the bottom cover and the supports for the six radial straps, which will also be fixed to this bottom cover, are being positioned and made to size.

The consumption of liquid helium has been reduced by the modification from about 75 l/h to about 30 l/h.

Because of the high heat load on the 1.8-K space, the magnet was originally equipped with two 1250 m³/h Beach-Russ vacuum pumps. The two are still needed to cool down from 4.2 K to 1.8 K within a reasonable time (which is important in our mode of operation, as we do not keep the system cold permanently). Although not really needed during operation of the system, the large pumping capacity serves us well in establishing an extra temperature margin: during normal experimental conditions a coil-tank temperature of between 1.5 K and 1.6 K can be maintained.

As the magnet is not kept at liquid helium temperature between runs it is important that helium transfer and cooldown to 1.8 K be done as efficiently as possible. We have reached an overall efficiency of 75% by starting the cooldown simultaneously with the transfer. Thus we precool the Joule-Thompson heat exchanger and adjacent valves and tubing with cold gas and can use the refrigeration power of the pumps at the earliest stage possible.

4. BURN-OUT PREVENTION

All of our 15-T and 20-T resistive magnets have shown gradual degradation at the end of their life span and regular calibrations and impedance checks indicate when a particular magnet should be overhauled. In our hybrid magnets the water-cooled insert consists of two nested Bitter coils, either axially-cooled, as in Nijmegen-I, or radially-cooled, as in Nijmegen-II. Here, failure of the innermost coil occurs very abruptly in a timespan of about 600 ms, resulting in the loss of most of the inner coil, and usually some damage to the outer coil. In radially-cooled inserts a failure is expected to result in the loss of most of both inner and outer coil, due to arcing between the two sections. Various schemes have been thought of to cut the power before the damage gets too big. A few have been tested with limited or no success. Close observations however have taught us recently, that in the last 30 seconds preceding the onset of a burn-out the impedance of the inner coil tends to change by an amount that can reliably be detected by computer guardance. We therefore have set up a computer-system that compares the actual impedance of the inner coil with the value estimated and calculated from the coil parameters, the momentary current and the inlet-temperature of the cooling water. The first tests have been very promising, even to the extent that, instead of opening the circuit-breakers, we may rely on a "slow switch-down" (about 10 s). This would prevent erraneous triggering of the quench-detector and so we would possibly avoid the loss of helium in case of the shut-down.

5. CONCLUSION

With all these modifications successfully implemented, Nijmegen-II has now become available for scientific research rather more routinely. Control of the fieldstrength is very convenient and sweep rates are no different from any of the other magnets whereas very little effort is required to handle and control the cryogenics of the system.

REFERENCES

1. Leupold, M.J., Iwasa, Y. and Weggel, R.J., 32 Tesla Hybrid Magnet System. J. Physique 1984, **45**, C1-41–44.

2. Leupold, M.J. and Iwasa, Y., Subcooled Superfluid Helium Cryostat for a Hybrid Magnet System. Cryogenics 1986, **26**, 579–85.

3. Leupold, M.J., Iwasa, Y., Hale, J.R., Weggel, R.J. and van Hulst, K., Testing a 1.8 K Hybrid Magnet System. In Magnet Technology, MT-9, eds. C. Marinucci and P. Weymuth, SIN, Villigen, 1985, pp. 215–18.

4. van Hulst, K. and Perenboom, J.A.A.J., Status and Development at the High Field Magnet Laboratory of the University of Nijmegen. IEEE Trans. Magn. 1988, **MAG-24**, 1397–1400.

5. Perenboom, J.A.A.J. and van Hulst K., The Nijmegen High Field Magnet Laboratory. Physica B 1989, **155**, 74–77.

PLASTIC BENDING LARGE DISPLACEMENT ANALYSIS AND SPRING-BACK OF A CONDUCTOR JACKET OF A SUPERCONDUCTING MAGNET FOR FUSION REACTORS

Roberto E. Gori
Istituto di Scienza e Tecnica delle Costruzioni, Padua University.
Via Marzolo n.9, 35131 Padova, Italy.
Pierluigi Zaccaria
Progetto RFX, Istituto Gas Ionizzati del CNR.
Corso Stati Uniti n.4, 35020 Padova, Italy.

ABSTRACT

In this paper a finite element large displacement elastoplastic analysis is used to model the relevant plastic forming steps and the residual strains and stresses produced by conductor bending during reel winding, reel unwinding and coil winding. The original proposed approach allows simply to analyse a part of conductor contained between two cross-sections and can easily take into account the nonhomogeneities of the composite.

INTRODUCTION

Recently proposed preliminary designs of future A15 force cooled superconductors for toroidal and poloidal field coils for tokamaks, have a flat rectangular or a square cross section.

As it is well known, relevant plastic deformations occur during the manifacturing and winding of cables [1,2]. In the manufacturing process, either in reaction before storage on reel, or in the case of reaction on reel, springback effect (of opposite sense) occurs after unwinding; furthermore, during the winding on the bobbin, the conductor is deformed from a straight configuration to the shape of the coil, which can present regions with a small radius. This causes a deformation of the conductor cross-section (keystoning) which needs to be controlled. Furthermore keystoning may apply a pressure on the conductor and produce a degradation of the superconducting properties. One way to simulate such behaviour is a numerical analysis of the different forming steps. The results of such an analysis are here reported. Previous works on this subject were carried out by Zehlein [3,4] where only an homogeneous rectangular cross section of a full conductor was analitically investigated, and a "homogenized" cross section was proposed. A complete hollow conductor was numerically analysed by the authors [5,6].

THE FINITE ELEMENT MODEL

The structural analysis of the proposed NET superconductor has been carried out by means of ANSYS 4.3 code [7].

Two cases have been studied. The first regards the single steel jacket of the conductor. The second one deals with a complete conductor: steel jacket plus internal superconducting material. The cross section of jacket is a hollow rectangle; external dimensions are 48.0x31.3 mm, internal 38.0x21.3 mm. The adopted meshes, shown in **Figure 1**, represent one half of a

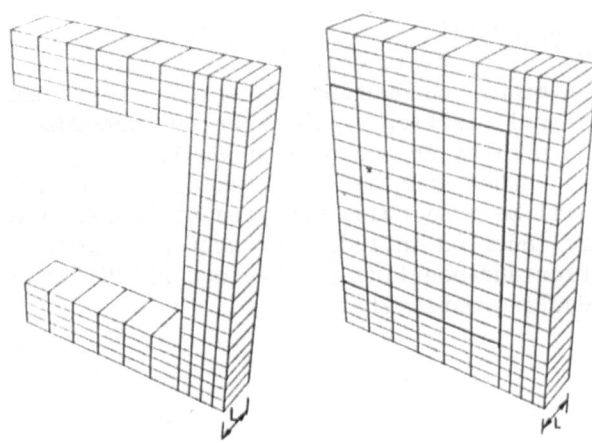

FIGURE 1 - 3D meshes of the analyzed proposed NET superconductor

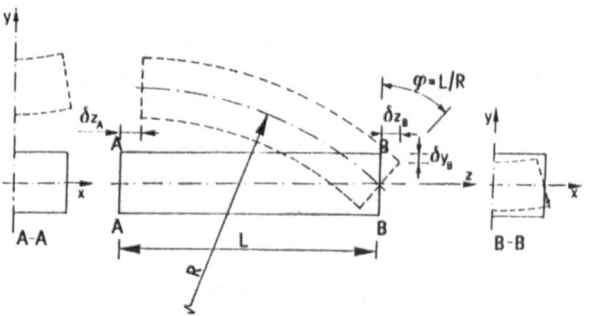

FIGURE 2 - Geometric scheme of the deformed shape

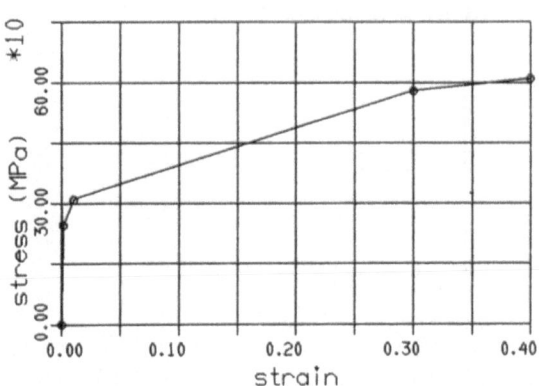

FIGURE 3 - Jacket material stress-strain relationship (316LN stainless steel)

conductor. Suitable boundary conditions are taken into account in accordance with the symmetry plane (yz).

The main object of the present approach is to simulate a series of loading conditions to reproduce the conductor for-

ming during the manifacturing process of the coil. In this procedure, a pure bending deformation is assumed to occur during forming process, in absence of axial winding forces [4].

The elements used to model the case and the conductor are 8 noded 3D isoparametric solid elements (STIF45) which include modified extra displacement shapes.

In the analysis of a complete conductor, three-dimensional interface elements (STIF52) have been used as gap elements to model the contact between internal conductor and case. The friction value is assumed to be zero. But it must be noted that friction does not influence significantly the longitudinal stresses. In fact the adopted boundary conditions force the cross-sections to remain plane, as described in the following.

Geometrical Constraint Conditions

Due to the cyclic symmetry of the pure bending deformation, the Bernoulli hypothesis can be assumed. Therefore the analysis can be limited to the part of conductor contained between two cross-sections; the corresponding meshes, shown in **Figure 1**, have a single element in the longitudinal (z) direction. The adopted hypothesis is still valid when a nonlinear analysis is carried out.

The nonlinear analysis is carried out for consecutive loading conditions corresponding to different curvature radii of the centroidal axis (**Figure 2**).

For each radius R, a set of equations (constraint equations and imposed displacements) has to be satisfied:

$$\delta z_{i,A} = L - R \sin(L/R) \qquad (1)$$

$$\delta x_{i,A} = \delta x_{i,B} \qquad (2)$$

$$y_{i,A} + \delta y_{i,A} - R[1 - \cos(L/R)] = \frac{\delta z_{i,B}}{\sin(L/R)} \qquad (3)$$

where $i = 1...N$; δ_x, δ_y, δ_z are the updated nodal large displacements, $y_{i,A}$ and $y_{i,B}$ are the y coordinates of the nodes of sections A and B. Equations (1), (2) and (3) force sections A and B to remain plane and to have the same deformation in their planes. These cyclic symmetric equations have been used for the analysis of the steel jacket.

When the complete conductor has to be analysed, for the nodes of the internal conducting material, equations (1) and (2) are still valid, and the following further equations have to be satisfied:

$$\delta z_{i,B} = (y_{i,B} + \delta y_{i,B}) \tan (L/R) \qquad (4)$$

$$\frac{\delta y_{i1,B} - \delta y_{i2,B}}{\cos(L/R)} = \delta y_{i3,A} - \delta y_{i4,A} \qquad (5)$$

Equation (4) forces a rotation L/R to section B, while equation (5) prescribes equal relative displacements, in the sections A and B, between the nodes of the corresponding gap elements (i_1 and i_2 of section B and i_3 and i_4 of section A).

Elastoplastic Material Behaviour

The stress-strain relationship of the steel of case is shown in **Figure** 3 (E=190 GPa).Multilinear kinematic hardening, with Von Mises yield criteria, is assumed. For internal conducting material, a linear elastic isotropic behaviour is considered as a first approach (E=40 GPa).

When further experimental results will be available, the present model may be adopted including orthotropic properties and non-linear behaviour of the conducting material.

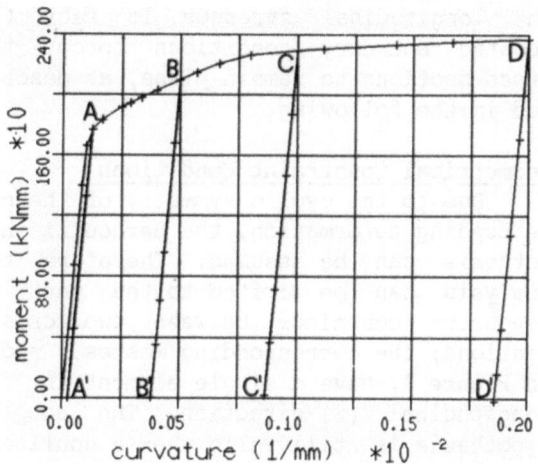

FIGURE 4 - Moment-curvature diagram, for **load history** I on the jacket alone: monotonic bending to A) R=8000mm, B) R=2000mm, C) R=1000mm, D) R=500mm; and the corresponding unloading to A') R=41700mm, B') R=2630mm, C') R=1150mm, D') R=538mm

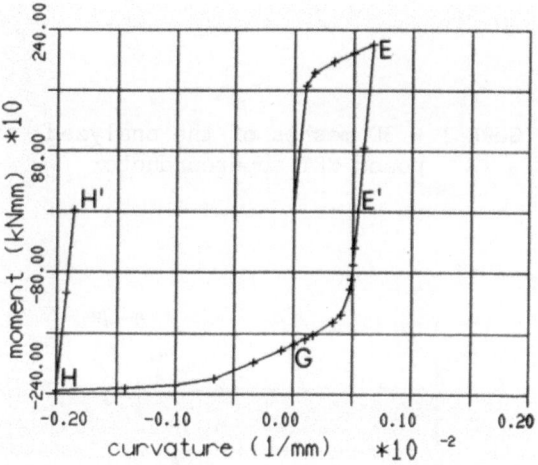

FIGURE 6 - Moment-curvature diagram, for **load history** III on the jacket alone: monotonic bending to E) R=1500 mm, unloading to E') R=1850 mm, reverse bending to G) R=∞, then to H) R=-500 mm, finally springback, unloading to H') R=-538 mm

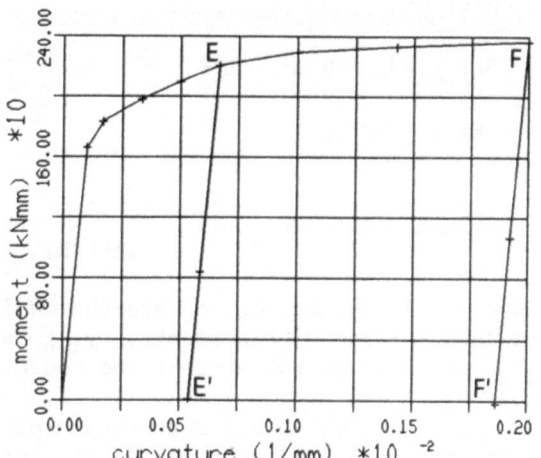

FIGURE 5 - Moment-curvature diagram, for **load history** II on the jacket alone: monotonic bending to E) R=1500mm, unloading to E') R=1850mm, rebending to F) R=500mm, finally springback, unloading to F') R=538mm

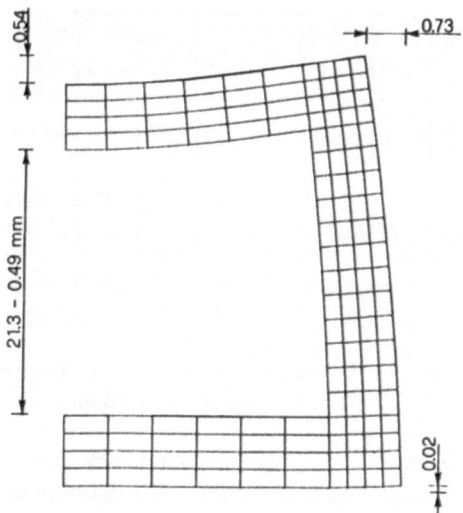

FIGURE 7 - Keystoning for maximum bending (R=500mm) of jacket, **load history** II (F)

THE LARGE DISPLACEMENT NONLINEAR ANALYSIS

A large displacement procedure has been adopted based on an updated tangent stiffness matrix and the direct integration method. The alternate restart function of ANSYS has been used to change the applied constraint conditions, at each lo-

ading step. **Figures 4, 5** and **6** show the different load histories to which the conductor has been subjected. Load history I considers four different monotonic bendings and unloadings. Load histories II and III concern with bending and reverse bending.

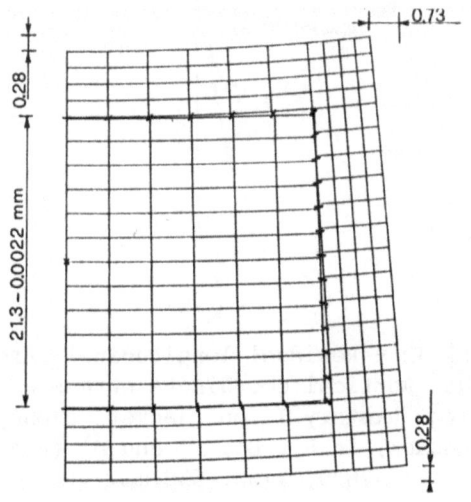

FIGURE 8 - Keystoning for maximum bending (R=500mm) of complete conductor, **load history II (F)**

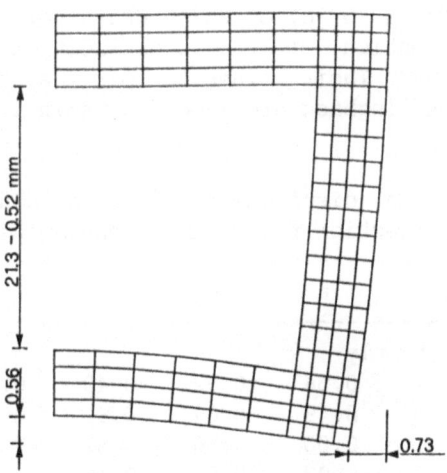

FIGURE 9 - Keystoning for **maximum** reverse bending (R=-500 mm) of jacket alone, **load history III (H)**.

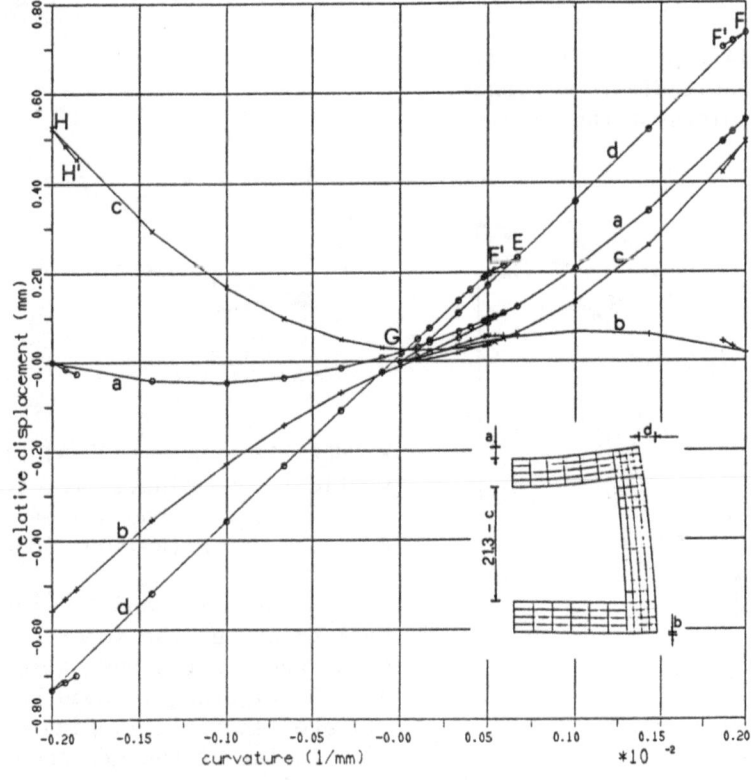

FIGURE 10
Relative displacements vs curvature for jacket alone for **load histories II and III**

RESULTS

Maximum Curvature Keystoning

Figures 7 and 8 show the keystoning of the jacket and of the complete conductor, for rebending to R = 500 mm, while **Figure** 9 represents the maximum reverse bending to R = 500 mm.

In **Figure 10**, the relative displacements a, b, c, d, which characterize the deformation of the cross-section, are plotted. Their values for maximum bending and springback are shown in **Table 1**.

TABLE 1.

Relative displacements (mm) in conductor cross-section for maximum bending(R=500mm)

case	a	b	c	d
Jacket only				
F	0.54	0.02	0.49	0.73
F'	0.49	0.04	0.42	0.70
H	0.00	-0.56	0.52	-0.73
H'	-0.02	-0.51	0.46	-0.70
Complete conductor				
F	0.28	0.28	0.00	0.73
F'	0.12	0.24	-0.12	0.51

Stress Distributions

Figures 11 to 14 represent the distributions of the longitudinal stresses, in the middle fiber points of the vertical

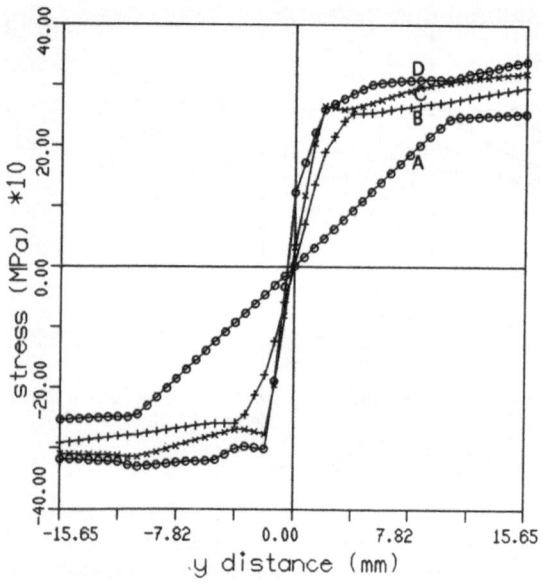

FIGURE 11 - Longitudinal stresses versus vertical coordinate, for **bending history** I on jacket alone, in A, B, C and D (R=8000, 2000, 1000, 500 mm)

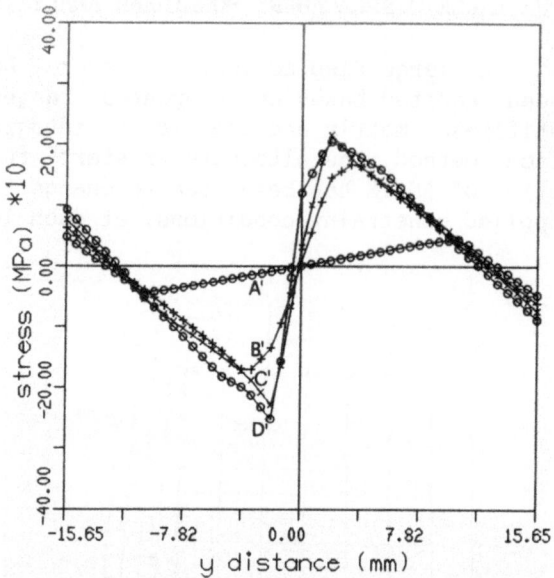

FIGURE 12 - Residual longitudinal stresses versus vertical coordinate stresses, for **bending history** I on jacket alone, · at springback, in A', B', C' and D' (R=41700, 2630, 1150, 538 mm)

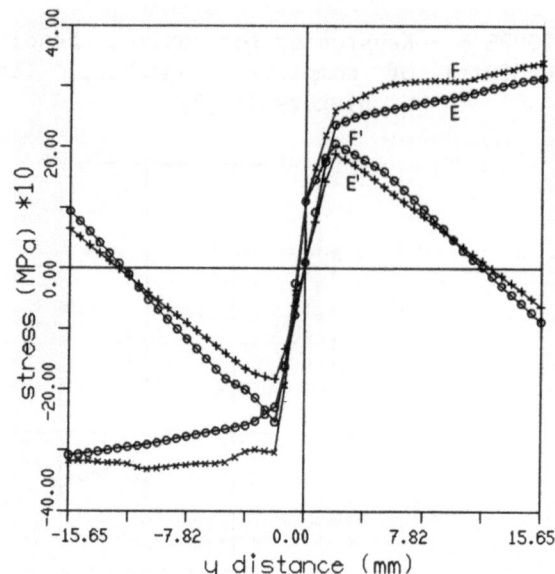

FIGURE 13 - Longitudinal stresses versus vertical coordinate, for **load history** II (bending and rebending) on jacket alone in E, E', F and F'(R=1500, 1850, 500, 538 mm)

web. **Figures 11.** and **12.** (residual stresses) point out the importance of the plastic zone corresponding to bending B, C and D (**Figure 4.**). Load histories II and III are described in **Figures 13.** and **14.**.

As far as the stresses are concerned, the difference between bending and rebending seems to be negligible, so as the difference between rebending and reverse bending (F, F' and H, H'). On the other

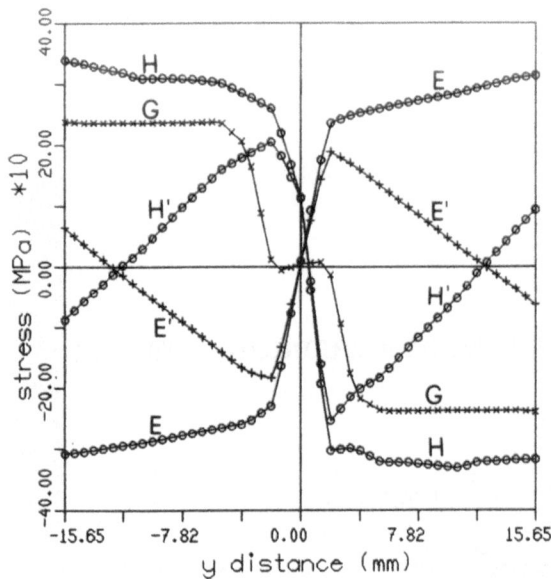

FIGURE 14 - Longitudinal stresses versus vertical coordinate for **load history** III (bending and reverse bending) on jacket alone, in E, E', G, H, H' (R=1500, 1850, ∞, -500, -538 mm)

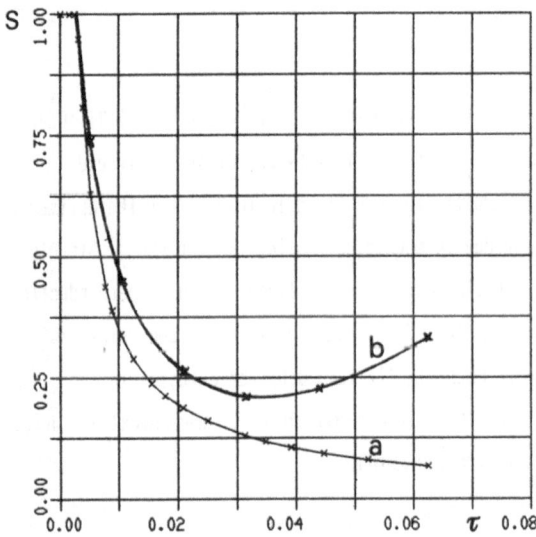

FIGURE 15 - Springback ratio (s) [4] versus thickness to radius ratio (τ): a) jacket alone, b) complete conductor

hand reverse bending must be avoided **[4]** in order to exclude change of the sign of plastic strain.

Full Springback

Referring to several unloadings the springback ratio has been evaluated and is shown as a function of τ (= t/R) in **Figure 15.** (t = 31.3 mm).

CONCLUSIONS

The procedure used presents several advantages with respect to an analysis on a complete 3D model, of a suitable length such as [5]: a) The dimensions of the problem are dramatically reduced, allowing a finer discretization; b) The stress perturbations at the ends of the model, due to the applied forces or displacements, are avoided; c) Computational time and memory request for non-linear analyses are largely reduced.

The authors are currently working to improve the present procedure in order to take into account also axial forces during coil manufacturing.

Acknowledgements
This research has been carried out under NET contract N.288/88-1/FU-I/NET.

REFERENCES

1. <u>Fusion Engineering and Design</u>, Special Issue on The IEA Large Coil Task, Vol.7 (1988),7 September 1988.

2. Nyilas, A., and Zehlein, H., Theoretical and Experimental Studies of Forming and Failure of Forced Flow Cooled Superconductor Bends, <u>Proceed. 11th Symp. on Fusion Techn.</u>, Austin, USA, 1985.

3. Zehlein, H., Recent inprovements of residual stress analysis of composite superconductor cables for fusion magnet windings, <u>Proceedings of the 8th SMIRT</u>, Brussels, 1985.

4. Zehlein, H., Residual stresses and strains in some new superconductors with composite cross sections, <u>Proceedings of the 9th Int. Conf. on Magnet Technology</u>, Zurich, 1985, 665-668.

5. Gori, R., Schrefler, B. and Zaccaria, P., NET Conductor Case Nonlinear Material and Large Displacement Analysis, Ist. Scienza e Tecnica delle Costruzioni, Report 9/88, Padova, 1988.

6. Mitchell, N., Gori, R. and Collier, D., Structural Analysis of the NET Toroidal Field Coils and Conductor. <u>Proceedings of the 15th Symp. on Fusion Technology</u>, Utrecht, The Netherlands, 1988.

7. ANSYS Engin. Analysis Systems, User's Manual, Rev.4.3, Houston, Pa, 1986.

STRUCTURAL CHARACTERISTICS AND THE INFLUENCE OF MECHANICAL DISTURBANCE IN SUPERCONDUCTING TOROIDAL COILS

Y. Kannoto[*], H. Hashizume[**], M. Minami[*], K. Hayakawa[*],

T. Takagi[**] and K. Miya[**]

[*] Mitsubishi Heavy Industries, LTD.
2-1-1 Shinhama, Arai-cho, Takasago, Hyogo Pref. Japan
[**] Nuclear Engineering Research Laboratory, University of Tokyo
Tokai-mura, Ibaraki Pref. Japan

Superconducting toroidal coils for SMES were fabricated.   The coils are composed of superconductor (Nb Ti/Cu), structural component, stabilizer and insulator.   The coil components are bounded mainly by solder and epoxybond. Cracking or slipping may occur in the components due to large electromagnetic force or seismic load.   For the optimum design of coils, evaluation of mechanical behavior and heat generation due to mechanical motion is required.   Structural analysis and model test plays an important role on establishment of a design guideline for large scaled superconducting toroidal coils.   In this study the mechanical behavior of the coils were tested and numerically evaluated using finite element analysis based on the composite theory.   The model coils were mechanically loaded by the actuator in the cryogenic vesse with 7 kA current. Strain, temperature and AE signals of the coil were measured.   Mechanical and quench behaviors measured in the tests were predicted very well by the numerical calculations.

1. Introduction

It is well known that one of the most critical issues in utilizing superconducting magnet is how to protect the magnet from quench.   To develop a method for the protection, however, it should be clarified how the quench initiates in the magnet.   From the viewpoint of the cost problem it is also important to optimize the coil design after adequately evaluating the safety factor against quench.   The object of this study is, therefore, to establish a method for predicting mechanical behavior of toroidal magnet and quench occurrence caused by the mechanical loading.

Figure 1 shows the flowchart to optimize the coil design with consideration for the protection from quench due to mechanical disturbance.   The points of this optimiza-

tion flowchart are prediction of slip occurrence components, non-linearity behavior of the coil due to the slip and the determination of presence or absence of quench caused by the heat generation due to slip.
In this study, the behavior of the toroidal coil subjected to mechanical disturbance was clarified by evaluating the influence of slip, frictional heat generation and the increase of the conductor temperature through the test of model toroidal coils and numerical analysis based on the composite theory.

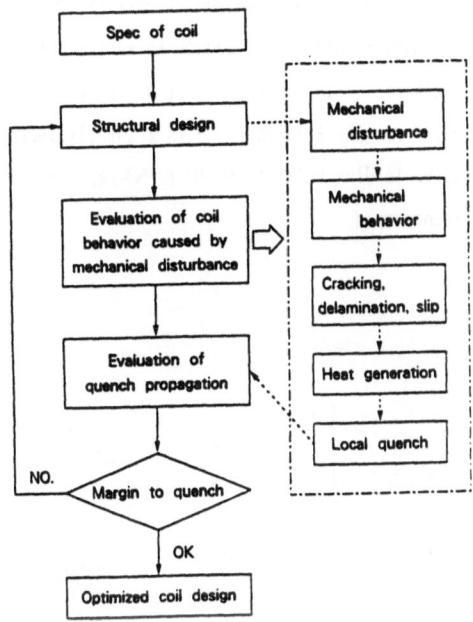

Fig. 1 The flowchart of optimum coil design

2. Model test

Figure 2 illustrates the cross section of the toroidal magnet for a large-sized SMES.   In this type of magnet, the superconductor is embedded in the structural material with the stabilizing material to suppress the quench.   For testing the influence of mechanical disturbance, the loading test with existence of current was executed with two types of toroidal coils simulating the SMES magnets shown in figure 2.

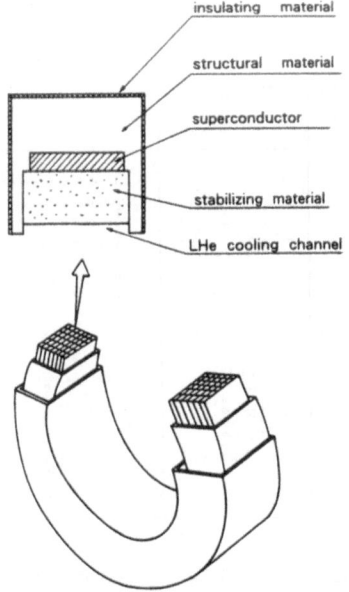

Fig. 2 Cross section of the superconducting coil
for large-sized SMES

Both of them are toroidal coils, composed of Nb Ti/Cu conductor and the other members, having an inside diameter of 350 mm and consisting of 3 layers x 3 lines. Model 1 coil is a fully stabilized coil for the 100 kWh class, while Model 2 is a partially stabilized coil for the quench test.   In figure 3, the cross section of these two two magnets are shown as well as the characteristics of the magnets.

As shown in Figure 4, each coil is installed in the loading frame and inserted into the cryogenic tank, then mechanical load was applied with existence of current flowin the LHe coolant.   The current of both magnets is 7kA and mechanical load is applied up to 4.6 ton in the Model 1 coil, while up to 2 tons in the Model 2 coil.

The following quantities are measured through the loading in the LHe circumstance ;

- Coil current
- Magnetic field of coil center
- Voltage of coil
- Loading force

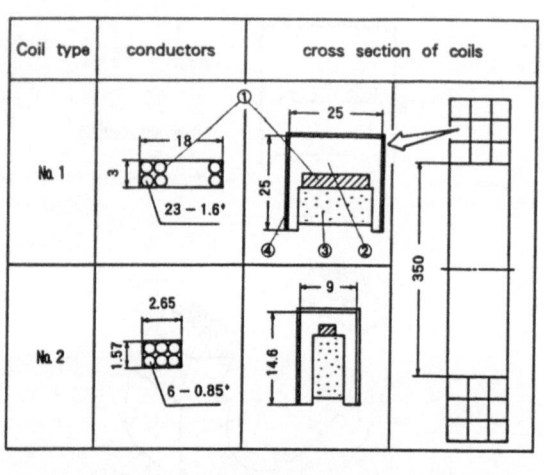

① Superconductors : NbTi／Cu   ② Structural material : aiminum alloy

③ Stabilizing material : aluminum   ④ Insulating material : GFRP

Fig. 3 Cross section of model coils

- Displacement of loading point
- Strain of coil
- AE signal from coil
- Temperature of coil

Figure 5 shows the details of where measurement for strain, temperature and AE signal is carried out. Experiment of the model 1 coil, a fully stabilized coil against quench, is carried out to examine the strain behavior of the coil to the mechanical load, while that of the Model 2 coil to observe the effects of the mechanical disturbance on the quench in addition to the strain behavior. In the experiment of the Model 2 coil, therefore, the loading force is gradually increassed with the loading velocity of 10 to 50 kg／s until quench is occurred.

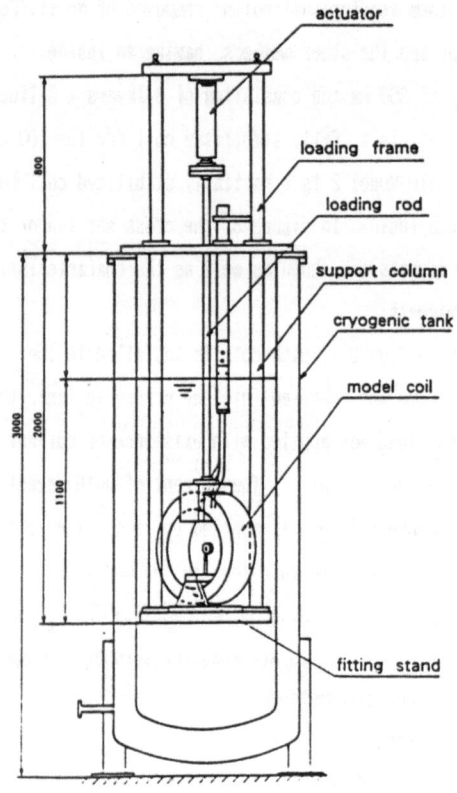

Fig. 4 Test facility and loading concept

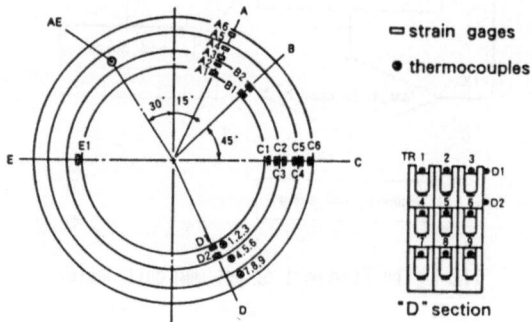

Fig. 5 Measuring points of strain, temperature and AE

3. Numerical analysis based on FEM

In this section, the method of numerical analysis for the toroidal coils, composed of the composite structure as is seen in Figure is described.
Some of the authors have developed the method for quantitative evaluation of the frictional heat generation due to slip between composite elements based on the composite theory for an attempt to determine whether

quench will be cause by the increase of the super-conductor temperature. [2] [3]

In this analysis theory, the composite structure is modeled as multi-layered composite laminate elements, using the Voigt's hypothesis, Reuss's hypothesis and the law of mixture. For the slip behavior between the composite elements, strain increment $d\varepsilon$ is defined as the

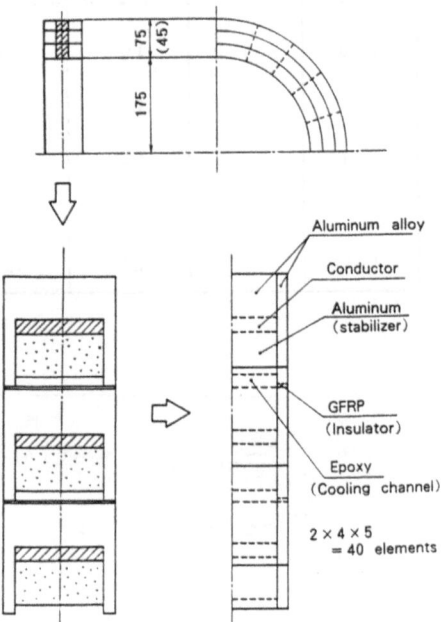

Fig. 6　Calculation model

the sum of elastic strain $d\varepsilon^e$, plastic strain $d\varepsilon^p$, and slip strain $d\varepsilon^s$ as follows :

$$\{dc\} = \{d\varepsilon^e\} + \{d\varepsilon^p\} + \{d\varepsilon^s\} \quad (1)$$

It is assumed that the magnitude of the slip is proportional to the shearing stress between layers, and that the slip occurs in the following stress condition :

$$fs = \frac{\sigma_N}{\sigma_0} + \frac{\tau^2}{\tau_0^2} - 1 = 0 \quad (2)$$

Above, τ_0 and σ_0 indicate the stress to delamination, assuming that the shearing stress or normal stress has an independent effect. The following equation is used to evaluate the heat generation W due to the slip behavior, and the heat transfer analysis is adopted to calculate the increase of the conductor temperature to determine whether quench will occur :

$$q_s = \{d\overset{s}{\varepsilon}\}\{\tau\} / dt \quad (3)$$

Figure 6 shows the model for the numerical analysis where the composite theory is adopted.

4. Experimental and numerical results

The comparison of the numerical and experimental results in case of the Model 1 coil is shown in table 1

Table. 1　Numerical and experimental test results of No. 1 coil

	Experimental result	Numerical results	
		$\sigma_s = \tau_0 = 15\text{MPa}$	$\sigma_s = \tau_s = 10^{11}\text{Pa}$
Displacement　δ		4.6mm	
Load　Fmax	4.6ton	5.1ton	13.0ton
Strain　ε min	$-2200\,\mu$	$-1990\,\mu$	$-3810\,\mu$
Residual Strain	$-600\,\mu$	$-1050\,\mu$	$-860\,\mu$

in terms of the maximum load, maximum strain and residual strain.

Here numerical analysis is carried out under the displacement control condition.

Figure 7 illustrates the behavior of the strain on the inner layer at the point A1 (see fig. 5).

As shown in Figure 7, the load vs strain curve provides non-linearity between 1 and 4 tons.

This is caused by the expansion of the slip region due to bending force on each layer caused by delamination or slip between the layers.

This is explained by the fact that the curve assuming the slip limitation τ_0 =15 MPa almost agrees with the experimental result while the curve with the assumption of $\tau_0 = \infty$ (no slip) is straying out of the experimental curve.

In both analyses, material non-linearity and plastic deformation, are considered.　These results indicate that slip effect should be taken into account in evaluation of the behavior of laminated coil.

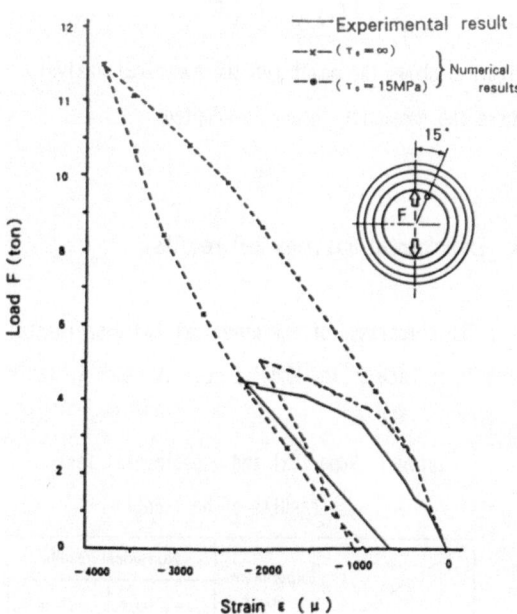

Fig. 7 Structural behavior of No. 1 coil

The experimental results about quench in the Model 2 coil is summarized in the table 2.   The loading tests with 7kA current were repeated nine times (T1-T9) and quench had occurred seven times in the test.

The first quench occurred at F=0.9 ton in T-1, and the second occurred at F=1.05 ton in T-3.   Thereafter, the load was increased to F=2.0 ton in T-4 without quench. However, in T-5 and subsequent tests, quench had occurred five times within the range of 1.66 to 1.95 ton in load F.   This range can be considered as the region where quench is be easily caused by mechanical slip. Figure 8 shows the relationship between the load and displacement at the loading points in the T-1 and T-3 tests.

As shown in Figure 8, the initial slip was observed on the inner layer at around 0.2 ton in load F, thereafter the slip region was expanded in a hoop direction.

The slope of the curve for T-3 experiment is modest in comparison with that of T-1 curve, which probably is caused by the change of the slip limitation character- istics after the coil has experienced slip.

Table. 2 Experimental results of No. 2 coil

Test No.	Current (kA)	Quench load Fq (ton)	Quench disp. ΔD (mm)
T-1		0.9	3.1
T-2		(~1ton)	No quench
T-3		1.05	4.2
T-4		(~2ton)	No quench
T-5	7	1.66	9.9
T-6		1.95	12.4
T-7		1.84	12.0
T-8		1.88	12.0
T-9		1.94	12.5

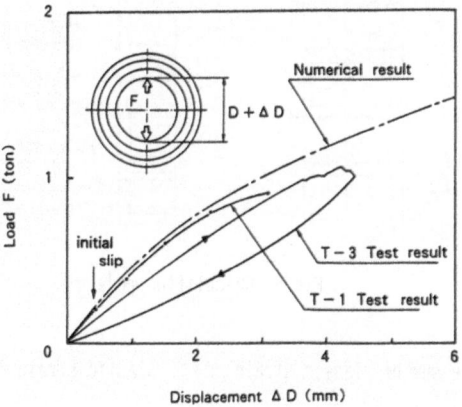

Fig. 8 Load vs displacement behavior of No. 2 coil

Figure 9 shows an example of the schematic diagram of the current, voltage, temperature, etc., which apparently change when quench occured.

It is difficult to discuss the details of the quench phenomenon simply based on this result. However, the quench prediction method based on the AE signal or strain behavior should be studied in the future.

Figure 10 shows the comparison between the experimental and numerical results of the load and the displacement at the occurrence of quench.   In the experiment, the initial quench had occurred around 1 ton in load F.

However, the load can be increased due to training and quench is repeated around 2 tons in F.

Numerical result show that quench could occur at 1.94 ton (displacement of loading point= 10mm).

This result almost agrees with the experimental result.

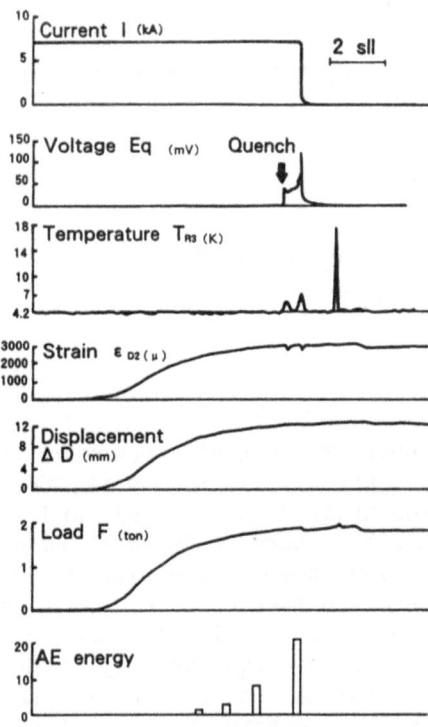

Fig. 9 Schematic diagrams of current, voltage, etc. during T5 test

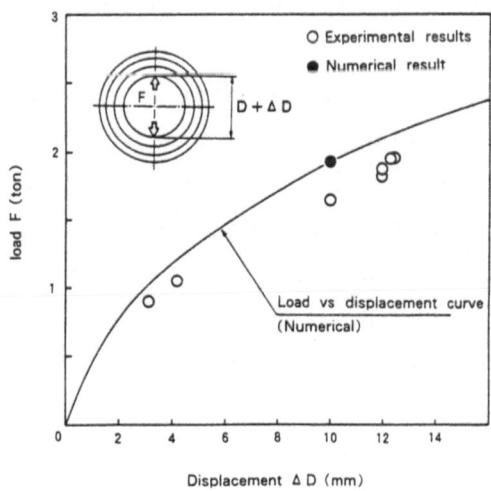

Fig. 10 Quench load comparison of experimental and numerical results

5. Conclusion

As described above, in order to evaluate the effects of the mechanical disturbance on the composite magnet for a large-sized SMES, the mechanical loading test for model coils have been performed with the existence of the current.   Numerical analysis based on the composite element method has also been carried out. The following conclusions were obtained through this study ;

(1) Non-linearity due to slip was observed even in relatively low loading stage when no plastic deformation occurs.   It indicates that the slip between layers affects the thermomechanical characteristics of the coil.

(2) It was verified by the model coil test that quench was caused by the mechanical disturbance.   Prediction for quench occurrence by the numerical analysis agreed with the experimental results for the model 2 coil.

(3) Relationship between the AE signal and quench should be studied in the future.

References

[1] T. Nakano, K. Hayakawa and M. Shimizu, Design study of a 5GWh SMES. Proceedings of 23rd ICEC, 1988, pp. 547-551.

[2] K. Miya, H. Yanagi, H. Hashizume and Y. Hamaoka, Study on the prediction of quench occurrence in superconducting coils, IEEE Trans. on Mag. Vol. 24, No. 2 (1988) 1548-1551

[3] T. Takagi, K. Miya, Y. Takeuchi and J. Tani, Finite element slip analysis of multilayer beam plate based on a composite theory, Fusion Eng. Des. 7 (1988) 1-13

[4] S. W. Tsai and E. M. Wu, A general theory of strength for anisotropic materials, J. Composite materials 5 (1971) 58-80

[5] T. Y. Chang and H. Aoki, A constitutive model for structural analysis of fusion magnets, Nucl. Eng. Des. 58 (1980) 237—245

TEMPERATURE RISE DUE TO FRICTIONAL SLIDING OF SUS316L VS SUS316L AND SUS316L VS POLYIMIDE AT 4 K

Akira IWABUCHI and Tomomi HONDA
Department of Mechanical Engineering, Iwate University
4-3-5 Ueda, Morioka 020, Japan

ABSTRACT

Frictional heat is one of the sources for the quench of a magnet. However, it is uncertain how much temperature rises due to sliding in a magnet. In this work, frictional properties were obtained under a fretting condition using SUS316L steel and polyimide at 4.2 K in liquid helium. The temperature rise was measured using a thermocouple inserted into a steel specimen during sliding. The maximum temperature rise was 9.3 K during fretting for the metal-metal contact. The temperature peak appeared twice during one fretting cycle corresponding to the slip in a cycle, and the time lag was 30 to 50 ms after the sliding velocity reached maximum. Temperature rise was linearly proportional to the product of a maximum sliding velocity and frictional force, related to the frictional rate. The temperature rise resulted from the balance between the heat generation during the slip and the cooling during the stick in one cycle.

INTRODUCTION

The development of new superconductors has been popular worldwide, and some had the transition temperature above 77 K. However, these new ceramic superconductors have a rather low critical current density and are brittle. Therefore, it is expected that the superconducting wire used in a large scale magnet is still metallic compound superconductor.

One of the most serious problems which should be solved in the design of a large scale magnet is to avoid the quench for the safety and reliability. Frictional heat due to micro-sliding at the contact points, especially between wires and spacers, is the source of the heat for the quench [1,2]. However, it is uncertain exactly how much temperature rises during this sliding although some attempts have been made [3,4]. For the optimum design of a magnet, the proper choice of materials should be done on consideration of the tribological point of view. Therefore, the tribological investigation is needed in the magnet technology.

In this work temperature rise was measured using thermocouple in a fretting condition, i.e., the oscillating slip motion with a small amplitude, and the relationships between the temperature rise and sliding properties were discussed.

EXPERIMENTAL DETAIL

Apparatus

The experimental apparatus used in this work was the same one used in the previous work [5]. A cylindrical specimen with two curved ends (30 mm in radius) was clamped by two cylindrical specimens with a flat end surface. The clamped specimen was driven oscillately up and down by using an eccentric device and an elastic beam. Frictional force and normal load were measured by strain gauges attached to rings respectively. Sliding distance was also measured by strain gauges attached to the spring plate. Specimens were put in liquid helium. Details were described in Reference 5.

Figure 1 shows the schematic diagram of the specimen with a thermocouple. The hole (0.25 mm in diameter) was made 0.4 mm below the surface by EDM, and Au-0.07at% Fe vs Chromel thermocouple (75 μm diameter each) was inserted into the hole, after which grease for cryogenic temperature (Apiezon N) was filled. The cold junction was placed in liquid helium. As the thermoelectomotive force was about 13 mV/K at 4 K, a DC amplifier was used to record the data.

Frictional force, slip distance and thermoelectromotive force were recorded by a digital recorder (NEC-Sanei, Omnilight 8M36). Sampling time was 50 to 200 μs.

Materials

Specimen materials used were SUS316L austenitic stainless steel and polyimide (Capton 125 μm thick, Du Pont-Toray). The former was used as a structural material and the latter as the material for insu-

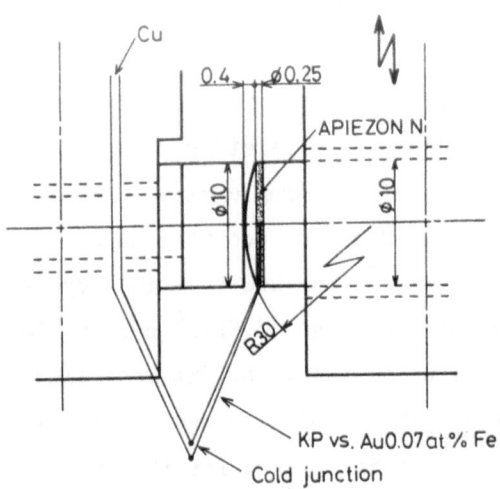

FIGURE 1 Schematic diagram of the specimen with a thermocouple.

lating tape in a magnet. The stainless steel was used in this work because thermal conductivity of the steel is lower than that of copper, which is a stabilizer around the superconducting wires, and, therefore, it was easy to measure the temperature rise.

After being turned, the metallic specimen was polished by emery papers and buffed. Surface roughness was 0.05 μm Rmax and hardness was 328 Hv. Thin polyimide was cut in size and adhered on the metallic flat specimen.

Procedure

The combinations of specimen were SUS316L against the same metal and SUS316L against polyimide.

Normal load applied was 20 N, the slip amplitude of the fretting was about 100 μm and the frequency was mainly 8.3 Hz. Fretting cycles were 5×10^4 cycles. When the temperature rise was measured at low sliding speed, the specimen was moved by rotating the pulley at the eccentric device by hand.

RESULTS AND DISCUSSION

Figure 2 shows the coefficient of friction and temperature rise against the number of fretting cycles at a frequency of 8.3 Hz. The coefficient of friction increased after 200 fretting cycles and reached 2.2 at 3.5×10^4 cycles for the metal-metal pair, but it kept almost constant at 0.3 or decreased slightly over the fretting cycles for the metal-polymer pair.

The temperature rise was coincident with the coefficient of friction qulitatively, and the temperature rise was higher for the metal-metal pair than for the metal-polymer pair. The maximum temperature rise during fretting was about 9.3 K for the metal-metal pair. The temperature rise for the metal-polymer pair reached the maximum value of 4 K at 50 fretting cycles and was almost constant at 2.8 K after 10^3 cycles.

The significant increase in coefficient of friction for the metal-metal pair was caused by the disruption of surface oxide layer by the fretting motion and the occurrence of the strong adhesion between contacting surfaces in liquid helium [5]. The constant coefficient of friction for the metal-polymer pair resulted from the increase in the mechanical strength of polyimide at 4 K [5].

In order to examine the temperature rise in one cycle under the fretting condition precisely, the driving pulley was rotated by hand after the continuous fretting test of 5×10^4 cycles. Figure 3 represents the measured results of displacement, frictional force, sliding velocity and temperature rise for SUS316L against itself. It can be seen that there are two sliding regimes: slip and stick in one cycle. The period of this oscillation was

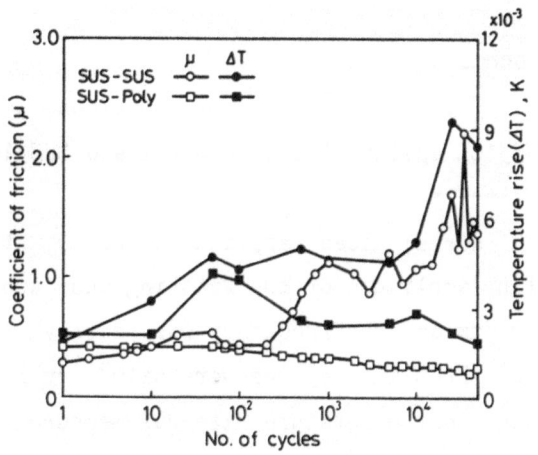

FIGURE 2 Coefficient of friction and temperature rise in the fretting condition.

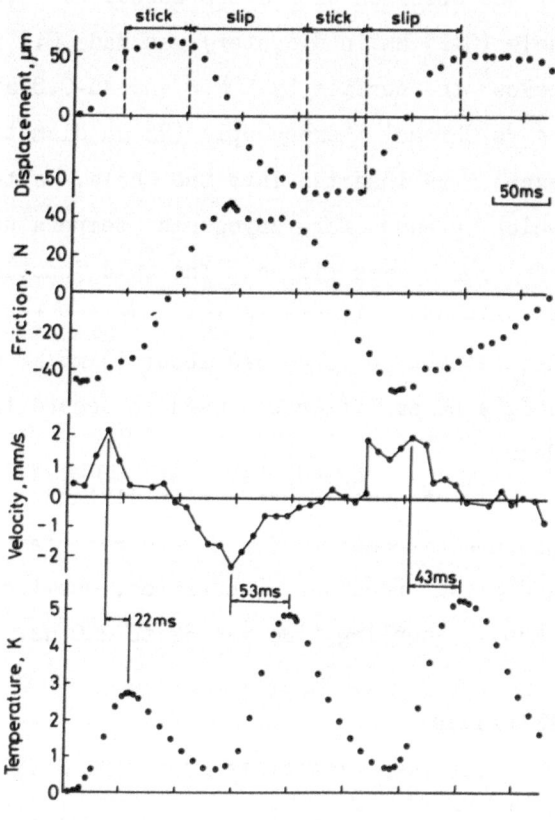

FIGURE 3 Displacement, frictional force, velocity and temperature rise for SUS316L-SUS316L pair at low speed driven by hand.

304 ms, and the mean sliding velocity in the cycle (2x(peak-to-peak slip amplitude)/period) was 0.40 mm/s. The mean sliding velocity in the slip (peak-to-peak slip amplitude/slip duration) was 1.12 mm/s and the maximum sliding velocity in the slip was 2.3 mm/s, which was as high as twice of the mean sliding velocity in the slip. There were two temperature peaks a cycle, corresponding to the slip. The time to the peak was about 56 ms and the rate of the temperature rise was about 64 K/s in this figure. The temperature peak appeared about 20 to 50 ms after reaching the maximum sliding velocity. This time lag was related to the heat transfer from the contacting surface to the site where the thermocouple was located. The temperature peak increased gradually. It is thought that the temperature rise results from the heat balance of the heating during the slip and the cooling during the stick.

The relationships between temperature rise ΔT and maximum velocity Vmax in the

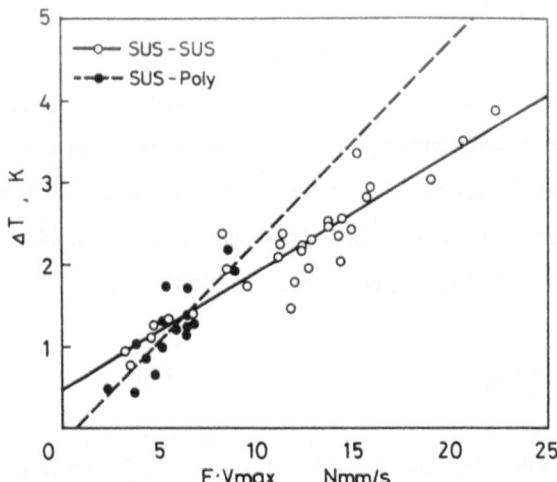

FIGURE 5 Relationships between temperature rise ΔT and the product of the maximum velocity and frictional force FVmax for two combinations.

slip are revealed in Fig. 4 for two combinations. It was apparent that the temperature rise was proportional linearly to the maximum velocity for two sliding pairs. The temperature rise for the metal-metal pair was greater than that for the metal-polymer pair, and the gradient was a little steeper for the metal-metal pair than for the metal-polymer pair. This result was explained by that the frictional force for the metal-metal pair was higher than that for the metal-polymer, as shown in Fig. 2.

Hence, the relationship between ΔT and the product of the maximum velocity and the frictinal force was examined, where the product was regarded as the frictional power, as shown in Fig. 5. Different lines were drawn for different specimen combinations.

It is apparent that the different gradient is caused by the change in thermal conductivity of the material and heat distribution rate in the pair. The thermal

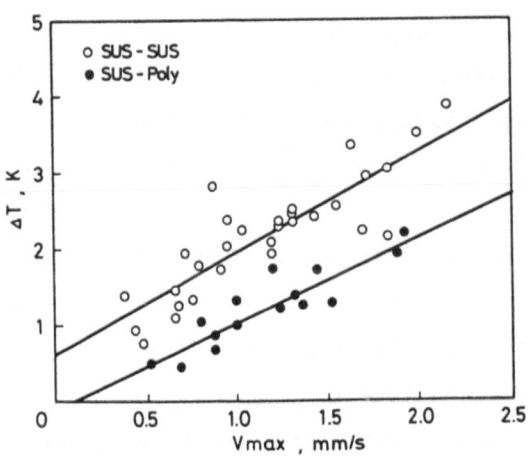

FIGURE 4 Relationships between temperature rise ΔT and the maximum velocity Vmax in the slip for two combination.

conductivity is 0.28 W/mK and about 0.03 W/mK for SUS316L and polyimide at 4 K, respectively. When two solid bodies contact and slide each other, the frictional heat generated, Q, is transmitted into the two bodies. If it is assumed that all the heat is conducted into the solid bodies, the following equation is derived:

$$Q = Q_1 + Q_2 \qquad (1)$$

where Q_1 and Q_2 are the heat distributed into body 1 and body 2, respectively. The heat distribution rate for body 1, s_1 is defined as

$$s_1 = Q_1/Q = k_1/(k_1 + k_2) \qquad (2)$$

where k_1 and k_2 are the thermal conductivity of body 1 and body 2 respectively [6]. The distribution rate is 0.5 for the metal-metal pair, and 0.90 for the metal specimen of the metal-polymer pair in this work. The ratio of the rate of the former to the latter was 0.56, which means that the temperature rise of the metal-polymer pair is greater than that of the metal-metal pair, if the product of the velocity and frictional force is equal. The ratio of the gradients in Fig. 5 is 0.58, which is almost the same as the theoretical one.

The results shown in Fig. 4 were obtained for rather slow speed by rotating a pulley by hand. The results for higher speeds were obtained by using the starting acceleration of a DC motor. Figure 6 shows these results at four different rotating speeds, i.e, frequencies of the fretting. The temperature peak increased with the fretting cycles, which was caused

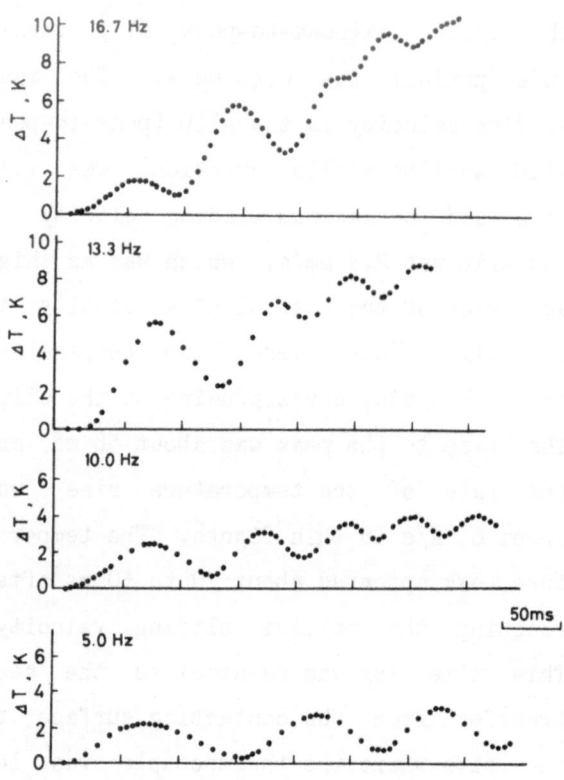

FIGURE 6 Temperature rise ΔT at the beginning of the fretting with different frequencies.

by the increase in the sliding velocity in the accelerating process. With an increase in frequency, the increasing rate in temperature peak became high. The large temperature drop was observed between the temperature peaks at 5 Hz; while the temperature drop was small and eventually disappeared at the third peak at 16.7 Hz. At lower speeds the stick time, that is, the cooling time was long between the slips in a cycle, so that the temperature drop was great because the cooling rate of liquid helium was maintained constant, and vice versa.

The temperature peaks were plotted against the maximum velocity in Fig. 7. The lines at 5 and 10 Hz were coincident with the line obtained by hand in Fig. 4.

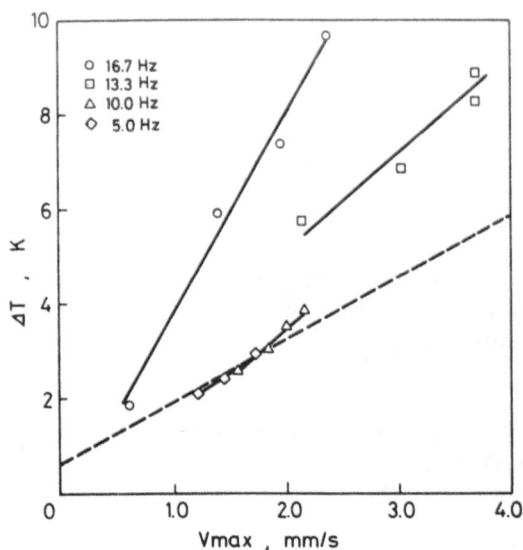

FIGURE 7 Relationships between temperature rise ΔT and the maximum velocity Vmax under the fretting with different frequencies.

At higher frequencies of 13.3 and 16.7 Hz the lines were away from this line and the gradients were steeper. These results were caused by shortening the cooling time, as the frictional heat depended on the product of the maximum velocity in the slip and the frictional force, as shown in Fig. 5, and the frictional force was constant.

CONCLUSIONS

1. The temperature rise reached 9.3 K in the fretting condition for SUS316L-SUS316L pair. It was about 3 K for SUS-316L-polyimide pair.

2. The temperature rise was linearly proportional to the product of the maximum velocity and the frictional force, i.e., the frictional power.

3. The temperature rise against the frictional power for SUS316L-polyimide pair was steeper than that of SUS316L-SUS316L pair due to the low thermal conductivity of polyimide and the higher heat distribution rate of metal specimen.

4. The temperature rise resulted from the heat balance between the heating during the slip and the cooling during the stick in one cycle. The gradient of the temperature peaks against the maximum sliding velocity increased with the frequency due to the reduction in cooling time.

ACKNOWLEDGMENTS

This work was supported by Grant-in-Aid (No. 63055002 and 01050022) from Ministry of Education, Japan.

REFERENCES

1 Edwards, V.W. and Wilson, M.N., The effect of adhesion between turns on the training of superconducting magnet. Cryogenics, 1978,18, 423-425.
2 Iwasa,Y., Experimental and theoretical investigation of mechanical disturbances in epoxy-impregnated superconducting coils. 1. General introduction. Cryogenics, 1985, 25, 304-306.
3 Kensley, R.S. and Iwasa, Y., Transient slip behaviour of metal/insulator pairs at 4.2 K. Cryogenics, 1981, 21,479-489.
4 Tsukamoto, O. Takao, T. and Honjoh, S., Estimating the size of disturbances due to conductor motion in superconducting windings. IEEE Trans. on Magnetics, 1988, 24, 1182-85.
5 Iwabuchi, A., Honda, T., and Tani, J., Tribological properties at temperatures of 293, 77 and 4 K in fretting. Cryogenics, 1989, 29, 124-131.
6 Spure, R.T., Temperatures reached by sliding thermocouples. Wear, 1980, 61, 175-182.

FRACTURE TOUGHNESS OF 304 STAINLESS STEEL IN HIGH MAGNETIC FIELDS AT CRYOGENIC TEMPERATURE

Eiji Fukushima, Sadao Kobatake, Minoru Tanaka
and Hiroyasu Ogiwara
Research and Development Center,
Toshiba Corporation,
4-1 Ukishima-cho, Kawasaki-ku, Kawasaki, 210, Japan

ABSTRACT

Fracture toughness tests were made on 304 stainless steel at 4 K, varying the precracking temperature (300 K and 77 K) and magnetic field (0 T and 8 T). The obtained results are discussed, taking the TRIP* effect and the promotion of martensite formation in magnetic fields into consideration.

Fracture toughness values (J_{IC}) of 43 - 62 kJ/m^2 and tearing modulus values (T_{mat}) of 11 - 20 were obtained under various test conditions. The highest values of J_{IC} and T_{mat} were obtained for cases in which a remarkable TRIP effect was expected. Low values of T_{mat} were obtained in cases of applying a magnetic field (8 T). The acceleration of martensite formation in the magnetic field might be considered to result in such low values of T_{mat} in these cases.

INTRODUCTION

Various different results have recently been reported on the fracture toughness of austenitic steels in high magnetic fields at cryogenic temperatures [1 - 3]. To solve this problem, fracture toughness tests were made on 304 stainless steel at 4 K, varying precracking temperatures (300 K and 77 K) and magnetic fields (0 T and 8 T) as parameters. The obtained results were discussed, taking the TRIP effect and the promotion of martensite formation in magnetic fields into consideration.

MATERIALS AND METHODS

Materials

A commercially available 304 stainless steel plate was used for the tests. The analytical values for the chemical components and the mechanical properties at room temperature of the material are shown in TABLE 1.

The shape and size of the specimen is shown in FIGURE 1. The specimens were precracked at 300 K (room temperature) and 77 K (liquid nitrogen temperature) before the tests.

---

* TRIP: transformation induced
plasticity

TABLE 1 Analytical values for chemical components and mechanical properties at room temperature of material

C	Si	Mn	P	S	Ni	Cr
0.05	0.27	0.78	0.033	0.013	8.08	18.13

Yield strength (MPa)	Tensile strength (MPa)	Elongation (%)
323	627	63

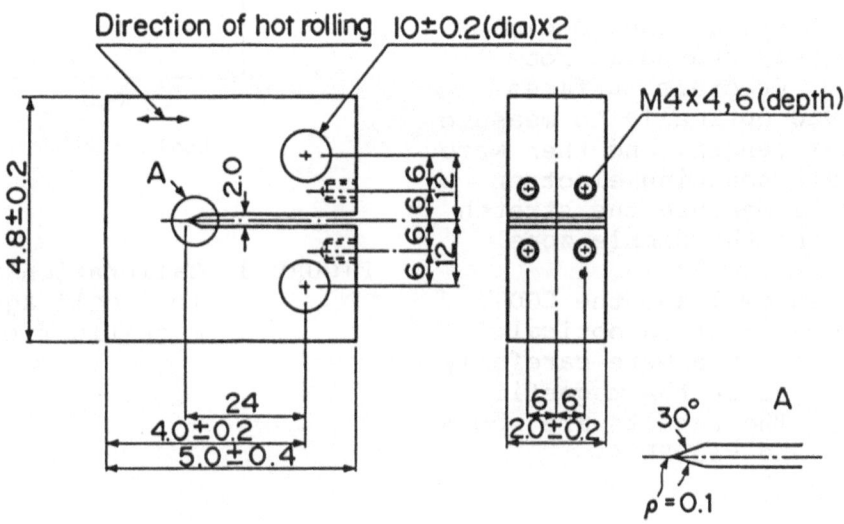

FIGURE 1 Shape and size of specimen

Methods

Tests were made using a test facility that had especially been made for fracture toughness tests in high magnetic fields at cryogenic temperatures. A scheme of the facility is shown in FIGURE 2. The maximum load capacity of the facility was 10 tons, and the bore diameter of the superconducting solenoid was about 200 mm.

Tests were made employing the multi-specimen method. The values of the J integral were calculated using the equation:

$$J = \frac{A}{Bb_0} f(a_0/W) \qquad (1)$$

where A is the area under the load-COD (crack opening displacement) curve, B is the specimen thickness (20 mm), b_0 is the ligament width (W-a_0, W: specimen width (40 mm), a_0: precrack length) and f (a_0/W) is given by the following equations.

$$F(a_0/W) = \frac{1 + \beta}{1 + \beta^2} \qquad (2)$$

$$\beta = \left[\left(\frac{2a_0}{b_0} \right)^2 + 2 \left(\frac{2a_0}{b_0} \right) + 2 \right]^{1/2}$$

$$- \left(\frac{2a_0}{b_0} + 1 \right) \qquad (3)$$

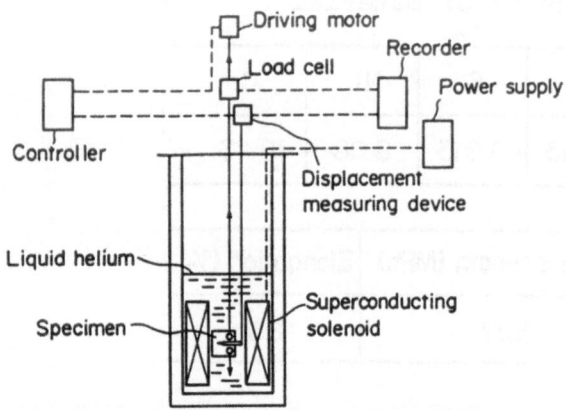

FIGURE 2   Scheme of test facility

After testing, the specimens were fatigue-fractured at room temperature. Fracture surfaces were observed optically to measure the precrack length, and then were observed by a scanning electron microscope to measure the stretch zone width and the dimple zone width [4].

The load cell and the COD measuring device of an optical interferometer type were carefully calibrated against the magnetic field (8 T). The results are shown in FIGURE 3 and FIGURE 4, respectively.

RESULTS AND DISCUSSION

The obtained results are shown in FIGURE 5 and TABLE 2. FIGURE 5 shows the stretch zone width and R curves for each test condition. The fracture toughness values (J_{IC}) are the J integral values at the cross points of the stretch zone width line and R curves. The tearing modulus values (T_{mat}) correspond to the slopes of the R curves [5]. TABLE 2 is a list of values for J_{IC} and T_{mat} under each test condition.

The following facts are taken into consideration to explain the obtained results.

(1) Little amount of martensite is formed around the precrack when it is introduced at 300 K, while a large amount of martensite is formed in front and both sides of the precrack when it is introduced at 77 K [6].

(2) The application of a high

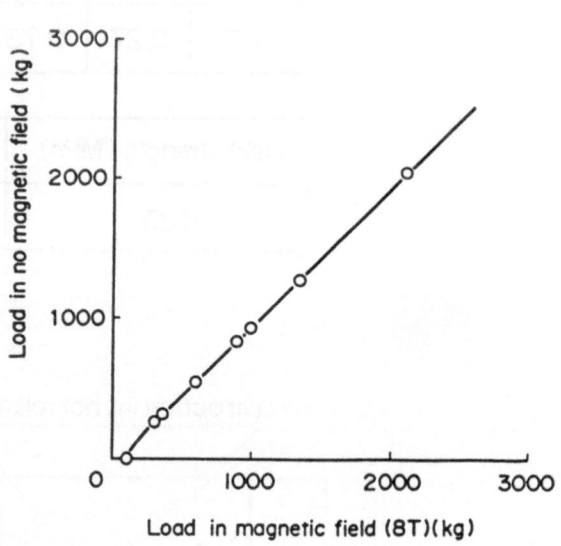

FIGURE 3   Calibration results of load cell against magnetic field (8 T)

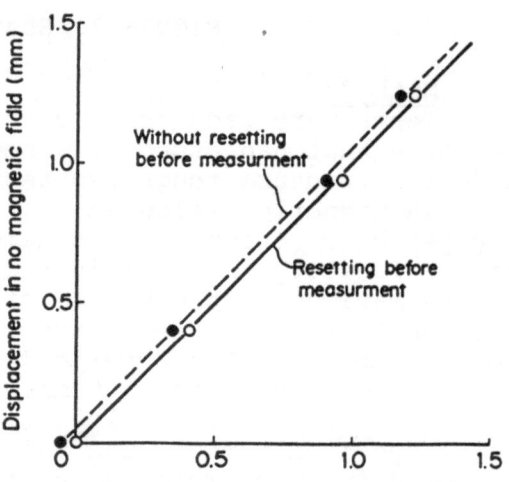

FIGURE 4   Calibration results of displacement measuring device against magnetic field (8 T)

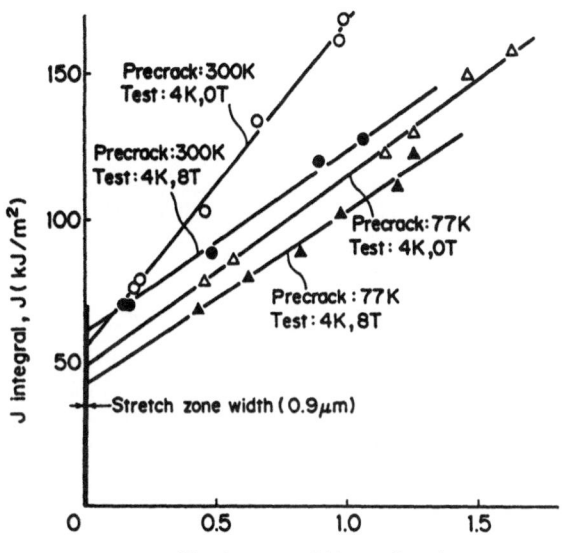

FIGURE 5   Test results (R curves at various test conditions and stretch zone width)

magnetic field promotes martensitic transformation [7].

(3) Martensite (α') itself is a brittle phase. But when crack growth accompanies martensitic transformation, it may be expected that the J integral value will increase according to the TRIP effect.

Among the various test conditions shown in TABLE 2, the case applying a magnetic field (8 T) coupled with precracking at room temperature gave the highest J_{IC} value. The most remarkable TRIP effect may be expected in this case. The TRIP effect also seems to have a dominant effect on determining the J_{IC} values under other test conditions.

As for T_{mat}, the highest value was obtained in the case of applying no magnetic field coupled with precracking at room temperature. This is also the case in which a remarkable TRIP effect is expected. Low values of T_{mat} were obtained for cases of applying a magnetic field. In these cases, a large amount of martensite might have been formed in a relatively wide area at the very beginning of crack growth according to the acceleration of martensite formation in the magnetic field. This might be the reason for the low values of T_{mat}.

These explanations should be supported experimentally. The authors are now observing the effect of martensitic transformation on the above-mentioned issues by means of X ray analyses and magnetization measurement on the fracture surfaces of the specimens.

CONCLUSIONS

A summary of the results is as follows.

(1) J_{IC} values of 43 - 62 kJ/m^2 and T_{mat} values of 11 - 20 were obtained for 304 stainless steel at test conditions varying the precracking temperature (300 K and 77 K) and the magnetic field (0 T and 8 T) as parameters.

(2) To explain the obtained results, it seems to be necessary to take into account the TRIP effect and the promotion of martensite formation in the magnetic field.

(3) The highest values of J_{IC} and T_{mat} were obtained for cases in which a remarkable TRIP effect was expected.

(4) Low values of T_{mat} were obtained in cases of applying a magnetic field. The acceleration of martensite formation in the

TABLE 2   Values of fracture toughness (J_{IC}) and tearing modulus (T_{mat}) at various test conditions

Precracking temperature	Test conditions	Fracture toughness, J_{IC} (kJ/m^2)	Tearing modulus, T_{max}
300K	4 K , 0tesla	56	20
	4 K , 8teslas	62	11
77K	4 K , 0tesla	49	12
	4 K , 8teslas	43	11

magnetic field might be considered to result in such low values of T_{mat}.

This work was performed through Special Coordination Funds from the Science and Technology Agency of the Japanese Government.

REFERENCES

[1] J. W. Morris, Jr.: U. S.-Japan low temperature structural materials and standards workshop, JAERI, Naka, Ibaraki, Japan (May, 1988)

[2] H. Yanagi and A. Nyilas: VAMAS technical working party meeting, superconducting and cryogenic structural materials, Tokyo, Japan (May, 1988)

[3] E. Fukushima, S. Kobatake, M. Tanaka and H. Ogiwara: Adv. Cryog. Engng-Mater., 1988, 34, p36

[4] JSME standard method of test for elasto-plastic fracture toughness J_{IC}, JSME (1981)

[5] P. C. Paris: Elastic-plastic fracture, STP 668, American Society of testing and materials, Philadelphia, (1979), p5

[6] E. Fukushima: J. of the Iron and Steel Inst. of Japan, 1989, 75, p879

[7] M. A. Krivoglaz and V. D. Sadovskiy: Fiz. Metal. Metalloved. (1964), 18, p502

SUPERCONDUCTING MAGNETS IN HIGH RADIATION ENVIRONMENTS: DESIGN PROBLEMS AND SOLUTIONS*

S. J. ST. LORANT and E. TILLMANN
Stanford Linear Accelerator Center
Stanford University, Stanford, CA 94309

ABSTRACT

As part of the Stanford Linear Collider Project, three high-field superconducting solenoid magnets are used to rotate the spin direction of a polarized electron beam. The magnets are installed in a high-radiation environment, where they will receive a dose of approximately 10^3 rad per hour, or 10^8 rad over their lifetimes. This level of radiation and the location in which the magnets are installed, some 10 meters below ground in contiguous tunnels, required careful selection of materials for the construction of the solenoids and their ancillary cryogenic equipment, as well as the development of compatible component designs. This paper describes the materials used and the design of the equipment appropriate for the application. Included are summaries of the physical and mechanical properties of the materials and how they behave when irradiated.

INTRODUCTION

Three superconducting solenoid magnets and their associated systems are part of the polarized electron beam facility to be installed at the Stanford Linear Collider. As an intense electron beam passes axially through each solenoid, high magnetic field quality, precise mechanical alignment and long-term stability are essential requirements. The magnets are located some ten meters below ground, in the immediate proximity of other beam lines and in a radiation field that may exceed 10^3 rad per hour at each magnet. All services, monitoring and control must therefore be remote and located above ground, as access to the accelerator housing and magnets is prohibited during operation.

From the beginning of this project, we were much concerned with the high radiation levels and the remoteness of the magnets from occupied service buildings. A closed refrigeration cycle was initially not included in the specifications; nonintrusive weekly batch refilling of local liquid helium storage dewars was mandated, with the option of adding a refrigerator at a later date.

Figure 1 is a schematic representation of one of the solenoid installations. The magnet is supplied with cryogens through a composite transfer line containing both the liquid nitrogen and liquid helium circuits in a common vacuum enclosure, connected to storage vessels above ground.

The environmental parameters of the facility are as follows:

- *Ionizing radiation:* 10^3 rad per hour, or 10^8 rad over the lifetime of the system.

- *Seismic Loads:* California earthquake loads of 0.75 G in any direction, without internal failures or personnel hazards.

* Work supported by Department of Energy contract DE–AC03–76SF00515.

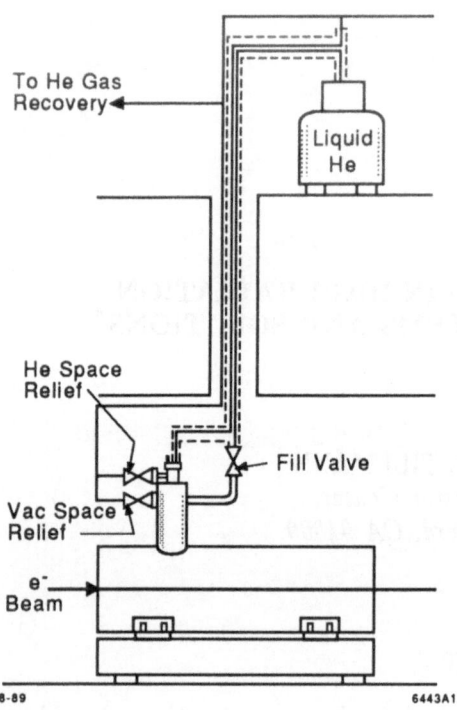

FIGURE 1. Schematic representation of one magnet installation.

- *Access:* Severely limited in space by contiguous beam lines, and restricted in time to a few hours of access every month.
- *Duty Cycle:* 5000 charge and discharge cycles over the lifetime of the system.
- *Thermal Cycles:* One cooldown and one warmup to room temperature per month.
- *Operation:* Remote and automatic for all phases except the periodic refilling of the cryogen storage dewars.

The solenoids themselves were fabricated by a commercial company [1] to exacting specifications determined by the beam optics and the environment described above. The design maximum magnetic field is the same for each solenoid, 4.5 T, with an azimuthal variation of the integrated transverse magnetic field at any radius less than 1.25 cm limited to less than ±0.1%. The current is restricted to 150 A at maximum field so as not to exceed the specified heat inleak of 1 W. This thermal load precluded the use of separate field trim windings or multiple joints in the conductor, so that the field accuracy had to be achieved by precision winding of a single, 18.3 km length of rectangular conductor.

The interplay of these seemingly unrelated requirements led to interesting engineering compromises, some of which are discussed here.

GENERAL CONSIDERATIONS

Initially, we considered the ionizing radiation environment and the limited access to be separate issues in the system concept, but as the design proceeded, we developed a design philosophy to minimize the impact of either constraint:

- Organic materials, unless absolutely essential, to be avoided.
- Organic materials demanded by the construction to have physical and mechanical properties which do not substantially degrade after a lifetime exposure to 10^8 rad.
- The properties of all materials used to be supported by a substantial data base of experimental measurements.
- The peripheral components usually associated with superconducting magnets to be certified for use in the environment or redesigned to have the capability or be relocated to unrestricted areas.
- Cooling of the conductor to be by pool boiling helium, with the added requirement of an *in-situ* reservoir of liquid helium sufficient for one refill every 24 hours.

During this process, we realized that while there exists a considerable body of information on the subject of radiation damage, there is a commensurate unevenness in the level of understanding of the effect of radiation on materials, particularly on magnet components. We therefore ranked our magnet subsystems according to the perceived severity of the problems induced in each by the expected radiation damage:

- Insulation, thermal and electrical.
- Superconductor, stabilizer.
- Instrumentation and control.
- Venting, vacuum and safety.
- Structural components and installation supports.

As the last two items either are constructed from materials not affected by ionizing radiation at dose levels of 10^{10} rad, and at the rather modest implied neutron fluences [2], or else are located in unrestricted access areas, we shall not discuss them further here.

TABLE 1. Industrial insulating materials of interest for cryogenic applications involving radiation.

Commercial Standard or Chemical Name	Trade Name® [3]	Description
NEMA/ASTM G–10	Spauldite G–10 G–10CR G–10–773	Glass fabric, epoxy resin [diglycidyl ether of bisphenol A (DGEBA) cured with dicyanodiamide (DCD)].
NEMA/ASTM G–11	Spauldite G–11 G–11CR G–11–963	Glass fabric, high-temperature epoxy resin [diglycidyl ether of bisphenol A (DGEBA) cured with diamino diphenyl sulphone (DDS)].
Polyvinyl formal	Formvar	
Polyimide	H-film, Kapton	$(C_{22} H_{10} N_2 O_4)_n$.
S-glass	S–2, S–901 S–994	Silica-alumina-magnesia fiber (high strength and modulus).
E-glass		Lime-alumina-borosilicate fiber (low modulus, good electrical properties).

INSULATION

Table 1 lists the insulating materials selected for the construction. They are materials whose radiation resistance has been well documented in accordance with our design guidelines, and whose behavior at cryogenic temperatures was well known to us.

Several reviews [4 (and references cited therein), 5] have noted that buried in the wealth of experimental data relating to radiation damage is evidence on secondary phenomena, such as localized nuclear heating exacerbated by the reduction in the thermal conductivity of the material. Studies which reported thermal and electrical properties jointly with measurements of the mechanical properties therefore received special attention. Excellent data on the stability of organic insulations were obtained at Oak Ridge National Laboratory [6]. At doses of 2×10^8 rad (γ) the volume resistivities of a bisphenol A epoxy, an inorganic-filled epoxy, a polyvinyl formal wire coating and an FR–5 type glass-epoxy laminate were found to be virtually unchanged. At higher-dosage levels, the mechanical and electrical properties of glass-reinforced epoxy laminates deteriorated significantly [7]. This is shown in Figs. 2 and 3.

The flexural and compressive strengths of G–10CR and G–11CR laminates fall to 10–15% of their original values after γ irradiation to 2.4×10^9 rad, and even lower after 10^{10} rad. As both irradiations were accompanied by fast neutrons at high fluences, we made this data the reference

base for our worst-case irradiation scenario, particularly as the measurements were made following a warmup to ambient temperature. Figure 4 shows data obtained at CERN [8], used to validate our design parameters.

Polyvinyl formal is frequently used where thin, uniform and tough, abrasion-resistant wire insulation is required. Its mechanical properties, as a function of radiation dose, are shown in Fig. 5 [9–11].

Polyimide film developed to compete with mylar, a material unstable in a radiation environment, and has the interesting property that its ultimate tensile strength increases to about 5×10^8 rad while its electrical properties remain essentially unchanged [12].

In the solenoid windings, the turn-to-turn insulation is provided by a 12 μm-thick polyvinyl formal coating applied to the superconductor during manufacture, all ground insulation is made from S-glass reinforced DCD-cured G–10CR, while the layer-to-layer insulation is 50 μm-thick polyimide H-film sheet. No impregnation or overbanding is used to hold the superconductor; the precision winding technique and the tight dimensional specifications ensure that the winding tolerances of ±100 μm are maintained. A byproduct of this method of construction is adequate percolation cooling of the innermost layer.

Each coil-containing helium vessel is supported in its vaccum tank by six race-track-

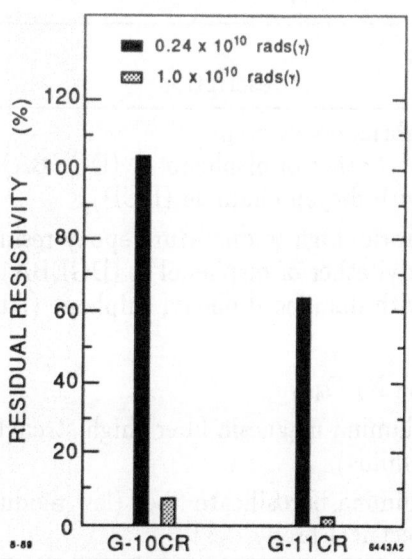

FIGURE 2. *Residual electrical resistivity of G-10CR and G-11CR after irradiation at 5 K and warmup to ambient temperature.*

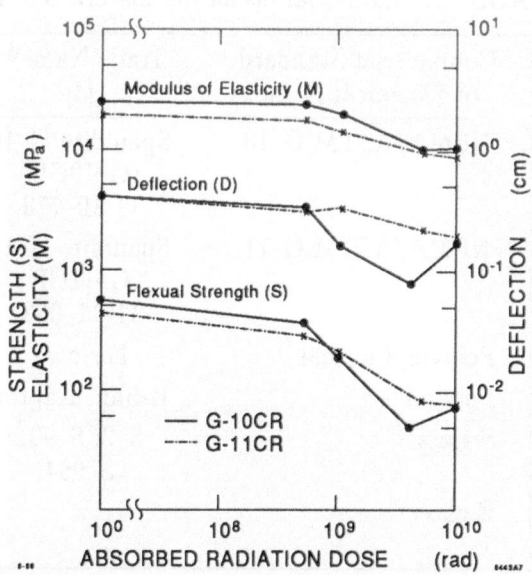

FIGURE 4. *The modulus of elasticity, flexural strength and the deflection characteristics of G-10CR and G-11CR at various radiation doses.*

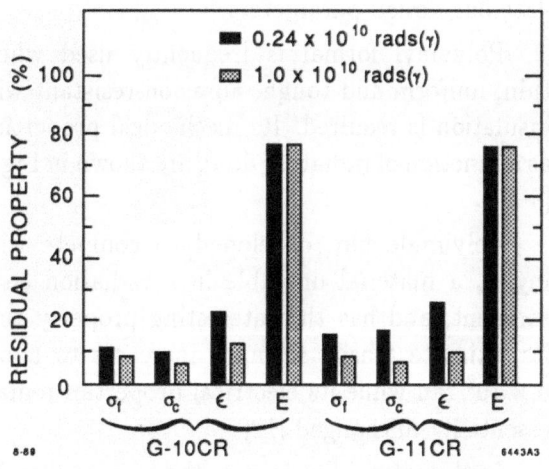

FIGURE 3. *Residual mechanical properties of G-10CR and G-11CR (flexural strength σ_f, strain ϵ, and modulus E, and compressive strength σ_c) after irradiation at 4 K and warmup to ambient temperature.*

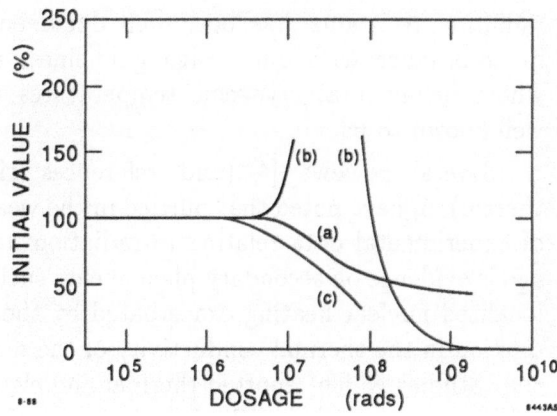

FIGURE 5. *Effect of radiation on (a) tensile strength, (b) elongation, and (c) elastic modulus of polyvinyl formal. The initial values are 51 MPa, 2%, and 3.45 GPa, respectively.*

unwoven glass fibers. The transfer lines accessing the solenoids are likewise insulated.

SUPERCONDUCTOR AND STABILIZER

There is considerable evidence that the bulk properties of NbTi alloys, T_c and H_{c2} in particular, are decreased very little by irradiation [5]. It also appears that irradiation can enhance the J_c in low J_{c0} materials while in high J_{c0} materials J_c decreases linearly for fluences up to about 4×10^{22} neutrons per m^2, at which point J_c is

shaped links of unidirectional filament-wound S-glass fibers in a matrix of DDS-cured DGEBA resin. The radiation shields are separately suspended with polydirectionally wound G-10CR rods.

The multilayer insulation blankets thermally insulating the various components inside the cryostat are made from 10–30 layers of 12 μm-thick aluminum foil separated by mats of

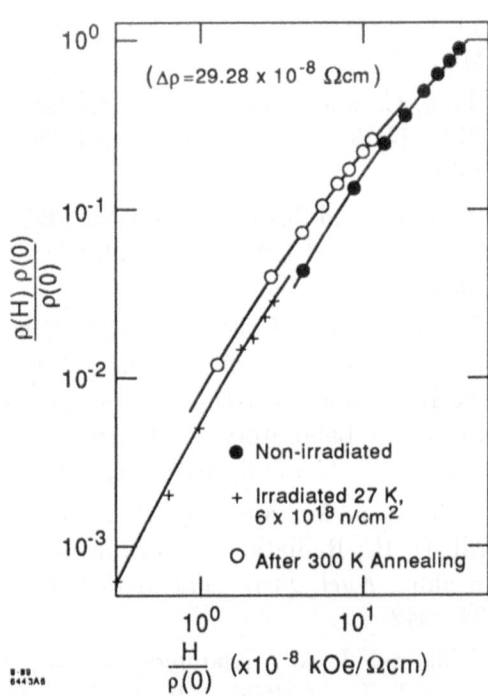

FIGURE 6. *Kohler plot for the copper stabilizer in Vacryflux superconducting wire (reference resistivity of the copper = 2.68 × 10⁻⁸ ohm-cm).*

degraded by about 10%. As the conductor for the solenoids was specified to operate at 60% of its short sample value at maximum field, its performance is not affected.

The magnetoresistance of the copper stabilizer is modified by radiation: the changes due to radiation damage and subsequent treatment are well documented [13]. Kohler's rule is closely obeyed by copper in the magnetic field range of 0–10 T and over a purity range of $200 \leq RRR \leq 7000$ [14]. Figure 6 illustrates the behavior of the copper stabilizer in our Vacryflux® superconducting wire [15,16].

INSTRUMENTATION AND CONTROL

All sensors installed in the magnets are radiation resistant:

- *Temperature:* The surface thermometer gauges are foils of resistance-compensating alloys of nickel and manganin.
- *Pressure:* The magnet-mounted pressure transducers have 17-4PH stainless steel diaphragms

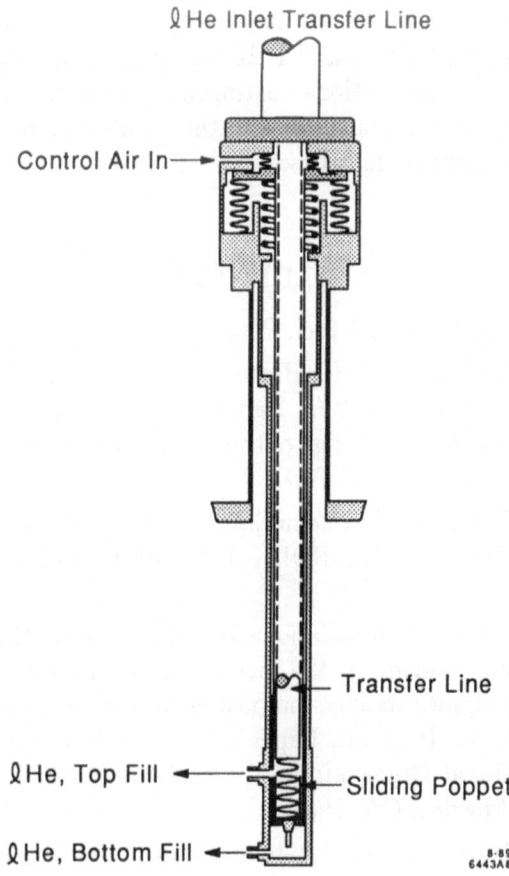

FIGURE 7. *The inline three-way cooldown and fill valve.*

to which compensated foil strain gauges are bonded.

- *Vacuum:* The all-metal cold cathode ionization gauges and thermocouple gauges which monitor the vacuum are rated to 10^{10} rad.
- *Control:* Two inline, bellows sealed, three-way valves control the cooldown and fill sequences in each magnet (Fig. 7). No elastomers or composites of any kind are used. The valves are remotely actuated by compressed air.

CONCLUSIONS

Three magnet systems were designed and built according to the philosophy described above. The fully configured solenoids have been tested successfully for several hundred hours each, and are now awaiting installation in the accelerator.

ACKNOWLEDGMENTS

We would like to thank our many colleagues at SLAC, CERN, and BNL for their advice and the many suggestions unstintingly given, as well as for the generous access to their materials radiation damage data bases.

REFERENCES

1. WANG NMR Inc., Pleasanton, CA.
2. Kulcinski, G. L., Brown, R. G., Lott, R. G., and Sanger, P. A., Radiation damage limitations in the design of the Wisconsin Tokamak fusion reactor. *Nucl. Tech.*, 1974, **22**, 20–35.
3. Registered trademarks. Military designations in MIL-HDBK-17A, 1971, and updates.
4. Egusa, S., Irradiation effects and degradation mechanism on the mechanical properties of polymer matrix composites at low temperature. Preprint, Paper FZ–05, 1989 International Cryogenic Materials Conference, Los Angeles, CA, 1989.
5. Brown, B. S., Low-temperature radiation effects in superconducting fusion-magnet materials. *J. Nucl. Matl.*, 1980, **97**, 1.
6. Kernohan, R. H., Coltman, R. R., and Long, C. J., Radiation effects on organic insulators for superconducting magnets. ORNL/TM–7077, Oak Ridge National Laboratory, Oak Ridge, TN, 1979.
7. Coltman, R. R., Klabunde, C. E., Kernohan, R. H., and Long, C. J., ibid.
8. Liptak, G., Schuler, R., Maier, P., Schönbacher, H., Haberthur, B., Müller, H., and Zeier, W., Radiation tests on selected electrical insulating materials for high power and high voltage applications. CERN 85–02, Technical Safety and Inspection Commission, CERN European Laboratory for Particle Physics, Geneva, Switzerland, March 1985.
9. Bopp, C., and Sisman, O., ORNL 928, Oak Ridge National Laboratory, Oak Ridge, TN, 1951.
10. Bopp, C., and Sisman, O., ORNL 1373, Oak Ridge National Laboratory, Oak Ridge, TN, 1954.
11. Calkins, C. and Collins, C., APEX 261, in Radiation Damage of Materials: Engineering Handbook, MPS/Int. CO 66–25, CERN European Laboratory for Particle Physics, Geneva, Switzerland, November 1966.
12. Koehler, A. M., Measday, D. F., and Morrill, D. H., Radiation damage in mylar and H-film. *Nucl. Instr. and Meth.*, 1965, **33**, 341–342.
13. Williams, J. M., Klabunde, C. E., Redman, J. K., Coltman, R. R., and Chaplin, R. L., The effects of radiation on the copper normal metal of a composite superconductor. *IEEE Trans. Magn.*, 1979, **MAG–15**, 731–734.
14. Fickett, F., Magnetoresistivity of copper and aluminum at cryogenic temperatures. Proceedings of the Fourth International Conference on Magnet Technology, Brookhaven National Laboratory, Upton, NY, 1972, 539.
15. Private communication. Vacryflux is the registered trade mark of Vacuumschmelze GMBH, Hanau, FRG.
16. Bonjour, E., Brauns, P., Lagnier, R., and Van de Voorde, M., Low-temperature behaviour in organic materials in a radiation field. CERN 77–03, CERN European Laboratory for Particle Physics, Geneva, Switzerland, February 1977.

INVESTIGATION OF IMPREGNANTS FOR SUPERCONDUCTING MAGNETS[*]

D.L. Zha[**], S. Han, L.Z. Lin
Institute of Electrical Engineering, Academia Sinica
P.O. Box 2703, Beijing 100080, China

ABSTRACT

A kind of powder filled epoxy composite by suitable choice of fillers and epoxy resins was developed. The composite shows features as small thermal contraction, high tensile strength, good handling properties and relatively high thermal conductivity. Composition and properties of the composite are presented. Meanwhile some other impregnants in common use in China as paraffin wax and beeswax are also investigated and compared with each other. The data is useful for fabricating superconducting magnets and evaluating their stability.

INTRODUCTION

Lots of practical superconducting magnets are adiabatically stabilized. Mechanical disturbances is main factor to affect the stability of these magnets. The aim of various stabilizing measures is avoiding the happening of these disturbances or eliminating the energy released in case that the disturbance occurred. Four techniques are usually used to stabilize the wires of a superconducting magnet: pretensioning and unimpregnating method, wet winding technique, wax impregnation and resin impregnation. Impregnant used at low temperature must be able to restrain the motion of conductors when the magnet works. It should possess good low temperature properties. Its thermal contraction must be similar to that of metals. It should have enough strength as a constituent material and be able to undergo thermal cycling. With good handling properties, it must be easy to permeate into magnet before curing. Furthermore the impregnant's high thermal conductivity is useful for the stability of magnets. The research aimed at improving the low temperature properties of epoxy composite by reasonably selecting fillers and epoxy resins. We will study systematically thermal contraction, viscosity, strength and thermal conductivity of epoxy EHD1 and EHD2. As a comparison, the properties of paraffin wax and beeswax is also investigated.

----

* The work supported by National Natural Science Foundation of China
** Present address: Kejian Corp. Ltd., Shekou, Guangdong 518067, China.

SELECTION OF COMPONENT PARTS
AND EXPERIMENTS

The physical properties of epoxy composites depend mainly on their composition. We must have careful selection of various components to meet the demands for use at low temperature. There are lots of choices of epoxy resins and curing agents. There are also many kinds of fillers to be selected. We will optimize the composition through tests and seek the best match.

Taking account of the experience of others and actual conditions in China, we select bisphenol-A epoxy resin as the main part of the composite. Finally we take E-51 resin with low molecular weight on account of its fluidity. Furthermore, we choose liquid acid anhydride (HK-021) as first curing agent to reduce the viscosity of the mixture and thermal contraction of final composite. To improve its tenacity at low temperature, we choose plasticizer HSA-1 as the second curing agent. To make its thermal contraction smaller further, we had better mix inorganic filler with the resin. The mixing should not make mixture's viscosity increase too much but be able to improve the composite's thermal conductivity. So we choose surface-treated crystal silica powder as our filler. The model HGH-600 is a kind of quartz powder with diameter of 20 micrometers. The quartz's thermal conductivity at low temperature is very high but thermal contraction extremely small [1]. We think it can meet our demands.

We mix E-51, HK-021, HSA-1 and HGH-600 in certain compositions and make them cured above 373 K. There was sediment found in all cured resins. The reason is that quartz's specific gravity is large and resin's viscosity small. If curing speed is low and the mixture is laid still for some time, heavy powder will sink and become sediment. We used catalyst to speed curing reaction and sediment problem was solved. The catalyst DMP-30 can make HGH-600 and resin mixed very well. The curing reaction is much faster with catalyst. The catalyst of 0.2% (0.2g DMP-30 per 100g E-51) is enough to prevent filler powder sinking and keep the mixture possess good fluidity for some time so as to impregnating. As for curing temperature and curing time, we found that graded temperature curing was more effective. Relatively low temperature at first made the resin easy to permeate into magnet and air bubbles easy to spill out.

To seek the composite with the best properties at low temperature, we optimize its composition through measuring its viscosity before curing and its thermal contraction. We also test its ability to undergo thermal cycling by soaking it in liquid nitrogen for several times. We found the fact that the cured resin composite didn't crack as long as the technique was proper and there was no sediment found. Furthermore we had metals (such as screw and bolt) cured in resin composite and tested in liquid nitrogen. We found there would not be crack if there was no lump resin. And if there was glass fiber between metal and resin (the case same as in actual Nb_3Sn magnets), there would not be any crack found. This indicates that the epoxy resin was suitable for impregnating superconducting magnets.

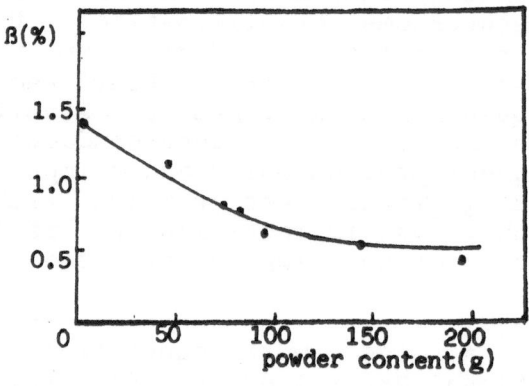

FIGURE 1. Filler's affect on
thermal contraction

The proportion of filler had main effect on its thermal contraction and viscosity. Thermal contraction(β) is just relative contraction from room temperature to liquid nitrogen. That is $\beta = (l_{293} - l_{77} / l_{293})$. Viscosity was expressed with conditional viscosity. For instances, water's viscosity is 11.5+0.5 seconds at 293K. Figure 1 and Table 1 indicates respectively affects of filler density (filler weight per 100g E-51 resin) on thermal contraction and viscosity

TABLE 1   The relation of filler and viscosity(seconds,393K)

Filler (g)	0	50	100	150	200
Viscosity	14.0	17.5	19.4	31.0	45.5

Considering impregnating technique and the mixture's viscosity in Table 1, we think filler content per 100g E-51 had better not exceed 100g. In addition, the proportion of two curing agents has affects on thermal contraction. We find that there would be the minimum contraction when HK-021 and HSA-1 are 50g and 30g relatively. Besides, curing time has some affect on thermal contraction of cured resin, as Fig. 2 indicated. Graded temperature is used to cure the resin. At first, 373K for 2 hours, 393K for 2 hours, and finally we raise the temperature to

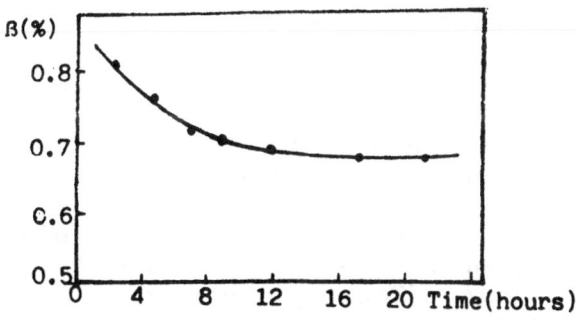

FIGURE 2. Curing time's affect on thermal contraction

403K. Test proves that thermal contraction tends to be consistent when curing time at 403K exceeds 15 hours. We chose curing condition as 373K/2h+393K/2h+403/17h.

The composition is chosen as: E-51/100g+HGH-600/100g+HK-021/50g+HSA-1/30g+DMP-30/0.2g, and it is named EHD2. As a comparison, EHD1 is the composition without quartz powder filler.

PHYSICAL PROPERTIES OF EPOXY RESIN

We has got lots of data about thermal contraction and viscosity during optimizing the composition of epoxy resin. The thermal contraction of EHD1 is 1.25%, and EHD2 0.67%. The viscosity of EHD1 is 19s at 373K for several hours, but the viscosity of EHD2 varies as Fig. 3 shows.

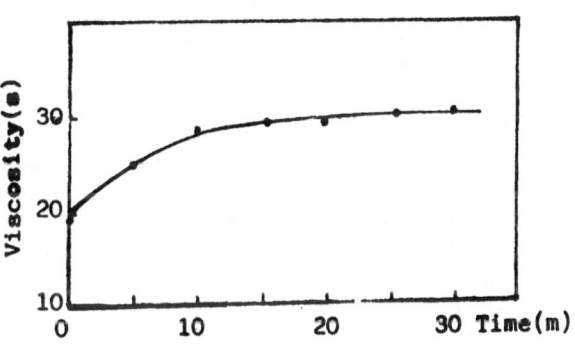

FIGURE 3. Viscosity of EHD2 resin

We find that its viscosity has some increment at first several minutes, this is just why catalyst could prevent filler sinking. Fig. 3 shows that viscosity remains below 30 seconds in 30 minutes. It is suitable for impregnation.

Thermal conductivity of epoxy composite has main affect on the transient stability of superconducting magnets[2]. We measured thermal conductivity k of EHD1 and EHD2 (as Fig. 4 shows). Upper points are the values for EHD2. Obviously

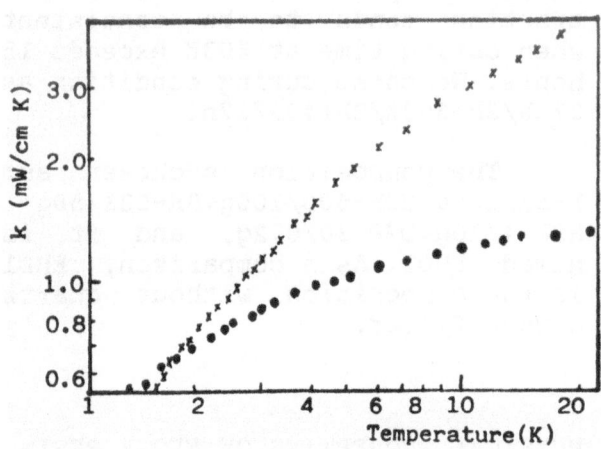

FIGURE 4. Thermal conductivity of epoxy composites

there are some increment, compared with EHD1. The value increases to 1.65 and 2.70 times of EHD1 at 4.2K and 10K. At the temperature from 2 to 15K, thermal conductivity of EHD2 is proportional to temperature approximately. This is due to filling of quartz powder. The improvement of thermal conductivity will result in the improvement of magnets' stability.

Tensile strength of epoxy composite had great influence on the stability of superconducting magnets. It is demanded that its strength at low temperature should be large enough to bear electromagnetic stress. We measured tensile strength of EHD1 and EHD2 at room temperature and liquid nitrogen temperature (Seeing Table 2). There are also other researcher's data present in Table 2. Our results are similar with FMNA and able to meet the demand for use at low temperature and for stability of magnets.

TABLE 2. Tensile strength of epoxy resin

Epoxy		EHD1	EHD2	EP0 [3]	FMNA [1]
Strength kg/mm2	293K	7.27	6.97	1.02	7.13
	77K	12.38	11.91	10.6	10.6

As stated above, the epoxy EHD2 possesses following features: low thermal contraction, good penetration properties before curing, high strength and thermal conductivity. We wounded several windings in diameter about 50mm, some of which glass fiber were padded in layers. The windings were impregnated in EHD2 in vacuum technique and sawed in vertical section. It was found that epoxy resin could permeate into the innermost layer of the windings uniformly. They were soaked in liquid nitrogen for many times and no cracks could be found with magnifying glass. The epoxy resin EHD2 was proven suitable to impregnate magnets and effective to use at low temperature.

SOME PROPERTIES OF PARAFFIN WAX AND BEESWAX

Wax impregnation is one of the earliest methods to stabilize superconducting magnets. Some researchers are still using these technique. Wax has some tenacity at room temperature, but it becomes extremely brittle at cryogenic temperature and is easy to break to powder when subjected to stress. So its main function is volume filling effect and to prevent conductor from moving. Thermal conductivity of paraffin wax is very low. There is a experiment formula to express its thermal conductivity between 93 K and 303 K [5]:

$$k=0.005(1+0.0016(t-273)) \text{ W/m.K}$$

Even if boundary resistance at low temperature is not accounted, its thermal conductivity is much lower than epoxy resin. So transverse thermal conducting capacity of wax-filling magnets is extremely poor and transient stability is bad.

Wax possesses excellent fluidity properties and its viscosity is very low (Tab. 3). Even for large windings, it is easy for wax to permeate into inner layers as long as vacuum impregnating and vacuum-pressure technique is performed.

TABLE 3. Viscosity of wax (s)

Temperature(C)	70	80	90	100
Para. wax	12.0	11.5	11.0	10.5
Beeswax	14.5	13.0	12.5	12.0

Thermal contraction of paraffin wax and beeswax determine their filling effect as well as stabilizing effect of the magnet. Contraction values were measured and given in table 4. It shows that wax's contraction was much larger than epoxy resin. Especially paraffin wax's is nearly one fifth. Therefore wax-filling magnets is just filled partially at low temperature.

TABLE 4.   Thermal contraction of paraffin wax and beeswax

	linear contraction 77K -293K	volume contraction 293K -373K	77K 373K
Paraffin wax	2.0%	13.8%	19.8%
Beeswax	1.4%	9.2%	13.4%

Table 4 shows that beeswax has relatively small thermal contraction. So beeswax is more effective to stabilize superconducting magnets. It was used in some laboratories, such as a 19.3T magnet at ITP [4].

CONCLUSIONS

Epoxy resin and wax are impregnants used most frequently to stabilize superconducting magnets. Their physical and chemical properties are utterly different, so their stabilizing mechanism are not identical.

Epoxy resin impregnation exhibited excellent long term stability of superconducting magnets. Because once the epoxy resin was cured, it had no tendency to flow. But the set epoxy resin might crack during energization, thus there would be training quenches. By optimizing resin composition, EHD2 with good properties at low temperature was developed. The new epoxy resin EHD2 exhibits following features: small thermal contraction, high tensile strength, good fluidity property and relatively high thermal conductivity. It is suitable for practical superconducting magnets impregnation.

Paraffin wax and beeswax just perform filling effect. They have very large thermal contraction and are easy to become powder at low temperature. Beeswax has better effect as an impregnant for its relatively small thermal contraction.

ACKNOWLEDGMENTS

The authors would like to thank Z.X.Li, X.C.Wu in Cryogenic Lab. for their activities during measuring thermal conductivity of epoxy resin composites and L.Z.Chao for measuring tensile strength.

REFERENCES

[1] H.Brechna,   Superconducting Magnet System, Springer-Verlag (1973), Berlin, Heide
[2] M.N. Wilson, Superconducting Magnets, Clanendon Press, Oxford, 1983
[3] C.Y. Hua, P.W. Xu, J.Y. Wang Cryogenic Physics (in Chinese) 4 (1982) 4, 284
[4] P. Turowski, Th. Schneider, IEEE Trans. on Magnetics, MAG-24 (1988) 2, 1063
[5] M.Freund, Paraffin Products, Elsevier Scientific Publishing Company (1982)

CREEP TEST OF COMPOSITE MATERIALS UNDER IRRADIATION CONDITION

T. Nishiura, S. Nishijima, K. Katagiri, T. Okada, J. Yasuda*, T. Hirokawa*
ISIR Osaka University, Osaka 567, Japan
*Shikishima Canvas Co. Ltd., Ohmihachiman, Shiga 523, Japan

ABSTRACT

In order to simulate fusion magnet conditions for insulating composite materials, creep tests on the epoxy based FRP were carried out under γ-ray irradiation. The creep deformation under irradiation during the test was much larger than those tested on the non-irradiated and post-irradiated specimens. This result suggests that the radiation damage of FRP in mechanical properties is enhanced by stress and that, therefore, the evaluation of radiation damage using irradiated samples obtained by conventional method can lead to significant underestimation. The mechanism of this enhancement of deformation as well as microscopic deformation mechanisms are discussed on the view point of the change of activation energy in the process of the chemical reaction induced by resin - γ-ray interaction and the separation between matrix resin and reinforcing fibers.

INTRODUCTION

In order to develop of the radiation resistant insulators for a fusion magnet, radiation damage of composite materials has been studied[1-4]. In a fusion reactor, the insulator is to be subjected to radiation and stress at cryogenic temperature, simultaneously. However, the evaluation of radiation induced change in mechanical properties on organic materials has, so far, been carried out with tests using the post-irradiated samples[1-7], and few experiments concerning to the simultaneous effects of stress and irradiation on the damage are conducted [8]. In this work, in order to clarify the simultaneous effect of stress and irradiation on composite materials (FRP), creep tests are performed in irradiation environment (creep under irradiation).

The creep of FRP is presumed to be controlled by both the characters of matrix resin and interface between fiber and matrix. The matrix resin and the interface in FRP have been known to be susceptible to the damage in the irradiation condition. According to the recent report on the simultaneous effect of stress and irradiation on fracture of rubber films by Dickinson et al, one shot of electron beam on the rubber films in tensile stress state makes remarkable extension of crack, and they concluded that the damage of films is promoted by such simultaneous conditions[8]. Although epoxy resin have superior resistivity to irradiation than such rubbers, the extent of damage promoted by combined effects in epoxy resin or epoxy based FRP can not be presumed theoretically at present.

Experimental evaluation is made to assess the combined effect of stress and irradiation on mechanical properties of epoxy based FRP. The data obtained are analyzed to elucidate the deformation mechanism taking place.

EXPERIMENTAL

Specimens were cut from the epoxy-based glass FRP plates of 2mm thickness(Lamivelle-A, Nittohdenkoh Co. Ltd.) to the rectangular shape of 60x10 mm as shown in Fig.1.

Creep test was conducted using the three point bending test machine which was designed for creep test. The distance of span was 40 mm. Creep deformation was measured using a dial gauge or a differential transformer. Irradiation was made using Co-60 source at the dose rate 0.006-0.019 MGy/h in a large part of this work and exceptly made by electron beam with 20 MeV and 240 mA current of LINAC at Radiation Laboratory of ISIR in Osaka University. Tests were made at room temperature in air and at liquid nitrogen temperature in liquid nitrogen.

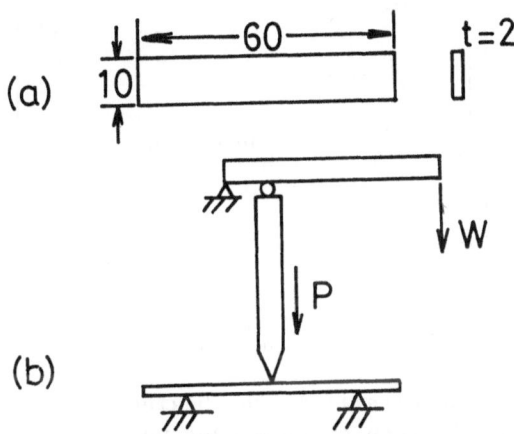

Fig.1. Dimension of specimen(a) and schematic diagram of creep test equipment(b).

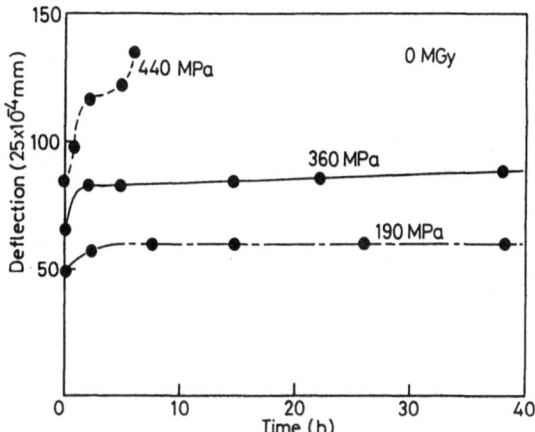

Fig.2. Creep curves of non-irradiated specimen at RT corresponding to various stress.

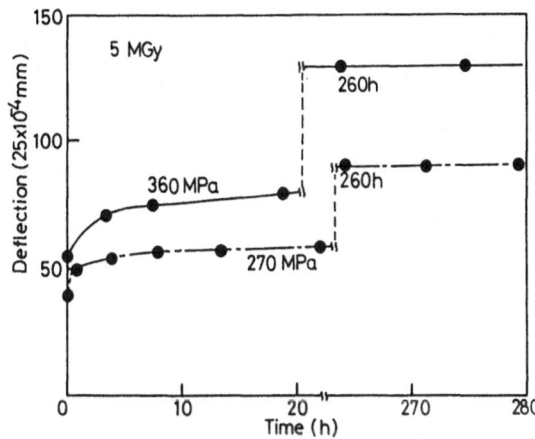

Fig.3. Creep curves of post-irradiated (5MGy) specimen at RT corresponding to various stress.

RESULTS

Room Temperature (RT) Creep

The creep curves of unirradiated specimens at RT is shown in Fig. 2. The results were obtained by a dial gauge and hence the measurement was made intermittently. There were found large instantaneous deflection as compared to the creep deflection. The beginning point of deflection in the figure does not indicate relative magnitude of the instantaneous deflection. Therefore, the creep curves except the beginning point exhibit the true data point. The creep curve at the stress of 440 MPa (the tensile strength of the specimen is 400

MPa) shows the 1st (transition), the 2nd (steady) stage although that stage is difficult to be identified since the stage is too short in this case and the 3rd (acceleration) stage. At the stress of 190 and 360 MPa creep curves exhibit a faint increase in the 2nd stage while the creep deformation is considerably large in the 1st stage.

The creep curves in 5 MGy irradiated specimens are shown in Fig. 3. Both at 270 and 360 MPa creep deformation is remarkable in the 1st stage, and faint in the 2nd stage like that of non-irradiated ones. The creep deformation in the 1st stage in 10 MGy irradiated specimens were larger than those of 5 MGy. The similar faint increase at the 2nd stage was

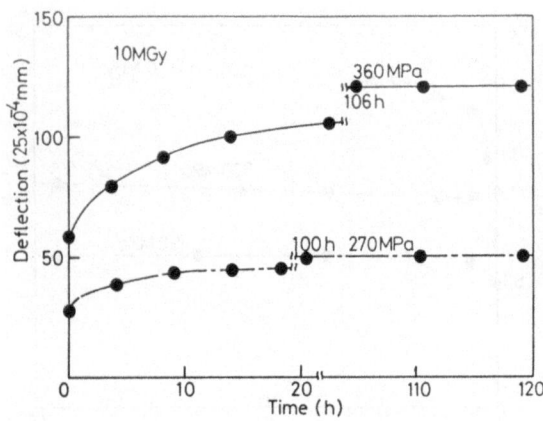

Fig.4. Creep curves of post-irradiated (10MGy) specimen at RT corresponding to various stress.

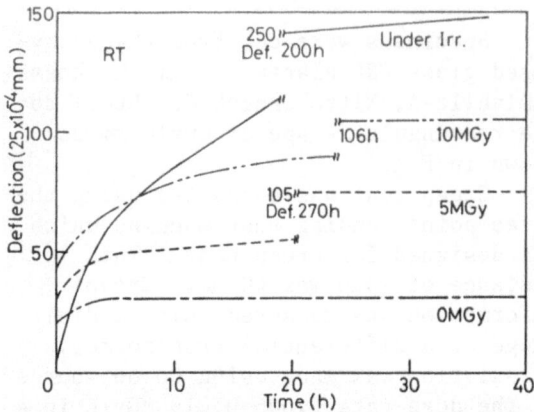

Fig.6. Creep curves for various specimens at stress of 360 MPa at RT.

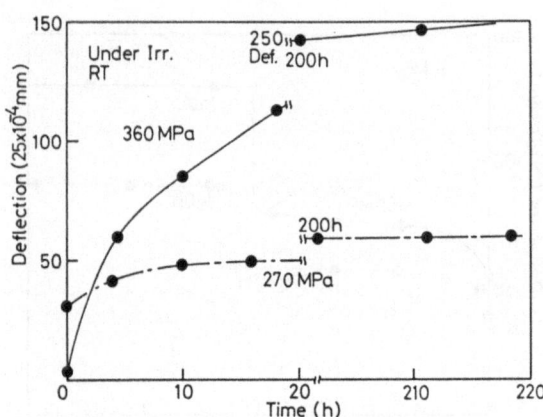

Fig.5. Creep curves of specimen under irradiation condition with the dose rate of 0.019 MGy/h at RT.

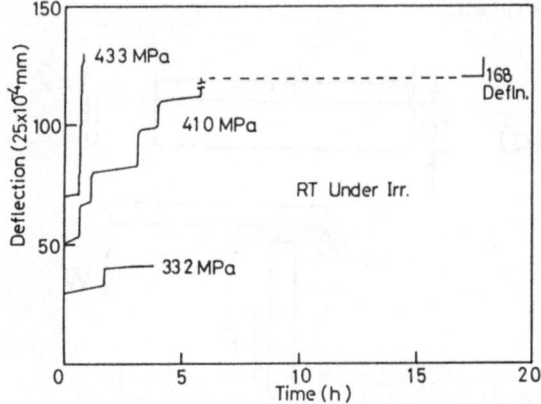

Fig.7. The details in the beginning region of creep on specimens under irradiation condition.

observed(Fig. 4).

On the other hand, the creep under irradiation at dose rate of 0.019 MGy/h at the stress level of 360 MPa shows large increase at both the 1st and the 2nd stages in comparison with those of non-irradiated and post-irradiated one(Fig. 5). The considerably large creep deformation in the 1st /and the early 2nd stage demonstrates that the GFRP is susceptible to simultaneous effect of the stress and the radiation. Creep curves for various specimens at the stress of 360 MPa are shown in Fig. 6. The larger amount of creep in specimen under irradiation is clearly noticed among these curves.

It appears worthy of note that the

creep deformation is large even at lower irradiation dose level, that is, the combined effect of stress and irradiation enhances the damage. The damage should not be neglected even at the low irradiation dose level where any damage can not be observed if the stress and the irradiation were applied to the sample, separately.

The creep deformation under irradiation is observed in details which was measured by a transformer continuously(Fig. 7). At 410 MPa, the stepwise increase of deformation is observed intermittently and the total creep deformation comes to be large.

The temperature rise due to absorption of γ-ray is measured by the thermocouple embedded in FRP, and is found to be about 15°C (and hence the temperature of specimen is 35°C during the irradiation) in the steady state. In order to examine the effect of temperature rise on creep, the specimen is heated up to 48°C by an electric lamp during the creep test. The creep amount is not increased by this temperature rise. The authors, therefore, conclude that resulting increase in creep amount under irradiation is not due to an increase in temperature of FRP.

Liquid nitrogen temperature(LNT) Creep

The results of creep for non-irradiated specimen at LNT are shown in

Fig. 8. The applied stress at LNT are nearly twice large to observe the same amount of creep at RT, due to the increase of strength (the tensile strength of the specimen at LNT is 1000 MPa) and rigidity. In non-irradiated specimens, the creep amount at 726 and 979 MPa is faint at first and comes to be small markedly at the 2nd region. The creep amount at 1105 MPa increases stepwisely at the first region but after that the creep rate is small.

The specimens under irradiation(Fig. 9) with a dose rate of 0.006 MGy/h show the stepwise creep deformation in the 1st stage at 927 MPa. The stepwise deformation at 1068 MPa is larger than that in non-irradiated one at 1105 MPa.

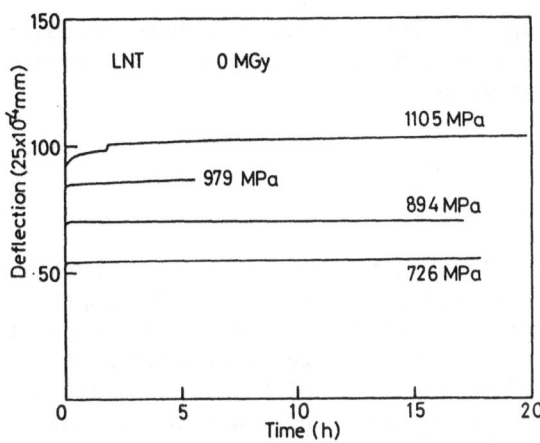

Fig.8. Creep curves of non-irradiated specimen at LNT corresponding to various stresses.

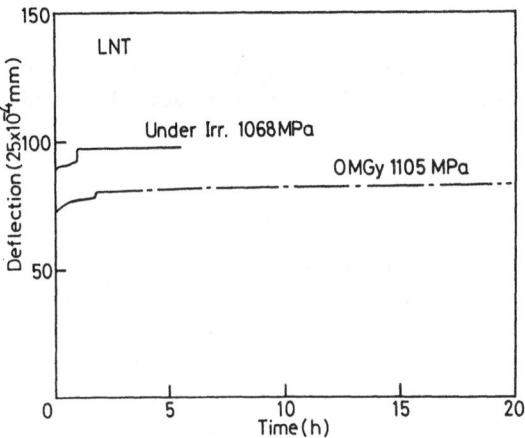

Fig.10. Creep curves for various specimens with stress of about 1000 Mpa at LNT.

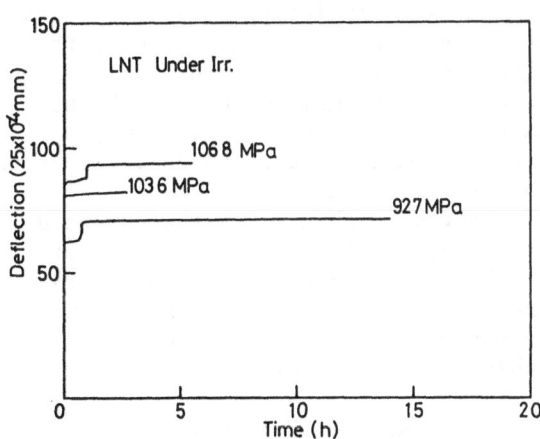

Fig.9. Creep curves of specimen under irradiation condition at LNT corresponding to various stresses.

Hence, creep deformation under irradiation at LNT is also larger compared with that of the non-irradiated.

Creep curves for various specimens with stress of about 1000 MPa at LNT are shown in Fig. 10. It is noticed that the larger step of creep deformation under irradiation than that of non-irradiated specimen occurs.

The surface of specimen in the vicinity of the compressed area by a loading wedge is shown in Fig. 11. The array of whitening spots are observed along the vertical weaves of glass fiber. These spots are estimated to be formed by debonding and/or microcracks which are induced with stress concentration at the beneath of the wedge. The intermittent generation of these debonding and microcracks makes the intermittent deformation and, therefore, lead to the

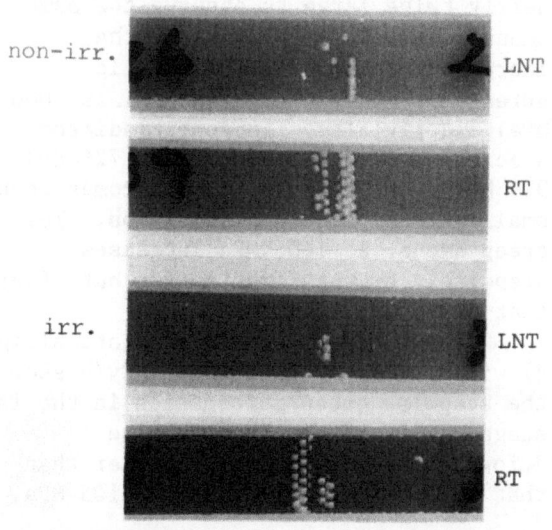

Fig.11. Array of whitening spots on the specimens surface just under the loading wedge.

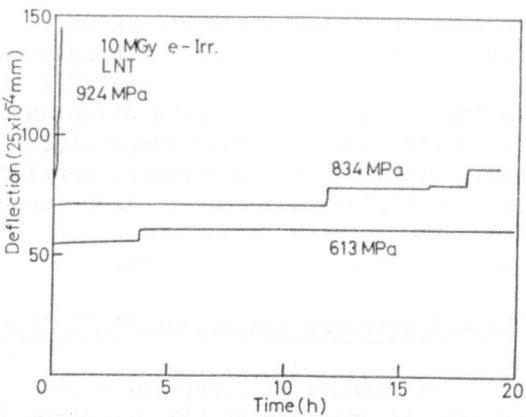

Fig.12. Creep curves of e-beam irradiated specimens at RT corresponding to various stresses.

stepwise creep deformation.

Creep curves at LNT using e-beam post-irradiated specimens in the dose of 10 MGy at RT are shown in Fig. 12. At relatively lower stress in comparison with those of γ-irradiated specimen, some large step wise creep are exhibited. This difference is estimated depending on the different quality of sources, and is not fully understood now.

DISCUSSION

The damage in FRP proceeds with radiation dose. In the creep test under irradiation, the dose to the sample increases with the time and hence the creep rate is expected to be increased corresponding to the damage proceeding. The creep phenomenon, however, can be monitored at the lower dose level where the creep of the irradiated sample was not be recognized.

The reason of this enhanced creep in the FRP is presumed due to the combined effect of stress and irradiation, that is, the debonding and microcrack formation is enhanced by the combined effect. The enhancement of crack growth by the combined effect of stress and irradiation has been reported in the test

of electron induced fracture of rubbers by Dickinson et al[8] as followings, that while the electron-induced scissions at zero stress tend to react quickly to reform the original bond or produce a crosslink, when the load-bearing chains in the crack tip region are under stress, electron induced bond scissions would tend to lead to the separation of newly created chain ends, thereby partially suppressing reattachment and favoring crack growth. Moreover, they observed the results similar to ours that fracture proceeds stepwise in accordance with electron beam and fracture acceleration occurs at the small dose (though the sample differs from ours). Thus, irradiation induced scission might occur in our experiments.

Another mechanism of the enhanced creep deformation by the combined effect may be explained by the Arrhenius' rate theory. The rate constant K for the formation of debonding and cracks is exhibited by the following formula.

$$K = A\exp[-(E-f)/kT]$$

where A is constant, E is activation energy, f is energy by stress effect, k is Boltzmann's constant and T is temperature. According to the formula, an apparent activation energy is reduced by f, and, therefore, at the larger load level the creep rate is enhanced. Under the irradiation, molecules can be excited directly and go over the potential and hence the creep deformation could be taken place even at the cryogenic temperatures.

The mechanism of stepwise creep is not understood at present in this experiment, and this is future task.

In order to suppress creep deformation under irradiation, the reduction of the deformation in matrix resin should be made. There are two methods to accomplish the reduction of deformation in matrix, one is to select the superior resin to the irradiation damage(like BT resin), and another is an advancement of construction of fiber weave. From the interlaminar strength, the resistance of irradiation damage in BT is suggested to be better than that of used epoxy resin. Creep test using BT resin based FRP under irradiation is now planned.

CONCLUSION

In order to investigate the simultaneous effect of stress and irradiation on the mechanical property in FRP, creep test under irradiation has been carried out. The results are concluded as followings.
(1)The simultaneous effect of stress and irradiation yields the enhanced damage on epoxy based FRP compared with the damage induced by the combination of the stress and the irradiation applied separately.
(2)Therefore, the evaluation of radiation induced damage on the test using post-irradiated sample would lead to some risks caused by overestimation.

ACKNOWLEDGMENTS

The authors are grateful to the members of Radiation Laboratory in ISIR for irradiation. This work is partly supported by Grant in Aid for Scientific Research No.0105002, Ministry of Education in Japan.

REFERENCES

1. Evans D. and Morgan T., Adv. Cryog. Eng., 1981, 28, 147-64.
2. Weber H.W., Kubasta E., Steiner W., Benz H. and Nylund K., J. Nucl. Mat., 1983, 115, 11-15.
3. Hagiwara M., Udagawa A., Kawanishi S., Egusa S. and Takeda N., J. Nucl. Mat., 1985, 133&134, 810-4.
4. Klabunde C.E. and Coltman Jr. R.R., J. Nucl. Mat., 1983, 117, 345-50.
5. Okada T., Nishijima S. and Yamaoka H., Adv. Cryog. Eng., 1983, 32, 145-51.
6. Nishijima S., Okada T., Miyata K. and Yamaoka H., Adv. Cryog. Eng., 1988, 34, 35-42.
7. Nishiura T., Katagiri K., Nishijima S., Okada T. and Nakahara S., Cryog. Eng., 1988, 34, 43-50.
8. Dickinson J.T., Klakken M.L., Miles M.H. and Jensen L.C., J. Polymer Sci.: Polymer Phys. Ed., 1985, 23, 2273-93.

MECHANICAL BEHAVIOR OF INSULATING MATERIAL
OF
FORCED FLOW SUPERCONDUCTING MAGNETS

S.Nishijima, T.Okada and M.Kawakami*
ISIR Osaka University,Ibaraki, Osaka, Japan
* Arisawa Mgf. Co., Ltd., Joetu, Niigata, Japan

ABSTRACT

The mechanical behavior of forced flow superconducting magnets has been simulated using small sized insulator-stainless steel sample. The insulation was made using glass clothes impregnated with B-stage epoxy on the stainless steel which simulated the cable conduit . After the insulated stainless steels were stacked, the assemble was cured to form the model sample. The flexural tests were performed on the model to investigate the mechanical behavior of the system. The shear modulus of the insulating layers were also examined.

INTRODUCTION

One of the most important problems to be solved in the large sized superconducting magnets such as fusion magnets is the degradation of the rigidity of the magnets [1]. As the size of the magnet increases, the thermal and/or electromagnetic forces increases and hence the mechanical behavior comes to be more important compared with other problems in large magnets.

It is recognized that the forced flow system is desirable to increase the rigidity of the large magnets. In this system the prepreg materials which have two roles as insulating material and mechanical binder are arranged around the conduit. After the winding the prepreg is fully cured applying heat to make the coil in one body. The rigidity of the such magnets can be increased compared with that of pool boiling system because the forced flow coils need not to have cooling channels formed by composite materials and the windings are bound firmly by the prepregs.

Even the prepreg method is adopted in the magnets the rigidity of the magnet is still lower than estimated by the law of mixture [2,3]. The estimation of the strength or rigidity appears to be problem. The windings can be considered as the composite containing the stainless steel couduit and the fiber reinforced plastics (FRP). For this reason the failure mode is ought to be complicated. Since the windings contain the FRP of which rigidity and strength is considerably small compared with those of stainless steel, the boundary failure between couduit and FRP and/or the FRP failure should decrease the strength and/or rigidity of the windings.

In this work the specimen which simulates the part of the forced flow windings was prepared and the strength and the rigidity of the winding were studied from the view point of the FRP and boundary behavior.

STRENGTH OF WINDINGS

The strength of the windings of forced flow cables is estimated first in this work. The forced flow cables are usually insulated by the prepreg (FRP with B-stage resin) and then wound to form a magnet. After the winding the whole coil is warmed to cure the prepreg and to make the winding one body. The winding is considered to be a composite composed with conduit and FRP. It means that the strength of the winding is not controlled only by the strength of the each component (conduit and FRP) but also by the interface strength between conduit and FRP. The strength of the windings is ought to be estimated considering the winding as the composite. In this work the strength of the windings is estimated by means of 3 point bending tests using the small sized insulator-stainless steel system.

In Fig.1 the schematic illustration of the specimen was shown. The 304 stainless steel 100mm in length, 15mm in width and 2mm in thickness were wrapped with the prepreg of which size were 0.2 mm in thickness and 15 mm in width. The two wrapped stainless steel strips are piled up and cured at the pressure of 1MPa, 423K for 1.5 hr. and 423 K for 15 hr. without pressure. The prepreg was made by plain woven E-glass with epoxy.

The 3 point flexural tests were performed at liquid nitrogen (LNT) and room temperature (RT) changing the span. The loading tip radius was 5 mm. The deformation speed was set at 1mm/min. The span length was changed form 30 to 90 mm.

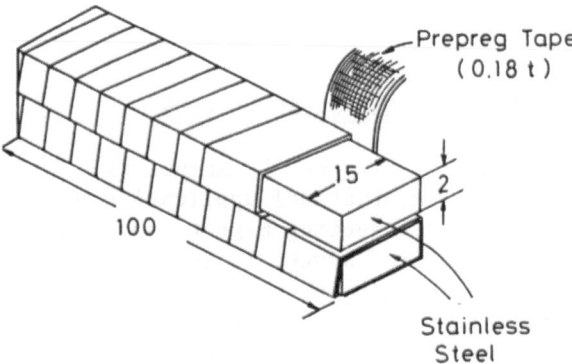

Fig.1 Schematic illustration of the tested specimen.

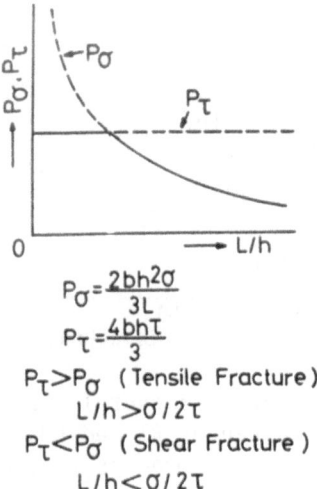

$$P_\sigma = \frac{2bh^2\sigma}{3L}$$

$$P_\tau = \frac{4bh\tau}{3}$$

$P_\tau > P_\sigma$ (Tensile Fracture)

$L/h > \sigma/2\tau$

$P_\tau < P_\sigma$ (Shear Fracture)

$L/h < \sigma/2\tau$

Fig.2. Stress distribution in flexural test and span dependence of flexural breaking load and shear breaking load.

In 3 point flexural test the failure mode can be changed with the span length [4,5]. In smaller span region the interlaminar shear failure is apt to be introduced and in larger span region the flexural failure occurs. It means that the various failure mode can be realized in 3 point bending tests with changing the span length.

In 3 point bending tests when the tensile stress reaches the tensile strength, the tensile fracture occurs. When the shear stress reaches the shear strength (in this case interlaminar shear strength), the specimen breaks in shear mode. The loads where the tensile P_σ and shear failure P_τ occurs are also plotted against the span length L in Fig.2. To get the flexural strength the L should be large enough being P_τ larger than P_σ . It means that when the complicated stresses apply to the windings in the magnets the failure mode varies with the stress condition.

The flexural strength obtained in 3 point bending tests was presented in Fig.3. In this figure the flexural strength was calculated by the simple beam theory without considering the change of failure mode. At the smaller span length the flexural strength appears to be small because the failure mode is different from the flexural failure. As mentioned above the nominal flexural

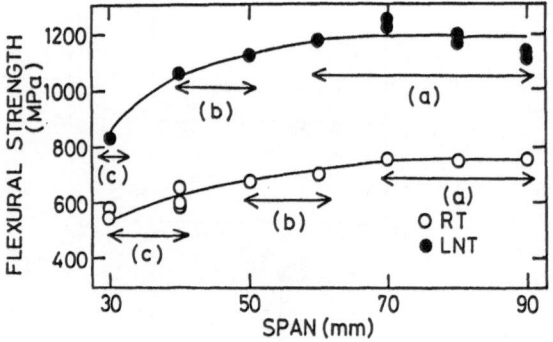

Fig.3 Change of flexural strength of the model obtained at RT and LNT. The calculation was made using the simple beam theory without considering the failure mode.

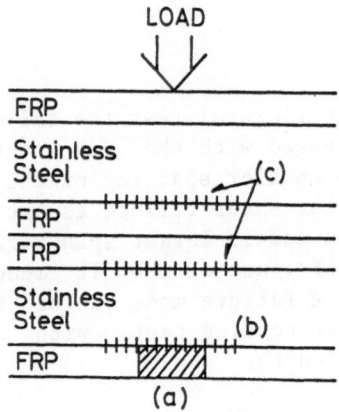

Fig.4 Schematic illustration of the specimen and the failure mode.

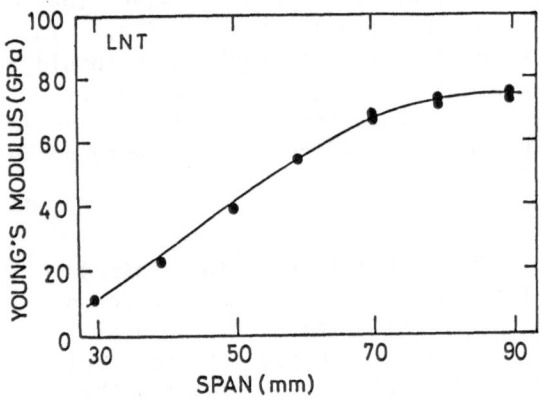

Fig.5 Change of Young's modulus of the model obtained at LNT. The calculation was made based on simple beam theory.

strength is changed by the span length and hence the mechanical behavior of the windings varies with the stress condition even the identical load is applied to the windings.

Figure 4 shows the schematic illustration of the cross section. At RT when the span length is larger than 70mm, the FRP arranged at the outer area was broken in tensile mode (a). When the span length ranges from 50 to 60 mm, the interface failure was taken place between the stainless steel and FRP in the tensile region (b). If the span length L is smaller than 40 mm the interface failure occurred at the center region (c). The interface strength between stainless steel and FRP is thought to be smaller than that between FRPs. Same change of failure mode was taken place at LNT and the failure mode was also shown in Fig.4.

Figure 5 shows the span dependence of Young's modulus obtained at LNT based on the simple beam theory. The Young's modulus at small span length is found to be small due to the shear deformation at the FRP region.

The specimen can be thought to be the composite beam and the Young's moduli of the components are different each others. Consequently the discontinuity of the tensile (or compressive)stress in the specimen is brought and results in the shear stress. The results show that the induced shear stress causes the interface failure and shear deformation. In considering the rigidity of the magnets, the shear deformation has to be taken into account.

INTERLAMINAR SHEAR MODULUS

The rigidity of the magnet is important in practical application of the large superconducting magnets and hence the forced flow system is usually employed to increase the rigidity. In estimating the rigidity of the magnet the shear modulus of the windings is to be measured or calculated. The shear behavior of the winding was found to be controlled by the shear deformation of the insulating material in the previous chapter. In this work the interlaminar shear modulus of the insulating material is measured.

The interlaminar shear modulus was measured by the same method to obtain the in plain shear modulus. The in plain shear modulus is obtained by the two Young's moduli in principal axes, that in 45 degree off-axis direction and Poisson's ratio considering the composite orthotropic.

The Young's modulus in theta off-axis direction is given by following equation.

$$1/E(\theta)=(1/E(0))\cos^4\theta+(1/E(\theta))\sin^4\theta$$
$$+(1/G(0)-\nu_0/E(\theta))\sin^2\theta\cos^2\theta \quad (1)$$

where $E(\theta)$ Young's modulus in theta direction, ν_0 Poisson's ratio in 0 degree direction and $G(0)$ shear modulus. Rearranging equation (1) by substituting Young's modulus in 45 degree direction, $E(45)$, into, the equation (2) is obtained.

$$1/G(0)=4/E(45)-(1/E(0)+1/E(90)$$
$$-2\nu_0/E(0)). \quad (2)$$

Then the shear modulus $G(0)$ can be obtained. Therefore, four independent parameters are needed to calculate the shear modulus that is Young's moduli in two principal axes and 45 degree direction and Poisson's ratio. The off-axis angle dependence of shear modulus is calculates as

$$1/G(\theta)=1/G(0)\cos^2\theta\times(1/E(\theta)+1/E(90)$$
$$+2\nu_0/E(0))\sin^2 2\theta \quad (3)$$

The interlaminar shear modulus was obtained in the same manner as mentioned above. The calculation was made assuming the plane stress condition and hence it is important to prepare the specimen properly. The specimens were cut out as shown in Fig.5. The specimen was glass cloths reinforced epoxy of which glass content was approximately 70% by weight. The reinforcement was plane woven cloth with silane finish. The thick plate of 40 mm thickness was sliced to make 2mm thick sheet as presented in Fig.6. From the sheet seven specimens, of which direction were different each other, were cut. The

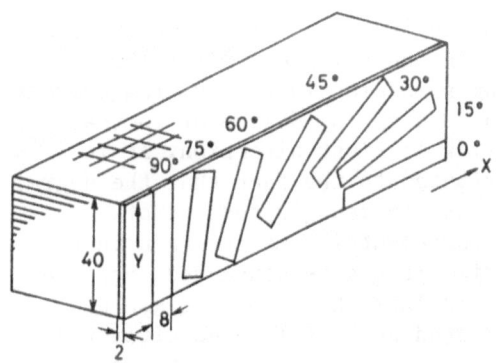

Fig.6 Sample preparation for interlaminar shear modulus measurement.

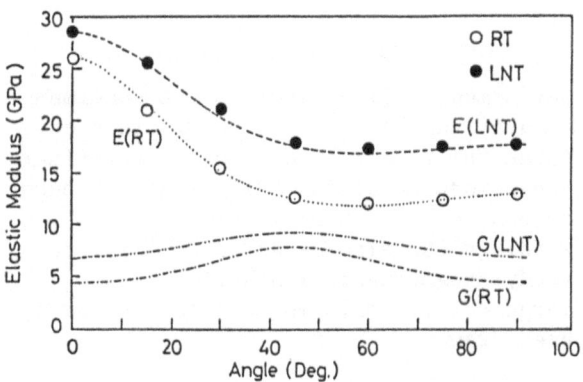

Fig.7 Off-axis angle dependence of Young's modulus and shear modulus.

Young's moduli of these specimens were measured and the interlaminar shear modulus was calculated. The Young's modulus was measured by means of the dynamic measurement. In this method the Young's modulus was derived from the resonance frequency of the specimen [6-8].

The off-axis angle dependence of Young's modulus was presented in Fig.7. The plotted points present the measured values. The measured temperature is room (RT) and liquid nitrogen temperature (LNT). The curves shown in the figure are the results obtained by the equation 1. The angle dependence of shear modulus is also presented in this figure. The interlaminar shear modulus was obtained as $G(0)$ and was 4.40 and 6.79 GPa at RT and LNT, respectively. The measured values coincide with the calculation. The good agreement in $E(0)$, $E(45)$ and $E(90)$ is natural because the calculation was made using these values. Beside these angles the good agreement was found and it means the rightfulness of the analysis even in the thickness direction. The Young's modulus increases with decreasing temperature. The degree of increase is smaller in fiber direction. It is attributed to the fact that the degree of increase in Young's modulus of reinforcement is smaller than that of the matrix. It can be concluded that the interlaminar shear modulus can be evaluated accurately even after the mechanical defects are introduced in the windings.

CONCLUSION

The 3 point flexural tests were performed at RT and LNT on the specimen simulating the forced flow windings to check the failure mode. The interlaminar shear modulus of the FRP which is thought to control the shear modulus of the windings was also measured in order to study the mechanical behavior of the large coils. Following conclusions have been drawn.

(1) The flexural strength of the stainless steel-FRP system varied with the stress conditions because the fracture mode were changed. The observed fracture modes in this work were the tensile failure of FRP, the boundary failure between FRP and conduit in tensile stress region and the boundary failure in the shear stress region. Another possible mode is the interlaminar failure between the FRPs. Although the failure does not bring the magnet failure, causes the degradation of the rigidity of the magnet because the discontinuous deformation is taken place at the failure area.

(2) The interlaminar shear modulus were successfully measured considering the composite materials orthotropic. Using the obtained data the Young's modulus in various direction can be calculated with considerable accuracy. The interlaminar shear modulus was 4.4 and 6.79 GPa at RT and LNT, respectively.

ACKNOWLEDGMENT

This work is partly supported by Grant in Aid for Scientific Research NO.0105002, Ministry of Education in Japan.

REFERENCES

[1] Yoshida, K., Nishi, M.F., Takahashi, Y., Tsuji, H., Koizumi, K., Okuno, K. and Ando, T., Design of the proto-type conductors for the fusion experimental reactor. IEEE Trans., 1989, Mag-25, 1488-91.
[2] Shimamoto, S., Ando, T., Hiyama, Y., Tsuji, H., Takahashi, Y., Tada, E., Nishi, M., Yoshida, K., Okuno, K., Koizumi, K., Kato, T., Nakajima, H., Takahashi, O., Shimada, M., Sanada, Y., Iida, F. and Yasukochi, K., Domestic test results of the Japanese LCT coil. IEEE Trans.,1983, Mag-19, 851-58.
[3] Hattori, Y., Yoshida, Y., Nakajima, H., Koizumi, K., Oshikiri, M. and Shimamoto, s., Measurement of winding rigidity on superconducting coil for fusion magnet design, Proc. MT-9 Zurichi, 1984, 371-374.
[4] Okada, T., Nishijima, S., Yamaoka, H., Miyata, K., Tsuchida, Y., Mizobuchi, K., Kuraoka, Y. and Namba, S., Mechanical properties of unidirectionally reinforced materials, In Nonmetallic Materials and Composites at Low Temperatures 3, ed. G. Hartwig and D. Evans, Plenum Publishing corporation, 1986, pp. 127-142.
[5]Okada, T., Nishijima, S., Yamaoka, H., Miyata, K., Fujioka, K., Kuraoka, Y. and Namba, S., Mechanical properties of unidirectionally reinforced composite materials, Adv. Cryog. Eng., 1986, 32, 203-208.
[6] Okada, T., Nishijimam, S., Mastusita, K., Okamoto, T., Yamaoka, H. and Miyata, K., Dynamic Young's modulus and internal friction in a composite material at low temperatures, Adv. Cryog. Eng., 1984, 30, 9-16.
[7] Nishijima, S., Mastusita, K., Okada, T., Okamoto, T. and Hagihara, T., Dynamic Young's modulus and internal friction in composite materials, In Nonmetallic Materials and Composites at Low Temperatures 3, ed. G. Hartwig and D. Evans, Plenum Publishing corporation, 1986, pp. 143-151.
[8] T.Okada, S.Nishijima, K.Mastusita and Okamoto, T., Effect of interlaminar failure on dynamic Young's modulus and internal friction on composite materials, Adv. Cryog. Eng., 1988, 34, 115-122.

COMPARISON OF TEST RESULTS WITH ANALYSIS OF GLASS FIBRE REINFORCED EPOXY SHEAR TEST SAMPLES

F. Fardi, N. Mitchell, R. Poehlchen,
NET Team Garching, Boltzmannstr. 2, 8046 Garching, FRG

ABSTRACT

In large coils the mechanical strength of the winding is important in maintaining electrical insulation, preventing heat generation through relative movement and giving overall rigidity to support the magnetic forces. The insulation system that gives a good combination of these three factors is glass fibre reinforced epoxy resin with vacuum/pressure impregnation. The usual failure mode of this insulation is in shear, and tests have been performed at 4 K as part of the establishment of a database for the NET and ITER projects which utilize large superconducting coils under nuclear conditions. To allow the shear test specimens to be irradiated at 4 K, the geometry has to be more compact than is conventional for shear test specimens. These have been analysed with a non-linear elastic-plastic finite element model to confirm their suitability for measuring the insulation shear strength and to attempt deductions about the failure mode.

INTRODUCTION

There are several potential applications for epoxy bonded magnets, both normal and superconducting, where the insulation is subject to nuclear radiation (i.e. high energy physics, nuclear fusion research).

As part of the development of radiation tolerant insulation systems for the superconducting toroidal field coil system of the NET Tokamak fusion experiment (2), an extensive programme of insulation testing is underway between the NET Team and ABB Zuerich (1). A range of insulation samples using different epoxies and different reinforcing glasses are being manufactured with subsequent irradiation and testing at 4 K. To allow the shear test specimens to fit into the reactor in which the irradiation is performed, together with the associated cryogenic cooling, much smaller specimens have to be used than is normal for lap shear specimens (3). The test specimen geometry is shown in Figure 1. At this size, edge effects can dominate the stress system in the insulation, and the primary purpose of this analysis is the confirmation that the specimens can give an acceptable approximation to uniform shear stress under the applied loading. A further objective is to try to identify if the degradation of any "bulk" epoxy properties (such as yield stress) can be the cause of shear failure.

The test results as they are available at present are fully reported in (1) and will not be redescribed here. In summary, the use of an "R" glass insulation (boron free) with "Olitherm" ABB resin with flexibiliser shows no degradation after a dose of $5 \cdot 10^8$ rads (higher doses are not yet tested). This is in contrast to the use of "E" glass (with boron) which had lost 60-80 % of its strength. Average shear strengths (based on applied load and insulation area) are well over 130 MPa. For testing, the shear samples (figure 1 and 2) are placed in a close fitting tube at 4 K. They are loaded in compression by a piston pressing on the projecting ends of the sample. The load is transmitted through a rod outside the 4 K cryostat, giving the necessary thermal insulation. Displacements are measured outside the cryostat and thus include a substantial length of steel. Typical load - deflection curves are shown in Fig. 3. A "normal" shear specimen with full strength insulation fails at about 45 KN (average shear stress over 130 MPa).

ANALYSIS PROCEDURE

The model has been constructed using 8 noded brick elements as shown in Fig. 2. The epoxy layer is two elements thick. The load is applied on the specimen ends, simulating the effect of the piston by requiring a uniform vertical displacement from the end nodes. The effect of the

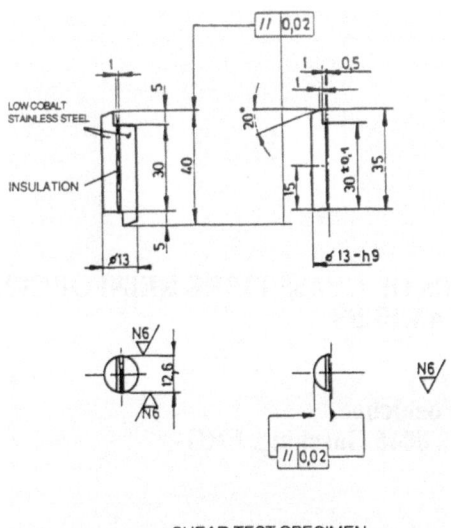

Fig. 1

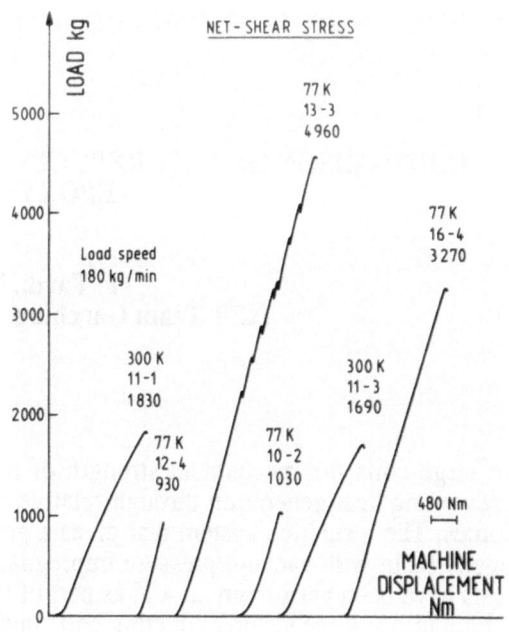

Fig. 3   Test Results

containing   tube is important and is modelled by constraining the radial motion of the outer surface nodes - only vertical slip is allowed.

Initial calculations with linear elastic material properties showed that both insulation and steel contained stresses above the yield point, and so non-linear elasto-plastic materials models have been included, as shown in Fig. 4.   (5).

The epoxy was modelled isotropically although it will certainly have strong orthotropic behaviour. However in this loading case most of the significant loads act normal to the insulation layers, and inexact representation of the properties in the plane of the glass reinforcing layers should not have much

effect. Since the onset of plastic behaviour and the subsequent elasticity is difficult to define, a sensitivity analysis has been performed with the materials properties as also shown in Fig. 4.

The specimen is loaded with successive load steps up to 45 kN, which is the maximum indicated from the test results. The first load step is chosen at 15 kN and is near the limit of the elastic range. The subsequent 6 load steps are in the non-linear range.

RESULTS

The displacement of the load transmission surfaces at the ends of the specimen are shown in Fig. 5 and the displaced shape at 45 kN load in Fig. 6. It is evident that substantial plastic deformation of the specimen ends is occuring and producing the highly non-linear curve shown in Fig. 5. This plastic deformation dominates the load-displacement plot. The total displacement at 45 kN is about 3 mm, compared with the measured 4-5 mm in Fig. 3. However the non-linearity in the tests is markedly less. This suggests that the long piston length in the tests adds appreciably to the displacements (as expected) and that the yield stress of the steel is somewhat higher than has been taken here.

Fig. 7 shows the insulation layer shear stress distribution, Fig. 8 the direct stresses, Fig. 9 the friction factor and Fig. 10 the Von Mises equivalent stress. There is some tendancy for a peaking of the

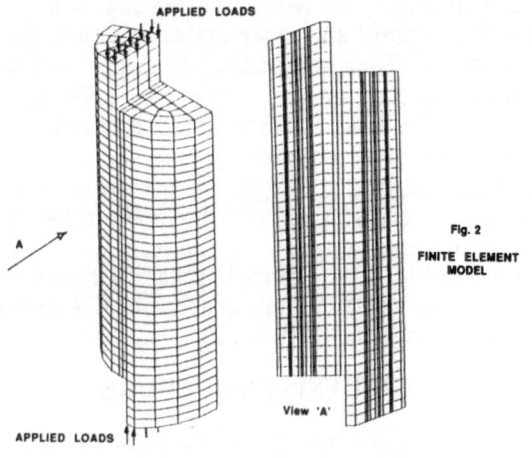

Fig. 2   Finite Element Model

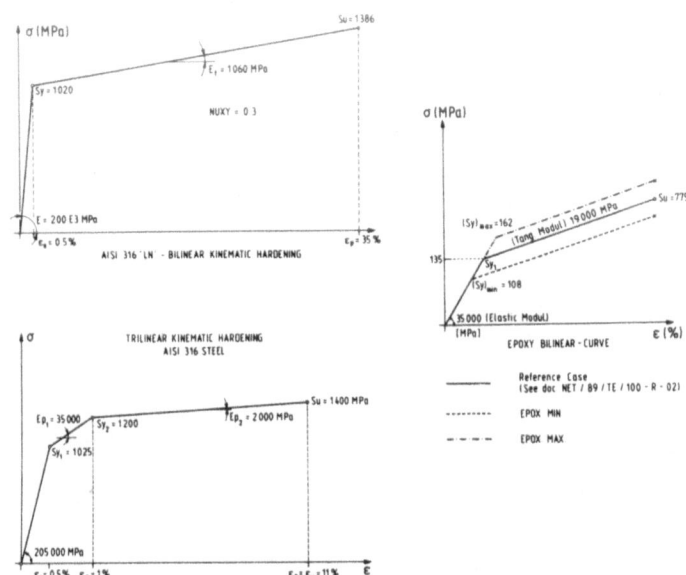

Fig. 4   Material Properties

shear stress at the ends of the insulation and the overall normal compression is quite high (always over 50 MPa). The friction factor shows some peaking along the edges and particularly at the corners. This suggests that local failure may occur initially at the edges leading to a load redistribution. The Von Mises stress lies above the "bulk" yield stress of the insulation suggesting that this is not a particularly good failure indicator.

Fig. 11 shows comparisons of the distribution of the shear stress in the insulation by a linear elastic finite element calculation, an analytic expression (4) and the elasto-plastic method. The plasticity of the steel in the test specimen, although limited to the ends, is significant in reducing the shear stress peaking.

SENSITIVITY TO MATERIAL PROPERTIES

A sensitivity study of the influence of both steel and insulation elastic-plastic properties (Fig. 4) has been performed. Generally, the influence of the variation in mechanical properties of the epoxy (in practice, Sy) on the results of the reference analysis is negligeable.

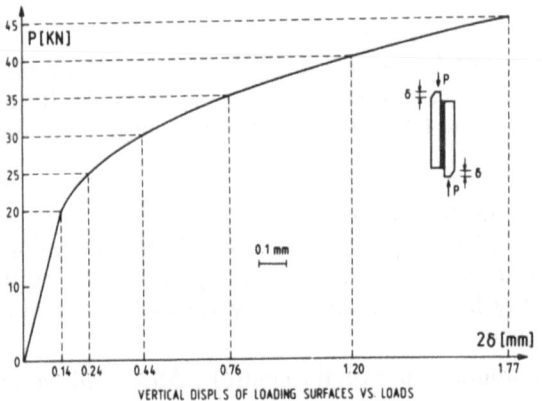

Fig. 5   Calculated Load-Deflection Curve

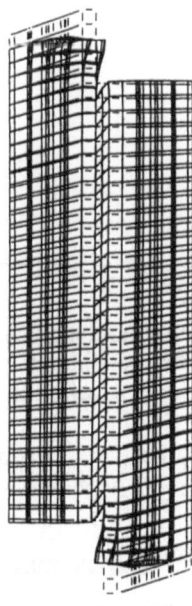

Fig. 6   Deflected Shape

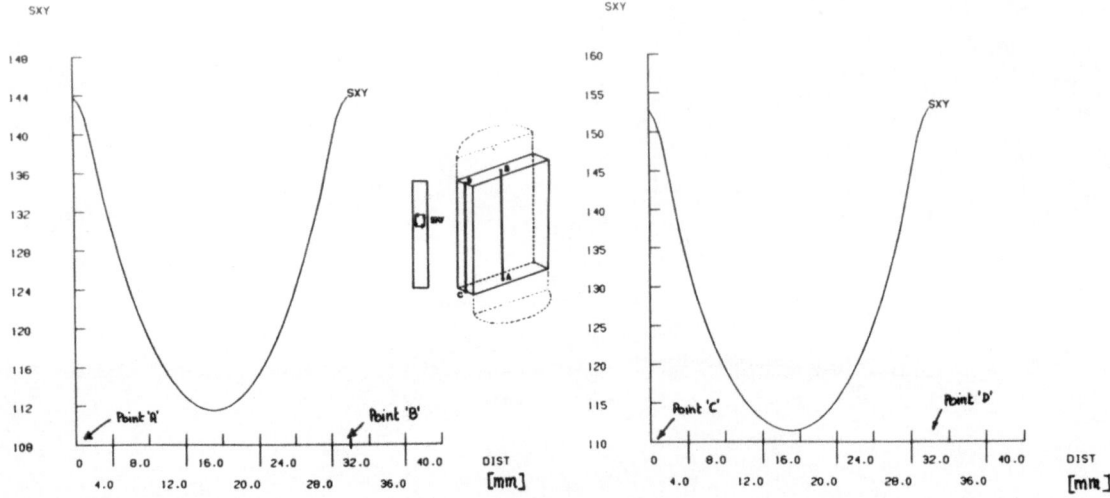

Fig. 7 Shear in Insulation (45 KN Load)

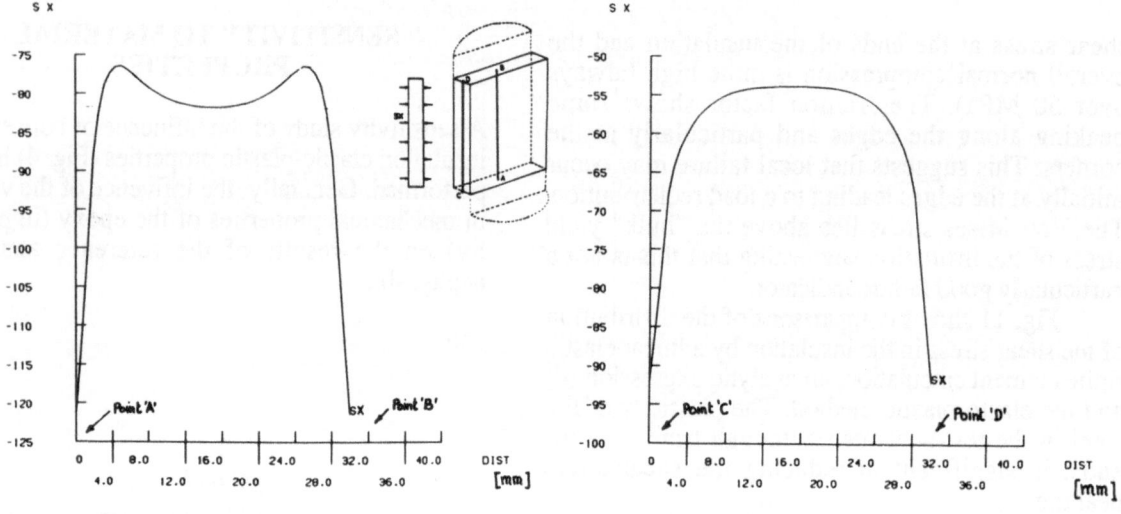

Fig. 8 Compression on Insulation (45 KN Load)

A total variation of:

$$\Delta Sy = \frac{162 - 108}{135} \rightarrow 40\ \%$$

for the two extreme analyses cases (epox. min & epox. max) has been considered.

The percentage of variation in stress values is defined by:

$$\Delta \sigma = \frac{|(\sigma)^{max.}| - |(\sigma)^{min.}|}{|(\sigma)ref.|} \times 100$$

in which $(\sigma)^{max,min} \rightarrow$ epox. max. and min. cases and $(\sigma)^{ref} \rightarrow$ reference case

1. Shear Stresses
    The figures [3] through [10] show that when considering a variation of $\pm$ 20 % in the Sy for the epoxy, the peak stresses do not change more than about 1 %.

2. Direct Stresses
    More significantly but still slight are the variation on the direct (normal) stresses along the epoxy layer. For the same variation in insulation properties.

minimum values (in the centre) $\Delta S_x \leq 3.8 - 4.5\ \%$

peaks (at the extremes) $\Delta S_x \leq 8\ \%$

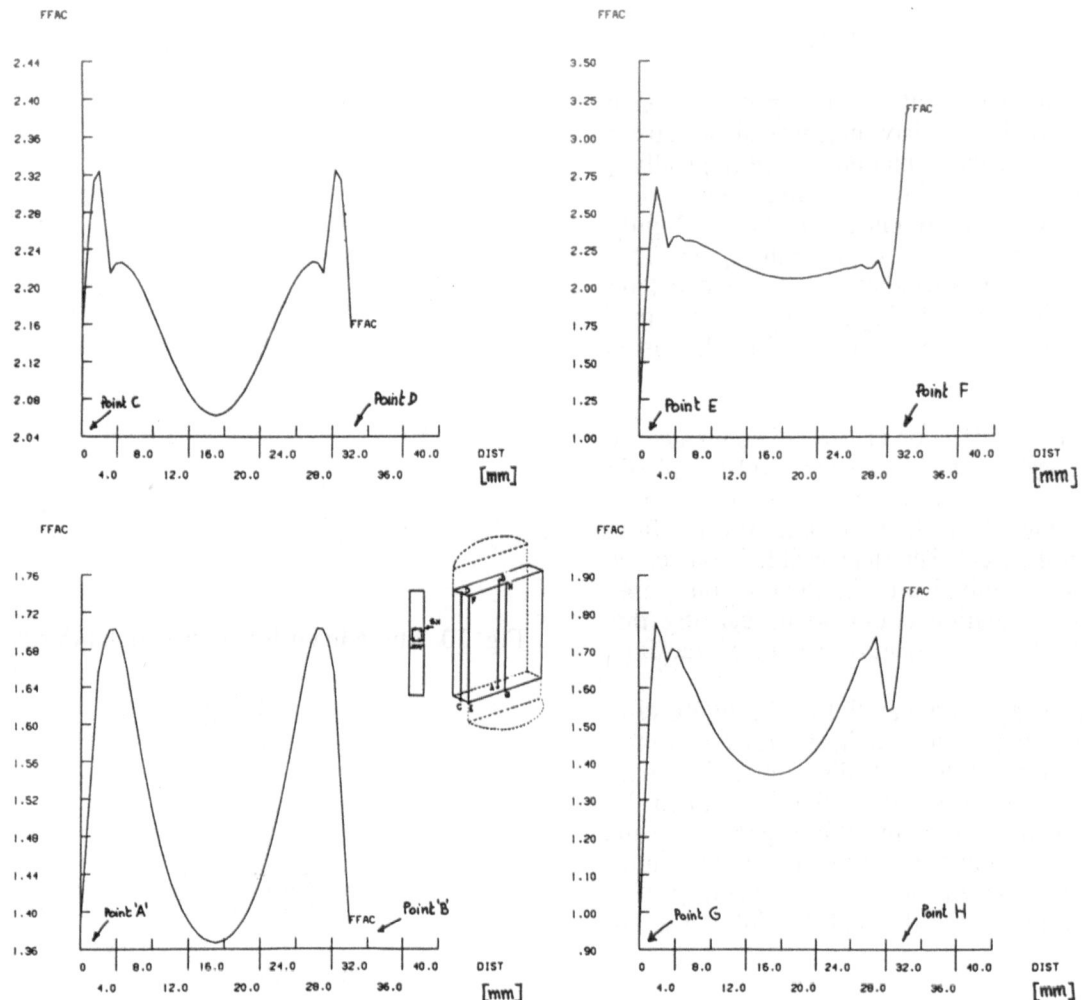

Fig. 9   Friction Factor in Insulation (45 KN Load)

In addition, the direct stress values obtained from the 3 different analysis cases tend to even out as the external loads increase (P = 40 kN . 45 kN). This is true also for the shear stresses and consequently for their ratio (i.e. friction factor).

3.   Friction Factors
The following variations occur:

-middle of the epoxy (in the centre): Δ (f.f.) ≤ 1-2%

-middle of the epoxy layer (at one edge): Δ (f.f.)
$$\leq 7 - 8\ \%$$

-int. epoxy-steel (in the centre): Δ (f.f.) ≤ 8-9 %

-int. epoxy-steel (at one edge): Δ (f.f.) ≤ 5 - 6 %
The above (maximum) values are referred to the P = 40 kN external load. They decrease on increasing the load from 40 to 50 kN.

4.   Vertical Displacements of the Load Support Position vs. Load
The relative vertical displacements of the load support position vs. load do not differ significantly

(1 - 2 %) from those evaluated by the reference case. This means that they are essentially produced by plastic deformations occuring in the region (steel) near the loading points while the effects due to the shear deformations in the epoxy affect only slightly these values.

5.   Stress Intensity
There are no significant variations in the equivalent S.I. (Tresca's criterion) evaluated along one of the most stressed paths of the epoxy layer.

Changing the steel properties (by making it stiffer in plastic yielding) has rather more influence. The friction factor peaks increase and decrease by about 50 % on the surface at the ends and by 10 % in the centre of layer at the ends. The shear stress is rather less affected (an increase of about 4 %) and the direct normal stress then accounts for the changes in friction factor, increasing and decreasing at the edges by up to 50 %.
The overall displacement of the end pistons are greatly reduced to about 1mm at 45 kN with

much less non-linearity.

CONCLUSIONS

1) Despite their small size, the specimens seem able to provide a fairly uniform shear stress distribution over the insulation surface (typically ± 15 % of the average). Unwanted direct compressive normal stresses are present and could be slightly increasing the apparent strength of the epoxy. This compression is an unavoidable result of the method of support (in a tube) during testing, since the tube walls must react the moment caused by the piston loads.

2) The load-deflection curves are dominated by the steel properties and in particular its elastic-plastic transition. Some of the experimental curves seem to show non-linear behaviour which must come from yielding of the steel. The steps visible in one curve are almost certainly due to the steel or connecting rods - the insulation could no be causing this amount of deflection by partial yielding or cracking.

3) The shear stress parallel to the reinforcing glass seems to be a good indicator for the failure. The insensitivity of the stress distribution (including concentration factors and friction factors) to the yield stress of the bulk material suggests that the radiation is directly affecting the interlaminar bonding. In some of the stronger specimens Von Mises stresses above the bulk "yield" stress are occuring.

REFERENCES

[1] The Mechanical Strength of Irradiated Electric Insulation of Superconducting Magnets, R. Poehlchen et al., paper presented at Inter. Conf. on Materials and Cryogenics, Los Angeles, 1989.

[2] The NET Project, Fusion Technology, July 1988, Vol. 14, No. 1.

[3] Mechanical Tests on Insulation Systems for the JET Poloidal Coils, J. Last, A. Bond, E. Salpietro, 10 Symp. of Fus. Tech., Padua, 1978.

[4] Festigkeitsverhalten und Ingenieurmäßige Berechnung von einschnittig-Überlappten Metallverbindungen, Dr. Ing. Ortwin Hahn, Aachen, 11. Nov. 1975.

[5] Fabricability Study of the NET Toroidal Field Coil System, ABB Report HISM20378, Zürich, 1988, for NET Contracts 723 and 726.

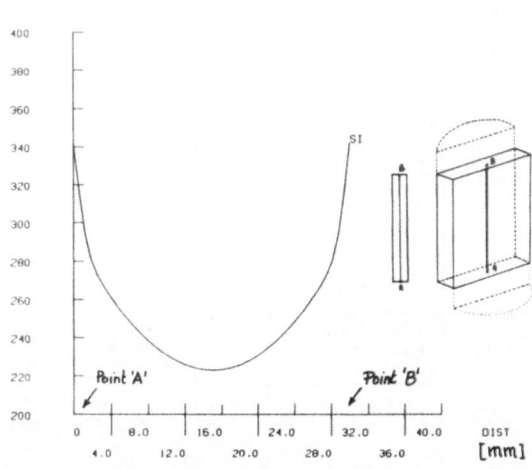

Fig. 10 Stress Intensity in Insulation (45 KN)

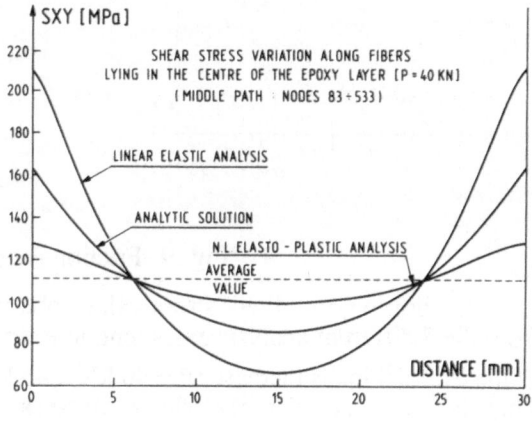

Fig. 11 Comparison of Calculations of Shear in Insulation (40 KN Load)

DEVELOPMENT OF RADIATION-RESISTANT MAGNETS FOR HIGH INTENSITY BEAM LINES

K. H. Tanaka, Y. Yamanoi, M. Takasaki
Beam Channel Group, Physics Department Ⅲ,
KEK, National Laboratory for High Energy Physics,
1-1 Oho, Tsukuba-shi, Ibaraki-ken, 305, Japan

T. Suzuki
Radiation and Safety Control Center,
KEK, National Laboratory for High Energy Physics,
1-1 Oho, Tsukuba-shi, Ibaraki-ken, 305, Japan

K. Kato
Electromagnet division, TOKIN Corporation
6-7-1 Koriyama, Taihaku-ku, Sendai-shi, Miyagi-ken, 982, Japan

ABSTRACT

We developed two types of radiation-resistant magnets for a new external proton beam line of the KEK 12 GeV Proton Synchrotron (KEK-PS). One was designed with the polyimide resin pre-impregnated (PRP) glass cloth to insulate the magnet coil. Most of the magnets for the new external proton beam line were manufactured using the PRP and no other organic material. The radiation life of the insulator was tested at the existing proton beam of the KEK-PS and no serious damage was observed up to 4×10^8Gy. The other type was designed for a higher radiation dose of 10^{11}Gy which often occurs at just downstream of production targets. The magnet coil was insulated with inorganic materials only, such as high-alumina cement and asbestos tape.

INTRODUCTION

A new counter experimental hall[1] is being constructed at the KEK 12 GeV Proton Synchrotron (KEK-PS) and a new external proton beam line will be installed in order to supply primary protons from a new extraction point of PS to the new hall. The completion of the building of the new hall will be by the end of this year and immediately followed by the magnet installation. All the magnets for the new line have already been manufactured and are now ready for installation. A schematic illustration of the new beam line is shown in Fig. 1. The total length of the new line is about 100 m and the number of magnets is approximately 40 including small steering magnets. Three target stations are prepared in the new hall to supply secondary particles for experiments.

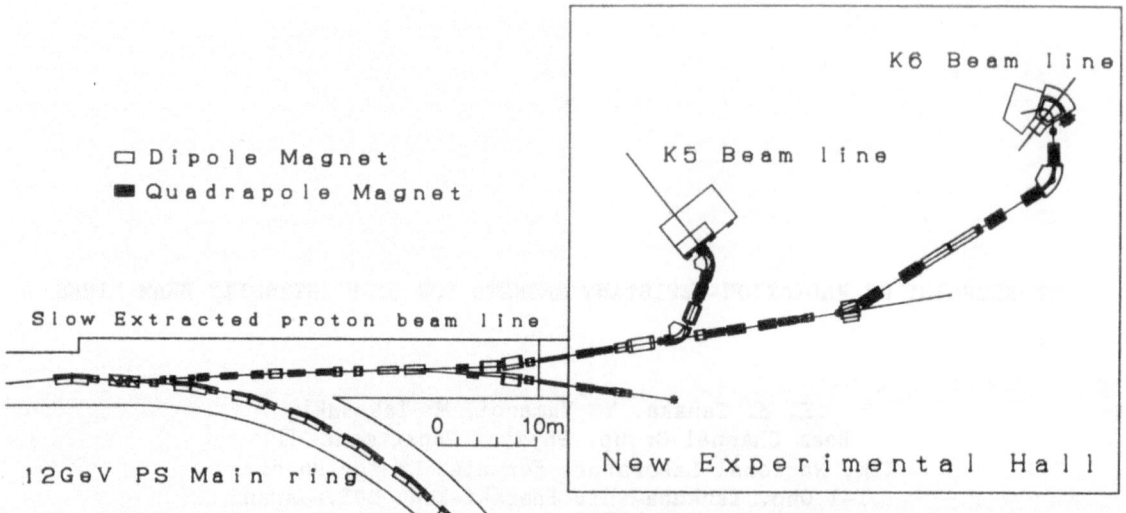

Fig. 1. Schematic illustration of the new counter experimental hall and beam lines of KEK-PS.

The criterion of design of the new line was to realize a ten-year stable operation with the beam up to 1×10^{13} protons per second (pps). This intensity was approximately ten times higher than the present intensity of 1×10^{12} pps, and was expected to be realized in near future by the upgraded KEK-PS in connection with the Japanese Hadron Project[2]. Thus the life time of the beam line magnets should be over the radiation dose of some 10^8Gy, at least a factor of ten more than the conventional ones presently used in the KEK-PS[3]. The magnets placed just downstream of the target station require higher radiation hardness up to 10^{11}Gy. For such a case the magnet must be assembled without any organic materials. Therefore, we developed two new types of radiation-resistant magnets for 10^8 Gy and for 10^{11}Gy.

POLYIMIDE MAGNETS

For magnets for the radiation dose of 10^8Gy, we employed the polyimide resin pre-impregnated (PRP) glass cloth to insulate the magnet coil. The radiation life of PRP is expected to be several ten times longer than conventional epoxy resin pre-impregnated insulators [4]. The radiation life of the PRP glass cloth was tested at the existing external proton beam line of the KEK-PS. The polyimide we used was BT (bismaleimide triazine) resin prepared by Mitubishi Gas Chemical Company, Inc[5]. The test samples were strips of 5mm wide, 50mm long and 0.25mm thick. The 12GeV protons were focused on the samples with $1cm^2$ spot. Approximately 1.5×10^{12} protons were irradiated in every 2.5 seconds and 10^8Gy was achieved in 10 days (one experimental cycle). The tensile strength was measured at the absorbed dose of 10^7, 10^8, 4×10^8 and 10^9 Gy.

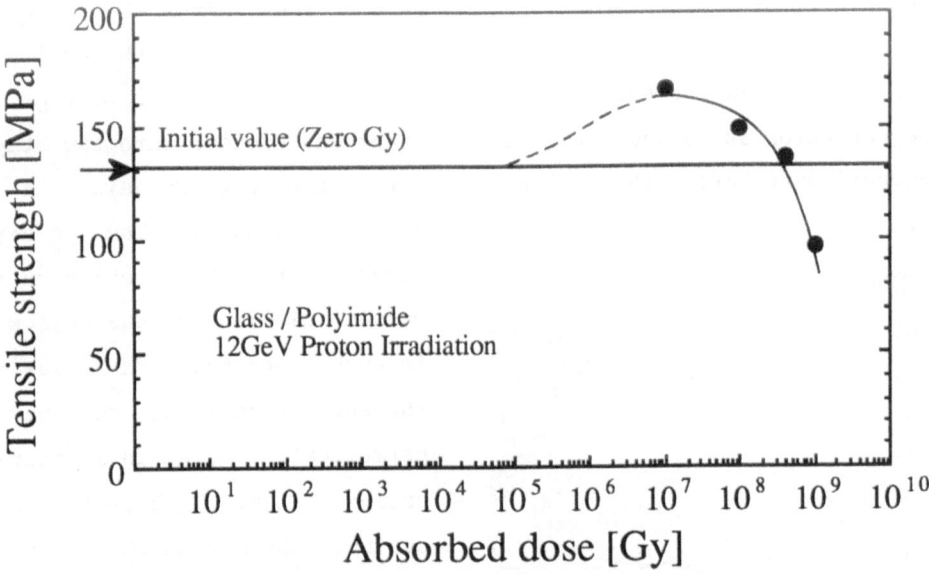

Fig. 2. Tensile strength of cured PRP glass cloth plotted against the absorbed dose by 12GeV proton irradiations.

No serious fall-off of the strength was observed up to 4 x 10⁸Gy. However, at 10⁹Gy, the strength became approximately two thirds of the initial value. The results are shown in Fig. 2 and ensure the radiation life of PRP insulated coil over 10⁸Gy. The break down voltage of the PRP insulator irradiated to 10⁸Gy was also measured by a single-needle-plane test. Using DC voltages at the increase rate of 1kV/s and a needle with 1⁻³mm of radius of curvature, the dielectric strength was found to be more than 400kV/mm. This value corresponds to 4kV/mm of standard AC break down test. This fact ensures also the radiation resistivity of PRP glass cloth over 10⁸Gy. Most magnets of the new line were manufactured with the PRP glass cloth insulator. The glass cloth tape we used was 0.25mm in thickness and was wrapped in a double layer on the conductor so that the insulation thickness between

conductors was 1mm. No other organic material was used in the magnet except for the PRP. Ceramic tubes were used for the electric insulation between the water manifold and the coil. Copper tubes were welded to both ends of the ceramic tube.

CEMENT MAGNETS

For magnets for the radiation dose of 10¹¹Gy, even the PRP must be eliminated. We assembled a small Q magnet (25cm in bore diameter and 40 cm long) from completely inorganic materials. The magnet coil was insulated by high-alumina cement (HAC) and asbestos tapes. First, the hollow conductor was wound into a final Q-magnet coil shape without any insulators. The distance between conductors was temporarily kept by copper spacers 2mm in thickness as shown in Fig. 3. Second, the asbestos

728

tape was wrapped by hand after the
spacers were removed. Inorganic ceramic
bond [6] was used to fix the asbestos
tape on the conductor as needed. Figure
4 shows the coil just before the filling
with cement.

Fig. 3. Coil winding with copper spacers.

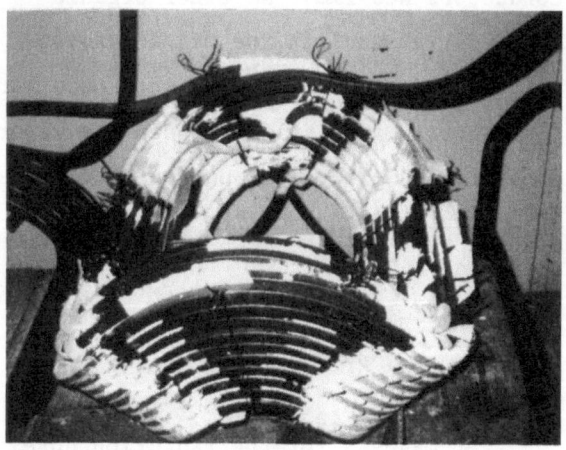

Fig. 4. Coil just before cement filling

The cement we used was mixture of 27% HAC,
55% natural Al_2O_3 and 18% water by weight.
The most crucial problem was how to
eliminate excess water efficiently from
the cement in the curing stage. In order
to get mechanically hard cement with high
electric resistance, some suitable
fraction of water must be eliminated from
the cement surface. The solution we
found was to make a cement bath of wood

planks. The cement was poured into the
wooden box in which the coil was placed,
and left 48 hours for curing. The drying
stage was divided into two stages. The
first stage was two days in a 120 °C oven
and the second was also two days but in a
vacuum tank. The electric resistance
between windings of the coil was measured
in each stage and $10^3 M \Omega$ was achieved at
the end. Finally the coil was
hermetically sealed in a stainless-steel
casing. Figure 5 shows the coil before
casing. The air in the casing was pumped
down for additional two days and replaced
by dry nitrogen as shown in Fig. 6. The
pumping hole was sealed by metallic
gaskets.

Fig. 5. Coil cured by the cement.

Fig. 6. Final pumping down.

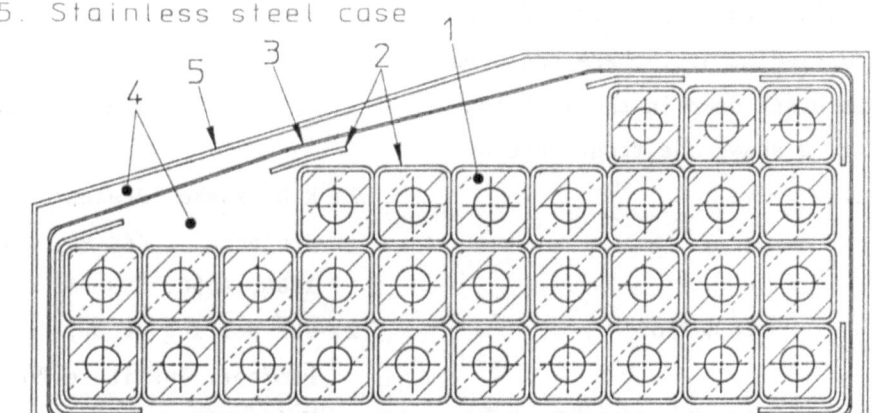

1. Hollow conductor
2. Asbestos tape
3. Stainless steel band
4. Cement 27%HAC
5. Stainless steel case

Fig. 7. Cross section of the cement insulated magnet coil.

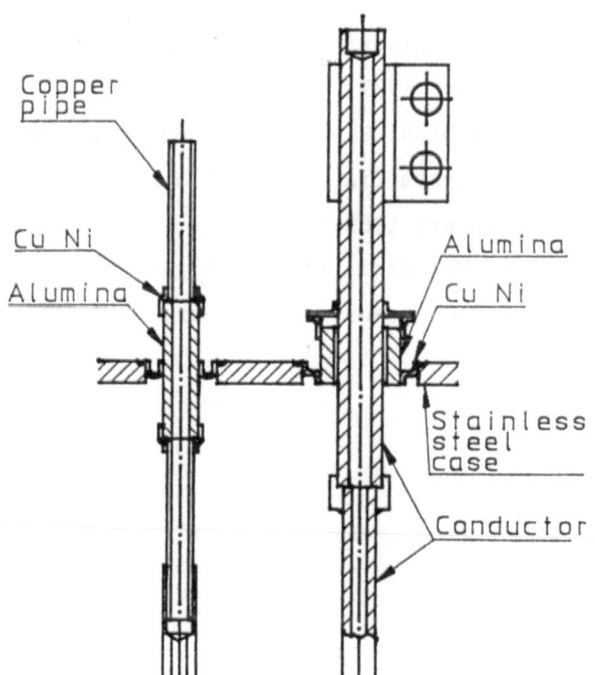

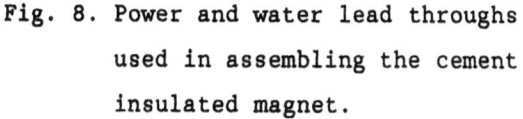

Fig. 8. Power and water lead throughs used in assembling the cement insulated magnet.

Fig. 9. Front surface of cement insulated magnets. The end plate of the magnet is temporarily removed.

The cross section of the coil is shown schematically in Fig. 7. The metalized ceramic tubes indicated in Fig. 8 were used as power and water lead-throughs to the coil. Since the completion of the cement insulated magnet, there has been no electric break down in the last two years. Fig. 9 shows the front face of the magnet.

CONCLUSION

We successfully developed two new types of radiation resistant magnets for 10^8Gy and 10^{11}Gy, respectively. The magnets for the new external proton beam line of the KEK-PS were manufactured with this new technology developed as described above and are now ready for installation. The new beam line will be able to accept the full extracted beam from the upgraded KEK-PS and for a mini kaon factory in the new experimental hall.

The authors would like to express their thanks to Prof. H. Hirabayashi and Prof. K. Nakai for their encouragement throughout the present study. One of

authors (Tanaka) was provided with support of Grant-in Aid for Encouragement of Young Scientists, the Ministry of Education, Science and Culture.

References

1. K.H.Tanaka, Present Status and Future Projects of KEK-PS, Nuclear Physics, 1986, A450, 533c-537c.

2. JHP Working Group, Japanese Hadron Project, Institute for Nuclear Study, University of Tokyo, March 1989.

3. R.L.Keizer and M.Mottier, Radiation-Resistant Magnets, CERN 82-05, June 1982.

4. H. Schoenbacher and A.Stolarz-lzycka, Compilation of Radiation Damage Test Data Part II : Thermosetting and thermoplastic resins, CERN 79-08, August 1979

5. Mitubishi Gas Chemical Company, Inc., BT Resin (in Japanese) 4th Ed., January 1989

6. Nissan Chemical Industries, Ltd., Bond X (in Japanese), February 1984

EFFECT OF COLD WORK AND HEAT TREATMENT ON THE 4°K TENSILE, FATIGUE AND FRACTURE TOUGHNESS PROPERTIES OF INCOLOY 908

M. M. Morra, I. S. Hwang, R. G. Ballinger, M. M. Steeves and M. O. Hoenig

Massachusetts Institute of Technology
Department of Materials Science and Engineering
Plasma Fusion Center
Cambridge, Massachusetts 02139

ABSTRACT

The influence of prior cold work on tensile properties, fatigue crack growth rates and fracture toughness of Incoloy 908 was examined at 293°K and 4.2°K. Fatigue crack growth rates were measured using a constant ΔK technique. Crack length was determined using compliance. Fracture toughness was determined by the J-integral technique. Properties following heat treatment were compared for two starting conditions: (1) solution annealed at 980°C for 1, hour and (2) solution annealed followed by 20% cold work. Three heat treatments were studied: (1) aging at 650°C for 200 hours, (2) aging at 700°C for 100 hours, and (3) aging at 750°C for 50 hours.

The results of this investigation show that the fracture toughness of Incoloy 908 is greater than $230 MPa\sqrt{m}$ at 4.2°K for all heat treatments. The fatigue response at 4.2°K is almost independent of heat treatment and is comparable to stainless steel. The yield strength of the solution annealed and aged material ranged from 823 to 961 MPa at 293°K and from 980 to 1070 MPa at 4.2°K depending on heat treatment. The cold worked and aged material exhibited significantly higher yield strength, ranging from 1240 to 1280MPa and 1320 to 1490MPa at 293°K and 4.2°K, respectively. Ductility is only slightly reduced (3-6% as measured by total elongation (25mm gage)) by cold working.

The results of this investigation show that Incoloy has excellent mechanical properties. These properties, coupled with a low thermal expansion coefficient, make Incoloy 908 an attractive alternative for superconducting magnet structural applications.

INTRODUCTION

High-field superconducting magnets exert large Lorentz forces during operation. These forces lead to stresses that can exceed those allowable for most materials. Advances in magnet design are now pushing the property limits of available structural materials.

One advanced super-conducting magnet design is the cable-in-conduit conductor (CICC) [1,2]. In this design, Nb$_3$Sn superconductor is enclosed in a suitable sheathing material through which liquid helium is pumped for coding. However, when Nb$_3$Sn is used, a wind-and-react fabrication procedure is required to overcome problems associated with the brittle behavior of this material. With this technique the superconductor is formed by a high temperature heat treatment after coil fabrication. This fabrication process results in a close mechanical coupling between wire and sheath. The strain dependence of the superconductor is such that a large differential in thermal expansion between the conduit structural material and Nb$_3$Sn can cause a significant degradation in performance. This is especially true after cooling the magnet from a reaction temperature of 650°C or greater to the 4.2°K operation temperature. In response to this need a new alloy, designated as Incoloy 908, has been developed. Incoloy 908 exhibits high strength along with a coefficient of thermal expansion that is compatible with that of Nb$_3$Sn. The material is being used as part of the US Demonstration Poloidal Coil (US-DPC) program at M.I.T. [3].

The composition of Incoloy 908 is shown in Table 1. The composition has been optimized to insure a low thermal expansion coefficient, good mechanical properties, phase stability, workability and weldability. Strengthening is achieved by γ', [Ni$_3$(Al, Ti)] precipitation.

In this paper we present the initial results of mechanical property evaluation for solution-anneal-then-age and cold-worked-then-aged Incoloy alloy 908. The 20% cold-worked-then-aged condition was chosen due to its being representative of material used in CICC fabrication. The properties reported in this paper are: (1) tensile, (2) fatigue, and (3) fracture toughness (J based).

MATERIALS

The Incoloy 908 material was supplied by Inco Alloys International, Huntington, West Virginia, in the form of 27mm and 16mm plate in the solution annealed and 20% cold worked conditions, respectively. The starting material was then subjected to three aging heat treatments: (1) 650°C for 200 hours, (2) 700°C for 100 hours, and (3) 750°C for 50 hours. Specimens were heat treated in vacuum after machining. The elastic modulus was determined by acoustic techniques and found to be 175GPa at room temperature [4].

EXPERIMENTAL PROCEDURES

Tensile, fatigue and fracture toughness tests were performed using a cryogenic test facility developed at M.I.T. [4]. The test facility consists of a servohydraulic fatigue machine equipped with a specially designed load frame. The loading system,

TABLE 1
Chemical Composition of Incoloy 908 Plates
Used in Mechanical Tests (Weight percent).

Element	SA	CW
Fe	40.6	40.83
Ni	49.7	48.74
Cr	3.83	4.12
Mn	0.04	0.09
S	-	0.001
Si	0.14	0.17
P	-	0.0018
B	-	0.004
Nb	2.99	3.04
Al	1.04	1.10
Ti	1.58	1.54
C	0.01	0.013

with a maximum capacity of 90kN, is designed to be operated using a cryogenic Dewar. Total load-train compliance was estimated to be 2.6×10^{-5}mm/N. The load-cell used was calibrated to an accuracy of 0.25% of full scale (90kN). Specimen strain or crack opening displacement were measured using extensometers designed for use at temperatures as low as 4°K. The extensometers were calibrated on a weekly-basis against a dial gauge with a precision of 2.5 micrometers.

Tensile test specimens were fabricated by wire electrodischarge machining (EDM). Tests specimens were fabricated in accordance with ASTM E8 (latest revision). The specimens were oriented with the loading axis parallel to the plate rolling direction. All tests were strain-controlled at 2×10^{-4}sec^{-1} using an extensometer with 12.7mm gauge-length.

Fatigue crack growth rate test specimens were cut by wire EDM. The 12.7mm thick compact tension specimen design was in accordance with ASTM E647 (latest revision). Specimens were oriented with the loading axis parallel to the rolling direction and the crack path in the transverse (LT) direction. Crack length was determined using an unloading compliance technique with a crack opening displacement (COD) extensometer mounted at the front surface of the specimen. Constant ΔK tests were performed using computer control. Testing was performed with a 10Hz sine wave and stress ratio (R) of 0.1.

The J-integral specimen was of a 25.4mm thick compact tension design in accordance with ASTM E813 (latest revision). Specimens were precracked at room temperature with a final ΔK of 33MPa$\sqrt{m}$. Initial crack length ranged between 0.65-0.70a/W. A COD extensometer mounted at load-line was used for measurement of both the load-line displacement and unloading compliance. Crack length was determined from the unloading compliance after the method of Saxena and Hudak [6]. Specimen rotation and moving crack effects were taken into account in the crack length determination. Tests were run in COD-control with a typical rate of 0.1mm/min. All testing was in accordance with ASTM E813-87. Data analysis was in accordance with ASTM E1152-87.

RESULTS AND DISCUSSION

Tensile Properties

Tensile properties of 20% cold-worked-then-aged (CW) and solution-annealed-then-aged (SA) material are compared in Tables 2 and 3. Prior cold-work results in a 30%-60% increase in yield strength with only moderate increase in the ultimate tensile strength at room temperature. At 4.2°K, the effect of cold-work on the yield strength and the ultimate

TABLE 2
Comparison Between the 293°K Tensile Properties
for Solution-Annealed (SA)-Then-Aged Material
and Cold-Worked (CW)-Then-Aged Material.

Heat Treatment		σ_y (MPa)	σ_{uts} (MPa)	ε %
CW	650°C/200 h	1279	1499	11
SA	650°C/200 h	961	1354	13
CW	700°C/100 h	1241	1451	13
SA	700°C/100 h	970	1361	14
CW	750°C/50 h	1248	1413	16
SA	750°C/50 h	823	1219	13

TABLE 3
Comparison Between the 4°K Tensile Properties
for Solution-Annealed (SA)-Then-Aged Material
and Cold-Worked (CW)-Then-Aged Material.

Heat Treatment		σ_y (MPa)	σ_{uts} (MPa)	ε %
CW	650°C/200 h	1489	1906	17
SA	650°C/200 h	1070	1780	20
CW	700°C/100 h	1434	1882	17
SA	700°C/100 h	1150	1770	23
CW	750°C/50 h	1320	1800	20
SA	750°C/50 h	980	1610	24

tensile strength is similar to that at room temperature. Cold work results in a slight decrease in total elongation.

The weak temperature dependence of the tensile properties is in contrast to that for stainless steels [7,8]. For these materials a factor of three difference is often observed. This weak temperature dependence is a significant advantage for applications where temperature transients are expected.

Fatigue Crack Growth Rate
Fatigue crack growth rates were determined for the 20% cold worked then aged condition both at room temperature and 4.2°K. Results are shown in Figures 1-3.

All three heat treatment conditions exhibit similar crack growth rate, within a factor of two. The effect of temperature is appreciable for each heat treatment. About a factor of three decrease in the crack growth rate is observed when the temperature is decreased from room temperature to 4.2°K.

Incoloy 908 shows very similar crack growth rate behavior at 4.2°K to that of stainless steel Type 316 [9]. However the temperature dependence of fatigue behavior makes the slope of da/dN vs. ΔK curve much lower than that of stainless steel. As with the tensile properties the weak temperature dependence of crack growth rate of Incoloy 908 can be attributed to precipitation hardening as the primary strengthening mechanism. The results of fits to a Paris law relationship are summarized in Table 4.

Fracture Toughness
J-integral fracture toughness data were obtained for the solution annealed and aged conditions only. Results are summarized in Table 5. Typical J-resistance curves are shown in Figures 4 and 5.

TABLE 4
Paris Law Constants of 20% Cold Worked Incoloy
908 at Room Temperature and 4.2°K

$$da/dN = c(\Delta K)^m, R=1$$

	c (10^{-11}mm/cycle)		m	
	R.T.	4.2°K	R.T.	4.2°K
650°C 200h	2.0814	7.0346	4.5812	4.0618
700°C 100h	297.13	199.45	3.1889	3.0398
750°C 50h	348.73	771.95	3.2518	3.4473

TABLE 5
Comparison of J-Integral Fracture Toughness
at 4.2 K.

Alloy	σ_y (MPa)	K_{IC} (MPa√m)		
		E813-87	E813-81	Ref.
316LN	894	200±0.5	183±1	191±1*
908 650°C 200 h	1070	235±5	232±14	
908 750°C 50 h	980	240±4	235±5	
JBK-75	1398			108‡
High Cr-Ni SS	1336	224±11	213±5	213±11

*[7,8] ‡[10]

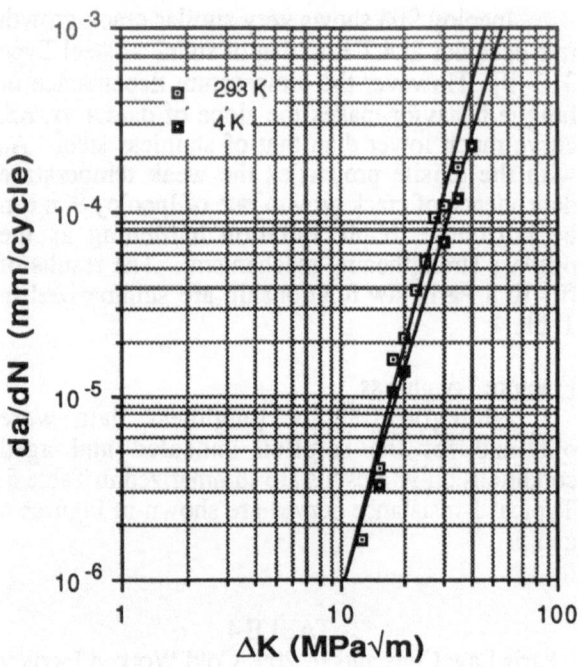

Figure 1. Fatigue crack growth rate of Incoloy 908, 20% cold worked 650°C for 200 hours condition.

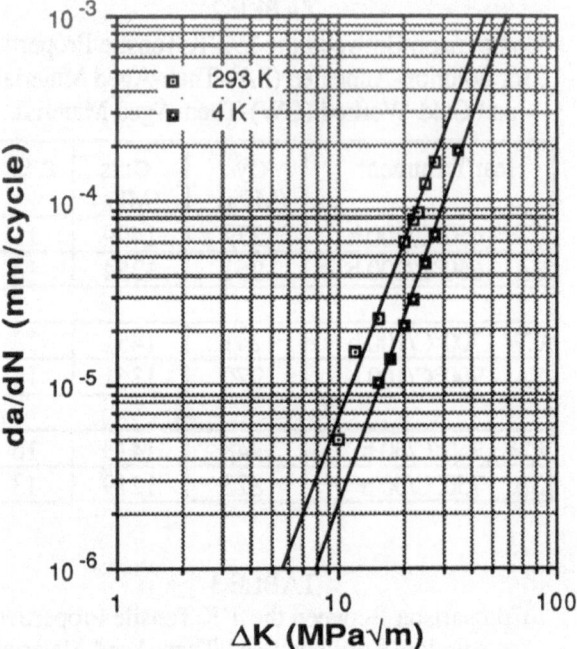

Figure 3. Fatigue crack growth rate of Incoloy 908, 20% cold worked 750°C for 50 hours condition.

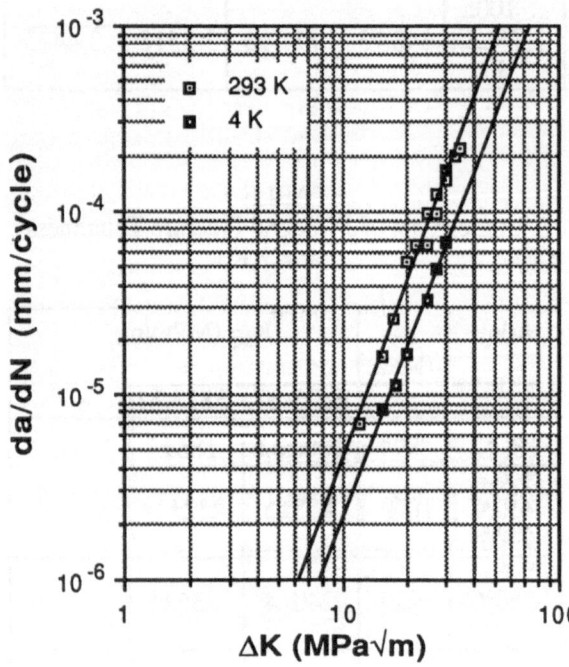

Figure 2. Fatigue crack growth rate of Incoloy 908, 20% cold worked 700°C for 100 hours condition.

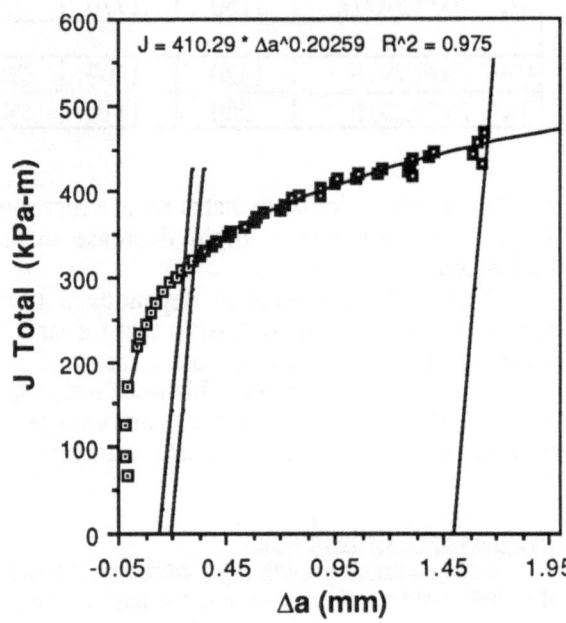

Figure 4. J-resistance curve of Incoloy 908 at 4°K, solution annealed then aged at 650°C for 200 hours.

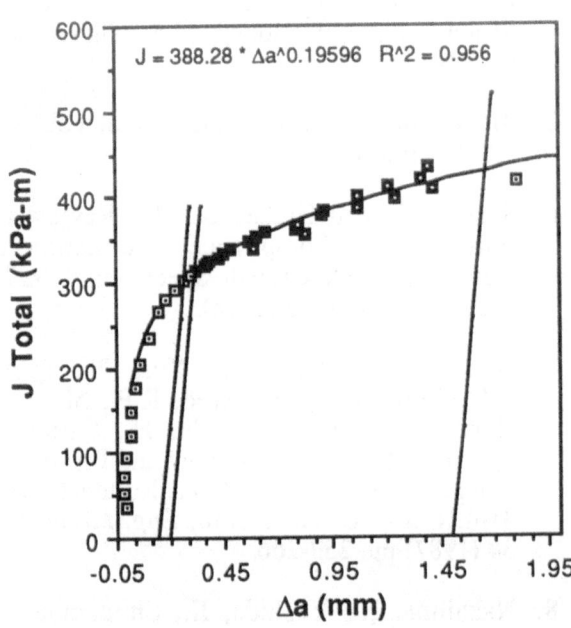

$J = 388.28 \cdot \Delta a^{0.19596}$ R^2 = 0.956

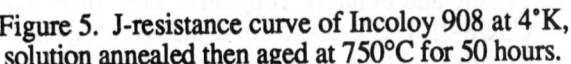

Figure 5. J-resistance curve of Incoloy 908 at 4°K, solution annealed then aged at 750°C for 50 hours.

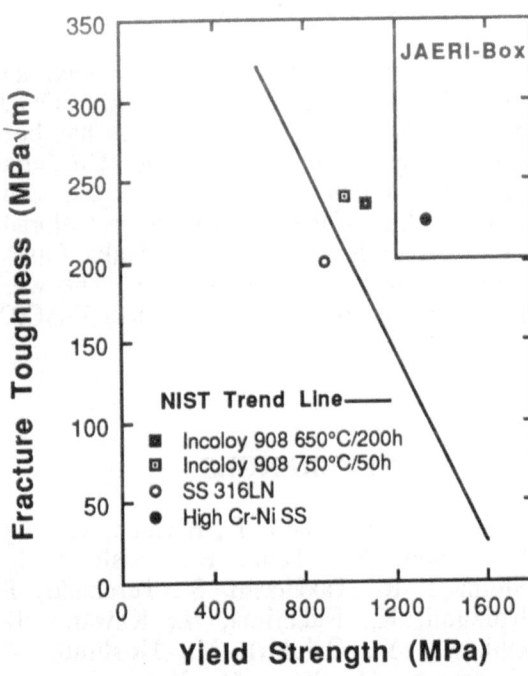

Figure 6. Comparison of 4°K Mechanical Properties of Some Cryogenic Structural Material

Due to the complexity of the J-integral technique especially at 4.2°K, our test system including software was verified using materials with known fracture toughness values. Type 316LN stainless steel and a high Cr-Ni stainless steel were obtained from the U.S. National Institute of Standards and Technology. Two specimens of each materials were made as a part of round-robin test programs between United States and Japan [7,8] Results of our tests on these materials and round-robin data averages are compared in Table 5. For both materials excellent agreement was observed. The difference is considered to be small enough compared with large uncertainty in most of round-robin data. Excellent data reproducibility was also observed

Also shown in Table 5 is the fracture toughness of other well-known cryogenic materials. Incoloy 908 in the solution annealed then aged condition shows the highest fracture toughness at 4.2°K among these alloys. Performance of these alloys are further compared on a yield strength-fracture toughness diagram, as shown in Figure 6. While the high Cr-Ni stainless steel meets the Japanese target characterized by the JAERI box, solution annealed then aged Incoloy 908 falls outside the target region due to a lower yield strength. However, the yield strength of the 20% cold worked material is squarely in this region. Fracture toughness testing of the 20% cold worked material is in progress, but the relatively small decrease in

ductility suggests that toughness will also not be significantly reduced.

CONCLUSIONS

1. Low coefficient of expansion alloy, Incoloy 908, has been shown to exhibit adequate yield strength and high fracture toughness based on initial results at 4.2°K. The properties combined with its thermal expansion coefficient compatible with Nb_3Sn, makes it an attractive material for CICC magnets.

2. By 20 percent cold work prior to ageing Incoloy 908 is shown to achieve the target yield strength for high field superconducting magnet materials defined by 1200 MPa yield strength. Fracture toughness of cold worked then aged material is expected, based on tensile data, not to decrease significantly from 235 MPa√m obtained with solution annealed then aged material at 4.2 K.

3. Tensile and fatigue crack growth rate behavior shows only a weak temperature dependence compared with those of austenitic stainless steel.

ACKNOWLEDGEMENTS

Authors would like to thank Scott Nicol and Salim Mathew for experimental assistance. Our J-integral test system development work has been benefited significantly from the US-Japan collaboration on round robin program organized by Dr. Ralph Tobler of the NIST, Boulder, Colorado and Mr. Hideo Nakajima of the Naka Fusion Research Establishment, JAERI, Japan. This work has been supported by US DOE Contract DE-AC02-78ET-51013.

REFERENCES

1. Shimamoto, S., Ando, T., Hiyama, T, Tsuji, H., Nishi, M., Tada, E., Yoshida, K., Shimada, R., Takahashi, S., Terakado, T., Koizumi, K., Nakajima, H., Kawano, K., Ohkawa, Y., Oshikiri, M., Hoshino, M., Yamamura, H., Satou, M., Yaguchi, E. and Ohgane, Y., "Design and Fabrication Status of the Demo. Poloidal Coils," Proc. 12th Fusion Eng. Symp., IEEE (1988) pp.7-10.

2. Henning, C., Editor,"Magnet Design Technical Report-ITER Definition Phase," Lawrence Livermore National Laboratory, Livermore, CA, U.S.A. (1989)

3. Steeves, M.M., Painter, T.A., Tracey, J.E., Hoenig, M.O., Takayasu, M., Randall, R.N., Morra, M.M., Hwang, I.S. and Marti, P., "The Further Progress in the Manufacture of the US-DPC Test Coil," 11th Int'l Conf. Mag. Tech., Tsukuba, Japan (1989) paper NC-04.

4. Morra, M.M.,"Incoloy 908-New High Strength Low Thermal Expansion Alloy for Cryogenic Structural Applications," M.S. Thesis, MIT (1989).

5. Annual Book of ASTM Standards, vol. 3.01, ASTM (1987).

6. Saxena, A. and Hudak, S.J. Jr.,"Review and Extension of Compliance Information for Common Crack Growth Specimens," Int'l J. Frac., vol. 14, no. 5 (1978).

7. Ogata, T., Ishikawa, K., Yuri, T., Tobler, R.L., Purtscher, P.T., Reed, R.P., Shoji, T., Nakaro, K. and Takahashi, H., "Effect of Specimen Size, Side-grooving and Precracking Temperature on J-integral Test Results for AISI 316LN at 4°K," Adv. Cryo. Eng. Mater., vol. 34 (1987) pp. 259-266.

8. Nakajima, H., Yoshida, K., Shimamoto, S., Tobler, R.L. and Reed, R.P.,"Round Robin Tensile and Fracture Toughness Test Results for CSUS-JN1 (Fe-25Cr-15Ni-0.35N) Austenitic Stainless Steel at 4°K," Int'l Cryo. Mater. Conf., Los Angeles, U.S.A. (1989) paper BZ-03.

9. Tobler, R.L.,"Near-Threshold Fatigue Crack Growth Behavior of AISI 316 Stainless Steel," Adv. Cryo. Eng. Mater., vol. 32 (1986) pp. 321-327.

10 Summers, L.T. and Dalder, E.N.C.,"An Investigation of the Cryogenic Mechanical Properties of Low Thermal Expansion Superalloys," Adv. Cry. Eng. Mater., vol. 32 (1986) pp. 73-80

EFFECT OF STIFFNESS ON SERRATED DEFORMATION AT VERY LOW TEMPERATURES UNDER CONSTANT LOADING RATE CONDITIONS AND ITS COMPUTER SIMULATION

K.Shibata and K.Fujita,

Department of Metallurgy and Materials Science,
The Universuty of Tokyo,
Hongo 7-3-1, Bunkyo-ku, Tokyo 113, Japan

SYNOPSIS : Serrated deformation behavior at very low temperatures such as 4K has been studied by many researchers. Details of this discontinuous unstable deformation, however, have remained unclarified, especially the deformation behavior under constant loading rate conditions which has been examined by only a few researchers. The objective of the present paper is to investigate the deformation under constant loading rate conditions, especially the effects of stiffness, that is, back-stress due to the constraint of surrounding structures on the rapid discontinuous deformation. The onset stress of the discontinuous deformation was lower under constant loading rate conditions than under constant cross-head velocity conditions. It was shown by computer simulation that the amount of rapid deformation under constant loading rate conditions could be effectively decreased and the onset of the first discontinuous deformation delayed by small back-stress.

INTRODUCTION

Almost all metallic materials exhibit discontinuous serrated deformation behavior in the load-elongation curve under constant cross-head velocity conditions at temperatures below 30K. As the load drops during the serration, rapid unstable deformation occurs in a local region in the specimen, together with a noticeable rise in temperature. Therefore, in tensile tests at these low temperatures conventional testing methods are not available and the new standard for testing is necessary. On the other hand, such unstable characteristics of the deformation and the temperature rise due to the discontinuous deformation cannot be neglected for sound and safe operation of machinery used at very low temperatures.

The serrated deformation at very low temperatures is influenced by many factors. These factors can be classified into two groups, one concerned with material properties and the other with the testing conditions. As for the former, effects of following factors have been examined: strength level[1], temperature dependence on strength[2], degree of work hardening[3], thermal conductivity [2] and specific heat[4]. As for the latter, following factors have been studied: strain rate[3,5,6], specimen size[3,4], specimen geometry [3], cooling method (in liquid He or gaseous He) [6,7], machine stiffness [3,8,9] and loading method[10]. The last factor means that deformation behavior under constant loading rate (CLR) condition is different from that

under constant cross-head velocity(CCV) condition. AS for as deformation under CLR condition is concerned, only little work[10,11] has been performed.

The present paper investigates the effect of machine stiffness on deformation behavior at very low temperatures, especially under CLR condition, using computer simulation.

EXPERIMENTAL PROCEDURES

Specimen

A vacuum melted Fe-42%Ni steel was supplied by Kawasaki Steel Corporation and subjected to investigation. This steel is convenient, because no martensite is induced by plastic deformation and its thermal properties at very low temperatures have been examined relatively well. The chemical composition in wt % is 41.7Ni, 0.001C, 0.0007N and 0.052Al. Hot rolled plate was solution treated at 1373K for 1 h followed by water quenching. Blanks were cut from the solution treated plate and machined to round tension test specimens of 5mm

diameter with 10 mm gage length. The diameter of the screwed end section was 12mm.

Tensile Tests

Tensile tests were carried out with a closed loop electro-hydraulically actuated machine with a cryostat. Variation of the length of the fillet and reduced section with deformation was measured by using a clip-on gage mounted on the specimen and a differential transducer mounted on the actuator.

Computer Simulation

Detail of the procedures of the computer simulation have been described in a previous paper[4,11] for deformation under CCV conditions.

In the practical case, it is conceivable that machinery can be elastically back-stressed due to the constraint of the surrounding structures. In the present work, simulation was performed for this case as shown schematically in FIGURE 1. The mechanical model is shown in FIGURE 2, the specimen being loaded at a constantly increasing rate but subjected to the back force (Pb), which was assumed to be linearly proportional to total elongation of the specimen. A symbol E' represents a proportional constant as shown below.

$$Pb=E'(\Delta L(e)+\Delta L(p))So/Lo$$

Where, $\Delta L(e)$: elastic elongation of specimen

$\Delta L(p)$: plastic elongation of

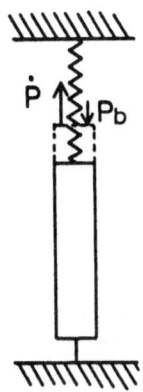

FIGURE 1. Schematic drawing of deformation of machinery under constant loading rate conditions with back-stress.

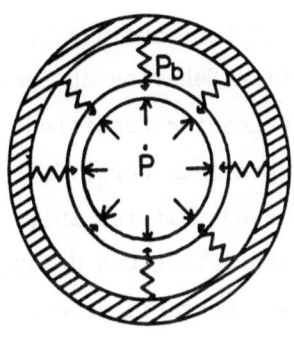

FIGURE 2. Mechanical model of deformation under constant loading rate conditions with back-stress.

specimen (summation of plastic elongation of elements)

So, S(R)o: initial sectional area of specimen and pull-rod

Lo, L(R)o: initial length of specimen and pull-rod

RESULTS AND DISCUSSION

Initiation of Discontinuous Deformation under Constant Crosshead Velocity and Constant Loading Rate Conditions

Engineering stress-strain curves obtained by actual tensile tests are shown in FIGURE 3 under (a)CCV and (b) CLR conditions. Strain in FIGURE 3 was measured using a clip-on gage. Under both conditions, initial elastic strain rate, that is, the loading rate was controled to be the same. Under CLR conditions, rapid deformation occurs intermittently to produce a stair-like stress-strain curve. Comparing (a) and (b) of

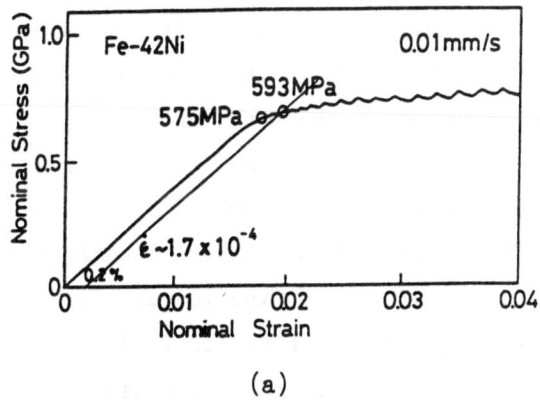

(a)

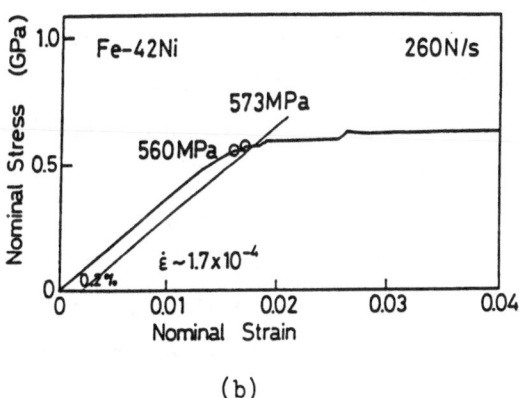

(b)

FIGURE 3. Stress-strain curves obtained by actual tensile tests under constant cross-head velocity (a) and under constant loading rate (b) conditions. Strain was measured using a clip-on gage.

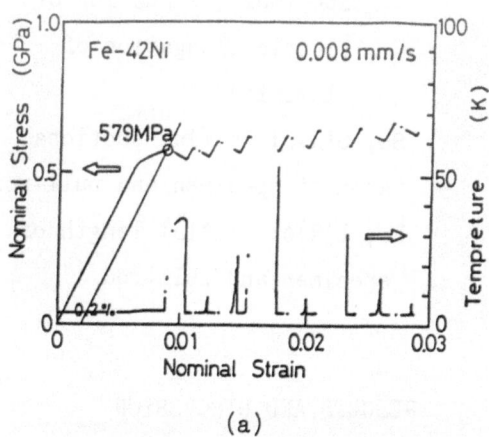 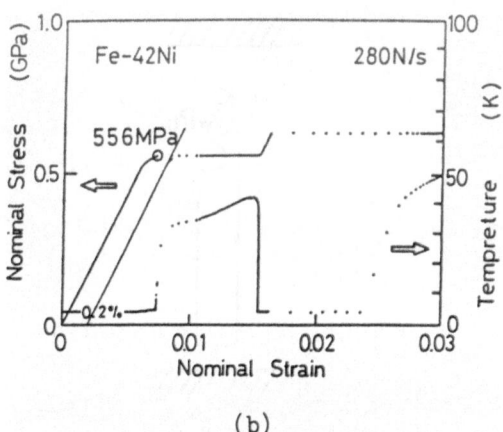

(a)　　　　　　　　　　　　　　(b)

FIGURE 4. Stress- and temperature-strain curves obtained by calculation under constant cross-head velocity (a) and under constant loading rate (b) conditions.

FIGURE 3, the first rapid discontinuous
deformation initiates at lower stress
levels under CLR conditions.　This
result is of importance from a practical
viewpoint.

　　Calculated results are shown in
FIGURE 4: (a) for CCV and (b) for CLR
conditions.　The initiation of rapid
discontinuous deformation at lower
stress is reproduced under CLR condition
as in the actual tensile tests.　In the
case of FIGURE 4, the specimen was deformed by 0.3% only in 0.5ms from the
initiation of the rapid deformation.
Therefore, a mean strain rate during
this interval amounts to an appreciably
high value of around 6 s^{-1}.

　　It was observed in actual tests and
also by computer simulations that
neither serrated deformation under CCV
condition nor stair-like deformation
under CLR condition took place in liquid
nitrogen.

Effects of Surroundings Stiffness on
Deformation Behavior of Specimen under
Constant Loading Condition

　　In previous work[10,11], it
was observed that the initiation of
rapid deformation was delayed and the
amount of the rapid deformation increased with a decrease in the loading rate.
Furthermore, the amount of the rapid
deformation decreased with an increase
in workhardening[11].　The present work
investigated the effects of back stress
on deformation under CLR condition using
computer simulation.　The procedures
and assumption in the calculation were
mentioned in the previous section.

　　Stress-strain and temperature-
strain curves are shown in FIGURE 5
in the case of a constant loading rate
of 550 N/s.　Curves (a) are for the
case without back stress while curves
(b) and (c) are under conditions of
back-stress.　It was assumed that
a proportional constant for (b) and (c)
was one twentieth and one tenth of

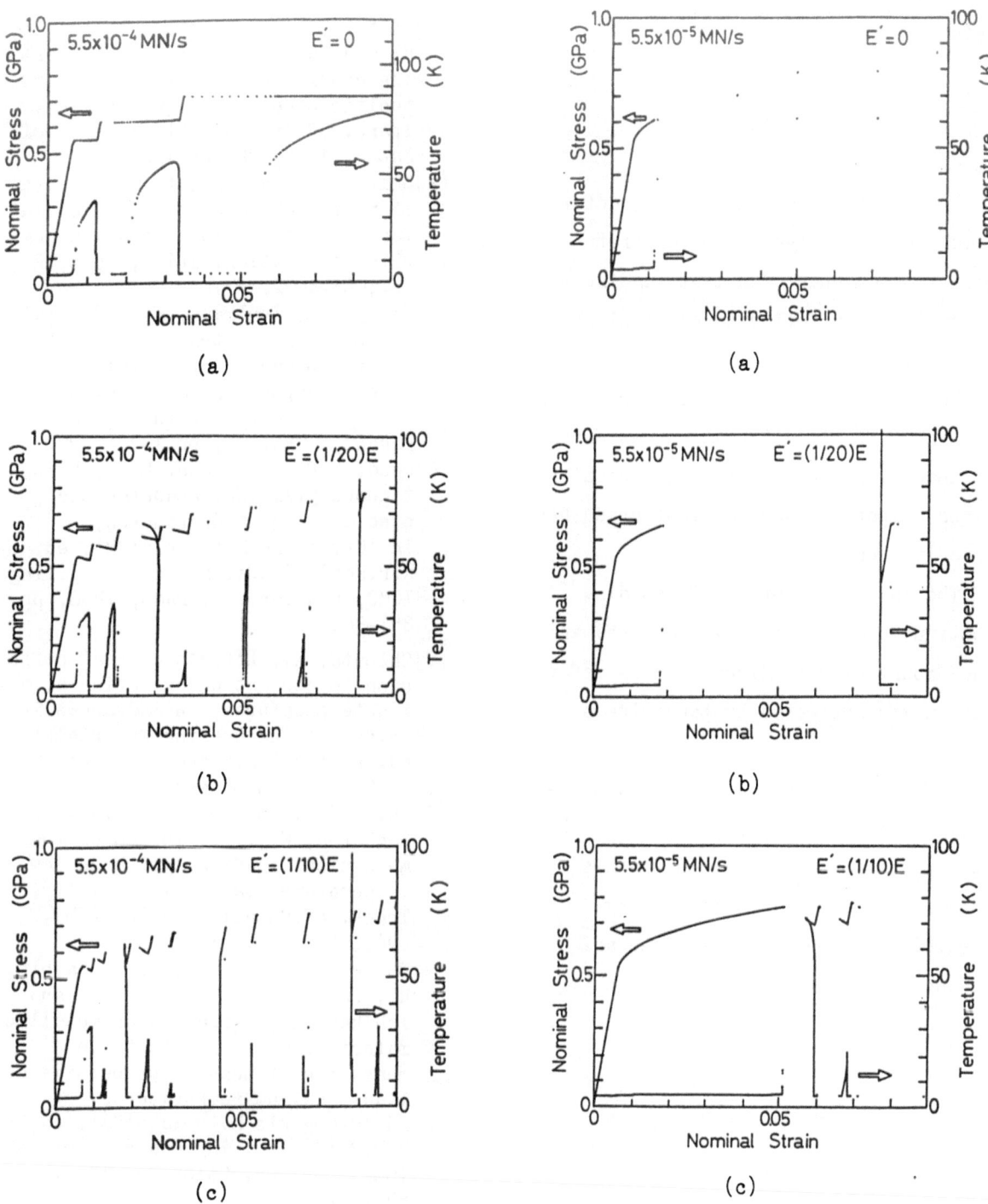

FIGURE 5. Stress- and temperature-
strain curves obtained under constant
loading rate (550 N/s) condition with
back stress.

FIGURE 6. Stress- and temperature-
strain curves obtained under constant
loading rate (55 N/s) condition with
back stress.

Young's modulus for the specimen 142GPa
respectively.   As the back-stress increases, the initiation of the first
discontinuous deformation was delayed

and the amount of each rapid deformation
decreased.   Calculated results at the
loading rate of 55 N/s are exhibited in
FIGURE 6.   The effects of back stress

are similar to a case of the loading rate of 550 N/s.

CONCLUSIONS

Using computer simulation, deformation behavior at very low temperatures was examined under constant loading rate conditions and following conclusions were obtained.

(1)The initiation of the first discontinuous rapid deformation was enhanced under constant loading rate conditions compared under constant cross-head velocity conditions.

(2)The initiation of the first discontinuous deformation was delayed and the amount of each discontinuous deformation was decreased by back stress.

The authors are grateful to Cryogenic Center of The University of Tokyo for experimantal assistance. This Research financially supported by Special Coordination of Science and Technology Agency of the Japanese Government.

REFERENCES

[1]Shibata, K., Martensitic transformation and serration of Fe-Ni binary alloys at 4.2K, Proc. of International Conference on Martensitic Transformations, The Japan Institute of Metals, Sendai, 1986, pp.509-514.

[2]Shibata, K., Discussion about Basinski's model for serration at very low temperatures with computer, in 'Cryogenic Materials'88', ed. by R.P.Reed, Z.S.Xing and E.W.Collings, ICMC, Boulder, Colorado, 1988, pp. 865-872.

[3]Shibata, K., Sakamoto, H., Fujita, K. and Fujita, T., Effect of testing conditions on serration of austenitic steels in liquid helium, Trans. of Iron and Steel Inst. of Japan, 1988, 28, pp.136-142.

[4]Shibata, K. and Fujita, T., Serration of Fe-Ni austenitic Steels at very low temperatures and its computer simulation, ibid., 1986, 26, pp.1065-1072.

[5]Ogata, T., Ishikawa, K. and Nagai, K., Effects of strain rate on the tensile behavior of stainless steels, copper and an aluminum alloy at cryogenic temperatures, Tetsu-to-Hagane, 1985, 71, pp.1390-1397.

[6]Reed, R.P. and Simon, N.J., Discontinuous yielding in austenitic steels at low temperatures, in 'Cryogenic Materials'88', ed. by R.P.Reed, Z.S.Xing and E.W.Collings, ICMC, Boulder, Colorado, 1988, pp. 851-863.

[7]Shibata, K., Effects of heat diffusion to coolant on serration at very low temperatures, in 'Advances in Cryogenic Engineering Materials', ed. by A.F.Clark and R.P. Reed, 1990, 36, (in press).

[8]Chin, G.Y., Hosford, W.F.,Jr. and Backofen, W.A., Influence of the mechanical loading system on low-temperature plastic instability, Trans. of Metall. Soc. of AIME, 1964, 230, pp.1043-1048.

[9]Shibata, K., Fujita, K., Sakamoto, H., Fukushima, E., Nagai, K. and Ishikawa K., Serration of metallic materials and its effects on measuring of tensile properties at liquid helium temperature, in 'Cryogenic Materials'88', ed. by R.P.Reed, Z.S.Xing and E.W.Collings, ICMC, Boulder, Colorado, 1988, pp. 873-882.

[10]Ogata, T., Ishikawa, K., Reed, R.P. and Walsh, R.P., Loading rate effects on discontinuous deformation in load-control tensile tests, in 'Advances in Cryogenic Engineering Materials', ed. by Clark, A.F. and Reed, R.P., 1988, 34, 233-240.

[11]Shibata, K., Computer simulation of deformation under load controling conditions, in 'Cryogenic Materials' 88', ed. by R.P.Reed, Z.S.Xing and E.W.Collings, ICMC, Boulder, Colorado, 1988, pp.883-892.

MECHANICAL PROPERTIES OF SUS304 STAINLESS STEEL UNDER COLD THERMAL CYCLES

Yoshihiko Mukai and Arata Nishinura
Faculty of Engineering, Osaka University
2-1 Yamada Oka, Suita, Osaka, Japan 565

ABSTRACT

A cold thermal cyclic test between a room and a cryogenic temperature under a thermal deformation restraint was conducted on an austenitic stainless steel (SUS304) which is popularly used for the cryogenic equipment or structures. SUS304 stainless steel was yielded by reducing the temperature to about 50K, and reverse yielded by the warming process. These hysteresis loops in the stress-strain curves obtained under the cold thermal cycles suggest that SUS304 stainless steel has the potential to cause cold thermal fatigue failure. Therefore thermal fatigue must be considered in the design of cryogenic structures especially in the case where a stress concentration exists and uniform cooling is applied.

INTRODUCTION

As the needs for cryogenic environment and the development of cryogenic equipment or structures have increased and advanced many investigations of mechanical behavior under constant temperatures and the development of new materials have been carried out [1][2]. A study of material properties under cold thermal cycling, however, has not been reported although the materials used in such structures are exposed to cold thermal cycles. Therefore, it is possible that the cold thermal cycles have not been taken into account or have not been concerned in previous designs of cryogenic equipment.

In this study SUS304 stainless steel a very popular material for cryogenic structures was used, and its cold thermal cyclic properties under deformation restraint investigated. Also the potential for cold thermal fatigue is discussed.

MATERIALS AND METHODS

A round bar of SUS304 with a chemical composition (wt%) of 0.05C, 0.25Si, 1.41Mn 0.032P, 0.026S, 8.06Ni and 18.32Cr, was heat-treated at 1323K, 3.6ks, followed by a water quench. Its initial diameter was 20mm, and its mechanical properties are shown in TABLE 1.

TABLE 1 Mechanical properties of SUS304

Temp. (K)	E(GPa)	$\sigma_{0.2}$(MPa)	σ_u(MPa)
293	188.6	277.6	715.8
77	197.0	388.8	1655.8
4.2	187.3	491.6	1869.7

An example of the stress-strain curve, which was measured by a strain gage, is shown in FIG.1. This data was obtained by a repetition of loading and unloading, and the test temperature was changed from 4.2K to 293K and down to 4.2K again successively.

As the non-linearity was caused by

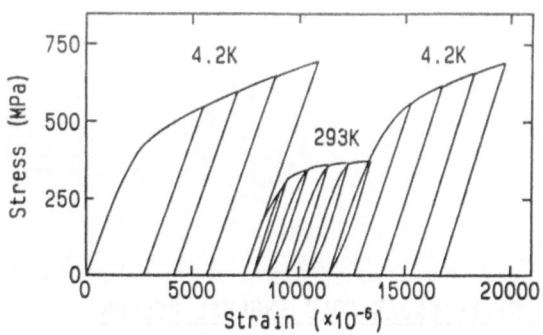

FIGURE 1 Stress-strain curves measured
successively at each temperature

the unloading process at 293K, a hysteresis occurred in the reloading process. This hysteresis curve was observed also at 77K but not at 4.2K. The range of cyclic plastic strain became larger, as the strain at unloading became large.

Also when the tensile test was performed at 4.2K following the 293K test, the propotional limit did not return to the unloaded stress level of the original test at 4.2K, as shown in FIG.1.

These two behaviors are supposed to be due to the occurrence of the structural transformation from an austenite to a martensite and its reverse transformation.

The thermal expansion of the material tested, the driving force causing the thermal yielding, was measured and the

result is shown in FIG.2, together with reference data [3]. The yielding is caused when the material is cooled down to cryogenic temperature under a thermal deformation restraint.

An experimental apparatus for cold thermal cycle testing is shown in FIG.3. The specimen was fixed to the restraint plates with FRP nuts to reduce heat transmission, and belt heaters were attached to both restraint plates to keep their temperatures constant. A feed was attached to the pipe specimen as shown in FIG.4, and cold He gas or warm N_2 gas was flowed into the specimen during a cooling or a warming process. The restraint jig with the specimen was placed in a vacuum chamber to prevent heat convection.

The thermal stress measurement was carried out by strain gages attached to 4 posts, which were connected in bridge circuit to cancel the bending strains and to self-compensate for small temperature changes.

The axial strain (ε_z) and the tangential strain (ε_θ) was measured with a bi-axial strain gage (G.L.= 1mm) mounted on the parallel part of the specimen. The mechanical strain due to the thermal shrinkage under the deformation restraint was evaluated by adding up the apparent strains obtained under the free shrinkage and the deformation restraint.

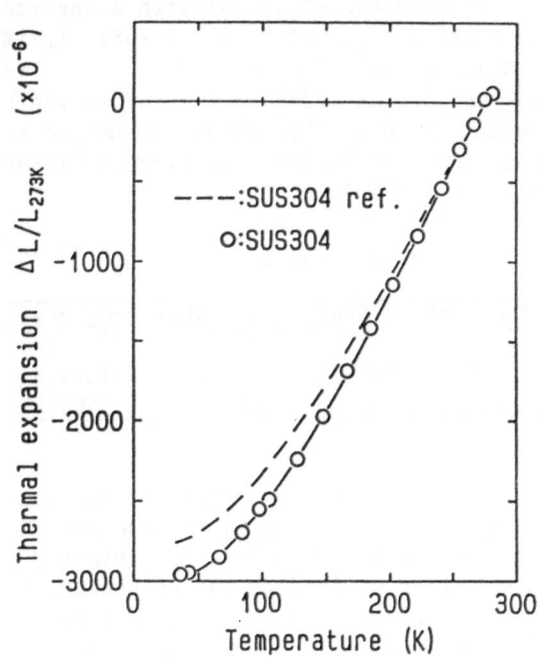

FIGURE 2 Thermal expansion of SUS304

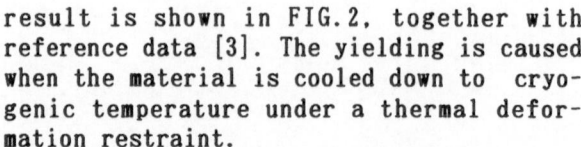
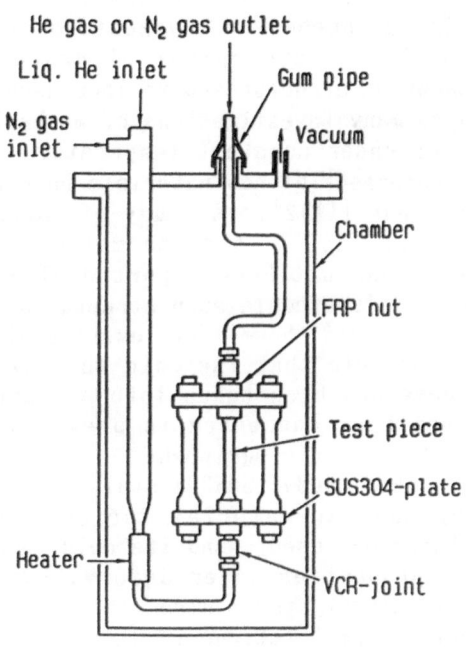

FIGURE 3 Apparatus for cold thermal
cycle tests

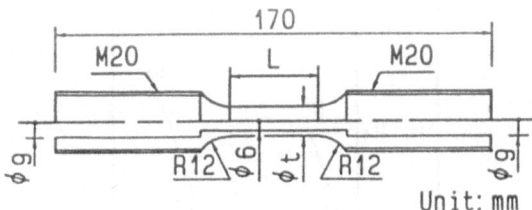

FIGURE 4 Specimen configuration

Temperatures at a parallel part of the specimen, restraint plates and posts were measured by Au·0.07Fe-chromel thermocouples and recorded continuously.

On the mounting the specimen on the restraint jig with FRP nuts, an initial stress below 100MPa was applied and the cold thermal cyclic tests performed up to 10 cycles. The parameters L and t, shown in FIG.4, are given in TABLE 2 for each specimen, together with K_e, the thermal stress concentration factor to be described later.

TABLE 2 Summary of L and t sizes

T.P.code	L(mm)	t(mm)	K_e
SUS I	30.0	9.5	1.89
SUS II	20.0	8.5	2.87

RESULTS

The variatin of the thermal stress, temperature and ε_θ against ε_z of specimen SUS I is shown in FIG.5.

In the first cycle, both the thermal stress and the strain increase and obvious yielding appears over the proportional limit as the temperature was reduced. Poisson's ratio is about 0.3 in this elastic region.

On the warming to room temperature, non-linear behavior was observed after an elastic deformation, and a compressive stress and a tensile strain remained at room temperature. This stress-strain condition is present because of the smaller cross section of the parallel part in comparison with that of the threaded part of the specimen.

After the second cycle, a visible change in the stress-strain curve was no longer observed. The reason for this is as follows: The driving force to produce the mechanical strain under the thermal deformation restraint is the thermal

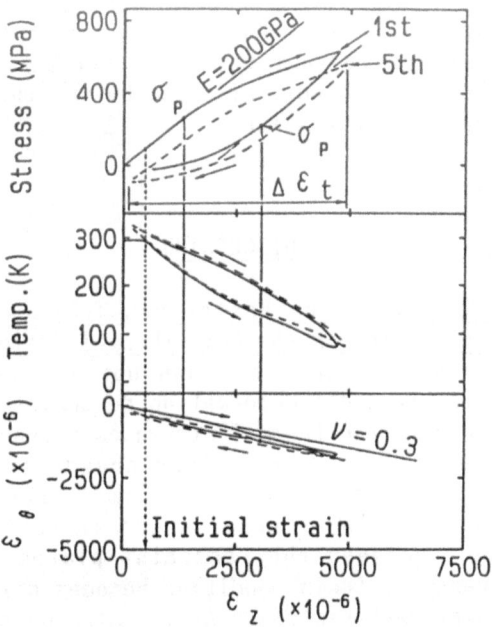

FIGURE 5 Change of thermal stress, temperature and ε_θ against ε_z (SUS I)

stress, therefore the cyclic strain condition becomes stationary, as the thermal strain range remains constant under the applied temperature changes.

The result of SUS II, where thermal stress concentration was made higher, is

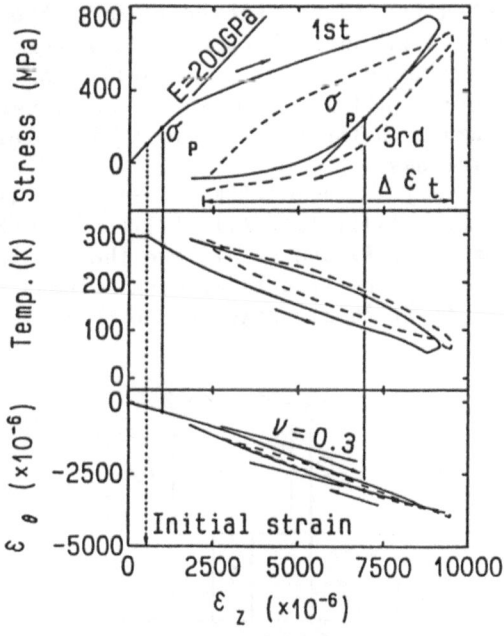

FIGURE 6 Change of thermal stress, temperature and ε_θ against ε_z (SUS II)

shown in FIG.6. A larger stress-strain hysteresis was obtained in spite of the same temperature change. From this result, we recognize that the cold thermal cycle behavior is hardly affected by the thermal stress concentration.

DISCUSSION

Effect of Shape Factor and Temperature Distribution on the Cold Thermal Cycles

When a smooth round bar is cooled uniformly under deformation restraint, the change in the thermal stress as a function of a certain temperature change is the same at every cross section. When the cross sectional area varies in the distance between the restraint plates, the stress or strain condition becomes different at each section. As the strains shown above were measured on the parallel part, the thermal stress concentrations were calculated.

In the following discussion, theoretical thermal stress in the elastic condition under uniform cooling is discussed, assuming the thermal expansion coefficient (α) and the elastic modulus (E) are independent on a temperature.

As shown in FIG.7, column 1 and column 2, which are the same material and have a different length and a cross sectional area ($A_2 > A_1$) respectively, are connected, and the deformation of the connected rod is restricted at both the top and the bottom. When a certain temperature change (ΔT) is applied to both columns uniformly, the stress change of column 1 in the elastic limit is obtained as follows:

$$\Delta \sigma_1 = - E \alpha \Delta T K_e \quad \text{-------- (1)}$$
$$K_e = A_2 (l_1 + l_2)/(A_1 l_2 + A_2 l_1) \quad \text{---- (2)}$$

where A_1, A_2 and l_1, l_2 are the sectional areas and the lengths of each column.

As K_e is a parameter constructed with

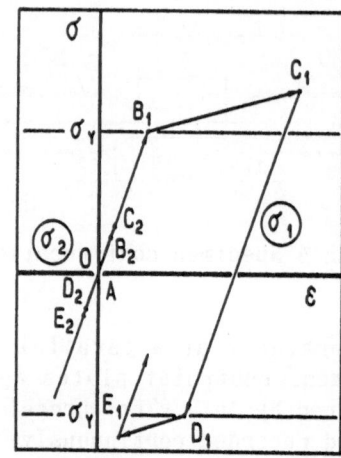

FIGURE 8 Illustration of the thermal stress-strain relationship

shape factors, it is considered to give a measure of the stress concentration using as basis the smooth bar. K_e of each specimen was given in TABLE 2.

An illustration of the thermal stress-strain relationship obtained theoretically is shown in FIG.8. From this figure, it is seen that column 1 produced a large plastic strain hysteresis, and column 2 moved only on the elastic line, and that the compressive stress and the tensile strain remained in column 1 after the thermal cycle. Therfore the initial high stress in the elastic region scarcely affected the stress-strain relationships of the later cycles.

The relationship between the thermal stress and the temperature is shown in FIG.9. The stress increment of column 1 becomes larger than that of the uniform

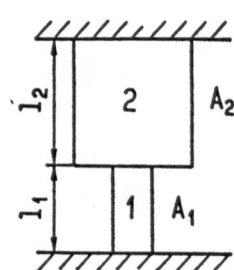

FIGURE 7 A model for thermal stress concentration

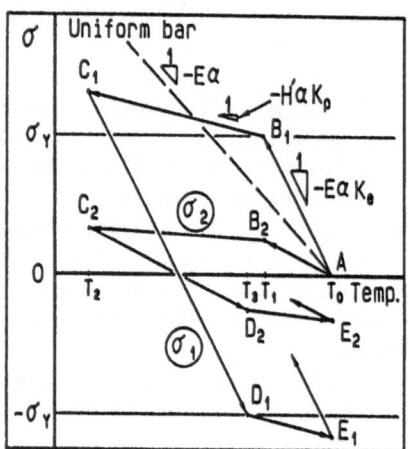

Figure 9 Illustration of the thermal stress-temperature relationship

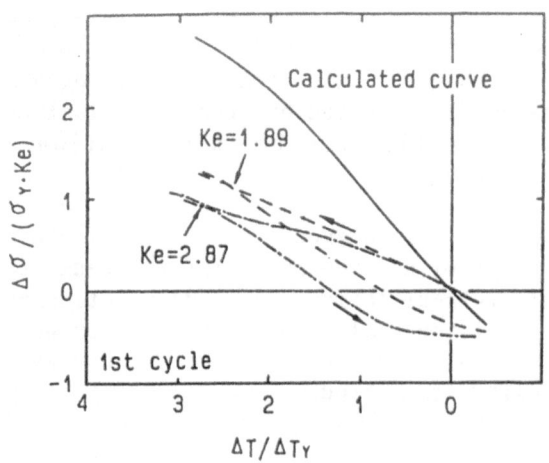

FIGURE 10 $\Delta\sigma/(\sigma_Y \cdot K_e)$ vs $\Delta T / \Delta T_Y$

bar, and after the yielding of column 1, the non-linear hysteresis curve is present in column 2.

The comparison of the test results with the calculated ones under ideal uniform cooling is shown in FIG.10. The longitudinal axis is the non-dimensional thermal stress divided by $(\sigma_Y \cdot K_e)$, and the horizontal axis shows the non-dimensional temperature divided by ΔT_Y, which is the temperature change required to yield the smooth bar under a deformation restraint. ($\Delta T_Y = \sigma_Y/E\alpha$) An estimate of ΔT_Y, α, σ_Y and E is 14×10^{-6}/K, 200MPa and 200GPa respectively. The calculated curve obtained from the thermal expansion is also shown in FIG.10. From this result, it is clear that the elastic behavior observed during the cooling pro-

cess is hardly dependent on K_e, and that the thermal stress measured is smaller than the calculated value for uniform cooling. The reason why the smaller stress was obtained is considered to be due to the thermal distribution in the longitudinal direction of the specimen.

When the relation between the temperature change of column 2 (ΔT_2) and that of column 1 (ΔT_1) satisfies the following equation,

$$\Delta T_2 = k \cdot \Delta T_1 \quad\text{-------- (3)}$$

eq.(1) and (2) become :

$$\Delta\sigma_1 = - E\alpha\Delta T K_e^* \quad\text{-------- (4)}$$
$$K_e^* = A_2(l_1+kl_2)/(A_1l_2+A_2l_1) \quad\text{--- (5)}$$

As k becomes small (<1), K_e^* becomes smaller than K_e.

The result of the calculation with K_e^* (k is supposed to be 0.25.) is shown in FIG.11. The experimental results in the elastic region shown in FIG.10 ($\Delta T / \Delta T_Y < 1.0$) come close to the calculated curve. From this fact, it is clear that the longitudinal distribution of temperature reduces the thermal stress, and it is suggested that if uniform cooling is established, the larger stress or strain would be produced by the cold thermal cycles.

<u>The Potential for Cold Thermal Fatigue</u>

On both specimens, the stress-strain relationships became stationary. The total strain ranges in the hysteresis curves were considered to become constant after the second cycle, because the mechanical strain due to the thermal shrinkage is expected to be constant, assuming that the temperature change would not be varied and the material properties such as the thermal expansion would not be changed by the thermal cycles. Therefore the cold thermal cycles under constant change of tem-

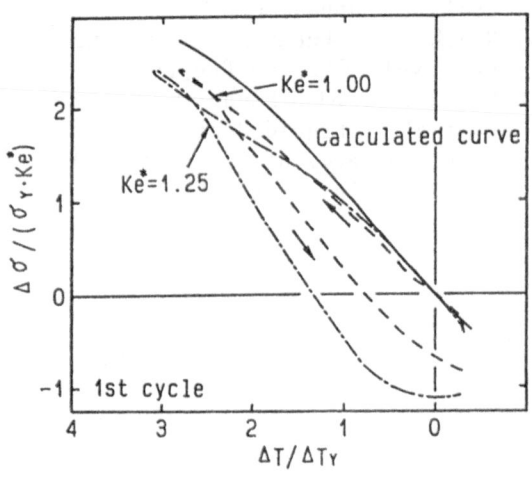

FIGURE 11 $\Delta\sigma/(\sigma_Y \cdot K_e^*)$ vs $\Delta T / \Delta T_Y$
(k = 0.25)

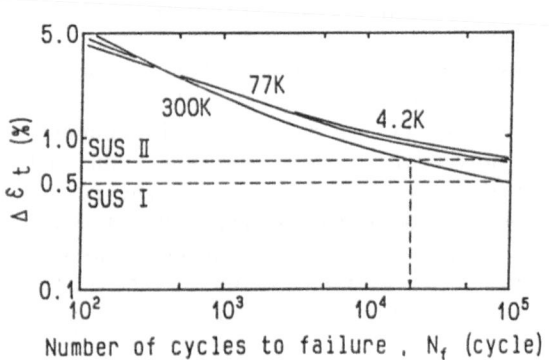

FIGURE 12 Comparison of test results with fatigue data according to [4]

perature can be considered to be a phenomenon similar to the strain controlled fatigue.

The comparison of fatigue data under the total strain range controlled test [4] with the total strain range obtained in this study is shown in FIG.12.

Although the regression curves of the total strain range $(\Delta \varepsilon_t)$ as a function of the fatigue life (N_f) are different depending on the test temperature, and while it is not clear whether the strain controlled test results correspond to those of cold thermal fatigue, the fatigue lives of SUS I and SUS II are expected to be about 10^5 and 2×10^4 cycles respectively. This assumes that the 300K results, which give the shortest fatigue life, can be used as reference. As mentioned above, the total strain range is considered to become larger under uniform cooling, while the fatigue lives of the specimens tested are supposed to become shorter if uniform cooling is applied.

As the cryogenic structures become larger and more complex, the stress or strain fields concentrated by thermal deformation restraints, such as welds, are expected to increase. On such structures, the design philosophy for cold thermal fatigue will become more important.

CONCLUSIONS

In this study, the cold thermal cyclic properties of the austenitic stainless steel (SUS304) were investigated under temperature changes between a room and a cryogenic temperatures (about 50K), and the potential for cold thermal fatigue discussed. The main results obtained are summarized as follows:

(1) When a repetition of loading and unloading was applied to SUS304 stainless steel, and the tensile test temperature was changed from 4.2K, 293K and to 4.2K successively, a non-linearity occurred in the unloading process at 293K, and a hysteresis loop was present in the reloading process. Also the proportional limit (or 0.2% off-set stress) in the subsequent test at 4.2K did not reach the unloaded stress level of the prior test at 4.2K.

(2) A cyclic plastic strain was produced by the cold thermal cycles under the deformation restraint. As the driving force to produce the mechanical strain was the thermal strain, the total strain range could not be varied under constant temperature change. This strain range, however, was hardly affected by the thermal stress concentration and the distribution of temperature.

(3) It is suggested that SUS304 stainless steel has the potential to cause the cold thermal fatigue under deformation restraint, specially in an application where a stress concentration exists and uniform cooling is applied.

ACKNOWLEDGEMENTS

Authors wish to thank the members of the Low Temperature Center, Osaka University where this work has been performed.

REFERENCES

[1] Fultz, B, Chang, G.M. and Morris, Jr., J.W., Effects of Magnetic Fields on Martensite Transformations and Mechanical Properties of Steels at Low Temperatures. In Austenitic Steels at Low Temperatures, ed. R.P. Reed and T. Horiuchi, Plenum Press, New York, 1983, pp. 199-209

[2] McHenry, H.I., Structural Alloys. In Materials at Low Temperatures, ed. R.P. Reed and A.F. Clark, American Society for Metals, 1983, pp. 371-412

[3] The Cryogenic Association of Japan, Data Book of Cryogenic Metal Materials, 1981, (in Japanese)

[4] Suzuki, K., Fukakura, J. and Mori, T., Low-Cycle Fatigue Properties of 304L Stainless Steel under Axial-Strain Control at Liquid Herium Temperature. J. of the Soc. of Mat. Sci., 1985, 385 pp.80-84 (in Japanese)

CREEP AND DISCONTINUOUS DEFORMATION BEHAVIOR
OF AUSTENITIC STEELS AT CRYOGENIC TEMPERATURES

T. Ogata, K.Ishikawa, T. Yuri, and O. Umezawa
National Research Institute for Metals, Tsukuba Labs.
Tsukuba, Ibaraki 305, JAPAN

ABSTRACT

We studied the behavior of austenitic steels to evaluate the accurate properties of structural materials for superconducting magnet systems at low temperatures. We carried out the load- and displacement-controlled tensile tests, and creep tests at 293, 77, and 4 K. A catastrophic discontinuous deformation occurs in load-controlled tensile tests. This phenomenon indicates that if a material at low temperatures is in the freely deforming condition and a deforming force is kept constant during a deformation, it will fail at a much lower stress than the stress obtained in the usual displacement-controlled tests. A creep rate of 10^{-10} s^{-1} after 200 hours test at a stress level of the 0.2 % yield strength was detected even at 4 K for SUS 310S. A creep rate of SUS 304L and 316LN at 4K at the proof stress of each steel became almost zero after tens of hours testing.

INTRODUCTION

To evaluate the practical mechanical properties of structural materials for high-field superconducting magnet applications, where constant stress is applied to the materials, we studied the creep and the discontinuous deformation as a time-dependent unique deformation of austenitic stainless steels in liquid helium. In the previous paper [1], we performed short-term creep tests and primitive load-controlled tensile tests and showed the effect of work-hardening rate of the steels on the deformation. Recently, cryogenic creep behavior has been recognized as one of the important properties of structural materials used for superconducting magnets. Long-term creep behavior for copper alloy was reviewed by Tien and Yen [2], the behavior for the 304LN plate and 316L weld metal was reported by Roth et al.[3], and further studies were conducted by Reed[4], McDonald [5], and Ogata [6]. The magnitude of creep strain in the long-term creep test is still of great concern for the design of applications. On the other hand, the discontinuous deformation occurs during mechanical tests at temperatures near 4 K. In load-controlled tensile tests, the discontinuous deformation is large and abrupt and could lead directly to catastrophic fracture [7,8]. This deformation differs from that reported in the displacement-controlled tests,

because the amount of displacement in load-controlled tests is not limited by the testing machine, and no load drop occurs during the deformation. Discontinuous deformation in load-controlled tests results in a lower ultimate strength than that measured with displacement-controlled tests and might affect the design codes for pressure vessels. So it is important to clarify this deformation behavior and to predict the mechanical properties when a non-restricted deformation force is applied to structural materials at low temperatures.

In this paper, we discuss the results of long-term creep tests on the austenitic stainless steels for more than 200 hours and the effects of loading rate on the discontinuous deformation behavior and tensile properties with updated data.

EXPERIMENTAL PROCEDURES

Materials

Materials used for creep tests were typical austenitic stainless steels for cryogenic applications; SUS 304L, 310S, and 316LN. Among the steels, SUS 304L has a lowest yield strength and the highest work-hardening rate due to strain-induced martensitic transformations, 310S has a higher yield strength and

Table 1. Chemical composition of the steels tested in this study (wt.%).

	C	Si	Mn	P	S	Ni	Cr	Mo	N
SUS 304L	0.016	0.67	1.52	0.027	0.009	10.03	18.24		
SUS 310S	0.04	0.79	0.93	0.020	0.001	19.20	25.18		
SUS 316LN	0.019	0.50	0.84	0.025	0.001	11.16	17.88	2.62	0.18
AISI 304L	0.02	0.6	1.4	0.02	0.01	9.7	18.4		
AISI 310	0.08	0.7	1.7	0.02	0.02	20.8	24.8	0.1	

Table 2. Tensile properties of the steels.

Material	Temperature (K)	Yield Strength (MPa)	Tensile Strength (MPa)	Elongation (%)	Reduction in Area (%)
SUS 304L	293	236	590	82.4	79.0
	77	431	1285	80.0	69.5
	4	505	1476	48.0	61.1
SUS 310S	293	248	587	61.6	64.9
	77	562	1103	80.0	53.3
	4	783	1269	54.8	43.9
SUS 316LN	293	342	716	71.9	85.0
	77	859	1517	75.5	70.1
	4	1072	1697	54.7	60.1

a lower work-hardening rate, and 316LN has a highest yield strength and an in-between work-hardening rate. Chemical composition and tensile properties are given in Table 1 and Table 2, respectively. Materials used for tensile tests were AISI 304L and 310 austenitic stainless steels and the chemical composition is also listed in Table 1.

Creep test

Round bar specimens with a 6 or 6.25 mm in diameter and a 32 mm in gauge length were machined from hot-rolled plates. Creep tests were conducted at liquid helium temperature (4 K), liquid nitrogen temperature (77 K), and room temperature (293 K). A servo-hydraulic testing machine in the load-controlled mode was used in this study. Tests were performed at selected stress levels; 60 %(0.6), 80 %(0.8), 100 %(1.0) and 110 %(1.1) of the 0.2 % yield strength at each temperature. The specimen stress was ramped up to the testing stress in 10 s and held at the level for more than 200 h. Creep strain was monitored with the 25-mm gage-length extensometer. A vacuum insulated cryostat with a 25 litter liquid helium reservoir was used for the tests at each temperature. The evaporation rate of liquid helium in the cryostat was about 0.27 litters per hour and liquid helium was refilled every 48 h manually. The evaporation rate of liquid nitrogen was 7 kg per 200 h. Details of this testing procedure and the system were described elsewhere[6].

Load-controlled Tensile Tests

The round specimens had gauge lengths of 38

mm and reduced-section diameters of 6.35 mm. Specimen strain was measured with two pairs of clip-on strain gage extensometers; each pair included one normal and one high-range extensometer. Loading rates were varied from 0.5 to 5000 N/s. The strain rate in the linear-elastic region of the stress-strain curves was calculated to be 9×10^{-8} s^{-1} to 9×10^{-4} s^{-1}, corresponding to the loading rates of 0.5 N/s to 5000 N/s. Displacement-controlled tensile tests were performed in the stroke-control mode. All these tensile tests were carried out with specimens submerged in liquid helium. Details of this testing procedure was described elsewhere[7].

RESULTS AND DISCUSSION

Creep curves

The strain-time curves for SUS 310S at 293 K, 77 K, and 4 K are shown in Figure 1. The creep strain increased with an increase of creep stress and the curves appear to be logarithmic creep behavior. At 4 K, the creep rate at the stress level of 80 % of yield strength was almost 0 (below the resolution limit of the strain indicating system) and the rate at the stress level of yield strength at 200 h was 8.3×10^{-11} s^{-1}. Figure 2 a), b) shows the creep curves for 304L and 316LN, respectively. The creep behavior of 316LN was similar to that of 310S. The creep strain of 304L was smaller than that of 310S at the same stress ratio to yield strength. At $T/T_m = 0.05$-0.3 and within about 0.2 % creep strain, creep behavior is considered to be logarithmic creep as follows:

$$\varepsilon = \alpha \ln t + C$$

where ε is creep strain, t is time, α and C are constant independent of time. Figure 4 shows the creep curves as a function of log time at 4 K. The creep strain yields a straight line in the figure, however, the creep curve for 310S at 110 % yield strength is curved concave upward in the steady-state region. This phenomenon was also found in 310S at 77 K at the yield strength and in 316LN at 293 K and 77 K. The creep rate for 304L at a stress level of above 80 % yield strength was lower than that for 310S or 316LN. The reason for the lower creep rate for 304L is considered to be its lower yield strength and higher work-hardening rate. The results of creep rates at 200 h are summarized in Table 3.

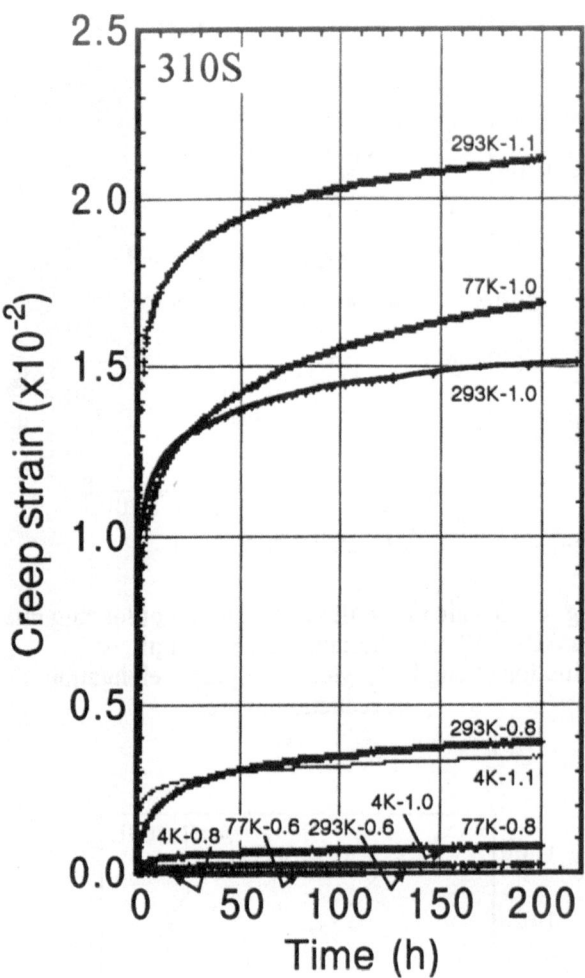

Fig. 1 Creep curves for SUS 310S at 293, 77, and 4 K.

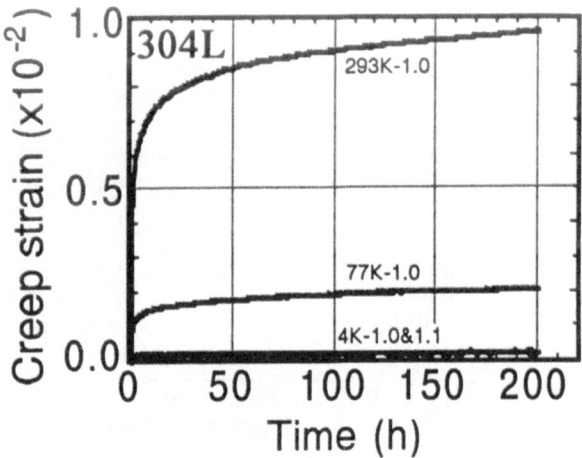

Fig. 2 Creep curves for SUS 304L.

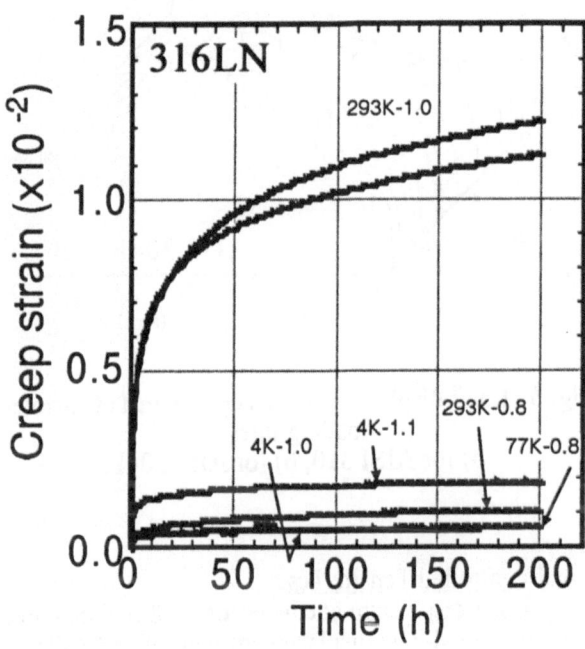

Fig. 3 Creep curves for SUS 316LN.

Table 3. Creep rate at 200 h.

Creep stress/	Creep rate ($\times 10^{-10}$ s^{-1})		
0.2 % Yield stress	304L	310S	316LN
293 K			
1.1		16	
1.0	18	14	34
0.8		9.8	2.6
0.6		0	
77 K			
1.0	2.0	25	24
0.8		2.4	1.3
0.6		0	
4 K			
1.1	0.15	4.6	1.3
1.0	0	0.83	0
0.8		0	
0.6		0	

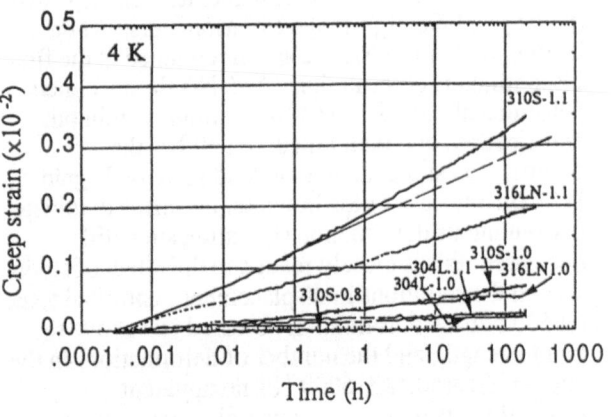

Fig. 4 Creep strain at 4 K as a function of log time for SUS 310S, SUS 304L and 316LN.

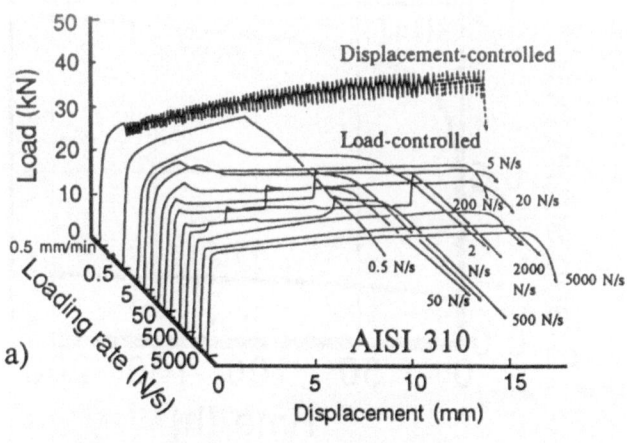

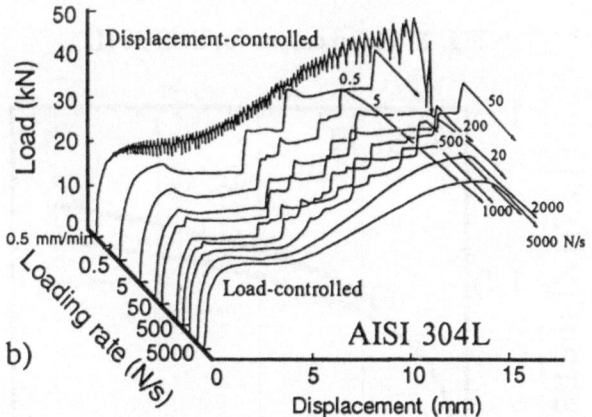

Fig. 5 Load-displacement curves obtained at various
loading rates.
a) for AISI 310, b) for AISI 304L.

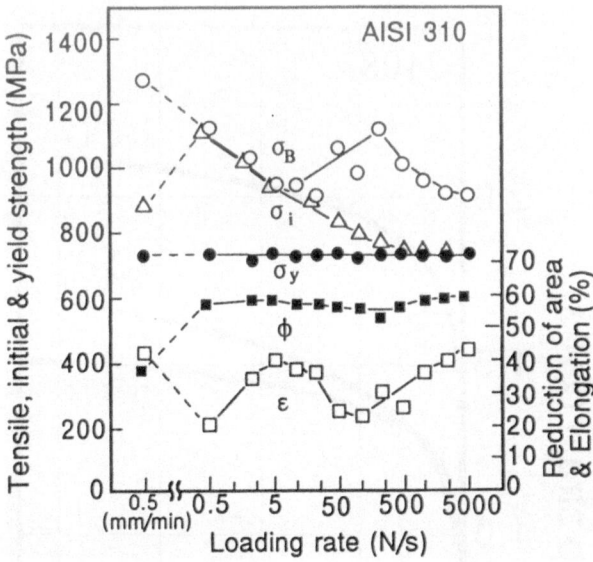

Fig. 6 Tensile properties as a function of loading rate
for AISI 310. σ_u: ultimate tensile strength, σ_i:
initiation strength, σ_y: yield strength, ε: elongation,
ϕ: reduction of area.

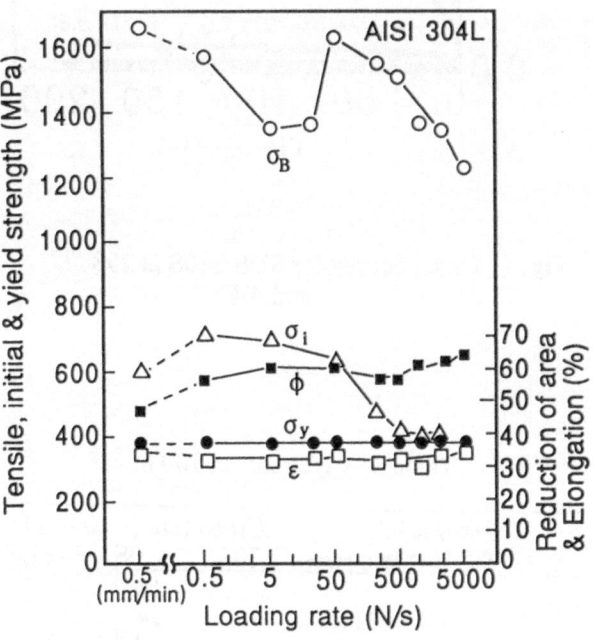

Fig. 7 Tensile properties as a function of loading rate
for AISI 304L. σ_u: ultimate tensile strength, σ_i:
initiation strength, σ_y: yield strength, ε: elongation,
ϕ: reduction of area.

Load-controlled Tensile Tests

Load-displacement curves for AISI 310 obtained
from both load- and displacement-controlled tensile
tests are shown in Figure 5 a). In the figure, 0.5
mm/min represents the displacement-controlled test; the
broken line indicates that the curve was calculated from
a stroke signal because specimen extension exceeded
the range of the high-range extensometers. At the
loading rate of 0.5 N/s the specimen failed at the first
discontinuous deformation. At 5 N/s the specimen
deformed about 15 mm (40% in strain) within one
deformation, and then failed. At 50 N/s the
deformation stopped, and the load increased again.
With the increase of loading rate, the stress to initiate
discontinuous deformation (initiation strength)
decreased; this is closely related to the effect of strain
rate on the serrations in displacement-controlled tests.
At 500 N/s the deformation occurred just above the
yield strength, and the number of deformations in the
curve increased. At 5000 N/s no apparent
discontinuous deformation was observed owing to
specimen heating.

The serrations that occurred near yield strength in
the displacement-controlled test seemed to be

negligible, however, large amount of discontinuous
deformation is characteristic of load-control tests. The
change in length during the first deformation for AISI
310 at loading rates of 500, 50, and 5 N/s was about
2, 7, and 15 mm. The deformation locally heats the

specimen and stops when a balance occurs between the local strength and the work-hardening at the elevated temperature. Subsequent discontinuous deformations occur in other areas of the specimen where less local strain hardening has occurred. In displacement-controlled tests, the deformation is suppressed by the testing machine, and when the deforming stress decreases, this suppression results in serrations of smaller strain increments. Figure 5 b) shows the load-displacement curves for AISI 304L steel. The curves for AISI 316LN were similar to the curves for AISI 310. [8] The curves for 304L were different from those of AISI 310, and the deformation stopped frequently, which is due to its lower initiation strength and also to its high work-hardening rate accompanied by martensitic transformation.

The mechanical properties of AISI 310 and 304L are shown in Figures 6 and 7 as a function of loading rate. In the figure, σ_u, σ_i, σ_y, ε, and ϕ represent ultimate tensile strength, initiation strength, yield strength, elongation, and reduction of area, respectively. Yield strength was not affected by the loading rate, because the strain rate in the elastic range is lower than 10^{-3} s^{-1}, and specimen heating occurs after the plastic deformation begins. Ultimate tensile strength and initiation strength decreased with an increase in loading rate. At the loading rate 5000 N/s, the ultimate strengths of AISI 310 and 304L were 71 and 65%, respectively, of those obtained from displacement-controlled tests. This is due to the discontinuous deformation that leads to specimen failure at higher temperature before the specimen is fully work-hardened at 4 K. In addition, the ultimate strength decreased to 71 or 80 % around the loading rate of 5 - 10 N/s, where the strain rate was about 2×10^{-4} s^{-1}. Above phenomenon indicates that if a material at low temperatures is in the freely deforming condition, it will fail at a much lower stress than the stress obtained in the usual displacement-controlled tensile tests. This change of ultimate strength in load-controlled tests may control the code for material selection, because 1/3 σ_u is lower than 1/2 to 1/3 σ_y for the higher yield strength materials.

CONCLUSION

1. The steady-state type behavior and logarithmic type creep were found in these steels at low temperatures.

2. Even at 4 K, a creep rate of 10^{-10} s^{-1} after 200 hours of test at a stress level of the yield strength was detected for the alloy SUS 310S.
3. At 293 K at a stress level of yield strength for each steel, the creep strain at 200 h was beyond 1 %.
4. Materials under load-controlled mode could fail at a much lower stress than the stress obtained in the usual displacement-control tensile tests.

ACKNOWLEDGMENTS

The authors wish to thank the members of the 1st Research Group for supplying the liquid helium. The work for load-controlled tensile tests were mainly carried out at the National Institute of Standards and Technology with Dr. R.P. Reed and Mr R.P. Walsh.

REFERENCES

1. Ogata, T., and Ishikawa, K., Time-dependent deformation of Austenitic Stainless Steels at Cryogenic Temperatures, Cryogenics, 1986, 26, 365-369
2. Tien, J.K. and Yen, C.T., Cryogenic Creep of Metals, Adv. Cryog. Eng., 1984, 30, 319-338
3. Roth, L. D., Manhardt, A. E., Dalder, E. N. C., and Kershaw, Jr., R. P., Creep of 304 LN and 316L Stainless steels at Cryogenic Temperatures, Adv. Cryog. Eng., 1986, 32, 369-376
4. Reed, R.P., Simon, N.J., and Walsh, R.P., Creep of Copper: 4 to 295 K, Adv. Cryog. Eng., 1990, 36,
5. McDonald, L.C. and Hartwig, K.T., Creep of Aluminum at Cryogenic Temperature, Adv. Cryog. Eng., 1990, 36,
6. Ogata, T., Ishikawa, K., and Umezawa, O., Low Temperature Creep Behavior of Stainless Steels, Adv. Cryog. Eng., 1990, 36,
7. Ogata, T., Ishikawa, K., Reed, R. P., and Walsh, R. P., Loading Rate Effects on Discontinuous Deformation in Load-control Tensile Tests, Adv. Cryog. Eng., 1988, 34, 233-240
8. Lee, H.M., Reed, R.P., and Han J.K., Load-controlled Tensile Tests of Austenitic Steels at 4 K, Adv. Cryog. Eng., 1990, 36,

FATIGUE AND FRACTURE OF Ti ALLOYS AT CRYOGENIC TEMPERATURES

K. Nagai, T. Yuri, O. Umezawa and K. Ishikawa
National Research Institute for Metals, Tsukuba Labs.
1-2-1, Sengen, Tsukuba, Ibaraki 305, Japan

ABSTRACT

Tensile properties, Charpy absorbed energy, and fracture toughness at room temperature, 77 K, and 4 K were determined for unalloyed titanium and titanium alloys. The titanium alloys with reduced content of oxygen and iron showed an excellent combination of high yield strength and high toughness at 4 K. Fracture behavior at cryogenic temperatures is also discussed. Fracture mode transition at low temperature was observed for higher oxygen unalloyed titanium and b.c.c. type titanium alloys. High cycle fatigue data were also determined at 4 K for some titanium alloys. Sub-surface fatigue crack initiation prevailed at lower cyclic stress and influenced the longer life property.

INTRODUCTION

High strength-to-weight ratio and nonmagnetism are a feature of titanium(Ti) alloys. In addition, the Ti alloys have higher electric resistivity and smaller thermal conductivity than austenitic stainless steels. All these properties are favorable for cryogenic structural use. However, comparatively lower fracture toughness as well as high material cost is thought to retard the wide use of Ti alloys at cryogenic temperatures. Hence, few studies have focussed on the cryogenic mechanical properties of Ti alloys[1,2,3,4].

Future application of superconducting magnet technology may require varieties of properties, not only mechanical properties but also physical properties, for structural components. In order to get a more complet data base on cryogenic mechanical properties of metals and their alloys, we have determined tensile properties, Charpy absorbed energy, fracture toughness, and S-N curves at cryogenic temperatures for Ti and its alloys. In the present paper, the results and the fracture behavior are briefly summarized[5,6,7].

MATERIALS AND METHODS

Materials

The chemical compositions of the materials tested in the present study are shown in Table 1. The testing materials were unalloyed Ti, alpha alloys (Ti-5Ta, Ti-5Al-2.5Sn ELI and Ti-5Al-2.5Sn Normal : ELI and Normal designate an extra-low-interstitial level and a normal interstitial level, respectively), alpha-beta alloys(Ti-6Al-4V Normal, ELI, and SPELI: SPELI means a SuPer ELI level), and beta alloys(They are designated as 15-3-3-3 and 15-5-3). The beta alloys were solution-heat-treated 3.6 ks at 1050 K and not aged. Others were conventionally mill-annealed.

Mechanical Tests

Tensile tests were done to determine yield strength (YS), ultimate tensile strength (UTS), elongation (ELN) and reduction of area (RA) using round bar type specimens of 3.5 mm in diameter. Charpy absorbed energy (CVN) was obtained using a 10 mm square and 2mm V-notched standard specimen. Fracture toughness, K_{IC} was evaluated with compact tension type specimens. A fatigue tester with a helium recondansation system [8] was utilized for LHT high cycle fatigue tests.

Table 1 Chemical compositions of titanium alloys tested
in the present study in weight percent.

Material	O	H	Fe	Al	V	Others
Unalloyed Ti #1	0.088	0.0029	0.066	-	-	Ti:bal.
Unalloyed Ti #2	0.143	0.0008	0.110	-	-	Ti:bal.
Unalloyed Ti #3	0.203	0.0040	0.030	-	-	Ti:bal.
Unalloyed Ti #4	0.068	0.0006	0.04	-	-	Ti:bal.
Unalloyed Ti #5	0.113	0.0013	0.04	-	-	Ti:bal.
Ti-5Ta	0.091	0.0028	0.028	-	-	Ta:4.75, Ti:bal.
Ti-5Al-2.5Sn Normal	0.165	0.0086	0.232	5.15	-	Sn:2.67, Ti:bal.
Ti-5Al-2.5Sn ELI	0.057	0.0058	0.19	5.15	-	Sn:2.66, Ti:bal.
Ti-6Al-4V Normal	0.135	0.0053	0.20	6.34	4.23	Ti:bal.
Ti-6Al-4V ELI #1	0.108	0.0029	0.209	6.19	4.14	Ti:bal.
Ti-6Al-4V ELI #2	0.104	0.0032	0.20	6.23	4.25	Ti:bal.
Ti-6Al-4V SPELI	0.054	0.0055	0.03	5.97	4.12	Ti:bal.
15-3-3-3	0.171	<0.001	0.100	2.92	14.89	Sn:2.81, Cr:2.87, Ti:bal.
15-5-3	0.088	0.0064	0.039	3.08	-	Mo:15.44, Zr:5.31, Ti:bal.

Test temperatures were room temperature (RT), 77
K (LNT), and 4 K (LHT).

RESULTS AND DISCUSSION

Load-Displacement Curves

Load-displacement curve at LHT is characterized by its serrated form or discontinuous deformation behavior as seen in Fig. 1. This discontinuous
deformation is called "serration"[9].

For the Ti alloys studied here, the load-
displacement curves at LHT were classified into
three groups as follows:

A) Low strength and good ductility, small load
drop by serration, and comparatively high strain
hardening rate (unalloyed Ti and Ti-5Ta),

B) High strength and relatively poor ductility,
large load drop by serration, and little strain
hardening (Ti-5Al-2.5Sn and Ti-6Al-4V), and

C) High strength and poor ductility, high strain
hardening rate, premature failure, and no large load
drop by serration (beta Ti alloys).

Yield Strength and Ductility

Yield strength increased with decreasing test
temperature as shown in Fig. 2. The YS at LHT
ranged between 500 and 1900 MPa, and seems to
be higher in the order of unalloyed Ti, alpha Ti
alloy, and alpha-beta (beta) Ti alloy.

Figure 3 describes the change of elongation as a
function of test temperature. In the unalloyed Ti and
Ti-5Ta, the ELN was larger at LNT than at RT, and
decreased a little at LHT. In Ti-5Al-2.5Sn and Ti-
6Al-4V alloys, a nearly equal ELN was observed at
RT and LNT, but a large decrease by half occurred
at LHT. In beta alloys the ELN decreased with a
decrease in temperature.

The reduction of area decreased with decreasing

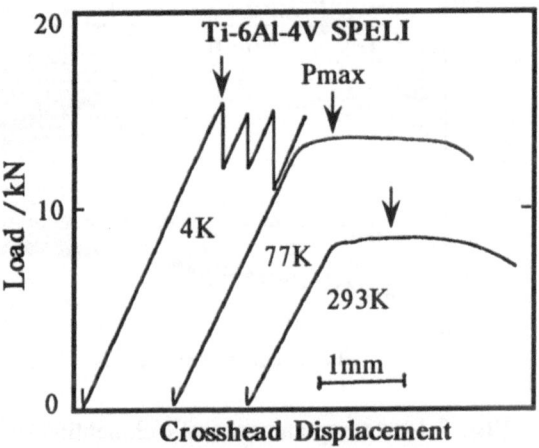

Fig. 1 Load-displacement curves of Ti-6Al-4V
SPELI at 293, 77 and 4 K.

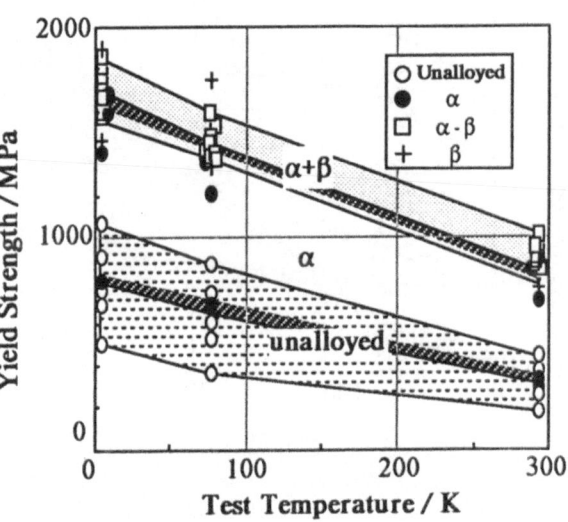

Fig. 2 Low temperature yield strength of Ti
alloys as a function of temperature.

test temperature similar to the elongation as seen in Fig. 4. However a drop of RA at LHT was not observed for Ti-6Al-4V alloys. The RA of the beta alloys was high at RT, but became nearly zero at cryogenic temperatures corresponding to the occurrence of premature failure.

The ELN and RA at LHT are plotted as a function of YS at LHT in Fig. 5. The ELN has a strong negative correlation between YS, irrespective of alloy types. This trend was seen at all the temperatures. Namely, higher yield strength results in lower ductility for Ti alloys. Although the similar correlation was observed between RA and YS at RT and LNT, the deviation from linearity appeared at

LHT as shown in Fig. 5(b). In other words, the beta alloys have lowest RA due to premature failure, and the alpha-beta alloys seems to have larger RA than the alpha alloys at the given YS level.

Charpy Absorbed Energy and Fracture Behavior

Changes of CVN as a function of test temperature are shown in Fig. 6. This shows that low oxygen and low strength alloys have high CVN over 50 J in the test temperature range, however higher oxygen or higher strength results in lower CVN at cryogenic temperatures as well as at RT.

The change of CVN by test temperature may be described in terms of fractography as follows:
A) In ELI level unalloyed Ti, only equiaxed dimples of several ten microns in diameter were seen irrespective of temperature. The CVN was over 50 J and sometimes exceeded 200 J. On the other hand, in a high oxygen unalloyed Ti cleavage-like failure occurred even at RT. The CVN was very low in that case.

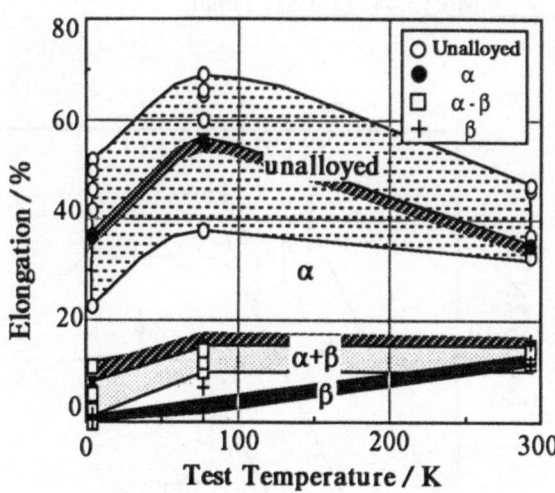

Fig. 3 Low temperature tensile elongation of Ti alloys as a function of temperature.

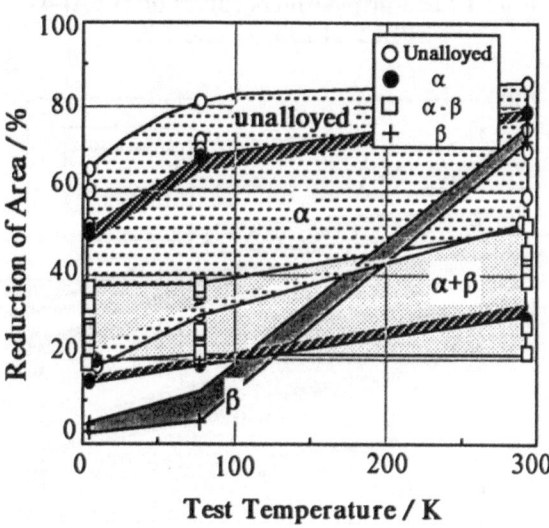

Fig. 4 Low temperature reduction of area of Ti alloys as a function of temperature.

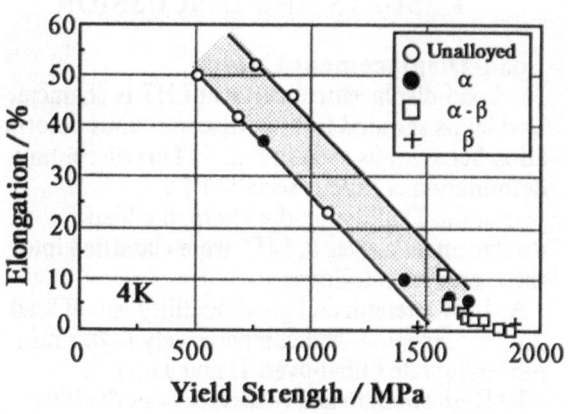

(a) elongation

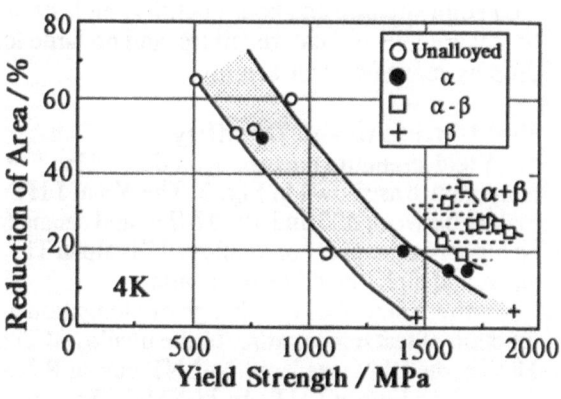

(b) reduction of area

Fig. 5 Yield strength and ductility at liquid helium temperature for Ti alloys:
a) elongation and b) reduction of area.

B) In the Ti-5Ta, the equiaxed dimples appeared at room temperature, but the cleavage-like failure occurred at LNT and LHT and decreased the CVN.
C) In the Ti-5Al-2.5Sn alloys, the equiaxed dimples also appeared at RT and the "groove"[5] fracture surface characterized the fracture at LNT and LHT. The appearance of the groove corre-sponded to the decrease in CVN from 50 J to 10 J.
D) In the Ti-6Al-4V alloys, a macroscopically flat fracture surface covered with so-called microdimples of several microns in diameter was observed at all the temperatures. The temperature dependence of CVN was similar to that for the Ti-5Al-2.5Sn alloy.

E) In the beta alloys, the equiaxed dimples were seen at room temperature and the cleavage-like failure at LNT. The room temperature CVN was about 50 J but the LNT CVN less than 10 J. This degradation phenomenon is quite similar to low temperature brittleness in ferrous alloys. However, further work is necessary in order to clarify the causal mechanism of this failure.

Figure 7 shows the combination of YS and CVN at LHT. For comparison, similar data of various kinds of austenitic steels measured by the authors are given as a trend band. The impact toughness of Ti alloy is believed to be very low compared with that of austenitic stainless steels. That may be true for high strength alloys, but is not always true for low strength materials.

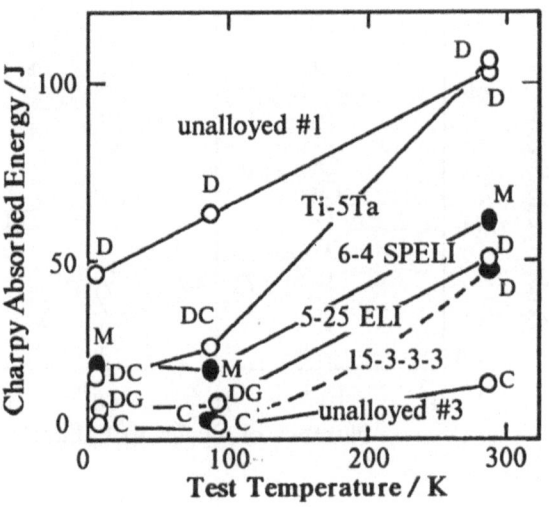

Fig. 6 Charpy absorbed energy transition curves of Ti alloys. Fracture modes are shown by D: dimple, C: cleavage-like, M: microdimple, DC: D+C, and DG: D+"groove"[5].

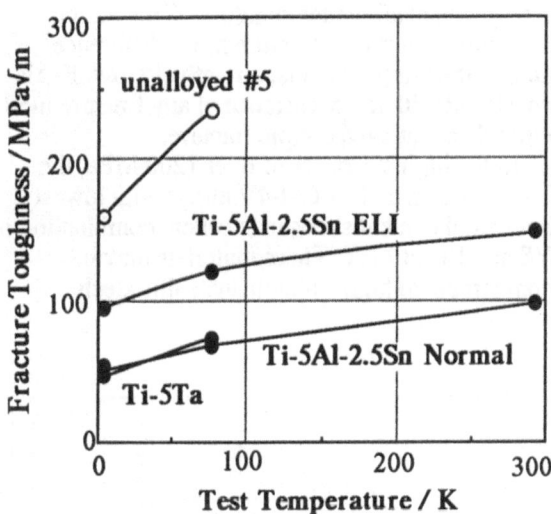

(a) unalloyed and α-Ti

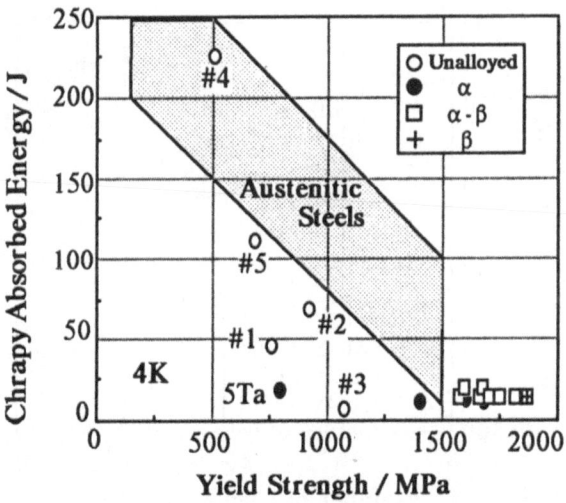

Fig. 7 Yield strength and Charpy absorbed energy at liquid helium temperature for Ti alloys. The trend band represents austenitic steels' data at liquid helium temperature.

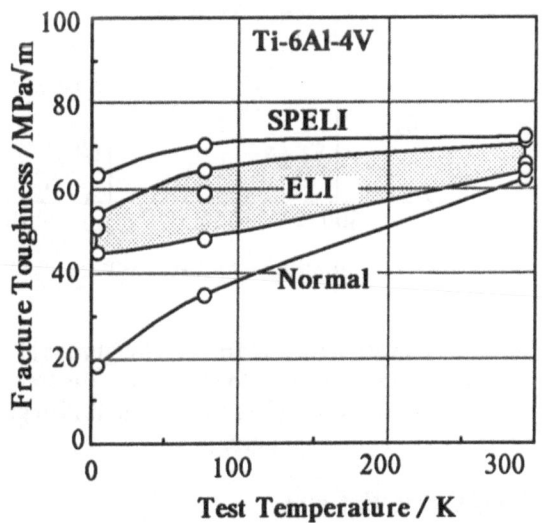

(b) α-β Ti

Fig. 8 Fracture toughness transition curves for Ti alloys: a) unalloyed and alpha Ti alloys and b) Ti-6Al-4V alloys.

Fracture Toughness

The KIC was less dependent on temperature than CVN as seen Fig. 8. The most important result was that reducing the oxygen content of Ti-5Al-2.5Sn and Ti-6Al-4V alloys brought about an increase in KIC especially at cryogenic temperatures, and consequently yielded an LHT KIC nearly equal to the room temperature value.

In addition, low strength unalloyed Ti revealed fairly high fracture toughness. Unalloyed Ti #4 was too tough to determine KIC. The combination of YS and KIC at LHT may be comparable to that of SUS 316L. Unalloyed Ti #5 yielded the KIC of about 150 MPa√m, which is believed to the maximum among the reported KIC values for Ti at LHT.

The KIC is plotted as a function of YS in Fig. 9 . No simple correlation between YS and KIC was recognized at all the temperatures. This can be explained in terms of fracture mode difference among the alloys. That is, a relatively low KIC is consistent with the occurrence of any fracture mode other than equiaxed dimple fracture.

In the high YS range of over 1200 MPa, the Ti-5Al-2.5Sn and Ti-6Al-4V alloys with lowest interstitial content show an excellent combination of YS and KIC at LHT. These materials may be comparable to high strength austenitic steels.

strength at given fatigue life was generally proportional to the material strength (YS or UTS).

Interestingly, the fatigue crack initiation site differed depending on the cyclic stress level. At higher cyclic stress, the crack initiated at the specimen surface. On the other hand, subsurface crack initiation prevailed at lower cyclic stress, irrespective of alloy type, chemical composition, and microstructure.

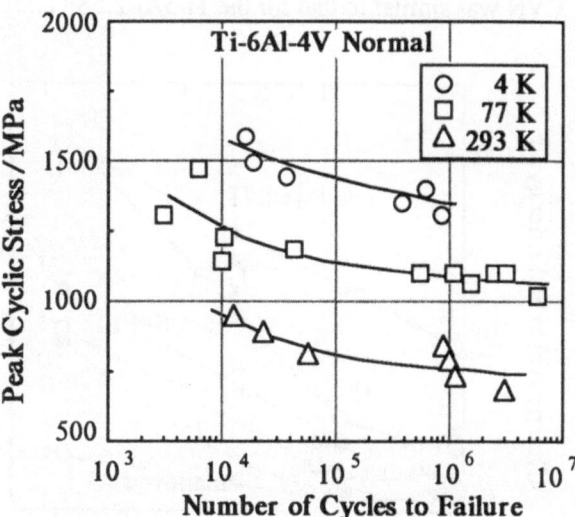

Fig. 10 The S-N curves of Ti-6Al-4V ELI alloy at RT, liquid nitrogen temperature, and liquid helium temperature.

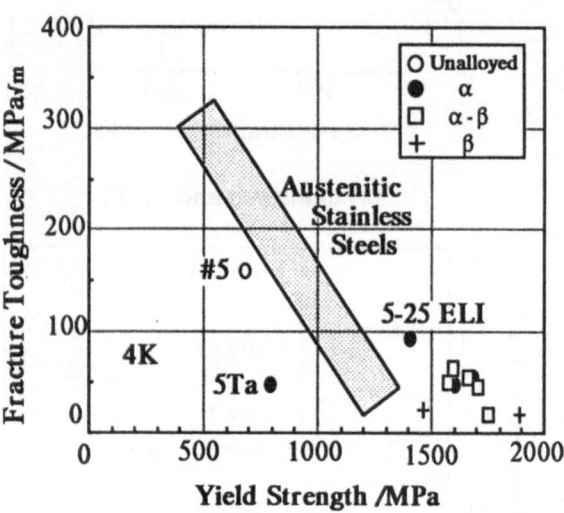

Fig. 9 Yield strength and fracture toughness at liquid helium temperature for Ti alloys. The trend band for austenitic stainless steels' data is also given here.

High Cycle Fatigue

The fatigue strength increased as the temperature decreased as shown in Fig.10. This corresponds to an increase in strength at lower temperature and the absence of brittle failure in this alloy.

The S-N curves at LHT are given for Ti-5Al-2.5Sn and Ti-6Al-4V alloys in Fig. 11. The fatigue

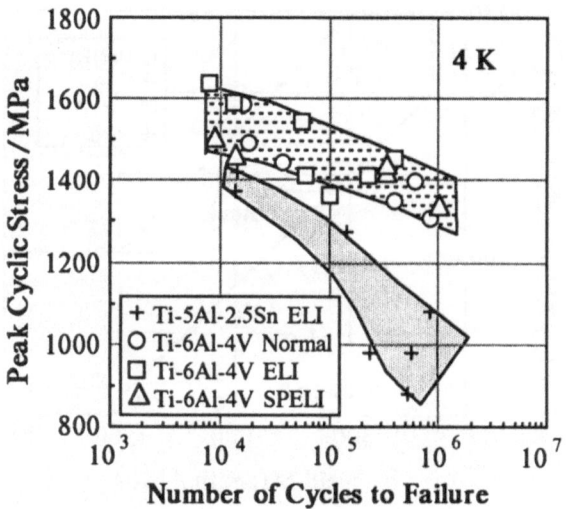

Fig. 11 The S-N curves at liquid helium temperature for Ti-5Al-2.5Sn ELI and Ti-6Al-4V alloys.

The subsurface crack initiation had microstructural origins; low Al small alpha grains in Ti-5Al-2.5Sn ELI[10], and the alpha plate in Ti-6Al-4V alloys[11]. In the latter, the finer and more homogeneous equiaxed alpha grain structure yielded higher fatigue strength in longer fatigue life range.

SUMMARY

Toughness

The strength and toughness at cryogenic temperatures are summarized as follows:
• The strength of unalloyed Ti is low. Low oxygen unalloyed Ti shows equiaxed dimple failure which brings about high toughness. High oxygen unalloyed Ti is brittle due to cleavage-like failure.
• The strength of Ti-5Ta is also low. The toughness is high at room temperature, but low at cryogenic temperatures also due to cleavage-like failure.
• The strength and toughness of Ti-5Al-2.5Sn is excellent. Reduction of oxygen content improves the toughness.
• The Ti-6Al-4V has a higher strength and a lower toughness than the Ti-5Al-2.5Sn, but also has a good combination of both. This alloy seems to have a little higher toughness than the Ti-5Al-2.5Sn when compared at the same strength level. Reduction of interstitials is also helpful in improving the toughness.
• The beta alloys have high strength. They show a cleavage-like fracture surface at cryogenic temperatures. Therefore, the toughness is low.

This review demonstrates some remedies for higher toughness at cryogenic temperatures.
1)From the viewpoint of fracture mode, the cleavage-like failure should be eliminated.
2)Reduction of interstitials effectively improves the cryogenic temperature toughness.
3)When a high strength over 900 MPa is required at LHT, alloying is inevitable. However, what alloy type or what alloying element is best is still disputable. Further work is also necessary to clarify the mechanism of the cleavage-like failure.

Fatigue

The high cycle fatigue strength depends mainly on the material strength. Ti alloys have higher fatigue strength at lower temperature. However, longer life strength is influenced by the fatigue crack initiation mechanism. At lower cyclic stress, subsurface crack initiation occurs predominantly reflecting the microstructural imhomogeneity.

REFERENCES

1. Van-Stone,R.H., Shannon,Jr.,J.L., Pierce, W.S. and Low,Jr., J.R., Influence of composition, annealing treatment, and texture on the fracture toughness of Ti-5Al-2.5Sn plate at cryogenic temperatures. ASTM STP 651, 1978, 154-179.

2. Scwatzberg, F.R., Cryogenic Materials Data Handbook, 1970.

3. Fawlkes, C.W. and Tobler, R.L., Fracture testing and results for a Ti-6Al-4V alloy at liquid helium temperature. Eng. Frac. Mech., 1976, 8, 487-500.

4. Tobler, R.L., Fatigue crack growth and J-integral fracture parameters of Ti-6Al-4V at ambient and cryogenic temperature. ASTM STP 601, 1976, 346-370.

5. Nagai,K, Ishikawa,K, Mizoguchi,T and Ito,Y, Strength and fracture toughness of Ti-5Al-2.5Sn ELI alloy at cryogenic temperatures. Cryogenics, 1986, 26, 19-23.

6. Nagai,K, Hiraga,K, Ogata,T and Ishikawa,K, Cryogenic temperature mechanical properties of beta-annealed Ti-6Al-4V alloys. Trans. JIM, 1985, 26, 405-413.

7. Nagai,K, Ogata,T, Yuri,T, Ishikawa,K, Mizoguchi,T and Ito,Y, Fatigue fracture of Ti-5Al-2.5Sn ELI alloy at liquid helium temperature. Trans. ISIJ, 1987, 27, 377-382.

8. Nagai,K, Ogata,T, Yuri,T and Ishikawa,K, Fatigue testing at 4K with a helium recondensation system. Adv. Cryog. Eng. Mater. , 1986, 32, 329-338.

9. Shibata,K and Fujita,T, Serration of Fe-Ni austenitic steels at very low temperatures and its computer simulation. Trans. ISIJ., 1986, 26, 1065-1073.

10.Umezawa,O, Nagai,K and Ishikawa,K, Internal crack intiation in high cycle fatigue for Ti-5Al-2.5Sn alloy at cryogenic temperatures. Tetsu-to-Hagane, 1989, 75, 159-166.

11.Umezawa,O, Nagai,K and Ishikawa,K, unpublished data.